HERBICIDES AND ENVIRONMENT

Edited by **Andreas Kortekamp**

Herbicides and Environment
Edited by Andreas Kortekamp

Published by InTech
Janeza Trdine 9, 51000 Rijeka, Croatia

Publishing Process Manager Ana Nikolic

Technical Editor Teodora Smiljanic

Cover Designer InTech Design Team

Additional hard copies can be obtained from orders@intechopen.com

Herbicides and Environment, Edited by Andreas Kortekamp
p. cm.
ISBN 978-953-307-476-4

For EU product safety concerns:
IN TECH d.o.o., Prolaz Marije Krucifikse Kozulić 3, 51000 Rijeka, Croatia,
info@intechopen.com or visit our website at intechopen.com.

Meet the editor

Andreas Kortekamp, PhD, was born in 1970 in Germany. After the first education steps and civilian service he studied biology at the University of Karlsruhe and got his PhD in natural sciences in 2001. He has served as a visiting scientist at the University of Kaiserslautern, University of Halle, and at the Federal Centre for Breeding Research on Cultivated plants (now Julius-Kühn-Institute) during his time as a postdoctoral researcher and assistant professor at the University of Hohenheim.

Contents

Preface

Maybe you think the use of herbicides is something simple. Herbicides are used to kill weeds to prevent their excessive growth. Thus, the herbicide selected is just sprayed onto the respective weed and, if you took the right one, the weed will die within a couple of days. Maybe you think this is the end of the story. Out of your sprayer, out of your mind! You are fatally wrong. In fact, this is just the beginning of the story. Once released into the environment, herbicides may cause unexpected side-effects on non-target organisms and may interfere with biological or chemical processes in soil or aquatic habitats before they fade away due to (bio)degradation.

Since weeds reduce crop yield and quality and interfere with cultivation and harvest operations, herbicides are without any doubt important agrochemicals, especially where conservation tillage is adopted. With respect to sales of pesticides worldwide, herbicides seem to be the most important agrochemicals. However, as mentioned before, herbicides are more than just anti-weed agents. In the following chapters of this book, several authors summarise different features and aspects of herbicides, especially their impact on environmental processes, such as nitrogen cycling, their adverse and unwanted effects on non-target organisms or, in contrary, their beneficial effects on other pests and diseases, their persistence or turn-over in soil and sediments, their toxicological relevance, and at least, their impact on human health. All these important parts of the herbicide story are each treated in several chapters of this book.

A notable feature of herbicides is the fact that their biological activity extends beyond the effect on target organisms. Herbicides may affect organisms in the same ecosystem or in other habitats where herbicides are transmitted mainly by wind currents during application or by rain. Furthermore, herbicides can affect non-target organisms directly or indirectly by altering the composition of plants or other organisms and by changing microclimates in an given ecosystem such as crop land. This may lead to an altered turnover of micro- and macroelements in soil which in turn may favour adapted (micro)organisms. Once applied to the field, herbicides are translocated sooner or later into the soil. There, they may just stick to soil particles waiting for a physical or chemical degradation or they may be absorbed by (micro)organisms and reduced via biochemical processes. The important question of persistence or degradation of herbicidal compounds, the fate of herbicides in soil and how to avoid residues is treated in several chapters of this book.

However, (eco)toxicological relevant amounts of herbicides may run off into surface water before completely metabolised or adsorbed to soil particles. These herbicide residues may interfere with aquatic (micro)organisms or ecosystems. Thus, this topic is also handled by some authors of this book.

Organisms living in soil or water and affected by herbicides do not represent the only non-target organisms. An improper and excessive use of herbicides may also interfere with human health. The use of herbicides as bioweapons may cause human pesticide poisoning, especially when the products used are contaminated with unwanted by-products as it was the case during the Vietnam War. Even though this inadmissible use represents an extreme incidence, contaminated water or residues of herbicides due to repeated applications may cause an impact on human health and is still an issue, especially in developing countries.

Pesticides are only released after passing several tests regarding their putative effects on non-target organisms including mammals. Therefore, the development and ability of test systems as useful tools to investigate unwanted properties of herbicides is also described in this book.

Beside the (eco)toxicological relevance of at least some herbicides, their use may also be restricted due to the emergence of weed populations resistant to distinct herbicidal compounds. Since resistance of weeds to herbicides is an increasing problem worldwide, this topic is also presented by several authors.

Finally, the last two chapters give us an outlook regarding alternative sources that may be used to produce new and environmental-friendly herbicides. It seems that byproducts, generated as unavoidable residues during fermentation and other bio-chemical processes, or eukaryotic organisms such as microalgae are appropriate sources of herbicidal compounds.

In the end, we can conclude that herbicides are much more than just weed killers. They may exhibit beneficial or adverse effects on other organisms. Given their toxicological, environmental but also agricultural relevance, herbicides are an interesting field of activity not only for scientists working in the field of agriculture. It seems that the investigation of herbicide-induced effects on weeds, crop plants, ecosystems, microorganisms, and higher organism requires a multidisciplinary approach. Some important aspects regarding the multisided impacts of herbicides on the living world are highlighted in this book. I am sure that the readers will find a lot of helpful information, even if they are just interested in a specific topic.

Andreas Kortekamp

DLR Agricultural Research Station,
Department of Phytomedicine Neustadt/Weinstrasse
Germany

Part 1

Effects on Non-target Organisms

1

Herbicides: The Face and the Reverse of the Coin. An *in vitro* Approach to the Toxicity of Herbicides in Non-Target Organisms

Amália S. Jurado[1], Maria A. S. Fernandes[2], Romeu A. Videira[3], Francisco P. Peixoto[3] and Joaquim A.F. Vicente[2]
[1]*CNC, Department of Life Sciences, Coimbra University, Coimbra,*
[2]*IMAR, Department of Life Sciences, Coimbra University, Coimbra*
[3]*CECAV, University of Trás-os-Montes e Alto Douro, Vila Real*
[1,2,3]*Portugal*

1. Introduction

I. Historical aspects, benefits and disadvantages of herbicide use

During thousands of years up to about a hundred years ago, man expended most of required energy in arable farming with mechanical operations aiming to remove weeds and providing suitable conditions for the efficient growth of crop plants, considering that weeds compete with beneficial and desired vegetation, which means that weeds are plants growing where man does not which them to grow. With the dawn of industrialization, labor to the factories decreased manpower on the farms, which forced to think of more efficient mechanical means of weed control.

The need of weed management is as old as agriculture itself. Six stages in the evolution of weed control practices can be considered: 1) 10,000 B.C.— removing weeds by hand; 2) 6,000 B.C.— the use of primitive hand tools to till the land and destroy weeds; 3) 1,000 B.C.— animal-powered implements like harrows; 4) 1920 A.D.— mechanically-powered implements like cultivators, blades, harrows, finger-weeders, rotary-hoes, rod-weeders, etc.; 5) 1930 A.D.— biological control and; 6) 1947 A.D.— chemical control, with the commercial development of organic herbicides such as 2,4-D and MCPA (Hay, 1974).

Especially in the last century, for various reasons among which the population explosion, in his effort to produce adequate supplies of food, man needed to combat efficiently the attacks of various pests on agricultural and horticultural crops. Pesticides, falling into three major classes: insecticides, fungicides, and herbicides (or weed killers), are required. Herbicides, specifically, are used for control of weeds.

At the end of the twentieth century, with an estimated world population of 6 billion people, some 700 million were undernourished and 1.3 billion exist on an inadequate diet. In 2009 FAO says that 1.02 billion people are undernourished, corresponding to 15 percent of the estimated world population of 6.8 billion. Undoubtedly, the first problem of Humankind is the lack of food, which affects especially underdeveloped countries. So, the urgent need for much greater application of herbicides and other agrochemicals is essential to increase food supply. Crops can duplicate or increase even more at the expenses of the agrochemicals use.

Another advantage of the greater efficiency of modern agricultural practice is the liberation of land. This can be of major importance in countries where the space dedicated to agriculture is limited, and where there is a very high population density or soil is inadequate for crop yields.

For more than a century chemicals have been employed for total weed control. For the removal of all plants from railway tracks, timber yards and car roads, crude chemicals such as rock salts, crushed arsenical ores, creosote, oil wastes, sulfuric acid, and copper salts were used (Whitten, 1966; Green et al., 1987). Under these conditions, all plants were killed and these compounds all function as total herbicides and treated areas remain toxic to plants for months or even years. However, for agricultural purposes, it was preferred that chemicals would selectively kill the weeds, but not harm the crop plants.

Selectivity can be based, for instance, in the fact that the larger rougher surfaces of weed leaves are more effectively wetted by the spray than the narrow, smooth cereal leaves in which there was much greater run-off of the toxicant. In a few cases selectivity is based on biochemical differences between weeds and crop plants, such that the latter have a unique defense mechanism. The absence of β-oxidases from leguminous plants enables 2,4-DB to be applied safely for selective weed control in these crops. Maize detoxifies the triazine herbicides (e.g. atrazine and simazine) by enzymic hydrolysis, but weeds do not possess this enzyme. Physicochemical properties may be important for selectivity. The soil-applied herbicide, triazine, have low aqueous solubility meaning that the chemical only penetrates some 5 cm downwords in the soil. Hence, germinating weed seeds are killed but more deep-rooted crops are unaffected. The phenoxyacetic acids were the first really effective selective herbicides, being the product of war-direct research. They came into use at a time when the maximum food production with the minimum labor force was a vital factor in the war effort (Cremlyn, 1991).

Herbicides and research related to their biochemical mechanisms of action have helped to unravel details of several of the biochemical pathways in plants. Historically, effects of various chemicals on photosynthesis and respiration have been studied since around 1878. Then, chloroform was described as reversibly inhibiting photosynthesis at concentrations that did not affect respiration (Moreland, 1993).

In France, in 1933, the first important discovery in the field of selective weed control was the use of DNOC, a contact herbicide that killed the majority of annual weeds infesting the crop without causing appreciable damage to the cereals. Its selectivity is explained by the fact that DNOC is not translocated in plants and perennial weeds were not killed because, although their top growth was desiccated, their extensive root systems survived and in due course sent up further shoots (Cremlyn, 1991). DNOC was developed for use in Europe and dinoseb (DNBP) for use in USA (Moreland, 1993). DNOC and other dinitro compounds played a big part in increasing food production during World War II as herbicides. They can also be used as insecticides, especially as winter wash for fruit trees (Cremlyn, 1991).

On field observations conducted in 1940 and 1941 by Templeman and Sexton, phenoxyacetic acids were toxic to dicotyledonous, but not monocotyledonous plants. Eventually, 2,4-D was commercially developed for use in U.S.A. and MCPA for use in Europe from 1947 (Rao, 2000). The discovery of phenoxyalcanoates revolutionized practices and the success achieved in their use probably stimulated industry to invest in research that led to the discovery of the large variety of herbicides that has been used untill now (Moreland, 1993).

Templeman and Sexton also reported, on field observations made in 1940 with esters of arylcarbamates and thiocarbamates, that they were toxic to monocotyledonous but not to dicotyledonous species (Moreland, 1993).

In 1949, Macdowall reported inhibition of the Hill reaction, in illuminated chloroplasts, by phenylurethane (I_{50}=2 mM) and dinitrophenol (DNP) (I_{50}=0.63 mM). However, based on I_{50}, arylcarbamates and phenylureas were up to 2500 times more active than phenylurethane, as was detected by Wessels and van der Veen in 1956 (Moreland, 1993).

Herbicides are the most widely used class of pesticides accounting for more than 60% of all pesticides applied in agriculture (Zimdahl, 2002). One of the main concerns about the use of herbicides is their effects on non-target organisms, especially mammalian toxicity.

Most modern herbicides have low mammalian toxicity, because research for new herbicides often rejects chemicals affecting metabolic pathways that are shared by mammals. In many cases, the low mammalian toxicity has to do with the fact that these chemicals interfere with biochemical pathways that do not exist in mammals, such as photosynthesis, essential amino acids biosynthesis or chlorophyll biosynthesis (Shaner, 2003).

There are about 20 mechanisms of action that have been elucidated for herbicides. Of these, some do share common target sites with mammalians. However, the consequence of inhibiting a common target site in plants can be quite different than in animals. What may be a lethal event in plants, e.g. inhibition of 4-hydroxyphenylpyruvate dioxygenase (HPPD), can even have a beneficial effect in mammals, e.g. treatment for tyrosinemia type I (Shaner, 2003). Also a common enzyme existing in plants and mammals, e. g. acetyl-CoA carboxilase (ACCase), can be inhibited by the herbicides cyclohexanediones and the aryloxyphenoxypropionates, killing plants, and not affecting ACCase in mammals. ACCase plays the same role in the fatty acid biosynthesis of the different organisms, despite vast differences in its regulation and protein structure (Incledon & Hall, 1997; Shaner, 2003). These herbicides suppress de novo synthesis of fatty acids in sensitive species by inhibiting the activity of plastid-associated ACCase. Aryloxyphenoxypropionates were reported by Hoechst AG in 1971 (Nester, 1982), and cyclohexanediones (alloxydim and sethoxydim) were introduced by Nippon Soda in 1977 and 1978 (Iwataki & Hirono, 1979; Ishikawa et al., 1985), being used as postmergence herbicides for the control of annual and perennial grasses in broadleaf crops.

Other commercially available herbicides that affect similar target sites in plants and mammals, with the exception of the bipyridiliums, have minimal toxicity to mammals because they are rapidly metabolized and excreted by mammals. As the herbicides do not accumulate in mammalian tissue, they cannot affect the biosynthetic pathways (Shaner, 2003).

On the other hand, the effects of inhibiting a particular enzyme in plants can be quite different from those observed in mammals. For example, considering that protoporphyrin biosynthesis is critical in plants and animals for the production of chlorophyll and heme, PROTOX (protoporphyrinogen IX oxidase) inhibitors cause the accumulation of Proto IX (protoporphyrin IX), which then absorbs light energy and produces reactive oxygen species that disrupt membranes by lipid peroxidation. In mammals, the accumulation of Proto IX also occurs in the liver and bile, but these organs are not normally exposed to light intensities like plant leaves. So, there is little opportunity for the Proto IX to produce degrading reactive oxygen species in animals (Shaner, 2003). Peroxidizing herbicides include diphenylether herbicides and oxodiazoles causing accumulation of large amounts of Proto IX, which generates singlet oxygen by light. If this oxygen free radical is not quenched, lipid peroxidation can be initiated (Lydon & Duke, 1988; Matringe & Scalla, 1988; Camadro et al., 1991).

It was observed in 1980 that glyphosate was a highly specific inhibitor of shikimate pathway enzyme 5-enolpyruvylshikimate 3-phosphate synthase (Steinrucken & Amrhein, 1980). Afterwards, sulfonylureas and imidazolinones were introduced as herbicides interfering with biosynthesis of branched-chain amino acids (leucine, isoleucine, and valine). It was shown that both herbicides inhibited the activity of acetolactate synthase (Ray, 1984; Shaner et al., 1984)

Herbicides can also inhibit carotenoid synthesis. As carotenoids are free radical scavengers, they protect chlorophyll from photooxidation (bleaching herbicides). One of the first compounds shown to function in this way was the pyridinazinone SAN 6706 (Bartels & McCullough, 1972). Afterwards, other compounds identified with similar functions include fluridone, difunon, dichlormate, and aminotriazole (Moreland, 1993).

Some times, e.g. glufosinate in suicide attempts, much of the toxicity has been related to the associated surfactant that is included in the formulation of the herbicide rather than to the active herbicide (Koyama et al., 1997).

From all the information here considered, it is obvious that the use of herbicides implies the existence of advantages and disadvantages resulting therefrom, which, in general, are listed in Table 1.

ADVANTAGES	DISADVANTAGES
– They kill unwanted plants. – They help the crops grow by destroying the weed that is robbing the crops water, nutrients and sunlight. – They can be safely used whereas in some cases manually or mechanically removing weeds can destroy the crop. – They can be used on closely planted crops where other methods cannot be used. – Most of the time one application of the herbicide is enough whereas other methods have to be continually used. – They are easy to use. – They work fast. – Herbicides are relatively cheap, and cheaper than manual weeding. – Non-selective herbicides can effectively clear fields, where houses and roads can then be built. – They can destroy plants bearing diseases. – Some are biodegradable, and become relatively harmless after decay.	– Some herbicides are non-biodegradable being harmful for a long period of time. – They are all at least slightly toxic. – They can cause illnesses, and even cause accidental or suicidal death (like paraquat). – They can be carried into streams by runoff rainwater or leached into underground water supplies polluting them. – Herbivores may eat the plants treated with herbicides and then carnivores eat the herbivores. The toxic herbicide would be passed up the food chain increasing in concentration, being harmful for those animals and man.

Table 1. Advantages and disadvantages of herbicides.

Scientists have made phenomenal progresses in understanding the selective action of hundreds of herbicides by studying their absorption and translocation patterns, mechanisms of action in plants, degradative and detoxification mechanisms in plant and soil, interactions with other pesticides and chemicals, etc. All this has helped in making more effective, economical and safe recommendations for control of numerous weeds (Rao, 2000).

Lately, scientists have already researched on natural product-based herbicides, which are generally considered safer than synthetic herbicides. However, relatively few commercial herbicidal agents have been derived from naturally occurring compounds. One of the indirect and important benefits of the chemical composition and structural characteristics of natural products is that most of these compounds are rapidly degraded in the natural environment. Thus, this accounts for the perception that most natural products are environmentally benign (Dayan et al., 1999).

An alternative approach of natural products-based weed management consists of having crops producing their own phytotoxins to prevent or suppress the growth of competing weeds in its immediate surroundings (allelopathy). Although some species strongly repress the development of other plants, allelopatic crops have had little success so far (Dayan et al., 1999).

Weed scientists are now facing new challenges, particularly in the light of emergence of weeds resistant to herbicides and concerns and questions about herbicide residues in food, soil, groundwater and atmosphere (Rao, 2000).

Finally, when scientists were capable of getting genes that encode target site enzymes cloned, modified and expressed in transgenic plants, they could confer herbicide tolerance. The possibility of developing herbicide-resistant crops began a new era in the use of herbicides, stimulating work on the genetics of herbicide resistance.

II. Classification of herbicides

By 1976, the herbicides that interfered with electron transport and phosphorylation in both chloroplasts and mitochondria were separated into five classes: a) electron transport inhibitors, b) uncouplers, c) energy transfer inhibitors, d) inhibitory uncouplers (multiple types of inhibition), and e) electron acceptors (Moreland, 1993).

Electron transport inhibitors include the Hill reaction inhibitors. Uncouplers dissociate electron transport from ATP formation by permeabilization to protons collapsing ΔpH. Energy transfer inhibitors interact with the coupling factor complex. Inhibitory uncouplers are those herbicides that act both as electron transport inhibitors and uncouplers. Electron acceptors are herbicides like bipiridilliums that intercept electron flow from photosystem I.

With isolated mitochondria, the inhibitory uncouplers uncouple phosphorylation at low molecular concentrations and inhibit electron transport at higher concentrations (Moreland, 1993).

The pure electron transport inhibitors also have been called diuron-type and include chlorinated phenylureas, pyridazones, s-triazines, triazinones, uracils, and ureacarbamates. This type of herbicides inhibits the Hill reaction but do not inhibit mitochondrial electron transport. Hence, light was known to be required for the expression of toxicity, i.e., plants maintained in the dark revealed no signs of toxicity (Moreland, 1993).

Inhibitory uncouplers were divided into two groups: dinoseb-type and dicryl-type. Included as dinoseb-type are dinitrophenols, benzimidazoles, benzonitriles, bromophenoxim, and thiadiazoles. Dicryl-type includes arylanilides, dinitroanilines, diphenylethers, bis-carbamates and perfluidone (Moreland & Novitzky, 1988).

Herbicides are designated by common names approved by the Weed Science Society of America or the British Standards Institution (Moreland, 1980), and organic herbicides are classified on the basis of: a) method of application; b) chemical affinity and structure similarity, and; c) mode of action (Rao, 2000).

On the basis of the method of application, herbicides are distributed in two groups: 1) soil-applied and 2) foliage-applied. All herbicides applied at preplanting (surface or incorporation) and preemergence (to crop, weeds, or both) are included in the soil-applied group; those applied at postemergence on the plant parts are included in the foliage-applied group.

Here we also present two general classifications of herbicides according to the mode of action and according to the chemical composition.

According to the mode of action, we present in Table 2 the classification extracted and adapted from a review (Moreland, 1980).

Classes of herbicides	**Examples of herbicides**
*CHLOROPLAST-ASSOCIATED REACTIONS	
TYPES OF INHIBITORS	
Electron transport inhibitors	Diuron, atrazine, pyrazon, etc.
Uncouplers	Perfluidone.
Energy transfer inhibitors	1,2,3-thiadiazolyl-phenylureas.
Inhibitory uncouplers	Dinitrophenols, acylanilines, etc.
Electron acceptors	Paraquat, diquat.
Carotenoid synthesis inhibitors	Amitrole, dichlormate, etc.
*MITOCHONDRIAL ELECTRON TRANSPORT AND PHOSPHORYLATION	
TYPES OF INHIBITORES	
Electron transport inhibitors	Pyriclor.
Uncouplers	Isipropyl ester of glyphosate.
Energy transfer inhibitors	No herbicide.
Inhibitory uncouplers	Dinitrophenols, acylanilines, etc.
*MEMBRANE INTERACTIONS	
Composition alterations	Dinoben, perfluidone, ioxynil.
Permeability and integrity actions	Paraquat, diclofop-methyl.
*CELL DIVISION	
Energy relations	Chlorpropham, trifluralin.
Microtubules action	Prophan, amiprofos-methyl.
*NUCLEIC ACID METABOLISM	
DNA and RNA action	Trifluralin.
PROTEIN SYNTHESIS	
Protein synthesis inhibitors	Glyphosate.

Table 2. Classification of herbicides according to the mode of action at the level of the main metabolic systems in the cell.

The other classification, distributing herbicides by groups with chemical affinities (Cremlyn, 1991; Rao, 2000), is presented in Table 3.

Classes of herbicides	Examples of herbicides	Mode of action
ALIPHATICS	[1]Dalapon, [2]TCA.	[1]Interference with RNA synthesis; [1,2]modifications of protein structure.
AMIDES Acylanilides	Propanil, pentanochlor.	Inhibition of Hill reation in photosynthetic electron transport.
Chloroacetanilides	[1]Allidochlor, [2]propachlor, [2]alachlor, [2]butachlor, [2]pretilachlor, [2]diethatylether, [2]metolachlor, [2]metazochlor.	[1] Inhibition of protein synthesis, respiration; [2] Inhibition of cell division and protein synthesis.
Benzamides	Benzoylpropethyl, flamprop-isopropyl, fampropmethyl, isoxaben.	Unknown.
Other amides	[1]Naptalam, diphenamid, [2]propyzamide, butam, [3]diflufenican.	[1]Alteration of auxin transport; [2]interference with cell division; [3]inhibition of carotenoid biosynthesis.
AROMATIC CARBAMATES AND THIOCARBAMATES Arylcarbamates	Propham, chloropropham, barban.	Interference with photosynthesis; alteration of meristematic activity; desorganization of microtubular assembly, inhibiting mitosis.
Thiocarbamates	EPTC, butylate, cycloate, pebulate, diallate, triallate, benthiocarb, asulam, nisulam.	Interference with lipid biosynthesis.
ARSENICALS	DSMA, MSMA, cacodylic acid.	They are uncouplers and inhibit respiration.
BENZOTHIADIAZOLES	Bentazon.	Inhibition of photosynthesis and induces lipid peroxation.
BIPYRIDILIUMS	Diquat, paraquat.	Induce lipid peroxidation by reactive oxygen species production.
CARBOXYLIC ACIDS Phenoxyacetic acids	[1]2,4-D, MCPA, 2,4,5-T, etc.	Inhibition of normal plant growth mimicking auxins (IAA); [1] Inhibition of oxidative phosphorylation.
Phenoxybutiric acids	2,4-DB, MCPB.	Inhibition of normal plant growth mimicking auxins (IAA).

Classes of herbicides	Examples of herbicides	Mode of action
Benzoic acids	TBA, dicamba.	Inhibition of normal plant growth mimicking auxins (IAA)
Pyridine derivatives	Picloram, clopyralid, triclopyr.	Inhibition of normal plant growth mimicking auxins (IAA).
Propionic acid derivatives	Dchlorprop, mecoprop, fenoprop, haloxyfop, fenoxaprop, fluazifop-P.	Inhibition of normal plant growth mimicking auxins (IAA).
CHLOROALIPHATIC ACIDS	[1]Dalapon, [2]TCA.	[1]Interference with RNA synthesis; [1,2]modifications of protein structure.
CINEOLES	*Cinmethylin.	Inhibition of asparagine synthetase.
CYCLOHEXANEDIONES	Clethodim, cycloxidim, sethoxydim, tralkoxydim.	Interference with lipid metabolism.
DINITROANILINES	Trifluralin, benfluralin, profluralin, ethalfluralin, fluchloralin, oryzalin, nitralin, butralin, pendimethalin, dipropalin.	Inhibition of cell division.
DIPHENYL ETHERS	Nitrofen, fluorodifen, bifenox, [1]oxyfluorfen, fomesafen.	Inhibition of Hill reation in photosynthesis and photophosphorylation; [1] inhibition of protoporphyrinogen oxidase.
IMIDAZOLIDINONES	Buthidazole.	Inhibition of r espiration and photosynthesis.
ISOXAZOLIDINONES	Clomazone.	– –
IMIDAZOLINES	Imazaquin, imazethapyr.	Inhibition of cell division by action on the enzyme acetolactate synthase (ALS) involved in valine and isoleucine biosynthesis.
NITRILES	[1]Dichlobenil, [2]bromoxynil, [2]ioxynil.	[1]Inhibition of cellulose biosynthesis; [2]uncoupling of oxidative phosphorylation and Hill reation inhibition.
ORGANOPHOSPHORUS COMPOUNDS	[1]Glyphosate, bensulide, **tributyl phosphorotrithiolate.	[1]Inhibition of biosynthesis of phenylalanine.

Classes of herbicides	Examples of herbicides	Mode of action
OXADIAZOLES	Oxadiazon.	Causes rapid peroxidative damage.
PHENOLS	Dinoseb, DNOC.	Uncouplers of oxidative phosphorylation.
N-PHENYLPHTHALAMIDES	Flumiclorac.	Inhibition of protoporphyrinogen oxidase.
PHENYL TRIAZINONES	Sulfentrazone.	Inhibition of protoporphyrinogen oxidase.
PHTHALAMATES	Naptalam.	It is an auxin inhibitor.
PYRAZOLIUMS	Difenzoquat.	It affects nucleic acid biosynthesis, photostynthesis, ATP production, K+ absorption and P incorporation in phospholipids and DNA.
PYRIDAZINES	[1]Chloridazon, [2]metflurazon, [2]norflurazon, [1]pyridate.	[1]Inhibition of Hill reation in photosynthesis; [2]Inhibition of carotenoid biosynthesis.
PYRIDINONES	Fluridone.	Bloks carotenoid biosynthesis.
PYRIMIDINYL THIOBENZOATES	Pyrithiobac.	Inhibition of cell division by action on the enzyme acetolactate synthase (ALS) involved in valine and isoleucine biosynthesis.
QUINOLINE CARBOXYLIC ACIDS	Quinclorac.	Auxin-type herbicide. Accumulates ACC, cyanide and ethylene.
SULPHONYLUREAS	Sulfometuron methyl, DPX-F 6025, DPX-F 5384, chlorsulfuron, DPX-L 5300, bensulfuron methyl.	Inhibition of cell division by action on the enzyme acetolactate synthase (ALS) involved in valine and isoleucine biosynthesis.
TRIAZINONES	Metribuzin.	– –
TRIAZOLOPYRIMIDINE SULFONANILIDES	Flumetsulam.	Inhibition of cell division by action on the enzyme acetolactate synthase (ALS) involved in valine and isoleucine biosynthesis.

Classes of herbicides	Examples of herbicides	Mode of action
TRIAZINES	Simazine, atrazine, prometryne, desmetryn, methoprotryne, terbutryn, cyanazine, eglinazine ethyl, aziprotryne, metrybuzin, metamitron, prometron, terbuthylazine, methroprotryne, aziprotryne.	Inhibition of Hill reation in photosynthetic electron transport.
TRIAZOLES	***Amitrole.	Probable interference with carotenoid biosynthesis leading to photooxidation of chlorophyll.
URACILS (PYRIMIDINES)	Bromacil, terbacil, lenacil, UCC-C4243.	Inhibition of Hill reation in photosynthetic electron transport.
UREAS	Monuron, diuron, linuron, monolinuron, chlorotoluron, isoproturon, difenoxuron, thiazafluron, terbuthiuron, ethidimuron, fluometuron, S3552.	Inhibition of Hill reation in photosynthetic electron transport.
OTHER HETEROCYCLIC HERBICIDES	[1]Endothal, [2]ethofumesate, [3]benzolin, [4]oxadiazon, [4,5]difenzoquat.	[1]Inhibition of lipid and protein biosynthesis; [2]interference with plant cuticular wax; [3]interence with cell growth and development; [4]inhibition of carbon dioxide fixation and electron transport in photosynthesis; [5]alteration of DNA and RNA synthesis.

*Is a member of the herbicide group called cineoles (monoterpenic cyclic ethers) (Romagni et al., 2000)
**Defoliant, induces early leaf abscission through changes in the levels of plant hormones (Matolcsy, 1988)
***Its use in fodder and food crops has been banned worldwide (suspicious of carcinogenicity)

Table 3. Classification of herbicides according to chemical classes.

2. *In vitro* toxicological assays: some biological model-systems

In this section, we will address some of the experimental strategies used by our group to evaluate the toxic potential of pesticides upon interaction with biological systems, using *in vitro* assays. The authors are aware that many other models have been successfully used with the same objectives, some of which will be discussed in confront with authors results in the next section. However, an exhaustive description of *in vitro* model systems used to herbicide toxicological assessment is out of the scope of this chapter.

I. Mitochondria, a key organelle to push cells towards survival or death

Mitochondria are efficient power plants of energy generation for eukaryotic cells. Moreover, mitochondria play a crucial role in several physiological processes and are involved in cell replication, differentiation and apoptosis (Boelsterli, 2003). In this way, mitochondria endowed eukaryotic cells with a broad spectrum of functionalities that prokaryotic cells do not have, but, concomitantly, they brought with them a multiplicity of vulnerable points to host cells. In fact, these organelles are a key target for xenobiotics to induce dysfunction at a cell, organ or organism level and the impairment of energy production is not the only harm of mitochondrial poisoning.

There are several means by which xenobiotics interfere with mitochondrial function. However, as most xenobiotics considered in this chapter are lipophilic molecules, their effects on the activities of mitochondrial matrix soluble enzymes involved in the fatty acid β–oxidation and tricarboxylic acid oxidation pathways will not be directly addressed. The focus of this section will be mainly the xenobiotic-induced alterations of mitochondrial membrane-related processes, such as electron transport, transmembrane potential ($\Delta\Psi$) generation, oxidative phosphorylation, oxygen reactive species generation, permeability transition and the release of pro-apoptotic proteins.

Three important features make mitochondria vulnerable to membrane-active compounds. The first one is the unique lipid composition of the inner membrane (Daum, 1985), with virtually no cholesterol and a high content of cardiolipin (CL), a phospholipid that is common in bacterial membranes but, in eukaryotic cells, only exists in membranes of mitochondria and chloroplasts (Gennis, 1989). Several xenobiotics have a high affinity for CL, which explains their mitochondriotropic effects (Boelsterli, 2003). A very well known example is given by adryamicin-like anthraquinones (Wallace & Starkov, 2000). The accumulation of these xenobiotics at the inner mitochondrial membrane may have different consequences. In the case of adryamicin, for example, it leads to the generation of oxidative stress due to redox cycling of quinone-semiquinone moiety (Boelsterli, 2003). Considering the important roles CL plays in mitochondria (see below), xenobiotic incorporation in specific CL-enriched domains of the inner mitochondrial membrane may predictably have disastrous outcomes for mitochondrial bioenergetics and also for the fate of cells towards survival or death. Cardiolipin is included in the quaternary structure of the complexes II, IV and V (Eble et al., 1990; Fry & Green, 1981; Sedlak et al., 2006) of the mitochondrial electron transport system. Besides those complexes, the complex I also requires cardiolipin for exhibiting optimal activity (Fry & Green, 1981). Furthermore, cytochrome c (cyt c) establishes specific interactions with CL (Rytomaa et al., 1992), which have been suggested to participate in a well coordinated mechanism of programmed cell death, involving cyt c-catalyzed peroxidation of CL, permeabilization of the outer mitochondrial membrane and release of cyt c and other apoptogenic factors from the mitochondria (Gonzalvez & Gottlieb, 2007; Basova et al., 2007). As proposed by Basova et al. (2007), CL association with cyt c could be a decisive step in the early stages of apoptosis, turning off the normal functions of this hemoprotein in mitochondrial electron transport and turning on its peroxidase activity. Therefore, xenobiotics, directly interacting with CL or modulating the bilayer properties of the mitochondrial membranes, may have drastic consequences in mitochondrial bioenergetics and/or apoptosis (Gonzalvez & Gottlieb, 2007).

The second feature which makes mitochondria vulnerable to xenobiotics is related with the high potential across the inner mitochondrial membrane, typically in the range of -180 to -220 mV. The anisotropic distribution of both charges and pH (the matrix side being negatively charged and slightly alkaline in opposition to the outer side of the membrane, facing the intermembrane space) confers to mitochondria a high susceptibility to accumulate large amounts of positively charged lipophilic compounds. The accumulation of these compounds into mitochondria, at concentrations that may exceed those of the cytosol by several orders of magnitude, exposes these organelles to potentially toxic compounds at much higher levels than any other cellular compartment.

The third reason why mitochondria are prone to be poisoned by lipophilic compounds concerns with the protein-regulated permeability of the inner mitochondrial membrane, which has a crucial importance for the functional integrity of mitochondria. Therefore, drastic consequences for mitochondrial intactness and bioenergetic functioning can be predicted whenever compounds promote an increase of the passive permeability of mitochondrial inner membrane, impair the activity of transport proteins embedded within it or interfere with the assembling of protein components of the mitochondrial permeability transition pore (MPTP). Xenobiotics that cause the opening of MPTP may also promote the release of apoptosis mediators, inducing apoptosis.

On the other hand, xenobiotics that interact with one or several complexes of the mitochondrial electron transport system, impairing the normal electron flow, may enhance the generation of reactive oxygen species (ROS), leading to an imbalance between prooxidant species and the cellular antioxidants. The complex I (NADH: ubiquinone oxidoreductase) is often a site of interference of xenobiotics acting as electron acceptors in reductase-catalyzed reactions. Redox cycling compounds, as the herbicide paraquat, may be particularly harmful. The generation of ROS, amplified by the cycling process, may lead to oxidative DNA damage and lipid peroxidation (Boelsterli, 2003). Additionally, xenobiotics that promote oxidative stress as well as those that induce a sustained increase of mitochondrial calcium concentration may disrupt mitochondrial function by opening the MPTP. The consequent uncoupling of oxidative phosphorylation, due to the collapse of mitochondrial membrane potential, and the release of pro-apoptotic factors are, in this case, the main cause of cellular damage (Boelsterli, 2003).

Although rat liver mitochondria constitute a common model for toxicity assessment, plant mitochondria can be an alternative model, offering several advantages, namely high activities preserved for much longer periods of time, easier availability and low cost (Vicente & Madeira, 2000). In the context of herbicide toxicity evaluation, the use of plant mitochondria is particularly important since useful plants (crops) are often target of the non-selective toxic effects of those compounds.

II. Unicellular microorganisms, a simple model system to study the complex routes of cell/membrane toxicity

A great number of compounds produced by agrochemical industries, including herbicides, are lipophilic molecules. In the case of biocide compounds, this feature is important to increase their ability to overcome cell membrane barriers. In a toxicokinetic context, the disposition of a foreign molecule in a biological system necessarily involves the passage across cell membranes. Thus, lipophilicity is a major determinant of compound capacity to exert noxious effects on target organisms. However, cellular compartmentalization by phospholipid-based membranes is a feature common to all living organisms and, then, all share a high vulnerability to lipophilic compounds. The selective effects of these compounds on specific organisms may arise from factors that affect compound metabolism, distribution or excretion (toxicokinetics) and from differences in the existence, accessibility or structural singularities of the molecular target of these compounds (toxicodynamics). The level of concern increases when the interaction of lipophilic compounds with membranes results in unspecific alterations of the physical properties of the lipid bilayer, compromising cell normal functioning. In fact, the lipid bilayer, besides defining a permeability barrier and providing a matrix for embedded proteins, endows the membrane with the capacity of regulating protein function (Lundbaek et al., 2010). Bilayer material properties such as thickness, intrinsic curvature, and the elastic compression and bending moduli, have been shown as having a key role on protein folding, trafficking, distribution in the plane of the membrane and function (McIntosh & Simon, 2006). Therefore xenobiotic-induced bilayer deformations or perturbation of their physical properties may cause drastic effects on a great diversity of vital cell functions. In this case, the interactivity with membrane lipid bilayer is involved in both toxicokinetics and toxicodynamics of the compounds.

In microbial cells, most vital functions are ascribed to the plasma membrane, e.g. signal and energy transductions, regulation of the intracellular environment and transport of solutes (Kaback, 1972; Stock et al., 1990; Trumpower & Gennis, 1994). Not surprisingly, perturbations of membrane structure and function by pollutants will end in more or less severe disturbance of cell growth and viability. Therefore, these simple systems can offer the opportunity to screen the cytotoxic potential of membrane-active compounds. Measurements of growth parameters provided by bacterial cell cultures in liquid media (e.g. specific growth rate, final bacterial yield), the counting of colony forming units in solid growth media and the evaluation of bacterial respiratory activity or other metabolic parameters have been extensively used as short-term screening tests for *in vitro* assessment of the toxicity of several environmental pollutants (Hernando et al., 2003; Sikkema et al., 1995; Monteiro et al., 2005).

On the other hand, bacteria have one, in maximal two, membranes surrounding the cellular content, which can be easily isolated providing a pure membrane fraction. Two experimental approaches may be developed by using bacterial membranes as a biological model system: a) evaluation of the influence of membrane biophysical properties, perturbed by the accumulation of alien lipophilic molecules, on cellular toxic effects; b) identification of bacterial mechanisms of regulation of the membrane lipid composition triggered by xenobiotic incorporation and its impact on counteracting cell toxicity. In fact, alterations of membrane lipid content (the relative proportions of different lipid classes and the acyl chain composition of phospho- and glycolipids) are common strategies for adaptation and responses to stress in bacteria (Jurado et al., 1991; Sikkema et al., 1995; Weber et al., 1996). Since bacterial lipid regulatory system is highly precise and sensitive, studies of the molecular mechanisms of membrane adaptation, together with determinations of growth inhibition and impairment of bacterial metabolic activities, can provide an alternative methodology for toxicity assessment. Thus, microorganisms offer the advantage of allowing the assessment of xenobiotic toxic effects ranging from the molecular to the cell/whole organism level. These data, together with the knowledge of physicochemical characteristics of the compounds, can be useful to establish structure-activity relationships.

In the field of ecotoxicology, two main reasons explain the considerable attention effects of lipophilic pollutants on microorganisms have drawn: firstly, microorganisms are responsible for recycling organic material in soil ecosystems and, then, are valuable indicators of pollutant ecological impact. Thus, toxicological data concerning soil bacteria may be used to define upper limits for concentration of pollutants and to predict environmental toxicity risks; secondly, several of those microorganisms have shown to be able to thrive in high concentrations of organic pollutants converting them into non-toxic metabolites (Dejonghe et al., 2003; Sorensen et al., 2003; Neumann et al., 2004). The interest here is to use these microorganisms for the removal of pollutants from the environment (bioremediation).

However, in the present work, the emphasis will be on the use of microorganisms as models to correlate xenobiotic-induced perturbations at a molecular level with the impairment of physiological processes studied *in vitro*. Different microorganisms have been used for this purpose such as the yeast *Sacharomyces cerevisiae* (Cardoso & Leão, 1992) and *E. coli* (Heipieper et al., 1991) as the most studied bacterial model. Microorganisms offer over multicellular organisms several advantages: (i) yeast/bacterial growth in laboratory controlled conditions is easy and economic; (ii) these models avoid ethical issues associated with animal use, providing alternative *in vitro* toxicity tests; (iii) microorganisms allow to correlating in the same system the toxic effects exerted by xenobiotics on the organism as a whole, assessed *in vivo* (e.g. growth disturbances), and their toxic action on cell functions assayed *in vitro* (e.g. protein/channel activities or passive membrane permeability).

In our laboratory, a strain of the thermophile *Bacillus stearothermophilus*, grown in a liquid medium (L-Broth) at 65ºC (within the optimal temperature range), has been extensively used as a model system for assessing the cytotoxicity of different groups of environmental pollutants, as herbicides (Pereira et al., 2009; 2010), insecticides (Donato et al., 1997a,b; Martins et al., 2003; Monteiro et al., 2005) and organometals (Martins et al., 2005). Toxicity indicators have been taken from parameters of growth (lag time, specific growth rate and bacterial yield in the stationary phase) when growth is followed by turbidity measurements or by counting the colony-forming units (c.f.u.) after plating serial dilutions of the liquid cultures on agar plates. Other indicators of toxicity have been provided by monitoring the respiratory activity (oxygen consumption rate supported by different respiratory substrates). Studies of growth or respiration performed in a broad concentration range have allowed the determination of toxicity indices as EC50, NOAEL and/or LOEC. A very good correlation has been observed between toxicity data provided by *B. stearothermophilus* and by other model systems, namely rat liver mitochondria (Monteiro et al., 2008; Pereira et al., 2009).

B. stearothermophilus presents several advantages to be used as a toxicological model system, in addition to the general ones of bacterial cells. Firstly, cultures of this bacterium with high cell densities are rapidly obtained at optimal conditions (temperature and pH) of growth (doubling times around 15 min in the absence of toxicants). Thus, a significant amount of cells (around 10^8 colony forming units per ml of growth medium) may be obtained in a few hours. Hence, in comparison with other time-consuming assays for toxicological assessment, *B. stearothermophilus* growth experiments are very advantageous. Moreover, it is also possible in a short time to isolate homogeneous membrane preparations for lipid extraction or for monitoring enzymatic reactions. Additionally, *B. stearothermophilus* growth showed to be very sensitive to perturbations of membrane physical properties and efficient molecular mechanisms of membrane adaptation (involving changes of membrane lipid composition) are triggered by adverse growth conditions (Jurado et al., 1991; Donato et al., 1997a and 2000; Luxo et al., 1998; Rosa et al., 2000). For these reasons this model is particularly useful for drug toxicity assessment at the membrane level.

On the other hand, the growth of this bacterium occurs at high temperatures (optimally at 60-65ºC), preventing contamination, simplifying sterilization procedures and also assuring safety for the operator. Concerning the respiration assays, protoplasts are advantageous in comparison with other systems, as animal mitochondria, for several reasons: economics, easier manipulation and high performance during much longer periods (several days). Additionally, the respiratory system of this bacterium shows a great similarity with that of mitochondria. However, an accurate determination of the levels at which xenobiotics act in the bacterial oxidative phosphorylation system is technically difficult, since there are not known inhibitors for the respiratory complexes, except for the cytochrome oxidase (CN^- as inhibitor). In spite of this limitation and the restricted number of substrates that efficiently support the respiration of protoplasts, very good correlations have been obtained regarding the concentration range of xenobiotics that impair the electron transfer system in both bacterial and eukaryotic models (Donato et al., 1997b; Monteiro et al., 2005; 2008; Pereira et al., 2009).

Taking together growth, respiration and membrane physical data collected with *B. stearothermophilus*, a remarkable parallelism may be established with results from studies carried out with other toxicological model systems, revealing differential toxic effects exerted by xenobiotics and their metabolites (Donato et al., 1997b and 2000), xenobiotic isomers (Martins et al., 2003; Monteiro et al., 2003) or different molecules of the same family of compounds (Pereira et al., 2009 and 2010).

III. Cell cultures of plant tissues, an economical model to evaluate phytotoxicity of chemical agents of stress

Plant and animal cell cultures have proven to be a useful tool for assessing the toxicity of chemical compounds *in vitro* (Steward, 1983; Babich et al., 1986). Diverse cytotoxicity endpoints can be considered. The most used are the cell viability, membrane integrity, changes in the content of vital macromolecules (DNA, proteins) or small molecules involved in cell metabolism (ATP, NADH), and antioxidant defenses (e.g. GSH).

Studies with plant tissue cultures are particularly meaningful when the goal is to screen the general toxicity of herbicides in non-target organisms. Useful plants (crops) can be susceptible organisms to the toxic action of those compounds, not only because they coexist with the target-plants (weeds), but also because weeds and crops do share a great number of features putatively affected by the toxic action of the herbicides.

Potato (*Solanum tuberosum* L.) is one of the most important crops in the temperate climate regions of the world (Zuba & Binding, 1989). It has been cultivated in Europe, but its production has been also increasing in China, India and Indonesia, as this crop tolerates a great variety of climates, from temperate to tropic, and from humid to arid. As a food crop, its value results from the accumulation of high amounts of starch in the tuber in combination with proteins, lipids and minerals. Therefore, the interest in studying herbicide toxic effects on potato tissue cultures *in vitro* is dual: a) to appraise the susceptibility or eventual resistance of this important food crop to herbicide toxicity and b) to use these accessible cultures as model systems for screening the toxicity of herbicides *in vitro*.

Tubers, roots, leaves, and other parts of the *Solanum tuberosum* plant have been extensively used as explants for initiating callus (friable, pale-brown lumps) formation in the presence of adequate culture media. The fully mature cells of potato tubers preserve, even after a long period of quiescence, the ability to behave as growing cells when appropriately stimulated, showing activities such as respiration supported by the carbohydrates from starch hydrolysis, protein synthesis, uptake of water and inorganic salts harnessing its metabolism and available energy (Steward, 1983). Plant cell cultures constitute a true clonal growth of the original plant, retaining not only the inheritance and totipotency of their nuclei but also the competence of the cytoplasm to regulate gene expression (Steward, 1983).

The use of plant tissue cultures has several advantages over other methods for toxicity assessment: a) high sensitivity, versatility, speed and simplicity; b) minimal costs for the equipment; c) technical expertise is easy to obtain; d) growth conditions are easily controlled in laboratory. Additionally, in callus tissue cultures, the simplicity of the structural organization and the absence of cellular wall barrier amenable to an effective herbicide penetration and translocation (Mumma & Davidonis, 1983; Magalhães et al., 1989; Nellessen & Fletcher, 1993) make this sensitive model a useful tool to study herbicide toxicity and metabolism in plants (Sandermann et al., 1984; Smeda & Weller, 1991). The growth of non-green callus tissue in the dark precludes the analysis of herbicide effects on the photosynthetic system, but allows to address studies to other putative targets of the toxic action of those compounds, such as mitochondria which have a crucial role in ATP synthesis in these cells devoid of chlorophyl. Thus, potato callus tissue has been used as a relevant material source for *in vivo* and *ex vivo* studies of herbicide toxicity at the mitochondrial level (Peixoto et al., 2008; 2009a).

In summary, plant cell cultures have been established as a good alternative model for ecotoxicological evaluation and also as a means to study herbicide mechanisms of action and metabolism in plants (Harms, 1992).

IV. Fish as an economical vertebrate model for toxicity assessment

Fish are currently used as toxicological models with a major importance in hazard identification, environmental risk assessment, and biomedical research in human development and diseases (Hill et al., 2005; Schmale et al., 2007). Fish are the most numerous, the most diverse and the oldest group of vertebrates, allowing the investigation of fundamental biological principles that can be extrapolated to humans. For example, studies on renal physiology of fishes led to diagnostic methods remaining in use in medicine today (Beyenbach, 2004). In a toxicological context, fish are also potential target for action of toxic compounds, such as herbicides, since aquatic ecosystems are the final reservoir of numerous chemicals produced and used by man (Pritchard, 1993). Thus, they are used in acute and chronic studies to evaluate the susceptibility of individual species to a wide variety of pollutants (*e.g.* herbicides).

Acute toxicity assessment, the first step of chemical toxicological evaluation, allows to determining the LC_{50} (96 h) value (e.g., according to OECD guideline 203, *i.e., as the concentration of the test substance resulting in 50% mortality of the experimental fish over a period of 96 hours*; OECD203, 1992). Although death is the endpoint in acute toxicological assays and it represents an unambiguous parameter, its environmental significance is questionable. The information concerning the way how the chemical compounds exert their acute toxicity in fish is scarce (Nagel, 2002). On the other hand, the chronic effects resulting from long-term exposure to low concentrations are environmentally relevant, allowing the comprehension of the molecular mechanisms underlying the toxic effects and the establishment of structure-activity relationships of several classes of xenobiotics (Pritchard, 1993). Therefore, it is important to understand chronic effects that arise from chemical exposure of fish at all life stages, i.e. embryonic, larval, juvenile and adult (Lawrence & Hemingway, 2003).

The research strategies used to assess chronic toxicity depend on the research team experience, the fish species used and also the physico-chemical properties of the compounds to be evaluated. The main goals are: firstly, identification of the main route of entry of the toxic substance in the body (skin, gills, food), determination of toxic distribution between various organs and the pathways of biotransformation and elimination (toxicokinetics); secondly, identification of the organs and/or tissues more sensitive to the toxic action, and the cellular and biochemical pathways underlying the toxicity of the chemical compound (toxicodynamics).

Economic advantages together with the general properties shared by many species of fish such as high fecundity, external fertilization of transparent eggs and relatively brief generation times make fish a valuable model for toxicological research. Small size fish species, such us medaka (*Oryzias latipes*) and zebrafish (*Danio rerio*) have transparent externally developed embryos, facilitating experiments in developmental toxicology, carcinogenesis, mutagenesis and organ-specific toxicity assays (Wakamatsu et al., 2001; Wiegand et al., 2001; Nagel, 2002; Hill et al., 2005). Other species like rainbow trout (*Oncorhynchus mykiss*), feathead (*Pimephales promeals*) and tilapia (*Oreochromis niloticus*) having large size are more appropriate for the experiments requiring large amounts of tissue, to pursue gene expression profiles, proteomics and lipidomics. Their size also allows better assessment for the histological and anatomical investigation. However, the main focus of the toxicological studies is still the biochemical changes associated with toxic exposure, namely, in proteins involved in xenobiotic detoxification, metabolism and excretion (Lawrence & Hemingway, 2003). The endpoint test includes the measurement of the quantities/activities of cytochrome P450 (CYP) monooxygenase system, UDP glucuronosyl, glutathione transferases and metallothioneins. To assess the possible correlation between oxidative stress induced by xenobiotics and the toxic effects, the presence of intracellular oxidative reactive species are usually investigated, quantifying the lipid peroxidation, the reduced glutathione and α-tocopherol intracellular stores and the activity of enzymes like glutathione reductase, glutathione peroxidase, catalase and superoxide dismutase (Figueiredo-Fernandes, 2006; Peixoto et al., 2006). Additionally, direct damage to DNA, including chromosomal aberrations such as micronucleus formation, DNA adducts (covalent attachment of a chemical to DNA) and strand breakage are also evaluated to indentify genotoxic compounds (Çavas & Könen, 2007).

3. Results

I. Studies with plant and animal mitochondria

There are only a small number of toxicological studies using isolated plant mitochondria. However, to understand the mechanism of action of the herbicides on plants, it will be important to study their effects on plant mitochondria considered as the main target for many of them. Some plant mitochondrial fractions exhibit activities higher than animal preparations, which are preserved over several days, especially when obtained by purification with a Percoll gradient (Neuburger et al., 1982; Vicente & Madeira, 2000). These fractions exhibit higher degrees of purity, as compared with mitochondria of animal origin, namely rat liver mitochondria, avoiding interferences with contaminants. In spite of this, the toxicological evaluation aiming environmental security for man, comprehensively prefers the use of mammalian mitochondria, considering the differences between plant and animal mitochondrial structure and function.

Mitochondria are known as the main intracellular organelle responsible for cell ROS production and we should bear in mind that compounds interacting with mitochondrial membranes putatively disturb the coupling efficiency between oxidation and

phosphorylation and also exacerbate mitochondrial ROS production. A range of biotic and abiotic stresses also raise ROS levels in plants due to both defense responses and perturbations of chloroplastic and mitochondrial metabolism (Van Camp et al., 1998).
Several studies were performed in our laboratories, with isolated plant and animal mitochondria, showing that this organelle could represent a target for several well known herbicides. Here we describe and discuss the main results obtained in our studies, considering other studies presented in literature.

1. DNOC

The compound DNOC (4,6-Dinitro-*o*-cresol) is an herbicide included in the phenols, being also used as fungicide and insecticide. DNOC in the concentration range of 10-50 μM, acted as a classical uncoupler of oxidative phosphorylation in rat liver mitochondria, promoting an increase in succinate-supported mitochondrial respiration in state 4 and a dissipation of $\Delta\Psi$. The protonophoric activity of DNOC was evidenced by the induction of mitochondrial swelling in hyposmotic K^+-acetate medium, in the presence of valinomycin. At higher concentrations (≥ 50 μM), DNOC induced an inhibition of succinate-supported respiration, and a decrease in the activity of the succinate-dehydrogenase. Treatment of Ca^{2+}-loaded mitochondria with uncoupling concentrations of DNOC resulted in mitochondrial permeability transition (MPT), associated with membrane protein thiol oxidation by ROS, as evidenced by mitochondrial swelling in isosmotic sucrose medium (Castilho et al., 1997). Similarly to rat liver mitochondria, DNOC acts as a classical uncoupler of oxidative phosphorylation in potato tuber mitochondria at low concentrations (≤ 100), and as an inhibitor of mitochondrial respiration at high concentrations (≥ 100 μM) (Vicente et al., 1998). This mode of action is in agreement with a type of herbicides named inhibitory uncouplers, according to a general classification (Moreland, 1980; 1993). Thus, low concentrations (up to about 100 μM) only induce uncoupling effect and higher concentrations also inhibit the mitochondrial electron transport chain.
In conclusion, DNOC affected similarly plant and animal mitochondria, with a higher sensitivity for rat liver mitochondria, requiring lower concentrations to produce the same results.

2. Dinoseb; 2,4-D; 2,4,5-T; 2,4,5-TP and MCPA

Studies on phenoxyacetic acid herbicides are here considered and compared with the phenol herbicide, dinoseb. The herbicide dinoseb (2-*sec*-butyl-4,6-dinitrophenol) partially inhibited both FCCP (carbonylcyanide p-trifluoromethoxyphenylhydrazone)-uncoupled respiration and state 3 respiration, indicating its limited interaction with the mitochondrial redox chain at the level of the complexes II and III, as reflected by the low inhibition induced by 500 nM dinoseb on succinate dehydrogenase (10%) and cytochrome *c* reductase (20%), respectively. Additionally, it increased the rate of state 4 oxygen consumption, stimulated ATPase activity, induced mitochondrial membrane permeabilization to protons (H^+), and depressed $\Delta\Psi$. Thus, dinoseb has a limited effect on the activities of the redox chain complexes at the same time it acts as an uncoupler of oxidative phosphorylation in rat liver mitochondria (Palmeira et al., 1994a). In fact, dinitrophenols, e.g., dinoseb, were previously described as uncouplers of oxidative phosphorylation in mung bean mitochondria (Moreland & Novitzky, 1988). Differently to dinoseb, the herbicide 2,4-D (2,4-dichlorophenoxyacetic acid) has a more pronounced inhibitory action on the redox chain and acts as uncoupler only at relatively high concentrations. Thus, 2,4-D, in the concentration range of up to 800 μM, decreased ΔΨ as a function of concentration. State 3 and FCCP-uncoupled respiration were depressed by approximately the same extent (60% at 700 μM), ruling out direct interactions on phosphorylation assembly independent of the redox chain. In fact, the herbicide, at 600 μM, strongly inhibited succinate dehydrogenase (50%) and cytochrome *c* reductase (75%). Therefore, 2,4-D specifically affects the redox chain at the mitochondrial complexes II and III, as opposed to a limited effect of dinoseb whose action is essentially related with uncoupling. Additionally, 2,4-D also uncoupled oxidative phosphorylation in rat liver mitochondria at concentrations 1000-fold higher than those required for a similar dinoseb effect (≥150 μM) (Palmeira et al., 1994a). Just as 2,4-D, 2,4,5-T (2,4,5-trichlorophenoxyacetic acid), MCPA (4-chloro-2-methylphenoxyacetic acid) and 2,4,5-TP [2-(2,4,5-trichlorophenoxy)propionic acid], at high concentrations, alter energy metabolism in rat liver mitochondria by uncoupling oxidative phosphorylation. The herbicide 2,4,5-TP possesses the strongest uncoupling properties followed by 2,4,5-T, MCPA and 2,4-D (Zychlinski & Zolnierowicz, 1990). *In vivo* treatment of rats with 2,4,5-T and 2,4-D also causes damage to mitochondrial and cellular membranes, leading to cellular toxicity. This implies that, *in vivo*, these herbicides also have the potential to inhibit the oxidative phosphorylation of mitochondria, affecting energy production in cells (Sulik et al., 1998). Similarly to rat liver mitochondria, in potato tuber mitochondria 2,4-D has an uncoupling effect which causes the complete loss of phosphorylation by permeabilizing the inner mitochondrial membrane to H^+. In plant mitochondria, simultaneously with uncoupling, 2,4-D also inhibited the electron transport chain by blocking the activity of dehydrogenases (succinate dehydrogenase, malate dehydrogenase and NAD-dependent malic enzyme) linked to the respiratory chain (Pireaux et al., 1992). As with DNOC, inhibitory action in plant mitochondria with 2,4-D also required higher concentrations (>3mM) to obtain the same range of inhibition, as compared with animal mitochondria.

3. Dicamba

Dicamba (2-methoxy-3,6-dichlorobenzoic acid) is included in the group of benzoic acid herbicides (classe of carboxylic acids). In the concentration range of 4-30 μmol/mg protein (2-15 mM), promoted an increase in state 4 succinate-supported respiration in rat liver mitochondria. Nevertheless, at the concentration of 50 μmol/mg protein (25 mM), it strongly inhibitied succinate-supported mitochondrial respiration (state 4). These results indicate not only an uncoupling effect, as a consequence of an increase on the permeability of inner mitochondrial membranes to protons, but also a strong inhibitory effect on the redox complexes of the mitochondrial respiratory chain. The protonophoric activity of dicamba was evidenced by the induction of mitochondrial swelling in hyposmotic K^+-acetate medium, in the presence of valinomycin. The inhibitory effect of dicamba on the respiratory complexes was evidenced by: 1) the inhibition, at the same extent, of state 3 respiration and FCCP uncoupled respiration with parallel dissipation of the ΔΨ, and 2) the strong inhibition of the mitochondrial complexes II and III, as reflected from the decrease in the enzymatic activity of the succinate dehydrogenase and cytochrome c reductase. Dicamba, in the concentration range of 4-30 μmol/mg protein, also inhibited the the activities of the ATPase and ATP synthase. From these results it was concluded that dicamba decreases oxidative phosphorylation by a dual effect on the redox chain (inhibition of redox complexes, and stimulation of proton leakage through the mitochondrial inner membrane) (Peixoto et al., 2003a). Similarly to rat liver mitochondria, dicamba interfered with potato tuber mitochondria by inhibiting the activities of the respiratory complexes II and III, of ATP synthase, and also stimulating the proton leakage through the mitochondrial inner membrane (Peixoto et al., 2003b). In turn, comparative effects of dicamba and a related compound, 2-chlorobenzoic acid, on potato tuber mitochondria showed that dicamba is a stronger mitochondrial respiratory chain inhibitor and uncoupler. So, considering the results, it was suggested that differences in the lipophilicity due to their chemical structures are related to the different sensitivities of mitochondrial bioenergetics to the referred compounds (Peixoto et al., 2003b).

4. Metolachlor and Alachlor

We also studied chloroacetanilide herbicides included in the larger group of amides for their effects on mitochondria. Metolachlor [2-chloro-N-(2-ethyl-6-methyl-phenyl)-N-(2-methoxy-1-methylethyl) acetamida], in the concentration range of 400-1000 nmol/mg protein (0.4-1.0 mM), interferes with rat liver mitochondrial bioenergetics, inhibiting state 3 and FCCP-uncoupled respirations supported by either malate/glutamate or succinate as the respiratory substrates. These results demonstrated that both complex I and complex II are sensitive to metolachlor. Accordingly, metolachlor-induced $\Delta\Psi$ dissipation and depression of the phosphorylation rates of rat liver mitochondria can be explained by the inhibitory action of the herbicide exerted in complex I- and complex II-dependent respirations. This assumption was supported by the parallel depressive effect on state 3 and FCCP-uncoupled respirations dependent on both malate/glutamate and succinate and the absence of effects when a mixture of ascorbate+TMPD was used as substrate (complex IV). The low stimulatory effect of state 4 respiration suggest a low ability of the compound to induce inner membrane permeabilization. Therefore, the toxicological effects of metolachlor on rat liver mitochondrial bioenergetics, expressed by its inhibitory action on mitochondrial respiratory chain for complexes I and II, may be relevant to understand the mechanism responsible for its toxic action (Pereira et al., 2009). Just as metolachlor, the herbicide alachlor [2-chloro-N-(2,6-diethyphenyl)-N-(methoxymethyl acetamide)], in the concentration range of 400-1200 nmol/mg protein (0.4-1.2 mM), interferes with rat liver mitochondrial bioenergetics by inhibiting mitochondrial respiratory chain activity at complexes I and II levels (Pereira et al., 2010).

5. Linuron

Linuron [N´- (3,4-dichlorophenyl)-N-metoxy-N-methylurea] is an herbicide included in the group of ureas. In the concentration range of up to 160 μM, linuron interferes with rat liver mitochondrial bioenergetics, inhibiting state 3 respiration supported by either malate/glutamate (65%) and succinate (8%) as the respiratory substrates. The FCCP-uncoupled respiration supported by malate/glutamate was inhibited by 40%. Nevertheless, succinate-supported FCCP-uncoupled respiration was not affected by the tested linuron concentrations. These results indicate that complex I of the mitochondrial respiratory chain is more sensitive to linuron than complex II. This assumption was also supported by a higher linuron-induced dissipation of $\Delta\Psi$ (11%) and depression of the phosphorylation rate (70%) of rat liver mitochondria respiring malate/glutamate as compared with succinate-dependent respiration (7% and 33%, respectively). The same concentration range of linuron progressively stimulated state 4 respiration using either malate/glutamate or succinate as respiratory substrates. However, at 120 μM linuron, a higher stimulation of state 4 supported by succinate (100%) was detected, as compared with that supported by malate/glutamate (20%) reinforcing the idea that the inhibitory effect of linuron is mainly exerted at the level of complex I. On the other hand, this result indicates that linuron has also some ability to induce inner membrane permeabilization to cations. This explains the strong inhibitory effect of succinate-supported state 3 respiration, taking into account that complex II is note very sensitive to linuron.

The toxicological effects of linuron on rat liver mitochondrial bioenergetics, expressed by its inhibitory action on mitochondrial respiratory chain for complex I and membrane permeabilization, may be relevant to understand the mechanism responsible for its toxic action (unpublished results).

6. Paraquat

Paraquat (1,1´-dimethyl-4,4´-bipyridylium dichloride), included in the group of bipyridiliums, in the concentration range of up to 10 mM, promoted an increase in succinate-supported mitochondrial respiration (state 4) on rat liver mitochondria, indicating some energy uncoupling due to some permeabilization of mitochondrial membrane to H^+. The deleterious effect of the herbicide on membrane organization was confirmed by its ability to induce lipid peroxidation (Palmeira et al., 1995a). So, this must be the main cause of permeabilization. State 3 respiration is a little depressed by about 15% at 10 mM paraquat, whereas CCCP-uncoupled respiration is depressed by about 30%. The inhibitory effect of paraquat on uncoupled respiration reflects its interaction with the mitochondrial respiratory chain. In fact, succinate cytochrome-*c* reductase and cytochrome-c oxidase are inhibited by 35 and 39% respectively, at 10 mM paraquat, whereas succinate dehydrogenase is not significantly affected, indicating that paraquat partially inhibits the redox chain at mitochondrial complexes III and IV. Furthermore, paraquat partially inhibits the ATPase activity for concentrations in the range of 1-2 mM, suggesting a direct effect on this enzyme complex. However, at higher concentrations (5-10 mM), the ATPase activity is stimulated, probably as a consequence of the described uncoupling effect. Paraquat depresses $\Delta\Psi$ of rat liver mitochondria as a function of herbicide concentration. In addition, the depolarization induced by ADP is decreased and repolarization is biphasic suggesting a double effect. Repolarization resumes at a level consistently higher than the initial level before ADP addition, for concentrations up to 10 mM. This particular effect is clear at 1 mM paraquat and tends to fade out with increasing concentrations of the herbicide. Based on these results, it was concluded that paraquat uncouples oxidative phosphorylation of rat liver mitochondria by inducing lipid peroxidation and, also, by inhibiting redox chain and ATPase/synthase activity (Palmeira et al., 1995a). Comparative studies of the effect of paraquat on mitochondrial bioenergetics of rat liver and potato mitochondria confirmed the protonophoric action of herbicide on both mitochondrial membranes. However, the sensitivity of potato tuber mitochondria to paraquat was lower than that of rat liver mitochondria (Vicente et al., 2001). Differences between rat liver and potato tuber mitochondria were also observed on the effects of paraquat on both $\Delta\Psi$ and oxygen consumption of complex-I-dependent respiration. Thus, paraquat (20 mM) considerably dissipated complex-I-dependent $\Delta\Psi$ and stimulated state 4 oxygen consumption of liver mitochondria, but not potato tuber mitochondria, even at 40 mM paraquat (Vicente et al., 2001). According to these results potato tuber mitochondria, in contrast to rat liver mitochondria, are protected against paraquat radical (PQ^+) afforded by complex I redox activity (Vicente et al., 2001). Interestingly, the complex I may afford protection for the paraquat effects on plant mitochondria, explained by Nagata et al. (1987). Working with

bovine heart and yeast mitochondria, the reduction of NAD^+ to NADH can be catalyzed at the expense of the paraquat radical, avoiding its degradative effects. Studies performed to understand the different sensitivity of plant and animal mitochondria to paraquat toxicity revealed that it may also be related with antioxidant agents (Vicente et al., 2001; Peixoto et al., 2004) and different native aliphatic contents of the mitochondria membrane phospholipids (Peixoto et al., 2004). In fact, the levels of superoxide dismutase (SOD), glutathione reductase and α-tocopheral content in potato tuber mitochondria were significantly higher than in rat liver mitochondria, which, in turn, revealed higher values of lipid peroxidation and protein oxidation induced by paraquat (Vicente et al., 2001; Peixoto et al., 2004). Also, the total number of double bonds (unsaturated index) in rat liver mitochondrial membranes was higher than in potato tuber mitochondrial membranes (Peixoto et al., 2004). From these studies it was also concluded that peroxides are not major intermediates in paraquat toxicity at the mitochondrial level since the activity of the glutathione peroxidase and catalase are much higher in rat liver than in potato tuber mitochondria (Peixoto et al., 2004).
A summary of the results of the effects of some herbicides on plant mitochondria is depicted in Figure 1.

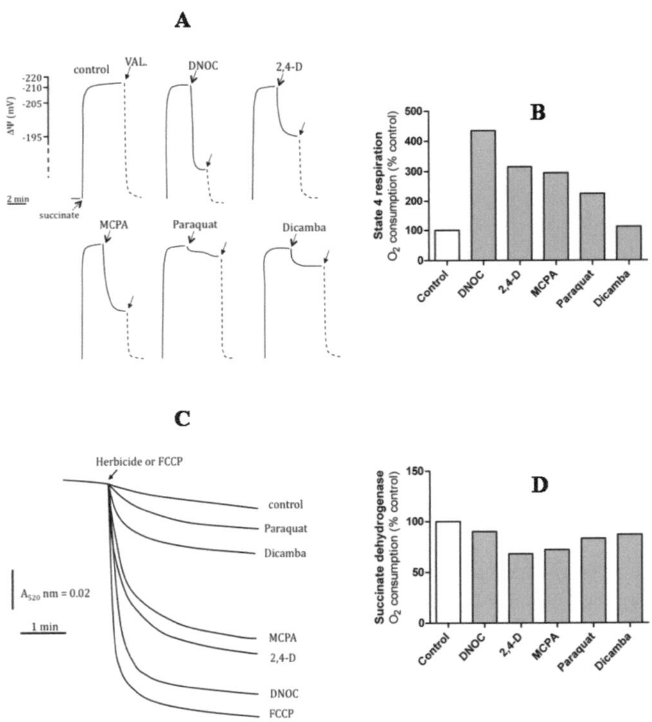

Fig. 1. Effects of DNOC (80 μM), 2,4-D (3 mM), MCPA (3 mM), paraquat (40 mM) and dicamba (2 mM) on succinate-supported transmembrane potencial (ΔΨ) (**A**), state 4 O_2 consumption (**B**), swelling in hypo-osmotic K^+-acetate medium (**C**), and succinate dehydrogenase activity (**D**) of potato tuber mitochondria. Val., valinomycin; FCCP, carbonyl cyanide *p*-trifluoromethoxyphenyl-hydrazone. The traces represent typical recordings from several experiments with different mitochondrial preparations.

Herbicide	Log Kp	LD_{50}	Chemical structure
DNOC	2.39	7.0	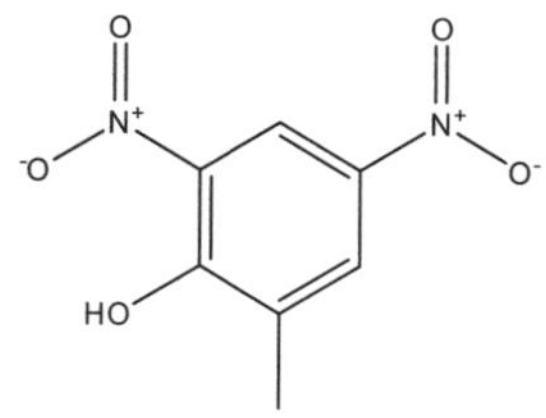
Dinoseb	2.29	25	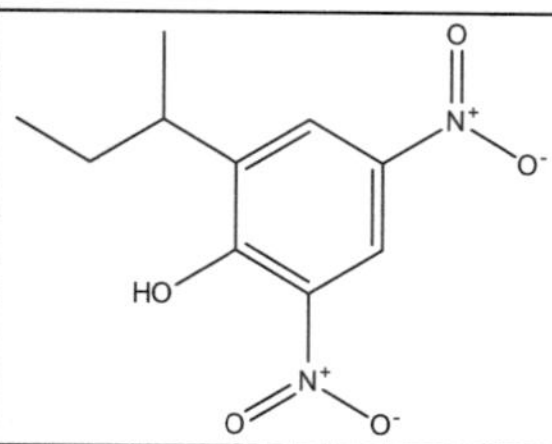
2,4-D	-0.83	469	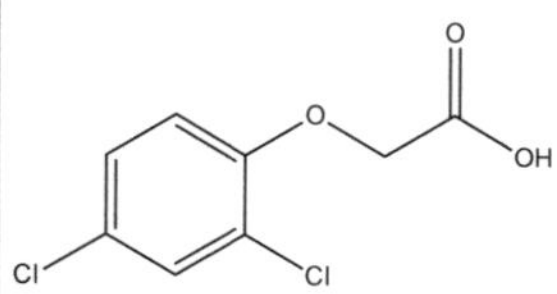
MCPA	-0.81	962	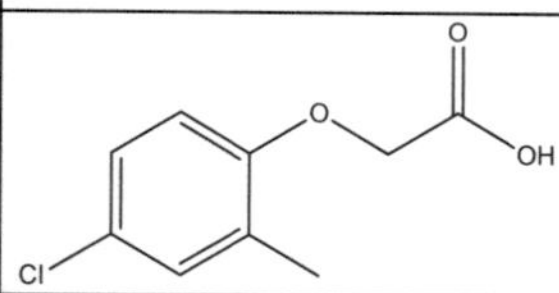
2,4,5-T	3.72	820	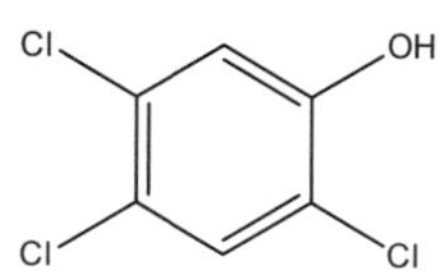
2,4,5-TP	4.0	500	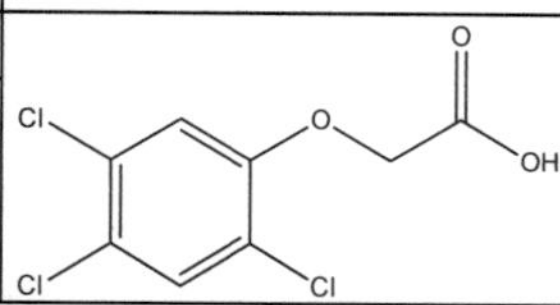
Dicamba	-1.88	1581	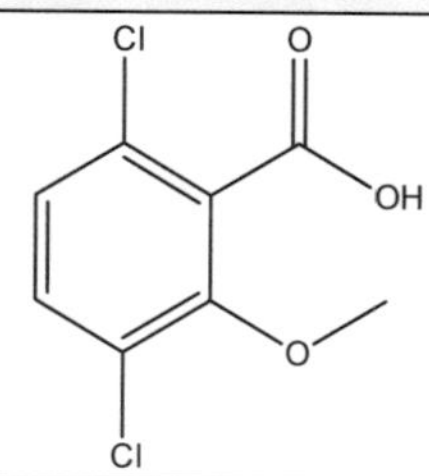
Metolachlor	3.4	1200	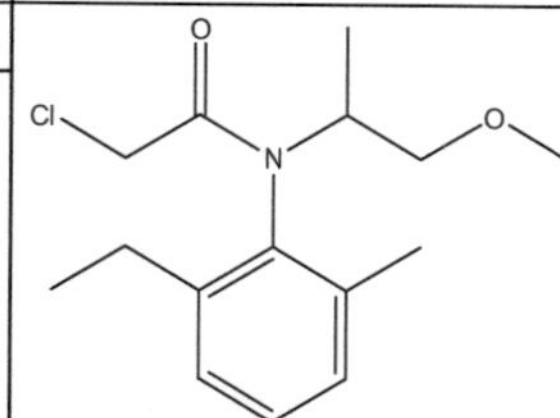
Alachlor	3.09	1200	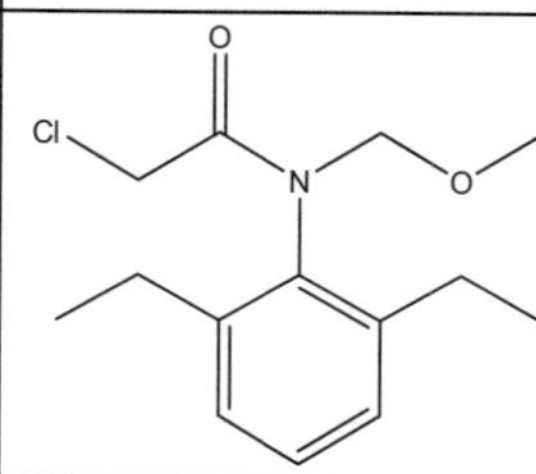
Linuron	3.0	1146	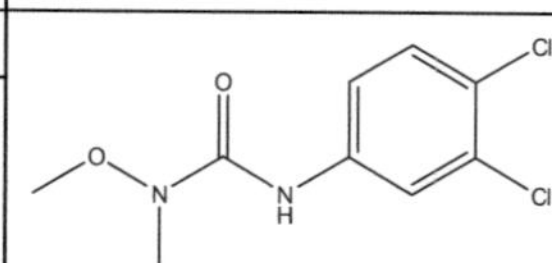
Paraquat	-4.5	110	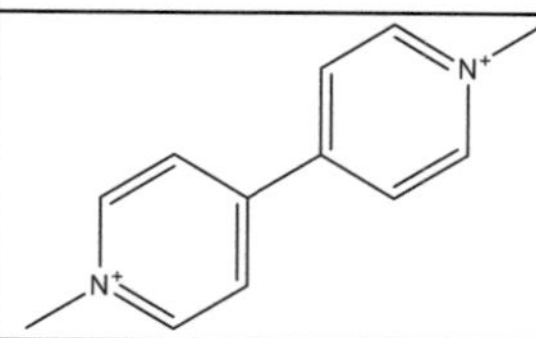
Oxyfluorfen	4.86	> 5000	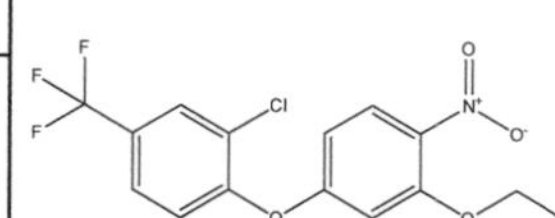

Table 4. Chemical structure, Log Kp (octanol-water partition coefficient at pH 7, 20 °C) and Rat-Acute oral **LD_{50}** (mg Kg^{-1}) values of the referred herbicides. (Data of Log Kp and LD_{50} obtained from: IUPAC global availability of information on agrochemicals, http://sitem.herts.ac.uk/aeru/iupac/index.htm, July 2010).

Comparing the results presented in Figure 1 with the Log Kp values of the herbicides studied (Table 4), efforts were made to establish a relationship between lipophylicity and their mechanisms of action.

Exception for paraquat, a good correlation between herbicide Log Kp values (Table 4) and the transmembrane potencial dissipation, the stimulation of state 4 respiration and the proton-dependent swelling induced by them can be established, hence, corroborating their effects on mitochondrial membrane permeabilization inducing uncoupling of mitochondrial phosphorylation.

These toxicological effects of the herbicides on mitochondria, putatively related to their incorporation into the mitochondrial membrane, cause an alteration on the surface charge density and a disturbance in the physicochemical and structural properties of the inner membrane, resulting in an unequivocal disturbance of electron delivery between redox complexes and, consequently, affecting the phosphorylative system. Obviously paraquat acts in a completely different way.

The mechanism of action of paraquat has been related to reactive oxygen species produced by paraquat redox cycling. Nevertheless, paraquat affects mitochondrial functions and induces a slight permeabilization of mitochondrial membrane probably due to lipid peroxidation (Peixoto et al., 2004). Mitochondrial protein thiol oxidation was also observed by paraquat treatment (Peixoto et al., 2004). Respiratory complexes from electron respiratory chain can also be negatively affected by all these herbicides as revealed from succinate dehydrogenase activity assays (Figure 1D). In this case, correlation with lipophylicity is not so obvious.

II. Effects on Prokaryotic and Eukaryotic cell cultures

In this section, studies of the effects of herbicides on bacteria and plant cell cultures will be described. If the use of plant cell cultures is easily accepted as an important means to estimate the potential toxicity of those compounds in target and non-target organisms, the interest in performing herbicide studies by using bacterial cells as toxicological model systems is apparently less obvious. However, considering the crucial role microorganisms play in soil as well as in aquatic environments, a significant ecological damage is predictable if these microbial communities are affected by the accumulation of herbicides in soil, upon their application in agriculture fields, or in aquatic systems, as a consequence of herbicide runoff. In fact, microorganisms are of fundamental importance for the vital functioning of any ecosystem. The microflora of the soil, namely in the top soil layer (0-15 cm) where an intense microbiological activity takes place, has a high responsibility for soil fertility because of its capacity to convert potential plant nutrients in active or available forms. On the other hand, microorganisms densely colonizing fresh water and marine sediments promote the transformation of organic matter and constitute food source for organisms of higher trophic levels. Therefore, the adverse effects of chemical agents such as herbicides upon microorganisms may constitute an early warning indicator of ecosystem perturbation.

In the literature, we can find a significant amount of studies showing herbicide-sensitive bacteria (Cserháti et al., 1992; Hernando et al., 2003) and micro-algae (Nystrom et al., 1999; Li et al., 2008), as well as microorganisms that not only tolerate the presence of high concentrations of herbicides but also are able to degrade them (Dejonghe et al., 2003; Sorensen et al., 2003; Neumann et al., 2004). If, on the one hand, the biodegradation constitutes an important factor favoring the removal of herbicides from the environment, on the other hand, the elimination of sensitive microorganisms and the proliferation of the most tolerant ones, due to the selection pressure exerted by those compounds, may cause changes to the existing microbiological balances with potential adverse consequences for the ecosystem. An example of differential sensitivity to herbicides has been found for sulfonylurea herbicides among fresh-water and marine micro-algae (Nystrom et al., 1999), microbial aquatic communities (Thompson et al., 1993) and soil bacteria (Forlani et al., 1995).

As previously referred, adverse effects may be caused by herbicides on beneficial microorganisms and their associated transformations in soil, affecting plant-nutrient availability. Therefore, the evaluation of herbicide toxicity exerted in soil microorganisms is recommendable to define upper limit concentrations. With the purpose of contributing to collect data concerning herbicide effects on soil microorganisms, three herbicides (the chloroacetamides metolachlor and alachlor and the arylurea linuron) were studied using *B. stearothermophilus* as a bacterial model system.

1. Metolachlor and alachlor toxic effects on bacterial cells

Metolachlor (Pereira et al., 2009) and alachlor (Pereira et al., 2010) inhibit the growth of a strain of *B. stearothermophilus* in the concentration range of 100 to 600 μM. The herbicides added to cultures of this Gram-positive bacterium, grown in a complex liquid growth medium (L-Broth) at 65 °C, affect the bacterial growth in the dependence of the concentration, inducing an increase of the lag time, decrease of the specific growth rate and reduction of the maximal optical density attained in the stationary phase of growth. The effective concentrations which alter these growth parameters by 50% compared to the control (EC_{50}) are around 400 to 450 μM for metolachlor and slightly higher than 500 μM for alachlor. Comparing these values of EC_{50} with those determined in identical conditions for other pesticides, such as DDT (Donato et al., 1997b), methoprene (Monteiro et al, 2005) or tributyltin (TBT) (Martins et al., 2005), it is clear that the herbicides metolachlor and alachlor have a much lower toxicity. Accordingly, the herbicide effects on the bacterial respiratory activity, evaluated in terms of the oxygen consumption rate in protoplasts, also denote a moderate toxicity. Thus, for oxygen consumption supported by NADH, metolachlor and alachlor EC_{50} of 1.2 μmol/mg protein or 1.2 mM (Pereira et al., 2009) and 2.2 μmol/mg protein or 2.2 mM (Pereira et al., 2010), respectively, are two order of magnitude higher than those found for DDT (Donato et al., 1997b) or the organotin TBT (Martins et al., 2005). The respiratory activity elicited by ascorbate/TMPD showed to be insensitive to both herbicides in the range of 0.5 to 5.0 μmol/mg protein (0.5 to 5.0 mM), suggesting that the herbicides act upstream the cytochrome c oxidase segment. These results are in accordance to those found in rat liver mitochondria for both herbicides at the same concentration range (Pereira et al., 2009; 2010).

Therefore, data obtained for the bacterial model, in terms of growth and respiratory activity, and for rat liver mitochondria, concur with each other and with other results in the literature (Stevens & Sumner, 1991) indicating that the herbicides alachlor and metolachlor exhibit a reduced toxicity in comparison with other tested pesticides or pollutants.

2. Linuron toxic effects on bacterial cells

The toxic effects of linuron on the growth and respiratory activity of *B. stearothermophilus* were also evaluated. Although concentration-dependent effects of linuron on bacterial growth follow the same trend of the tested chloroacetamides, resulting in longer lag times, lower growth rates in the exponential phase and lower bacterial yields, as compared to the control, they revealed to be much more severe. Thus, the inhibition of bacterial growth by linuron occurred at concentrations one order of magnitude lower than those required by the other tested herbicides to promote the same effect. An identical behavior was observed when herbicide effects on the respiratory activity of protoplasts, elicited by NADH, were monitored. The effective concentration of linuron of 98 nmol/mg protein or 98 μM (unpublished data) to reduce for a half the protoplast oxygen consumption rate relatively to the control (protoplast preparation without herbicide, containing a correspondent amount of DMSO, *i.e.* the herbicide vehicle solvent) is also one order of magnitude lower than the equivalent concentrations of alachlor and metolachlor (see above) and one order of magnitude higher than the concentration of TBT to produce the same effect (Martins et al., 2005). These results are also in accordance with those obtained using the mitochondrial model system (see above).

These findings confirm previous data (Donato et al., 1997b; Monteiro et al., 2005; 2008) showing a good parallelism between the results obtained with the bacterial model and rat liver mitochondria, suggesting a similar sensitivity of both models for chemical toxicity assessment.

3. Paraquat effects on plant cell cultures

Paraquat is a non-selective contact herbicide of the bipyridilium class, whose incorrect use may cause pulmonary injury in humans (Boelsterli, 2003). This organ-selective toxicity results from parquat structural similarity with diamines, such as putrescine, which are actively transported by type I and type II alveolar cells. The toxicodynamics of this compound is related with its high redox-cycling activity (Boelsterli, 2003). In plants, the mechanism of action of this herbicide consists of affecting photosynthesis, being reduced by light reaction I (Cremlyn, 1991). Since this herbicide is absorbed from the foliage, but not from the roots, it is predictable those effects do not occur on non-green cells.

In order to appraise paraquat toxicity in non-target plants, studies were performed using cell cultures from non-green potato (*S. tuberous* L.) tuber calli (Peixoto et al., 2008). Several endpoints were used to evaluate the cytotoxic effects of this herbicide: cell growth, total synthesised protein; cellular integrity; adenine nucleotide content and the activity of several antioxidant enzymes. Paraquat concentration of 50 µM promoted a complete inhibition of the growth of callus tissue cells. Studies of fluorescein release showed that paraquat concentrations which inhibited cellular growth by 50% (5 µM) did not affect the integrity of the cell membrane. Accordingly, the protein/cellular weight ratio did not significantly alter upon incubation of plant cells with this concentration of paraquat (5 µM). This finding is in agreement with studies performed in potato tuber-isolated mitochondria (Vicente et al., 2001) showing that concentrations of paraquat up to 30 mM do not induce swelling depending on permeabilization to H^+.

Since previous studies using plant and animal mitochondrial fractions had shown paraquat exerts deleterious effects on animal mitochondria bioenergetics due to oxidative stress and also disturbs, although less severely, plant mitochondria functioning (Vicente et al., 2001; Peixoto et al., 2004), the adenosine nucleotides in callus tissue cultures exposed to this herbicide were quantified in order to obtain information regarding paraquat effects on oxidative phosphorylation, glycolisis and pentose pathways (Peixoto et al., 2008). Paraquat induced ATP and ADP depletion, together with an increase of AMP, and strongly increased the NAD^+/NADH ratio, revealing decrease of the cellular redox state. The depletion of intracellular ATP and NADH together with the increase of the levels of ADP, AMP and NAD^+, due to paraquat effects, was also reported in isolated rat-hepatocytes (Palmeira et al., 1994b; Palmeira, 1999). The ATP depletion has been related to the interference with the mitochondrial or glycolytic energy pathways in isolated rat hepatocytes (Palmeira, 1999) as well as in non-green callus tissue (Peixoto et al., 2008). On the other hand, the redox cycling of paraquat may justify the depletion of cellular reducing equivalents, in both biological systems. Both events, depletion of ATP and decrease of the cellular redox state, may lead to the impairment of cellular functions, hence promoting inhibition of potato tuber calli cell growth (Peixoto et al., 2008) and decrease of hepatocyte viability (Palmeira et al., 1995b). It is also suggested that lipid peroxidation induced by superoxide anions generated by paraquat redox cycling, plays an important role in the toxicity mechanism of this herbicide (Krall et al., 1988; Peter et al., 1992). In isolated rat hepatocytes it was demonstrated that paraquat promoted an increase in thiobarbituric acid (TBA)-reactive species as a function of concentration and incubation time (Palmeira et al., 1995b). In parallel, it was shown that at the same concentration range paraquat induced loss of intracellular glutathione (GSH) with a concomitant increase of the oxidised form (GSSG), decrease of protein thiols and cell death (Palmeira et al., 1995b). These events have been suggested as being interrelated.

Antioxidant enzymes can be up-regulated during oxidative stress (adaptive response induced by H_2O_2 as a second messenger) and their activities have been proposed as being involved in mechanisms of herbicide toxicity or resistance (Allen et al., 1997).

In order to clarify the level of protection plant cells exhibit against paraquat-induced oxidative stress, the activity of antioxidant enzymes was determined in crude extracts of potato callus tissue previously exposed to different concentrations of the herbicide (Peixoto et al., 2008). It was observed that paraquat induced a high stimulation of superoxide dismutase (SOD) and glutathione reductase (GR) but did not affect glutathione transferase (GST). Catalase (CAT) was stimulated only at low paraquat concentrations (1μM) and was inhibited at higher concentrations of the herbicide. The high increase of hydrogen peroxide resulting from the dismutation of superoxide anions generated by paraquat redox cycling, which is favoured by the high herbicide-induced stimulation of SOD, likely justifies why CAT is inhibited above 1μM paraquat, since potato CAT is very sensitive to H_2O_2 (Beaumont et al., 1990). These data agree with those reported for tobacco plants, which also showed an increase of SOD activity and CAT inhibition (Furusawa et al., 1984).

GR activity is responsible for the reduction of GSSG back to GSH. Since paraquat promotes a decrease of GSH in potato calli cells (Peixoto et al., 2008), its stimulatory effect on GR activity may result from a cellular response to GSH level decrease. Finally, the GST which has been shown to promote detoxification of several herbicides, as dicamba and 2,4-D (Hatton et al., 1998), by conjugation of the electrophilic herbicides with GSH, seems to have no effect on paraquat, since its activity was not stimulated by the latter herbicide in contrast with the formers.

In conclusion, studies of paraquat toxicity in plant cell cultures concur with those reported in animal cells, showing that the most probable cause of cell death in non-target organisms involves decrease of the cell energy charge, decrease of GSH content and depletion of cellular reducing equivalents needed to assure the normal cell functioning, including energy metabolism and antioxidant defence mechanisms.

4. Dicamba, 2,4-D and MCPA effects on plant cell cultures

The cytotoxic effects of the herbicides dicamba, 2,4-D and MCPA on cell cultures of potato tuber calli (*S. tuberousum* L.) were evaluated upon exposition to the herbicides for a period of 4 weeks, by monitoring the changes in cell growth, membrane integrity, energy charge, cell redox state and activity of several antioxidant enzymes, in studies parallel to those previously referred for paraquat (Peixoto et al., 2008; 2009a).

Regarding the cell growth, the concentrations at which herbicides inhibited growth by 50% (IC_{50}) were 59, 50 and 20 μM for MCPA, dicamba and 2,4-D, respectively. Studies of fluorescein release to monitor membrane integrity showed that, in contrast to paraquat (see above), the three herbicides tested affected membrane integrity at concentrations at which growth was inhibited by 50%. Unexpectedly, MCPA at the concentration of 120 μM, at which growth was completely inhibited, induced a decrease of fluorescein absorbance at 458 nm, apparently reflecting a decrease of fluorescein release (Peixoto et al., 2009). These results were clarified by using fluorescence microscopy. Cells were loaded with fluorescein diacetate, a non-polar and non-fluorescent molecule, which easily permeates the plasma membrane of intact cells (Guilbaut & Kramer, 1964). Into the cell, the esterified molecules undergo hydrolysis by intracellular esterases, generating fluorescein, which fluoresces by exposure to short wavelength light (Prosperi, 1990). Therefore, viable cells containing fluorescein can be observed by fluorescence microscopy (Peixoto et al., 2009a). When the membrane is disrupted, fluorescein is released occurring an increase of extracellular fluorescence, detected by microscopy or by the spectrometric method. In the case of cells incubated with 120 μM MCPA, fluorescence microscopy showed a strong decrease of fluorescence (Peixoto et al., 2009a). This unexpected result is due to intracellular esterase's inhibition, which in turn avoids fluorescein production. This study served to show that the fluorescein assay, as a method to monitor membrane integrity, should be carefully interpreted. The effects of the three herbicides on membrane integrity are consistent with their capacity to induce H^+ permeabilization and $\Delta\Psi$ dissipation in isolated mitochondria from plant as well as animal origin (Zychlinski & Zolnierowicz, 1990; Peixoto et al., 2004).

The protein/cellular weight ratio showed an increase by action of the three herbicides tested (Peixoto et al., 2008; 2009a). This parameter should reflect herbicide effects on cell intactness. This was supported by concentrations of paraquat that did not affect membrane integrity and, concomitantly, did not induce significant alteration of protein/cellular weight ratio (Peixoto et al., 2008). In the case of MCPA, the increase in the protein/cellular weight ratio is more likely to be a consequence of cellular water and electrolyte loss with a decrease of cell volume and weight, rather than the increase of protein synthesis. This conclusion is supported by the observation of increased number of cells per gram of callus and cells becoming harder and dark-brownish with increasing herbicide concentrations (Peixoto et al., 2009a).

The effect of the three herbicides on the intracellular content of adenosine nucleotides, reflected by a decrease of ATP content with a concomitant increase of the levels of ADP and AMP, indicates an inhibitory action in cell metabolism. Since non-green callus tissue was used, ATP generation depends on glycolisis and oxidative phosphorylation. Considering the herbicides of the chlorophenoxyacetic acid group (2,4-D and MCPA), the MCPA seems to have been less efficient in reducing the cell energy charge (taken as [ATP]+0.5 [ADP]/[ATP]+[ADP]+[AMP]) than 2,4-D (Peixoto et al., 2008; 2009a). Dicamba is probably somewhat less efficient yet.

In summary, the inhibition of cell growth promoted by these herbicides may be correlated with the decreased availability of ATP, which is needed for ion translocation, nutrient import and metabolite export. Interestingly, the same order of efficiency (2,4-D>MCPA≈dicamba) was observed in terms of herbicide-induced growth inhibition.

Cell redox state as influenced by the herbicides tested was also evaluated in terms of $NAD^+/NADH$ ratio. All the herbicides tested increased $NAD^+/NADH$ ratio, revealing a decrease in the cellular redox state. The herbicide 2,4-D promoted a stronger effect as compared with MCPA or dicamba. Paraquat and 2,4-D showed similar efficiency in promoting the decrease of cell energy charge, but paraquat effect on the cell redox state was the highest of all the herbicides tested in potato callus cells. Consistently, paraquat revealed to be the most potent growth inhibitor, too (Peixoto et al., 2008).

As previously referred for paraquat, the activity of the antioxidant enzymes CAT, GST and GR was evaluated upon incubation of the potato callus tissue with different concentrations of the herbicides 2,4-D, MCPA and dicamba (Peixoto et al., 2008; 2009a). Considering the two herbicides of the phenoxyacetic acid group, it was observed that 2,4-D (Peixoto et al., 2008) strongly stimulated the catalase (150%) and superoxide dismutase (about 50%) activities at the highest concentration assayed (50 μM), whereas MCPA, up to 120 μM, did not induce any significant change (Peixoto et al., 2009a). Apparently, MCPA does not stimulate production of the superoxide anion and hydrogen peroxide, in contrast with 2,4-D and paraquat (Peixoto et al., 2008). Dicamba promoted an increase of CAT and SOD activities, although at a smaller extent than 2,4-D (Peixoto et al., 2008).

Concerning the GST activity, an enzyme involved in the detoxification of a great deal of pollutants, it was stimulated by both phenoxyacetic acid herbicides (2,4-D and MCPA) and by dicamba, in contrast with paraquat, as previously referred. As GST uses GSH for conjugation with electrophilic compounds, it would be expected GR activity to be also stimulated by the herbicides. In fact, it was observed a stimulation of GR activity in callus tissue cultures incubated with all the herbicides tested, paraquat included (Peixoto et al., 2008; 2009a). Although the experimental approach used in these studies do not enable authors to conclude that GST is involved in 2,4-D, MCPA and dicamba bioinactivation in potato callus tissue, their results concur with data in the literature regarding different herbicides and several plant species for which GST activity plays an important role in detoxification (Hatton et al., 1998). In the case of paraquat, GR activity stimulation was probably induced as a cellular response to an intracellular decrease of GSH used to detoxify the reactive oxygen species generated as a consequence of the herbicide redox cycling (see above). Additionally, GSH oxidation could be also related to other cytotoxic events such as lipid peroxidation and decrease in protein thiols, which have been demonstrated in isolated rat hepatocytes treated with 2,4-D and paraquat (Palmeira et al., 1995b).

In conclusion, data obtained in plant and animal cell cultures concur to suggest similar mechanisms of toxicity by 2,4-D and paraquat in non-target organisms, although with different relative potencies. The decrease of GSH/GSSG ratio is probably the primary event in the cytotoxic process leading to death. In the case of MCPA and 2,4-D, the lack of studies in animal cells precludes any comparison. However, it is probable that MCPA cytoxicity results primarily from disturbance of membrane physicochemical properties leading to loss of electrolytes and metabolites, decrease in ATP availability and decrease of the cellular redox state. Dicamba biological activity shows some similarities with both 2,4-D and MCPA. It disrupted the membrane integrity as 2,4-D and MCPA but induced an increase of catalase and SOD activities, in contrast with MCPA, although in a lower extent than 2,4-D. At the mitochondrial level, dicamba uncoupled oxidative phosphorilation by inducing H^+ permeabilization, as MCPA, but also inhibited the electron transfer in the mitochondrial redox system, as 2,4-D (see above).

Since all the herbicides tested in potato callus tissue promote significant alterations in plant as well as in animal mitochondria functioning, the bioenergetic impairment may be suggested as one of the mechanisms underlying the cytotoxicity of those compounds in non-target organisms.

III. Effects of paraquat and oxifluorfen on tilapia fish (*Oreochromis niloticus*)

Hundreds of herbicides of different chemical structures, extensively used to control a wide variety of agricultural pests, contaminate aquatic habitats due to leaching and runoff water from treated areas. Fish are among the non-target organisms that can be seriously affected by the ecological imbalance imposed through this excessive charge of chemicals. In freshwater, the presence or absence of fish could be used as a bioindicator of the degree of water pollution. Considering that fisheries are an important source of food provided with protein and lipid of high nutritional value, representing an economic benefit for many countries, herbicide contamination of surface waters derived from agricultural practices is a problem of worldwide importance. On the other hand, fish constitute the vertebrate model most used in the field of ecotoxicology.

Mortality data, although being the definitive demonstration of toxicity for the organism in study, are often left out of ecotoxicological studies as being unable to provide reliable information about the environmental hazard resulting from pollutant contamination. Therefore, rather than mortality, physiological parameters (Tortorelli et al., 1990) and biochemical alterations (Stephensen et al., 2002) have been extensively used to perform an integrated evaluation of the risk pollutants may represent for the wildlife.

Toxicity exerted by many herbicides in aquatic organisms has been associated to an increased generation of ROS (Di Giulio et al., 1989; Livingstone et al., 1990). These toxic effects can be promoted via redox cycling, paraquat being the paradigmatic example, or result from herbicide metabolism by phase I, with production of ROS as by-products. Under physiological conditions, antioxidant defenses are available in fish, like in mammals, to scavenge ROS or to prevent its production (Figueiredo-Fernandes et al., 2006; Peixoto et al., 2006). In the presence of pollutants promoting oxidative stress, organisms can adapt by up-regulation of antioxidant defenses such as GSH-related enzymes, which constitute non-specific biomarkers of exposure to pro-oxidative xenobiotics. Thus, alterations in the activity of antioxidant enzymes have been proposed as early indicators of exposure to pollutants, including herbicides (Di Giulio et al., 1989).

Several field studies for biomonitoring environmental pollutants have shown that alterations of GSH-dependent/producing enzymes as well as GSH content may be useful tools to evaluate the ecotoxicological effects exerted by herbicides in fish. Stephensen et al. (2002) showed that the levels of GSH and GSH-related enzymes in the liver of rainbow trout (*Oncorhynchus mykiss*) constitute suitable biomarkers for oxidative stress in fish developed in laboratory. In this context, tilapia (*Oreochromis niloticus*) has been used as a biological model for toxicological assessment, showing several advantages: high growth rates, easy reproduction, effciency in adapting to diverse diets, great resistance to diseases and handling practices, as well as good tolerance to a wide variety of husbandry conditions (Fontaínhas-Fernandes, 1998). On the other hand, tilapia constitutes an indicator species in biomonitoring programs, since an increased level of its biotransformation enzymes may reflect the existence of environmental pollutants.

Figueiredo-Fernandes et al. (2006) carried out studies of the effects of paraquat on the hepatic levels of antioxidant enzymes in tilapia. Studies regarding the toxic action of xenobiotics in animals are often performed in liver, since this organ is a main target of chemical injury. This fact results from its role in xenobiotic biotransformation and its unique position within the circulatory system. Therefore, biochemical and histological alterations in fish hepatocytes have been extensively reported as important biomarkers of exposure to toxic compounds, in laboratory as well as in field studies.

The effects of paraquat on oxidative stress enzymes of tilapia hepatocytes were studied in the dependence of temperature and gender (Figueiredo-Fernandes et al., 2006). Males and females of tilapia were kept in tanks at 17 and 27 °C (breeding non-compatible vs breeding compatible temperatures). A qualitative analysis showed that fish mortality induced by paraquat was accompanied by the appearance of extensive areas of hepatocytic necrosis. Preliminary experiments were carried out to select test concentrations of paraquat corresponding to 25% of the LC50, at which hepatocytes showed no signals of necrosis. SOD and GST activities showed sex-dependence, since males present higher values than females at both temperatures assayed. No significant alterations were found in the dependence of temperature. SOD, GR and GST activities were increased by exposition to paraquat. An increase of GR activity induced by paraquat was also demonstrated (Stephensen et al., 2002) in rainbow trout. The authors justify this fact due to the ability of GR to reduce paraquat to its cation radical, causing a chain reaction with paraquat radical promoting oxidation of GSH, which in turn induced an increased activity of GR.

We have also observed alterations of the condition factor (CF) and the hepatosomatic index (HSI) of tilapia exposed to paraquat (Figueiredo-Fernandes et al., 2006). These parameters serve as indicators of the general organism well being. A high condition factor corresponding to a high body weight/length relationship generally means good environmental conditions. Also a high HSI, defined as the ratio of liver weight to body weight, normally reflects a high reserve of energy. However, an increase of both parameters, CF and HSI, was promoted by paraquat in tilapia. An increase of HSI was also registered in different laboratories upon exposition of fish to different pollutants. This alteration was interpreted (Stephensen et al., 2002) as being due to an increased activity of xenobiotic biotransformation enzymes.

Other studies were conducted in tilapia exposed to the herbicide oxifluorfen (Peixoto et al., 2006). The exposure of tilapia for a period of 14 to 21 days to sub-lethal concentrations (0.3 to 0.6 mg/L) of oxyfluorfen [2-chloro-1-(3-ethoxy-4-nitrophenoxy)-4(trifluoromethyl) benzene], a diphenyl ether herbicide commonly used in agriculture to control broadleaf and grassy weeds with specific recommendations (EPA 1992), resulted in a progressive increase of the activity of catalase and, to a lower extent, of glutathione reductase. These results are consistent with those reported in the literature concerning different aquatic organisms in the presence of pollutants such as polycyclic aromatic hydrocarbons (PAHs) and polychlorinated biphenyls (PCBs) (Otto & Moon, 1995). Regarding glutathione S-transferase (GST), an enzyme engaged in the detoxification of many xenobiotics, by conjugation of their phase I metabolites, it was observed an increase of its activity induced by 7 days exposition to oxyfluorfen, but thereafter the activity decreased to levels lower than the control. Data in literature are also variable, reporting an increase (Otto & Moon, 1995) or a decrease (Pedrajas et al., 1995) of GST activity in fishes exposed to PAHs and PCBs. In contrast to the other enzymes, SOD activity showed a progressive decrease along the 21 days of exposition to the herbicide. These data are supported by other studies using rainbow trout injected with paraquat (Pedrajas et al., 1995) but are opposite to results obtained with other species exposed to other pollutants. In conclusion, fish antioxidant enzymes have a limited potential to be used as pollution biomarkers, since very different responses can be observed depending both on the xenobiotic and on the organism under study.

It was also observed that tilapia exposure to oxyfluorfen promoted changes in the hepatic fatty acid profile, with special incidence on the unsaturated acids (Peixoto et al., 2006), decreasing the oleic acid content and increasing the nervonic acid (24:1). It was shown that oxyfluorfen also induced alterations of the fatty acid composition in plants (Watanabe et al., 2001), promoting an increase of the monounsaturated chains (C16:1 and C18:1). Disturbance of the fatty acid biosynthesis regulatory network by a potential oxyfluorfen-induced oxidative stress may explain those alterations.

Oxyfluorfen (0.3 and 0.6 mg/L) was also evaluated on the mitochondrial hepatopancreas bioenergetics of tilapia (Peixoto et al., 2009b). Although no significant change was registered in respiration supported by piruvate/malate (complex I) or succinate (complex II) as detected by oxygen consumption, the mitochondrial $\Delta\Psi$ was dissipated in some extent. Furthermore, the phosphorylation rate decreased, indicating a perturbation of the phosphorylative mitochondrial system, potentially affecting the ATP availability. The controversial results in herbicide-induced oxygen consumption (no effect), $\Delta\Psi$ dissipation and decrease of the phosphorylation rate can putatively be explained as a limited direct effect on the respiratory complexes, referred in literature as "Biochemical Threshold Effect" (Rossignol et al., 2003). The big size of the fishes used in the study could also explain small effects on the mitochondrial respiration, since bigger and fatter fishes could present higher resistance to xenobiotic, due to its accumulation in fat tissue, particularly when the xenobiotic is lipophilic.

In conclusion, oxyfluorfen induces biochemical changes in fish hepatopancreas regarding the antioxidant enzymatic system, fatty acid metabolism and mitochondrial bioenergetics. However, the relationships between those toxicity parameters have not been completely clarified, yet.

4. Concluding remarks

A great deal of work has been done to assess the environmental risks and health hazards resulting from the extensive and intensive use of herbicides. As other pesticides, the herbicides are common contaminants of surface water and soils. When applied to the agricultural areas, they may have different fates: microbial or non-biological degradation, plant uptake or adsorption and transport by surface water far from the site of application. Therefore, organisms in soil and water niches, from microorganisms to plants and animals, may be exposed to a large number of these pollutants and their metabolites. Additionally, the persistence of these compounds in the environment, associated with their lipophilicity, renders them amenable to progressive accumulation in biological tissues (bioaccumulation) and, consequently, prone to be concentrated along the food chains, increasing the concentration towards the top of the chain (biomagnification). On the other hand, water contamination by agricultural run-off and aerial spraying of herbicides may also have dangerous repercussions on public health even in areas far from those primarily affected. Considering all these risks, including those resulting from a daily consume of eventually contaminated food, it is obvious the general public concern with an unrestricted use of herbicides. The acute and chronic toxic effects of herbicides on useful animals and humans have implemented the research focused on the herbicide mechanisms of action.

A great body of evidence has shown that many environmental contaminants affect transversely the most of the organisms of an ecosystem including man, affecting basic cell functioning and structural properties common to all living cells. In the case of lipophilic compounds, where a great number of herbicides are included, membrane structure and function constitute the first potential target for the toxic effects of the incoming compound, predictably reflected by perturbations in the bioenergetic process, since this physiological event is on the strict dependence of membrane intactness and stability. On the other hand, this process is a basic mechanism required to drive the activity of all living systems. In prokaryotic cells energy transduction, as well as most vital cell functions, is ascribed to the cytoplasmic membrane. Not surprisingly, perturbations of membrane structure and function by toxic compounds will end in severe perturbation and inhibition of cell growth and viability. On the other hand, mitochondria have a crucial role as energy suppliers in eukaryotic cells, being their functioning strictly dependent on membrane intactness and the proper activity of membrane protein complexes. Thus, these organelles are common targets for the toxic action of lipophilic compounds that, upon incorporation into mitochondrial membranes, can act as uncoupling agents, electron-transport inhibitors or energy-transfer inhibitors.

Huge efforts have been done by our research group to develop predictive tests of the ecological hazards resulting from the exposure to pesticides, assessing in laboratory their toxic effects on model biological systems provided by prokaryotic, animal and plant cells. On the other hand, aiming at elucidating the mechanisms underlying the toxic action of pesticides at a cellular or subcellular level, different experimental approaches have been improved. Three kinds of studies have been implemented in our laboratory to address xenobiotic cytotoxicity *in vitro*: 1) a biophysical approach to lipophilic compound/membrane interactions using membrane lipid models, 2) the use of prokaryotic/eukaryotic cells and 3) the use of mitochondria from plant/animal origin, to evaluate and identify the mechanisms underlying the toxic action of xenobiotics *in vitro*.

A great deal of work has been carried out in our laboratory showing that pesticide membrane incorporation leads to alterations of the structural and physical properties of membrane lipid bilayer, induction of lateral phase separation and generation of lipid microdomains, with predictable impact on protein distribution and sorting in different regions of the membrane (Donato et al., 2000; Videira et al., 2002). Consequent repercussions in membrane permeability and the activity of membrane-associated proteins have been also reported (Videira et al., 2001). One of the biochemical mechanisms that have shown a strong correlation with xenobiotic-induced disturbance of membrane lipid structure and organization is the oxido-reductive systems of bacteria and mitochondria of eukaryotic cells, with obvious impact in all cellular energetic processes (Donato et al., 1997b; Videira et al., 2001; Monteiro et al., 2005 and 2008; Pereira et al., 2009).

A remarkable parallelism has been established between pesticide effects on prokaryotic models (namely the micro-organisms *Bacillus stearothermophilus* and *B. subtilis*) and the other model systems. Thus, partitioning and physical effects have been estimated with bacterial membrane lipids similar to those described for synthetic lipid membrane models and native eukaryotic membranes (Donato et al., 1997c; Donato et al., 2000; Martins et al., 2003; 2005); the impairment of the electron transfer in the bacterial respiratory system has been also demonstrated in rat liver mitochondria (Monteiro et al., 2008; Pereira et al., 2009); the inhibition of bacterial growth and loss of animal cell cultures viability have shown the same dependence on the structural characteristics of chemical related compounds, such as organotin compounds (unpublished data). Mitochondrial fractions of eukaryotic cells have also shown to be plausible alternative *in vitro* model systems for an evaluation of pesticide toxicity, which can simultaneously clarify its mechanisms of action. Data obtained from mitochondria assays have shown to be strongly correlated with cytotoxic parameters

provided by more complex biological systems as whole organisms, like fish (Peixoto et al., 2006). These ones have been widely used as models to evaluate the health of aquatic systems and in studies of xenobiotic-induced pathological effects on vertebrates.
The good correlations established with data provided by these different model systems in studies of herbicide toxicity, as described in this chapter, contribute to validate the use of our models in predictive tests of environmental hazards and toxicity assessment. As also emphasized along this chapter, the most remarkable advantage in using models such as bacteria and plant cells or organelles regards the use of a biological material easily and economically obtained, minimizing or avoiding ethical issues associated with the use of mammals.

5. References

Allen, R.D.; Webb, R.P. & Schake, S.A. (1997). Use of transgenic plants to study antioxidant defenses. *Free radical Biol Med*, 23, 473-479

Babich, H.; Shopsis, C. & Borenfreud, E. (1986). In vitro cytotoxicty testing of aquatic pollutants (cadmium, copper, zinc, nickel) using established fish cell lines. *Ecotoxicol Environ Saf*, 11, 91-99

Bartels, P. G. & McCullough, C. (1972). A new inhibitor of carotenoid synthesis in higher plants: 4-chloro-5-(dimethylamino)-2-(*a,a,a*-trifluoro-*m*-tolyl-3(2H)-pyridazinone. *Biochem Biophys Res Commun*, 48, 16-22

Basova, L.V.; Kurnikov, I.V.; Wang, L.; Ritov, V.B.; Belikova, N.A.; Vlasova, I.I.; Pacheco, A.A.; Winnica, D.E.; Peterson, J.; Bayir, H.; Waldeck, D.H. & Kagan, V.E. (2007). Cardiolipin switch in mitochondria: shutting off the reduction of cytochrome *c* and turning on the peroxidase activity. *Biochemistry*, 46, 3423-3434

Beaumont, F.; Jouve, H.M.; Gagnon, J.; Gaillard, J. & Pelmont, J. (1990). Purification and properties of catalase from potato tubers (*Solanum tuberosum*). *Plant Sci*, 72, 19-26

Beyenbach, K.W. (2004). Kidney sans glomeruli. *Am J Physiol Renal Physiol*, 286, F811-F827

Boelsterli, U.A. (2003). *Mechanistic Toxicology: The Molecular Basis of How Chemicals Disrupt Biological Targets*, Taylor & Francis, London, New York

Camadro, J-M; Matringe, M.; Scalla, R. &Labbe, P. (1991). Kinetic studies on protoporphyrinogen oxidase inhibition by diphenyl ether herbicides. *Biochem J*, 277, 17-21

Cardoso, H. & Leão, C. (1992). Mechanisms underlying the low and high enthalpy death induced by short-chain monocarboxylic acids and ethanol in *Saccharomyces cerevisiae*. *Appl Microbiol Biotechnol*, 38, 388-392

Castilho, R.F.; Vicente, J.A.F.; Kowaltowski, A.J. & Vercesi, A.E. (1997). 4,6-dinitro-*o*-cresol uncouples oxidative phosphorylation and induces membrane permeability transition in rat liver mitochondria. *Int J Biochem Cell Biol*, 29,1005-1011

Çavas, T. & Könen, S. (2007). Detection of cytogenetic and DNA damage in peripheral erythrocytes of goldfish (*Carassius auratus*) exposed to a glyphosate formulation using the micronucleus test and the comet assay. *Mutagenesis*, 22, 263–268

Cremlyn, R. J. (1991). *Agrochemicals. Preparation and Mode of Action*, Wiley, Great Britain

Cserháti, T.; Illés, Z. & Nemes-Kósa, S. (1992). Effect of benzonitrile ester herbicides on the growth of some soil bacteria. *J Appl Bacteriol*, 72, 523-528

Daum, G. (1985). Lipids of mitochondria, *Biochim Biophys Acta*, 822, 1-42

Dayan, F.; Romagni, J.; Tellez, M.; Rimando, A. & Duke, S. (1999). Managing weeds with natural products, *Pesticide Outlook*, October, pp.185-188

Dejonghe, W.; Berteloot, E.; Goris, J.; Boon, N.; Crul, K.; Maertens, S.; Hofte, M.; de Vos, P.; Verstraete, W. & Top, E.M. (2003). Synergistic degradation of linuron by a bacterial consortium and isolation of a single linuron-degrading *Variovorax* strain. *Appl Environm Microbiol* 69, 1532-1541

Di Giulio, R.T.; Washburn, P.C.; Wenning, R.J.; Winston, G.W. & Jewell, C.S. (1989). Biochemical responses in aquatic animals: a review of oxidative stress. *Environ Toxicol Chem*, 8, 1103-1123

Donato, M.M.; Jurado, A.S.; Antunes-Madeira, M.C. & Madeira, V.M.C. (1997a). Effects of a lipophilic environmental pollutant (DDT) on the phospholipid and fatty acid contents of *Bacillus stearothermophilus. Arch Environ Contam Toxicol*, 33, 341-349.

Donato, M.M.; Jurado, A.S.; Antunes-Madeira, M.C. & Madeira, V.M.C. (1997b). Comparative study of the toxic actions of DDT and DDE on the growth and respiratory activity of a microorganism used as a model. *Appl Environ Microbiol*, 63, 4948-4951.

Donato, M.M.; Antunes-Madeira, M.C.; Jurado, A.S. & Madeira, V.M.C. (1997c). Partition of DDT and DDE into membranes and extracted lipids of *Bacillus stearothermophilus. Bull Environ Contam Toxicol*, 59, 696-701

Donato, M.M.; Jurado, A.S.; Antunes-Madeira, M.C. & Madeira, V.M.C. (2000). Membrane lipid composition of *Bacillus stearothermophilus* as affected by lipophilic environmental pollutants: an approach to membrane toxicity assessment. *Arch Environ Contam Toxicol*, 39, 145-153

Eble, K.S.; Coleman, W.B.; Hantgan, R.R. & Cunningham, C.C. (1990). Tightly associated cardiolipin in the bovine heart mitochondrial ATP synthase as analysed by 31P nuclear magnetic resonance spectroscopy. *J Biol Chem*, 265, 19434-19440

Figueiredo-Fernandes, A.; Fontaínhas-Fernandes, A.; Peixoto, F.; Rocha, E. & Reis-Henriques, M.A. (2006). Effects of gender and temperature on oxidative stress enzymes in Nile tilapia *Oreochromis niloticus* exposed to paraquat. *Pest Biochem Physiol*, 85, 97–103

Fontaínhas-Fernandes, A.A. (1998). Tilapia production, In: *Aquaculture Handbook*, Reis-Henriques, M.A. (Ed.), pp. 135-150

Forlani, G.; Mantelli, M.; Branzoni, M.; Nielsen, E. & Favilli, F. (1995). Differential sensitivity of plant-associated bacteria to sulfonylurea and imidazolinone herbicides. *Plant Soil*, 176, 243-253

Fry, M. & Green, D.E. (1981). Cardiolipin requirement for electron transfer in complex I and III of the mitochondrial respiratory chain. *J Biol Chem*, 256, 1874-1880

Furusawa, I.; Tanaka, K.; Thanutong, P.; Mizuguchi, A.; Yazaki, M. & Asada, K. (1984). Paraquat resistant tobacco calluses with enhanced superoxide dismutase activity. *Plant Cell Physiol*, 25, 1247-1254

Gennis, R.B. (1989). *Biomembranes: Molecular Structure and Function*. Springer-Verlag, New York

Gonzalvez, F. & Gottlieb, E. (2007). Cardiolipin: Setting the beat of apoptosis. *Apoptosis*, 12, 877-885.

Green, M.B., Hartley, G.S., & West, T.F. (1987) *Chemicals for Crop Improvement and Pest Management*, 3rd Edn, Pergamon Press, Oxford, p. 207

Guilbaut, G.G. & Kramer, D.N. (1964). Fluorometric determination of lipase acylase a- and b-chymotrypsin and inhibitors of these enzymes. *Anal Chem*, 36, 409-412

Harms, H. (1992). In-vitro systems for studying phytotoxicity and metabolic fate of pesticides and herbicides in plants. *Pestic Sci*, 35, 277–281

Hatton, P.J.; Cummins, I.; Price, L.J.; Cole, D.J. & Edwards, R. (1998). Glutathione transferases and herbicide detoxification in suspension-cultured cells of giant foxtail (Setaria faberi). *Pestic Sci*, 53, 209-216

Hay, J. R. (1974). Gains to the grower from weed science. *Weed Sci*, 22, 439-442

Heipieper, H.-J.; Keweloh, H. & Rehm, H.-J. (1991). Influence of phenols on growth and membrane permeability of free and immobilized *Escherichia coli*. *Appl Environ Microbiol*, 57, 1213-1217

Hernando, M.D.; Ejerhoon, M.; Fernández-Alba, A.R. & Chisti, Y. (2003). Combined toxicity effects of MTBE and pesticides measured with *Vibrio fischeri* and *Daphnia magna* bioassays. *Water Res*, 37, 4091-4098

Hill, A.J.; Teraoka, H.; Heideman, W. & Peterson, R.E. (2005). Zebrafish as a model vertebrate for investigating chemical toxicity. *Toxicol Sci*, 86, 6-19

Incledon, B. J. & Hall, J. C. (1997). Acetyl-coenzyme A carboxylase: quarternary structure and inhibition by graminicidal herbicides. *Pestic Biochem Physiol*, 57, 74-83

Ishikawa, I.; Yamada, S.; Hosaka, H.; Kawana, T.; Okunuki, S. & Kohara, K. (1985). Herbicidal properties of sethoxydim for the control of gramineous weeds. *J Pestic Sci*, 10, 187-193

Iwataki, I. & Hirono, Y. (1979). The chemical structure and herbicidal activity of alloxydin-sodium and related compounds. In: *Advances in Pesticide Science*, Geissbuhler, H.; Brooks, G.T. & P.C. Kearney, (Eds.), Part 2, pp. 235-243, Pergamon Press, Oxford

Jurado, A.S.; Pinheiro, T.J.T. & Madeira, V.M.C (1991). Physical studies on membrane lipids of *Bacillus stearothermophilus*. Temperature and calcium effects. *Arch Biochem Biophys*, 289, 167-179

Kaback, H.R. (1972). Transport across isolated bacterial cytoplasmic membranes. *Biochim Biophys Acta*, 265, 367-416

Koyama, K.; Koyama, K. & Goto, K. (1997). Cardiovascular effects of a herbicide containing glufosinate and a surfactant: in vitro and in vivo analysis in rats. *Toxicol Appl Pharmacol*, 145, 409-414

Krall, J.; Bagley, A.C.; Mullenbach, G.T.; Hallewell, R.A. & Lynch, R.E. (1988). Superoxide mediates the toxicity of paraquat for cultured mammalian cells. *J Biol Chem*, 263, 1910-1914

Lawrence, A.J. & Hemingway K.L. (2003). Introduction and Conceptual Models, In: *Effects of Pollution on Fish - Molecular Effects and Population Responses.* Lawrence, A. & Hemingway, K., (Eds.), pp. 1–13, Blackwell Science Ltd, Oxford

Li, X.-Q.; Ng, A.; King, R. & Durnford, D.G. (2008). A rapid and simple bioassay method for herbicide detection. *Biomarker Insights*, 3, 287-291

Livingstone, D.R.; Garcia-Martinez, P.; Michel, X.; Narbonne, J.F.; O'Hara, S.; Ribera, D. & Winston, G.W. (1990). Oxyradical generation as a pollution-mediated mechanism of toxicity in the common mussel, *Mytilus edulis* L. and other molluscs. *Funct Ecol*, 4, 415-424

Lundbaek, J.A.; Collingwood, S.A.; Ingólfsson, H.I.; Kapoor, R. & Andersen, O.S. (2010). Lipid bilayer regulation of membrane protein function: gramicidin channels as molecular force probes. *J R Soc Interface*, 7, 373-395

Luxo, P.C.; Jurado, A.S. & Madeira, V.M.C. (1998). Lipid composition changes induced by tamoxifen in a bacterial model system. *Biochim Biophys Acta*, 1369, 71-84

Lydon, L. & Duke, S.O. (1988). Porphyrin synthesis is required for photobleaching activity of the p-nitrosubstituted diphenyl ether herbicides. *Pestic Biochem Physiol*, 31, 74-83

Magalhães, M.J.A.; Ferreira, J.R.; Frutuoso, L. & Taínha, A. (1989). Study of the disappearance of endosulfan, parathion, trichlorfon and pirimicard from broccoli and Portuguese cabbage. *Pestic Sci*, 27, 23–31

Martins, J.D.; Monteiro, J.P.; Antunes-Madeira, M.C.; Jurado, A.S. & Madeira, V.M.C. (2003). Use of the microorganism *Bacillus stearothermophilus* as a model to evaluate toxicity of the lipophilic environmental pollutant endosulfan. *Toxicol In Vitro*,17, 595-601

Martins, J.D.; Jurado, A.S.; Moreno, A.J.M. & Madeira, V.M.C. (2005). Comparative study of tributyltin toxicity on two bacteria of the genus *Bacillus*. *Toxicol In Vitro*, 19, 943-949

Matolcsy, Gy.; Nádasy, M. & Andriska, V. (1988). *Pesticide Chemistry*, Elsevier, Amsterdam

Matringe, M. & Scalla, R. (1988). Studies on the mode of action of acifluorfen-methyl in nonchlorophyllous soybean cells. *Plant Physiol*, 86, 619-622

McIntosh, T.J. & Simon, S.A. (2006). Roles of bilayer material properties in function and distribution of membrane proteins. *Annu Rev Biophys Biomol Struct*, 35, 177-198

Monteiro, J.P.; Martins, J.D.; Luxo, P.C.; Jurado, A. S. & Madeira, V.M.C. (2003). Molecular mechanisms of the metabolite 4-hydroxytamoxifen of the anticancer drug tamoxifen: use of a model microorganism *Toxicol In Vitro*, 17, 629-634

Monteiro, J.P.; Jurado, A.S.; Moreno, A.J.M. & Madeira, V.M.C. (2005). Toxicity of methoprene as assessed by the use of a model microorganism. *Toxicol In Vitro*, 19, 951-956

Monteiro, J.P.; Oliveira, P.J.; Moreno, A.J.M. & Jurado, A.S. (2008). Disruption of hepatic mitochondrial bioenergetics is not a primary mechanism for the toxicity of methoprene. Relevance for toxicological assessment. *Chemosphere*, 72, 1347-1354

Moreland, D.E. (1980). Mechanisms of action of herbicides. *Ann Rev Plant Physiol*, 31, 597–638

Moreland, D. E. & Novitzky, W. P. (1988). Effects of inhibitors and herbicides on membrane potential of mung bean mitochondria. *Pestic Biochem Physiol*, 31, 247-260

Moreland, D. E. (1993). Research on biochemistry of herbicides: An historical overview. *Z. Naturforch*, 48c, 121-131

Mumma, R.O. & Davidonis, G.H. 1983. Progress in pesticide biochemistry and toxicology, In: *Plant Tissue Culture and Pesticide Metabolism*, Hutson, D.H. & Roberts, R., (Eds.), Wiley, pp. 255–274, Chichester

Nagata, S.; Günther, H.; Bader, J. & Simon, H. (1987). Mitochondria catalyze the reduction of NAD by reduced methylviologen, FEBS Lett, 210, 66-70

Nagel, R. (2002). DarT: the embryo test with the zebrafish *Danio rerio* - a general model in Ecotoxicology and Toxicology. *Altex, 19,* Suppl 1, 38-48

Nellessen, J.E. & Fletcher, J.S. (1993). Assessment of published literature pertaining to the uptake/accumulation, translocation, adhesion and biotransformation of organic chemicals by vascular plants. *Environ Toxicol Chem*, 12, 2045–2052

Neuburger, M. ; Journet, E.-P. ; Bligny, R. ; Carde, J.-P. & Douce, R. (1982). Purification of plant mitochondria by isopycnic centrifugation in density gradients of percoll. *Arch Biochem Biophys*, 217, 312–323

Neumann, G.; Teras, R.; Monson, L.; Kivisaar, M.; Schauer, F. & Heipieper, H.J. (2004). Simultaneous degradation of atrazine and phenol by *Pseudomonas* sp. strain ADP: effects of toxicity and adaptation *Appl Environm Microbiol*, 70, 1907-1912

Nystrom, B.; Bjornsater, B. & Blanck, H. (1999). Effects of sulfonylurea herbicides on non-target aquatic microorganisms. *Aquat Toxicol*, 47, 9-22

OECD 203 (1992). OECD Guideline for Testing of Chemicals. Test N° 203: Fish, Acute Toxicity Test. www.oecd.org/dataoecd/17/20/1948241.pdf

Otto, D.M.E. & Moon, T.W. (1995). 3,3',4,4'-tetrachlorobiphenyl effects on antioxidant enzymes and glutathione status in different tissues of rainbow trout. *Pharmacol Toxicol*, 77, 281-287

Palmeira, C.M.; Moreno, A.J.M. & Madeira, V.M.C. (1994a). Interactions of the Herbicides 2,4-D and dinoseb with liver mitochondrial bioenergetics. *Toxicol Appl Pharmacol*, 127, 50-57

Palmeira, C.M.; Moreno, A.J.M. & Madeira, V.M.C. (1994b). Metabolic alterations in hepatocytes pomoted by the herbicides paraquat, dinseb and 2,4-D. *Arch Toxicol*, 68, 24-31

Palmeira, C.M.; Moreno, A.J. & Madeira, V.M.C. (1995a). Mitochondrial bioenergetics is affected by the herbicide paraquat. *Biochim Biophys Acta*, 1229, 187-192

Palmeira, C.M.; Moreno, A.J. & Madeira, V.M.C. (1995b). Thiols metabolism is altered by the herbicides paraquat, dinoseb and 2,4-D: a study in isolated hepatocytes. *Toxicol Lett*, 81, 115-123

Palmeira, C.M (1999). Herbicide-induced mitochondrial and cellular liver toxicity: a review of paraquat, dinoseb, and 2,4-D effects. *Toxic Subst Mech*, 18, 187-204

Pedrajas, J.R.; Peinado, J. & Lopez-Barea, J. (1995). Oxidative stress in fish exposed to model xenobiotics; oxidatively modified forms of Cu,Zn-superoxide dismutase as potential biomarkers. *Chem Biol Interact*, 98, 267-282

Peixoto, F.; Vicente, J.A.F. & Madeira, V.M.C. (2003a). The herbicide dicamba (2-methoxy-3,6-dichlorobenzoic acid) interacts with mitochondrial bioenergetics function. *Arch Toxicol*, 77, 403-409

Peixoto, F.; Vicente, J.A.F. & Madeira, V.M.C. (2003b). Comparative effects of herbicide dicamba and related compound on plant mitochondrial bioenergetics. *J Biochem Mol Toxicol*, 17, 185-192

Peixoto, F.; Vicente, J.A.F. & Madeira V.M.C. (2004). A comparative study between plant and animal mitochondria with paraquat reveal that hydrogen peroxide is not related to the observed toxicity. *Toxicol in Vitro*, 18, 733-739

Peixoto, F.; Alves-Fernandes, D.; Santos, D. & Fontaínhas-Fernandes, A. (2006). Toxicological effects of oxyfluorfen on oxidative stress enzymes in tilapia *Oreochromis niloticus. Pest. Biochem. Physiol, 85: 91–96*

Peixoto, F.; Gomes-Laranjo, J.; Vicente, J.A.F. & Madeira, V.M.C. (2008). Comparative effects of herbicide dicamba, 2,4-D and paraquat on non-green potato tuber calli. *J Plant Physiol*, 165,1125-1133

Peixoto, F.; Lopes, M.L.; Madeira, V.M.C. & Vicente, J.A.F. (2009a). Toxicity of MCPA on non-green potato tuber calli. *Acta Physiol Plant*, 31, 103-109

Peixoto, F.; Santos, D.L.; Vilela, S. & Fontaínhas-Fernandes, A. (2009b). Characterization of liver Nile tilapia (*Oreochromis niloticus*) mitochondria and mitochondrial bioenergetics alterations by exposure to oxyfluorfen herbicide (Goal). *Arquivo Brasileiro de Medicina Veterinária e Zootecnia*, 61, 386-392

Pereira, S.P.; Fernandes, M.A.S.; Martins, J.D.; Santos, M.S.; Moreno, A.J.M.; Vicente, J.A.F.; Videira, R.A. & Jurado, A.S. (2009). Toxicity assessment of the herbicide metolachlor: comparative effects on bacterial and mitochondrial model systems. *Toxicol In Vitro*, 23, 1585-1590

Pereira, S.P.; Fernandes, M.A.S.; Martins, J.D.; Moreno, A.J.M.; Vicente, J.A.F.; Videira, R.A. & Jurado, A.S. (2010). Toxicity assessment of the herbicide Alachlor using in vitro model systems. International Courses on Toxicology, May 11-14, p.38, University of Coimbra. Portugal

Peter, B.; Wartena, M.; Kampinga, H.H. & Konings, W.T. (1992). Role of lipid peroxidation and DNA damage in paraquat toxicity and the interaction of paraquat with ionizing radiation. *Biochem Pharmacol*, 43, 705- 715

Pireaux, J.-C.; Dizengremel, P. & Botton, B. (1992). Effects of pesticides on oxidative phosphorylation and electron transfer in potato tuber mitochondria. *Plant Sci*, 83: 23-30

Pritchard, J.B. (1993). Aquatic Toxicology: Past, present and Prospects. *Environ Health Perspect*, 100, 249-257

Prosperi, E. (1990). Intracellular turnover of fluorescein diacetate. Influence of membrane ionic gradients on fluorescein efflux. *Histochem J*, 22, 227-233

Rao, V. S. (2000). *Principles of Weed Science*, 2nd Edn, CRC Press, New Hampshire, USA

Ray, T.B. (1984). Site of action of chlorsulfuron. Inhibition of valine and isoleucine biosynthesis in plants. *Plant Physiol*, 75, 827-831

Romagni, J.G.; Duke, S.O. & Dayan, F.E. (2000). Inhibition of plant asparagine synthetase by monoterpene cineoles. *Plant Physiol*, 123, 725-732

Rosa, S.M.L.J.; Antunes-Madeira, M.C.; Matos,M.J.; Jurado, A.S. & V.M.C. (2000). Lipid composition and dynamics of cell membranes of *Bacillus stearothermophilus* adapted to amiodarone. *Biochim Biophys Acta, 1487*, 286-295

Rossignol, R.; Faustin, B.; Rocher, C.; Malgat, M. & Letellier T. (2003). Mitochondrial threshold effects. *Biochem J*, 370, 751-762

Rytomaa, M.; Mustonen, P. & Kinnunen, P.K. (1992). Reversible, non-ionic, and pH-dependent association of cytochrome c with cardiolipin-phosphatidylcholine liposomes. *J Biol Chem*, 267, 22243-22248

Sandermann, H.; Scheel, D. & Trenck, T.V.D. (1984). Use of plant cell cultures to study the metabolism of environmental chemicals. *Ecotoxicol Environ Saf*, 8, 167–82

Schmale, M.C; Nairn, R.S. & Winn, R.N. (2007). Aquatic Animal Models of Human *Disease. Comp Biochem Physiol C Toxicol Pharmacol*, 145, 1–4

Sedlak, E.; Panda, M.; Dale, M.P.; Weintraub, S.T. & Robinson, N.C. (2006). Photolabeling of cardiolipin binding subunits within bovine heart cytochrome c oxidase. *Biochemistry*, 45, 746-754

Shaner, D.L.; Anderson, P.C. & Stidham, M.A. (1984). Potent inhibitors of acetohydroxyacid synthase *Plant Physiol*, 76, 545-546

Shaner, D. L. (2003). Herbicide safety relative to common targets in plants and mammals. *Pest Manag Sci*, 60, 17-24

Sikkema, J.; de Bont, J.A.M. & Poolman, B. (1995). Mechanisms of membrane toxicity of hydrocarbons. *Microbiol Rev*, 59, 201-222

Smeda, R.J. & Weller, S.C. (1991). Plant-cell and tissue-culture techniques for weed science research. *Weed Sci*, 39, 497–504

Sorensen, S.; Bending, G.D.; Jacobsen, C.S.; Walker, A. & Aamand, J. (2003). Microbial degradation of isoproturon and related phenylurea herbicides in and below agricultural fields. *FEMS Microbiol Ecol*, 45, 1-11

Steinrucken, H.C. & Amrhein, J. (1980). The herbicide glyphosate is a potent inhibitor of 5-enolpyruvylshikimate 3-phosphate synthase. *Biochem Biophys Res Commun*, 94, 1207-1212

Stephensen, E.; Sturve, J. & Forlin, L. (2002). Effects of redox cycling compounds on glutathione content and activity of glutathione-related enzymes in rainbow trout liver. *Comp Biochem Physiol Part C*, 133, 435-442

Stevens, J.T. & Sumner, D.D. (1991). Herbicides, In: *Handbook of Pesticide Toxicology*, Hayes, W.J. & Laws, E.R. (Eds.), pp. 1317-1408, Academic Press, New York

Steward, F.C. (1983). Reflections on aseptic culture, In: *Handbook of Plant Cell Culture*, Evans, D.A.; Sharp, W.R.; Ammirato, P.V. & Yamada, Y., (Eds.), pp. 1-10, Macmillan Publishing Company, NewYork

Stock, J.B.; Stock AM & Mottonen, J.M. (1990). Signal transduction in bacteria. *Nature*, 344, 395-400

Sulik, M.; Kisielewski, W.; Szynaka, B.; Kemona, A.; Sulik, A.; Sulkowska, M. & Baltaziak, M. (1998). Morphological changes in mitochondria and lysosomes of hepatocytes in acute intoxication with 2,4-dichlorophenoxyacetic acid (2,4-D). *Materia Medica Polonia*, 30, 16-19

Thompson, D.G.; Holmes, S.B.; Thomas, D.; MacDonald, L. & Solomon, K.R. (1993). Impact of hexazinone and metsulfuron methyl on the phytoplankton community of a mixed wood/boreal forest lake. *Environ Toxicol Chem*, 12, 1695-1707

Tortorelli, M.C.; Hernández, D.A.; Vásquez, G.R. & Salibián, A. (1990). Effects of paraquat on mortality and cardiorespiratory function of catfish fry *Plecostomus commersoni*. *Arch Environ Contam Toxicol*, 19, 523-529

Trumpower, B.L. & Gennis, R.B. (1994). Energy transduction by cytochrome complexes in mitochondrial and bacterial respiration: the enzymology of coupling electron transfer reactions to transmembrane proton translocation. *Annu Rev Biochem*, 63, 675-716

Van Camp, W. ; Van Montagu, M. & Inze D. (1998). H_2O_2 and NO: redox signals in disease resistance *Trends Plant Sci*, 3, 330–334

Vicente, J.A.F.; Santos, M.S.; Vercesi, A.E. & Madeira, V.M.C. (1998). Comparative effects of the herbicide dinitro-*o*-cresol on mitochondrial bioenergetics. *Pestic Sci*, 54, 43-51

Vicente, J.A.F. & Madeira, V.M.C. (2000). The excellente of turnip mitochondrial fractions. *Biochem Educ*, 28, 104-106

Vicente, J.A.F.; Peixoto, F.; Lopes, M.L. & Madeira, V.M.C. (2001). Differential sensitivities of plant and animal mitochondria to the herbicide paraquat. *J Biochem Mol Toxicol*, 15, 322-330

Videira, R.A.; Antunes-Madeira, M.C. &. Madeira, V.M.C (2001). Ethylazinphos interaction with membrane lipid organization induzes increase of proton permeability and impairment of mitochondrial bioenergetic functions. *Toxicol Appl Pharmacol*, 175, 209-216

Videira, R.A.; Antunes-Madeira, M.C. &. Madeira, V.M.C. (2002). Differential effects induced by alpha- and beta-endosulfan in lipid bilayer organization are reflected in proton permeability. *Biochim Biophys Acta*, 1564, 140-148.

Wakamatsu, Y.; Pristyazhnyuk, S.; Kinoshita, M.; Tanaka, M. & Ozato, K. (2001). The see-through medaka: a fish model that is transparent throughout life. *Proc Natl Acad Sci USA*, 98, 10046–10050

Wallace, K.B. & Starkov, A.A. (2000). Mitochondrial targets of drug toxicity. *Annu Rev Pharmacol Toxicol*, 40, 353-388

Watanabe, K.; Ohori, Y.; Sato, Y. Boger, P. & Wakabayashi, K. (2001). Changes in fatty acid composition of neutral lipid in mung bean cotyledons by oxyfluorfen-induced peroxidation. *Pestic Biochem Physiol*, 69, 166-173

Weber, F.J. & de Bont, J.A.M. (1996). Adaptation mechanisms of microorganisms to the toxic effects of organic solvents on membranes. *Biochim Biophys Acta*, 1286, 225-245

Whitten, J. L. (1966) *That we may live*, Van Nostrand, Princeton, N.J., p. 31

Wiegand, C., Krause, E., Steinberg, C. & Pflugmacher, S. (2001). Toxicokinetics of atrazine in embryos of the zebrafish (*Danio rerio*). *Ecotoxicol Environ Saf*, 49, 199–205

Zimdahl, R. L. (2002). My view. *Weed Sci*, 50, 687

Zuba, M. & Binding, H. (1989). Isolation and culture of potato protoplasts, In: *Biotechnology in Agriculture and Forestry 8. Plant Protoplasts and Genetic Engineering I*. Bajaj, Y.P.S., (Ed.), pp. 124-146, Springer-Verlag, New York

Zychlinski, L. & Zolnierowicz, S. (1990). Comparison of uncoupling activities of chlorophenoxy herbicides in rat liver mitochondria. *Toxicol Lett*, 52, 25-34

2

Impact of Herbicides on Non-Target Organisms in Sustainable Irrigated Rice Production Systems: State of Knowledge and Future Prospects

Victor Galhano[1], José Gomes-Laranjo[1], Eduardo Fernández-Valiente[2], Romeu Videira[3] and Francisco Peixoto[3]

[1]*CITAB – Centre for Research and Technology of Agro-Environment and Biological Sciences, University of Trás-os-Montes and Alto Douro (UTAD), Vila Real*

[2]*Department of Biology, Faculty of Sciences, Autonomous University of Madrid*

[3]*CECAV – Centre of Animal Sciences and Veterinary, UTAD, Vila Real*

[1,3]*Portugal*

[2]*Spain*

1. Introduction

Feeding the 9 billion people expected to inhabit our planet by 2050 will be an unprecedented challenge for all the mankind (Ash et al., 2010). Nevertheless, producing enough food for the world's population in 2050 will be easy, but doing it at an acceptable cost to the planet will depend on research into everything, from high-tech seeds to low-tech farming practices (Anonymous, 2010). Research into rice (*Oryza sativa* L.), the most important food crop in the developing world and a staple food for more than half of the world's population, is crucial for the development of strategies that will increase global food security (Normile, 2008). Appropriate integrated management of parasitic rice weeds is thus expected to increase in importance due to their general invasive nature and their abilities to adapt to changing conditions such as those imposed by more and more unpredicted global climate changes. Modern sustainable paddy cultivation worldwide involves extensive use of agrochemicals such as insecticides, fungicides but especially herbicides. Herbicide demand has unique characteristics compared with other common productive inputs in rice culture systems such as land, labour, seeds and chemical fertilizers (Yamamoto & Nakamura, 2003). The goal of herbicide use is to kill or stunt weed infestation allowing the rice to grow and gain a competitive advantage (Monaco et al., 2002). The use of rice herbicides has been expanding enormously worldwide over the past 20-40 years. However, herbicides are considered a "two-edged sword" (Kudsk & Streibig, 2003) or the "reverse of the coin" (Jurado *et al.*, 2010; chapter 1, this book), since the subsequent dispersion of herbicide compounds and their degradation products in rice fields and adjacent areas with strong ecological value still threatens the integrity of ecosystems, thus resulting in serious global environmental concern (Olofsdotter et al., 1998). One of the major issues about environmental herbicide contamination in wetland rice fields is its bioaccumulation in ecosystem primary producers and its subsequent propagation through trophic chain. Therefore, reliable legislation and risk assessment tools are needed to carry out the monitoring of herbicide residues in autochthonous living organisms inhabiting rice fields.

The main objective of this review is to compile and summarize relevant update information on herbicide weed control and consequent impacts of these agrochemicals on rice field organisms as well as on the environment, mainly in European and particularly Mediterranean countries. For additional information, though some of them are being not necessarily updated, the readers are reported to the pioneer works of Padhy (1985), Pingali & Roger (1995) and Roger (1996). A considerable amount of research has been published since these previous reviews and this chapter summarizes only the main findings related to impacts of herbicides on the ecology of sustainable rice field agroecosystems in the new policy context. Moreover, we have chosen examples of herbicide effects on non-target rice field organisms that, in our view, best illustrate different effects and environmental impacts. It is not our intention to suggest that these are the sole examples or the most suitable herbicides. Clearly, there are numerous and elegant examples of herbicide effects on non-target rice field organisms that fall outside of the small, but in our opinion acceptable, number of cases we have chosen to present.

2. Herbicide use and predominant weeds in irrigated rice fields

According to the immense worldwide literature database quested, herbicides are the most frequently detected chemical pollutants in water. Moreover, herbicides account for more than 80% of the total consumption of pesticides utilized for crop protection, with a total spending of about € 110 million.year^{-1} (Ferrero & Tinarelli, 2007). Herbicides[1] are chemicals used to manage unwanted plants in the ecosystem, plants that are commonly referred to as "weeds". Weeds are the most important biological constraint to increasing yields wherever rice is grown and every rice researcher and farmer must be provided with a great amount of useful scientific information on weed control. Unlike the periodic outbreaks of insect pests and plant diseases, rice weeds are ever-present and threatening (Olofsdotter et al., 1998).

The application of different rice field herbicides varies with the region. For instance, in North America and Western Europe, due to high costs of labour, the chemical control of weeds is heavily done with herbicides, contrasting to East Asia and Latin America where herbicides are much less used (Carvalho, 2006). In European rice fields, weeds are considered the most noxious organisms affecting rice production. It has been estimated that without weed control, at a yield level of 7 to 8 t.ha^{-1}, yield loss can be as high as about 90% (Ferrero & Tinarelli, 2007 and references therein).

At present, rice is grown in all continents under various environmental conditions which can be separated into four main ecosystems: irrigated lowland, rainfed lowland, upland and flood prone (De Datta, 1981; Ferrero & Tinarelli, 2007). The cultural practices, weed species present, and control methods are somewhat different in these systems (Monaco et al., 2002). In this review we are only concerned about sustainable irrigated lowland rice production systems (both seeded and transplanted), which account for about half of the world's rice harvested area and provide 75% of total rice production (Ferrero & Tinarelli, 2007).

[1] Herbicides were preferably named by their common names in the text, but sometimes registered trademarks are cited and their use is not free for everyone. In view of the vast number of trademarks, it was not possible to indicate each particular case and contribution. The authors accepted no liability for this.

Sustainable irrigated rice should be perceived here as a prolonged existence and functioning of several important interrelated elements of irrigated rice culture (Sutawan, 2005). In direct-seeded fields flooding cannot be used until crop establishment, so post-emergence weed control is essential.

3. Impact of herbicides on soil and water non-target organisms in wetland rice fields

Besides weeds, herbicides can act upon other species, causing serious side effects on non-target rice field inhabiting organisms. Moreover, herbicide residues contaminate soils and water, remain in the rice crop, enter the food chain, and finally are ingested by humans with rice foodstuffs and water (Liebman, 2001). Traditionally, paddy fields are home-ecosystems to many species. In early 1991 Kenmore (as cited in Clay, 2004) wrote that "*Rice ecosystems often have more than 700 animal species per hectare in highly intensified fields in the Philippines and over 1000 so far described in Asian species of higher trophic level predators and parasitoids.*" Besides the application of ever-increasing quantities of synthetic fertilizers, the increasing application of pesticides, mainly herbicides, has led to the disappearance of much of this biodiversity. In addition, non-target survivors have been continuously threatened by these xenobiotics.

3.1 Effects on non-photosynthetic microorganisms

Frequently, non-photosynthetic microorganisms undergo stress conditions caused by herbicide application. For instance, the metabolism of the Gram-negative bacteria *Stenotrophomonas maltophilia*, sometimes present in rice field irrigation channels (Reche & Fiuza, 2005), could be negativelly affected by some rice field herbicides (Lü et al., 2009). These authors showed that a mixture of quinclorac and bensulfuron-methyl (BSM) induced the activity of the antioxidant enzymes superoxide dismutase and catalase of a *S. maltophilia* strain (WZ2), thus demonstrating the induced oxidative stress caused by the herbicides. The effect of BSM on a soil microbial community in a model paddy microcosm was studied by Saeki & Toyota (2004). BSM did not affect bacterial numbers remarkably, either in the overlying water or in the surface paddy soil, but the nitrification potential was significantly suppressed was significantly suppressed.

The composition of culture-independent microbial communities and the change of nitrogenase activities under butachlor application to paddy soil were investigated by Chen et al. (2009). The results of their work showed that the nitrogen-fixing ability was suppressed shortly after butachlor application but was augmented after 37 days both in upper and lower soil layers. A significant variation on microbial community shift was also demonstrated, favouring the diazotrophic microorganisms within the general bacterial communities imposed by butachlor, which may be a reason for the boosting nitrogen-fixation ability in paddy soils. Other effects of butachlor applied at 5.5 $\mu g.g^{-1}$ to 22.0 $\mu g.g^{-1}$ to microbial populations inhabiting dried paddy soil resulted in a decline of actinomycetes number and and an increase of bacteria and fungi, but fungi were easily affected by butachlor compared to bacteria, particularly at higher butachlor concentrations (Min et al., 2001).

3.2 Effects on photosynthetic microorganisms

3.2.1 Effects on microalgae

Besides nonphotosynthetic organisms, on the first level of rice field trophic chain we can find the phytoplankton, microscopic algae which are the food for the next steps in trophic chain.

Planktonic green algae are the ones generally used as test species for the first tier aquatic risk assessment of herbicides (Ishihara, 2009). Wetland contamination could result in a die-off of most algal species present, causing a severe decline in this food source; alternatively, certain species or groups of algae could be selectively inhibited (Ferraz et al., 2004).

An *in vitro* toxicity bioassay screening conducted by Marques et al. (2008) on surface waters and sediment elutriates proceeding from River Pranto waters, which irrigate Western Portugal's rice fields, as well as on the irrigated rice fields in "Quinta do Seminário" (Soure, Portugal), revealed that *Pseudokirchneriella subcapitata* was more sensitive to the overall physico-chemical conditions in natural samples than *Chlorella vulgaris*, its growth being inhibited under water samples from both sites. In addition, these authors found that water samples, mainly those from the main irrigation/drainage channel of the rice fields, were more deleterious to microalgae than those from River Pranto or any of the elutriates. Surprisingly, the qualitative chemical analysis done by these researchers did not reveal the presence of the herbicides applied in the field like molinate and propanil. These two herbicides plus MCPA were used in microbiotests in order to assess the comparative toxicity effects of herbicide active ingredients (a.i.) *versus* commercial products (c.p.) on the green algae *Raphidocelis subcapitata* (Pereira et al., 2000). It was demonstrated in laboratory that water samples fortified with each active ingredient caused higher toxic effects than the respective formulated product, mainly with molinate and propanil. A study carried out by Leitão et al. (2007) in an irrigated rice field ecosystem in the river Sado basin, a characteristic Mediterranean lowland basin located in southwestern Portugal, a highly acute toxicity (80%) was detected on *P. subcapitata* in June after propanil (0.2 µg a.i.L^{-1}) application to paddy fields. Sabater & Carrasco (1998) found that species of isolated microalgae responded differently to molinate concentrations tested in laboratory: at 44.6, 50.2, 3.2, and 1.12 ppm, the growth of *Chlorella saccharophila*, *C. vulgaris*, *Scenedesmus acutus*, and *Scenedesmus subspicatus* was strongly inhibited after 96 h, respectively, whereas no growth was observed at 69.8 ppm for *C. vulgaris* and 2.2 ppm *S. subspicatus*. This study showed also that the two *Chlorella* species were considerably more tolerant than the two *Scenedesmus* species isolated from Spanish rice fields. Vendrell et al. (2009) tested the acute toxicity of glyphosate on *S. acutus*, *S. subspicatus*, *C. vulgaris* and *C. saccharophila* isolated from samples collected at Albufera lake in Valencia (Spain), one of the most important rice areas in Europe with a very rich flora and fauna. Although glyphosate is not applied to Valencian rice fields, its massive spraying in other agricultural crops surrounding the Albufera National Park (ANP), a protected ecosystem where rice is cultivated in harmony with local fauna and flora, is still a common practice. In a microplate bioassay, the authors demonstrated the acute toxicity induced by glyphosate on the four *Chlorophyceae* and the herbicide concentrations eliciting a 50% growth reduction over 72 h (EC_{50}) ranged from 24.5 to 41.7 mg.L^{-1}, while a 10% growth inhibition was achieved with 1.6 to 3.0 mg.L^{-1}, difficult to find both in the paddy field and in the ANP lake. Therefore, it could be concluded from this study that glyphosate is not a dangerous herbicide for the ANP ecosystem due to its low microalgae toxicity at low glyphosate concentrations. Also in the ANP ecosystem, Ferraz et al. (2004) assessed the sensitivity of the same aforementioned four algal species to propanil and mefenacet using single species short-termed (72 h) toxicity tests. The 72 h-EC_{50} of propanil and mefenacet ranged from 0.29 to 5.98 mg.L^{-1}, and from 0.25 to 0.67 mg.L^{-1}, respectively for the four algal species.

The relative toxicity of hydroquinone on submerged aquatic weed green musk chara (*Chara zeylanica* Willd.) was investigated by Pandey et al. (2005) to explore possible use of the phytotoxin as herbicide management of the weed. It was found that hydroquinone was phytotoxic to *C. zeylanica* at 0.01 mM and lethal at 0.075-0.10 mM, resulting in death after 3-12 days. Moreover, it is important to highlight that this phytotoxin has a short life in the environment and a promissory potential for weed management.

3.2.2 Effects on cyanobacteria

The nitrogen-fixing cyanobacteria form a prominent component of microbial population in rice paddy fields, since they significantly contribute to fertility as natural biofertilizers (Fernández Valiente et al., 2000; Singh & Datta, 2007). The influence of herbicides on cyanobacteria has been extensively reviewed in many studies (Padhy, 1985; Pingali & Roger, 1995), though most have been restricted to laboratory cultures (Whitton, 2000). Furthermore, one of us, together with the research team colleagues of the Biology Department of the Autonomous University of Madrid, has contributed significantly to the comprehension of herbicide action mechanisms on Mediterranean rice field cyanobacteria (Leganés & Fernández-Valiente, 1992; González Tomé, 1996). Generally, cyanobacteria are quite sensitive to herbicides, because they share many of the physiological features of higher plants, which form the site of herbicide action (Whitton, 2000). Leganés & Fernández-Valiente (1992) showed that N_2-fixing cyanobacteria were mostly relatively tolerant to 2,4-D, at least under field conditions. Differences have been found between the tolerance to herbicides of cyanobacteria and the one of other organisms (Whitton, 2000). For example, a study performed in liquid culture showed that hexazinone was more toxic to green algae, diatoms and duckweed than to cyanobacteria, whereas green algae were more tolerant to diquat than cyanobacteria and diatoms (Peterson et al., 1997).

Certain rice field herbicide resistant strains of cyanobacteria have been isolated and characterized in laboratory studies, but their outdoor survival, competence and biofertilizer potentials have only recently been characterized. In a research undertaken by Singh & Datta (2007), four natural strains of *Anabaena variabilis* that showed multiple herbicide resistance to arozin, alachlor, butachlor and 2,4-D, were inoculated in growing rice plants, being demonstrated that mutant strains had stable resistance to herbicides under outdoor conditions in flooded soils.

Butachlor can boost *Anabaena sphaerica* biomass and accelerate the amount of nitrogen fixation (Suseela, 2001). Working with butachlor and also with arozin, alachlor and 2,4-D, (Singh & Datta, 2005) developed an immobilization technique with Ca-alginate that provides protection to diazotrophic cyanobacteria inoculants against the growth-toxic action of the herbicides. These authors tested the effect of graded concentrations (2 to 25 $mg.L^{-1}$) of the four common rice field herbicides on immobilized free-living isolates of *Nostoc punctiforme, Nostoc calcicola, A. variabilis, Gloeocapsa* sp., *Aphanocapsa* sp. and on a laboratory strain of *Nostoc muscorum*. It was demonstrated that: 1) *A. variabilis* showed maximum natural tolerance towards all the four tested herbicides; 2) all cyanobacterial isolates showed progressive inhibition of growth with increasing dosage of herbicides in both free and immobilized states; 3) arozin was more toxic to cyanobacterial growth compared to the other three herbicides; and 4) at herbicidal lethal concentrations, Ca-immobilized cells showed prolonged survival times when compared to their free-living counterparts, suggesting that immobilized N_2-fixing cyanobacteria could be used as better inocula delivery system for enhancing rice agriculture.

High bensulfuron-methyl concentrations (8-10 ppm) inhibited the growth and photosynthesis of over 50% in *A. variabilis* and *Nostoc commune* rice field isolated;

nitrogenase activity decreased by 94-98% in *A. variabilis* and by 85-86% in *N. commune* after 24 hours' incubation with 10 ppm and 20 ppm of the herbicide, respectively (Kim & Lee, 2006). Ahluwalia et al. (2002) proved that the incorporation of relatively higher doses (> 5 $\mu g.mL^{-1}$) of diquat into *N. muscorum* and *Cylindrospermum* sp. cultures could be highly toxic, thereby reducing their chlorophyll *a* (Chl*a*) content and contributing to a progressive decrease in growth which culminates in complete lysis of the cells with the increasing level of the herbicide. The highest concentration tested (15 $\mu g.mL^{-1}$) has been found to be algicidal for both cyanobacteria. At this concentration, the same authors demonstrated that paraquat supplemented into culture medium containing *Cylindrospermum* sp. also had an algicidal effect (Kaur et al., 2002).

3.3 Effects on invertebrates

Herbicides are responsible for a general reduction in the numbers of invertebrates within agricultural landscapes, which in turn compromises food supply for higher taxa (Stoate et al., 2009). Populations of copepods, cladocerans and ostracods fluctuate during the paddy-growing season in response to flooding, field drainage, ploughing and other practices (Tarazona & Dohmen, 2007 and references therein). In order to better understand the toxic results obtained in field conditions, particularly the influence of herbicide formulations on the toxic effects on immobilized forms of the crustaceans *Daphnia magna, Thamnocephalus platyurus* and *Artemia salina,* Pereira et al. (2000) performed different laboratory microbiotests with water samples fortified with the active ingredient and the respective formulated or commercial product of some rice field herbicides usually applied in Portuguese paddies *viz.* molinate, MCPA and propanil. The results obtained after the bioassays suggested that: 1) the organic solvents and surfactants present in some of the tested formulations affected the toxicity of the sample; 2) except for propanil, water samples fortified with formulated herbicides seemed to be more toxic than the respective active ingredient solutions on the tested crustaceans.

In a 6-day bioassay carried out by Faria et al. (2007) with the macroinvertebrate *Chironomus riparius* larvae, both in laboratory and *in situ* (from high and low contaminated rice fields and in the adjacent wetland channel) during molinate and propanil treatments, the larvae were not affected by the respective herbicide concentrations in water and sediments. In laboratory experiments, Burdett et al. (2001) used *Chironomus tepperi* midge larvae to assess the relative toxicity of formulated molinate, clomazone and thiobencarb applied to Australian rice crops. Whereas clomazone had no effect on *C. tepperi* at concentrations up to 0.288 $mg.L^{-1}$, molinate significantly increased its development time at antecipated field concentrations (AFC; 3.6 $mg.L^{-1}$) and above. Thiobencarb reduced emergence success of the adults at 0.0625 times the AFC and it decreased male adult size and increased development time for males and females at 0.125 times the AFC. The authors of this study also assessed the non-target effects of the herbicides on the aquatic invertebrate communities through shallow experimental ponds using commercial application rates and, one week after treatment, they saw that only thiobencarb had a significat effect, supressing populations of chironomids, calanoids and cyclopids. Besides *C. tepperi*, Wilson et al. (2000) conducted some laboratory acute and chronic toxicity tests on the herbicide benzofenap using adults of the aquatic snail *Isidorella newcombi*. Whereas in 24-h acute bioassays the midge larvae did not show significant mortality at 1.2 mg a.i. L^{-1} [double the maximum recommended levels for field application (RLFA) expected in rice fields at the permitted rate of 2 $L.ha^{-1}$], no significant snail mortality was recorded in the acute bioassays at concentrations up to 76 mg a.i. L^{-1} (120 times the maximum RLFA); it was concluded that benzofenap does not represent a significant risk to these invertebrates in downstream environments when applied to rice fields at the permitted rate of 2 $L.ha^{-1}$.

The high acute toxicity effects (100%) of propanil (0.2 µg a.i.L^{-1}) on *D. magna* were demonstrated by Leitão et al. (2007) in rice field floodwaters from river Sado basin. These authors also study the relationship between native fauna composition and the ecotoxicological variables throughout rice crop season. The analysis of variance partition with the accepted RDA (redundancy analysis) model showed that the macroinvetebrate distribution was strongly correlated with the ecotoxicological parameters (higher toxicity) in 27.5% of total variation and 17.7% with floodwater characteristics. It was demonstrated that macroinvertebrates' assemblages inhabiting rice fields tend to be different in their richness and abundance according to the paddy sediment and water characteristics.

In a study performed by Uno et al. (2001), the accumulative characteristics of herbicide residues in the organs of two bivalve mollusks, *Corbicula leana* and *Anodonta woodiana*, were examined during rice planting seasons of 1992 and 1994, using a small artificial stream under natural conditions. It was shown that thiobencarb accumulated in *C. leana* at extremely high levels in the midgut gland (12.45 and 15.70 µg.g^{-1}, in 1992 and 1994, respectively) and in the gonad (15.80 and 16.40 µg.g^{-1}, in 1992 and 1994, respectively); these levels were about 100 times as high as those of *A. woodiana*.

3.4 Effects on vertebrates

As we saw before, during the aquatic phase of rice plant growth, many organisms other than rice colonize the paddy fields, but fish from source rivers, in particular, are sometimes seen as detrimental to the rice harvest (Williams, 2006) and could be affected by herbicide pollution as described next.

3.4.1 Effects on fish

The effect of herbicides on estuaries, rivers, and fragile coastal zones are all reflected on the reduction of fish catch (Clay, 2004). In a recent 90-day experiment carried out by Moraes et al. (2009) with the objective of evaluating the effects of herbicide commercial formulations on acetylcholinesterase (AChE), thiobarbituric acid-reactive substances (TBARS), catalase (CAT) and metabolic parameters in teleost fish (*Leporinus obtusidens*) exposed to field concentrations of clomazone (376 µg.L^{-1}) and propanil (1644 µg.L^{-1}) on Brazilian rice paddy waters, it was shown that AChE activity decreased in the brain and muscle, whereas TBARS levels decreased in brain, muscle and liver tissues and liver CAT decreased after exposure to both herbicides. The results obtained by these authors suggest that environmentally relevant rice field herbicide concentrations are toxic to *L. obtusidens*. Crestani et al. (2006) suggested that alanine aminotransferase and aspartate aminotransferase activities could be used as early biomarkers of clomazone fish toxicity, since enzyme activities were significantly elevated in silver catfish's (*Rhamdia quelen*) liver after 12 to 24 hours' clomazone exposure to nominal concentrations used in Brazilian paddy fields (0.4-0.7 mg.L^{-1}). Working also with *R. quelen*, Miron et al. (2005) demonstrated significantly short-term (96 h) effects of exposure to environmentally relevant nominal concentrations of clomazone (LC_{50} = 7.32 mg.L^{-1}) and quinclorac (LC_{50} = 395 mg.L^{-1}) on AChE activity in brain and muscle tissues of silver catfish, but the fish fingerlings survived at metsulfuron-methyl concentrations as high as 1200 mg.L-

[1]. They suggested that AChE activity could be used as an early biomarker for studies on fish toxicity, since enzyme's activity increased by 98 and 179% in fish's brain after quinclorac and metsulfuron methyl exposure, respectively.

3.4.2 Effects on amphibians

In a recent study performed by Kang et al. (2009) about *Bombina orientalis*, a worldwide common amphibian that frequently spawns in rice fields where massive application of herbicides occurs, the deleterious effects of molinate on embryonic survival and development abnormality in *B. orientalis* embryos were demonstrated. A statistical significant decrease in embryonic survival was detected at mouth open stage after exposure to 100 μM molinate (46.8% *vs.* 81.1% in control) and then in the tadpole stage at 50 μM molinate (35.9% *vs.* 68.9% in control). The authors thus concluded that molinate at 50 μM (the lowest observed effective dose; LOED) was detrimental for survival and development following zygote transcription after midblastula transition in frog embryos, causing severe development abnormities like bent trunk, neurula with yolk plug, bent tail, tail dysplasia, ventral blister, eye dysplasia, thick-set body and cephalic dysplasia. Acute toxicity tests were carried out by Saka (1999), on five species of Japanese amphibian larvae to assess the risk posed by thiobencarb. The 24, 48, 72 and 96 h-LC_{50} (median lethal concentration) values ranged from 0.9 to 6.5 $mg.L^{-1}$ of thiobencarb. Moreover, in all tested species, the newly hatched larvae seemed to be slightly more resistant to the herbicide than well-developed larvae, which led the authors to conclude that thiobencarb residues in paddy water can be lethal to amphibians through larval development.

4. Legislation and assessment of contamination risks by rice field herbicides

Concerns about environmental protection have increased over the years from a global viewpoint. As the criteria especially for standard risk assessments are well harmonized amongst OECD member states (Streloke, 2007), there is only a need to give a short overview here. Needless to say, there is a raft of regulations and standards to negotiate in testing any herbicide. Diffuse pesticide pollution of water bodies and potentially adverse effects on aquatic communities gave rise to current legal formulations of the European Union (EU) such as the Council Directive 91/414/EEC concerning the placing of plant protection products on the market (EEC, 1991) or the Water Framework Directive (WFD) (EC, 2000) that presents the concept for the sustainable use of water resources by integrated river basin management (Schriever & Liess, 2007). Where appropriate, these Directives call on other European legislation in related areas such as test methods, classification and labelling, and maximum residue levels (MRLs) (Hussey & Bell, 2004). The WFD intends to provide an overall framework for a cleaner and safer water ecosystem, particularly regarding surface freshwater and ground water bodies (*i.e.* lakes, streams, rivers, estuaries, coastal waters etc.). The Directive 91/414/EEC defines the principles and procedures to be used for authorization of plant protection products, and its annexes outline the basis for coordination or harmonization of data requirements and regulatory decisions. The implementation of this Directive has led to an EU-wide regulatory process for evaluating the safety of herbicides to humans and the environment, whilst leaving the responsibility for approval of plant protection products in individual countries to member states. The Directive has established a positive list (Annex 1) that lists those herbicides that have been judged to be "without unacceptable risk" to people or the environment, and all new active molecules proposed for use as herbicides within the EU must be deemed acceptable according to the Directive. Article 5 of the Directive requires that the use of plant protection products and their residues should not have any harmful effects on human or animal health or on groundwater, or have any unacceptable influence on the environment. In this respect, all existing active ingredients (about 800) introduced into the market prior to 2000 had transitional approval, pending their re-evaluation using modern toxicological and environmental protocols with a view to inclusion in Annex 1. Identification of products to be supported for re-registration was sought by 2002, and evaluation of those using modern toxicological and environmental protocols is required by 2012. Approved substances will be listed in Annex 1 of the Directive (EEC, 1991; Vogelezang-Stoute, 2003; Hussey & Bell, 2004; Carlile, 2006).

The risk to the environment covers the fate and behaviour of an active ingredient (*i.e.*, exposure) as well as its possible effects on non-target organisms (EEC, 1991; Benfenati et al., 2007). It should be stressed that the Directive 91/414/EEC and associated technical annexes are constantly under revision.

For reasons of preventive health protection and protection of the environment, the use of plant protection products has to be limited to the minimum level compatible with effective crop protection. The MRLs (in the EU) or tolerances (in the USA) are established for crops and food commodities (Siebers & Hänel, 2003). In countries with no national legislation, the MRLs are set by the Codex Alimentarius Commission, an international body that aims to protect the health of consumers (Granby et al., 2008). Groundwater contamination has also received increasing attention over the last few years as most of the drinking water is drawn from wells (Vidotto et al., 2004). In the last few years, great activity in regulating the level of herbicide in water has been carried out in the EU. The maximum allowable concentrations in groundwaters (for drinking and any other use) are set to 0.1 $\mu g.L^{-1}$ for any individual chemical and 0.5 $\mu g.L^{-1}$ for total herbicide load (Gan & Bondarenko, 2008).

In Europe, regulations and regulatory methods to assess and control the impact of herbicides in the aquatic environment aim to protect the ecosystem and public health, while monitoring contamination levels and any potential negative effects; in addition to the REACH (Registration, Evaluation, Authorization, and Restriction of Chemicals) law (EC 1907/2006) (EU, 2006), regarding chemicals and their safe use, which entered into force on 1st June 2007, there are specific regulations towards protecting health and ensure quality of all water resources such as the Drinking Water Directive (DWD, Council Directive 98/83/EC) (EC, 2007; Nielsen et al., 2008; Adrián et al., 2009). Environment quality standards for 33 priority substances in surface aquatic bodies have been recently established in the Directive 2008/105/EC (EC, 2008), offspring of the WFD (EC, 2000), and about a third of the priority substances covered by this directive are pesticides (Köck et al., 2010). For additional aspects relating to regulatory authorities and legislation in many other parts of the world besides the EU, readers are reported to the works of Marrs & Ballantyne (2004), Carlile (2006), Racke (2007), Zimdahl (2007), and Nielsen et al. (2008).

It is important to stress that the ecotoxicological assessment of rice herbicides is particularly complex due to the specificities of this crop where aquatic and terrestrial communities are mixed in a human-managed wetland-type agro-bio-ecosystem. According to Tarazona & Dohmen (2007), any assessment must cover at least three main aspects: 1) potential effects on the paddy community; 2) potential effects on associated wetlands and water bodies, and 3) potential risks for vertebrates feeding on the paddy. The initial low tier assessment may be conducted using the generic approach used for other crops. For higher tier risk assessment it is recommended to develop a specific conceptual model for identifying the key elements that should be addressed (Tarazona & Dohmen, 2007).

5. Environmental exposure and contamination

Since rice is an irrigated crop, the use of pesticides during cereal growth may affect the quality of the surrounding aquatic environment (Pereira et al., 2000). Irrigation control after the application of herbicides could be the most important approach to avoiding environmental impact (Yamamoto & Nakamura, 2003). Several herbicides may enter aquatic environments through spray drift or runoff events, drainage or leaching, resulting in contamination of surface waters and groundwaters (Cerejeira et al., 2003; Faria et al., 2007), thus affecting phytoplankton and delicate autochthonous living organisms and also human health (Srivastava et al., 2010).

In a recent review about the ecological status of agricultural systems across the European Union, Stoate et al. (2009) reported to various studies about the presence of herbicide contamination, some exceeding the 0.1 $\mu g.L^{-1}$ EU limit, in both surface waters and groundwaters in areas occupied by intensive agriculture, rice fields included. Silva et al. (2006) detected residues of chlorfenvinphos, cycloxydim, 3,4-dichloroaniline, MCPA, molinate, oxidiazon, profoxydim, and propanil in 62% of 171 water samples collected from 22 groundwater wells used for public supply, domestic supply, and irrigation purposes in oriziculture areas of "Baixo Sado" region. In this study, 6% of total samples presented maximum concentration levels of at least one of the compounds above 0.1 $\mu g.L^{-1}$ and molinate was the most frequently detected (55%), particularly with maximum concentration levels above the maximum limit.

5.1 Surface water bodies

During three years (2000 to 2003), Marchesan et al. (2007) detected clomazone, propanil and quinclorac in water samples during the rice growing season in the Vacacaí and Vacacaí-Mirim Rivers, located in Rio Grande do Sul State (Brazil) and concluded that river water contamination by rice herbicides was probably caused by rice water management used in the fields. In a study designed by Castro et al. (2005) to monitor molinate accumulation in surface waters and underground waters during herbicide application in rice fields of central Portugal (valley of River Pranto), it was concluded that the thiocarbamate was rapidly dissipated in the environment, reaching levels as high as 3.9 $\mu g.L^{-1}$ and 15.8 $\mu g.L^{-1}$ in underground and in the river receiving tail waters, respectively (the legally recommended limits are bellow 5 $\mu g.L^{-1}$). Some common rice field herbicides, *e.g.* propanil and molinate have been detected in surface waters of Portugal's river basins (Tejo, Sado and Guadiana) and in groundwater samples collected from wells in seven different areas in the Tejo and Sado basins (Cerejeira et al., 2003). Clomazone dissipation in water samples from irrigated rice cultures was analyzed by Carlomagno et al. (2010) in a water management study implemented in Uruguay during the 2008-2009 harvest season. As expected, early irrigation of rice plots was accompanied by increasing concentrations of clomazone in water layers: 77 $ng.mL^{-1}$ in the day of flooding to 129 $ng.mL^{-1}$ after 6 day flooding. The high levels of clomazone attained in this study are a matter of concern, since contamination of drainages and rivers near rice paddies holds the potential for unacceptable levels of clomazone in drinking waters.

5.2 Groundwater systems

Rice fields are considered a flood control and a groundwater-charging area. Therefore, the risk of rice herbicides contaminating the groundwater cannot be neglected. Depending on the soil type and metereological conditions, among other abiotic and biotic factors and application rates, thiobencarb poses less risk for groundwater when applied to flooded rice fields than other herbicides. In a 2-year study done by Phong et al. (2009), the concentrations of thiobencarb were in the hundreds of $\mu g.kg^{-1}$ in the top soil layer (0-5 cm) and became significantly lower in tens of $\mu g.kg^{-1}$ in deeper soil layers (5-10 and 10-15 cm) with a short life time (17 days) in the top layer, 69 days in the 5-10 cm layer and 165 days in the 10-15 cm layer. In another study, it was found that bentazon is one of the most frequently detected herbicide in groundwater in Northern Italy, even though its use is either prohibited or reduced in many districts in this country (Garagna et al., 2005).

6. Herbicide effects on non-target cyanobacteria: the case study of Lower Mondego River Valley, Portugal

In the European Union (EU), rice is cultivated in about 410 000 ha, mostly located in the Mediterranean countries. It is mostly grown in concentrated areas such as the Po valley in Italy, the Rhône delta in France, and the Thessaloniki area in Greece. In Spain, rice cultivation is scattered in several areas such as the Aragon area, the Guadalquivir valley, the Ebro delta and Valencia Albufera (Ferrero & Tinarelli, 2007). In Portugal, the approximately 25 000 ha of rice cultivation area is concentrated mainly in three regions: the Tagus and Sorraia valleys (10 000 ha), the Mondego (8000 ha), and the Sado and Caia (7000 ha) river valleys (Calha et al., 1999; Silva et al., 2006). In all European countries rice is cultivated in permanent flooded conditions, with short periods during which soil is dried to favour rice rooting (in the early stages) or weed control treatments. The conventional irrigation system is also known as a "flow-through" system, because water is usually supplied in a series from the topmost to the bottommost basin and is regulated by floodgates by means of removable boards. Water circulation implies that effects of the herbicide application can also be observed in areas different from that of direct application. Moreover, paddy flooding favours the flow of solutes from the surface to the subsoil waters and may result in significant exchanges between these two aqueous bodies (Vidotto et al., 2004). Because of the high water demand for maintaining flooded conditions in the rice fields, Mediterranean rice-growing areas are restricted to the lower parts of river basins and near or directly on river deltas (Comoretto et al., 2008).

The management of Portuguese rice ecosystems has adapted to Mediterranean climate conditions: they are dry in the winter, serving as cattle grazing prairies, while in the summer they are flooded up to 15-30 cm, with water coming from nearby dams or rivers, thus becoming temporary aquatic habitats managed with a variable degree of intensity; the paddies of rectangular or similar shaped flooded fields comprised mostly rice plants, surrounded by dry bunds (levees) rich in weeds, connected by irrigation canals and ditches that serve as contiguous aquatic habitats (Lima, 1997; Leitão et al., 2007).

The Lower Mondego River Valley is located in the central-western region of Portugal (40° 10′ N, 8° 41′ W, 5 m). The valley consists of about 15 000 ha where the main agricultural crop is "paddy" rice, which occupies about 60% of the usable area (Cabral et al., 2001). Non cultivated areas, such as swamps, are usually located in the perimeter of the valley and exhibit a flourished fauna and flora, whereas drainage canals, which are widespread across the whole valley, also constitute a biological reservoir (Anastácio et al., 1999). Additional geologic, hidrogeologic and climate details of the valley were described by Andrade & Stigter (2009). Rice cultivation starts normally with seeding from the beginning of April to the end of May, which depends on water availability for field flooding, the choice of rice variety (early or late) and climatic conditions. The growing cycle continues until September-October, when rice is harvested. The rice cropping is supported by an organized system of irrigation-drainage canals around the paddies and the water flow is controlled by small and simple dams, constructed in strategic points to provide the entrance or discharge of water. Local rice fields are abundantly fertilized and a variety of pesticides, including herbicides, are intensively and sometimes indiscriminately used in weed control programs. The application of agrochemicals occurs mainly during the end of April up to June, but additional amounts of fertilizers or pesticides are added along the whole cropping season, depending on the type of culture demands and on the rice crop regional conditions (Marques et al., 2008).

6.1 Herbicides: overall use and selection

Almost all the major weeds in the Mondego Valley are grasses (family *Poaceae*) and, given the efficiency of the herbicides used, red rice (*O. sativa* L.) is undoubtedly the adventitious plant which poses more problems, followed by *Leersia oryzoides* (L.) Sw., which is spreading fast. Other important weeds found in the area are the *Poaceae* barnyward grass [*E. crus-galli* (L.) Beauv.] and early watergrass [*Echinochloa oryzoides* (Ard.) Fritsch]; the *Ponteridaceae* roundleaf mudplantain [*Heteranthera reniformis* Ruiz & Pavón] and ducksalad [*Heteranthera limosa* (Sw.) Willd.]; the *Cyperaceae* triangle (*Scirpus maritimus* L.) and dirty dora (*C. difformis* L.); the *Alismataceae* water-plantain (*A. plantago-aquatica* L.); and the algae *Sphaeroplea annulina* (class *Chlorophyceae*) and *Chara foetida* (class *Charophyceae*) (CACMV, 2010).

Most of the herbicides used in rice are selective, and the time of application relative to the growth stage of the crop and the weeds, along with proper water management, is critical for crop safety and good weed control (Monaco et al., 2002). The recommended herbicides for integrated weed management (IWM) are depicted in Table 1. Only a short overview of the more important characteristics of herbicides is described. For more detailed and specific information about herbicide properties and modes of action, the reader is reported to more extensive and specialized literature such as Tomlin (2000), Aizawa (2001), Stenersen (2004), Mackay et al. (2006), and Krieger (2010). It is important to mention that some active ingredients were already excluded from the *EU Pesticides database* (http://ec.europa.eu/sanco_pesticides/public/index.cfm; last updated on 03/08/2010; last accessed on 3 November 2010) but we still include them here just for information purposes, since some local farmers still use them and because of the holistic approach of this review.

6.2 Ecotoxicological impacts of Basagran® (*a.i.* bentazon) and Ordram® (*a.i.* molinate) on non-target rice field cyanobacteria

Since 2005 our research group has been developing various relevant studies about the effects of two most commonly used rice field herbicides - Basagran® and Ordram® - on native non-target rice field cyanobacteria. In one of those studies, the effects of the two selective herbicides recommended for IWM on rice were laboratory-assessed in *Anabaena cylindrica* during a short-term experiment of 72 h (Galhano et al., 2009). The results obtained in this work revealed that both herbicides had a pleiotropic effect on the cyanobacterium at the range of concentrations tested (0.75-2 mM). Cyanobacterial growth (expressed as dry weight, D.W.) as well as photosynthetic pigments [Chl*a*, carotenoids (Car), phycobiliproteins (PBP)] were more adversely affected by molinate than bentazon commercial formulations. *A. cylindrica* growth of over 50% was observed soon after 48 h with 1.5-2 mM of Ordram. The protein content increased with both herbicides although the

Weeds/ Active Ingredient	Form.	Concent. (g a.i. ha^{-1})	TC	Recommended application conditions
Monocotyledonous				
Cycloxydim (authorized only at national level)	EC	200	Xi	Only for red rice control. Rice preplant and with red rice in the 1-2 leaves stage
Cyalofop-butyl	EC	300	Xn; N	In rice post-emergence and grass weeds with 1-3 leaves
Molinate (cancelled at national level in 28/04/2010) (*)	FG	4500	Xn; N	In rice preplant, pre-emergence or early post-emergence, with grass weeds in the 2-leaves stage
Oxadiazon	EC	300 - 400	Xn; N	Apply after soil preparation and flood the soil after application. In rice preplant (5–8 days after application)
Profoxydim(1) (pending chemical in the *EU Pesticides database*)	EC	100 – 150	C; N	In rice post-emergence; apply 100 - 120 g from the 4-leaf stage to tillering, and 100 - 150 g from the beginning to the middle of tillering
Monocotyledonous (families *Poaceae, Alismataceae, Cyperaceae, Pontederiaceae*) and Dicotyledonous				
Azimsulfuron	WG	20 - 25	N	In rice post-emergence, from the 2-3 leaves stage to tillering; with grass weeds from 2-leaves stage to the beginning of tillering and other weeds to the 4-6 leaves stage
Bensulfuron-methyl	WG	51 - 60	Is	In early rice post-emergence and weeds with 2-3 leaves
Bensulfuron-methyl + Metsulfuron-methyl	WG	40 + 1.6 - 50 + 2	Is	In early rice post-emergence and weeds with 2-3 leaves; it does not control grassy weeds
Bentazon (cancelled at national level in 16/04/2008) (*)	SL	1440 - 2400	Xn; N	In rice post-emergence from the 3-leaves stage to tillering, and weeds with 3-5 leaves
Bispyribac-sodium	SC	25 - 30	Xi; N	In rice post-emergence, from the 3-leaves stage to tillering; with grass weeds from 2-leaves stage to the beginning of tillering, and other weeds with 4-6 leaves
MCPA (ester)	SL	630 - 1050	Xn	In rice post-emergence and with weeds in active growth (4-6 leaves)

MCPA (potassium salt)	SL	800 - 1200	Xn	In rice post-emergence and with weeds in active growth (4-6 leaves)
Penoxsulam	OD	40	Xi; N	In rice post-emergence from the 2-3 leaves stage to stalk stage, and weeds with 1-4 leaves
Propanil (not included in the *EU Pesticides database*); (cancelled at national level in 30/09/2010) (*)	EC SC WG	3600 - 9000 3600 - 8640 3600 - 9000	Xn; N Xn; N Xn; N	In rice post-emergence and grass weeds with 3-4 leaves; apply lower and higher doses in nurseries and sowed or planted rice, respectively
Ridges (Monocotyledonous and Dicotyledonous)				
Glyphosate (ammonium salt) (some formulations were cancelled at national level in 14/08/2008) (*)	SG	272 - 3600	Is; N	Apply when weeds are well developed by using a bell glass during crop cycle or after harvest; apply lower and higher doses to control annual and perennial weeds, respectively
Glyphosate (isopropylammonium salt) (some formulations were cancelled at national level in 26/08/2010) (*)	SL	540 – 3600		
Algae				
Copper sulphate (some formulations will be cancelled at national level in 31/02/2012) (*)	XX	750 - 1875	Xn; N	As soon as algae became perceptible; apply by spraying or by placing jute bags in the entrance of irrigations ditches
Copper and calcium sulphate (some formulations will be cancelled at national level in 30/04/2011) (*)	WP	1700		

Table 1. Herbicide active ingredients and chemicals recommended in plant protection to control rice field weeds (adapted from: DGADR, 2009). Abbreviations: Form., formulation [EC, emulsifiable concentrate; FG, granules; OD, oil dispersion; SC, irritant; SG, water soluble granules; SL, aqueous liquid solution; WG, exempt of classification; XX, others; WP, wettable powder]; Concent., concentration; TC, toxicological classification [C, corrosive; Is, exempt; N, nocive and dangerous to the environment; Xi, irritating or sensitizing; Xn, nocive]. Observations: (1)It must be applied in a mixture with 0.75 $L.ha^{-1}$ of the commercially available wettable adjuvant Dash HC® (adjuvant concentration should not exceed 0.5% in low volume applications); (2)Treated rice for human consumption should be cleaned; (*) DGADR, 2010. *Cancelamento de AVs e APVs (Circular 16/2004) – Exhaustive Cancellation List of All Active Ingredients in Portugal Since 01/01/2001 (Last Review: 10 June, 2010).* Direcção-Geral de Agricultura e Desenvolvimento Rural; Ministério da Agricultura, do Desenvolvimento Rural e das Pescas, 23 pp. (in Portuguese).

effect was more remarkable with the highest concentration of Ordram. Concerning to carbohydrate content, it was shown that Ordram increased this organic fraction whereas Basagran decreases it. Photosynthesis (P_m^{Chl}) and dark respiration (R_d^{Chl}) normalized to Chl*a* were inhibited by both herbicide formulations in a time- and dose-response manner within the experiment time, and higher Ordram concentrations full stopped O_2 evolution after 48 h. Pointing to safety environmental precautions, the findings obtained with our study suggest the reduction or even ban of molinate from the agro-ecosystem rice field because of its strong inhibitory action on soil autochthonous microflora, mainly on the important diazotrophic primary producers, the cyanobacteria. Together with pulse-amplitude-modulation (PAM) fluorimetric routine parameters, namely maximum photosystem II (PS II) quantum efficiency (Fv/Fm), effective quantum efficiency of PS II for a light-acclimated sample (Φ_{PSII}), and photochemical quenching (qP; proportion of light excitation energy converted to photochemical act by the active PS II reaction centres), part of the results published by Galhano et al. (2009) were previously presented to the overall scientific community in the 15th European Bioenergetics Conference held in Dublin, from 19 to 24 July 2008 (Galhano et al., 2008). Relevant findings related to Basagran and discussed in the meeting are displayed in Figure 1 carpet-plot.

But we wanted to go further on and test the effects of the herbicides on other cyanobacteria strain isolated and subsequently identified by suitable molecular biology tools (Galhano et al., 2010a). So, the next step was the assessment of toxicity of Basagran and Ordram on *Nostoc muscorum*, an abundant and well characterized cyanobacterium from Mondego River Valley rice fields. Once again, in a short-term exposure experiments during 72 h with a concentration range from 0.75 to 2 mM, the toxicity of commercial formulations on growth and some biochemical and physiological parameters cited before were studied (Galhano et

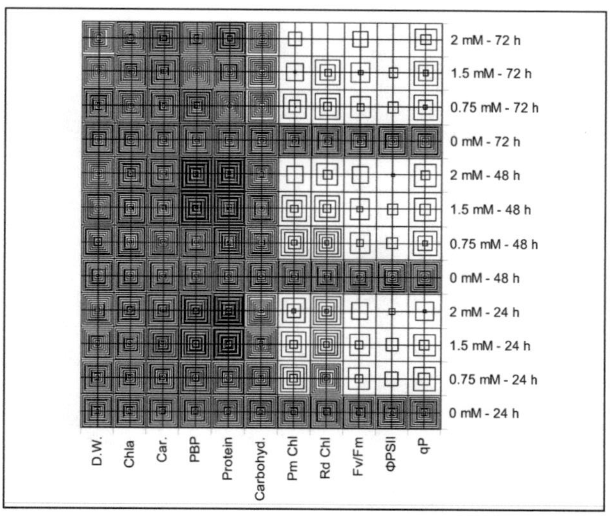

Fig. 1. Effect of Basagran on dry weight (D.W.), chlorophyll *a* (Chl*a*), carotenoids (Car), phycobiliproteins (PBP), protein, carbohydrates, photosynthetic rate (P_m^{Chl}), dark respiration rate (R_d^{Chl}) and fluorescence parameters of *A. cylindrica* after exposure for 24, 48 and 72 h. Relative values are means ± SE of at least three independent experiments. The plotted values are visualised by the number of the contour lines, with successive lines corresponding to values differering by 0.05 (here above zero, *i.e.* bigger than the control).

al., 2010d). The results almost entirely obtained in this study confirmed the mode of action of both herbicides on cyanobacteria: 1) molinate was more toxic than bentazon to growth, respiration, Chl*a*, Car, and PBP contents; 2) protein content increased with both herbicides but the effect was mostly evident at higher molinate concentrations (1.5-2 mM); 3) the herbicides had contrasting effects on carbohydrates content - molinate increased it whereas bentazon caused a decrease of this organic fraction; 4) both photosynthesis and respiration were inhibited by Ordram and Basagran.

At this point of our research we were intrigued about the insufficient information on the biochemical mode of action of both herbicides at cellular level, particularly concerning bentazon. We knew from previous works that bentazon acts as an inhibitor of photosynthesis by blocking the electron transfer flow in the PS II and CO_2 fixation. The blockage of PS II induced by bentazon in the presence of light induces secondary effects on several metabolic pathways, such as the production of singlet and triplet Chl energized states, as well as various reactive oxygen species (ROS) like *e.g.* the singlet oxygen (Macedo et al., 2008). The very recent reviews of Latifi et al. (2009) and Pospíšil (2009) on oxidative stress science that came to us almost at the same time were very exciting and inspiring, thus contributing to the next follow up step. Therefore, by interlinking these reviews, we hypothesized that bentazon, like most environmental stresses *e.g.* heavy metals, high light, UV-B, heat, salinity and drought, induced the production of ROS in cyanobacteria, causing

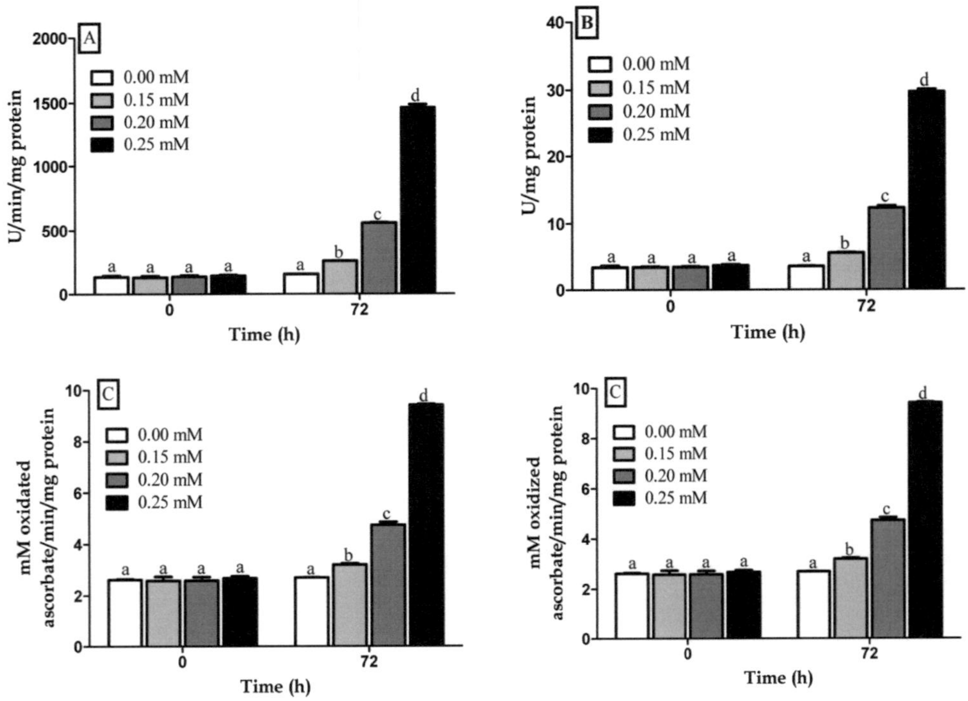

Fig. 2. Effect of molinate on the activities of (A) SOD, (B) CAT, (C) APX, and (D) GST of *A. flos-aquae* after exposure to bentazon for 0 and 72 h. The values (%) relative to controls are means ± SE of three to six independent experiments. Results of the one-way ANOVA factorial analysis. Values with a common letter are not significantly different according to Tukey's test (P<0.05).

oxidative damage. The obtained results confirmed the formulated hypothesis, since the activities of antioxidant enzymes such as superoxide dismutase (SOD), catalase (CAT), ascorbate peroxidase (APX), and glutathione *S*-transferase (GST) increased in a time- and Basagran dose-response manner and were higher than those in the control samples after 72 h for a concentration range of 0.75 to 2 mM. On the basis of these data and information available elsewhere we proposed a hypothetical mechanistic model for Basagran-induced oxidative stress on *A. cylindrica* (Galhano et al., 2010c). A similar trend was attained with the active ingredient of Basagran *i.e.* bentazon in an *Aphanizomenon flos-aquae* strain isolated from "Quinta do Seminário" rice fields, also in the Baixo Mondego River Valley region (Fig. 2). Acute and short-term (72 h) experiments also demonstrated that the active ingredient caused a time- and concentration (0.15-0.25 mM) dependent decrease of *A. flos-aquae* growth, photosynthetic pigments, total protein, carbohydrates, photosynthesis and respiration (Galhano, V., et al., 2010; unpublished results).
Meanwile, we did not forget the antioxidant system machinery response of cyanobacteria exposed to the formulation Ordram in acute and short-termed experiments (72 h). Contrary to Basagran, the thiocarbamate dramatically decreased the activity of all antioxidative stress enzymes tested in a time- and concentration-dependent manner within a concentration range of 0.75 to 2 mM. The collapse of both enzymatic (*e.g.* SOD, CAT, APX, GST) and non-enzymatic (*e.g.* glutathione pool and total carotenoids) antioxidants was evident after 72 hours' exposure to the highest Ordram concentrations. A radical decline in saturated fatty acids was also observed in this study (Galhano et al., 2010b).

7. Conclusions and future prospects

The present environmental concerns about rice field herbicide residues in water, soil and rice foodstuffs will probably not vanish in the next years. Not enough has been done to reduce the risks caused by herbicide pollution in rice agriculture. Even though the switch to low-dose agents has significantly reduced herbicide consumption, most surface water and groundwater samples still contain herbicides, sometimes at levels harmful for human health and for environment. At present, it is not possible to assess the future risks to human beings and to environment caused by the hazardous properties of herbicides used today and the trends will vary from country to country (OECD, 2008). Notwithstanding, to guarantee minimal negative side-effects on rice field ecosystems other than the soil-plant systems, herbicides should have no or low toxicity, except for the target weeds. Improved formulations will be needed to reduce off-target deposition, improve retention on target, and enhance uptake and translocation (Arias-Estévez et al., 2008). As many European rice fields are located in natural parks or environmentally protected areas, the future of rice in the European countries will most likely be interlinked to the development of new varieties and environmentally and economically sustainable methods of production, which allow to increase rice yield and quality by improving water and fertilizer efficiency and minimizing the use of crop protection products like herbicides (Ferrero, 2007; Ferrero & Tinarelli, 2007). At present there is a very considerable literature related to herbicide effects on the flora and fauna of rice fields but a conspicuous lacuna related to rice field biodiversity affected by this kind of xenobiotics still exists. Ecological studies contrasting intensive irrigated rice systems reported in this review with the traditional rainfed rice lands have not been carried out so far. As a result of a heavy dependence on selective herbicides for weed control in rice fields worldwide, a serious problem has emerged in the last years: several weed species have evolved resistance to herbicides, including the most pernicious grass weed, *Echinochloa* spp., becoming resistant to rice gramicidines (Olofsdotter et al., 1998; Vidotto et al., 2007; Gressel & Valverde, 2009). So far, some mechanisms of evolved resistance of rice weedy species are still unknown (Dyer & Weller, 2005). Therefore, in future weed management programs, Gressel & Valverde (2009) recommended the use of transgenic herbicide-resistant rice cultivars to better achieve the control of weeds that have evolved herbicide resistance.

Interestingly, some organisms that have evolved in natural ponds near rice fields are now being used as biological control agents in rice culture, as for example in Japan, where the problem of annual weeds infesting paddy fields has been countered by the introduction of several species of tadpole shrimp (*Triops* spp.) which agitates the soil surface, uproots weed seedlings, creates turbid water (which compromises photosynthesis) and consumes weed buds (Williams, 2006). Therefore, nowadays, as in the future, the commonly accepted best approach to manage rice field weeds is to follow an IWM strategy that includes good land preparation, good water management, a competitive crop, and a judicious herbicide use (Monaco et al., 2002; Upadhyaya & Blackshaw, 2007; Demont et al., 2009).

The use of biological control agents (natural enemies and pathogens) (Castle et al., 2006) and/or biologically-based products (allelochemicals) (Belz, 2007) should also have to be implemented. The use of bioherbicides are also interesting alternatives for use in rice IWM programs (Charudattan et al., 2002; Kendig et al., 2003). A suitable and periodic water quality monitoring together with the improvement of good agricultural practices will be advisable in future IWM programs.

More important yet, rice researchers and farmers must go one step further and rapidly to an era of precision farming, which helps to reduce the cost of production and improve productivity on an ecologically sustainable basis. They should launch a movement for achieving an evergreen revolution in rice farming systems based on ecologically sustainable and location-specific precision farming technologies (Pretty, 2005; Swaminathan & Rao, 2009). Dr. Norman E. Borlaug, The Nobel Prize in Peace of 1970, called it the "Blue Revolution" of the 21st century to complement the so-called "Green Revolution" of the 20th century in order to feed the growing world population (Borlaug, 2004).

In conclusion, we must say that nowadays, not only the scientific community, but also the general public, including rice farmers and extension workers, should be aware of the need for a continual review of rice field herbicides once they have been authorized to be lunched into the market, mainly due to their unpredictable effects on both the environment and human health. We think that this timely and up-to-date review can significantly improve the information in this research area and contribute to a better understanding of the effects of rice field herbicides on non-target organisms, which inevitably will lead to a rationalization of their use in future integrated weed management programs.

8. References

Adrián, J.; Fernández, F.; Muriano, A.; Obregon, R.; Ramón-Azcon, J.; Tort, N. & Marco, M.-P. (2009). Biosensors for pharmaceuticals and emerging contaminants based on novel micro and nanotechnology approaches. In: *Biosensors for Environmental Monitoring of Aquatic Systems: Bioanalytical and Chemical Methods for Endocrine Disruptors*, Vol. 5: Water Pollution, Part J, Barceló, D. & Hansen, P.-D. (Eds.), pp. 47-68, Springer-Verlag, ISBN: 978-3-540-00278-9, Berlin/Heidelberg

Ahluwalia, A.S.; Kaur, M. & Dahuja, S. (2002). Toxicity of a rice field herbicide in some nitrogen-fixing algae. *Indian Journal of Environmental Health,* Vol. 44, No. 4, October 2002, 298-302, ISSN: 0367-827X

Aizawa, H. (2001). *Metabolic Maps: Pesticides, Environmentally Relevant Molecules and Biologically Active Molecules,* Academic Press, ISBN: 0-12-045605-2, San Diego, California

Anastácio, P.M.; Nielsen, S.N.; Frias, A.F. & Marques, J.C. (1999). CRISP (crayfish and rice integrated system of production): 4. Modelling water, algae and oxygen dynamics. *Ecological Modelling,* Vol. 123, No. 1, November 1999, 29-40, ISSN: 0304-3800

Andrade, A.I.A.S.S. & Stigter, T.Y. (2009). Multi-method assessment of nitrate and pesticide contamination in shallow alluvial groundwater as a function of hydrogeological setting and land use. *Agricultural Water Management,* Vol. 96, No. 12, December 2009, 1751-1765, ISSN: 0378-3774

Anonymous (2010). Editorial - How to feed a hungry world. *Nature,* Vol. 466, No. 7306, July 2010, 531-532, ISSN: 0028-0836

Arias-Estévez, M.; López-Periago, E.; Martínez-Carballo, E.; Simal-Gándara, J.; Mejuto, J.-C. & García-Río, L. (2008). The mobility and degradation of pesticides in soils and the pollution of groundwater resources. *Agriculture Ecosystems & Environment,* Vol. 123, No. 4, February 2008, 247-260, ISSN: 0167-8809

Ash, C.; Jasny, B.R.; Malakoff, D.A. & Sugden, A.M. (2010). Feeding the future. *Science,* Vol. 327, No. 5967, February 2010, 797-797, ISSN: 0036-8075

Belz, R.G. (2007). Allelopathy in crop/weed interactions - an update. *Pest Management Science,* Vol. 63, No. 4, April 2007, 308-326, ISSN: 1526-498X

Benfenati, E.; Clook, M.; Fryday, S. & Hart, A. (2007). QSARs for regulatory purposes: the case for pesticide authorization. In: *Quantitative Structure-Activity Relationships (QSAR) for Pesticide Regulatory Purposes,* Benfenati, E. (Ed.), pp. 1-57, Elsevier, ISBN: 978-0-444-52710-3, Amsterdam, The Netherlands; Oxford, UK

Borlaug, N.E. (2004). Prospects for world agriculture in the twenty-first century. In: *Sustainable Agriculture and the International Rice-Wheat System* Lal, R.; Hobbs, P.R.; Uphoff, N. & Hansen, D.O. (Eds.), pp. 1-18, Marcel Dekker, ISBN: 0-8247-5491-3, New York

Burdett, A.S.; Stevens, M.M. & MacMillan, D.L. (2001). Laboratory and field studies on the effect of molinate, clomazone, and thiobencarb on nontarget aquatic invertebrates. *Environmental Toxicology and Chemistry,* Vol. 20, No. 10, October 2001, 2229-2236, ISSN: 0730-7268

CACMV (2010). *Cooperativa Agrícola do Concelho de Montemor-o-Velho. Protecção Integrada do Arroz ("Integrated Pest Management in Rice"),* Secção de Protecção Integrada, Cooperativa Agrícola do Concelho de Montemor-o-Velho (CACMV), Montemor-o-Velho (in Portuguese), 66 pp.

Cabral, J.A.; Marques, J.C. & Nielsen, S.N. (2001). Modeling mosquitofish (*Gambusia holbrooki*) responses to Genapol OXD-080, a non-ionic surfactant, in rice fields. *Ecological Engineering,* Vol. 16, No. 4, February 2001, 537-544, ISSN: 0925-8574

Calha, I.M.; Machado, C. & Rocha, F. (1999). Resistance of *Alisma plantago aquatica* to sulfonylurea herbicides in Portuguese rice fields. *Hydrobiologia,* Vol. 415, November 1999, 289-293, ISSN: 0018-8158

Carlile, W.R. (2006). *Pesticide Selectivity, Health and the Environment,* Cambridge University Press, ISBN: 978-0-521-81194-1, Cambridge, UK; New York

Carlomagno, M.; Mathó, C.; Cantou, G.; Sanborn, J.R.; Last, J.A.; Hammock, B.D.; Roel, A.; González, D. & González-Sapienza, G. (2010). A clomazone immunoassay to study the environmental fate of the herbicide in rice (*Oryza sativa*) agriculture. *Journal of Agricultural and Food Chemistry,* Vol. 58, No. 7, April 2010, 4367-4371, ISSN: 0021-8561

Carvalho, F.P. (2006). Agriculture, pesticides, food security and food safety. *Environmental Science & Policy,* Vol. 9, No. 7-8, November-December 2006, 685-692, ISSN: 1462-9011

Castle, L.A.; Wu, G.S. & McElroy, D. (2006). Agricultural input traits: past, present and future. *Current Opinion in Biotechnology,* Vol. 17, No. 2, April 2006, 105-112, ISSN: 0958-1669

Castro, M.; Silva-Ferreira, A.C.; Manaia, C.M. & Nunes, O.C. (2005). A case study of molinate application in a Portuguese rice field: herbicide dissipation and proposal of a clean-up methodology. *Chemosphere,* Vol. 59, No. 7, May 2005, 1059-1065, ISSN: 0045-6535

Cerejeira, M.J.; Viana, P.; Batista, S.; Pereira, T.; Silva, E.; Valério, M.J.; Silva, A.; Ferreira, M. & Silva-Fernandes, A.M. (2003). Pesticides in Portuguese surface and ground waters. *Water Research,* Vol. 37, No. 5, March 2003, 1055-1063, ISSN: 0043-1354

Charudattan, R.; Chandramohan, S. & Wyss, G.S. (2002). Biological control. In: *Pesticides in Agriculture and the Environment,* Wheeler, W.B. (Ed.), pp. 25-58, Marcel Dekker, ISBN: 0-8247-0809-1, New York

Chen, W.-C.; Yen, J.-H.; Chang, C.-S. & Wang, Y.-S. (2009). Effects of herbicide butachlor on soil microorganisms and on nitrogen-fixing abilities in paddy soil. *Ecotoxicology and Environmental Safety,* Vol. 72, No. 1, January 2009, 120-127, ISSN: 0147-6513

Clay, J.W. (2004). *World Agriculture and the Environment: A Commodity-by-Commodity Guide to Impacts and Practices,* Island Press, ISBN: 1-55963-367-0, Washington, D.C.

Comoretto, L.; Arfib, B.; Talva, R.; Chauvelon, P.; Pichaud, M.; Chiron, S. & Höhener, P. (2008). Runoff of pesticides from rice fields in the Ile de Camargue (Rhône river delta, France): Field study and modeling. *Environmental Pollution,* Vol. 151, No. 3, February 2008, 486-493, ISSN: 0269-7491

Crestani, M.; Menezes, C.; Glusczak, L.; Miron, D.d.S.; Lazzari, R.; Duarte, M.F.; Morsch, V.M.; Pippi, A.L. & Vieira, V.P. (2006). Effects of clomazone herbicide on hematological and some parameters of protein and carbohydrate metabolism of silver catfish *Rhamdia quelen. Ecotoxicology and Environmental Safety,* Vol. 65, No. 1, September 2006, 48-55, ISSN: 0147-6513

DGADR (2009). *Actualização dos Produtos Fitofarmacêuticos Aconselhados em Protecção Integrada: Arroz* - Actualizado a 28 de Outubro de 2009 ("*Integrated Pest Management in Rice*"), Direcção-Geral de Agricultura e Desenvolvimento Rural; Ministério da Agricultura, do Desenvolvimento Rural e das Pescas (in Portuguese; last update: 28/10/2009), 4 pp.

De Datta, S.K. (1981). *Principles and Practices of Rice Production*, John Wiley & Sons, ISBN: 0-471-08074-8, New York

Demont, M.; Rodenburg, J.; Diagne, M. & Diallo, S. (2009). *Ex ante* impact assessment of herbicide resistant rice in the Sahel. *Crop Protection,* Vol. 28, No. 9, September 2009, 728-736, ISSN: 0261-2194

Dyer, W.E. & Weller, S.C. (2005). Plant response to herbicides. In: *Plant Abiotic Stress*, Jenks, M.A. & Hasegawa, P.M. (Eds.), pp. 171-214, Blackwell Publishing, ISBN: 978-1-4051-2238-2, Oxford, UK; Ames, Iowa, USA

EC (2000). Council of the European Communities: Directive 2000/60/EC Establishing a Framework for Community Action in the Field of Water Policy. *Official Journal of European Community,* L 327, 1

EC (2000). Directive 2000/60/EC of the European Parliament and of the Council of 23 October 2000 establishing a framework for Community action in the field of water policy. *Official Journal of European Community,* L 327: 77, 22.12.2000

EC (2010). *European Commission. Environment Policies Water website* [Online]. Available: http://ec.europa.eu/environment/water/index_en.htm [last updated on 22/09/2010; last accessed on 3 November 2010)].

EC (2008). Council of the European Communities: Directive 2008/105/EC on Environmental Quality Standards in the Field of Water Policy. *Official Journal of European Community,* L 348, 84

EEC (1991). Council Directive of 15 July 1991 concerning the placing of plant protection products on the market. No. 91/414/EEC. *Official Journal of European Community,* L 230: 137, European Union, Brussels

EU (2006). *The DG Environment REACH website* [Online]. Available: http://ec.europa.eu/environment/chemicals/reach/reach_intro.htm [last updated on 20/05/2010; last accessed on 3 November 2010)].

Faria, M.S.; Nogueira, A.J.A. & Soares, A.M.V.M. (2007). The use of *Chironomus riparius* larvae to assess effects of pesticides from rice fields in adjacent freshwater ecosystems. *Ecotoxicology and Environmental Safety,* Vol. 67, No. 2, June 2007, 218-226, ISSN: 0147-6513

Fernández Valiente, E.; Ucha, A.; Quesada, A.; Leganés, F. & Carreres, R. (2000). Contribution of N_2 fixing cyanobacteria to rice production: availability of nitrogen from ^{15}N-labelled cyanobacteria and ammonium sulphate to rice. *Plant and Soil,* Vol. 221, No. 1, 107-112, ISSN: 0032-079X

Ferraz, D.G.d.B.; Sabater, C. & Carrasco, J.M. (2004). Effects of propanil, tebufenozide and mefenacet on growth of four freshwater species of phytoplankton: a microplate bioassay. *Chemosphere,* Vol. 56, No. 4, July 2004, 315-320, ISSN: 0045-6535

Ferrero, A. (2007). Rice scenario in the European Union. *Cahiers Agricultures,* Vol. 16, No. 4, July-August 2007, 272-277, DOI: 10.1684/agr.2007.0108

Ferrero, A. & Tinarelli, A. (2007). Rice cultivation in the E.U. ecological conditions and agronomical practices. In: *Pesticide Risk Assessment in Rice Paddies: Theory and*

Practice, Capri, E. & Karpouzas, D.G. (Eds.), pp. 1-24, Elsevier, ISBN: 978-0-444-53087-5, Amsterdam

Galhano, V.; de Figueiredo, D.R.; Alves, A., Correia, A.; Pereira, M.J.; Gomes-Laranjo, J. & Peixoto, F. (2010a). Morphological, biochemical and molecular characterization of *Anabaena, Aphanizomenon* and *Nostoc* strains (Cyanobacteria, Nostocales) isolated from Portuguese freshwater habitats. *Hydrobiologia* (accepted for publication)

Galhano, V.; Gomes-Laranjo, J. & Peixoto, F. (2010b). Exposure of the Portuguese rice field cyanobacterium *Nostoc muscorum* to molinate (Ordram®): Effects on the antioxidant system and fatty acid profile. *Aquatic Toxicology* (accepted for publication)

Galhano, V.; Peixoto, F. & Gomes-Laranjo, J. (2010c). Bentazon triggers the promotion of oxidative damage in the portuguese ricefield cyanobacterium *Anabaena cylindrica*: Response of the antioxidant system. *Environmental Toxicology,* Vol. 25, No. 5, October 2010, 517-526, ISSN: 1520-4081

Galhano, V.; Peixoto, F.; Gomes-Laranjo, J. & Fernández-Valiente, E. (2009). Differential effects of bentazon and molinate on *Anabaena cylindrica,* an autochthonous cyanobacterium of Portuguese rice field agro-ecosystems. *Water Air and Soil Pollution,* Vol. 197, No. 1-4, February 2009, 211-222, ISSN: 0049-6979

Galhano, V.; Peixoto, F.; Gomes-Laranjo, J. & Fernández-Valiente, E. (2010d). Comparative toxicity of bentazon and molinate on growth, photosynthetic pigments, photosynthesis, and respiration of the Portuguese ricefield cyanobacterium *Nostoc muscorum*. *Environmental Toxicology,* Vol. 25, No. 2, April 2010, 147-156, ISSN: 1520-4081

Galhano, V.M.F.; Gomes-Laranjo, J.C.E. & Peixoto, F.M.P. (2008). Growth and photosynthetic performance of the ricefield cyanobacterium *Anabaena cylindrica* to the herbicide bentazon. *Biochimica Et Biophysica Acta-Bioenergetics,* Vol. 1777, July 2008, S24-S24, ISSN: 0005-2728

Gan, J. & Bondarenko, S. (2008). Determination of pesticides in water. In: *Analysis of Pesticides in Food and Environmental Samples,* Tadeo, J.L. (Ed.), pp. 231-256, CRC Press, ISBN: 978-0-8493-7552-1, Boca Raton

Garagna, S.; Vasco, C.; Merico, V.; Esposito, A.; Zuccotti, M. & Redi, C.A. (2005). Effects of a low dose of bentazon on spermatogenesis of mice exposed during foetal, postnatal and adult life. *Toxicology,* Vol. 212, No. 2-3, September 2005, 165-174, ISSN: 0300-483X

González Tomé, R., (1996). *Estudio de Diversas Técnicas Agronómicas en Relación con las Cianobacterias Fixadoras de N_2 en el Cultivo del Arroz*. Ph.D Thesis. Facultad de Ciencias Biológicas, Universidad de Valencia, Valencia [co-supervised by Prof. Eduardo Fernández-Valiente (UAM, Madrid)] (in Spanish)

Granby, K.; Petersen, A.; Herrmann, S.S. & Poulsen, M.E. (2008). Levels of pesticides in food and food safety aspects. In: *Analysis of Pesticides in Food and Environmental Samples,* Tadeo, J.L. (Ed.), pp. 287-318, CRC Press, ISBN: 978-0-8493-7552-1, Boca Raton

Gressel, J. & Valverde, B.E. (2009). A strategy to provide long-term control of weedy rice while mitigating herbicide resistance transgene flow, and its potential use for other crops with related weeds. *Pest Management Science,* Vol. 65, No. 7, July 2009, 723-731, ISSN: 1526-498X

Hussey, D.J. & Bell, G.M. (2004). Regulation of pesticides and biocides in the European Union. In: *Pesticide Toxicology and International Regulation,* Marrs, T.C. & Ballantyne, B. (Eds.), pp. 501-512, John Wiley & Sons, ISBN: 0-471-49644-8, Chichester, West Sussex, England; Hoboken, New Jersey, USA

Ishihara, S. (2009). Studies on risk assessment procedure of rice herbicides for microalgae in river ecosystem. *Journal of Pesticide Science,* Vol. 34, No. 3, 184-186, ISSN: 1348-589X

Kang, H.S.; Park, C.J. & Gye, M.C. (2009). Effects of molinate on survival and development of *Bombina orientalis* (Boulenger) embryos. *Bulletin of Environmental Contamination and Toxicology,* Vol. 82, No. 3, March 2009, 305-309, ISSN: 0007-4861

Kaur, M.; Ahluwalia, A.S. & Dahuja, S. (2002). Toxicity of a rice field herbicide in a nitrogen fixing alga, *Cylindrospermum* sp. *Journal of Environmental Biology,* Vol. 23, No. 4, 359-363, ISSN: 0254-8704

Kendig, A.; Williams, B. & Smith, C.W. (2003). Rice weed control. In: *Rice: Origin, History, Technology, and Production,* Smith, C.W. & Dilday, R.H. (Eds.), pp. 457-472, John Wiley & Sons, ISBN: 0-471-34516-4, Hoboken, New Jersey

Kim, J.-D. & Lee, C.-G. (2006). Differential responses of two freshwater cyanobacteria, *Anabaena variabillis* and *Nostoc commune,* to sulfonylurea herbicide bensulfulron-methyl. *Journal of Microbiology and Biotechnology,* Vol. 16, No. 1, January 2006, 52-56, ISSN: 1017-7825

Köck, M.; Farré, M.; Martínez, E.; Gajda-Schrantz, K.; Ginebreda, A.; Navarro, A.; de Alda, M.L. & Barceló, D. (2010). Integrated ecotoxicological and chemical approach for the assessment of pesticide pollution in the Ebro River delta (Spain). *Journal of Hydrology,* Vol. 383, No. 1-2, March 2010, 73-82, ISSN: 0022-1694

Krieger, R. (Ed.) (2010). *Hayes' Handbook of Pesticide Toxicology, Volumes 1 & 2,* 3rd Ed., Academic Press, ISBN: 978-0-12-374367-1, San Diego, California

Kudsk, P. & Streibig, J.C. (2003). Herbicides - a two-edged sword. *Weed Research,* Vol. 43, No. 2, April 2003, 90-102, ISSN: 0043-1737

Latifi, A.; Ruiz, M. & Zhang, C.-C. (2009). Oxidative stress in cyanobacteria. *FEMS Microbiology Reviews,* Vol. 33, No. 2, March 2009, 258-278, ISSN: 0168-6445

Leganés, F. & Fernández-Valiente, E. (1992). Effects of phenoxy acetic herbicides on growth, photosynthesis, and nitrogenase activity in cyanobacteria from rice fields. *Archives of Environmental Contamination and Toxicology,* Vol. 22, No. 1, January 1992, 130-134, ISSN: 0090-4341

Leitão, S.; Pinto, P.; Pereira, T. & Brito, M.F. (2007). Spatial and temporal variability of macroinvertebrate communities in two farmed Mediterranean rice fields. *Aquatic Ecology,* Vol. 41, No. 3, September 2007, 373-386, ISSN: 1386-2588

Liebman, M. (2001). Weed management: a need for ecological approaches. In: *Ecological Management of Agricultural Weeds,* Liebman, M.; Mohler, C.L. & Staver, C.P. (Eds.), pp. 1-39, Cambridge University Press, ISBN: 0-521-56068-3, Cambridge, UK; New York, USA

Lima, A.S. (1997). Current situation of rice production in Portugal and the main diseases that occur. In: *Rice Diseases in the Mediterranean Region and Breeding for Resistance,*

Vol. 15 (3), Chataigner, J. (Ed.), pp. 29-40, Cahiers Options Méditerranéennes (CIHEAM), Montpellier

Lü, Z.; Sang, L.; Li, Z. & Min, H. (2009). Catalase and superoxide dismutase activities in a *Stenotrophomonas maltophilia* WZ2 resistant to herbicide pollution. *Ecotoxicology and Environmental Safety,* Vol. 72, No. 1, January 2009, 136-143, ISSN: 0147-6513

Macedo, R.S.; Lombardi, A.T.; Omachi, C.Y. & Rörig, L.R. (2008). Effects of the herbicide bentazon on growth and photosystem II maximum quantum yield of the marine diatom *Skeletonema costatum. Toxicology in Vitro,* Vol. 22, No. 3, April 2008, 716-722, ISSN: 0887-2333

Mackay, D.; Shiu, W.Y.; Ma, K.-C. & Lee, S.C. (2006). *Handbook of Physical-Chemical Properties and Environmental Fate for Organic Chemicals, Volumes 1-4,* 2nd Ed., CRC Press/Taylor & Francis, ISBN: 978-1-56670-687-2, Boca Raton, Florida

Marchesan, E.; Zanella, R.; de Avila, L.A.; Camargo, E.R.; Machado, S.L.d.O. & Macedo, V.R.M. (2007). Rice herbicide monitoring in two Brazilian rivers during the rice growing season. *Scientia Agricola,* Vol. 64, No. 2, March-April 2007, 131-137, ISSN: 0103-9016

Marques, C.R.; Abrantes, N.; de Figueiredo, D.R.; Pereira, M.J. & Gonçalves, F. (2008). Are *Pseudokirchneriella subcapitata* and *Chlorella vulgaris* affected by environmental samples from a rice field? *Water Air and Soil Pollution,* Vol. 189, No. 1-4, March 2008, 49-59, ISSN: 0049-6979

Marrs, T.C. & Ballantyne, B. (Eds.) (2004). *Pesticide Toxicology and International Regulation,* John Wiley & Sons, ISBN: 0-471-49644-8, Chichester, West Sussex, England; Hoboken, New Jersey, USA

Min, H.; Ye, Y.-F.; Chen, Z.-Y.; Wu, W.-X. & Du, Y. (2001). Effects of butachlor on microbial populations and enzyme activities in paddy soil. *Journal of Environmental Science and Health Part B-Pesticides Food Contaminants and Agricultural Wastes,* Vol. 36, No. 5, July 2001, 581-595, ISSN: 0360-1234

Miron, D.d.S.; Crestani, M.; Shettinger, M.R.; Morsch, V.M.; Baldisserotto, B.; Tierno, M.A.; Moraes, G. & Vieira, V.L.P. (2005). Effects of the herbicides clomazone, quinclorac, and metsulfuron methyl on acetylcholinesterase activity in the silver catfish (*Rhamdia quelen*) (Heptapteridae). *Ecotoxicology and Environmental Safety,* Vol. 61, No. 3, July 2005, 398-403, ISSN: 0147-6513

Monaco, T.J.; Weller, S.C. & Ashton, F.M. (2002). *Weed Science: Principles and Practices,* 4th Ed., John Wiley & Sons, Inc., ISBN: 0-471-37051-7, New York

Moraes, B.S.; Loro, V.L.; Pretto, A.; da Fonseca, M.B.; Menezes, C.; Marchesan, E.; Reimche, G.B. & de Avila, L.A. (2009). Toxicological and metabolic parameters of the teleost fish (*Leporinus obtusidens*) in response to commercial herbicides containing clomazone and propanil. *Pesticide Biochemistry and Physiology,* Vol. 95, No. 2, October 2009, 57-62, ISSN: 0048-3575

Nielsen, E.; Østergaard, G. & Larsen, J.C. (2008). *Toxicological Risk Assessment of Chemicals: A Practical Guide,* Informa Healthcare, ISBN: 978-0-8493-7265-0, New York

Normile, D. (2008). Reinventing rice to feed the world. *Science,* Vol. 321, No. 5887, July 2008, 330-333, ISSN: 0036-8075

OECD (2008). *OECD Environmental Outlook to 2030*, OECD Publishing, ISBN: 978-92-64-04048-9, Paris

Olofsdotter, M.; Watson, A. & Piggin, C. (1998). Weeds: a looming problem in modern rice production. In: *Sustainability of Rice in the Global Food System*, Dowling, N.G.; Greenfield, S.M. & Fisher, K.S. (Eds.), pp. 165-173, Pacific Basin Study Center; International Rice Research Institute, ISBN: 971-22-0107-4, Davis, California, USA; Manila, Philippines

Padhy, R.N. (1985). Cyanobacteria and pesticides. *Residue Reviews*, Vol. 95, 1-44, ISSN: 0080-181X

Pandey, D.K.; Mishra, N. & Singh, P. (2005). Relative phytotoxicity of hydroquinone on rice (*Oryza sativa* L.) and associated aquatic weed green musk chary (*Chara zeylanica* Willd.). *Pesticide Biochemistry and Physiology*, Vol. 83, No. 2-3, October-November 2005, 82-96, ISSN: 0048-3575

Pereira, T.; Cerejeira, M.J. & Espírito-Santo, J. (2000). Use of microbiotests to compare the toxicity of water samples fortified with active ingredients and formulated pesticides. *Environmental Toxicology*, Vol. 15, No. 5, December 2000, 401-405, ISSN: 1520-4081

Peterson, H.G.; Boutin, C.; Freemark, K.E. & Martin, P.A. (1997). Toxicity of hexazinone and diquat to green algae, diatoms, cyanobacteria and duckweed. *Aquatic Toxicology*, Vol. 39, No. 2, September 1997, 111-134, ISSN: 0166-445X

Phong, T.K.; Nhung, D.T.T.; Motobayashi, T. & Watanabe, H. (2009). Behavior of simetryn and thiobencarb in the plough zone of rice fields. *Bulletin of Environmental Contamination and Toxicology*, Vol. 83, No. 6, December 2009, 794-798, ISSN: 0007-4861

Pingali, P.L. & Roger, P.A. (Eds.) (1995). *Impact of Pesticides on Farmer Health and the Rice Environment*, Kluwer Academic Publishers, ISBN: 0-7923-9522-0, Massachusetts/ Dordrecht

Pospíšil, P. (2009). Production of reactive oxygen species by photosystem II. *Biochimica Et Biophysica Acta-Bioenergetics*, Vol. 1787, No. 10, October 2009, 1151-1160, ISSN: 0005-2728

Pretty, J. (2005). Sustainability in agriculture: Recent progress and emergent challenges. In: *Sustainability in Agriculture*, Hester, R.E. & Harrison, R.M. (Eds.), pp. 1-15, RSC Publishing, ISBN: 0-85404-201-6, Cambridge

Racke, K.D. (2007). Pesticide residues in food and international trade: Regulation and safety considerations. In: *Pesticide Chemistry: Crop Protection, Public Health, Environmental Safety*, Ohkawa, H.; Miyagawa, H. & Lee, P.W. (Eds.), pp. 29-41, Wiley-VCH, ISBN: 978-3-527-31663-2, Weinheim

Reche, M.H.L.R. & Fiuza, L.M. (2005). Bacterial diversity in rice-field water in Rio Grande do Sul. *Brazilian Journal of Microbiology*, Vol. 36, No. 3, July-September 2005, 253-257, ISSN: 1517-8382

Roger, P.A. (1996). *Biology and Management of the Floodwater Ecosystem in Ricefields*, International Rice Research Institute, ISBN: 971-22-0068-X, Manila, Philippines

Sabater, C. & Carrasco, J.M. (1998). Effects of molinate on growth of five freshwater species of phytoplankton. *Bulletin of Environmental Contamination and Toxicology,* Vol. 61, No. 4, October 1998, 534-540, ISSN: 0007-4861

Saeki, M. & Toyota, K. (2004). Effect of bensulfuron-methyl (a sulfonylurea herbicide) on the soil bacterial community of a paddy soil microcosm. *Biology and Fertility of Soils,* Vol. 40, No. 2, July 2004, 110-118, ISSN: 0178-2762

Saka, M. (1999). Acute toxicity tests on Japanese amphibian larvae using thiobencarb, a component of rice paddy herbicides. *Herpetological Journal,* Vol. 9, No. 2, April 1999, 73-81, ISSN: 0268-0130

Schriever, C.A. & Liess, M. (2007). Mapping ecological risk of agricultural pesticide runoff. *Science of the Total Environment,* Vol. 384, No. 1-3, October 2007, 264-279, ISSN: 0048-9697

Siebers, J. & Hänel, R. (2003). Assessment of residue analytical methods for crops, food, feed, and environmental samples: the approach of the European Union. In: *Handbook of Residue Analytical Methods for Agrochemicals, Volumes 1 & 2,* Lee, P.W.; Aizawa, H.; Barefoot, A.C. & Murphy, J.J. (Eds.), pp. 13-37, Wiley, ISBN: 0-4714-9194-2, Chichester, West Sussex, England / Hoboken, NJ, USA

Silva, E.; Batista, S.; Viana, P.; Antunes, P.; Serôdio, L.; Cardoso, A.T. & Cerejeira, M.J. (2006). Pesticides and nitrates in groundwater from oriziculture areas of the 'Baixo Sado' region (Portugal). *International Journal of Environmental Analytical Chemistry,* Vol. 86, No. 13, November 2006, 955-972, ISSN: 0306-7319

Singh, S. & Datta, P. (2005). Growth and survival potentials of immobilized diazotrophic cyanobacterial isolates exposed to common ricefield herbicides. *World Journal of Microbiology & Biotechnology,* Vol. 21, No. 4, June 2005, 441-446, ISSN: 0959-3993

Singh, S. & Datta, P. (2007). Outdoor evaluation of herbicide resistant strains of *Anabaena variabilis* as biofertilizer for rice plants. *Plant and Soil,* Vol. 296, No. 1-2, July 2007, 95-102, ISSN: 0032-079X

Srivastava, S.; Goyal, P. & Mohan Srivastava, M. (2010). Pesticides: Past, present, and future. In: *Handbook of Pesticides: Methods of Pesticide Residues Analysis,* Nollet, L.M.L. & Rathore, H.S. (Eds.), pp. 47-65, Taylor & Francis, ISBN: 978-1-4200-8245-6, Boca Raton, Florida

Stenersen, J. (2004). *Chemical Pesticides: Mode of Action and Toxicology,* CRC Press, ISBN: 0-7484-0910-6, Boca Raton, Florida

Stoate, C.; Báldi, A.; Beja, P.; Boatman, N.D.; Herzon, I.; van Doorn, A.; de Snoo, G.R.; Rakosy, L. & Ramwell, C. (2009). Ecological impacts of early 21st century agricultural change in Europe - A review. *Journal of Environmental Management,* Vol. 91, No. 1, October 2009, 22-46, ISSN: 0301-4797

Streloke, M. (2007). Current EU regulation in the field of ecotoxicology. In: *Pesticide Chemistry: Crop Protection, Public Health, Environmental Safety,* Ohkawa, H.; Miyagawa, H. & Lee, P.W. (Eds.), pp. 403-414, Wiley-VCH, ISBN: 978-3-527-31663-2, Weinheim

Suseela, M.R. (2001). Effect of butachlor on growth and nitrogen fixation by *Anabaena sphaerica*. *Journal of Environmental Biology,* Vol. 22, No. 3, July 2001, 201-203, ISSN: 0254-8704

Sutawan, N. (2005). The need to keep irrigated rice culture sustainable. In: *Rice is Life: Scientific Perspectives for the 21st Century. Proceedings of the World Rice Research Conference held in Tokyo and Tsukuba, Japan, 4-7 November 2004 [CD-ROM]*, Toriyama, K.; Heong, K.L. & Hardy, B. (Eds.), pp. 329-331, IRRI; Japan International Research Center for Agricultural Sciences, ISBN: 971-220-204-6 Manila, Philippines; Tsukuba, Japan

Swaminathan, M.S. & Rao, S.A. (2009). Rice for sustainable food and nutritions security. In: *Rice Improvement in the Genomics Era*, Datta, S.K. (Ed.), pp. 375-415, CRC Press, ISBN: 978-1-56022-952-0, Boca Raton, Florida

Tarazona, J.V. & Dohmen, G.P. (2007). Ecotoxicology of rice pesticides. In: *Pesticide Risk Assessment in Rice Paddies: Theory and Practice*, Capri, E. & Karpouzas, D.G. (Eds.), pp. 69-90, Elsevier, ISBN: 978-0-444-53087-5, Amsterdam

Tomlin, C.D.S. (2000). *The Pesticide Manual: A World Compendium*, 12th Ed., British Crop Protection Council, ISBN: 978-1901396140, Farnham, Surrey, UK

Uno, S.; Shiraishi, H.; Hatakeyama, S.; Otsuki, A. & Koyama, J. (2001). Accumulative characteristics of pesticide residues in organs of bivalves (*Anodonta woodiana* and *Corbicula leana*) under natural conditions. *Archives of Environmental Contamination and Toxicology*, Vol. 40, No. 1, January 2001, 35-47, ISSN: 0090-4341

Upadhyaya, M.K. & Blackshaw, R.E. (2007). Non-chemical weed management: Synopsis, integration and the future. In: *Non-chemical Weed Management: Principles, Concepts and Technology*, Upadhyaya, M.K. & Blackshaw, R.E. (Eds.), pp. 201-209, CABI, ISBN: 978-1-84593-290-9, Wallingford, UK; Cambridge, MA

Vendrell, E.; Ferraz, D.G.d.B.; Sabater, C. & Carrasco, J.M. (2009). Effect of glyphosate on growth of four freshwater species of phytoplankton: a microplate bioassay. *Bulletin of Environmental Contamination and Toxicology*, Vol. 82, No. 5, May 2009, 538-542, ISSN: 0007-4861

Vidotto, F.; Ferrero, A.; Bertoia, O.; Gennari, M. & Cignetti, A. (2004). Dissipation of pretilachlor in paddy water and sediment. *Agronomie*, Vol. 24, No. 8, December 2004, 473-479, ISSN: 0249-5627

Vidotto, F.; Tesio, F.; Tabacchi, M. & Ferrero, A. (2007). Herbicide sensitivity of *Echinochloa* spp. accessions in Italian rice fields. *Crop Protection*, Vol. 26, No. 3, March 2007, 285-293, ISSN: 0261-2194

Vogelezang-Stoute, E. (2003). The authorisation of pesticides in the light of sustainability. In: *Pesticides: Problems, Improvements, Alternatives*, den Hond, F.; Groenewegen, P. & van Straalen, N.M. (Eds.), pp. 31-52, Blackwell Science, ISBN: 0-632-05659-2, Oxford, UK; Malden, MA, USA

Whitton, B.A. (2000). Soils and rice-fields. In: *The Ecology of Cyanobacteria: Their Diversity in Time and Space*, Whitton, B.A. & Potts, M. (Eds.), pp. 233-255, Kluwer Academic Publishers, ISBN: 0-7923-4735-8, Dordrecht

Williams, D.D. (2006). *The Biology of Temporary Waters*, Oxford University Press, ISBN: 0-19-852812-4, Oxford; New York

Wilson, A.L.; Stevens, M.M. & Watts, R.J. (2000). Acute and chronic toxicity of the herbicide benzofenap (Taipan 300) to *Chironomus tepperi* Skuse (Diptera: Chironomidae) and *Isidorella newcombi* (Adams and Angas) (Gastropoda: Planorbidae). *Archives of*

Environmental Contamination and Toxicology, Vol. 38, No. 2, February 2000, 176-181, ISSN: 0090-4341

Yamamoto, H. & Nakamura, K. (2003). Sampling sediment and water in rice paddy fields and adjacent water bodies. In: *Handbook of Residue Analytical Methods for Agrochemicals, Volumes 1 & 2,* Lee, P.W.; Aizawa, H.; Barefoot, A.C. & Murphy, J.J. (Eds.), pp. 892-907, Wiley, ISBN: 0-4714-9194-2, Chichester, West Sussex, England / Hoboken, NJ, USA

Zimdahl, R.L. (2007). *Fundamentals of Weed Science,* 3rd Ed., Elsevier, Inc./Academic Press, ISBN: 978-0-12-372518-9, San Diego, California

3

Transient Effect of the Herbicide Flumioxazin on Physiology of *Vitis vinifera* L. cv. Pinot Meunier

Bigot Aurélie[1], Clément Christophe[2] and Vaillant-Gaveau Nathalie[2]
[1]*Laboratoire d'Eco-Toxicologie, URVVC-SE, EA 2069, Université de Reims Champagne-Ardenne, UFR Sciences Exactes et Naturelles, Bâtiment 18, Moulin de la Housse – BP 1039, F-51687 REIMS Cedex 2*
[2]*Laboratoire de Stress, Défenses et Reproduction des Plantes, URVVC-SE, EA 2069, Université de Reims Champagne-Ardenne, UFR Sciences Exactes et Naturelles, Bâtiment 18, Moulin de la Housse – BP 1039, F-51687 REIMS Cedex 2*
France

1. Introduction

Pesticides are widely used to control pests and diseases in crop production. Flumioxazin (fmx), or 2-[7-fluoro-3,4-dihydro-3-oxo-4-(2-propynyl)-2H-1,4-benzoxazin-6-yl]-4,5,6,7-tetrahydro-1H-isoindole-1,3(2H)-dione, is a N-phenylphthalimide herbicide registered for pre-emergence control of broadleaved weeds in peanut (*Arachis hypogaea* L.), soybean (*Glycine max* L.), sorghum (Grichar, 2005) and as an early pre-plant burndown treatment in cotton (*Gossypium hirsutum* L.) (Main et al., 2003).

Fmx inhibits protoporphyrinogen oxidase (protox) in the chlorophyll biosynthetic pathway, resulting in light-induced membrane lipid peroxidation (Scott et al., 2001). In the presence of protox inhibitors, tetrapyrroles accumulate, especially protoporphyrin IX (proto IX). Protox inhibition leads to the accumulation of its substrate protoporphyrinogen, which is readily oxidized to proto IX by oxidative enzymes. Proto IX is a quite effective photosensitizer that transfers absorbed light energy to molecular oxygen to form singlet oxygen. The singlet oxygen peroxidizes lipids leading to the destruction of cellular membranes (Moreland, 1999).

Fmx is a pre-emergence herbicide applied on soil at the end of winter at a concentration of 5 mM. Fmx inhibits the development of redroot pigweed (*Amaranthus retroflexus*), lambsquaeter (*Chenopodium album*), jimsonweed (*Datura stramonium*), morningglory (*Ipomoea spp.*), nutsedge (*Cyperus spp.*) and prickly sida (*Sida spinosa* L.) (Nagano, 1999; Niekamp et al., 1999).

Fmx enters plants mainly throughout root and tolerant crop species avoid its injury by rapid detoxication metabolism (Yoshida et al., 1991). Fmx is applied to control adventive plants but the presence of such molecules in the foliage of non-target crops and in the soil has been reported (Jame et al., 1999). Little information is available on the effect of fmx on crop physiology, especially in grapevine, although it is one of the most frequently used herbicides in vineyards. We showed that fmx dramatically affects grapevine physiology *in vitro* (Saladin et al., 2003a, b, c, d). Various concentrations of this herbicide have a negative impact on vine plantlet leaf growth, as revealed by tissue dehydration and cell membrane alteration, decrease in osmotic potential and accumulation of proline. Moreover, fmx treatment results in a reduction of plantlet growth and photosynthesis and further induces some perturbation in leaf carbohydrate partitioning (Saladin et al., 2003b). Proteomic analysis of grapevine under fmx stress suggested that photosynthesis-related proteins, enzymes involved in photorespiration and enzymes of sugar metabolism were impaired (Castro et al., 2005). However, these results were obtained with juvenile plantlets grown *in vitro* and thus have to be considered cautiously before being extended to the whole plant cultivated in vineyards.

The aim of this study is to further determine the effects of fmx treatment on the photosynthetic characteristics of grapevine cutting leaves. The combined measurements of chlorophyll fluorescence and gas-exchange rates were proved to be a useful approach for distinguishing stomatal *versus* nonstomatal effects, as well as for estimating the importance of various types of energy use, such as thermal dissipation and photorespiration (Hendrickson et al., 2004).

2. Materials and methods

2.1 Plant material, growth conditions

Canes of *Vitis vinifera* L. cv. Pinot Meunier were collected in winter, treated with cryptonol (2% v/v) to prevent contamination and stored in the dark at 4 °C for a minimum of two weeks. They were then cut into fragments to obtain two consecutive fertile buds and one sterile bud (Mullins, 1966; Mullins & Rajasekaran, 1981). The sterile bud was soaked for three min in a 0.1% (w/v) 3-indol butyric acid aqueous solution in order to stimulate rhizogenesis. Then, the cuttings were placed in 300 ml pots containing perlite: sand (1: 2) at 25 °C and 75% relative humidity in the greenhouse, with a 16 h photoperiod at a photosynthetic photon flux density of 400 µmol $m^{-2}s^{-1}$ (Lebon et al., 2005).

2.2 Fmx treatments

Plants were daily irrigated with a nutrient solution optimized for grapevine culture (Coïc & Lesaint, 1971). After eight weeks, when the cuttings had eight leaves, the fmx solution (commercial herbicide Pledge®) was sprayed only one time on the soil with different solutions of fmx in water: 0.5 mM, 5 mM (concentration recommended by the manufacturer) or 50 mM. Simultaneously, the soil of control cuttings was sprayed with water.

2.3 Growth measurements

At the end of the experimentation, ten plants per treatment were harvested, separated into shoot and root parts, and their fresh weights were determined.

2.4 Measure of gas exchanges

The net photosynthetic rate (Pn), the stomatal conductance (gs), the intercellular CO_2 concentration (Ci) and the transpiration rate (T) were measured using a portable infrared gas analyser (LI-Cor Model 6400, Lincoln, NE, USA). The infrared gas analysis system was equipped with a clamp-on leaf cuvette that exposed 6 cm^2 of leaf area. Light, temperature and humidity were 400 µmol m^{-2} s^{-1}, 25±1 °C and 30% respectively.

Photosynthetic light response curves: Response of Pn to photosynthetic photon flux (PPF) was measured by illuminating the leaf at decreasing PPF (from 2000 to 0 µmol m^{-2} s^{-1}) until Pn was constant. The apparent quantum yield of CO_2 fixation (ΦCO_2) was calculated as the slope of the linear portion of the response curves between 0 and 100 µmol m^{-2} s^{-1} PPF. CO_2 was maintained at a constant level of 360 mmol l^{-1} using an LI-6400-01 CO_2 injector (LI-Cor 6400 Lincoln, NE, USA) with a high pressure liquefied CO_2 cartridge source.

2.5 Chlorophyll fluorescence measurements

The chlorophyll *a* fluorescence of the leaves was quantified on attached leaves with an IMAGING-PAM Chlorophyll Fluorometer (Walz, Effeltrich, Germany). The measuring system applies array of blue light-emitting diodes (LEDs) (peak wavelength, 470 nm) for saturating light pulses. The frequency of the pulses was adjusted to 10 Hz. Measurements were carried out at maximal distance between the camera and the leaf, corresponding to a 25 x 34 mm area. The image captured by the charge-coupled device (CCD) camera was composed of 640 x 480 pixels.

During the whole experiment, the measurements were systematically performed on the adaxial side on the central parts of the young leaves. The leaves used for measurements were pre-conditioned in the dark. The initial fluorescence (F_o) was obtained after 0.5 hour of dark adaptation. Maximal fluorescence (F_m) was obtained with a saturating flash (1 s, 13 000 µmol m^{-2} s^{-1}). The ratio of variable to maximal fluorescence (F_v/F_m) was calculated. The protocol for fluorescence measurement was similar to the one described by Genty *et al.* (1989), but the measurements were performed on attached leaves. The relative quantum yield of PSII (Φ_{PSII}) at steady state is defined as $(F'_m-F_s)/F'_m$ where F_s and F'_m are respectively steady-state fluorescence and maximum fluorescence in the light. ΦPSII represents the number of electrons transported by a PSII reaction centre per mole of quanta absorbed by PSII. Both Photochemical (q_P) and total non-photochemical quenching (q_{NP}) were calculated according to van Kooten & Snel (1990). The Stern-Volmer equation (NPQ) was used as an indicator of the activity of energy dissipation in the pigment bed of PSII. NPQ was proportional to the effective rate constant for energy dissipation in the antennae as well as in the concentration of quenching centres (Demmig-Adams *et al.*, 1996).

Fluorescence light response curves: Response of F_v/F_m, Q_P and Q_{NP} to PPF (the light response curve) were measured by illuminating the leaf with actinic light at increasing PPF (0 to 1200 µmol m^{-2} s^{-1}).

2.6 Chlorophyll assay

Chlorophyll contents were determined at the end of the experiment. Leaf slices were dissected and pigments were extracted under overnight continuous agitation in 80% (v/v) acetone amended with 0.5% (w/v) $MgCO_3$ to prevent chlorophyll acidification at 4°C. Crude extract was centrifuged at 10,000 g for 10 min at 4°C, and the supernatant was used to estimate spectrophotometrically pigment concentrations according to the absorbance coefficients determined by Lichtenthaler (1987). Results were expressed in mg g^{-1} fresh weight (FW).

2.7 Statistical analysis

Five replicates plants per treatment and three replicate measurements per plant were carried out. All data were analysed using the Mann & Whitney test at the 0.05 probability level.

3. Results

3.1 Growth

In the event of fmx excess (5 and 50 mM), the plant growth is inhibited (Fig. 1). The leaves growth is more affected than the root growth.

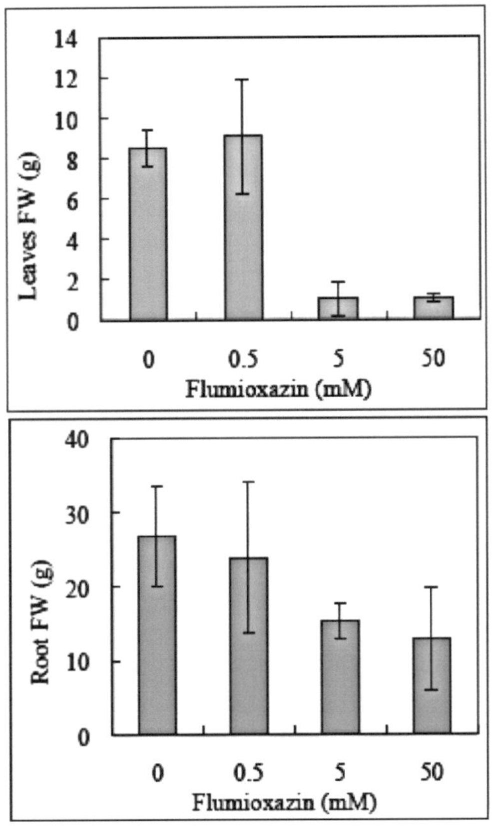

Fig. 1. Leaves and root fresh weight of grapevine after the spraying of various fmx concentrations. Each value is the mean of 15 measurements ± SD.

3.2 Gas exchanges

Photosynthetic responses of grapevine grown with various herbicide concentrations were analysed to determine whether fmx modifies Pn, gs, T and Ci. The Pn decreased significantly after one day at 0.5, 5 and 50 mM fmx (Fig. 2A). Fifteen days after spraying with 5 and 50 mM fmx, Pn was steady and equal to zero. gs was significantly affected by fmx (Fig. 2B). Detectable inhibition of gs occurred after one day using 5 and 50 mM fmx and after three days at 0.5 mM fmx. Whatever fmx concentration, the treatment leads to the closure of stomata in grapevine after three days and 15 days after, gs was equal to 0 for 5 and 50 mM. Transpiration of grapevine following fmx treatment was also reduced significantly at 0.5, 5 and 50 mM after two days (Fig. 2D). Ci was not affected with 0.5 mM fmx, but Ci increased at 50 mM after 3 days and at 5 mM fmx after 5 days (Fig. 2C). At 0.5 mM fmx and after 5 days, Pn and T increased and after 45 days, they were equal to control.

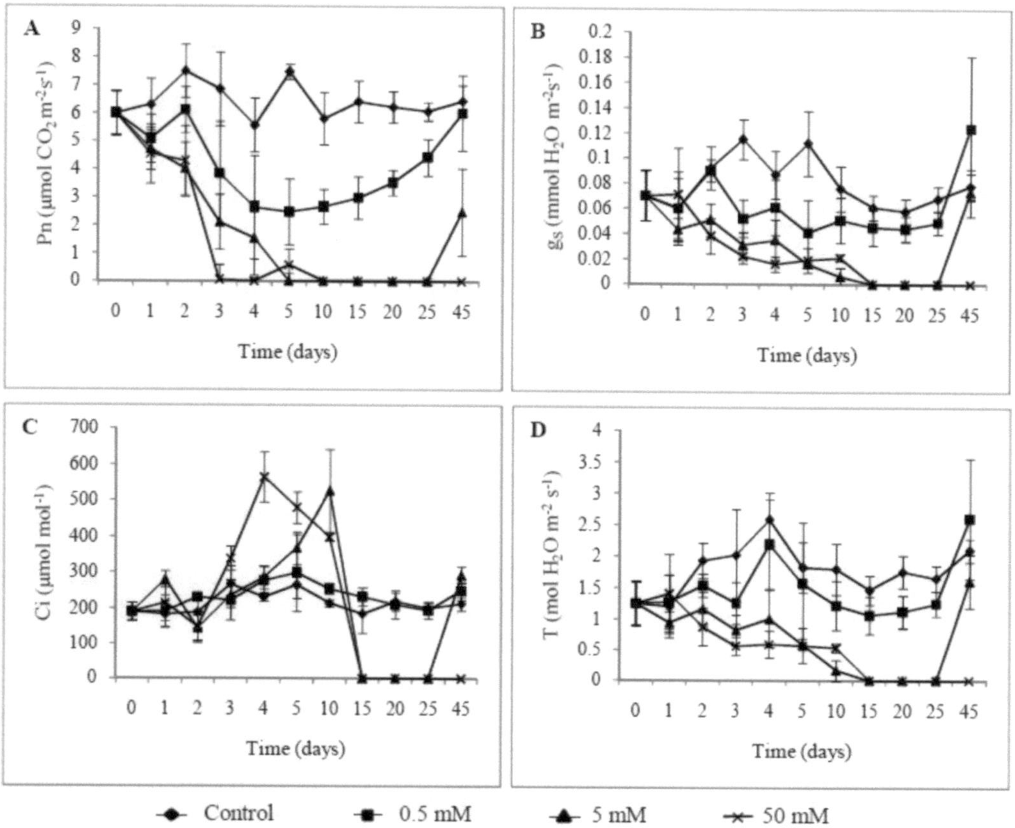

Fig. 2. Changes in net photosynthesis (P_n) (A), stomatal conductance (g_S) (B), intercellular CO_2 concentration (Ci) (C), and transpiration rate (T) (D) in leaves of grapevine after the spraying of various fmx concentrations. Each value is the mean of 15 measurements ± SD.

Photosynthetic light response curves

To clarify the nature of the mechanisms involved in plant adaptation to the treatment, both CO_2 distribution in the leaf and the capacities of mesophyll to assimilate atmospheric CO_2 were analyzed after ten days. The study of these processes was performed by analysing the light response curve and by measuring Pn variations in response to the increase of CO_2 concentration after ten days of herbicide stress application. The leaves of the treated plants showed a drastically lower photosynthetic capacity than the control leaves. Similarly, saturating PPF and apparent quantum yield of CO_2 fixation (ΦCO_2) showed a significant decrease with increasing fmx concentrations (Table 1). Cuttings treated with 50 mM fmx did not respond to the PPF because the plants died. Also, dark respiration and compensation point decreased when the fmx stress increased. In addition, photosynthesis was saturated at 10.86 µmol m^{-2} s^{-1} in the control and light-saturated net CO_2 assimilation rate (A_{sat}) decreased by 15% and 98% using 0.5 and 5 mM fmx respectively. Using the high PPF, the slope of curves was null (Table 1), meaning that there was no photoinhibition. The ratio $\Phi PSII/\Phi CO_2$ was inversely correlated to the efficiency of light involvement for carbon fixation. It was higher in leaves grown at 5 mM fmx concentration (Table 1), indicating that the light was less efficient in treated cuttings.

	Fmx concentrations			
	Control	0.5 mM	5 mM	50 mM
Apparent quantum yield of CO_2 fixation (ΦCO_2)	0.044±0.002[a]	0.041±0.006[a]	0.006±0.007[b]	-
Dark respiration Rd (μmol m^{-2} s^{-1})	1.10±0.54 [a]	1.31±0.34 [a]	0.622.±0.710[b]	-
Compensation point Γ (μmol)	26.32±1.30 [a]	31.67±3.01 [b]	101.42±10.01 [c]	-
A_{sat} (μmol m^{-2} s^{-1})	10.86±1.28 [a]	9.14±0.93 [b]	0.139±0.055 [c]	-
PPF (μmol m^{-2} s^{-1}) for A_{sat}	1000 [a]	500 [b]	250 [c]	-
Slope with PPF > 1500 μmol m^{-2} s^{-1}	-0.0003±0.0008 [a]	0.0003±0.0001 [a]	0.0005±0.0003 [a]	-
$\Phi PSII/\Phi CO_2$	15.43 [a]	12.92 [b]	50.33[c]	-

Table 1. Analyses of photosynthetic light response curves: the apparent quantum yield of CO_2 fixation (ΦCO_2), dark respiration (Rd), compensation point (Γ), light-saturated net CO_2 assimilation rate (A_{sat}), PPF (μmol m^{-2} s^{-1}) for A_{sat}, slope with PPF > 1500 μmol m^{-2} s^{-1} and ratio $\Phi PSII/\Phi CO_2$ of grapevine with various flumioxazin concentrations after ten days. The grapevine treated with 50 mM of fmx were dead. Statistical analyses were carried out using the Mann and Whitney test. Means for a considered parameter were not significantly different when followed by the same letter ($P \geq 0.05$).

3.3 Chlorophyll fluorescence

All chlorophyll fluorescence parameters strongly dropped as the fmx concentration increased (Fig. 3). The Fv/Fm ratio used as the means of maximal photochemical efficiency of PSII was not modified in the controls nor in the 0.5 mM fmx treated cuttings, whereas it dropped down to zero after four days using 5 and 50 mM fmx (Fig. 3A). Similarly ΦPSII slowed down significantly after four days of 5 and 50 mM fmx treatments (Fig 3B). Quenching was also affected in the same way: q_P and q_{NP} decreased significantly after ten days using 5 and 50 mM fmx (Fig. 3C, D). There was no PSII activity after 10 days using 50 mM fmx and after 15 days using 5 mM.

Identification of fmx damage in the leaf

Picture of the fluorescence showed a marked decrease in fluorescence emission when the cuttings were exposed to the highest concentrations of fmx (Fig. 4). Fm images allowed early detection of fluorescence variations than Fv/Fm images. 0.5 mM fmx treatment induced only a slight fluorescence decline in the leaves during the whole treatment (Fig. 3). More drastic modifications appeared in the veins of leaves from four days using 5 mM fmx (Fig. 4). Using 50 mM fmx, the fluorescence decline appeared significant after four days in the veins and next spread rapidly throughout the entire leaf, the damages spread throughout the mesophyll (Fig. 4).

Fluorescence light response curves

Figure 5 presents the changes in light response curves of chlorophyll fluorescence in leaves ten days after fmx treatment. The responses of Fv/Fm, Q_P and Q_{NP} to PPF were measured by illuminating the leaf with actinic light at increasing PPF 0 to 1200 μmol m^{-2} s^{-1}. Treated plants responded less strongly to the light than the control. Fv/Fm and Q_P decreased with

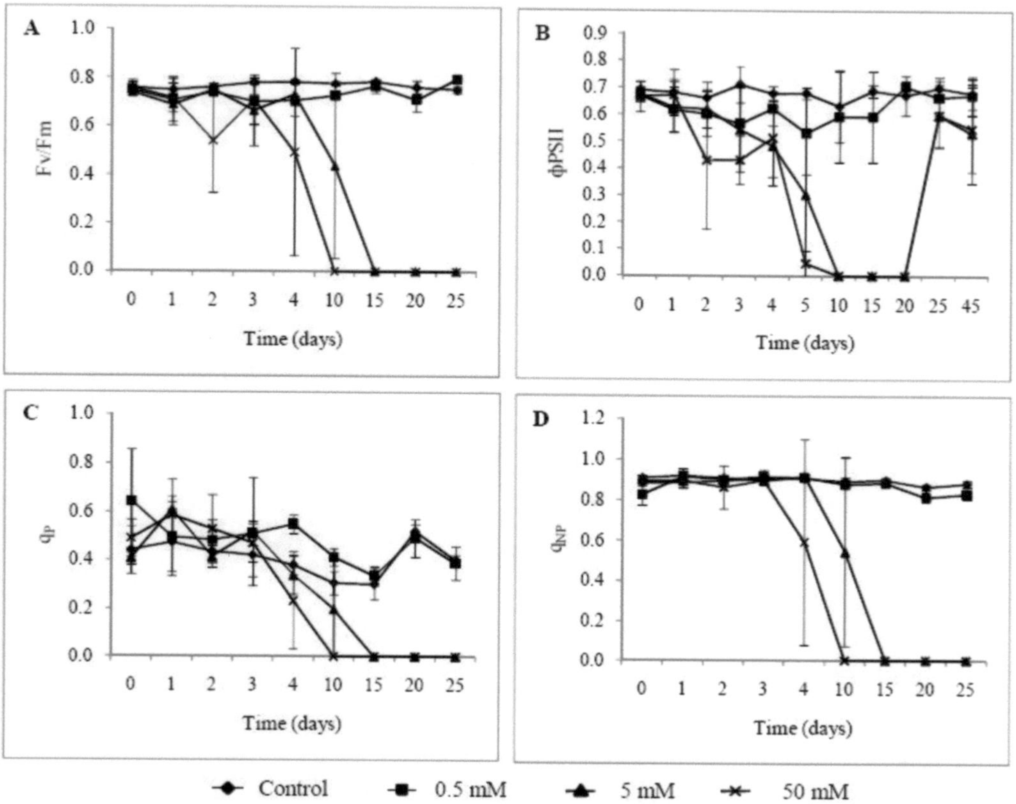

Fig. 3. The Chl fluorescence parameters of PSII (relative units) in leaves of grapevine after the spraying of various fmx concentrations. Each value is the mean of 15 measurements ± SD.

increasing light intensity while Q_{NP} increased. The fluorescence kinetics showed that the increase of fmx concentration led to a decrease in the maximal efficiency of PSII photochemistry and a decrease in the coefficients of photochemical and non-photochemical quenchings.

3.4 Chlorophyll contents

Fmx leads a decrease in the total chlorophyll, chlorophyll a and b concentration and in the carotenoid concentration. We measured a decline in the ratio chlorophyll a/chlorophyll b (Table 2).

	Fmx concentration			
	Control	0.5 mM	5 mM	50 mM
Chl tot (mg.g FW)	1.728±0.259	0.995±0.302	0.593±0.198	0.021±0.016
Chl a (mg.g FW)	1.368±0.168	0.749±0.211	0.464±0.136	0.018±0.014
Chl b (mg.g FW)	0.359±0.097	0.245±0.112	0.128±0.076	0.002±0.008
Carot (mg.g FW)	0.628±0.057	0.348±0.083	0.263±0.055	0.037±0.006
Chl a / Chl b	3.927±0.594	3.378±0.976	2.972±0.644	0

Table 2. Chlorophyll total, a, b and carotenoid concentration and chlorophyll a / chlorophyll b ratio with various fmx concentrations. Each value is the mean of 15 measurements ± SD.

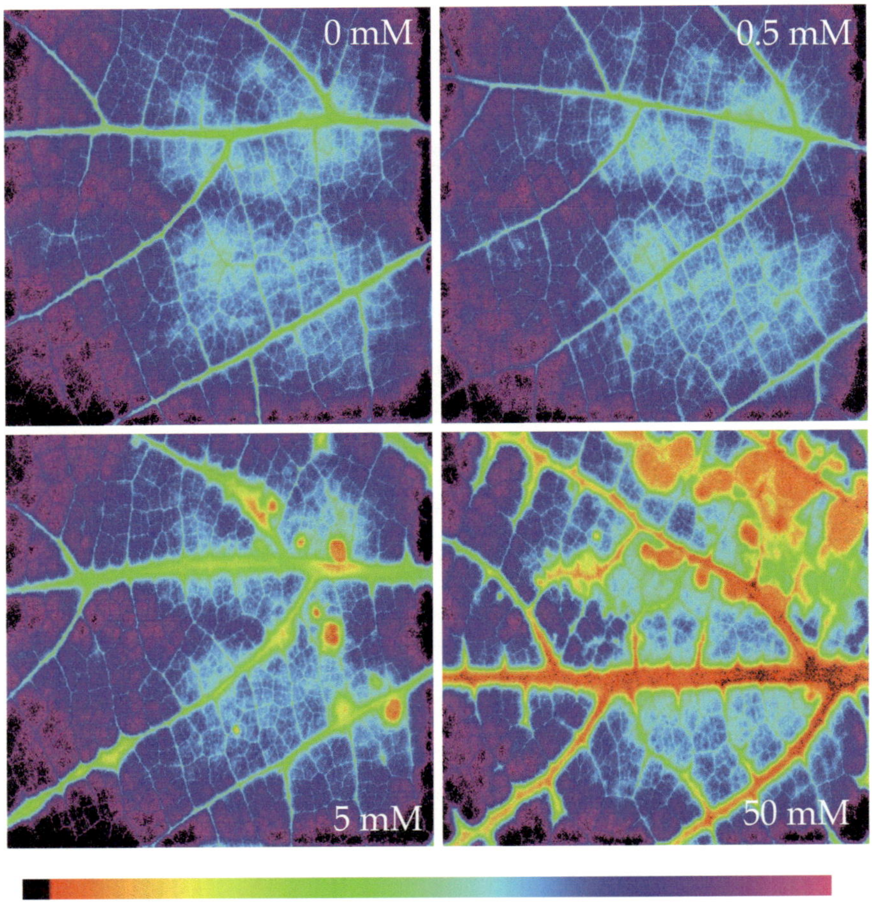

Fig. 4. Fluorescence imaging of the dynamic evolution of abiotic stress induced 4 days after flumioxazin treatment. A grapevine leaf was 30 min dark-adapted and submitted to saturation pulse. A photograph of maximum fluorescence (Fm) was captured. Data have been mapped to the colour palette. The false colour code ranges from black (0.000) to pink (1.000), as shown at the bottom of the images.

4. Discussion

These results provide new insights into the effects of fmx herbicide on grapevine physiology through the analysis of numerous parameters. We have demonstrated a transient fmx effect on Pinot Meunier physiology. The answer of this cultivar was different that observed with the Chardonnay (Bigot et al, 2007). They complement preliminary information on the stress effects of this herbicide on plant physiology *in vitro* (Saladin et al., 2003a, b, c, d; Castro et al., 2005) and help to further understand how action of herbicide acts on non-target grapevine. The soil-applied herbicide is known to be a peroxidizing agent, through the inhibition of protoporphyrinogen IX oxidase in the chlorophyll biosynthetic pathway (Scott et al., 2001). It appears that fmx affects other metabolic functions i.e. all the photosynthetic parameters we evaluated. It induces a strong net photosynthesis inhibition and a parallel decrease of stomatal conductance and transpiration. The photosystem II activity is also affected.

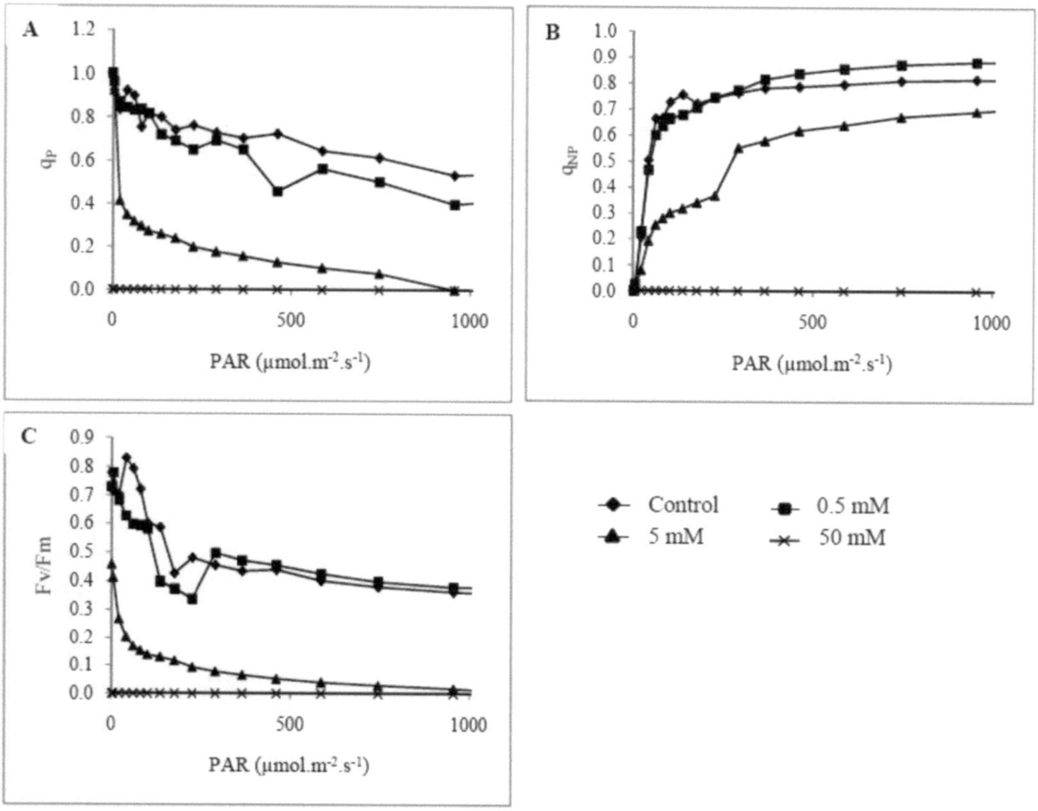

Fig. 5. Fluorescence light response curves: Q_P (A), Q_{NP} (B), Fv/Fm (C), to PPF (0 to 1200 µmol m^{-2} s^{-1}) in leaves of grapevine after ten days with various fmx concentrations.

Fmx inhibits CO_2 assimilation

All the photosynthetic parameters of grapevine cutting leaves were significantly reduced after 25 days of fmx treatment. Net photosynthesis and transpiration reduction were associated with decline of stomatal conductance. The photosynthesis decrease in leaves may be caused by stomatal closure. However, the reduction of Fv/Fm and the quantum yield of CO_2 assimilation indicate that the efficiency of photochemistry is also impaired in grapevine treated with 5 and 50 mM fmx.

There is a strong relationship between photosynthetic electron transport and carbon fixed by plants (Genty et al., 1989). $\Phi PSII/\Phi CO_2$ is an estimate of the relationship between the rate of electron transport and carbon fixation. If four electrons are consumed per mol CO_2 fixed and if the light is equally distributed between the two photosystems, $\Phi PSII/\Phi CO_2$ should theoretically be 8 minimum. Experimentally, $\Phi PSII/\Phi CO_2$ greater than 8 is obtained, meaning that electrons are also used for other processes than photosynthesis, such as photorespiration, N assimilation, or pseudocyclic electron transport (Genty et al., 1989).

Fmx affects chlorophyll a fluorescence

The fluorescence arising from chlorophyll is almost exclusively associated with PSII (Schreiber et al., 1994). Since PSII functioning is sensitive to a wide range of environmental variations, chlorophyll fluorescence provides numerous information on the effects of stresses on plants (Schreiber et al., 1994). Our results clearly show that fmx significantly inhibits the quantum yield of PSII electron transport (ΦPSII) in grapevine cuttings. We also demonstrate that such a decrease in ΦPSII was associated to the alteration of q_P and q_{NP}. A decrease in q_P induced by fmx indicates a higher proportion of closed PSII reaction centres, i.e., an increase in the proportion of the reduced state of Q_A, (Genty et al., 1989), which probably generates a decrease in the proportion of available excitation energy used for photochemistry (Havaux et al., 1991). Concomitant with the reduction of Fv/Fm, we observed that q_{NP} decreased drastically with increasing fmx concentrations, suggesting that the fmx treatment does not involve non-radiative energy dissipation.

In leaves of grapevine cuttings, the capacity for CO_2 assimilation decreases to almost zero after five days whatever fmx concentration. However, after ten days of 0.5 mM fmx treatment, while there is negligible CO_2 assimilatory capacity, ΦPSII remains at approximately 12% when compared to control leaves. This suggests that a certain rate of non-cyclic electron transport is required to maintain CO_2 assimilation. An alternative way to CO_2 assimilation for electrons would be oxygen reduction by photorespiration and/or a Mehler reaction (Brestic et al., 1995). Changes in fluorescence yield in grapevine leaves are also associated with modifications in the antenna pigments, in the efficiency of excitation trapping at the active centres of PSII, or in changes in the thylakoid membrane (Calatayud & Barreno, 2001). The *in vitro* application of fmx to grapevine induced, on the one hand, disorganization of internal photosynthetic membranes (Saladin et al., 2003a) and affected, on the other hand, an oxygen-evolving enhancer protein and a LHCII type III chlorophyll a/b binding protein, a major component of light-harvesting antennae complex of PSII (Castro et al., 2005).

Fmx treatment on the soil, provokes leaf fluorescence damages that first occur in the veins and next spread throughout the mesophyll. Fmx treatment induces depression of photosynthesis in grapevine and involves also heterogeneity of leaf photosynthesis. Such heterogeneity may be the consequence of patchy stomatal closure and/or collapse of part of the mesophyll due to loss of turgor, associated with a low lateral CO_2 diffusion capacity (Cornic & Massacci, 1996). It also results in decreases in the photosynthetic efficiency and capacity of leaves. These observations further suggest that either fmx or a by-product penetrate the plant throughout the roots and are thus distributed in the whole plant through the veins. These results are consistent with Castro et al. (2005) who found significant changes in root and shoot proteome and who suggest that the herbicide could act systemically in grapevine tissues, probably *via* root uptake.

5. Conclusion

We have demonstrated a transient fmx effect on grapevine physiology characterized by strong increases of Pn, gs and ΦPSII at 0.5 and 5 mM fmx after 45 days. The grapevine was able to partially overcome the damages caused by herbicides (Saladin & Clément, 2005). In the vineyard the herbicide caused mild stress (Saladin et al., 2003c, d). It may be explained by a detoxification of the herbicide in the rootstock and/or a low fmx uptake by the roots, which is due to a deeper root system or different soil adsorption characteristics (Saladin et al., 2003d). Moreover, in the vineyard fmx was applied at the end of the winter, when canes have no leaf and when the sap flow is low.

6. References

Bigot, A., Fontaine, F., Clément, C. & Vaillant-Gaveau, N. (2007). Effect of the herbicide flumioxazin on photosynthesis performance of Grapevine *(Vitis vinifera L.). Chemosphere*, 67 : 1243-1251.

Brestic, M.; Cornic, G.; Fryer, M.J. & Baker, N.R. (1995). Does photorespiration protect the photosynthetic apparatus in French bean leaves from photoinhibition during drought stress? *Planta*, 196: 450-457, ISSN 0032-0935

Calatayud, A. & Barreno, E. (2001). Chlorophyll *a* fluorescence, antioxidant enzymes and lipid peroxidation in tomato in response to ozone and benomyl. *Environmental Pollution*, 115: 283-289, ISSN 0269-7491

Castro, A.J.; Carapito, C.; Zorn, N.; Magné, C.; Leize, E.; Van Dorsselaer, A. & Clément, C. (2005). Proteomic analysis of grapevine (*Vitis vinifera* L.) tissues subjected to herbicide stress. *Journal of Experimental Botany*, 56: 2783-2795, ISSN 0022-0957

Coïc, E.C. & Lesaint, C. (1971). Comment asssurer une bonne nutrition en eau et élements minéraux en horticulture ? Horticulture. France. 8: 11-14

Cornic, G. & Massacci, A. (1996). Leaf photosynthesis under drought stress. In: Baker NR, ed. Photosynthesis and the environment. The Netherlands: Kluwer Academic Publishers, pp. 347-366

Demming-Adams, B.; Adams III, W.W.; Baker, D.H.; Logan, B.A.; Bowling, D.R. & Verhoeven, A.S. (1996). Using chlorophyll fluorescence to assess the fraction of absorbed light allocated to thermal dissipation of excess excitation. *Physiologia Plantarum*, 98: 253-264, ISSN 0031-9317

Genty, B.; Briantais, J.M. & Baker, N.R. (1989). The relationships between the quantum yield of photosynthetic electron transport and quenching of chlorophyll fluorescence. *Biochimica et Biophysica Acta*, 990: 87-92, ISSN 0005-2736

Grichar, W.J. (2005). Weed control and grain sorghum tolerance to flumioxazin. *Crop Protection*, 25(2): 174-177, ISSN: 0261-2194

Havaux, M.; Strassera, R.J. & Greppinb, H. (1991). Effects of incident light intensity on the yield of steady-state chlorophyll fluorescence in intact leaves. An example of bioenergetic homeostasis. *Environmental and Experimental Botany*, 31(1): 23-32, ISSN: 0098-8472

Hendrickson, L.; Ball, M.C.; Wood, J.T.; Chow, W.S. & Furbank, R.T. (2004). Low temperature effects on photosynthesis and growth of grapevine. *Plant Cell and Environment*, 27(7): 795-809, ISSN: 0140-7791

Jame, Y.W.; Cessna, A.J.; Biederbeck, V.O.; Grover, R.; Smith, A.E. & Korven, H.C. (1999). Herbicide residues and yield effects from repeated flood-irrigations of alfalfa with water containing monouron or simazine. *Canadian Journal of Plant Science*, 79: 639-645, ISSN 1918-1833

Lebon, G.; Duchêne, E.; Brun, O. & Clément, C. (2005). Phenology of flowering and starch accumulation in grape (*Vitis vinifera* L.) cuttings and vines. *Annals of Botany London*, 95: 943-948, ISSN 0305-7364

Lichtenthaler, H.K. (1987). Chlorophylls and carotenoids, the pigments of photosynthetic biomembranes. *Methods in Enzymology*, 148: 350-382, ISBN: 978-0-12-182048-0

Main, C.R.; Ducar, J.T.; Whitty, E.B. & MacDonald, G.E. (2003). Response of three runner-type peanut cultivars of flumioxazin. *Weed Technology*, 17: 89-93, ISSN 0890-037X

Moreland, D.E. (1999). Biochemical mechanisms of action of herbicides and the impact of biotechnology on the development of herbicides. *Journal of Pesticide Science*, 24: 299-307, ISSN 1349-0923

Mullins, M.G. & Rajasekaran, K. (1981). Fruiting cuttings: revised method for producing test plants of grapevine cultivars. *American Journal of Enology and Viticulture*, 32(1): 35-40, ISSN 0002-9254

Mullins, M.G. (1966). Test-plants for investigations of the physiology of fruiting in *Vitis vinifera* L. *Nature*, 209: 419-420, ISSN: 0028-0836

Nagano, E. (1999). Herbicidal efficacy of protoporphyrinogen oxidase inhibitors. In: Böger, P. and Wakabayashi, K., Editors. Peroxidizing Herbicides, Springer-Verlag, Berlin/Heidelberg, pp. 294-302

Niekamp, J.W.; Johnson, W.G. & Smeda, R.J. (1999). Broadleaf weed control with sulfentrazone and flumioxazin in no-tillage soybean (*Glycine max*). *Weed Technology*, 13: 233-238, ISSN 0890-037X

Saladin, G. & Clément, C. (2005). Physiological effects of pesticides on cultivated crops. In "Agriculture and Soil Pollution: New Research", Nova Science Publishers, ISBN 1-59454-310-0 Inc., pp 53-86

Saladin, G.; Clément, C. & Magné, C. (2003)a. Stress responses of grapevine to flumioxazin herbicide involves dehydration, cell membrane, alteration and osmoregulation process. *Plant Cell Report*, 21: 1221-1227, ISSN 0721-7714

Saladin, G.; Magné, C. & Clément, C. (2003)b. Impact of flumioxazin herbicide on growth and carbohydrate physiology in *Vitis vinifera* L. *Plant Cell Report*, 21: 821-827, ISSN 0721-7714

Saladin, G.; Magné, C. & Clément, C. (2003)c. Effects of flumioxazin herbicide on carbon nutrition of *Vitis vinifera* L. *Journal of Agricultural and Food Chemistry*, 51(14): 4017-4022, ISSN 0021-8561

Saladin, G.; Magné, C. & Clément, C. (2003)d. Stress reactions in *Vitis vinifera* L. following soil application of the herbicide flumioxazin. *Chemosphere* 53: 199-206, ISSN 0045-6535

Schreiber, U.; Bilger, W. & Neubauer, C. (1994). Chlorophyll fluorescence as a non-intrusive indicator for rapid assessment of *in vivo* photosynthesis. In: Schulze ED, Caldwell MM, eds. Ecophysiology of photosynthesis. Berlin: Springer-Verlag, 49-70

Scott, G.H.; Askew, S.D. & Wilcut, J.W. (2001). Economic evaluation of diclosulam and flumioxazin systems in Peanut (*Arachis hypogaea*). *Weed Technology*, 15: 360-364, ISSN 0890-037X

van Kooten, O. & Snel, J.F.H. (1990). The use of chlorophyll fluorescence nomenclature in plant stress physiology. *Photosynthesis Research*, 25: 147-150, ISSN 0166-8595

Yoshida, R.M.; Sakaki, R.; Sato, R.; Haga, T.; Nagano, E.; Oshio, H. & Kamoshita, K. (1991). S-53482 - a new N-phenyl phthalimide herbicide. Proc. Brighton Crop Prot. Conf. Weeds, 1: 69-75

4

Unexpected Side Effects of Herbicides: Modulation of Plant-Pathogen Interactions

Andreas Kortekamp
DLR Agricultural Research Station, Department of Phytomedicine
Neustadt/Weinstrasse
Germany

1. Introduction

Herbicides are widely used as an important alternative to prevent excessive growth of weeds in agricultural crop land, particularly where conservation tillage is adopted. Weeds reduce crop yield and quality, interfere with cultivation and harvest operations and seem to be the most economically important of all pests with respect to sales of pesticides worldwide. As an interesting side effect, the biological activity of herbicides extends beyond their effect on target organisms and herbicides may influence plant-pathogen interactions through their effect on the pathogen, the plant, or on the surrounding soil organisms including symbiotic interactions. This phenomenon was first observed in the early 1940s by Smith et al. (1946) and described in more detail since 1960. Several studies examining the direct effects of various herbicides on plant pathogens and disease development have been published.

The objective of this chapter is to summarise publications in which herbicide applications have resulted in a direct effect on fungal plant pathogens *in vitro* or in the field or on disease development by influencing the metabolism of the plant. The question that arises is if the direct or indirect effect of herbicides on microorganisms is a general feature of these agrochemicals, also provoking some kind of stress that leads to a reprogramming of the plant's physiology, or if the observed effect relates to a distinct mode of action within a given plant-pathogen relationship. In some cases, herbicide-resistant plants were used to specify the effect on pathogens and plants when herbicides are applied at field rates. Even though the application of field doses may represent the true situation in the field where herbicide-resistant crops are planted, the effect of sublethal doses on non-transformed plants and pathogens remains more or less obscure. Thus, some hormetic effects, although defined as side effects of putatively toxic compounds but playing an important role regarding plant health, plant growth, or even harvest, can not be explained. Some, if not many herbicides seem to provoke hormetic effects (Duke et al., 2006). Hormesis refers to stimulatory effects caused by toxic compounds. Paracelsus, an ancient leader in toxicology, declared that all things are poison and are not poison. Only the dose matters and it is only the dose that makes a thing not to be a poison. He considered that substances, although toxic at higher concentrations or doses, can be stimulatory or even beneficial when used at low doses. Even though this phenomenon was recognised a long time ago, hormesis was mainly discussed in the biomedical literature, especially in toxicology and radiation biology. Sublethal doses of

toxic compounds or radiation, for instance, were found to provoke stimulatory responses instead of reducing the vitality of human cells (Calabrese & Baldwin, 2002). However, hormesis is not restricted to mammals and can be found within all groups of organisms, from higher plants and animals to bacteria and fungi (Calabrese, 2005). Interestingly, herbicides seem to induce hormesis in both, plants and pathogens (Duke et al., 2006). Stimulations have been shown for various physiological and biochemical parameters such as gene expression and enzyme activity (Ahn, 2008), growth, biomass, and protein content (reviewed in Duke et al., 2006), and chlorophyll content in plants (Kortekamp, 2010).
Several highly informative reports considering side effects of herbicides have been published in recent years (e.g. Duke et al., 2007; Sanyal & Shrestha, 2008). However, a big part of the work was done with herbicide-resistant plants, not representing biochemical processes in non-transformed plants as mentioned above. Furthermore, mainly pathogens able to grow on artificial media were used. Even though the direct effect of the active compound, the additive(s) and/or the formulated product can be tested *in vitro* very easily in most cases, only a few details about the mechanisms that are involved when the pathogen enters its host are known. Moreover, especially the underlying mechanisms playing an important role in case for biotrophic pathogens (that can not be investigated without the respective host) are still far from being well known. Therefore, some examples, e.g. the grapevine-downy mildew interaction representing a biotrophic plant-pathogen relationship, are picked out to throw some light on these highly sophisticated plant-pathogen-herbicide interactions.

2. Herbicides with antifungal properties

2.1 Glyphosate

Glyphosate (N-[phosphonomethyl]glycine), mainly sold under the trade name Roundup, is a systemic broad-spectrum herbicide that inhibits 5-enolpyruvyl shikimate 3-phosphate synthase (EPSPS), a key enzyme in the biosynthesis of aromatic acids and secondary metabolites. Blockage of this pathway results in massive accumulation of shikimate in affected plant tissues leading to a deficiency of significant end-products such as lignins, alkaloids, and flavonoids and a decrease in CO_2 fixation and biomass production in a dose dependant manner (Olesen & Cedergreen, 2010). EPSPS is also present in fungi and bacteria, but not in animals, and organisms with glyphosate-sensitive EPSPS may be affected by glyphosate. In plants, glyphosate is readily translocated throughout the plant within a few days after treatment and thus affects roots or rhizomes even after foliar application. One reason for the popularity of glyphosate is that glyphosate-resistant plants had been developed. In such a case, glyphosate can be applied over the top to glyphosate-resistant plants as a postemergence herbicide to kill unwanted weeds without affecting the crop. In most cases, the use of glyphosate on resistant crops reduces the need for pre-emergence herbicides and other postemergence herbicides.
Some studies have shown that the application of glyphosate on glyphosate-resistant plants alters the susceptibility of such plants towards plant pathogens. Reports of both enhanced and reduced disease severity have been published and glyphosate seem to have preventative and curative properties. Furthermore, the formulation and adjuvants used to enhance the efficiency of the active compound can dramatically affect germination, growth, and propagation of fungal plant pathogens (Smith & Hallett, 2006; Weaver et al., 2006; Weaver, et al., 2009; Wyss et al., 2004). Interestingly, glyphosate has also been reported to

control mammalian pathogenic fungi (Nosanschuk et al., 2001) and was active against apicomplexan parasites that cause diseases such as malaria and toxoplasmosis (Roberts et al., 2002).

2.1.1 Soil pathogens

There are several reports indicating that glyphosate inhibits fungal species involved in soilborne diseases. *Sclerotium rolfsii*, for instance, is a common soilborne plant pathogen known to persist on crop residues. Banana growers noted that rotting residues inadvertently sprayed with glyphosate had little mycelial growth and fewer sclerotia than those not sprayed with the herbicide. Growth of *Sclerotium rolfsii* was retarded on culture plates amended with benomyl or glyphosate, each at the commercial rate of application. Both amended media reduced the radial growth of *S. rolfsii* compared to the control; however, glyphosate-amended medium had the greater inhibitory effect (Westerhuis et al., 2007). Radial growth of other pathogens such as *Pythium ultimum* and *Fusarium solani* f.sp. *pisi* was also retarded with increasing concentrations of the herbicide (Kawate et al., 1992), which also referred to conidial germination and sporulation in *F. solani* f.sp. *glycines* (Sanogo et al., 2000). In contrast to the results described above, Harikrishnan and Yang (2001) found no negative effect of glyphosate on vegetative growth of several *Rhizoctonia solani* isolates and anastomosis groups. However, the herbicide influenced the production of fruiting bodies of this pathogen. The number of sclerotia produced was higher but these sclerotia remained smaller in the presence of the herbicide compared to the untreated control.

Even though inhibitory effects of glyphosate on several plant diseases have been reported, some pathogens were unaffected and/or glyphosate increased disease severity of host plants. In some cases, glyphosate affected growth and reproduction of a given pathogen *in vitro* but showed an adverse effect in the field. Glyphosate inhibited, for instance, the development of *Nectria galligena* mycelium *in vitro* but increased the number of lesions when apple shoots were inoculated with a mycelium derived from a medium containing glyphosate (Burgiel & Grabowski, 1996). Thus, even though glyphosate exhibit a negative effect towards distinct pathogens in some test systems, this herbicide may show other effects *in vivo*. In greenhouse studies using glyphosate-resistant sugar beet, increased disease severity was observed following glyphosate application and inoculation with *Rhizoctonia solani* and *Fusarium oxysporum* (Larson et al., 2006). This increase in disease was not fungal mediated, since there was no direct effect of glyphosate on both fungal species as tested in *in vitro* studies. Thus, the herbicide seems to reduce the plant's ability to protect itself against pathogens. Glyphosate was also shown to be phytotoxic to sugarcane and herbicide treatment resulted in increased disease severity caused by *Pythium arrenomanes* (Dissanayake et al. 1998). Furthermore, glyphosate application caused injury and death of *Lolium multiflorum* as a result of increased Pythium root rot (Kawate and Appleby, 1987). Even sublethal doses of glyphosate inhibited the expression of resistance in soybean to *Phytophthora megasperma* f. sp. *glycinea* (Keen et al., 1982), in bean to *Colletotrichum lindemuthianum* (Johal & Rahe, 1990), and in tomato to *Fusarium* spp. (Brammal & Higgins, 1988). Furthermore, glyphosate applied to the soil increases the disease symptoms caused by *Cylindrocarpon* sp. in grapevine (Whitelaw-Weckert, 2010).

Despite the fact that glyphosate may have a direct effect on a crop plant and the respective pathogens, repeated glyphosate use has also an impact on the microbial community composition. Repeated applications favour species belonging to the group of Proteobacteria in glyphosate-treated soils than occurring in untreated control soils (Lancaster et al., 2010);

glyphosate mineralisation was reduced when glyphosate was applied several times. Gimsing et al. (2004) found that glyphosate mineralisation rates are positively correlated with *Pseudomonas* spp. population size. However, results of Lancaster et al. (2010) indicate that a repeated application of glyphosate is associated with an increase of those soil microorganisms capable of metabolising the herbicide. Altered microbial community may repress *Pseudomonas* species such as the beneficial species *P. fluorescens* and may modulate plant-pathogen interactions as well.

2.1.2 Leaf pathogens

There are several cases of inhibitory effects of glyphosate on certain leaf diseases in various crops. Transgenically modified wheat with tolerance to glyphosate showed very low infection rates regarding leaf rust caused by *Puccinia triticina* and stem rust caused by *P. graminis* f.sp. *tritici* when treated with field doses one day prior to inoculation with the pathogen (Anderson & Kolmer, 2005). The leaf rust control by glyphosate decreased with reduced application rates and longer periods of time between herbicide application and rust inoculation indicating a direct toxic effect. However, control of leaf rust in wheat conditioned by glyphosate is effective for at least 21 days (Anderson & Kolmer, 2005), but how glyphosate inhibits rust infection was not investigated. The herbicide may act as a systemic fungitoxic compound itself or may induce a systemic resistance, since also non-treated leaves were protected after herbicide application. In wheat straw, Sharma et al. (1989) reported an inhibition of *Pyrenophora tritici-repentis* pseudothecia production by glyphosate. Glyphosate has been shown to reduce sporulation, growth, and disease development caused by other cereal fungal pathogen such as *Septoria nodorum* on wheat (Harris & Grossbard, 1979), Rhizoctonia root rot (Wong et al., 1993), and take-all of wheat caused by *Gaeumannomyces graminis*, as well as *Rhynchosporium secalis* and *Drechslera teres* on barley (Toubia-Rahme et al., 1995; Turkington et al., 2001).
Feng et al. (2005) showed by using glyphosate-resistant wheat and soybeans that rust infections and symptoms caused by *Puccinia striiformis* f.sp. *tritici*, *Puccinia triticina*, and *Phakopsora pachyrhizi*, respectively, can be suppressed when plants had been sprayed with formulated glyphosate. The authors proposed that when rust spores became exposed to the herbicide, glyphosate was able to inhibit fungal EPSPS, thus, through the same mechanism described for its herbicidal activity. Their studies with glyphosate-resistant wheat revealed that rust control activity of glyphosate is not mediated through the induction of SAR (systemic acquired resistance) genes, but that glyphosate provided both preventative and curative activities in greenhouse experiments and in the field. However, rust control seemed to depend on the systemic glyphosate concentration in the host plant during germination of rust spores and the first infection events. Thus, rust spores just entering the plant in order to receive nutrients have to be exposed to a lethal concentration of glyphosate. Furthermore, field data obtained from glyphosate-resistant soybeans suggest that rust control by glyphosate is influenced by environmental conditions, and rust races may differ in glyphosate sensitivity (Feng et al., 2008). Also species-specific differences in glyphosate sensitivity seem to exist, so that rust control in soybean requires higher doses than rust control in wheat (Feng et al., 2008). There are also intra-specific variations in *R. solani* as shown by Verma and McKenzie (1985).
Since glyphosate is originally used as an herbicide to prevent growth of unwanted weeds, the use of fungi and bacteria as biological control agents was tested as an alternative to chemical herbicides or, much more interesting, in combination with herbicides. In many

cases, weed control and disease incidence were enhanced when the biocontrol agent was applied after glyphosate treatment (Boyette et al., 2006; Boyette et al., 2008a; Boyette et al. 2008b). The authors demonstrated that an application of glyphosate prior to *Myrothecium verrucaria* provided better weed control in kudzu (*Pueraria lobata*), redvine (*Brunnichia ovata*), and trumpetcreeper (*Campis radicans*). This was also the case for green foxtail, which was sufficiently controlled when treated with glyphosate prior to *Pyricularia setariae* inoculation (Peng & Byer, 2005). These results suggest that timing of glyphosate application in relation to combined treatment with a bioherbicide is important. Wyss et al. (2004) reported that certain pesticides and their adjuvants affected spore germination and growth of *Phomopsis amaranthicola*, an effective bioherbicides against *Amaranthus* species. Several herbicides such as glyphosate had also negative effects on spore germination of *P. setariae* (Peng & Byer, 2005). Thus, one strategy to overcome direct toxic effects of herbicides is a sequential rather than simultaneous application of the synthetic herbicide and the bioherbicides. Applying glyphosate prior to pathogen application would allow the absorption, translocation, and the full action of the herbicide (with minimised degradation) and reduces its possible toxicity to the biocontrol agent. Furthermore, glyphosate interactions with bioherbicides were found to be synergistic. Sharon et al. (1992) showed that glyphosate suppressed the plant's defence by lowering phytoalexin production and biosynthesis of other phenolics. Even a sublethal dose of glyphosate suppressed the shikimate pathway in sicklepod (*Cassia abtusifolia*) infected with *Alternaria cassiae*, thus reducing the resistance of this weed (Sharon et al., 1992). Numerous examples in the literature have correlated production or transformation of preformed phenolic compounds and plant defence. In most cases, an activation of the enzyme phenylalanine ammonia-lyase (PAL) plays a pivotal role, and compounds that inhibit PAL activity have caused increased susceptibility to disease (Hoagland, 2000). This seems also to refer to crop plants such as soybean. Glyphosate was able to block resistance to *Phytophthora megasperma*, even in an incompatible interaction by lowering the glyceollin production, an important phytoalexin and part of the resistance machinery in soybean (Keen et al., 1982).

2.2 Glufosinate ammonium

The non-selective herbicide glufosinate ammonium is an ammonium salt of phosphinothricin and used as a postemergence contact herbicide currently being marketed under the trade name Basta® or Liberty®. Glufosinate ammonium efficiently kills various kinds of plants, since it is a glutamic acid analog that inhibits glutamine synthetase by irreversible binding (Hoerlein, 1994). Glutamine synthetase is also a target to control bacteria pathogenic to humans such as *Mycobacterium tuberculosis* (Nilsson et al., 2009). The inhibition of glutamine synthetase in plants results in an accumulation of toxic ammonium that disturbs electron transport systems and induces production of free radicals (Krogmann et al., 1959). Free radicals in turn cause lipid peroxidation and cell death (Devine et al., 1993; Hess, 2000).

Glufosinate ammonium resistant plants have been produced successfully by introducing a *bar* gene from the soilborne microbe *Streptomyces hygroscopicus* or the *pat* gene from *S. viridochromogenes*. The genes encode the phosphinothricin acetyl transferase, which converts glufosinate ammonium to a nonphytotoxic acetylated metabolite (Murakami et al., 1986; Thomson et al., 1987). Glufosinate ammonium-resistant crops allow the control of weeds through an application of suitable amounts of the herbicide. As a beneficial side effect, glufosinate ammonium can also reduce fungal diseases in plants. Wang et al. (2003) had shown that glufosinate significantly reduced disease development of *Rhizoctonia solani* and *Sclerotinia homoeocarpa* on transgenic bentgrasses expression the *bar* gene under controlled conditions. Glufosinate ammonium also reduced Pythium blight caused by *Pythium aphanidermatum* in transgenic bentgrasses expressing the *bar* gene, even though the herbicide did not alter mycelial growth *in vitro* (Liu et al., 1998).

In rice, two important diseases, blast and brown leaf spot, were also diminished in transgenic rice when treated with glufosinate ammonium (Ahn, 2008). The herbicide inhibited the formation of appressoria of the two pathogens *Magnaporthe grisea* and *Cochliobolus miyabeanus* in a dose-dependant manner; but the same treatment did not affect conidial germination of both pathogens. However, glufosinate ammonium almost completely inhibited mycelial growth of both fungi *in vitro* and triggered the transcription of pathogen related (PR) genes and hydrogen peroxide accumulation in rice and *Arabidopsis thaliana* (Ahn, 2008). Furthermore, a pretreatment with glufosinate ammonium 24 h prior to infection greatly increased blast protection. These results indicate that the herbicide is able to activate the resistance response of the plant, and that the induced mechanisms are more effective after a time lag between application and infection. Thus, both direct inhibition of pathogen infection and activation of the defence system by glufosinate ammonium seem to be responsible for disease protection in transgenic rice. Other reports also showed that treatments with glufosinate ammonium enhanced resistance against rice sheath blight caused by *R. solani* on *bar*-transgenic rice (Uchimiya et al., 1993). In that case, herbicide treatment led to a substantial suppression of blight symptoms, even when applied two days after inoculation, indicating a curative capacity of glufosinate ammonium.

Even though the direct effects of glufosinate ammonium on plant pathogens are not well understood in most cases, this herbicide may inhibit glutamine synthetase activity in fungi or fungal like organisms similar to inhibition of glutamine synthetase in plants. Consistent with amino acid biosynthesis being the primary target of glufosinate ammonium, inhibition of glutamine synthetase in the presence of the herbicide leads to a reduced nitrogen metabolism, a reduced hyphal protein content, and thus to a restricted growth and biomass yield of various *Trichoderma* species (Ahmad et al., 1995). Furthermore, especially the expression of those genes involved in protein biosynthesis and energy production seem to be important for oomycetous pathogens during germination and at the onset of a biotrophic or hemibiotrophic infection. Whereas the amino acid biosynthetic genes are expressed at basal levels during release of zoospores (that are produced in sporangiospores or sporocysts), they are upregulated in germinated cysts of *Phytophthora* species, indicating a requirement of elevated amino acid production and metabolism at the early infection events. Among those genes expressed at early infection stages, glutamine synthetase was also upregulated in germinated cysts of *P. nicotianae* (Shan et al., 2004). During the biotrophic phase of the *P. infestans*-potato interaction, the free amino acid pool within the plant leaf increases, that corresponds with the expression of host amino acid biosynthesis genes (Grenville-Briggs & Van West, 2005; Grenville-Briggs et al., 2005). As infection progresses, there is a decrease in the level of free amino acids within the infected plant tissue and a corresponding increase in the expression of both host and pathogen amino acid biosynthesis genes (Grenville-Briggs & Van West, 2005; Grenville-Briggs et al., 2005) which attests the need of high levels of amino acids for an sufficient growth of the pathogen. Interestingly, growth and propagation of oomycetous pathogens such as *P. infestans* and *Pythium ultimum* were also inhibited by glufosinate ammonium *in vitro*, especially when cultivated on media containing low amounts of nutrients (Kortekamp, 2008).

Direct inhibition of mycelial growth was also observed for several fungal species putatively pathogenic to grapevine. *Botrytis cinerea, Guignardia bidwellii, Penicillium expansum,* and

Phomopsis viticola were exposed to various concentrations of glufosinate ammonium in an *in vitro* assay (Albrecht & Kortekamp, 2009). The herbicidal compound caused reduction of mycelial growth in a dose-dependant manner as it was shown for other phytopathogenic fungi. However, *G. bidwellii* seem to be extremely sensitive, since mycelial growth was reduced about of 80%, even when the pathogen was exposed to a 500fold diluted solution of glufosinate ammonium normally applied to the field. Even though the pathogen *P. expansum* seemed to be less sensitive towards this herbicidal compound with regard to mycelial growth, spore production of this fungus was nearly completely inhibited when exposed to the same low concentration used to suppress growth of *G. bidwellii*, maybe allowing an effective control of this challenging pathogen late in the growing season. An application of glufosinate ammonium also caused severe effects on growth and development of the obligate biotrophic grapevine pathogen *Plasmopara viticola* in a dose-dependant manner (Kortekamp, 2008; Kortekamp, 2010). High doses were unacceptable phytotoxic, but low doses did not cause any visible negative effect on grapevine leaf samples. Moreover, low doses increased chlorophyll concentrations as a result of a hormetic-stimulatory response.

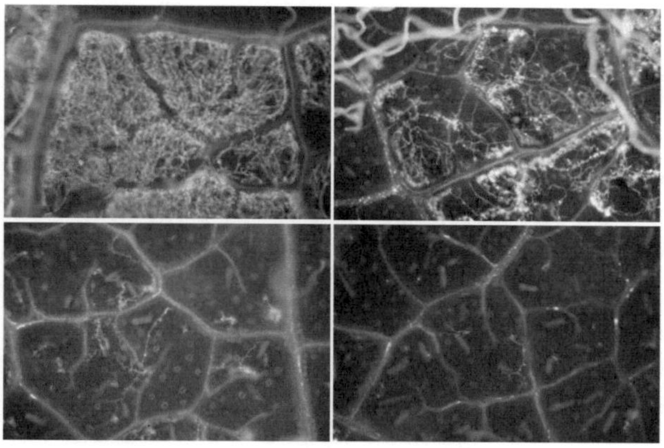

Fig. 1. Mycelial growth of *P. viticola* 7 day post inoculation. Incubation of leaf discs on glufosinate ammonium led to a retarded hyphal growth in a dose-dependant manner.

Even though germination of sporangiospores and zoospore release of the pathogen was not effected when exposed to low concentrations, spreading of the intercellular mycelium was reduced also leading to a dramatically reduced sporulation (Kortekamp, 2010). However, higher doses up to the rate normally applied to the filed completely inhibited each developmental step of the disease cycle. Interestingly, glufosinate ammonium exhibited preventative and curative features. Pre- and postinfectional treatments resulted in significant reduced sporulation rates. The inhibitoric effect of glufosinate ammonium on spore production decreased with increasing time intervals between inoculation and treatment, since the pathogen was able to establish a dense network of hyphae within the infected tissue and started to sporulate after few days. However, if the herbicide was applied prior to inoculation, the preventative effect increased with increasing time intervals between treatment and inoculation. This suggests an activation of defence mechanisms in

the plant. Alternatively or in addition, an application of the herbicide might cause an uncomfortable and improper environment due to a reduced level of amino acids, reduced nitrogen availability in general, an altered pH and/or an accumulation of ammonium. Especially changes in the pH, nitrogen availability, and ammonium concentrations have been suggested as a regulatory factor for colonisation of pathogenic fungi.
Ammonification (the active secretion of ammonium) of the host tissue leading to an alkalinisation of the host environment has been suggested to be a key factor in the enhancement of pathogenicity of several fungi such as *Alternaria* and *Colletotrichum* (Duan et al., 2010; Eshel et al., 2002; Prusky et al., 2001). Both fungi are necrotrophic pathogens that are able to degrade cells or cell components to receive small fragments suitable for their own nutrition. This degradation of host cells seem to depend on suitable pH values, since most lytic enzymes are pH-sensitive regarding their maximum activity. Furthermore, changes in host pH are signals activating the production of pathogenicity factors via the regulation of gene expression (Kramer-Haimovich et al., 2006). It was recently shown that ammonium secretion and accumulation plays a key role as a pathogenicity factor during infection of tomato by *Colletotrichum* species and induces the transformation of the biotrophic to a necrotrophic infection (Alkan et al., 2008). Furthermore, addition of ammonium to a plant-pathogen system induces appressorium formation in *Alternaria alternata* and enables the pathogen to overcome defence mechanisms even in a resistant tobacco cultivar (Duan et al., 2010). Ammonium accumulation in plants is also associated with senescence promotion due to a decrease of glutamine synthetase activity (Chen and Kao, 1996; Chen et al., 1997). Plant tissues undergoing senescence are suitable resources for necrotrophic pathogens and saprophytes but do not represent adequate environments for biotrophic pathogens which rely on living host cells. Thus, high ammonia levels seem to favour necrotrophic fungi but maybe suppress the growth of biotrophic pathogens such as *P. viticola* on grapevine.

2.3 Triazine herbicides

The principle mode of action of triazine herbicides is the inhibition of photosynthesis. The triazines were shown to inhibit PSII but have no effect on PSI (Trebst, 2008). Several effects of triazines on soil organisms, especially on fungi causing soilborne diseases, were reported in the 1960s and 1970s. Especially atrazine had high inhibitory effects on *Fusarium moniliforme*, *F. oxysporum*, and *Aspergillus* species (Curl et al., 1968; Bozarth and Tweedy; 1971; Kabana and Curl, 1970; Rattanakreetakul et al., 1990). Several *Aspergillus* species were also repressed in soil by cyanazine, an herbicide that inhibits the growth of at least six other important soil fungi at field doses (Abdel-Fattah et al, 1983). Interestingly, this effect was not observed in artificial media. However, atrazine and other triazine herbicides seem to have an impact on growth and the production or viability of spores and fruiting bodies in soil cultures and on artificial media. Beam et al. (1977) demonstrated that enzyme activities and mycelial growth of *Rhizoctonia solani* were significantly reduced by prometryn and sclerotium production of *Sclerotium rolfsii* was reduced or even repressed by atrazine. This was also the case for *S. sclerotiorum* when triazine herbicides were applied to soil or media. Atrazine, simazine, and metribuzin inhibited mycelial growth or the development of normal apothecia and sclerotia at low concentrations (Casale and Hart, 1986). In another study, sclerotia germination was stimulated by triazine herbicides (Radke and Grau, 1986). Simazine and atrazine enhanced stipe formation but stipes and apothecia were malformed, whereas metribuzin enhanced stipe and mycelial growth without malformations. These herbicides also induced the germination of *Cochliobolus sativus* spores which resulted in a loss of viability (Isakeit and Lockwood, 1989). *C. sativus* (*Bipolaris sorokiniana*) is the causal agent of a wide variety of cereal diseases. This pathogen can infect roots, leaves, stems, flowers, and head tissues just like other *Cochliobolus* species. Even though *C. sativus* was greatly affected by triazine herbicides, these herbicides had no influence on germination and viability of conidia of other *Cochliobolus* species such as *C. heterostrophus*, *C. carbonum*, and *C. victoriae* (Isakeit and Lockwood, 1989). There may be species differences among the genus *Cochliobolus*. Russin et al. (1995) reported that atrazine did also not reduce the production and germination of microsclerotia of *Macrophomina phaseolina* in sorghum but reduced fungal growth. Despite the fact that atrazine and other triazines could have a direct effect on fungal pathogens, they are able to modulate plant-pathogen interactions due to changes in the physiology of the plant. Atrazine applications to sugarcane plants growing in soils infested with *Pythium arrenomanes* resulted in increased root and shoot growth, even though root colonization by *P. arrenomanes* was unaffected by the herbicide. Furthermore, root rot symptom severity was not reduced. However, atrazine inhibited mycelial growth of *P. arrenomanes in vitro* when applied at the label rate (Dissanayake et al., 1998). The mechanism of root and shoot growth stimulation of triazine herbicides was shown to be an increase in the activity of nitrite reductase and transaminase (Ries et al., 1967) and seem also to refer to pea and sweet corn (Wu et al., 1972). Even though triazine and maybe other triazine herbicides are able to inhibit *P. arrenomanes in vitro*, such an effect was not observed in the field. If both, the herbicide and the pathogen are present, growth stimulation by atrazine seems to be greater than growth reduction induced by *P. arrenomanes*. Other data recently published indicate that herbicide treatments, especially when applied at field rates, may lower the effect of the fungicide. Heydari et al. (2007) conducted two field experiments to investigate the impact of three preemergence herbicides on the efficacy of commonly fungicides against *Rhizoctonia solani*. In one trial, the effectiveness of fungicides on fungal pathogenicity was reduced in the presence of prometryn and two dinitroaniline herbicides. The authors suggested that the herbicide-mediated suppression of fungicidal activity occurred perhaps because herbicides concentrations in the soil were high shortly after application but diminished gradually due to inactivation (Heydari et al., 2007). However, the fact that herbicides interfere with fungicidal activity of other pesticides may also be due to the presence of variable soil factors including texture, pH, temperature, moisture, and organic matter, which all might have in influence on microbial activity in soil.

Hill and Stratton (1991) tested the antifungal capacity of metribuzin towards *Alternaria solani*. Metribuzin was used for both preemergence and postemergence control of weeds in potatoes that can be affected by *A. solani*. The results presented for metribuzin indicated that this herbicide is relatively nontoxic towards *A. solani in vitro*. Interestingly, the herbicide interacted in an additive manner when applied at low doses together with a fungicide but antagonistically at higher doses. Thus, the type of interaction between triazine herbicides and fungicides seems to depend on the concentration of the components in mixtures. Reasons for this are still far from being well understood.

2.4 Dinitroaniline herbicides

Dinitroaniline herbicides are selective, wide-spectrum herbicides, which are used extensively in vegetable and field crops. The herbicidal effect results from an uptake by roots and the negative effect on root development. Dinitroaniline herbicides disrupt mitosis by binding to plant tubulin to form a complex, thus, inhibiting the formation of microtubules (Strachan & Hess, 1983).

Dinitroanilines have been reported to reduce disease incidence by different pathogens in various crops such as cherry, which can be affected by several *Phytophthora* species leading to a crown rot (Wilcox, 1996). Dissanayake et al. (1998) and Canady et al. (1986) reported that pendimethalin and trifluralin inhibited mycelial growth of *Pythium arrhenomones* and root colonization of *Macrophomia phaseolina*, respectively. However, both herbicides seem to be able to increase disease incidence of seedling damping-off caused by *Rhizoctonia solani* in cotton (Neubauer & Avizohar-Hershenson, 1973) and germination of sclerotia produced by *Sclerotinia sclerotiorum* (Radke & Grau, 1986). Trifluralin is also able to increase the severity of Fusarium root rot, since it induces hypocotyl swelling in soybean, which allows a more successful penetration of the pathogen *Fusarium oxysporum* (Carson et al., 1991). Even though trifluralin showed no effect on damping-off of cotton seedlings, pendimethalin lowered the effectiveness of fungicides applied in combination with the herbicide (Heydari et al., 2007).

In contrast to the results mentioned above, dinitroaniline herbicides may provoke a remarkable increase in resistance of pretreated plants to soil-borne pathogens, such as *Fusarium* species, even if applied at very low concentrations (Grinstein et al., 1976). Thus, this effect can not surely be attributed to a direct fungitoxic mode of action. However, the increase in resistance correlates with the amount of herbicide applied and correlates negatively with the production of ethylene. Ethylene seems to play an important role in inducing certain disease symptoms of wilt diseases (Cronshow & Pegg, 1979; Cohen et al., 1986) by predisposing plant tissues to the damage of lytic enzymes or other fungal-derived pathogenicity factors. Even though a dose-dependent suppression of ethylene production and an induction of resistance in *Fusarium*-infected plants may be a result of different mechanisms, dinitroanilines are capable to induce the production of antifungal compounds leading to an occlusion of the pathogen (Grinstein et al., 1984).

Beside the effects of dinitroanilines on soil-borne pathogens, these herbicides seem also to interfere with the phyllosphere microflora. Population and species composition of microbial communities on leaf surfaces are mainly influenced by physico-chemical characteristics of the leaves. However, specific environmental conditions and agrochemicals can modify the leaf surface and, thus, its microflora. Shukla et al. (1988) showed that potato leaves treated with herbicides harbored lower population compared to the untreated control. Especially *Penicillium brevicompactum*, *Fusarium oxysporum*, *Mucor racemosus*, and *Rhizopus species* were repressed after fluchloralin (Basaline®) application, whereas other species such as *Oidiodendron echinulatum* were isolated only from herbicide treated plants. This indicates that some fungal species were directly affected by selected herbicides, but others are favored and find a more convenient habitat when the population of (most) other fungi diminished due to the effect of the herbicide. Interestingly, also bacterial populations seem to be affected by herbicides. In case of untreated potatoes, the population increased with time whereas the population decreased initially after an application of fluchloralin and other herbicides but recovered from herbicide treatment rapidly (Shukla et al., 1988).

2.5 Quaternary ammonium herbicides

The site of action for quaternary ammonium herbicides such as paraquat and diquat is in the chloroplast. Paraquat is known to act on the PS I within the photosynthetic membrane. The free electrons from the PS I react with the paraquat ion to give a free radical form that interferes with oxygen leading to superoxides. The production of reactive oxygen species (ROS) in turn results in lipid peroxidation and photobleaching (Duke, 1990). Thus, paraquat acts in the presence of light and the herbicidal activity increases with increased light intensity. Inhibitory effects of paraquat on mycelial growth of pathogens were reported in *Rhizoctonia solani* (Black et al., 1996), *Rhizopus stolonifer* (Wilkinson and Lukas, 1969), *Sclerotium rolfsii* (Kabana et al., 1966), *Septoria nodorum* and *S. tritici* (Harris & Grossbard, 1979; Jones & Williams, 1971), and to a lesser extend in *Fusarium moniliforme* (Rattanakreetakul et al., 1990). Paraquat seems to enhance the toxicity of fungicides as reported by Awadalla & El-Refaie (1994). In pot tests, damping-off caused by *R. solani* was better controlled by fungicides when the soil was treated with paraquat or simazine. Both herbicides increased the toxicity of fungicides against mycelial growth of the pathogen maybe due to an increased concentration of ROS in the plant.

2.6 Protoporphyrinogen oxidase (PPO) inhibitors

This herbicide group consists of a large number of compounds that cause an uncontrolled autooxidation of protoporphyrinogen and a rapid lipid peroxidation (Sandmann & Böger, 1982; Duke, 1990). Therefore, these compounds were termed as peroxidising herbicides. They have a contact action and cause leaf burn, desiccation, cell death, and therefore also growth inhibition (Matringe et al., 1992). Several studies have reported that PPO inhibitors enhance the defence mechanisms in plants leading to a decrease in disease severity. Some of these results have been recently reviewed by Sanyal and Shrestha (2008). Nelson et al. (2002) conducted some experiments to determine the response in soybean after an inoculation with *Sclerotinia sclerotiorum* and an application with several PPO inhibitor herbicides. Lesions caused by *S. sclerotiorum* exhibited smaller sizes when treated with PPO inhibitors. Furthermore, some of these herbicides induced an increase in phytoalexin production, but only in leaves and not in stems (Nelson et al., 2002). Furthermore, even though these experiments include glyphosate resistant plants that should not differ in their response regarding a PPO inhibitor application, these plants produced more phytoalexins than near-isogenic glyphosate susceptible cultivars. Lesion size was not only reduced by all PPO inhibitors on the treated leaf but also on non-treated leaves of the same plant. The authors suggest that the herbicides induced a systemic resistance response and that these herbicides mimic a hypersensitive response due to an increased production of reactive oxygen species (ROS). The generation of ROS in turn can result in lipid peroxidation and cell wall lignification leading to a reinforcement of cell walls.

2.7 Other herbicides

Antifungal effects or effects on disease development have been reported for several other classes of herbicides including amide herbicides such as propyzamide (Burgiel & Grabowski, 1996), carbanilate herbicides such as desmedipham (Pakdaman et al., 2002), chloroacetanilide herbicides such as acetochlor,, alachlor, and metolachlor (Cohen et al., 1996; Russin et al., 1995), diphenyl ether herbicides such as lactofen (Dann et al., 1999), and phenoxy herbicides such as clodinafop and 2,4-D (Pakdaman et al., 2002). Most of them showed broad antifungal effects and were able to inhibit the growth of fungal pathogens belonging to different taxonomical groups. Thus, their activity against phytopathogens or symbiotic organisms does not depend on their specific mode of action, even though the toxicity towards fungi may differ with regard to a given pathogen or distinct plant-pathogen interactions.

3. Herbicide-bacteria interactions

Once herbicides are released into the environment, mainly to affect weeds as their primary targets, they have to be degraded and eliminated during time to avoid long-lasting negative effects regarding soil microbiology or groundwater safety. Since a large number of herbicides have been introduced during the past four decades, the fate of these compounds is becoming increasingly important. Thus, several results describing the metabolism of herbicides by microorganisms in soil and water have been published. Especially *Bacillus* and *Pseudomonas* species showed high capacities to degrade various herbicides (Wang et al., 2008; Moneke et al., 2010). However, herbicides are known to change the microbial community in soils (Sapundjieva et al., 2003), including those species relevant for symbiotic interactions with plants (Khan et al., 2004), and will surely affect phytopathogenic bacteria. This topic was excluded from this review and has to be considered in more detail elsewhere.

4. Conclusion

The mechanisms of herbicide-pathogen interactions are not well understood in most cases. Some herbicides seem to have fungitoxic or at least fungistatic properties and affect mycelial growth, production of spores or fruiting bodies, or spore germination, whereas others provoke indirect effects on soil and leaf organisms that are antagonistic to pathogens. In some cases, herbicides showed no effect *in vitro* but lowered disease incidence on the respective host plant. Thus, herbicides may also stimulate the physiology of the plant, e.g. by altering phytoalexin production, mineral and nutrient composition, or source-sink relationships. These alterations may lead to a reduced susceptibility due to physiological changes not favourable for a given pathogen or an induction of resistance and, thus, may affect the incidence of disease. On the other side, herbicides may cause an increase in diseases due to direct stimulatory effects on growth and reproduction of the pathogen, effects on the virulence of the pathogen (Ware, 1980) or by inactivating parts of the defence battery of the host plant.

Effects of herbicides described in this review are not restricted to distinct fungal pathogens, since effects have been observed in necrotrophic, hemibiotrophic, and biotrophic species, and many fungal pathogens are affected by various herbicides applied to different crops. Furthermore, antifungal capacities of the active compound and/or the adjuvants or the modulation of the physiology of the plant leading to increased or decreased disease severity do not depend on the plant tissue affected. Both effects, lowered or enhanced disease incidence, can be observed in case for phytopathogens infecting leaves, stems or roots. However, in some cases, results obtained from *in vitro* experiments differ from those generated in the field or on the host plant. Thus, future research may also include high throughput methods, such as chip based technologies, to illuminate all mechanisms involved in plant-pathogen interactions that are modulated by herbicides. This trilateral communication has to be considered as a molecular and biochemical crosstalk between the plant and the pathogen, the plant and the herbicide, and the pathogen and the herbicide. New information about mechanisms can be obtained by the generation of gene expression profiles, the observation of physiological and morphological changes at tissue level or even in single cells, and an analysis of all relevant compounds such as phenolics, phytoalexins, and proteins (metabolomic approach). In some cases the plant itself and its herbicide-modulated physiology play the predominant role within a given plant-pathogen interaction.

However, depending on the compound used the pathogen may represent the main target that will be arrested or even killed by the herbicide.
There are only few reports about additive or even synergistic effects of combined applications of herbicides together with fungicides (Hill & Stratton, 1991; Schuster & Schroder, 1990), even though these effects can be expected. With regard to the data presented by Hill and Stratton (1991) and Heydari et al. (2007), the simultaneous use of an herbicide and a fungicide to control diseases and weeds could lead to antagonistic interactions between these two kinds of pesticides. This could cause a reduction in the efficacy of both the fungicide and the herbicide. It would be useful to determine the potential herbicide-fungicide interactions in distinct plant-pathogen combinations and to use herbicides that interact synergistically with fungicides, thus they can be used to lower the amount of the fungicides necessary to prevent diseases. Unfortunately, only few data on the ecotoxic effects of pesticide combinations exist, even though considerable data have been published on the effects of individual agrochemicals towards non-target organisms and ecological processes. Thus, the investigation of herbicide-induced effects on plant-pathogen interactions, regardless if applied alone or in combination with other pesticides, requires a multidisciplinary approach combining plant physiology, plant pathology, biochemistry, microbiology, and weeds science and represents a highly interesting field in plant science.

5. References

Abdel-Fattah, H.M.; Abdel-Kader, M.I.A. & Hamida, S. (1983). Selective effects of two triazine herbicides on Egyptian soil fungi. *Mycopathologia*, 82, 143-151, ISSN 0301-486X.

Ahmad, I.; Bisset, J. & Malloch, D. (1995). Effect of phosphinothricin on nitrogen metabolism of *Trichoderma* species and its implications for their control of phytopathogenic fungi. *Pesticide Biochemistry and Physiology*, 53, 49-59, ISSN 0048-3575.

Ahn, I.P. (2008). Glufosinate ammonium-induced pathogen inhibition and defense responses culminate in disease protection in *bar*-transgenic rice. *Plant Physiology*, 146, 213-227, ISSN 0032-0889.

Albrecht, M. & Kortekamp, A. (2009). The *in vitro* effect of the herbicide Basta® (Glufosinate ammonium) on potential fungal grapevine pathogens. *European Journal of Horticultural Science*, 74, 112-117, ISSN 1611-4426.

Alkan, N.; Fluhr, R.; Sherman, A. & Prusky, D. (2008). Role of ammonia secretion and pH modulation on pathogenicity of *Colletotrichum coccodes* on tomato fruit. *Molecular Plant-Microbe Interactions*, 21, 1058-1066, ISSN 0894-0282.

Anderson, J.A. & Kolmer, J.A. (2005). Rust control in glyphosate tolerant wheat following application of the herbicide glyphosate. *Plant Disease*, 89, 1136-1142, ISSN 0191-2917.

Awadalla, O.A. & El-Refaie, I.M. (1994). Effect of herbicides on toxicity of fungicides against *Rhizoctonia solani* causing damping-off of cotton. *Journal of Phytopathology*, 140, 187-192, ISSN 0931-1785.

Beam, H.W.; Curl, E.A. & Rodriguwz-Kabana, R. (1977). Effects of the herbicides fluometuron and prometryn on *Rhizoctonia solani* in soil cultures. *Canadian Journal of Microbiology*, 23, 617-623, ISSN 1480-3275.

Black, B.D.; Russin, J.S.; Griffin, J.L. & Snow, J.P. (1996). Herbicide effects on *Rhizoctonia solani in vitro* and Rhizoctonia foliar blight of soybean (*Glycine max*). *Weed Science*, 44, 711-716, ISSN 0043-1745.

Boyette, C.D.; Reddy, K.N. & Hoagland, R.E. (2006). Glyphosate and bioherbicides interaction for controlling kudzu (*Pueraria lobata*), redvine (*Brunnichia ovata*), and trumpetcreeper (*Campis radicans*). *Biocontrol Science and Technology*, 16, 1067-1077, ISSN 0958-3157.

Boyette, C.D.; Hoagland, R.E. & Weaver, M. (2008a). Interaction of a bioherbicide and glyphosate for controlling hemp sesbania in glyphosate-resistant soybean. *Weed Biology and Management*, 8, 18-24, ISSN 1444-6162.

Boyette, C.D.; Hoagland, R.E.; Weaver, M.A. & Reddy, K.N. (2008b). Redvine (*Brunnichia ovata*) and trumpetcreeper (*Campis radicans*) controlled under field conditions by a synergistic interaction of the bioherbicide, *Myrothecium verrucaria*, with glyphosate. *Weed Biology and Management*, 8, 39-45, ISSN 1444-6162.

Bozarth, G.A. & Tweedy, B.G. (1971). Effect of pesticides on growth and sclerotial production of *Scerotium rolfsii*. *Phytopathology*, 61, 11140-1142, ISSN 0031-949X.

Brammal, R.A. & Higgins, V.J. (1988). The effects of glyphosate on resistance of tomato to Fusarium crown rot and root diseases and on the formation of host structural defence barriers. *Canadian Journal of Botany*, 66, 1547-1555, ISSN 0008-4026.

Burgiel, Z. & Grabowski, M. (1996). The effect of selected herbicides on pathogenicity and growth of *Nectria galligena* Bres. *Folia Horticulturae*, 8, 13-19, ISSN 0867-1761.

Calabrese, E.J. & Baldwin, L.A. (2002). Defining hormesis. *Human and Experimental Toxicology*, 21, 91-97, ISSN 0960-3271.

Calabrese, E.J. (2005). Paradigm lost, paradigm found: The re-emergence of hormesis as a fundamental dose response model in the toxicological sciences. *Environmental Pollution*, 138, 378-411, ISSN 0269-7491.

Canady, C.H.; Helsel, D.G. & Wyllie, T.D. (1986). Effects of herbicide-induced stress on root colonization of soybeans by *Macrophomia phaseolina*. *Plant Disease*, 70, 863-866, ISSN 0191-2917.

Carson, M.L.; Arnold, W.E. & Todt, P.E. (1991). Predisposition of soybean seedlings to Fusarium root rot with trifluralin. *Plant Disease*, 75, 342-347, ISSN 0191-2917.

Casale, W.L. & Hart, L.P. (1986). Influence of four herbicides on carpogenic germination and apothecium development of *Sclerotinia sclerotiorum*. *Phytopathology*, 76, 980-984, ISSN 0031-949X.

Chen, S.J. & Kao, C.H. (1996). Ammonium accumulation in relation to senescence of detached maize leaves. *Botanical Bulletin of Academia Sinica*, 37, 255-259, ISSN 0006-8063.

Chen, S.J.; Hung, K.T. & Kao, C.H. (1997). Ammonium accumulation is associated with senescence of rice leaves. *Plant Growth Regulation*, 21, 195-201, ISSN 0167-6903.

Cohen, R.; Riov, J.; Lisker, N. & Katan, J. (1986). Involvement of ethylene in herbicide-induced resistance to *Fusarium oxysporum* f.sp. *melonis*. *Phytopathology*, 76, 1281-1285, ISSN 0031-949X.

Cohen, R.; Blaier, B.; Schaffe, A.A. & Katan, J. (1996). Effect of acetochlor treatment on Fusarium wilt and sugar content in melon seedlings. *European Journal of Plant Pathology*, 102, 45-50, ISSN 0929-1873.

Cronshow, D.K. & Pegg, G.F. (1976). Ethylene as a toxin synergist in Verticillium wilt of tomato. *Physiological Plant Pathology*, 9, 33-44, ISSN 0885-5765.

Curl, E.A.; Rodriguez-Kabana, R. & Funderburg, H.H. (1968). Influence of atrazine and varied carbon and nitrogen amendments on growth of *Sclerotium rolfsii* and *Trichoderma viride* in soil. *Phytopathology*, 58, 323-328, ISSN 0031-949X.

Dann, E.K.; Diers, B.W. & Hammerschmidt, R. (1999). Suppression of Sclerotinia stem rot of soybean by lactofen herbicide treatment. *Phytopathology*, 89, 598-602, ISSN 0031-949X.

Devine, M.D.; Duke, S.O. & Fedtke, C. (1993). *Physiology of Herbicide Action*, Englewood Cliffs, ISBN 0133690679, Prentice Hall.

Dissanayake, N.; Hoy, J.W. & Griffin, J.L. (1998). Herbicide effects on sugarcane growth, Pythium root rot, and *Pythium arrenomanes*. *Phytopathology*, 88, 530-535, ISSN 0031-949X.

Duan, W.J.; Zhang, X.Q.; Yang, T.Z.; Dou, X.W.; Chen, T.G.; Li, S.J.; Jiang, S.J.; Huang, Y.J. & Yin, Q.Y. (2010). A novel role of ammonia in appressorium formation of *Alternaria alternata* (Fries) Keissler, a tobacco pathogenic fungus. *Journal of Plant Diseases and Protection*, 117, 112-116, ISSN 1861-3829.

Duke, S.O. (1990). Overview of herbicide mechanisms of action. *Environmental Health Perspectives*, 87, 263-271, ISSN 0091-6765.

Duke, S.O.; Cedergreen, N.; Velini, E.D. Belz, R.G. (2006). Hormesis: Is it an important factor in herbicide use and allelopathy? *Outlooks in Pest Management*, 17, 29-33, ISSN 1743-1026.

Duke, S.O.; Wedge, D.E.; Cerdeira, A.L. & Matallo, M.B. (2007). Herbicide effects on plant disease. *Outlooks on Pest Management*, 18, 36-40, ISSN 1743-1026.

Eshel, D.; Miyara, I.; Ailing, T.; Dinoor, A. & Prusky, D. (2002). pH regulates endoglucanase expression and virulence of *Alternaria alternata* in persimmon fruits. *Molecular Plant-Microbe Interactions*, 15, 774-779, ISSN 0894-0282.

Feng, P.C.C; Baley, G.J.; Clinton, W.P.; Bunkers, G.J.; Alibhai, M.F.; Paulitz, T.C. & Kidwell, K.K. (2005). Glyphosate inhibits rust disease in glyphosate-resistant wheat and soybean. *Proceedings of the National Academy of Sciences* (USA), 102, 17290-17295, ISSN 0027-8424.

Feng, , P.C.C.; Clark, C.; Andrade, G.; Balbi, M.C. & Caldwell, P. (2008). The control of Asian rust by glyphosate in glyphosate-resistant soybeans. *Pest Management Science*, 64, 353-359, ISSN 1526-498X.

Gimsing, A.L., Borggaard, O.K., Jacobsen, O.S., Aamand, J. & Sørensen, J. (2004). Chemical and microbiological soil characteristics controlling glyphosate mineralization in Danish surface soils. *Applied Soil Ecology*, 27, 233-242, ISSN 0929-1393.

Grenville-Briggs, L.J & Van West, P. (2005). The biotrophic stages of oomycete-plant interactions. *Advances in Applied Microbiology*, 57, 217-243, ISBN 978-0-12-374788-4.

Grenville-Briggs, L.J.; Avrora, A.; Bruce, C.R.; Williams, A.; Birch, P.R.J. & Van West, P. (2005). Elevated amino acid biosynthesis in *Phytophthora infestans* during appressorium formation and potato infection. *Fungal Genetics and Biology*, 42, 244-256, ISSN 1087-1845.

Grinstein, A.; Katan, J. & EShel, Y. (1976). Effect of dinitroaniline herbicides on plant resistance to soilborne pathogens. *Phytopathology*, 66, 517-522, ISSN 0031-949X.

Grinstein, A.; Lisker, N.; Katan, J. & Eshel, Y. (1984). Herbicide-induced resistance to plant wilt diseases. *Physiological Plant pathology*, 24, 347-356, ISSN 0885-5765.

Harikrishnan, R. & Yang, X.B. (2001). Influence of herbicides on growth and sclerotia production in *Rhizoctonia solani*. *Weed Science*, 49, 241-247, ISSN 0043-1745.

Harris, D. & Grossbard, E. (1979). Effects of the herbicides gramoxone W and roundup on *Septoria nodorum*. *Transactions of the British Mycological Society*, 73, 27-34, ISSN 0007-1536.

Hess, F.D. (2000). Light-dependent herbicides: an overview. *Weed Science*, 48, 160-170, ISSN 0043-1745.

Heydari, A.; Misaghi, I.J. & Balestra, G.M. (2007). Pre-emergence herbicides influence the efficacy of fungicides in controlling cotton seedling damping off in the field. *International Journal of Agricultural Research*, 2, 1049-1053, ISSN 1816-4897.

Hill, T.L. & Stratton, G.W. (1991). Interactive effects of the fungicide chlorothalonil and the herbicide metribuzin towards the fungal pathogen *Alternaria solani*. *Bulletin of Environmental Contamination and Toxicology*, 47, 97-103, ISSN 0007-4861.

Hoagland, R.E. (2000). Plant pathogens and microbial products as agents for biological weed control. In: *Advances in Microbial Biotechnology*, Tewari, J.P.; Lakhanpal, T.N.; Singh, J.; Gupta, R. & Chamola, V.P. (Eds.), 213-255, APH Publishing Corp., ISBN 81-7648-078-9, New Dehli, India.

Hoerlein, G. (1994). Glufosinate (Phosphinothricin), a natural amino acid with unexpected herbicidal properties. *Reviews of Environmental Contamination and Toxicology*, 138, 73-145, ISSN 0179-5953.

Isakeit, T. & Lockwood, J.L. (1989). Lethal effect of triazine and other triazine herbicides on ungerminated conidia of *Cochliobolus sativus* in soil. *Soil Biology and Biochemistry*, 21, 809-817, ISSN 0038-0717.

Johal, G.S. & Rahe, J.E. (1990). Role of phytoalexins in the suppression of resistance of *Phaseolus vulgaris* to *Colletotrichum lindemuthianum* by glyphosate. *Canadian Journal of Plant Pathology*, 12, 225-235, ISSN 1715-2992.

Jones, D.G. & Williams, J.R. (1971). Effect of paraquat on growth and sporulation of *Septoria nodorum* and *Septoria tritici*. *Transactions of the British Mycological Society*, 57, 351-357, ISSN 0007-1536.

Kabana, R.R.; Curl, E.A. & Funderburk, H.H. (1966). Effect of herbicides on growth of *Rhizoctonia solani*. *Phytopathology*, 56, 1332-1333, ISSN 0031-949X.

Kabana, R.R. & Curl, E.A. (1970). Nontarget effects of pesticides on soilborne pathogens and diseases. *Annual Review of Phytopathology*, 18, 311-321, ISSN 00664286.

Khan, W.S.; Zaidi, A. & Aamil, M. (2004). Influence of herbicides on chickpea-*Mesorhizobium* symbiosis. *Agronomie*, 24, 123-127,

Kawate, M.K. & Appleby, A.P. (1987). Effect of soil pH on availability of glyphosate in soil to germinating ryegrass seedlings. *Journal of Applied Seed Production*, 5, 45-49, ISSN 8755-8750.

Kawate, M.K.; Kawate, S.C.; Ogg, A.G. & Kraft, J.M. (1992). Response of *Fusarium solani* f.sp. *pisi* and *Pythium ultimum* to glyphosate. *Weed Science*, 40, 497-502, ISSN 0043-1745.

Keen, N.T.; Holliday, M.J. & Yoshikawa, M. (1982). Effects of glyphosate on glyceollin production and the expression of resistance to *Phytophthora megasperma* f. sp. *glycinea* in soybean. *Phytopathology*, 72, 1468-1470, ISSN 0031-949X.

Kortekamp, A. (2008). Knocked out with Basta®! - Are herbicides effective against downy mildew of grapevine? *Journal of Plant Diseases and Protection, Special Issue* XXI, 107-112, ISSN 1861-4051.

Kortekamp, A. (2010). Side effects of the herbicide glufosinate ammonium on *Plasmopara viticola* and other fungal pathogens. In: *Proceedings of the 6th International Workshop of Grapevine Downy and Powdery Mildew*, Calonnec, A.; Delmotte, F., Emmet, B.; Gadoury, D.; Gessler, C.; Gubler, D.; Kassemeyer, H.-H.; Magarey, P.; Raynal. M.; Seem, R. (Eds.), 13-15, ISBN 978-2-7380-1279-1, Bordeaux (France).

Kramer-Haimovich, H.; Servi, E.; katan, T.; Rollins, J.; Okon, Y. & Prusky, D. (2006). Effect of ammonia production by *Colletotrichum gloeosporioides* on *pelB* activation, pectate lyase secretion, and fruit pathogenicity. *Applied and Environmental Microbiology* 71, 1034-1039, ISSN 0099-2240.

Krogmann, D.W., Jagendorf, A.T. & Avron, M. (1959). Uncouplers of spinach chloroplast photosynthesis phosphorylation. *Plant Physiology*, 34, 272-277, ISSN 0032-0889.

Lancaster, S.H., Hollister, E.B., Senseman, S.A. & Gentry, T.J. (2010). Effects of repeated glyphosate applications on soil microbial community composition and the mineralization of glyphosate. *Pest Management Science*, 66, 59-64, ISSN 1526-498X.

Larson, R.L., Hill, A., Fenwick, A., Kniss, A.R., Hanson, L.E. & Miller, S.D. (2006). Influence of glyphosate on Rhizoctonia and Fusarium root rot in sugar beet. *Pest Management Science*, 62, 1182-1192, ISSN 1526-498X.

Liu, C.A.; Zhong, H.; Vargas, J.; Penner, D. & Sticklen, M. (1998). Prevention of fungal diseases in transgenic, bialaphos- and glufosinate-resistant creeping bentgras (*Agrostis palustris*). *Weed Science*, 46, 139-146, ISSN 0043-1745.

Matringe, M.; Camadro, J.M.; Block, M.A.; Joyard, J.; Scalla, R.; Labbe, P. & Douce, R. (1992). Localization within the chloroplast of protoporphyrinogen oxidase, the target enzyme for diphenylether-like herbicides. *Journal of Biological Chemistry*, 267, 4646-4651, ISSN 0021-9258.

Moneke, A.N.; Okpala, G.N. & Anyanwu, C.U. (2010). Biodegradation of glyphosate herbicide *in vitro* using bacterial isolates from four rice fields. *African Journal of Biotechnology*, 9, 4067-4074, ISSN 1684-5315.

Murakami, T.; Anzai, H.; Imai, S.; Satoh, A.; Nagaoka, K. & Thompson, C.J. (1986). The bialaphos biosynthetic genes of *Streptomyces hygroscopicus*: molecular cloning and characterization of the gene cluster. *Molecular and General Genetics*, 205, 42-50, ISSN 0026-8925.

Nelson, K.A.; Renner, K.A. & Hammerschmidt, R. (2002). Effects of protoporphyrinogen oxidase inhibitors on soybean (*Glycine max* L.) response, *Sclerotinia sclerotiorum* disease development, and phytoalexin production by soybean. *Weed Technology*, 16, 353-359, ISSN 0890-037X.

Neubauer, R. & Avizohar-Hershenson, Z. (1973). Effect of herbicide, trifluralin, on Rhizoctonia disease in cotton. *Phytopathology*, 63, 651-652, ISSN 0031-949X.

Nilsson, M.T.; Krajewski, W.W.; Yellagunda, S.; Prabhumurthy, S.; Chamarahally, G.N.; Siddamadappa, C. Srinivasa, B.R.; Yahiaoui, S.; Larhed, M.; Karlén, A.; Jones, T.A & Mowbray, S.L. (2009). Structural basis for the inhibition of *Mycobacterium tuberculosis* glutamine synthetase by novel ATP-competitive inhibitors. *Journal of Molecular Biology*, 393, 504-513, ISSN 0022-2836.

Nosanschuk, J.D., Ovalle, R. & Casadevall, A. (2001). Glyphosate inhibits melanization of *Cryptococcus neoformans* and prolongs survival of mice after systemic infection. *Journal of Infectious Diseases*, 183, 1093-1099, ISSN 0022-1899.

Olesen, C.F. & Cedergreen, N. (2010). Glyphosate uncouples gas exchange and chlorophyll fluorescence. *Pest Management Science*, 66, 536-542, ISSN 1526-498X.

Pakdaman, B.S.; Khabbaz, H.; Goltapeh, E.M. & Afshari, H.A. (2002). *In vitro* studies on the effects of sugar beet fields prevalent herbicides on the beneficial and deleterious fungal species. *Pakistan Journal of Plant Pathology*, 1, 23-24, ISSN 1812-5425.

Peng, G. & Byer, K.N. (2005). Interactions of *Pyricularia setariae* with herbicides for control of green foxtail (*Setaria viridis*). *Weed Technology*, 19, 589-598, ISSN 0890-037X.

Prusky, D.; McEvoy, J.L.; Leverentz, B. & Conway, W.S. (2001). Local modulation of host pH by *Colletotrichum* species as a mechanism to increase virulence. *Molecular Plant-Microbe Interactions*, 14, 1105-1113, ISSN 0894-0282.

Radke, V.L. & Grau, C.R. (1986). Effects of herbicides on carpogenic germination of *Sclerotinia sclerotiorum*. *Plant Disease*, 70, 19-23, ISSN 0191-2917.

Rattanakreetakul, C.; Korpraditskul, V. & Chamsawarng, C. (1990). Side effects of some herbicides to *Fusarium moniliforme* and its antagonistic microorganisms. *Kasetsart Journal (Natural Science Suppl.)*, 24, 41-48, ISSN 0075-5192.

Ries, S.K.; Chmiel, H.; Dilly, D.R. & Filner, P. (1967). The increase in nitrate reductase activity and protein content of plants treated with simazine. *Proceeding of the national Academy of Sciences* USA, 58, 526-532, ISSN 0027-8424.

Roberts, C.W., Roberts, F., Lyons, R.E., Kirisits, M.J., Mui, E.J., Finnerty, J., Johnson, J.J., Ferguson, D.J.P., Coggins, J.R., Krell, T., Coombs, G.H., Milhous, W.K., Kyle, D.E., Tziposi, S., Barnwell, J., Dame, J.B., Carlton, J. & McLeod, R. (2002). The shikimate pathway and its branches in apicomplexan parasites. *Journal of Infectious Diseases*, 185 (Suppl. 1), 25-36, ISSN 0022-1899.

Russin, J.S.; Carter, C.H. & Griffin, J.L. (1995). Effects of grain sorghum (*Sorghum bicolor*) herbicides on charcoal rot fungus. *Weed Technology*, 9, 343-351, ISSN 0890-037X.

Sandmann, G & Böger, P. (1982). Mode of action of herbicidal bleaching. *ACS Symposium Series*, 181, 111-130, ISBN 9780841206991.

Sanogo, S.; Yang, X.B. & Scherm, H. (2000). Effects of the herbicides on *Fusarium solani* f.sp. *glycines* and development of sudden death syndrome in glyphosate-tolerant soybean. *Phytopathology*, 90, 57-66, ISSN 0031-949X.

Sanyal, D. & Shrestha, A. (2008). Direct effect of herbicides on plant pathogens and disease development in various cropping systems. *Weed Science*, 56, 155-160, ISSN 0043-1745.

Schuster, E. & Schröder, D. (1990). Side-effects of sequentially-applied pesticides on non-target soil microorganisms: field experiments. *Soil Biology and Biochemistry*, 22, 367-373, ISSN 0038-0717.

Shan, W.; Marshall, J.S. & Hardham, A.R. (2004). Gene expression in germinated cysts of *Phytophthora nicotianae*. *Molecular Plant Pathology*, 5, 317-330, ISSN 1464-6722.

Sharma, U.; Adee, E.A. & Pfender, W.F. (1989). Effect of glyphosate herbicide on pseudothecia formation by *Pyrenophora tritici-repentis* in infested wheat straw. *Plant Disease*, 73, 647-650, ISSN 0191-2917.

Sharon, A.; Ghirlando, R. & Gressel, J. (1992). Glyphosate suppression of an elicited defense response. Increased susceptibility of *Cassia obtusifolia* to a mycoherbicide. *Plant Physiology*, 98, 654-659, ISSN 0032-0889.

Shukla, A.K.; Tiwari, B.K. & Mishra, R.R. (1988). Effect of foliar application of herbicides, benthiocarb, 2,4-D and fluchloralin on phyllosphere microflora of potato. *Plant and Soil*, 106, 277-280, ISSN 0032-079X.

Smith, N.R.; Dawson, V.T. & Wenzel, M.E. (1946). The effect of certain herbicides on soil microorganisms. *Proceedings – Soil Science Society of America*, 10, 197-201, ISSN 0038-0776.

Smith, D.A. & Hallett, S.G. (2006). Interactions between chemical herbicides and the candidate bioherbicides *Microsphaeropsis amaranthi*. *Weed Science*, 54, 197-201, ISSN 0043-1745.

Strachan, S.D. & Hess, F.D. (1983). The biochemical mechanism of action of the dinitroaniline oryzalin. *Pesticide Biochemistry and Physiology*, 20, 141-150, ISSN 0048-3575.

Thompson, C.J.; Movva, N.R.; Tizard, R.; Crameri, R.; Davies, J.E. & Lauwereys, M. (1987). Characterization of the herbicide resistance gene *bar* from *Streptomyces hygroscopicus*. *EMBO Journal*, 6, 2519-2523, ISSN 0261-4189.

Toubia-Rahme, H.; Ali-Haimoud, H.; Barault, G. & Albertini, L. (1995). Inhibition of *Drechslera teres* scleroid formation in barley straw by application of glyphosate or paraquat. *Plant Disease*, 79, 595-598, ISSN 0191-2917.

Trebst, A. (2008). The mode of action of triazine herbicides, In: *The triazine herbicides: 50 years revolutionizing agriculture*, LeBaron, H.M., McFarland, J.E. & Burnside, O. (Eds.), 101-110, Elsevier, ISBN 978-0-444-51167-6, Oxford (UK).

Turkington, T.K.; Orr, D.D. & Xi, K. (2001). The influence of roundup® on *in vitro* growth and sporulation of *Rhynchosporium secalis* and *Pyrenophora teres*. *Canadian Journal of Plant Pathology*, 23, 307-311.

Uchimiya, H.; Iwate, M.; Nojiri, C.; Samarajeewa, P.K.; Takamatsu, S.; Ooba, S.; Anzai, H.; Christensen, A.H.; Quail, P.H. & Toki, S. (1993). Bialaphos treatment of transgenic rice plants expressing a *bar* gene prevents infection by the sheath blight pathogen (*Rhizoctonia solani*). *Nature Biotechnology*, 11, 835-836, ISSN 1087-0156.

Verma, P.R. & McKenzie, D.L. (1985). *In vitro* effects of herbicides on mycelial growth of AG2-1 and AG4 *Rhizoctonia isolates* from canola/rapeseed. *Phytopatholgy*, 75, 1363, ISSN 0031-949X.

Wang, Y.; Browning, M.; Ruemmle, B.A.; Chandlee, J.M. & Kausch, A.P. (2003). Glufosinate reduces fungal diseases in transgenic glufosinate-resistant bentgrasses (*Agrostis* spp.). *Weed Science*, 51, 130-137, ISSN 0043-1745.

Wang, Y.S.; Liu, J.C.; Chen, W.C. & Yen, J.H. (2008). Characterization of acetanilide herbicides degrading bacteria isolated from tea garden soil. *Microbial Ecology*, 55, 435-443, ISSN 0095-3628.

Ware, G.W. (1980). Effects of pesticides on nontarget organisms. *Residue Reviews*, 76, 173-210, ISSN 0080-181X.

Weaver, M.A.; Boyette, C.D.& Hoagland, R.E. (2006). Compatibility of the bioherbicides *Myrothecium verrucaria* with selected pesticides. *Phytopathology*, 96, S121, ISSN 0031-949X.

Weaver, M.A.; Jin, X.; Hoagland, R.E. & Boyette, C.D. (2009). Improved bioherbicidal efficacy by *Myrothecium verrucaria* via spray adjuvants or herbicide mixtures. *Biological Control*, 50, 150-156, ISSN 1049-9644.

Westerhuis, D., Vawdrey, L.L., & Piper, R. (2007). An *in vitro* study into the effect of glyphosate on *Sclerotium rolfsii*. *Australasian Plant Disease Notes*, 2, 23-24, eISSN 1833-928X.

Whitelaw-Weckert, M.A. (2010). Interaction between *Cylindrocarpon* and glyphosate in young vine decline. *Phytopathologia Mediterranea*, 49, 117-118 (Abstract), ISSN 0031-9465.

Wilkinson, V. & Lukas, R.L. (1969). Gramoxone W: Its Effects on spores and mycelia of *Rhizopus stolonifer*. *Transactions of the British Mycological Society*, 53, 297-299, ISSN 0007-1536.

Wilcox, W.F. (1996). Influence of dinitroaniline herbicides on growth, sporulation, and infectivity of four *Phytophthora* ssp. pathogenic to deciduous fruit trees. *Phytopathology*, 86, 906-913, ISSN 0031-949X.

Wong, P.T.W.; Dowling, P.M.; Tesoriero, L.A. & Nicol, H.I. (1993). Influence of preseason weed management and in-crop treatments on 2 successive wheat crops. 2. take-all severity and incidence of Rhizoctonia root rot. *Australian Journal of Experimental Agriculture*, 33, 173-177, 0816-1089.

Wu, M.T.; Singh, B. & Salunkhe, D.K. (1972). Influence of foliar application of s-triazine compounds on fresh weight, dry weight, chemical composition, and enzymatic activity of pea and sweet corn seedlings. *Journal of Experimental Botany*, 23, 793-800, ISSN 0022-0957.

Wyss, G.S.; Charudattan, R.; Rosskopf, E.N. & Littell, R.C. (2004). Effects of selected pesticides and adjuvants on germination and vegetative growth of *Phomopsis amaranthicola*, a biocontrol agent for *Amaranthus* spp. *Weed Research*, 44, 469-482, ISSN 0043-1737.

Part 2

Effects on Agro Ecosystems

5

Herbicides Effect on Nitrogen Cycling in Agroecosystems

Damin, V[1]. and Trivelin P.C.O.[2]
[1]*Universidade Federal de Goiás, Escola de Agronomia e Engenharia de Alimentos, Dep. de Ciência do Solo, Goiânia, GO*
[2]*Centro de Energia Nuclear na Agricultura, Lab. Isótopos Estáveis. Av. Centenário, nº 303, Caixa Postal 96, CEP 13416-000, Piracicaba, SP*
Brazil

1. Introduction

With the widespread use of post-emergent herbicides and their increased use in areas under conservation management systems (no-till and minimum tillage), concerns about environmental and public health problems due to herbicidal molecules has been growing. The use of herbicides has been justified by the resultant reduction in production costs because expenses regarding application and product price are lower with these treatments compared with existing alternatives. However, studies in the literature have demonstrated that there can be an increasing need for greater doses of nitrogen (N) fertilizers and pesticides after herbicides have been applied (Cakmak, 2007; Damin et al., 2008, 2009).
The use of nitrogen fertilizers in doses that are enough to supply the demand of crops is one of the main practices associated with high productivity. However, these fertilizers are expensive, and their indiscriminate use can cause the emission of green-house gases (Sherlock et al., 1989), the contamination of superficial waters with nitrate and the destruction of ozone in the stratosphere, with N_2O as an intermediary (Groffman, 2000). In addition, fossil fuels, which are non-renewable resource, are consumed in the manufacture of N fertilizers.
Recent research has shown that about 15 to 20% of the N introduced by fertilization can be lost after plants have been desiccated by herbicides. This losses can be even greater when the N that was already in the system is considered. Moreover, some laboratory studies have shown an increase in the emission of N_2O, a gas that has a global heating potential 298 times greater than that of CO_2, in areas using desiccants.
Despite the importance of awareness about the effects of herbicides in the nitrogen cycle, this subject has been seldom discussed and studied in the literature. Within this context, this chapter will discuss the effects of herbicides in the nitrogen cycle processes that determine its availability to plants in agricultural systems.

2. Nitrogen in the agroecosystems

In an attempt to minimize the production costs and environmental impacts that occur because of the use of N fertilizers, the adoption of management strategies that minimize the

N losses in agroecosystems has been proposed. Such strategies include the efficient application of N at the moment the crop needs it most. N presents a great difficulty, however, because it is a highly reactive element that is subject to various changes in the soil and in plants (Moreira & Siqueira, 2006). Most of the changes of N in soil are intermediated by the microbial community, which fluctuates in both time and space. Therefore, besides properties such as pH, organic matter and CTC content, N dynamic is influenced by factors that affect the microbiota. These factors include aeration, water availability and the presence of toxic or stimulants substances to microorganisms.

The effects of herbicides on soil microbiota depend on the mechanism and other features of the product; type of metabolism of the microorganism (e.g., photolithotrophic, photoorganotrophic, chemolithotrophic, chemoorganotrophic), which determines if the metabolic routes that are affected by the herbicide are present in the microorganism; and also edaphic and climatic factors. Some herbicides, such as the paraquat, are toxic to biota in general. Others, such as glyphosate and glufosinate ammonium show low toxicity to non-target organisms. In fact, an increase in microbial biomass after the application of these products has been observed in some situations. This result has been attributed to the products serving as a source of nutrients. It is worth stressing, however, that some studies have shown that glyphosate is toxic to some organisms, such as certain strains of nitrogen-fixing bacteria (NFB).

Besides affecting processes of the N cycle that are intermediated by the microbiota, the herbicides can change the metabolism of N inside the plant, depending on the mechanism of the product. The herbicide glufosinate ammonium, for example, inhibits the incorporation of NH_4^+ in organic compounds. This effect can cause a build-up of NH_4^+ in vegetal tissue. Because its ions are toxic to plants, part of the accumulated NH_4^+ can be exuded through the roots and/or converted into NH_3 and lost in the atmosphere, reducing the N availability in the system. The exit of N from the plant can increase the C/N relation of the formed stubble, thus reducing its mineralization rate.

Some of the processes of the N cycle and the effect desiccant herbicides have over them are presented below.

2.1 Nitrogen losses

Among the main forms of nitrogen loss in agroecosystems are removal by crop, the volatilization of soil ammonium, the gas loss of nitrogen oxides (NO_2, N_2O, NO) and elemental nitrogen (N_2), the lixiviation of soil nitrate leaving the root system's target range and NH_3 loss by the aerial parts of plants.

Sutton et al. (1993) reported that losses of nitrogen by the aerial parts of vegetables can increase ammonium levels in the atmosphere, noting emission values ranging from 5-15 kg ha^{-1} $year^{-1}$ for N-NH_3. This nitrogen output may be responsible for 15-20% of the total NH_3 gas emissions (Asman et al., 1998).

The magnitude of nitrogen losses through the aerial parts of crops varies according to each species. Wetselaar & Farquhar (1980) estimated from research available in the literature that the total N losses through the aerial parts of wheat, rice, grassy sorghum, cotton and rye crops were within a range of 16 to 40, 48, 48, 21 and 42 kg ha^{-1} N, respectively. For barley crops, Schjoerring et al. (1993) found losses of 1-2 kg ha^{-1} $year^{-1}$ of N-NH_3 under three fertilization regimes with N. For corn, Francis et al. (1993) found losses of 45 to 81 kg ha^{-1} N, with application rates ranging from 50 to 300 kg ha^{-1} N. NH_3 losses through the aerial parts of sugar cane were indirectly estimated by Trivelin (2000) at around 50 kg ha^{-1} $year^{-1}$. This number represents half of the dose of N fertilizer applied to the crop sprouts.

Aside from its effects on various species, N losses can be linked to the developmental stages of plants, N availability in the soil, stomatal conductance, leaf temperature, NH_3 concentrations in the atmosphere, ammonium concentrations in the plant, the activity of the glutamine sinthetase enzyme and environmental stresses (Schjoerring et al., 1998; Parton et al., 1988). Another important factor that affects not only the intensity but also the way in which ammonium gas is exchanged among leaves and the atmosphere is the ammonium compensation point, which was defined by Farquhar et al. (1980). When NH_3 concentrations in the atmosphere are below the ammonium compensation point of the plant, emission will occur through the leaves; for concentrations above, it is absorbed (Farquhar et al., 1980; Holtan-Hartwig & Bockman, 1994).

Senescence is the developmental stage that shows the highest NH_3 emission potential. During this stage, there is an increase in protein degradation, and the liberated N is transferred to the glutamate. Afterward, the glutamate N is converted to NH_4^+ by the glutamate dehydrogenase (GDH) enzyme. GDH is an enzyme of N metabolism that usually reaches a higher degree of activity during senescence (Ragster & Chrispeels, 1981; Laurière & Daussand, 1983). NH_4^+ is available for transformation in the composts of N transport, such as glutamine and asparagine (Ghosh et al., 1995; Nakasathien et al., 2000), and some of the ions are converted into NH_3 according to the following balanced reaction:

$$NH_4^+ \leftrightarrow NH_3 + H^+ \qquad pKa = 9.2$$

When the pH is 9.2, 50% of the molecules are dissociated (NH_4^+), and 50% are in the molecular form (NH_3). In the cytoplasm, the pH is usually around 7.2. At this pH value, only 1% of the molecules are in the NH_3 form. However, it is important to stress that the NH_4^+ build-up in the cytoplasm can promote H^+ exit to the apoplast, transiently increasing the interior pH of the cell and, consequently, the NH_3 concentration (Britto & Kronzucker, 2002). There is evidence that NH_3 transport is easier through membranes (Kleiner, 1981), and that it can be stomatally lost.

2.1.1 Effect of herbicides

Herbicides are used in agricultural areas to induce the senescence of plants. The duration of the senescence phase deriving from herbicide application is very short, and the intervention may occur at any developmental stage. Therefore, it is possible that the behavioral standards of N are different from those observed in natural senescence conditions. Besides inducing senescence, glyphosate and glufosinate ammonium herbicides also affect the plant's N metabolism.

The herbicidal action of glyphosate is attributed to its inhibition of the EPSPs (enolpyruvylshikimate phosphate synthase) enzyme that is responsible for one of the synthesis phases of the tryptophan amino acids phenylalanine and tyrosine. However, as a secondary effect, there is an inhibition in the synthesis of phenolic compounds deriving from those amino acids, which increases the activity of the PAL (phenylalamine ammonia lyase) enzyme due to a response effect (Duke & Hoagland, 1985). The PAL acts in the lyase of phenylalanine and tyrosine amino acids, resulting in the formation of phenolic acids and NH_4^+ (Hoagland et al., 1979).

The glufosinate ammonium herbicide inhibits the action of the glutamine sinthetase, which is responsible for the glutamate synthesis from ammonium and glutamine. This metabolic route is considered the main form of ammonium assimilation in organic composts (Schjoerring et al., 1998). Consequently, the application of glufosinate ammonium results in endogenic NH_4^+ in the plant (Manderscheid et al., 2005).

Therefore, the increase of NH_4^+ in plants desiccated with glyphosate and glufosinate ammonium should not be connected to the induction senescence alone. Damin et al. (2010b) quantified N losses by *Brachiaria decumbens* after desiccation with glyphosate, glufosinate ammonium and paraquat, all of which promoted senescence. However, only the first two increased the losses of N in the system.

Manderscheid et al. (2005) assessed the ammonium concentration in five weed species after applying glufosinate ammonium. They observed an NH_4^+ increase in studied species. Damin et al. (2008; 2010a) observed losses of around 10 to 20% of applied N fertilizer after the application of glyphosate and glufosinate ammonium in *Brachiaria decumbens* and *Pennicetum glaucum* (Table 1). In those same studies, a reduction of around 40% of the total-N in the plant was observed after the application of herbicides.

From a practical point of view, the effect of herbicides on the total N of plants used for soil covering is important because it can affect the mineralization rate due to a raise in the C:N ratio, and it can effect the immobilization of N by soil microorganisms, a process that reduces the nutrient's availability to plants.

3. N mineralization/immobilization

When organic residue is added to the soil, extra-cellular enzymes are liberated by the biota, which degrade the residue into substances with smaller molecular weights (monomers). These monomers are absorbed and metabolized by microbial cells, which transform them into inorganic forms, a process called mineralization. Mineralization occurs at the same time as immobilization, which refers to the incorporation of mineral nutrients to the microbial cells to meet the nutritional demand of the soil microbiota.

Treatments	Above ground part		Roots		Entire plant		Soil		Losses on soil-plant system	
	P. g.[1]	B. d.[2]	P. g.[1]	B. d.[2]	P. g.[1]	B. d.[2]	P. g.[1]	B. d.[2]	P. g.[1]	B. d.[2]
	$g\ m^{-2}$									
Control	13.8 **a**	11.5	3.4 **a**	6.1 **a**	17.2 **a**	17.6 **a**	2.4 **b**	2.1 **b**	5.4 **b**	9.9 **b**
Glyphosate	10.8 **b**	8.1	2.7 **b**	2.9 **b**	13.6 **b**	11.1 **b**	3.4 **a**	3.4 **a**	8.0 **a**	15.1 **a**
Glufosinate	11.0 **b**	8.1	2.0 **b**	2.4 **b**	12.9 **b**	10.5 **b**	4.0 **a**	3.1 **a**	8.0 **a**	16.0 **a**
LDS (0,05)	2.1	3.8	0.5	1.8	3.0	4.9	0.7	0.8	2.3	4.9
P	**	ns	*	**	**	**	*	**	*	*
C.V. (%)	19.2	26.3	31.7	32.3	8.9	25.1	28.3	18.9	14.9	24.2

[1]Measurement of ten replicates. [2]Measurement of six replicates. *Means followed by different letters in columns differ from each other by Tukey's Test (α = 0.05). **Means followed by different letters in columns differ from each other by Tukey's Test (α = 0.01). The total dose of N was 25 g m^{-2} in pearl millet and 15 g m^{-2} in signal grass.

Table 1. Data of N recovered and lost from ^{15}N-labeled fertilizer (Ndff) in the soil-plant system after herbicide application in *Brachiaria decumbens* (B.d.) and *Pennicetum glaucum* (P.g.). Retrieved from Damin et al. (2010a).

From the incorporation of vegetable residues in soil to their mineralization, several living organisms act in this process. In the initial phase, the vegetable residues are fragmented into smaller litter by the soil's macro-fauna. Subsequently, a diverse microbial community acts to liberate extra-cellular enzymes, which turn the substances that are easily decomposed into monomers that can then be absorbed by microbial cells. Substances that are difficult to degrade, such as cellulose and lignin, are degraded by a more specialized group of micro-organisms, which compose the secondary and tertiary mineralizers.

Residues that are rich in lignin and polyphenols have slower mineralization rates in the soil due to the resistance these substances have to degradation (Moreira & Siqueira, 2006; Oliveira et al., 2002). The C/N ratio also has a major influence on the mineralization rate of vegetable residues, as it determines the N availability in the microbiota. Crop residues with a C/N ratio around 15:1 to 20:1 (e.g., soybeans, beans, lupine, forage turnips) tend to degrade faster than those with a C/N ratio higher than 30:1 (e.g., black oat, pearl millet, corn, sorghum). Heinzmann (1985) studied the N dynamics after winter green fertilization with black oat, forage turnip, hairy vetch, lupine and wheat and their effects on soy, corn and bean crops. He observed that the N liberation from legume and forage turnip residues occurred in the first weeks, whereas the black oat liberated more N in the blooming and graining stages.

3.1 Effect of herbicides

In addition to chemical composition, the mineralization of crop residues added to the soil depends on factors that affect the composition and activity of the decomposing microbial community. These factors include temperature, humidity, aeration and the presence of toxic or nurturing substances. Concerning the effect of herbicides on mineralization rates, some observations should be made: 1) Herbicides can increase vegetable residue decomposition rates due to a greater detachment of leaves and roots caused by senescence. Thus, they may have a beneficial physical effect on stubble mineralization (Snapp & Borner, 2005). 2) However, the increased C/N ratio in the stubble caused by the application of these products may decrease the mineralization rate. 3) Furthermore, these herbicidal molecules may alter the soil's microbial community and cause an increase or a decrease in mineralization, depending on the affected microorganism populations.

Argenta et al. (2001) observed an increase in the C:N ratio on the aerial parts of black oats 16 days after the application of glyphosate, glufosinate and paraquat. Although the paraquat did not decrease the N content of the plant during senescence, there was a remobilization of the nutrient to the reproductive structures and to the root system (Marschner, 1995). With the increase in the C/N ratio, the stubble mineralization rate can decrease. In fact, Damin et al. (2009; 2010c) assessed the N decomposition rate present in pearl millet and black oat stubble after the application of glyphosate and glufosinate ammonium. They observed an increase in the stubble dry matter remaining on the soil surface and a decrease in N liquid mineralization (Tables 2 and 3). In their studies, the absorption of N in stubble (enriched with ^{15}N) by corn plants was also assessed, and the herbicide-treated stubble was observed to be less effective in supplying N to plants.

The mineralization of organic substances is mainly mediated by chemoorganotrophic micro-organisms, which should not be damaged by the application of glyphosate and glufosinate ammonium because these herbicides act on metabolic routes that are typical of lithotrophic organisms (chemolithotrophic or photolithotrophic). Therefore, the reduction in mineralization observed by Damin et al. (2009; 2010c) was attributed to the increase in the stubble C/N ratio.

In fact, Acinelli et al. (2002) observed that the application of the recommended doses of glyphosate and glufosinate ammonium did not interfere in the activity or in the soil microbial carbon biomass. The application of higher doses than recommended resulted in greater microbial activity, which can be connected to the death of lithotrophic micro-organisms, ensuring a competitive advantage to heterotrophs. Similar results were found for glyphosate by Haney et al. (2000), Ratcliffe et al. (2006) and Zabaloy & Gómez (2008).
Some studies have shown that the N mineralization that is already in the soil is increased by the application of glyphosate. Grossbard (1985) assessed the effect of glyphosate on soil N mineralization and observed an increase in mineralization when the herbicide was used. Haney et al. (2002) evaluated under laboratory conditions the effect of applying atrazine a mix of atrazine and glyphosate on the mineralization of total-C and total-N present in the soil. These authors observed a greater C and N mineralization in the soil treated with the mixture. The results showed that the glyphosate favored the chemoorganotrophic microbiota, causing an increase in the soil edaphic organic matter mineralization.
In tillage systems in Brazil's Cerrado, a reduction in deposited stubble mineralization is desirable because it increases the permanence time of residue over the soil. However, the lower effectiveness of the residue in supplying N to plants and microorganisms can reduce productivity when the N supply through fertilization is not appropriate because the addition of low N content organic residues to the soil results in the immobilization of the nutrient by the microbiota, reducing its availability to plants. Therefore, the use of herbicides for covering crop desiccation in tillage system should be considered in the decision making processes of N fertilization management.

Treatments	DM	Carbon	Nitrogen	C:N	ΔDM	ΔC	ΔN
	g per pot		mg per pot			%	
Control	6.5 b	2.6 b	117.2 b	22.6 b	49.9 a	49.4 a	69.6 a
Glyphosate	8.5 a	3.7 a	159.9 a	24.3 b	34.1 b	22.1 b	51.8 b
Glufosinate	9.3 a	3.8 a	160.2 a	28.6 a	28.3 b	23.2 b	49.7 b
F-test	**	**	**	*	**	**	**
CV (%)	15.5	13.6	16.2	12.4	25.6	30.3	12.3

DM = Dry mass; C = carbon; N = nitrogen; C:N = Carbon to Nitrogen ratio; Δ = (Non-recovered fraction/ total amount added)*100. Means followed by the same letter in a given column indicates nonsignificant differences at the 5% level by Tukey test. **, * = Significant at 1% and 5% levels, respectively.

Table 2. Dry mass, carbon, nitrogen, and C:N ratio in above-ground parts of black oat treated with herbicides at 101-days-old grown in soil.

4. N exuding through the roots

The rhizosphere may be defined as the soil zone that is influenced by the roots. The deposition of exuded matter and root fragments in this region favors the growth and activity of the soil microbial community, which has important functions in cycling and in the supply of nutrients to plants, especially N. About 15% of the total CO_2 set by plants is deposited in the rhizosphere in a cereal crop cycle (Nguyen, 2003). Sugars, amino acids, organic acids, phenols, exoenzymes and ions such as NH_4^+ are the main compounds exuded by the roots (Kraffczyk et al., 1984; Marschner, 1995; Dakora & Phillips, 2002; Nguyen, 2003; Paterson, 2003).

Exudation may be an important way for N to exit the plant. Klein et al. (1988) estimated that the C:N ratio of exuded compounds for the *Poaceae* family species ranged from 2 to 2.7. However, other studies have demonstrated that the exuding of compounds containing N is very low, and the exuded compounds are rich in sugars (Deubel et al., 2000; Paterson, 2003; Merbach et al., 2003). The main factors affecting the amount and composition of material exuded from the roots are vegetable species, plant development stage, environmental stress, nutritional state, injuries, application of chemical products and activity and composition of the soil microbial community (Marschner, 1995; Holland et al., 1996).

Treatments	Net N mineralization	Non-recovered	Soil – Plant
		mg per pot	
Control	201.2 a	52.9	418.0 a
Glyphosate	150.0 b	43.9	383.9 ab
Glufosinate	148.5 b	74.6	368.3 b
F-test	**	NS	*
CV (%)	8.8	54.6	8.9
		%	
Control	42.3 a	11.2	88.8
Glyphosate	35.3 b	10.3	90.2
Glufosinate	33.5 b	16.8	83.1
F-test	**	NS	NS
CV (%)	8.7	55.4	8.8

Means followed by the same letter in a given column indicates nonsignificant differences at the 5% level by Tukey's test; **, * = Significant at 1% and 5% levels, respectively; NS = nonsignificant difference.

Table 3. Net N mineralization, non-recovered fraction and soil-plant recovery of the nitrogen arising from black oat residue with or without herbicide previous application.

4.1 The effect of herbicides in exudated N

Herbicides may affect some of the mentioned factors and, consequently, root exudation. Kremer et al. (2005) evaluated amino acid and carbohydrate exudation after glyphosate was applied to conventional crops and crops genetically modified for herbicide tolerance. They concluded that the exudation was stronger in both crops after the product was applied. Similar results were obtained in bean plants that were desiccated with glyphosate (Liu et al., 1997). These authors found aspartic acid, glutamic acid, serine, glycine, amino butyric acid, valine, isoleucine, leucine, tyrosine, phenylalanine and proline in the material exuded from the roots. They observed that the valine, isoleucine and glycine concentrations were higher in the plants desiccated with the product.

The herbicides that were used as a replacement for the glyphosate also increased root exudation. Damin et al. (2010b) evaluated the compound exudation containing N and the root detachment in *Brachiaria decumbens* plants after the application of glyphosate, glufosinate ammonium and paraquat. They observed an increase in exudation (Figure 1) and the detachment of roots after the application of three products. Paraquat, unlike glyphosate and glufosinate ammonium, does not have its action connected to the N metabolism of the plant. As all herbicides increase N exudation, this process is probably not connected exclusively to the product's active mechanism. Physiologic events characteristic of senescence could justify the results obtained in this research. However, Kremer et al. (2005) observed that the exudation of sugars and amino acids was higher in soybeans genetically modified for tolerance to glyphosate than in the conventional soybeans. In this case, the increase in exudation was not connected to the development of senescence after the application of the herbicide.

It is worth stressing that the losses from exudation and root detachment together represented less than 4% of the N applied to the entire plant and, therefore, they should not affect in a relevant way the N content of the same plant (Damin et al., 2010b). The compound N exudation, notwithstanding, may modify the composition and activity of the soil microbial community due to the supply of carbon sources that are highly soluble and readily available.

There is a great diversity of microorganisms in the soil, which, according to the kind of nutrition, may be grouped in photolithotrophic, photoorganotrophic, chemolithotrophic and chemoorganotrophic. Phototrophic microorganisms (photolithotrophic and photoorganotrophic) are rare, and chemotrophic microorganisms are abundant in the soil. All kinds of fungi are chemoorganotrophic, i.e., they use organic molecules as a source of carbon and energy, whereas bacteria and actinomycetes may come from a variety of groups. Processes such as nitrification and N biological fixation are mainly mediated by chemolithotrophic organisms, which use CO_2 as a carbon source and obtain energy from inorganic compound oxidation (Moreira & Siqueira, 2006).

Due to the nutritional type, the supply of easily decomposable organic carbon sources caused by exudation offers a competitive advantage to chemoorganotrophic microorganisms, modifying the proportion of microorganism populations in the soil microbial community. In addition, chemoorganotrophs may be favored by the use of herbicides such as glyphosate and glufosinate, which act in metabolic routes that are present only in lithotrophic organisms (photolithotrophic and chemolithotrophic).

Among the chemoorganotrophic microorganisms that are important for agriculture are the decomposers of organic residues and most of the plants' pathogens. Kremer et al. (2005) observed an increase in fungi biomass of the kind *Fusarium* sp. after the application of glyphosate to soybeans. Liu et al. (1997) observed an increase in the number of colonies and a higher development of the *Pytium* sp. fungi after addition to the environment through the bean roots cultivated with glyphosate desiccated or exuded herbicide. As the glyphosate can also be exuded by the roots (Kremer et al., 2005; Tuffi Santos et al., 2008), it is possible that the population increase of pathogenic microorganisms after the application of the herbicide is connected to the N compound exudation as well as to the presence of the herbicide.

Aside from those processes, a higher colonization of roots by pathogenic microorganisms may be associated with the effects the herbicide has on the plant's defense system. Liu et al. (1997) also observed that bean plantlets growing in the middle of *Pytium* sp. showed a higher lignin content when glyphosate was not added to the environment. Studies in sterile environments have shown that the increase in *Pythium* and *Fusarium* sp. is connected to the effectiveness of the glyphosate herbicide (Johal & Rahe, 1984; Levesque & Rahe, 1992; Levesque et al., 1992; Levesque et al., 1993; Descalzo et al., 1996).

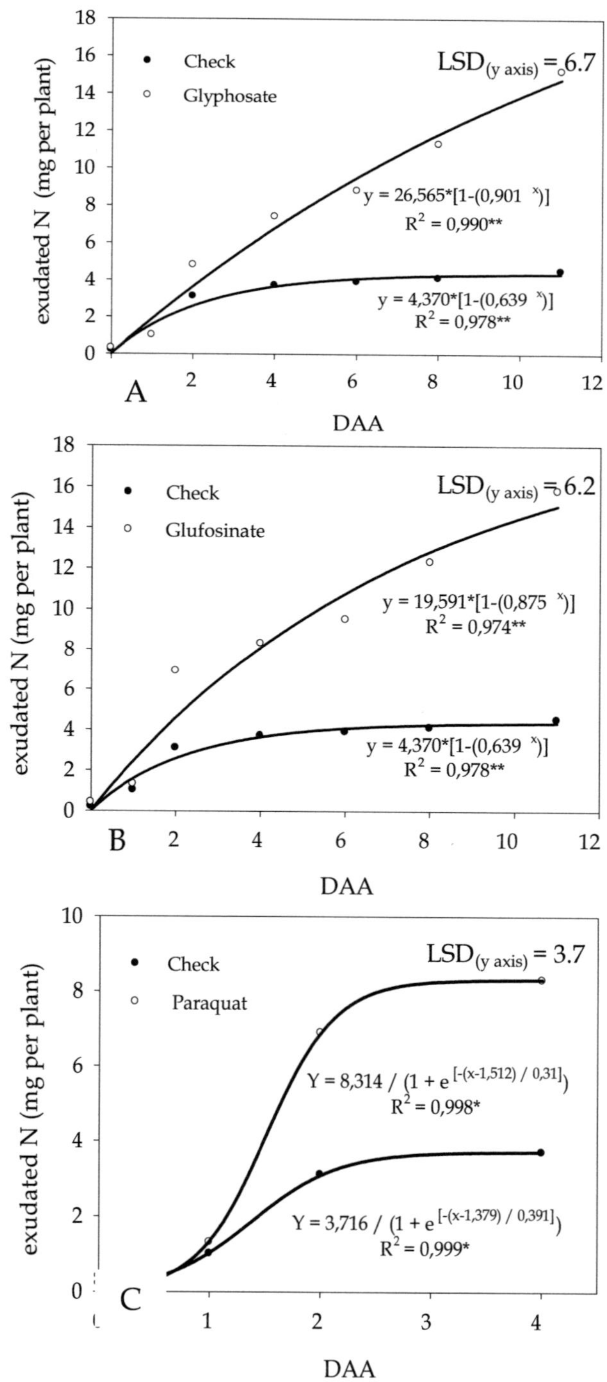

Fig. 1. Time accumulated exudation of nitrogen after herbicide application on *Brachiaria decumbens* Stapf. A – Glyphosate (F_{herb}=2.8*/F_{DAA}=13.5**/$F_{Herb \times DAA}$=4.3**); B – Glufosinate-ammonium (F_{herb}=4.9*/F_{DAA}=15.4**/$F_{Herb \times DAA}$=4.9**); C – Paraquat F_{herb}=6.7*/F_{DAA}=15.2**/$F_{Herb \times DAA}$=3.1*). DAA = days after herbicide application

It is likely that other chemoorganotrophic microorganisms, such as denitrificants, are favored by the use of glyphosate and glufosinate, as the highest exudation results in increased C and N availability in both the soil and these herbicidal molecules (unlike paraquat) do not affect the soil's chemoorganotrophic microbiota.

5. Nitrification/denitrification

In addition to reducing the agronomic effectiveness of N fertilization, N losses because of the denitrification-nitrification processes in soil may have a negative impact on the environment because considerable amounts of N_2O are generated in these processes. This gas is considered the fourth-greatest contributor to the greenhouse effect, although it is present in the atmosphere and in the stratosphere in small amounts. Moreover, N_2O is an intermediary in the ozone layer destruction process (Tabatabai et al., 1981; Griffith, 2005). It is estimated that approximately 20% of the global gas emissions of nitrogen oxides and elemental N are connected to agricultural activity. However, in countries such as Australia, agriculture may be responsible for 80% of these emissions (Australian Greenhouse Office, 2001).

The main processes that result in the gas loss of N oxides and elemental N are nitrification and denitrification. During nitrification (in the oxidation of NH_4 into NO_2), there may be slight production of nitrous oxide by a chemical dismutation of nitroxyl (NOH) or by the action of nitrite reductase (Schimdt, 1982; Bremner, 1997). In this case, the formation of N_2O may occur under aerobic conditions, during the nitrification. Although the N_2O production by nitrification might, in some situations, have an emission potential similar to that of the denitrification process (Granli & Bockman, 1994; Wang et al., 1997), denitrification seems to be the main N_2O source in the soil (Tiedje, 1994).

Biological denitrification is defined as an anaerobic breathing process mediated by microorganisms that are able to use NO_3 or NO_2 as final electron acceptors. The main products of denitrification are N_2 and N_2O, and the proportion of production of each gas varies according to the environmental conditions. There are more than 125 bacterial species that are capable of performing denitrification, including phototrophs, lithotrophs and organotrophs (the most important group).

The main environmental factors that interfere in denitrification rates are O_2 supply, water content, temperature, organic matter, the presence of organic substances and ammonium (NH_4^+) and nitrate (NO_3^-) concentration (Firestone & Davidson, 1989). A number of studies have demonstrated that an increase in carbon content results in higher N_2 and N_2O emissions. Furthermore, pH may interfere in N_2O and N_2 emissions, as it affects nitrification, when it is reduced below 3.5 (Focht & Verstraete, 1977).

5.1 The effect of herbicides in nitrification/denitrification

As discussed in section 4.1, chemoorganotrophic microorganisms may be favored by the application of some herbicides, and such is the case for glyphosate and glufosinate ammonium. Laboratory studies have demonstrated the increase of N_2O emission in soils treated with glyphosate (Bollag & Henninger, 1976; Yeomans & Bremner, 1985; Carlisle & Trevors, 1986), which might be attributable to the herbicide serving as a nutrient source to organisms and offering a competitive advantage to denitrificants through the killing of chemolithotrophic microorganisms. However, higher N_2O emissions were also observed for soils with plants desiccated by these molecules (Robertson et al., 1987; Tenuta & Beauchamp, 1996). This observation may be attributed to the indirect effects of the herbicide in the rizosphere due to the killing of the plants, such as the increase in promptly available carbon content and a reduction in the environment's NO_3 absorption, leaving it more susceptible to microbial attack.

Most nitrifying microorganisms are chemolithotrophic and, therefore, they can show the metabolic routes present in the plants, including those that were affected by glyphosate or glufosinate ammonium. Damin et al. (2009) evaluated the NO_3^- content of two soils (haplustox and quartpsament) 90 days after the permanence of millet stubble desiccated with these herbicides. The N-NO_3^- content in the glyphosate and glufosinate treatments was reduced in the quartpsament, but not in the haplustox. The haplustox soil had a higher sorting capacity than the quatzpsament, as it could be noticed in the contents of the clay, the CTC and the organic matter of soils (Table 4). In fact, the NH_4^+ availability to nitrifying microorganisms was lower in the haplustox than in the quatzpsament due to the higher ion absorption to the colloidal fraction of the soil. Moreover, the glyphosate showed a great affinity toward the iron and aluminum oxides and the hydroxides (Prata et al., 2000) common in the oxisoils. The strong connection between the herbicide and these fractions reduced the bioavailability of the product to the microorganisms.

Some authors have observed a rise in NO_3^- content in the soil after the plants are killed. This result can be attributed to a higher availability of NH_4^+ in the soil as it stops being absorbed by the plant. Although the decrease in nitrification indicates a damaging effect to the microbiota, it is important to highlight that the conversion of NH_4^+ into NO_3^- in the soil may promote a higher risk of superficial water contamination by nitrate and a lower availability of the nutrient to future crops.

	NH_4^+		NO_3^-	
Treatments	**Quartpsament**	**Haplustox**	**Quartpsament**	**Haplustox**
		mg per pot		
Witness+	229.7	297.7	117.2baA	31.5 aB
Glyphosate	197.7	331.6	59.3 bA	34.2 aA
Glufosinate	212.6	314.5	82.6 bA	21.4 aB
Average	213.3 b	314.6 a	86.4	29.1
Herbicide F	0.0 ns		3.8*	
Soil F	24.3**		43.7**	
Interaction F	0.9 ns		4.1*	

+Witness (stubble with no application of herbicides); Means followed by the same letter in a given column indicates nonsignificant differences at the 5% level by Tukey's test. **, * = Significant at 1% and 5% levels, respectively

Table 4. NH_4^+ and NO_3^- content in two soils 90 days after the permanence on the soil of *Pennicetum glaucum* stubble desiccated with herbicides (taken from Damin et al., 2009)

6. Biological N fixation

Biological N_2 fixation (BNF) is an important entrance path for N in agroecosystems, and it is the main source of the nutrient in leguminous farming. In these plants, BNF is performed through the symbiosis between N fixing bacteria (NFB), which break the N_2 triple bond through the action of the nitrogenase enzyme, reducing it to NH_3, while the plants supply photosynthates to the bacteria.

Only a small percentage of prokaryotes are able to accomplish BNF. Among these prokaryotes, however, there is high morphological, physiological and philogenetic diversity. In addition, N fixing micro-organisms are found among the heterotrophs, the anoxygenic phototrophs, the archaebacteria and the cyanobacteria. The main kinds of fixing bacteria, which are economically important, are *Rhizobium* and *Bradyrhizobium*, which are predominantly chemoorganotrophic metabolic proteobacteria, although there are some chemolitotrophic strains of *Bradyrhizobium japonicum*.

6.1 Effect of herbicides

Due to the diverse kinds of metabolism of N fixing bacteria, it is difficult to establish which conditions are favorable or unfavorable to NFB, as the effects of a determined factor will depend on the bacteria community present in the soil. This difficulty may be one of the reasons for the contradictory results in the literature concerning the effect of herbicides on nodulation.

The growth, survival and nitrogenase activity results vary according to the herbicide doses applied, the species/strains studied and the experimental conditions (Faizah et al., 1980; Mallik & Tesfai, 1983; Eberbach & Douglas, 1989; Martensson, 1992; Moorman et al., 1992; Mallik & Tesfai, 1985; Moorman, 1986). Recently, Santos et al. (2004) observed differentiated responses in the growth of strains of native Brazilian *Bradyrhizobium* spp. because of the application of different commercial formulations of glyphosate. This differentiation reveals the importance of considering the adjuvant components of those herbicides in impact evaluations. With soybeans that are tolerant to glyphosate (Roundup Ready, RR), King et al. (2001) reported that the herbicide increased the number and decreased the weight of nodes. Jaworski (1972) observed a reduction of BNF after the application of glyphosate.

Some authors have attributed the BNF reduction after glyphosate application to the inhibition of the EPSPs enzymes in microorganisms (Malkores, 2000). Moorman (1986) evaluated the effect of glyphosate on the development of *Bradyrhizobium* spp. in the environment with and without the addition of amino acids. It was observed that the herbicide reduced the *Bradyrhizobium* spp. population only when no amino acids were added. It is worth stressing, however, that only chemolithotrophic microorganisms, such as some strains of *Bradyrhizobium* spp., should show this metabolic route.

Some authors have attributed the BNF reduction after glyphosate application to the physiological changes caused to the plants. It is common for nutritional disorders to happen in transgenic soybeans after glyphosate application, and two of them are particularly important: manganese (Mn) and iron (Fe). The lack of manganese in the tissue may increase the ureide concentration in the aerial parts, signaling to the BNF to stop fixation (Gordon, 2007). Moreover, the application of the herbicide reduces the production of flavonoids, substances that stimulate the genetic expression of nodulation.

7. Final consideration

The effect of herbicides on a soil-plant system's N loss and on the mineralization rates of vegetable residues should be considered when predicting the N availability to plants, in nutrient-caused environmental impact studies, in production cost estimations and in decision making concerning the appropriate time to apply N doses in areas under a tillage system. This kind of information is important for creating managing strategies that increase the efficiency of N fertilization and reduce the environmental impacts caused by the use of fertilizers and herbicides.

8. References

Accinelli, C., Screpanti, C., Dinelli & G.; Vicari, A. (2002) Short-time effects of pure and formulated herbicides on soil microbial activity and biomass. *International Journal of Environmental and Analytical Chemistry*, v.82, p. 519–527.

Argenta, G., Silva, P.R.F., Fleck, N. G., Bortoline, C. G., Neves, R. & Agostinetto, D. (2001) Efeitos do manejo mecânico e químico da aveia-preta no milho em sucessão e no controle do capim-papuã. *Pesquisa Agropecuária Brasileira*, v. 36, p. 851-860.

Asman, W.Ah., Sutton, M.A. & Schjorring, J.K. (1998) Ammonia: emission, atmospheric transport and deposition. *New Phytologist*, v.139, p.27-48.

Australian Greenhouse Office, (2001). National Greenhouse Gas Inventory. 1999. With Methodology Supplements. Australian Greenhouse Office, Commonwealth of Australia, Canberra.

Bremner, J.M. Sources of nitrous oxide in soils. (1997) *Nutrient Cycling in Agroecosystems*, v.49, p.7-16.

Britto D.T. & Kronzucker, H.J. (2001). NH_4^+ toxicity in higher plants: a critical review. *Journal of Plant Physiology*, v.159, p.567-584.

Bollag, J. M. & Henninger, N. M. (1976). Influence of pesticides on denitrification in soil and with an isolated bacterium. *Journal of Environmental Quality*, v.5, p. 15–19.

Cakmak, I. 2007. Efeitos do glyphosate na nutrição de micronutrientes de plantas. *Informações Agronômicas*, Piracicaba, n.119, p.5-8, set.

Carlisle, S. M. & Trevors, J. T. (1986). Effect of the herbicide glyphosate on nitrification, denitrification and acetylene reduction in soil. *Water Air Soil Pollut.* v.29, p.391–401.

Dakora, F.D. & Phillips, D.A. (2002) Roots exudates as mediators of mineral acquisition in low-nutrient environments. *Plant and Soil*, v.245, p.35-47.

Damin, V., Franco, H.C.J, Moraes, M.F, Franco, A., Trivelin, P.C.O. (2008) Nitrogen loss in *Brachiaria decumbens* after application of glyphosate or glufosinate-amonnium. *Scientia Agricola*, Piracicaba, v.65, n.4, p.402-407.

Damin, V., Trivelin, P. C. O. & Barbosa, T. G. (2009). Mineralização do nitrogênio da stubble de milheto dessecado com herbicides. *Revista Brasileira de Ciência do Solo*, v. 33, p. 925-934.

Damin, V.; Trivelin, P.C.O.; Franco, H.C.J.F. & Barbosa, T.G. (2010a)Nitrogen (15N) loss in soil-plant system after herbicides application in *Penissetum glaucum*. *Plant and Soil*, v. 328, p. 245-252.

Damin, V., Trivelin, P. C. O., Carvalho, S. J. P., Moraes, M. F. & Barbosa, T. G. (2010b) Herbicides application increases nitrogen (15N) exudation and root detachment of *Brachiaria decumbens* Stapf. *Plant and Soil* . DOI: 10.1007/s11104-010-0402-6, on-line first.

Damin, V., Trivelin, P. C. O., Barbosa, T. G., Carvalho, S. J. P & Moraes, M.F. (2010, no prelo). mineralization and corn recovery of 15Nitrogen from black oats residues treated with herbicides. *Journal of Plant Nutrition*.

Descalzo, R.D., Punja, Z.K., Levesque, C.A. & Rahe J.E. (1996). Identification and role of *Pythium* species as glyphosate synergists on bean (*Phaseolus vulgaris* L. Topcrop) grown in different soils. *Mycological Research*, v.100, p.971-978.

Deubel, A., Gransee, A. & Merbach, W. (2000) Transformation of organic rhizodephositions by rhizosfere bacteria and its influence on the availability of tertiary calcium phosphate. *Journal of Plant Nutrition and Soil Science*, v.163, p.387-392.

Duke, S.O. & Hoagland, R.E. (1995) Effects of glyphosate on metabolism of phenolic compounds. In: Grossbard, E.; Atkinson, D. (Ed.). *The herbicide glyphosate*. London: Butterworths. p. 75-91.

Eberbach, P.L.; Douglas, L.A. Herbicide effects on the growth and nodulation potential of *Rhizobium trifolii* with *Trifolium subterraneum* L. Plant and Soil, v.119, p.15-23, 1989.

Faizah, A.W.; Broughton, W.J.; John, C.K. Rhizobia in tropical legumes. XI. Survival in the seed environment. Soil Biology and Biochemistry, v.12, p.219-227, 1980.

Farquhar, G.D.; Firth, P. M.; Wetselaar, R.; Weir, B. On the gaseous exchange of ammonia between leaves and the environmental: determination of the ammonia compensation point. Plant Physiology, Rockville, v.66, n.4, p. 710-714, 1980.

Focht, D.D., Verstraete. W. Biochemical ecology of nitrification and denitrification. In 'Advances in microbial ecology'. (Ed. M. Alexander) pp. 135-214. (Plenum Press: New York), 1977.

Francis, D. D.; Schepers, J. S.; Vigil, M.F. Post-anthesis nitrogen loss from corn. Agronomy Journal, Madison, n.85, n.3, p. 659-663, 1993.

Ghosh, S.; Paliyath, P.; Peirson, D.; Fletcher, R.A. Nitrogen mobilization during senescence. In: SRIVASTAVA, H.S.; SINGH, R.P. (Ed.). Nitrogen nutrition in higher plants. New Delhi: Associated Publishing, 1995. p. 337–365.

Gordon, B. Adubação com manganês em soja convencional e soja resistente ao glifosato. Informações Agronômicas, Piracicaba, n.117, p.6-7, 2007.

Grossbard, E. Effects of glyphosate on the microflora: with reference to the decomposition of treated vegetation and interaction with some plant pathogens. In: Grossbard, E.; Atkinson, D. (Ed.). The herbicide glyphosate. Londres: Butterworths, 1985. p.159-185.

Granli, T.; Bockman, O.C. Nitrous oxide from agriculture. Journal of Agricultural. Science. v.12, p.7-128, 1994.

Griffith, 2005 GRIFFITH, J. Nitrogen pollution turn bays and rivers into dead zones. Ithaca: College of Agriculture and Life Sciences News, Cornel University, 2005. p.19-21.

Groffman, P.M. Nitrogen in the environment. In: SUMMER, M.E. (Ed.) Handbook of soil science. Boca Raton: CRC Press, 2000. p.190-200.

Haney R.L.; Senseman, S.A.; Hons, R.M.; Zuberer, D.A. Effect of glyphosate on soil microbial activity and biomass. Weed Science, Lawrence, v.48, n.1, p.89–93, 2000.

Haney, R. L.; Senseman, S. A.; Hons, F. M. Effect of Roundup Ultra on microbial activity and biomass from selected soils. Journal of Environmental Quality, Madison, v. 31, n. 3, p.730-735, 2002.

Heinzmann, F.X. Resíduos culturais de inverno e assimilação de nitrogênio por culturas de verão. Pesquisa Agropecuária Brasileira, Brasília, v.20, n.9, p.1021-1030, 1985.

Holtan-Hartwig, L.; Bockman, O.C. Ammonia exchange between crops and air. Norwegian Journal of Agricultural Sciences, Norwegian, v.14, Supplement, p.1-41, 1994.

Hoagland, R.E.; Duke, S.O.; Elmore, C.D. Effects of glyphosate on metabolism of phenolic compounds. III. Phenylalanine ammonia-lyase activity, free amino acids, soluble protein and hydroxiphenolic compounds in axes of dark-grown soybeans. Physiologia Plantarum, Kobenhavn, v. 46, n.4, p. 357-366, 1979.

Holland, J.N.; Cheng, W.; Crossley, D.A. J. Herbivore induced changes in plant carbon allocation assestment of below-ground C fluxes using carbon-14. Oecologia, Berlim, v.107, n.1, p.87-94, 1996.

Jaworski, E.G. Mode of action of N-phosphonomethylglycine: inhibition of aromatic amino acid biosynthesis. Journal of Agricultural and Food Chemistry, Easton, v. 20, n. 6, p. 1195-1198, 1972.

Johal, G.S.; Rahe J.E. Effect of soilborne plant-pathogenic fungi on the herbicidel action of glyphosate on bean seedlings. Phytopathology, Lancaster, v.74, n.7, p. 950-955, 1984.

King, C.A.; Purcell, L.C.; Vories, E.D. Plant growth and nitrogenase activity of glyphosate-tolerant soybean in response to foliar glyphosate applications. Agronomy Journal, v.93, p.179-186, 2001.

Klein, D.A.; Frederick, B.A.; Biondini, M.J.T. Rhizosphere microorganisms effects on soluble amino acids, sugars and organic acids in the roor zone of *Agropyron cristatum, A. smithii* and *Bouteloua gracilis*. Plant and Soil, Dordrecht, v.110, n.1, p.19-25, 1988.

Kraffczyk, I.; Trolldenier, G.; Beringer, H. Soluble root exudates of maize (*Zea mays* L.): influence of potassium supply and rhizosfere microorganisms. Soil Biology and Biochemistry, Amsterdam, v.16, n.4, p.315-322, 1984.

Kremer, R.J.; Means, N.E.; Kim, S. Glyphosate affects soybean root exudation and rhizosphere microorganisms. International Journal of Environmental Analytical Chemistry, Oxfordshire, v. 85, n.15, p. 1165-1174, 2005.

Laurière, C.; Daussand, J. Identification of the ammonium dependent isoenzymes of glutamine dehidrogenase as the form induced by senescence or darkness stress in the first leaf of wheat. Physiologia Plantarum, Kobenhavn, v.58, n.1, p. 89-92, 1983.

Levesque, C.A.; Rahe, J.E. Herbicide interactions with fungal root pathogens, with special reference to glyphosate. Annual Review of Phytopathology, Palo Alto, v.30, n.1, p.579-602, 1992.

Levesque, C.A.; Rahe, J.E.; Eaves, D.M. Fungal colonization of glyphosate-treated seedlings using a new root plating technique. Mycological Research, Cambridge, v.97, n.3, p.299-306, 1993.

Levesque, C.A.; Rahe, J.E.; Eaves, D.M. The effect of soil heat treatment and microflora on the efficacy of glyphosate in seedlings. Weed Research, Oxford, v. 32, n.5, p.363-373, 1992.

Liu, L.; Punja, Z. K.; Rahe, J. E. Altered root exudation and suppression of induced lignification as mechanisms of predisposition by glyphosate of bean roots (*Phaseolus vulgaris* L.) to colonization by *Pythium* spp. Physiological and Molecular Plant Pathology, London, v.51, n.2, p. 111-127, 1997.

Malkones, H.P. Comparason of the effects of differently formulated herbicides on soil microbial activities - a review. Journal of Plant Diseases and Protection, Sttutgart, v.8, p.781-789, 2000.

Mallik, M.A.B.; Tesfai, K. Compatibility of *Rhizobium japonicum* with commercial pesticides in vitro. Bulletin of Environmental Contamination and Toxicology, v.31, p.432-437, 1983.

Mallik, M.A.B.; Tesfai, K. Pesticidal effect on soybean-rhizobia symbiosis. Plant and Soil, v.85, p.33-41, 1985.

Manderscheid, R.; Schaaf, S.; Mattsson, M.; Schjoerring, J.K. Glufosinate treatment of weeds results in ammonia emission by plants. Agriculture, Ecosystems and Environment, Amsterdam, v.109, n.1/2, p. 129-140, 2005.

Marschner, H. Mineral nutrition of higher plants. 2nd .ed. London, UK: Academic Press, 1995. 889p.

Merbach, W.; Schlze, J.; Richert, M.; Rrocco, E.; Mengel, K. A comparison of different ^{15}N application techniques to study the net rhizodeposition in the plant-soil system. Journal of Plant Nutrition and Soil Science, Weinhein, v.163, n.3, p.375-379, 2003.

Moorman, T.B. Effects of herbicides on the survival of *Rhizobium japonicum* strains. Weed Science, v.34, p.628-633, 1986.

Moorman, T.B.; Becerril, J.M.; Lydon, J.; Duke, S.O. Production of hydroxybenzoic acids by *Bradyrhizobium japonicum* strains after treatment with glyphosate. Journal of Agricultural and Food Chemistry, v.40, p.289-293, 1992.

Moreira, F.M.S.; Siqueira, J.O. Microbiologia e bioquímica do solo. Lavras: UFLA, 2006. 729p.

Nakasathien, S.; Israel, D. W.; Wilson, R. F.; Kwanyuen, P. Regulation of seed protein concentration in soybean by supra-optimal nitrogen supply. Crop Science, Madison, v.40, n.5, p. 1277-1284, 2000.

Nguyen, C. Rhizodeposition of organic C by plant, mechanisms and controls. Agronomie, Paris, v.23, n.5/6, p.375-396, 2003.

Oliveira, F.H.T.; Novais, R.F.; Alvarez V.V.H.; Cantarutti, R.B.; Barros, N.F. Fertilidade do solo no sistema plantio direto. In: Alvarez, V.V.H.; Schaefer, C.E.G.R.; Barros, N.F.; Mello, J.W.V.; Costa, L.M. (Ed.) Tópicos em ciência do solo. v.2. Viçosa : Sociedade Brasileira de Ciência do Solo, 2002. p.393-486.

Paterson, E. Importance of rhizodeposition in the coupling of plant and microbial productivity. European Journal of Soil Science, Oxford, v.54, n.4, p.741-750, 2003.

Parton, W.J.; Morgan, J.A.; Altenhofen, J.M.; Harper, L.A. Ammonia volatilization from spring wheat plants. Agronomy Journal, Madison, v.80, n.3, p.419-425, 1988.

Prata, F.; Lavorenti, A.; Regitano, J.B.; Tornisielo, V.L. Influência da matéria orgânica na sorção e dessorção do glifosato em solos com diferentes atributos mineralógicos. Revista Brasileira de Ciência do Solo, Viçosa, v.24, n.4, p.947-951, 2000.

Ratcliffe, A.W.; Busse, M.D.; Shestak, C.J. Changes in microbial community structure following herbicide (glyphosate) additions to forest soils. Applied Soil Ecology, Amsterdam, v.34, n.2/3, p.114–124, 2006.

Ragster, L.E.; Chrispeels, M.J. Autodigestion in crude extracts of soybean leaves and isolated chloroplasts as a measure of proteolytic activity. Plant Physiology, Rockville, v.67, n.1, p.104-109, 1981.

Robertson, G. P., Vitousek, P. M., Matson, P. A. and Tiedje, J. M. 1987. Denitrification in a clear-cut Loblolly pine (*Pinus taeda* L.) plantation in the southeastern U.S. Plant Soil 99: 119–129.

Santos, J.B.; Jacques, R.J.S.; Procópio, S.O.; Kasuya, M.C.M.; Silva, A.A.; Santos, E.A. Efeitos de diferentes formulações comerciais de glyphosate sobre estirpes de *Bradyrhizobium*. Planta Daninha, v.22, p.293-299, 2004.

Sherlock, R.R.; Freney, J.R.; Smith, N.P.; Cameron, K.C. Evaluation of a sampler for assessing ammonia losses from fertilized fields. Fertilizer Research, Springer, v.21, n.2, p. 61-67, 1989.

Schimidt, E. Nitrification in soil. In: STEVENSON, F.J.; BREMNER, J.M.; HAUCK, R.D.; KEENEY, D.R. (Ed.) Nitrogen in agricultural soils. Madison: American Society of Agronomy, 1982. 940p. (Agronomy Series, 22)

Schjoerring, J.K.; Kyllingsbaek, A.; Mortnsen, J.V.; Byskovi-Nielsen, S. Field investigations of ammonia exchange between barley plants and the atmosphere. I. Concentration profiles and flux densities of ammonia. Plant Cell Environment, Heidelberg, v.16, n.2, p. 161-167, 1993.

Schjoerring, J. K.; Husted, S.; Mattsson, M. Physiological parameters controlling plant-atmospheric ammonia exchange. Atmosferic Environmental, Amsterdam, v.32, n.3, p.491-498, 1998.

Snapp, S.S.; Borden, H. Enhanced nitrogen mineralization in mowed or glyphosate treated cover crops compared to direct incorporation Plant and Soil, Dordrecht, v.270, n.1/2, p.101-112, 2005.

Sutton, M.A.; Pitcairn, C.E.R.; Fowler, D. The exchange of ammonia between the atmosphere and plant communities. In: Begon, M.; Fitter, A.H. Advances in Ecological Research, New York, v.24. p.301-393, 1993.

Tabatabai, M.A.; Burwell, R.E.; Keeney, D.R.; Logan, T.J.; Nelson, E.W.; Olson, R.A.; Randall, G.W.; Timmons D.R.; Verry, E.S.; White, E.M. Nutrient concentrations and accumulations in precipitation over the North Central Region. Ames: Iowa State University, 1981. 142p. (Research Bulletin, 594)

Tenuta, M. and Beauchamp, E. G. Denitrification following herbicide application to a grass sward. Canadian Journal of Soil Science 76:15–22, 1996

Tiedje, J.M. Denitrifiers. In: Weaver, W.; Angel, J.S.; Bottomley, P.S. (Ed.) Methods of soil analysis. Part 2. Microbiological and biochemical properties. Madison: Soil Science Society of America, 1994. p. 245-267.

Trivelin, P. C. O. Utilização do nitrogênio pela cana-de-açúcar: três casos estudados com uso do traçador ^{15}N. 2000. 143p. Tese (Livre Docência) – Escola Superior de Agricultura "Luiz de Queiroz", Universidade de São Paulo, Piracicaba, 2000.

Tuffi Santos, L.D.; Santos, J.B.; Ferreira, F.A.; Oliveira, J.A.; Bentivenha, S.; Machado, A.F.L. Exsudação radicular de glyphosate por *Brachiaria decumbens* e seus efeitos em plantas de eucalipto. Planta Daninha, Viçosa, v.26, n.2, p.369-374, 2008.

Zabaloy, M. C.; Gómez, M. A. Microbial respiration in soils of the Argentine pampas after metsulfuron-methyl, 2,4-D, and glyphosate treatments. Communications in Soil Science and Plant Analysis, Philadelphia, v.39, n.3/4, p.370-385, 2008.

Wang, M.Y.; Siddiqi, M.Y.; Ruth, T.J.; Glass, A.D.M. Ammonium uptake by rice roots. I. Fluxes and subcellular distribution of $^{15}NH_4^+$. Plant Physiology, Rockville, v.103, n.4, p.1249-1258, 1993

Wetselaar, R.; Farquhar, G. D. Nitrogen losses from tops of plants. Advances in Agronomy, New York, v.33, n.1, p. 263-302, 1980.

Yeomans J.C.; Bremner, J.M. Denitrification in soil: Effects of herbicides. Soil Biology and Biochemistry, v.17, p.447-452, 1985.

6

Critical Revision and Development Perspectives of Herbicide Residues Analysis in Agro Ecosystems

Verónica Cesio[1], Lucía Pareja[1], Silvina Niell[1], Lucia Geis Asteggiante[1], Bernardo Böcking[2], Claudio García[3], Grisel Fernández[4], Amadeo R. Fernández-Alba[5] and Horacio Heinzen[1]
[1]Farmacognosia y Productos Naturales, Facultad de Química UDELAR General Flores 2124, Montevideo 11800,
[2]Donistar S.C, Col. Rubio, Salto,
[3]INIA Las Brujas, Las Brujas, Canelones,
[4]Malherbología EEMAC, Facultad de Agronomía, UDELAR Km366 Ruta 3 Paysandú,
[5] Grupo de Investigación de Laboratorio de Residuos de Plaguicidas, Almería, Universidad de Almería, Cañada de San Urbano s/n, Almeria
[1,2,3,4]Uruguay
[5]Spain

1. Introduction

Weeds are defined as plants growing in the inadequate place at the inadequate moment. They interfere with the normal crop development and can drop down yields dramatically. Winter cereals show lower losses in average, due to its dense stands and competitive abilities. But, in summer crops, particularly corn that has low densities per area and lower initial growth rates than weeds, very high yield reductions are possible. Rice is very susceptible to weed interferences. There are evidences of production losses from 20 to 96% (Table 1). Weed control is performed in 100% of rice cultures either manually, mechanically or chemically. In average, registered losses are around 10% of production even when weed control has been done. Weed control cost is around 5%. Consequently, the global cost of this factor is around 15% of production. In sum, this 15% is the negative impact weeds have in agricultural and food production.
To control losses in the crop production and as support of the new agronomical techniques for intensive cropping, like no tillage or direct sewing techniques, weed chemical control has been adopted as a usual technique to improve yields.

2. Herbicide application

Herbicide application timing depends on various factors such as class of herbicide, weed species aiming to control, mechanic labor, climate, conditions and soil type. This latter factor is very important, particularly soil preparation, previous and present vegetation, humidity content on soil surface and temperature that makes probable a quick weed germination. A

Crop	% Production Decrease	Cause	Reference
Rice	20-30	Competition, inadequate weed control. (Brazil)	(Harri, L., 1994)
Rice	25 (11pl/m^2) 49 (54 pl/m^2) 79(269 pl/m^2)	*Echinocloa colonum*	(Smith, R. J., 1968)
Rice	40	Competition, weed control 45 d post crop emergence.	(Chebatarrof N., 2007)
Corn	10-15	Weed resistant to control	(Cepeda, S. et al., 1995)
Corn	27-40 97 (if the weed is *Sorghum halepense* or *Cynodon dactylon*)	No weed control at all	(Rossi, A. et al., 1996)

Table 1. Losses caused by weeds in rice and corn.

general concept is that the herbicide should be applied when the crop has the maximum resistance and weeds are in the most susceptible period (Genta and Villamil, 1992). Herbicide application at pre-emergence stage is closely related with the culture labors of the soil. Once the crop has been sewed, herbicide application is related with crop development and how infested are the present weeds.

2.1 Preplant treatment

In rice crops, an herbicide application (i.e. glyphosate) is usually applied in autumn to control existent weeds when machinery labor is needed. This is a non selective treatment and it is done in order to prepare the surface for future work such as levees for irrigation. Probability of weed germination increases with earthwork, favorable conditions of humidity and crop absence.

After land planting, due to humidity conditions of soil in autumn and winter, often great amounts of weeds appear and a herbicide is applied to control them (i.e. glyphosate).

2.2 Preemergence treatment

This treatment is done after sowing but before weed emergence. The control consists in combating the seeds that could be in the first soil layer or seeds that have recently germinated when a good control can be achieved.

The main advantage of pre-emergence and preplant treatments is eliminating weed competition in the initial steps of crop development. In general, these treatments are safer and more consistent than post-emergence ones because in crops that irrigation is needed or when it has rained it is difficult to apply the product or hoe. Some disadvantages are that perennial weeds are hard to control and the effectiveness of the products depends on the kind of soil (organic matter content and pH). Also these products require to be mobilized by water in soil, so they are not effective under drought conditions.

2.3 Postemergence treatment

Selective products must be used in order to eliminate weeds without damaging the crop. The class of weed that can be controlled depends on the tolerance of the crop to the herbicide. The risk of damaging the crop is high; temperature is one of the factors that affects the most the effectiveness and volatilization of many post-emergent herbicides, like clomazone. In general when they are applied weeds have already born. An effective weed control is sometimes difficult to achieve because there are many factors that must be taken into account (right climatic conditions, doses, soil humidity level, etc.)

Herbicides are the most employed agrochemicals in agriculture nowadays. Pre or post emergence, root or leaf absorbed, they are applied in huge amounts to hamper weed development worldwide. There are many possible molecular targets to exert the herbicidal action and therefore, herbicides are the vastest group of agrochemicals as there are 53 different chemical families of weed controllers. The most representative herbicide chemical classes are listed in Table 2.

GROUP OF HERBICIDES	EXAMPLES	MODE OF ACTION	PERSISTANCE IN SOIL (DT50)*
Amide	Florasulam	Branched chain amino acid synthesis (AHAS or ALS) inhibitors	2 - 18 d
	Propanil	Photosynthetic electron transport inhibitor	N/A
Aromatic Acid	Dicamba Quinclorac	Auxin growth regulator	14 d N/A
	Bispyribac-sodium	Branched chain amino acid synthesis (AHAS or ALS) inhibitors	< 10 d
	Aminopyralid Picloram	Auxin growth regulator	8 – 35 d 30 – 90 d
Bipyridiles	Paraquat	Promotes autoxidation	
Dinitroaniline	Pendimethalin	Microtubule assembly inhibition	3 - 4 mo
Diphenyl Ether	Oxyfluorfen	Protoporphyrinogen oxidase inhibitor	5 – 55 d
Imidazolinone	Imazapyr	Branched chain fatty acid biosynthesis inhibitors	30-150 d
organphosphate	Glyphosate	Inhibits 5-enolpyruvylshikimate-3-phosphate synthase.	1 - 130 d
Oxazole Herbicides	Topramezone	p-Hydroxyphenyl pyruvate dioxygenase inhibitor	9 – 81 d
Phenoxy Herbicides	MCPA 2,4-D Mecoprop-p Dichlorprop-P 2,4-DB	Auxin growth regulator	< 7 d < 7 d 3 - 13 d < 7 d
	Fenoxaprop-P-ethyl Cyhalofop-butyl Diclofop-methyl	Lipid biosynthesis inhibitors	1 – 10 d 2 – 10 h 1 – 57 d

GROUP OF HERBICIDES	EXAMPLES	MODE OF ACTION	PERSISTANCE IN SOIL (DT50)*
Pyrazole Herbicides	Pinoxaden	Lipid biosynthesis inhibitors	< 1 d
Pyridine Herbicides	Fluroxypyr Diflufenican	Auxin growth regulator	5 - 9 d 36,3 d
	Pyroxsulam	Branched chain amino acid synthesis (AHAS or ALS) inhibitors	< 15 d
Triazine Herbicides	Atrazine	Photosynthetic electron transport inhibitor	16 - 77 d
Triazolone Herbicides	Carfentrazone-ethyl	Branched chain amino acid synthesis (AHAS or ALS) inhibitors	2,5 - 4 d
	Propoxycarbazone-sodium Flucarbazone-sodium	Branched chain amino acid synthesis (AHAS or ALS) inhibitors	12 - 56 d 17 d
Triazolopyrimidine Herbicides	Flumetsulam	Branched chain amino acid synthesis (AHAS or ALS) inhibitors	≤ 1 mo
Urea	Isoproturon	Photosynthetic electron transport inhibitor	6-28 d
Sulphonylureas	Amidosulfuron Bensulfuron-methyl Flupyrsulfuron-methyl-sodium Iodosulfuron-methyl-sodium Metsulfuron-methyl Pyrazosulfuron-ethyl Sulfosulfuron Thifensulfuron-methyl Triasulfuron Tribenuron-methyl Foramsulfuron Chlorsulfuron	Branched chain amino acid synthesis (AHAS or ALS) inhibitors	3 - 29 d 14 d 1 - 5 d 52 d 10 - 21 d 11 - 47 d 1 - 7 d 19 d 3,5 - 5,1 d 1,5 - 9,4 d 4 - 6 w
Unclassified Herbicides	Bentazone Clomazone	Photosynthesis inhibition Carotenoid biosythesis inhibition	12 d 30 - 135 d

* Degradation half-life: d – days, w – weeks, mo – months; N/A: not available

Table 2. Most representative chemical families of herbicides

Economical aspects

Higher land, use intensity and short crop rotations (or sometimes no rotation at all) plus no tillage technology leads to an increase in the usage and dependency of agrochemicals. Pests and weeds problem increases when land use is more and more intensive. Weeds negatively affect grain production and quality, which declines prices. Also, harvesting becomes complex and more expensive. Resources competition and allelopathy interferences, like light, water, CO_2 and soil nutrient competition provokes important or complete economic losses. On top of that, weeds are pest hosts and disseminate crop infections.

2.4 Herbicides trade and usage

In 2009, herbicides represented 75% of agrochemical imports in kg and 59.5% in U$S/CIF, showing they continue to be the most employed agrochemicals in Uruguay. As shown in Figure 1, herbicides prevail in agrochemicals world trade. In Uruguay, this tendency is even stronger when compared to Brazil although Argentina has bigger differences than Uruguay (63%, 20% and 9% herbicides, insecticides and fungicides respectively) (casafe website).

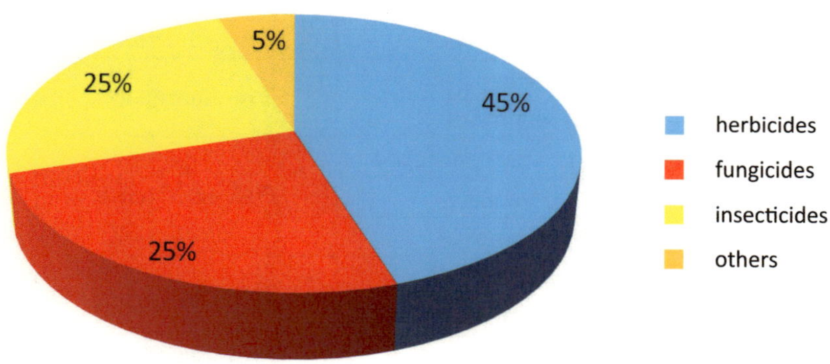

Fig. 1. Agrochemicals world trade (Rana, S. AGROW, 2010)

Herbicides show the higher relative increase when compared to fungicides and insecticides between 2003 and 2009. Glyphosate accounts for 57% of herbicides total.
As stated above, rice production needs intensive herbicide use. From 141.500 ha of rice crop analyzed and registered, the 98% (138.000 ha) required the appliance of herbicides with repeated control treatments in about the 10% of the treated area (Molina et al., 2010). Fields never used for rice crops were included in the area under treatment, along with pasture, which despite of doing the adequate rotation, showed the presence of weeds in quantities above the threshold for economic damage requiring control with herbicides. The active ingredients used are shown in Table 3.
As shown in Table 4, the most common herbicide treatment is atrazine for corn and sorghum in mixture with a chloroacetamide in pre-emergence. Post-emergent options such as imidazolinones (Imazethapyr + imazapyr or imazapic + imazapyr) are preferred for corn with zero tillage because of straw interactions with pre-emergent herbicides. However, at present there are only limited numbers of cultivars with tolerance for these herbicides and so very few hectares had been treated with imidazolinones in the last years. Other post-emergent options as growth regulators herbicides, the mixture of iodosulfuron plus foramsulfuron or topramezone had not been extensively adopted. They are more expensive

Products or mixtures applied	Area of appliance (ha) T	%
Quinclorac+Clomazone+Propanil	45660	30
Clomazone+Glyphosate	18198	12
Clomazone+Quinclorac	8975	6
Bispyribac-sodium+Clomazone	7521	5
Quinclorac+Clomazone+Propanil+Pyrazosulfuron-ethyl	6964	5
Bispyribac-sodium	5342	4
Clomazone+Propanil	4977	3
Others	52667	35
Total	150.304	100
Clomazone alone or as mixture	118017	79
Quinclorac alone or as mixture	77683	52

Table 3. Active ingredients used in rice production in Uruguay

Crop	Herbicide	Application
Corn	Atrazine	Pre-emergence
	chloroacetamides	
	Iodosulfuron + foramsulfuron	Post-emergence
	Topramezone	
	imidazolinonas	
Sorghum	Atrazine	Pre-emergence
	chloroacetamide	
Soybean	Glyphosate	Post-emergence
	Imazethapyr	pre/ post-emergence
	Diclosulam	pre/ post-emergence
Wheat and Barley	Pyroxsulam (only wheat)	Post-emergence
	Flucarbazone (only wheat)	Post-emergence
	Iodosulfuron	Post-emergence
	Pinoxaden	Post-emergence
	Diclofop-methyl	Post-emergence
	Fenoxaprop	Post-emergence
	Metsulfuron	Post-emergence
	Metsulfuron + Chlorsulfuron	Post-emergence
	growth regulators (2,4D amine, Dicamba, Picloram)	Post-emergence
Artificial pastures	Flumetsulam	Post-emergence
	2,4 DB ester	Post-emergence

Table 4. Herbicides most commonly used in extensive crops and artificial pastures in Uruguay

and their performances had not demonstrated clear advantages compared with the classic pre-emergent mixture of atrazine+chloroacetamides except in specific weed situations.

There are potential risks of persistence with this treatment including atrazine. In production systems alternating crops and pastures where corn or sorghum crops grown for silage are followed by pastures with legumes and grasses or by oats, plant-back period may resulted insufficient for atrazine dissipation and damage occurrence.

Soybean area, entirely GMO planted crop, is basically treated with glyphosate. Nevertheless, It has been registered an increased trend to complement glyphosate with residuals herbicides like imazethapyr or diclosulam looking to reduce number of glyphosate applications and to broad control spectrum. There is an increasing movement of the big breeders companies towards the inclusion in the new varieties, some resistant genes against particular herbicide mode of action. Following the trend initiated by the glyphosate resistant RR soybean, crop varieties resistant to imidazolinones (rice, corn) glufosinate (rice) sulphonylureas (sunflower) has been registered. The use of these varieties will boost the use of the specific herbicides and therefore new challenges on environmental risk assessment on soil, water and food can be foreseen.

Herbicide options for winter cereals, wheat and barley are various and especially in case of weed grasses. Number and total use of graminicides has increased markedly. The most used are pinoxaden, pyroxsulam, iodosulfuron, flucarbazone, diclofop methyl and fenoxaprop. The use of these herbicides has five-folded since 2006 even though winter crop area has just doubled during this period, in association with the widespread infestations of grasses in agriculture lands.

Other herbicides widely used for winter cereals are the sulphonylureas, metsulfuron and chlorsulfuron as they have a broad spectrum of weed control, may be sprayed early and have important residual effects, so satisfactory controls are reached in species with continued emergences fluxes as commonly happens with winter weed species in our country. However, application of these herbicides implies a risk of persistence as commonly happens when used in fallow seed-bed preparation without considering the plant-back guidelines for crop rotation. See for example Bradford et al, 2008. Its low cost and wide spectrum makes them highly attractive for use in mixtures with glyphosate in fallow seed-bed preparation for summer and winter crops, being frequent situations of persistence problems.

These problems, occasional phytotoxicity effects and risk of resistance, have been promoting a return to traditional growth regulators treatments such as 2,4-D amine in mixtures with dicamba, picloram, etc. or new options like aminopyralid. Also in artificial pastures use of herbicides is incremental. The most common treatment in this areas is flumetsulam alone or in mixtures with 2,4-DB ester.

In this introduction it is shown that herbicide usage has been integrated systematically in cropping systems, that there is no universal procedure to perform weed control and different herbicidal combinations are employed in different stages of crop production. Although some herbicides are used in very low doses, and many of them are designed to inhibit specific pathways of plant metabolism, their interference with metabolic pathways of other organisms is a growing concern. Some herbicides are persistent between two cropping systems and can affect the new crop, notably atrazine and the sulphonylureas. Therefore, for public health reasons and systems sustainability the knowledge of the level of remaining herbicide residues has to be ascertained. In the past, sometimes the low dosage of these compounds hampered a clear determination of their residual level in the different environmental compartments. Nowadays the development of new sample treatment procedures either instrumental, like ASE or MASE or not, like QuEChERS, coupled to hyphenated MS/MS techniques allow the detection of very low levels of these compounds in different matrices. The present chapter aims to present some characteristics of herbicide analysis, to explore the reasons why many currently used herbicides have not been included in the most common Multi Residue Methods developed and to give some insights on the different aspects that should be considered when trying to include herbicides in a multiresidue method.

3. Herbicide residues analysis

The analysis of herbicides has been confined to single residue methods, or class herbicide residue analysis (phenoxyacid herbicides, imidazolinones, sulphonylureas, triazines). Nevertheless, the real situation is that cationic, anionic, basic or acid herbicides together with fungicides and insecticides can be applied successively over a crop looking for different protecting effects. They accumulate in the crop and the environment and when evaluating their presence in the different compartments they can seldom be determined in one single analytical procedure. To define the scope of a multirresidue method that will allow the simultaneous determination of many herbicides belonging to different chemical classes, the physicochemical properties of the agrochemicals must be carefully evaluated. The physicochemical properties of some representative herbicides are listed in Table 5.

Herbicide	Koc	log Kow	log Ws (mg L^{-1}) 20-25°C	H (Pa m^3 mol^{-1})	Vapor Pressure (mPa)
2,4-D	60	2.58 -2.83	4.37	1.32 x 10-5	1.86 x 10-2
2,4-DB	-	-	1.66	-	-
Amidosulfuron	-	1.63	0.95	5.34 x 10-4	2.20 x 10-2
Aminopyralid	-	0.201	3.39	-	9.52 x 10-6
Atrazine	39 – 173	2.50	1.52	1.5o x 10-4	3.85 x 10-2
Bensulfuron-methyl	-	0.79	1.83	2.00 x 10-11	2.80 x 10-9
Bentazone	13.3 – 176	-0.46	2.76	-	5.4 x 10-3
Bispyribac-sodium	-	-1.03	4.87	3.12 x 10-11	5.05 x 10-6
Carfentrazone-ethyl	15 – 35	3.36	1.08	2.47 x 10-4	7.2 x 10-3
Chlorsulfuron	40	-0.99	4.50	3.50 x 10-11	3.00 x 10-6
Clomazone	150 – 526	2.50	3.04	4.19 x 10-3	19.2
Cyhalofop-butyl	5247	3.31	-0.36	9.51 x 10-4	5.30 x 10-2
Dicamba	2	-1.88	> 5.40	6.10 x 10-5	1.67
Dichlorprop-P	-	-0.25	2.77	2.47 x 10-5	0.06
Diclofop-methyl	14000 - 24400	4.58	-0.10	2.19 x 10-1	0.25
Diflufenican	-	4.90	-1.30	1.18 x 10-2	4.25 x 10-3
Fenoxaprop-P-ethyl	-	4.58	-0.15	2.74 x 10-4	1.80 x 10-1

Florasulam	2-069	-1.22	3.80	4.35 x 10-7	1.00 x 10-2
Flucarbazone-sodium	-	-1.84	4.64	< 1 x 10-11	< 1 x 10-6
Flumetsulam	5-182	-0.68	1.69	-	3.7 x 10-7
Flupyrsulfuron-methyl-sodium	-	0.10	2.78	1 x 10-8	1 x 10-6
Fluroxypyr	-	-1.24	3.76	1.06 x 10-8	3.78 x 10-6
Foramsulfuron	38 – 151	-0.78	3.52	5.80 x 10-12	4.20 x 10-8
Glyphosate	-	< -3.20	4.02	< 2.10 x 10-7	1.31 x 10-2
Imazapic	-	0.39	3.33	-	< 1 x 10-2
Imazapyr	-	0.11	4.05	-	< 0.013
Imazaquin	-	0.34	1.78-2.08	3.70 x 10-12	< 0.013
Iodosulfuron-methyl-sodium	0.80 – 152	-0.70	4.40	2.29 x 10-11	6.70 x 10-6
Isoproturon	-	2.50	1.81	1.46 x 10-5	3.15 x 10-3
MCPA	-	2.75	2.47	5.50 x 10-5	2.30 x 10-2
Mecoprop-p	12-025	0.1004	2.94	2.18 x 10-4	1.60
Metsulfuron-methyl	-	0.018	3.45	4.50 x 10-11	3.3 x 10-7
Oxyfluorfen	2891 (sand)	4.47	-0.94	-	0.0267
Pendimethalin	-	5.20	-0.48	2.728	1.94
Picloram	-	1.90	-0.25	-	8 x 10-11
Pinoxaden	-	3.20	2.30	-	2 x 10-9
Propanil	239 – 800	3.30	2.11	1.70 x 10-4	0.02
Propoxycarbazone-sodium	28,8	-1.55	4.62	< 1 x 10-10	< 1 x 10-5
Pyrazosulfuron-ethyl	-	3.16	0.99	-	4.20 x 10-5
Quinclorac	-	-0.74	-1.19	-	< 0.01
Sulfosulfuron	33	-0.77	3.21	8.83 x 10-9	3.1 x 10-5
Thifensulfuron-methyl	-	0.02	3.35	9.70 x 10-16	1.70 x 10-5
Topramezone	-	-1.52	2.71	-	< 1 x 10 -12
Triasulfuron	-	-0.59	2.91	< 8 x 10 -5	< 2 x 10 -3
Tribenuron-methyl	-	0.78	3.31	1.03 x 10-8	5.20 x 10-5

Table 5. Physicochemical and environmental properties of some selected herbicides

As an example, when looking for a procedure that can include the analysis of imidazolinone residues their peculiar pH behavior has to be considered. The Imidazolinones are amphoteric molecules that can exist in anionic, neutral or cationic states depending upon the pH of the environment. The pKa values of imidazolinone herbicides range from 1.9 to 3.9. When the pH of the extracting solvent is greater than their pKa, these herbicides are usually present mainly in an anionic state as carboxylates. On the other hand, the acidic functional group is completely in a non-ionized state when the pH is two units lower than the most

acidic pKa value. Therefore factors such as pH, organic carbon content, and ionic strength may affect their extractability to perform the analysis. All these factors also affect the fate of the imidazolinone herbicides in the environment which are relatively persistent in soil with half-lives ranging from 30 to 150 days and therefore, may have carryover effects on the rotation system of winter-summer crops. Moreover, it has been reported that the imidazolinone herbicides show high potential for leaching because of their relatively low pKa values (1.9–3.9). (Ramezani et al., 2009)
But not only have these factors operated. The matrix characteristics are crucial for the precision and accuracy of the analysis. Differences in pesticide absorption and recoveries were found when comparing results from contrasting soils (Boivin et al., 2005).

Environmental samples

The major public health concern on herbicide residues in the environment is due to the fact that many of them have low Koc and relatively low Kow and therefore, can either lixiviate, leach or run off and contaminate ground and surface water. Many herbicides have been identified as endocrine disrupters and they threaten wild life as Atrazine (Barchanska & Baranowska, 2009). The usual methodology for contaminant analysis in water is the solid phase extraction either off line or on line coupled with the analytical instrument (GC or HPLC). The adsorbed compounds are eluted from the stationary phase with a polar organic solvent and the extract is analyzed directly. This simple procedure allows sample concentration increasing the limit of detection of the method to ngkg^{-1} level or ppts, fulfilling the requirements of the European legislation on drinking water, where pesticide limits of quantification (LOQs) of 0.025μgL^{-1}, four times lower than the maximum allowed (0.1μgL^{-1}), are required. To perform the SPE extraction RP C18 silica gel cartridges have been employed, but the most widely extraction phases employed are polymeric styrene, polystyrene divinil-benzene polymers or N-vinylpyrrolidone. These cartridges are commercially available with different brand names. This solid phase has a medium polarity that allows the selective retention of relatively polar compounds. In this way many different herbicides belonging to various chemical families (sulphonylureas, anilides, imidazolinones, phenoxy acids) have been selectively absorbed and quantitatively analyzed. Solid phase microextraction (SPME) is a very attractive analytical technique for herbicide residue analysis in water. Serôdio (Serôdio et al., 2004) analyzed 13 herbicides which are endocrine disruptors performing a micro extraction using a poliacrilate coated sorptive stirring bar for pesticide extraction, performing a back extraction procedure with acetonitrile to recover all analytes with high selectivity. A miscellaneous procedure for the selective analysis of triazine herbicides in water was presented by (Nelson et al., 2004). A portable system based on immunoextraction and reversed-phase HPLC was developed for the field analysis of herbicides in groundwater and surface water. Using inmunotemplates to concentrate atrazine, simazine, and cyanazine and analyzing them through HPLC, a LOQ of a 0.2mg L^{-1}up to 100mg L^{-1} were achieved. This procedure can be applied to the high throughput analysis of water samples as a single determination took less than 8 min. Another type of class analysis of herbicide residues is the use of molecular imprinted polymers to selectively extract sulphonylureas in water (Bastide et al., 2005). The highly selective polymer concentrates the herbicide residues allowing their determination at the sub ppt level.
MIP technique can be coupled to chemiluminscence sensing, taking advantage of the reaction of many sulphonylureas with luminal+H_2O_2 as reported by (Xie et al., 2010) reaching detection levels of 1-10 nM L^{-1} of sulphonylureas in water. MIP methodology with

its different applications off line, on line or in line has been recently reviewed. Due to the low application doses of herbicides (Table 6) the development of such polymers offers the possibility of sub trace analysis for different herbicide classes. Nevertheless, the improvement of the new LC-MS/MS equipment, allow the detection of many agrochemicals in water, just by filtrating and injecting the sample.

3.1 Herbicide residue analysis in water

In the last decades, liquid-liquid extraction (LLE) has been largely replaced by solid phase extraction (SPE). In 2009 Pinxteren et al., compared the performance of SPE and microwave assisted-solvent extraction (MASE or MAE), both in combination with LC–MS/MS, for the analysis of 10 pesticides in water. The main conclusion of the study was that SPE has the potential of larger sensitivity whereas MASE is faster, provides slightly better recoveries, and represents a promising alternative to conventional off-line SPE concerning low to medium polar compounds (Pinxteren et al., 2009)

One of the advantages of SPE compared to other techniques is the variety of sorbents available. As an example, Geiss et al., compared the performance of 21 cartridges packed with different octadecyl silica, polymeric, modified silica, ion exchange and carbon materials for the extraction of highly polar organophosphorus pesticides from water. In this study the polymeric cartridges were found to have the best performance, and the recovery rates obtained with this material and with the octadecyl solid phases were observed to correlate with the octanol–water partition coefficient of the pesticides (Geiss & Gebert, 2006). Similar results were obtained when testing the extraction of nine herbicides from rice paddy fields water by (Roehrs et al., 2009). Another advantage of SPE is the possibility of automation (Petrovic et al., 2010).

Herbicides in environmental samples such as river, sea or agricultural water are generally highly diluted; therefore a high level of enrichment is required for their detection. As the instruments improve their sensitivity, the pre concentration factor required previous to the analysis is smaller. In reported literature concerning water analysis, as time pass by, the amount of sample required for the analysis is smaller with the clear trend to reduce the sample clean up and pre concentration of the sample.

Ten years ago, Jeannot et al., described two different methods for the analysis of triazines and sulphonylureas in surface waters. They used 0.5-10 L of sample for a liquid-liquid extraction with dichloromethane and two different solid phase extractions (SPE) using the Carbopack B (graphitized carbon black) and the C-18 bonded silica cartridge. The recoveries for each of these methods varied from 83-93% in liquid-liquid mode with RSD between 2-10%, 60-96% in SPE mode on Carbopack B with RSD 3-17% and 67-100% in SPE C-18-bonded silica mode with RSD 2-7% (Jeannot et al., 2000).

In 2003 Zanella et al., validated a method for the determination of 2,4-D, quinclorac, bentazone, clomazone and propanil in surface and agriculture waters. The method was applied for levels between 0.1 and 0.5 mgL^{-1} after a 500-fold-pre-concentration with recoveries ranging from 85.7 to 109.8 % and RSD of 1.8 to 13.4% analyzed by HPLC with UV detection (Zanella et al., 2003). HPLC-DAD was also employed for the determination of these herbicides plus four sulphonylureas, cyhalofop-butyl and bispyribac-sodium but using an OASIS HLB cartridge in tap and rice paddy field waters (Roehrs et al., 2009). The concentration level achieved was 1000 fold detecting herbicides at concentration as low as 40 $ngkg^{-1}$ for tap water and 0.3 μgkg^{-1} for paddy field water.

Ayano et al., in 2004, described a multianalyte method for the determination of five sulphonylureas and three ureas in water. The analysis consisted on a SPE extraction with a polystyrene polymer cartridge (PS2), and ODS C-18-bonded silica cartridge (C-18) and an N-vinylpyrrolidone polymer cartridge (OASIS). The analytes determination and quantitation were performed by liquid chromatography with mass spectrometry (LC-MS). Average recoveries of the eight analytes from water samples were in the range 70-120% with relative standard deviations (RSD) below 20% and the LOQs were between 10 and 100ngL^{-1} (Ayano et al., 2004).

Carabias-Martínez et al., in 2004, developed a method for the simultaneous determination of 10 sulphonylurea and phenylurea herbicides and one of their most common degradation products (3-chloro-4-methyl-phenyl urea) in water. In this procedure LC with diode array UV detection and electrospray mass spectrometry (LC-ESI/MS) in the positive mode were used for the separation, identification and quantification of the selected analytes. A 1000-fold-pre-concentration step based on solid phase extraction was applied for the simultaneous extraction of sulphonylureas and phenylureas from water samples. Three different types of sorbents were compared, silica-based C-18, Oasis HLB and LiChrolut EN obtaining the highest recoveries (70-95%) with the Oasis HLB cartridges (Carabias-Martínez et al., 2004). In 2006, Polati et al., presented the determination of 6 sulphonylureas in surface water. The methodology consisted on a pre concentration/SPE step using a Strata RP-18 E and two polymeric phases, Strata-X and Lichrolut EN, followed by HPLC-UV (240nm)-MS^n analysis. After a 1000/1 pre-concentration the LODs were lower than 26.9ngL^{-1} with recoveries around 81-113% and RSD in the range 10-22 % (Polati et al., 2006).

Kuster et al., in 2007, described the analysis of 14 polar herbicides in the Ebro river delta, during the main growing season of rice, by an automated on-line solid phase extraction followed by liquid chromatography tandem mass spectrometry.

The extraction of the herbicides was performed with polymeric cartridges Hysphere Resin GP (polydivinilbenzene), 10 mL samples were loaded onto the cartridge and the target analytes were eluted directly onto the chromatography column. The detection was performed using a triple quadrupole (QqQ) mass spectrometer and the electrospray interface was operated in both positive and negative mode. Some of the selected herbicides presented difficulties in their recoveries such as alachlor, molinate, propanil, diuron, chlorotoluron, with recoveries below 60% and atrazine, metolachlor, cyanazine with recoveries higher than 120% (Kuster et al., 2007). When analyzing complex matrixes, with minimal work up, matrix effects are noticeable in LC-MS/MS methods and can influence notably the results (identification, quantification and confirmation). The major matrix interferences are due to co-eluting compounds from the sample matrix that can affect the analyte ionization process leading to a signal enhancement or signal suppression. Different approaches have been proposed to overcome matrix effects: sample clean-up, the standard additions method, use of matrix-matched standards, a simple sample dilution if the instrument provides enough sensitivity, but the most widely method employed involves the use of appropriate internal standards. Environmental waters have high sample composition variability and it is difficult to find representative water samples that can be employed for blank determinations and method validation. The problem is still present when using the same type of water (surface, ground) because of their different origin. Therefore, the use of matrix-matched standards, widely applied in analysis of fruits and vegetables, does not provide a straightforward solution in the environmental field. If it is not possible to apply the "dilute and shoot" methodology, the use of analyte isotope-labelled internal standard is the preferred method to avoid matrix effects influence. This approach has some drawbacks, as only a limited number of reference standards of all possible contaminants are available which are also very expensive. Sometimes other compounds with similar chromatographic properties or structural analogues of the analytes are used but the obtained data is no always good enough (Marín et al., 2009). These authors compared the matrix effects from seven different types of water either using HPLC or UHPLC and the use of labeled internal standards to quantitate 37 pesticides using nine labelled internal standards (I.S). They concluded that in environmental waters, matrix effect was generally a negative effect but no general rules can be applied. Only the labelled pesticide as I.S assured a good correction. Nevertheless for not too loaded waters, the uncorrected values were satisfactory.

In 2009, Mazzella et al. developed a method for the simultaneous determination of 30 triazines, phenylurea and chloroacetanilide herbicides in fresh and estuarine waters, this work addresses two objectives; the development of an accurate method based on ESI-MS/MS detection and the investigation of matrix effects. The pre concentration of the analytes was accomplished by using SPE with Oasis HLB cartridges, with recoveries between 73 to 122% and RSD ranged from 6 to 22% (Mazzella et al., 2009).

Ouyang et al., evaluated the performance of different SPE sorbents for the analysis of 10 sulphonylureas herbicides. The sorbents studied were: silica-based ODS-C18 and two polymeric sorbents, Oasis HLB and Cleanert HXN. Analytes determination and quantitation was carried out with liquid chromatography with electrospray mass spectrometry equipped with ion trap analyzer. The recovery rates range from 76.6 to 109.1% with RSD between 0.3 to 13.8% with the HLB cartridges which was the best one of the three cartridges evaluated (Ouyang et al., 2009).

Also gas chromatography, mainly with MS detection, has been extensively employed for the measurement of herbicide residues in water; different methods had been reported in the literature.

A GC-MS method was developed for the detection of triazine herbicides (atrazine, cyanazine, simazine) and their decomposition products (deethylatrazine, deisopropylatrazine) in environmental waters. The water samples were extracted using an octadecylsilica SPE cartridge with recoveries and RSD in their acceptable ranges (Ma et al., 2003).

SPME combined with GC-MS was developed and employed for the determination of 10 herbicides (alachlor, atrazine, chlorotoluron, diclofop, diflufenicam, ethofumesata, isoproturon, linuron, terbutryn and trifluralin) in surface and ground water. Microextraction was performed with a polyacrylate fiber. It was found that thermally unstable phenylurea herbicides decompose and the resulting anilines can be used for their identification. The recovery of herbicides varied between 94±16 and 107±12% and the detection limit was below 1 $\mu g\ L^{-1}$. It was stated that the method is sensitive, reproducible, easy to perform, and can be applied for the quantitative determination of these herbicides in water (Carabias-Martínez et al., 2003a; Carabias-Martínez et al., 2003b).

In 2008 Crespo-corral et al., describe the determination of carbamate, phenylurea and phenoxy acid herbicide residues by gas chromatography after a potassium ter-butoxide/dimethyl sulphoxide/ethyl iodide derivatization reaction. The method consisted on the pre concentration of the sample using a C-18 sep-Pack cartridge followed by the derivatization of the extract for the analysis with GC-FID and GC/MS in electron impact (EI) and selected ion monitoring mode. The recoveries were in the range 81-99% with RSD 0.9-20.6% (Crespo-Corral et al., 2008).

As a curiosity, phenoxy herbicides have been detected even in rainfall waters in Canada (Hill et al 2002).

3. Herbicide residues analysis in soil

Because of the possible strong binding of herbicides to the polar or apolar components of soil and sediment, the use of an appropriate extraction and pre concentration method is the pre requisite of a reliable chromatographic analysis (Cserháti, 2004). The analysis of pesticides in solid environmental matrices, such as soil, sewage, sludge and sediments, has been addressed in much fewer occasions than in water, probably because of the comparatively greater complexity of the matrices and the absence of environmental quality standards (EQS).

The first and probably the most important step in pesticide residue analysis in soils is soil characterization. There are many reports in the bibliography showing that the efficiency of the extraction method is highly dependent not only on the nature of the soil but also the amount of organic matter in it, being of paramount importance (Merini et al., 2008, Niell et al., 2010). Many nonpolar herbicides are adsorbed to soil organic matter and the cationic herbicides bind tightly to humic acids.

3.3 Sample conservation

In case that the soil sample will not be analyzed immediately, freezing to -20°C is a reasonable measure. Nevertheless, care should be taken if the moist sample is frozen. Sample defrosting takes several hours and during this period microbial growth is exponential and biodegradation of pesticides can occur. Therefore, to prevent the microbial degradation of herbicides it is advisable to freeze-dry the sample and store it properly afterwards. Soil samples without water are easy to handle and to subsample to perform the analysis. To develop the method, the blank soil sample employed has to be treated as the real samples will be handled. Soils can be either reconstructed adding water or extracted directly with the extraction solvent. Nevertheless, it has to be considered that the dryness of the soil increase the clay portion of the soil sorption capacity (Haouari et al., 2006), (Merini et al., 2008), (He et al., 2006) but when water is added the interaction of the herbicide with the humic matter of the soil is increased. These authors also found that the recoveries of 2,4DCP were higher as the water amount of the soil was increased. It is desirable therefore to standardize the water content to the soil by adding a precise amount of water prior of conducting the extraction of the herbicides trapped in the soil.

3.4 Spiking procedure

During the development and validation of the analytical method for soil analysis, the spiking procedure is a crucial step for a successful appropriate analytical method. Pesticides interact in different ways with soils making the reconstruction of that situation in vitro a difficult task. Given that is almost impossible to overcome handicap, there are other points that have to be considered: Spiking solvent selection, volume of the spiking solution, sample homogenization, temperature for solvent evaporation. Merini et al., in 2008, evaluated such parameters for Argentinean loamy soils. They have found that >80% recovery of 2,4 D was obtained using 2000 μl of Methanol as spiking solvent but recoveries were lower using lower volumes. When water was used for spiking, the recoveries dropped down to <50%.

3.5 Extraction solvents

Neutral analytes can be easily extracted using polar organic solvents (MeOH, ACN) or aqueous (MeOH:H_2O, $(CH_3)_2CO$:H_2O) solutions of organic solvents. Ionic herbicides need water solutions to extract them from soils. In many cases, in order to minimize Van der Waals interactions between clay and humic acids, electrolytes like KCl can be employed to give enough ionic force to the solution to displace the herbicides from the soil. In some cases, the electrolyte also gives an adequate pH to get the most soluble form of the herbicide and to break the interaction with charged soils particles. Particularly sulphonylureas can be easily extracted with MeOH: Aq.0.1M $(NH_4)_2CO_3$ (1:9) from different types of soils having either low or high organic matter content. Nevertheless, the method failed to yield good recoveries from other compounds like cyhalofop-butyl or bentazone. Also ionic compounds like bispyribac sodium cannot be recovered properly (Niell et al., 2010). The most employed herbicides nowadays cannot stand a thermally conducted extraction using methodologies such as MASE or Soxhlet. Pressurized Solvent Extraction and supercritical CO_2 extraction with polar solvents as modifiers are methodologies usually reported for herbicide analysis, but room temperature extractions either ultrasonic or shaker stirred still are the preferred methodologies to perform herbicide extraction from soils. (Baugros et al., 2009; García-Valcarcel & Tadeo, 2009). An example of the influence of the extraction solvents and extraction methodology is shown in the following example. Six chemically different herbicides residues (three sulphonylureas, bentazone, bispyribac-sodium and cyhalofop-butyl) were extracted from four soils of different compositions commonly employed for rice cultivation in Uruguay. The results are shown in Figure 2

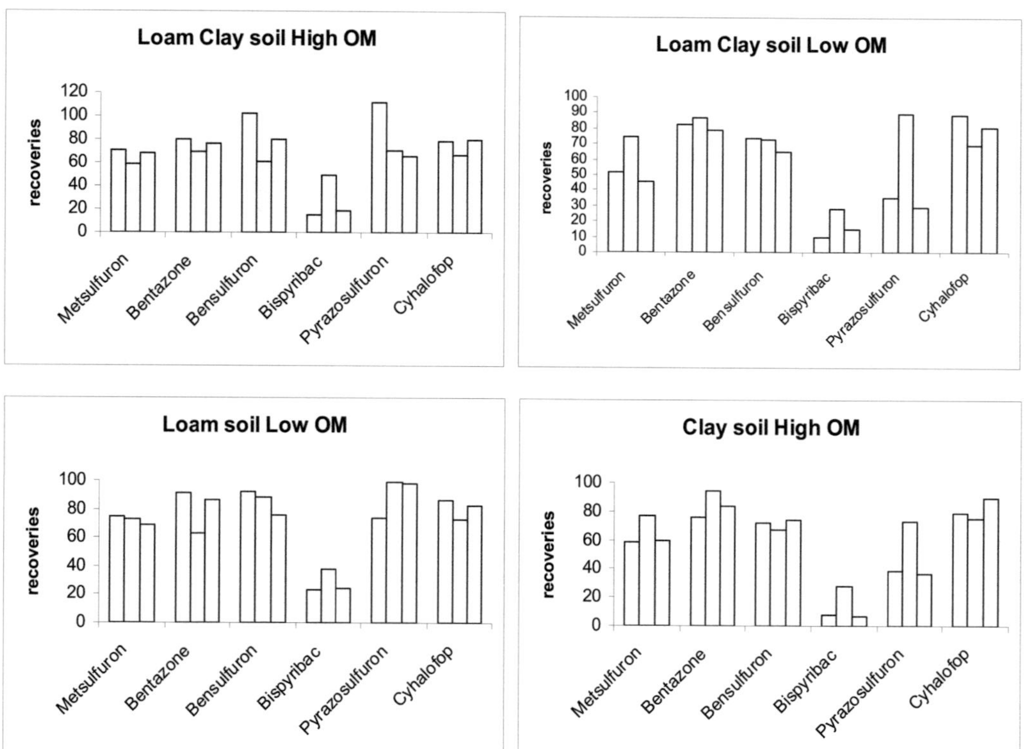

Fig. 2. Recoveries of herbicides from different soil types using two different extraction solvents and two different sitrring procedures (Heinzen et al, unpublished)

□ MeOH: AcOEt (7:3),Ultrasonic bath for 15 minutes

▩ MeOH: 0.1 M $(NH_4)_2CO_3$ (1:9; v/v), magnetic stirred.

▧ MeOH: AcOEt (7:3), magnetic stirred.

Herbicide recoveries depend on the soil type and the extraction procedure. Different recoveries were observed for the herbicides depending On the type of soil. As a general trend it can be observed that the amount of organic matter usually decreases herbicide recoveries either in clay or loamy soils and the recoveries from loam soils are better than those obtained from clay soils. Aqueous solution of high ionic strength with 10% Methanol gave better recoveries when extracting herbicides from clay soils. Even if the same solvent is used, the stirring method, either ultrasound or magnetic stirring, influences the overall herbicide recovery.

Synthetic herbicides are active at very low doses. Table 6 shows the usual application doses of imidazolinones, sulfonylureas and some individual compounds that are employed worldwide.

Herbicide	**kg ha^{-1}**	**kgm^{-2}(x10^{-6})**
Sulphonylureas	0.3-0.5	3-5
Imidazolinones	0.1-0.35	1-3.5
Alachlor	0.5-2.0	5-20
Clomazone	1.1	11
Metribuzin	0.57	6
Quinclorac	5-10	50-100

Table 6. Application dose of herbicides per area unit

The concentration of the herbicide applied per m^2 is within the mg range. Taking into account that usually a 1 kg soil sample is made from mixing many subsamples that covered approximately a 10x20 cm of the superficial soil, the amount of herbicide to be detected is in the sub milligram range or ppbs.

In a miniaturized protocol, usually less than 20 g of soil are employed for residue analysis. The detection and quantitation procedure must be sensitive enough to detect the compounds under investigation at such level. The physicochemical properties like the Henry´s law constant and Kow shows that most of the herbicides are either polar or not volatile enough to be GC-analyzed. Nevertheless, in the past, GC/ECD, NPD or sQuad MS operating in the SIM mode detectors were the method of choice to quantify the agrochemicals searched. In many cases, a derivatization step was mandatory as for phenoxy auxin herbicides sometimes indirect detection of the analytes was the only option: thermolabile herbicides like ureas are detected as the corresponding anilines (Berrada et al., 2004). Phenylureas, can be determined directly by GC but some derivatization reactions have been reported to obtain less polar and more volatile compounds. Some of the reagents commonly used are the alkyl iodide and heptafluorobutyric anhydride (HFBA), where the latest one gives very stable derivatization products with a high ECD response. This family of pesticides can also be determined as their thermal decomposition products (isocyanates) that may be obtained quantitatively under closely controlled conditions (Tadeo et al., 2000). Phenylureas also decompose to the corresponding anilines in the GC injection port. This property was applied (Berrada et al., 2004) for the GC determination of phenylureas after MSPE extraction from vegetables with a polyacrilate fiber.

The other old instrumental approach was to use HPLC coupled to UV and fluorescence detectors and later on sQuadMS. The fluorescence detector is very sensitive and selective, and applying pre or post column derivatization, very complicated analytes like glyphosate and AMPA have been routinely analyzed worldwide since twenty years ago.

On the other hand UV detection has many interferences from the soil matrix as it is neither as sensitive nor as specific towards analytes in complex mixtures. Nevertheless, they have been used for years. (Patsias et al., 2002, Niell et al., 2010) Normally, the detection level is Not lower than $0.1 mgkg^{-1}L^{-1}$ in the injected sample. Therefore, sample size must be increased and the extracting solution concentrated in order to get enough analyte to be detected. The main disadvantage is that the other matrix components also concentrate, yielding noisy chromatograms and the herbicide peaks are difficult to identify and in many cases impossible to quantify if a S/N 10:1 has to be reached. HPLC/MS operating in the SIM mode is a more powerful analytical tool and many problems had been solved using this approach, but mainly focused on single residue analysis. The MS detector is also a universal one and interferences from matrices are common, driving to false positives in many cases.

LC-MS/MS techniques are the most powerful methods for pesticide trace analysis particularly herbicides. They offer the possibility of selectively analyze the fragmentation of an ion which was ionized in the first MS cycle. The selectivity achieved and the sensitivity is also remarkable. This advance in the detection and quantitation of trace compounds boosted the strict EU-regulations on contaminant levels allowed in different food matrices. Screening of contaminants is performed mainly with Time of flight instruments (ToF), in MS or MS/MS modes. ToF instruments can determine exact masses for the compounds detected and therefore, unequivocal identification of the analyte is performed. Sorbents used for on-line SPE or MSPE extraction have included both traditional (alkylbonded silica and polymers) (Djozan & Ebrahimi, 2008) and novel molecular imprinted polymers (MIP) materials. MIPs have been proven to be valuable materials for the selective extraction of pesticides (the template molecule and structurally related compounds). The inherent selectivity of the molecular recognition of these materials allows a high degree of sample clean-up to be achieved (Cacho et al., 2009). Zhang et al., reports the preparation of a new non-covalently bonded MIP and its evaluation for pre concentration of metribuzin in soil samples, this MIP was prepared by in-situ polymerization using methacrylic acid (MAA) as the functional monomer and ethylene glycol dimethacrylate (EDMA) as cross-linker. An on-line procedure was, furthermore, employed for the quantitative determination of metribuzin with pre concentration on the monolithic polymer pre-column. MIP can be combined with stir bar extraction for the determination of nicosulfuron (Yang et al., 2010), achieving a sensitivity for water and soils at the nM level. The sample preparation consisted on the maceration of 25 g of soil sample with acetone: water (4:1), followed by filtration and later extraction with dichloromethane in the presence of 20% sodium chloride solution. This sample solution was concentrated and analyzed directly with the on-line procedure. The effectiveness of the MIP was evaluated by LC (Zhang et al., 2009).

In 2008, Lesueur et al., compared four different extraction methods for the analysis of 12 herbicides and two transformation products. The methods were a) ultrasonic solvent extraction (USE) consisting on the extraction of 20g of sample with a 1:2 solution of water/acetonitrile, b) a PLE experiment where 5g of sample were mixed with 1g of silica gel and extracted with acetonitrile/acetone (1:1), c) QuEChERS method and d) European Norm DIN 12393. The analyses were performed by using a LC tandem mass spectrometry system equipped with an electrospray ionization interface operated in positive mode. The lowest recoveries were obtained with the USE method; on the contrary the higher recoveries were obtained with QuEChERS method, between 27-120%. Nevertheless, the lowest limits of detection were obtained with the European Norm DIN 12393 and the highest with QuEChERS and PLE methods (Lesueur et al., 2008).

In 2009, Wu et al., analyzed four sulphonylureas by dispersive solid-phase extraction (dSPE) followed by dispersive liquid-liquid microextraction (DLLME) and HPLC. The dSPE-DLLME procedure consisted on the extraction of 10 g of soil with 20 mL acetone/0.15 M $NaHCO_3$ (2:8) vigorously shaken for 30 min. For dSPE 0.15g of C-18 per 10mL extract was added and shaken. For the DLLME 5mL aliquot of the extract were mixed with 60µL chlorobenzene (as extraction solvent), after vortexing and centrifugation the chlorobenzene residue was dissolved in acetonitrile and analyzed by HPLC. The recovery rates varied from 76-93% with RSD from 5 to 7%. The authors noticed that compared with other conventional sample preparation methods, this analytical technique offers advantages such as simplicity, ease operation, relatively short analysis time, and lower consume of solvent (Wu et al., 2009). Ionic fluids, one of the most popular class of green solvents, are interesting solvents for performing DLLME of herbicides either in water or soil and food aqueous extracts. A room temperature ionic liquid, 1-hexyl-3-methylimidazolium hexafluorophosphate ($[C_6MIM][PF_6]$), was used as extraction solvent and Triton X 114 was used as dispersant. A mixture of 175µL of ($[C_6MIM][PF_6]$) and 50µL 10% Triton X 114 was rapidly injected into the 20mL honey sample by syringe. Herbicides (chlortoluron, prometon, propazine, linuron and prebane) residues were quantified below the 0.01 $mgkg^{-1}$ range (Wang, 2010).
Sulphonylureas have been detected in Canadian wetlands sediment where no agriculture activities had been performed (Degenhart et al., 2010) and in a real situation, a Multirresidue method by Pressurized Liquid Extraction and GC of vineyard soils allowed the simultaneous and GC allowed the simultaneous determination of two herbicides, three fungicides and two Insecticides at ppb level (Schreck et al., 2008).
Diez et al., studied the soil dissipation kinetics of 12 herbicides used on a rain-fed barley crop, the extraction of the soils was carried out with a mixture of acetone, water, and acetic acid (30:7.5:0.3) followed by the analysis in a GC/MS. The extraction method was previously validated for the extraction of 40 herbicides in soils, with recoveries between 71-108% and RSD in the range 0.6-8% (Díez et al., 2008; Díez & Barrado, 2010). The determination by GC-(ITD)MS/MS of triazines, alachlor and metabolites from soils was also performed after Microwave Assisted Extraction (Vryzas et al., 2007).

3.6 Herbicide residues analysis in cereals.

Analysis of herbicide residue involves different steps such as extraction, interference removal, determination of the herbicide residues and their confirmation. All these analysis are generally performed by using several analytical techniques.
The determination of herbicides was initially carried out by colorimetric and spectrophotometric methods but the sensitivity was not enough to meet the regulations and therefore more sensitive methods have been created in order to reach the limits of detection demanded.
The extraction techniques used for the analysis of herbicide residue in cereals depends on the characteristic of the matrix and the polarity of the herbicides.
In recent years, the development of multiresidue methods (MRM) has taken a high profile, however, there are several pesticides that cannot be included in MRM thus single residue methods (SRM) has to be implemented, which means more workload and expenses for the laboratories (Poulsen et al., 2009).
Some examples of herbicides which need these kinds of SRM are glyphosate and their metabolites, acidic herbicides or trazines. Table 7 summarizes different MRM which involve

the determination of herbicides in cereals used during cereals cropping, while in Table 9 the main single extraction methods for some important herbicides are also presented.
Another important factor that the analyst has to take into account is whether the sample suffered an industrial process or if it is raw material. During cereal processing specially during the milling step a big part of the fats components, which are generally found in the bran, are lost. This is an advantage from an analytical point of view, because the sample treatment will not be so exhaustive.

3.7 Extraction and clean-up

Extraction of herbicides from food depends on their polarity and also on the type of matrix. Generally it consists on the homogenization of the sample with an organic solvent alone or mixed with water or pH adjusted, using an ultrasonic bath, a blender or a homogenizer (Lambropoulou & Albanis, 2007).
After the extraction process generally a clean-up procedure is carried out in order to remove the co-extracted compounds that may act as interference during the chromatographic analysis, causing problems in the detection and quantitation of the analytes.
As discussed before, extraction of residues from food depends strongly on the polarity of the selected herbicides, but also on the selectivity and sensitivity of the detection technique employed in their determination.
Nowadays, one of the most used MRM method is QuEChERS (Quick, Easy, Cheap, Effective, Rugged and Safe) but for certain herbicides the use of PSA during the clean-up, or the pH during the extraction causes low recoveries due to poor extraction or chemical interaction with the PSA sorbent. Therefore, several modifications have been implemented in order to improve the recoveries for these kinds of herbicides (CRL website; Anastassiades et al., 2003; Lehotay et al., 2005).
Despite the fact that in routine analysis laboratories prefer the use of few methods with a wide scope, some pesticides, especially some kinds of herbicides require particular extraction conditions and / or clean-up.
The most used herbicides in cereal cropping belong to the phenoxy, phenyl and sulphonylurea families of herbicides. Nevertheless, there are some others like glyphosate, propanil, bentazone (a banned herbicide within the EU) that are also widely used (Herrmann et al., 2006).
This section is focused on the analytical techniques for the determination of most of these herbicides.
The phenoxy herbicides are widely applied as salts or esters but they decompose by hydrolysis, in the plant, to their respective acids thus the extraction of these herbicides is generally carried out with high polarity organic solvents such as MeOH, EtOH or mixtures with water. However these herbicides are best analyzed when they are extracted as their free acids. In general, both acid and basic hydrolysis followed two different strategies. On one hand, samples are hydrolyzed and the free acids extracted with an organic solvent. On the other hand, the parent herbicide and its conjugates are extracted by organic solvents and subsequently hydrolyzed.
Table 7 summarizes the main methodologies used for the analysis of these herbicides in cereals.

Matrix	Extraction Procedure	Clean-up	Ref.	Analytical Technique
Unpolished Rice	Pressurized Liquid Extraction	LLE/SPE:ENVI-Carb/LC-NH_2	(Itoh et al., 2009)	Liquid Chromatography Atmospheric Pressure Photoionization mass spectrometry
Polished rice	QuEChERS (1% HAc in MeCN)	---	(Niell et al., 2010)	Liquid chromatography DAD detection
Barley	QuEChERS	MSPD: PSA, $MgSO_4$	(Díez et al., 2006)	Liquid Chromatography tandem mass spectrometry Gas Chromatography-time of flight-mass spectrometry
Unpolished Rice	Supercritical Fluid Extraction	SPE: BondElut® PSA over Sep-Pak® Florisil column connected in tandem	(Kaihara et al., 2002)	Liquid Chromatography-Electrospray Ionization mass spectrometry
Rice	CH_2Cl_2 ultrasonic bath	SPE: Florisil	(Pengyan et al., 2006)	Gas Chromatography mass spectrometry by selected ion monitoring mode
Cooked polished rice, wheat	MeCN, shaker, ultrasonic bath	---	(Lee et al., 2009)	Liquid Chromatography-Electrospray Ionization mass spectrometry
Rice	QuEChERS	MSPD: PSA, $MgSO_4$/ GPC	(Liu et al., 2007)	Gas Chromatography mass spectrometry
Rice	QuEChERS	MSPD: PSA, MgSO4	(Libin et al., 2006)	Gas Chromatography mass spectrometry
Rice	QuEChERS	Supelclean ENVI-18 cartridge/Supelclean ENVI-carb II/PSA cartridge.	(Takatori et al., 2008)	Liquid Chromatography-Electrospray Ionization mass spectrometry
Rice	QuEChERS	MSPD: PSA, $MgSO_4$, GCB	(Nguyen et al., 2007)	Gas Chromatography mass spectrometry by selected ion monitoring mode
Unpolished rice	Ethylacetate, homogenizer	GPC/ SPE Florisil	(Zhang et al., 2006)	Gas Chromatography mass spectrometry by selected ion monitoring mode
Rice	QuEChERS	MSPD: PSA, $MgSO_4$, GCB	(Nguyen et al., 2008)	Gas Chromatography mass spectrometry by selected ion monitoring mode

Matrix	Extraction Procedure	Clean-up	Ref.	Analytical Technique
Rice, wheat, maize, barley, oats	Accelerated Solvent Extraction	a. ENVI-18 cartridges/ENVIcarb cartridges/Sep-Pak NH_2 cartridge b. Sep-Pak Alumina N cartridge/Sep-Pak NH_2 cartridge	(Pang et al., 2006)	a. Gas Chromatography mass spectrometry, b. Liquid Chromatography-Electrospray Ionization mass spectrometry
Maize	a.Water/1% formic acid in MeCN b. Water/1% formic acid in MeOH c. Water/1% formic acid in acetone	---	(Mol et al., 2008)	Ultraperformance liquid Chromatography electrospay ionization mass spectrometry
Soybean	MeCN:acetone:light petroleum (1:1:1), homogenizer	GPC	(Pizzutti et al., 2007)	Liquid Chromatography-Electrospray Ionization mass spectrometry
Soybean	LLE (MeCN, ethylacetate, 1M phosphate buffer pH 7	SPE: SAX/PSA	(Hirahara et al., 2005)	Gas Chromatography. Detectors FPD, ECD, MSD
Wheat, Rye, Rice, Maize	Buffered QuEChERS	MSPD: PSA, $MgSO_4$	(Herrmann et al., 2006)	Gas Chromatography Ion trap Detection
Wheat	Citrate Buffered QuEChERS	MSPD: PSA, $MgSO_4$, C-18	(Walorckyk, 2007)	Gas Chromatography mass spectrometry
Wheat flour	Water/MeOH, Ultra Turrax	NaCl/ChemElut	(Klein & Alder, 2003)	Liquid Chromatography Electrospray Ionization/atmospheric pressure ionization mass spectrometry
Wheat Flour	Citrate buffered QuEChERS Basic Hydrolysis (optional)	Freezing (optional)	CRL webpage	Liquid Chromatography Electrospray mass spectrometry
Wheat	Citrate Buffered QuEChERS	MSPD: PSA, $MgSO_4$, C-18	(Walorckyk, 2008)	Gas Chromatography mass spectrometry

Table 7. Main methodologies used for the multiresidue analysis of herbicides in cereals.

Herbicides like molinate, atrazine, diuron are normally included and determined in many of the MRM methods listed above along with insecticides and funguicides (Cervera et al., 2010). The following example (Figure 3) shows the multiresidue analysis using HPLC-MS/MS of fifteen herbicides currently employed in rice production. Table 8 summarizes the most relevant analytical features of the analyzed pesticides.

Pesticide	Mode of action	tR (min)	Quantitation MRM1	Confirmation MRM2	Fragmentor (V)	CE1 (eV)	CE 2 (eV)
Azimsulfuron	Sulphonylureas	17.3	447.1/178.1	425.0 / 182.1	120 / 90	10	15
Bensulfuron-methyl	Sulphonylureas	18.7	411.1 / 182.0	411.1 / 149.0	150	20	15
Bispyribac sodium	pyrimidinyloxybenzoic acid	19.2	453.1 / 297.1	453.1 / 179.1	150	15	20
Bromacil	uracyl	14.2	261.0 / 205.0	261.0 / 188.0	90	10	20
Clomazone	unclassified	19.4	240.1 / 125.0	240.1 / 89.0	150	20	60
Cyhalofop-butyl	aryloxyphenoxypropionic	10.4	357.8 / 256.0	357.8 / 302.0	150 / 165	40	12
Fluroxypyr	pyridine	14.5	255.0 / 181.0	255.0 / 209.0	120	20	15
Imazapic	imidazolinone	9.6	276.1 / 163.2	276.1 / 145.0	150	30	40
Imazapyr	imidazolinone	6.5	262.1 / 149.1	262.1 / 217.0	150	30	30
Imazaquin	imidazolinone	14.3	312.1 / 199.0	312.1 / 153.2	90/150	30	50
Imazosulfuron	Sulphonylureas	18.6	413.1 / 156	413.1 / 153.0	150	20	5
Molinate	thiocarbamate	21.4	188.2 / 126.1	188.2 / 55.1	80	10	20
Pyrazosulfuron-ethyl	Sulphonylureas	20.3	415.1 / 182.1	415.1 / 139.1	120 / 90	25	50
Propaquizafop	aryloxyphenoxypropionic	26.3	444.1 / 371.1	444.1 / 100.1	140	15	15
Quinclorac	quinolinecarboxylic acid	14.4	242.0 / 223.0	242.0 / 161.0	90	10	40

Table 8. Instrument acquisition data used for the analysis of the selected herbicides by LC-QqQ/MS tR, retention time, CE, collision energy.

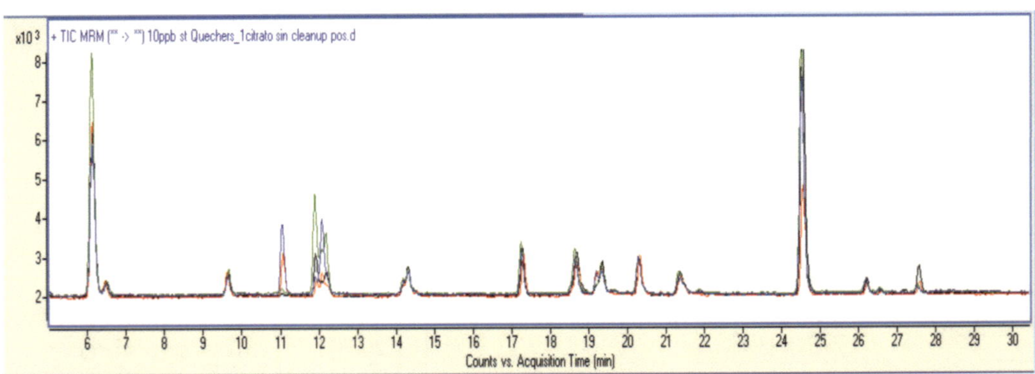

Fig. 3. Analysis of the selected herbicides by LC-QqQ/MS after different QuEChERS procedures at $5 \mu g kg^{-1}$ (Pareja et al, unpublished)

For conditions see Table 8.

Different QueChERS based procedures allowed a good extraction and clean up of the complex rice matrix at very low levels, with recoveries ranging between 60-120% (Pareja et al, 2009)

4. Single residue methods

Analyte	Matrix	Extraction Procedure: SRM	%Recovery (% RSD)	Ref.
Glyphosate, chlormequa, mepiquat	Wheat	Glyphosate 3 g of wheat with 25 mL water were extracted twice by ultrasonication. The extract was centrifuged and filtered. The clean up was performed by using a polystyrene reverse phase column. Chlormequat, mepiquat 10 g of sample was extracted with MeOH/water/acetic acid by ultraturrax. The centrifuged extracts were clean-up on a SPE-C-18 column with MeOH/water/ammonium acetate buffer.	89, 93, 90	(Andersen et al., 2007)
Glyphosate	Cereals	1 g of grain flour with 20 mL water were vortexed for 1 min. Overnight standing extraction and then centrifuged for 10 min. Clean up: 1 mL of the extract was loaded into a preconditioned 100 mg C-18-bonded silica cartridge. The cartridge was placed on top of a calibrated tube. 1.5 mL of the supernatant was transferred into the cartridge and by means of over pressure passed through the cartridge and collected into the tube.	88-99 (5.3-7.1)	(Hogendoorn et al., 1999)
Fenoxaprop-p-ethyl, isoxadifen-ethyl	Rice	25 g of sample were extracted with 40 mL acetonitrile/0.1M hydrochloric acid (80:20) and mixed for 10 min on a shaker. The extract was filtered and the paper washed twice with acetone. The filtrate was loaded onto the Chemelut CE250 column and eluted with 150 mL cyclohexane/ethylacetate, the eluate evaporated to dryness.	74.1-98.3 83.3-100.9	(Lucini & Molinari, 2010)
Chlormequat, Mepiquat	Wheat and other cereals flour	3 g of flour were packed into a 11 mL extraction cell and extracted with EtOH at 100 atm at 120 °C with an extraction time of 15 min divided as follows: preheating (1min), heating (8 min directly set by microprocessor on the grounds of the selected temperature value), static step (5 min) and purging (1 min). Subsequent to extraction 50 µL of IS solution were added to each sample.	83-99	(Marchese et al., 2009)
Triazines	Cereal-based foods	PMAE. 2 g of sample with 20 mL MeOH were placed in the extraction vessel. The magnetron power was set at 100% (600 W). The extraction temperature increased gradually until it reached 105 °C in that moment the extraction was performed for 10 min. After the completion of the extraction the vessels were allowed to cool at room temperature for 20 min. The extract was filtered and the sediments were rinsed 3 times with MeOH. The mixture was dried by rotatory evaporator at 45 °C and redissolved in acetonitrile. AMAE. 2 g of sample were refluxed with 40 mL MeOH for 20 min in a modified household microwave oven with 30 % output maximum power of 800 W and a distilling flask fitted with water cooling condenser tube. The extract was filtered and the sediments were	71.9-83.7 (5.3-10.6) 71.5-80.6 (6.9-12.9) 38.9-74.9 (3.8-9.4) 64.5-74.2 (3.3-5.6)	(You et al., 2007)

Analyte	Matrix	Extraction Procedure: SRM	%Recovery (% RSD)	Ref.
		rinsed three times with MeOH. The mixture was dried by rotatory evaporator at 45 °C and redissolved in acetonitrile. UE 2 g of sample with 40 mL MeOH were placed into a 100 mL flask and extracted in a water bath and sonicated for 90 min. After completing extraction the extract was filtered and the flask rinsed three times with MeOH. The mixture was dried by rotatory evaporator at 45 °C and reconstituted in acetonitrile. SE 2 g of sample were placed in a glass soxhlet thimble and 40 mL MeOH were added in a 100 mL flask, The flask was fitted with water cooling condenser tube and immersed in a water bath. SE was carried out for 2 h at 100 °C. After that the extract was filtered and the sediments were rinsed three times with MeOH. The mixture was dried by rotatory evaporator at 45 °C and rsuspeded in acetonitrile.		
13 phenoxy acids herbicides	Rice	QuEChERS. 10 g of rice with 5 mL water were vortexed for 1 min. 10 ml 0.5 % acetic acid in MeCN were added and shaken two times in vortex. 4 g $MgSO_4$, 1.0 g of tri-Na, 0.5 g di-Na citrate and 1g of NaCl were added, vortexed and centrifuged. Clean up: an aliquot of the extract with 250 mg C-18, 100 mg of alumina neutral and 1.5 g $MgSO_4$, were shaken and centrifuged. 5 mL of the solution were concentrated to dryness and reconstituted for analysis.	45-104 (< 13.3)	(Koesukwiwat et al., 2008)
15 phenylureas herbicides	Rice	Solvent Extraction/SPE. 10 g of rice were homogenized with 50 mL MeCN for 2 min, and filtered into a mixing cylinder and sealed with a stopper, 10 g of NaCl were added and shaked, and the phases were allowed to stand for 20 min. A 25 mL portion of the supernatant was evaporated to dryness, and dissolved in 2 mL n-hexane. Clean up: the extract was transferred to a Florisil SPE column and eluted with 5 mL acetone:n-hexane (40:60), then concentrated to dryness and dissolved in MeCN: water (1:1) for analysis.	75.3-104.3 (1.5-9.6)	(Mou et al., 2008)
Isoproturon	Cereal grains and pasta	Solvent Extraction. 15 g of sample with 4 mL of water and 30 mL acetone were placed in a centrifuged bottle homogenized using an ultra turrax for 30 s at 13500 rpm. Then 30 mL hexane and 30 mL dichloromethane were added to the bottle and homogenized for further 1 min at 13500 rpm. The matrix and extract were separated by centrifugation for 2 min and the extract decanted into a 200 mL volumetric flask. The matrix was resuspended in 30 mL acetone, 30 mL dichloromethane and 30 mL hexane and extracted again for 1 min. The 2nd extract was mixed with the previous one and made up to 200 mL with dichloromethane. A 4 mL aliquot was transferred into a borosilicate glass culture tube, 50 µL of 50%	94-100 (0.8-5.8)	(Winrow et al., 2003)

Analyte	Matrix	Extraction Procedure: SRM	%Recovery (% RSD)	Ref.
		propylene glycol in acetone was added and the mixture evaporated under a nitrogen stream. The residue was reconstituted with 1 mL dichloromethane and placed onto a SPR cartridge containing 200 mg aminopropyl sorbent. The sample tube was rinsed successively with 0.5 mL dichloromethane and 2 mL dichloromethane/MeOH (99:1). The combined eluates were evaporated to nearly dryness and the residue reconstituted in 1 mL MeOH for analysis.		
Phenoxy acids herbicides	Cereals (plant samples)	5 g of sample were extracted with 0.1M NaOH (2x25mL) in a sorvall homogenizer, the extract was filtered and the filter cake washed twice with 5 mL of the basic aqueous solution. The extract was mixed with 25 mL saturated sodium chloride solution and the pH lowered to near 5 by the addition of 2 M H_2SO_4, the solution let stand for 15 min and the liquid decanted. Then the pH of the solution was lowered again to 1 the solution was transferred to a separatory funnel and extracted with diethyl ether (2 x 50 mL). The organic phase was extracted with 0.5 M $NaHCO_3$ (2 x 25 mL), the combined aqueous solution acidified to pH 1 by adding carefully 3M H_2SO_4 (10 mL) and extracted with $CHCl_3$ (2 x 25 mL). The organic phase was filtered through Na_2SO_4 and solvent concentrated to dryness under vacuum.	95.5-104.0 (1.4-10.3)	(Sánchez-Brunete et al., 1994)
Glyphosate, Gluphosinate and their metabolites	Rice, soybean	Rice 5 g of milled rice and 40 mL water were sonicated for 3 min, after resting for 30 min it was centrifuged for the AG1-X8 anion exchange chromatography. Soybean 5 g of soybean sprouts with 25 mL water were mixed in a shaker for 10 min, after resting for 1 min 15 mL acetone were added and centrifuged for the AG1-X8 anion exchange chromatography. 30 mL of soybean extract or rice extract were applied onto a preconditioned AG1-X8/ Dowex 1-X2 column which was first washed with 15 mL 40 % acetone and eluted by repeated elution with 15 mL of 0.5 N HCl five times. Each pooled eluate was dried under reduced pressure at a temperature lower than 55 °C for the derivatization reaction and Florisil clean-up. Florisil clean-up of the derivatization products: derivatives in mixed standard solution were dissolved in 1 mL ethylacetate and applied to a Florisil cartridge (500 mg, 6 mL) previously conditioned with 5 mL ethylacetate. The cartridge was eluted with 10 mL ethylacetate followed by 10 mL of acetone, 10 mL acetone / MeOH (1:1) mixture, 10 mL acetone / MeOH (1:2) mixture and 10 mL MeOH. Each pooled fraction of mix standard solution was dried under reduced pressure and then dissolved in 1 mL ethylacetate for analysis.	72-119 (6.7-9.6) 86-101 (4.7-6.5)	(Tseng et al., 2004)

Analyte	Matrix	Extraction Procedure: SRM	%Recovery (% RSD)	Ref.
Trazine herbicides	Wheat	PLE 7 g of sample were mixed with 4.5 g of hydromatrix and placed in a 34 mL extraction cell and extracted with dichloromethane/n-hexane (1:4) solution. The PLE conditions were as follows: pressure 1500 psi, heating time 5 min, purge volume 60%, purge time 100 s, 2 static cycles, static time 6 min. The total extraction time was 21 min, the extract were evaporated to dryness in a rotavapor at 40-50 °C and the dry residue reconstituted with 5 mL dichloromethane/n-hexane/acetone (1:1:1). The clean-up was performed using OASIS MCX cartridges conditioned with 10 mL dichloromethane, once the retention step has been completed the cartridges were dried under vacuum for 10 min. The elution of the retain compounds was accomplished with 4 mL of 25% ammonia solution /MeOH (15:85). The organic phase was then dried under a stream of nitrogen at 40-50 °C and reconstituted in 0.5 mL of MeOH/water (1:1) for analysis.	106-125 (6-18)	(Carabias -Martínez et al., 2007)

Table 9. Main methodologies used for the single residue analysis of herbicides in cereals. (Pareja el al unpublished)

Glyphosate is a highly polar herbicide, very soluble in water and insoluble in most organic solvents. For this reason its extraction is generally performed with water or water/chloroform, sometimes at acidic pH, but in this process different components of the matrix are co-extracted thus a clean-up procedure is required (Tadeo et al., 2000).

In Table 9 three different analytical methodologies for the determination of glyphosate in cereals are described. Hogendoorn et al., in 1999 described the rapid determination of glyphosate by means of pre-column derivatization with 9-fluorenylmethyl chloroformate and coupled column liquid chromatography with fluorescence detection. The overall recovery of this herbicide was 86% with a RSD of 9.5%. This procedure implies the extraction with water followed by a clean-up step in a C-18 bonded silica cartridge. The extract was then subjected to derivatization and analysis (Hogendoorn et al., 1999).

In 2004 Tseng et al., reported the analysis of this herbicide and its metabolites in rice and soybean by extraction with water or acetone followed by a clean-up in an AG1-X8/ Dowex 1-X2 anion exchange column and then a single derivatization with trimethylortoacetate in the presence of HAc. The derivative products were then purified using a Florisil cartridge. The detection of these analytes was made by using a gas chromatograph with flame photometric detector. The recoveries and RSDs were 72% and 6.5% for rice and 86% and 6.5 % for soybean respectively with a limit of detection of 20 $\mu g kg^{-1}$ (Tseng et al., 2004).

In 2007, Andersen et al., organized an intercomparison study for the determination of glyphosate and other herbicides in wheat. The authors described the procedure used by their laboratory for the analysis of these herbicides as follows; the extraction of the analytes with water and then a clean-up step online on a polystyrene based reverse phase column and separated by ion chromatography-HPLC tandem mass spectrometry. The average recovery was 96% with a coefficient of variation of 4% (Andersen et al., 2007).

The EU-CRL for SRM in Stuttgart used a straightforward methanol extraction at pH<2 of wheat flour to validate a new multi class residue method for traditionally analyzed herbicides and metabolites using SRM like glyphosate, AMPA and gluphosinate and MPPA. This procedure was used to analyze some fruits and vegetables matrices.

Glyphosate shares structural similarities with amino acids and therefore has low UV absorption and fluorescence is low and also presents the disadvantage that vaporizes easily upon heating. Thus, it presents difficulties in the quantification by high performance liquid chromatography (LC-UV), LC fluorescence detection and gas chromatography. The analysis of this herbicide by gas chromatography requires its derivatization, this procedure involves the use of trifluoroacetic anhydride (TFAA) and trifluoroethanol, TFAA and diazomethane or HBFA and 2-choroethanol (Tadeo et al., 2000).

Gas and liquid chromatography with classical detectors.

Gas chromatography is widely used in the analysis of herbicide residues, due to the high selectivity and sensitivity of the detection systems.

The classical detectors most often used are the flame ionization detector (FID), nitrogen-phosphorus detector (NPD), Thermo ionic detection, electro capture detection (ECD), but in the last decade MS coupled to GC has been the choice for the analysis of herbicides, especially in MRM methods (Tadeo et al., 2000).

The NPD is employed for herbicides containing nitrogen, such as triazines, dinitroanilines or chloroacetamides, this detection system allows limits of detection in the range of μg-mg kg^{-1}.

In the case of ECD, this detector has high sensitivity for halogenated compounds, although its linear range is narrow. It has been used in the analysis of halogenated phenoxyacids, benzonitriles, dinitroanilines, glyphosate and multiresidue analysis, frequently after derivatization.

Liquid Chromatography (LC) is very useful for polar, thermally labile and low-volatility pesticides which in general cannot be directly analyzed by GC, so LC is preferred instead of the use of derivatization techniques.

Most pesticides, including those not easily analyzed by GC, can also be separated by high-performance liquid chromatography without the need of chemical derivatization. phenylureas, organophosphorus pesticides, triazines, quaternary ammonium compounds and chlorinated phenoxy acids are examples of pesticides submitted to LC analysis (Thurman et al., 2005).

Conventional LC detectors such as the UV detector are, however, not selective enough for pesticide analysis in complex matrices. Moreover, selective detectors such as fluorescence detection can only be applied after derivatization.

GC-MS/MS and LC-MS/MS

Over the last 20 years liquid chromatography–mass spectrometry (LC–MS) techniques have advanced dramatically in their sensitivity, specificity and reliability. Detection of sub-ppt concentrations is becoming routine for many organic analytes and methods achieving detection of a few hundred femtograms of some analytes have been reported. Such progress Is mostly due to the development of hyphenated LC–tandem MS techniques, which are today the methods of choice for the determination of trace organic analytes in food and environmental samples. Such growth in the use of LC–MS/MS for the analysis of organic contaminants in environmental matrices has been compelled by the need for high-quality data on their occurrence in the environment at very low concentration levels (Petrovic et al., 2010).

5. Conclusion

Herbicides are integrated to the general cropping systems. Routine herbicide application in intensive crop production is performed following a more or less strict calendar but the number of applications can be more frequent if a weed suddenly threatens the culture. The rotational no-tillage based productive system winter crop-summer crop uses many herbicides like atrazine, sulphonylureas and imidazolinones that can accumulate in soils and in a carry-over phenomenon injure the newly planted crop. Herbicides can be found in every environmental compartment but their occurrence in water, soils and crops are of paramount importance from a toxicological point of view. Herbicides that have low Kow and Koc, can be found in water, either ground or surface through run off, leaching or lixiviation. Therefore herbicides can be found not only in the application site as they migrate through water flows. On the other hand, high Koc and /or Kow herbicides remain trapped in soils. Many different analytical methodologies have been developed to determine pesticide residue concentrations. Herbicide residue analysis is not only a valuable tool to evaluate threshold damage limits for the carry over phenomenon but also their residual levels knowledge are fundamental to evaluate the sustainability of the global agroecosystem. The development of chromatographic hyphenated MS/MS techniques allow the broadening of MRMs scope and therefore, lower detections limits with high accuracy have been reached for a high number of pesticides, but the most commonly developed MRM include only few herbicides. The challenges of herbicide residue analysis are the low application dose of the active substances and therefore, sample handling and clean-up procedures are still the bottle neck of the analytical methodologies. Selective clean up procedures have been developed to isolate and concentrate the searched compounds during the last years. In this context, an increasing amount of methods for different food and environmental matrices have been published. New analytical procedures based on selective polymer extraction (MIP, SPME) are interesting possibilities. As general procedure, SPE extraction is the method of choice for water. Nevertheless, for herbicide residue analysis in soil, the type of soil determined the scope and clean up methodology to follow. After the success of the GMO soybean RR glyphosate-resistant variety, the trend for new crop varieties introduction is based on their selective resistance to specific herbicides like imidazolinones or gluphosinate and sulphonylureas. If these herbicides will be applied in no-tillage production systems at the same ratio as glyphosate is nowadays, environmental issues will probably arise, as imidazolinones and sulphonylureas are semi persistent molecules.

6. References

Ampong-Nyarko, K. & De Datta, S.K. (1991) *A handbook for weed control in Rice,* Manila Philippines.

Anastassiades, M.; Lehotay, S. J.; Štajnbaher, D.; Schenck, F. J. (2003) *J. AOAC Int.* 86, 412-431.

Anderson, R.L.; Barrett, M.R. (1985) *J. Environ. Qual.* 14, 1, 111-114.

Andersen, J. H.; Bille, R. L.; Granby, K. (2007) *Food Add. Contam.* 24, 140-148.

Ayano, E.; Kanazawa, H.; Ando, M.; Nishimura, T. (2004) *Anal. Chim. Acta* 507, 211-218.
Barchanska, H.; Baranowska, I. (2009) *Rev. Environ. Contam. and Toxicol.* 200, 53-84.
Bastide, J.; Cambon, J.; Breton, F.; Piletsky, S.; Rouillon, R. (2005) *Anal. Chim. Acta* 542, 97-103.
Baugros, J.B.; Olive, C.C.; Giroud, B.; Gauvrit, J. Y.; Lanteri, P.; Grenier-Loustalot, M. F.; (2009) *J. Chromatogr. A* 1216: 4941-4949.
Boivin, A.; Cherrier, R.; Schiavon, M. (2005) *Chemosphere* 61, 668–676.
Bradford, W.; Minton, M. A.; Matocha, S. A. (2008) *Weed Technol.* 22, 3, 425-430.
Cacho, C.; Turiel, E.; Perez-Conde, C. (2009) *Talanta* 78, 1029-1035.
Carabias-Martínez, R.; García-Hermida, C.; Rodríguez-Gonzalo, E.; Soriano-Bravo, F. E.; Hernández-Méndez. J. (2003) *J. Chromatogr. A* 1002, 1–12.
Carabias-Martínez, R.; Rodríguez-Gonzalo, R.; Fernández-Laespada, E.; Calvo-Seroner, L.; Sánchez San Roman, F. J. (2003) *Water Research* 37, 928–938.
Carabias-Martínez, R.; Rodríguez-Gonzalo, E.; Herrero-Hernández, E.; Hernández-Méndez, J. (2004) *Anal. Chim. Acta* 517, 71-79.
Caseley, J.J. (1996) *Serie de Estudios FAO: Producción y protección vegetal* -Manejo de Herbicidas para países en desarrollo, Montevideo, Uruguay.
Cepeda, S.; Rossi, A. R.; Ponsa, J. (1995) *Tolerancia o resistencia de malezas a los herbicidas,* In: Carpeta de Producción Vegetal, Tomo XIII. Generalidades, Información, N.125, 5p. EEA Pergamino.
Cervera, M.I.; Medina, C.; Portolés, T.; Pitarch, E.; Beltrán, J.; Serrahima, E.; Pineda, L.; Muñoz, G.; Centrich, F.; Hernández, F. (2010) *Anal. Bioanal. Chem.* 397, 7, 2873-2891.
Chebatarrof N. (2007) *Conexión Tecnológica* 202, 1-6.
Crespo-Corral, E.; Santos-Delgado, M. J.; Polo-Díez, L. M.; Soria, A. C. (2008) *J. Chromatogr. A* 1209, 22-28.
Cserháti, T. (2004) *Biomed. Chromatogr.* 18, 350-359.
Degenhardt, D.; Cessna, A.J.; Raina, R.; Pennock, D.J.; Farenhorst, A. (2010) *J. Environ. Sci. Heal. B* 45, 1, 11-24.
Díez, C.; Barrado, E.; Marinero, P.; Sanz, M. (2008) *J. Chromatogr A* 1180, 10–23
Díez, C.; Barrado, E. (2010) *Anal. Bioanal. Chem.* 397, 1617-1626.
Díez, C.; Traag, W. A.; Zommer, P.; Marinero, P.; Atienza, J. (2006) J. Chromatogr. A 1131, 11-23.
Djozan, D.; Ebrahimi, B. (2008) Anal. Chim. Acta 616, 2, 2, 152-159
García-Reyes, J.F.; Jackson, A.U.; Molina-Díaz, A.; Cooks, R.G. (2009) *Anal. Chem.* 81, 2, 820-829.
García Valcarcel, A.I.; Tadeo, J.L. (1999) *J. Agric. Food Chem.* 47, 3895–3900.
García-Valcarcel, A. I.; Tadeo, J. L. (2009) *Anal. Chim. Acta* 641, 117-126.
Geiss, S.; Gebert, S. (2006) *Acta Hydrochim. Hydrobiol.* 34, 464.
Genta, H.; Villamil, J. (1992) *Serie Técnica* INIA ISBN: 9974-556-26-0,141, Montevideo, Uruguay
Granby, K.; Johannesen, S.; Vahl, M. (2003) *Food Add. Contam.* 20, 8, 692-698
Haouari, J.; Dahchour, A.; Peña-Heras, A.; Louchard, X.; Lennartz, B.; Alaoui, M.E.; Satrallah A. (2006) *J. Environ. Sci. Heal. B* 41, 6, 883-893.
Harri, L. (1994) *Manual de identificação e controle de plantas danhinas: plantio direto e convencional,* 6 ed. Nova Odessa, SP (Brazil). Editora Plantarum.
He, Y.; Xu, J.; Wang, H.; Zhang, Q.; Muhammad, A. (2006) *Chemosphere* 65, 497–505.

Herrmann, S.S; Christensen, H.B.; Poulsen, M. E. (2006) *Technical report for Cereals*, National Food Institue, the Danish Technical University. CRL for Cereals and Feeding Stuff.

Hill, B.D.; Harker, K.N.; Hasselback, P.; Moyer, J.R.; Inaba, D.J.; Byers, S.D. (2002) *Can. J. Plant Sci.* 82, 2, 481-484.

Hogendoorn, E. A.; Ossendrijver, F. M.; Dijkman, E.; Baumann, R. A. (1999) *J. Cromatogr. A* 833, 67-73.

Howatt, K.A.; Endres, G.J. (2006) *Weed Technol.* 20, 1, 67-73. http://www.crlpesticides.eu/docs/public/tmplt_article.asp?CntID=748&LabID=200&Lang=EN

http://www.casafe.org/estad/m2009.htm

Itoh, N.; Otake, Y.; Aoyagi, Y.; Matsuo, M.; Yarita, T. (2009) *Chromatographia* 70, 1073-1078.

Jeannot, R.; Sabik, H.; Sauvard, E.; Genin, E. (2000) *J. Chromatogr. A* 879, 51-71.

Kaihara, A.; Yoshii, K.; Tsumura, Y.; Ishimitsu, S.; Tonogai, Y. (2002) *J. Health Sci.* 48, 173-178.

Klein, J. & Alder, L. (2003) *J. AOAC Int.* 86, 1015-1036.

Koesukwiwat, U.; Sanguankaew, K.; Leepipatpiboon, N. (2008) *Anal. Chim. Acta* 626, 10-20.

Kumar, A.; Malik, A.K.; Picó, Y. (2010) *Electrophoresis* 31,13, 2115-2125.

Kuster, M.; López de Alda, M. J.; Barata, C.; Raldua, D.; Barceló, D. (2007) *Talanta* 75, 390-401.

Lambropoulou, D. A. & Albanis, T. A. (2007) *Anal. Bioanal. Chem.* 389, 1663-1683.

Lee, S. J.; Park, H. J.; Kim, W.; Jin, J. S.; El-Aty, A.M.A.; Shim, J. H.; Shin, S. C. (2009) *Biomed. Chromatogr.* 23, 434-442.

Lehotay, S. J.; Maštovská, K.; Lightfield, A. R. (2005) *J. AOAC Int.* 88, 615-629.

Lesueur, C.; Gartner, M.; Mentler, A.; Fuerhacker, M. (2008) *Talanta* 75, 284-293.

Libin, L.; Hashi, Y.; Yaping, Q.; Haixia, Z.; Jinming, L. Chin. (2006) *J. Anal. Chem.* 34, 783-786.

Lieqing, Y.; Xiaomin, Z.; Jie Z. (2010) *Anal. Chim. Acta* 670, 72–77.

Liu, L. B.; Hashi, Y.; Qin, Y. P.; Zhou, H. X.; Lin, J. M. (2007) *J. Chromatogr. B* 845, 61-68.

Lucini, L. & Molinari, G. P. (2010) *Pest Manag. Sci.* 66, 621-626.

Hirahara, Y.; Kimura, M.; Inoue, T.; Uchikaea, S.; Otani, S.; Haganuma, A.; Matsumoto, N.; Hirata, A.; Maruyama, S.; Iizuka, T.; Ukyo, M.; Ota, M.; Hirose, H.; Suzuki, S.; Uchida, Y. (2005) *J. Health Sci.* 51, 617-627.

Ma, W. T.; Fu, K. K.; Cai, Z.; Jiang, G. B. (2003) *Chemosphere* 52, 1627–1632.

Marchesan, E.; Sarzi S.; de Avila G.M.; Machado,L.A.; Zanella, R.; Primel, E.G.; Mussoi Macedo, V.R.; Marchezan, M.G. (2010) *Ciencia Rural* 40, 5,1053-1059.

Marchese, S.; Perret, D.; Bafile, E.; Gentili, A.; Caretti, F.; Berardino, M. (2009) *Cromatographia* 70, 761-767.

Marín, J.M.; Gracia-Lor,E.; Sancho,J.M.; López,F.J:; Hernández, F. (2009) *J. Chromatogr. A* 1216, 1410–1420.

Mazzella, N.; Delmas, F.; Delest, B.; Méchin, B.; Madigou, C.; Allenou, J. P.; Gabellec, R.; Caquet, T. (2009) *J. Environ. Monit.* 11, 108-115.

Merini, L. J.; Cuadrado, V.; Giuliette, A. M. (2008) *Chemosphere* 71, 11, 2168-2172

MGAP/Servicios Agrícolas, in http://www.mgap.gub.uy/dgssaa/last accessed July 2010.

Mol, H. G. J.; Bolaños, P.; Zomer, P.; de Rijk, T. C.; Stolker, A.A.M.; Mulder, P. P. J. (2008) *Anal. Chem.* 80, 9450-9459.

Molina, F.; Cantou, G.; Roel, A. (2010) last accessed August 2010

<http://www.inia.org.uy/estaciones/ttres/actividades/2010/Resumen 2009_10.pdf>.

Mou, R. X.; Chen, M.; Zhi, J. (2008) *J. Chromatogr. B* 875, 437-443.

Mullen, A.K.;,Clench, M.R.; Crosland, S.; Sharples, K.R. (2005) *Rapid Communications in Mass Spectrometry* 19, 18, 2507-2516.

Nanita, S.C.; Pentz, A.M.; Bramble, F.Q. (2009) *Anal. Chem.* 81, 8, 3134-3142.

Nelson, M.A.; Gates, A.; Dodlinger, M.; Hage, D.S. (2004) *Anal. Chem.* 76, 805-813.

Nguyen, T. D.; Han, E. M.; Seo, M. S.; Kim, S. R.; Yun, M. Y.; Lee, D. M.; Lee, G. H. (2008) *Anal. Chim. Acta* 619, 67-74.

Nguyen, T. D.; Lee, B. S.; Lee, B. R.; Lee, D. M.; Lee, G. (2007) *Rapid Commun. Mass Spectrom.* 21, 3115-3122.

Niell, S.; Pareja, L.; Geis Asteggiante, L.; Cesio, M. V.; Heinzen, H. (2010) *Food Add. Contam. A* 27, 206-211.

Oppenhuizen, M.E.; Cowell, J.E. (1991) *J AOAC Int* 74, 2, 317-323.

Ouyang, X.; Zhang, W.; Chang, N, Pan, C.; Zhang, J.; Niu, W. (2009) *J. Anal. Chem.* 64, 935-940.

Pang, G. F.; Liu, Y. M.; Fan, C. L.; Zhang, J. J.; Cao, Y. Z.; Li, X. M., Li, Z. Y.; Wu, Y. P.; Guo, T. T. (2006) *Anal. Bioanal. Chem.* 384, 1366-1408.

Pareja,L.; Cesio,V.; Heinzen, H.; Fernandez-Alba, A.; (2009) LAPRW , 128.

Patsias,J.; Papadakis,N.E:; Papadopoulou-Mourkidou, E. (2002) *J. Chromatogr. A* 959, 153–161.

Pengyan, L.; Qingxue, L.; Yusong, M.; Jinwei, L.; Xuan, J. Chin. (2006) *J. Chromatogr.* 24, 228-234.

Pizzutti, I. R.; de Kok, A.; Zanella, R., Adaime, M. B.; Hiemstra, M.; Wickert, C.; Prestes, O. D. (2007) *J. Chromatogr. A* 1142, 123-136.

Petrovic, M.; Farré, M.; López de Alda, M.; Pérez, S.; Postigo, C.; Köck, M.; Radjenovic, J.; Gros, M.; Barceló, D. (2010) *J. Chromatogr. A* 1217, 4004–4017.

Pinxteren, M. V.; Bauer, C.; Popp, P. (2009) *J. Chromatogr. A* 1216, 5800-5806.

Polati, S.; Bottaro, M.; Frascarolo, P.; Goseti, F.; Gianotti, V.; Gennaro, M. C. (2006) *Anal. Chim. Acta* 579, 146-151.

Poulsen, M. E.; Christensen, H. B.; Herrmann, S. S. (2009) *Accred. Qual. Assur.* 14, 477-485.

Ramezani, M.; Simpson,N.; Oliver,D.; Kookana,R.; Gill,G.; (2009) *J. Chromatogr. A* 1216, 5092–5100.

Roehrs, R.; Zanella, R.; Pizzuti, I.; Adaime, M. B.; Pareja, L.; Niell, S.; Cesio, M. V.; Heinzen, H. (2009) *J AOAC Int.* 92, 4, 1191-1195.

Rodríguez-Gonzalo, E.; Carabias-Martínez, R.; Cruz, E.M.; Domínguez-Álvarez, J.; Hernández-Méndez, J. (2009) *J Sep. Sci.* 32, 4, 575-584.

Rossi, A. R.; Ponsa, J.; Cepeda, S. (1996) *Maíz: las malezas, su manejo y control*, In: Publicación Miscelánea, N.77, cap.5. EEA Rafaela.

Pareja, L, Cesio, M.V., Heinzen, H., Fernández-Alba A.R. (2010)unpublished results

Sánchez-Brunete, C.; Martínez, L.; Tadeo, J. (1994) *J Agric Food Chem* 42, 2210-2214.

Schoeneberger, P.J.; Wysocki, D.A.; Benham, E.C.; Broderson, W.D. (1998) Field book for describing and sampling soils. Natural Resources Conservation Service, USDA, National Soil Survey Center, Lincoln, NE.

Schreck,E.; Geret,F.; Gontier,L.; Treilhou, M. (2008) *Talanta* 77, 298–303.

Serodio, P.; Nogueira, J. M. (2004) Anal. Chim. Acta 517, 147-153.

Smith, R. J. (1968) *Weed Science*, 16, 2, 252-255
Smith, S.; Cooper, C.M.; Lizotte, R.E. (2007) *Int. J. Ecol. Environ. Sci.* 33, 4, 223-231.
Tadeo, J. L.; Sánchez-Brunette, C.; García-Valcárcel, A. I.; Martínez, L.; Pérez, R. A. (1996) *J. Chromatogr. A* 754, 347-365.
Tadeo, J. L.; Sánchez-Brunette, C.; Pérez, R. A.; Fernández, M. D. (2000) *J. Chromatogr. A* 882, 175-191.
Takatori, S.; Okihashi, M.; Okamoto, Y.; Kitagawa, Y.; Kakimoto, S.; Murata, H.; Sumimoto, T.; Tanaka, Y. (2008) *J AOAC Int.* 91, 871-883.
Tan, S.; Evans, R.R.; Dahmer, M.L.; Singh, B.K.; Shaner, D.L. (2005) *Pest Manag. Sci.* 61, 3, 246-257.
Thurman, E. M.; Ferrer I.; Fernández-Alba A.R. (2005) *Chapter 8 LC-MS. I: Basic principles and technical aspects of LC-MS for pesticide analysis,* In: Chromatographic-Mass Spectrometric Food Analysis for Trace Determination of Pesticide Residues in series Comprehensive Analytical Chemistry, Fernández-Alba A. R. (Ed.), 369-401, Elsevier B.V., ISBN 13: 978-0-444-50943-7
Tomlin, C. (1996) *The Pesticide Manual*, 14th Ed. British Crop Protection Council, Farnahan, UK.
Tseng, S. H.; Lo, Y. W.; Chang, P. C.; Chou, S. S.; Chang, H. M. (2004) *J. Agric. Food Chem.* 52, 4057-4063.
Turiel, E.; Martín, A. (2010) *Anal. Chim. Acta* 668, 87–99.
Vryzas, Z.; Tsaboula, A.; Papadopoulou-Mourkidou, E. (2007) *J. Sep. Sci.* 30, 15, 2529-2538.
Vryzas, Z. & Papadopoulou-Mourkidou, E. (2002) *J. Agric. Food Chem.* 50, 18, 5026-5033.
Wang, Y.; You, J.; Ru, R.; Ciao, Y.; Gao, S.; Zhang, Y.; You, A. (2010) *J Chromatogr A* 1217, 26, 4241-4246.
Walorckyk, S. (2007) *J. Chromatogr. A* 1165, 200-212.
Walorckyk, S. (2008) *J. Chromatogr. A* 1208, 202-214.
Wells, M. J. M.; Yu, L. Z. (2000) *J. Chromatogr. A* 885, 237-250.
Winrow, S.; Hird, S.; Brown, P. (2003) *Food Add. Contam.* 20, 1064-1071.
Woudneh, M.B.; Ou, Z.; Sekela, M.; Tuominen, T.; Gledhill, M. (2009) *J. Environ. Qual.* 38, 3, 948-954.
Wood, A.; In: Alan Wood's Web Site last accessed august 2010, http://www.alanwood.net/pesticides/class_herbicides.html
Wu, Q.; Wang, C.; Liu, Z.; Wu, C.; Zeng, X.; Wen, J.; Wang, Z. (2009) *J. Chromatogr. A* 1216, 5504-5510.
Xie, C.; Gao, S.; Zhou, H.K.; Li, H.F. (2010) *Luminescence* Jul 13. [Epub ahead of print]
Ying, W.; Jingyan, Y.; Ruibing, R.; Yao, X.; Shiqian, G.; Hanqi, Z.; Aimin; Y. (2010) *J. Chromatogr. A* 1217, 4241–4246.
You, J.; Zhang, H. Zhang, H.; Yu, A.; Xiao, T.; Wang, Y.; Song, D. (2007) *J. Chromatogr. B* 856, 278-284.
Yu, Z.G.; Ding, W.M.; Chen, Y.H.; You, J.; Liu, Y.Y.; Wang, H.; Yang, Z.D. (2010) *Chromatographia* 72, 1-2, 33-38.
Zanella, R.; Primel, E. G.; Gonçalves, F. F.; Kurz, M. H. S.; Mistura, C. M. (2003) *J. Sep. Sci.* 26, 935-938.
Zhang, W. G.; Chu, X. G.; Cai, H. X.; An, J.; Li, C. J. (2006) *Rapid Commun. Mass Spectrom.* 20, 609-617.
Zhang, Z.; Yang, G.; Zheng, Z.; Chen, Y. (2009) *Cromatographia* 69, 615-619.

7

Prevention of Herbicides Pollution Using Sorbents in Controlled Release Formulations

Fernández-Pérez, M., Villafranca-Sánchez, M.,
Flores-Céspedes, F. and Daza-Fernández, I.
University of Almería
Spain

1. Introduction

At present, the consume of great quantities of herbicides for crop protection is supposing an important challenge for the maintenance of the welfare state. The correct use of such substances will contribute to protect the natural resources avoiding environment pollution and harm on the public health.

In spite of the great effort made by the agrochemical industry in the last decades, developing new biodegradable herbicides, and producing new types of formulations such as concentrated in suspension, granules, soluble liquids, and so, there are still important problems derived from the immediate release of the active principles which compose them. Over 95% of sprayed conventional formulations of herbicides reach another destination than their target species, including nontarget species, air, water, bottom sediments, and food (Mogul et al., 1996). This situation leads to toxicity risks near the crops as a result of the high concentrations of active applied mass, as well as by the losses of active ingredients as a consequence of processes such as biodegradation, chemical degradation, photolysis, evaporation, surface runoff, and percolating groundwater, without considering the specific danger of these formulations to the applier by inhalation or absorption through skin.

The result of trying to compensate such losses is a tendency showing the use of excessive quantities of these dangerous chemical substances. This situation is an important economic loss and, at the same time, it is perilous for human health as well as for environment (Mogul et al., 1996).

2. Important

In the agrochemical industry the use of controlled release formulations (CRFs) could be a potential solution for low efficacy and environmental pollution derived from the use of conventional pesticide formulations (Scher, 1999). The aims of CRFs are to protect the supply of the agent, to allow the release of the agent to the target at a controlled rate, and to maintain its concentration in the system within the optimum limits, over a specified period of time, thereby providing great specificity and persistence (Flores Céspedes et al., 2007).

The parameters that affect the properties of CRFs depend on the nature and type of polymer used. Despite of several polymers employed in CRFs natural polymers such as starch, ethylcellulose, lignin, chitosan and alginate, are preferred to synthetic polymers because of their non-toxic, low cost, free availability, and biodegradability characteristics (Fernández Pérez et al., 2004; 2010).

Alginate gel has been applied to produce an effective CR carrier of drugs (Liew et al, 2006) and pesticides (Fernández Pérez et al., 2000; 2004; 2005) because it forms strong gels in aqueous media and is bioerodible.

In this chapter, discussion is focused on the use of several sorbents as modifying agents in the preparation of alginate-based controlled release formulations of herbicides. The addition of sorbents such as natural and activated clays, humic acid, activated carbon or mineral carbon in CRFs shows that the encapsulation efficiency increases and a better control on release profiles of active ingredient is reached. Therefore, they reduce the environmental pollution by eliminating the need for widespread distribution of a large amount of herbicides at one time.

3. Information

3.1 Herbicides

Herbicides decrease growth, seed production and competitiveness of susceptible weeds, are an integral part of management systems for agricultural practices. Systemic widely soil-applied herbicides have been selected in this study. IUPAC name, applications and action mode are summarized in Table 1.

Herbicide	IUPAC name	Applications/action mode
Diuron	[3-(3,4-dichlorophenyl)-1,1-dimethylurea]	Widely soil-applied herbicide for general weed control
Atrazine	[2-chloro-4-(ethylamine)-6-(isopropylamine)-s-triazine]	Selective triazine herbicide used to control broadleaf and grassy weeds. It is also used as a non-selective herbicide on non-cropped industrial lands and on fallow lands.
Isoproturon	[3-[4-isopropylphenyl)- 1,1-dimethylurea]	Selective, systemic herbicide used in the control of annual grasses and broadleaved.
Chloridazon	[5-amino-4-chloro-2-phenyl-3(2H)-pyridazinone]	Selective herbicides, to be absorbed predominantly by the roots, with translocation to all plant parts. They are used for the control of annual broad-leaved weeds in field and vegetable crops.
Metribuzin	[4-amino-3-methylthio-6-ter-butyl-1,2,4-triazin-5(4H)-one]	

Table 1. Selected herbicides.

Herbicides applied in the field have several fates, including uptake by plants, degradation in plants or soil, off-target, movement, immobilization and unintended deposition in the environment. In a well-planed management system, herbicide should effectively control weeds with little or no adverse environmental effects. Nevertheless, high amounts of herbicides used in agronomic practices lead to the existence of polluted groundwater sources, mainly through leaching process. The most important elements that condition a herbicide to be susceptible to leach, being a potential pollutant in groundwater, are the properties of the herbicide, taking sorption and persistence in soil as the most relevance ones. Soil persistence of an herbicide is measured through half-life time in soils (DT_{50}, time required for the herbicide to be degraded to a 50% of the initial amount of the herbicide in soil). Herbicides are ranked on the basis of DT_{50} as non-persistent when DT_{50} is under 30 days, slightly persistent when DT_{50} is between 30 and 100 days and persistent when DT_{50} is over 100 days. In this way, high half-life herbicides are mainly persistent and thus they possess great lixiviation potential to groundwaters. In addition, as previously mentioned, the sorption capacity of soil by a determined herbicide is an important factor that also affects potential to polluted groundwaters. The different models used to predict herbicide movement in soil take the constants describing the sorption process (K_d and K_f) as key input in the aforementioned models (Jury et al., 1987). Therefore a specific constant can be calculated for each herbicide, known as K_{oc}, which is a partition coefficient in relation to organic carbon in soil (O.C.) according to the equation.

$$K_{oc} = \frac{K_f}{O.C.} \times 100 \qquad (1)$$

Some physical-chemical properties of the herbicides selected in this study are shown in Table 2 (PPDB, 2009).

Herbicide	Molecular weight (g mol^{-1})	Watersolubility (mg L^{-1})	Log K_{ow}	K_{oc} (mL g^{-1})	Soil DT_{50} (days)	Vapor pressure (mPa)
Diuron	233.1	36.4 (25 °C)	2.84	1067	75.5	$1.10 \cdot 10^{-3}$
Atrazine	215.7	33 (22 °C)	2.50	100	75	$3.85 \cdot 10^{-2}$
Isoproturon	206.3	65 (22 °C)	2.50	122	12	$3.15 \cdot 10^{-3}$
Chloridazon	221.6	340 (20 °C)	1.19	199	31	< 0.01
Metribuzin	214.3	1050 (20 °C)	1.60	38	11.5	0.06

Table 2. Physical-chemical properties of selected herbicides.

To evaluate leaching potential (LP) of a non-ionic herbicide, the K_{oc} and DT_{50} values can be used in the following equation 2 (Guftason, 1989).

$$LP = log(DT_{50}\ soil) \times (4 - log(K_{oc})) \qquad (2)$$

According to the values obtained through the application of this equation, herbicides can be classified as it follows: Non-leachers (LP < 1.8), transitional/intermediate leachers (1.8 < LP < 2.8) and leachers (LP > 2.8). LP parameters for diuron, isoproturon, chloridazon, metribuzin and atrazine have been obtained using the aforementioned equation. These values were 1.83, 2.07, 2.54, 2.57 and 3.75 for diuron, isoproturon, chloridazon, metribuzin and atrazine, respectively. It is clear that atrazine is a leacher due to its high LP value and diuron, isoproturon, chloridazon, metribuzin are right between transitional/intermediate leachers. Consequently diuron, isoproturon, chloridazon, metribuzin have been found to leach, and atrazine has been widely confirmed by the widespread detections in surface water and groundwater (Buchanan et al., 2009).
The risk of pollution resulting from rapid runoff and leaching of these relatively highly soluble herbicides can be minimized through their application in controlled release formulations (CRFs).

3.2 Alginate

Alginates constitute a family of linear binary copolymers, consisting of (1 / 4) linked β-D-mannuronic acid (M) and α-L-guluronic acid (G) residues (see Fig. 1a and b). Chemical composition and sequence may vary widely between algae species and even between different parts of the algae and the time of year when it is harvested. (Gombotz & Wee, 1998). Alginate has been determined to be a true block copolymer with no regular repeating unit; i.e. the alginate molecule consisted of M-blocks, G-blocks and stretches of blocks of a predominantly alternating structure.

The capability of this copolymer to form stable biodegradable gels in the presence of divalent cations has been known and studied extensively since the seventies. These gelation properties can be attributed to the simultaneous binding of the divalent cations such as Ca^{2+} to different chains of α-l-guluronate blocks (G-blocks) (Figure 2). As a result of their configuration, these chains form electronegative cavities, capable of holding the cations via ionic interactions, resulting in the cross-linking of the chains into a structure resembling an "egg box" (Papageorgiou et al., 2006).

Alginate has been used successfully for many years in the food and beverage industry as a thickening agent, a gelling agent and a colloidal stabilizer. Alginate also has several unique properties that have enabled it to be used as a matrix for the entrapment and/or delivery of a variety of products on many fields including agriculture, food, cosmetics, construction, pharmacology, analytics, biotechnology, and medicine (Liew et al., 2006).

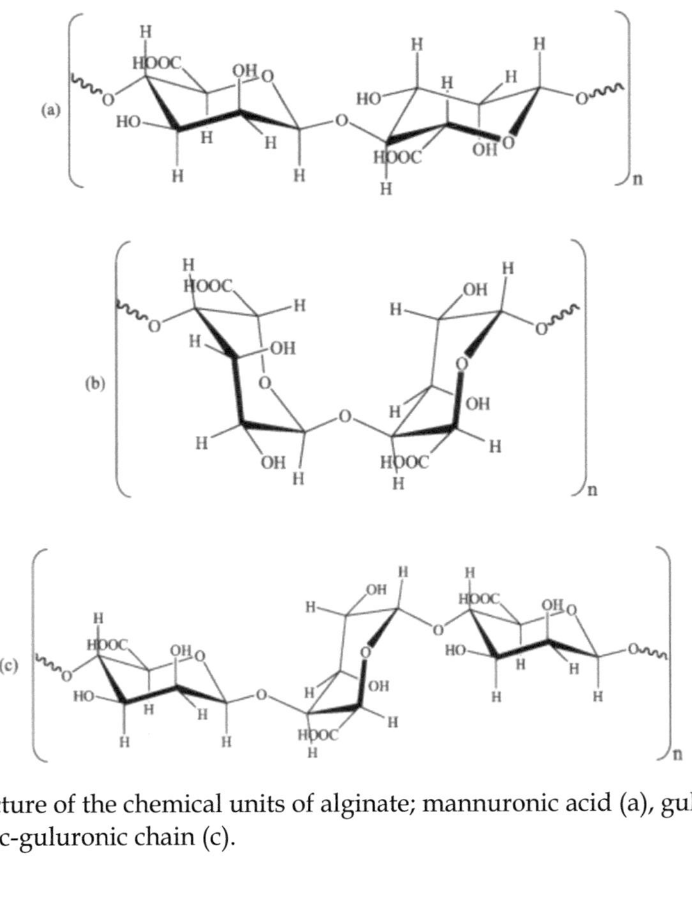

Fig. 1. The structure of the chemical units of alginate; mannuronic acid (a), guluronic acid (b) and mannuronic-guluronic chain (c).

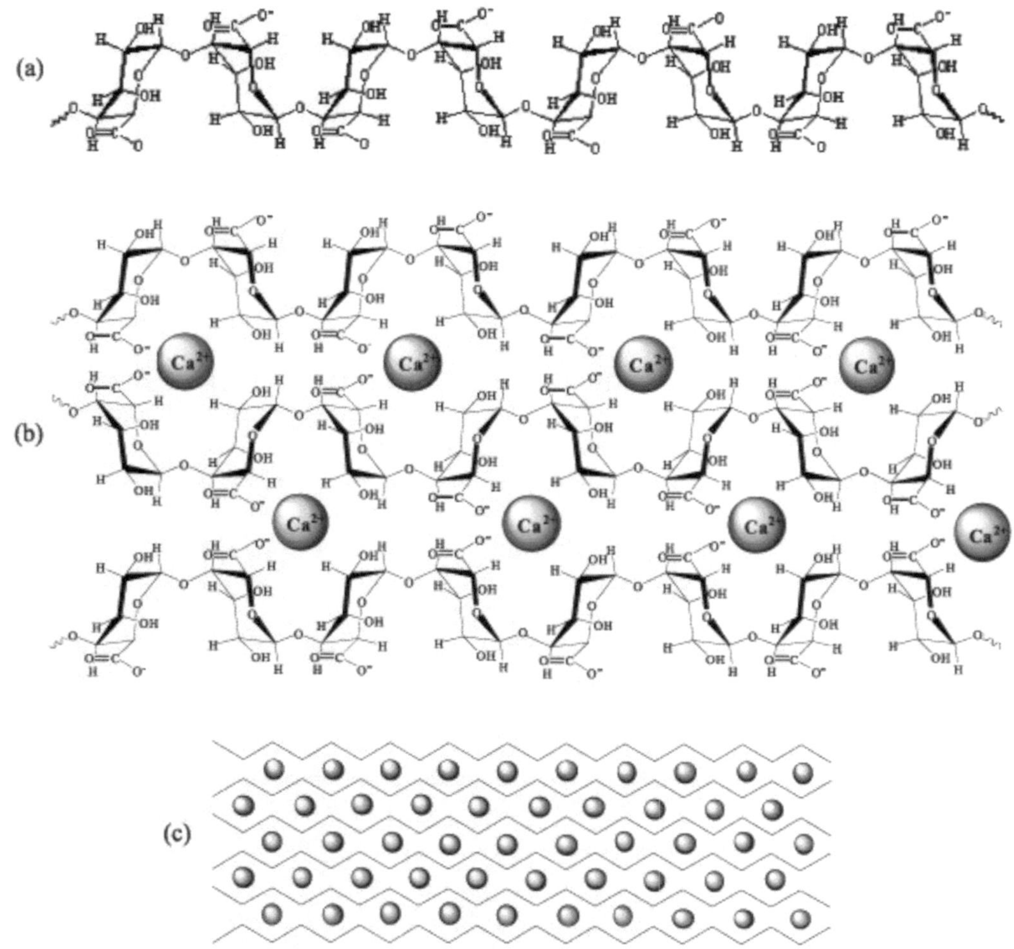

Fig. 2. Alginate gel formation, guluronic acid homopolymers in solution (a), binding between homopolymers chains by Ca^{2+} ions (b), gel formation by binging homopolymeric chains by Ca^{2+} ions (c).

3.3 Sorption and sorbents

Sorption is one of the main retention phenomena that determine the transport, transformation, and biological effects of herbicides in the environment. Besides, sorption is one of the methods which have been applied for removing organic and inorganic pollutants from waters using sorbents with highly developed surface properties. These types of sorbents can be also used as modifying agents in controlled release formulations to obtain different release profiles. The potential use of sorbents such as natural and acid-trated bentonite, a mineral carbon such as anthracite, and an activated carbon as modifying agents in alginate-based controlled release formulations of herbicides has been evaluated (Villafranca-Sánchez et al., 2000; Garrido-Herrera et al., 2006; Flores-Céspedes, et al., 2007).

Bentonites is a well-known silicate which include a significant group of natural materials mostly composed from fine-grained particles of minerals from the group dioctahedral smectite-montmorillonite. Viraraghavan and Kapoor (1994) noted that the abundance and low cost of bentonite make it a strong candidate as a sorbent for the removal of heavy metals and organic contaminants from wastewaters. The composition of bentonite varies, though it consists mainly of montmorillonite. Montmorillonite clays have the smallest crystals, the largest surface area and the highest cation exchange capacity. Thus montmorillonite clays would be expected to have the highest sorptive capacity. Crystallochemical formula of bentonite:

$$[Si^{4+}_{7.36}Al^{3+}_{0.64}]_T\,[Al^{3+}_{3.18}Fe^{3+}_{0.32}\,Mg^{2+}_{0.687}]_O O_{20}(OH)_4X^{+}_{0.76}$$

Previous studies have shown that acid treatments given to the clays produced an increase in the specific surface area of the samples (González-Pradas et al., 1991) and also improve their adsorptive characteristics (Jovanovic & Janackovic, 1991; González-Pradas et al., 1993).

Activated carbon in its broadest sense includes a wide range of processed amorphous carbon-based materials. It is not truly an amorphous material but has a microcrystalline structure. The activated carbons in general have a strongly developed internal surface and are usually characterized by a polydisperse porous structure consisting of pores of different sizes and shapes. These structural properties make activated carbons excellent adsorbents. The adsorption capacity of an activated carbon is determined by the physical or porous structure but strongly influenced by the chemical structure of the carbon surface. Carbon-oxygen surface groups are by far the most important surface groups that influence the surface characteristics such as the wettability, polarity, and acidity, and the physico-chemical properties such as catalytic, electrical, and chemical reactivity of these materials.

Anthracite is an intriguing feedstock for premium carbon materials; indeed, anthracite is essentially a carbon material. Most anthracites contain 92–98% carbon, virtually all of which is present as aromatic carbon in large polycyclic sheets. These sheets may contain 30 or more fused aromatic rings, resulting in extraordinary properties such as highly ordered carbon that also exhibits a high ultra-microporosity pore volume. The microporosity of anthracites makes them useful in, e.g., water filtration applications and sorption process.

	Surfacearea ($m^2\ g^{-1}$)	**MicroporeVolume ($mL\ g^{-1}$)**	**C.E.C meq/100g**	**pH**
B	65.46	15.04	65.63	9.02
$B_{0.5}$	182.97	42.03	76.88	1.87
$B_{1.0}$	298.84	68.65	73.75	1.84
$B_{2.5}$	484.78	111.36	54.38	2.19
An	23.40	5.37		8.87
C	858.07	197.11		9.11

B: natural bentonite; **$B_{0.5}$**: bentonite treated with H_2SO_4 0.5 M; **$B_{1.0}$**: bentonite treated with H_2SO_4 1.0 M; **$B_{2.5}$**: bentonite treated with H_2SO_4 2.5 M; **An**: anthracite; **C**: activated carbon

Table 3. Physical-chemical characteristics of sorbents.

3.4 Controlled release formulations of herbicides

3.4.1 Preparation of controlled release formulations (CRFs)

Herbicides CRFs were prepared based on the gelling properties of the alginate in the presence of divalent cations. It was made up of formulations in water containing different percentages of technical grade herbicide, sodium alginate (A), bentonite (B) or acid-treated bentonite ($B_{0.5}$, $B_{1.0}$,$B_{2.5}$), anthracite (An) and activated carbon (C) (shown in Table 4).

Herbicide	Formulation	A.I. (%)	A (%)	B (%)	$B_{0.5}$ (%)	$B_{1.0}$ (%)	$B_{2.5}$ (%)	An (%)	C (%)
Atrazine	AA	0.60	1.40						
	AAB	1.20	1.40	5.00					
	$AAB_{0.5}$	1.20	1.40		5.00				
	$AAB_{2.5}$	1.20	1.40				5.00		
Diuron	DA	0.60	1.40						
	DAB	1.20	1.40	5.00					
	$DAB_{0.5}$	1.20	1.40		5.00				
	$DAB_{2.5}$	1.20	1.40				5.00		
Isoproturon	IA	0.67	1.87						
	IAB	1.19	1.87	3.28					
	$IAB_{0.5}$	1.19	1.87		3.28				
	$IAB_{1.0}$	1.19	1.87			3.28			
	$IAB_{2.5}$	1.19	1.87				3.28		
Chloridazon	ClA	0.28	1.47						
	ClAB	1.17	1.39	4.75					
	ClAAn	1.17	1.39					4.75	
	ClAC	1.17	1.39						4.75
	$ClABC_{20}$	1.17	1.39	3.29					1.46
Metribuzin	MA	0.28	1.47						
	MAB	1.17	1.39	4.75					
	MAAn	1.17	1.39					4.75	
	MAC	1.17	1.39						4.75
	$MABC_{20}$	1.17	1.39	3.29					1.46

Table 4. Percentage (by weight) of the components of CR formulations containing herbicides (water to 100%).

The alginate-sorbent mixtures were vigorously stirred and then were dropwise added to a gellant bath of 0.25 M $CaCl_2$ using the apparatus described by Connick (1982). The resulting beads were filtered and dried first at room temperature and then in an oven to constant weight.

Characteristics of dry granules such as active ingredient content, calcium content, granule size, average weight and encapsulation efficiency are shown in Table 5.

Herbicide	Formulation	Active ingredient (%)	Ca^{2+} (%)	Average weight (mg/granule)	Average diameter (mm/granule)	Encapsulation Efficiency[a] (%)
Atrazine	AA	19.34	-	0.93	0.78	90.91
	AAB	11.60	10.90	2.49	1.52	98.74
	$AAB_{0.5}$	12.33	10.85	2.62	1.62	96.32
	$AAB_{2.5}$	13.01	10.54	2.92	1.82	97.30
Diuron	DA	21.53	-	0.90	1.08	98.85
	DAB	12.75	8.30	2.66	1.38	98.56
	$DAB_{0.5}$	13.09	9.25	2.77	1.54	98.33
	$DAB_{2.5}$	13.57	13.36	2.77	1.80	98.56
Isoproturon	IA	15.85	10.10	1.49	1.22	98.46
	IAB	13.77	20.50	2.34	1.58	97.21
	$IAB_{0.5}$	13.80	23.10	2.48	1.60	91.92
	$IAB_{1.0}$	13.65	22.70	2.66	1.73	98.04
	$IAB_{2.5}$	14.06	25.00	2.68	1.79	99.27
Chloridazon	ClA	7.18	16.37	0.94	0.99	74.91
	ClAB	12.13	3.99	2.28	1.17	92.66
	ClAAn	13.17	4.77	2.54	1.25	92.44
	ClAC	12.72	5.32	3.02	1.73	99.67
	$ClABC_{20}$	11.93	5.76	2.57	1.30	93.61
Metribuzin	MA	4.00	16.84	0.72	0.95	38.12
	MAB	9.78	6.93	2.22	1.17	75.98
	MAAn	11.31	10.55	2.16	1.23	79.86
	MAC	11.84	6.88	2.69	1.66	90.24
	$MABC_{20}$	10.95	7.08	2.02	1.19	80.14

[a] Encapsulation efficiency= (amount of herbicide in dry product / amount of herbicide in formulation processed) x 100

Table 5. Characteristics of CR granules (dry products) containing atrazine, diuron, isoproturon, chloridazon and metribuzin.

3.4.2 Water uptake test

Water uptake behavior, including swelling ratio and water uptake kinetics, is a very important property for an alginate-herbicide delivery vehicle, because it has a great influence on controlled delivery behaviour. The swelling behavior of the CR granules was studied by using the method of Franson and Peppas (1983). As an example, the water uptake of the granules containing diuron and chloridazon vs. time is shown in Figure 3.

The presence of a higher amount of microporosity in granules increases the amount of water uptake in formulations containing sorbents as modifying agents (Fernández Pérez et al., 2005). The extent of swelling of the anthracite, bentonites and activated carbon determines the volume occupied in the matrix and the area that can interact with the diffusing molecules of herbicide. Thus, water uptake and intensity of interactions of the herbicide with anthracite, bentonites and activated carbon could affect the diffusion through the granules, and hence, the release of the active ingredient. For granules without sorbents, it is necessary to consider that there are not additional bonds with other components of matrix on drying; thus expansion by swelling process is possible (Flores-Céspedes et al., 2007).

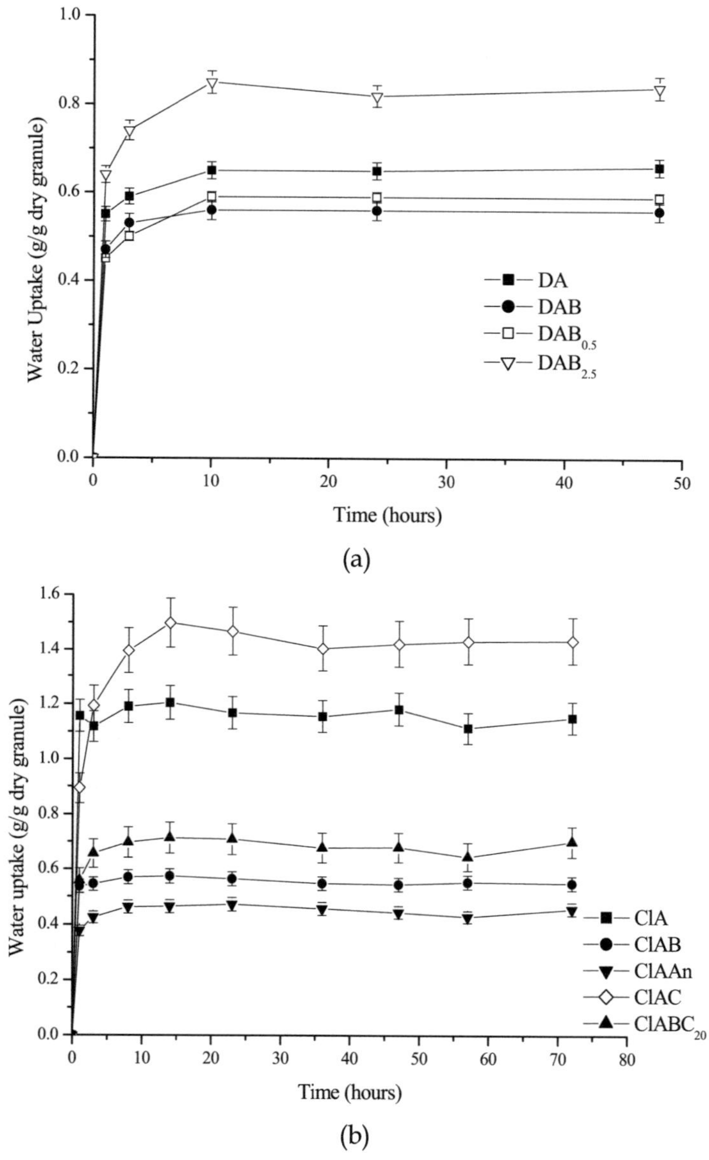

Fig. 3. Water uptake of diuron (a) and chloridazon (b) granules over time (error bars represent the standard deviation of three replicates).

3.4.3 Sorption studies

The effect of incorporation of different sorbents into alginate-based formulations on the rate of herbicide release can be evaluated through the study of the interactions between the active ingredients and the sorbents. Sorption isotherms of herbicides with natural bentonite, acid-treated bentonite, anthracite and activated carbon were compared using the K_f parameter of the Freundlich equation. The K_f and n values were calculated from the least-square method applied to the linear form of the Freundlich equation and their values are summarized in Table 6.

Herbicide	Sorbent	K_f ($mg\ kg^{-1}$)	n	r
Atrazine	B	0.28	1.15	0.998
	$B_{0.5}$	80.86	0.62	0.996
	$B_{2.5}$	46.34	0.54	0.994
Diuron	B	2.58	0.69	0.970
	$B_{0.5}$	11.08	0.50	0.987
	$B_{2.5}$	13.00	0.74	0.991
Isoproturon	B	0.48	0.72	0.979
	$B_{0.5}$	8.53	0.75	0.987
	$B_{1.0}$	8.50	0.78	0.998
	$B_{2.5}$	13.43	0.79	0.999
Chloridazon	B	4.09	0.92	0.999
	An	$1.41 \cdot 10^3$	0.26	0.960
	C	$2.03 \cdot 10^5$	0.09	0.974
Metribuzin	B	0.53	1.17	0.983
	An	$0.55 \cdot 10^3$	0.20	0.996
	C	$1.16 \cdot 10^5$	0.14	0.964

Table 6. Freundlich coefficients, K_f and n_f, for the sorption of herbicides on the bentonite and acid-treated bentonites, anthracite and activated carbon samples.

Sorption capacities (K_f) of the acid-treated bentonite samples were higher than that obtained with natural bentonite for diuron, isoproturon and atrazine. Sulfuric acid treatments to the bentonite resulted in higher efficiencies of sorption due to the increase of the specific surface of the acid treated samples. A slightly increase of K_f values for acid-treated bentonites is observed for diuron and isoproturon as acid treatment increases. For atrazine, the lower pH generated and the higher cation exchange capacity of $B_{0.5}$ sample compared with these observed for $B_{2.5}$ lead to a higher sorption capacity for the first. The decrease of the pH leads to a greater protonation of atrazine and then a greater sorption takes place by a cation exchange mechanism.

Sorption of chloridazon and metribuzin on bentonite produce the lowest K_f values and the higher was obtained for the activated carbon, being intermediate the values obtained of K_f for the sorption of the herbicides on anthracite. The K_f values for sorption of chloridazon and metribuzin on activated carbon are approximately 10^2 times higher than those obtained for sorption on anthracite and approximately 10^5 times higher than those obtained for sorption of herbicides on bentonite. The highest porosity and surface area of the activated carbon compared with those obtained for the anthracite and bentonite samples could explain the higher sorption capacity of activated carbon for the herbicides. The higher value of K_f obtained for anthracite, with regarding to bentonite, seems to be related to the higher affinity of the surface of anthracite by herbicides (Andrésen et al., 2004).

3.4.4 Water release kinetics

Water release experiments, made as a static immersion water test, let to obtain a deeper understanding of the release mechanism of the herbicides from CRFs.

Figures 4, 5 and 6 showed the cumulative release of atrazine, chloridazon and metribuzin, respectively, from alginate-bentonite based CR granules and the solubility profile for technical grade product.

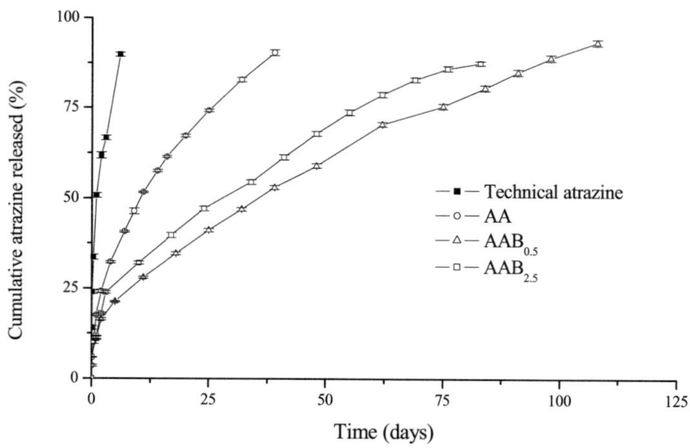

Fig. 4. Cumulative release of atrazine from alginate-bentonite CR granules into static water.

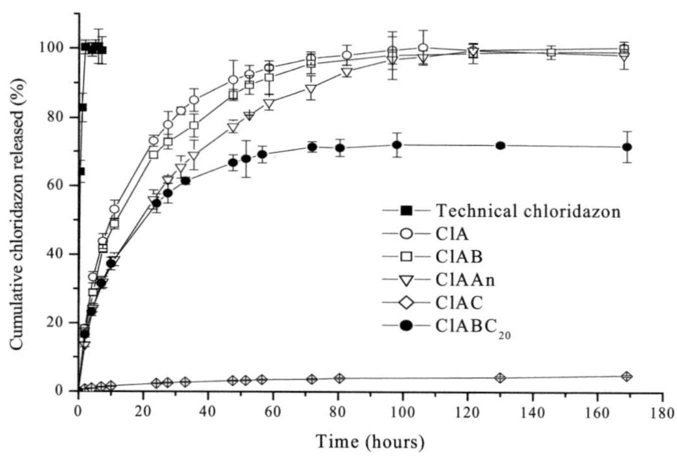

Fig. 5. Cumulative release of chloridazon from alginate CR granules into static water.

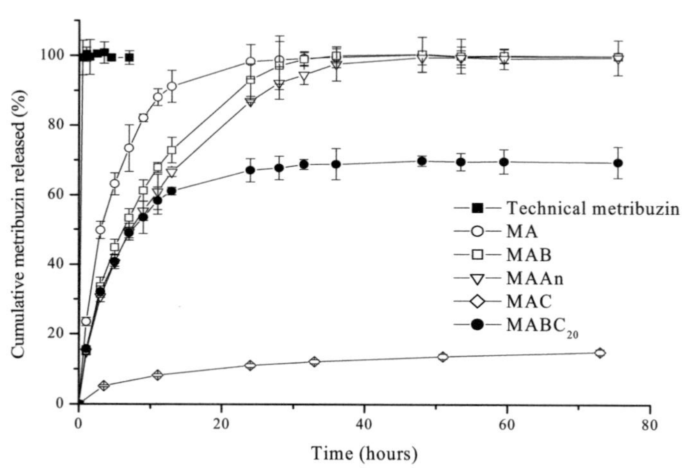

Fig. 6. Cumulative release of metribuzin from alginate CR granules in water.

The influence of the modifying agents such as natural and acid-treated bentonite, anthracite and activated carbon, appears clearly defined for all CR systems prepared. The presence of the sorbents produce a slower release rate in comparison to the CRFs prepared without these modifiers, being higher the decrease in the release rate of active ingredient for the sorbent with higher sorption capacity.

To evaluate the influence of the modifiers on the release rate of herbicides from CR granules, the release data must be analysed using diffusion-controlled models. The decline in the release of herbicide with time observed in Figures 4, 5 and 6 is probably due to an increase in the distance through which dissolved molecules have to diffuse as the depleted zone advances to the center of the matrix. In diffusion-controlled matrix systems this usually means that the release is proportional to the square root of time. Alginate-sorbents formulations could be described as systems containing finely divided solute particles, which are uniformly dispersed within the matrix phase. Higuchi (1963) originally analyzed analogous systems, such as drugs dispersed in a stationary matrix, e.g. semisolid ointment. The application of the model proposed by this author, involves that, (1) a pseudo-steady state exists, (2) the active ingredient particles are small compared to the average distance of diffusion, (3) the diffusion coefficient is constant, and (4) a perfect sink condition exists in the external media. In these conditions, the following equation was derived for spherical monolith systems, assuming Fickian diffusion ($y = kt^{1/2}$) (Higuchi, 1963):

$$\left[\frac{1-\left(1-\frac{M_t}{M_0}\right)^{2/3}-\frac{2}{3}\frac{M_t}{M_0}}{2}\right]^{1/2}=K_H t^{1/2} \qquad K_H=\left(\frac{1}{C_0 r^2}P\right)^{1/2} \tag{3}$$

M_t/M_0 is the fraction of active ingredient released at time t and K_H is a constant that depends on the radius of the sphere (r), the initial concentration of the active ingredient (C_0), and the permeability of the matrix (P). K_H values and correlation coefficients were obtained

Herbicide	Formulation	$K_H \times 10^2$ ($h^{-1/2}$)	r	$P \times 10^4$ ($mg\ h^{-1} mm^{-1}$)
	ClA	4.90	0.999	0.65
	ClAB	4.37	0.999	1.50
Chloridazon	ClAAn	3.78	0.999	1.30
	ClAC	0.09	0.985	0.00061
	$ClABC_{20}$	2.92	0.996	0.66
	MA	10.00	0.999	1.33
	MAB	7.34	0.998	3.36
Metribuzin	MAAn	6.36	0.998	2.71
	MAC	0.36	0.986	0.0081
	$MABC_{20}$	5.20	0.998	1.57

Table 6. Constants from fitting Higuchi equation to the release data of chloridazon and metribuzin in water and matrix permeability parameter.

by applying the model proposed by Higuchi to release data. As examples, the values for the formulations containing chloridazon and metribuzin are presented in Table 6 together with values of matrix permeability.

The values of P for the granules that contain modifiers, in general, decrease as the herbicide sorption capacity of the modifiers increases. This variation order for P values is in agreement with those observed when the release rate was evaluated from the CR granules. This coincidence suggests that the factors which affect and control the permeability and the release of active ingredient from the different granules studied should be the same, and therefore produced by the presence of modifying agents as anthracite, bentonite, acid-treated bentonite and activated carbon.

Ritger and Peppas equation was used to obtain the time taken for 50% of the herbicides to be released (T_{50}). This equation is used to relate the amount of active ingredient released as an exponential function of the release time (Ritger and Peppas, 1987).

$$\frac{M_t}{M_0} = Kt^n \tag{4}$$

M_t/M_0 is the percentage of active ingredient released at time t, K is a constant that incorporates characteristics of the macromolecular network system and the active ingredient, and n is a diffusional parameter, which is indicative of the transport mechanism.

The values of K and n and the correlations coefficients for atrazine, chloridazon and metribuzin are presented in Table 7. Values of n close to 0.43 are indicative of Fickian diffusion in spherical monolithic matrixes (Ritger and Peppas, 1987). Slightly different values than 0.43 of n could be explained by the complexity of the heterogeneous system involved together with the capacity of the sorbent to interact with the active ingredient. According to n values, the release of herbicide from the alginate-based CR formulations into water is due to diffusion where the sorption capacity of sorbent for herbicide and formulation permeability are the most influential factors.

T_{50} values, calculated from K and n constants, are presented in Table 7. The order of variation in T_{50} parameter for the different groups of prepared system is explained if we consider that the extent of the interactions between herbicides and modifying agents will affect the diffusing process and so on the release rate of herbicides from the alginate-based granules. This aspect has been quantified with sorption experiments of herbicides with bentonite, acid treated bentonite, anthracite and activated carbon samples. A higher sorption capacity result in a slower release of herbicide.

Taking into account the variation shown for T_{50} values by the different systems prepared, this study might be useful for selecting the most appropriate formulation, depending on the environmental factor that affect the herbicide mobility. This can be useful to prevent the polluting environmental risk derived from the use of herbicides mainly in soils with a low sorption capacity, where the herbicides shows a greater potential to cause groundwater pollution. In this sense, herbicides mobility from CRFs can be evaluates through leaching experiments in soil columns. Mobility of diuron, isoproturon and atrazine from an alginate-bentonite controlled release formulation in layered greenhouse soil (i.e., native soil, amended soil, peat, and sand) was researched. These studies indicate that the use of CR granules reduces the vertical mobility of herbicides into the soil layer columns and also diminish its presence in the leachate compared to the technical and commercial products (Fernández-Pérez et al., 1999; Villafranca-Sánchez et al., 2000).

Herbicide	Formulation	K (days)$^{-n}$	n	r	T_{50} (days)
Atrazine	AA	0.15	0.53	0.999*	9.69
	$AAB_{0.5}$	0.11	0.42	0.997*	36.78
	$AAB_{2.5}$	0.13	0.40	0.996*	29.01
		$K \times 10^2$ (h^{-n})	n	r	T_{50} (h)
Chloridazon	ClA	14.47	0.51	0.992	11.23
	ClAB	14.01	0.49	0.991	13.59
	ClAAn	10.75	0.51	0.996	19.81
	ClAC	0.50	0.46	0.987	$2 \cdot 10^4$
	$ClABC_{20}$	12.66	0.45	0.996	21.03
Metribuzin	MA	24.96	0.55	0.993	3.54
	MAB	16.65	0.58	0.992	6.71
	MAAn	15.91	0.56	0.996	7.77
	MAC	3.64	0.33	0.991	$2 \cdot 10^3$
	$MABC_{20}$	16.77	0.53	0.994	7.89

Table 7. Constants from fitting the empirical equation $M_t/M_o = Kt^n$ to the release data of herbicides in water.

4. Conclusions

Different sorbents have been applied as modifying agents to produce alginate-based CRFs of potential pollutant herbicides. The encapsulation efficiency increases and a better control on release profiles of active ingredient is reached with the use of these sorbents. The release of herbicides from the CRFs into water may be due to diffusion where the sorption capacities of sorbent for herbicide, water uptake and permeability of the formulations are the most influential factors. Concerning to the sorption capacities, the release is affected not only by the addition of natural bentonite and anthracite to the CRFs but also, and in a greater extent, by the addition of sorbents with higher sorption capacity such as acid-treated bentonites and activated carbon. The wide range of T_{50} values obtained was due to the different physical-chemical properties of the herbicides and also by the addition of sorbents. Taking into account the variation of T_{50} values, it is possible to select the most appropriate formulation, depending on the environmental factor that affected herbicide mobility. Therefore, the risk of pollution resulting from rapid runoff and leaching of relatively highly mobile herbicides can be minimized through their application in sorbents-alginate-based CRFs.

Significant potential exists to improve conventional methods of herbicide applications by the use of sorbent mixtures with different sorption capacity to modulate the release profiles of herbicides in alginate-based CRFs. Besides, low-cost lignin-sorbents mixtures coated with biodegradable polymers could be developed to obtain new CRFs of herbicides.

5. References

Andrésen, J.M.; Burgess, C. E.; Pappano, P.J. & Schobert, H. H. (2004). New directions for non-fuel uses of anthracites. *Fuel Process. Technol.* 85:1373-1392. ISSN:0378-3820.

Buchanan, I.; Liang, H. C.; Khan, W.; Liu, Z.; Singh, R.; Ikehata, K. & Chemel-Ayala, P. (2009). Pesticides and herbicides. *Water Enviroment Research*. 81:1731-1816. ISSN: 1554-7531.

Connick, W.J. Jr. (1982). Controlled release of the herbicides 2,4-D and dichlobenil from alginate gels. *J. Appl. Polym. Sci.* 27: 3341-3348. ISSN: 0021-8995.

Fernández-Pérez, M.; Flores-Céspedes, F.; González-Pradas, E.; Villafranca-Sánchez, M.; Pérez-García, S. & Garrido-Herrera, F.J. (2004). Use of activated bentonites in controlled release formulations of atrazine. *J. Agr. Food Chem.* 52:3888-3893. ISSN: 1520-5118.

Fernández-Pérez, M.; González-Pradas, E.; Villafranca-Sánchez, M.; Flores-Céspedes, F. & Ureña-Amate, M. D. (2000). Bentonite and humic acid as modifying agents in controlled release formulations of diuron and atrazine. *J. Environ. Qual.* 29:304-3010. ISSN: 0047-2425.

Fernández-Pérez, M.; Villafranca-Sánchez, M.; Flores-Céspedes, F.; Pérez-García, S. & Daza-Férnández, I. (2010). Prevention of chloridazon and metribuzin pollution using lignin-based formulations. *Environ. Pollut.* 158:1412-1419. ISSN: 0269-7491.

Fernández-Pérez, M.; Villafranca-Sánchez, M.; Flores-Céspedes, F.; Garrido-Herrera, F.J. & Pérez-García, S. (2005). Use of bentonite and activated carbón in controlled release formulations of carbofuran. *J. Agric. Food Chem.* 53: 6697-6703.ISSN: 1520-5118.

Fernández-Pérez, M.; Villafranca-Sánchez, M.; González-Pradas, E. & Flores-Céspedes, F. (1999). Controlled Release of Diuron from an Alginate-Bentonite Formulation: Water Release Kinetics and Soil Mobility Study. *J. Agric. Food Chem.* 47:791-798. ISSN: 1520-5118.

Flores-Céspedes, F.; Villafranca-Sánchez, M.; Pérez-García, S. & Fernández-Pérez, M. (2007). Modifying sorbents in controlled release formulations to prevent herbicides pollution. *Chemosphere* 69:785-794. ISSN: 0045-6535.

Franson, N. M. & Peppas, N.A. (1983) Influence of copolymer composition on non-fickian water transport through glassy copolymers. *J. Appl. Polym. Sci.* 28:1299-1310. ISSN: 0021-8895.

Garrido-Herrera, F. J.; González-Pradas, E. & Fernández-Pérez, M. (2006). Controlled release of isoproturon, imidacloprid, and cyromazine from alginate-bentonite-activated carbon formulations. *J. Agric. Food Chem.* 54:10053-10060. ISSN: 1520-5118.

Gombotz, W. R. & Wee, S. F. (1998) Protein release from alginate matrices. *Advances Drug Delivery Reviews*. 31:267-285. ISSN: 0169-409X

González-Pradas, E.; Villafranca Sánchez, M.; Socías-Viciana, M.; Castillo-Sánchez, J. & Fernández-Pérez, M. (1993). Removal of 3-(3,4-dichlorophenyl)-1,1dimethyl urea from aqueous solution by natural and activated Bentonite. *J. Chem. Tech. Biotechnol.* 56: 67-71. ISSN: 1097-4660.

González-Pradas, E.; Villafranca-Sánchez, M.; Villafranca-Sánchez, E.; Del Rey-Bueno, F.; Valverde-García, A. & García-Rodríguez, A. (1991). Evolution of surface properties in bentonites as a function of acid an heat treatments. *J. Chem. Tech. Biothechnol.* 52: 211-248. ISSN: 1097-4660.

Gustafson, D.I. (1989) Groundwater ubiquity score: a simple method for assessing pesticide leachability. *Environ. Toxic. Chem.* 8:339-357. ISSN: 0730-7268.

Higuchi, T. (1963). Mechanism of sustained-action medication. Theoretical analysis of rate of release of solid drugs dispersed in solid matrices. *J. Pharm. Sci.* 52:1145-1149.

Jovanovic, N. & Janackovic, J. (1991) Pore structure and adsorption properties of an acid-activated bentonite. *Appl. Clay Sci.* 6: 59-68. ISSN: 0169-1317

Jury, W.A.; Focht, D. D. & Farmer, W. J. (1987). Evaluation of pesticide groundwater potential from standard indices of soil chemical adsorption and biodegradation. *J. Environ. Qual.* 16:422-428. ISSN:0047-2425.

Liew, C.V.; Chan, L.W.; Ching, A.L. & Heng, P.W.S. (2006). Evaluation of sodium alginate as drug release modifier in matrix tablets. *Inter. J. Pharm.* 309: 25-37. ISSN: 0378-5173

Mogul, M.G.; Akin, H.; Hasirci, N.; Trantolo, D.J.; Gresser, J.D. & Wise, D.L. (1996). Controlled release of biologically active agents for purposes of agricultural crop management. *Resources, Conservation and Recycling* 16:289-320. ISSN: 0921-3449.

Papageorgiou, S. K.; Katsaros, F. K.; Kouvelos, E. P.; Nolan, J. W.; Le Deit, H. & Kanellopoulos, N. K. (2006). Heavy metal sorption by calcium alginate beads from *Laminariadigitata. J. Hazard. Mater.* 137: 1765–1772, ISSN: 0304-3894.

PPDB (2009). The Pesticide Properties Database (PPDB) developed by the Agriculture & Environment Research Unit (AERU), University of Hertfordshire, funded by UK national sources and the EU-funded FOOTPRINT project (FP6-SSP-022704).

Ritger, P.L. & Peppas, N.A. (1987). A simple equation for description of solute release I. Fickian and non-Fickian release from non-swellable devices in the form of slabs, spheres, cylinders or discs. *J. Control. Rel.* 5:23-36. ISSN: 0168-3659.

Scher, H.B., 1999. *Controlled-Release Delivery Systems for Pesticides*. Marcel Dekker Inc., ISBN: NY. ISBN-13: 978-0824719883.

Villafranca-Sánchez, M.; González-Pradas, E.; Fernández-Pérez, M.; Martínez-López, F.; Flores-Céspedes, F. & Ureña-Amate, M.D. (2000). Controlled release of isoproturon from an alginate-bentonite formulation: water release kinetics and soil mobility. *Pest. Manag. Sci.* 56: 749-756. ISSN: 1526-4998.

Viraraghavan, T. & Kapoor A. (1994). Adsorption of mercury from wastewater by bentonite *Appl. Clay Sci.* 9: 31-49. ISSN: 0169-1317.

Part 3

Herbicides in Soil

8

The Fate of Herbicides in Soil

Sonia Blasioli, Ilaria Braschi and Carlo E. Gessa
Department of Agroenvironmental Sciences and Technologies, University of Bologna
Italy

1. Introduction

The agrochemical spreading is a common and essential agricoltural practice to obtain high quality, large harvests.
Agrochemicals are classified according to the target organisms designed to be controlled (insects, weeds, fungi). Of all the target organisms, weeds cause by far the greatest economic loss due to their interference in crop production. It is not surprising therefore, that herbicides are the most common class of agrochemicals in the world (48% of the total expenditure) and in Europe (43%) outstripping fungicides (35%) and insecticides (14%). Europe, Asia, and the United States are the largest consumers of agrochemicals; in Europe, France has the biggest agricoltural areas, and is the highest-ranking country for pesticide consumption followed by Germany and Italy (see http://www.croplife.org/ and http://www.ecpa.be).
Bad agricoltural practice and accidental spreading of high doses of agrochemicals can determine toxic effects in humans and the environment; pesticides can accumulate in organisms and achieve critical concentrations for the human and ecosystem health.
Agrochemicals were used for the treatment of human diseases like malaria and typhus. However, high doses of some pesticides can be highly toxic to humans. Laboratory experiments have shown that the administration of high doses of pesticides to animals can cause cancer, mutagenesis, and even death; moreover, exposure to low doses can cause skin irritation and breathing problems. In the "infamous" case of DDT, for instance, which was introduced onto the market in 1940 for the malaria and typhus control, the central nervous system was attacked causing loss of memory, tremblings, and personality changes. Paraquat, a dipyridylic herbicide, is an extremely toxic systemic pesticide; it can enter in the body by inhalation, ingestion or direct contact. It is expecially toxic to the lungs, but can cause gastrointestinal apparatus, kidney, liver, and heart disorders and the weakening of other organs with vital functions.
Plants that are sensitive to pesticide molecules may show signs of growth inhibition and loss in biomass even as far as necrosis, but may be able to develop resistance to certain pesticides (see http://www.weedscience.org; Yuan et al, 2007). Agrochemicals may also have a toxic effect on nontarget plants (Madhun & Freed, 1990) when transported away from the treated site (soluble herbicides or surface erosion).
Soil and aquatic ecosystems contain a multitude of microorganisms. After pesticide spreading, microbic activity may be reduced. However, in some situations an enhancement in microbial activity may occur (Lewis et al., 1978; Pozo et al., 1994).

The leaching of soluble and highly mobile molecules, wilful discharge in underground wells and accidental dumping in water bodies contribute to water contamination. Carabias Martinez et al. (2000) monitored the concentration of fifteen herbicides selected owing to their frequency of use, the amounts used, their toxicity and their persistence in river basins in the provinces of Zamora and Salamanca (Spain). After six months, the presence of six out of the fifteen herbicides monitored, was detected at levels ranging from the detection limit to 1.2 µg/L. The presence of these herbicides was related to agricultural activities as well as the kind of crop and its treatment period.

The prediction of herbicide movement and fate in soils represents an important strategy in limiting their environmental impact (Figure 1). Physical, chemical, and biological processes regulate herbicide mobility and degradation in soil: rainfall and irrigation water can move herbicides along the soil profile; sites negatively charged of clay mineral surfaces and/or organic matter can adsorb herbicides in their cationic form at soil pH; microbial activity can promote herbicide transformation. Different transfer and degradation processes which control the movement and the fate of pesticides in the environment are reported in the Table 1. Except physical processes, other processes depend on soil characteristics.

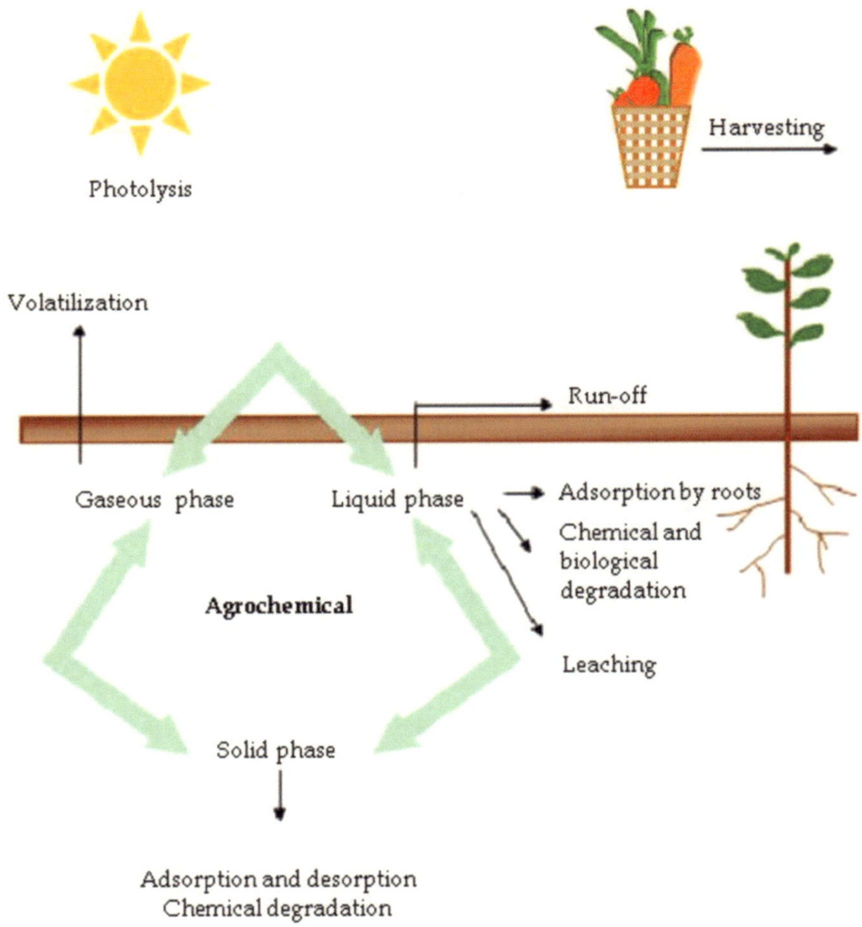

Fig. 1. The fate of agrochemicals in the environment.

Process	Consequence	Factors
Movement (processes that relocate agrochemicals without changing their structure)		
Physical drift	Movement due to wind action	Wind speed, drop sizes
Volatilization	Loss due to evaporation from soils, plants, and waters	Vapor pressure, wind speed, temperature
Adsorption	Removal due to interaction with soils, plants, and sediments	Clay content, organic matter, moisture
Absorption	Uptake by plant roots or animal ingestion	Cell membrane transport, contact time
Leaching	Horizontal and vertical movement downward through the soil	Water content, soil texture, clay and organic matter contents
Erosion	Wind and water action	Rainfall, wind speed, sizes of clay and organic matter
Degradation (processes that modify the chemical structure)		
Photochemical	Assorption of sunlight (i.e., ultraviolet radiation)	Chemical structure, intensity and duration of exposure
Microbial	Degradation by microorganisms	Environmental factors (pH, moisture, temperature) organic matter content
Chemical	Hydrolysis and redox reactions	pH modifications, same factors as microbial degradation
Metabolism	Adsorption by plants or animals	Adsorption capacity, metabolism, interactions with microorganisms

Table 1. Movement and degradation processes of agrochemicals in the environment (Pierzynski et al., 2000).

2. Chemico-physical parameters affecting the fate of herbicides in soil

The fate of herbicides such as that of any organic molecule released into the environment is determined by their chemico-physical characteristics.

Solubility. The solubility of an herbicide is important in predicting its behaviour in water and its mobility in soil. Agrochemical water solubility is a function of temperature, pH, and ionic strength and is affected by the presence of other organic substances such as the dissolved organic matter (DOM) (Pierzynsky et al., 2000). Two methods are frequently used to estimate organic molecule solubility based on i) chemical structure (*Kps*) and ii) the *n*-octanol/water partition coefficient (K_{OW}). n-Octanol/water coefficients are determined by the following equation which highlights that there is an inverse relationship between solubility and K_{OW}:

$$K_{OW} = \frac{concentration\ of\ organic\ chemical\ in\ octanol \left(\frac{mg}{L}\right)}{concentration\ of\ organic\ chemical\ in\ water \left(\frac{mg}{L}\right)} \quad (1)$$

Persistence. The persistence of an herbicide is defined as the time in which the molecule remains in the soil and is usually expressed as half-live. Half-live ($t_{1/2}$) refers to the time required to halve the organic molecule concentration compared with its initial level.

Half-life values are important in understanding the potential environmental impact of a chemical; in fact, a molecule which degrades quickly, has a low $t_{1/2}$ value and thus the impact of this species on the environment is reduced if the degradation products are harmless. On the contrary, the environmental impact of species with a high $t_{1/2}$ value can be substantial even if the molecule is only moderately toxic.

The prediction of herbicide half-life and thus, its persistence in the environment is an important parameter in agronomic practice because it supplies information on the residual activity of agrochemicals which could cause damages to the successive crops.

For a first order reaction, the half-life is determined by the following equation:

$$t_{1/2} = \frac{0.693}{k} \quad (2)$$

where k is the kinetic constant of the degradation reaction involving the agrochemical. *Volatilization.* Volatilization of organic molecules is responsible for the transfer of molecules from aquatic and soil environments into the atmosphere. As with the solubility, it is important to know the contribution of agrochemical volatilization in predicting its residual amount and thus, its persistence in the environment.

The volatilization of herbicides from waters depends on the chemical and physical properties of the molecules in question (e.g., vapour pressure and solubility), their interaction with suspended materials and sediments, the physical properties of the water bodies (depth, turbulence, and velocity) and any water-atmosphere interface properties.

The solubility of a gas dissolved in an aqueous solution is well defined by the Henry constant, calculated using the homonymous equation:

$$K_H = \frac{P_{gas}}{C_{aq}} \quad (3)$$

where K_H is the Henry constant, P_{gas} is the gas partial pressure and C_{aq} is its concentration in the aqueous phase. For high K_H values, the molecule prefers to leave the liquid phase in order to pass into the atmosphere. This constant is useful to describe the agrochemical fugacity from a water body but also from soil solid components which are always surrounded by water in adsorbed form.

The rate of volatilization can be indicated as half-life, which is the time required to halve the organic molecule concentration in water compared with its initial value. The volatilization half-lives of different molecules are reported in the table 2.

Factors that influence the volatility of organic molecules from soils include the chemical and physiochemical properties of the pollutant (i.e., vapour pressure, solubility, the structure and nature of the functional groups, and adsorption-desorption characteristics), concentration, soil properties (soil moisture content, porosity, density, and organic matter and clay contents) and environmental factors like temperature, humidity, and wind speed.

Volatilization	Agrochemical	$t_{1/2}$
Low	Dieldrin	327 d
	3-bromo-1-propanol	390 d
Medium	Phenantrene	31 h
	Pentachlorophenol	17 d
	DDT	45 h
	Aldrin	68 h
	Lindane	115 d
High	Benzene	2.7 h
	Toluene	2.9 h
	O-xylene	3.2 h
	Carbon tetrachloride	3.7 h

Table 2. Volatilization rates of some organic molecules (Pierzynsky et al., 2000).

Photolysis. Photochemical reactions involve sunlight radiation and play an important role in the degradation of molecules on soil surfaces and in aquatic environments. Photolysis in the soil is difficult to determine because of the heterogeneous nature of soils and low sunlight penetration. Nevertheless, it is an important herbicide degradation process in soil since it is always active.

In water as well as in soil, photolysis can occur either by direct or indirect processes. In direct photolysis, sunlight is absorbed directly by organic molecules which alter its chemical structure. The indirect process occurs in the presence of natural photosensitive species such as nitrates or humic acids which can absorb the light and subsequently transfer excitation energy to the organic molecule.

Biodegradation. Herbicide biodegradation is due to microorganism activity and is a function of those properties which influence microbial activity such as temperature and pH: a temperature or pH decrease slows down the biotic degradation rate since under such conditions microbial activity is reduced. This could explain the presence of certain molecules such as antibiotics, in the deeper layers of soils and waters (Gavalchin & Katz, 1994; Van Dijk & Keukens, 2000).

Adsorption-desorption. The ability of herbicides to adsorb on soils and sediments and their tendency to desorb are the most important factors affecting soil and water contamination. Adsorption depends on both molecule and soil chemico-physical properties. In soil, the surfaces responsible for adsorption are colloidal particles and among these, organic matter and clays. Organic matter, due to its chemical affinity with agrochemical molecules, has the greatest adsorption strength towards these species; high surface area and the interlayer charge of clays, such as expandable phyllosilicates, make these sorbents good for organic molecules.

Adsorption on clays or organic matter can occur with the following interactions: van der Waals forces, hydrogen bondings, dipole-dipole interactions, ionic exchange, covalent bondings, protonation, ligand exchange, cationic and H_2O bridging, and/or hydrophobic interactions.
Cationic species adsorb on soil by electrostatic attraction while anionic molecules can adsorb on positively charged soil colloids, even if the adsorption of negative species is less strong than on the negatively charged clay surfaces. In acidic soils, herbicides which have amino groups can protonate to quaternary ammonium ions ($-NH_3^+$) and form H-bonds with the oxygen atoms of the phyllosilicate surfaces or with nitrogen atoms of organic matter; molecules which have acidic functional groups remain in a neutral form (COOH).
Nonpolar molecules can interact with the hydrophobic moieties of soil organic matter: this hydrophobic bond is responsible for the strong adsorption of DDT and organochloride insecticides on soil organic matter.
The strength of adsorption affects molecule mobility along the soil profile and thus, its bioactivity, persistence, biodegradation, leaching, and the volatilization process. The adsorption of an agrochemical onto the soil components can be considered as the first step towards its chemical degradation.
Organic molecule adsorption modeling by soils is frequently done using adsorption isotherms.
Adsorption isotherms are built by measuring the residual concentrations of pollutant in aqueous solution at the equilibrium point, after the adsorption on soil of different initial concentrations. For each concentration point, the adsorbed molecule concentrations are determined by the difference between initial and equilibrium concentrations. Adsorption data are commonly fitted using two different models described by the Langmuir and Freundlich equations.
The Langmuir equation is:

$$\frac{x}{m} = \frac{K_l bC}{1 + K_l C} \qquad (4)$$

where x/m is the mass of organic molecule adsorbed per unit of soil weight; C is the equilibrium concentration of the organic molecule; K_l is the Langmuir constant that is related to binding strength. The linear form of the Langmuir equation is:

$$\frac{C}{x/m} = \frac{1}{K_l b} + \frac{C}{b} \qquad (5)$$

If a plot of C/(x/m) vs C is a straight line, then the adsorption data satisfy the Langmuir equation, and b can be calculated from the slope and K_l from the intercept.
The Freundlich equation is:

$$\frac{x}{m} = K_f C^{1/n} \qquad (6)$$

where x/m and C are the same as above; K_f and n are empirical constants.
The K_f value is a measure of the extent of adsorption whereas the 1/n value indicates the affinity of organic molecule for the sorbent surface. If the 1/n value is lower than 1, there is high affinity between the adsorbate and the adsorbent. If 1/n is equal to 1, the solute is equally

distributed between the solution and adsorbent surface. If the 1/n value is higher than 1, the adsorption is called "cooperative" because there is cooperation between the adsorbed molecules and the new molecules approaching the surface to promote the adsorption.
The linear form of the Freundlich equation is obtained by logarithmic transformation:

$$\log\left(\frac{x}{m}\right) = \log K_f + \frac{1}{n}\log C \tag{7}$$

A plot of log (*x/m*) vs log C should produce a straight line, with 1/n being equal to the slope and K_f the intercept.
For many organic molecules, especially nonpolar species, adsorption can be constant, that is the adsorbed concentration is proportional to the equilibrium concentration. The Freundlich equation can be simplified:

$$K_d = \frac{x/m}{C} \tag{8}$$

where K_d is the distribution constant. The adsorption constant can be normalized to the organic carbon content of the soil (f_{OC} = OC%/100): the new constant is know as K_{OC} which is independent of soil type and specific to a given pollutant. The constant is calculated using the following equation:

$$K_{OC} = \frac{K_d}{f_{OC}} \tag{9}$$

Adsorption isotherms. Adsorption isotherms of organic molecules are divided into four classes, according to the nature of the initial curve portion (Giles et al., 1960). The four classes are know as H (high affinity), L (Langmuir type), C (constant partition), and S (sigmoidal or with an "s" form) isotherms (Figure 2). The L curves are the best known: the initial curvature shows that as more sites in the substrate are filled, it becomes increasingly difficult for solute to find an available vacant site. The H isotherm is a special case of L curve, where the solute has a high affinity for the surface especially at low concentrations. The C curves are characterized by the constant partition of solute between the liquid and solid phase; the constant partition is independent of concentration right up to the maximum possible adsorption, where an abrupt change in the slope to a horizontal plateau occurs. The initial part of the S curves describes contrary conditions in comparison with the other isotherms: the more solute has already been adsorbed, the easier it is for additional amounts to become fixed. This implies a side-by-side association between adsorbed molecules, helping to hold them to the surface. This has been called "cooperative adsorption".
Abiotic and biotic transformations. Both abiotic and biotic reactions are responsible for the transformation of herbicides in soils and waters. One of the two processes may be dominant, but usually both of these participate simultaneously in molecule degradation. The principal abiotic reactions that occur in water are hydrolysis, oxidation-reduction, and photolysis; in sediments, hydrolysis and redox reactions may prevail. Redox reactions in aquatic environments can be mediated by direct or indirect photolysis or catalyzed by metal species. In soil, abiotic reactions occur in the liquid phase (i.e. soil solution) and at the solid-liquid interface. In soil solution, hydrolysis and redox reactions are the most common abiotic transformations; these reactions are catalyzed by clays, organic matter and metal oxides.

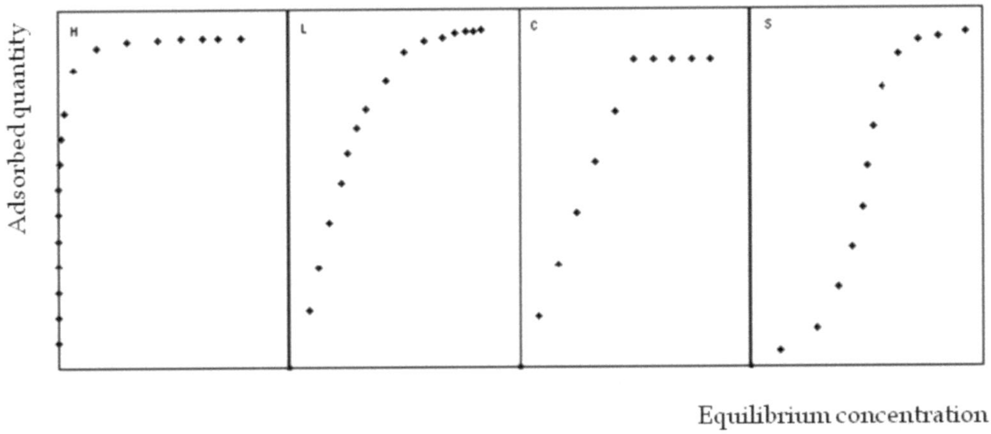

Fig. 2. The four main classes of isotherms. From left to right: high affinity (H), Langmuir (L), constant partition (C) and S curves.

Biotic reactions are classified as: i) biodegradation (contaminants are used as a substrate for microorganism growth); ii) cometabolism (contaminants are transformed by metabolic reactions without being used as an energy source); iii) accumulation (contaminants are accumulated in microorganisms); iv) polymerization or conjugation (contaminants are bonded to other organic molecules); v) secondary effects of microbial activity (contaminants are transformed by secondary microbial effects such as pH and redox changes) (Bollag & Liu, 1990).

Biodegradation is considered the principal mechanism for the conversion of organic molecules into CO_2, H_2O and mineral salts.

Although these reactions are mediated by microorganisms, abiotic processes are also involved, especially in transformations related to categories iv) and v).

Braschi et al. (2000) have investigated the degradation of primisulfuron, a sulfonylurea herbicide, in microbial communities enriched with soils polluted by herbicide. The authors find that the degradation reaction of primisulfuron firstly occurs by means of hydrolysis and photolysis processes and that the role of microorganisms subsequently becomes important in the degradation of the herbicide metabolites.

3. Soil inorganic phase: clay minerals

Soil solid phases are almost totally characterized by inorganic components (fragments of rocks, primary and secondary minerals, amorphous materials); the organic component is only a small fraction.

Minerals are the most diffuse inorganic species in the lithosphere. From a chemical point of view, they are classified as: i) silicates formed by oxygen and silicon and ii) nonsilicates, such as oxides, carbonates, phosphates, sulphates.

Silicon tetrahedron is the building unit of silicates: different classes of silicates are obtained by the polymerization of building units.

Layered aluminosilicate minerals, known as clay minerals, have a profound influence on many soil chemical reactions because of their high active surface area. They have regular layers of tetrahedral and octahedral sheets: tetrahedral sheets are comprised of silicon and oxygen atoms with three out of every four oxygen atoms shared between adjacent tetrahedra. There are two types of octahedral sheets: dioctahedral and trioctahedral. Dioctahedral sheets have two out of every three octahedral sites occupied, most often by the trivalent Al cation. Trioctahedral sheets have all octahedral sites occupied by divalent cations, which are commonly Mg ions. Clays have structures that are either 1:1, 2:1, or 2:1:1 layers of tetrahedral and octahedral sheets. 1:1 clay minerals have one tetrahedral and one octahedral sheet held together by sharing an apical tetrahedral oxygen. 2:1 clay minerals have an octahedral sheet posed between two tetrahedral sheets. 2:1:1 layered clays are similar to 2:1 clays with an additional dioctahedral or trioctahedral sheet between the 2:1 layers (Pierzynski et al., 2000).

The expansion property after water adsorption is typical to the 2:1 arrangement. In 1:1 clays, the oxygens of octahedral sheet bond with the oxygens of the tetrahedral sheet by means of H-bondings: this arrangement does not permit good expansion in water for this clay. In 2:1 clays, no hydrogen bonds are formed between the oxygens of tetrahedral and octahedral sheets; a weak repulsion develops and distributes over the surface, allowing the clay minerals to expand easily in water (Figure 3).

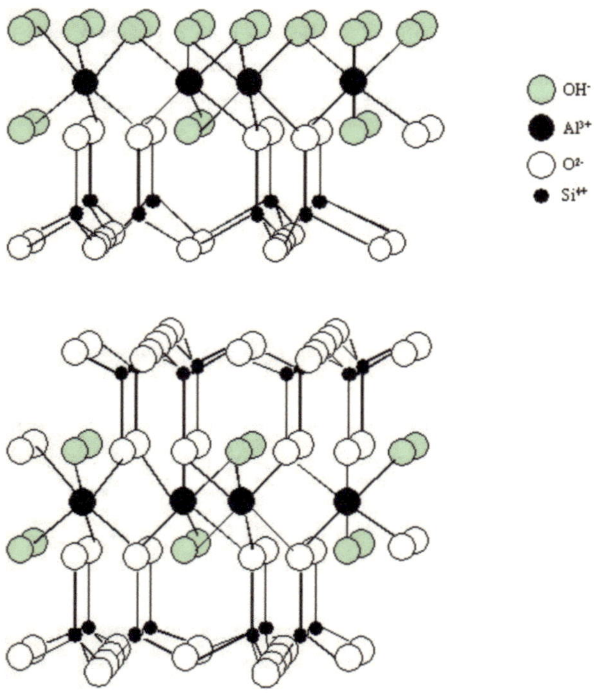

Fig. 3. Different arrangements of clay minerals: 1:1 on the top, 2:1 on the bottom.

Clay minerals contain active sites for adsorption which are localized in the tetrahedral sheet, and are formed by siloxanes (siloxane cavity). This cavity is bordered by six sets of lone-pair electrons (from the oxygen atoms) that give a nucleophilic character to the surface: polar or positively charged species can interact with this highly negative surface (Sposito, 1984). Water can occupy the siloxane cavity by pointing the hydroxylic group inside the cavity; the polarization inducted by the surface, makes the water molecules strongly polarized and then reduces their mobility. The siloxane cavity extends along the whole tetrahedral sheet giving high adsorption properties to clays.

Clays can have a permanent negative charge which comes from *isomorphous substitutions* and a variable pH dependent charge. Isomorphous substitution occurs when an element substitutes for another in the mineral structure, such as Al^{3+} substituting Si^{4+}. If an element of a lower charge is substituted for an element of a higher charge, a permanent negative charge develops in the clay mineral: the net negative charge electrostatically attracts positive and polar species. The variable charge is due to the presence of surface hydroxylic groups that can lose or accept protons as a function of soil solution pH.
Cations in the soil solution are bonded to the surface of clay minerals by electrostatic interactions and can return in solution by the substitution of other cations or by dilution. The most representative exchange cations are K^+, Ca^{2+}, Mg^{2+}, and Na^+. Other cations, such as organic cations, are bonded onto clay minerals as a function of their affinity for the surface and their concentration in the soil solution.
Cations adsorbed onto clay surfaces are surrounded by hydration water. This water is more acidic than the water in the soil solution because cations attract the hydroxylic groups of water and move the water dissociation equilibrium towards higher values. This results in a decrease of 2-3 units of hydration sphere pH in comparison with the soil solution. The surface acidity effect is an important factor in the adsorption processes of herbicides which have nitrogen groups: in fact, the nitrogens contained in some agrochemical structures can be protonated by acidic surfaces and agrochemicals can be adsorbed by cationic exchange.

4. Soil organic matter

The organic components of soils are characterized by:
- vegetable and animal residues which are partially degraded and *in transformation;*
- the biomass of living organisms;
- materials of the neogenesis.

Vegetable and animal residues are slowly decomposed by microbial attack on molecular and ionic compounds which can be transformed by polycondensation in macromolecules with complex and unknown chemical structures: these are known as humic substances.
Humic substances have colloidal dimensions, high specific areas and are able to adsorb molecules or ions. The dark colour of humic compounds promotes the sunlight radiation absorption and thus, the increase of soil temperature.
Organic matter plays an important role in the chemistry of soils: it covers the pores created by roots or pedofauna action by stabilizing the soil structure. Organic matter affects the water flow into the pores (capillary porosity): in fact, the coexistence of hydrophilic and hydrophobic properties in the same structure makes organic matter a material which is able to retain moisture or to repel the water by decreasing its flow along the pores. Moreover, organic matter forms macroscopic aggregates ("cements") with inorganic species (i.e. Fe and Al oxides and hydroxides) which stabilize the soil structure.
Finally, the organic matter can interact with agrochemicals by H-bondings, van der Waals forces, H_2O bridgings, and hydrophobic bondings.

5. Dissolved organic matter, DOM

Dissolved organic matter (DOM) is defined as "the amount of organic matter that is able to dissolve in the field conditions". DOM plays an important role in the biogeochemistry of carbon, nitrogen, and phosphorous, in pedogenesis and in the transport of pollutants in soils (Kalbitz et al., 2000).

The source of virtually all DOM in soils is photosynthesis; this includes both recent photosynthate (throughfall, leaf litter, root exudates, decaying fine roots) as weel as the leaching and decomposition of older, microbially processed soil organic matter (McDowell, 2003). DOM ranges in age from hours to days, to decades and even up to thousands of years.

Sinks of DOM include microbial transformation and immobilization, mineralization (to CO_2, inorganic N, etc.), precipitation, and adsorption on mineral surfaces.

Microbial soil communities are the most important agents in DOM formation. Guggenberger et al. (1994) studied DOM fractionation and structure and demonstrated that microbial metabolites constitute a significant portion of DOM. According to these studies, the carbohydrate fraction of DOM is chemically different from that in plant residues or bulk humus, in that DOM carbohydrates have a higher proportion of hexose- and deoxysugars than pentose sugars. Since pentose sugars are rarely found in microbial cells, DOM may be predominantly of microbial origin.

Zsolnay (1996) and Tipping (1998) have supposed that the DOM can be partitioned into mobile and an immobile fractions according to the pore sizes of the soil matrix. Only the mobile DOM fraction in macro- and mesopores is subjected to convective transport by seepage. DOM in micropores is immobile and interacts with the mobile fraction by diffusion.

Several field studies (Jardine et al., 1989; Michalzik et al., 2000) have shown that the DOM concentration and flux in soil solutions decrease significantly with soil depth because DOM is adsorbed along the soil profile. High molecular weight fractions are preferentially adsorbed when compared with low molecular weight components (Gu et al., 1995). The presence of aromatic rings, carboxylic acids, N- and S-containing groups, and amino acid residues in organic molecules increases the adsorption capacity (McKnight et al., 1992). Adsorption involves the free surfaces of colloidal minerals and the presence of organic matter which has already been adsorbed further reduces DOM adsorption (Kalbitz et al., 2000).

Anions in soil solutions, such as sulphate and phosphate, compete with DOM for adsorption sites (Tipping, 1981). Kaiser & Zech (1997) and Beck et al. (1999) confirmed the role of phosphate in DOM removal from adsorption sites: this behaviour is also observed when the phosphate concentration is lower than the concentrations of other anions. Competition between sulphate and DOM for adsorption sites is evident when the sulphate concentration is higher than 10 mM (Kaiser & Zech, 1998).

Polyvalent (Al^{3+}, Fe^{3+}, Ca^{2+} and Mg^{2+}) or monovalent (Na^{+} e K^{+}) cation activity affects the solubility of organic matter (Baham & Sposito, 1994). Chemical reactions between anionic functional groups of organic molecules and solution cations can reduce the surface charge density, alter the structural conformation of the adsorbed species, and consequently reduce solubility. At high concentrations of ions in solution, these processes increase and the solubility of organic matter is reduced by flocculation (Tipping & Woof, 1990).

Various environmental factors influence DOM concentrations and fluxes in soils. Temperature is always a factor regulating DOM microbial production (Mulholland et al., 1990).

One of the most consistent findings in both field and laboratory studies is that DOM concentrations increase following rewetting after dry periods (Lundquist et al., 1999; Tipping et al., 1999; Zsolnay et al., 1999). It is likely that reduced rates of decomposition in dry soils cause microbial products to accumulate. This, together with cell death and lysis, can contribute to high DOM concentrations in the soil leachate after dry periods.

The significant effect of precipitation and water fluxes on DOM is the DOM release at the beginning of large rainfall events. Storm events can alter DOM concentration and flux: high pore water velocity leads to low contact times between the soil solution and the solid matrix and creates chemical and physical nonequilibrium conditions. These conditions increase the DOM concentration in the soil solution leaving the soil poor in organic matter.
Land use changes, such as afforestation, liming, and fertilization, converting forests into arable sites, and other management activities influence DOM dynamics by i) changing the input of organic matter, ii) changing the substrate quality and iii) altering the rates, extent, and pathways of microbial degradation and the synthesis of organic matter (Cronan et al., 1992).
The understanding of the dynamics and fluxes of DOM in soils is important in limiting the loss of organic matter from the soil, improving agronomic practices, and reducing the environmental impact of substances adsorbed on the DOM, such as agrochemicals.

5.1 The effect of DOM on the fate of herbicides

The water solubility of herbicides is one of the most important physical properties controlling the transport and fate of chemicals in aquatic systems (Chiou et al., 1986). The formation of soluble complexes between agrochemicals and DOM can be considered responsible for the transport of pollutants towards water bodies. Previous studies have indicated that low concentrations of dissolved and/or suspended particulate-bound natural organic matter in water can significantly enhance the solubility and stability of many hydrophobic organic compounds, notably DDT and some polychlorobiphenyls (PCBs) (Wershaw et al., 1969; Hassett and Anderson, 1979; Caron et al., 1985).
Chiou et al. (1986) observed the water solubility enhancement of solutes characterized by low water solubility such as DDT, 2,4,5,2',5'-PCB, trichlorobenzene, and lindane, due to their interaction with the dissolved humic and fulvic acids extracted from soil and aquatic sediments. The effectiveness of DOM in enhancing solute solubility appears to be largely controlled by DOM molecular size and polarity.
In the presence of soil, contradictory results have been obtained in studies concerning the effect of DOM on the behaviour of cationic or positively ionisable pesticides in soil and waters probably owing to different experimental conditions (Pennington et al., 1991; Barriuso et al., 1992; Klaus et al., 1998; Seol & Lee, 2000).
The nature of DOM (exogenous or endogenous) influences the adsorption and desorption of dimefuron, atrazine and carbetamine (Barriuso et al., 1992). The authors observed that DOM chemico-physical properties, like organic carbon content, pH, and conductivity, strongly affect herbicide adsorption. Moreover, different DOM additions to soils (pretreatment with DOM solution before herbicide adsorption or preincubation of DOM solution with herbicide before soil addition) influences adsorption as a function of herbicide solubility. Increased adsorption of less soluble atrazine and dimefuron, after soil pretreatement with DOM solution, can be explained by an increase in soil adsorption capacity related to the increase of soil C content via adsorption of some organic compounds from DOM solutions. The fate of the highly soluble carbetamide is different: its adsorption decrease can be explained by the coverage of soil hydrophilic sites by DOM organic compounds adsorbed during the preincubation.
Pennington et al. (1991) observed an opposite behaviour of the DOM: no interactions between the DOM extracted from different soils and the herbicides bromacil, metribuzin, alachlor, were observed. Diquat and paraquat are weakly adsorbed on DOM and their adsorption onto soils is not affected by the presence of soluble organic matter.

Despite the negative charge of negatively ionisable pesticides, which constrains them in the soil liquid phase and subsequently, in water courses, only a few studies on their behaviour in soil in the presence of DOM have been conducted (Spark & Swift, 2002; Said-Pullicino et al. 2004).

Although the interactions of pesticides with DOM are affected by the ionic strength of the solutions, DOM is often used in laboratory trials, after the removal of salts, i.e. as a purified organic fraction, or further fractionated in humic and fulvic acids (Spark & Swift, 2002; Chiou et al., 1986). Metals bound to DOM constituents or contained in free form are lost during the purification procedure, and rarely purified DOM has been studied as a function of both the ionic strength and saline composition (Carter & Suffet, 1982).

5.2 Case study: The fate of cyhalofop herbicide in soils treated with DOM from composts

Cyhalofop-butyl (butyl(R)-2-[4-(4-cyano-2-fluorophenoxy)phenoxy]propionate, ClincherTM), is an acetyl CoA carboxylase inhibitor for post-emergence control of barnyard grass (Echinochloa spp) and silver top (Lepthochloa fusca) in rice (Buendia et al., 1998; APVMA, 2005). The esters of aryloxyphenoxyalkanoic acids act as pro-herbicides. The formulation as esters facilitate the uptake through the plant cuticle and, once inside the plant, are transformed within a few hours into their acidic form, i.e. the active herbicide (Hendly et al., 1985; Ferreira et al., 1995). In soils, cyhalofop-butyl is quickly transformed into its more soluble negatively ionisable acidic form (Jackson & Douglas, 1999).

The potential for contamination of water bodies is high in areas where rice is cultivated in flooded conditions (Celis et al., 1998; Charizopoulos & Papadopoulou, 1999; Cerejeira et al., 2000, Miao et al., 2003). Agrochemicals applied to aquatic environments such as paddy-fields are matter of concern due to their potential leaching (Boesten & van der Linden, 1991; Müller et al., 2007) and persistence in soil and waters (Braschi et al., 2003).

The addition of organic amendments to the soil is an agricultural practice that is considered to potentially affect the fate of pesticides in soil (Cox et al., 2001; Hesketh et al., 2001), by introducing a remarkable amount of exogenous soluble organic matter. The extent to which exogenous DOM is involved in transportation through the soil is yet to be understood.

The effect of exogenous DOM from two composts on the behaviour of cyhalofop-butyl (CB) and cyhalofop-acid (CA) (Figure 2) was studied in two different soils (a paddy-field sediment, and a forest soil, Table 3) by means of solubility tests, determination of adsorption-desorption isotherms in soils, leaching experiments on soil columns (for experimental details, see Blasioli et al., 2008). To study the effect of the saline component on herbicide behaviour in soils, DOMs were used without any desalting treatment or pH modification.

Cyhalofop-butyl degradation is slow in the paddy-field sediment and leads to the cyhalofop-acid formation; in the forest soil, the degradation is faster and three byproducts were detected (cyhalofop-acid, -amide, and –diacid, Figure 2). The degradation is mediated by microorganisms as confirmed by the cyhalofop-butyl stability in autoclaved soils.

The water solubility of cyhalofop-butyl is unchanged in DOM solution at pH 6.0. This probably results from the highly hydrophilicity of these DOMs in contrast with the completly hydrophobic character of CB molecule. On the contrary, the solubility of CA doubles suggesting interaction with the DOMs. Since at working pH (about 6.0) cyhalofop-

Soil	pH		TOC[a] (g kg^{-1})	CEC[b] (cmol$_{(+)}$ kg^{-1})
	H_2O	$CaCl_2$		
Paddy-field sediment	6.00	5.50	14.0	2.00
Forest soil	7.10	6.80	73.0	9.90

Table 3. Summary of the characteristics of investigated soils. [a] TOC: Total Organic Carbon; [b] CEC: Cationic Exchange Capacity (modified from Blasioli et al., 2008).

acid, characterized by a pK_a of 3.8 (APVMA, 2005), mostly exists in anionic form, a molecular interaction, responsible for this increase in solubility, must occur despite the repulsion between DOM and CA negative charges. Similarly to water and cation bridgings formed between the carboxylate groups of humic substances and the soil phases in the presence of base metals (Sposito, 1984), polar and/or ionic interactions between metal cations-rich DOM and CA anionic moiety may be assumed to be responsible for the increase of CA solubility.

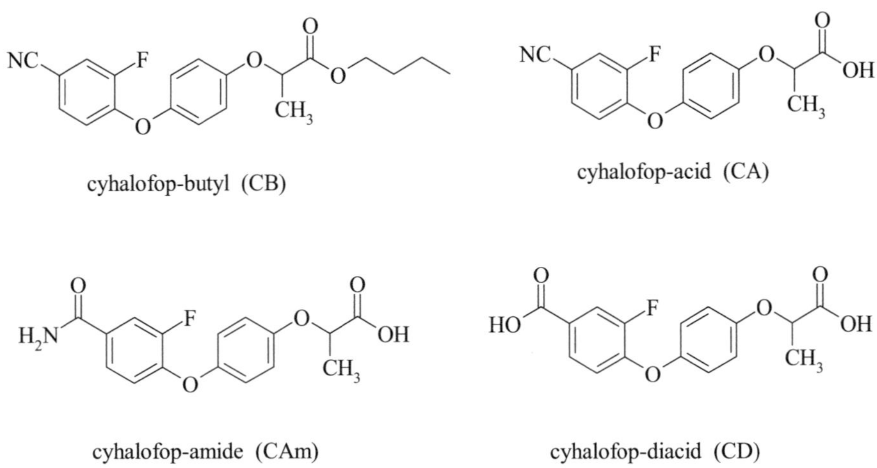

Fig. 2. Structure of cyhalofop-butyl and its byproducts.

The adsorption of cyhalofop-butyl and cyhalofop-acid is higher on the forest soil than on the sediment (Table 4). The pretreatment of soils with DOM solutions significantly decreases the cyhalofop-butyl adsorption but increases that of the acid. Since the DOM does not interact with the cyhalofop-butyl, the adsorption decrease may be due to the occupation by the DOM of soil sites available for the pro-herbicide.

The adsorption of acid is reversible on the paddy-field sediment but not reversible on the forest soil before and after soil pretreatment with DOM solutions. Reversible adsorption can be explained by two mechanisms: *i)* DOM which was previously adsorbed on soil surfaces increases the number of organic sites available to interact with CA by polar/ionic mechanisms; *ii)* the addition of base cations increases the base metal saturation degree of the sediment, hence linking CA carboxylate moiety to a higher extent via cation and water bridges (Sposito, 1984).

On the forest soil, the amount of adsorption sites available for hydrophobic interaction with CA does not vary as a consequence of DOM adsorption on the soil. This is reasonably supported by the very high content of organic matter in forest soil (TOC 73.0 g kg^{-1}, Table 3).

Sample	Cyhalofop-butyl	Cyhalofop-acid
	K_d L kg^{-1}	K_{Fads} $\mu g^{(1-1/n)}$ $mL^{1/n}$ g^{-1}
Paddy-field sediment	1.30	0.52
Paddy-field sediment + DOM_A	n.a.	4.70
Paddy-field sediment + DOM_M	n.a.	2.60
Forest soil	3.34	3.00
Forest soil + DOM_A	0.85	4.90
Forest soil + DOM_M	0.61	3.50

Table 4. Distribution coefficient of cyhalofop-butyl and Freundlich constants for cyhalofo-acid adsorbed in soils. n.a.: not available because no adsorption was observed (modified from Blasioli et al., 2008). DOM_A and DOM_M extracted from compost A (blend of winery byproducts) and M (blend of municipal solid waste), respectively.

The increases of the adsorption extent after pretreatment of forest soil with DOMs may be due to the ionic content of DOM solutions whose cationic component, complexing the negatively charged groups, lowers the repulsion between the negative charges and hence favours hydrophobic interactions of both soil organic matter and CA.

Adsorption data interpretation has been confirmed by CA leaching experiments on soil colums (Figure 3). The leaching of column with DOM solutions reduces the mobility of cyhalofop-acid in both soils. In fact, on the paddy-field sediment, which is poor in organic matter, the increase in base metal saturation degree due to DOM addition, promotes cation and water brigding formation with the carboxylate CA group. On the forest soil, which is rich in organic matter, the metals contained in the DOM, decrease the electrostatic repulsion between the soil organic matter and the carboxylate groups of the cyhalofop in acid, amide, and diacid forms, by reducing their mobility along the soil profile.

Further confirmation for the adsorption data was obtained by eluting the soil columns with K^+ and Ca^{2+} solutions (the most representative monovalent and divalent cations found in DOM_M solution). In both soils, the K^+ ion reduces the mobility of cyhalofop–acid whereas the Ca^{2+} ion decreases the mobility in the paddy-field sediment but not in the forest soil. The difference in the binding activity of the two cations in the adsorbed form towards the carboxylates can be explained. In fact, while in adsorbed form K^+ may give rise to a cation bridge with carboxylates due to the formation of an inner-sphere complex, Ca^{2+} forms a water bridge, a weaker outer-sphere complex mediated by water molecules (Sposito, 1984; Theng, 1982). The different strength between the K- and Ca-complexes may be responsible for the best CA retention in soils eluted with K ions in comparison with Ca ions.

In conclusion, the saline component of compost extracts seems to account to a large extent for the increased adsorption of the anionic herbicide cyhalofop-acid in both paddy-field and forest soils, generating highly reversible complexes with the former and non-reversible complexes with the latter, both involving the CA carboxylate group.

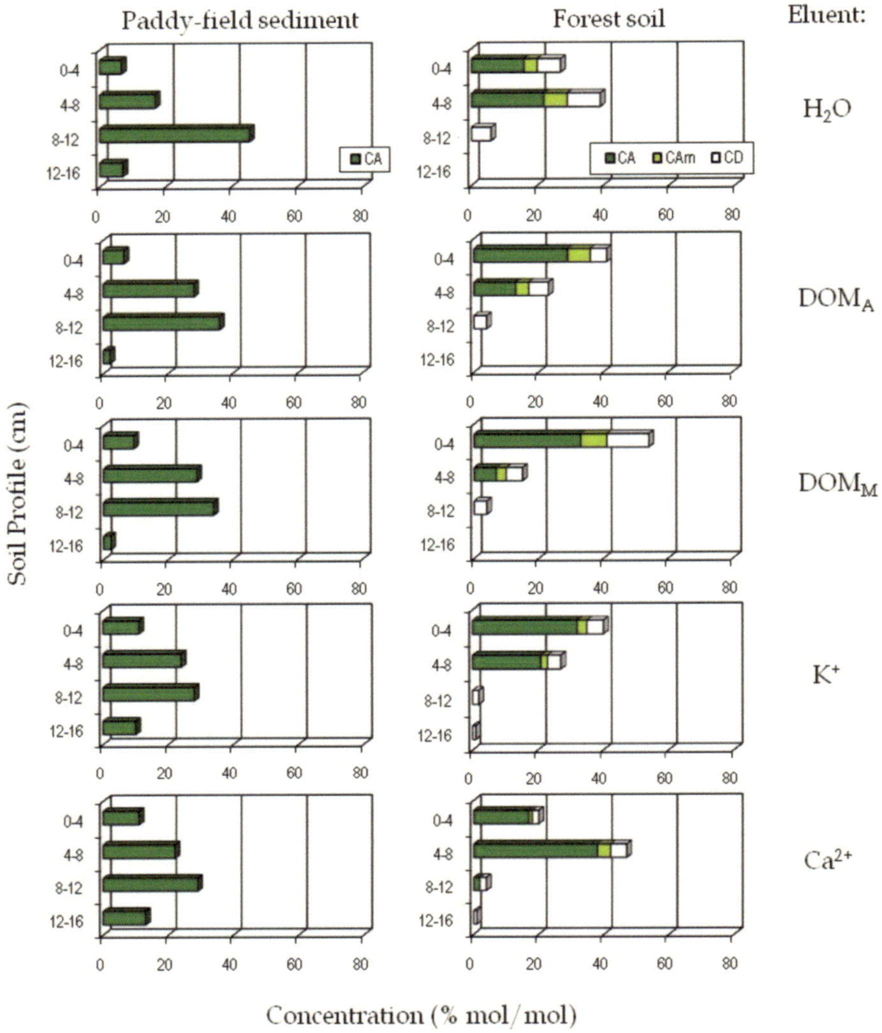

Fig. 3. Mobility of cyhalofop-acid and byproducts along the paddy-field sediment and forest soil column profiles leached with exogenous dissolved organic matter (DOM_A or DOM_M) or metal (K^+ or Ca^{2+}) solutions (mean of two replicates) (modified from Blasioli et al., 2008).

6. Conclusions

The prediction of the movement and the fate of herbicides in soils represents an important strategy in limiting their environmental impact. The chemico-physical properties of herbicides affect their behaviour in soil and regulate their interaction mechanisms with organic and inorganic soil phases. Among these, dissolved organic matter plays an important role: DOM influences the mobility of herbicides by complex interactions that can facilitate or reduce the movement of chemicals along the soil profile.

The knowledge of soil phase characteristics and the mechanisms involved in herbicide transformation can help to understand the fate of herbicides in soil.

7. References

APVMA. (2005). Evaluation of the new active cyhalofop-butyl in the product barnstorm herbicide. *Australian Pesticide and Veterinary Medicines Authority,* July 2005, Camberra, Australia, 1-39, ISSN1443-1335

Baham, J. & Sposito, G. (1994). Adsorption of dissolved organic carbon extracted from sewage sludge on montmorillonite and kaolinite in the presence of metal ions. *Journal of Environmental Quality,* Vol. 23, No. 1, 147-153, doi: 10.2134/jeq1994.00472425002300 010023x

Barriuso, E., Baer, U. & Calvet, R. (1992). Dissolved organic matter and adsorption-desorption of dimefuron, atrazine and carbetamide by soils. *Journal of Environmental Quality,*Vol. 21, No. 3, 359-367, doi: 10.2134/jeq1992.00472425002100030009x

Beck, M. A., Robarge, W. P. & Buol, S. W. (1999). Phosphorus retention and release of anions and organic carbon by two Andisols. *European Journal of Soil Science,* Vol. 50, 157-164

Blasioli, S., Braschi, I., Pinna M. V., Pusino A. & Gessa, C. E. (2008). Effect of undesalted dissolved organic matter from composts on persistence, adsorption and mobility of cyhalofop herbicide in soils. *Journal of Agrochemical and Food Chemistry,* Vol. 56, No. 11, 4102-4111, doi: 10.1021/jf7036906

Boesten, J. J. T. I. & van der Linden, A. M. A. (1991). Modeling the influence of sorption and transformation on pesticide leaching and persistence. *Journal of Environmental Quality,* Vol. 20, No. 2, 425–435, doi: 10.2134/jeq1991.00472425002000020015x

Bollag, J. M. & Liu, S. Y. (1990). Biological transformation processes of pesticides, In: *Pesticides in the Soil Environment: Processes, Impacts, and Modeling,* Cheng, H. H., (Ed. SSSA Book Series No. 2), 169-211, Soil Science Society of America, Madison, WI

Braschi, I., Pusino, A., Gessa, C., & Bollag, J. M. (2000). Degradation of primisulfuron by combination of chemical and microbiological processes. *Journal of Agricultural and Food Chemistry,Vol.* 48, No. 6, 2565-2571, doi: 10.1021/jf990604q

Braschi, I., Pusino, A. & Gessa, C. E. (2003). Adsorption-desorption of pretilachlor and cinosulfuron on paddy field sediment. *Proceedings of XII Symposium Pesticide Chemistry,* Piacenza, Giugno 2003, 161-167

Buendía, J., Barotti, R., Fullick, K., Thompson, A. & Velilla, J. (1998). Cyhalofop, a new postemergence herbicide for grass control in rice. *Proceedings of 6th EWRS Mediterranean Symposium,* Montpellier, France, May 1998, Oxford, UK, Blackwell

Carabias Martínez, R., Rodríguez Gonzalo, E., Fernández Laespada, M. E. & Sánchez San Román, F. J. (2000). Evaluation of surface- and ground-water pollution due to herbicides in agricultural areas of Zamora and Salamanca (Spain). *Journal of Chromatography A,* Vol. 869, No. 1-2, (February 2000) 471-480, ISSN 0021-9673/00/

Caron, G., Suffet, I. H. & Belton, T. (1985). Effect of dissolved organic carbon on the environmental distribution of nonpolar organic compounds. *Chemosphere,* Vol. 14, No. 8, (November 1984) 993-1000, doi:10.1016/0045-6535(85)90020-7

Carter, C. W. & Suffet, I. H. (1982). Binding of DDT to dissolved humic materials. *Environmental Science and Technology,* Vol. 16, No. 11, (November 1982) 735-740, doi: 10.1021/es00105a003

Celis, R., Barriuso, E. & Houot, S. (1998). Sorption and desorption of atrazine by sludge-amended soil: dissolved organic matter effects. *Journal of Environmental Quality,* Vol. 27, No. 6, 1348-1356, doi: 10.2134/jeq1998.00472425002700060011x

Cerejeira, M. J., Silva, E., Batista, S., Trancoso, A., Centeno, M. S. L. & Silva-Fernandes, A. (2000). Simazine, metribuzine and nitrates in ground water of agricultural areas of Portugal. *Toxicological & Environmental Chemistry*, Vol. 75, No. 3, 245–253, ISSN: 1029-0486

Charizopoulos, E. & Papadopoulou-Mourkidou, E. (1999). Occurence of pesticide in rain of the Axio stream basin, Greece. *Journal Environmental Science and Technology*, Vol. 33, No. 14, (June 1999) 2363–2368, doi: 10.1021/es980992x

Chiou, C. T., Malcolm, R. L., Brinton, T. I. & Kile, D. E. (1986). Water solubility enhancement of some organic pollutants and pesticides by dissolved humic and fulvic acids. *Environmental Science and Technology*, Vol. 20, No. 5, (May 1986) 502-508, doi: 10.1021/es00147a010

Cox, L., Cecchi, A., Celis, R., Hermosín, M. C., Koskinen, W. C. & Cornejo, J. (2001). Effect of exogenous carbon on movement of simazine and 2,4-D in soils. *Soil Science Society of America Journal*, Vol. 65, No. 6, 1688-1695, doi: 10.2136/sssaj2001.1688

Cronan, C. S., Lakshman, S. & Patterson, H. H. (1992). Effects of disturbance and soil amendments on dissolved organic carbon and organic acidity in red pine forest floors. *Journal of Environmental Quality*, Vol. 21, No. 3, 457-463, doi: 10.2134/jeq1992.00 472425002100030025x

Ferreira, K. L., Burton, J. D. & Coble, H. D. (1995). Physiological basis for antagonism of fluazifop-P by DPX-PE350. *Weed Science*, Vol. 43, 184-191

Gavalchin, J. & Katz, S. E. (1994). The persistence of fecal-borne antibiotics in soil. *Journal of AOAC International*, Vol. 77, 481-485

Giles, C. H., MacEwan, T. H., Nakhwa, S. N. & Smith, D. (1960). Studies in adsorption: Part XI. A system of classification of solution adsorption isotherms, and its use in diagnosis of adsorption mechanisms and in measurement of specific surface areas of solids. *Journal of Chemical Society*, Vol. 111, 3973-3993

Gu, B., Schmitt, J., Chen, Z., Liang, L. & McCarthy, J. F. (1995). Adsorption and desorption of different organic matter fractions on iron oxide. *Geochimica and Cosmochimica Acta*, Vol. 59, 219-229, ISSN: 0016-7037

Guggenberger, G., Zech, W. & Schulten, H.-R. (1994). Formation and mobilization pathways of dissolved organic matter: evidence from chemical structural studies of organic matter fractions in acid forest floor solutions. *Organic Geochemistry*, Vol. 21, (January 1994) 51-66, doi:10.1016/0146-6380(94)90087-6

Hassett, J. P. & Anderson, M. A. (1979). Association of hydrophobic organic compounds with dissolved organic matter in aquatic systems. *Environmental Science and Technology*, Vol. 13, No. 12, (December 1979) 1526-1529, doi:10.1021/es60160a017

Hendly, P., Dicks, J. W., Monaco, T. J., Slyfield, S. M., Tummon, O. J., & Barret, J. C. (1985). Translocation and metabolism of pyridinyloxyphenoxypropionate herbicides in rhizomatous quackgrass (*Agropyron repens*). *Weed Science*, Vol. 33, 11-24

Hesketh, N., Brookes, P. C. & Addiscott, T. M. (2001). Effect of suspended soil material and pig slurry on the facilitated transport of pesticides, phosphate and bromide in sandy soil. *European Journal of Soil Science*, Vol. 52, 287-296

Jackson, R. & Douglas, M. (1999). An aquatic risk assessment for cyhalofop-butyl: a new herbicide for control of barnyard grass in rice. *Proocedings of XI Symposium Pesticide Chemistry: Human and Environmental Exposure to Xenobiotics*, Cremona, Italy, 1999, 345-354

Jardine, P. M., Weber, N. L. & McCarthy, J. F. (1989). Mechanism of dissolved organic carbon adsorption on soil. *Soil Science Society of America Journal*, Vol. 53, No. 5, 1378-1385, doi: 10.2136/sssaj1989.03615995005300050013x

Kaiser, K. & Zech, W. (1997). Competitive sorption of dissolved organic matter fractions to soils and related mineral phases. *Soil Science Society of America Journal,* Vol. 61, No. 1, 64-69, doi: 10.2136/sssaj1997.03615995006100010011x

Kaiser K. & Zech W.(1998). Soil dissolved organic matter sorption as influenced by organic and sesquioxide coatings and sorbed sulfate. *Soil Science Society of America Journal,* Vol. 62, No. 1, 129-136, doi: 10.2136/sssaj1998.03615995006200010017x

Kalbitz, K., Solinger, S., Park, J.-H., Michalzik, B. & Matzner, E. (2000). Controls on the dynamics of dissolved organic matter in soils: a review. *Soil Science,* Vol. 165, No. 4, (April 2000) 277-304

Klaus, U. Mohamed, S., Volk, M. & Spiteller, M. (1998). Interaction of aquatic humic substances with anilazine and its derivatives: the nature of the bound residues. *Chemosphere,* Vol. 37, No. 2, (July 1998) 341-361, doi: 10.1016/S0045-6535(98)00050-2

Lewis, J. A., Papavizas, G. C. & Hora, T. S. (1978). Effect of some herbicides on microbial activity in soil. *Soil Biology and Biochemistry,* Vol. 10, No. 2, (June 1977) 137-141, doi: 10.1016/0038-0717(78)90084-6

Lundquist, E. J., Jackson, L. E. & Scow, K. M. (1999). Wet-dry cycles affect dissolved organic carbon in two California agricultural soils. *Soil Biology Biochemistry,* Vol. 31, No. 7, (July 1999) 1031-1038, doi: 10.1016/S0038-0717(99)00017-6

Madhun, Y. A. & Freed, V. H. (1990). Impact of pesticides on the environment, In: *Pesticides in the Soil Environment: Processes, Impacts, and Modeling,* Cheng, H. H., (Ed. SSSA Book Series No. 2), 429-466, Soil Science Society of America, Madison, WI

McDowell, W. H. (2003). Dissolved organic matter in soils-future directions and unanswered questions. *Geoderma.* Vol. 113, 179-186, ISSN: 0016-7061

McKnight, D. M., Bencala, K. E., Zellweger, G. W., Aiken, G. R., Feder, G. L. & Thorn, K. A. (1992). Sorption of dissolved organic carbon by hydrous aluminium and iron oxides occurring at the confluence of Deer Creek with the Snake river, Summit country, Colorado. *Environmental Science and Technology,* Vol. 26, No. 7, (July 1992) 1388-1396, doi: 10.1021/es00031a017

Miao, Z., Padovani, L., Riparbelli, C., Ritter, A. M. & Trevisan, M. (2003). Prediction of the environmental concentration of pesticide in paddy field and surrounding surface water bodies. *Paddy and Water Environmental,* Vol. 1, No. 3, (August 2003) 121–132, doi: 10.1007/s10333-003-0021-4

Michalzik, B., Kalbitz, K., Park, J. H., Solinger S. & Matzner. (2001). Fluxes and concentration of dissolved organic matter-a synthesis for temperate forests. *Biogeochemistry,* Vol. 52, No. 2, 173-205, doi: 10.1023/A:1006441620810

Mulholland, P. J., Dahm, C. N., David, M. B., Di Toro, D. M., Fisher, T. R., Kögel-Knabner, I., Meybeck, M. H., Meyer, J. L. & Sedell, J. R.(1990). What are the temporal and spatial variations of organic acids at the ecosystem level?. In: *Organic Acids in Aquatic Ecosystems.* E. M. Perdue & E. T. Gjessing (Eds.). Life Sciences Research Report 48, 315-329, John Wiley & Sons, Chichester, ISBN 0-471-92631-0

Müller, K., Magesan, G. N. & Bolan, N. S. (2007). A critical review of the influence of effluent irrigation on the fate of pesticides in soil. *Agriculture, Ecosystem and Environment,* Vol. 120, No. 2-4, (May 2007) 93–116, doi: 10.1016/j.agee.2006.08.016

Pennington, K. L., Harper, S. S. & Koskinen, W. C. (1991). Interactions of herbicides with water-soluble soil organic matter. *Weed Science,* Vol. 39, No. 4 (Oct.-Dec., 1991) 667-672

Pierzynski, G. M., Sims, J. T. & Vance, G. F. (2000). Organic Chemicals in the Environment, In: *Soils and environmental quality,* CRC Press (2nd Ed.), Lewis Publishers, ISBN 0-87371-680-9, USA.

Pozo, C., Salmeron, V., Rodelas, B., Martinez-Toledo, M. V. & Gonzalez-Lopez, J. (1994). Effects of the herbicide alachlor on soil microbial activities. *Ecotoxicology*, Vol. 3, No. 1, (March 1994) 4-10, ISSN 0963-9292 (Print) 1573-3017 (Online)

Said-Pullicino, D., Gigliotti, G. & Vella, A. J. (2004). Environmental fate of triasulfuron in soils amended with municipal waste compost. *Journal of Environmental Quality,* Vol. 33, No. 5, 1743-1751, doi: 10.2134/jeq2004.1743

Seol, Y. & Lee, L. S. (2000). Effect of dissolved organic matter in treated effluents on sorption of atrazine and prometryn by soils. *Soil Science Society of America Journal,*Vol. 64, No. 6, 1976-1983, doi: 10.2136/sssaj2000.6461976x

Spark, K. M. & Swift, R. S. (2002). Effect of soil composition and dissolved organic matter on pesticide sorption. *Science of the Total Environment*, Vol. 298, No. 1-3, (October 2002) 147-161, doi: 10.1016/S0048-9697(02)00213-9

Sposito, G. (1984). The reactive solid surface in soils, In *The surface chemistry of soils*. Oxford University Press, 143-147, Claredon Press, ISBN: 0-19-503421-X, New York

Theng, B. G. K. (1982). Clay polymer interactions: Summary and perspectives. *Clays and Clay Minerals,* Vol. 30, No. 1, 1-10

Tipping, E. (1981). The adsorption of aquatic humic substances by iron oxides. *Geochimica and Cosmochimica Acta*, Vol. 45, No. 2, (February 1981) 191-199, doi: 10.1016/0016-7037(81)90162-9

Tipping, E. & Woof, C. (1990). Humic substances in acid organic soils: Modelling their release to the soil solution in terms of humic charge. *Journal of Soil Science*, Vol. 41, No. 4, (December 1990) 573-586, doi: 10.1111/j.1365-2389.1990.tb00227.x

Tipping, E. (1998). Modelling of the proprieties and behaviour of dissolved organic matter in soils. *Mitteilungen Deutsche Bodenkundliche Gesellschaft,* Vol. 87, 237-252

Tipping, E., Woof, C., Rigg, E., Harrison, A. F., Inneson, P., Taylor, K., Benham, D., Poskitt, J., Rowland, A. P., Bol, R. & Harkness, D. D. (1999). Climatic influences on the leaching of dissolved organic matter from upland UK moorland soils, investigated by a field manipulation experiment. *Environmental International,* Vol. 25, No. 1, (January 1999) 83-95, doi :10.1016/S0160-4120(98)00098-1

Van Dijk, J. & Keukens, H. J. (2000). The stability of some veterinary drugs and cocciostats during composting and storage of laying hen and broiler feces, In: *Residues of veterinary drugs in food, Proceedings of the Euroresidue IV Conference,* Veldhoven, The Netherlands, May 2000, L. A. van Ginkel and A. Ruiter Eds.

Wershaw, R. L., Burcar, P. J. & Goldberg, M. C. (1969). Interaction of pesticides with natural organic matter. *Environmental Science and Technology,* Vol. 3, No. 3, (March 1969) 271-273, doi: 10.1021/es60026a010

Yuan, J. S., Tranel, P. J. & Stewart, C. N. (2007). Non-target-site herbicide resistance: a family business. *Trends in Plant Science,* Vol. 12, No. 1, (January 2007) 6-13, doi: 10.1016/j.tplants.2006.11.001

Zsolnay, A. (1996). Dissolved humus in soil waters. In *Humic Substances in Terrestrial Ecosystem*. A. Piccolo (Ed. Elsevier), 171-223, Amsterdam

Zsolnay, A., Baigar, E., Jimenez, M., Steinweg, B. & Saccomandi, F. (1999). Differentiating with fluorescence spectroscopy the sources of dissolved organic matter in soils subjected to drying. *Chemosphere,* Vol. 38, No. 1, (January 1999) 45-50, doi: 10.1016/S0045-6535(98)00166-0

9

Environment Behavior and Fate of a Novel Pyrimidynyloxybenzoic Herbicide ZJ0273 in Aerobic Soil

Haiyan Wang, Yanfei Zhang, Man Yu, Juying Li, and Qingfu Ye
Zhejiang University, Institute of Nuclear Agricultural Sciences, Hangzhou, 310029
China

1. Introduction

Herbicides are universally used in crop production all over the world. Although herbicides are developed to be biologically active to weeds, their residues are frequently found in plants, soils, and surface and underground waters, where they may bring about a risk concern for human health, ecosystems, or the productivity of subsequent crops (Walker et al., 1997; Vithala & White, 1996).

ZJ0273, propyl 4-(2-(4,6-dimethoxypyrimidin-2-yloxy)benzylamino) benzoate (shown in Fig. 1), is a new broad-spectrum herbicide (Lu et al., 2004; Yang et al., 2008). It is marketed under the trade names of Youli (EC) and Youli II (SC) in China, and both products were registered for use in oilseed rape in 2003. In recent years, the herbicide has been rapidly adopted for use in China with the cumulative application area reaching 666,667 hectares in 2009. Applications of the herbicide at the dosage of 40-60 g a.i. ha^{-1} are commonly used to achieve effective control (80-90%) in rape fields against pre- and post-emergence weeds, including equal alopecurus (*Alopecurus aequalis Sobol.*), crickweed (*Malachium aquaticum L.*), chickweed (*Stellaria media L.*), redroot amaranth (*Amaranthus retroflexus L.*), japanese alopecurus (*Alopecurus japonicus Steud.*), annual bluegrass (*Poa annua L.*), common polypogon (*Polypogon fugax Nees ex Steud.*), and spinefruit buttercup (*Ranunculus muricatus L.*) (Tang et al., 2005). ZJ0273 greatly inhibits acetolactate synthase (ALS) in vivo and it has little inhibitory effect on ALS in vitro, which differs from the typical commercial ALS inhibitors, such as sulfonylurea and pyrimidine salicylic acid herbicides.

Study on environmental behavior and fate for new agrochemicals in soil is an indispensable requirement for safe use of pesticides on food crops (Langenbach et al., 2000, 2001; Fent et al., 2003; Mordaunt et al., 2005). A series of issues, such as extractable residues, bound residue formation and mineralization of herbicides, as well as persistence time of parent compound, are of the most important concern. Many studies showed that soil-bound residues of herbicides, such as sulfonylurea herbicides, may cause damage to rotation or substitution crops (Moyer et al., 1990; Ye et al., 2003, 2004). In Europe, the Uniform Principles (CEC, 1997) state that in laboratory tests, if non-extractable soil residues are formed at > 70% of the initial dose after 100 days with mineralization to CO_2 at < 5%, there will be no authorization unless it is demonstrated that no accumulation of residues occurs in soil under field conditions (Craven, 2000; Craven and Hoy, 2005). In addition, it is important to understand if mineralization will detoxify agrochemicals with polyaromatic rings. However, up to now, relative information on ZJ0273 is still not well documented.

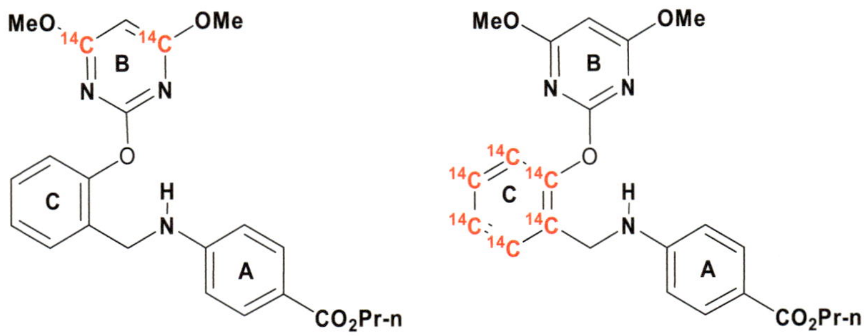

[*pyrimidine*-4,6-^{14}C] ZJ0273 [*benzyl*-U-^{14}C] ZJ0273

Fig. 1. Structures of radiotracers, ZJ0273 with asterisks marking the position of ^{14}C.

Therefore, from the above, the aim of the present study, by using multi-^{14}C-labeled propyl 4-(2-(4,6-dimethoxypyrimidin-2-yloxy)benzylamino)benzoate, was to investigate environmental behavior and fate of the herbicide, characterize the fate processes, persistence, degradation pathways, bound residues formation, mineralization of ZJ0273, as well as plant availability and phytotoxicity of bound residues of ZJ0273 in well-characterized soils under aerobic conditions.

2. Materials and methods

2.1 Chemicals

Propyl 4-(2-(4,6-dimethoxy[4,6-^{14}C]pyrimidin-2-yloxy)benzylamino)benzoate ([*pyrimidine*-4,6-^{14}C] ZJ0273; radiochemical purity 99.4%; chemical purity 98.1%; specific activity 3.77×10^7 Bq/mmol) and propyl 4-(2-(4,6-dimethoxypyrimidin-2-yloxy)[*phenyl*-U-^{14}C]benzylamino)benzoate ([*benzyl*-U-^{14}C] ZJ0273; radiochemical purity 98.9%; chemical purity 98.1%; specific activity 3.74×10^7 Bq/mmol), were synthesized according to the methods of Yang et al. (2008 & 2009). The chemical purity was analyzed by high performance liquid chromatography (HPLC), and the radiochemical purity was determined by HPLC-liquid scintillation counting (LSC), and thin layer chromatography-isotope imaging analysis (TLC-IIA) (Yang et al., 2006 & 2009).

2.2 Soils

Three different agricultural soils were used in the incubation experiment. The soils included a Red clayey soil (S_1), a Fluvio-marine yellow loamy soil (S_2) and a Coastal saline soil (S_3). The soil samples were taken from the surface layer (0-15 cm) in fields from different quarters of Zhejiang Province, China. The bulk soil samples were air dried, mixed, and passed through a 1-mm sieve before use. Some basic physicochemical characteristics of the soils were determined using standard methods (Nelson and Sommers, 1982; Gee and Bauder, 1986) and are given in Table 1.

2.3 Incubation experiments

To characterize the fate processes of the herbicide ZJ0273 in aerobic soils, two multi-position ^{14}C labelings, [*pyrimidine*-4,6-^{14}C] ZJ0273 and [*benzyl*-U-^{14}C] ZJ0273 were employed in this study. Test vessels for the degradation study were 500-mL glass flasks. For each soil, a test system of three replicates was used. Prior to the application of [*pyrimidine*-4,6-^{14}C] ZJ0273, 3×300 g of soil (dry weight equivalent) was weighed for each soil and placed in sealed polypropylene bags, and the soil moisture content was adjusted to about 30% of the field water holding capacity by addition of distilled water. The test soils were then pre-incubated at 25±1 °C for 10 d to allow the microorganisms to acclimatize. After the initial acclimation, 15.0 mg of [*pyrimidine*-4,6-^{14}C] ZJ0273 (3.77×10^7 Bq/mmol) and 15.0 mg of non-labeled analogue were dissolved in methanol, and 5.0 mL (1.340×10^5 Bq) was added to each soil sample. After the applied herbicide was thoroughly mixed, the treated soil samples were left in a fume hood to allow the evaporation of methanol and then transferred into incubation vessels. The water content of the spiked soil samples was adjusted to 60% of the water holding capacity by adding distilled water. Each incubation flask was connected to a series of air-tight test tubes, which allowed for the scrubbing of CO_2 from the inlet air (with 5.0 M NaOH, two traps) and for maintaining constant soil moisture, and entrapment of volatiles (with 1.0M H_2SO_4/glycol, 5/10 v/v, one trap) and $^{14}CO_2$ (with 10mL of 0.5M NaOH, two traps) (Fig. 2; EPA, 2002). During incubation, a slow and continuous air flow was maintained in all systems at 25±2 °C. At different time intervals (5, 10, 20, 30, 45, 60, 75, 90 and 100 d after treatment), the traps and a subsample of the treated soils (10.0 g, air-dried weight equivalent) were removed to determine the radioactivity associated with extractable, non-extractable residue, and $^{14}CO_2$, as well as the fraction of [*pyrimidine*-4,6-^{14}C] ZJ0273 remaining in the parent molecule form. The treatment and sampling for [*benzyl*-U-^{14}C] ZJ0273 followed the same procedures as given above for [*pyrimidine*-4,6-^{14}C] ZJ0273, except that 1.326×10^5 Bq of radioactivity was amended in each 300 g soil sample.

Property	Soil type		
	S_1 Red clayey soil	S_2 Fluvio-marine yellow loamy soil	S_3 Coastal saline soil
pH (H_2O)	4.20	7.02	8.84
OM[a] (g kg^{-1})	8.40	30.50	9.50
Total N (%)	0.34	2.90	1.80
CEC[b] (cmol kg^{-1})	6.62	10.83	10.17
Clay (%)	39.0	8.0	24.3
Silt (%)	41.1	71.2	71.1
Sand (%)	19.9	20.8	4.6
P (mg kg^{-1})	3.21	25.20	10.80

[a] Organic matter;
[b] Cation Exchange Capacity.

Table 1. Basic physical-chemical properties of the soils.

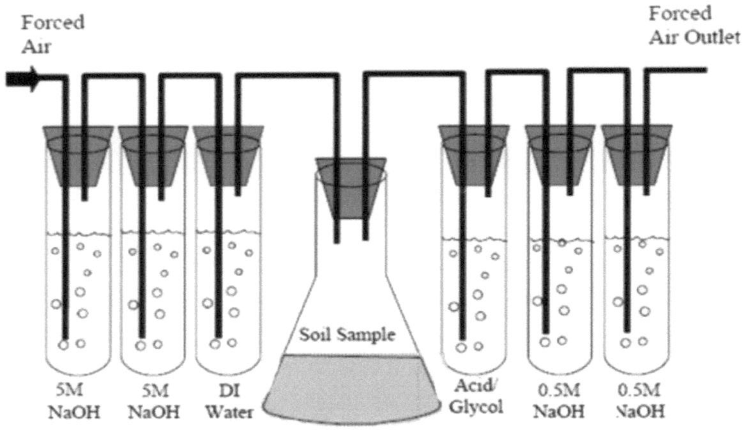

Fig. 2. Experimental set-up for monitoring bound residues of [*pyrimidine*-4,6-^{14}C] ZJ0273 and [*benzyl*-U-^{14}C] ZJ0273 during incubation with fresh soil.

2.4 Extractable residue and parent compound

The soil subsamples were transferred to 100-mL polypropylene centrifuge tubes, and consecutively extracted with solvents with decreasing polarity by following a protocol similar to Mordaunt et al. (2005). The soil samples were shaken in each solvent (50 mL) for 24 h, using the following solvents in a successive order: 0.01 mol/L $CaCl_2$, acetonitrile/water (9:1, v/v), methanol, and dichloromethane. At the end of the extraction period, the sample was centrifugated, supernatant collected and the final volume brought up to 50.0 mL. From the extracts obtained from the first three extraction steps, an aliquot of 1.0 mL was withdrawn and mixed in 10.0 mL of scintillation cocktail A and analyzed for ^{14}C activity (dpm) using LSC. The dichloromethane extract from the final extraction step was condensed to dryness and then the residue was dissolved in 50 mL methanol. An aliquot of 1.0 mL was withdrawn for analysis of radioactivity. The radioactivity of total ER was the sum of radioactivity determined after each extraction step.

The $CaCl_2$ extract was further adjusted to pH 3.0 and partitioned with dichloromethane (1:1, v/v) for three consecutive times. The dichloromethane phase was combined and condensed on a vacuum rotary evaporator at 40 °C until dry and the residue was pooled with those from the other three extraction steps. The combined supernatants were then condensed to about 1.0 mL on a vacuum rotary evaporator at 40 °C and centrifuged at 18,000 g for 15 min. A 20-µL aliquot was injected into a Waters HPLC and the eluted fraction corresponding to the parent compound peak was manually collected at the outlet for measurement of ^{14}C. The HPLC system was composed of a Waters 600 multi-solvent delivery unit, a Waters 996 photodiode array (PDA) detector operating at 301 nm and 254 nm, a Diamonsil C_{18} column (5-µm, 250 × 4.6 mm, Dikma Technologies, Beijing, China), and a C_{18} guard column. The column temperature was maintained at 30 °C. The mobile phase A and B were made of H_2O (HPLC grade) and methanol, respectively, both containing 0.1% acetic acid. The elution was achieved at the flow rate of 1.0 mL min^{-1} using a gradient program (minutes/%A: 0/80, 0-40/25, 40-80/25, 80-90/0) and was collected and mixed with 10.0 mL of scintillation cocktail A. Their ^{14}C radioactivity was determined by LSC.

2.5 Bound residue

After the sequential extraction, residual soils were left in a fume hood overnight to allow the evaporation of organic solvents and a 1.0-g aliquot of the air-dried soil was combusted to $^{14}CO_2$ on a biological oxidizer (OX-600, R.J. Harvey Instrument Corp., Hilldale, NJ, USA), followed by measurement of ^{14}C radioactivity by LSC to determine the amount of bound residue formed in soils. The total time of oxidation was 4.0 min, the combustion temperature was 900 °C and the catalyzer temperature was 680 °C. The $^{14}CO_2$ evolved from combustion was trapped in 15 mL of scintillation cocktail B and the radioactivity was determined by Quantulus 1220 ultra-low Level liquid scintillation spectrometer (ULLLSS; Quantulus 1220, Wallac, Turku, Finland). The derived ^{14}C activity was defined as non-extractable or bound residue in this study. The ^{14}C recovery of the combustion method was determined to be ≥ 92.5% from analysis of samples spiked with known radioactivity of ^{14}C labeled ZJ0273.

2.6 Measurement of mineralization to ^{14}C-CO_2

At each sampling interval, $^{14}CO_2$ in the incubation flask was swept into the traps by purging each system with air for two hours to ensure that $^{14}CO_2$ was thoroughly trapped by the NaOH solution. The NaOH traps were exchanged with new solution, and the contents from the two used traps were combined and diluted into the volume of 50 mL with H_2O, after which 2.0 mL of the solution was mixed in 10 mL of scintillation cocktail A. After the samples were kept in dark for 24 h (to remove chemiluminescence), ^{14}C measurement was carried out on the ULLLSS, due to the relatively low radioactivity, which was calibrated with ^{14}C-standards to correct for any quenching effect.

2.7 MS analyses

To elucidate the structures of the intermediates, LC-MS/MS analysis was carried out on a Micromass Quattro micro API™ with a HPLC detector and triple quadrupole mass analyzer for determining mass-to-charge ratio (m/z) for a wide variety of analytes (Waters, Milford, MA, USA). Control of the instruments and calculation was made using MassLynx V4.1 software (Waters). The instrument was operated in positive ESI ionization mode. Operating conditions were optimized by constantly introducing a standard solution of the above-mentioned compounds to the HPLC flow via a T-connector with a split ratio of 1:1. The signal was optimized on the total ion current in MS mode by changing cone-, capillary-, extractor- and RF lens voltages in the source and resolution and ion-energy in the analyzer. At the same time the collision voltages and resolution in the second quadrupole were optimized. The structures of the intermediates were determined by comparing with mass spectral data of intermediate standards.

2.8 Plant availability experiment

Rice and corn were used for the plant availability assay. The soils containing BR from the above were mixed with fresh soils at three different ratios so that the initial contents of ^{14}C-BR of ZJ0273 were 0.6, 1.2 and 1.8 nmol g^{-1} dried soil corresponding to 11.2, 22.4, and 33.7 Bq g^{-1} dried soil. The mixed soils (130 g dry weight equivalent) were placed in 200-mL plastic pots for cultivation of corn and Petri dishes (d = 12 cm) for rice cultivation. After the soil moisture was adjusted to 60% and 80% of the maximum field water holding capacity for corn and rice, respectively, five germinated corn seeds were sown into each pot and 30 germinated rice seeds were sown into each Petri dish. The pots and Petri dishes without amendment of ^{14}C-BR soil were similarly prepared and used as the control. Five replicates were set up for each treatment. Corn and rice seedlings were cultivated under the same greenhouse conditions (25/20 °C, day/night; humidity, 80%; light, 12 h/ 12h), with irrigation every day. Visual inspection and photo taking were carried out at 5, 7 and 14 d after sowing. All seedlings were harvested at 14 d after sowing. The height of shoots was measured. Each seedling was divided into shoots and roots. The roots were washed with tap water. All the plant parts were dried at 60 °C to a constant weight. Aliquots of five dried rice plants and individual corn plant were combusted on the biological oxidizer and the released $^{14}CO_2$ was absorbed in 15 mL liquid scintillation cocktail B. The radioactivity was measured by ULLLSS to estimate the amount of BR that was accumulated by the plant. The recovery efficiency of the above combustion procedure was > 95%.

2.9 Characterization of released residues

After cultivation of rice and corn seedlings, the soils that were amended with BR at 1.8 nmol g^{-1} were extracted using the procedure given in Section 2.4 and the extracts were further characterized for the composition of the released residues by the method of Section 2.7.

2.10 Phytotoxicity assays

The phytotoxicity of the released compounds from BR was evaluated using rice as the indicator species. The stock solutions of pure M1(38.1 mg), M3 (27.6 mg) and ZJ0273 (42.3 mg) were prepared in 1.0 mL dimethyl formamide (1%), and diluted to a final volume of 100 mL with distilled water after addition of 1.0 mL isopropyl alcohol (1%), 0.5 mL TW-80 (0.5%) and 1.0 mL methanol (1%). Five treatment solutions (0.5, 1, 2.5, 5 and 10 μM) were prepared by diluting the stock solution with distilled water. After germination at 25 °C in Petri dishes containing filter papers moistened with distilled water, 20 seeds of rice were placed in the Petri dishes and treated with 9 mL of the compounds derived from the BR at the five concentrations. After cultivation in an illuminated incubator (25/20 °C, day/night; humidity, 80%; light, 12 h /12 h) for 7 d, shoot height of rice seedlings was measured. Two different blanks were used. In blank 1, rice seeds were exposed to solutions containing carrier-solvents but no ZJ0273 or its metabolites. In control 2, rice seeds were exposed to water only. Each treatment was replicated 5 times. The inhibition ratio (IR) was expressed as IR = 1-(shoot height of treated rice / shoot height of control). A three parameter logarithm equation ($y = a - b*\ln(x + c)$) was fitted to the data of inhibition ratio as a function of herbicide concentration by nonlinear regression using Origin 7.5 (Microcal Software, Northampton, MA).

2.11 Statistical analysis

All measurements were in three replicates and the arithmetic means and standard errors of means (means ± SEM) were calculated from the repeated measurements. Significance between treatments was determined by a one-way analysis of variance (ANOVA) using Origin 6.0 (Microcal Software, Northampton, MA, USA).

3. Results and discussion

3.1 Mass balance

Mass balance was conducted to calculate the amount of $^{14}CO_2$ (mineralization), extractable residue, and bound residue for both ^{14}C labels. Throughout the entire incubation, mass

Incubation Days	Soil	$CaCl_2$	$CH_3CN:H_2O$	CH_3OH	CH_2Cl_2
5d	S_1	13.8±5.3	75.6±5.8	8.2±0.9	1.7±0.4
	S_2	11.9±0.9	76.4±1.9	7.2±0.8	2.6±0.5
	S_3	16.7±3.2	66.6±1.3	9.97±2.5	2.8±1
10d	S_1	11.4±3.9	76.6±3.8	9.6±0.3	1.3±0.3
	S_2	11.8±2	74.7±1.6	8.5±0.6	1.8±0.3
	S_3	14.1±1.4	70.6±1.2	8.4±1.4	2.9±0.4
20d	S_1	12.6±3	73.8±3.6	9.5±0.9	1.7±0.2
	S_2	19.2±3	61.7±2.5	0.8	2.6±0.4
	S_3	23.5±9.8	57.6±10.1	8±1.2	2.7±0.5
30d	S_1	15.5±4.9	75.1±5.6	6.4±0.3	0.9±0.8
	S_2	36.6±5.3	48.6±5.4	5±0.9	1.1±0.2
	S_3	35.8±7.5	47.5±7.9	6.2±0.9	1.5±0.4
45d	S_1	20.2±7.8	67.4±11.1	5.2±0.7	1.4±0.2
	S_2	34.6±7.9	40.6±1	4.2±0.1	1.1±0
	S_3	39.8±2.7	36.9±8.2	5.5±1.1	2±1.4
60d	S_1	30.4±7.4	54.9±9.9	3.8±1.3	4±0.3
	S_2	34.9±0.6	36±1.2	4.2±0.5	1.5±0.5
	S_3	32.6±1.9	30.4±6.2	3.8±0.5	1.1±0.3
75d	S_1	24.4±7.3	61±10.6	4.6±1.5	1.3±2.9
	S_2	39.3±2.3	26.7±1.2	4.1±0.1	1.5±0.2
	S_3	35.6±2.2	24.5±5	3.6±0.9	2±1.3
100d	S_1	36.3±2.7	35.9±6.3	2.8±0.7	0.9±0.4
	S_2	33.8±2.4	18±2.9	3.6±0.6	2.5±1.6
	S_3	30.8±.38	15.7±0.7	3±0.2	0.7±0.2

Table 2. Dissipation of extractable residue of [*pyrimidine*-4,6-^{14}C] ZJ0273 and its metabolites after four-step extraction in soils

balances as percentage of the initially added ^{14}C radioactivity were from 90.0±5.8% to 104.4±4.6%, which indicated good mass recoveries for the procedures utilized in this study.

3.1.1 Extractable residue (including parent compound and metabolites)

It is manifested that extraction efficiency of ^{14}C activity immediately after application of ^{14}C-labeled ZJ0273 ranged from 91.1 to 100.1% for [*pyrimidine*-4,6-^{14}C] ZJ0273 and from 92.3 to 99.0% for [*benzyl*-U-^{14}C] ZJ0273 in the three tested soils. During the first 30-d incubation, the fraction of total extractable residues (ER) from all extraction steps in the three soils was over 90% of the initially applied ^{14}C for [*pyrimidine*-4,6-^{14}C] ZJ0273. Nevertheless, total ER experienced a more dramatic downward trend during the next 30 d period, especially in S_2 and S_3 (Table 2). The lag phase observed was similar to the phenomenon which was found in other studies (Roeth, 1986; Smith and Lafond, 1990). Pesticide degradation in soils was generally mediated by microorganisms and microbial degradation was commonly characterized by an initial lag phase due to adaptation of the microbes (Gevao et al., 2000). At 100 d after treatment, total ER declined to 75.9% for S_1, 57.9% for S_2 and 50.2% for S_3. As a consequence, the level of total ER of [*pyrimidine*-4,6-^{14}C] ZJ0273 in the acidic S_1 soil was significantly higher than that in the neutral S_2 and alkaline S_3 soils ($p < 0.05$). Throughout the experiment, however, , there was no significant discrepancy in ER dissipation trends between the two ^{14}C labels ($p \geq 0.05$) according to the comparison of data.

Extractable ^{14}C residue was further analyzed for the fraction of parent compound. Data analysis demonstrated that the fall of the parent compound in both [*pyrimidine*-4,6-^{14}C] ZJ0273 and [*benzyl*-U-^{14}C] ZJ0273 treatments complied with a first order decay model, with the regression coefficient R^2 being 0.98 to 0.99 ($p < 0.01$, data not shown). The estimated first-order half-lives were 51.0, 20.1 and 17.6 d for S_1, S_2 and S_3, respectively, for the [*pyrimidine*-4,6-^{14}C] ZJ0273 treatment. Almost identical half-lives (53.7 d, 20.4 d and 17.2 d for S_1, S_2 and S_3, respectively) were obtained for the [*pyrimidine*-4,6-^{14}C] ZJ0273 treatment ($p < 0.05$). However, other scientists discovered diverse results in 2,4-D degradation by two multiple-position ^{14}C-labels. For instance, Gaultier et al. (2008) used [ring-U-^{14}C] 2,4-D (R2,4-D) and [carboxyl-^{14}C] 2,4-D(C2,4-D) to measure degradation rates of 2,4-D, finding that rates of R2,4-D degradation in soils were limited more by sorption than rates of C2,4-D degradation. Smith and Muir (1980) and Xie et al. (1997) presented 15–20% more [ring-U-^{14}C] 2,4-D bound to soil compared with [carboxyl-^{14}C] 2,4-D. In this study, for both labels, the dissipation of the parent compound was positively correlated with soil pH ($p < 0.001$). Accordingly, ZJ0273 was more stable and resistant to transformation under acidic conditions. The similarities observed for the two labels offered evidence that the behavior of the parent compound was independent from the labeling position, which, as expected, was a direct validation of the rigorousness of the experimental procedures used in this study. The slower degradation in the acidic soil may attribute to the following factors. First and foremost, the propyl group of ZJ0273 is prone to break and thereby forming 4-(2-(4,6-dimethoxypyrimidin-2-yloxy)benzylamino)benzoic acid under alkaline condition, while it is probably impossible to take place in acidic soil. Second of all, the acidic soil S_1 had a lower OM content than soils S_2 and S_3, and it is likely that the lower OM resulted in a reduced formation of bound residue between OM and the parent molecule, or a higher level of free parent compound. Last but not least, it is also likely that there were differences in microbial populations and makeup because of various soil pH conditions, which generated slower transformations in the acidic soil.

At the first few sampling intervals, the majority of ER was extracted by acetonitrile-water, whereas relatively small fractions were from other extractions. Mordaunt et al. (2005) depicted that a significant large portion of ER was derived from $CaCl_2$ extraction for atrazine, dicamba, and isoproturon, but from acetonitrile-water extraction for lindane and trifluralin. The difference was attributed to the differences in pesticide aqueous solubility. Specifically, the water solubility is 33 mg L^{-1} for atrazine, 6500 mg L^{-1} for dicamba, and 65 mg L^{-1} for isoproturon, but much lower for lindane (7.3 mg L^{-1}) and trifluralin (0.2 mg L^{-1}). Compared with these pesticides, the aqueous solubility for ZJ0273 is substantially lower at 1.5mg L^{-1} and the very low solubility would probably implicate the association of most ER with the acetonitrile-water extract during the initial phase of the incubation experiment. On the contrary, the fraction from the $CaCl_2$ extraction step gradually rose as incubation time further increased. For instance, from 5 d to 100 d after the treatment, the fraction of $CaCl_2$ extracted residue rocketed from 13.8% to 36.3% for [*pyrimidine*-4,6-^{14}C] ZJ0273 (Table 2). Correspondingly, a similar phenomenon was found in the sequential extractions for the [*benzyl*-U-^{14}C] ZJ0273 treatment. In our previous study, several metabolites, such as 2-(4,6-dimethoxypyrimidin-2-yloxy)benzoic acid (M1), 4-(2-(4,6-dimethoxypyrimidin-2-yloxy)benzylamino)benzoic acid (M2), 4-(2-(4,6-dimethoxypyrimidin-2-yloxy)benzamido)benzoic acid (M3) and 4,6-dimethoxypyrimidin-2-ol and/or 4,6-dimethoxypyrimidin-2(1*H*)-one (M4), were identified. All of these metabolites (log P=1.40-4.11) were more polar than the parent compound (log P=5.04), which induced the conclusion that the increased recoveries of ^{14}C by $CaCl_2$ extraction suggested accumulation of polar metabolites in the soils.

3.1.2 Mineralization

The cumulative mineralization rates of [*pyrimidine*-4,6-^{14}C] ZJ0273 and [*benzyl*-U-^{14}C] ZJ0273 in three aerobic soil over time was present in Fig. 3. As can be seen from the graph that mineralization of both labels displayed a lag phase lasting from the beginning to about 20 d after the treatment. This lag phase coincided with what was observed in dissipation of extractable residues, suggesting that time was in need for the acclimatization of microbial degraders of this compound. After 20 d, mineralization increased with incubation time, and appeared to accelerate towards the end of the incubation. Mineralized fractions of [*pyrimidine*-4,6-^{14}C] ZJ0273 or [*benzyl*-U-^{14}C] ZJ0273 were substantially higher in S_2 and S_3 than in S_1 at the same interval time. At the end of 100-d incubation, the total fraction mineralized in S_1 came least at less than 1.2% of the applied activity for [*benzyl*-U-^{14}C] ZJ0273, with S_2 at 7.7% and S_3 at 9.9%. In soils treated with [*pyrimidine*-4,6-^{14}C] ZJ0273, at the end of 100-d incubation, only 0.5% mineralization occurred in S_1, while the cumulative mineralization rate was 6.6% for S_2 and 5.0% for S_3. The overall mineralization rates of [*pyrimidine*-4,6-^{14}C] ZJ0273 and [*benzyl*-U-^{14}C] ZJ0273 were correlated with soil pH ($p < 0.05$) and organic matter content ($p < 0.05$). The soil S_1 was more acidic (pH 4.20) and also had less organic matter than the other two soils (Table 1), which may have brought about limited microbial activity in this soil and hence a limited mineralization potential (Boivin et al., 2005). The slow mineralization of ZJ0273 in acidic soil and implications for risk assessment and management merit were further investigated. Mineralization to $^{14}CO_2$ was found to be consistently greater with [*benzyl*-U-^{14}C] ZJ0273 than with [*pyrimidine*-4,6-^{14}C] ZJ0273 in the same soil throughout the incubation. For instance, at the end of the 100-d incubation, the mineralized fractions of [*pyrimidine*-4,6-^{14}C] ZJ0273 reached 6.6% for S_2, and 5.0% for S_3, while the corresponding fractions of [*benzyl*-U-^{14}C] ZJ0273 were 7.7% for S_2, and 9.9% for S_3. These differences suggested that, under the same conditions, mineralization of ZJ0273 involved more active ring cleavage at the benzyl ring than that at the pyrimidine ring.

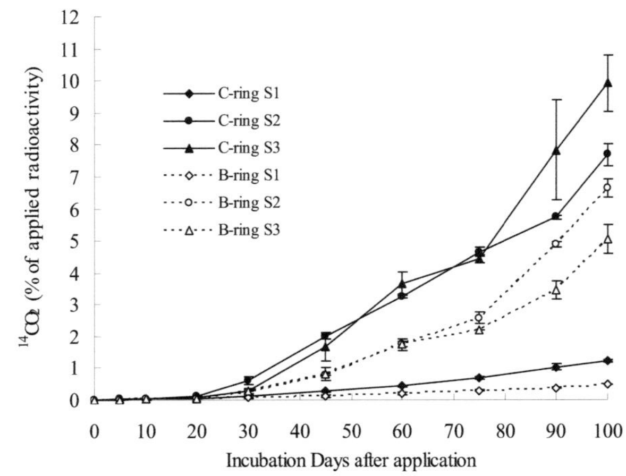

Fig. 3. Mineralization of [*pyrimidine*-4,6-^{14}C] ZJ0273 and [*benzyl*-U-^{14}C] ZJ0273 in three soils

3.1.3 Bound residue

The time-dependent accumulation of bound residue in soils for the two labels was presented in Fig. 4. The fraction of BR following treatment of [*pyrimidine*-4,6-^{14}C] ZJ0273 increased rapidly in all three tested soils. The fraction of BR in S_1 was found to be significantly smaller than in the other two soils ($p < 0.01$) and BR formation was correlated with soil pH ($p < 0.05$). At the end of 100-d incubation, BR represented around 35.3% of the initially applied ^{14}C in

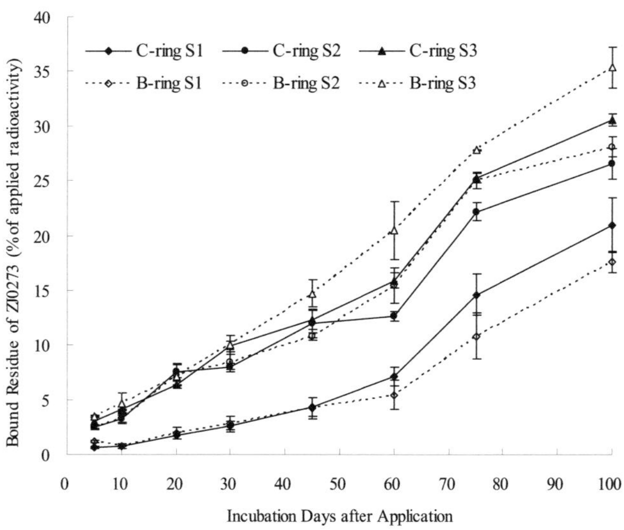

Fig. 4. Formation of bound residue after treatment of [*pyrimidine*-4,6-^{14}C] ZJ0273 and [*benzyl*-U-^{14}C] ZJ0273 in aerobic soils

S_3, 28.1% in S_2, and 17.5% in S_1, which were far lower than the 70% as described as the non-accumulative criteria in the directive by CEC (1997). The formation of soil-bound residues for many pesticides has been reported to be mediated by the activities of soil microorganisms (Smith & Philips, 1975; Haider, 1983; Krause et al., 1985). During the entire incubation, ANOVA analysis showed that there was no significant difference in BR formation between the two ^{14}C labels ($p \geq 0.05$), which implies that formation of BR in soils involved parent compound and/or metabolites with both pyrimidine and benzyl rings, rather than metabolites with a single ring.

3.2 Identification of degradation intermediates

The highest radioactivity measured by HPLC and ULLLSS was found in the samples at 60 DAT. For the [*pyrimidine*-4,6-^{14}C] ZJ0273 treatment, as shown in Fig. 5A and 5C, radioactivity was detected in five peaks at the retention time (t_R) of 10, 40, 47, 55, and 70 min. Other components that were not radio-labeled were not further identified. For [*benzyl*-U-^{14}C] ZJ0273 treatment, only four radioactive peaks at the retention times (t_R) of 40, 47, 55, and 70 min were detected while the peak at 9.80 min was absent (Fig. 5D). The radioactive intermediate peaks were found in both [*pyrimidine*-4,6-^{14}C] ZJ0273 and [*benzyl*-U-^{14}C] ZJ0273 treatments, suggesting that they could be possible degradation products of ZJ0273. Further structure analysis was conducted with the assistance of LC-MS/MS.

The parent ZJ0273 had a retention time of 70 min, and its mass spectrum included ion fragments at m/z 245($C_{13}H_{13}N_2O_3$), 424$[M+H]^+$, and 446$[M+Na]^+$. The selected ion chromatography of m/z 424 included the major ions (m/z, %) of 364 (85), 245(100), and 185(10), which were consistent with the structure of standard ZJ0273 (Yang et al., 2009; Han et al., 2009a).

The peak at 47 min appeared in the HPLC trace and in the total ion chromatogram of degradation products of ZJ0273 (Fig. 5A and 5B), and was identified in the LC-MS/MS chromatogram. The mass spectrum of the intermediate at 47 min revealed m/z 245($C_{13}H_{13}N_2O_3$), 382$[M+H]^+$, and 404$[M+Na]^+$, which was consistent with the structure of compound M1 with a molecular weight of 381. The selected ion chromatography of M1 included the major ions (m/z, %) of 107 (35), 160 (30), 185 (20), 214 (15), 245 (100), and 319 (10) (Table 3). The fragment ion (m/z 245) and two ion adducts (m/z 382($C_{20}H_{20}N_3O_5$), 404($C_{20}H_{19}N_3O_5Na$)) were simultaneously monitored, and the extracted ion chromatogram showed three peaks (m/z 382, 404 and 245), which rose and fell synchronously. The extracted ion chromatogram showed that this compound appeared to be ^{14}C-4-(2-(4,6-dimethoxypyrimidin-2-yloxy)benzylamino)benzoic acid. This was confirmed by comparing the retention time of this compound in HPLC with that of a standard of 4-(2-(4,6-dimethoxypyrimidin-2-yloxy)benzylamino)benzoic acid.

Compound M2 that appeared at 55 min in the LC-MS chromatogram was identified. The mass spectrum of the metabolic intermediate at 55 min reveals m/z 396$[M+H]^+$, and 418$[M+Na]^+$, which is consistent with the structure of compound M2. The selected ion of m/z 396 included the major ions (m/z, %) of 107 (20), 139 (20), 185 (30), 230 (20), 245 (100), 300 (20), and 364 (80) (Table 3). The fragment ion (m/z 245) and three ion adducts (m/z 107(C_7H_7O), 185($C_{11}H_9N_2O$), 364($C_{20}H_{18}N_3O_4Na$)) were monitored simultaneously at the optimal conditions, and the extracted ion chromatogram showed two peaks (m/z 396 and 418) consistent with the standard of M2. In addition, this compound was found radioactive in both treatments of [*pyrimidine*-4,6-^{14}C] ZJ0273 and [*benzyl*-U-^{14}C] ZJ0273, indicating that both the pyrimidine and benzyl rings were contained in the structure of M2. As M2 had a t_R of 55 min and matched the mass spectra of the standard, the intermediate compound was thus confirmed as ^{14}C-4-(2-(4,6-dimethoxypyrimidin-2-yloxy)benzamido)benzoic acid.

Proposed degradation compound	FW	t_R(min)	MS spectrum ions of the selected ion (m/z) (%, abundance)
M1	381	47	107(35),160(30),185(20),14(15),245(100)
M2	395	55	107(20), 139(20), 185(30), 230(20), 245(100), 300(20), 364(80)
M3	276	40	107(45), 139(100), 153(75), 170(35), 259(35)
M4	156	10	100(55), 125(30), 139(100), 141(30)

Table 3. HPLC-retention times (t_R), mass spectra of ZJ0273 and its major degradation products in aerobic soils.

The peak for compound M3 appeared at 40 min in the chromatogram and in the total ion chromatogram of the LC-MS of ZJ0273 metabolites (Fig. 5A and 5B). The mass spectrum of the metabolite at 40 min revealed m/z 259($C_{13}H_{11}N_2O_4$), 277[M+H]$^+$, and 299[M+Na]$^+$ (Table 2). The selected ion of m/z 277 included the major ions (m/z, %) of 107 (45), 139 (100), 153 (75), 170 (35), and 259 (35). The fragment ion (m/z 259) and two ion adducts (m/z 277 ($C_{13}H_{13}N_2O_5$), 299($C_{13}H_{13}N_2O_5Na$)) were monitored simultaneously. The intermediate was found radioactive in both [*pyrimidine*-4,6-^{14}C] ZJ0273 and [*benzyl*-U-^{14}C] ZJ0273 treatments, which showed that the structure of M3 contained both pyrimidine and benzyl rings. The HPLC retention time of this component was in accordance with the standard of 2-(4,6-dimethoxypyrimidin-2-yloxy)benzoic acid. Comparison of the extracted ion chromatogram with the mass spectrum of the authentic standard further verified M3 as ^{14}C-2-(4,6-dimethoxypyrimidin-2-yloxy)benzoic acid with a molecular weight of 276.

The peak of compound M4 appeared at 10 min in the chromatogram and in the total ion chromatogram of LC-MS of ZJ0273 degradation products (Fig. 5A and 5B). The mass spectrum of this compound at 10 min revealed m/z 157[M+H]$^+$, which was consistent with the structure of compound M4 with a molecular weight of 156. The selected ion chromatograph of 157 included the major ions (m/z, %) of 100 (55), 125 (30), 139 (100), and 141 (30) (Table 3). Since the compound was found only in the [*pyrimidine*-4,6-^{14}C] ZJ0273 treatment, it should contain the pyrimidine ring. The extracted ion chromatogram, when compared with mass spectrum of an authentic standard confirmed the degradation product M4 to be 4,6-dimethoxypyrimidin-2-ol, a keto-enol tautomer of 4,6-dimethoxypyrimidin-2(1*H*)-one, as described by Chang et al. (2007).

3.3 Proposed degradation pathways

As illustrated in Fig. 6, there were two possible metabolic pathways resulting in the degradation of ZJ0273. In all three aerobic soils, the parent compound was initially hydrolyzed and de-esterified by cleavage of the propyl group to form M1. Since the side chain carbon atom of the benzene ring tended to be oxidized abiotically or biotically, carbonylation occurred. After that, conjugation between the π orbital of the phenyl group or

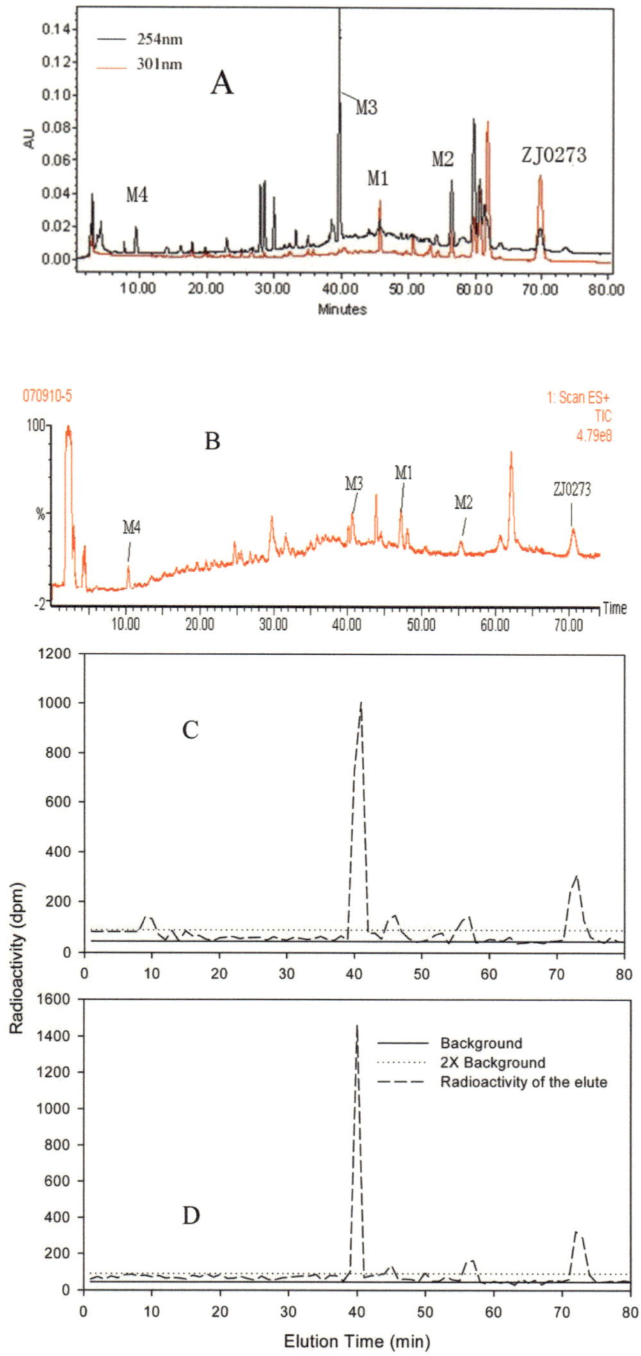

Fig. 5. Charts of HPLC, TIC of LC-MS, and radioactive chromatogram obtained from an extract of ^{14}C-ZJ0273 degradation products. (A) HPLC chromatogram at the wavelength of 254 nm and 301 nm; (B) LC-MS ESI$^+$; (C) Radioactive chromatogram of elution of [*pyrimidine*-4,6-^{14}C] ZJ0273 degradation products at 60 DAT; (D) Radioactive chromatogram of elution of [*benzyl*-U-^{14}C] ZJ0273 degradation products at 60 DAT.

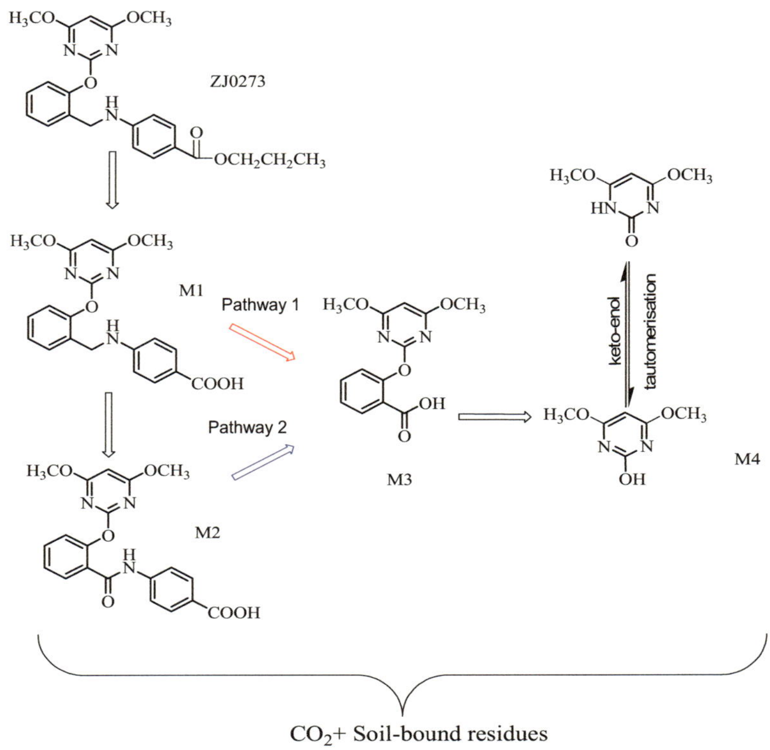

Fig. 6. Degradation pathways of ZJ0273 in aerobic soils. Pathway 1 showed the route of ZJ0273→M1→M3→M4; Pathway 2 followed the other route of ZJ0273→M1→M2→M3→M4.

the pyrimidine ring and the p orbital of the oxygen atom in the compound M3 was the dominant reaction. As the conjugation force of the pyrimidine ring was stronger than that of the phenyl ring, the fragments were supposed to be M4 and salicylic acid.Then M4 was ultimately decomposed to carbon dioxide. The second degradation pathway was also initiated by the hydrolysis to form M1. The carbon atom of the benzylamine group with higher activity was readily acylated to generate M2 with an amide group. In the intermediate M2, the amino group was an easy-to-leave group and was hydrolyzed with acid and/or base catalysts, which was followed by carboxylated with the phenyl ring and cleaved to form the main degradation product M3. The M3 was transformed as described in the first step to yield M4, and finally mineralized to CO_2.

3.4 Bioavailability and phytotoxicity of bound residue

3.4.1 Plant availability and phytotoxicity

In order to explore plant availability and potential phytotoxicity of BR from ZJ0273, visual characterization and quantitative analysis including measurement of plant height, dry weight of shoot, root, and total plant of rice and corn seedlings were employed after exposure to soil BR fromZJ0273. Compared with the control, growth depression of rice seedlings by the BR was observed at the concentration of 1.8 nmol g^{-1}, while no inhibition occurred at the concentration of 0.6 or 1.2 nmol g^{-1} in all soils at 5 d after treatment (DAT). After growing in BR amended soils for 7 d, growth of rice was significantly impeded at all concentration levels in the test soils and the suppression aggravated with growing concentrations. At 14 DAT, leaf chlorosis and growth inhibition of rice seedlings were recorded (Fig. 7a), which indicated a more pronounced inhibition compared with that at 7 DAT. Plant height, dry weights of shoot, root and the whole plant of rice seedlings decreased dramatically as the BR amendment rate increased in all the three soils at 14 DAT (Table 4), supporting the visual observations. These results demonstrated that ZJ0273 and/or its degradation products released from BR in the soils imposed severe phytotoxic impacts on the rice seedlings. Analysis of ^{14}C radioactivity in rice seedlings 14 DAT witnessed a rise in ^{14}C content in shoot, root and the whole plant with increasing amendment rates, coinciding with the drop in the plant height and dry weight (Table 5). It is suggested that the accumulation of the released chemicals from BR of ZJ0273 in the soils made great contribution in inhibiting plant growth. The total ^{14}C activity in the root reached a maximum in the soils amended with BR at 1.2 nmol g^{-1}, indicating that absorption of BR-derived chemicals was suppressed beyond this amendment rate (Table 5).

Over the whole process of cultivation, no significant visual growth depression of corn seedlings was observed 14 DAT in any treatment (Fig. 7b). Determination of plant height, dry weights of shoot, root and whole plant of corn seedlings further justified the visual observations (Table 5). Compared with the control, no significant inhibition was found in the plant height, dry weight of shoot, root and whole plant ($p > 0.05$, Table 5). Quantification of ^{14}C radioactivity in different parts of corn seedling revealed that the total amount of ^{14}C in shoot and root increased with the rise of the BR amendment rate in the test soils (Table 6). The radioactivity of ^{14}C on a dry biomass basis was higher for the shoot of corn at a higher BR amendment rate, but no significant increase was gained for the root when the amendment rate increased from 1.2 to 1.8 nmol g^{-1} (Table 6).

Soil	Conc. (nmol g^{-1})	height(cm)	shoot (mg)	Root (mg)	Total weight (mg)
S_1	0	40.5±1.6	152±13	72±2	227±13
	0.6	43.3±1.3	166±15	79±8	243±21
	1.2	40.2±1.1	155±13	77±5	232±17
	1.8	40.4±1.5	162±13	80±5	238±14
S_2	0	34.3±1.7	120±14	82±7	197±10
	0.6	34.8±1.1	123±11	88±6	210±14
	1.2	33.8±1.6	116±11	84±4	212±28
	1.8	32.1±1.3	110±10	86±7	208±16
S_3	0	38.2±2.0	176±21	74±6	252±26
	0.6	38.9±2.3	179±20	68±6	259±30
	1.2	37.6±1.8	169±19	70±5	240±25
	1.8	37.1±2.0	168±22	72±7	248±21

Table 4. Bioavailability and phytotoxicity of bound residue derived from ZJ0273 on corn seedlings.

Soil	Conc. (nmol g^{-1})	Height (cm)	Shoot (mg)	Root (mg)	Total Weight (mg)
S_1	0.0	26.0±0.2	37.2±0.7	7.3±0.2	44.4±0.8
	0.6	18.5±0.2	25.4±0.3	6.0±0.1	31.5±0.6
	1.2	17.4±0.2	22.6±0.2	5.3±0.1	29.0±0.4
	1.8	15.8±0.2	19.2±0.3	4.6±0.1	24.5±0.4
S_2	0.0	26.5±0.3	35.4±0.7	7.7±0.2	42.9±1.0
	0.6	17.3±0.2	19.1±0.5	5.9±0.3	25.2±0.8
	1.2	14.9±0.1	15.3±0.5	5.2±0.1	20.1±0.7
	1.8	11.9±0.2	12.6±0.7	3.4±0.1	16.0±0.7
S_3	0.0	25.5±0.3	38.9±1.0	6.9±0.2	45.9±1.2
	0.6	20.5±0.4	29.3±0.6	5.6±0.1	32.0±1.4
	1.2	14.8±0.2	17.5±0.6	4.9±0.1	22.6±0.8
	1.8	12.4±0.2	15.5±0.6	3.0±0.1	18.4±0.7

Table 5. Bioavailability and phytotoxicity of ZJ0273 bound residue on rice seedlings.

After plant cultivation and the subsequent extraction in Section 2.4, decreases in the radioactivity of ^{14}C in the whole soil indicated the fraction of BR released. The fractions of ^{14}C released after rice planting were 68.3%, 57.0%, and 61.1% of the applied activity, respectively, for S_1, S_2 and S_3 amended with BR at 1.8 nmol g^{-1}. The released amounts were considerably greater than those in the corresponding soils grown with corn, which ranged from 32.6% to 38.9%. The inhibition caused by the BR on rice seedlings in the different soils followed the order $S_1 > S_3 > S_2$, similar to the order for the fractions of ^{14}C released. The interactions between pesticides, along with their metabolites, and soils are extremely complicated and may change with soil pH, microbial community and soil enzyme activity (Khan, 1982; Dec & Bollag, 1988). During the cultivation, different soil conditions for corn and rice may have contributed to the different release rates of BR. Furthermore, much larger amounts of the released ^{14}C were absorbed into the corn plants than rice seedlings, despite the amount of ^{14}C per dry biomass was smaller for corn. All in all, the lower accumulation rates of BR-derived residues on a biomass basis may have led to the insignificant injury in corn seedlings.

3.4.2 Characterization of released residues

Extractable products from soils amended with BR after planting of rice or corn were resolved on HPLC and the radioactivity was measured in the eluted fractions. The retention times of the three radioactive peaks coincided with those for the authentic standards for parent ZJ0273 (70 min), M1 (47 min) and M3 (40 min). Both M1 and M3 are degradation products that were previously identified upon degradation of ZJ0273 in soil. The metabolite M1 was a hydrolysis product of ZJ0273 and its further metabolism may lead to the formation of M3, which belongs to the first degradation pathway mentioned above. ZJ0273 and its degradation products may bind to the soil through physical and/or chemical interactions such as van der Walls forces, ligand exchange, charge-transfer complexes, hydrophobic partitioning, and covalent bonding. The formed soil BR of ZJ0273 could be

Soil	Conc. (nmol g-1)	Shoot (Bq g-1)	Root (Bq g-1)	Shoot (Bq)	Shoot (Bq)
S_1	0	0	0	0	0
	0.6	6.8±0.4	18.0±1.6	1.2±0.1	1.3±0.1
	1.2	12.7±0.8	38.7±2.6	2.0±0.2	2.8±0.2
	1.8	16.2±0.9	38.4±2.6	2.7±0.2	3.5±0.4
S_2	0	0	0	0	0
	0.6	24.4±3.4	31.9±2.3	2.3±0.3	3.8±0.4
	1.2	45.6±4.1	62.9±2.8	3.3±0.4	6.3±0.6
	1.8	75.5±6.6	62.8±6.3	6.9±0.6	8.1±1.2
S_3	0	0	0	0	0
	0.6	11.5±0.7	30.0±1.7	2.2±0.2	2.0±0.2
	1.2	16.5±1.0	45.1±2.5	2.7±0.2	2.8±0.3
	1.8	20.3±2.7	48.1±1.9	2.8±0.3	4.2±0.4

Table 6. ^{14}C radioactivity in corn seedlings.

Fig. 7. (a). Effects of bound residues of ^{14}C-ZJ0273 on growth of rice seedlings in soils at 14 d after application; (b). Effects of bound residues of ^{14}C-ZJ0273 on growth of corn seedlings in soils at 14 d after application.

released by physico-chemical mechanisms or through biochemical processes. However, in this study it was found that only M1, M3 and ZJ0273 were released from the soil BR upon planting and the relative ratios of the three compounds varied with soil types and plant types. In the extractable residues, ZJ0273 and its degradation products M1 and M3 constituted 6.9-10.7%, 14.8-17.7%, and 68.0-77.1%, respectively (see Fig. 8), for soils grown with corn, as compared with 15.1-31.0%, 25.6-39.8%, and 38.3-45.1% in rice growing soils. In all soils, M3 was the dominant compound released from soil BR derived from ZJ0273.

Soil	Conc. (nmol g^{-1})	Shoot (Bq g^{-1})	Root (Bq g^{-1})	Shoot (Bq)	Root (Bq)
S_1	0	0	0	0	0
	0.6	35.7±0.9	61.9±2.6	0.52±0.03	0.23±0.01
	1.2	71.5±2.6	100.1±5.3	1.03±0.08	0.31±0.02
	1.8	110.7±4.4	129.1±2.6	1.28±0.08	0.32±0.01
S_2	0	0	0	0	0
	0.6	61.5±1.6	214.0±6.3	1.35±0.07	1.33±0.07
	1.2	130.3±2.0	373.7±15.1	1.97±0.07	2.15±0.08
	1.8	203.5±8.2	387.9±8.0	2.59±0.10	1.36±0.07
S_3	0	0	0	0	0
	0.6	27.4±1.2	63.3±2.3	0.75±0.04	0.36±0.02
	1.2	66.1±2.6	101.1±2.0	1.31±0.07	0.51±0.02
	1.8	114.3±5.2	121.2±3.5	1.62±0.12	0.40±0.02

Table 7. ^{14}C radioactivity in rice seedlings.

3.4.3 Phytotoxicity of the released compounds

Phytotoxicity of ZJ0273 and the two identified compounds 4-(2-(4,6-dimethoxypyrimidin-2-yloxy)benzylamino)benzoic acid and 2-(4,6-dimethoxypyrimidin-2-yloxy)benzoic acid was further evaluated by measurement of the shoot height of rice seedlings exposed to solutions containing these compounds. No significant differences were found in the shoot height of rice seedlings exposed to control 1 and control 2 solutions. Thus, the inhibition ratio was calculated based on control 2. No significant inhibition effect was observed on rice by ZJ0273 at 0.5 and 1 μM (see Fig. 8). However, the plant height decreased at 2.5 μM of ZJ0273 and the inhibition reach 28.9% at 10 μM. The inhibition effect caused by M1 and M3 both increased with the growth of concentration. At the same concentration, the inhibition rate followed the order of 2-(4,6-dimethoxypyrimidin-2-yloxy)benzoic acid (M3) > 4-(2-(4,6-dimethoxypyrimidin-2-yloxy)benzylamino)benzoic acid (M1) > ZJ0273, with the inhibition rate at 10 μM reaching 75.5%, 86.0%, and 28.9% for M1, M3, and ZJ0273, respectively. The IC50 values of M1, M3 and ZJ0273, defined by nonlinear regression of rice shoot height over the treated doses, were assessed to be 1.93, 0.49 and 33.16 μM, respectively (see Fig. 8). Results of the phytotoxicity assay suggested that M1 and M3 were more biologically active than the parent ZJ0273 against rice. It may be further concluded that it was the degradation products of ZJ0273, M3 in particular, that possibly imposed the inhibition effect on the growth of rice seedlings.

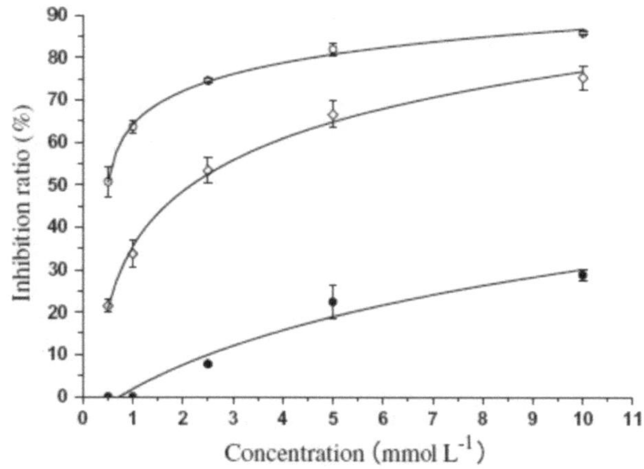

Fig. 8. Inhibition of ZJ0273 (- d -, y = -23.58 + 21.45 * ln(x + 2.27), R2 = 0.9643), M1 (- e -, y = 37.69 + 17.13 * ln(x + 0.12), R^2 = 0.9954) and M3 (-s-, y = 68.12 + 8.30 * ln(x - 0.38), R^2 = 0.9969) on growth of rice seedling measured as reductions in shoot height.

Previous studies have shown that ZJ0273 inhibits the synthesis of the branched-chain amino acids in a susceptible plant in vivo and the inhibition effect is counteracted by the addition of branched-chain amino acids, while no inhibition is detected under in vitro conditions. These results suggest that ZJ0273 is a pro-herbicide and it is the metabolites of ZJ0273 in the plant that contribute to the inhibition on ALS (Chen et al., 2005). Far more severe inhibition effect of M3 on ALS was studied in preliminary studies (unpublished). The IC50 of M3 on the ALS of the etiolated pea shoots was estimated to be 0.39 μM (Yukio et al., 1996). Therefore, the cause of phytotoxicity to rice from soil BR of ZJ0273 may be attributed to the release of ZJ0273, M1, and M3 from the soil BR and the subsequent inhibition on the biosynthesis of amino acids in plants due to the inhibition of ALS primarily by M3.

4. Conclusion

As a novel acetolactate synthase potential inhibitor, ZJ0273 had short to intermediate persistence in aerobic soils, with half-lives ranging from 17.2 to 53.7 d for the parent molecule. The disappearance of the parent compound was accompanied with production of metabolites, formation of BR and mineralization to CO_2. The overall transformation of ZJ0273 appeared to closely rely on soil properties such as pH and organic matter content. In an acidic soil with low organic matter content, ZJ0273 was degraded much slower with limited mineralization and reduced formation of BR. Use of ^{14}C labels at different positions provided similar and hence confirmatory information on the behavior of the parent compound and formation of BR. However, mineralization of [*benzyl*-U-^{14}C] ZJ0273 was consistently greater than that of [*pyrimidine*-4,6-^{14}C] ZJ0273under the same conditions, suggesting that ring cleavage at the benzyl ring was more active than the pyrimidine ring. Judging from the information of BR and mineralization, ZJ0273 met the non-accumulative criteria as stated in the directives by the Commission of the European Communities (BR < 70% of the initial dose after 100 d with mineralization to CO_2 at > 5%).

In this study, four aromatic intermediates of ZJ0273, 4-(2-(4,6-dimethoxypyrimidin-2-yloxy)benzylamino)benzoic acid, 4-(2-(4,6-dimethoxypyrimidin-2-yloxy)benzamido)benzoic acid, 2-(4,6-dimethoxypyrimidin-2-yloxy)benzoic acid, and 4,6-dimethoxypyrimidin-2-ol, were identified. Two possible metabolic pathways could lead to the degradation of ZJ0273.On the one hand, in all three aerobic soils, the parent compound was initially hydrolyzed and de-esterified by cleavage of the propyl group to form 4-(2-(4,6-dimethoxypyrimidin-2-yloxy)benzylamino)benzoic acid. Since the side chain carbon atom of the benzene ring tended to be oxidized abiotically or biotically, carbonylation occurred. Conjugation was the dominant reaction in the compound 2-(4,6-dimethoxypyrimidin-2-yloxy)benzoic acid, which was ultimately decomposed to carbon dioxide. On the other hand, the degradation pathway was also initiated by the hydrolysis to 4-(2-(4,6-dimethoxypyrimidin-2-yloxy)benzylamino)benzoic acid, which was readily acylated to generate 4-(2-(4,6-dimethoxypyrimidin-2-yloxy)benzamido)benzoic acid, then, to form the main degradation product 2-(4,6-dimethoxypyrimidin-2-yloxy)benzoic acid, which then was transformed to yield 4,6-dimethoxypyrimidin-2-ol, and finally mineralized to CO_2.

Meanwhile, plant availability and phytotoxicity of soil bound residues of herbicide ZJ0273 showed that only ZJ0273 and its two degradation products M1 and M3 were identified in the extract of the released residues from BR after planting of corn or rice. Phytotoxicity assay of the three compounds revealed that M3 played a dominant role in the inhibition effect on the growth of rice seedlings. In the extractable residues released from BR, the most biologically active M3 made up of the largest fraction in all soils. Therefore, it may be concluded that the main cause of phytotoxicity from exposure to soil BR of ZJ0273 is related to the release of ZJ0273 and its degradation products and the subsequent inhibition on ALS by M3. The elevated accumulation of ^{14}C on a biomass basis in rice as compared with corn provided an explanation to the occurrence of phytotoxicity only in the former. Results from this study clearly showed that when a field previously treated with ZJ0273 is subsequently used for planting of rice, there would be an increased likelihood for herbicide induced phytotoxicity. This hypothesis is valuable since it helps to select the succeeding rotation crops to prevent yield losses.

All these findings provided the basic information that might be useful for assessing the factors related to the environmental fate and behavior of this commonly used herbicide. However, absorption, translocation, distribution, residue in the oilseed rape metabolism, and mode of action of the herbicide still remain unknown, which are well worth studying in the near future.

5. References

Boivin, A.; Amellal, S.; Schiavon, M.; Genuchten, M.T. (2005). 2,4-Dichlorophenoxyacetic acid (2,4-D) sorption and degradation dynamics in three agricultural soils. *Environ. Pollut.*, 138, 92-99.

Chen, J.; Yuan, J.; Liu, J.D.; Fu, Q.M.; Wu, J. (2005). Mechanism of action of the novel herbicide ZJ0273. *Acta Phytophylacica Sinica*, 32, 48-52.

Commission of the European Communities. (1997). European Commission Directive 97/57, establishing Annex VI (Uniform Principles) to Directive 91/414. Commission of the European Communities, Directorate-General for Agriculture, DG VI B II-1, Brussels, Belgium.

Craven, A. (2000). Bound residues of organic compounds in the soil: the significance of pesticide persistence in soil and water: a European regulatory view. *Environ. Pollut.*, 108, 15-18.

Craven, A.; Hoy, S. (2005). Pesticide persistence and bound residues in soil-regulatory significance. *Environ. Pollut.*, 133, 5-9.

Dec, J.; Bollag, J.M. (1988). Microbial release and degradation of cathecol and chlorophenols bound to synthetic humic acid. *Soil Sci. Soc. Am. J.*, 52, 1366-1371.

Environmental Protection Agency. Final Report to the U.S. Environmental Protection Agency Washington, D.C. (Topic Area: Environmental Chemistry 98-NCERQA-A3), 2002. p. 93.

Fent, G.; Hein, W.J.; Moendel, M.J.; Kubiak, R. (2003). Fate of ^{14}C-bisphenol A in soils. *Chemosphere*, 51,735-746.

Gaultier, J.; Farenhorst, A.; Cathcart, J.; Goddard, T. (2008). Degradation of [carboxyl-^{14}C] 2,4-D and [ring-U-^{14}C] 2,4-D in 114 agricultural soils as affected by soil organic carbon content. *Soil Biol. Biochem.*, 40, 217-227.

Gee, G.W.; Bauder, J.W. (1986). Particle-size analysis. In *Methods of Soil Analysis*, Part 1, Physical and Mineralogical Methods; Klute, A., Ed.; Soil Science Society of America: Madison, WI, pp. 383-412.

Gevao, B.; Semple, K.T.; Jones, K.C. (2000). Bound pesticide residues in soils: a review. *Environ. Pollut.*, 108, 3-14.

Haider, K. (1983). Abstracts. IUPAC Pesticide Chemistry. Miyamoto et al. (Eds.). p. 177–182, Pergamon Press, New York.

Khan, S.U. (1982). Bound pesticide residues in soil and plants. *Residue Rev.*, 81, 1-25.

Krause, A.; Hancock, W.G.; Minard, R.D.; Freyer, A.J.; Honeycutt, R.C.; LeBaron, H.M.; Paulson, D.L.; Lui, S.Y.; Bollag, J.M. (1985). Microbial transformation of the herbicide metolachlor by a soil actinomycete. *J. Agri. Food Chem.*, 33, 584-589.

Langenbach, T.; Schroll, R.; Paim, S. (2000). Fate and distribution of ^{14}C-atrazine in a tropical oxisol. *Chemosphere*, 40, 449-455.

Langenbach, T.; Schroll, R.; Scheunert, I. (2001). Fate of the herbicide ^{14}C-terbuthylazine in Brazilian soils under various climatic conditions. *Chemosphere*, 45, 387-398.

Lu, L.; Chen, J.; Wu, J.; Ling, W.; Mao, L.S.; Li, M.Z.; Cai, X.; Peng, W.L.; Wu, Y,; Wu, S.G.; Wang, H.J.; Wang, G.C.; Cui, H.; Han, S.D.; Qiu, W.L.; Wang, Y.H. (2004). 2-Pyrimidinyloxy-N-arylbenzylamine derivatives, their processes and uses. U.S. patent 6800590 B2.

Mordaunt, C.J.; Gevao, B.; Jones, K.C.; Semple, K.T. (2005). Formation of non-extractable pesticide residues: observations on compound differences, measurement and regulatory issues. *Soil Biol. Biochem.*, 133, 25-34.

Moyer, S.R.; Esau, R.; Kozub, G.C. (1990). Chlorsulfuron persistence and response of nine rotational crop in alkaline soils of Southern Alberta. *Weed Technol.*, 4, 543-548.

Nelson, D.W.; Sommers, L.E. (1982). Total carbon, organic carbon, and organic matter. In: *Methods of Soil Analysis*, Page, A.L., Miller, R.H., Keeney, D.R. (Eds.), Part 2, Chemical and Microbiological Properties, pp. 539-579, American Society of Agronomy, Madison, Wisconsin.

Roeth, F.W. (1986). Enhanced herbicide degradation in soil with repeat application. *Rev. Weed Sci.*, 2, 45-65.

Smith, A.E.; Lafond, G.P. (1990). Effects of long-term phenoxyalk-anoic acid field applications on the rate of microbial degradation. *ACS Symposium Series*, 426, 14-22.

Smith, A.E.; Muir, D.C.G. (1980). Determination of extractable and nonextractable radioactivity from prairie soils treated with carboxyl- and ring-labelled [^{14}C]2,4-D. *Weed Research*, 20, 123-129.

Smith, A.E.; Philips, D.V. (1975). Degradation of alachlor by Rhizoctonia solani. *Agron. J.*, 7, 347-349.

Tang, Q.H.; Chen, J.; Lu, L. (2005). An innovative research for novel rape herbicide ZJ0273. *Chin. J. Pestic. Sci.*, 44, 496-502.

Vithala, R.; White, C.K. (1996). Aerobic metabolism of flupropacil in sandy loam soil. *Pestic. Sci.*, 47, 235-240.

Walker, S.S.; Robinson, G.R.; Hargreaves, P.A. (1997). Weed control with atrazine and chlorsulfuron is determined by herbicide availability and persistence in soils. *Aust. J. Agric. Res.*, 48, 1003-1009.

Xie, H.; Guetzloff, T.F.; Rice, J.A. (1997). Fractionation of pesticide residues bound to humin. *Soil Sci.*, 162, 421-429.

Yang, Z.M.; Wang, W.; Han, A.L.; Ye, Q.F.; Lu, L. (2009). Determination of herbicide ZJ0273 residue in rapeseed by radioisotopic tracing method. *Food Chem.*, 114, 300-305.

Yang, Z.M.; Ye, Q.F.; Lu, L. (2008). Synthesis of herbicidal ZJ0273 labeled with tritium and carbon-14. *J. Label. Compd. Radiopharm.*, 51, 182-186.

Yang, Z.M.; Ye, Q.F.; Tang, Q.H.; Lu, L. (2006). Microsynthesis of a novel herbicide, propyl 4-[2-[(4,6-dimethoxy-2-pyrimidinyl)oxy]benzylamino]benzoate. *Acta Agriculturae Nucleatae Sinica*, 20(5), 423-428 [in Chinese].

Ye, Q.F.; Sun, J.H.; Wu, J.M. (2003). Causes of phytotoxicity of metsulfuron-methyl bound residues in soil. *Environ. Pollut.*, 126, 417-423.

Ye, Q.F.; Wu, J.M.; Sun, J.H. (2004). Bioavailability of bound residue derived from C-14-labeled chlorsulfuron in soil and its mechanism of phytotoxicity. *J. Environ. Sci.*, 16, 262-267.

Yukio, N.; Masahiro, M.; Kazuhiko, S., Nobuhide, W. (1996). Dimethoxypyrimidines as novel herbicides. Part 2. Synthesis and herbicidal activity of O-pyrimidinylsalicylates and analogues. *Pestic. Sci.*, 47, 115-124.

10

Application of a Laboratory Bioassay for Assessment of Bioactivity of ALS-inhibiting Herbicides in Soil

Anna M. Szmigielski[1], Jeff J. Schoenau[1], Bryce G.L. Geisel[1], Frederick A. Holm[2] and Eric N. Johnson[3]
[1]*Soil Science Department, University of Saskatchewan, Saskatoon, SK*
[2]*Plant Sciences Department, University of Saskatchewan, Saskatoon, SK*
[3]*Agriculture and AgriFood Canada, Scott, SK*
Canada

1. Introduction

Acetolactate synthase (ALS) herbicides inhibit the biosynthesis of branched chain amino acids (valine, leucine and isoleucine) in sensitive plants. The ALS-inhibitor group of herbicides includes sulfonylureas, imidazolinones, triazolopyrimidines, pyrimidinyl oxybenzoates , and sulfonylamino carbonyl trizolinones. They control a wide spectrum of broadleaf weeds and grasses and are commonly used in cereal and pulse crops, soybean and rice. Tolerant plants rapidly metabolize ALS-herbicides to an inactive product while sensitive plants show little or no metabolism of ALS-herbicides (Sweetser et al., 1981). ALS inhibition is a biological pathway that exists only in plants and not in animals, and therefore the ALS-inhibiting herbicides are considered to be safe (Colborn & Short, 1999). Because of the very high plant toxicity of ALS-inhibiting herbicides to susceptible plants, the application rates of these herbicides are remarkably low, typically between 3 to 150 g ai/ha (Senseman, 2007) making these herbicides environmentally attractive. The bioavailability of ALS-herbicides to plants is soil dependent, and the efficacy in weed control may decrease in soils of high organic matter and clay content and low pH. Dissipation of ALS-herbicides varies greatly with environmental conditions, soil characteristics and type of herbicide. Although the half-lives are relatively short, the small residual quantities remaining in soil may be of agronomic concern due to the high potency of these herbicides at low concentrations. The expected levels of soil residual ALS-inhibiting herbicides one year after application are at or below one part per billion concentrations. Addressing concerns regarding possible damage to successive crops requires the ability to detect extremely low concentrations of these herbicides in soil.

2. Plant bioassay techniques

Herbicide residues in soils can be determined by plant bioassays or by chemical methods. Chemical methods are specific, sensitive and quantify the total amount of herbicide residue in soil (Klaffenbach & Holland, 1993; Galletti et al., 1995; Stout et al., 1997; Szmigielska et al.,

1998; Smith, 1995). However, they may be costly, requiring extraction solvents and sophisticated analytical equipment, and can be time consuming as well. Plant bioassays are simple, inexpensive, and measure a phytotoxic portion of soil residual herbicide which typically varies with soil type and plant species. Also, because bioassays are non-specific, the effect of all residual herbicides present in soil is measured by bioassays (Johnson et al., 2005). Parameters that are frequently assessed in plant bioassays are root or shoot length, fresh or dry weight of roots or shoots, leaf area or plant height, visual estimation of plant injury, physiological and morphological effects such as photosynthetic activity, water consumption, or chlorosis (Horovitz, 1976). These measurements are assessed relative to a control sample which is needed because of the variation in plant growth in soils of different properties. Therefore, having a control soil that is identical in properties to the treated soil is considered essential for accuracy of a bioassay.

Various plant species have been used in bioassays for the determination of ALS- herbicides in soil, primarily using root measurements because of the inhibiting effect of ALS-herbicides on cell division at the root tips of susceptible plants. Some of the crops that have been used are corn for chlorsulfuron (Anderson & Humburg, 1987; Groves & Foster, 1985; Hsiao & Smith, 1983; Morishita et al., 1985), sunflower for MON-37500 and triasulfuron (Hernández-Sevillano et al., 2001), lentil for metsulfuron (Szmigielska et. al., 1998), and canola for imazethapyr (Szmigielska & Schoenau, 1999). Eliason et al. (2004) reported a sensitive bioassay using oriental mustard (*Brassica juncea* L.) as an indicator plant for flucarbazone. The mustard root bioassay was further improved by Szmigielski et al. (2008) and was used in Canadian prairie soils for investigation of other ALS-inhibiting herbicides that included imazamox-imazethapyr, sulfosufuron, florasulam, pyroxsulam and thiencarbazone. In this bioassay, oriental mustard plants are grown in 50 g soil. Soil is wetted to 100% field capacity, hand-mixed and transferred to a 2-oz Whirl-Pak bag. Subsequently, the soil in the Whirl-Pak bag is gently packed to form a layer that is approximately 8 cm deep and 1 cm thick (Fig. 1).

Fig. 1. Mustard root length bioassay performed in Whirl-Pak bags.

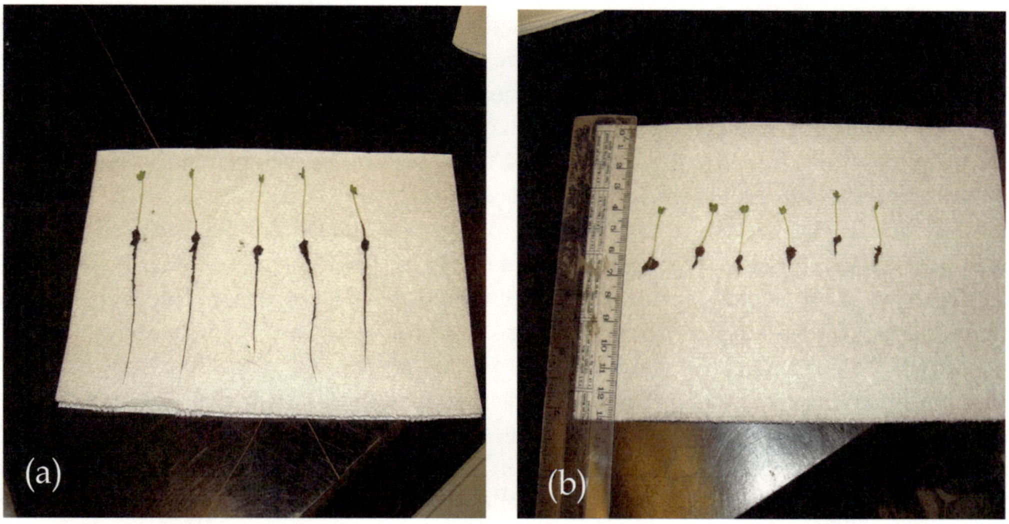

Fig. 2. (a) Mustard plants grown in control (uncontaminated) soil; (b) Mustard plants grown in treated (contaminated) soil.

Six oriental mustard seeds are planted per bag and plants are grown for three days in a fluorescent canopy. Plants are harvested after opening the bag and washing the soil away from the roots with water, and the length of roots is measured with a ruler. Root lengths in soils free of ALS-inhibiting herbicides are consistent among soils, and are in the approximate range of 7 cm ± 1 cm. A root length of 6 cm or less is considered to be indicative of herbicide residue present in the soil (Fig. 2). Because this bioassay is completed in three days, the reduction of root length is primarily due to the herbicide presence in soil as the effect of the nutrient status of the soil is minimized.

3. Effect of soil properties on herbicide phytotoxicity

Soil characteristics have a major influence on the bioavailability of ALS-inhibiting herbicides mainly because they affect sorption of herbicides to soil constituents. Most important properties in this regard are soil organic matter content, pH and texture. For studies of the relationships between herbicides, soil properties and plant species, dose-response curves based on a log-logistic regression model are frequently used (Seefeldt et al., 1995):

$$y = C + (D - C)/(1 + (x/I_{50})^{b}) \qquad (1)$$

where y = plant response, x = herbicide concentration, C = lower limit of log-logistic curve, D = upper limit of log-logistic curve, I_{50} = concentration corresponding to 50% plant growth inhibition, and b = slope of the curve around the I_{50} value. I_{50} values can be estimated from the dose-response curves and can be used for examining relationships of herbicide phytotoxicity with factors such as soil characteristics, plant species or type of herbicide.

Organic matter in the form of humus is colloidal in nature with a highly reactive surface containing many functional groups capable of binding herbicide molecules. Reduced phytotoxicity of flucarbazone (Eliason et al., 2004; Geisel et al., 2008), imazethapyr (Szmigielska & Schoenau, 1999), metsulfuron (Szmigielska et al., 1998), pyroxsulam and thiencarbazone (Szmigielski A.M., unpublished) in Canadian prairie soils of high organic matter content was explained by increased herbicide sorption. Using mustard root bioassay, dose-response curves were constructed for these herbicides and the I_{50} values were estimated. An example of varying dose-response curves with soil organic carbon content for pyroxsulam in Canadian prairie soils is shown in Fig. 3. Correlations of the I_{50} values with soil properties revealed that phytotoxicity of these herbicides was primarily related to soil organic carbon. However, soils used in the above studies had a broad range of organic carbon content but a relatively narrow range of soil pH. The narrow range of soil pH might have limited an assessment of the effect of soil pH on phytotoxicity of ALS-herbicides, as herbicide sorption to soil surfaces has been reported to be also pH-dependent. ALS-inhibiting herbicides are weak acids with pKa values in an approximate range of 3 to 5 (Senseman, 2007); therefore in soil solution at pH lower than the pKa value, these herbicides exist predominantly in nonionic form, whereas at pH values higher than the pKa, they are ionized. When these herbicides are in anionic form, their solubility in water increases and higher herbicide concentration is present in soil solution resulting in higher herbicide phytotoxicity. Soil pH also affects the ionic charges of the organic matter and clay colloids, with higher pH increasing the negative charge. Therefore, soil adsorption of weakly acidic herbicides generally decreases as soil pH increases due to repulsion of the herbicide anions from the negatively charged organic and clay surfaces. The relationship of herbicide adsorption and soil pH has been shown for many ALS-inhibiting herbicides including chlorsulfuron (Mersie & Foy, 1985), imazaquin and imazethapyr (Renner et al., 1988; Che et al., 1992; Loux & Reese, 1992; Goetz et al., 1986), imazapyr (Wang & Liu, 1999; Wehtje et al., 1987), and chlorimuron (Goetz et al., 1989). The effect of clay content on herbicide bioavailability is similar to the effect of organic matter in that the high surface area of clay can increase herbicide sorption and may further reduce herbicide bioavailability.

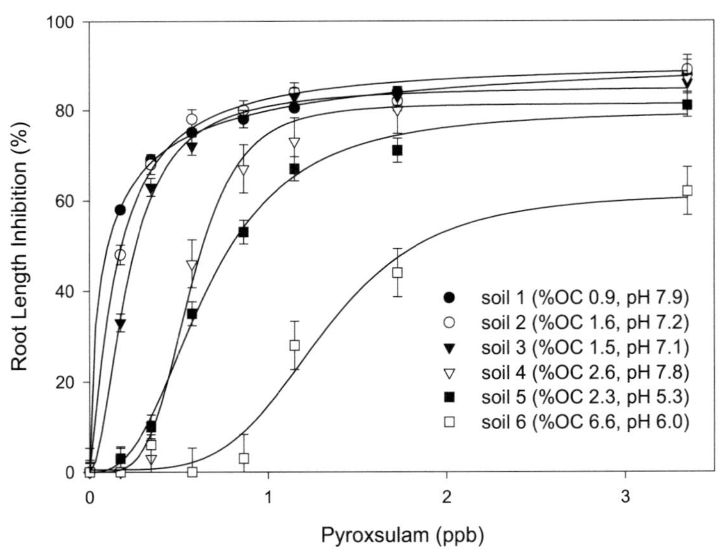

Fig. 3. Dose-response curves for pyroxsulam in Saskatchewan (Canada) soils as determined by the mustard root length bioassay under laboratory conditions.

The reduced phytotoxicity associated with high organic carbon and clay content and low soil pH may result in decreased efficacy of ALS-inhibiting herbicides. However, these soil characteristics may contribute to minimizing the injury to rotational crops by lowering the phytotoxicity of ALS-herbicide residues remaining in soil one year after application.

Because soil properties vary within the farm field landscape, the bioavailability of ALS-herbicides is affected by field topography (Schoenau et al., 2005). Studies of metsulfuron (Szmigielska et al., 1998), imazethapyr (Szmigielska & Schoenau, 1999) and flucarbazone (Eliason, 2003;) in Canadian prairie soils revealed that in lower slope positions, herbicide phytotoxicity was decreased as compared to the upper slope positions in the same farm field (Fig. 4). Reduced phytotoxicity in the lower slope soil is explained by the higher organic matter and clay content and lower pH compared to the upper slope soil. Thus, potential landscape effects on the phytotoxicity of ALS-herbicides should be taken into consideration when herbicides are applied to fields of variable topography.

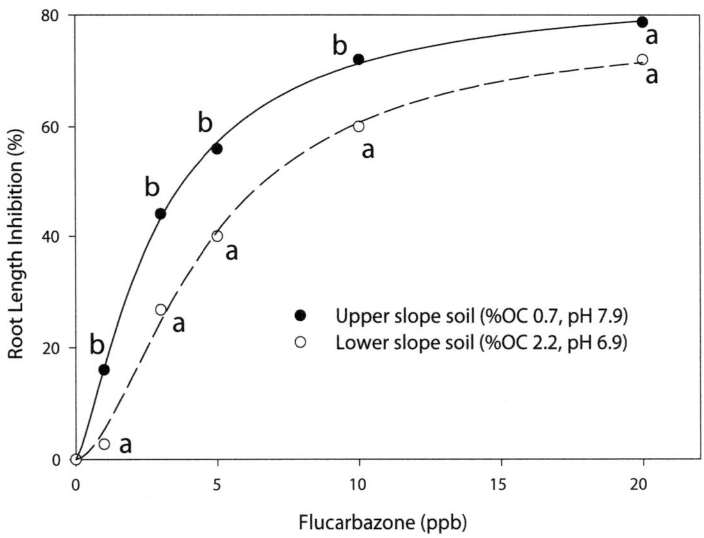

Fig. 4. Effect of landscape positions on phytotoxicity of flucarbazone in a soil from southern Saskatchewan (Canada); means for each flucarbazone concentration followed by the same letter are not significantly different at $p \leq 0.05$.

4. Herbicide dissipation under laboratory and field conditions

Herbicide dissipation in soils is influenced by environmental conditions and soil properties (Walker, 1991). The two primary mechanisms of ALS-herbicide dissipation are microbial degradation and chemical hydrolysis. Both these processes are dependent on soil water and temperature with faster dissipation occurring in moist and warm soils, particularly for herbicides that dissipate mainly through microbial processes (Beckie & McKercher, 1989; Joshi et al., 1985; Walker & Brown, 1983; Brown, 1990). Furthermore, experiments showed that degradation of ALS-herbicides is faster in non-sterile, microbially active soils than in sterile soils. The sterile degradation is slower and is due solely to chemical hydrolysis (Joshi et al., 1985).

Soil properties such as organic matter content, soil pH and texture play an important role in the dissipation of ALS-herbicides. High organic matter and clay content decrease the dissipation rate by limiting the amount of herbicide available in soil solution for biodegradation because of the adsorption process. A low soil pH tends to increase the persistence of ALS-herbicides as herbicide's adsorption to soil particles is enhanced under acidic conditions. Also, at low soil pH the ALS-herbicides are likely to exist as neutral molecules which are less soluble in water than ionized molecules, thus further reducing the amount of herbicide available for degradation.

Generally, the pattern of dissipation of ALS-herbicides is biphasic both under laboratory and field conditions (Brown, 1990; Loux et al., 1989; Hill & Schaalje, 1985; LaFleur, 1980; Eliason et al., 2004). In the biphasic dissipation process, initial rapid dissipation is followed by a slower dissipation rate at lower residual concentrations. In the rapid stage of dissipation, the readily available portion of the herbicide is degraded, whereas in the slow stage, the remaining molecules are tightly adsorbed to the soil particles and are less available for dissipation. (Zimdahl & Gwynn, 1977). A two-compartment (bi-exponential) regression model is frequently used to describe dissipation of ALS-herbicides in soil (Hill & Schaalje, 1985):

$$C = C_0 \exp[-(k_s + k_r)t] + C_0 \frac{k_r}{k_s + k_r - k_d} \{(\exp[-k_s t] - \exp[-(k_s + k_r)t]\} \qquad (2)$$

where C = herbicide concentration remaining in soil after time t, C_0 = initial herbicide concentration, k_d = dissipation rate constant, k_s = surface loss rate constant, and k_r = retention rate constant.

Herbicide half-lives can be estimated from the dissipation curves and their relationships with parameters such as environmental conditions or soil characteristics can be examined, as the half-lives of ALS-herbicides vary greatly with temperature, moisture, soil type, and also with herbicide type. Goetz et al. (1990) found half-lives for imazethapyr to range from 192 to 318 days in a silty-clay soil and from 78 to 270 days in a silty-loam soil both incubated at different soil moisture and temperature conditions. Chlorsulfuron half-life was longer at lower temperature (229 days at 10 C and 62.5 days at 40 C) and varied with soil pH (88.5 days at pH 6.2 and 144 days at pH 8.1) (Thirunarayanan et al., 1985). For amidosulfuron half-life values ranged from 14 days in a loamy sand incubated at 30 C to 231 days in a clay incubated at 10 C (Smith & Aubin, 1992). Beckie & McKercher (1989) reported half-lives for DPX-A7881 herbicide to increase from 33 to 214 days in a soil with pH adjusted from 5.5 to 8.1. Using the mustard root bioassay, Eliason et al. (2004) studied flucarbazone dissipation in Canadian prairie soils of contrasting properties under laboratory conditions of 25 C and moisture content of 85% field capacity. Flucarbazone half-lives ranged from 6 to 110 days and half-lives were significantly correlated with soil organic carbon with longer half-lives in soils of higher organic carbon. Johnson E.N (unpublished) found that the half-life for flucarbazone dissipation under field conditions in a loamy textured Dark Brown Chernozem was approximately 11 days and was not dependent on flucarbazone application rate (Fig. 5). Although half-lives of ALS-herbicides are generally short under optimal conditions of moisture and temperature, under conditions of drought and/or cold weather the ALS-inhibiting herbicides may persist in soil and may carry over to the next growing season at levels that cause injury to rotational crops.

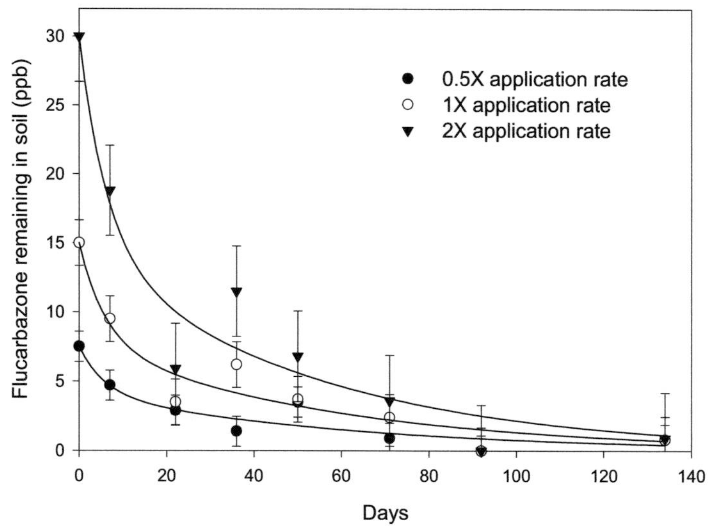

Fig. 5. Field dissipation of flucarbazone in a loamy textured Dark Brown Chernozem soil (organic carbon 2.3%, pH 5.3) in central Saskatchewan (Canada) at different application rates.

5. Predicting potential carry-over injury to subsequent crops

For rotational crop damage to occur, a herbicide must have sufficient persistence in soil, must be available to plants at a phytotoxic level, and the rotational crop must be susceptible to the residual concentration of the herbicide remaining in soil at the time of planting (Hartzler et al., 1989). Because of the high potency of the ALS-herbicides, the minuscule residual levels may remain active and may exert an effect on sensitive crops. Moyer et al. (1990) reported that when chlorsulfuron was applied in wheat on alkaline soils of relatively low organic matter in Alberta (Canada) at the recommended level of 40 g ai/ha, successive crops were affected; the required time for recropping barley, canola, pea, bean, flax, potato, alfalfa, sugar beet and lentil ranged from 2 to 7 years. In another study in Alberta (Moyer & Esau, 1996), imazamethabenz reduced the yield of sugar beet seeded 1 year after application, while imazethapyr application increased the risk of yield loss of flax, corn, meadow bromegrass, mustard, sunflower, timothy, wheat, canola, sugar beet and potato seeded between 1 to 3 years later. Crops that were reported as very sensitive to sulfonylurea herbicides are lentil, sugar beet and turnip, and as sensitive are alfalfa, canola, corn, flax, garden cress, lettuce, mustard and sunflower (Smith, 1995; Moyer & Hamman, 2001).

Typically, ALS-herbicides remaining in soil one year after application are at very low concentrations and are difficult to detect by analytical methods. Therefore plant bioassays are frequently used for the determination of bioactive herbicide residues. For the results of a bioassay to be most reliably interpreted, plant response in contaminated soil should be compared to the plant response in a control (uncontaminated) soil. However, having a control soil sample identical in properties to treated soil may be a problem especially in testing farm field soils. To overcome this difficulty, a designated non-contaminated soil may be used as a control as described by Watson & Checkel (2005). In their method the target crop, a check crop and a sensitive crop are planted in both the soil submitted for testing and a check soil, and the symptoms consistent with herbicide damage are reported.

In the mustard root length bioassay reported by Szmigielski et al. (2008), root lengths in non-contaminated soils are uniform among soils, thus the need for a control sample is eliminated. This method is not intended to determine ALS-herbicides quantitatively but rather to "red-flag" the soils for potential presence of herbicide residues. For flucarbazone that was applied in replicated field trials in western Canada, comparison of the results of the mustard root bioassay and chemical analysis with the yield of subsequent crops revealed that the bioassay was a better predictive tool for yield reduction than chemical analysis (89% and 27% agreement, respectively). While these results showed that the mustard root bioassay provides a good level of accuracy in predicting injury, 6% of the results were false positive and 5% were false negative. False positives (flucarbazone detected by the bioassay but no crop injury observed in the field) pose no risk of crop damage; however they would restrict re-cropping options. False negatives (no flucarbazone detected by the bioassay but crop injury observed in the field) could represent significant crop damage and loss of income for the grower.

Interpreting bioassay results and making re-cropping recommendations is a complex task and should be approached with caution (Watson & Checkel, 2005). Soil field sampling is critical for the bioassays because a single sample may not be representative of the whole field unless a sample is carefully obtained either by using a composite sample from different locations in the field or by sampling different parts of the field separately. Factors such as soil characteristics (organic matter content, pH and texture), farm field topography, previous herbicide use, crop history and weather conditions should be considered together with the bioassay results when determining which crops to grow in the following year.

6. Herbicide interactions after successive field applications

Repeated applications of ALS-inhibiting herbicides may result in interactions of the residues existing in soil from previous years with a herbicide that is applied in succession. Combined effect of two or more herbicides may be additive if actual and expected responses are similar, may be synergistic if the actual response is greater than the expected, and may be antagonistic if the actual response is less than the expected (Colby, 1967). Because many ALS-herbicides have residual properties, a potential for interactions exists with successive applications.

Limited research has been conducted on the crop response from repeated applications of the same ALS-herbicide or different ALS-herbicides. Moyer & Hamman (2001) reported that residues of MON 37500 herbicide combined with either imazethapyr or metsulfuron, or trisulfuron resulted in additive injury to sugar beet. Johnson et al. (2005) reported that the application of ALS-herbicide can predispose the following crop to higher levels of phytotoxicity from postemergence ALS-herbicide application.

Interactions of imazamethabenz, flucarbazone, sulfosulfuron and florasulam in combination with imazamox/imazethapyr in western Canadian soils were investigated in laboratory experiments (Geisel, 2007) and field trials (Geisel et al., 2008). In the laboratory experiments, soils were amended with individual herbicides and combinations of herbicides, and their effect on mustard root length inhibition was measured. The interactions between the investigated herbicide combinations were additive: the expected (calculated) responses and the actual (observed) responses to each pair of herbicides were the same, as seen in Fig. 6 for the imazamox/imazethapyr and florasulam combination.

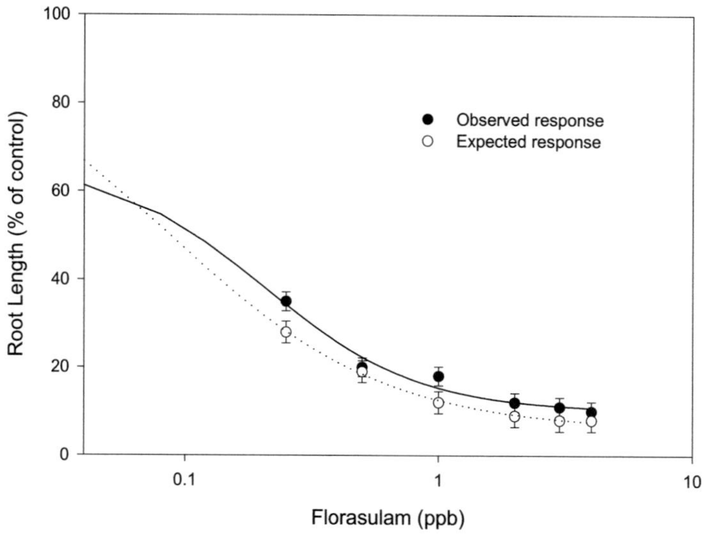

Fig. 6. Mustard root response (as % of control) for florasulam in combination with imazamox/imazethapyr in a Dark Brown clay textured Chernozem soil from southern Saskatchewan (Canada).

In the field trials, herbicides were applied sequentially over the course of 2 years; in the first year only imazamox/imazethapyr was applied and in the second year imazamethabenz, flucarbazone, sulfosulfuron or florasulam was added to the plots. All plots were sampled in the third year before the next crop was seeded and the herbicide residues were determined with the mustard root bioassay. Similar to observations for the laboratory experiments, in the field trials herbicide residue combinations showed additive injury.

As application of ALS-herbicides in successive years is becoming a frequent practice, producers need to be aware of the fact that the bioactivity of the herbicide residues persisting in soil from previous years may add to the bioactivity of the applied ALS-herbicide and that this practice may result in increased risk of injury to subsequent crops that are sensitive to both herbicides.

For some herbicides, repeated applications may also lead to enhanced degradation. Enhanced degradation occurs when a herbicide is applied to a field that received a prior treatment of the same herbicide (Roeth et al., 1989; Walker & Welch, 1991). It is believed that enhanced degradation is a result of microbial adaptation which consequently increases the rate of microbial activity. Enhanced degradation helps in minimizing the concentration of residual herbicide that may persist in soil to the following season. However, it may also result in reduced weed control in the year of application (Johnson et al., 2005).

7. Conclusions

Plant bioassays are an effective tool in research and in soil testing because they detect the phytotoxic amount of herbicide present in soil. A laboratory bioassay based on the root length inhibition of oriental mustard for detection of ALS-herbicides is simple and quick; it is completed in three days and uses only 50 g of soil per replication. With measurements generally requiring 4 replications, the total amount of soil needed to perform the bioassay is 200 g. Root development in a Whirl-Pak bag is not restricted as the bioassay is completed before the roots grow to the bottom of the bag, therefore root reduction is only due to the presence of an ALS-herbicide. Recovery of roots from soil is very easy because soil is removed from roots by a gentle stream of water after the bags are cut open, and roots do not get damaged or broken before being measured. Consequently, the results of the bioassay are reproducible (coefficient of variation of approximately 6% based of 4 replications) and sensitive (detection limit of approximately 1 ppb).

The mustard root bioassay was successfully used to examine activity and behavior of several ALS-inhibiting herbicides in soils of the Canadian prairies. Phytotoxicity and persistence of ALS-herbicides was found to be mainly affected by organic matter content in prairie soils: higher organic matter content results in decreased phytotoxicity and in slower dissipation. Thus efficacy in weed control in the season of application will be lower and potential for herbicide carry-over to the next season will higher in soils of high organic carbon. A study of carry-over injury showed that a mustard root bioassay is a useful technique for predicting potential crop damage due to residual ALS-herbicides. Laboratory and field evaluations of the effects of combinations of different ALS-inhibiting herbicides in soil showed that interactions among residues in prairie soils are additive.

8. References

Anderson, R.L. & Humburg, N.E. (1987). Field duration of chlorsulfuron bioactivity in the central Great Plains. *J. Environ. Qual.* 16:263-266

Beckie, H.J. & McKercher, R.B. (1989). Soil residual properties of DPX-A7881 under laboratory conditions. *Weed Sci.* 37:412-418.

Brown, H.M. (1990). Mode of action, crop selectivity and soil relations of the sulfonylurea herbicides. *Pestic. Sci.* 29:263-281.

Che, M.; Loux, M.M.; Traina, S.J. & Logan, T.J. (1992). Effect of pH on sorption and desorption of imazaquin and imazethapyr on clays and humic acid. *J. Environ. Qual.* 21:698-703.

Colborn, T. & Short, P. (1999). Pesticide use in the U.S. and policy implications: A focus on herbicides. *Toxicol. Ind. Health.* 15:241-276.

Colby, S.R. (1967). Calculating synergistic and antagonistic responses of herbicide combinations. *Weed Sci.* 15:20-22.

Eliason, R. (2003). Assessing the phytotoxicity and persistence of flucarbazone in prairie soils. MSc. Thesis. University of Saskatchewan, Saskatoon, SK, Canada.

Eliason, R.; Schoenau, J.J.; Szmigielski, A.M. & Laverty, W.M. (2004). Phytotoxicity and persistence of flucarbazone-sodium in soil. *Weed Sci.* 52:857-862.

Galletti, G.C.; Bonnetti, A. & Dinelli, G. (1995). High performance liquid chromatographic determination of sulfonylureas in soil and water. *J. Chromatogr.* 692:27-37.

Geisel, B.G.L. (2007). The phytotoxic effect of ALS-inhibiting herbicide combinations in prairie soils. MSc. Thesis. University of Saskatchewan, Saskatoon, SK, Canada.

Geisel, B.G.L.; Schoenau, J.J.; Holm, F.A. & Johnson, E.N. (2008). Interactions of ALS-inhibiting herbicide residues in three prairie soils. *Weed Sci.* 56:624-627.

Goetz, A.J.; Lavy, T.L. & Gbur, E.E. (1990). Degradation and field persistence of imazethapyr. *Weed Sci.* 38:421-428.

Goetz, A.J.; Walker, R.H.; Wehtje, G. & Hajek, B.F. (1989). Sorption and mobility of chlorimuron in Alabama soils. *Weed Sci.* 37:428-433.

Goetz, A.J.; Wehtje, G.; Walker, R.H & Hajek, B. (1986). Soil solution and mobility characterization of imazaquin. *Weed Sci.* 34:788-793.

Groves, K.E.M. & Foster, R.K. (1985). A corn (*Zea mays* L.) bioassay technique for measuring chlorsulfuron levels in three Saskatchewan soils. *Weed Sci.* 33:825-828.

Hartlzler, R.G.; Fawcett, R.S. & Owen, M.D. (1989). Effects of tillage on trifluralin residue carryover injury to corn. *Weed Sci.* 37:609-615.

Hernández-Sevillano, E.; M. Villarroya, M.; Alonso-Prados, J.L & J.M. García-Baudín, J.M. (2001). Bioassay to detect sulfosulfuron and triasulfuron residues in soil. 15: 447-452.

Hill, B.D. & Schaalje, G.B. (1985). A two-compartment model for the dissipation of deltamethrin on soil. *J. Agric. Food Chem.* 33:1001-1006.

Horowitz, M. (1976). Application of bioassay techniques to herbicide investigations. *Weed Res.* 16:209-215.

Hsiao, A.I. & Simth, A.E. (1983). A root bioassay procedure for the determination of chlorsulfuron, diclofop acid and sethoxydim residues in soils. *Weed Res.* 23:231-236.

Johnson, E.N.; Moyer, J.R; Thomas, A.G.; Leeson, J.Y.; Holm, F.A.; Sapsford, K.L.; Schoenau, J.J.; Szmigielski, A.M.; Hall, L.M.; Kuchuran, M.E. & Hornford, R.G. (2005). Do repeated applications of residual herbicides result in herbicide stacking? In *Soil Residual Herbicides: Science and Management. Topics in Canadian Weed Science, ed.* R.C. Van Acker, 53-70, Volume 3. Sainte-Anne-de Bellevue, Québec: Canadian Weed Science Society - Sociéte canadienne de malherbologie.

Joshi, M.M.; Brown, H.M. & Romesser, J.A. (1985). Degradation of chlorsulfuron by soil microorganisms. *Weed Sci.* 33:888-893.

Klaffenbach, P. & Holland, P.T. (1993). Analysis of sulfonylurea herbicides by gas-liquid chromatography. Determination of chlorsulfuron and metsulfuron-methyl in soil and water samples. *J. Agric. Food Chem.* 41:396-401.

Lafleur, K.S. (1980). Loss of pesticides from congaree sandy loam with time: characterization. *Soil Sci.* 130:83-87.

Loux, M.M.; Liebl, R.A. & Slife, F.W. (1989). Availability and persistence of imazaquin, imazethapyr, and clomazone in soil. *Weed Sci.* 37:259-267.

Loux, M.M. & Reese, K.D. (1992). Effect of soil pH on adsorption and persistence of imazaquin. *Weed Sci.* 40:490-496.

Mersie, W. & Foy, C.L. (1985). Phytotoxicity and adsorption of chlorsulfuron as affected by soil properties. *Weed Sci.* 33:564-568.

Morishita, D.W.; Thill, D.C.; Flom, D.G.; Campbell, T.C. & Lee, G.A. (1985). Method for bioassaying chlorsulfuron in soil and water. *Weed Sci.* 33:420-425.

Moyer, J.R. & Esau, R. (1996). Imidazolinone herbicide effects on following rotational crops in southern Alberta. *Weed Technol.* 10:100-106.

Moyer, J.R.; Esau, R. & Kozub, G.C. (1990). Chlorsulfuron persistence and response of nine rotational crops in alkaline soils of southern Alberta. *Weed Technol.* 4:543-548.

Moyer, J.R. & Hamman, W.H. (2001). Factors affecting the toxicity of MON 37500 residues to following crops. *Weed Technol.* 15:42-47.

Renner, K.A.; Meggitt, W.F. & Penner, D. (1988). Effect of soil pH on imazaquin and imazethapyr adsorption to soil and phytotoxicity to corn (*Zea mays*). *Weed Sci.* 36:78-83.

Roeth, F.W.; Wilson, R.G.; Martin, A.R & Shea, P.J. (1989). Enhanced soil degradation of butylate and EPTC in Nebraska fields. *Weed. Technol.* 2:45-65.

Schoenau, J.J.; Szmigielski, A.M. & Eliason, R.C. (2005). The effect of landscape position on residual herbicide activity in prairie soils. In *Soil Residual Herbicides: Science and Management. Topics in Canadian Weed Science, ed.* R.C. Van Acker, 45-52, Volume 3. Sainte-Anne-de Bellevue, Québec: Canadian Weed Science Society - Société canadienne de malherbologie.

Seefeldt, S. S.; Jensen, J.E. & Fuerst, E.P. (1995). Log-logistic analysis of herbicide dose-response relationships. *Weed. Technol.* 9:218-227.

Senseman, S.A. (2007). Herbicide Handbook, ninth ed. Weed Science Society of America, Lawerence, KS.

Smith, A.E. (1995). A review of analytical methods for sulfonylurea herbicides in soil. *Intern. J. Environ. Anal. Chem.* 59:97-106.

Smith, A.E. & Aubin, A.J. (1992). Degradation of the sulfonylurea herbicide [^{14}C]amidosulfuron (HOE075032) in Saskatchewan soils under laboratory conditions. *J. Agric. Food Chem.* 40:2500-2504.

Stout, S.J.; daCunha, A.R. & Safarpour, M.M. (1997). Simplified determination of imidazolinone herbicides in soil at parts-per-billion level by liquid chromatography/electrospray ionization tandem mass spectrometry. *J. Assoc. Off. Anal. Chem.* 80:426-432.

Szmigielska, A.M. & Schoenau, J.J. (1999). Analysis of imazethapyr in agricultural soils by ion exchange membranes and a canola bioassay. *Commun. Soil Sci. Plant Anal.* 30:1831- 1846.

Szmigielska, A.M.; Schoenau, J.J. & Greer, K. (1998). Comparison of chemical extraction and bioassay for measurement of metsulfuron in soil. *Weed Sci.* 46:487-493.

Szmigielski, A.M.; Schoenau, J.J.; Irvine, A. & Schilling, B. (2008). Evaluating a mustard root-length bioassay for predicting crop injury from soil residual flucarbazone. *Commun. Soil Sci. Plant Anal.* 39:413-420.

Sweetser, P.B.; Schow, G.S. & Hutchison, J.M. (1981). Metabolism of chlorsulfuron by plants: biological basis for selectivity of a new herbicide for cereals. *Pestic. Biochem. Physiol.* 17:18-23.

Thirunarayanan, K.; Zimdahl, R.L. & Smika, S.E. (1985). Chlorulfuron adsorption and degradation in soil. *Weed Sci.* 33:558-563.

Walker, A. (1991). Influence of soil and weather factors on the persistence of soil-applied herbicides. *Appl. Plant Sci.* 5:94-98.

Walker, A. & Brown, P.A. (1983). Measurement and prediction of chlorsulfuron persistence in soil. *Bull. Environ. Contam. Toxicol.* 30:365-372.

Walker, A. & Welch, S.J. (1991). Enhanced degradation of some soil-applied herbicides. *Weed Res.* 31:49-57.

Wang, Q. & Liu, W. (1999). Correlation of imazapyr adsorption and desorption with soil properties. *Soil Sci.* 164:411-416.

Watson, P.R. & Checkel, S. (2005). Soil residual herbicide bioassays: Science and practice. In *Soil Residual Herbicides: Science and Management. Topics in Canadian Weed Science, ed.* R.C. Van Acker, 71-79, Volume 3. Sainte-Anne-de Bellevue, Québec: Canadian Weed Science Society - Société canadienne de malherbologie.

Wehtje, G.; Dickens, R.; Wilcut, J.W. & Hajek, B.F. (1987). Sorption and mobility of sulfometuron and imazapyr in five Alabama soils. *Weed Sci.* 35:858-864.

Zimdahl, R. L. & Gwynn, S.M. (1977). Soil degradation of three dinitroanilines. *Weed Sci.* 25:247-251.

11

An Electrochemical Approach to Quantitative Analysis of Herbicides and to the Study of Their Interactions with Soils Components

Ana Valeria Juarez, Julieta Soledad Riva and Lidia Mabel Yudi
Departamento de Fisicoquímica, Facultad de Ciencias Químicas, Universidad Nacional de Córdoba, Instituto de Investigaciones en Fisicoquímica de Córdoba (INFIQC)
Argentina

1. Introduction

Triazine herbicides are an important group of pesticides. Most of these compounds derived from the heterocyclic shown in Figure 1. It has three nitrogen atoms in positions 1, 3 and 5 and different substituents in position 2, 4 and 6. These triazines have a protonation - deprotonation site on the nitrogen atom labeled 5 in Figure 1. Substituents of different compounds are shown in table 2.

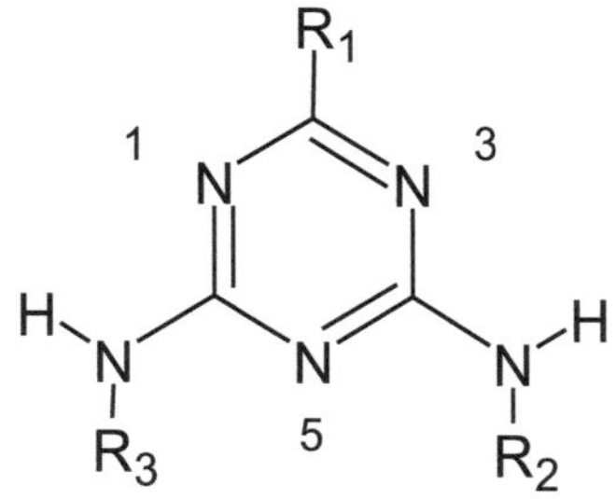

Fig. 1. Chemical structure of s-triazine core, substituent in R_1 : Cl, SCH_3 or OCH_3; R_2 and R_3: hydrocarbons chains.

These herbicides have been extensively applied to pre and post-emergence weed control. Many studies were focused on ecological and health hazards of these compounds and their toxic effects are very well known. For this reason, the use of some triazine pesticides has been banned in some countries or their permitted levels in drinking water is very low, so that analytical procedures for quantitative determination of several triazines, as well as their degradation products, at low levels are often requested.
In this sense, several analytical techniques have been developed, like HPLC (Katsumata et al., 2006), CG-MS, capillary electrophoresis (Frías et al., 2004), solid-phase micro-extraction coupling with GC, LC, ion mobility spectrometry (Garcia Galan et al., 2010; Mohammadi et al., 2009; Sanchez Ortega et al., 2009; Quintana et al., 2001) and with HPLC (Zhou et al.,

2009; See et al., 2010), immunosensors (Bahnd et al., 2005) and multi-biosensor based on immobilized Photosystem II (Touloupakisa et al., 2005; Dong et al., 2009), micellar electrokinetic chromatography (Zhang et al., 2008), tandem techniques (Beale et al., 2009; Tsang et al.,2009; Lacina et al., 2010) cyclic voltammetry (Fuchiwaki et al, 2009; Zapardiel et al., 2000) and differential-pulse polarography (Ignjatovic et al., 1993; Kubo et al., 2008; Vaz et al., 1996) on solid electrodes, photosynthetic electron transport (PET) electrochemical biosensors (Campàs, et al., 2008; Preuss & Hall, 1995), PET colorimetric detection (Brewster & Lightfield, 1993; Shao et al., 2002) and adsorptive stripping voltammetry in dispersed media (Pedredo et al., 1995).

In the last years, the environmental pollution by pesticides has become in a serious problem especially in aquatic ecosystems, due to their heavy use in agriculture and to their persistence. The half-lives of herbicides vary from weeks to several months and, under environmental conditions, they are usually degraded to compounds with better water solubility. Indeed, the most important physicochemical properties of these pesticides and their degradations products are the solubility in water and the capacity to be retained by soils (Aelion & Mathur, 2001; Besse-Hogan et al., 2009). So that, the use of agricultural chemicals requires knowledge of their stability and transformation in the environment as well as their influence on micro-organisms. These s-triazine herbicides and some of their degradation products are used by water and soil microbes as a source of energy (alkyl fragments) and nitrogen (amine fragments) (Lyapchenko et al., 2004). For this reason, not only the development of new sensitive and selective analytical techniques for the determination of s-triazine herbicides and their metabolites in the environment, but also the recognition of their interactions with different elements, especially with heavy metals cations and organic compounds present in soils, are important problems in modern s-triazine chemistry. The study of complex formation or adsorption behavior between herbicides and cations or organic molecules contained in soils is an important topic because it determines pesticide mobility, its bioavailability and its effectiveness. Regarding to the interaction with inorganic species, Al(III) is a cation present in most soils, and several authors have studied its complexes with different herbicides in aqueous solutions or in complex model systems, which closely simulate those found in soils by using pure montmorillonite or montmorillonite covered by different amounts of OH-Al species (chlorite-like complexes) as adsorbents (Sannino et al., 1999). Several methods were employed in these investigations: macroscopic and molecular scale techniques, potentiometric titration data combined with EXAFS, ATR-FTIR and NMR, as well as spectroscopic data (Jonsson, 2007).

The present chapter is focused in the progress made in the s-triazine quantification as well as in the study of their interactions with inorganic compounds of soils employing electrochemical methods applied at the interface between two immiscible electrolyte solutions (ITIES) (Juarez & Yudi, 2003; Juarez & Yudi 2008; Juarez & Yudi 2009).

The interface between two immiscible electrolyte solutions and the transport of different ions across it are an important branch of electrochemistry because of their importance in the examination of heterogeneous kinetics and potential analytical applications (Reymond et al., 2000). This methodology is used as an appropriate electroanalytical technique for quantitative determination of organic ions. The possibility of working in an oil/water system overcomes problems such as the low solubility of many organic compounds, like the case of s-triazines, in water. Moreover, the traces quantification of pesticides in different kind of samples requires pre-concentration techniques. In the past few years, new

techniques were developed like liquid-liquid extraction, solid phase extraction, molecular imprinted polymers and carbon nanotubes, among others. These pre-concentration procedures were employed before the quantification of the pesticide, coupled to different techniques like GC-MS, capillary electrophoresis, non-aqueous capillary electrophoresis and micellar electrokinetic capillary chromatography (Katsumata et al., 2006; Sambe et al, 2007; Zhou et al., 2006; Carabias-Martínez et al, 2006; Hu et al., 2009; Pinto et al., 2010; See et al., 2010). In this sense, the use of a combined procedure consisting in a previous pre-concentration stage, followed by square wave voltammetry at a water/1,2-dichloroethane interface has achieved to improve the detection limit for s-triazines quantification (Juarez & Yudi, 2009). The pre-concentration of the analyte in the organic phase is possible due to its high solubility and partition coefficient in this solvent.
On the other hand, voltammetry at ITIES has proven to be a valuable tool to elucidate the stoichiometry of complex formation (Reymond et al., 1998; O'Dwyer & Cunnane et al., 2005; Azcurra et al., 2003; Caçote et al., 2004; Rahman et al., 2001; Yudi et al., 1992) and to identify and evaluate successive complex formation at the interface (Kakiuchi & Senda, 1991; Kakiuchi, 1993; Reymond et al., 1998). With the purpose of contributing to the knowledge of the interaction between Prometryne and soils components, the complex formation of the herbicide PROM with Al(III) cation at the water / 1,2 - dichloroethane interface, has been studied and the results are presented in this chapter (Juarez & Yudi, 2008).

2. Methodology

2.1 Liquid-liquid Interfaces

The interface between two immiscible electrolytes is one with its own dynamics. The structure of double layer in these interfaces has been studied since 1939. Over the years, the information obtained was used to develop a clearly defined model of the liquid-liquid interface (Girault, 1987). Surface tension measurements (Girault & Schiffrin, 1983) and capacitance (Samec at al., 1983) provides access to important interfacial parameters. The model proposed from these results, considers an interfacial area on which the molecules of both solvents are mixed. The penetration of ions in the solvent mixture zone (ZMS) depends on their hydrophobicity or hydrophilicity (Girault & Schiffrin, 1984). Figure 2 schematizes two immiscible solutions (α and β) with an ion X^{z+} in common. Under these conditions a potential difference is generated determined by the Nernst-Donnan equation (Koryta, 1979):

$$\Delta_{\alpha}^{\beta}\varphi = \left[\mu_{X^{z+}}^{o}(\alpha) - \mu_{X^{z+}}^{o}(\beta)\right] / z_X F + \frac{RT}{z_X F}\ln\left[\frac{a_{X^{z+}}^{\alpha}}{a_{X^{z+}}^{\beta}}\right] \tag{1}$$

$$\Delta_{\alpha}^{\beta}\varphi = -\Delta G_{tr,X^{z+}}^{o,\alpha\rightarrow\beta} / z_X F + \frac{RT}{z_X F}\ln\left[\frac{a_{X^{z+}}^{\alpha}}{a_{X^{z+}}^{\beta}}\right] \tag{2}$$

where φ is the internal potential, $\mu_{X^{z+}}^{0}(\alpha)$ and $\mu_{X^{z+}}^{0}(\beta)$ are the standard chemical potentials, z_X is the ion charge and $a_{X^{z+}}^{\alpha}$ and $a_{X^{z+}}^{\beta}$ are the X^{z+} activities in both phases α and β, respectively. $\Delta G_{tr,X^{z+}}^{0,\alpha\rightarrow\beta}$ is the Free Energy difference of solvatation for the X^{z+} ion in α and β, respectively, which is related to the standard transfer potential of X^{z+} ion according to equation 3.

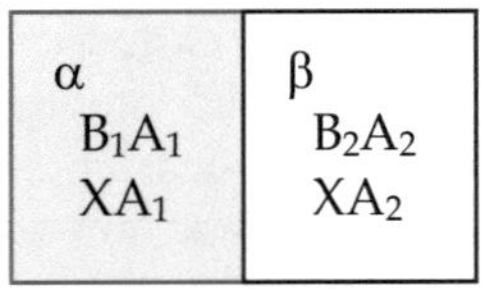

Fig. 2. B_1A_1 and B_2A_2: base electrolytes in α and β phases, respectively. X^{z+}: transferable cation between both phases.

$$\Delta_{\alpha}^{\beta}\varphi_{X^{z+}}^{o} = -\frac{\Delta G_{tr,X^{z+}}^{o,\alpha\rightarrow\beta}}{z_x F} \tag{3}$$

In absence of current flux, the equilibrium condition is established and the potential difference, $\Delta_{\alpha}^{\beta}\varphi_{eq}$, is determined by equation 1. If a potential difference greater than the equilibrium potential is applied, by an external source, to the system, two processes occur: the charging of the double layer and the ion transfer trough the interface. Figure 3 shows the voltammetric profiles obtained when a linear potential sweep is applied at the aqueous/organic interface in presence and absence of a semi hydrophobic X^{z+} cation. The transfer of base electrolyte, between the two phases, limits the potential window (grey line). Within this potential window, the interface behaves as an ideal polarizable electrode. The presence of semi hydrophobic X^{z+} ions, in the system, gives rise to positive and negative current peaks when their transfer from aqueous to organic phase, and vice versa respectively, occurs at potential values within the limits of potential window, as can be seen in black line in Figure 3.

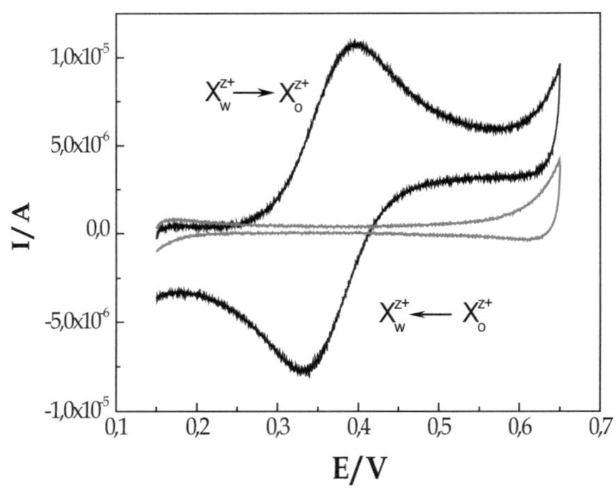

Fig. 3. Voltammetric profiles obtained when a linear potential sweep is applied at the interface between two immiscible solutions: (grey) Base electrolytes present in both phases; (black) base electrolytes and a semi hydrophobic cation. *w*: aqueous phase, *o*: organic phase.

2.2 Charge transfer reactions between liquid / liquid interfaces

One of the characteristics of the ITIES is the diversity of charge transfer reactions which can be studied by electrochemical methodologies (Girault, 1993). These charge transfer reactions can be classified into three main categories:

- Direct ion transfer,
- Assisted ion transfer,
- Electron transfer.

2.2.1 Direct ion transfer

In this case, an ion present in the aqueous phase is transferred to the organic phase (or vice versa) according to equation 4, favored by polarizing the interface.

$$X_{(w)}^{z+} \rightleftarrows X_{(o)}^{z+} \quad (4)$$

From a phenomenological point of view, an ion transfer reaction includes three major steps (Reymond, 2000):

1. mass transfer from the bulk of one phase to the interface (mainly diffusion),
2. electrochemical ion transfer reaction,
3. mass transfer in the other phase away from the interface.

When a linear potential sweep is applied and a reversible diffusion controlled ion transfer from the aqueous to the organic phase occurs, the current - time dependence is given by (Nicholson & Shane, 1964; Bard & Wiley, 1980):

$$I(\tau) = zFAc_X^{w*} D_X^{w1/2} \left(\frac{zFv}{RT} \right)^{1/2} \pi^{1/2} \chi(\tau) \quad (5)$$

where c_X^{w*} is X^{z+} concentration in the bulk of aqueous solution, A is the interfacial area in cm^2, D_X^w is the diffusion coefficient in aqueous phase, $\tau = (zF/RT)vt$, t is time, v is the sweep rate in Vs^{-1}, z is the ion charge F, R and T are the Faraday constant, the universal constant and the temperature, respectively, and $\chi(\tau)$ is the current function. When the current signal reaches a maximum value, $\pi^{1/2}\chi(\tau)$ is equal to 0.4463 and the peak current is proportional to X^{z+} concentration, to the square root of D_x andv, according Randles-Sevcik equation (Bard & Wiley, 1980):

$$I_p = 0.4463 zFAc_X^{w*} D_X^{w1/2} \left(\frac{zFv}{RT} \right)^{1/2} \quad 25°C \quad (6)$$

The peak potential, $\Delta\phi_P$, is related to standard transfer potential according to equation 7:

$$\Delta\phi_P - \Delta\phi_{1/2} = \Delta\phi_P - \Delta\phi^{o'} + \frac{RT}{zF} \ln \left(\frac{D_w}{D_o} \right)^{1/2} \quad (7)$$

Where $\Delta\phi_{1/2}$ is the half-wave polarographic potential and D_w and D_o are the diffusion coefficients of X^{z+} in both phases.

2.2.2 Facilitated ion transfer

In the case of a highly hydrophilic ion, its transfer from the aqueous to the organic phase does not occur within the potential window due to the high positive value of free energy for this process. However, this transfer can occur when a ligand present in the organic phase acts as a complexing agent for the ion. In this way, the ion-ligand complex formation in organic phase, decreases the free energy for ion transfer.

Depending on the different concentration ratios and association constants, different types of facilitated ion transfer reactions can be envisaged, as described below (Girault, 1993).

ACT (aqueous complexation followed by transfer): Complex formation occurs in aqueous phase, previous to charge transfer.

$$X^{z+}_{(w)} + sL_{(w)} \rightleftarrows XL^{z+}_{s(w)}$$
$$XL^{z+}_{s(w)} \rightleftarrows XL^{z+}_{s(o)} \qquad (8)$$

TOC (transfer followed by complexation in the organic phase): Complex formation occurs in organic phase after the ion transfer.

$$X^{z+}_{(w)} \rightleftarrows X^{z+}_{(o)}$$
$$X^{z+}_{(o)} + sL_{(o)} \rightleftarrows XL^{z+}_{s(o)} \qquad (9)$$

TIC/TID (transfer by interfacial complexation/dissociation): Heterogeneous charge transfer occurs by interfacial complex formation or dissociation.

$$X^{z+}_{(w)} + sL_{(o)} \rightleftarrows XL^{z+}_{s(o)} \qquad (10)$$

Equation 11 describes the reversible half-wave potential for a facilitated transfer process when the following conditions are fulfilled:

a. $c_L < c_{X^{z+}}$ (the ligand concentration is lower than cation concentration).
b. High Partition coefficient for the ligand, so that its aqueous phase concentration is negligible.
c. High complex formation constant, so that the free ion concentration in organic phase is negligible (Girault, 1993).

$$\Delta_o^w \phi_{1/2}^{tr} = \Delta_o^w \phi_{X^{z+}}^0 + \left(\frac{RT}{z_X F}\right)\ln\left[\frac{D_{L(o)}}{D_{XL_s^{z+}(o)}}\right]^{1/2} + \left(\frac{RT}{z_X F}\right)\ln\left[\frac{K^o_{XL_s^{z+}}}{c^w_{X^{z+}}}\right] \qquad (11)$$

Where $\Delta_o^w \phi_{X^{z+}}^0$ is the standard transfer potential of free X^{z+} ion, D_L and $D_{XL_s^z}$ are the diffusion coefficients of ligand and complex, respectively in organic phase, $K^o_{XL_s^{z+}}$ is the complex constant formation in organic phase and $c^w_{X^{z+}}$ is the ion free concentration in aqueous phase. This expression allows obtaining $K^o_{XL_s^{z+}}$ from the variation of $\Delta_o^w \phi_{1/2}^{tr}$ with X^{z+} concentration in water.

Moreover, equation 12 relates the half-wave potential with cation and ligand concentration, in aqueous and organic phase respectively. From the slope values of $\Delta\phi_{1/2}$ vs ln c_L plots, it is possible to obtain the complex stoichiometry, providing $c^w_{X^{z+}}$ is constant (Homolka et al., 1982; Samec et al., 1982):

$$-\frac{RT}{zF}(s-1)\ln c_L^o = \Delta\phi_{1/2} - \Delta\phi^o + \frac{RT}{zF}(\ln c^w_{X^z} + \ln s) + \frac{1}{2}\ln\frac{D_{L(o)}}{D_{X^{z+}(w)}} - (s-1)\ln 2 \qquad (12)$$

Where *s* is the complex stoichiometry and the other parameters have the same meaning.
The i / time variation for a facilitated transfer process is given by (Homolka et al., 1982):

$$I(\tau) = s^{-1} zFAc_L^o D_L^{1/2} \left(\frac{zFv}{RT} \right)^{1/2} \chi(\tau) \tag{13}$$

The current function takes different values depending on the stoichiometry of the complex formed. Table 1 resumes the parameters of this function at the peak potential.

Parameter	Stoichiometry (cation: ligand)		
	1:1	1:2	1:3
$z(\Delta\phi_{P1} - \Delta\phi_{1/2})$/ V	0.02825	0.0710	0.0988
$z(\Delta\phi_{P2} - \Delta\phi_{1/2})$/ V	-0.03075	-0.0161	-0.0135
$z\ \Delta\phi_P$ / V	0.0590	0.0871	0.1123
χ_{P1}	0.4463	0.3533	0.3033
χ_{P2}	-0.2760	-0.2206	-0.1910
$\lvert\chi_{P2}\rvert / \chi_{P1}$	0.618	0.625	0.630

Table 1. Parameters of adimensional current function, $\chi(\tau)$ at the peak potential calculated for a reversible charge transfer and different complex stoichiometry, P1 and P2 correspond to positive and negative peaks respectively.

2.2.3 Electron transfer

The first evidence of electron transfer at liquid/liquid interfaces were found by Guainazzi *et al* (Girault, 1993; Guainazzi, 1975). They could reduce Cu(III) in aqueous phase to Cu(0) using hexacarbonylvanadate tetrabutylammonium in 1,2- dichloroethane. Moreover, Samec *et al.* (Samec et al., 1977) measured the oxidation current of ferrocene in nitrobenzene by the ferricianure reduction in aqueous phase. Also, electron interfacial transfer has been applied for the electropolymerization in liquid/liquid interfaces, with the purpose of studying the polymerization mechanism, permeability of polymers to ions and new synthetic routes. (Gorgy et al., 2002; Cheng & Corn, 1999; Rieger et al., 2006; Johans et al., 2002; Marecek et al., 2005; Maeda et al., 2001; Cheng et al., 2000; Fantini et al., 2003)
The equation describing the electron transfer process when hemicouples are present at each phase (α and β) , is given by:

$$O_1^{\alpha} + R_2^{\beta} \rightleftarrows R_1^{\alpha} + O_2^{\beta} \tag{14}$$

2.3 Voltammetric techniques

A briefly explanation of the basic voltammetric techniques, cyclic voltammetry (CV) and Square Wave Voltammetry (SWV), applied to liquid/liquid interfaces is carried out below.
The polarization of liquid/liquid interfaces requires the use of four electrodes system similar to the electrochemical cell shown in Figure 4.

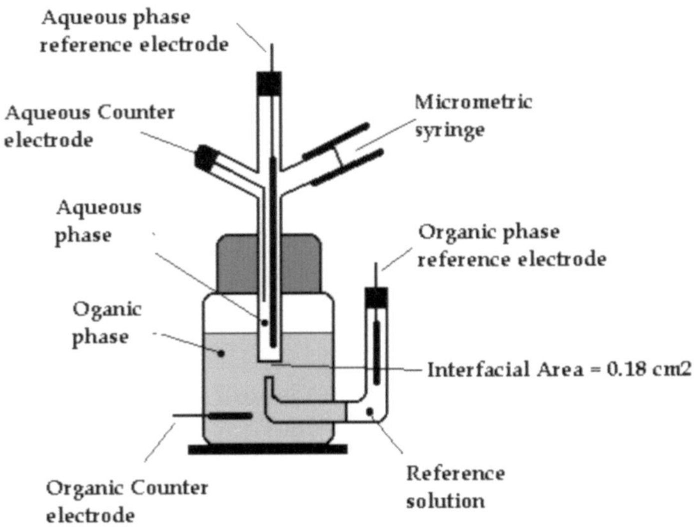

Fig. 4. Glass cell employed for electrochemical measurements.

In this system two Ag/AgCl reference electrodes are immersed in each phase to control the applied potential, ΔE, and two Pt counter electrodes allow the current flowing along the system.
In this way, the electrochemical cell containing only the base electrolytes, can be schematized as follows:

Ag	**AgCl**	**TPhAsCl** **1x10^{-2} M (w)**	**TPhAsDCC** **1x10^{-2} M (*o*)**	**LiCl** **1x10^{-2} M (*w*)**	**AgCl**	**Ag**
	1	**2**		**3**		

TPhAsDCC is tetraphenyl arsonium dicarbolylcobaltate salt employed as organic base electrolyte in 1,2-dichloroethane and TPhAsCl is tetraphenyl arsonium chloride salt dissolved in water and employed as reference solution for the Ag/AgCl electrode corresponding to the organic phase. The contact between these two solutions generates an ideal non-polarizable interface (labeled 2 in the scheme) and a potential difference ($\Delta\phi_{INP}$) dependent of $TPhAs^+$ concentration in both phases is established.
The Galvani potential difference ($\Delta_o^w\phi$) at the o/w interface is related to the applied potential (ΔE) by the following equation:

$$\Delta_o^w\phi = \Delta E + \Delta E_{ref}^w - \Delta E_{ref}^O + \Delta\phi_{inp} \qquad (16)$$

Where ΔE_{ref}^w and ΔE_{ref}^o are the potential differences at interfaces 3 and 1 in the cell scheme and $\Delta\phi_{INP}$ is the difference potential at the non-polarizable interface 2.
The potential-time profiles for the electrochemical techniques employed in this case are shown in Figure 5.
Cyclic and SW voltammograms were recorded using an Autolab (Eco-Chemie, Utrecht, Netherlands) equipped with a PSTAT 30 potentiostat and the GPES 4.3 software package. Typical SW instrumental parameters, unless otherwise stated, were: square-wave frequency f=8–40Hz, square-wave amplitude E_{SW}=35 mV and scan increment dE=3mV.

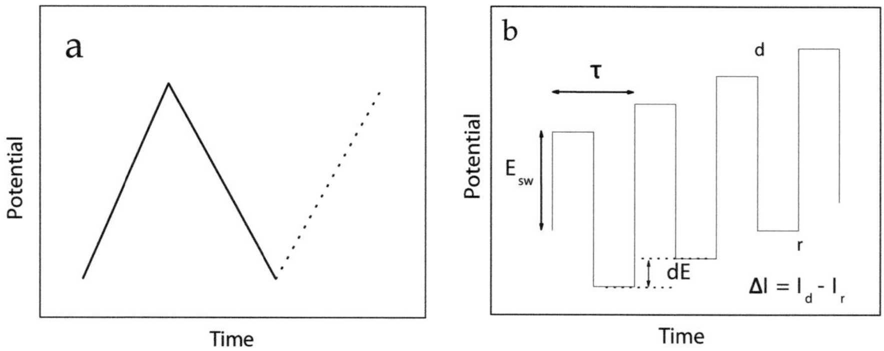

Fig. 5. a- Potential –time profile for Cyclic Voltammetry. b- Potential-time profile for Square Wave Voltammetry.

The base electrolyte solutions were 1.0×10^{-2} M LiCl (Merck p.a.) in ultrapure water and 1.0×10^{-2} M tetraphenyl arsonium dicarbollyl cobaltate (TPhAsDCC) or tetrapentyl ammonium tetraclorophenyl borate (TPnATPhB) in 1,2-dichloroethane (DCE, Dorwill p.a.). The pH of the aqueous phase was adjusted within the range of 1.50 – 8.00 by addition of HCl (Merck p.a.) and LiOH (Merck p.a.) respectively.
All the herbicides employed were of analytical grade. Table 2 resumes the physicochemical properties of the s-triazines studies.

Compound	Substituent			MW g.mol^{-1}	pK_a	Water solubility mg.L^{-1}	log P
	R1	R2	R3				
ATR	Cl	C_2H_5	C_3H_7	215.68	1.68	30	2.70
PRO	Cl	C_3H_7	C_3H_7	230.09	1.85	5	2.91
PROM	SCH_3	C_3H_7	C_3H_7	241.37	4.05	33	3.34

Table 2. Physicochemical parameters for s - triazines. ATR: Atrazine. PRO: Propazine. PROM: Prometryne

3. Results

3.1 Electrochemical behavior of Triazines

As first step, the study of electrochemical behavior of s-triazines herbicides at a water/DCE interface was performed. Figure 6 shows the voltammetric response for the three s-triazines studied: Atrazine (ATR), Propazine (PRO) and Prometryne (PROM). In all cases a reversible ion transfer is observed. The positive peak potential, E_p^+ and the difference peak potential, $\Delta E_p = E_p^+ - E_p^- = 0.060$ V, were constant with sweep rate, ν. The positive peak current, I_p^+, is proportional to $\nu^{1/2}$, as expected for a reversible diffusion controlled mechanism. The differences in currents values observed in the figure would be arising from different equilibrium concentrations of each species at the present pH conditions. Taking into account the partition coefficient (P) and acid constant (K_a^w) values for the herbicides (table 2) and pH conditions, it is possible to obtain the fraction of the protonated, HX^+, and neutral species, X, of the herbicides ($\alpha_{HX^+}^w, \alpha_X^w$) in water and the fraction of neutral species into the organic phase (α_X^o). For the present pH values (0.80 and 0.88), α_X^o is equal to 0.983 and 0.987 for ATR and PRO respectively. In the case of PROM α_X^o = 0.686 and $\alpha_{HX^+}^w$ = 0.313 at pH 1.05. In this case, both species coexist at this pH and different transfer mechanisms can take place which determine the current values.

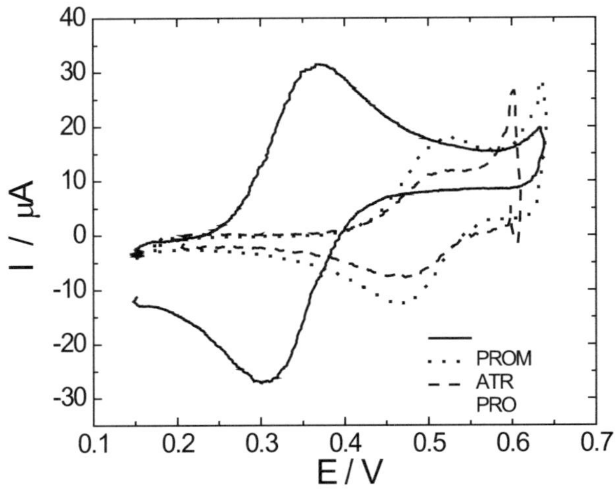

Fig. 6. Cyclic voltammogram obtained at v = 0.050 Vs^{-1} for ATR, PRO and PROM. Aqueous phase composition (apc): 1.00 x 10^{-2} M LiCl + 5.00 x 10^{-4} M s-triazine (—— PROM, pH = 1.05; – – – PRO, pH = 0.88; ······· ATR, pH = 0.80). Organic phase composition (opc): 1.00x10^{-2} M TPhAsDCC. Reprinted from Electroanalysis 15(2003)1481, A.V. Juarez and L.M.Yudi, Copyright (2003) with permission from Wiley.

There are two possible mechanisms that could be responsible for the voltammetric response observed as described by eq. 4 and 10: direct transfer of the protonated herbicide or H^+ transfer from aqueous phase, facilitated by the herbicide present in the organic phase. As can be deduced from the eq. mentioned above, the dependence of peak potential and current with experimental conditions (such as herbicide concentration, pH) allows the determination of the mechanism. In that sense, the dependence of $\Delta_o^w\phi_{1/2}^{tr}$ and I^+_p with both, herbicide concentration and pH was analyzed.

A facilitated proton transfer mechanism (reaction 10) is favored when the species X predominates over HX^+ (i.e. at pH >> pK_a^w conditions) and when this neutral species is highly soluble in organic phase. In this way, the necessary condition under which the facilitated proton transfer occurs can be written as (Homolka et al., 1984):

$$K_a^w \cdot P_X / c_{H+}^w >> 1 \tag{16}$$

where c_{H+}^w is the bulk concentration of the proton in the aqueous solution. If c_{H+}^w is higher than that of the neutral triazine (X) in the organic phase, the charge transfer process is controlled by the diffusion of X towards the interface. In this case, the reversible half- wave transfer potential, $\Delta_o^w\phi_{1/2}^{tr}$, is given by equation 11 (Homolka et al., 1984). Then, a linear variation of $\Delta_o^w\phi_{1/2}^{tr}$ with pH (slope = 0.059 V) is predicted for a facilitated proton transfer mechanism provided the condition $c_{H+}^w >> c_X^o$ is fulfilled.

Figure 7 shows the variation of $\Delta_o^w \phi_{1/2}^{tr}$ with pH for ATR (a), PRO (b) and PROM (c) at several sweep rates. In all systems, the herbicide was added to the aqueous phase. In the case of ATR and PRO, a linear dependence is observed in the whole range of pH analyzed. This is an indication that facilitated proton transfer is occurring in both cases, although the s-triazines were dissolved in aqueous phase, at pH < pK_a^w. Under these experimental conditions, a direct transfer of HX^+ species from the aqueous to the organic phase would be expected. Nevertheless, according to equation 16, the condition is fulfilled in the whole pH range studied. So, whatever the phase in which ATR or PRO are dissolved, the partition equilibrium favors the transfer of the neutral species to the organic phase. From this phase, they act as proton acceptor and facilitate the transfer to the organic phase (reaction 10). From the intercept value in Fig. 7 and equation 11, the dissociation constant of protonated herbicide in the organic solvent, $K^o_{XL_s^z}$, can be calculated. In this way, $K_a^o = 4.60 \times 10^{-9}$ and 7.03×10^{-9} were obtained for ATR and PRO, respectively. These values are approximately 10^6 folds lower than those for aqueous phase, as expected due to the low permittivity of organic media.

In the case of PROM, the behavior is quite different, because it has a K_a^w value lower than the other herbicides. For pH > 2.00 a facilitated proton transfer is observed, similar to the case of ATR and PRO. While an inflection is observed in Figure 7 at pH 2.00, below this value, $\Delta_o^w \phi_{1/2}^{tr}$ is independent of pH demonstrating a change in the transfer mechanism.

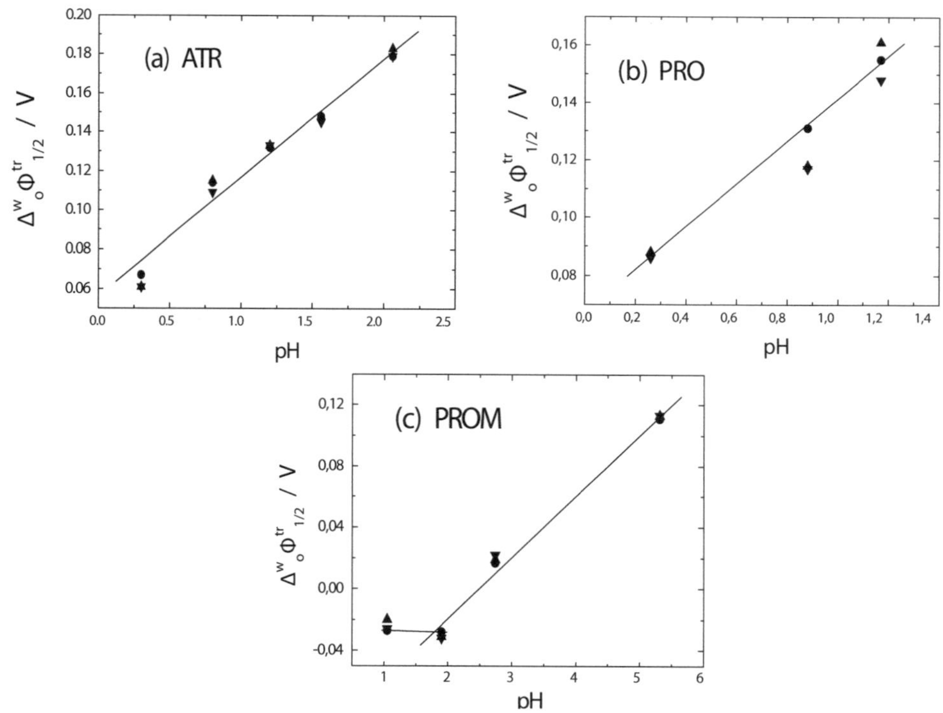

Fig. 7. Plot of $\Delta_o^w \phi_{1/2}^{tr}$ vs. pH for s-triazines: (a) ATR; (b) PRO; (c) PROM. $\Delta_o^w \phi_{1/2}^{tr}$ values were obtained at different sweep rate: (▼) 0.010 Vs^{-1}; (●) 0.050 $V.s^{-1}$; (▲) 0.100 Vs^{-1}. apc: 1.00×10^{-2} M LiCl + 5.00×10^{-4} M s-triazine. opc: 1.00×10^{-2} M TPAsDCC. Reprinted from Electroanalysis 15(2003) 1481, A.V. Juarez and L.M.Yudi, Copyright (2003) with permission from Wiley.

On the other hand, from plots of I_p^+ vs triazine concentration is possible to obtain the diffusion coefficients of the herbicides in the organic phase, as predicted by eq. 13, if a linear relationship is obtained. For this purpose, the experimental conditions were selected to ensure that the facilitated proton transfer is the occurring mechanism, so that, in all cases the pH values ensure the condition of $\alpha_X^o \cong 1$. Figure 8 shows the plots of $I_p^{+\cdot}$ vs concentration obtained for the three herbicides. As can be seen, a linear relationship is obtained with correlation coefficients 0.999, 0.992 and 0.996 for ATR, PRO and PROM respectively, in the concentration ranges 2.50×10^{-5}M < c_{Triaz} < 2.50×10^{-4} M (PRO and PROM) and 2.50×10^{-5}M < c_{Triaz} < 5.00×10^{-4} M (ATR). From the slope values obtained from these curves, D_X values equal to 1.56×10^{-6}, 8.83×10^{-7} and 2.90×10^{-6} $cm^2\ s^{-1}$ were calculated for ATR, PRO and PROM respectively (equation 13).

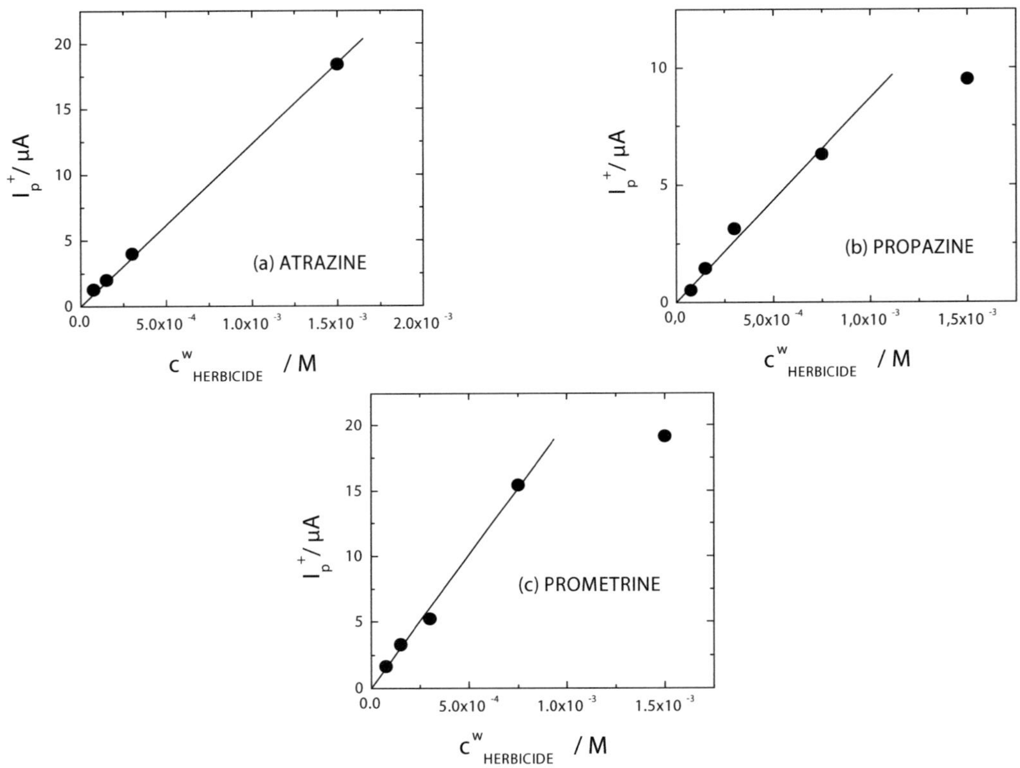

Fig. 8. Plot of I_p^+ vs. triazine concentration: (a) ATR, pH 1.20; (b) PRO, pH 1.20; (c) PROM, pH 2. 00. v = 0.050 V s^{-1}. Reprinted from Electroanalysis 15(2003) 1481, A.V. Juarez and L.M.Yudi, Copyright (2003) with permission from Wiley.

The results obtained demonstrate that quantitative analysis of ATR and PRO can be carried out at a liquid - liquid interface provided these species partition in the organic phase and the pH of aqueous phase satisfy the condition $c^w_{H+} >> c^o_X$. In this case, facilitated proton transfer is the electrochemical process responsible for the faradaic current, which is proportional to s - triazine concentration in organic phase. For PROM quantification two experimental conditions could be used: at pH ≥ 2.00, peak current is proportional to PROM concentration in organic phase as described above for ATR and PRO, while at pH < 2.00

protonated PROM in aqueous phase and neutral PROM in organic phase coexist, as can be expected for the fraction values calculated. Then, both processes could be occurring and determining the peak current value. For this reason, the latter experimental condition is not the best to carry out PROM quantification.

3.2 Prometryne quantification

From the results shown in the previous section, employing cyclic voltammetry technique, the detection limit (DL) obtained for the herbicides studied was 2.50x10^{-5} M. This value is not lower enough for trace quantification. In that sense a more sensitive electrochemical technique is required. For these purpose, square wave voltammetry (SWV) experiments were carried out at liquid/liquid interfaces (Juarez et al., 2005).
As mentioned above, PROM facilitates proton transfer from the aqueous to the organic phase under the condition pH ≥ 2.00. PROM has a high partition coefficient, log P = 3.34, which promotes the partition equilibrium to the organic phase where it acts as a proton acceptor.

3.2.1 Calibration curves before and after pre-concentration procedure

The calibration curve of PROM before pre-concentration experiments was plotted employing ΔI_p values from SWV experiments at f = 8 Hz and pH 2.50. From this curve it was possible to determine a linear range from 1.00x10^{-6} M to 5.00x10^{-4} M with a correlation coefficient of 0.9999. The detection limit (DL) reached under these experimental conditions was 1.50x10^{-6} M, determined from standard deviation of a set of three replicates.
The DL value obtained is lower than that employing I_p values from cyclic voltammetry experiments, but it is not lower enough to determine trace levels of PROM. For this reason, a pre-concentration treatment was developed. This procedure consists in concentrate the herbicide in the organic phase before the electrochemical measurements were performed. For this purpose, 500.00mL of aqueous phase containing different concentration of PROM was stirred with three aliquots of 10.00 mL of DCE. The resulting 30.00 mL of organic solution was evaporated at room temperature in a rotary evaporator to reduce the volume to 2.0 or 3.0 mL. Then, 5.00 mL of solution was prepared with this extract and TPhAsDCC adding DCE up to the final volume. The pH of herbicide aqueous solution was fixed at 8.00 to ensure that the neutral form of the analyte predominates, and to favor a complete extraction to the organic phase. The concentration range was 1.0x10^{-8} M to 2.40x10^{-7} M in aqueous solution (corresponding to a range of 1.00x10^{-6} M to 2.40x10^{-5} M in organic phase after the extraction). The variation of ΔI_p with the herbicide concentration in aqueous phase is shown in Figure 9. A linear range from 8.0x10^{-8} M to 2.4x10^{-7} M was obtained. The correlation coefficient was 0.9858 and DL = 1.0x10^{-7} M was determined from standard deviation of a set of three replicates. As can be seen, the pre-concentration treatment allows an increase of the analytical signal.
It is important to remark that the extraction procedure of the herbicide not only decreases the detection limit, but also allows the possibility of purification of the samples. This advantage can be very useful in the analysis of real samples, because they could have interferents or other hydrophilic contaminants which would not transfer to the organic phase. Thus, it is possible to isolate the herbicide from the real aqueous matrix to the organic phase for its quantification.

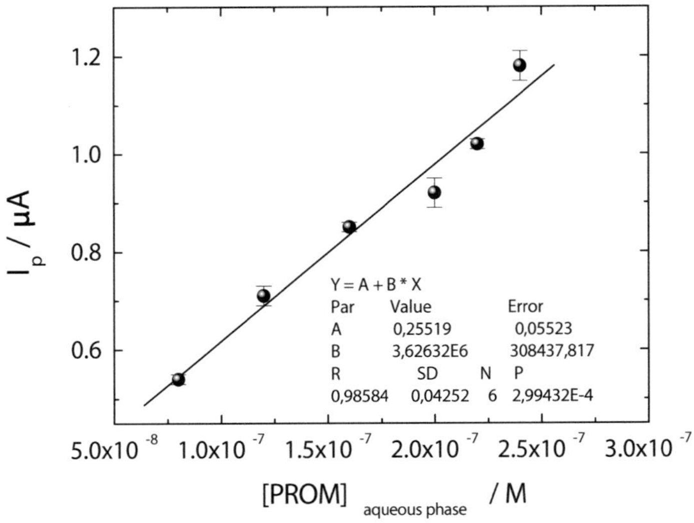

Fig. 9. Calibration curve of PROM employing pre-concentration treatment. apc: LiCl 1.00×10^{-2} M, pH 2.00; opc: TPhAsDCC 1.00×10^{-2} M + x PROM. This organic phase was obtained from pre-concentration of PROM aqueous solutions in the concentration range: 1.0×10^{-8} M - 2.4×10^{-7} M (showed in x-axis). SWV parameters: f = 8 Hz, E_{SW} = 35 mV, dE = 3mV. Reprinted from Electroanalysis 21(2009)767, A.V. Juarez and L.M.Yudi, Copyright (2009) with permission from Wiley.

3.2.3 Standard addition method

The addition of standards was carried out on a pre-concentrated sample obtained from an aqueous solution containing 1.0×10^{-8} M PROM. Different aliquots of a 2.00×10^{-3} M PROM solution to the organic pre-concentrated phase were made. Figure 10 shows the square wave voltammograms obtained after the additions of the standard solution. The ΔI_p values obtained after each addition are shown in Figure 11 as a function of PROM added concentration. The correlation coefficient obtained from linear regression was 0.997. From x-axis intercept, a 1.0×10^{-6} M initial concentration in the organic pre-concentrated phase was determined. Certainly, this concentration value corresponds to PROM 1.0×10^{-8} M in the original aqueous phase. So that, these results allow concluding that standard addition method is appropriated for the quantification of PROM in pre-concentrated samples.

3.3 Prometryne-Al(III) interactions

The interaction between herbicides and soils compounds affect the adsorption, transport and degradation processes and, in consequence the fate of the herbicides in the environment. Al(III) is a cation present as important inorganic component of soils, so that its interaction with PROM was analyzed employing cyclic voltammetry at liquid/liquid interfaces.

As mentioned in methodology section, equations 11 and 12 relate the transfer potential with stoichiometry and complex formation constant. The analysis of transfer potential and current allows elucidating the charge and stoichiometry of the complex formed (Homolka et al., 1982; Samec et al., 1982; Reymond et al., 1998; Iglesias et al., 1998; O'Dwyer & Cunnane, 2005; Katano et al., 2000; Azcurra et al., 2003; Dassie et al., 1999; Caçote et al., 2004; Rahman et al., 2001; Yudi et al., 1992) as well as to obtain thermodynamic (Samec et al., 1982;

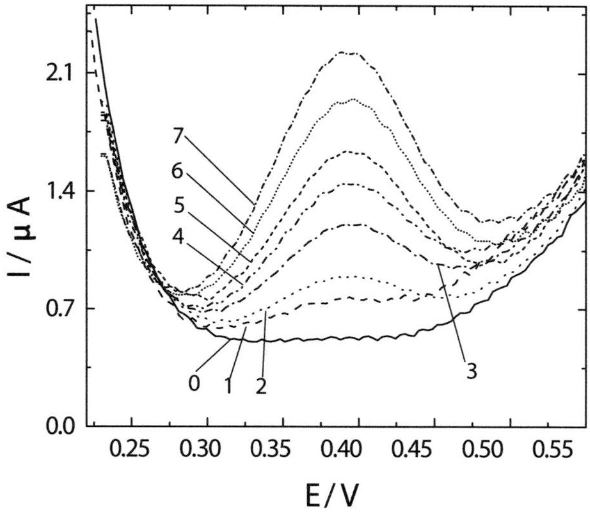

Fig. 10. SWV voltammetric profiles obtained after standard addition of a PROM 2.00×10^{-3} M solution to the organic pre-concentrated phase. apc: LiCl 1.00×10^{-2} M, pH 2.00, *f* 8 Hz. opc: TPhAsDCC 1.00×10^{-2} M + PROM (obtained by pre-concentrating a 1.0×10^{-8} M PROM aqueous solution), with the following aggregates: **0:** 0 μL; **1:** 10 μL; **2:** 20 μL; **3:** 30 μL; **4:** 40 μL; **5:** 50 μL; **6:** 60 μL; **7:** 70 μL. SWV parameters: f = 8 Hz, E_{sw} = 35 mV, dE = 3mV. Reprinted from Electroanalysis 21(2009) 767, A.V. Juarez and L.M.Yudi, Copyright (2009) with permission from Wiley.

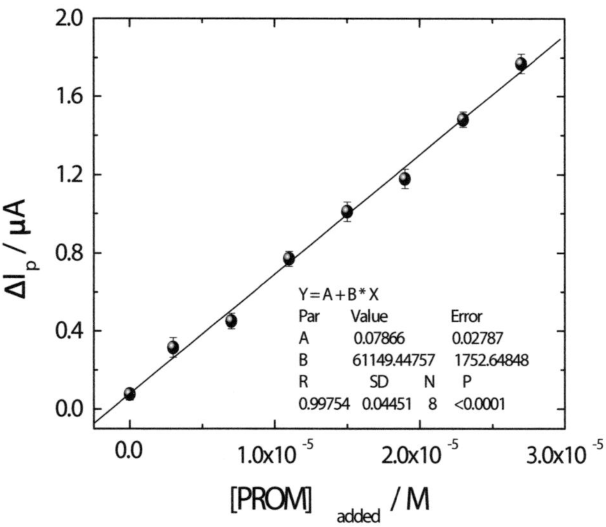

Fig. 11. Variation of ΔI_p with PROM concentration for standard addition experiments. The experimental conditions are the same than those in Fig. 8. Reprinted from Electroanalysis 21(2009) 767, A.V. Juarez and L.M.Yudi, Copyright (2009) with permission from Wiley.

Reymond et al., 1998; Iglesias et al., 1998; Caçote et al., 2004; Yudi et al., 1992; Koryta, 1979; Ferreira et al., 2006; Dassie & Baruzzi, 2002) and kinetic data (Samec et al., 1982; Seno et al., 1990; Sabela et al., 1994; Beatti et al., 1995; Shao & Mirkin, 1997, Homolka et al., 1984) of facilitated ion transfer. The global process of the X^{z+} ion transfer is given by equation 10. In this way, experiments changing ion and ligand concentration can be used for studing transfer mechanism as well as charge and stoichiometry of the complex and also, the complex formation constant.

With the aim of elucidating the complex formed between Al and PROM, experiments at different pH values: 1.50, 3.60, 4.50 and 5.30 were carried out. The relation Al:PROM was 3:1 in all cases. From the analysis of voltammetric response obtained at these pH values it was concluded that a competition between Al(III) and H^+ takes place: at low pH values, H^+ transfer prevails while at high pH values the facilitated transfer of Al(III) is observed.

When the Al:PROM relation changes, the complex formed in the organic phase depends on the cation and PROM concentration. Thus, depending on the PROM concentration, two different processes were observed: when the ligand concentration is lower than the concentration of the both cations, H^+ transfer is the only process observed at pH values between 1.50 and 4.50. It is worthwhile to note that even when the Al(III) concentration is higher than the H^+ concentration, H^+ transfer prevails. This fact indicates a higher formation constant value for $HPROM^+$ with respect to Al(III)-PROM. On the other hand, if the ligand concentration is higher than the H^+ and Al(III) concentrations, the transfer of both cations is observed. In this case, the voltammograms obtained shows two overlapped waves. One of the process is due to Al(III) transfer facilitated by PROM.

3.3.1 Determination of the Al(III) transfer mechanism

Scheme 1 shows one of the experiments carried out with the purpose to establish the transfer mechanism of the Al:PROM complex.

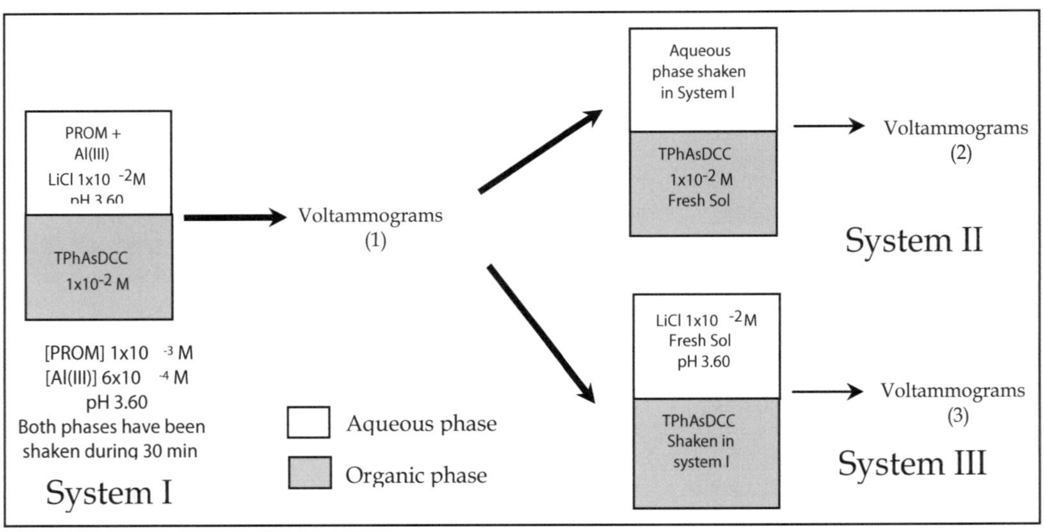

Scheme 1. An experiment carried out for determination of transfer mechanism. Reprinted from Electrochim. Acta, 54 (2008) 530, A.V. Juarez and L.M.Yudi, Copyright (2008) with permission from Elsevier.

Both solutions of system I were shaken to establish the partition equilibrium and then, the electrochemical measurement was taken. After this, both phases were separated and electrochemically analyzed. Figure 12 compares the voltammograms obtained for system I (see scheme 1) with those obtained for system II (Figure 12 (a)) and III (Figure 12 (b)). From these results, it is possible to conclude that the total amount of PROM in the system prevails in the organic phase after agitation, even in the presence of Al(III) in the aqueous phase, and that cation transfer occurs by a facilitated mechanism. This mechanism is confirmed by the analysis of voltammetric parameters (I_p^+, E_p^+ and ΔE_p).

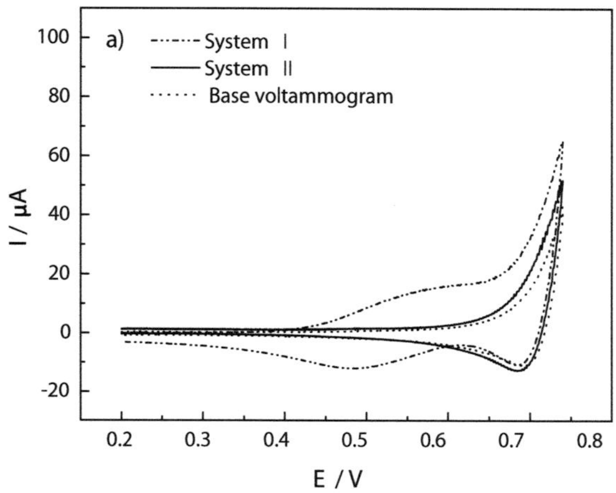

a) (····) Voltammogram for base solutions: apc: 1.00x10^{-2} M LiCl, pH = 3.60. opc: 1.00x10^{-2} M TPhAsDCC. (··—··—··) Voltammogram for System I: apc: 1.00x10^{-2} M LiCl + 6.00x10^{-4} M $Al(NO_3)_3$ + 1.00x10^{-3} M PROM, pH = 3.60. opc: 1.00x10^{-2} M TPhAsDCC. (—) Voltammogram for System II: apc resulting from the agitation of system I, opc: 1.00x10^{-2} M TPhAsDCC.(fresh solution).

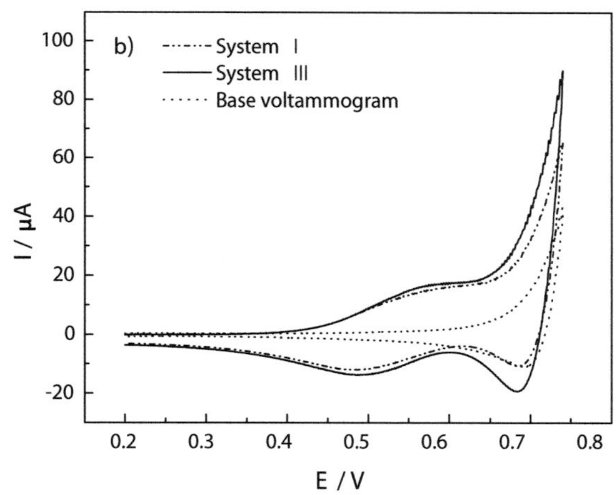

b)(····) Voltammogram for base solutions: apc: 1.00x10^{-2} M LiCl, pH = 3.60. opc: 1.00x10^{-2} M TPhAsDCC. (··—··—··) Voltammogram for System I: apc: 1.00x10^{-2} M LiCl + 6.00x10^{-4} M $Al(NO_3)_3$ + 1.00x10^{-3} M PROM, pH = 3.60. opc: 1.00x10^{-2} M TPhAsDCC. (—) Voltammogram for System III: apc: 1.00x10^{-2} M LiCl, pH = 3.60 (fresh solution). opc: resulting from the agitation of system I. Sweep rate= 0.050 Vs^{-1}. Reprinted from Electrochim. Acta, 54 (2008) 530, A.V. Juarez and L.M.Yudi, Copyright (2008) with permission from Elsevier.

Fig. 12. Voltammetric profiles corresponding to the experiment described in scheme 1.

3.3.2 Determination of the complex stoichiometry

With the aim to determine the stoichiometry of Al:PROM complex, experiments changing the ligand concentration were performed. Figure 13 shows the voltammetric response obtained. The pH value was fixed at 4.50, and the cation concentration was higher than ligand concentration employed to avoid H^+ competition. Under these experimental conditions, the dependence of current with ligand concentration is given by eq. 13, whenever, a reversible diffusion controlled transfer occurs. As can be seen, the peak current increased with PROM concentration and E_p^+ shifts to more negatives values as predicted by equations 12 and 13. To carry out these experiments it was necessary to change the organic electrolyte, and TPnATClPhB was employed because it allows increasing the positive limit of the potential window. In this way, a better determination of the peak potential of Al(III) transfer could be done. As a consequence of the increase in the positive limit, a second process was observed around E = 0.800 V as the concentration of ligand increased. This process could be likely due to H^+ facilitated transfer.

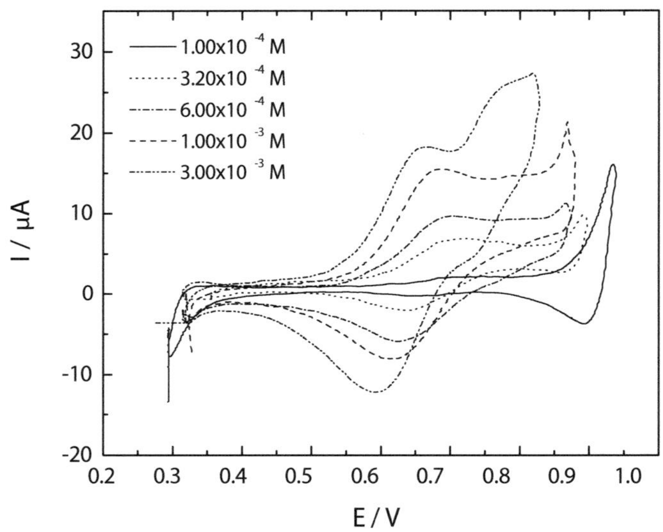

Fig. 13. Voltammetric profiles corresponding to Al(III) facilitated transfer by PROM at different concentrations of ligand. Apc: $1.00x10^{-2}$ M LiCl + $1.00x10^{-2}$ M $Al(NO_3)_3$, pH = 4.50. opc: $1.00x10^{-2}$ M TPATPhClB + x M PROM, x = (—) $1.00x10^{-4}$ M; (····) $3.20x10^{-4}$ M; (-·-·-) $6.00x10^{-4}$ M; (— —) $1.00x10^{-3}$ M; (··—··—··) $3.00x10^{-3}$ M. Sweep rate= 0.050 Vs^{-1}. Reprinted from Electrochim. Acta, 54 (2008) 530, A.V. Juarez and L.M.Yudi, Copyright (2008) with permission from Elsevier.

Plots of I_p *vs* square root of sweep rate are linear as predicted by eq. 13 for this kind of ion transfer. With the purpose to calculate the stoichiometry and the charge of the complexes, the slopes of these plots at different ligand concentrations were obtained. Figure 14 compares the theoretical slope values obtained for different complexes and the experimental values obtained in this case. From the analysis of this figure, it is possible to conclude that PROM:Al(III) complex stoichiometry changes with the ligand concentration, while the charge of the transferred species is constant and equal to 2. The charge and stoichiometry complex can also be deduced from eq 12. From the analysis of E_p vs log [PROM] variation

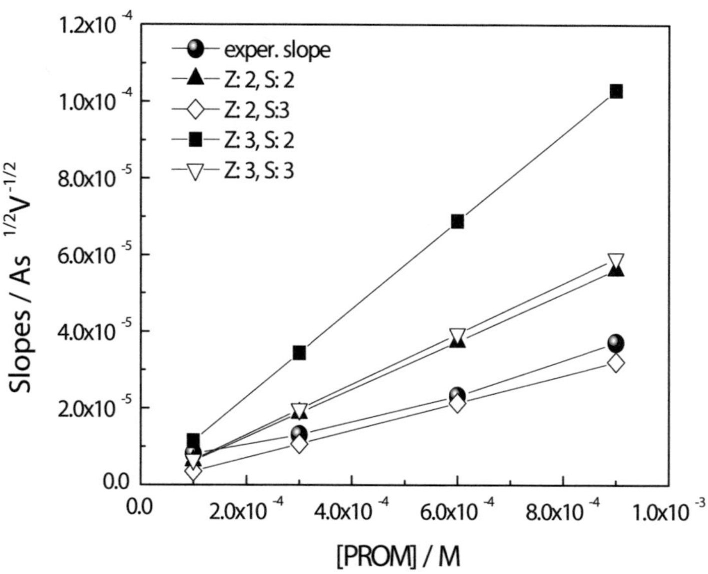

Fig. 14. Plot of experimental and theoretical slope values $I_p/v^{1/2}$ vs. [PROM]. Experimental slope: (•). Theoretical slopes for the following values of z and s: (▲) z: 2, s: 2; (◊) z: 2, s: 3; (■) z: 3, s: 2; (∇) z: 3, s: 3. Aqueous phase composition: 1.00x10^{-2} M LiCl + 1.00x10^{-3} M $Al(NO_3)_3$, pH = 4.50. Organic phase composition: TPhAsDCC 1.00x10^{-2} M + PROM n M. Sweep rate= 0.050 Vs^{-1}. Reprinted from Electrochim. Acta, 54 (2008) 530, A.V. Juarez and L.M.Yudi, Copyright (2008) with permission from Elsevier.

(not shown), a slope value equal to 0.062 V/dec was obtained, in a concentration range between 3.20x10^{-4} M to 1.00x10^{-3} M. This value approximates to the theoretical value corresponding to a complex with a charge +2 and a stoichiometry of 1:3 (Homolka et al., 1982).

4. Conclusions

This chapter resumes the results obtained from the electrochemical study of *s*-triazines herbicides at liquid-liquid interfaces. These herbicide compounds (Atrazine, Propazine and Pometryne) can be quantitatively analyzed from the electrochemical transfer current at the water/1,2-dichloroethane interface.

Regarding to the transfer of ATR and PRO, a facilitate proton transfer from aqueous phase trough the interface to form the protonated species in the organic phase was observed. In these cases, transfer potential varies with pH as predicted by theory and peak current is proportional to triazine concentration in organic phase and independent of the pH (for $c_{H+}^{w} > c_{L}^{o}$ conditions). For these two species it was possible to calculate the acid dissociation constant in organic phase, K_a^o, from the intercepts in the graph of $\Delta_o^w \phi_{1/2}^{tr}$ vs pH. The values obtained were 4.60 x 10^{-9} and 7.03 x 10^{-9} for ATR and PRO, respectively.

PROM behavior is quite different and depends on the pH values. At low pH values, the transfer of protonated PROM from aqueous to organic phase can occur. Under these conditions, the peak potential is independent on the pH and peak current is proportional to protonated fraction of PROM, which decreases as pH increases up to pH = 2.00. Above this value, the process change to a facilitated proton transfer, showing a behavior similar to that observed for ATR and PRO.

Calibration curves for these species, employing cyclic voltammetry, show detection limits of 2.50 x 10^{-5} M. This limit must be lowered to be able to apply this electrochemical methodology to the analytical determination of ATR, PRO and PROM in waste water or soil samples. However, the experimental conditions found so far were used to apply square - wave voltammetry technique at a liquid – liquid interface.

In that sense, the electrochemical response of PROM, employing SWV technique at liquid-liquid interface, was studied. The detection limit found was 1.50×10^{-6} M. This DL value is still very high for trace determinations of PROM required in environmental studies compared with other available techniques. However, one of the advantages of this system is the possibility of pre-concentrate the herbicide in the organic phase. Extraction procedures to the organic phase are possible due to the high partition coefficient of PROM and high solubility in 1,2-DCE. The aqueous:organic volume ratio equal to 500:30 and the later reduction of the volume of the organic phase enriched with PROM, by a factor of 6, yield an overall pre-concentration factor of 100. The detection limit could be lowered to 1.0×10^{-7} M under these experimental conditions. The concentration ranges from 1.0×10^{-6} to 5.0×10^{-5} M, without pretreatment, and between 8.0×10^{-8} to 2.4×10^{-7} M, carrying out the pre-concentration procedure previous to electrochemical measurement, were used for the calibration curves.

On the other hand, the standard addition method is highly efficient in this kind of systems and presents several advantages like less use of reactive and easy sample manipulation. A linear response in the concentration range between 1.0×10^{-6} to 2.7×10^{-5} M with correlation coefficient of 0.997 was obtained. The lower concentration value, in this case, corresponds to an aqueous PROM solution 1.0×10^{-8} M.

It is worthwhile to discuss the practical aspect of the procedure here proposed. In this sense, the extraction and pre-concentration methods developed by other authors (Herzog et al., 2008; Berduque & Arrigan, 2006; Berduque et al., 2005; Collins et al., 2008; Kim & Amemiya, 2008) have several practical advantages over the present procedure, as the use of low organic phase volume, among others. Nevertheless, the results obtained in this study justify evaluating the possibility of carrying out PROM pre-concentration following the thin film approach (Kim & Amemiya, 2008) or the electrochemistry modulated liquid - liquid extraction procedure (Berduque & Arrigan, 2006; Berduque et al., 2005; Collins et al., 2008; Kim & Amemiya, 2008).

Regarding to the study of the interaction between PROM and Al(III) at the water /1,2-DCE interface, it was possible to determine the charge and stoichiometry of the complex formed. A facilitated Al(III) transfer through the liquid - liquid interface takes place, depending on pH and PROM concentration. From the analysis of the experimental results, a competition of H^+ and Al(III) for the ligand is observed. At low pH values (pH<2.00), only H^+ transfer occurs, a pH higher than 4.50 only Al(III) is transferred, and at intermediate pH values, the transfer of both cations takes place. To determine the stoichiometry of the Al:PROM complex, the experiments were carried out at pH 4.50 because no H^+ transfer was observed under these conditions. From the experimental results, we conclude that the stoichiometry depends on PROM concentration: 1:3 at $c_{PROM} \geq 3.00 \times 10^{-4}$ M and 1:2 for $c_{PROM} < 3.00 \times 10^{-4}$ M, while the charge of the transferred species is 2+. Therefore, the ion forming the complex with the herbicide, at pH 4.50, is $Al(OH)^{2+}$. This statement is supported by the fact that $Al(OH)^{2+}$ is one of the predominant species at this pH value.

5. References

Aelion, C.M.; Mathur, P.P. (2001). *Environ. Toxicol. Chem.* 20, 11 (November 2001) (2411-2419).

Azcurra, A.I.; Yudi, L.M.; Baruzzi, A.M. (2003). *J. Electroanal. Chem.*, 560, 1, (December 2003), (35-42).

Bahnd, S.; Surugiu, I.; Dzgoev, A.; Ramanathan, K.; Sundaram, P.V.; Danielsson, B. (2005). *Talanta,* 65, 2 (January 2005) (331-336).

Bard, A. J.; Faulkner, L. R.. Wiley, Ed., Electrochemical Methods: Fundamentals and applications. (New York, 1980).

Beale, D.J.; Porter, N.A.; Roddick, F.A. (2009). *Talanta*, 78, 2 (April 2009) (342-347).

Beattie, P.D.; Delay, A.; Girault, H.H. (1995) *J.Electroanal. Chem.*, 380, 1-2, (January 1995), (167-175)

Berduque, A. Sherburn, M. Ghita, A.W. Dryfe, D.W.M. Arrigan. (2005). *Anal. Chem. 77*, 22, (October, 2005), (7310–7318).

Berduque, A.; Arrigan, D.W.M. (2006). *Anal. Chem. 78*, 8, (March 16), (2717–2725).

Besse-Hogan, P.; Alekseeva, T.; Sancelme, M.; Delort, A.M.; Forano, C. (2009). *Environmental Pollution*, 157, 10 (October 2009) (2837-2844).

Brewster, J. D.; Lightfield, A. R. (1993). *Anal. Chem.*, 65, 18, (September 1993) (2415-2419).

Caçote, M.H.M.; Pereira, C.M.; Tomaszewski, L.; Girault, H.H.; Silva, F. (2004) . *Electrochim. Acta*, 49, 2, (January 2004), (263-270).

Campàs, M.; Carpentier, R. ; Rouillon, R. (2008). Biotechnology Advances, 26, 4 (July-August 2008) (370-378).

Carabias-Martínez, R; Rodríguez-Gonzalo, E.; Miranda-Cruz, E.; Domínguez-Álvarez, J.; Hernández-Méndez, J. (2006). *J. Chromatogr. A*, 1122, 1-2, (July 2006), (194-201).

Cheng, Y.; Corn, R. M. (1999). *J. Phys. Chem. B*. 103, (September, 1999), (8726-8731).

Cheng, Y.; Murtomäki, L.; Corn R. M. (2000). *J. Electroanal. Chem.* 483,1-2, (March 2000), (88-94).

Collins, C. J.; Berduque, A.; Arrigan, D.W.M. (2008). *Anal. Chem.* 80, 21, (October, 2008), (8102–8108).

Dassie, S. A.; Baruzzi A. M. (2002) *J.Electroanal. Chem.*, 522, 2, (April 2002), (158-166)

Dong, X.; Zhu, L.; Wang, J.; Wang, J.; Xie, H.; Hou,X.; Jia, W. (2009). *Chemosphere,* 77, 3 (October 2009) (404-412).

Fantini, S.; Clohessy, J.; Gorgy, K.; Fusalba, F.; Johans, C.; Kontturi, K.; Cunnane, V.J. (2003). *European Journal of Pharmaceutical Sciences*. 18, 3-4, (March 2003), (251-257).

Ferreira, E.S.; Garau, A.; Lippolis, V.; Pereira, C.M.; Silva, F. (2006). *J.Electroanal. Chem.*, 587, 1, (February 2006), (155-160)

Frías, S.; Sánchez, M. J.; Rodríguez, M.A. (2004). *Anal. Chim. Acta,* 503, 2 (February 2004) (271-278).

Fuchiwaki, Y.; Sasaki, N.; Kubo, I. (2009). *Journal of Sensors,* Vol 2009 (2009), Article ID 503464, 6 pages. doi:10.1155/2009/503464

García Galán, M.J.; Díaz Cruz, M.S.; Barceló, D. (2010). *J. Hydrology*, 383, 1-2, (March 2010) (30-38).

Girault H. H.; Schiffrin, D. J. (1983). *J. Electroanal. Chem.*, 150, 1-2, (July 1983), (43-49).

Girault, H. H. (1987). *Electrochim. Acta*, 32, 3, (March 1987), (383-385).

Girault, H.H. (1993). *In Modern Aspects of Electrochemistry*. B. C. J.O'M. Bockirs, R. White, (Plenum Press),vol. 25, (1-62). New York.

Girault, H.H.; Schiffrin, D. J. (1984). *J. Electroanal. Chem.*, 170, 1-2, (July 1984), (127-141).

Gorgy, K.; Fusalba, F.; Evans, U.; Kontturi, K.; Cunnane, V.J. 125, 365-373 (2002). *Synthetic metals*. 125, 3, (December 2001), (365-373).

Guainazzi, M.; Silvestry, G.; Survalle, G. (1975). *J. Chem. Soc. Chem. Commun.*, 200 (1975).

Herzog, G.; Kam, V.; Berduque, A.; Arrigan, D.W.M. (2008). *J. Agric. Food Chem.* 56, 12, (May, 2008), (4304–4310).

Homolka, D.; Holub, K.; Marecek ,V. (1982). *J. Electroanal. Chem.* 138, 1, (August 1982), (29-36).

Homolka, D.; Marecek, V.; Samec, Z.; Base, K.; Wendt, H. (1984). *J. Electroanal. Chem.*, 163, 159, (March 1984), (159-170).

Hu, Y.; Wang, Y.; Hu, Y.; Li, G.(2009). *J. Chromatogr. A*, 1216, 47, (November 2009), (8304-8311).

Iglesias, R.; Dassie, S. A.; Yudi, L. M.; Baruzzi A. M. (1998) *Anal. Scien.* 14, 1, (February 1998), (231-236).

Ignjatović, L.M.; Marković, D. A.; Veselinović, D. S.; Bešić, B. R. (1993). *Electroanalysis*, 5, 5 (June-July 1993) (529-533).

Johans, C.; Clohessy, J.; Fantini, S.; Kontturi, K.; Cunnane, V.J. (2002). *Electrochemistry communications*, 4, 3, (March 2002), (227-230).

Jonsson, C. "Modelling of glyphosate and metal-glyphosate speciation in solution and at solution-mineral interfaces", Doctoral Thesis, Umeå University, Faculty of Science and Technology, Chemistry, 2007, Sweden.

Juarez, A. V.; Yudi, L.M. (2003). *Electroanal.*, 15, 18, (October 2003), (1481-1487).

Juarez, A. V.; Yudi, L.M. (2008). *Electrochim. Acta*, 54, 2, (December 2008), (530-534).

Juarez, A. V.; Yudi, L.M. (2009). *Electroanal.*, 21, 6, (March 2009), (767-771).

Juarez, A.V.; Baruzzi, A.M.; Yudi, L.M. (2005). *J. Electroanal. Chem.*. 577, 2, (April 2005), (281-286).

Kakiuchi, T.; Senda, M. (1991). *J. Electroanal. Chem.*, 300, 1-2, (February 1991), (431-445).

Kakiuchi,T. (1993). *J. Electroanal. Chem.*, 345, 1-2, (February 1993), (191-203).

Katano, H.; Kuboyama, H.; Senda, M. (2000). *J.Electroanal. Chem.*, 483, 1-2, (March 2000), (117-123)

Katsumata, H.; Kaneco, S.; Suzuki, T.; Ohta, K. (2006). *Anal. Chim. Acta*, 577, 2, (September 2006), (214-219).

Kim, Y.; Amemiya, S. (2008). *Anal. Chem. 80*, 15, (July, 2008), (6056–6065).

Koryta, J. (1979). *Electrochim. Acta*, 24, 3, (March 1979), (293-300).

Kubo, I.; Shoji, K.; Fuchiwaki, Y.; Suzuki, H. (2008). Electrochemistry, 76, 8 (July 2008), (541-544).

Lacina, O.; Urbanova, J.; Poustka, J.; Hajslova, J. (2010). *J. Chromatography A* 1217, 5 (January 2010), (648-659).

Lyapchenko, N.; Eitner, K.; Schroeder, G.; Brzezinski, B. (2004). *J. Molec. Struc.*, 690, 1-3, (March 2004), (45-51).

Maeda, K. *et al.* (2001). *J. Electroanal. Chem.*, 516, 1-2, (December 2001), (103-109).

Mareček, V.; Jänchenová, H.; Stibor, I.; Budka, J. (2005). *J.Electroanal.Chem.*, 575, 2, (February 2005), (293-299).

Mirčeski, V.; Lovrić, M. (2001). *J. Electroanal. Chem.* 497, 1-2, (February 2001),(114-124).

Mohammadi, A.; Ameli, A.; Alizadeh, N. (2009). *Talanta*, 78, 3 (May 2009), (1107-1114).

Nicholson, R. S.; Shain, I. (1964). *Anal. Chem.*, 36, 4, (April 1964), (706 -723).

O'Dwyer, P.; Cunnane, V.J. (2005). *J. Electroanal. Chem.*, 581, 1, (July 2005), (16-21).

Pedrero, M.; Rosa Alonso, M.; de Villena, F. JM.; Pingarrón, J.M. (1995). *Electroanalysis*, 7, 7 (July 1995), (644-648).

Pinto, M. I.; Sontag, G.; Bernardino, R.J.; Noronha, J.P. (2010). A review, *Microchemical Journal, Available online 21 June 2010* Microchem. J. (2010), Doi:10.1016/j.microc.2010.06.010

Preuss, M.; Hall, E. A. H. (1995). *Anal. Chem.* 67, 13 (July 1995), (1940- 1949).

Quintana, J.; Martí, I.; Ventura, F. (2001). *J. Chromatogr. A*, 938, 1-2 (December 2001), (3-13).

Rahman, A.; Doe, H.; Sakurada, N.; Arakawa, R. (2001). *Electrochim. Acta*, 47, 4, (November 2001), (623-631).

Reymond, F.; Carrupt, P.; Girault, H. H. (1998). *J. Electroanal. Chem.*, 449, 1-2, (June 1998), (49-65).

Reymond, F.; Fermin, D. J.; Lee, H. J.; Girault, H. H. (2000). *Electrochimica Acta*, 45, 15-16, (May 2000), (2647-2662).

Reymond, F.; Lagger, J.P; Carrupt, P.A.; Girault, H.H. (1998). *J. Electroanal. Chem.*, 451, 1-2, (July 1998), (59-76).

Rieger, J.; Dubois, P.; Jérôme, R.; Jerome, C. (2006). *Langmuir*, 22, 18, (August, 2006), (7471-7479).

Sabela, A.; Mareček, V.; Koryta, J.; Samec, Z. (1994). *Collect. Czech. Chem. Commun*, 59, 6, (June 1994), (1287-1295).

Sambe, H.; Hoshina, K.; Haginaka, J. (2007). *J. Chromatogr. A*, 1152, 1-2, (June 2007), (130-137).

Samec, Z.; Marecek, V.; Homolka, D. (1982). *J. Electroanal. Chem.* 135, 2, (May 1982), (265-283).

Samec, Z.; Marecek, V.; Homolka, D. (1983). *J. Electroanal. Chem.*, 126, 1-3, (September 1981), (121-129).

Samec, Z.; Marecek, V.; Weber, J. (1979). *J. Electroanal. Chem.* 96, 2, (February 1979), (245-247).

Sanchez Ortega, A.; Unceta, N.; Gomez Caballero, A.; Sampedro, M.C.; Akesolo, U.; Goicolea, M.A.; Barrio, R. J. (2009). *Anal. Chim. Acta*, 641, 1-2 (May 2009), (110-116).

Sannino, F.; Filazzola, M.T.; Violante, A.; Gianfreda, L. (1999). *Chemosphere*, 39, 2, (July 1999), (333-341).

See, H.H.; Sanagi, M.M.; Ibrahim, W.A.W.; Naim, A.A. (2010). *J. Chromatography A*, 1217, 11 (March 2010), (1767-1772).

Senō, M.; Iwamoto, K.; Chen, Q.Z. (1990). *Electrochim. Acta*, 35, 1, (January 1990), (127-134).

Shao, C.Y.; Howe C.J.; Porter, A.J.; Glover, L.A. (2002). *Appl. Environ. Microbiol.*, 68, 10, (October 2002), (5026-5033).

Shao, Y.; Mirkin, M.V. (1997) *J. Am. Chem. Soc.* 119, 34, (August 1997), (8103–8104).

Touloupakisa, E.; Giannoudib, L.; Piletskyb, S.A.; Guzzellac, L.; Pozzonic, V.; Giardia, M.T. (2005). *Biosensors and Bioelectronics*, 20, 10 (April 2005), (1984-1992).

Tsang, V. W.H.; Lei, N.Y.; Lam, M.H.W. (2009). *Marine Pollution Bulletin*, 58, 10 (October 2009), (1462- 1471).

Vaz, C.M.P.; Crestana, S.; Machado, S.A.S.; Mazo, L. H.; Avaca, L. A. (1996). *International Journal of Environmental Analytical Chemistry*, 62, 1 (January 1996), (65-76).

Yudi, L.M.; Baruzzi, A.M.; Solis, V.M. (1992). *J. Electroanal. Chem.*, 328, 1-2, (July 1992), (153-164).

Zapardiel, A.; Bermejo, E.; Perez, J. A.; Chicharro, M. (2000). *Fresenius J. Anal. Chem.* 367, 5 (June 2000), (461- 466).

Zhang, S.H.; Yang, Y. Y.; Han, D. D.; Wang, C.; Zhou, X.; Huan Zang, X.; Wang, Z. (2008). *Chinese Chem. Let.*, 19, 12 (December 2008), (1487-1490).

Zhou, J.; Chen, J.; Cheng, Y.; Li, D.; Hu, F.; Li, H. (2009). *Talanta*, 79, 2 (July 2009), (189-193).

Zhou, Q.; Xiao, J.; Wang, W.; Liu, G.; Shi, Q.; Wang, J. (2006). *Talanta*, 68, 4, (February 2006), (1309-1315).

12

Application of Bioassays in Studies on Phytotoxic Herbicide Residues in the Soil Environment

Tomasz Sekutowski
Institute of Soil Science and Plant Cultivation – National Research Institute, Pulawy Department of Weed Science and Tillage Systems, Wroclaw Poland

Science is food for the mind
War and Peace, L. Tolstoy 1869

1. Introduction

The primary aim of the application of herbicides is to protect plantations against the competitive action of many weed species found in the field of a given crop. Herbicides may be used both directly to the soil and in foliar applications. In relation to the type and method of application (single vs. split dose) a portion (in foliar applications) or the entire amount (in soil-applied agents) of herbicide reaches the soil (Praczyk & Skrzypczak, 2004; Woźnica, 2008). Each active ingredient in a herbicide which penetrates the soil medium, undergoes certain biophysical and biochemical processes. At the time the herbicide active ingredient enters the soil it is separated between the solid phase (soil particles) and the aqueous phase (soil solution). In the soil medium only this portion of the active ingredient is available to plants, which is found in the liquid phase. However, herbicide molecules adsorbed or chemically bound with the solid phase are not absorbed by plants. Under field conditions this balance is constantly disturbed as a result of the action of the edaphone and through changes in temperature and moisture content of soil, which affects the availability of herbicide to weeds and crops (Vicari et al., 1994; Sadowski, 2001).
Depending on the applied cultivation regime and climatic and soil conditions observed in a given vegetation season only a portion of herbicide active ingredient residue found in the soil is available to plants and under advantageous conditions may exhibit phytotoxic action. Thus the determination of the level of residue, degradation rate and translocation of herbicide active ingredients in the soil is so significant both for the agricultural practice and for the protection of the agricultural environment (Sadowski et al., 2002; Sadowski & Kucharski, 2004).
At the selection of a detection technique the most important criterion in the evaluation is the concentration, at which a given analyte may be found in the tested sample. Instrumental methods, such as gas chromatography (GC) or liquid high performance chromatography (HPLC), make it possible to determine the total content of active ingredients in the soil at the time of the application or several weeks after the application of herbicides (Ahmad & Crawford, 1990; Sadowski, 2001; Sadowski et al., 2001a; Kucharski & Sadowski, 2006). This

problem appears when the herbicide is used once or several times in the vegetation season in small doses of <50 g/ha, since already at the moment of application the level of herbicide active ingredients is slight and does not exceed 10^{-2} mg/kg (Sadowski et al., 2002). The evaluation of risk resulting from the occurrence of herbicide residue in the soil medium until recently was based only on the results of chemical analyses, which supplied information on the presence, content and type of the chemical substance, preventing an evaluation of harmful ecological effects of the herbicide residue. Thus the traditional, chemical approach to the assessment of the level of herbicide residue in the agrophytocenosis for well over a decade has been supplemented by ecotoxicological analyses. In such analyses the level of herbicide residue is evaluated on the basis of a specific, comprehensive response of standard indicator organisms to the active ingredient varying both chemically and in terms of its concentration, contained in the tested soil sample. In such analyses the biotest methods is used with the application of e.g. a plant biodetector. This method is to determine a biologically effective level of the herbicide active ingredient residue immediately after application, as well as to follow the dynamics of decline for this substance in the soil environment in the course of several months or even for more than a year. Biotests also facilitate an objective evaluation of the level of residue, due to the fact that all higher plants have a certain sensitivity to different xenobiotics (e.g. herbicides) found in the soil environment. The phytotoxic effect of active ingredients originating from herbicides may be observed on the basis of the reduction of dry or fresh weight of roots or aboveground parts (stems, leaves) of test plants (Günther et al., 1993; Stork & Hannah, 1996; Sarmah et al., 1999; Sadowski et al., 2002; Demczuk et al., 2004; Sekutowski & Sadowski, 2005; 2006; 2009). Thanks to the wide-scale application of the bioindication method using plants it is possible to evaluate the degree of contamination not only for the soil, but also for the entire agrophytocenosis (Dećkowska et al., 2008).

2. The behavior of herbicides in the soil environment

Active ingredients of herbicides, after penetrating to the soil, are separated between the solid phase (soil particles) and the liquid phase (soil solution). In the soil only this portion of the active ingredient is available to plants, which is found in the soil solution within the rhizosphere. In turn, herbicide molecules adsorbed or chemically bound with the solid phase are not available to plants. They may constitute a certain reserve, which under advantageous climatic conditions may become available to plants. Availability of these active ingredients in the soil fluctuates constantly, since they are removed from the soil solution as a result of immobilization, elution or diffusion. Bounding of these chemical substances by the soil sorption complex is a factor determining their occurrence and through accumulation may significantly alter their deposition time in the soil environment. Under field conditions this equilibrium is constantly disturbed by changes in temperature, moisture content, cultivation measures and the soil entomofauna, which has a crucial effect on the amount of the herbicide active ingredient which is available to weeds or crops within a specified period of time. On the one hand, the process of binding reduces mobility of these residues in the soil profile and their penetration to aquatic zones, while on the other hand, it reduces the possibility of their removal from soil using plants themselves and soil microorganisms (Sadowski et al., 2001a, 2002; Praczyk & Skrzypczak, 2004; Woźnica, 2008).
The process of degradation and translocation of herbicide active ingredients depends on many environmental and soil factors, which determine their adsorption and absorption in

the soil. Such factors include the type of soil, mechanical composition (particularly the content of clays and zeolites), temperature, moisture content, content of organic matter (humus), pH of soil as well as the content of soil entomofauna biomass (Walker & Welch, 1989; Vicari et al., 1994; James et al., 1999; Sarmah et al., 1999; Sadowski & Kucharski, 2004).

The sorption capacity of herbicides is defined by the index of soil sorption of the herbicide (K_d) and sorption of the herbicide to organic carbon (K_{oc}). Active ingredients of herbicides characterized by very high mobility in the soil environment have $K_{oc} < 100$ ml/g (clopyralid, nicosulfuron, sulfosulfuron, sulcotrione, dicamba), while herbicides with $K_{oc} > 2000$ ml/g exhibit poor mobility (trifluralin, diquat, pendimethalin, diclofop, fenoxaprop-P) (Praczyk & Skrzypczak, 2004; Woźnica, 2008).

The translocation of herbicides in the soil profile is frequently disturbed by crops themselves or by weeds, absorbing water from the soil solution. Since a portion of the root system of plants frequently reaches a depth of 50 - 60 cm and has a very big suction power we may often observe the process of leaching or even the movement of residue of certain herbicides (e.g. chlorsulfuron) from deeper soil layers towards the rhizosphere (Walker et al., 1989; Sadowski et al., 2001b). In the opinion of Sadowski & Kucharski (2004), elution of herbicide active ingredients being derivatives of sulfonylurea (chlorsulfuron, sulfosulfuron) and phenoxyacetic acid (2.4 D, MCPA) from the soil profiles is strongly dependent on the initial moisture content and the absorbing capacity of soil. They showed in their studies that with an increase in the initial moisture content of soil, the degree of leaching for these substances increased markedly, reaching a certain maximum. When soil reached the maximum water capacity (under field conditions this process is observed during heavy rains), then the percentage of leached active ingredient of a herbicide is markedly reduced. Also Beckie & McKercher (1990), Oppong & Sagar (1992) and Günter et al. (1993) were of an opinion that apart from moisture content, also absorbing capacity of soil has a decisive effect on herbicide mobility. In their studies Günter et al. (1993) showed that in soil with poor absorbing capacity metsulfuron and triasulfuron were subjected to elution much faster than in soil with a high absorbing capacity. Mobility was also dependent on the active ingredient itself, with metsulfuron being much more active than triasulfuron.

Also the depth to which herbicide active ingredients penetrate under field conditions is not specifically defined, since it depends on many factors (e.g. absorbing capacity, granulometric composition, cultivation measures). On the basis of studies concerning the translocation of herbicide active ingredients Helling & Turner (1968) determined the relative mobility index of herbicides (R_f), dividing them into five classes. In another study Walker & Welch (1989) showed that chlorsulfuron (ALS group) was capable of penetrating to a depth of 50 cm 63 days after application, despite the fact that a bigger part of its residue was detected in a layer up to 25 cm deep. In turn, another active ingredient from the same chemical group, i.e. triasulfuron, did not penetrate deeper than 10 cm, and its residue remained at that depth throughout the entire period of the experiment (125 days).

Stability of active ingredients of herbicides in the soil is also dependent on its physico-chemical properties and on the course of degradation dynamics. A very important indicator, which defines potential persistence of the herbicide active ingredient in the soil environment, is the half-life period (DT_{50}). It is a time period required for the degradation of the active ingredient to half its initial concentration in soil. The value of DT_{50} is a characteristic feature of individual active ingredients of herbicides and it may range from several days (e.g. quizalofop-P, mesotrione, MCPA) to as long as several months (e.g. trifluralin, ethofumesate, pendimethalin). Most active ingredients of herbicides used in agricultural plantations has DT_{50} of less than 60 days (e.g. florasulam, clomazone, clopyralid, bentazone), while in vegetable growing it is below 20 days (e.g. clethodim, cycloxdim, metazachlor, pyridate) (Praczyk, 2004; Praczyk & Skrzypczak, 2004; Woźnica, 2008). Half-life (DT_{50}) is only a rough indication of the potential persistence of herbicide active ingredients in soil. Under field conditions degradation of a herbicide and its translocation may occur faster or much slower, since it is a result of interactions between chemical properties of the active ingredient itself and moisture content, temperature, absorbing capacity of soil, pH and soil microorganisms. Thus the risk of persistence and translocation of herbicide active ingredients in soil may not be considered only on the basis of one of the above mentioned parameters (e.g. DT_{50}, K_{oc}, R_f), as under field conditions the interactions of all these factors affect the rate of chemical and biological processes, which in turn determine the behavior of active ingredients of herbicides in the soil environment. Table 1 presents characteristics of selected active ingredients of herbicides, which have a decisive effect on their behavior in the soil environment.

Active ingredient	Group HRAC	Solubility in water [mg/l]	DT_{50} [days]	K_{OC} [ml/g]	R_f movement index in soil environment
quizalofop-P	A	0.4	<1	1024	small (R_f = 0.0-0.34)
florasulam	B	6360 (pH 7)	2-18	4-54	
trifluralina	K1	0.22	60-132	2500-13700	
diquat	D	700000	1000	>32000	
pendimethalin	K1	0.3	30-150	6700-29400	
amidosulfuron	B	9 (pH 5.8)	3-29	33.7	medium (R_f = 0.35-0.64)
clomazone	F3	1100	15-45	104-608	
ethofumesate	N	50	15-250	97-245	
alachlor	K3	242	15-30	170-200	
MCPA	O	734	5-6	25-157	
sulfosulfuron	B	1627 (pH 7)	11-47	5-89	large (R_f = 0.65-1.0)
metamitron	C1	1700	7-70	91-392	
bentazone	C3	570	12-45	13-176	
mesotrione	F2	2200	3-7	19-390	
clopyralid	O	143	14-56	4.6	

Source: Helling & Turner, (1968); Praczyk & Skrzypczak, (2004); Woźnica, (2008); modified

Table 1. Examples of active ingredient of herbicides and selected physico-chemical properties affecting their behavior in soil

Annually repeated application of herbicides in the same field may affect the dynamics of degradation and translocation, as well as the level of residue of their active ingredients. After penetrating into the soil the action of a herbicide on a crop or weeds is determined within the rhizosphere by the degree of availability and the sensitivity of the plant to the active ingredient. Strong vertical translocation of certain active ingredients of herbicides several days after application, particularly in lessive soils may be dangerous for the soil environment due to the possible penetration into the ground waters causing their contamination (Beckie & McKercher, 1990; Sadowski & Kucharski, 2003). Thus studies are necessary which would facilitate an evaluation of a threat posed by the application of herbicides in relation to agrophytocenosis. In ecotoxicology the adopted methods for the determination of the levels of bioavailable phytotoxic residue of herbicide active ingredients in soil include biotests, due to their high efficiency, relatively very high sensitivity and limited testing costs in comparison to instrumental methods (Fahl et al., 1995; Hollaway et al., 1999; James et al., 1999; Sadowski et al., 2002; Sadowski & Kucharski, 2004; Sekutowski & Sadowski, 2006). Plant species exhibiting high sensitivity to the action of selected active ingredients of herbicide, such as *Sinapis alba, Fagopyrum esculentum, Sorghum saccharatum, Lepidium sativum, Helianthus annuus, Zea mays* or *Cucumis sativus* are used as detectors. In particular cases biotests may also provide information on transport and on the situation of applied active ingredients (Günther et al., 1993; Sadowski & Kucharski, 2004). We may find numerous examples in literature concerning applications of plant biodetectors in studies on herbicide active ingredient residue (Günther et al., 1993; Stork & Hannah, 1996; Sarmah et al., 1999; Sekutowski & Sadowski 2005; 2006; 2009).

3. Division of biological methods used in studies on the soil environment

Analytical methods using biological material are becoming promising alternatives for conventional analytical methods and in certain cases they may even replace them (Hollaway et al., 1999). They are commonly applied mainly due to their specificity and low unit costs. In toxicological analyses we may distinguish two groups of applications for biological methods in the assessment of the effect of xenobiotics (e.g. herbicides) on the soil environment:

a. bioanalytical tests, which are connected with the use of biological organisms as receptors of specific chemical substances, e.g. herbicides. Due to the method of the utilization of the biological component we distinguish:
- biosensors, in which the biological component is the active element (e.g. an enzyme, antibodies - ELISA test),
- biotests, in which a whole plant organism or its part (e.g. seeds, roots) are the control and measuring element (Hollaway et al., 1999; van Wyk & Reinhardt, 2001).

b. biomonitoring, which may be conducted in two ways:
- through the formation of passive accumulation samplers based on typical analytical tests of biological samples,
- through observation of plant or animal bioindicators (Fahl et al., 1995; Alonso-Prados et al., 2002).

4. Bioassay

Bioassay or biotest (Greek *bios* – life + Latin *testari* - indicate) may be defined as an experimental biological sample (the whole organism or its part), which aim is to detect a toxic substance found in the environment or to identify its harmful action, by quantitative determination of the effect of the tested substance in relation to the control object.

In studies conducted using biotests three methods are typically applied, with the first two being conducted under controlled (laboratory) conditions, while the third being run using a population of organisms living under natural conditions (*in situ*).

a. phytotoxicity tests conducted in a laboratory, during which the substance exhibiting phytotoxic action is artificially introduced to the tested object (e.g. soil). Next the test is performed with an appropriately selected indicator organism e.g. a plant (a phytotest). Thus collected results are a source of information on toxicity of a given substance under controlled conditions. The main aim of such a test is to calibrate the biotest, which will next be used to estimate phytotoxicity of tested samples (e.g. collected from contaminated areas).
b. phytotoxicity tests conducted at a laboratory on the basis of respective samples (e.g. soil) collected from contaminated areas. Phytotoxicity of such samples is compared with the phytotoxicity of reference samples (biotests). On this basis the interval is determined, within which residue e.g. of herbicides may have an adverse effect on crops (e.g. residual effect).
c. phytotoxicity tests conducted on the site in which a population of sensitive organisms is living (conditions of their natural occurrence) (Kuczyńska et al., 2005, Namieśnik & Szefer 2009).

Moreover, biotests may be classified in terms of the used organism (e.g. bacteria, plants, animals), which constitute the active element of the test. In ecotoxicology in studies on the residue of different xenobiotics (e.g. herbicides) the most frequently applied include plants and their seeds, due to the specific action of the tested preparations and in view of the humane, economic and practical aspects.

On the basis of the dose ↔ final effect dependence, which may be expressed e.g. by the reduction of fresh or dry weight of the test plant in comparison to the control object, we may determine values of indicators being a quantitative measure of phytotoxicity of the tested substance. Phytotoxic action of active ingredients contained in herbicides may be determined using such indicators as ED_{10}, ED_{50} or ED_{90} (effective dose), i.e. determining the concentration of the active ingredient causing a specific biological effect at 10%, 50% or 90% its maximum value. Another applied indicator is index IC_{50} or IC_{90} (inhibition concentration), i.e. the concentration of e.g. herbicide in the soil environment, which causes a reduction of fresh or dry weight of the test plant (roots, stems, leaves) by 50% or 90% in comparison to the control (not treated with this herbicide).

The dose ↔ final effect dependence may also be used to predict risk, i.e. to determine the dose and persistence of herbicide residue, at which the probability of phytotoxic effects is high or small. An example exhibiting this dependence may be here a study conducted by Sadowski et al. (2007) or Sadowski & Sekutowski (2008), referring to the phytotoxic action of herbicide active ingredient residue on successive crops. Those authors using biotests showed that herbicide residue in soil may be hazardous for successive crops at two critical moments. The first refers to resowing, i.e. situations when for different reasons, most frequently independent of the farmer, the plantation is eliminated. In turn, the other moment refers to residue persisting in soil and exhibiting phytotoxic action immediately after the crop is harvested or even for the next several months. A similar phenomenon on fields in which chlorsulfuron and metsulfuron were applied, observed in the form of extensive damage to sugar beet or rape plantations found in the period of 2 successive years, was reported by Walker & Welch (1989) and Walker et al. (1989). The above mentioned effect is manifested only because crops (e.g. beet, rape) exhibit very high sensitivity to herbicides from the ALS group. Plant species with a narrow range of tolerance (stenobionts) characterized by high sensitivity to specific chemical groups or active ingredients of herbicides are referred to as indicator species or bioindicators. Thus biotests are very often used in biomonitoring, to evaluate the consequences potentially caused by herbicides on individual elements of agrophytocenosis (e.g. crops, soil or water).

4.1 Criteria for the selection of a bioindicator

Species of indicator plants should be characterized by a narrow range of responses and exhibit high sensitivity to specific chemical substances, with their response being specific and adequate to the concentration of the chemical substance and easily observable (e.g. strong inhibition of root growth).

Bioindicators should meet the following requirements:

- common occurrence,
- a wide range of distribution,
- a long life cycle or several generations within a year,
- being easily recognizable,
- genetic homogeneity,
- high sensitivity to specific chemical substances,
- stability and repeatability of responses,
- low unit costs and easy laboratory culture.

In turn, plant bioindicators used in phytotests should have the following characteristics:

- small and even seeds,
- uniform germination power and energy of seeds,
- a short emergence period (1-2 days),
- a short vegetation period,
- high biomass of stems, leaves or roots,
- high sensitivity in relation to one chemical group (e.g. phenoxy acids, sulfonylurea).

When selecting a bioindicator for a test it is also necessary to take into consideration the age and sensitivity of individual tissues to the tested herbicide. A similar opinion was also expressed by Shim et al. (2003) and Demczuk et al. (2004), who in their studies conducted using different weed species and *Cucumis sativus* plants observed a diverse response of individual plant tissues to tested active ingredients of herbicides. They showed that sensitivity to residue of sulfonylurea herbicide depended to a considerable degree on the age of tissues and their location. The youngest roots and leaves of test plants turned out to be most sensitive.

Thus one of the basic guarantees of an appropriately conducted biotest is the selection of an appropriate test plant. An example of a dependence between the phytoindicator and the response to the herbicide active ingredient is presented in Fig. 1-2. In the analyses 3 test plants were used, i.e. *Sinapis alba, Fagopyrum esculentum* and *Cucumis sativus*, as well as 2 active ingredients of herbicides belonging to different chemical groups (phenoxy acids - 2.4 D and sulfonylurea - nicosulfuron). For the detection of 2.4 D residue *Cucumis sativus* proved to be most suitable, since root growth inhibition by 50% (IC_{50}) occurred already at a concentration of 0.18 mg/kg. For the two other species, i.e. *Sinapis alba* and *Fagopyrum esculentum*, IC_{50} ranged from 0.4 to 0.5 mg/kg (Fig. 1).

In turn, in the detection of nicosulfuron residue the highest sensitivity was found for the test with the use of *Sinapis alba*. Root length reduction by 50% (IC_{50}) occurred already at a concentration of 0.125 mg/kg. Sensitivity of the test (IC_{50}) with the use of *Fagopyrum esculentum* and *Cucumis sativus* was markedly lower and amounted to 0.25 mg/kg for *Fagopyrum esculentum* and 0.55 mg/kg for *Cucumis sativus*, respectively (Fig. 2).

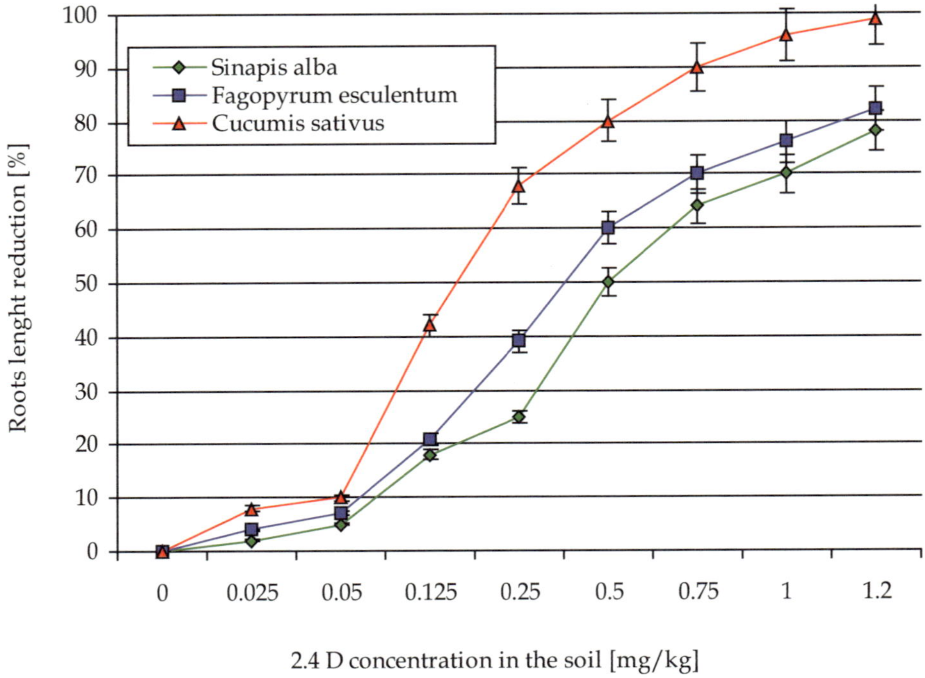

Fig. 1. 2.4 D effect on the tested plant in terms of roots lenght reduction

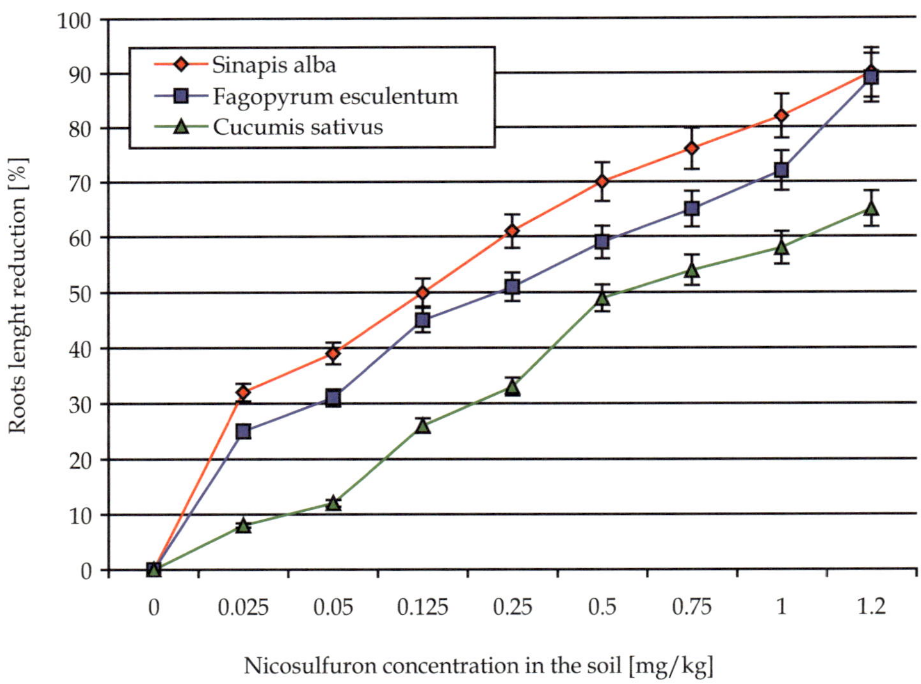

Fig. 2. Nicosulfuron effect on the tested plant in terms of roots lenght reduction

4.2. Conventional bioassays

A conventional bioassay, used in the detection of herbicide active ingredients in soil, consists in the sowing of seeds of a test plant (adequately sensitive to the tested substance or chemical group) into the soil sample containing the residue. Examples of procedures required for the establishment of such a bioassay are presented in a diagram in Fig. 3.

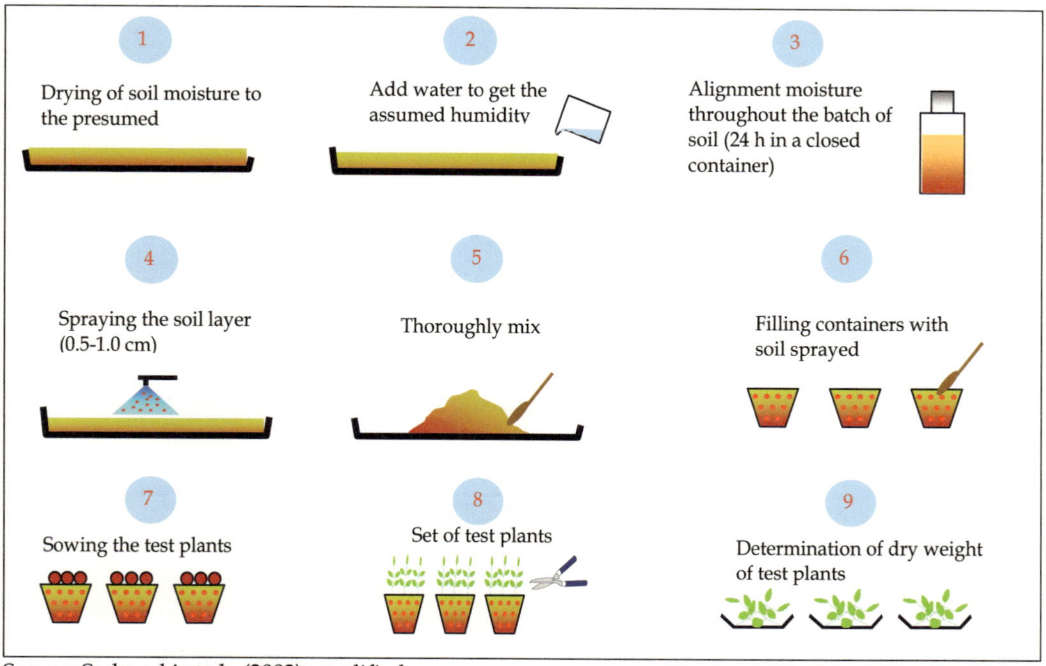

Source: Sadowski et al., (2002); modified

Fig. 3. Example diagram of a conventional bioassay setting

4.2.1 Availability of herbicide active ingredients to plants

The bioassay method is also used in the determination of values of ED_{50}, ED_{90} or IC_{50}, IC_{90}. The duration of a conventional bioassay depends to a considerable degree on the test plant, or rather on the tested part of the bioindicator (roots, leaves) and the active ingredient of a given herbicide, and it may range from 7 days (roots) to 14 days (leaves, stems). After a period of 7 or 14 days from the establishment of the test fresh and then dry weight of roots or leaves and stems is determined (by cutting and drying at a temperature of 105^{0}C, and weighing on an analytical scale). Next the percentage loss of fresh and dry weight is calculated in relation to the control plants (sown into the soil containing no herbicide), while thus collected results for the dependence between weight loss in the phytotest and the concentration of the herbicide active ingredient in the soil are used in the graphic presentation of this dependence (Fig. 4).

Figure 5 presents an example of a phytotest using *Cucumis sativus* established in soil containing different concentrations of chlorsulfuron. Results recorded from the bioassay constitute a source of information on the toxicity of chlorsulfuron under controlled conditions. The theoretical objective of such a test is to determine IC_{50} for chlorsulfuron and to calibrate the phytotest, which will next be used in the estimation of phytotoxicity of

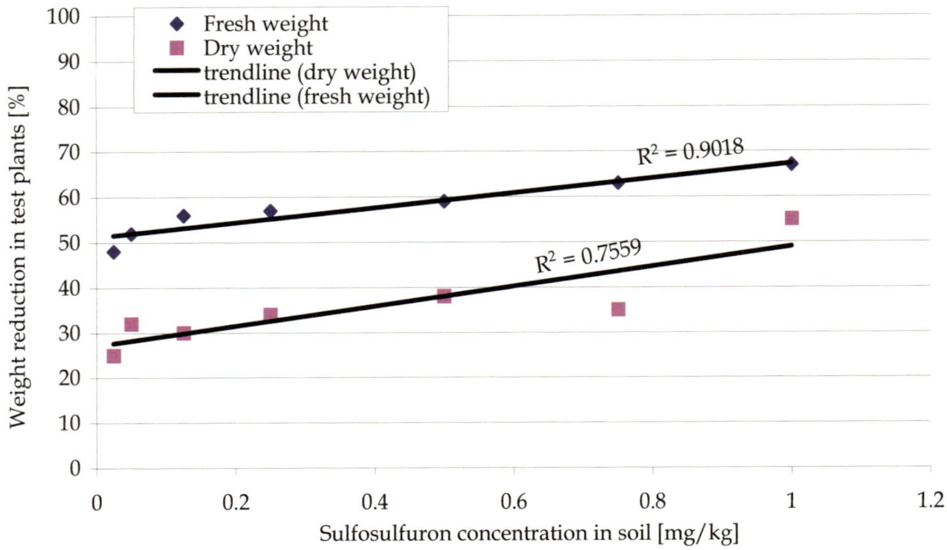

Fig. 4. Changes fresh and dry weight of *Sinapis alba* under the influence of different sulfosulfuron concentrations in soil

Fig. 5. Te effect of chlorsulfuron on fresh weight reduction of *Cucumis sativus* (determination of IC_{50})

samples of soil collected from a field containing residue of chlorsulfuron (the practical objective). Thanks to this test it will al be possible to determine whether in that field plants from family *Cucurbitaceae* will be exposed to the phytotoxic action of chlorsulfuron residue.

4.2.2 Distribution of herbicide active ingredients in the soil profile

Knowledge on the translocation and distribution of active ingredients in the soil profile and factors affecting this process is required both for the protection of the soil environment and a more efficient use of herbicides. Most studies in this field have been conducted mainly using lysimeters. Unfortunately, the primary drawback of the lysimeter model is connected with the high cost of one assay and limitations related with the collection of soil samples, resulting from the disruption of the soil profile in the lysimeter column. After several samplings the lysimeter column has to be refilled with a new undisturbed soil profile. In turn, analyses conducted under laboratory conditions using bioassays do not have such limitations. Moreover, they are more efficient and provide the experimenter with more flexibility and control over a much bigger number of parameters observed during the process of herbicide translocation. Soil collected from such a model is used as a substrate for bioassays and the filtrate may be used in chemical analyses. This method makes it possible to determine in a very precise way the distribution of phytotoxic residue of herbicide active ingredients in the soil profile. Another advantage of this model is the possibility of arbitrary modeling of irrigation in the soil profile, which facilitates a comprehensive evaluation of the residue balance in the soil - water system. Figure 6 presents an example diagram of such a model in action, in which the bioassay method was used to determine the distribution of herbicide residue.

In the opinion of Sadowski & Kucharski (2004) the degree of leaching and as a consequence the distribution of a portion of herbicide active ingredients depends on the initial soil moisture content. Figure 7-8 presents the distribution of certain active ingredients of

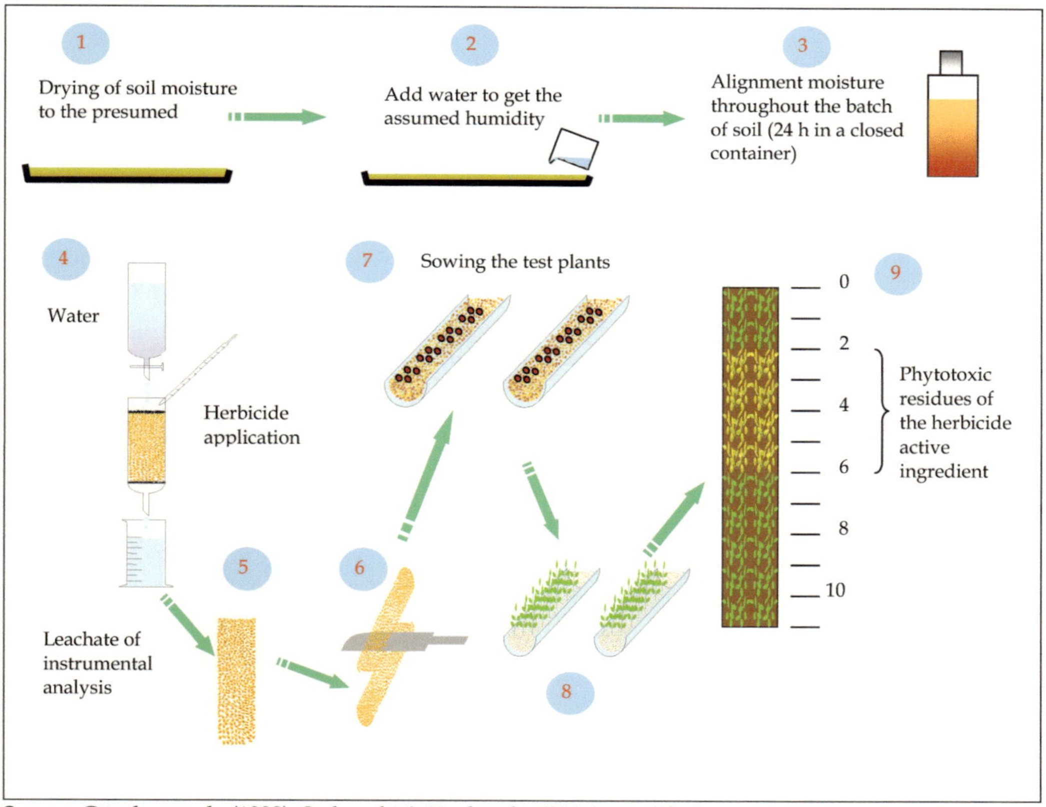

Source: Günther et al., (1993); Sadowski & Kucharski, (2004); modified

Fig. 6. The application of the phytotest method to determine residue of herbicide active ingredients in the soil profile

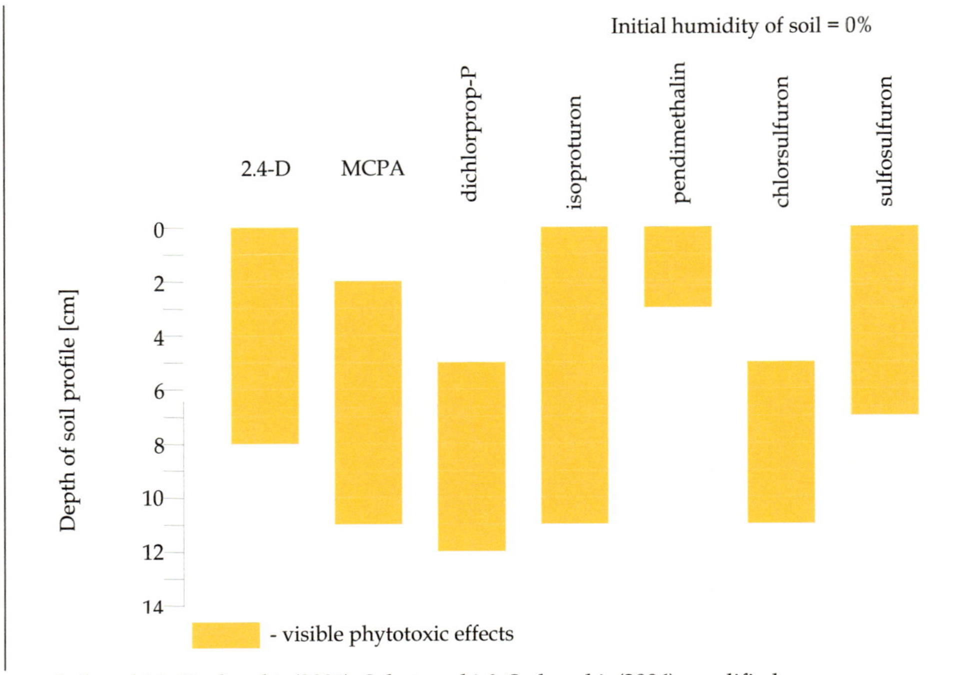

Source: Sadowski & Kucharski, (2004); Sekutowski & Sadowski, (2006); modified

Fig. 7. Rate of translocation of active ingredients depending on initial soil moisture content

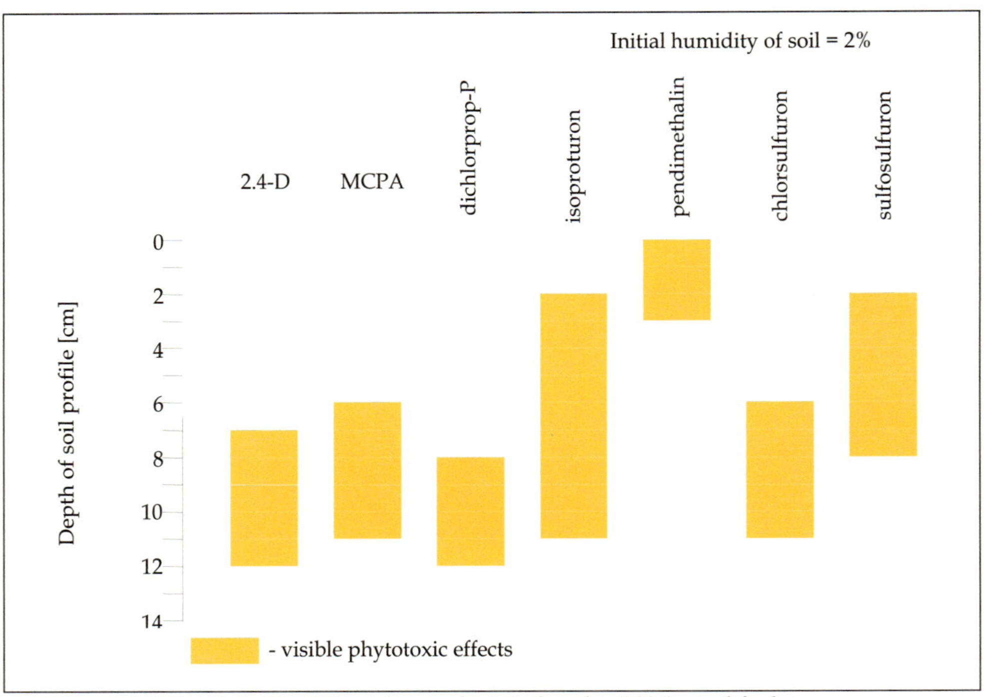

Source: Sadowski & Kucharski, (2004); Sekutowski & Sadowski, (2006); modified

Fig. 8. Rate of translocation of active ingredients depending on initial soil moisture content herbicides depending on changes in the initial soil moisture content. Most active ingredients, which were transferred on air dry soil (0% moisture content) were detected by test plants (*Sinapis alba*) mainly in the surface soil layer (Fig. 7). The highest leaching level was found for isoproturon (0-11 cm), while the least leached substance turned out to be pendimethalin (0-3 cm). An increase in the initial soil moisture content by 2% caused a marked shift of residue deeper within the soil profile practically for all the tested active ingredients. Only pendimethalin residue remained at the same level (Fig. 8).

From the practical point of view the distribution of the main portion of herbicide active ingredients, as well as the degree of their leaching to deeper soil layers are highly significant, since they determine the effectiveness of herbicides (particularly those soil-applied). Moreover, they also determine the degree of herbicide translocation outside the root zone, which may increase the risk of their being transferred to ground waters (Sadowski & Kucharski, 2004).

4.2.3 Dynamics of degradation of herbicide active ingredients in soil

Dynamics of degradation occurs most intensively in the surface soil layer (0-20 cm) and it is closely related with the processes of degradation and translocation of herbicide active ingredients. In this layer the intensity of biological and chemical processes is dependent to a high degree on the temperature and soil moisture content, as well as the tillage systems (Sadowski, 2001; Sadowski & Kucharski, 2004; Rola & Sekutowski, 2005). In order to determine the dynamics of degradation and translocation of herbicide active ingredients in the soil, at specified time intervals samples are collected, onto which test plants (phytoindicators) are sown. An example of a phytotest for different herbicide active ingredients is presented in Fig. 9.

An example given here presents an experiment conducted using the bioassay method (with *Sinapis alba* as a phytodetector) under field conditions referring to the dynamics of translocation and degradation of rimsulfuron depending on the tillage method applied. In the first 6 weeks after application no marked differences were observed in the course of the dynamics of rimsulfuron degradation depending on the tillage method used. Accelerated

Test plant - *Sinapis alba*

1 = Control
2 = foramsulfuron + iodosulfuron (45 g/ha + 1.5 g/ha)
3 = nicosulfuron (60 g/ha)
4 = rimsulfuron (12 g/ha)
5 = rimsulfuron (15 g/ha)
6 = tifensulfuron (11.5 g/ha)
7 = dicamba + rimsulfuron (240 g/ha + 7.5 g/ha)

Fig. 9. Degradation of different active ingredients of herbicides in the 0-20 cm soil layer (biotest method)

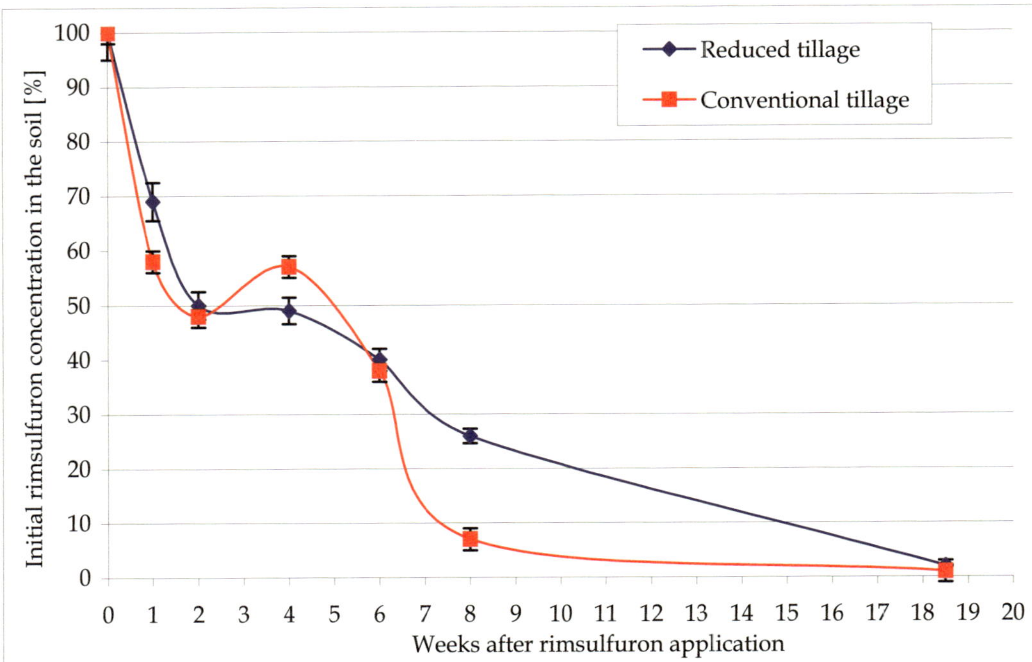

Source: Sekutowski, (2009)

Fig. 10. Rimsulfuron degradation rates in the 0-20 cm soil layer (mean in the years 2005-2007)

translocation and dynamics of degradation was found as late as 7 weeks after application and it was markedly diversified depending on the tillage systems (Fig. 10).

The presented examples of the application of plants as phytodetectors in the bioassay method more precisely illustrate the phytotoxic action of herbicide active ingredients (even those found in trace amounts) for agrophytocenosis than their concentration in soil determined using chemical analyses. A similar opinion was presented by Hollaway et al. (1999), who in their studies concerning the detection of sulfonylurea herbicide residue in soil using three methods, i.e. bioassay, ELISA and HPLC, stated that a bioassay using *Pisum sativum* and *Lens culinaris* plants as bioindicators was most sensitive. Biotests detected residue of sulfonylurea herbicides at 0.1 - 1.0 mg/ha soil, ELISA at 0.1 - 10 mg/ha, while HPLC at 3 - 10 mg/ha, respectively.

Depending on soil and climatic conditions only a portion of residue contained in soil is available to plants. Biotests used in biomonitoring make it possible to evaluate whether this part of residue may exhibit phytotoxicity towards agrophytocenosis.

The presented examples of conventional phytotests using different plants and their seeds as phytodetectors, conducted according to standardized national procedures, frequently happen to be complicated, they require considerable laboratory space and are time-consuming (BN-83 9180-25, 1983; BN-83 9180-27, 1983; BN-84 9180-30, 1984; PN-ISO 17616, 2010). For several years now ready-to-use tests (toxkits) have been commercially available, sold in the form of packages, allowing the evaluation of phytotoxicity of tested samples within a short time (1-3 days). They contain cryptobiotic forms of bioindicators (e.g. seeds of plants - Phytotoxkit™), coming from standard breeding, which may be stored for 6 months and when needed prepared for the test within a very brief time (Phytotoxkit, 2004).

4.3 Phytotoxkit microbiotest

The necessity to conduct analyses of many soil samples within a relatively short time has led to the introduction of miniature phytotoxicity tests, called microbiotests or second generation tests, as alternatives for conventional phytotests. An example of such a microbiotest is a rapid (72 h) test - Phytotoxkit™ (Phytotoxkit, 2004). Professor Guido Persoone (with a team of co-workers) from the University of Ghent in Belgium was the creator of the toxkit tests (Persoone, 2005). The principle of such a phytotest is based on germinating seeds of *Sorghum saccharatum*, *Lepidium sativum* and *Sinapis alba*, which as a result of contact with the tested herbicide active ingredient found in soil exhibit a specific reaction (a lack of germination or reduced root length). The use of standard seeds facilitates test standardization and maintenance of reproducible results irrespective of the laboratory, at which analyses are being conducted. The specific nature of Phytotoxkit™ results in the omission of all labor-consuming activities connected with conventional biotests, thus considerably reducing the time required to obtain the reading (from 14 to 3-5 days). Moreover, this test makes it possible to obtain a direct measurement of root length using image tools, thanks to which a graphic presentation of the dependence between root length reduction in phytodetectors and the phytotoxic concentration of tested herbicide active ingredients is faster and much easier in comparison to a conventional biotest. This test makes it also possible to more comprehensively estimate the phytotoxic effect of herbicide residue not only on the soil environment, but also on the entire agrophytocenosis. An example of a Phytotoxkit™ tst conducted using a standard set of plants is presented in Fig. 11.

The diverse chemical character of herbicides prevents the use of only one type of a Phytotoxkit™ containing standard phytodetectors supplied in the kit. Due to the specific response of different plant species to the presence of herbicide active ingredients belonging to different chemical groups it is necessary to supplement knowledge on the applicability of other plants. Thus the test is very often modified, which consists in the replacement of standard test plants with other plant species, such as e.g. *Helianthus annuus, Cucumis sativus* or *Fagopyrum esculentum*. Thanks to the modification of Phytotoxkit™ it was possible to extend the collection of plants potentially applicable in the determination of herbicide residue, e.g. derivatives of benzoic acid, phenoxy acids and sulfonylourea (Fig. 12).

Similarly as in case of conventional biotests, Phytotoxkit™ may be used in the determination of values of ED_{50} and IC_{50} and the determination of the level of residue, rates of degradation and translocation of herbicide active ingredients in soil. An example in this respect may be an experiment conducted using a modified Phytotoxkit™ under laboratory conditions, consisting in the determination of ED_{10} and ED_{50} for dicamba. The run biotest

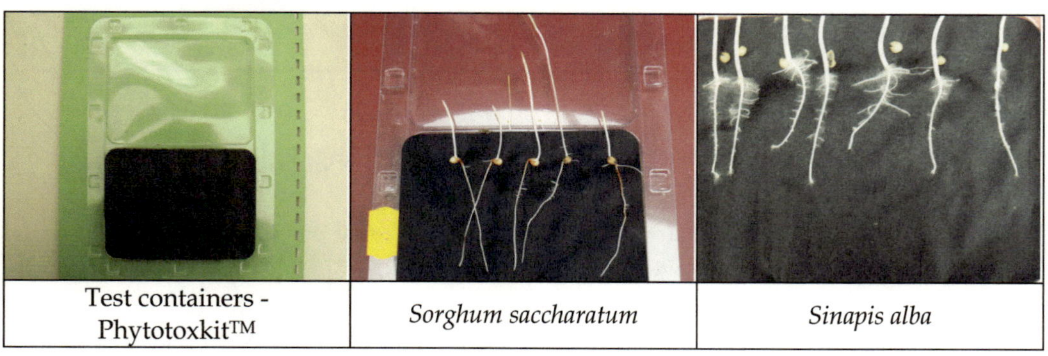

Test containers - Phytotoxkit™	*Sorghum saccharatum*	*Sinapis alba*

Fig. 11. Phytotoxkit™ with standard test plants.

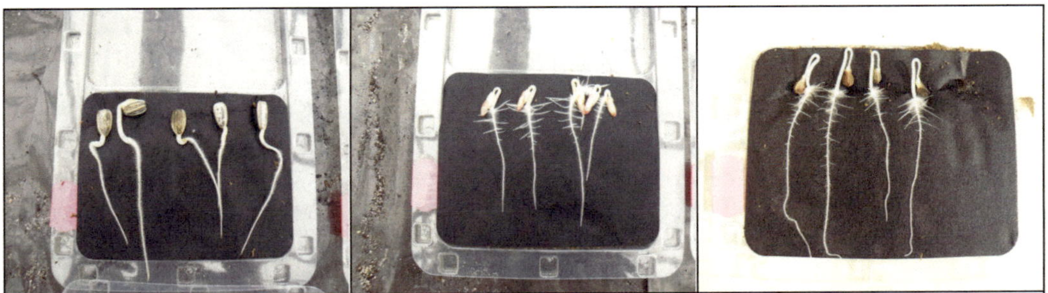

Helianthus annuus	*Cucumis sativus*	*Fagopyrum esculentum*

Fig. 12. Phytotoxkit™ with alternative test plants.

showed that significant differences in the reduction of root length in *Fagopyrum esculentum* and *Cucumis sativus* were obtained for concentrations ranging from 0.025 mg/kg to 0.25 mg/kg. The strongest response to the tested substance was recorded for *Fagopyrum esculentum*, while it was weakest in case of *Sinapis alba*. The detoxication capacity in relation to dicamba in *Fagopyrum esculentum* (ED_{50}) was eliminated already at a concentration of 0.125 mg/kg, while a further increase in the concentration of the tested substance in soil (1.2 mg/kg) resulted in root length reduction by 99%. In turn, ED_{50} for the other two species, i.e. *Cucumis sativus* and *Sinapis alba* fell within the range of 0.25 - 0.5 mg/kg soil (Fig. 13).

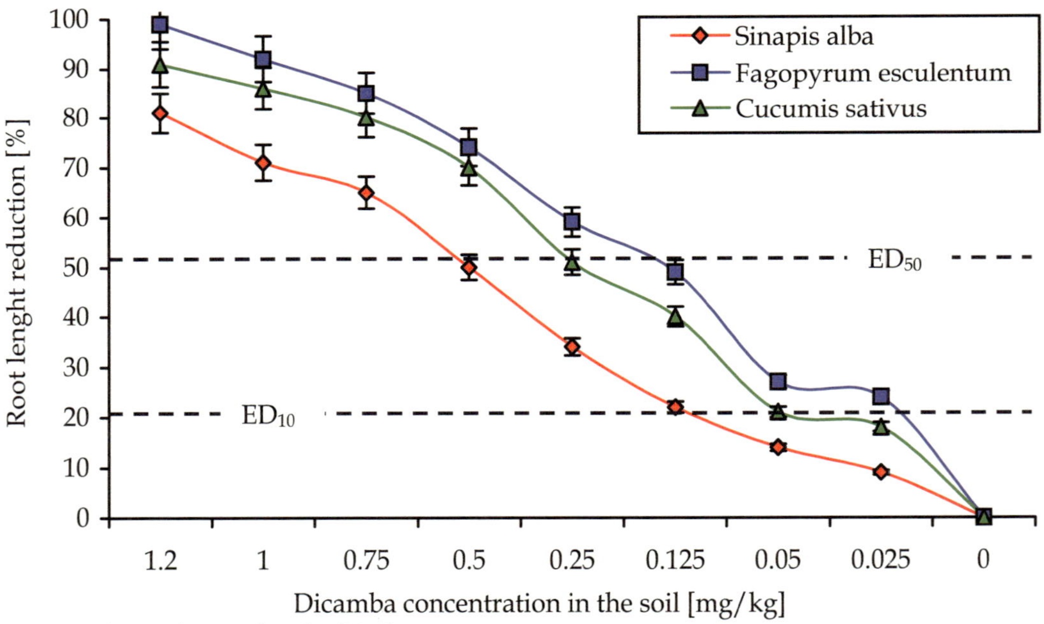

Source: Sekutowski & Sadowski, (2009)

Fig. 13. The effect of dicamba on root length reduction in tested plant

The above example very well shows the response (sensitivity) of the phytodetector to the tested active ingredient of the herbicide. In analyses using plants as detectors, it is crucial to select an appropriate plant for the tested herbicide active ingredient. A sufficiently sensitive plant detector makes it possible to conduct tests on microresidue of 0.01 mg/kg soil.

4.4 Sets of biotests (batteries)

The selection of an appropriate biotest in studies on agrophytocenosis depends on the type of required information, the concentration of herbicide active ingredient residue in the analyzed sample of soil (water), as well as the species-specific sensitivity of the tested plant. In case of the use of only one phytodetector species the estimated phytotoxicity reflects the sensitivity of only this one tested species. Such a procedure may result in an error connected with an underestimation of phytotoxicity of the analyzed herbicide active ingredient in relation to the entire agrophytocenosis. This risk may be minimized thanks to the application of a battery of biotests, which action is based on the use of plant species of different sensitivities to active ingredients of herbicides belonging to one chemical group. Batteries of tests may be formed within one test (e.g. Phytotoxkit™), which may include several species of test plants exhibiting different sensitivity to a given chemical group. Moreover, sets of batteries may be established within several tests using different biodetectors of varying sensitivity to the same chemical group, e.g. Phytotoxkit™ → ELISA → HPLC (Hollaway et al., 1999).

5. Conclusion

Bioassays are methods commonly applied in ecotoxicology in the determination of the levels of bioavailable phytotoxic residue of herbicide active ingredients in soil. Tests with the use of rapidly germinating seeds have several very important advantages, as they are cheap and easy to perform, they do not require expensive laboratory equipment and they yield reproducible results. The phytotoxic effect of herbicide active ingredient may be stated on the basis of the dynamics of germination, seedling growth, reduction of dry or fresh weight of roots or aboveground parts (stems, leaves) of test plants. On the basis of selected parameters, such as the reduction of root length, the toxic effect of herbicide active ingredients may be determined already after approx. 24 h, while the dynamics of root growth - after 3-5 days from the onset of the test (Phytotoxkit™). In turn, the reduction in fresh or dry weight of aboveground parts of plants may be established after approx. 10-14 days (a conventional biotest).

Unfortunately, drawbacks of such a method include first of all the fact that it is impossible to identify the tested active ingredient. This problem may be solved by using different biological factors forming a set of biotests (Phytotoxkit™ → ELISA → HPLC), which will make it possible to precisely determine the herbicide active ingredient. It also needs to be stressed that biotests with the application of rapidly germinating seeds of selected plant species may be a good supplementation or even an alternative to classical instrumental measurements, used in the detection of phytotoxic residue of herbicide active ingredients in soil.

Probably the scope of bioassay application within the next few years will be increasing and thus collected information will constitute the basis for the initiation of analyses using classical analytical methods.

6. Acknowledgement

I would like to express my gratitude to Prof. J. Sadowski for providing me with literature and for his valuable remarks and suggestions in the course of preparation of this manuscript.

7. References

Ahmad I., Crawford G. (1990). Trace residua analysis of the sulfonylurea herbicide chlorsulfuron in soil by gas chromatography-electron capture detection. Journal Agric. Food Chem., 38: 138-141.

Alonso-Prados J.L., Hernandez-Sevillano E., Llanos S., Villarroya M., Garcia-Baudin J.M. (2002). Effects of sulfosulfuron soil residues on barley (*Hordeum vulgare*), sunflower (*Helianthus annuus*) and common vetch (*Vicia sativa*). Crop Protect., 21: 1061-1066.

Beckie H.J., McKercher R.B. (1990). Mobility of two sulfonylurea herbicides in soil. J. Agric. Food Chem., 38: 310-315.

BN-83 9180-25. (1983). Testy biologiczne. Ocena aktywności biologicznej herbicydów metodą testu hamowania wzrostu korzeni ogórka., pp. 3.

BN-83 9180-27. (1983). Testy biologiczne. Ocena aktywności biologicznej herbicydów i ich makropozostałości w glebie metodą testu gorczycy., pp. 6.

BN-84 9180-30. (1984). Testy biologiczne. Ocena aktywności biologicznej herbicydów i ich makropozostałości w glebie metodą testu gryki., pp. 5.

Czerniawska-Kulesza I., Cieszelczuk T., Kusza G., Cichoń A. (2006). Comparison of the Phytotoxkit Microbiotest and Chemical Variables for Toxicity Evaluation of Sediments. Wiley Periodicals, Inc. Environ Toxicol., 21: 367-372.

Dećkowska A., Pierścieniak M., Gworek B., Maciaszek D. (2008). Wybrane gatunki roślin jako wskaźniki zmian w środowisku. Ochr. Środ. Zasob. Natural., 37: 128-138.

Demczuk A., Sacała E., Grzyś E. (2004). Zmiany aktywności syntazy acetylomleczanowej (ALS) pod wpływem herbicydu Titus 25 DF u różnych odmian ogórka. Progr. Plant Protect./Post. Ochr. Roślin 44(2): 645-647.

Fahl G.M., Kreft L., Altenburger R., Faust M., Boedeker W., Grimme L.H. (1995). pH-Dependent sorption, bioconcentration and algal toxicity of sulfonylurea herbicides. Aquat. Toxicol. 31: 175-187.

Günther P., Pestemer W., Rahman A., Nordmeyer H. (1993). A bioassay technique to study the leaching behaviour of sulfonylurea herbicides in different soils. Weed Res., 33: 177-185.

Helling C.S., Turner B.C., (1968). Pesticide mobility: determination by soil thin-layer chromatography. Science, 162: 562-563.

Hollaway K .L., Kookana R.S., McQuinn D.J., Moerkerk M.R., Noy D.M., Smal M.A. (1999). Comparison of sulfonylurea herbicide residue detection in soil by bioassay, enzyme-linked immunosorbent assay and hplc. Weed Res., 39: 383-397.

James T.K., Holland P.T., Rahman A., Lu Y.R. (1999). Degradation of the sulfonylurea herbicides chlorsulfuron and triasulfuron in a high-organic-matter volcanic soil. Weed Res., 39, 137-148.

Kucharski M., Sadowski J. (2006). Wpływ wilgotności gleby na rozkład herbicydu - badania laboratoryjne. Prog. Plant Protection/Post. Ochr. Roślin 46(2): 750-753.

Kuczyńska A., Wolska L., Namieśnik J. (2005). Application of biotests in environmental research. Crit. Rev. Anal. Chem., 35(2): 135-154.

Namieśnik J., Szefer P. (2009). Analytical Measurements in Aquatic Environments. Publisher. CRC Press, pp. 503. ISBN: 978-1-4200-8268-5.

Oppong F. K., Sagar R. (1992). The activity and mobility of triasulfuron in soil as influence by organic matter, duration, amount and frequency of rain. Weed Res., 32: 157-165.

Persoone G. (2005). Nowe metody w ekotoksykologii osadów i gleb. Mat. Konf. pt. Nowe metody w ekotoksykologii. Mikrobiotesty do szybkiej oceny toksyczności. Warszawa, 15-16 września 2005: 23-25.

Phytotoxkit. (2004). Seed germination and early growth microbiotest with higher plants. Standard Operational Procedure. Nazareth, Belgium: MicroBioTest Inc., pp. 24.

Polska Norma - PN-ISO 17616. (2010). Jakość gleby - Zasady wyboru i oceny biotestów do ekotoksykologicznej charakterystyki gleby i materiałów glebowych., pp. 16.

Praczyk T. (2004). Diagnostyka uszkodzeń herbicydowych roślin rolniczych. Wyd. PWRiL, Poznań, pp. 144.

Praczyk T., Skrzypczak G. (2004). Herbicydy. Wyd. PWRiL, Poznań, pp. 274.

Rola H., Sekutowski T. (2005). Wpływ systemów uprawy na dynamikę rozkładu wybranych herbicydów sulfonylomocznikowych stosowanych w kukurydzy. Pam Puł., 140: 239-243.

Sadowski J. (2001). Wpływ terminu stosowania na dynamikę rozkładu herbicydów w glebie. Prog. Plant Protect./Post. Ochr. Roślin 41(1): 134-139.

Sadowski J., Kucharski M., Rola H. (2001a). Pozostałości herbicydów w środowisku glebowo-wodnym. Biul. Nauk., 12: 23-32.

Sadowski J., Rola H., Domaradzki K. (2001b). Wpływ stanu i stopnia zachwaszczenia na dynamikę rozkładu herbicydów w glebie. Prog. Plant Protect./Post. Ochr. Roślin 41(1): 140-144.

Sadowski J., Rola H., Kucharski M. (2002). Zastosowanie biotestów do oceny poziomu pozostałości herbicydów w glebie. Prog. Plant Protect./Post. Ochr. Roślin 42(1): 152-158.

Sadowski J., Kucharski M. (2003). Monitoring of herbicidal pollution in ground and surface water on arable land of South-West Poland. J. Plant Prot. Res., 43(3): 241-245.

Sadowski J., Kucharski M. (2004). Wpływ czynników agrometeorologicznych na pobieranie i fitotoksyczność pozostałości herbicydów zawartych w glebie. Prog. Plant Protect./Post. Ochr. Roślin 44(1): 355-363.

Sadowski J., Kucharski M., Sekutowski T. (2007). Ocena zagrożeń upraw następczych przez pozostałości wybranych herbicydów stosowanych w uprawach rzepaku. Prog. Plant Protect./Post. Ochr. Roślin 47(3): 246-253.

Sadowski J., Sekutowski T. (2008). Ocena ryzyka fitotoksycznego działania pozostałości herbicydów na uprawy następcze. W: Ekotoksykologia w ochronie środowiska. Wyd. PZITS, Wrocław (884): 349-354.

Sarmah A.K., Kookana R.S., Alston A.M. (1999). Degradation of chlorsulfuron and triasulfuron in alkaline soil under laboratory conditions. Weed Res., 39: 83-92.

Sekutowski T., Sadowski J. (2005). Wpływ technologii uprawy na dynamikę rozkładu pozostałości nikosulfuronu. Prog. Plant Protection/Post. Ochr. Roślin 45(2): 1065-1068.

Sekutowski T., Sadowski J. (2006). Use of bioassays for assessment of residues level of herbicides active ingredients in soil. Pesticides/Pestycydy (1-2): 59-64.

Sekutowski T. (2009). Effect of different tillage systems on maize yield and degradation dynamics of rimsulfuron in soil. Acta Sci. Pol., Agricul., 8(2): 9-20.

Sekutowski T., Sadowski J. (2009). PHYTOTOXKIT™ microbiotest used in detecting herbicide residue in soil. Environ. Prot. Eng., 35(1): 105-110.

Shim S.I., Lee B.M., Ryu E.J., Kang B.H. (2003). Response of leaf acetolactate synthase from different leaf positions and seedling ages to sulfonylurea herbicide. Pesticide Biochem. Physiol., 75: 39-46.

Stork P., Hannah M.C. (1996). A bioassay method for formulation testing and residue studies of sulfonylurea and sulfonanylide herbicides. Weed Res., 36: 271-280.

Vicari A., Catizone P., Zimdahl R.L. (1994). Persistence and mobility of chlorsulfuron and metsulfuron under different soil and climatic conditions. Weed Res., 34: 147-156.

Walker A., Welch S.J. (1989). The relative movement and persistence in soil of chlorsulfuron, metsulfuron-methyl and triasulfuron. Weed Sci., 29: 375-383.

Walker A., Cotterill E.G., Welch S.J. (1989). Adsorption and degradation of chlorsulfuron and metsulfuron-methyl and triasulfuron. Weed Sci., 29: 281-287.

Woźnica Z. (2008). Herbologia. Podstawy biologii, ekologii i zwalczania chwastów. Wyd. PWRiL, Poznań, pp. 430.

Van Wyk L.J., Reinhardt C.F. (2001). A bioassay technique detects imazethapyr leaching and liming-dependent activity. Weed Technol., 15(1): 1-6.

13

Adsorption and Photochemical Behaviour of the Herbicide Monuron on Clay Surfaces

Hafida Mountacer[1], Laila Tajeddine[1] and Mohamed Sarakha[2]
[1]*Laboratoire des Sciences de L'Environnement et Développement, Université Hassan 1er, Settat*
[2]*Laboratoire de Photochimie Moléculaire et Macromoléculaire, UMR CNRS 6505, Université Blaise Pascal, F-63177 Aubière Cede*
[1]*Morocco*
[2]*France*

1. Introduction

The intensive use of pesticides on large areas of agricultural soil has given rise to concern about their fate in the environment. For many years, their presence as well as their environmental impact has attracted much attention due to the wide use of such compounds in agriculture and household applications (Beck et al., 1993; Serrano et al., 2005; Mansour, 1993). Less than 1% of total applied pesticides reach the target pests, the vast majority being dispersed in the different environmental compartments (water, air and soil) but mainly in aquatic environment via agricultural runoff or leaching (Pimentel, 1995). The application of herbicides on agricultural soils is a well-established and effective practice to control weed growth. They represent about 50% of the demand for agricultural chemicals; their prolonged use involves the risk of their retention in crops and soils. Soil is an ultimate reservoir for these products, whether applied directly or received indirectly from spray drift and residues of treated sites (Barbash & Resek, 1996). These products can be the subject of various transformations and they can react with different fractions of soil minerals or organic.

Mineral soil colloids play an important role in adsorption of polar organic compounds, mainly due to the high surface areas associated to their small particle size and, in the case of Smectites, because they have internal (expandable) surfaces accessible to water and polar organic molecules such as pesticides (Lagaly, 1994; Cox et al., 1995; Lee & Kim, 2002). These clay minerals possess layered structures and the presence of interlayer space thus gives a sterically constrained reaction environment for pesticide molecules when intercalated (Caine et al., 1999). Iron is one of the most abundant transition metals in soil and is considered to play a large role in photoinduced redox reactions generating active oxygen species such as $\cdot OH$ (Sherman, 1989). The clay surface is also covered with humic substances that represent 60%–70% of soil organic matter, probably via formation of clay-metal-organic complexes. These complexes are characterized by the presence of stable radical species detected by ESR (Senesi & Schnitzer, 1977).

Regarding the presence of these substances in soil, they may either contribute to the degradation of the pesticide through the formation of reactive species (Albanis et al., 2002) or be involved as inhibitors by a screen effect phenomenon (Hebert & Miller, 1990).

The present work reports the results of monuron (3-(4-chlorophenyl)-1,1-dimethylurea) adsorption onto some selected clays followed by its degradation induced by light solar excitation.

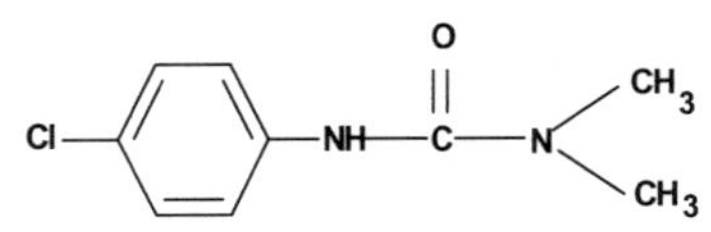

Monuron

Monuron is an herbicide from the group of phenylurea derivatives widely used as an inhibitor of photosynthesis and it is used mainly for the total weed control of non-crop areas. This chemical is characterized by long life times in the environment; its persistence is rather high about 10 months (Khan, 1980) which means that there is a potential risk of waters contamination. Not much information can be found on monuron adsorption and photoreactivity at the surfaces clays and soils.

The aims of the work are first to assess the influence of different components of soils clay of Moroccan region and commercial clays, montmorollonite and kaolin, in adsorption of a monuron herbicide and second to study the photodegradation process of monuron at the surface of some montmorillonite clays.

2. Part I: Adsorption studies of monuron on surface clays

2.1 Introduction

The sorption of pesticides by soil colloidal particles is of great interest in the transport of these compounds in runoff, surface and ground waters, because this particulate matter can act as a carrier of organic contaminants from point sources (Hermosin et al., 1993), and that is also affected by the solid/solution ratio (Cox et al., 1996). Batch sorption experiments were carried out under various conditions to determine the influence of solution chemistry (herbicide concentration, pH and solid-liquid ratio).

2.2 Materials and methods

2.2.1 Adsorptive molecule

Monuron was purchased from Aldrich standards with a purity of 99%. It is a crystalline solid with a vapor pressure of 0.067mPa (25°C) and water solubility 230mg/l (Zhou et al., 2001) Monuron concentrations were determined by high performance liquid chromatography HPLC Water equipped with two pumps type 510, UV- VIS detector with standard diode arrays 996 and a sampler type 717. The unit is controlled by Millennium software. The column used was C_8 phase reverse grafted by silica of granulometry 5µm, length 250m and internal diameter of 4.5 mm. The mobile phase was Methanol and water 60/40 at flow rate of 1mL/min. Under these experimental conditions, the retention time of monuron was 6.5 min.

The herbicide solutions were prepared with ultra pure water (Millipore MilliQ) of resistivity 18.2 MΩcm^{-1} and the pH has been measured using a pH-meter JENWAY 3310 equipped with a combined electrode of Orion type.

2.2.2 Mineral sorbents

Clays extracted from soils: The clays were extracted from the first 20cm of tow Moroccan soils from Settat and Berrechid. The soil samples were air dried and sieved to pass 2 mm mesh. Physicochemical properties are given in table1.

Origin of soil	pH	CEC(10^{-2}meq/g)	OM%	Clay %	Sand %	Limon%	Silt%
Settat	7.8	27.17	3.7	30.4	26.3	35.5	15.4
Berrechid	7.95	23.10	2.15	19.5	37.4	21.8	19.5

O.M: Organic Matter
CEC: Cation Exchange Capacity

Table 1. Selected physical and chemical characteristics of soil

The clay fraction of the soils (< 2μm particle size fraction) was obtained by sedimentation using method developed by Tessier and Collaborators (Robert & Tessier, 1974; Robert, 1975). In the followings, the clays extracted from Settat and Berrechid are named CESS and CEBS respectively.

Pure clays: Adsorption experiments with pure clays are also carried out for monuron, the minerals used as model adsorbents were potassic montmorillonite: K-M and kaolin: Ka. Both were purchased from Fluka chemistry AG and they were used as received.

Major elements were analyzed by X-ray fluorescence using AXIOS-P analytical equipment. The chemical composition of all studied clays fraction is summarized in table 2.

Clay	K-M	Ka	CEBS	CESS
% SiO_2	71,42	45,41	32,743	41,445
% Al_2O_3	14,362	35,066	13,139	16,22
% Fe_2O_3	2,992	0,652	3,95	4,612
% MnO	0,016	0,012	0,069	0,078
% MgO	1,412	0,211	1,513	2,133
% CaO	0,207	0,052	16,937	10,558
% Na_2O	0,372	0,13	0,228	0,326
% K_2O	1,722	1,453	2,257	3,032
% TiO_2	0,518	0,054	0,426	0,644
% P_2O_5	0,049	0,108	0,334	0,782

Table 2. Chemical composition of the studied clays

2.2.3 Adsorption studies

Adsorption of monuron on clays was carried out by the batch equilibration technique (Sukul & Spiteller, 2000) using various mass adsorbent-solution volume ratio and at the desired herbicide concentration, the suspensions were shaken in centrifuge tubes at a known time and then centrifuged at 3500rpm for roughly 10min. The resulting was then filtered through 0.45μm cellulose membranes and the filtrate was analyzed by HPLC Water under the condition mentioned in paragraph 2.1.

Difference between Ce and Ci were assumed to be due to adsorption. Adsorption isotherms were obtained by plotting the amount of herbicide adsorbed (Cs = mmolkg^{-1} clay, Ce (mM)).

2.3 Results and discussion

2.3.1 Characterization of studied clays

The Cation Exchange Capacity CEC and Surface Specific Area SSA are important characteristics of clays. CEC represents the overall quantity of charges available on the exchanger material. It is related to the surface density of charges through the SSA. For the

sorbents having no organic matter, SSA is anticipated to become a sorbent characteristic most likely to reflect sorption capacity.
Table 3 gathers the CEC and SSA values for different minerals used. These results indicate that montmorilonite represent the major fraction of soil clays and have been confirmed by some other characterization method.

Mineral	Ka	K-M	CESS	CEBS
CEC (meq/100g)	5,72	60,72	52,47	47,25
SSA ($m^2.g^{-1}$)	5	166	60	21

Table 3. Cation Exchange Capacity and Surface Specific Area for clay minerals

The Infra Red spectra were obtained with Perkin Elmer 2000 spectrophotometer. The clay pastilles were prepared with KBr (0.1%). Infra Red spectrums of clays are illustrated in Figure 1.

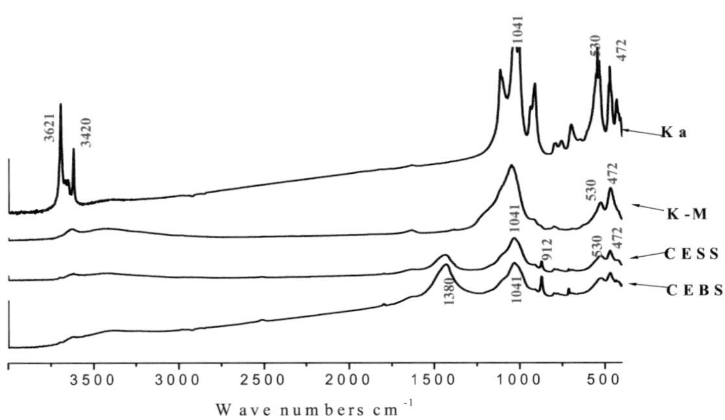

Fig. 1. IR spectrum of kaolin (Ka), K-montmorillonite (K-M), Clay extracted from Settat and Berrechid soil (CESS) and (CEBS).

The kaolin spectrum reveal an intense band at 3621 cm^{-1} corresponding to stretching vibration of OH groups coordinated to Al-Al pairs and another band located at 3420cm^{-1} corresponding to the H_2O-stretching vibrations (Russell & Fraser, 1994). However, CESS and CEBS spectra present a broad band located at 1382cm^{-1} that can be attributed to the presence of carbonate solid phases (Gueu et al., 2007). Such results are also confirmed by XRD analysed.
Moreover, all the spectra show a large band around 1041cm^{-1} equivalent to Si-O-stretching (Madejova & Komadel, 2001). The double band observed at 530 and 472cm^{-1} was related to Al-O-Si and Si-O-Si deformations (Madejova et al., 2002).
XRD measurements were performed using a Philips X-Ray diffractometer employing nickel filtered CuKά radiation. The X-Ray powder diffraction pattern of clays is given in Fig.2.
The principal equidistance of the most intense lines as well as the hkl plans relative to these rays are presented in table 4.
The comparison of experimental d_{hkl} and those taken from JCPDS chart show that CESS and CEBS are composed of several mineralogical phase; montmorillonite, saponite, chlorite, illite, quartz and calcite (table III and figure 2). These soils clays have similar mineralogical phases with different percentages. saponite and calcite represent the major phase of CEBS. However, CESS is rich of illite and montmorillonite.

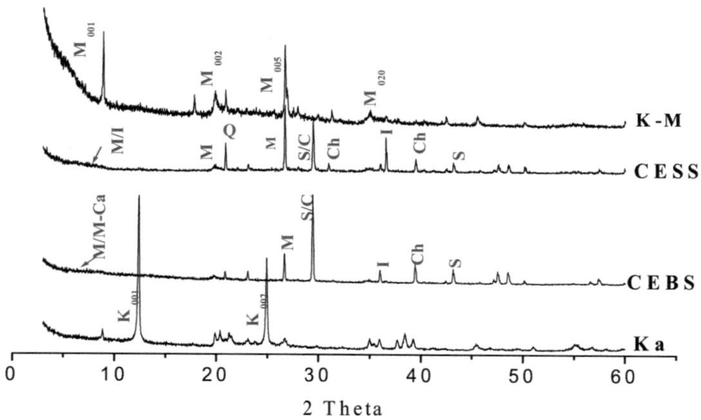

M: Montmorillonite, Q: Quartz, I: Illite, Ch: Chlorite, S: Saponite, C: Calcite and Ka: Kaolin

Fig. 2. XRD pattern of clays studies

K-M		Ka		CEBS and CESS		
d_{hkl}	hkl	d_{hkl}	Hkl	d_{hkl}	hkl	Mineralogical phase
9,94	001	7,15	001	10,1	001	Montmorillonite
4,48	002	3,58	002	4,48	002	
3,35	005	2,39	003	4,26	*	Quartz
2,56	020	1,79	004	3,04	005	Saponite + Calcite
2,13	004	1,49	060	2,13	134	Saponite
************		**************		2,89	005	Chlorite
				2,28	133	
				2,46	004	Illite

Table 4. Basal distances obtained for clays

2.3.2 Adsorption kinetics

In order to determine the optimum contact time for monuron adsorption onto clays, equilibrium concentrations were measured at definite times. As can be seen from Figure 3.

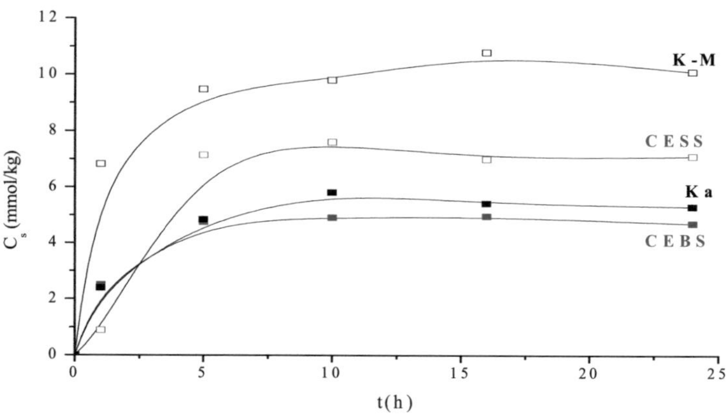

Fig. 3. Effect of contact time on adsorption

From figure 3, when the equilibrium time was increased, the amount of adsorption was also increased. Under our experimental conditions, the equilibrium was reached within 6h for K-montmorillonite (K-M), within 10h for kaolin (Ka), clay extracted from Settat soil (CESS), and clay extracted from Berrechid soil (CEBS).
Adsorption kinetics is usually controlled by different mechanisms. The most limiting ones are the diffusion mechanisms. The initial curved portion is attributed to a rapid external diffusion or boundary layer diffusion and surface adsorption, while the linear portion represents a gradual adsorption stage due to intraparticle diffusion. This starts to decrease due to the low concentration in solution as well as fewer available adsorption sites (Guibal et al., 2003).

2.3.3 Adsorption isotherms

Figure 4 and 5 show monuron adsorption isotherms in commercial clays and soils clays. Most of the curves do not reach the plateau of saturation when the solid/liquid ratio was 5mg/ml. Only 11.97 mmol of monuron was adsorbed in one Kg of K-montmorillonite. This is equivalent to 9.4%. The amounts of monuron adsorption in CESS, CEBS and kaolin, were 6.8, 5.4 and 5.8mmol/Kg respectively with corresponding percent adsorptions; 5.1, 4.01 and 4.3.
In order to optimize the design of an adsorption system, it is important to establish the most appropriate correlation for the equilibrium curves. In this respect, the equilibrium experimental data for adsorbed monuron on clays were studies using Freundlich equation.
The Freundlich equation (Freundlich, 1906) is an empirical equation employed to describe heterogeneous systems; characterized by the heterogeneity factor nf, describes reversible adsorption, and is not restricted to the formation of the monolayer,

$$C_s = K_f C_e^{nf}$$

Where K_f is the adsorption capacity evaluated at 1mM C_e and n_f is the intensity factor, expressing the slope of the adsorption isotherm.
A linear form of the Freundlich expression can be obtained by taking logarithms of the equation:

$$\log C_s = \log K_f + n_f . \log C_e,$$

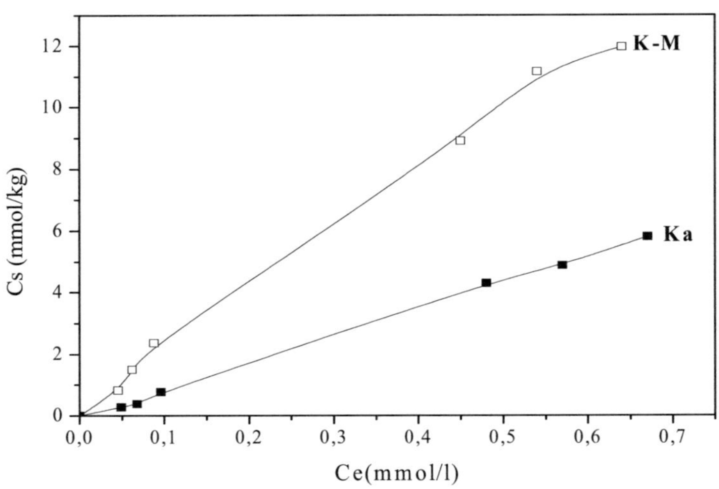

Fig. 4. Monuron adsorption isotherms on K-montmorillonite and kaolin

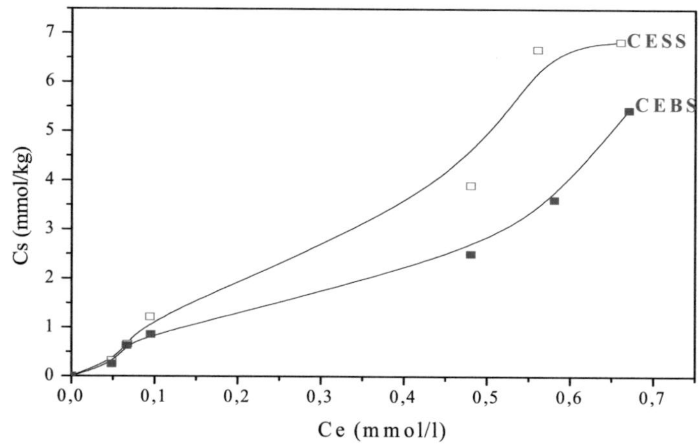

Fig. 5. Monuron adsorption isotherms on clays extracted from soils CESS and CEBS

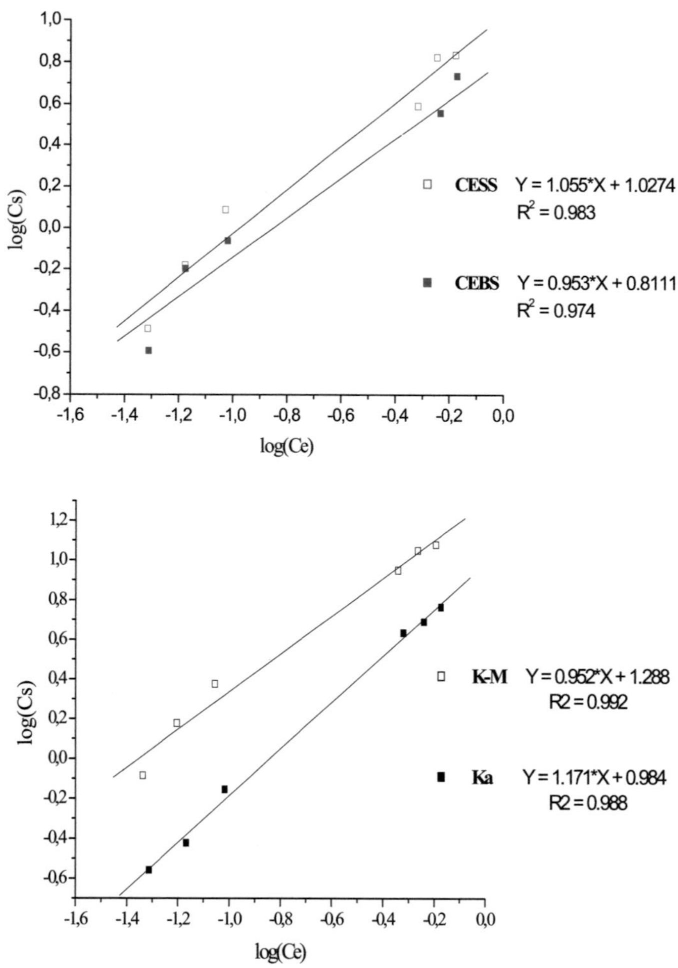

Fig. 6. Freundlich isotherms adsorption on K-M, Ka, CEBS and CESS

Therefore, the plots of logCs vs logCe for the adsorption of monuron onto clays were employed to generate the intercept value of K_f and the slope of n_f.
Figure 6 illustrate Freundlich isotherms for studied clays.
The adsorption isotherms of monuron followed the Freundlich equation with a high regression factor (R^2>0.98). Table 5 contains the adsorption coefficients values K_f and n_f extrapolated from Fig.6

Clay	K_f	n_f	R^2
K-M	19,4	0,952	0,992
Ka	9,64	1,171	0,983
CESS	10,6	1,055	0,974
CEBS	6,47	0,953	0,988

Table 5. Freundlich sorption coefficients K_f and n_f of monuron adsorption onto clays

The adsorption capacity of monuron was in the following order:

K-M > CESS> CEBS > Ka

K_f values indicated that adsorption capacity of CESS is better than that of CEBS. This difference can be attributed to CEC and specific surface area; these parameters are the most key factors influencing adsorption on clays. High CEC of CESS suggests that this clay has strong ability for cation exchange. By consequent this clay has an important adsorption capacity when compared to the CEBS
The value of n_f is smaller than 1 for K-M and CEBS clays, it's reflecting favourable adsorption then the sorption capacity increases and new adsorption sites occur. For Ka and CESS clays the value of n_f is larger than 1, the adsorption bond becomes weak and unfavourable adsorption takes place, as a result of the decrease in adsorption capacity (Jiang et al., 2002; Tsai et al., 2003).
In order to determine the adsorption type, the Dubinin-Radushkevich equation can be expressed as follows (Dubinin, 1960):

$$\text{Ln } C_s = \ln X_m - k\varepsilon^2,$$

Where ε (Polanyi Potential) can be written as:

$$\varepsilon = RT\ln(1+1/C_e),$$

C_s is the amount of pesticide adsorbed per unit weight of adsorbent (mol/g), X_m is the adsorption capacity (mol/g), C_e is the equilibrium concentration of pesticide in solution (mol/L), k is the constant related to adsorption energy (mol^2/kj^2), R is the gas constant (kJ/mol K), T is temperature (K).
Radushkevich and Dubinin (Radushkevich, 1949; Dubinin, 1965) have reported that the characteristic sorption curve is related to the porous structure of the sorbent. The constant, *k*, is related to the mean free energy of sorption per mole of the sorbate as it is transferred to the surface of the solid from infinite distance in the solution (Hasany & Choudhary, 1996).
A plot of $\ln C_s$ vs ε^2 should be linear. The values of Xm and k were calculated from intercept and slope of this plot. The results are summarized in Table 6.

Clay	$X_m \times 10^4$ (mol/g)	k x 10^2	R^2	E(kJ/mol)
K-M	2.4	0.93	0.989	7.33
Ka	2.2	1.14	0.996	6.62
CESS	1.8	1.04	0.972	6.93
CEBS	0.85	0.91	0.952	7.41

Table 6. The sorption parameters of Dubinin-Radushkevich equation for monuron on clays

The values of k increased from 0.91 x 10^{-2} to 1.14 x 10^{-2} in clays. The mean free energy change of adsorption (E) can be considered as the free energy change during the transfer of one mole of molecule from infinity in solution to the surface of the solid particles. Its value can be evaluated from the following equation:

$$E = -(2k)^{-0.5}$$

It should be noted that magnitude of E is useful for the estimation the adsorption type. If this value is found within 8-16kj/mol, adsorption type can be explained by ion exchange (Singh & Pant, 2004; Mahramanlioglu et al., 2002). The values of E estimated in the present study are 6.62, 6.93, 7.33 and 7.41 kJ/mol. for kaolin, clay extracted from Settat soil, K-montmorillonite and clay extracted from Berrechid soil respectively. The adsorption type under our experimental conditions can be considered as a physical adsorption (Paul & Long, 1957). However, the forces involved are intermolecular forces (Van Der Waals forces), and which do not involve a significant change in the electronic orbital patterns of the involved species. The adsorbed molecule is not affixed to a specific site at the surface but is free to undergo translational movement within the interface. It is predominant at low temperature, and is characterized by a relatively low energy of adsorption.

2.3.4 Initial pH effect

The amount of herbicide adsorbed on each adsorbent as a function of pH was also investigated (Figure7).

It is apparent that adsorption is independent on the pH of the solution. The pKa of monuron is too weak, it is about -0.082 (Ozcan et al., 2004), and does not have ionisable functions.

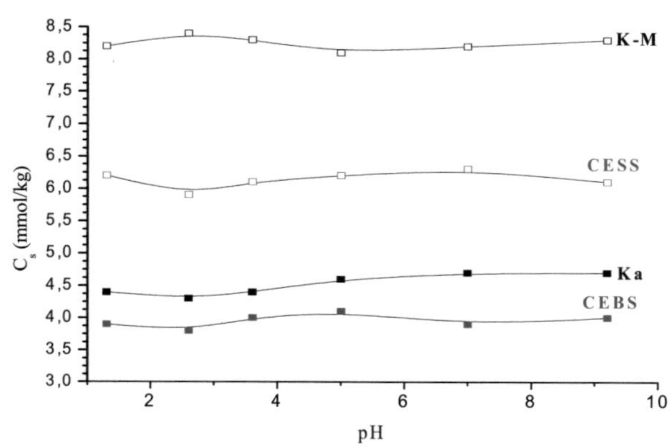

Fig. 7. pH effect on monuron adsorption by clays

Therefore the pH does not influence its sorption to the clays. However it's probable that pH affects the surface charge of clays; in the acidic medium, the H^+ ion can competitively exclude the adsorption of pollutant by exchanging with cation on the surface or in the interlayer region of the clay. When the pH value of the solution was increased, the surface of the adsorbent becomes negatively charged. According to the reported results (Yu et al., 2001), no significant attraction is observed between herbicide molecules and negatively surface.

2.3.5 Effect of the ratio solid/liquid

In order to select the best conditions of monuron adsorption in clays support, the influence of adsorbent/solution ratio on the capacity of adsorption were carried out at solid/liquid ratio ranging from 1 to 12 mg/ml. figure 8 presents the obtained results.

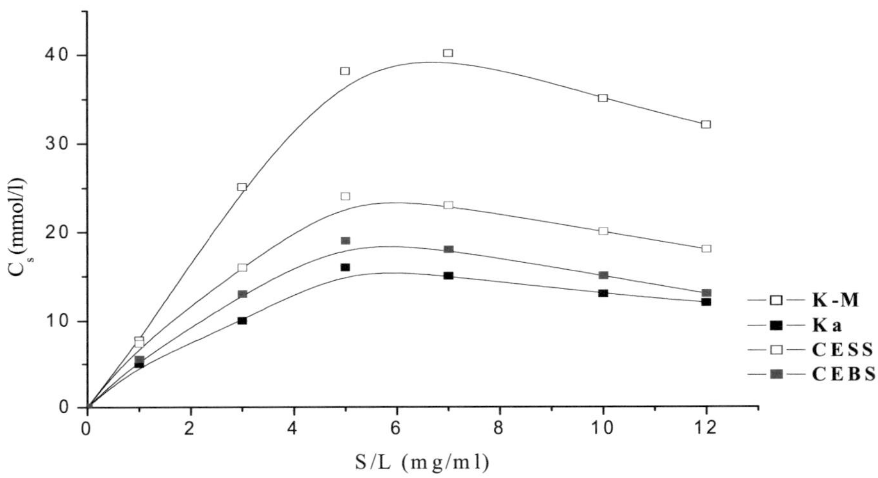

Fig. 8. Solid liquid ratio effect on monuron adsorption by clays

As clearly shown, the results indicate that the dispersion of the clays particles in suspension is modified according to the ratio S/L used. Clays have a better capacity of monuron adsorption in the range of 5-7mg/ml, the increase of ratio solid/liquid more than 7mg/ml allows the reduction of the rate of adsorption. Contact surface increases and the solute reaching the sorption sites of the different components when adsorbent/solution ratio increases. At a given mass, an aggregation phenomenon of the clay particles between them will be present that would decrease the total adsorbent surface and the rate of monuron adsorption (Mountacer et al. 2008).

2.3.6 Conclusion

Interaction between mineral sorbents and monuron herbicide is influenced by the dispersion conditions used in the preparation of the mixture. The favourable adsorption was realized at the following optimal conditions: contact time was 10 hours and sorbent/solution ratio was in the range of 5-7 mg/ml but pH does not have any effect on monuron adsorption onto clays studied.

The present work confirms that the mineralogical characteristics of clays play an important role in the retention of pollutants. A correlation between the maximum adsorption, CEC and

SSA of all sorbents was found. The adsorption capacity of monuron was in the following order: K-M > CESS> CEBS > Ka

3. Part II: Photodegradation process of monuron adsorbed on surface clays

3.1 Introduction

Photodegradation by solar light excitation is a process that pesticide may undergo once dispersed in the environment. The study of these abiotic transformations makes an appreciable contribution in determining the final fate of these xenobiotics. However, the phototransformation of herbicide adsorbed by clays in absence of water is an important question. Kinetic photodegradation of monuron was studied in various clay matrices: K-montmorillonite, Fe(III) exchanged montmorillonite (Fe(III)-montmorillonite) and montmorillonite complexed with humic acid (HA-montmorillonite) complexes. The nature of intermediate products during the photodegradation was investigated in order to compare the efficiency of different clays supports and to elucidate the mechanism of organic substrate photodegradation.

3.2 Materials and methods

3.2.1 Clay samples

The mineral used as a model adsorbent was K-montmorillonite (K-M), purchased from Fluka chemistry AG. Fe(III)-montmorillonite (Fe-M) and HA-montmorillonite (HA-M) were prepared by adsorption using the method developed by Cox and Koskinen (Cox & Koskinen, 1998). Clay minerals were analyzed by X-ray Fluorescence using AXIOS-P analytical equipment. The chemical composition of clay fractions is summarized in Table 7.

Composition (%)	Clay		
	K-M	Fe-M	HA-M
SiO_2	71,42	66,861	70,354
Al_2O_3	14,362	15,024	14,276
Fe_2O_3	2,992	5,342	2,999
MnO	0,016	0,019	0,017
MgO	1,412	1,609	1,426
CaO	0,207	0,102	0,2
Na_2O	0,372	0,322	0,434
K_2O	1,722	1,557	1,696
TiO_2	0,518	0,496	0,503
P_2O_5	0,049	0,052	0,047

Table 7. Chemical composition of clay fractions obtained by X-Ray fluorescence

Adsorption of monuron on clay samples was carried out by the batch equilibration technique, using 100 mg clay and 30 ml of pesticide solution at 0.5 mmol L^{-1}. The suspensions were shaken in centrifuge tubes for 10 hours and then centrifuged at 3500 rpm for 10 min. The solid was recovered and transferred to desiccator for vacuum drying during 24 h.

3.2.2 Irradiation experiments

30 mg of pesticide-clay was irradiated in Suntest set up at different times, the pesticide was extracted from each irradiated sample by 1 ml of methanol, agitated for 10 min and filtered

through 0.45 μm cellulose membranes. The procedure was repeated for all clay complexes prepared.

3.2.3 Analytical procedure

The photodegradation of monuron and the formation of the photoproducts were followed by high performance liquid chromatography. LC/MS analyses were carried out with Q-TOF-Micro/water 2699 from CRMP centre at the Blaise Pascal University. It is equipped with an electrospray ionization source (ESI) and a Waters photodiode array detector. Each single experiment permitted the simultaneous recording of both UV chromatogram at a preselected wavelength and an ESI-MS full scan. Data acquisition and processing were performed by MassLynx NT 3.5 system. Chromatography was run using a Nucleosil column100-5 C18 ec (250 9 4.6 mm, 5 lm). Samples (5-10 lL) were injected either directly or after evaporation of the solvent for better detection

3.3 Results and discussion

3.3.1 Absorption spectrum of the studied pesticide

The absorption spectrum of monuron is given in Fig 9. It shows a maximum at 245 nm with a molar absorption coefficient of 17800 $mol^{-1}L\ cm^{-1}$ and a small shoulder at 280 nm. The comparison with the solar light emission (represented here by the emission spectrum from the suntest lamp) clearly shows a very weak not negligible overlap. Such overlap permits a direct excitation of monuron from the sunlight.

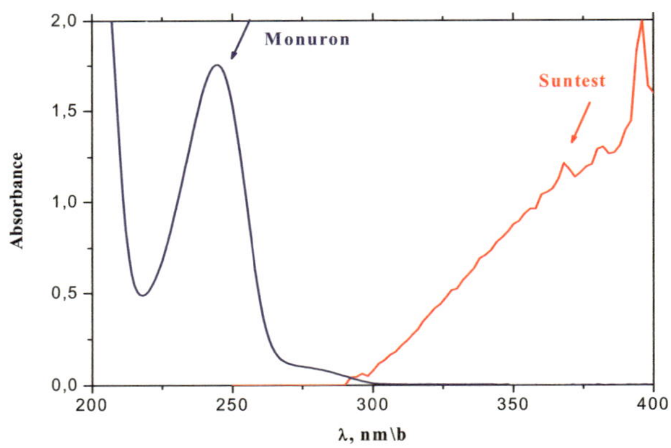

Fig. 9. The absorption spectrum of monuron in aqueous solution and the emission spectrum of suntest (xenon lamp)

3.3.2 Characteristics of clay fractions

The method used for determining CEC, which involves the complete exchange of the naturally-occurring cations by ammonium, is a standard method for CEC determination. Table 8 shows the values of CEC for the different studied minerals clay.

Mineral fraction	K-M	Fe-M	HA-M
CEC (meq/100g)	60.72	53.64	70.52

Table 8. Cation exchange capacity of clay minerals

The values consigned in Table 8 shows that the CEC of montmorillonite enriched by iron is low compared with that of K-montmorillonite. This result is due to the difficulty in moving the ferric cations by the ammonium cations. Moreover, the presence of the humic acid complexed with montmorillonite increases its capacity of cation exchange. This result is well correlated with those of many authors who showed that the CEC of the organic matters is higher than that of minerals clay (Baize, 1988).

The examination of IR spectra of clay fractions reveals the principal absorption bands; a slight band at 3631 cm^{-1} which is due to the stretching vibration of structural OH groups coordinated to Al-Al pairs. Adsorbed water gives a broad band at 3420 cm^{-1} corresponding to the H_2O-stretching vibrations (Russell & Fraser, 1994). The broad band around 1041 cm^{-1} corresponds to Si-O-stretching (Madejova & Komadel, 2001). The 530 and 472 cm^{-1} is related to the Al-O-Si and Si-O-Si deformations (Madejova et al., 2002). Results obtained are shown in Fig. 10.

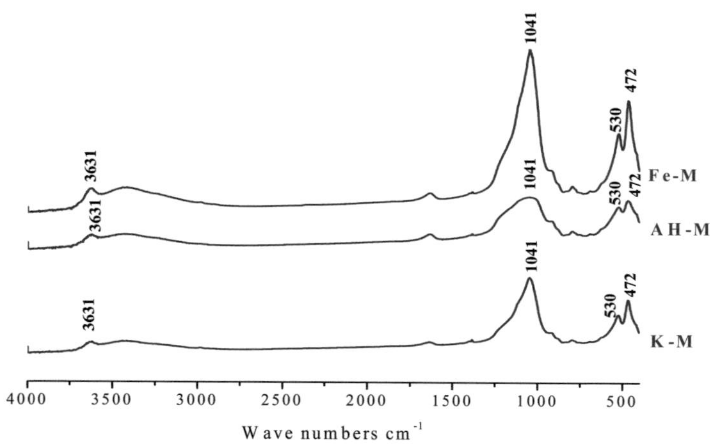

Fig. 10. IR spectra of K-M, Fe-M and HA-M.

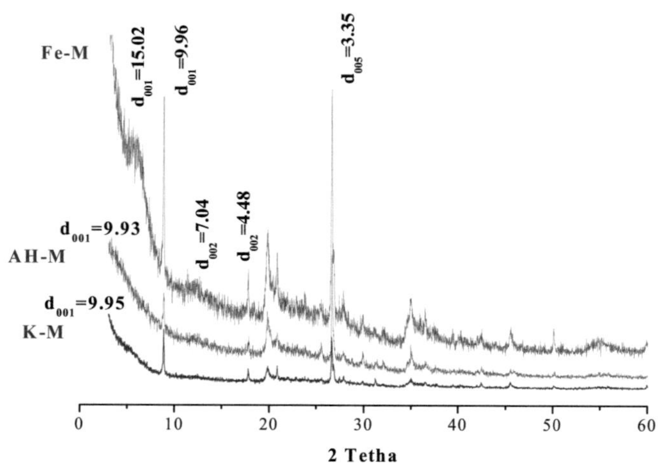

Fig. 11. XRD pattern of K-M, Fe-M and HA-M.

The spectra show also a widening of the band centred at 1041 cm^{-1} in AH-M and a stressing of the latter for Fe-M. Additionally, the band located at 472 cm^{-1} became more marked in the case of the complex Fe-M. This result confirms the exchange of potassium by iron, while this band weakened in the case of AH-M because of the interaction of humic acids in clay layer (Tajeddine et al., 2010).

The XRD measurements were performed using a Philips X-ray diffractometer employing nickel filtered CuKa radiation. The X-ray diffraction pattern of clay fractions is given in Fig.11. The equidistance appears to vary regularly according to the rate of hydration and the nature of exchanged cation. With the introduction of iron in montmorillonite, the reticular distance corresponding to plane 001 was evolved from 9.45 to 15.04. This indicates that the specific surface area was also increased. The presence of the equidistance 9.4 characteristic of K-M in the spectrum of Fe-M indicates that the substitution of K^+ by Fe^{3+} was not complete.

3.3.3 Photodegradation experiments of monuron adsorbed on surface clays

3.3.3-a Kinetics of monuron dispappearance in presence of humic acid (HA)

The kinetics of monuron disappearance on K-montmorillonite and HA-montmorillonite are shown in Fig. 12.

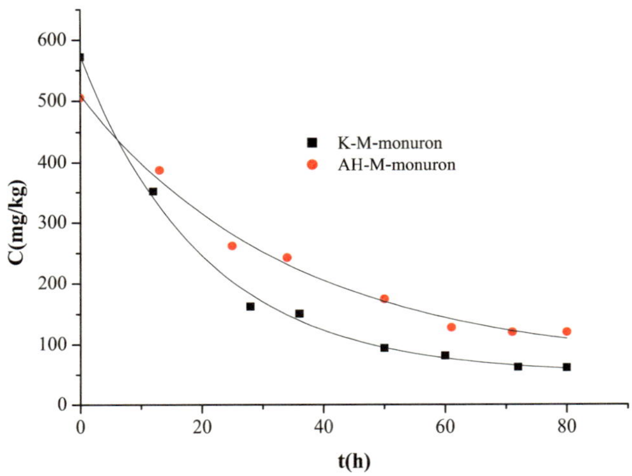

Fig. 12. Kinetics of disappearance of monuron irradiated on K-M and HA-M.

As clearly shown in Fig. 12, the photodegradation rate of monuron is slower for montmorillonite complexed with humic acid than that for K-montmorillonite alone. This is primarily due to the fact that light attenuation was more obvious by montmorillonite complexed with humic acid than K-montmorillonite. The absorption of the incidental light by the pesticide is quasi-total in the experiment of K-M. Kinetics and half lives of these experiences are given in Table 9.

The results show that the disappearance rate of monuron decreased from 18.4 to 9.09 mg kg^{-1} h^{-1} by the complexation of humic acid to K-montmorillonite. Its indicates that the presence of humic acid, complexed with K-montmorillonite, protects pesticide from photodegradation, due to the screen effect or light attenuation effect.

The phototransformation of monuron sensitized by the humic substances into aqueous medium was studied by Richard and Bengana (Richard et al., 1997). They showed that no

		k(mg kg^{-1}h^{-1})	k_{app}(h^{-1})	$t_{1/2}$
monuron	K-M	18.4	0.032	21h 35min
	HA-M	9.09	0.018	38h 54min

Table 9. Rate constants and half-life times of OPP and monuron irradiated onto K-M and HA-M

consumption of monuron was observed when it was irradiated alone at 365 nm with black light lamps. By contrast, monuron was degraded when it was irradiated in the presence of Fulvic acid or Humic Acid. It was observed that without oxygen no reaction takes place and that the rates of transformation are lower in oxygen-saturated than in air-saturated solutions. It was suggested that reactive excited triplet states of humic substances abstract an electron or a hydrogen atom (Canonica et al., 1995). The hydrogen atom is then transferred onto oxygen which is reduced into hydroperoxyl radical or superoxide anion; oxygen is thus needed for the chromophores to be regenerated. At high concentrations, oxygen quenches efficiently the triplets and inhibits the consumption of substrates. It is concluded that in this case a hydrogen atom transfer between oxidant excited states of humic acid and monuron occurs.

Under the current conditions, inhibitive processes, such as for example the auto-inhibition can also contribute in the reduction of the total process effectiveness (Vulliet et al., 2001).

$$AH^{*} + AH \rightarrow AH + AH^{*}$$

3.3.3-b Kinetics of monuron disppearance

The photodegradation kinetics of monuron adsorbed by Fe(III)-montmorillonite complexes are presented in Fig. 13.

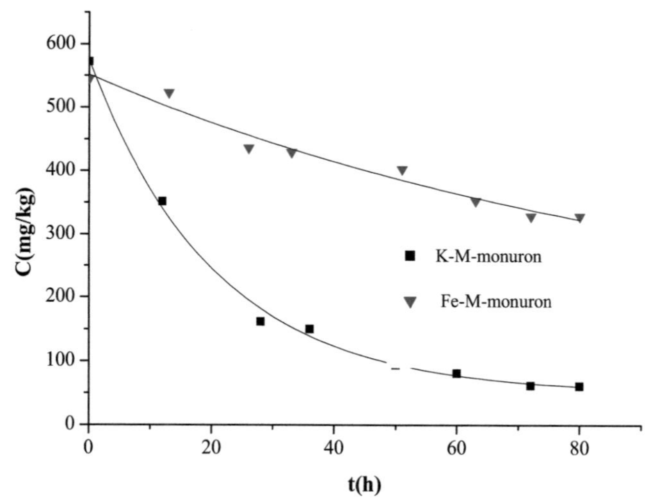

Fig. 13. Kinetics of disappearance of monuron irradiated on K-M and Fe-M

The presence of iron related to montmorillonite has a negative effect on the degradation rate of monuron. Table 10 illustrates the kinetics and the half lives of these experiences.

As shown in Table 10 the half-life times of monuron photodegradation were 43 and 78 hours for K-montmorillonite and Fe(III)-montmorillonite, respectively. The slow degradation can be explained by the absorption of light by Fe^{3+}, these ions cause a competition with the pesticide molecule.

		k(mg/kg.h)	$k_{app}(h^{-1})$	$t_{1/2}$
monuron	**K-M**	18.4	0.032	21h, 35min
	Fe-M	9.09	0.018	38h, 54min

Table 10. Rate constants and half-life time of monuron irradiated onto K-M and Fe-M

In aqueous medium Mest'ankova and collaborators have noted that the direct photolysis of Monuron was negligible: no photodegradation of monuron occurred in the absence of Fe(III) (Mest'ankova et al., 2004). The kinetics of monuron degradation are strongly dependent on Fe(III) concentration. The disappearance of monuron became faster when the concentration of Fe(III) increased from $1.0x10^{-4}$ to $1.0x10^{-3}mol.l^{-1}$. This effect can be correlated with the increase of the monomeric species present in Fe(III) solution. However at highest Fe(III) concentrations, $1.0x10^{-2}$ and $5.0x10^{-2}mol.l^{-1}$, a decrease of the rate of monuron disappearance was observed when compared to the kinetics of the reaction with Fe(III) concentration equal to $1.0x10^{-3}moll^{-1}$. These results illustrate the importance of the speciation of iron in aqueous solution. Indeed, the nature of the iron present in the environment is a parameter which can catalyse or induce pesticides photodegradation. Also, it should be noted that the method of introduction of iron into clay depends on the photochemical behaviour of pesticides; when it is added with clays, the pollutants photodegradation is accelerated. These results have been found in the studies carried out by Menager (Menager et al., 2009). However in our study iron was introduced between the layers of montmorillonte (Tajeddine et al., 2010).

3.3.4 Identification of photoproducts

The photoproducts formed in the irradiation of monuron adsorbed onto K-montmorillonite after 35 min irradiation were investigated by LC/MS analysis. Fig.14.

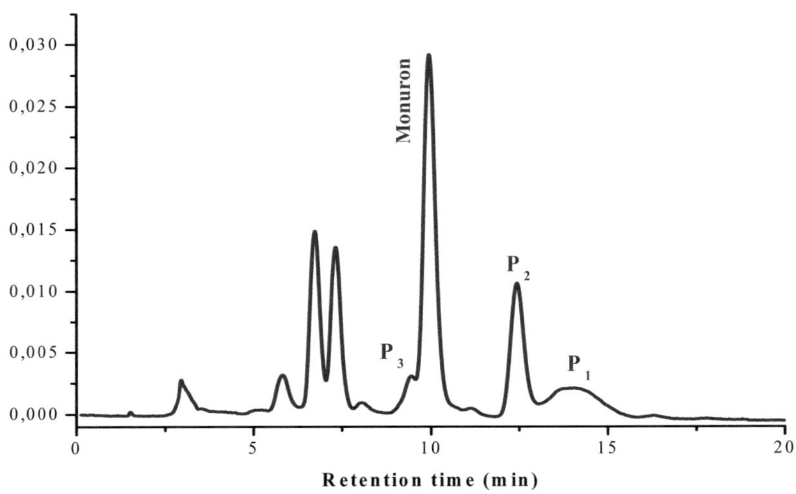

Fig. 14. Chromatogram of the methanol extract of the monuron irradiated on montmorillonite during 35 hours

Three photoproducts were identified by the molecular ion and mass fragment ions. The structures of the main photoproducts are presented in table 11. In addition to these compounds, some other products were detected but it was not possible to determine their structure due to their low concentration.

Photoproduct	Structure	Molecular weight (m/z)	Retention time (min)
P_1 N-(-chloro-4-chlorophenyl)-N',N-dimethylurea	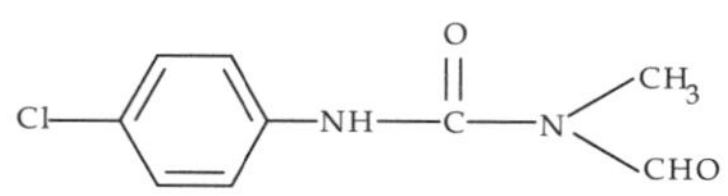	232	14
P_2 N-(4-chlorophenyl)-N'-formyl, N-dimethylurea	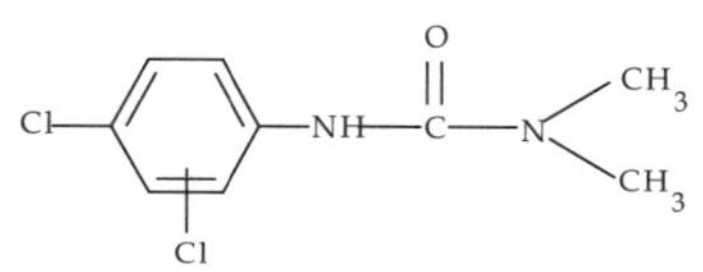	212	12.5
P_3 N-(-hydroxy-4-chlorophenyl)-N',N-dimethylurea	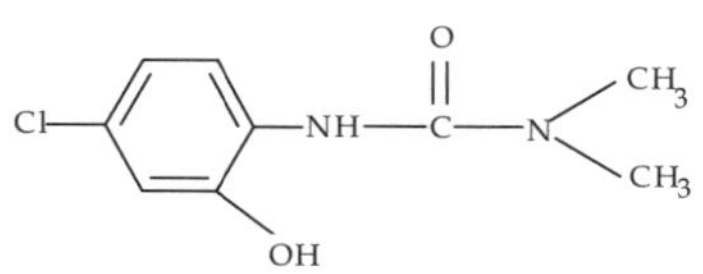	214	9.4

Table 11. Photoproducts obtained by irradiation of monuron adsorbed onto K montmorillonite

3.3.5 Photodegradation mechanism

Based on the intermediate products listed in table 11, the possible degradation pathway for monuron irradiated onto K-montmorillonite is proposed in schema 1.

Upon light excitation, the first step was initiated by the substitution of a hydrogen atom by a chlorine atom. This reaction occurs at the aromatic group. The formation of the two other products may attributed to the attack of the hydroxyl radicals, formed through the excitation of iron (III) species, to either the aromatic moiety or the methyl group of the amine moiety. In the latter case, the formation of the final product can be explained by oxidation of the methyl group to aldehyde function-N (CH3)2. This mechanism has been demonstrated in the most studied of phenylurea herbicides transformation (Macounova et al., 2003). It has been characterized as a majority path in the mechanism of photodegradation of aqueous monuron (Mest'ankova et al., 2004).

Some minor products may be formed from the photoproduct 2. The addition of OH on the aromatic cycle reacts rapidly with oxygen to form intermediates radical. These react with the radicals formed after the opening of the aromatic ring (Pramuro & Vicenti, 1993). The disappearance of the aromatic structure might be due to the formation of a carbene by removal of HCl as was proved in the case of linuron irradiated onto the sand and clay (Richard & Bengana, 1996) and the case of 4 chlorophenol-irradiated in water (Oudjehani & Boule, 1995).

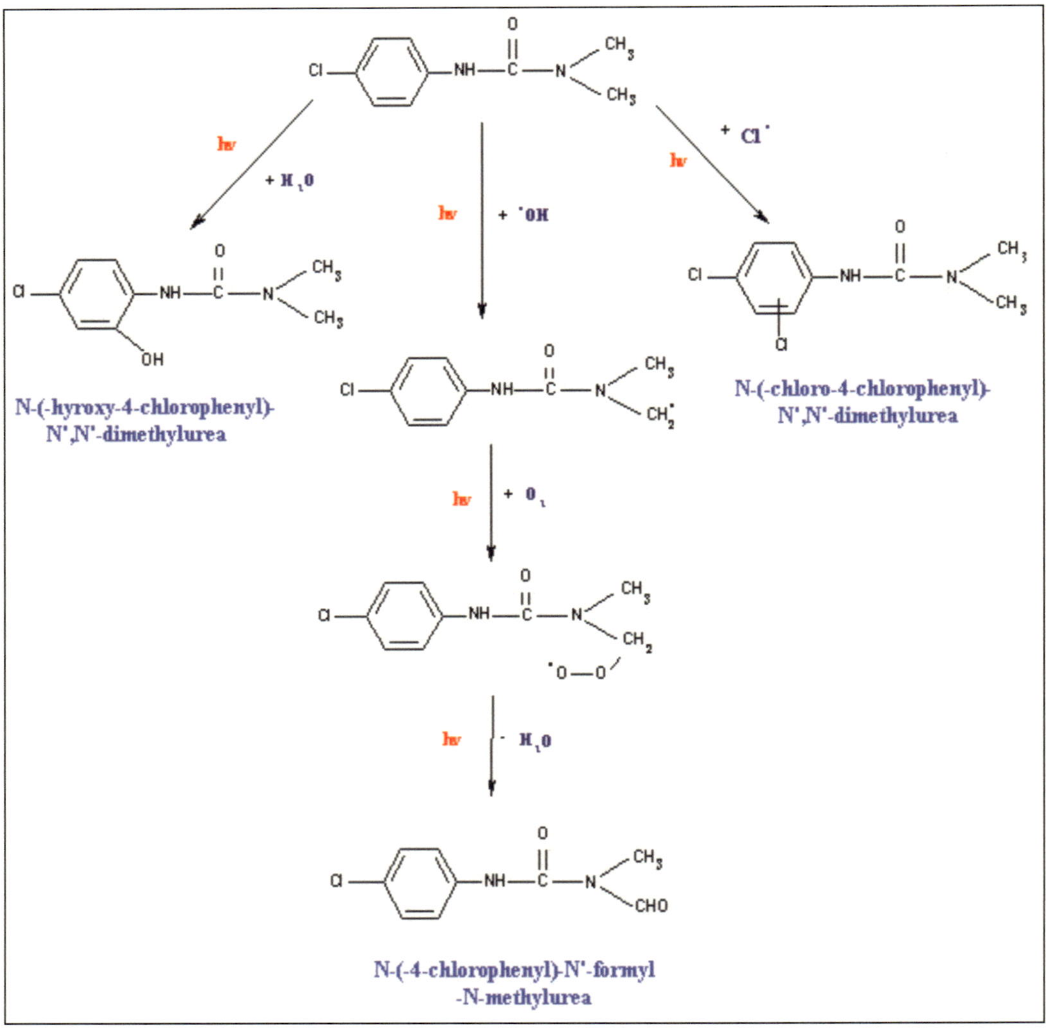

Schema 1. Phototransformation mechanism of monuron irradiated onto montmorillonite

4. Conclusion

The photo-transformation of monuron adsorbed on montmorillonite and irradiated under simulated sunlight is not sensitized by the humic acids and iron. On one hand, humic acid decreased the rate of monuron disappearance is slower owing to a screen effect phenomenon. On the other hand, the presence of iron cannot act an efficient photo inducer of monuron herbicide elimination from dry phase. However, in humid systems, Fe^{3+} appeared to be a good photocatalyst for the removal of monuron from water (Hana 2004).

From the analytical point of view, the Principal photoproducts identified by HPLC/MS/MS initially formed by irradiation of monuron adsorbed onto K-montmorillonite were; a product of oxidation of a methyl group (N-(4-chlorophenyl)-N'- formyl, N-dimethylurea), a product of hydroxylation on the level of the aromatic nucleus (N-(-hydroxy-4-chlorophenyl)-N',N-dimethylurea)and the 3rd product obtained by a photo-substitution of a hydrogen atom of the aromatic nucleus by a chlorine atom (N-(-chloro-4-chlorophenyl)-N',N-dimethylurea).

5. References

Albanis, T.A.; Bochicchio, D.; Bufo, S.A.; Cospito, I. D.; Auria, M.; Lekka, M. & Scrano, L. (2002). Surface adsorption and photoreactivity of sulfonylurea herbicides. *Int. J.Environ. Anal. Chem.*, Vol. 82, 561–569

Baize, D. (1988). *Guide des analyses courantes en pédologie: Choix-expression-presentation-interpretation.*, 172, ISBN 2-7380-0075-4, INRA Pub., Paris

Barbash, J.E. & Resek, E.A. (1996). *Pesticides in ground Water, Distribution, Trends and Governing Factors*, Ann Arbor, Michigan.

Beck, A.J.; Johnston, A. E. J. & Jones, K. C. (1993). Movement of non-ionic organic chemicals in agricultural soil. *Critical Rev. Environ. Sci. Technol.*, Vol. 23, 219-229

Caine, M.; Dyer, G.; Holder, J.V.; Osborne, B.N.; Matear, W.A.; McCabe, R.W.; Mobbs, D.; Richardson, S. & Wang, L. (1999). The use of clays as sorbents and catalysts. In: *Natural Microporous Materials in Environmental Technology* Misaelides, Maca´sˇek, P. F; Pinnavaia, T.J. and Colella, C. (eds)., 49–69, Kluwer, Dordrecht

Canonica, S.; Jans, U.; Stemmler, K. & Hoign , J. (1995). Transformation kinetics of phenols in water: photosensitization by dissolved natural organic material and aromatic ketones. *Environ. Sci. Technol.*, No. 29, 1822- 1831

Cox, L.; Hermosin, M.C. & Cornejo, J. (1995). Adsorption mechanisms of thiazafluron in mineral soil clay components. *European J. Soil Sci.*, Vol. 46, 431-442

Cox, L.; Hermosin, M.C. & Cornejo, J. (1996). Sorption of metamitron on soils of low organic matter content. *Chemosphere*, Vol.32, 1381-1389

Cox, L. & Koskinen, C. W. (1998). Sorption of imidacloprid on soil minera land organic components. *Soil Sci.soc. Am.*, Vol.62, 915-991

Dubinin, M. M. (1960). The potential theory of adsorption of gases and vapors for adsorbents with energetically non-uniform surface. *Chem. Rev.*, Vol. 60, 235–266.

Dubinin, M. M. (1965). Modern state of the theory of volume filling of micropore adsorbents during adsorption of gases and steams on carbon adsorbents. *Zhurnal Fizicheskoi Khimii.* Vol. 39, 1305–1317

Freundlich, H.M.F., (1906). Uber die adsorption in losungen. *Z. Phys. Chem.*, Vol. 57, 385-470

Gueu, S.; Yao, B.; Adouby, K. & Ado, G., (2007). Kinetics and thermodynamics study of lead adsorption on to activated carbons from coconut and seed hull of the palm tree. *International Journal of Environmental Science and Technology, Vol. 4, No. 1,* 11-17 .ISSN: 1735-1472

Guibal, E.; McCarrick, P. & Tobin, J. M. (2003). Comparison of the Sorption of Anionic Dyes on Activated Carbon and Chitosan Derivatives from Dilute Solutions. *Separation Science and Technology*, Vol. 38, No.12& 13, 3049 - 3073

Hasany, S. M. & Chaudhary, M. H.(1996). Sorption potential of Hare River sand for the removal of antimony from acidic aqueous solution. *Appl. Rad. Isot.,* Vol. 47, 467–471

Hebert, V.R. & Miller, G.C. (1990). Depth dependence of direct and indirect photolysis on soil surfaces. *J. Agric. Food Chem.,* Vol. 38, 913–918

Hermosin, M. C.; Martin-Olmedo, P. & Cornejo. (1993). Adsorption mechanisms of monobutyltin in clay minerals. Environ. *Sci. Technol.*, 27, 2606-2615

Jiang, J.Q., Cooper, C. & Ouki, S. (2002). Comparison of modified montmorillonite adsorbents. Part I: preparation, characterization and phenol adsorption. *Chemosphere* , Vol.. 47, No. 7, 711-716. ISSN: 0045-6535

Khan, S.U. (1980). *Pesticides in the Soil Environment.* Elsevier, Amsterdam

Lagaly, G. (1994). Layer charge determination by alkylammonuim ions. In: *Layer Charge Characteristics of 2:1 Silicate Clay Minerals.* Mermut, A.R. (Ed.), The Clay Minerals Society, Boulder, 46

Lee, S.Y. & Kim, S.J. (2002). Expansion of smectite by hexadecyltrimethylammonuim. *J. Clays Miner.*, Vol. 50, 435-442

Macounova, K.; Hrysova, H.; Ludvik, J. & Jirkovsky, J. (2003). Kinetics of photocatalytic degradation of diuron in aqueous colloidal solutions of Q-TiO_2 particles. *J. Photochem. Photobiol. A : Chemistry,* Vol. 156, 273-282

Mahramanlioglu, M.; Kizilcikh, I. & Biçer, I.O. (2002). Adsorption of fluoride from aqueous solution by acid treated spent bleaching earth *J. Fluorine Chem.,* Vol. 115, 41-49

Malato, S.; Caceres, J.; Fernandez-Alba, A.R.; Piedra, L.; Hernando, M.D.; Aguera, A. & Vial, A. (2003). Photocatalytic treatment of diuron by solar photocatalysis : Evaluation of main intermediates and toxicity. *Environ Sci Technol.,* 37, 2516 -2524

Mansour, M. (1993). *Fate and Prediction of Environmental Chemicals in Soils, Plants and Aquatic Systems.* Lewis Publishers, Boca-Raton, Ann Arbor, London, Tokyo

Menager, M.; Siampiringue, M. & Sarakha, M. (2009). Photochemical behaviour of phenylbenzoquinone at the surface of the clays: Kaolinite, bentonite and montmorillonite. *Journal of Photochemistry and Photobiology A: Chemistry,* Vol. 208, No. 2-3, 159-163

Mest'ankova, H.; Mailhot, G. ; Pilichowski, J.F. ; Kry´sa, J. ; Jirkovsky´, J. & Bolte, M. (2004). Mineralisation of monuron photoinduced by Fe(III) in aqueous solution. *Chemosphere,* 57, 1307–1315

Oudjehani, K. D. & Boule, P. (1995). Photolysis of halophenols in aqueous solution sensitised by hydroquinone or phenol. *New J.Chem.,* Vol. 19, 199–206.

Ozcan A. S., Erdem B., Ozcan A. (2004). Adsorption of Acid Blue 193 from aqueous solutions onto Na bentonite and DTMA-bentonite. *J Colloid Interface Sci.,* Vol. 280, No. 1, 44-54

Paul, M. A. & Long, F. A. (1957). HO and related indicator acidity functions. *Chem. Rev.,* 57, 1-45.

Pimentel, D. (1995). Amounts of pesticides reaching target pests: Environmental impacts and ethics. *Journal of Agricultural and Environmental Ethics,* Vol. 8, No.1, 17-29

Pramuro, E. and Vicenti, M. (1993). Photocatalytic degradation of monuron in aqueous TiO_2 dispesions. *Environ. Sci. Technol.*, 27, 1790-1795

Radushkevich, L. V. (1949). Potential theory of sorption and structure of carbons. *Zhurnal Fizicheskoi Khimii*, 23, 1410–1420

Richard, C. and Bengana, S.(1996). pH effect in the photocatalytic transformation of a phenylurea herbicide. *Chemosphere*, 33, 635-641

Richard, C.; Vialaton, D.; Aguer, J.P.; Andreux, L.F. (1997). Transformation of monuron photosensitized by soil extracted humic substances: energy or hydrogen transfer mechanism. *Journal of Photochemistry and Photobiology*, 3, 265-271

Robert, M. and Tessier, D. (1974). Méthode de préparation des argiles des sols pour des études minéralogiques. *An. Agro.*, 25, 859-882

Robert, M. (1975). Principes de détermination qualitative des minéraux argileux les plus fréquents dans les sols des régions tempérés. *An. Agro.*, 26, 363-399

Russell, J.D. & Fraser, A.R. (1994). Infrared methods. In: *Clay Mineralogy: Spectroscopic and chemical determinative methods* , M.J. Wilson Ed., 11-67, London,

Senesi, N. & Schnitzer, M. (1977). Effects of pH, reaction time, chemical reduction and irradiationon ESR spetra of fulvic. *Soil Sci.*, 123, 224–234

Serrano, S.; Garrido, F.; Campbell, C.G. & Garcia-Gonzalez, M.T. (2005). Competitive sorption of cadmium and lead in acidic soils of central Spain. *Geoderma.*, Vol.124, 91-99

Sherman, D.M. (1989). Crystal chemistry, electronic structures, and spectra of Fe sites in clay minerals: application to photochemistry and electron transport. In: *Spectroscopic Characterization of Minerals and Their Surfaces.* ACS Symposium Series 415. Coyne, L.M.; Blake, D.F. and McKeever, S.W.S. (Eds). American Chemical Society, Washington, DC, 284–309

Singh, T.S. & Pant, K.K. (2004). A comparative study for the removal of As(III) and As(V) by activated alumina.Natural Arsenic in *Groundwater: Occurrences, remediation and management.* Bundschuh, J.; Bhattacharya, P. &. Chandrashekharam, Eds., D., Taylor and Francis series, Balkema Publication The Netherlands, 139-145

Sukul, P. & Spiteller, M. (2000). Sorption study on metalaxyl in soils of different physicochemical properties. *Fresenius Envir. Bull.*, 701-711

Tajeddine, L.; Mountacer, H. & Sarrakha, M. (2008). Sorption of monuron in some selected mineral sorbents. Research Journal of Chemistry and Environment, Vol. 12, 83-89

Tajeddine, L.; Mountacer, H. & Sarrakha, M. (2010). Effect of iron and humic acid on photodegradation of some pesticides adsorbed on clay surfaces, *Arabian Journal of Chemistry*, Vol.. 3, 73–78

Tsai, W.T.; Lai, C.W. & Hsien, K.J. (2003). Effect of Particle Size of Activated Clay on the Adsorption of Paraquat from Aqueous Solution. *J. of Colloid and Interface Science*, Vol. 263, 29-34

Vulliet, E.; Emmelin, C.; Scrano, L.; Bufo, S. A.; Chovelon, J. M.; Meallier, P. & Grenier-Loustalot, M. F. (2001). Photochemical Degradation of Acifluorfen in Aqueous Solution. *J. Agric. Food Chem.*, Vol.49, No.10, 4795–4800

Yu, Y.; Zhuang, Y.Y. & Wang, Z.H. (2001). Adsorption of Water-Soluble Dye onto Functionalized Resin. *Journal of Colloid and Interface Science*, Vol. 242, No. 2, 288- 293
Zhou, L.H.; Li, F.S.and Zhu, X.J. (2001). Systems Evaluation Area. *J. Environ. Qual.*, 32, 84-95

14

Application of Diflufenican Herbicide on Soils Amended with Different Organic Wastes

Manuel Tejada[1], Marina del Toro[1], Isidoro Gómez[1] and Juan Parrado[2]
[1]*Grupo de investigación "Edafología ambiental". Departamento de Cristalografía, Mineralogía y Química Agrícola, E.U.I.T.A. Universidad de Sevilla, Crta de Utrera km. 1, 41013 Sevilla*
[2]*Departamento de Bioquímica y Biología Molecular. Facultad de Farmacia. Universidad de Sevilla, C/ Prof. García González 2, 41012 Sevilla, Spain*

1. Introduction

Pesticides are important tools in agriculture that help to minimize economic losses caused by weeds, insects and pathogens and have greatly improved food production (Farenhorst, 2006; Song et al., 2008; Romero et al., 2010). The ideal outcome of pesticide use occurs when a pesticide accomplishes the purposes for which it was applied and then rapidly breaks down into harmless components, such as carbon dioxide and water. However, extensive use of pesticides in conventional agricultural practices has increased dramatically over time and the potential negative effects on human health and the environment are now of concern (Renaud et al., 2004; Majumdar & Singh, 2007; Song et al., 2008).

According to El-Nahhal (2003), achieving proper weed control needs some incorporation usually via rainfall or irrigation and movement of the active ingredients through the soil profile. However, movement of the active ingredients may be associated with three problems: (1) dilution of herbicide in the zone of weed seed germination and accumulation in the root zone of the crop which may cause crop injury, (2) reduction of herbicidal activity due to distribution of the active ingredient below the root zone of the target weeds and (3) possible contamination of ground water. Thus only a small fraction of the applied herbicide contributes to the desired activity. These problems may result in repeated applications and/or increasing the applied rate, increasing cost and potential for ecological damage.

In Europe, pesticides are considered hazardous substances in accordance with current legislation regarding water (Directives 2000/60/EC, 2006/11/EC). Because of this, public concern over the residues of pesticides in environment, food and related commodities has increased over the last decades (Majumdar & Singh, 2007).

The presence of pesticides in groundwater has grown considerably in the last few years (Cox et al., 1997, Renaud et al., 2004; Carabias Martínez et al., 2000; Gomides Freitas et al., 2008). The relatively high water solubility and not readily degradable of these chemicals makes them potentially mobile in soil and surface and ground waters vulnerable to contamination (Carrizosa et al., 2001; El-Nahhal, 2003). For this reason it is of maximum interest that the development of agricultural strategies be directed to the decrease in pesticide movement.

However, pesticides pollution is not only limited to groundwater. Several studies on widely-used pesticides such as benomyl, chlorothalonil, dinocap, chlorpyrifos, metsulfuron-methyl, 2,4- D, glyphosate, MCPA, etc. have already shown that pesticide application leads to changes in soil nutrient levels and alterations to soil microbial activity, diversity and/or genetic structure (Edwards, 1989; Chen et al., 2001; Pandey & Singh, 2004; Zabaloy & Gómez, 2008; Černohlávková et al., 2009). Consequently, disturbances of microbial communities ensuring several key ecological processes in soil such as organic matter degradation and nutrient cycling, could harmfully altered soil fertility and sustainable agricultural productivity. Therefore, population size, enzymatic activity and biodiversity of certain systematic and physiological groups of microorganisms may serve as bioindicators of changes taking place in the soil following herbicide application.

Many soil enzymes can be used as potential indicators of soil quality for sustainable management because they are sensitive to ecological stress and land management practices (Tejada et al., 2009; 2010). Enzymes may react to changes in soil management more quickly than other variables and therefore may be useful as early indicators of biological changes (Bandick & Dick, 1999; Masciandaro et al., 2004).

Dehydrogenase activity has been proposed as a measure of overall microbial activity (Masciandaro et al., 2001), since it is an intracellular enzyme related to oxidative phosphorylation processes (Trevors, 1984). García et al. (1997) found that dehydrogenase activity is a good index of the soil microbial biomass in semiarid Mediterranean areas.

Respect to the hydrolases enzymes, urease is involved in the hydrolysis of urea to carbon dioxide and ammonia, which can be assimilated by microbes and plants (Kizilkaya & Bayrakli, 2005). β-glucosidase catalyzes the hydrolysis of β-D-glucopyranoside, an important source of energy for soil microorganisms (Turner et al., 2002; Kizilkaya & Bayrakli, 2005). Phosphatase is the enzyme involved in the hydrolysis of organic phosphorus to different forms of inorganic phosphorus (Amador et al., 1997; Kizilkaya & Bayrakli, 2005). Arylsulphatase is the enzyme involved in the mineralization of ester sulphate in soils (Pascual et al., 1998; Kizilkaya & Bayrakli, 2005).

There have been many investigations into the mechanisms by which herbicides are adsorbed by soil components of soil, particularly the clay minerals and organic matter. Sorption of herbicides is determined by the complex interrelationship between the physiochemical properties of the adsorbate, the nature of soil constituents and the experimental and environmental conditions under which sorption occurs (Ying & Williams, 2000).

An increasing number of studies conducted at the field scale, with lysimeters or in the laboratory have shown that organic matter is the main soil component contributing to the sorption of herbicides (Renaud et al., 2004) and sorption is one of the main processes that reducing the mobility of these chemicals in soils, the addition of exogenous organic matter to soil land has been suggested as a possible method to reduce herbicide movement (Sheng et al., 2001; Ben-Hur et al., 2003). Application of organic carbon to soils in the form of compost, sludge, efluent and crop residues is a common practice, because of its low costs and recycling of nutrients (Majumdar & Singh, 2007; Song et al., 2008). The use of organic wastes to enrich soils of low organic matter content and to aid as nutrient input can modify surfaces of soils and subsurface materials promoting sorption of herbicides and retarding their movement as a secondary effect.

However, the influence of organic matter on soil properties and sorption process depends upon the type, amount and dominant components of the added organic materials (Tejada et

al., 2007, 2008). This aspect is of great interest, since it supposes an important advance in the behavior of the herbicides in the soil after addition of different sources of organic matter.
Organic amendment can also affect the soil microbiota and soil enzymes production and therefore, biodegradation of pesticides by enhancing microbial activity and thus biodegradation, although the increase in sorption with the organic amendment can also protect the pesticides from biodegradation (Cox et al., 1997). Thus, organic wastes affecting both soil properties and microbial populations, generating a more complex herbicide behaviour in the soil system (Romero et al., 2010).
The objectives of this study were (1) to evaluate the sorption and mobility of the diflufenican herbicide in a soil amended with two organic wastes, and (2) to study the influence of these organic wastes on the diflufenican herbicide polluted soil, by analyzing their effects on soil enzymatic activities.

2. Materials and methods

2.1 Soils, organic wastes and herbicide characteristics

The soils used in this experiment were a Xerollic Calciorthid and a Typic Xerofluvent (Soil Survey Staff, 1987). The main characteristics of both soils are shown in Table 1.
Soil pH was determined in distilled water with a glass electrode (soil:H_2O ratio 1:2.5). Soil texture was determined by the Robinson's pipette method (SSEW, 1982). Soil total N was determined by the Kjeldhal method (MAPA, 1986). Soil total C was determined by the method of Yeomans & Bremner (1988). Cation exchange capacity was determined with 1 M ammonium chloride solution in ethanol/water (60:40, v/v) at pH 8.2 (Tucker, 1954

	Xerollic Calciorthid	Typic Xerofluvent
pH	7.7 ± 0.2	7.6 ± 0.1
Clay (g kg^{-1})	382 ± 15	180 ± 12
Silt (g kg^{-1})	332 ± 27	131 ± 15
Sand (g kg^{-1})	286 ± 39	689 ± 21
Textural class	Clay loam	Sandy loam
Total N (g kg^{-1})	0.83 ± 0.04	0.41 ± 0.03
Total C (g kg^{-1})	6.4 ± 0.5	2.6 ± 0.7
C/N ratio	7.7 ± 2.1	6.3 ± 1.2
Cation exchange capacity (cmol$_c$ kg^{-1})	11.3 ± 1.1	4.1 ± 0.6

Table 1. Main features of the experimental soils (mean ± standard error). Data are the means of four samples

The organic wastes applied were the organic fraction of a municipal solid waste (MSW) and cow manure (CM). The general properties of the organic wastes are shown in Table 2. Organic matter was determined by dry combustion, according to the official methods of the Spanish Ministry of Agriculture (MAPA, 1986). Humic and fulvic acids-like were extracted with 0.1 M sodium pyrophosphate and 0.1 M sodium hydroxide at pH 13 (Kononova, 1966). The supernatant was acidified to pH 2 with HCl and allowed to stand for 24 h at room temperature. To separate humic acids-like from fulvic acids-like, the solution was centrifuged and the precipitate containing humic acids-like was dissolved with sodium hydroxide (Yeomans & Bremner, 1988). After the removal of humic acids-like, the acidic filtrate containing the dissolved fulvic acid-like fraction was passed through a column of XAD-8 resin. The adsorbed fulvic was then recovered by elution with 0.1 M NaOH, desalted using Amberlyst 15-cation-exchange resin, and finally freeze-dried. The carbon content of humic and fulvic acids-like were determined by the method described. Total N was determined by the Kjeldhal method (MAPA, 1986). After nitric and perchloric acid digestion, total Ca, Mg, Fe, Cu, Mn, Zn, Cd, Pb, Ni and Cr concentrations were determined by atomic absorption spectrometer and K was determined by atomic emission spectrometer, according to MAPA methods (1986).

	MSW	CM
pH (H_2O)	6.2 ± 0.3	8.3 ± 0.2
Organic matter (g kg^{-1})	469 ± 15	764 ± 29
Humic acid-C (mg kg^{-1})	1030 ± 17	461 ± 13
Fulvic acid-C (mg kg^{-1})	711 ± 10	631 ± 24
Total N (g kg^{-1})	17.3 ± 1.3	29.2 ± 2.1
Fe (mg kg^{-1})	815 ± 38	407 ± 28
Cu (mg kg^{-1})	82.6 ± 9.8	24.2 ± 1.8
Mn (mg kg^{-1})	75.6 ± 8.1	14.1 ± 1.2
Zn (mg kg^{-1})	134 ± 13	10.3 ± 1.6
Cd (mg kg^{-1})	1.1 ± 0.3	0.28 ± 0.09
Pb (mg kg^{-1})	82.4 ± 3.6	5.3 ± 0.8
Ni (mg kg^{-1})	13.6 ± 1.5	2.4 ± 0.6
Cr (mg kg^{-1})	19.4 ± 1.7	0.29 ± 0.04

Table 2. Organic wastes chemical characteristics (mean ± standard error). Data are the means of four samples

Table 3 shows the acidic functional group contents of humic acids isolated from both organic wastes. The carboxyl group content was estimated by direct potentiometric titration at pH 8, the phenolic hydroxyl group content was estimated as two times the change in charge between pH 8 and pH 10, and the total acidity was calculated by addition (Ritchie & Perdue, 2003).
The herbicide used in this experiment was the diflufenican (Brodal®, Bayer CropScience). Diflufenican (2',4' - dichloro-2 - (α,α,α -trifluoro-m-toly-loxy) nicotinanilide, $C_{16}H_{11}F_5N_2O_2$) is a pre- and early post-emergence herbicide used for the selective control of broad leaf and grass weeds in winter cereals, with lower solubility in water than glyphosate and variable half-life in the environment (15 to 30 weeks) (Ashton et al., 1994). Brodal® physical and chemical properties are shown in Table 4.

	Total acidity	COOH	Phenolic-OH
		(mol kg^{-1})	
MSW	4.29 ± 0.04	3.19 ± 0.03	1.10 ± 0.03
CM	2.81 ± 0.02	2.00 ± 0.03	0.80 ± 0.01

Table 3. Acidic functional group contents (mean ± standard error) of humic acids isolated from MSW and CM. Data are the means of four samples

Ingredients:	Diflufenican	500 g l^{-1}
	Propane-1,2-diol	50 g l^{-1}
Appearance:		A viscous, light brown liquid suspension
Odour:		Characteristic slightly sweet
pH:		7.5 to 8.5 (undiluted)
Vapour pressure:		4.25 x 10^{-3} mPa (diflufenican) at 25 ° C
Vapour density:		Not available
Boiling point:		Greater than 100°C
Freezing/melting point:		Not available
Solubility:		Miscible with water
Density:		1.175 g ml^{-1} at 20° C
Flash Point:		> 100° C
Flammability (explosive) limits:		Not available
Auto-ignition temperature:		Not available
Partition coefficient (octanol/water):		Diflufenican: Log P_{OW}= 4.2

Table 4. Physical and chemical properties of Brodal®. The date are taken from Bayer CropScience

2.2 Experimental layout

2.2.1 Field experiments

For the field experiments and for each soil, the experimental layout was a randomized complete block design with four replications with a total amount of 24 plots, with each plot measuring 4 m × 3 m. Six treatments were used:

1. C, control soil (non organic amendment and without the application of herbicide)
2. C+H, plots non organic amendment and with application of herbicide
3. C+CM, plots organic amendment with CM and without the application of herbicide
4. C+MSW, plots organic amendment with MSW and without the application of herbicide
5. C+CM+H, plots organic amendment with CM and with the application of herbicide
6. C+MSW+H, plots organic amendment with MSW and with the application of herbicide

The CM was applied at a rate of 5 t ha^{-1} (fresh matter), whereas MSW was applied at a rate of 8.1 t ha^{-1} (fresh matter), in order to apply the same amount of organic matter to soil. The herbicide dose used was 200 ml ha^{-1}). The dose used was 200 ml ha^{-1}. Organic wastes were applied to the soil 1 month before applying the herbicide. Organic wastes were mechanically mixed with the soil, allowing the organic wastes to reach up to 25 cm from the soil surface

At 3, 15, 45, 90, 120, 150, 180, 210 and 250 days after applying the herbicide to soil, diflufenican was measured. To extract diflufenican we used the method described by Tejada (2009). Five grams of soil was shaken with 20 ml of 90:10 v/v methanol:H_2O for 1 h at 25 ± 1° C. After allowing the soil to settle, pesticide concentrations were determined by HPLC using a column Merck C18 ultrasphere (250 x 4.6 mm x 5 µm and 80 Å). The pesticide was eluted with a mobile phase of acetonitrile:H_2O:orthophosphoric acid of 85:15:0.25. The flow rate and wavelength were 1 ml min^{-1} and 220 nm. The temperature was constant at 25 °C using a Model 234 Column Oven Heater. The injected volume was 50 µl.

Also, the activity levels of four soil enzymes were measured at 3, 15, 45, 90, 120, 150, 180, 210 and 250 days after applying the herbicide to soil. Dehydrogenase activity was measured by reduction of 2-*p*-iodo-3-nitrophenyl 5-phenyl tetrazolium chloride to iodonitrophenyl formazan (García et al., 1993). Urease activity was determined by the buffered method of Kandeler & Gerver (1988) using urea as substrate. β-Glucosidase activity was determined using *p*-nitrophenyl-β-D-glucopyranoside as substrate (Masciandaro et al., 1994). Phosphatase activity was measured using *p*-nitrophenyl phosphate as substrate (Tabatabai & Bremner, 1969).

2.2.2 Laboratory experiments

For the laboratory experiments and for each soil, sorption studies and leaching experiments in methacrylate columns were made.

For sorption studies and for each experimental soil, the treatments used were:

1. C, non-organic amended control soil (10 g of soil)
2. C+CM, soil amended with CW at rate of 10% (10 g of soil + 1 g of CW)
3. C+MSW, soil amended with MSW at a rate of 16.3% (10 g of soil + 1.63 g of CW)

Diflufenican sorption was determined according to Cabrera et al. (2009) criteria. Triplicate samples (5 g) of the unamended and organic amended soil (C, C+CM and C+MSW) were treated with 10 ml of diflufenican (50%:50%, v/v) solution (initial concentrations, Ci, ranging from to 50 μM in 0.01 $CaCl_2$). Previously, it was determined that equilibrium was reached in less than 24 h, and that no measurable degradation occurred during this period. Equilibrium concentrations (Ce) in the supernatants were determined by HPLC. Sorption isotherms were fitted to Freundlich equation ($Cs=K_f \times Ce^{1/n_f}$) and sorption coefficients K_f and $1/n_f$ were calculated.

The diflufenican analysis were performed following the methodology previously described.

The system to determine diflufenican leaching in soils, consisted of 30-cm long X 3.1 cm i.d. methacrylate columns filled with 20 cm of soil. Glass wool plus 10 g of sea was placed on the bottom of the columns to prevent losses of soil and contamination of leachates with soil particles. The columns were hand-packed with 180 g of soil without and with organic wastes (10 g of MSW and 10.9 g of CM) with the intention of applied the same amount of organic matter and 10 g of sea sand was placed on the soil surface.

The columns were saturated with water, allowed to drain for 24 h, and then the amount of diflufenican corresponding to an application rate of 2 kg ha^{-1} was applied to the top of the columns. Daily, the columns were leached with 200 mm d^{-1} of distilled water and the leachates were collected and filtered. The diflufenican concentration was analyzed by HPLC at 1, 3, 7, 10, 15, 20 and 30 days after the application of herbicide to soil. The leaching experiment was conducted in triplicate.

2.3 Statistical analysis

Data were submitted to two-way ANOVA with treatment and sampling time as factors using the Statgraphics Plus 2.1 software package. The means were separated by the Tukey's test, considering a significance level of $P<0.05$ throughout the study. For the ANOVA, triplicate data were used for each treatment and every incubation day.

3. Results

3.1 Field experiments

During the experimental period and for both soils, the diflufenican soil contents applied as Brodal® presented higher values in clay loam texture soil than for sandy loam texture soil

(Figure 1). The slow diflufenican degradation makes think that the degradation of the diflufenican is not carried out in short time, indicating that this degradation happens superior at one time to the 250 days. For both soils, the diflufenican contents were decreasing progressively in soils amended with organic matter. However, this decrease depended on soil type and organic matter type applied. At the end of the experimental period and compared with the unamended soil, the diflufenican content significantly decreased 25.5% and 41.2% in the amended with CM and MSW, respectively, whereas for the sandy loam texture soil the diflufenican content significantly decreased 32.5% and 50.2% in the amended with CM and MSW, respectively.

Table 5 shows the evolution of soil dehydrogenase activity during the experimental period. The results indicated that at the end of the experiment and compared with the control soil, the dehydrogenase activity significantly decreased in the non-organic amended polluted soils (59.4% for the clay loam texture soil and 53.8% for the sandy loam texture soil), reflecting the adverse effects of the herbicide on this intracellular activity. Also and compared to the control soil, in non-polluted soil the application of organic matter increased significantly the dehydrogenase activity. However, this increase depended of the soil type and organic matter applied to the soil. For the Xerollic Calciorthid soil and at the end of the incubation period, the dehydrogenase activity increased 92.9% and 80% in soil amended with MSW and CM, respectively, whereas for the Typic Xerofluvent soil, the dehydrogenase activity increased 93.5% and 82.4% in soil amended with MSW and CM, respectively. The application of herbicide in organic-amended soils decreased dehydrogenase activity. However, this decrease was lower than for the non-amended herbicide polluted soil. At the end of the experiment and for the clay loam texture soil, the dehydrogenase activity significantly decreased 23.9% for the C+MSW+H treatment (compared to the C+MSW treatment), followed by 24.7% for the C+CM+H treatment (compared to the C+CM treatment), whereas for the sandy loam texture soil, the dehydrogenase significantly decreased 18.3% for the C+MSW+H treatment (compared to the C+MSW treatment), followed by 20.4% for the C+CM+H treatment (compared to the C+CM treatment).

Similar to the dehydrogenase activity, at the end of the experimental period the urease activity decreased in the non-organic polluted soils compared to the non-organic unpolluted soils (32.5% for the Xerollic Calciorthid soil and 28.6% for the Typic Xerofluvent soil) (Table 6). At the end of the incubation period, the application of organic matter to unpolluted soils significantly increased the soil urease activity. Again, this increase was higher for MSW-amended soils than for CM-amended soils. Also, the application of herbicide in organic-amended soils decreased urease activity. For the Xerollic Calciorthid soil, the urease activity decreased 15.4% for the C+MSW+H treatment (compared to the C+MSW treatment), followed by 21.9% for the C+CM+H treatment (compared to the C+CM treatment), whereas for the Typic Xerofluvent soil, the urease activity decreased 15% for the C+MSW+H treatment (compared to the C+MSW treatment), followed by 19% for the C+CM+H treatment (compared to the C+CM treatment).

The evolution of β-glucosidase and phosphatase activities was very similar to the enzymes described (Tables 7 and 8). The application of diflufenican in organic-amended soils decreased both enzymatic activities. Again, this decrease was higher in the sandy loam texture soil than the clay loam texture soil. Also, the application of organic matter to soils contaminated with the herbicide increased both enzyme activities compared with the non-organic amended polluted soils.

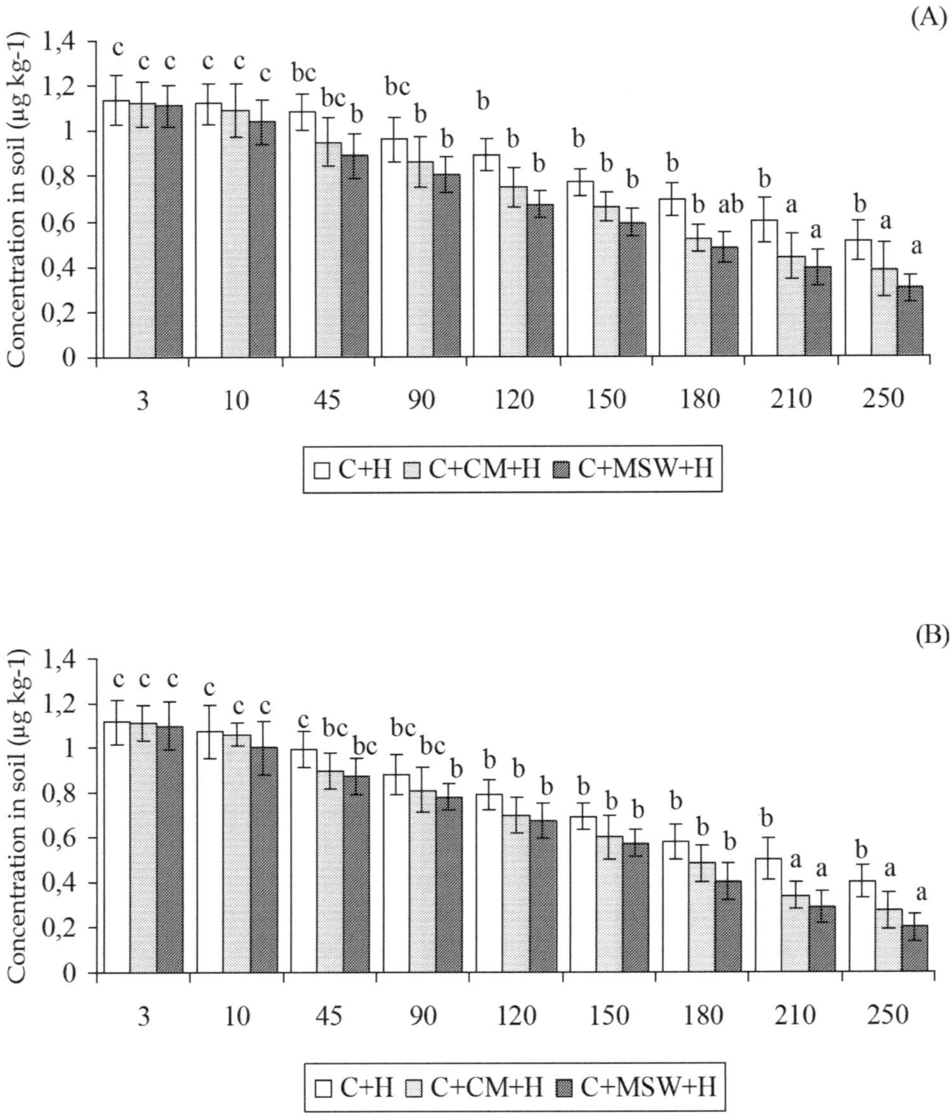

Fig. 1. Evolution of diflufenican (mean ± standard errors) in Xerollic Calciorthid (A) and Typic Xerofluvent (B) during the experimental for all experimental treatments
Column followed by the same letter(s) are not significantly different (p<0.05)

	3	15	45	90	120	150	180	210	250
	Xerollic Calciorthid								
C	2.5b† ± 0.4	2.4b ± 0.6	2.3b ± 0.3	2.1b ± 0.3	2.1b ± 0.4	2.0b ± 0.5	1.9b ± 0.4	1.8b ± 0.3	1.7b ± 0.5
C+H	2.2b ± 0.7	1.9b ± 0.3	1.7b ± 0.4	1.4a ± 0.5	1.1a ± 0.4	0.92a ± 0.15	0.88a ± 0.13	0.72a ± 0.17	0.69a ± 0.11
C+CM	2.8b ± 1.0	3.2bc ± 0.9	4.0bc ± 1.2	4.7c ± 1.1	5.4c ± 1.4	6.1c ± 1.7	7.0c ± 1.8	7.8c ± 1.9	8.5d ± 1.1
C+MSW	2.9b ± 0.8	3.3bc ± 1.1	4.3bc ± 1.0	5.0c ± 1.2	5.9c ± 1.1	6.5c ± 1.6	7.6c ± 1.6	8.5d ± 1.5	9.6d ± 1.5
C+CM+H	2.2b ± 0.5	2.6b ± 0.8	3.3bc ± 1.1	3.7bc ± 1.3	4.1bc ± 1.6	4.6c ± 1.0	5.3c ± 1.2	5.9c ± 1.1	6.4c ± 0.9
C+MSW+H	2.3b ± 0.6	2.8b ± 1.0	3.7bc ± 1.0	4.2bc ± 0.9	4.9c ± 1.5	5.5c ± 1.3	6.1c ± 1.0	6.9c ± 1.3	7.3c ± 1.2
	Typic Xerofluvent								
C	2.0b† ± 0.5	2.0b ± 0.6	1.8b ± 0.8	1.7b ± 0.6	1.6b ± 0.3	1.6b ± 0.4	1.5b ± 0.3	1.4b ± 0.4	1.3b ± 0.4
C+H	1.7b ± 0.3	1.5b ± 0.5	1.3b ± 0.2	1.1a ± 0.2	0.96a ± 0.11	0.89a ± 0.14	0.75a ± 0.13	0.67a ± 0.16	0.6a ± 0.11
C+CM	2.3b ± 0.4	2.6b ± 1.0	3.5bc ± 1.3	4.0bc ± 1.1	4.6c ± 1.2	5.3c ± 1.0	5.9c ± 1.3	6.5c ± 1.2	7.4c ± 0.9
C+MSW	2.4b ± 0.3	2.9b ± 0.9	3.9bc ± 1.2	4.8c ± 1.3	5.9c ± 1.0	6.9c ± 1.3	7.8d ± 1.6	8.6d ± 1.7	9.3d ± 1.5
C+CM+H	2.1b ± 0.7	2.3b ± 0.6	2.9b ± 1.5	3.4bc ± 1.1	4.0bc ± 1.0	4.4bc ± 1.2	4.9c ± 1.0	5.4c ± 1.2	5.9c ± 1.1
C+MSW+H	2.2b ± 0.5	2.5b ± 0.7	3.3bc ± 1.0	3.8bc ± 0.8	4.3bc ± 0.9	5.1c ± 1.1	5.6c ± 1.6	6.5c ± 1.5	7.6c ± 1.3

Table 5. Evolution of dehydrogenase activity (μINTF g^{-1} h^{-1}) (mean ± standard errors) in Xerollic Calciorthid and Typic Xerofluvent during the experimental for all experimental treatments. INTF: 2-*p*-iodo-3-nitrophenyl formazan

†Different letters following the numbers indicate a significant difference at $P<0.05$

	3	15	45	90	120	150	180	210	250
					Xerollic Calciorthid				
C	1.6† ± 0.3	1.5b ± 0.2	1.2ab ± 0.3	1.2ab ± 0.2	1.1ab ± 0.3	1.0ab ± 0.2	0.91a ± 0.11	0.84a ± 0.13	0.77a ± 0.11
C+H	1.5b ± 0.6	1.3b ± 0.5	1.1ab ± 0.2	0.95ab ± 0.19	0.88a ± 0.12	0.79a ± 0.18	0.68a ± 0.16	0.60a ± 0.16	0.52a ± 0.12
C+CM	1.7b ± 0.4	1.8b ± 0.4	2.3b ± 0.6	2.5bc ± 0.9	2.6bc ± 1.0	2.8b ± 0.6	2.9bc ± 0.9	3.0bc ± 0.7	3.2c ± 0.9
C+MSW	1.7b ± 0.2	1.8b ± 0.7	2.5bc ± 0.5	2.7bc ± 0.8	3.0bc ± 0.7	3.2c ± 0.7	3.4c ± 1.0	3.7c ± 0.9	3.9c ± 1.0
C+CM+H	1.7b ± 0.5	1.7b ± 0.5	2.0b ± 0.4	2.1b ± 0.3	2.1b ± 0.6	2.2bc ± 0.7	2.3b ± 0.5	2.4bc ± 0.6	2.5b ± 0.4
C+MSW+H	1.7b ± 0.4	1.6b ± 0.3	2.2b ± 0.6	2.4bc ± 0.5	2.5bc ± 0.5	2.7bc ± 0.4	2.9bc ± 0.8	3.2c ± 0.7	3.3c ± 0.6
					Typic Xerofluvent				
C	1.3b† ± 0.2	1.2b ± 0.3	0.98ab ± 0.12	0.88ab ± 0.15	0.76ab ± 0.11	0.68a ± 0.10	0.59a ± 0.12	0.50a ± 0.13	0.42a ± 0.11
C+H	1.2b ± 0.3	1.0b ± 0.1	0.90ab ± 0.15	0.82ab ± 0.16	0.74ab ± 0.17	0.62a ± 0.10	0.50a ± 0.11	0.44a ± 0.13	0.30a ± 0.09
C+CM	1.4b ± 0.3	1.5b ± 0.2	1.6b ± 0.4	1.7b ± 0.4	1.8b ± 0.5	1.9b ± 0.5	2.0b ± 0.8	2.0b ± 0.9	2.1bc ± 0.6
C+MSW	1.4b ± 0.2	1.6b ± 0.4	1.8b ± 0.4	2.2bc ± 0.8	2.5bc ± 0.4	2.9bc ± 0.7	3.2c ± 1.1	3.6c ± 1.2	4.0c ± 1.1
C+CM+H	1.3b ± 0.2	1.3b ± 0.2	1.4b ± 0.3	1.4b ± 0.3	1.3b ± 0.2	1.5b ± 0.2	1.6b ± 0.2	1.6b ± 0.3	1.7b ± 0.5
C+MSW+H	1.4b ± 0.3	1.6b ± 0.3	1.9b ± 0.5	2.2bc ± 0.9	2.5bc ± 0.6	2.7bc ± 0.8	2.9bc ± 0.9	3.2c ± 1.0	3.4c ± 0.9

Table 6. Evolution of urease activity (μmol NH_4^+ g^{-1} h^{-1}) (mean ± standard errors) in Xerollic Calciorthid and Typic Xerofluvent during the experimental for all experimental treatments
†Different letters following the numbers indicate a significant difference at $P<0.05$

	3	15	45	90	120	150	180	210	250
					Xerollic Calciorthid				
C	5.6b† ± 1.9	5.6b ± 1.8	5.5b ± 1.3	5.4b ± 1.1	5.3b ± 1.4	5.1b ± 1.2	4.9b ± 1.1	4.8b ± 1.3	4.7b ± 1.5
C+H	5.2b ± 1.5	5.0b ± 1.1	4.5ab ± 0.9	4.0ab ± 1.1	3.3a ± 0.8	2.9a ± 1.0	2.5a ± 0.8	2.3a ± 1.0	2.0a ± 1.2
C+CM	6.7b ± 1.8	8.0b ± 1.6	10.8c ± 1.1	12.5c ± 1.4	14.6c ± 1.5	16.6c ± 1.7	18.7c ± 1.5	20.5c ± 2.0	22.6d ± 2.6
C+MSW	6.8b ± 2.0	8.2b ± 2.0	11.9c ± 1.6	13.8c ± 1.7	15.4c ± 1.6	17.8c ± 1.6	20.2c ± 2.4	22.9d ± 1.8	25.0d ± 2.8
C+CM+H	6.5b ± 1.5	6.9b ± 1.6	8.8b ± 1.0	9.7bc ± 1.5	11.8c ± 1.1	14.1c ± 1.4	15.6c ± 1.6	16.7c ± 1.6	18.2c ± 2.1
C+MSW+H	6.6b ± 1.8	7.5b ± 2.1	10.0bc ± 1.8	12.1c ± 1.2	14.6c ± 1.3	15.7c ± 1.8	17.6c ± 2.2	18.8c ± 1.7	21.0c ± 1.9
					Typic Xerofluvent				
C	3.8bc† ± 1.1	3.7bc ± 1.3	3.4b ± 0.9	3.3b ± 1.0	3.2b ± 1.1	3.0b ± 0.6	2.9b ± 1.3	2.8b ± 1.1	2.6b ± 1.0
C+H	3.6bc ± 1.3	3.5b ± 1.1	3.1b ± 0.8	2.7b ± 0.8	2.4b ± 0.7	2.1a ± 0.4	1.7a ± 0.5	1.4a ± 0.5	1.2a ± 0.19
C+CM	4.0bc ± 1.6	4.5c ± 1.5	6.7c ± 1.5	8.0c ± 1.4	9.2c ± 1.5	10.3cd ± 1.9	11.5d ± 1.7	12.7d ± 1.6	14.8d ± 1.8
C+MSW	4.1bc ± 1.2	4.8c ± 1.6	7.2c ± 1.3	8.9c ± 2.0	10.1cd ± 1.9	11.5d ± 1.6	13.0d ± 1.6	14.8d ± 1.3	16.9d ± 1.5
C+CM+H	3.9bc ± 1.0	4.2c ± 1.2	5.4c ± 1.1	6.2c ± 1.6	7.5c ± 1.6	8.6c ± 1.1	9.7cd ± 1.8	11.0d ± 1.2	12.2d ± 1.2
C+MSW+H	3.9bc ± 1.3	4.5c ± 1.3	5.9c ± 1.9	7.0c ± 1.5	8.2c ± 1.8	9.9cd ± 1.3	11.0d ± 2.0	13.1d ± 1.1	14.3d ± 1.3

Table 7. Evolution of β-glucosidase activity (μmol PNP g^{-1} h^{-1}) (mean ± standard errors) in Xerollic Calciorthid and Typic Xerofluvent during the experimental for all experimental treatments. PNP: *p*-nitrophenol

†Different letters following the numbers indicate a significant difference at $P<0.05$

	3	15	45	90	120	150	180	210	250
Xerollic Calciorthid									
C	21.8b† ± 2.3	21.7b ± 2.0	21.2b ± 2.3	20.8b ± 2.2	20.5b ± 2.6	20.1b ± 2.4	19.7b ± 2.0	19.4b ± 1.7	19.0b ± 1.2
C+H	21.0b ± 1.9	19.8b ± 1.6	17.8a ± 1.6	16.7a ± 1.5	15.5a ± 1.6	14.7a ± 1.4	13.6a ± 1.7	12.4a ± 1.5	11.1a ± 1.6
C+CM	22.2b ± 2.0	23.1b ± 1.7	26.2bc ± 2.1	27.9c ± 2.2	29.1c ± 3.0	30.2c ± 2.6	31.4c ± 2.9	32.5c ± 3.0	34.1c ± 2.9
C+MSW	22.4b ± 2.4	23.7b ± 2.5	27.0bc ± 1.8	28.4c ± 2.3	30.2c ± 1.7	31.8c ± 2.6	33.0c ± 2.1	34.6c ± 2.5	35.9d ± 2.4
C+CM+H	21.7b ± 2.0	22.0b ± 1.8	22.9bc ± 2.4	24.0bc ± 2.4	25.0bc ± 2.8	25.8bc ± 1.9	26.5bc ± 2.6	27.4c ± 2.7	28.5c ± 2.0
C+MSW+H	21.7b ± 2.2	22.3b ± 2.4	24.4bc ± 2.2	25.4bc ± 2.1	26.1bc ± 1.4	27.5c ± 2.0	28.8c ± 1.7	29.7c ± 2.0	31.0c ± 2.6
Typic Xerofluvent									
C	18.7b† ± 1.2	18.5b ± 1.3	18.2b ± 1.4	17.9b ± 1.1	17.5b ± 1.9	16.6b ± 1.4	16.1b ± 1.5	15.7b ± 1.7	15.2b ± 1.4
C+H	18.2b ± 1.6	17.1b ± 1.5	15.1b ± 1.1	14.0a ± 1.4	13.2a ± 1.0	12.0a ± 1.8	11.1a ± 1.3	10.2a ± 1.6	9.4a ± 1.1
C+CM	19.2b ± 1.3	19.9b ± 1.1	22.5bc ± 1.4	23.7c ± 2.0	24.9c ± 2.1	26.6c ± 1.4	27.8c ± 2.7	28.9c ± 1.9	30.0c ± 2.6
C+MSW	19.4b ± 1.5	20.3b ± 1.8	23.6c ± 2.3	24.9c ± 1.8	26.5c ± 2.0	28.9c ± 1.7	30.6c ± 2.1	33.1c ± 2.7	35.0d ± 2.4
C+CM+H	19.1b ± 1.7	19.5b ± 1.6	21.1b ± 2.2	22.0bc ± 1.2	22.7bc ± 2.0	23.6c ± 2.2	24.6c ± 1.5	25.4c ± 2.6	26.1c ± 1.9
C+MSW+H	19.1b ± 1.8	19.8b ± 1.8	21.8b ± 1.8	22.9bc ± 1.6	24.2c ± 1.2	25.0c ± 1.9	26.6c ± 2.3	28.6c ± 2.2	30.5c ± 2.0

Table 8. Evolution of phosphatase activity (μmol PNP g^{-1} h^{-1}) (mean ± standard errors) in Xerollic Calciorthid and Typic Xerofluvent during the experimental for all experimental treatments. PNP: *p*-nitrophenol

†Different letters following the numbers indicate a significant difference at $P<0.05$

3.2 Laboratory experiments

For both soils, sorption isotherms of diflufenican on C, C+MSW, and C+CM treatments are shown in Figure 2. The results indicated that sorption of diflufenican on the clay loam texture soil was higher than the sandy loam texture soil. Also and for each experimental soil, the sorption of diflufenican on organic amended soils increased compared to non-organic amended soil. For each organic amended soil, the herbicide sorption with MSW was higher than with CM.

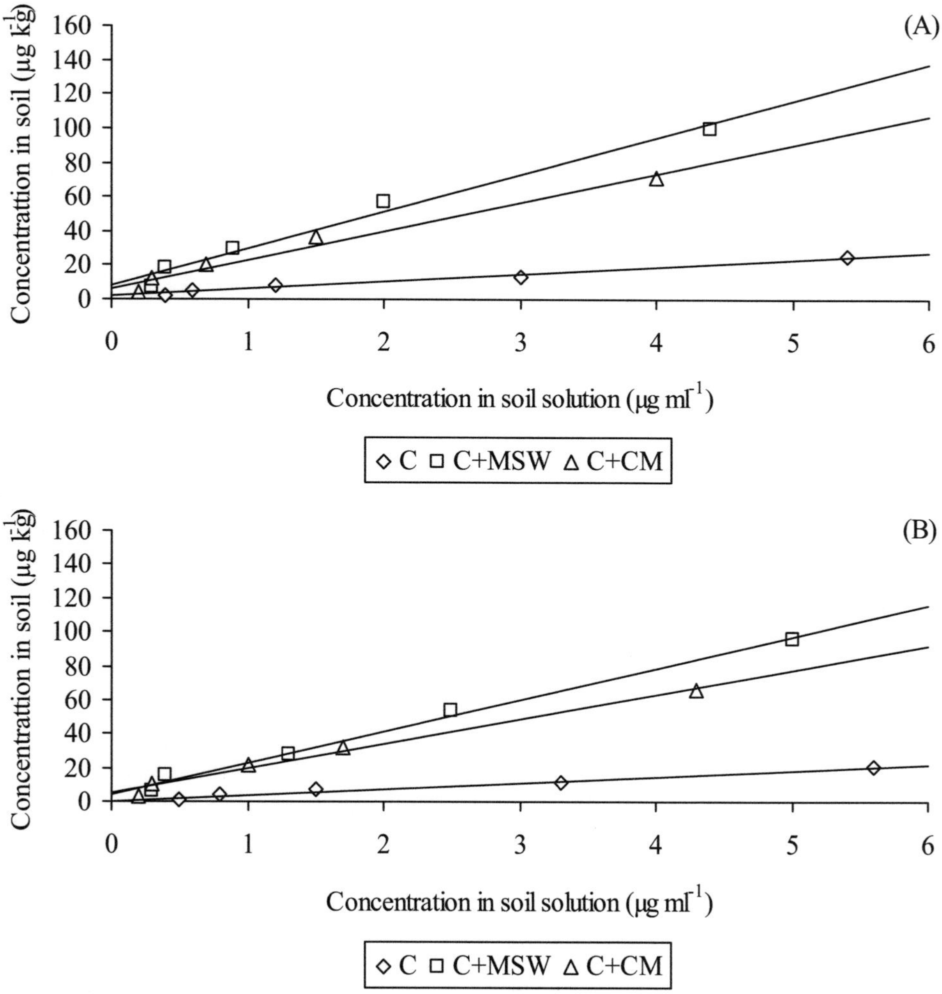

Fig. 2. Diflufenican sorption isotherms for Xerollic Calciorthid soil (A) and Typic Xerofluvent (B) soil. Symbols are experimental data points, whereas lines are the Freundlick-fit sorption isotherms

Sorption isotherms were fit to the Freundlich equation and sorption coefficients K_f and $1/n_f$ were calculated (Table 9). For non-organic amended soils, the K_f values were highest for Xerollic Calciorthid soil than for the Typic Xerofluvent soil. For both experimental soils, the results indicated that K_f values significantly increased in organic amended soils than for non-organic amended soils. However, diflufenican sorption was higher in soils amended with MSW than with CM. Also, the $1/n_f$ coefficients significantly decreased in organic amended soils than for non-organic amended soil. For organic amended soils, the $1/n_f$ coefficient was higher in the soil amended with MSW than for CM.

	Xerollic Calciorthid		
	K_f	$1/n_f$	R^2
C	7.94a† ± 1.2	0.89b ± 0.08	0.982
C+M	28.45b ± 2.3	0.77a ± 0.11	0.974
C+MSW	34.72b ± 2.8	0.81a ± 0.07	0.981
	Typic Xerofluvent		
C	5.76a† ± 0.96	0.93b ± 0.05	0.978
C+M	24.69b ± 2.8	0.81a ± 0.09	0.984
C+MSW	30.15b ± 2.7	0.87a ± 0.06	0.993

Table 9. Freundlich sorption coefficients K_f and $1/n_f$ (mean ± standard errors) for diflufenican in unamended and organic amended soils.
†Different letters following the numbers indicate a significant difference at $P<0.05$

Figure 3 shows the cumulative diflufenican leachates applied to unamended and organic-amended soil columns. At the end of the experiment, the diflufenican leachates were higher in the Xerollic Calciorthid soil than in the Typic Xerofluvent soil. Diflufenican leachates significantly decreased in organic-amended soils. However, this decrease depended of the organic matter type applied to the soil. For the Xerollic Calciorthid soil, the maximum concentratin of diflufenican in leachates was reduced from 7.4 μM for the unamended soil, to 4.1 μM or 4.9 μM for the MSW or CM-amended soil, whereas for the the Typic Xerofluvent soil the maximum concentratin of diflufenican in leachates was reduced from 8.9 μM for the unamended soil, to 5.8 μM or 6.2 μM for the MSW or CM-amended soil For both soils, no significant differences between the herbicide in leachates for organic-amended soils exits.

4. Discussion

Firstly, our results suggested that the herbicide time persistence depended of the soil texture type. The highest contents of diflufenican occurred in the clay loam texture soil probably due to adsorption of the herbicide with the clay, and therefore, the herbicide leachates were lowest in the clay loam texture soil than in sandy loam texture soil. These results are in agreement with those obtained by Singh et al., (2002), Renaud et al. (2004), Yen et al. (2003) and Flores et al. (2009), who suggested that when in low soil organic matter content, texture plays a fundamental role in the adsorption process of pesticides in soil. Also, these results suggest that the diflufenican persistence time was higher in the clay loam texture soil than in sandy loam texture soil. These results are in agreement with those obtained by Rouchaud et al. (1998) and Flores et al. (2009), who suggested that in soils with low organic matter content the herbicide adsorption, persistence and mobility depends mainly on soil texture.
Also, our results indicated that the diflufenican herbicide caused a toxic effect on soil enzymatic activity. These results are in agreement with those obtained by Tejada (2009), which observed the toxic effect of this herbicide in microcosm studies. Therefore, the study of soil enzymatic activities are an essential tool for knowing the state of soil pollution (Gianfreda et al., 2005; Tejada et al., 2007, 2008, Tejada, 2009). Since the diflufenican content

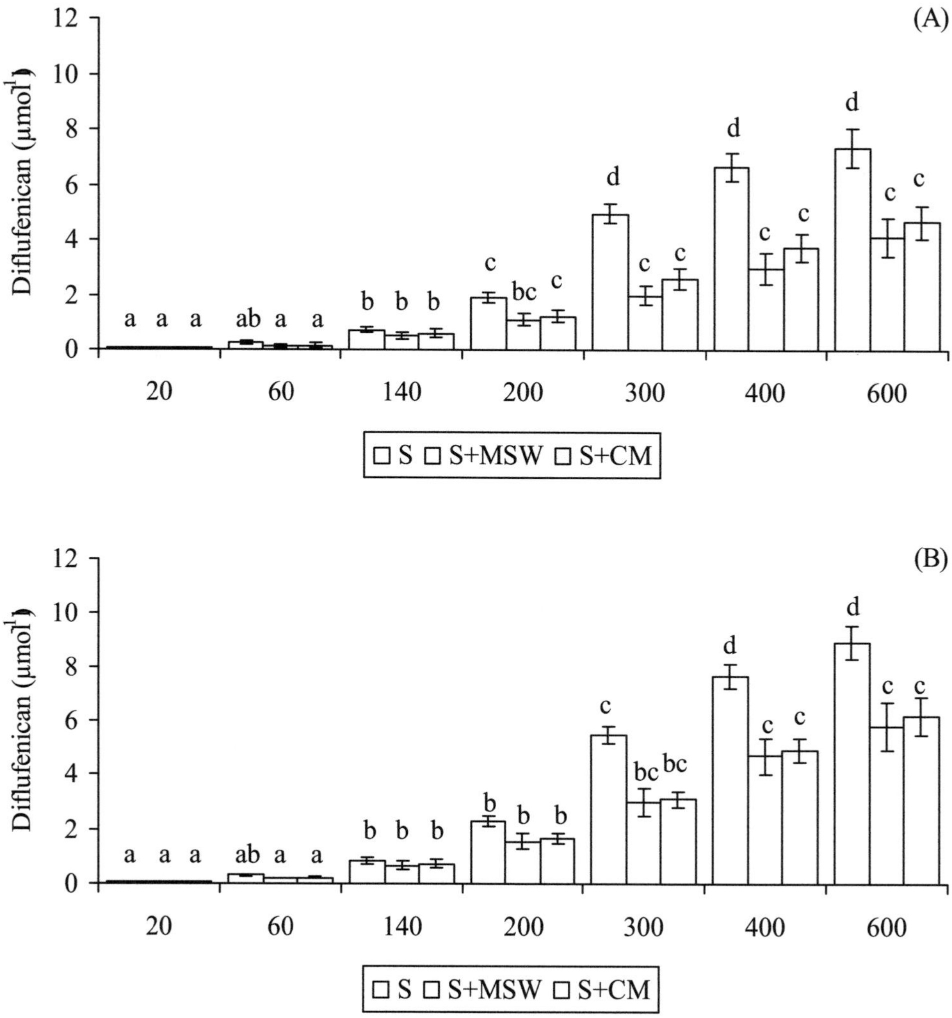

Fig. 3. Cumulative diflufenican leachates (mean ± standard errors) in unamended and organic amended treatments for Xerollic Calciorthid soil (A) and Typic Xerofluvent (B) soil. Column followed by the same letter(s) are not significantly different (p<0.05)

was higher the clay loam texture soil than in sandy loam texture soil, the inhibition of soil enzymatic activities in soil was higher in the Calciorthid Xerollic soil than in the Typic Xerofluvent soil.

The application of organic matter in unpolluted soils significantly increased enzymatic activities (Ros et al., 2003; Ferreras et al., 2006, Tejada et al., 2010). Soil microorganisms degrade organic matter through the production of a variety of extracellular enzymes, which explains the increase in enzymatic activities observed after the application of vermicomposts to the soil. However, this increase in soil enzyme activity was higher in the MSW than in the CM-amended soil. According to Tejada et al. (2010), the organic matter with higher fulvic than humic-acids contents is degraded more rapidly in the soil, promoting more positively the biological properties.

The application of the herbicide in MSW and CM-amended soils caused a decrease in the soil enzymatic activity inhibition. The diflufenican sorption isotherms and Freundlich sorption coefficients obtained in this study, suggested that organic matter play a fundamental role in the sorption of the herbicide in agricultural soils, probably as a result of the humic substances containing several major functional groups, such as carboxyl, phenolic, alcohol and carbonyl (Sluzny et al., 1999; Datta et al., 2001) According to Rouchaud et al. (1998) and Flores et al. (2009), when soils have a high content of organic matter, herbicide adsorption, persistence and mobility in soil depends mainly on the organic matter, whereas soil texture playing a more secondary role. However, our results also suggested that the chemical composition of the organic matter influenced in the diflufenican sorption.

Several studies of metal complexation with organic matter indicated that the sorption of heavy metals increased when the humic acid-like content increased in the organic matter, compared to the fulvic acid content, probably due to the humic acid-like possess a higher number of carboxylic groups than fulvic acid (Tejada et al., 2007, 2008).

The diflufenican sorption isotherms and Freundlich sorption coefficients indicated higher herbicide sorption in MSW than for CM-amended soils. Therefore, and similar to the heavy metals complexation, the sorption of herbicide increased with the humic acid content in the organic waste applied to the soil. The higher sorption probably caused a higher decrease of herbicide in the soil solution, and therefore, lowest availability of diflufenican availability for the soil microorganisms. This fact probably is the responsible of the increase in the soil enzymatic activities.

Also, the adsorption of the herbicide in organic-amended soils produced a decrease of the leachates herbicide. Since the diflufenican adsorption in soils amended with organic matter rich in humic acids was higher, herbicide losses were lower in soils amended with MSW than with CM.

5. Conclusions

It can be concluded that the diflufenican adsorption, persistence and mobility depends mainly on soil texture. Since the herbicide adsorption was higher in in the clay loam texture soil than in sandy loam texture soil, the leaching losses were highest in sandy loam texture soil than in clay loam texture soil. Also, the diflufenican herbicide caused a negative effect on soil enzymatic activities. The application of organic matter to the soil is a good environmental practice because decrease the diflufenican concentration in leachates and decreased the enzymatic activities inhibition. However, the beneficial effect depended of the organic matter chemical composition. The herbicide decrease was higher in MSW than for CM-amended soils. These results indicated that the addition of organic materials with a higher humic acid concentration may be considered a good strategy for decreased the diflufenican environmental pollution.

6. References

Amador, J.A., Glucksman, A.M., Lyons, J.B. & Gorres, J.H. (1997). Spatial distribution of soil phosphatase activity within a riparian forest. *Soil Science* 162(11), 808-825.

Ashton, I. A., Abulnaja, K. O., Pallett, K.E., Cole, D.J. & Harwood, J.L. (1994). The mechanism of inhibition of fatty acid synthase by the herbicide diflufenican. *Phytochemistry* 35(3), 587-590.

Bandick, A.K. & Dick, R.P. (1999). Field management effects on soil enzymes activities. *Soil Biology and Biochemistry* 31(11), 1471-1479.

Ben-Hur, M., Leteo, J., Farmer, W.J., Williams, C.F. & Nelson, S.D. (2003). Soluble and solid organic matter effects on atrazine adsorption in cultivated soils. *Soil Science Society of America Journal* 67(4), 1140-1146.

Cabrera, A., Cox, L., Fernández-Hernández, A., García-Ortiz Civantos, C. & Cornejo, J. (2009). Field appraisement of olive mills solid waste application in olive crops: Effect on herbicide retention. *Agriculture, Ecosystems & Environment* 132(3-4), 260-266

Carabias Martínez, R., Rodríguez Gonzalo, E., Fernández Laespada, M.E. & Sánchez San Román, F.J. (2000). Evaluation of surface- and ground-water pollution due to herbicides in agricultural areas of Zamora and Salamanca (Spain). *Journal of Chromatography A* 869(1-2), 471-480.

Carrizosa, M.J., Koskinen, W.C., Hermosin, M.C. & Cornejo, J. (2001). Dicamba adsorption-desorption on organoclays. *Applied Clay Science* 18(5-6), 223-231.

Černohlávková, J., Jarkovský, J. & Hofman, J. (2009). Effects of fungicides mancozeb and dinocap on carbon and nitrogen mineralization in soils. *Ecotoxicology and Environmental Safety* 72(1), 80-85.

Chen, S.K., Edwards, C.A. & Subler, S. (2001). Effects of the fungicides benomyl, captan and chlorothalonil on soil microbial activity and nitrogen dynamics in laboratory conditions. *Soil Biology & Biochemistry* 33(14), 1971-1980.

Cox, L., Celis, R., Hermosin, M.C., Becker, A. & Cornejo, J. (1997). Porosity and herbicide leaching in soils amended with olive-mill wastewater. *Agriculture, Ecosystems & Environment* 65(2), 151-161.

Datta, A., Sanyal, S.K. & Saha, S. (2001). A study on natural and synthetic humic acids and their complexing ability towards cadmium. *Plant and Soil* 235(1), 115-125.

Directive 2000/60/EC of 23 of October of 2000, related to Communitary Marc to Actuation about Water Politic. Diary Official of European Communities (DOCE), L 321, 19 of December of 2000.

Directive 2006/11/EC of the European Parliament and Council of 15 of February of 2006, related to Pollution Caused by Dangerous Substances tipped in the Aquatic Environment of the Community. Diary Official of European Communities (DOCE), L 64, 4 of March of 2006.

Edwards, C. (1989). Impact of herbicides on soil microcosms. *Critical Reviews in Plant Science* 8, 221-253.

El-Nahhal, Y. (2003). Persistence, mobility, efficacy and activity of chloroacetanilide herbicide formulation under greenhouse and field experiments. *Environmental Pollution* 124(1), 33-38.

Farenhorst, A. (2006). Importance of soil organic matter fractions in soil-landscape and regional assessments of pesticide sorption and leaching in soil. *Soil Science Society of America Journal* 70(3), 1005-1012

Ferreras, L., Gomez, E., Toresani, S., Firpo, L. & Rotondo, R. (2006). Effect of organic amendments on some physical, chemical and biological properties in a horticultural soil. *Bioresource Technology* 97(4), 635–640.

Flores, C., Morgante, V., González, M., Navia, R. & Seeger, M. (2009). Adsorption studies of the herbicide simazine in agricultural soils of the Aconcagua Valley, central Chile. *Chemosphere* 74(11), 1544-1549.

García, C., Hernández, T., Costa, F., Ceccanti B. & Masciandaro, G. (1993). The dehydrogenase activity of soil as an ecological marker in processes of perturbed system regeneration, *Proceedings of the XI International Symposium of Environmental Biogechemistry*, pp 89-100, Salamanca, Spain.

García, C., Hernández, T. & Costa, F. (1997). Potential use of dehydrogenase activity as an index of microbial activity in degraded soils. *Communications in Soil Science and Plant Analysis* 28 (1-2), 123-134.

Gianfreda, L., Rao, M.A., Piotrowska, A., Palumbo, G. & Colombo, C. (2005). Soil enzyme activities as affected by anthropogenenic alterations: intensive agricultural practices and organic pollution. *Science of Total Environment* 341(1-3), 265-279.

Gomides Freitas, L., Singer, H., Muller, S.R., Schwarzenbach, R.P. & Stamm, C. 2008. Source area effects on herbicide losses to surface waters- A case study in the Swiss Plateau. *Agriculture, Ecosystems & Environment* 128(3), 177-184

Kandeler, E. & Gerber, H. (1988). Short-term assay of soil urease activity using colorimetric determination of ammonium. *Biology and Fertility of Soils* 6(1), 68-72.

Kizilkaya, R. & Bayrakli, B. (2005). Effects of N-enriched sewage sludge on soil enzyme activities. *Applied Soil Ecology* 30(3), 195-202.

Kononova, M.M. (1966). *Soil Organic Matter: Its Nature, its Role in Soil Formation and in Soil Fertility*. (2nd ed.), Pergamon Press, Oxford.

Majumdar, K. & Singh, N. (2007). Effect of soil amendments on sorption and mobility of metribuzin in soils. *Chemosphere* 66(4), 630-637.

MAPA (1986). *Métodos oficiales de análisis. Tomo III (plantas, productos orgánicos fertilizantes, suelos, aguas, productos fitosanitarios, fertilizantes inorgánicos)*. Ministerio de Agricultura y Pesca, Madrid, 662 pp.

Masciandaro, G., Ceccanti, B. & García, C. (1994). Anaerobic digestion of straw and piggery waste waters. II. Optimization of the process. *Agrochimica* 38(3), 195-203.

Masciandaro G, Ceccanti, B., Benedicto, S. & Lee, H. (2001). Humic substances to reduce salt effect on plant germination and growth. *Communications in Soil Science and Plant Analysis*. 33(3-4), 365-378.

Masciandaro, G., Ceccanti, B., Benedicto, S., Lee, H.C. & Cook, F. (2004). Enzyme activity and C and N pools in soil following application of mulches. *Canadian Journal of Soil Science* 84, 19-30.

Pandey, S. & Singh, D.K. (2004). Total bacteria and fungal population after chlorpyrifos amd quinalphos treatments in groundnut (Arachis hypogaea L.) soil. *Chemosphere* 55(2), 197-205.

Pascual, J.A., Hernández, T., García, C. & Ayuso, M. (1998). Enzymatic activities in an arid soil amended with urban organic wastes: Laboratoty experiment. *Bioresource Technology* 64(2), 131-138.

Renaud, F.G., Brown, C.D., Fryer, C.J. & Walker, A. (2004). A lysimeter experiment to investigate temporal changes in the availability of pesticide residues for leaching. *Environmental Pollution* 131 (1), 81-91.

Ritchie, J.D. & Perdue, E.M. (2003). Proton-binding study of standard and reference fulvic acids, humic acids, and natural organic matter. *Geochimica et Cosmochimica Acta* 67(1), 85-96.

Romero, E., Fernández-Bayo, J., Castillo, J.M. & Nogales, R. (2010). Enzyme activities and diuron persistence in soil amended with vermicompost derived from spent grape marc and treated with urea. *Applied Soil Ecology* 44(3), 198-204.

Ros, M., Hernández, M.T. & García, C. (2003). Soil microbial activity after restoration of a semiarid soil by organic amendments. *Soil Biology and Biochemistry* 35(3), 463-469.

Rouchaud, J., Neus, O., Callens, D. & Bulcke, R. (1998). Isoxaflutol herbicide soil persistence and mobility in summer corn and winter wheat crops. *Bulletin of Environmental Contamination and Toxicology* 60(4), 577-584.

Sheng, G., Johnston, C.T., Teppen, B.J. & Boyd, S.A. (2001). Potential contributions of smectite clays and organic matter to pesticide retention in soils. *Journal of Agricultural and Food Chemistry* 49(6), 2899-2907.

Singh, N., Kloeppel, H. & Klein, W. (2002). Movement of metolachlor and terbuthylazine in core and packed soil columns. *Chemosphere* 47(4), 409-415.

Sluzny, C., Graber, E.R. & Gerstl, Z. (1999). Sorption of s-triazine herbicides in organic matter amended soils: Fresh and incubated systems. *Water, Air and Soil Pollution* 115(1-4), 395-410.

Soil Survey Staff. (1987). *Keys to Soil Taxonomy*. Soil Manage. Support Serv. Tech. Monogr. 6. 3rd ed. Cornell Univ., Ithaca, NY.

Song, N.H., Chen, L. & Yang, H. (2008). Effect of dissolved organic matter on mobility and activation of chlorotoluron in soil and wheat. *Geoderma* 146 (1-2), 344-352.

[SSEW], Soil Survey of England and Wales (1982). *Soil Survey Laboratory Methods*. Technical Monograph 6, SSEW, Harpenden, UK.

Tabatabai, M.A. & Bremner, J.M. (1969). Use of *p*-nitrophenol phosphate in assay of soil phosphatase activity. *Soil Biology and Biochemistry* 1(4), 301-307.

Tejada, M., Hernández, M.T. & García, C. (2007). Application of two organic wastes in a soil polluted by lead: Effects on the soil enzymatic activities. *Journal of Environmental Quality* 36(1), 216-225.

Tejada, M., Moreno, J.L., Hernández, M.T. & García, C. (2008). Soil amendments with organic wastes reduce the toxicity of nickel to soil enzyme activities. *European Journal of Soil Biology* 44(1), 129-140.

Tejada, M. (2009). Evolution of soil biological properties after addition of glyphosate, diflufenican and glyphosate+diflufenican herbicides. *Chemosphere* 76(3), 365-373.

Tejada, M., Hernández, M.T. & García, C. (2009). Soil restoration using composted plant residues: Effects on soil properties. *Soil & Tillage Research* 102(1), 109-117.

Tejada, M., Gómez, I., Hernández, T. & García, C. (2010). Utilization of vermicomposts in soil restoration: Effects on soil biological properties. *Soil Science Society of America Journal* 74(2), 525-532.

Trevors, J.T. (1984). Dehydrogenase activity in soil: A comparison between the INT and TTC assay. *Soil Biology and Biochemistry* 16(6), 673-674.

Tucker, B.M. (1954). The determination of exchangeable calcium and magnesium in carbonate soils. *Australian Journal of Agricultural Research* 5(4), 706-715.

Turner, B.L., Hopkins, D.W., Haygarth, P.M. & Ostle, N. (2002). β-Glucosidase activity in pasture soils. *Applied Soil Ecology* 20(2), 157-162.

Yen, J.-H., Sheu, W.-S. & Wang, Y.-S. (2003). Dissipation of the herbicide oxyfluorfen in subtropical soils and its potential to contaminate groundwater. *Ecotoxicology and Environmental Safety* 54(2), 151-156.

Yeomans, J.C. & Bremner, J.M. (1988). A rapid and precise method for routine determination of organic carbon in soil. *Communications in Soil Science and Plant Analysis* 19(13), 1467–1476.

Ying, G.G. & Williams, B. (2000). Laboratory study on the interaction between herbicides and sediments in water systems. Environmental Pollution 107(3), 399-405.

Zabaloy, M.C. & Gómez, M.A. (2008). Microbial respiration in soils of the Argentine Pampas after metsulfuron-methyl, 2,4-D and glyphosate treatments. *Communications in Soil Science and Plant Analysis* 39(3-4), 370-385.

15

Impacts of Biochar (Black Carbon) Additions on the Sorption and Efficacy of Herbicides

Alegría Cabrera Mesa[1] and Kurt A. Spokas[2]
[1]*IRNAS-CSIC*
[2]*United States Department of Agriculture – Agricultural Research Service*
[1]*Spain*
[2]*USA*

1. Introduction

The aim of this chapter is to review the effect of biochar on the fate and efficacy of herbicides. The increasing use and need of energy worldwide, together with the depletion of fossil fuels make the search and use of renewable energy sources a priority. Biomass is recognized as a potential renewable energy source, and pyrolysis is considered the most promising thermo-chemical conversion of biomass into bioenergy products (Özçimen & Karaosmanoğlu, 2004). Burning biomass in the absence of oxygen (pyrolysis) yields three products: a liquid (bio-oil), solid, and a gas (Bridgwater, 2003), with the traditional use of these products as renewable fuel and energy sources. Biochar currently lacks a universal definition, as can been seen in the range of definitions in the literature. According to Azargohar & Dalai (2006) *biochar can be considered the solid product of pyrolysis* and Sohi et al., (2010) define it as *biomass-derived char intended specifically for application to soil*. Warnock et al. (2007) defined biochar as the *term reserved for the plant biomass derived materials contained within the black carbon (BC) continuum*. We recommend that the term biochar be defined as the **solid residual remaining after the thermo-chemical transformation of biomass whose main intended purpose is as a means of C sequestration**. However, to retain the "biochar" classification there are two restrictions: 1) biochar itself cannot be used as a fuel source (although the utilization of the energy during production of the biochar is acceptable and encouraged) and 2) excludes those forms of black C derived from non-renewable (fossil fuel) resources [e.g. coal, petroleum, tires] (Lehmann et al., 2006).

The origin of charcoal in the environment can be natural or synthetic. In the first case, wildfire and volcanic processes are responsible for its formation (Scott, 2010; Scott & Damblon, 2010); meanwhile in the second case, thermal processes such as combustion and pyrolysis convert biomass into a char (residual solid product). Pyrolysis is described by Bridgwater (2003; 2006) as thermal decomposition in absence of oxygen, and is always the first step in the processes of gasification and combustion. Production of charcoal is favored by low temperatures and very long residence time conditions (Bridgwater, 1992). According to Goldberg (1985) black carbon is produced by the incomplete combustion of fossil fuels and vegetation that comprises the range of products of char, charcoal, graphite, ash, and soots, so biochar can be considered as a heterogeneous mixture of solid residues (Figure 1) (Hedges et al., 2000; Forbes et al., 2006; Jones et al., 1997; Spokas, 2010). In a study comparing oxidative stability of soot versus biochar, using a reference thermal stability method (OTC-375) (Gustafsson et al., 1997; 2001; Elmquist et al., 2004) it was observed that synthetic chars did not survive this analytical condition, with nearly 100% of the C oxidized at 375 °C. Thereby, the authors concluded that pyrolysis biochars are of a lower stability when compared to the natural chars (soot) in environmental sediments (Elmquist et al., 2006). In a recent review, this variability in the stability of biochars was placed in context of the spectrum of potential black C products with the molar ratio of oxygen: carbon (O:C) being suggested as a surrogate for biochar stability in soils (Spokas, 2010).

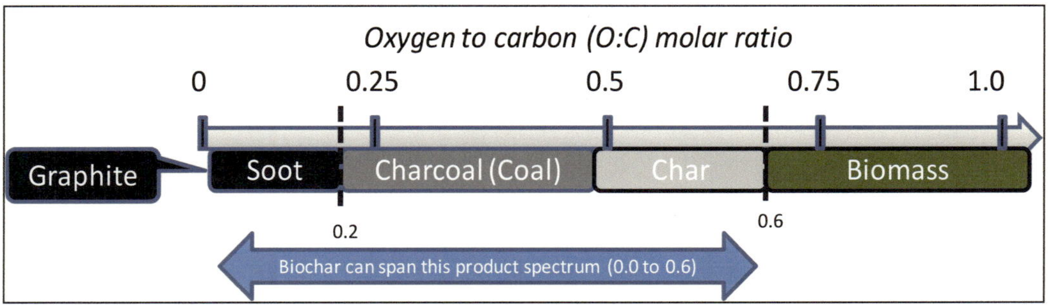

Fig. 1. Spectrum of black C products based on the oxygen to carbon ratio in the residual solid product. Adapted from Hedges et al. (2000), Bansal & Goyal (2005) and Spokas (2010).

For Bansal & Goyal (2005), activated carbon is a term that includes a wide range of processed amorphous carbonaceous materials with increased microcrystalline structure. Common parent materials for activated carbon are nutshells, peat, wood, coal (lignite or bituminous), and petroleum coke (Hassler, 1974). Activated carbon's superior adsorptive properties are due to their high surface area, micropore structure, high porosity, high sorption capacity, and high degree of surface reactivity (Pradhan & Sandle, 1999). Preparation of activated charcoal involves two steps: 1) the carbonization of the raw material (typically at temperatures below 800 °C) in an inert atmosphere, followed by 2) the activation by chemical or thermal methods of the carbonized product (Hassler, 1974; Yalcin & Sevinc, 2000). Chemical treatments both provide an opportunity to unify the surface behavior as well as to customize the overall surface area and porosity of the activated char (Hassler, 1974; Szymanski et al., 2002; Marsh & Rodríguez-Reinoso, 2006).

In summary, biochar properties are attributed to its characteristics, which depend on the pyrolysis conditions, temperature and time, as well as the feedstock used. There is renewed interest in the soil application of biochar as a means of increasing C sequestration and combating climate change. Biochar is a carbon-rich residue and the C is in a form that is more resistant to degradation than the original parent biomass whose primary purpose is as a vehicle for C sequestration, but the stability of the biochar spans the entire black C spectrum (e.g. Spokas, 2010). The land areas that are targeted for biochar applications are primarily agricultural land, due to the potential positive secondary impacts that biochar additions have on overall soil fertility as well as plant growth and yield (e.g. Busscher et al., 2010; Chan et al., 2007; Glaser et al., 2002; Novak et al., 2009; Sohi et al., 2010; Piccolo et al., 1996).

The application of organic materials to soils has been known for some time to increase herbicide sorption and reduce efficacy (e.g. Birk & Roadhouse, 1964; Weber et al., 1969).

Overall, there is the assumption that increased soil sorption could reduce leaching of agrochemicals to groundwater, thus improving overall water quality. On the other hand, increased sorption leads to decreased efficacy of soil applied herbicides thereby eliminating the residual action of these chemicals for weed management. The reader is encouraged to consult other chapters in this book, as well as Kookana et al. (1998), Gevao et al. (2000) and Laird & Koskinen (2008) for a review of herbicide sorption and fate in the soil environment. The impacts on the fate and dissipation of herbicides with biochar additions is still a developing research area, with only a limited number of biochars and soils that have been evaluated to date. To overcome this shortcoming, this chapter will review the existing literature and compile the results of the individual studies to examine if any overall conclusions can be gleamed from the data to date. In particular, production conditions (e.g. temperature, pyrolysis type, residency time) as well as parent material (biomass type) will be compiled to assess any dependency on these parameters as a function of herbicide efficacy, dissipation, and sorption. This review will provide insights for the future directions of herbicide-biochar research.

2. Charcoal, coal, ash and activated carbon: non-biochar additions

As mentioned in the introduction, biochar is the specialized name given to the spectrum of charred biomass products that are intended as a means of C sequestration. However, there has been some research conducted with non-biochar materials (e.g. activated charcoal applications and wood ashes). Activated charcoal starts as charred C prior to the activation process (e.g. steam, acid, or base activation) that optimizes sorption. Occasionally, the source of the C does not also fall within the biochar definition of a biomass (e.g. fossil fuels [coal], used tires, etc.).

There are minor differences in the overall chemical composition of activated charcoals compared to biochars, with the only significant difference being biochars possess higher oxygen contents (5-25%; e.g. Della Rocca et al., 1999; Brewer et al., 2009; Spokas & Reicosky, 2009; Zimmerman, 2010) than the corresponding activated charcoal (<10%; Yalcin & Sevinc 2000; Otowa et al., 1997) and natural black C forms (<7%; Hargrave et al., 1986; Ghetti, 1986; Hofrichter et al., 1997). The most striking difference is in surface area, with activated charcoals typically being >1000 $m^2\ g^{-1}$ and biochars are often two orders of magnitude smaller (<50 of $m^2\ g^{-1}$; e.g. Della Rocca et al., 1999; Brewer et al., 2009; Spokas & Reicosky, 2009; Zimmerman, 2010). Thereby, activated charcoals possess an increased potential for herbicide and other organic chemical sorption, as illustrated by activated charcoal's use as absorbent for analytical chemistry quantification methods (e.g. Betz et al., 1989).

2.1 Effect of carbonaceous materials on pesticide sorption

Sorption is one of the main processes that determine the fate of pesticides applied to soils. Soil sorption is characterized by a partition constant K_d, which is the concentration of pesticide in the solid phase divided into the pesticide concentration at the liquid phase at the equilibrium (Wauchope et al., 2002). Sorption isotherms of organic compounds on soils or amended soils are usually fitted to the Freundlich model, especially when the adsorption takes place on heterogeneous surfaces (Calvet, 1989):

$$C_s = K_f\, C_e^{1/n} \qquad (1)$$

where C_s is the concentration of pesticide sorbed, C_e is the equilibrium aqueous-phase concentration, K_f is the Freundlich capacity coefficient, and $1/n$ is the Freundlich exponent, which describes the degree of nonlinearity of the isotherm. When n=1, the Freundlich model reduces to a linear model.

Sorption isotherms can also be fitted to a Langmuir model, which assumes that the sorbent surface has a finite number of sorption sites and the sorptive energy is the same in all the sites (i.e. all sites are identical). Furthermore, each sorptive site is occupied only by a single molecule and there are no interactions between the sorbed molecules. Thereby, the maximum sorption a monolayer of sorbates on the surface of the sorbent can be given by:

$$q = QbC_e/1+bC_e \quad (2)$$

where q is the amount of sorbed pesticide per unit mass of soil, C_e is the equilibrium pesticide concentration in solution, Q is the Langmuir adsorption capacity, and b is the Langmuir affinity coefficient.

Among the carbonaceous materials more used as soil amendments are chars, especially wheat char, charcoal, ash, and activated carbon. Pesticide sorption increases in all cases when soils are amended with the carbonaceous materials (Table 1). Wang & Xing (2007) hypothesized that the sorption of organic compounds to un-charred biomass is dominated by absorption mechanisms, whereas adsorption becomes the dominant process with charred materials, largely due to the newly created atomic surfaces and micropores. In the case of wheat char and atrazine, one of the herbicides most used worldwide, the K_f value sorption on the amendment was 2012 (Loganathan et al., 2009), and the K_f increased by a factor of 5 in a sandy loam soil and of 4.3 in a clay loam soil, when both soils were amended with 1% (w/w) of wheat char. Sorption of pentachlorophenol, an organochlorine compound used as herbicide, insecticide, fungicide, algaecide, and disinfectant, was studied on three soils with different amendments (char, humic acids and peat), by Li et al. (2009). The highest sorption of pentachlorophenol in all the soils was observed when the soils were amended with char, followed by humic acids, and peat. Sorption was a function of organic carbon content (Li et al., 2009). The authors reported a 2.5-20 fold increase in the K_f values following the char addition. Zhang et al. (2006) studied the sorption of the pesticide benzonitrile in soil, wheat char and soil amended with the char at a rate of 1% (w/w) with the results of 40.5, >99, and 96.5% respectively, sorbed of the total of the pesticide initially applied. The wheat char was the dominant sorbent phase in the soil amended system, attributing approximately 90% of benzonitrile sorption to the wheat char and 10% to the soil. The sorption of benzonitrile on ash, 1% (w/w) amended soil and unamended soil was studied by Zhang et al. (2004), and a 10 fold increase was observed in the pesticide sorption on the silt loam soil amended with the wheat char. Sorption of the herbicide diuron on soil, char, ash, and amended soil has also been studied (Toth & Milham, 1975; Sheng et al., 2005; Yang et al., 2004; 2006; and Yang & Sheng, 2003a; 2003b). In the study by Toth & Milham (1975), ash was found to adsorb appreciable quantities of diuron from solution. Yang & Sheng (2003a) observed an increase in the sorption of diuron which ranged between 400-2500 times, when wheat and rice ash were used as sorbents and compared to the sorption on a silt loam soil. Increasing amounts of char added to soil, from 0.01 to 1% (w/w), resulted in an increase in sorption, directly proportion to the ash content.

Aged ash was compared as sorbent to fresh or non aged amendment by Yang & Sheng (2003b), and diuron sorption was found to decrease in the aged ash due to the competitive sorption of the dissolved organic matter present in the aged amendment. The competitive

Author	Pesticide	Amendment	Soil	Effect
Loganathan et al., 2009	Atrazine	Wheat char (1% w/w) 12.9% C, SA 10.1 $m^2 g^{-1}$	SL, 9% clay , 1.3% OC CL, 32% clay, 2.1% OC	↑ Sorption ↓ Desorption ↓ Mineralization slurries
Zhang et al., 2006	Benzonitrile	Wheat char (1% w/w)	SiL, 23% clay, 1.2% OC	↑ Sorption ↓ Degradation
Li et al., 2009	Pentachlorophenol	Willow woods shavings char 400ºC, 2h (2% w/w) SA 292 $m^2 g^{-1}$	SiL, 27% clay, 2.6% OC CL, 32% clay, 0. 9% OC SiL, 11% clay, 0.8% OC	↑ Sorption
Sheng et al., 2005	Diuron Bromoxynil Ametryne	Wheat char (1% w/w) SA 10.1 $m^2 g^{-1}$	SiL, 23% clay, 1.2% OC	↑ Sorption
Yang et al., 2006	Diuron	Wheat char SA 10.1 $m^2 g^{-1}$	SiL, 23% clay, 1.2% OC	↑ Sorption ↓ Degradation
Qiu et al., 2009a	Atrazine Dichlobenil	Wheat char (0.1-1% w/w)	SL	↑ Biodegradation
Zhang et al., 2005	Benzonitrile	Wheat char	SiL, 23% clay, 1.2% OC	↑ Biodegradation
Sobek et al., 2009	Fenuron Metoxuron Diuron Isoproturon Linuron Chlorotoluron Metobromuron Chlorbromuron Monolinuron	Black carbon (Fork-lift diesel soot)		Determination BC-water distribution coefficients
Qiu et al., 2009b	2,4-D Prometon Propanil	Black carbon (Rice straw) SA 1085 $m^2 g^{-1}$		2,4-D ↑ Sorption low pH Prometon ↑ Sorption high pH Propanil no pH effect
Yang et al., 2004	Diuron Bromoxynil Ametryne	Commercial AC SA 776 $m^2 g^{-1}$ Wheat carbon SA 310 $m^2 g^{-1}$		↑ Sorption Diur >AC no pH ffect ↑ Sorption Brom >AC <pH ↑ Sorption Ametryne >AC > pH
Xu et al., 2008	Clomazone	Burned rice straw (0.01-1% w/w)	SiL	↑ Sorption > ↑ Concentration ↓ Availability to barnyardgrass
Yamane and Green, 1972	Ametryne	Charcoal	SiC (subsoil)	↑ Sorption > high pH
Mukherjee, 2009	Atrazine	Charcoal (2.5% w/w)	SL, 17% clay, 0.4% OC	Slightly ↑ dissipation
Crepeau et al., 1991	Prometon, Prometryn 2,4-D, Carbofuan Dinoseb, Feanmiphos	Coal AC	SL, 10% clay, 0.4% OC	↑ pesticide retention > activated carbon ↓ leaching
Sudhakar & Dikshit, 1999	Endosulfan	Wood charcoal		Determination sorption kinetics
Sudhakar & Dikshit, 2010	Endosulfan Atrazine Monocrotophos	Wood charcoal		Competitive sorption Effect of pH, Cl^- fertilizer, DOM

Notes: AC: Activated Carbon; SA: Surface Area; SL: Sandy loam; CL: Clay loam,; SiL: Silt loam; SiC: Silty clay

Table 1. Summary of studies involving non-biochar materials (charcoals, coal, ash and activated carbons)

Author	Pesticide	Amendment	Soil	Effect
Keerthinarayana & Bandyopadhyay, 1997	Lindane	Wood charcoal		Sorption kinetics. Effect speed concentration, sorbent size,
Cui et al., 2009	Pentachlorophenol	Rice straw ash, 11.4% OC (1-10% w/w)	Sediment	↑ Sorption ↓ Genotoxicity, except high concentration
Hiller et al., 2007 2008	MCPA	Wheat ash, 11.8% C (1% w/w)	SL, 11% clay, 1.2% OC S, 0.6% clay, 0.5% OC	↑ Sorption ↓ Desorption
Hiller et al., 2009	MCPA	Wheat ash, 11.8% C (1% w/w)	SL, 11% clay, 1.2% OC S, 0.6% clay, 0.5% OC	↑ Sorption ↓ Desorption ↓ Degradation
Al Qodah et al., 2007	Deltamethrin Lambda-Cyhalothrin	Oil shale ash		Sorption > intermediate T Effect concentration, pH, particle size, speed, T
Yang and Sheng, 2003a	Diuron	Rice ash, 6%OC Wheat ash 14%OC (0.01- 1%/w)	SiL, 23% clay, 1.2% OC	↑ Sorption
Yang and Sheng, 2003b	Diuron Atrazine	Wheat ash 14% OC (1% w/w)	SiL, 23% clay, 1.2% OC	↑ Sorption Aging effect Herbicide competition
Wang et al., 2009	2,4,6-Trichlorophenol	Wheat ash 77% OC SA 410 $m^2 g^{-1}$		Metals slightly ↓ sorption and ↑ desorption
Zhang et al., 2004	Benzonitrile	Wheat ash 13% C (1% w/w) SA 10.1 $m^2 g^{-1}$	SiL, 23% clay, 1.2% OC	↑ Sorption ↓ Desorption ↓ Biodegradation
Gupta et al., 2006	2,4-D Carbofuran	CA SA 380 $m^2 g^{-1}$ AC SA 710 $m^2 g^{-1}$		Effect contact time, particle size, concentration, pH, temperature
Hilber et al., 2009	Dieldrin	AC SA 1200 $m^2 g^{-1}$ (0.02-0.08% w/w)	SiL, 9% clay, 2.2% OC	AC→dieldrin sequestration in soil. No effect on plant uptake except 0.08%
Domingues et al., 2007	α-cypermethrin	AC SA 972 $m^2 g^{-1}$		↑ Sorption
Hu et al., 1998	MCPB Imazalil Bentazone MCPP Dinoseb PCP Imidacloprid Linuron	AC SA 1340 $m^2 g^{-1}$		Effect pH on sorption Acidic ↓Sorption ↑pH Basic ↑Sorption ↑pH
Guo et al., 1993	Alachlor	AC 95.4% C (0.5-2.1 t C ha^{-1})	S, 5% clay, 0.9% OC	↑Sorption ↓Leaching (-) Degradation
Guo et al., 1991	Atrazine Alachlor	AC 95.4% C (0.5-2.1 t C ha^{-1})	S, 5% clay, 0.9% OC	↑ED_{50} with > AC concentration (-) Degradation bioactvity

Table 1. (Continued) (AC: Activated Carbon; CA: Carbonaceous adsorbent; SA: Surface Area; SL: Sandy loam; SiL: Silt loam; S: Sandy)

Author	Pesticide	Amendment	Soil	Effect
Hale et al., 2009	DDT DDD DDE DDMU	AC TOG SA 935 m^2 g^{-1} ACRS SA 900 m^2 g^{-1}	Water Sediment	SorptionDesorption Sequestration DOM ↓Sorption
Tomaszewski et al., 2007	DDT	AC TOG SA 935, F400 SA 1100 AC830 SA 900, ACRS SA 900	Sediment	Effective to sequester DDT from sediment
Lafrance et al., 1991	Aldicarb Lindane Pentachlorophenol	AC		Effect of pesticides and fulvic and humic acids on sorption
Bandala et al., 2006	Aldrin Dieldrin Heptacchlor Heptachlor Epoxide	Activated vegetal carbon		Efficient removal pesticides from water
Matsui et al., 2003	Simazine Simetryn Asulam	AC Coal based AC Wood based		Sorption kinetics
Ayranci and Hoda, 2005	Ametryn Aldicarb Dinoseb Diuron	AC-cloth SA 2500 m^2 g^{-1}		Effective to pesticide removal

Table 1. (Continued) (AC: Activated Carbon; SA: Surface Area)

sorption with atrazine was also studied, and a slight decrease in diuron sorption in presence of atrazine in the soil and in the amendment was also observed. The effect of the pH on pesticide sorption on the organic amendments was also studied by Li et al. (2009) with the result of an increase of sorption with the decrease in pH, which changes the pentachlorophenol species, and increases the static repulsion between the sorbent surface and the anionic form of the pesticide. However, it has been suggested that the presence of oxygenated groups on char could mask the impact of pH on the sorption of herbicides (Tessmer et al., 1997). Yang et al. (2004) did not observe any effect of pH on diuron sorption on activated carbon, due probably to the low density of functional groups on the surface of the activated carbon. Sheng et al. (2005) examined wheat char and 1% (w/w) char amended soil and hypothesized that the London forces between herbicide molecules and the electrically neutral char surface did not change with the pH. However, Yang et al. (2004) observed a decrease in diuron sorption with the pH increase on wheat char, because of the alteration of the surface charge properties by the deprotonation of the functional groups over the pH range. Contrary to the above charcoal studies, Sheng et al. (2005) also reported the same pH effect when diuron was sorbed only to soil, attributing the fact to the increase of the dissociation of the acidic groups with the increase in pH, increasing the hydrophilic of the organic matter, and reducing diuron sorption. Furthermore, Sheng et al. (2005) and Yang et al. (2004) also examined the pH effect on the sorption of the herbicides ametryne and bromoxynil. Both observed that sorption of bromoxynil on soil, wheat char, activated carbon and char-amended soil was higher at low rather than at high pH. The pK_a of this compound is 4.06, so at pH< 4.06 the herbicide is in the molecular form and at pH>4.06 in anionic form. The deprotonation of the phenolic hydroxyl group and the interactions and repulsion between the anion and the charge of the soil or amendment surfaces are the responsible for the sorption decrease with increasing pH of bromoxyinil. Ametryne is a basic herbicide, with pK_a of 4.1, at pH< 4.1 the compound is mostly in cationic form and at pH > 4.1 in molecular form. Sheng et al. (2005) studied the effect of 2 pH values, 3 and ~6.5, and found higher sorption of ametryne at lower pH on soil, wheat char and wheat char amended soil, in this case, the higher sorption occurs when the pesticide is in ionic form, due probably to electrostatic interactions of ametryne cations and the charges of the soil or char surface. Yang et al. (2004) included a value more of pH= 4 in the study and observed a different effect of pH on ametryne sorption. Ametryne sorption on activated carbon increased with the increase in pH, as result of the deprotonation of the herbicide molecule at low pH. However, when the sorbent was wheat char the highest sorption was observed at the intermediate pH, followed by the lower pH and the higher pH. The authors considered that the increase in sorption when the pH increased from 2 to 4 is due to the net result of the adsorption increase because of the ametryne deprotonation and the desorption decrease because of deprotonation of the functional groups of the sorbent surface. Yamane & Green (1972) also reported an increase on ametryne sorption on a silty clay soil when charcoal was added. Higher sorption of the herbicide at lower pH on the non amended soil was observed, but with the addition of charcoal to the soil the effect of pH on sorption of ametryne was the opposite, increasing the herbicide affinity for charcoal with protonation. For clomazone, a rice crops herbicide, Xu et al. (2008) reported an increase in sorption with the addition of burned rice straw to the soil, with an increase of 1.5 and 3 times in the K_f values when a silt loam soil was amended with the 0.1 and 0.5% (w/w) respectively. The acidic herbicide MCPA and the effect of wheat ash as soil amendment on its sorption were studied by Hiller et al. (2007; 2008; 2009). The authors observed that the isotherms of MCPA on wheat ash fitted to a Langmuir equation, but sorption on soil and ash amended soil did not fit to any model and calculated the K_d values, with a 15 and 10 fold increase when the sandy loam and the sandy soil were amended with 1% (w/w) of wheat ash. Activated carbon was used as amendment of a sandy soil for sorption of the herbicide alachlor (Guo et al., 1993). The soil was amended with the waste activated carbon at a rate of 0.5-2 t C ha^{-1}, increasing the value of K_f by a factor of 5-14 as compared with the non amended soil. Digested sewage sludge and animal manure were also studied as amendments, but waste activated carbon was more efficient in alachlor sorption. Konstantinou & Albanis (2000) examined the impact of coal fly-ash additions on the impact of 6 herbicides (atrazine, propazine, prometryne, propanil, molinate, propachlor). Their particular coal fly-ash had a carbon content of 2%, which is significantly lower than the typical charcoal. They showed that the adsorbed amounts of herbicides increased with the amount of fly ash addition, up to the amount sorbed by the "pure" fly ash. The adsorption coefficient (K_f) was exponentially related to the fly ash percent.

2.2 Effect of carbonaceous materials on herbicide desorption

The process of desorption has been less studied as compared to sorption. A pesticide that is sorbed to soil particles has to be desorbed to be available for transport and degradation. Desorption of atrazine in one step, successive steps and followed by solvent extraction for soil, char and char amended soil was performed by Loganathan et al. (2009). The multiple steps desorption was more efficient than the one step process, showing hysteresis in the sandy loam and clay loam soils and the amended soils. After the successive steps and solvent extraction desorption the amount of atrazine remaining sorbed on the char amended soils were higher than in nonamended soils. The successive desorption isotherms could be fitted to the Freundlich, Langmuir, and a combination of both equations. Desorption of MCPA is influenced by the presence of wheat ash in a sandy and sandy loam soil, decreasing MCPA desorption from 77 to 21% in the non amended and wheat ash amended sandy loam soil, and from 45 to 11% in the non amended and amended sandy soil (Hiller et al., 2007; 2008; 2009). MCPA desorption at high concentration (57 mg L^{-1}) the percentage of herbicide desorbed from the wheat ash was similar to the percentages desorbed from the amended soils, however, at the low concentration (5.7 mg L^{-1}) decreased the desorption of MCPA from the amendment. This fact could be propably due to the limited number of sorption sites of high energy binding available, which are occupied first at low concentration; to the extent of concentration is increasing, these sites are not available and herbicide molecules are retained to low energy sites and can be desorbed more easily. The same amendment was used for the desorption study of benzonitrile by Zhang et al. (2004). These authors evaluated the herbicide desorption from the nonamended soil and from the amendment, and observed a similar desorption of benzonitrile, but with different kinetics, slower desorption from wheat ash than from the silt loam soil, because of stronger sorption of benzonitrile on ash surface than on soil organic matter. Karanfil & Kilduff (1999) observed that the surface acidity (by activation) was a vital factor in the sorption of hydrophobic compounds. Increases in acidity caused a corresponding increase in surface polarity, which correspondingly resulted in reduced sorption. Konstantinou & Albanis (2000) also observed that the amount of herbicide desorbed with water decreased with increased fly-ash content. Thereby, indicating that the amount of potential available herbicide in the soil system decreases but fly-ash could be utilized to reduce the mobility of the herbicide in the soil system.

2.3 Effect of carbonaceous materials on pesticide degradation

The addition of organic amendments to soil can have contrasting effects on pesticide biodegradation; some authors have reported an increase as result of the microbial stimulation by the amendments, and others report a degradation decrease with the amendment, due to a lower pesticide bioavailability to microorganisms because of the increase in sorption.

For the herbicide atrazine, Loganathan et al. (2009) reported a decrease in mineralization of the herbicide in the 1% (w/w) wheat char amended soils as compared to the soils slurries, attributing the atrazine bioavailability decrease to the increase in sorption of the herbicide on the char amendment. Guo et al. (1991) suggested that atrazine and alachlor degradation could be inhibited by the presence of waste activated carbon, and stimulated by other uncharred amendments, such as municipal sewage sludge and manure. An increase on atrazine degradation by the addition of organic amendments to a sandy loam soil was also reported by Mukherjee (2009). Among the amendments studied, charcoal was included, although it was the amendment that had a lower decrease in atrazine degradation, as compared to rice straw, sawdust and farm yard manure. Qiu et al. (2009a) observed enhanced degradation of atrazine increasing the amount of wheat char added to the sandy loam soil from 0.1 to 1%. These increases were hypothesized to result from the nutrients present in char, and especially P, which could stimulate the microorganisms' activity thereby enhancing the biodegradation. Coapplication of another herbicide, dichlobenil resulted in a reduction of atrazine degradation in the soil and char amended soils, in comparison with the degradation of atrazine applied as the only herbicide. The suppression on the degradation of one herbicide by the metabolism of the other one applied could be due to the toxicity of the second compound or its metabolites, or by the competition for nutrients (Qiu et al., 2009a). Similar results were obtained for dichlobenil degradation as the only herbicide and coapplied with atrazine. Yang et al. (2006) observed a decrease and slower degradation of diuron in the 0.5% char amended soil as compared to the soil non-amended, with degradation lower than 40% and 55% respectively of the amount of diuron initially applied and after 10 weeks of study. Bioavailability of diuron to soil microorganisms was decreased by the presence of wheat char in the soil, due to the increase in herbicide sorption. For the weak acidic herbicide MCPA, a decrease in biodegradation because of wheat ash amendment was also reported by Hiller et al. (2009). The half-life of MCPA increased by a factor of 4.5 in the sandy loam soil with the 1% (w/w) wheat ash amendment, and correspondingly increased by 2.6 times in the sandy soil. The lower degradation of MCPA in the amended soils might be due to the increase in sorption, but also to the lower desorption observed.

Biodegradation of benzonitrile has been extensively studied by Zhang et al. (2004; 2005; 2006). In the first study a decrease in the biodegradation of benzonitrile in the 1% (w/w) wheat ash amended soil was reported. During the study the authors proved that there was no abiotical degradation of the organic compound and dissolved organic C did not affect its biodegradation. Reduction in the benzonitrile biodegradation was due to the combination of increased sorption and slow sorption from the ash and the acclimation period of the microorganisms. In the second study, initial degradation of benzonitrile was faster in the the soil amended than in the soil free of char, because of the dissolved nutrients of the wheat-residue derived char, which stimulated the cell growth and degradation in the amended soil. Degradation process was followed by a slower phase due to the adsorptive decrease by the presence of char. In the last work the authors studied the effect of initial benzonitrile concentration on the degradation. In the non-amended soil higher degradation was observed at lower benzonitrile concentration, however, in char amended soil, the higher initial concentration, the higher degradation. At the highest concentration a higher degradation was observed in the amended soil as compared to the non amended one, which could be attributed to the increase and stimulation of the microbial activity by the nutrients of the char. No significant pH effect was observed for benzonitrile degradation in char and char-amended soil, probably due to the high affinity sorption of benzonitrile by char that did not vary with pH.

2.4 Effect of carbonaceous materials on pesticide leaching

The addition of rich carbon amendments to soils can decrease pesticide leaching, as consequence of the increase in sorption. Crepeau et al. (1991) studied the mobility of the pesticides: 2,4-D, carbofuran, dinoseb, fenamiphos, prometon and prometryn in a sandy soil, two charcoals and the soil amended with three charcoals and activated carbon at three rates: 4:1, 7:1, and 10:1 (dry weight soil/weight coal carbon). The highest retention of the pesticides was observed with the activated carbon at the rates of 7:1 and 10:1. Leaching of pesticides also decreased with the addition of charcoal to the sandy soil. Leaching of alachlor was studied by Guo et al. (1993) in three different conditions, unsaturated flow, saturated flow and alternation of saturated and unsaturated flow in soil amended with activated carbon, municipal sewage sludge and animal manure at three rates of C content. In all cases the highest alachlor retention and recovery at the end of the study was observed in the waste activated carbon amended soil, which is in accordance with the higher sorption reported by the authors.

3. Impacts of biochar additions

The use of biochar as soil amendment is highly recommended because of the properties and benefits this amendment provides to soil and environment. As it was previously mentioned, among the properties of biochar are its high stability to degradation, and nutrient retention. Numerous works deals with the effect of biochar on soil properties (Warnock et al., 2007; Chan et al., 2007; 2008; Asai et al., 2009; Novak et al. 2009; 2010; Steinbeiss et al., 2009; Van Zwieten et al., 2010; Busscher et al., 2010; Sohi et al., 2010; Gaskin et al., 2010; Singh et al., 2010). There are limited studies (Table 2) detailing the effect of biochar soil amendments (fitting the definition of biochar) on pesticides and agrochemicals fate. Due to the fact that biochar is a black C the studies outlined in Section 2 (Table 1) are also applicable.

The use of biochar could contribute to the increase of pesticides sorption on soil, decreasing its mobility and reducing the contamination risks of surface and ground waters. The knowledge of biochar properties, which depends on the feedstock and pyrolysis conditions, is vital (Section 1). The characterization of biochar used in studies should be conducted and this data should be included in the manuscript, which would enable the comparison of the effects of the amendment on the pesticides behavior in soils across different studies. This comparison would allow the determination of the optimum pyrolysis conditions to yield biochar with the most appropriate properties to reduce the environmental contamination risks associated with the use of agrochemicals.

3.1 Effect of biochar on herbicide sorption

Triazine herbicides are among the more studied pesticides in the works about the effect of biochar on pesticide sorption. Spokas et al. (2009) observed an increase on atrazine sorption when a sandy loam soil was amended with 5% (w/w) biochar. The biochar was produced by fast pyrolysis at 500 °C of sawdust and had a surface area of 1.6 m^2 g^{-1}. The effect of biochar on atrazine sorption was greater at lower atrazine concentrations, and no effect of biochar addition was observed at the highest concentration of atrazine studied (17 μg mL^{-1}). These authors also studied the sorption of acetochlor, a chloroacetanilide herbicide, and reported a greater sorption of the herbicide in the soil amended with biochar, increasing the Freundlich K_f value from 4.1 to 6.6 $\mu g^{1-1/n}$ $mL^{1/n}$ g^{-1} with the amendment.

Wang et al. (2010) studied the effect of biosolids and two biochars on terbuthylazine sorption on sandy volcanic ash soil and forest topsoil. Pine (*Pinus* spp.) wood was the feedstock for both biochars, BC350 was produced at 350 °C and BC700 at 700 °C and were applied to soil at a rate of 1% (w/w). Sorption isotherms of terbuthylazine on soil and soil amended was described by a Freundlich model. The highest increase in sorption was observed when the soil was amended with BC700, followed by BC350. The authors found that the organic matter in the amendments had higher sorption capacity to the herbicide than endogenous organic matter in the landing soil, with lower content on organic carbon (OC). Sorption of atrazine to biochar was studied by Cao et al. (2009) and Cao & Harris (2010). The biochar feedstock was dairy manure and was produced at 200 and 350 °C in the first study and in the second were included the temperatures of 100 and 500 °C. Sorption of atrazine in the first study was higher on BC200 than on BC350, which was correlated to the organic carbon content (31 and 26% respectively) and the polymer aliphatic fraction. No significant differences were observed by the authors on the sorption of atrazine in single or binary solute (with Pb) systems when the adsorbent were the biochars, indicating limited sorption competition between Pb and atrazine. In the second study, the higher sorption of atrazine was observed with BC200, with higher OC content than BC350, but lower than BC100. It is thought that the lower sorption in BC100 with higher content on OC is due to the suppression of sorption by high dissolved organic C competing for sorption sites.

Zheng et al. (2010) have also studied the sorption of two triazine herbicides, atrazine and simazine, on biochar. The feedstock of this biochar was green waste which was a mixture of maple, elm and oak woodchips and bark. The pyrolysis temperature was set at 400 °C and biochar was separated in different fractions according to the particle size. Higher and faster sorption of atrazine and simazine occurred with the smaller particle size of biochar. The authors also reported an increase in sorption affinity of the biochar for the herbicides with the decrease in the solid/solution ratio and higher sorption at $pH < 7$, but, no pH effect at values ≥ 7 was observed. Sorption isotherms of the herbicides as single or double solute system were adjusted and fit to Freundlich models, and competitive sorption was observed for both herbicides. According to Zheng et al. (2010) biochar can be considered a heterogeneous sorbent with combined adsorption and partition mechanisms, which occurs on carbonized and non-carbonized fractions of the biochar. A biochar produced from red gum wood at two temperatures (450 and 850 °C) were used to study the sorption of the pesticides diuron and pyrimethanil on soil by Yu et al. (2006 and 2010) respectively. The sandy loam soil was amended with BC450 at rates from 0.1-5% (w/w) and from 0.1 to 1% for BC850. The sorption data were fitted to a Freundlich model and it was reported that there was an increase in the non-linearity behavior of the sorption with the increasing amendment rate, at the highest biochar rate data was better described by a Langmuir model.

Reference	Pesticide	Biochar	Soil	Effect
Yu et al., 2009	Carbofuran Clorpyrifos	Red gum wood chips 450 °C 2h (0.1-1% w/w) SA 27 $m^2 g^{-1}$ 850 °C 1h (0.1-1% w/w) SA 566 $m^2 g^{-1}$	SL, 11% clay	↓ Dissipation ↑DT_{50} > BC 850 ↓ Plant uptake ↓Bioavailability
Yu et al., 2006	Diuron	Red gum wood chips 450 °C 2h (0.1-5% w/w) SA 27 $m^2 g^{-1}$ 850 °C 1h (0.1-1% w/w) SA 566 $m^2 g^{-1}$	SL, 11% clay	↑ Sorption + > amount BC 850 ↑ Sorption irreversibility > amount BC 850 > microporosity
Spokas et al., 2009	Atrazine Acetochlor	Sawdust 500°C (5% w/w) SA 1.6 $m^2 g^{-1}$	SiL, 23% clay, 2.6% OC	↑ Sorption ↓ Dissipation ↓ Bioavailability
Jablonowski et al., 2010	Atrazine	Hardwood 450-500°C (0.1-5% w/w) SA 200 $m^2 g^{-1}$	C, 49.4% clay, 3.2%OC SiL, 7.8% clay, 1.3%OC	↑ Mineralization C Soil
Cao et al., 2009	Atrazine	Dairy manure, 200 °C 4h, 350 °C 4h		↑ Sorption BC 200
Wang et al., 2010	Terbuthylazine	Pinus radiata wood Commercial charcoal 350°C (1% w/w) Sawdust 700 °C (1% w/w)	S Landing soil, 1.2% OC Forest topsoil, 5.1% OC	↑ Sorption BC 700 ↓ Desorption
Yang et al., 2010	Chlorpyrifos Fipronil	Cotton straws 450°C 2h (0.1-1% w/w) SA 3.9 $m^2 g^{-1}$ 850 °C 1h (0.1-1% w/w) SA 159 $m^2 g^{-1}$	CL, 33% clay	↓ Dissipation> 1% BC 850 ↓ Bioavailability ↓ Plant uptake
Yu et al., 2010	Pyrimethanil	Red gum wood chips 450 °C 2h (0.1-1% w/w) SA 27 $m^2 g^{-1}$ 850 °C 1h (0.1-1% w/w) SA 566 $m^2 g^{-1}$	SL, 11% clay	↑ Sorption + > amount BC 850 ↑ Sorption irreversibility > amount BC 850
Sopeña et al., 2010	Isoproturon	Eucalyptus sp (0.1-1% w/w)	SL , 1.3% OC	↑ Sorption ↓ Desorption ↑Hysteresis ↓ Degradation↓ Mineralization
Cao & Harris, 2010	Atrazine	DM BC 100°C 4h SA~1.8 $m^2 g^{-1}$ BC 200°C 4h SA~2.7 $m^2 g^{-1}$ BC 350°C 4h SA~7 $m^2 g^{-1}$ BC 500°C 4h SA~13 $m^2 g^{-1}$		↑ Sorption BC 200
Zheng et al., 2010	Atrazine Simazine	Elm, oak, maple woodchips and barns 450°C SA 6.7-7.6 $m^2 g^{-1}$		> faster sorption < particle size BC Competitive sorption solid/solution , pH effect pH<7

Notes: SA: Surface area; BC: Biochar; DM: Dairy Manure; SL: Sandy loam; SiL: Silt loam; S: Sandy; C: Clay; and CL: Clay loam.

Table 2. Summary of biochar amendment studies on herbicide fate

The amount of diuron sorbed did increase with the biochar amount added to the soil. Biochar additions, even small additions, increased diuron sorption as compared with the non-amended soil. This clearly illustrates that presence of carbonaceous material, even in small amounts, can dominate sorption of organic compounds in soils. Similar results were obtained for the sorption of pyrimethanil on the same soil and using the same amendments rates and biochars (Yu et al., 2010). Higher sorption was observed in the soil amended with BC850, as compared to the amended BC450 soil, which is in accordance with the surface area of biochar, 566 and 27 $m^2 g^{-1}$, respectively. The authors suggested that sorption of the pesticide on the biochar amended soil was a combination of surface adsorption and absorption into biochar micropores. The increase of sorption and non linearity of the isotherms with the increase of biochar amendment was also observed by Sopeña et al. (2010) in the study of isoproturon sorption in a sandy loam soil. Biochar feedstock was red gum wood, but no further information about the biochar production and characteristics was reported thus limiting the comparison to other biochar results.

3.2 Effect of biochar on pesticide desorption

Studies of pesticide desorption have been reported by Yu et al. (2006; 2010), Wang et al. (2010), and Sopeña et al. (2010). In this first study of Yu et al. (2006), diuron desorption isotherms for a sandy loam soil amended with different rates of biochar produced at 450 and 850 °C were fitted to Freundlich models. Sorption-desorption hysteresis on the amended soils was observed, and irreversibility of the sorption increased with the increase in the amount of biochar applied to the soil. Sorption irreversibility was greater in the soil amended with the highest rate of BC850, which was attributed to the higher micro-porosity of this amendment as compared to BC450. In the second study pyrimethanil desorption was calculated as the percentage of the initial amount of herbicide adsorbed on the soil free of biochar and on the amended soil. Isotherms were not calculated and neither adjustment to any sorption model was performed. The authors observed a decrease on the percentage of pyrimethanil desorbed when the amount of biochar increased in the amended soils and lower herbicide desorption from the soil amended with BC850 than with BC450. The lower desorption reported for the BC850 could be due to the higher surface area (566 vs. 27 $m^2 g^{-1}$) and the higher number of micropores, maximum peak of pore size distribution at 0.49 vs. 1.1 nm, as compared to BC450. Desorption isotherms of terbuthylazine in unamended and amended soils were well described by the Freundlich model (Wang et al., 2010). The authors reported a lower desorption of terbuthylazine adsorbed on the soils treated with biochar, especially with BC700, than in the soils no amended. These biochars were not characterized, as to surface area and microporosity, but again the biochar produced at higher temperature shows a slower and lower desorption of the herbicide adsorbed. In case of isoproturon desorption the data also fitted well to Freundlich model and lower desorption of the herbicide with the increase in the biochar/soil ratio was observed by Sopeña et al. (2010).

3.3 Effect of biochar on pesticide degradation

The effect of biochar on the dissipation and degradation of pesticides in amended soils has been reported for the herbicides atrazine (Spokas et al., 2009; Jablonowski et al., 2010), acetochlor (Spokas et al., 2009), isoproturon (Sopeña et al., 2010), and the insecticides chlorpyrifos (Yu et al., 2009; Yang et al., 2010), fipronil (Yang et al., 2010), and carbofuran (Yu et al., 2009). Spokas et al. (2009) reported a lag phase of 11 days in the dissipation of atrazine in the non-amended and biochar amended silt loam soil. After that time dissipation was greater in the unamended soil. An increase in the degradation of atrazine was found by Jablonowski et al. (2010), in a clay soil adapted to atrazine, and amended with biochar from hardwood at 450-500 °C. The increase in mineralization and degradation of the herbicide is attributed to the stimulation of the soil microflora by the nutrients provided by biochar. Spokas et al. (2009) did not observe any lag phase for acetochlor dissipation, and the amendment of soil with biochar decreased the herbicide dissipation. The time of disappearance of the 50% of the initial amount of herbicide applied (DT_{50}) was calculated and increased from 9.7 days in the non amended soil to 34.5 d in the biochar amended soil. An increase in persistence of isoproturon in a sandy soil with biochar amendment was also reported by Sopeña et al. (2010) and DT_{50} increased from 2.2 d in the unamended soil to 5.6 d in the 2% (w/w) biochar amended soil. The amendment of a clay loam soil with biochar produced at 450 and 850 °C from cotton straws also caused a decrease in chlorpyrifos and fipronil dissipation (Yang et al., 2010). The dissipation decrease was higher with the higher content of biochar, and BC850 was more effective in reducing the loss of both insecticides. Under non sterilized conditions chlorpyrifos DT_{50} increased from 21 to 44 and 56 d with the 1% (w/w) BC450 and BC850 amendments, respectively. Furthermore, fipronil DT_{50} increased from 27 d in the non amended soil to 48 and 60 d with the 1% BC450 and 850, respectively. Under sterilized conditions DT_{50} were 2 and 3 times higher for chlorpyrifos and fipronil, which implies that degradation of the pesticides is mainly biotic. It was suggested that the decrease in pesticide dissipation in the soils was due to the lower bioavailability of chlorpyrifos and fipronil to the soil microorganisms, because of an increase in sorption and lower desorption promoted by the biochar. Yu et al. (2009) also reported a decrease in carbofuran and chlorpyrifos dissipation in a sandy loam soil when the soil was amended with biochar produced from red gum wood chips at 450 and 850 °C. These authors also found a decrease in pesticide dissipation with the increasing content of biochar, being BC850 more effective in the reduction of the insecticides dissipation. Carbofuran DT_{50} increased from 12 d in the non amended soil to 33 d in the 1% (w/w) BC850 amended soil and for chlorpyrifos the increase in DT_{50} was from 12 to 43 d with the addition of 1% BC850 to the soil. The effect of the presence of plants on the insecticides dissipation was also studied, with the result of faster loss of pesticide residues with the presence of plants in all soils, and especially in the soil amended with BC850. This could be due to the combination of effects of uptake by the plants and increase in degradation of the pesticides, because of the stimulation of microbial and biochemical activity.

3.4 Effect of biochar on pesticide plant uptake

Little research is reported on the effect of biochar on plant uptake of pesticides. Yu et al. (2009) used spring onion to study the effect of the biochar produced from red gum wood chips at 450 and 850 °C as amendment of a sandy loam soil on the plant uptake of chlorpyrifos and carbofuran. In the study of Yang et al. (2010) Chinese chives uptake of chlorpyrifos and fipronil from a clay loam soil amended with biochar prepared from cotton straw residue at 450 and 850 °C was determined. In both studies higher biomass production was observed in the cultivated soils amended with biochars than in the control soils, with the BC850 resulting in a larger increase than the BC450 amendment. In the soils spiked with carbofuran, greater biomass was produced from spring onion than in the soils spiked with chlorpyrifos. Soils cultivated with Chinese chives produced higher biomass in the soil spiked with chlorpyrifos as compared with fipronil. It was shown in the two studies that BC850 was more effective in reducing the uptake of the insecticides by the plants. The total amount of plant uptake of the pesticides in the whole plant or in parts of the plant decreased with the increasing content (0.1, 0.5, 1% w/w) of biochar in the soil, being the 1% BC850 treatment the most effective in the reduction of the total pesticide plant uptake (75% for carbofuran and 90% for chlorpyrifos in case of spring onion, and 52% for fipronil and 81% for chlorpyrifos in Chinese chives) as compared with the control or non amended soil.

4. Impact of biochar characteristics

There are several studies about pyrolysis conditions and biochar properties focused on soil quality improvement, but as mentioned above, reports on the study of biochar on the effect of pesticides fate are scarce. Thus, it is difficult to assess all the factors that predict biochar properties for the increase in sorption and retention, and decrease in leaching and mobility of herbicides to avoid environmental pollution associated to the use of agrochemicals. However, some trends can be elucidated from the available data.

4.1 Type of pyrolysis process

The first condition that appears to have an influence on herbicide sorption is the type of pyrolysis process. Biochar properties can be very varied and depend on the feedstock, the conditions of the thermal process used, and the changes post-pyrolysis caused by aging or activation treatments (Laird et al., 2009). The yield of gas, liquid or solid products also depends on the thermal process conditions. The yield of biochar follows the order: slow pyrolysis > intermediate pyrolysis > fast pyrolysis > gasification. The slow pyrolysis process implies low temperature, around 400 °C, and residence times of several minutes. The temperature of the fast and intermediate pyrolysis processes is moderate, around 500 °C, and the residence time is of ~1 second and 10-20 s, respectively. In gasification the temperature is higher, around 800 °C and the residence time is longer, more than 5 minutes (Bridgwater, 2006). Although the process with lower temperature yields the higher amount of char, temperatures lower than 400 °C produces biochar with low pH, CEC, and small surface area, which make it not suitable for improving soil quality (Lehmann, 2007), or increasing pesticide sorption on soil, decreasing its dissipation and availability (Yu et al., 2006; 2009; 2010; Wang et al., 2010; Yang et al., 2010). According to Sohi et al. (2010) feedstock is an important factor to determine the function of biochar in soil, but there is no consensus about the optimal feedstock in terms of soil use and energy production. The concentration of bases in biochar depends on the mineral content of the biomass used, and the ash content of the biochar on the type of feedstock, for example, softwood produces a biochar with lower content on ash than hardwood and corn or wheat wastes (Laird et al., 2009). Woody chars are considered to be more highly condensed, more aromatic and with higher surface area as cotton straw derived chars (Yang et al., 2010).

4.2 Pyrolysis temperature

Temperature is a key factor on biochar properties, such as surface area, microporosity, and stability. Brewer et al. (2009) reported different physical and chemical properties of the chars produced by fast pyrolysis and gasification than from the chars prepared by slow pyrolysis. The process temperature determines the type of carbon present in the biochar, being reaction time less decisive. In the studies, which deal with the effect of biochar on pesticide sorption, dissipation and bioavailability in soils, the authors report better properties to the biochar produced at higher temperature, as consequence of the higher surface area and microporosity. Yu et al (2006) observed higher hysteresis and sorption irreversibility of diuron with the biochar produced at 850 °C as compared to the one produced at 450 °C. The microporosity was greater in BC850 than in BC450, and diuron molecules sorbed could be retained in the micropores or have caused a slow and prolongued sorption phase, which may have led to the apparent hysteresis due to non equilibrium process. Yang et al. (2010) considered that temperature is more important than feedstock material on the biochar effectiveness in treatment of contaminated soils. Wang et al. (2010) observed a higher sorption of the herbicide terbuthylazine in the soil amended with biochar produced at 700 °C than at 350 °C, although the feedstock was the same. In Fig. 2 we have plotted the data of Freundlich K_f parameter for some pesticides sorbed to the biochar and temperature pyrolysis. Granted there are a limited number of studies with sufficient data to compare, but the hypothesis is that higher temperature biochars appear to possess a higher capacity for sorption than the lower temperature biochars. However, since there are limited studies that present sufficient characterization data on the biochar, this conclusion requires further investigation.

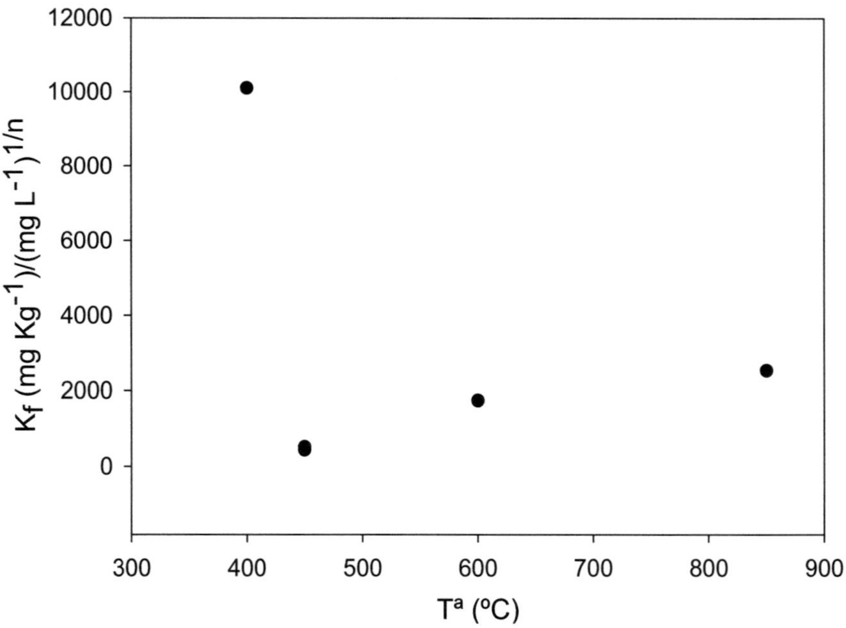

Fig. 2. Freundlich sorption parameter K_f correlated with the pyrolysis temperature (T^a). [Data of sorption on char from Zheng et al., 2010 (Table 2); Li et al., 2009, (Table 1); Aroguz, 2006 and Hameed et al., 2009].

4.3 Surface area

Fig. 3 displays the overall relationship between surface area and observed K_f sorption coefficient for various biochars. There is a relationship between surface area and sorption. However, due to the scatter in the graph, there is the suggestion that surface area is not the fundamental parameter and potentially the chemistry of the surface groups would be a better predictor variable as has been shown for activated charcoals (e.g. Bello et al., 2002).

5. Conclusions

Overall, the comparison of the impact of biochar on herbicide sorption is narrow due to the limited number of studies utilizing biochar. Furhermore, the lack of information within these studies to properly characterize the biochar material used further limits the ability to compare results. However, based on the analogous observations with non-biochar materials (Table 1), the overall conclusions are that biochar additions (Table 2) will:

- increase sorption,
- decrease dissipation rates,
- decrease leaching and movement of the herbicide in the soil, and
- reduce the bioavailability of the herbicide, which could lead to reduced efficacy.

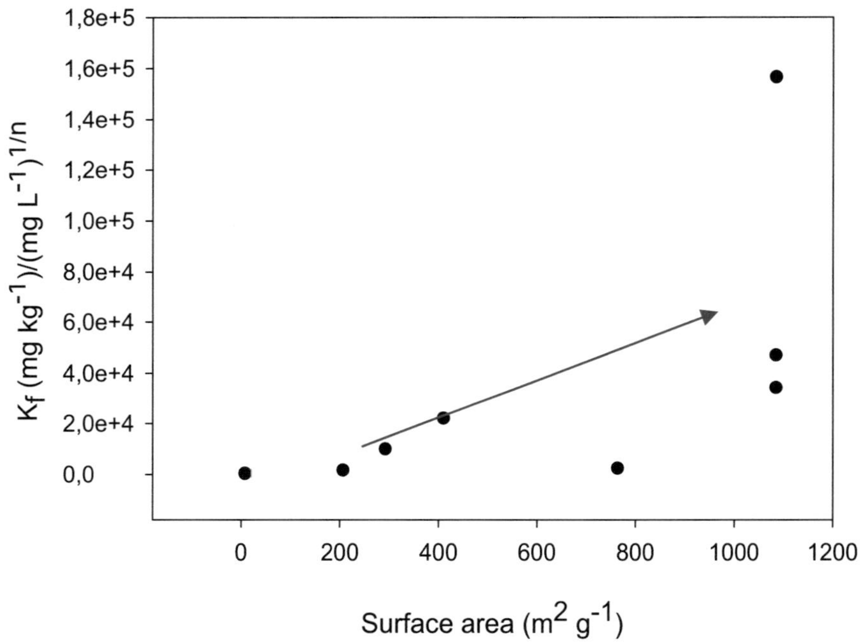

Fig. 3. Freundlich sorption parameter K_f correlated with the surface area determination of the biochar. [Data presented from Zheng et al., 2010 (Table 2); Li et al., 2009; Qiu et al., 2009b; Wang et al., 2009, (Table 1); Aroguz, 2006 and Hameed et al., 2009].

6. References

Al-Qodah, Z.; Shawaqfeh, A. T. & Lafi, W. K. (2007). Adsorption of pesticides from aqueous solution using oil shale ash. *Desalination*, Vol. 208, No. 1-3, (April 2007) page numbers (294-305), ISSN 0011-9164

Aroguz, A. Z. (2006). Kinetics and thermodynamics of adsorption of azinphosmethyl from aqueous solution onto pyrolyzed (at 600 °C) ocean peat moss (*Sphagnum sp.*).

Journal of Hazardous Materials, Vol. 135, No. 1-3, (July 2006) page numbers (100-105), ISSN 0304-3894

Asai, H.; Samson, B. K.; Stephan, H. M.; Songyikhangsuthor, K.; Homma, K.; Kiyono, Y.; Inoue, Y.; Shiraiwa, T. & Horie, T. (2009). Biochar amendment techniques for upland rice production in Northern Laos 1. Soil physical properties, leaf SPAD and grain yield. *Field Crops Research*, Vol. 111, No. 1-2, (March 2009) page numbers (81-84), ISSN 0378-4290

Ayranci, E. & Hoda, N. (2005). Adsorption kinetics and isotherms of pesticides onto activated carbon-cloth. *Chemosphere*, Vol. 60, No. 11, (September 2005) page numbers (1600-1607), ISSN 0045-6535

Azargohar, R. & Dalai A. K. (2006). Biochar as a precursor of activated carbon. *Applied Biochemistry and Biotechnology*, Vol. 131, No. 1-3, (March 2006) page numbers (762-773), ISSN 0273-2289

Bandala, E. R.; Andres-Octaviano, J.; Pastrana, P. & Torres, L. G. (2006). Removal of aldrin, dieldrin, heptachlor, and heptachlor epoxide using activated carbon and/or *Pseudomonas fluorescens* free cell cultures. *Journal of Environmental Science and Health part B*, Vol. 41, No. 5, page numbers (553-569), ISSN 0360-1234

Bansal, R. C. & Goyal, M. (2005). *Activated carbon adsorption*, CRC Press Taylor & Francis Group LCC, ISBN 0-8247-5344-5, Boca Raton, Florida, USA

Bello, G.; García, R.; Arriagada, R.; Sepúlveda-Escribano, A. & Rodríguez-Reinoso, F. (2002). Carbon molecular sieves from Eucalyptus globulus charcoal. *Microporous and mesoporous materials*, Vol. 56, No. 2, (November 2002) page numbers (139-145), ISSN 1387-1811

Betz, W. R.; Maroldo, S. G.; Wachob, G. D. & Firth, M. C. (1989). Characterization of carbon molecular sieves and activated charcoal for use in airborne contaminant sampling. *American Industrial Hygiene Association Journal*, Vol. 50, No. 4, (April, 1989) page numbers (181-187), ISSN: 1542-8117

Birk, L. A. & Roadhouse, F. E. B. (1964). Penetration of and persistence in soil of the herbicide atrazine. *Canadian Journal of Plant Science*, Vol. 44, No. 1, (January 1964), page numbers (21-27), ISSN 0008-4220

Brewer, C.; Schmidt-Rohr, K.; Satrio, J. A. & Brown, R. C. (2009). Characterization of biochar from fast pyrolysis and gasification systems. *Environmental Progress & Sustainable Energy*, Vol. 28, No. 3, (November 2009) page numbers (386–396), ISSN 1944-7442

Bridgwater, A.V. (1992). An overview of thermochemical conversion technologies, In: *Wood, Fuel for Thought*, Richards, G. E. (Ed.), page numbers (73-107), Harwell Laboratories Oxfordshire, ISBN 0705816516, Oxford, United Kingdom

Bridgwater, A. V. (2003). Renewable fuels and chemicals by thermal processing of biomass. *Chemical Engineering Journal*, Vol. 91, No. 2-3 (March 2003) page numbers (87-102), ISSN 1385-8947

Bridgwater, T. (2006). Biomass for energy. *Journal of the Science of Food and Agriculture*, Vol. 86, No. 12, (September 2006) page numbers (1755-1768), ISSN 0022-5142

Busscher, W. J.; Novak, J.M.; Evans, D. E.; Watts, D. W.; Niandou, M. A. S. & Ahmedna, M. (2010). Influence of pecan biochar on physical properties of a Norfolk loamy sand. *Soil Science*, Vol. 175, No. 1, (January 2010) page numbers (10-14), ISSN 0038-075X

Calvet, R. (1989). Adsorption of organic chemicals in soils. *Environmental Health Perspectives*, Vol. 83, (November 1989) page numbers (145-177), ISSN 0091-6765

Cao, X. & Harris, W. (2010). Properties of dairy-manure derived biochar pertinent to its potential use in remediation. *Bioresource Technology,* Vol. 101, No. 14, (July 2010) page numbers (5222-5228), ISSN 0960-8524

Cao, X.; Ma, L.; Gao, B. & Harris, W. (2009). Dairy-manure derived biochar effectively sorbs lead and atrazine. *Environmental Science & Technology,* Vol. 43, No. 9, (May 2009) page numbers (3285-3291), ISSN 0013-936X

Chan, K. Y.; Van Zwieten, L.; Meszaros, I.; Downie, A. & Joseph, S. (2007). Agronomic values of greenwaste biochar as soil amendment. *Australian Journal of Soil Research,* Vol. 45 No. 8, page numbers (629-634), ISSN 0004-9573

Chan K. Y.; Van Zwieten, L.; Meszaros, I.; Downie, A.; Downie, A. & Joseph, S. (2008). Using poultry litter biochars as soil amendments. *Australian Journal of Soil Research,* Vol. 46, No. 5, page numbers (437-444), ISSN 0004-9573

Crepeau, K. L.; Walker, G. & Winterlin, W. (1991). Use of coal to retard pesticide movement in soil. *Journal of Environmental Science and Health, part B,* Vol. 26, No. 5-6, page numbers (529-545), ISSN 0360-1234

Cui, XY.; Wang, HL.; Lou, LP.; Chen, YX.; Yu, YL.; Shi, JY.; Xu, L. & Khan, M. I. (2009). Sorption and genotoxicity of sediment-associated pentachlorophenol and pyrene influenced by crop residue ash. *Journal of Soil and Sediments,* Vol. 9, No. 6, (December 2009) page numbers (604-612), ISSN 1439-0108

Della-Rocca P. A.; Cerrella, E. G.; Bonelli, P. R. & Cukierman, A. L. (1999), Pyrolysis of hardwoods residues: on kinetics and chars characterization. *Biomass and Bioenergy,* Vol. 16, No. 1, (January 1999), page numbers (79-88), ISSN 0961-9534

Domingues, V. F.; Priolo, G.; Alves, A. C.; Cabral. M. F. & Delerue-Matos, C. (2007). Adsorption behavior of α-cypermethrin on cork and activated carbon. *Journal of Environmental Science and Health, part B,* Vol. 42, No. 6, (August 2007) page numbers (649-654), ISSN 0360-1234

Elmquist, M.; Gustafsson, O. & Andersson, P. (2004). Quantification of sedimentary black carbon using the chemothermal oxidation method: An evaluation of ex situ pretreatments and standard additions approaches. *Limnology and Oceanography-Methods,* Vol. 2, No. 2, (December 2004) page numbers (417- 427), ISSN 1541-5856

Elmquist, M.; Cornelissen, G.; Kukulska, Z. & Gustafsson, O. (2006). Distinct oxidative stabilities of char versus soot black carbon: Implications for quantification and environmental recalcitrance. *Global Biogeochemical Cycles,* Vol. 20, No. 2, (May 2006) article number GB2009, ISSN 0886-6236

Forbes, M. S.; Raison, R. J. & Skjemstad, J. O. (2006). Formation, transformation and transport of black carbon (charcoal) in terrestrial and aquatic ecosystems. *Science of the Total Environment,* Vol. 370, No. 1, (October 2006) page numbers (190-206), ISSN 0048-9697

Gaskin, J. W.; Speir, R. A.; Harris, K.; Das, K. C.; Lee, R. D.; Morris, L. A. & Fisher D. S. (2010). Effect of peanut hull and pine chip biochar on soil nutrients, corn nutrient status, and yield. *Agronomy Journal,* Vol. 102, No. 2, (March-April 2010) page numbers (623-633), ISSN 0002-1962

Gevao, B.; Semple, K. T. & Jones, K. C. (2000). Bound pesticide residues in soils: a review. *Environmental Pollution,* Vol. 108, No. 1, (April 2000), page numbers (3-14), ISSN: 0269-7491

Ghetti, P. (1986). DTG combustion behavior of coal-correlations with proximate and ultimate anlaysis data. *Fuel,* Vol. 65, No. 5, (May 1986) page numbers (636-639), ISSN 0016-2361

Glaser, B.; Lehmann, J. & Zech, W. (2002). Ameliorating physical and chemical properties of highly weathered soils in the tropics with charcoal-a review. *Biology and fertility of soils,* Vol. 33, No. 4, (June 2002) page numbers (219-230), ISSN 0178-2762

Goldberg, E. D. (1985). *Black carbon in the environment: properties and distribution,* John Wiley & Sons, ISBN 0471819794, New York, USA

Guo, L.; Bicki, T. J.; Felsot, A. S. & Hinesly, T. D. (1991). Phytotoxicity of atrazine and alachlor in soil amended with sludge, manure and activated carbon. *Journal of Environmental Science and Health part B,* Vol. 26, No. 5-6, page numbers (513-527), ISSN 0360-1234

Guo, L.; Bicki, T. J.; Felsot, A. S. & Hinesly, T. D. (1993). Sorption and movement of alachlor in soil modified by carbon-rich wastes. *Journal of Environmental Quality,* Vol. 22, No. 1, (January-March 1993) page numbers (186-194), ISSN 0047-2425

Gupta, V. K.; Ali, I.; Suhas, & Saini, V. K. (2006). Adsorption of 2,4-D and carbofuran pesticides using fertilizer and steel industry wastes. *Journal of Colloid and Interface Science,* Vol. 299, No. 2, (July 2006) page numbers (556-563), ISSN 0021-9797

Gustafsson, O.; Haghseta, F.; Chan, C.; Macfarlane, J. & Gschwend, P. M. (1997). Quantification of the dilute sedimentary soot phase: Implications for PAH speciation and bioavailability. *Environmental Science & Technology,* Vol. 31, No. 1, (January 1997) page numbers (203–209), ISSN 0013-936X

Gustafsson, O.; Bucheli, T. D.; Kukulska, Z.; Andersson, M.; Largeau, C.; Rouzaud, J. N.; Reddy, C. M. & Eglinton, T. I. (2001). Evaluation of a protocol for the quantification of black carbon in sediments. *Global Biogeochemical Cycles,* Vol. 15, No. 4, (December 2001) page numbers (881–890), ISSN 0886-6236

Hale, S. E.; Tomaszewski, J. E.; Luthy, R. G. & Werner, D. (2009). Sorption of dichlorodiphenyltrichloroethane (DDT) and its metabolites by activated carbon in clean water and sediment slurries. *Water Research,* Vol. 43, No. 7, (September 2009) page numbers (4336-4346), ISSN 0043-1354

Hameed, B. H.; Salman, J. M. & Ahmad, A. L. (2009). Adsorption isotherm and kinetic modeling of 2,4-D pesticide on activated carbon derived from date stones. *Journal of Hazardous Materials,* Vol. 163, No. 1, (April 2009) page numbers (121-126), ISSN 0304-3894

Hargrave, G.; Pourkashanian & Williams, A. (1986). The combustion and gasification of coke and coal tars. *Twenty-first Symposium (International) on Combustion/The Combustion Institute,* Vol. 21, No. 1 (January 1988), page numbers (221-230), ISSN 1540-7489

Hassler, J. W. (1974), *Purification with activated carbon, industrial, commercial and environmental,* Chemical Publishing Company, ISBN 0820603732 , New York

Hedges, J. I.; Eglinton, G.; Hatcher, P. G; Kirchman, D. L.; Arnosti, C.; Derenne, S.; Evershed, R. P.; Kogel-Knabner, I.; de Leeuw, J. W.; Littke, R.; Michaelis, W. & Rullkotter, J. (2000). The molecularly-uncharacterized component of nonliving organic matter in natural environments. *Organic Geochemistry,* Vol. 31, No. 10, page numbers (945-958), ISSN 0146-6380

Hilber, I.; Wyss, G. S.; Mäder, P.; Bucheli, T. D.; Meier, I.; Vogt, L. & Schulin, R. (2009). Influence of activated charcoal amendment to contaminated soil on dieldrin and nutrient uptake by cucumbers. *Environmental Pollution,* Vol. 157, No. 8-9, (August-September 2009) page numbers (2224-2230), ISSN 0269-7491

Hiller, E.; Fargasova, A.; Zemanova, L. & Bartal, M. (2007). Influence of wheat ash on the MCPA immobilization in agricultural soils. *Bulletin of Environmental Contamination and Toxicology,* Vol. 79, No. 4, (October 2007) page numbers (478-481), ISSN 0007-4861

Hiller, E.; Fargasova, A.; Zemanova, L. & Bartal, M. (2008). Influence of wheat ash on the MCPA immobilization in agricultural soils. *Bulletin of Environmental Contamination and Toxicology,* Vol. 81, No. 3, (September 2008) page numbers (285-288), ISSN 0007-4861

Hiller, E.; Bartal, M; Milicka, J. & Cernansky, S. (2009). Environmental fate of the herbicide MCPA in two soils as affected by the presence of wheat ash. *Water Air and Soil Pollution,* Vol. 197, No. 1-4, (February 2009) page numbers (395-402). ISSN 0049-6979

Hofrichter, M.; Bublitz, F. & Fritsche, W. (1997). Fungal attack on coal: I. Modification of hard coal by fungi. *Fuel Processing Technology,* Vol. 52, No. 1-3, (November 1997) page numbers (43-53), ISSN 0378-3820

Hu, JY.; Aizawa, T.; Ookubo, Y.; Morita, T. & Magara, Y. (1998). Adsorptive characteristics of ionogenic aromatic pesticides in water powdered activated carbon. *Water Research,* Vol. 32, No. 9, (September 1998) page numbers (2593-2600), ISSN 0043-1354

Jablonowski, N. D.; Hamacher, G.; Accinelli, C.; Berns, A. E.; Meng, F.; Martinazzo, R.; Langen, U.; Walraf, C.; Linden, A. & Burauel, P. (2010). Influence of biochar and activated char amendment on the biodegradation of ^{14}C-atrazine in atrazine adapted soils from Belgium and Brazil. In Setac Europe 20th Annual Meeting, Seville, May 2010

Jones, T. P.; Chaloner, W. G. & Kuhlbusch, T. A. J. (1997). Proposed bio-geological and chemical based terminology for fire-altered plant matter, In: *Sediment records of biomass burning and global change,* Clark, J. S.; Cachier, H.; Goldammer, J. G. & Stocks, B. (Eds.) NATO ASI Series, Serie I: Global Environmental Change. Vol. 51, page numbers (9-22), Publisher Springer-Verlag, ISBN 3-540-62434-1, Berlin, Germany

Karanfil, T. & Kilduff, J. E. (1999). Role of Granular Activated Carbon Surface Chemistry on the Adsorption of Organic Compounds. 1. Priority Pollutants. *Environmental Science & Technology,* Vol. 33, No. 18, (September 1999) page numbers (3217-3224), ISSN 0013-936X

Keerthinarayana, S. & Bandyopadhyay, M. (1997). Diffusion mechanism of lindane into wood charcoal. *Journal of Environmental Science and Health part B,* Vol. 32, No. 5, page numbers (673-700), ISSN 0360-1234

Konstantinou, I. K. & Albanis, T. A. (2000). Adsorption–desorption studies of selected herbicides in soil–fly ash mixtures. *Journal of Agricultural and Food Chemistry,* Vol. 48, No. 10, (September 2000) page numbers (4780-4790), ISSN 0021-8561

Kookana, R. S.; Baskaran, S. & Naidu, R. (1998). Pesticide fate and behaviour in Australian soils in relation to contamination and management of soil and water: a review. *Australian Journal of Soil Research,* Vol. 36, No. 5, (September 1998) page numbers (715-764), ISSN 0004-9573

Lafrance, P.; Villeneuve, J. P.; Mazet, M. ; Ayele, J. & Fabre, B. (1991). Organic compounds adsorption onto activated carbon : the effect of association between dissolved humic substances and pesticides. *Environmental Pollution,* Vol. 72., No. 4, page numbers (331-344), ISSN 0269-7491

Laird, D. A. & Kosikinen, W. C. (2008). Triazine soil interactions. In: *The triazine herbicides: 50 years revolutionizing agriculture,* LeBaron, H. M.; McFarland, J. E. & Burnside, O. (Eds.), page numbers (275-299), Elsevier, ISBN 978-0-444-51167-6, San Diego, CA

Laird, D. A.; Brown, R. C.; Amonette, J. E. & Lehmann, J. (2009). Review of the pyrolysis platform for coproducing bio-oil and biochar. *Biofuels, Bioproducts and Biorefining,* Vol. 3, No. 5, (September-October 2009) page numbers (547-562), ISSN 1932-104X

Lehmann, J.; Gaunt, J. & Rondon, M. (2006). Biochar sequestration in terrestrial ecosystems - a review. *Mitigation and Adaptation Strategies for Global Change,* Vol. 11 ,No. 2, (March 2006) page numbers (403-427), ISSN 1381-2386

Lehmann, J. (2007). Bio-energy in the black. *Frontiers in ecology and the environment,* Vol. 5, No. 7, (September 2007) page numbers (381-387), ISSN 1540-9295

Li, R.; Wen, B.; Zhang, S.; Pei, Z. & Shan, X. (2009). Influence of organic amendments on the sorption of pentachlorophenol on soils. *Journal of Environmental Sciences,* Vol 21, No. 4, page numbers (474-480), ISSN 1001-0742

Loganathan, V. A.; Feng, Y.; Sheng, G. D. & Clement, T. P. (2009). Crop-residue derived char influences sorption, desorption and bioavailability of atrazine in soils. *Soil Science Society of America Journal,* Vol. 73, No. 3, (May-June 2009) page numbers 967-974, ISSN 0361-5995

Marsh, H. & Rodríguez-Reinoso, F. (2006). *Activated carbon,* Elsevier Ltd, ISBN 0-08-044463-6, Oxford, Great Britain

Matsui, Y.; Fukuda, Y.; Inoue, T. & Matsushita, T. (2003). Effect of natural organic matter on powdered activated carbon adsorption of trace contaminants: characteristics and mechanism of competitive adsorption. *Water Research,* Vol. 37, No. 18, (November 2003) page numbers (4413-4424), ISSN 0043-1354

Mukherjee, I. (2009). Effect of organic amendments on degradation of atrazine. *Bulletin of Environmental Contamination and Toxicology,* Vol. 83, No. 6, (December 2009) page numbers (832-835), ISSN 0007-4861

Novak, J. M.; Busscher, W. J.; Laird, D. L.; Ahmedna, M.; Watts, D. W. & Niandou, M. A. S. (2009). Impact of biochar amendment on fertility of a southeastern coastal plain soil. *Soil Science,* Vol. 174, No. 2, (February 2009) page numbers (105-112), ISSN 0038-075X

Novak, J. M.; Busscher, W. J.; Watts, D. W.; Laird, D. A.; Ahmedna M. A. & Niandou, M. A. S. (2010). Short-term CO_2 mineralization after additions of biochar and switchgrass to a Typic Kandiudult. *Geoderma,* Vol. 154, No. 3-4, (January 2010) page numbers (281-288), ISSN 0016-7061

Otowa, T.; Nojima, Y. & Miyazaki, T. (1997). Development of KOH activated high surface area carbon and its application to drinking water purification. *Carbon,* Vol. 35, No. 9, page numbers (1315-1319), ISSN 0008-6223

Özçimen, D. & Karaosmanoğlu, F. (2004). Production and characterization of bio-oil and biochar from rapeseed cake. *Renewable Energy,* Vol 29, No. 5, (April 2004) page numbers (779-787), ISSN 0960-1841

Piccolo, A.; Pietramellara, G. & Mbagwu; J. S. C. (1996). Effects of coal derived humic substances on water retention and structural stability of Mediterranean soils. *Soil Use and Management,* Vol. 12, No. 4, (December, 1996) page numbers (209-213), ISSN 0266-0032

Pradhan, B. K. & Sandle, N. K., 1999. Effect of different oxidizing agent treatments on the surface properties of activated carbons. *Carbon,* Vol. 37, No. 8, (January 1999), page numbers (1323–1332), ISSN 0008-6223

Qiu, Y.; Pang, H.; Zhou, Z.; Zhang, P.; Feng, Y. & Sheng D. G. (2009a). Competitive biodegradation of dichlobenil and atrazine coexisting in soil amended with a char and citrate. *Environmental Pollution,* Vol. 157, No. 11, (November 2009) page numbers (2964-2969), ISSN 1873-6424

Qiu, Y.; Xiao, X.; Cheng, H.; Zhou, Z. & Sheng, G. D. (2009b). Influence of environmental factors on pesticide adsorption by black carbon: pH and model dissolved organic matter. *Environmental Science & Technology,* Vol. 43, No. 13, (July 2009) page numbers (4973-4978), ISSN 0013-936X

Scott, A. C. (2010). Charcoal recognition, taphonomy and uses in palaeoenvironmental analysis. *Palaeogeography, Palaeoclimatology, Palaeoecology,* Vol. 291, No. 1-2, Special Issue SI (May 2010) page numbers (11-39), ISSN 0031-0182

Scott, A. C. & Damblon, F. (2010). Charcoal: taphonomy and significance in geology, botany and archaeology. *Palaeogeography, Palaeoclimatology, Palaeoecology,* Vol. 291, No. 1-2, Special Issue SI (May 2010) page numbers (1-10), ISSN 0031-0182

Sheng, G.; Yang, Y.; Huang, M. & Yang, K. (2005). Influence of pH on pesticide sorption by soil containing wheat residue-derived char. *Environmental Pollution,* Vol. 134, No. 3, (April 2005) page numbers (457-463), ISSN 0269-7491

Singh, B. P.; Hatton, B. J.; Singh, B.; Cowie, A. L. & Kathuria, A. (2010). Influence of biochars on nitrous oxide emission and nitrogen leaching from two contrasting soils. *Journal of Environmental Quality,* Vol. 39, No. 4, (July-August 2010) page numbers (1224-1235), ISSN 0047-2425

Sobek, A.; Stamm, N. & Bucheli, T. D. (2009). Sorption of phenyl urea herbicides to black carbon. *Environmental Science & Technology,* Vol. 43, No. 21, (November 2009) page numbers (8147-8152), ISSN 0013-936X

Sohi, S.P.; Krull, E.; Lopez-Capel, E. & Bol, R. (2010). A review of biochar and its use and function in soil. In: *Advances in Agronomy,* page numbers (47-82), Publisher Elsevier Academic Press Inc., ISSN 0065-2213, San Diego, CA-92101-4495, USA

Sopeña, F.; Semple, K.; Sohi, S. & Bending, G. D. (2010). Isoproturon bioavailability in soil amended with biocharcoal. In Setac Europe 20th Annual Meeting, Seville, May 2010.

Spokas, K.; Koskinen, W. C.; Baker, J. M. & Reicosky, D. C. (2009). Impacts of woodchip biochar additions on greenhouse gas production and sorption/degradation of two herbicides in a Minnesota soil. *Chemosphere,* Vol. 77, No. 4, (October 2009) page numbers (574-581), ISSN 0045-6535

Spokas, K. A. & Reicosky, D. C. (2009). Impacts of sixteen different biochars on soil greenhouse gas production. *Annals of Environmental Science,* Vol. 3, (November 2009) page numbers (179-193), ISSN 1939-2621

Spokas, K. A. (2010). Review of the stability of biochar in soils: Predictability of O:C molar ratios. *Carbon Management.* accepted (October 2010) In Press.

Steinbeiss, S.; Gleixner, G. & Antonietti, M. (2009). Effect of biochar amendment on soil carbon balance and soil microbial activity. *Soil Biology & Biochemistry,* Vol. 41, No. 6, (June 2009) page numbers (1301-1310), ISSN 0038-0717

Sudhakar, Y. & Dikshit, A. K. (1999). Kinetics of endosulfan sorption on to wood charcoal. *Journal of Environmental Science and Health part B,* Vol. 34, No. 4, page numbers (587-615), ISSN 0360-1234

Sudhakar, Y. & Dikshit, A. K. (2010). Competitive sorption of pesticides onto treated wood charcoal and the effect of organic and inorganic parameters on adsorption capacity. *Journal of Environmental Engineering,* doi: 10.1061(ASCE)EE. 1943-7870.0000246

Szymanski, G. S., Karpinski, Z.; Biniak, S. & Swiatkowski, A. (2002). The effect of the gradual thermal decomposition of surface oxygen species on the chemical and catalytic properties of oxidized activated carbon. *Carbon,* Vol. 40, No. 14, (November, 2002) page numbers (2627-2639), ISSN 0008-6223

Tessmer, C. H.; Vidic, R. D. & Uranowski, L. J. (1997). Impact of oxygen-containing surface functional groups on activated carbon adsorption of phenols. *Environmental Science & Technology*, Vol. 31, No. 7, (July 1997) page numbers (1872-1878), ISSN 0013-936X

Tomaszewski, J. E.; Werner, D. & Luthy, R. G. (2007). Activated carbon amendment as a treatment for residual DDT in sediment from a superfund site in San Francisco Bay, Richmond, California, USA. *Environmental Toxicology and Chemistry*, Vol. 26, No. 10, (October 2007) page numbers (2143-2150), ISSN 0730-7268

Toth, J. & Milham, P. J. (1975). Activated-carbon and ash-carbon effects on the adsorption and phytotoxicity of diuron. *Weed Research*, Vol. 15, No. 3, page numbers (171-176), ISSN 0043-1737

Van Zwieten, L.; Kimber, S.; Morris, S.; Chan, K. Y.; Downie, A.; Rust, J.; Joseph, S. & Cowie, A. (2010). Effects of biochar from slow pyrolysis of papermill waste on agronomic performance and soil fertility. *Plant and Soil*, Vol. 327, No. 1-2, (February 2010) page numbers (235-246), ISSN 0032-079X

Wang, H.; Lin, K.; Hou, Z.; Richardson, B. & Gan, J. (2010). Sorption of the herbicide terbuthylazine in two New Zealand forest soils amended with biosolids and biochars. *Journal of Soil and Sediments*, Vol. 10, No. 2, (March 2010) page numbers (283-289), ISSN 1439-0108

Wang, X. & Xing, B. S. (2007). Sorption of organic contaminants by biopolymer-derived chars. *Environmental Science & Technology*, Vol. 41, No. 24 (November 2007) page numbers (8342-8348), ISSN 0013-936X

Wang, Y. S.; Shan, X. Q.; Feng, M. H.; Chen, G. C.; Pei, Z. G.; Wen, B.; Liu, T.; Xie, YN. & Owens, G. (2009). Effects of copper, lead, and cadmium on the sorption of 2,4,6-trichlorophenol onto and desorption from wheat ash and two commercial humic acids. *Environmental Science & Technology*, Vol. 43, No. 15, (August 2009), page numbers (5726-5731), ISSN 0013-936X

Warnock, D. D.; Lehmann, J.; Kuyper, T. W. & Rillig, M. C. (2007). Mycorrhizal responses to biochar in soil-concepts and mechanisms. *Plant and Soil*, Vol. 300, No. 1-2, (November 2007) page numbers (9-20), ISSN 0032-079X

Wauchope, R. D.; Yeh, S.; Linders, J.B.H.J.; Klokowski, R.; Tanaka, K.; Rubin, B.; Katayama, A.; Kördel, W.; Gerstl, Z.; Lane, M. & Unsworth, J. B. (2002). Review: pesticide soil sorption parameters: theory, measurement, uses, limitations and reliability. *Pest Management Science*, Vol. 58, No. 5, (May 2002) page numbers (419-445), ISSN 1526-498X

Weber, J. B.; Weed, S. B. & Ward, T. M. (1969). Adsorption of s-triazines by soil organic matter. *Weed Science*, Vol. 17, No. 4, (October 1969) page numbers (417-421), ISSN 0043-1745

Xu, C.; Liu, W. & Sheng, G. D. (2008). Burned rice straw reduces the availability of clomazone to barnyardgrass. *Science of the Total Environment*, Vol 392, No. 2-3, (March 2008) page numbers (284-289), ISSN 0048-9697

Yalcin, N. & Sevinc, V. (2000), Studies of the surface area and porosity of activated carbons prepared from rice husks, *Carbon*, Vol. 38, No. 14, (November 2000) page numbers (1943-1945), ISSN 0008-6223

Yamane, V. K. & Green, R. E. (1972). Adsorption of ametryne and atrazine on an oxisol, montmorillonite, and charchoal in relation to pH and solubility effects. *Soil Science Society of America Journal*, Vol. 36, No. 1, (January 1972) page numbers (58-64), ISSN 0361-5995

Yang, XB.; Ying GG.; Peng, PA.; Wang, L.; Zhao, JL.; Zhang, LJ. Yuan, P. & He, HP. (2010). Influence of biochars on plant uptake and dissipation of two pesticides in an

agricultural soil. *Journal of Agricultural and Food Chemistry,* Vol. 58, No. 13, (July 2010) page numbers (7915-7921), ISSN 0021-8561
Yang, Y. & Sheng, G. (2003a). Enhanced pesticide sorption by soils containing particulate matter from crop residue burns. *Environmental Science & Technology,* Vol. 37, No. 16, (August 2003) page numbers (3635-3639), ISSN 0013-936X
Yang, Y. & Sheng, G. (2003b). Pesticide adsorptivity of aged particulate matter arising from crop residue burns. *Journal of Agricultural and Food Chemistry,* Vol. 51, No. 17, (August 2003) page numbers (5047-5051), ISSN 0021-8561
Yang, Y.; Chun, Y.; Sheng, G. & Huang, M. (2004). pH-dependence of pesticide adsorption by wheat-residue-derived black carbon. *Langmuir,* Vol. 20, No. 16, (August 2004) page numbers (6736-6741), ISSN 0743-7463
Yang, Y.; Sheng, G. & Huang, M. S. (2006). Bioavailability of diuron in soil containing wheat-straw-derived char. *Science of the Total Environment,* Vol. 354, No. 2-3, (February 2006) page numbers (170-178), ISSN 0048-9697
Yu, XY.; Ying, GG. & Kookana, R. S. (2006). Sorption and desorption behaviors of diuron in soils amended with charcoal. *Journal of Agricultural and Food Chemistry,* Vol. 54, No. 22, (November 2006) page numbers (8545-8550), ISSN 0021-8561
Yu, XY.; Ying, GG. & Kookana, R. S. (2009). Reduced plant uptake of pesticides with biochar additions to soil. *Chemosphere,* Vol. 76, No. 5, (July 2009) page numbers (665-671), ISSN 0045-6535
Yu, XY.; Pan, LG.; Ying, GG. & Kookana, R. S. (2010). Enhanced and irreversible sorption of pesticide pyrimethanil by soil amended with biochars. *Journal of Environmental Sciences,* Vol. 22, No. 4, page numbers (615-620), ISSN 1001-0742
Zhang, P.; Sheng, G.; Wolf, D. C. & Feng, Y. (2004). Reduced biodegradation of benzonitrile in soil containing wheat-residue derived ash. *Journal of Environmental Quality,* Vol. 33, No. 3, (May-June 2004) page numbers (868-872), ISSN 0047-2425
Zhang, P.; Sheng, G.; Feng, Y. & Miller, D. M. (2005). Role of wheat-residue-derived char in the biodegradation of benzonitrile in soil: nutritional stimulation versus adsorptive inhibition. *Environmental Science & Technology,* Vol. 39, No. 14, (July 2005) page numbers (5442-5448), ISSN 0013-936X
Zhang, P.; Sheng, G.; Feng, Y. & Miller, D. M. (2006). Predominance of char sorption over substrate concentration and soil pH in influencing biodegradation of benzonitrile. *Biodegradation,* Vol. 17, No.1, (February 2006) page numbers 1-8, ISSN 0923-9820
Zheng, W.; Guo, M.; Chow, T.; Bennet, D. N. & Rajagopalan, N. (2010). Sorption properties of greenwaste biochar for two triazine pesticides. *Journal of Hazardous Materials,* Vol. 181, No. 1-3, (September 2010) page numbers (121-126), ISSN 0304-3894
Zimmerman, A. R. (2010). Abiotic and microbial oxidation of laboratory-produced black carbon (biochar). *Environmental Science & Technology,* Vol. 44, No. 4 (February 2010) page numbers (1295-1301), ISSN 0013-936X

Part 4

Herbicides in Aquatic Ecosystems

16

Effects of Herbicide Glyphosate and Glyphosate-Based Formulations on Aquatic Ecosystems

Gonzalo Luis Pérez[1], María Solange Vera[2] and Leandro Andrés Miranda[1]
[1]IIB-INTECH UNSAM CONICET
[2]UBA CONICET
Argentina

1. Introduction

Public awareness of worldwide increase herbicides use and their adverse effects on ecosystems has been growing over the past decades. Herbicides may reach water bodies via agricultural runoff and leaching processes, as well as by direct applications to control noxious aquatic weeds. Once in the aquatic ecosystems, herbicides may reduce environmental quality and influence essential ecosystem functioning by reducing species diversity and community structures, modifying food chains, changing patterns of energy flow and nutrient cycling and changing the stability and resilience of ecosystems. The aim of this chapter is to provide a general notion of the current knowledge concerning the direct and indirect effects of glyphosate and commercial formulations of glyphosate on aquatic ecosystems. Glyphosate based products are the leading post-emergent, systemic and non-selective herbicides for the control of annual and perennial weeds in the world. Here, we present a revision of their toxicity to non-target species of algae, aquatic plants, protozoa, crustaceans, molluscs, fish and amphibians. In addition, we describe the importance of each group of organisms in the functioning and health of aquatic ecosystems. With this information, a conceptual framework can be developed contributing to enhance our attention and concern about human impacts on ecosystems.

2. The scenario where glyphosate appeared on stage

The transition from biologically based to intensive-chemical based agricultural production systems advanced in North America and Europe soon after World War II. This change was supported by growing availability of inorganic fertilizers and organically synthesized pesticides. Afterwards, this type of agriculture has been adopted by other major crop production areas throughout the world during the 1960s and 1970s. The intensive cropping systems are characterized as large-scale production enterprises that utilize high inputs of chemical fertilizers and pesticides. Little emphasis is given to managing soil organic matter through use of traditional crop rotations, cover crops, or organic soil amendments that are central to maintaining the biological activity and allowing the long-term preservation of agroecological systems in biologically based cropping systems (Yamada et al., 2009). One of the most significant inputs necessary for successful intensive crop production are herbicides for management of the variety of weed infestations especially encountered in row cropping systems. This technology was rapidly adopted because most weeds could be controlled when matched with selective herbicides, which were compatible with the crop, and was considered more cost-effective than cultural methods of weed management. In this scenario the herbicide glyphosate appeared on stage.

Glyphosate under the trade name Roundup® was introduced in the marked by Monsanto Company during the 1970s. It was initially registered as a broad-spectrum, non selective, systemic herbicide for certain non-crop and plantation crop uses (fallowed fields, orchards, vineyards and timber plantations) and for the control of annual and perennial weeds before the emergence of agronomic crops (Folmar et al., 1979; Woodburn, 2000). The development of minimum and no-tillage cultivation systems (zero-till) for row-cropping systems greatly expanded the use of herbicides, such as glyphosate, as it became standard practice to apply herbicides to growing weeds in fields prior to planting. This "burndown" application eliminated the need for traditional tillage (such as plow tillage) and allowed farmers to plant crop seeds directly into soil beneath a mulch of dead plant residues. The no-till practice was rapidly adopted around the world and really booming in some countries of South America like Brazil, Argentina, Paraguay and Uruguay (Altieri & Pengue, 2006). The reasons for the rapid growth of this practice are manifold, but the most important aspects are mainly economical (less work for field preparation, few expenses on fuel and machinery and higher profits). Important ecological aspects have been also pointed out. Non-till practice improves soil quality avoiding organic matter lost (Bayer et al., 2006) and water evaporation, despite of an increment in the use of herbicides. In addition, this cultivation system protects soil from erosion. For example in the southern of Brazil, no-till practice was adopted to reduce extensive soil erosion resulting from intensive row-cropping (Bolliger et al., 2006).

However, glyphosate became the most widely used herbicide worldwide with the introduction of genetically modified (GM) glyphosate-resistant (GR) crops (Woodburn, 2000). Monsanto's glyphosate-tolerant Roundup Ready (RR) soybean was the first GR crops to be commercialized (Dill et al., 2008). In 1996, RR soybean was commercially available for the first time in the USA. These crops greatly improved conventional farmers' ability to control weeds, since glyphosate could be applied before seeding and sprayed several times during growth without damage the crop. Nowadays, glyphosate has established itself as the leading herbicide for the control of annual, perennial weeds and volunteer crops in a wide range of different situations (Woodburn, 2000). The arrival of GR soybean was followed by GR cotton, maize, canola, alfalfa and sugarbeet (Dill et al., 2008). These transgenic solutions (GR seeds + glyphosate) lead a sharp increase of worldwide areas under GR crops with concomitant increase of glyphosate use. The worldwide GR hectares planted during 1998 to 2008, increased from about 15 millions to more than 130 millions (Dill et al., 2008; James, 2008). Under these circumstances, only in USA, glyphosate usage increased from 3 10^6 kg of a.i. (active ingredient) in 1987 to more than 54 10^6 kg of a.i. in 2007 (Fig. 1). The two other countries with large areas under GR crops are Argentina and Brazil, however data set concerning glyphosate use in these countries are scarce.

The reported substantial increase in the global use of glyphosate has been also related with other items like herbicide price cuts and aggressive marketing, as well as the increased reliance on herbicides for weed control (Pengue, 2005). The latter issue is represented in the occurrence of weed population shifts toward less sensitive species and the evolution of

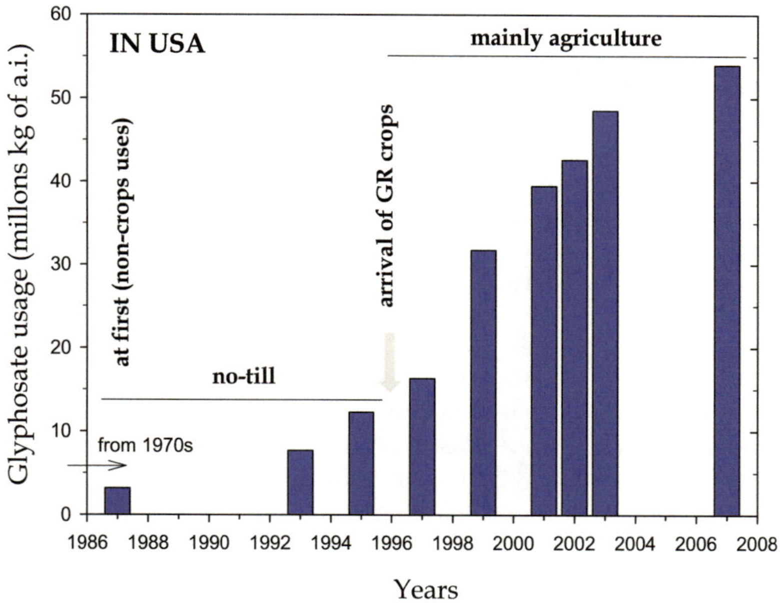

Fig. 1. Evolution of glyphosate usage in USA. Sources: USEPA, USDA.

herbicide-resistant weed populations. Glyphosate has been used worldwide since 1974 and, despite its widespread and long-term use, no case of evolved resistance to glyphosate under field conditions had been identified by 1993 (Holt et al., 1993). However, in 1996 the first case of weed resistance to glyphosate was documented in two accessions of the rigid ryegrass *Lolium rigidum,* from an orchard in Australia (Powles et al., 1998; Pratley et al., 1999). Since then, an increasing number of cases of glyphosate resistant biotypes have been reported. Currently, 14 GR weeds have been documented worldwide (Van Gessel, 2001; Pérez & Kogan, 2002; Powles, 2008; Binimelis et al., 2009; among others). Consequently, the average of glyphosate application per Ha showed a marked global increase associated with the appearance of a growing number of tolerant or resistant weeds. Bonny (2008) pointed out that the amount of glyphosate spread over the total US soybean land raised from less than 0.1 Kg/Ha in 1990 to more than 1.4 kg/Ha in 2006. Higher application rates (up to 5.6 a.i. kg/Ha) have been reported by Giesy et al. (2000). Regarding the use of other herbicides, at first the rapid growth in the use of glyphosate was accompanied by a decrease in the consumption of other former herbicides. However, for example in Argentina the consumption of the herbicides atrazine and 2.4-D, have risen again during the growing seasons of (2005-2006) (Binimelis et al., 2009). These observations coincide with Bonny (2008) who concluded that the total amount of herbicides applied per Ha in USA decreased initially between 1996 and 2001, but tended to rise afterwards.

3. Glyphosate (the molecule)

3.1 Chemistry

The chemical (technical-grade) name of glyphosate is N-(phosphonomethyl) glycine (IUPAC), an acid that belongs to chemical group of Phosphonoglycine or more generic: Organophosphonate herbicides (Fig. 2). Its main degradation product is the metabolite aminomethyl phosphonic acid (AMPA).

Glyphosate is an aminophosphonic analogue of the natural amino acid glycine and the name is a contraction of *glycine*, *phos-*, and *-ate*. The molecule has several dissociable hydrogens, especially the first hydrogen of the phosphate group. Technical-grade glyphosate has relatively low solubility in water (1.2 % at 25 ° C), and is insoluble in other solvents. Strong intermolecular hydrogen bonds stabilize the crystal lattice, causing the low water solubility. Various salts of glyphosate have much higher solubility, and do not lose any of the herbicidal properties of the parent compound (Franz, 1985). Glyphosate is commonly formulated in its form of isopropylamine salt (IPA salt), though other related chemical form are also commercialized. Glyphosate concentration is commonly expressed as mg a.i./L or mg a.e./L, where: a.e. (acid equivalents). Glyphosate is an unusual herbicide, in that essentially no structurally related compounds show any herbicidal activity (Hollander & Amrhein, 1980; Franz, 1985), whith the exception of glyphosine, which has reduced herbicidal effects but shows some interesting plant growth regulatory effects, such as enhancing ripening of sugar cane (Franz, 1985). The herbicidal properties of glyphosate were reported in 1971 (Baird et al., 1971). The compound was found during a study of the herbicidal effects of tertiary aminomethylphosphonic acids derived from various primary and secondary amines (Moedritzer & Irani, 1966). Only two of the compounds produced showed some herbicidal activity, but both had very low unit activities. Attempts to find other tertiary aminomethylphosphonic acids with improved herbicidal activity failed. As a last resort, it was suggested that degradation of the two compounds might give rise to a common, active metabolite (contrary to the general trend that metabolism reduces toxicity). Glyphosate was among the possible metabolites of the two compounds, and was found to have extremely high herbicidal activity (Franz, 1985).

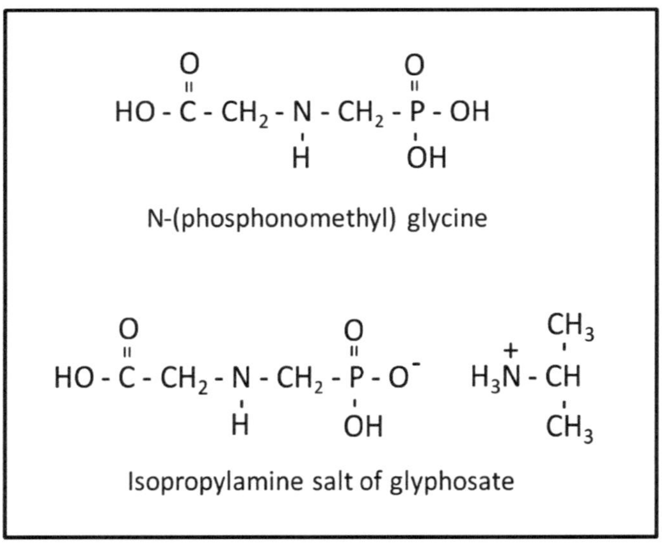

Fig. 2. Glyphosate molecule (as an acid and IPA salt).

3.2 Mode of action and biochemistry

Glyphosate is a systemic herbicide that is phloem mobile and is readily translocated throughout the plant. From the leaf surface, glyphosate molecules are absorbed into the plant cells were they are symplastically translocated to the meristems of growing plants

(Laerke, 1995). Unlike many contact herbicides, phytotoxic symptoms of glyphosate injury often develop slowly. Visible effects on most annual weeds occur within two to four days and may not occur for 7 days or more on most perennial weeds. Visual symptoms are a gradual wilting and yellowing of the plant which advances to complete browning and finally with the total deterioration and death of the plant.

Although glyphosate may ultimately perturb a variety of biochemical processes including protein synthesis, nucleic acid synthesis, photosynthesis and respiration, the primary mode of action of glyphosate was localized to the shikimate pathway of aromatic amino acid biosynthesis, a pathway that links primary and secondary metabolism. Its mode of action is the competitive inhibition of the enzyme 5-enolpyruvylshikimate 3-phosphate (EPSP) synthase, a chloroplast-localized enzyme in the shikimate pathway (Steinrücken & Amrhein, 1980; Bode et al., 1984a; Rubin et al., 1984). This inhibition prevents the production of chorismate, which is the last common precursor in the biosynthesis of numerous aromatic compounds in bacteria, fungi and plants. Essential aromatic amino acids are used by plants in protein synthesis and to produce many secondary plant products (e.g. growth promoters, growth inhibitors, lignin precursors, flavonoids, tannins, and other phenolic compounds). The major end products are the amino acids phenylalanine, tyrosine, and tryptophan.

Evidence for the involvement of this pathway in glyphosate toxicity has come from a variety of studies in a wide range of microorganisms, plant cell cultures and plants (e.g. Jaworsky, 1972; Haderlie et al., 1977; Gresshoff, 1979; Duke et al., 1980; Hollander & Amrhein, 1980; among others). Glyphosate caused a massive accumulation of shikimate in treated cells and tissues (Amrhein et al., 1980; Berlin & Witte, 1981; Bode et al., 1984; Rubin et al., 1984). These studies narrowed the possible site of action to three enzymes, involved in the conversion of shikimate to chorismate. The enzyme that was finally implicated was EPSP synthase; its inhibition by glyphosate is competitive with phosphoenolpyruvate (PEP), and non-competitive with respect to shikimate-3-phosphate with a single glyphosate binding site on the enzyme.

3.3 Commercial formulations of glyphosate

When the toxicity of herbicides is discussed, the focus is mostly on the active compound (in this case glyphosate acid or glyphosate salts). However, herbicides are formulated to increase their efficacy against target plants. Commercial glyphosate based herbicides contain other components, which are called inert ingredients. These inert ingredients are mainly surfactants, solvents and antifoam compounds. Shortly, surfactant refers to chemicals that have pronounced surface activity in aqueous solutions that can decrease surface tension and perturb membrane permeability or transport function of membranes including permeability to glyphosate (Riechers et al., 1994). Antifoam compounds are chemical additives that reduce and hinder the formation of foam in industrial process liquids. Numerous contributions have demonstrated that inert ingredients in glyphosate formulations have several folds higher toxicity on non-target organisms than glyphosate alone (e.g. Folmar et al. 1979; Wan et al. 1989; Cedergreen & Streibig, 2005; among others). Therefore, glyphosate formulations are chemical mixtures and must be considered as mixtures in toxicity assessments. In this context, studies regarding specific toxicity or generalization about toxicology of inert ingredients (e.g. surfactants) must to be conducted on glyphosate, inert ingredient and commercial formulation separately. The lack of such data will render any predictions about the effects of the formulations on glyphosate highly uncertain (Diamond & Durkin, 1997).

Glyphosate concentration, as well as nature and concentration of inert additives, depend on commercial formulations; though available information about inert ingredients in glyphosate products is commonly not listed on the label. The commercial formulation of glyphosate, Roundup®, is a most popular brand name for glyphosate herbicides. Roundup® contains IPA salt of glyphosate (35-50 %) plus inert ingredients. Roundup® concentration is commonly expressed as mg a.i./L or mg a.e./L of glyphosate, while also as mg of Roundup® (whole product) per L. We considered 1 mg a.i./L equal to 0.75 mg a.e./L (Relyea, 2006).

The surfactant in Roundup®, as well as in some other glyphosate based products, is the highly toxic polyethoxylated tallowamine compound (Smith & Oehme, 1992; Giesy et al., 2000). This material is referred to in the literature as MON0818, or polyoxyethyleneamine (POEA), present at about 15% in Roundup®. Presumably POEA is a derivative of tallow, a complex mixture of fat from the fatty tissue of cattle or sheep. Other trade names of glyphosate based herbicides include Aquamaster®, Filedmaster®, Touchdown®, Glyphos®, Duramax®, Durango®, Glyphomax®, Fosato®, Ron-do®, Vision®, Rodeo®, Sulfosato®, etc.

4. Glyphosate in aquatic environments

4.1 Offsite movement and direct applications

The use of herbicides and other chemical agents in agriculture may result in accidental introduction into waters. When it is applied as post emergence spray, herbicides may enter aquatic systems through accidental offsite movement in herbicide spray drift, or through transport in leaching and surface run-off.

Particularly, under field conditions, glyphosate is usually assumed to be rapidly and tightly adsorbed to soil. Consequently, glyphosate is unlike to leach into ground waters or runoff significantly into surface waters following application. In several laboratories (Rueppel et al., 1977; Crisanto et al., 1994; among others) and some field studies (Roy et al., 1989), the immobility of glyphosate in soils has been demonstrated. In contrast, other field studies showed detectable concentrations of glyphosate in flume and streams after applications. Even thought it was only restricted to a relatively brief window of time (about 1 day post application), due to the fast dissipation kinetics of glyphosate in the field. In natural waters, glyphosate dissipate rapidly (half lives < 4 days) being removed from water due to adsorption to suspended particulates followed by subsequent sedimentation and or biodegradation. However, longer half lives were reported in hard waters, where glyphosate residues could be measured after 11 days post application (Pérez et al., 2007).

Edwards et al. (1980) reported important glyphosate concentrations in runoff from natural rainfall following early springtime treatments in no-tillage agriculture soils. The highest concentration of glyphosate in runoff waters (5.2 mg/L) was found in runoff occurring 1 day after treatment at the highest rate (8.6 Kg/Ha of Roundup®) (Edwards et al., 1980). The maximum amount of glyphosate transport by runoff was 1.85% of the amount applied, most of which occurred during a single storm on the day after application. In addition, Feng et al. (1990) found in a treated watershed, a dramatic increase of glyphosate concentrations (about 1.1 and 1.5 mg/L) in two oversprayed streams in response to first rainfall event 27 h post application. Authors attributed their observations to several source of input mobilization of residues in ephemeral stream channels feeding the tributary; wash off of unabsorbed residues from overhanging vegetation, surface runoff and subsurface flow. Regarding POEA residues, based on adsorption and degradation data, leaching and runoff potential is

expected to be small. POEA strongly adsorb to soil (Giesy et al., 2000), although little information about POEA offsite movement is nowadays available.
Offsite movement of glyphosate is also possible through spray drift (e.g. Payne et al., 1990; Payne, 1992). Although the spray drift of pesticides is not compound specific, this is relevant when non-target effects of glyphosate based herbicides are considered, and several studies have specifically addressed the issue. Some studies reported that the spray deposition decreased to around 10 % of the application rate in the first 30 m and less to 5 % at a distance of 200 m (Payne et al., 1990; Riley et al., 1991). Other studies suggested that drift rates would be greater. For instance, residues have been measured 400 m downwind from applications sites (Yates et al., 1978; Payne & Thompson, 1992).
Considering offsite movement of glyphosate from treated soils through drift and run-off, Giesy et al. (2000) estimated an acute scenario considering worst-case exposure conditions. The estimate was based on two assumptions, *(a)* that runoff (2%) from 10 Ha field treated at the maximum single use rate of Roundup® entered to 1 Ha pond (2 m deep) and *(b)* that *10%* of maximum single application rate per hectare entered the pond through drift, assuming aerial application. Based on these assumptions, maximum concentrations of Roundup® in natural water would range from 0.27 to 0.41 mg/L (Giesy et al., 2000). However, clearly higher concentrations in surface waters could be expected if assumptions are changed. For instance, some authors have reported that glyphosate can be readily desorbed from soil and has the potential to be extensively mobile in the soil environment. Adsorption of glyphosate to soil particulates is determined by chemical and physical characteristic of soils, which in turn affect the potential for off-target movement of the herbicide through water runoff or subsurface flow. Interestingly, given that glyphosate is bound to soil through its phosphonic acid moiety, the addition of inorganic phosphorus could potentially release glyphosate from soil particles through competition for sorption sites (Franz et al. 1997; Pechlaner, 2002). Piccolo et al. (1994) reported in an experimental study with some European soils that desorption varied from around 15 to 80% of the absorbed herbicide according to the soil characteristic. These observations, as well as supposing higher rates of terrestrial uses and higher spray drift due to weather conditions, could elicit elevated off-target movements of glyphosate formulations in to water ecosystems. Particularly, these impacts will be more important in ponds, ephemeral streams and ditchbank areas of irrigation canals due to their low water volume, and higher perimeter and area /volume proportions.
On the other hand, some glyphosate based herbicides (e.g. Rodeo® and AquaMaster®) were specially formulated to be used as aquatic herbicides, and have been employed extensively to control noxious aquatic weeds and algal blooms (Seddon, 1981; Diamond & Durkin, 1997; Siemering et al., 2008). For this purpose, glyphosate based herbicides are directly applied in aquatic ecosystems and their residues can be expected to be higher than that resulting from agricultural and other non aquatic uses. Furthermore, glyphosate can move considerable distances in canal or stream waters affecting undesired places (Duke, 1988). Fifty-eight percent of applied glyphosate was detected at distances 8 and 14.4 Km downstream from sites of introduction (Comes et al., 1976). Regardless herbicide sources, it is very important to set up the amount of glyphosate that have been measured in the field. Unluckily, there are few relevant field data on the concentration of glyphosate in aquatic habitats. The highest concentrations that have been observed in nature were: 1.24 mg a.e./L (Newton et al., 1994); 1.54 mg a.e./L (Couture et al., 1995); 2.8 mg a.e./L (Legris & Couture, 1989) and 5.2 mg a.e. /L (Edwards et al., 1980).

4.2 Toxicity of glyphosate based herbicides in aquatic environments

4.2.1 Toxicity assessment

In this chapter, we extensively reviewed published contributions about glyphosate, glyphosate formulations and surfactants effects on non-target aquatic organisms. Different parameters (lethal and sublethal effects) were evaluated in reviewed studies to characterize the hazard of chemicals (e.g. mortality, growth, biomass, ^{14}C uptake, weight, density, length, pigments, mobility, reproduction, metabolism, etc). Results were expressed as LC (concentration lethal to 10% and 50% of test organisms), EC (effective concentration causing specified effects in 10% and 50% of test organisms) and IC (inhibition concentration to specified effects in 10% and 50% of test organisms). In addition, when available, values of NOEC (no observed effect concentration) and LOEC (lowest observed effect concentration) were pointed out.

When dose-response curves were not available or were not calculated due to experimental design (e.g. field studies carried out in micro- and mesocosms), contributions were described as concentrations of treatments and obtained outcomes (generally in % of control values). A complete resume of published outcomes in acute and chronic tests and field experiments were listed in tables (from Table 1 to Table 6). Concentrations were preferentially expressed as were originally reported in each reviewed contribution.

In order to relate the aquatic toxicity of the herbicide to realistic exposure levels, the expected environmental concentration (EEC) was taken as a reference value. We considered a EEC of 2.6 mg a.i./L, (following Relyea, 2005). Similar values were estimated by other authors: 1.87 mg a.i./L (Chen et al., 2004) and 3.73 mg a.i./L (Perkins et al., 2000); though higher values were also evaluated (e.g. 10. 13 mg a.i./L; Mann and Bidwell, 1999).

4.2.2 Effects on non-target aquatic plants and algae

Herbicides are manly designed to kill unwanted terrestrial plants. Consequently the most sensitive group of aquatic non-target organisms is expected to be aquatic plants and algae. Aquatic plants and algae play a pivotal role for the function of aquatic ecosystems (Scheffer et al., 1993). Aquatic plants aid in stabilizing the sediment both in lakes and running waters, and their presence affects sedimentation rates, flow velocity, nutrient uptake and recirculation. In addition, they provide refuges for insects, crustaceans and fish, and act as substrates for surface-living microorganisms, snails and other epiphyte grazers. Microalgae (phytoplankton and periphyton communities) provide the basis for a range of food-webs in the aquatic environment and are therefore fundamental to the functioning of aquatic ecosystems (Wetzel, 2001).

Single species test in algae and aquatic plants treated with glyphosate alone (i.e. technical grade acid or salts of glyphosate), showed a wide range of EC and IC values; indicating different sensibilities. Microalgae presented EC_{50} values for glyphosate treatments ranging from 0.68 mg a.e./L in the diatom *Skeletonema costatum* (Malcolm Pirnie, 1987) to around 600 mg a.e./L in the green algae *Chlorella pyrenoidosa* (Maule & Wright, 1984) (Table 1). It is important to clarify that these values indicate the concentration that elicited the 50 % of reduction in the evaluated parameter. Some works showed that 10 % of reduction (EC_{10}) could be reached between 3 to 16 folds lower concentrations than EC_{50}. For instance, 10% growth inhibition in the green algae *Scenedesmus subpicatus* was observed in treatments with 1.6 mg/L of glyphosate acid (Vedrell et al., 2009). In addition, Christy et al. (1981) reported a 10% growth inhibition in *Chlorella sorokiniana* at the concentration of 2 mg a.e./L. Regarding macrophytes, generally lower values of EC and IC were reported, indicating a higher sensibility. For example, IC_{50} and EC_{50} values ranged from 0.22 mg a.i./L for *Myriophyllum aquaticum* (Turgut & Fomin, 2002) to 46.9 mg/L for *Lemma minor* (Cedergreen & Streibig, 2005) (Table 1).

The relative toxicity of glyphosate itself vs. commercial formulations and surfactants only can be evaluated in studies specially designed to this purpose. In general, commercial formulations (e.g. Ron-do® and Roundup®) were more toxic than glyphosate alone. For example, Tsui & Chu (2003) observed a 7 folds higher toxicity of Roundup® than the IPA salt of glyphosate in the green algae *Selenastrum capricornutum* (Table 1). Alike results were reported for *Selenastrum capricornutum* and the macrophyte *Lemma minor*, showing 4 folds higher toxicity of Roundup® than glyphosate (Cedergreen & Streibig, 2005). Lower differences in toxicity were registered by other authors (e.g. Sáenz et al., 1997; Sobrero et al., 2007), reporting between 1.2 to 1.8 folds higher toxicity of commercial formulations than active ingredient (Table 1). POEA itself contributed to Roundup® toxicity with values ranged from about 45% for *Skeletonema costatum* to 85 % for *Selenastrum capricornutum* (Tsui & Chu, 2003).

Numerous studies have been published about pesticide toxicity assessment on microalgae, using single species tests. However, Bérard et al. (1999) demonstrated that single-species tests may fail to predict indirect or system responses to toxicants, such as changes in population competition or succession. According to these authors, studies focusing on the whole natural community provide more reliable predictions about herbicide safety in aquatic environments. In studies assessing communities, significant direct and indirect effects of commercial glyphosate formulations have been reported. For example, Schaffer & Sebetich (2004) reported an increment of 161 % in net primary production for phytoplankton community treated with 0.13 mg a.i./L of Rodeo® (commercial formulation without POEA). In contrast, Goldsborough & Brown (1988), registered a 50% of reduction in periphyton primary production at values varying from 35.4 to 69.7 mg a.i./L of Roundup® (Table 1). However, in this contribution, 4 of the 6 treated ponds showed a reduction in the mean values of primary production with much lower concentrations (a dosage of 0.89 mg a.i./L). In microcosms experiments with natural marine microbial community, significant effects in species number and relative abundance of phytoplankton were observed at 10 µg a.i./L of Roundup® (Stachowski-Haberkorn et al., 2008). Comparable results were obtained by Pesce et al. (2009), reporting changes in riverine algal communities exposed to about 10 µg/L of glyphosate alone, in a microcosms experiment. In addition, mesocosms studies showed remarkable results with a single pulse application of Roundup® at concentrations of 6 and 8 mg a.i./L (Pérez et al., 2007; Vera et al., 2010). Even if these two contributions assessed herbicide effects in worst case scenarios, glyphosate concentration at the end of the experiments (11 and 14 days respectively), were around 2 mg a.i./L and effects were still clearly observed. At day 11, significant differences were observed in chemical and biological variables (Pérez et al., 2007). For example, we observed changes in phytoplankton assemblage fractions, with a reduction of micro and nanophytoplankton densities (2.5 folds) and a concomitant increase of picocyanobacteria (PICY) densities (40 folds) accompanied by an increase of primary production. These results can be expected by either direct effect of herbicide (differences in sensibility among species) or indirect effects duo to interspecific competition. In addition, Vera et al. (2010) found that Roundup® produced a clear delay in periphytic colonization in treated mesocosms. The periphytic mass variables: dry weight (DW), ash-free dry weight (AFDW) and chlorophyll a, were always higher in control mesocosms. Despite the mortality of algae, (mainly diatoms), cyanobacteria was favoured in treated mesocosms. We also observed that Roundup® produced a long term shift in the typology of mesocosms, "clear" turning to "turbid" state due to an eutrophication process.

AQUATIC ALGAE & PLANTS	ASSESSED CHEMICAL	STUDY TYPE	ASSESSED PARAMETER	EFFECTS CONCENTRATION (mg/L) ##	REFERENCES
- Phytoplankton and Periphyton					
Chlorococcumhypnosporum	Gly. (acid)	SST (7 d.)	Growth	96h EC_{50} = **68.0**	Maule & Wright, 1984
Chlorella pyrenoidosa				96h EC_{50} = **590.0**	
Skeletonema costatum	Gly. (acid)	SST (7 d.)	Biomass	EC_{50} = **0.64**; NOEC = **0.28**	Malcolm Pirnie, 1987
(Periphyton community)	Roundup®	LTNC (4 h.)	NPP	4h EC_{50} = (between **35.4** to **69.7**)	Goldsborough & Brown, 1988
Scenedesmus acutus	Gly. (IPA salt)	SST (4d.)	Density	96h EC_{50} =**10.2**; NOEC = **2.0**	Sáenz et al., 1997
	Ron-do®			96h EC_{50} = **9**; NOEC = **3.2**	
Ankistrodesmus sp	Rodeo®	SST (10 d).	Density	96h EC_{50} = **74.0**	Gardner et al., 1997
(Phytoplankton community)	Rodeo®	LTNC (6 h.)	NPP	**0.13** mg/L elicited NPP increment	Schaffer & Sebetich, 2004
Selenastrum capricornutum	Gly. (IPA salt)	SST (4 d.)	Growth	96h EC_{50} =**41.0**	Tsui & Chu, 2003
	Roundup®			96h EC_{50} =**5.81**	
	POEA			96h EC_{50} =**3.91**	
Skeletonema costatum	Gly. (IPA salt)	SST (4 d.)	Growth	96h EC_{50} =**5.89**	Tsui & Chu, 2003
	Roundup®			96h EC_{50} = **1.85**	
	POEA			96h EC_{50} =**3.35**	
Selenastrum capricornutum	Gly. (n.c.)	SST (2 d.)	Growth	48h EC_{10} = **95.5**; 48h EC_{50} = **270.0**	Cedergreen & Streibig, 2005
	Roundup®			48h EC_{10} = **13.6**; 48h EC_{50} = **64.7**	
(Phytoplankton & periphyton community)	Roundup®	MES (11 d.)	Density #	**6** mg/L elicited changes in community structure	Pérez et al., 2007
(Microbial community)	Roundup®	MIS (7 d.)	Community structure #	**10** µg/L elicited a reduction in species number.	Stachowski-Haberkorn et al., 2008
(Microbial community)	Gly. (n.c.)	MIS (14 d.)	Community structure #	~ **10** µg/L elicited changes in algal community structure.	Pesce et al., 2009
Chlorella sacchorophila	Gly. (acid)	SST (3 d.)	Growth	72h EC_{10} = **3.0**;72h EC_{50} = **46.6**	Vedrell et al., 2009
Scenedesmus subspicatus				72h EC_{10} = **1.6**; 72h EC_{50} = **26.0**	
(Periphyton community)	Roundup®	MES (42 d.)	Density #	**8** mg/L elicited changes in community structure	Vera et al., 2010
- Macrophytes					
Myriophyllum sibiricum	Gly. (acid)	SST (14d.)	Root length #	14d IC_{10} = **0.59**;14d IC_{50} = **0.84**	Roshon et al., 1999
	Roundup®			14d IC_{50} = **1.22**	
Myriophyllum aquaticum	Gly. (n.c.)	SST (14d.)	Growth & chl *a* #	14d EC_{50} = **0.22** (for growth) 14d EC_{50} = **0.22** (for chl a)	Turgut & Fomin, 2002
Lemma minor	Gly. (n.c.)	SST (7 d.)	Growth	7d EC_{10} =**3.8**; 7d EC_{50} = **46.9**	Cedergreen & Streibig, 2005
	Roundup®			7d EC_{10} =**3.5**; 7d EC_{50} = **11.2**	
Myriophyllum spicatum	Roundup®	SST (21 d.).	Weight & length #	21d IC_{50} = **1.0** (for weight) 21d IC_{50} = **2.8** (for length)	Sánchez et al., 2007
Lemma gibba	Gly. (acid)	SST (10 d.).	Growth #	10d IC_{10} =**4.6**; 10d IC_{50} = **20.5**	Sobrero et al., 2007
	Roundup®			10d IC_{10} =**2.5**; 10d IC_{50} = **11.6**	

Abbreviations and acronyms: gly. (glyphosate); n.c. (not clarified); n.a (not available); d. (days); SST (single species laboratory tests); LTNC (laboratory tests with natural communities), MES (mesocosms studies); (MIS) microcosms studies; SSFE (single species field experiments); chl *a* (chlorophyll *a*); NPP (net primary production). Notes: (##) Effects concentrations were expressed as mg/L of formulation, mg a.i./L or mg a.e./L, see bibliographic references to clarify. (#) Several parameters were assessed in these contributions; remarkable examples of the reported outcomes were listed in tables.

Table 1. Effects of glyphosate, different commercial formulations of glyphosate and POEA on algae and aquatic plants

4.2.3 Effects on non-target aquatic bacteria and protozoa

The majority of the available pesticide data regarding aquatic microorganisms is for algae. Far fewer pesticide studies exist for aquatic bacteria and protozoa. Aquatic bacteria and protozoa (e.g. amoeboids, flagellates, ciliates and sporozoans) have key roles in the functioning of aquatics environments. Shortly aquatic bacteria occupy an important position in the aquatic food web since they are major actors in the decomposition of dead material, and thereby in the recycling of nutrients and carbon. They are extremely important in "Lake metabolism", being involved in mineralization processes and in the chemical transformation of elements between reduced and oxidized forms Protozoans are ecologically important as key links in food chains. Ubiquitous in aquatic environments, protozoans prey upon algae, bacteria, and other organisms and are themselves consumed by animals such as microinvertebrates. Thus, the ecological role of protozoa in the transfer of bacterial and algal production to successive trophic levels is very important (in the traditional food web and in the microbial loop). On the other hand, some protozoa are important as parasites and symbionts of multicellular animals.

Concentration effects of glyphosate itself on bacteria and protozoa varied widely and seem to indicate low sensibility (Table 2). For instance, EC_{50} values obtained in treatments with glyphosate ranging from 18.2 mg/L for the bacteria *Vibrio fischeri* (Bonnet et al., 2007) to 386.0 mg a.e./L for the ciliate *Tetrahymena pyriformis* (Tsui & Chu, 2003) (Table 2). However, lower concentrations have been reported to produce observable effects (Everett & Dickerson, 2003). These authors registered a LOEC value of 5 mg/L for the parasite ciliate *Ichthyophthirius multifiliis* treated with glyphosate acid.

Roundup® showed higher toxicity than glyphosate for bacteria and protozoa in the revised bibliography. Tsui & Chu (2003) reported 6 folds higher sensibility of *Vibrio fischeri* to

MICROORGANISMS	ASSESSED CHEMICAL	TYPE OF STUDY	ASSESSED PARAMETER	EFFECTS CONCENTRATION (mg/L) ##	REFERENCE
-Bacteria and Protozoa					
Ichthyophthirius multifiliis	Gly. (acid)	SST (5 h)	Mortality	NOEC = **2.5**, LOEC = **5.1**	Everett & Dickerson, 2003
	Roundup®			NOEC = n.a., LOEC = **0.07**	
Tetrahymena thermophila	Gly. (acid)	SST (24 h)	Mortality	NOEC = **10.1**, LOEC = n.a.	Everett & Dickerson, 2003
	Roundup®			NOEC = n. a., LOEC = **0.31**	
Brachionus calyciflorus	Gly. (n.c.)	SST (24 h)	Growth #	24h EC_{50} =**28.0**	Xi & Feng, 2004
Vibrio fischeri	Gly. (IPA salt)	SST (5 min.)	Growth	5min EC_{50} =**162.0**	Tsui & Chu, 2003
	Roundup®			5min EC_{50} =**24.9**	
	POEA			5min EC_{50} =**10.2**	
Tetrahymena pyriformis	Gly. (IPA salt)	SST (40 h.)	Growth	40h EC_{50} =**386.0**	Tsui & Chu, 2003
	Roundup®			40h EC_{50} =**29.5**	
	POEA			40h EC_{50} =**4.96**	
Euplotes vannus	Gly. (IPA salt)	SST (48 h.)	Growth	48h EC_{50} =**64.1**	Tsui & Chu, 2003
	Roundup®			48h EC_{50} =**23.5**	
	POEA			48h EC_{50} =**5.0**	
Euglena gracilis *	Roundup®	SST (7 d.)	Velocity #	NOEC = **0.05**, LOEC = **0.1**	Pettersson & Ekelund, 2006
	Avans®			NOEC = **0.05**, LOEC = **0.1**	
Vibrio fischeri	Gly. (acid)	SST (15 min.)	Bioluminescence	15min EC_{50} =**18.2**	Bonnet et al., 2007
	AMPA			15min EC_{50} =**53.4**	
Tetrahymena pyriformis	Gly. (acid)	SST (45 min.)	Enzyme activities #	45min EC_{50} =**87.9**	Bonnet et al., 2007
	AMPA			45min EC_{50} =**166.5**	

Abbreviations and acronyms: see Table 1. Notes: (*) *Euglena gracilis* is a mixotrophic_green flagellated; although in this resume was grouped with protozoa; see Table 1 for other notes.

Table 2. Effects of glyphosate, different commercial formulations of glyphosate, AMPA and POEA on bacteria and protozoa

Roundup® than to glyphosate acid. In addition, these authors observed 2.7 and 13 folds higher sensibility of the ciliates *Euplotes vannus* and *Tetrahymena pyriformis* to Roundup®, respectively. In addition, the ciliate parasite *Ichthyophthirius multifiliis* showed an accentuated response, being several times more sensible to Roundup® (Everett & Dickerson, 2003) (Table 3). Values of EC_{50} obtained in Roundup® treatments ranged from 23.5 to 29.5 mg a.e./L (Tsui & Chu, 2003); though lower values produced observable effects. For example, Everett & Dickerson (2003) registered LOEC values of 0.07 and 0.31 mg a.e./L for two ciliates. Besides, values of 0.1 mg a.i./L of Roundup® and Avans® (other glyphosate commercial formulation) elicited reduction of at least 50% in the swimming velocity of *Euglena gracilis* (Pettersson & Ekelund, 2006).

AQUATIC INVERTEBRATES	ASSESSED CHEMICAL	STUDY TYPE	ASSESSED PARAMETER	EFFECTS CONCENTRATION (m/L) ##	REFERENCES
-Crustaceans (copepods, cladocerans and amphipods)					
Daphnia magna	Roundup®	SST (48h)	Immobilization	48h EC_{50} = **3.0**	Folmar et al., 1979
Gammarus pseudolimnaeus	Roundup®	SST (48h)	Mortality	48h LC_{50} = **62.0**	Folmar et al., 1979
Daphnia pulex	Roundup®	SST (96 h.)	Immobilization	96h EC_{50} = **8.5**	Servizi et al., 1987
	POEA			96h EC_{50} =**2.0**	
Daphnia magna	Ron-Do®	SST (48h)	Immobilization	48h EC_{50} = **61.7**	Alberdi et al., 1996
Daphnia spinulata	Ron-Do®	SST (48h)	Immobilization	48h EC_{50} = **66.2**	Alberdi et al., 1996
Simocephalus vetulus	Vision®	SST (8 d)	Survival and reproduction #	**0.75** mg/L elicited survival and reproduction reduction	Chen et al., 2004
Ceriodaphnia dubia	Rodeo®	SST (48 h.)	Mortality#	48h LC_{50} = **415.0**	Tsui & Chu, 2004
	Roundup bioactive®			48h LC_{50} = **81.5**	
	Roundup®			48h LC_{50} = **5.7**	
Hyalella azteca	Rodeo®	SST (48 h.)	Mortality#	48h LC_{50} = **225.0**	Tsui &Chu, 2004
	Roundup bioactive®			48h LC_{50} = **120.0**	
	Roundup®			48h LC_{50} = **1.5**	
Ceriodaphnia dubia	Gly. (IPA salt)	SST (48 h.)	Mortality	48h LC_{50} = **415.0**	Tsui &Chu, 2003
	Roundup®			48h LC_{50} = **5.4**	
	POEA			48h LC_{50} = **1.2**	
Acartia tonsa	Gly. (IPA salt)	SST (48 h.)	Mortality	48h LC_{50} = **49.3**	Tsui & Chu, 2003
	Roundup®			48h LC_{50} = **1.77**	
	POEA			48h LC_{50} = **0.57**	
Daphnia magna	POEA (5:1)	SST (48 h.)	Mortality	48h LC_{50} = **0.18**	Brausch et al., 2007
	POEA (10:1)			48h LC_{50} = **0.097**	
	POEA (15:1)			48h LC_{50} = **0.85**	
-Molluscs (snails and mussels)					
Pseudosuccinea columella	Gly. (n. c.)	SST (12 d.)	Growth and hatching #	**1** mg/L elicited growth increment and **10** mg/L inhibited hatching.	Tate et al., 1997
Pseudosuccinea columella	Gly. (n. c.)	SST (12 d.)	Metabolism	**0.1** mg/L elicited an increment in free amino acids.	Tate et al., 2000
Utterbackia imbecillis	Roundup®	SST (24 h.)	Mortality #	48h LC_{50} = **18.3**	Conners & Black, 2004
-Others (insects and worms)					
Chironomous plumosus	Gly. (acid)	SST (48h)	Immobilization	48h EC_{50} = **55.0**	Folmar et al., 1979
	Roundup®			48h EC_{50} = **18.0**	
	POEA			48h EC_{50} = **13.0**	
Lumbriculus variegatus	Gly. (acid)	SST (2-4 d.)	Bioaccumulation & enzyme action	**0.05** mg/L elicited increase in sGST activity.	Contardo-Jara et al., 2009
	Roundup Ultra®			**0.05** mg/L elicited an increase in sGST activity.	

Abbreviations, acronyms and notes: see Table 1

Table 3. Effects of glyphosate, different commercial formulations of glyphosate and POEA on aquatic invertebrates

Surfactant POEA itself resulted more toxic to bacteria and ciliates (between 2.4 to 5.8 folds) than Roundup® (Tsui & Chu, 2003); though the degradation product of glyphosate (AMPA) resulted less toxic than glyphosate acid (Bonnet et al., 2007).

4.2.4 Effects on non-target aquatic invertebrates

Invertebrates comprise a large group of aquatic species with a wide variety in shape and size, and evolved to utilize different habitats and resources (e.g., insects, worms, snails, hydroids, crustacean, etc.). They can be practically divided in micro and macro groups. Micro-invertebrates (zooplankton) are keystone species (e.g. as food for predators and top-down controller of phytoplankton, periphyton and detritus) (Montenegro Rayo, 2004). The micro-invertebrates (rotifers and copepods) are usually more abundant in number of individuals, but their smaller size limit their impact to size discrimination of phytoplankton rather than a reduction of total algal biomass (Scheffer, 1998). Large cladocerans (e.g., *Daphnia spp.)* can feed efficiently on a wide range of particle types and sizes. Moreover, cladocerans are also important as a major food source for many fish species and predatory invertebrates. Some crustacean species (e.g., shrimps, crabs and crayfish) are important as food resource for men. Insects are foraging on zooplankton, periphytic algae, and detritus and themselves prey for fish and waterfowl. In addition, aquatic molluscs (i.e. snails and mussels) have also a significant importance in aquatic environments and in human life (e.g. parasite vectors; invasive species; top-down controller of phytoplankton, periphyton and detritus; source of food to higher trophic levels, as well as human food resources) (Brönmark & Hansson, 2005).

Aquatic invertebrates seem to have low sensibility to glyphosate itself (Table 3). Values of LC_{50} obtained in treatments with glyphosate ranged from 49.3 mg a.e./L for the marine copepod *Acartia tonsa* to 415 mg a.e./L for the cladoceran *Ceriodaphnia dubia* (Tsui & Chu, 2003). However, Tate et al. (1997 and 2000), reported remarkable results in the snail *Pseudosuccinea columella* (an intermediate snail host of *Fasciola hepatica*) treated with lower glyphosate concentrations (0.1, 1 and 10 mg/L). Concentrations of 1 mg/L elicited an increment in the growth in the third-generation of snails, as well as 0.1 and 10 mg/L elicited an inhibition of eggs hatching, abnormalities and polyembryony (Tate et al., 1997). Same authors observed significant differences in the metabolism of *P. columella.* Glyphosate concentrations of 0.1 mg/L induced about 2 folds increment in five free amino acids (Tate et al., 2000). In addition, Contardo-Jara et al. (2009) reported significant increment in the enzymes activities (sGST and SOD) of worm *Lumbriculus variegatus* at 0.05 and 0.1 mg a.i./L of glyphosate.

On the other hand, invertebrates showed higher sensibility to commercial formulations. For instance, Roundup® showed between 3 folds to 76 folds higher toxicity than glyphosate itself. Values of LC_{50} obtained in Roundup® treatments ranged from 1.5 mg a.e./L for the amphipod *Hyalella azteca* (Tsui & Chu, 2004) to 62.0 mg a.i./L for other amphipod *Gammarus pseudolimnaeu* (Folmar et al., 1979) (Table 3). In addition, 0.7 mg a.e./L of Vision® (a commercial formulation containing POEA) elicited a 100 % of mortality and more than 50 % reduction in total neonates per female in the cladoceran *Simocephalus vetulus* at values of pH = 7.5 (Chen et al., 2004).

Sublethal effects were observed at much lower concentration of Roundup (1.1 µg/L) in the Clam, *Ruditapes decussates,* showing histological alterations (Abdel-Nabi, et al., 2007; El-Shenawy, et al., 2009).

The surfactant POEA was several times more toxic for invertebrates than glyphosate itself and Roundup®. For example, Folmar et al. (1979) reported almost 1.4 folds higher toxicity of POEA relative to Roundup® in the midge larvae *Chironomous plumosus;* contributing with about 66 % of the Roundup® toxicity. Higher values were reported by Servizi et al. (1987) and Tsui & Chu (2003), being POEA 3 and 5 folds more toxic than Roundup®, respectively. POEA can contribute with more than the 90% of Roundup® toxicity (Tsui & Chu, 2003) On the other hand, commercial formulations without POEA (e.g. Ron-Do®, Rodeo® and Roundup bioactive®) showed lower toxicity (Alberdi et al., 1996; Tsui & Chu, 2004) than other formulations (Table 3).

4.2.5 Effects on non-target aquatic vertebrates

Fishes are well appreciated in human societies in many ways (e.g., economical, recreational, ecological). In many countries, commercial fishing has a large economic importance as national food supply and as an export product. Fishes are one of the most demanded pets and there are a lot of people who enjoy sport fishing. This vertebrate group has an amazing diversity in morphology, size and color, which reflects their life history adaptations (e.g., feeding behavior reproduction and habitat selection). It is well known that fish populations have both direct and indirect effects on ecosystem function and structure in general (e.g., nutrient dynamics and cycling, zooplanktonic community composition), and especially in freshwater ecosystems where they are top consumers on lower trophic levels (e.g., piscivore, planktivore, benthivore fish) (Scheffer, 1998; Montenegro Rayo, 2004; Brönmark & Hansson, 2005).

The major groups of amphibians found in lakes and ponds are frogs, toads and salamanders. Some species live their whole life in freshwater whereas other species are completely terrestrial and depended on water for their reproduction. Most tadpoles have a feeding apparatus that allows them to trap bacteria, phytoplankton, and other small particles suspended in the water. Many species also graze on periphytic algae and some species even have mouth parts adapted for a predatory feeding mode. Salamanders start to feed on zooplankton but as they grow they include larger invertebrates in their diet and some species even prey on tadpoles. Tadpoles, frogs, toads and salamanders are important source of food for fish and birds.

Aquatic fish and amphibians appear to have low sensibility to glyphosate itself (Table 4 and 5). Values of LC_{50} obtained in glyphosate treatments ranged from 130 mg a.i./L for the channel catfish *Ictalurus punctatus* (Folmar et al., 1979) to 620 mg a.i./L for the carp *Cyprinus carpio* (Neškovic et al., 1996) (Table 4). In amphibians, values of LC_{50} obtained with glyphosate (IPA salt) treatments, varied from 340 to 460 mg a.e./L in four tadpoles of Australian frogs (Table 5). However, lower values where obtained in treatments with glyphosate as an acid, reporting values of LC_{50} from 82 to 121 mg a.e./L (Mann & Bidwell, 1999). Similar toxicity was registered for Roundup® Biactive, a commercial formulation without surfactant POEA (Mann & Bidwell, 1999).

Wide differences were observed in the toxicity of glyphosate itself and commercial formulations. In fish, values of LC_{50} obtained with Roundup® treatments ranged from 2.3 mg a.i./L for the fathead minnows *Pimpehales promelas* (Folmar et al., 1979) to 14.5 mg /L for *Ictalurus punctatus* (Abdelghani et al., 1997). Treatments with Vision® showed middle LC_{50} (10.42 mg a.i./L) for the rainbow trout *Oncorhynchus mykiss* (Morgan & Kiceniuk 1992). In addition, values of 4 and 12 mg a.i./L of Eskoba III Max® (a commercial formulation with unknown surfactant) elicited the 25 % and 20% of mortality in juveniles of Pejerrey *Odontesthes bonariensis* and adults of Tosquero *Jenynsia lineata* respectively; though not lethal effects were observed in glyphosate treatments (Pérez & Miranda unpublished) (Fig. 3). However, much lower concentrations of Roundup® have shown to cause effects in the biometry, metabolism and enzyme activities of fish. For instance, Glusczak et al. (2007) reported a significant decrease in AChE activity (enzyme presents in cholinergic synapses and motor end plates) and

TBARS levels (a measure of oxidative stress) in the brain of silver catfish *Rhamdia quelen* exposed to 0.2 and 0.4 mg/L. Besides, significant reduction in the biometry of Piava *Leporinus obtusidens* was observed in treatments with 1 mg/L, eliciting a reduction of length (15%) and weight (50%) (Salbego et al., 2010). Other recent study, indicated molecular responses for the flounder *Platichthys flesus* treated with low doses of herbicide cocktail (< 10 µg/L of glyphosate) during a long-term contamination (62 days) (Evrard et al., 2010).

VERTEBRATES	ASSESSED CHEMICAL	STUDY TYPE	ASSESSED PARAMETER	EFFECTS CONCENTRATION (mg/L) ##	REFERENCE
Fish					
Oncorhynchus mykiss	Gly. (acid)	SST (96h)	Mortality	96h LC_{50} =**140.0**	Folmar et al., 1979
	Roundup®			96h LC_{50} =**8.3**	
	POEA			96h LC_{50} =**2.0**	
Pimpehales promelas	Gly. (acid)	SST (96h)	Mortality	96h LC_{50} =**97.0**	Folmar et al., 1979
	Roundup®			96h LC_{50} =**2.3**	
	POEA			96h LC_{50} =**1.0**	
Ictalurus punctatus	Gly. (acid)	SST (96h)	Mortality	96h LC_{50} =**130.0**	Folmar et al., 1979
	Roundup®			96h LC_{50} =**13.0**	
	POEA			96h LC_{50} =**13.0**	
Oncorhynchus mykiss	Roundup®	SSFE (96h)	Mortality	96h LC_{50} =**52.0**	Hildebrand et al., 1982
Oncorhynchus mykiss	Roundup®	SST (96h)	Mortality	96h LC_{50} =**8.5**	Servizi et al., 1987
	POEA			96h LC_{50} =**3.2**	
Oncorhynchus nerka	Roundup®	SST (96h)	Mortality	96h LC_{50} =**8.1**	Servizi et al., 1987
	POEA			96h LC_{50} =**2.6**	
Oncorhynchus mykiss	Vision®	SST (96h)	Mortality	96h LC_{50} =**10.2**	Morgan & Kiceniuk, 1992.
Cyprinus carpio	Gly. (acid)	SST (96h)	Mortality #	96h LC_{50} =**620.0**	Neškovic et al., 1996
Ictalurus punctatus	Roundup®	SST (96h)	Mortality	96h LC_{50} =**14.5**	Abdelghani et al., 1997
Lepomis macrochirus	Roundup®	SST (96h)	Mortality	96h LC_{50} =**13.0**	Abdelghani et al., 1997
Rhamdia quelen	Roundup®	SST (96h)	Metabolism & enzyme activity#	**0.2** mg/L elicited a decrease in AChE and TBARS.	Glusczak et al., 2007
Prochilodus lineatus	Roundup®	SST (96h)	Mortality & physiology #	96h LC_{50} =**13.7**	Carmo Langiano & Martinez, 2008
Leporinus obtusidens	Roundup®	SST (90 d.)	Biometry & enzyme activity#	**1** mg/L elicited a decrease in length and weight and AChE	Salbego et al., 2010
Platichthys flesus	Roundup® + AMPA	SST (62 d.)	Molecular and physiology	**0.16** µg/L elicited liver injury	Evrard et al., 2010
Odontesthes bonariensis	Gly. (IPA salt)	SST (96h.)	Mortality	Not observed lethal effect	Pérez & Miranda (unpublished)
	Eskoba III Max®			**4** mg/l elicited the 25% of mortality	
Jenynsia lineata	Gly. (IPA salt)	SST (96h.)	Mortality	Not observed lethal effect	Pérez & Miranda (unpublished)
	Eskoba III Max®			**12** mg/l elicited the 20% of mortality	

Abbreviations, acronyms and notes: see Table 1

Table 4. Effects of glyphosate, different commercial formulations of glyphosate and POEA on fish.

Comparable outcomes with treatments of Roundup® and Glyphos® (other commercial formulation containing POEA) were reported for amphibians (Table 5). In laboratory tests, values of LC_{50} varied from 2.6 mg/L of Glyphos® for tadpoles of the hylid *Scinax nasicus* (Lajmanovich et al., 2003) to 11.6 mg a.e./L of Roundup® for tadpoles of *Litoria moorei* (Mann & Bidwell, 1999). Middle concentrations (8 mg a.i./L) caused 100% of mortality in tadpoles of the toad *Rhinella arenarum* (Pérez & Miranda unpublished) exposed to Eskoba III Max® (Fig. 6), though LC values of 3.2 mg a.i./L were reported for this toad exposed to Roundup Ultra-Max (Lajmanovich et al., 2010). However, lower concentrations have shown significant effects in mortality and growth. For instance, Chen et al. (2004) reported 100 % of mortality in tadpoles of *Rana pipiens* treated with 0.75 mg a.e./L of Vision® at pH of 7.5. Cauble & Wagner (2005) observed 50% of mortality for tadpoles of *Rana cascadae* treated with 1.94 mg a.i./L of Roundup® and an earlier metamorphosis time with 1 mg a.i. /L (Table 5). In addition, 2 mg a.i./L significantly reduce the survival and growth in three of five tadpoles exposed to Roundup® (Relyea, 2004). The same author reported in a mesocosms experiment a 100% of mortality for the *Rana sylvatica* and *Hyla versicolor* tadpoles and around 98 % of mortality for *Rana pipiens* and *Bufo americanus* tadpoles due to a direct herbicide effect (Relyea, 2005). Tadpoles seem to be more sensible to commercial formulations than juveniles and adults.

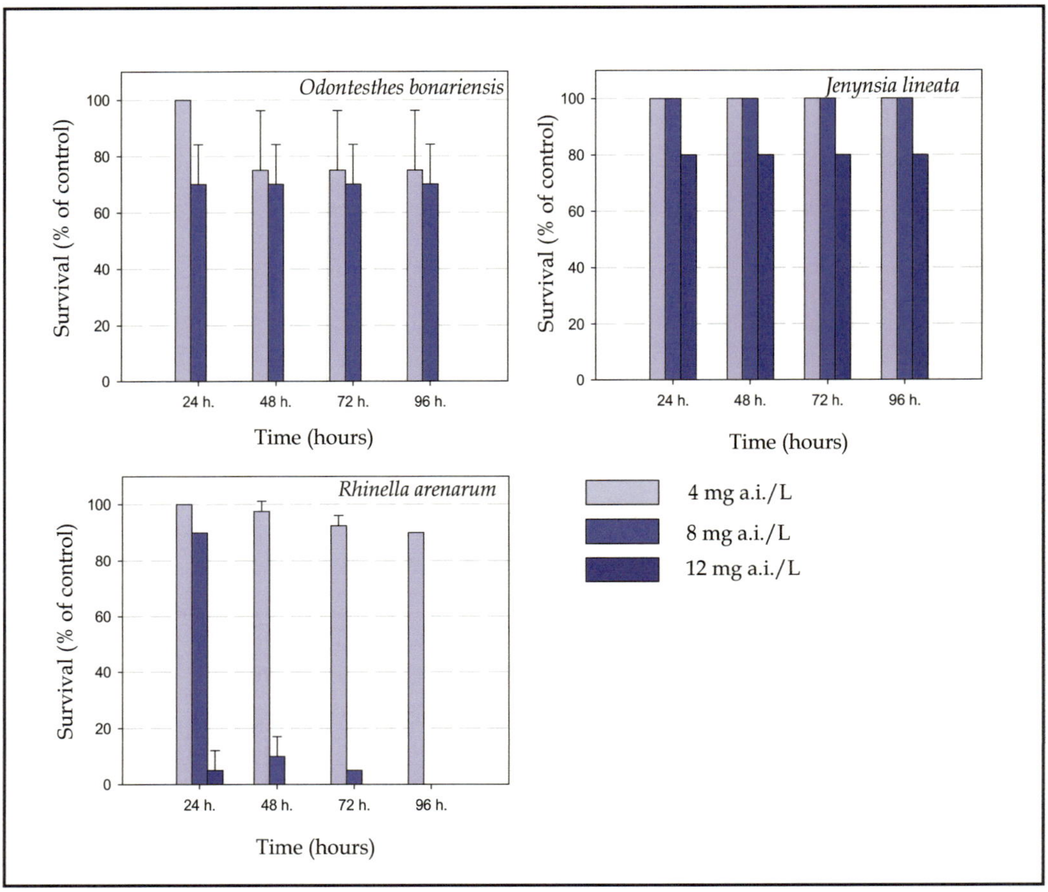

Fig. 3. Acute lethal effects of Eskoba III Max® on two fish species (*Odontesthes bonariensis* and *Jenynsia lineata*) and on tadpoles from *Rhinella arenarum*

POEA itself resulted more toxic than Roundup®, being this surfactant the more noxious component of several commercial formulations. Different authors concluded that the high mortality in fish and amphibian are actually due mainly to POEA surfactant and not to glyphosate itself (Folmar et al., 1979; Servizi et al., 1987; Perkins et al., 2000). In the fish *Pimephales promelas,* the relative contribution of glyphosate acid to the toxicity of Roundup® was about 30% (Folmar et al., 1979), while glyphosate (as IPA salt) was not toxic for 5 species of Australian frogs, and therefore without any contribution to Roundup® toxicity

(Mann & Bidwell, 1999). In fish, values of LC_{50} obtained in POEA treatments varied from 1 to 13 mg a.i./L (Folmar et al., 1979; Servizi et al., 1987), being these values up to 4 fold more toxic than Roundup®. Besides, Perkins et al. (2000), found LC_{50} values of 6.8 mg a.e./L for POEA treatments in African *Xenopus laevis* tadpoles, showing 1.8 fold higher toxicity than commercial formulation.

VERTEBRATES	ASSESSED CHEMICAL	STUDY TYPE	ASSESSED PARAMETER	EFFECTS CONCENTRATION (mg/L) ##	REFERENCES
-Amphibians (frogs and toads)					
Lymnodynastes dorsalis	Gly. (IPA salt)	SST (48 h.)	Morality	48h LC_{50} = > **400**	Mann & Bidwell, 1999
	Roundup®			48h LC_{50} = **3.0**	
	Roundup® Biactive			48h LC_{50} = > **400**	
	Touchdownt®			48h LC_{50} = **12.0**	
Heleioporus eyrei	Gly. (IPA salt)	SST (48 h.)	Morality	48h LC_{50} => **373**	Mann &Bidwell, 1999
	Roundup®			48h LC_{50} = **6.3**	
	Roundup® Biactive			48h LC_{50} => **427**	
	Touchdownt®			48h LC_{50} = **16.1**	
Crinia insignifera	Gly. (IPA salt)	SST (48 h.)	Morality	48h LC_{50} = > **466**	Mann & Bidwell, 1999
	Roundup®			48h LC_{50} = **3.6**	
	Roundup® Biactive			48h LC_{50} = > **494**	
	Touchdownt®			48h LC_{50} = **9.0**	
Xenopus laevis	Rodeo®	SST (96 h.)	Morality	96h LC_{50} = **5407**	Perkins et al. 2000
	Roundup®			96h LC_{50} = **9.4**	
	POEA			96h LC_{50} = **2.7**	
Scinax nasicus	Glyphos®	SST (96 h.)	Morality #	96h LC_{50} = **2.6**	Lajmanovich et al., 2003
Rana pipiens	Vision®	SST (8 d.)	Morality #	**0.75** mg/L elicited 100 % mortality	Chen et al., 2004
Rana pipiens	Roundup®	SST (16 d.)	Mortality & growth	Not observed significant effects	Relyea, 2004
Rana clamitans				**2** mg/L reduce survival & growth	
Rana catesbeiana				**2** mg/L reduce survival & growth	
Bufo americanus				**2** mg/L reduce growth	
Hyla versicolor				Not observed significant effects	
Rana sylvatica	Roundup®	MES (15 d.)	Mortality & biomass #	**3.8** mg/L elicited 100% mortality	Relyea, 2005
Rana pipiens				**3.8** mg/L elicited 98% mortality	
Bufo americanus				**3.8** mg/L elicited 98% mortality	
Hyla versicolor				**3.8** mg/L elicited 100% mortality	
Pseudacris crucifer				Not observed significant effects	
Rana cascadae	Roundup®	SST (42 d.)	Mortality & metamorphosis#	**1.94** mg/L elicited mortality and earlier metamorphosis times	Cauble & Wagner, 2005
Rhinella arenarum	Gly. (IPA salt)	SST (96h.)	Mortality	Not observed significant effects	Pérez & Miranda (unpublished)
	Eskoba III Max®			**8** mg/L elicited 100% mortality	

Abbreviations, acronyms and notes: see Table 1

Table 5. Effects of glyphosate, different commercial formulations of glyphosate and POEA on frogs and toads.

5. Conclusions

- Reviewing the available information on toxicity of glyphosate and its formulations on different groups of aquatic organisms, we have concluded that they are hazardous to the aquatic environment. Several contributions reviewed here reported significant effects of the herbicide at concentrations lower than EEC (2.6 mg a.i./L). Herbicide could be very noxious in standing waters like ponds, or in irrigation canals and impounded waters, where EEC can be reached. In these scenarios, toxicity could be exacerbated by other stressors and water characteristics (e.g. high temperature and pH, low O_2 concentration, presence of clay colloids, water hardness and other chemicals). Besides, toxicity also has showed to depend on organism life stage.

- Overall, ecotoxicological sub-lethal endpoints based on behavioral traits (e.g., predator avoidance, feeding, and locomotion) and other endpoints (e.g. growth, reproduction and metabolism) seem to be more sensitive indicators of effects (i.e. reporting lower effective concentrations) and give more insights into patterns of toxicity than survivorship tests (i.e. lethality). In addition, in doses dependent effects studies, commonly results are expressed as LC_{50} or EC_{50}. However, it is not possible to predict, for instance, if the 10 % of mortality or reduction in growth (i.e at lower herbicide concentration) do not have significant effects on population and eventually in the community. On the other hand, studies focused in natural or assembled communities (e.g. microcosm and mesocosms experiments) have provided interesting and significant outcomes regarding direct and indirect herbicide effects that could not be reached in single species laboratory tests. Although these laboratory tests are an essential protocol to rapidly identify the direct impacts of pesticides on organisms, they prevent an assessment of effects on organisms embedded in their natural ecological contexts.
- Glyphosate itself (as acid or salt) is generally considered to be slightly or moderately toxic to aquatic organisms (i.e., LC_{50} or EC_{50} between >1 to < 100 mg/L). However, some algae and aquatic plants showed higher sensibility, being glyphosate very toxic (EC_{50} between >0.1 to < 1 mg/L). Aquatic plants seem to be more sensitive to glyphosate than microalgae. The high toxicity of glyphosate in algae and aquatic plants is related with the mode of action of this compound (an herbicide) that interferes with plant metabolisms. On the other hand, much lower glyphosate toxicity was observed for other aquatic organisms (i.e. bacteria, protozoa, invertebrates, fish and amphibians). However, snails and worms seem to be exceptions; showing significant effects in growth, reproduction and metabolism at concentrations of < 1 mg/L of glyphosate.
- Commercial formulations and specially those containing the surfactant POEA, showed higher toxicity than the active ingredient itself for all the aquatic organisms studied. Roundup® showed to be up to 7 folds more noxious than glyphosate in algae and aquatic plants, up to 13 folds in protozoa, up to 42 folds in fish, up 70 folds in crustaceans, and up to 130 folds in frogs and toads. Algae and aquatic plants, showed significant effects with concentrations < 3 mg a.i./L. however, lower values were registered in studies of periphyton and micro plankton communities. Roundup® concentration of 10 µg a.i./L elicited changes in marine microbial community structure and 0.13 mg a.i./L of Rodeo® caused an increment in periphyton primary production. In addition, significant effects at concentrations relevant to environmental toxicity thresholds were also observed for other groups of aquatic organism. In protozoa, *Euglena gracilis* showed high sensibility, with 0.1 mg a.i. /L of Roundup® and Avans® eliciting significant sublethal effects. Different species of crustaceans showed lethal effects with values lower than 3 mg a.i./L of Roundup®. In Frogs and toads, relevant concentrations of glyphosate based products (< 2 mg a.i./L) elicited lethal and sublethal effects. Fish seems to be less sensitive to commercial

formulations, though some contributions showed significant sublethal effects in metabolism and enzyme activity at concentrations (< 2.5 mg a.i./L) of Roundup®

- The high toxicity observed for several commercial formulation of glyphosate was generally related with the content of POEA. In protozoa and invertebrates, POEA contributed with more than the 80% of Roundup® toxicity. Crustaceans showed values of EC_{50} and LC_{50} that ranged from 0.097 to 2 mg/L of POEA. In fish, glyphosate only contributed to the toxicity of Roundup® with around 30%, and values of EC_{50} ranged from 1 to 13 mg/L of POEA. In frogs and toads POEA seems to be the most toxic compound in commercial formulations. Glyphosate alone (as IPA salt) was not toxic for 5 species of Australian frogs ($LC_{50} > 343$ mg a.e./L) and therefore contributed little to Roundup® toxicity. In contrast, POEA could show lower contribution in algae and aquatic plants (as from about 46%).
- Stated the hazard of glyphosate and commercial formulations of glyphosate on aquatic environments and ecological implication of the effects reviewed here, we stressed the paucity of contributions studying the effects of glyphosate on several potential endangered aquatic species (e.g. hydroids, sponges, worms, flatworms, insects, and urodela species). We also emphasize the necessity of studies in natural communities or in assembled communities in order to evaluate direct and indirect effects upon different trophic levels.
- Finally we consider that glyphosate and commercial formulations of glyphosate could have particularly significant disruptive effects to waterbodies like ponds. Ponds have been widely recognized as very important freshwater habitats. These relative small and shallow still aquatic environments are very rich in genetic and taxonomic biodiversity; they are important refuges for amphibians and also for a bewildering variety of plants and animals, including many scarce and endangered species. In addition ponds are important places for insects hatching, fish larvae and juveniles refuges and net sites for wetland birds.

6. Acknowledgments

We thank Dr. Horacio Zagarese for help and support, as well as to CONICET (Consejo Nacional de Investigaciones Científicas y Técnicas) for financial assistant.

7. References

Abdel-Nabi, I.M.; El-Shenawy, N.S.; Taha I.A. & Moawad, T.I. (2007). Oxidative stress biomarkers and bioconcentration of Reldan and Roundup by the edible clam *Ruditapes decussates*. Acta Zool. Sinica, 53: 910-920

Abdelghani A. A., P. B. Tchounwou, A. C. Anderson, H. Sujono, L. R. Heyer, & Monkiedje, A. (1997). Toxicity evaluation of single and chemical mixtures of Roundup, Garlon-3A, 2,4-D, and Syndets surfactant to channel catfish (*Ictalurus punctatus*), bluegill sunfish (*Lepomis microchirus*), and crawfish (*Procambarus spp.*). *Environ. Toxicol. Water. Qual.* 12, 237-243

Alberdi, J. L.; Sáenz, M. E. Di Marzio, W. D. & Tortorelli, M. C. (1996). Comparative acute toxicity of two herbicides, Paraquat and Glyphosate, to *Daphnia magna* y *D. spinulata*. *Bull. Environ. Contam. Toxicol.*, 57, 229-235

Amrhein N., Deus B., Gehrke P. and Steinrucken H.C. (1980). The site of the inhibition of the shikimate pathway by glyphosate. II. Interference of glyphosate with chorismate formation in vivo and in vitro. *Plant Physiol.* 66, 830-834

Altieri, M.A. & Pengue, W. (2006). GM soybean: Latin America's new colonizer. *Seedling,* January issue

Baird, D. D.; Upchurch, R. P.; Homesley, W. B. & Franz, J. E. (1971). Toxicity of glyphosate. *Proc. North Cent. Weed Control Conf.,* 26, 64

Bayer, C.; Lovato, T.; Dieckow, J.; Zanatta, J. A. & Mielniczuk, J. (2006). A method for estimating coefficients of soil organic matter dynamics based on long-term experiments. *Soil and Tillage Research,* 91, 217-226

Bérard, A.,Leboulanger,C.,Pelte,T. (1999).Tolerance of *Oscillatoria limnetica* Lemmermann to atrazine in natural phytoplankton populations and in pure culture: influence of season and temperature. *Arch. Environ. Contam. Toxicol.,* 37,472–479

Berlin, J. & Witte, L. (1981). Effects of glyphosate on shikimic acid accumulation in tobacco cell cultures with low and high yields of cinnamoyl putrescines. *Z. Naturforsch Sect. C.,* 36, 210-214

Binimelis, R.; Pengue, W. & Monterroso, I. (2009). "Transgenic treadmill": Responses to the emergence and spread of glyphosate-resistant johnsongrass in Argentina. *Geoforum,* 40, 623–633

Bode, R.; Melo, C. & Birnbaum, D. (1984). Mode of action of glyphosate in Candida maltosa. *Arch. Microbiol.,* 140, 83-85

Bolliger, A.; Magrid, J.; Amado, T. J. C.; Neto, F. S.; Ribeiro, M. F. S.; Calegari, A.; Ralisch, R. & Neergaard, A. (2006). Taking stock of the Brazilian "zero-till revolution": a review of landmark research and farmers' practice. *Adv. Agron.,* 91, 47–110

Bonnet, J. L.; Bonnemoy, F.; Dusser, M. & Bohatier, J. (2007). Assessment of the potential toxicity of herbicides and their degradation products to non target cells using two microorganisms, the bacteria *Vibriof ischeri* and the ciliate *Tetrahymena pyriformis. Environ. Toxicol.,* 22, 78–91

Bonny, S. (2009). Genetically Modified Glyphosate-Tolerant Soybean in the USA: Adoption Factors, Impacts and Prospects - A Review. *Agron. Sustain. Dev.,* 28, 257-272

Brausch, J.; Beall, B. & Smith, P. (2007). Acute and Sub-Lethal Toxicity of Three POEA Surfactant Formulations to *Daphnia magna. Bull. Environ. Contam. Toxicol.,* 78, 510-514

Brönmark, C. & Hansson, L.-A. (2005). *The Biology of Lakes and Ponds,* Oxford University Press, New York.

Cauble, K. & Wagner, R. S. (2005). Sublethal effects of the herbicide glyphosate on amphibian metamorphosis and development. *Bull. Environ. Contam. Toxicol.,* 75, 429-435

Cedergreen, N. & Streibig, J. C. (2005). The toxicity of herbicides to nontarget aquatic plants and algae: Assessment of predictive factors and hazard. *Pest Manag. Sci.,* 61, 1152–1160

Chen, C. Y.; Hathaway, K. M. & Folt, C. L. (2004). Multiple stress effects of Vision® herbicide, pH, and food on zooplankton and larval amphibian species from forest wetlands. *Environ. Toxicol. Chem.,* 23, 823-831

Christy,S.L.; Karlander, E.P. & Parochetti, J.V. (1981). Effects of glyphosate on the growth rate of Chlorella. *Weed. Sci.,* 29, 5–7

Comes, R. D.; Bruns, V. F. & Kelly, A. D. (1976). Residues and persistence of glyphosate in irrigation water. *Weed Sci.,* 24, 47-

Conners D. E. & Black, M. C. (2004). Evaluation of lethality and genotoxicity in the freshwater mussel *Utterbackia imbecillis* (Bivalvia: Unionidae) exposed singly and in combination to chemicals used in lawn care. *Arch. Environ. Contam. Toxicol.*, 46, 362–371

Contardo-Jara,V.; Klingelmann,E. & Wiegand, C. (2009). Bioaccumulation of glyphosate and its formulation Roundup Ultra in *Lumbriculus variegatus* and its effects on biotransformation and antioxidant enzymes. *Environ. Pollut.*, 157, 57–63

Couture, G; Legris, J. & Langevin, R. (1995). Evaluation des impacts du glyphosate utilise dans le milieu forestier. Ministere des Ressources Naturelles, Direction de Penviron- ment forestier, Service du suivi environnemental, Charles- bourg, Quebec, Canada.

Crisanto T.; Sanchez-Martín, M. J. & Sanchez-Camazano, M. (1994). Mobility of pesticides in soil. Influence of soil properties and pesticide structure. *Toxicol. Environ. Chem.*, 45, 97-104

Diamond. G. & Durkin, P. (1997). Effects of Surfactants on the Toxicity of Glyphosate, with Specific Reference to Rodeo. Prepared under USDA FS Contract No. 53-3187-5- 12. Final Report.

Dill, G. M.; CaJacob, C. A. & Padgette, S. R. (2008). Glyphosate-resistant crops: adoption, use and future considerations. *Pest Manag. Sci.*, 64, 326–331

do Carmo Langiano, V. & Martinez, C. B. R., (2008). Toxicity and effects of a glyphosate-based herbicide on the Neotropical fish *Prochilodus lineatus*. *Comp. Biochem. Physiol.*, 147, 222 - 231

Duke, S. O.; Hoagland, R. E. & Elmore, C. D. (1980). Effects of glyphosate on metabolism of phenolic compounds. V. L-a-aminooxy-B-phenylpropionic acid and glyphosate effects on phenylalanine ammonia-lyase in soybean seedlings. *Plant Physiol.*, 65, 17-21

Duke, S. O. (1988). Glyphosate. In: *Herbicides, Chemistry, degradation and mode of action*, Kearney, P.C. & Kaufman, D.D. (Ed.), 1-70 pp, Marcel Dekker, New York

Edwards, W. M.; Triplett, G. B. & Kramer, R. M. (1980). A watershed study of glyphosate transport in runoff. *J. Environ. Qual.*, 9, 661-665

Everett, K. D. E. & Dickerson, H. W. (2003). *Ichthyophthirius multifiliis* and *Tetrahymena thermophila* tolerate glyphosate but not a commercial herbicidal formulation. *Bull. Environ. Contam. Toxicol.*, 70, 731–734

El-Shenawy, N.S.; Moawad, T.I.S.; Mohallal, M.E.; Abdel-Nabi , I.M. & Taha, I.A. (2009). Histopathologic biomarker response of clam *Ruditapes decussates* to Reldan and Roundup: A laboratory study. *Ocean Sci. J.*, 44: 27-34.

Evrard, E.; Marchand, J.; Theron, M.; Pichavant-Rafini, K.; Durand, G.; Quiniou, L. & Laroche, J. (2010). Impacts of mixtures of herbicides on molecular and physiological responses of the European flounder *Platichthys flesus*. *Comparative Biochemistry and Physiology Part C*, 152, 321-331

Feng, J. C.; Thompson, D. G. & Reynolds, P. E. (1990). Fate of Glyphosate in a Canadian Forest Watershed. 1. Aquatic Residues and Off Target Deposit Assessment. *J. Agric. Food Chem.*, 38, 1110-1118

Folmar, L. C.; Sanders, J. O. & Julin A. M. (1979). Toxicity of the Herbicide Glyphosate and Several of its Formulations to Fish and Aquatic Invertebrates. *Arch. Environ. Contain. Toxicol.*, 8, 269-278

Franz, J. E. (1985). Discovery, development and chemistry of glyphosate. In: *The Herbicide Glyphosate*, Grossbard, E. & Atkinson, D. (Eds.), 3-17. Butterworth & Co, London, UK

Franz, J. E.; Mao, M. K. & Sikorski, J. A. (1997). *Glyphosate: A Unique Global Herbicide.* ACS Monograph 189. American Chemical Society, 65-97, Washington, DC

Gardner,S. C.; Grue, C. E.; Grassley, J. M.; Lenz, L. A.; Lindenauer, J. M. & Seeley, M. E. (1997). Single species algal (*Ankistrodesmus*) toxicity tests with Rodeo® and Garlon® 3A. *Bull. Environ. Contam. Toxicol.*, 59, 492–499

Giesy, J.P.; Dobson, S. & Soloman, K. R. (2000). Ecotoxicological risk assessment for Roundup® herbicide. *Rev. Environ. Contam. Toxicol.*, 167, 35–120

Goldsborough, L. G. & Brown, D. J. (1988). Effect of Glyphosate (Roundup® Formulation) on Periphytic Algal Photosynthesis. *Bull. Environ. Contain. Toxicol.*, 41, 253- 260

Glusczak, L.; Miron, D. S.; Moraes, B. S.; Simões, R. R.; Schetinger, M. R. C.; Morsch, V. M. & Loro, V. L. (2007). Acute effects of glyphosate herbicide on metabolic and enzymatic parameters of silver catfish (*Rhamdia quelen*). *Comp. Biochem. Physiol. C.*, 146, 519–524

Gresshoff, P. M. (1979). Growth inhibition by glyphosate and reversal of its action by phenylalanine and tyrosine. *Aust. J. Plant. Physiol.*, 6, 177-185

Haderlie, L. C.; Widholm, J. & Slife, F. W. (1977). Effect of glyphosate on carrot and tobacco cells. *Plant Physiol.* 60, 40-3.

Hildebrand, L. D.; Sullivan, D. S. & Sullivan, T. P. (1982). Experimental studies of rainbow trout populations exposed to field applications of Roundup herbicide. *Arch. Environ. Contam. Toxicol.*, 11, 93-98

Hollander. H. & Amrhein, N. (1980). The site of action of the inhibition of the shikimate pathway by glyphosate. I. Inhibition by glyphosate of phenylpropanoid synthesis in buckwheat (*Fagopyrum esculentum* Moench). *Plant Physiol.*, 66, 823-829

Holt J.; Powles, S. & Holtum, J. (1993). Mechanisms and agronomic aspects of herbicide resistance. *Annual Review of Plant Physiology and Plant Molecular Biology,* 44, 203 – 229

James, C. (2008). Global status of commercialized biotech/GM crops: 2008. The first thirteen years, 1996 to 2008. ISAAA Brief No. 39. International Service for the Acquisition of Agri-biotech Associations, Ithaca, New York, USA

Jaworski, E. G. (1972). Mode of action of N-phosphonomethylglycine: inhibition of aromatic amino acid biosynthesis. J. *Agric. Food Chem.*, 20, 1195-1198

Laerke, P. E. (1995). Foliar absorption of some glyphosate formulation and their efficacy on plants. *Pesticide Science,* 44, 107-116

Lajmanovich, R. C.; Sandoval, M. T, & Peltzer, P. M. (2003). Induction of mortality and Malformation in *Scinax nasicus* tadpoles exposed to glyphosate formulations. *Bull. Environ. Contain. Toxicol.*,70, 612-618.

Lajmanovich, R. C.; Attademo, A.M.; Peltzer, P.M.; Junges C. & Cabagna, M. (2010). Toxicity of four herbicide formulations with glyphosate on *Rhinella arenarum* (Anura: Bufonidae) tadpoles: B-esterases and glutathione S-transferase inhibitions. *Arch. Environ. Contam. Toxicol.* doi:10.1007/s00244-010-9578-2

Legris, J. & Couture, G. (1989). Residus de glyphosate dans l'eau et les sediments suite a des pulverisations terrestres en milieu forestier en 1986. Publication #3322.

Gouvernement du Quebec, Ministere de l'Energie et des Ressources, Direction de la Conservation, Charlesbourg, Quebec, Canada

Malcolm, P., Inc. (1987) The toxicity of glyphosate technical to Skeletonema costatum. White Plains, New York, (Unpublished report submitted by Monsanto Ltd)

Mann, R. M. & Bidwell, J. R. (1999). The toxicity of glyphosate and several glyphosate formulations to four species of southwestern Australian frogs. *Arch. Environ. Contain. Toxicol.*, 26, 193–199

Maule, A. & Wright, S. J. L. (1984). Herbicide effects on the population growth of some green algae and cyanobacteria. *J. Appl. Bacteriol.*, 57, 369-376

Moedritzer, K. & Irani, R. R. (1966). The direct synthesis of aminomethyl phosphonic acid. *J. Org. Chem.*, 31, 1603-1607

Montenegro Rayo, K. (2004). *Ecotoxicological effects of DDT and glyphosate on aquatic organisms: a case study*, Lunds Universitet, Lund

Morgan, M. J. & Kiceniuk, J. W. (1992). Response of rainbow trout to a two month exposure to Vision®, a glyphosate herbicide. *Bull. Environ. Contam. Toxicol.*, 48, 772-780

Nešković, N. K.; Poleksic, V.; Elezovic, I.; Karan, V. & Budimir, M. (1996). Biochemical and histopathological effects of glyphosate on carp (*Cyprinus carpio*). *Bull. Environ. Contam. Toxicol.*, 56, 295–302

Newton, M.; Horner, L. H.; Cowell, J. E.; White, D. E. & Cole E. C. (1994). Dissipation of glyphosate and aminomethylphos- phonic acid in North American forests. Journal of Agricul- ture and Food Chemistry, 42, 1795-1802.

Payne N. J.; Feng J.C. & Reynolds, P. E. (1990). Off-target deposits and buffer zones required around water for aerial glyphosate applications. *Pestic. Science*, 30, 183-198

Payne, N. J. 1992. Off-target glyphosate from aerial silvicultural applications and buffer zones required around sensitive areas. *Pestic. Science*, 34, 1-8.

Payne, N. J. & Thompson, D. G. (1992). Off-target glyphosate deposits from Aerial silvicultural applications under various meteorological conditions. *Pestic. Science*, 34, 53-59

Pechlaner, R. (2002). Glyphosate in herbicides: an overlooked threat to microbial bottom-up processes in freshwater systems. *Verhand. Intern. Verien. Limnol.*, 28, 1831 - 1835.

Pengue, W. (2005). "Transgenic crops in Argentina: the ecological and social debt" *Bulletin of Science, Technology and Society*, 25, 314-322

Perez, A. & Kogan, M. (2003). Glyphosate-resistant Lolium multiflorum in Chilean orchards. *Weed Res.*, 43, 12–19

Pérez, G. L.; Torremorell, A.; Mugni, H.; Rodríguez, P.; Vera, M. S.; Do Nascimento, M.; Allende, L.; Bustingorry, J.; Escaray, R.; Ferraro, M.; Izaguirre, I.; Pizarro, H.; Bonetto, C.; Morris, D. P. & Zagarese, H. (2007). Effects of the herbicide Roundup on freshwater microbial communities: a mesocosm study. *Ecol. Appl.*, 17, 2310–2322

Perkins, P. J.; Boermans, H. J. & Stephenson, G. R. (2000). Toxicity of glyphosate and triclopyr using the frog embryo teratogenesis assay-Xenopus. *Environ. Toxicol. Chem.*, 19:940–945.

Pesce, S.; Batisson, I.; Bardot, C.; Fajon, C.; Portelli, C.; Montuelle, B. & Bohatier, J. (2009). Response of spring and summer riverine microbial communities following glyphosate exposure. Ecotoxicol. Environ. Saf., 72, 1905-1912

Pettersson, M. & Ekelund, N. G. (2006). Effects of the Herbicides Roundup and Avans on *Euglena gracilis*. *Arch. Environ. Contam. Toxicol.*, 50, 175–181

Piccolo, A.; Celano, G. & Pietramellara, G. (1992). Adsorption of the herbicide glyphosate on a metal-humic acid complex. *Sci. Total Environ.*, 123/124, 77-82

Powles, S. B; Lorraine-Colwill, D. F., Dellow, J. J. & Preston, C. (1998) Evolved resistance to glyphosate in rigid ryegrass (*Lolium rigidum*) in Australia. *Weed Sci.*, 46, 604–607

Powles, S. B. (2008). Evolved glyphosate-resistant weeds around the world: lessons to be learnt. *Pest. Manag. Scie.*, 64, 360–365

Pratley, J.; Urwin, N.; Stanton, R.; Baines, P.; Broster, J.; Cullis, K.; Schafer, D.; Bohn, J. & Krueger, R. (1999). Resistance to glyphosate in Lolium rigidum. I. Bioevaluation. *Weed Sci.*, 47, 405–411.

Relyea, R. A. (2004). Growth and survival of five amphibian species exposed to combinations of pesticides. *Environ. Toxicol. Chem.*, 23, 1737–1742

Relyea, R.A. (2005). The impact of insecticides and herbicides on the biodiversity and productivity of aquatic communities. *Ecol. Appl.*, 15, 618–627

Relyea, R. A. (2006). The impact of insecticides and herbicides on the biodiversity and productivity of aquatic communities: response. *Ecol. Appl.*, 16, 2027-2034.

Riechers, D. E.; Wax, L. M.; Liebl, R. A. & Bullock, D. G. (1995). Surfactant effects on glyphosate efficacy. Weed Technol., 9, 281 - 285

Riley, C. M.; Wisener, C. J.; Sexsmith, W. A. (1991). Estimating off-target spray deposition on the ground following the aerial application of glyphosate for conifer release in New Brunswick. *J. Environ. Sci. Health B.*, 26, 185-208

Roshon, R. D.; McCann, J. H.; Thompson, D. G. & Stephenson, G. R. (1999). Effects of seven forestry management herbicides on *Myriophyllum sibiricum* , as compared with other nontarget aquatic organisms. *Can. J. For. Res.*, 29, 1158-1169

Roy, D. N.; Konar, S. K.; Banerjee, S.; Charles, D. A.; Thompson, D. G. & Prasad, R. (1989). Persistence, movement and degradation of glyphosate in selected Canadian boreal forest soils. J. *Agric. Food Chem.*, 37, 437-440

Rubin, J. L.; Gaines, C. G. & Jensen, R. A. (1984). Glyphosate inhibition of 5-enolpyruvylshikimate 3-phosphate synthase from suspension-cultured cells of *Nicotiana silvestris*. *Plant Physiol.*, 75, 839-845.

Rueppel, M. L.; Brightwell, B. B.; Schaefer J. & Marvel J. T. (1977). Metabolism and degradation of glyphosate in soil and water. *J. Agric. Food Chem.*, 25, 517-528

Sáenz, M. E.; Di Marzio, W. D.; Alberdi, J. L. & del Carmen Tortorelli, M. (1997). Effects of technical grade and a commercial formulation of glyphosate on algal population growth. *Bull. Environ. Contam. Toxicol.*, 59, 638-344

Salbego, J.; Pretto, A.; Gioda, C.; de Menezes, C., Lazzari, R., Radünz Neto, J., Baldisserotto, B. & Loro, V. (2010). Herbicide Formulation with Glyphosate Affects Growth, Acetylcholinesterase Activity, and Metabolic and Hematological Parameters in Piava (*Leporinus obtusidens*). *Arch. Environ. Contam. Toxicol.*, 58, 740-745

Sánchez, D.; Graça, M. & Canhoto, J. (2007). Testing the Use of the Water Milfoil (*Myriophyllum spicatum* L.) in Laboratory Toxicity Assays. *Bull. Environ. Contam. Toxicol.*, 78, 421-426

Schaffer, J. D. & Sebetich, M. J. (2004). Effects of aquatic herbicides on primary productivity of phytoplankton in the laboratory. *Bull. Environ. Contam. Toxicol.*, 72:1032–1037

Scheffer, M.; Hosper, S. H.; Meijer, M. L.; Moss B. & Jeppesen, E. (1993). Alternative equilibria in shallow lakes. *Trends in Ecology and Evolution*, 8, 275–279

Scheffer, M. (1998). *Ecology of Shallow Lakes*, Chapman & Hall, London, UK.

Seddon, J. C. (1981). The control of aquatic weeds with the isopropylamine salt of N-phosphonomethyl glycine. Proceedings Aquatic Weeds and their Control, Assoc. Appl. Biol. Natl. Veg. Res. Stn., Wellesbourne, Warwick, England, 141-148

Servizi, J. A.; Gordon, R. W. & Martens, D. W. (1987). Acute toxicity of Garlon 4 and Roundup herbicides to salmon, *Daphnia*, and trout. *Bull. Environ. Contam. Toxicol.*, 39, 15-22

Siemering, G.; Hayworth, J. & Greenfield, B. (2008). Assessment of potential herbicide impacts to California aquatic ecosystems. *Arch. Environ. Contam. Toxicol.*, 55, 415–431

Smith, E. A. & Oehme, F. W. (1992). The biological activity of glyphosate to plants and animals: a literature review. *Vet. Hum. Toxicol.*, 34, 531-543

Sobrero, M.C., Rimoldi, F., Ronco, A.E., 2007. Effects of the glyphosate active ingredient and a formulation on *Lemma gibba* L. at different exoposure levels and assessment end-points. *Bull. Environ. Contam. Toxicol.*, 79, 537-543

Stachowski-Haberkorn, S.; Becker, B.; Marie, D.; Haberkorn, H.; Coroller, L. & de la Broise, D. (2008) .Impact of Roundup on the marine microbial community, as shown by an in situ microcosm experiment. *Aquat. Toxicol.*, 89, 232–241

Steinrücken, H. C. & Amrhein, N. (1980). The herbicide glyphosate is a potent inhibitor of 5-enolpyruvyl-shikimic acid-3-phosphate synthase. *Biochem. Biophys. Res. Commun.*, 94, 1207-1212

Tate, T. M.; Spurlock, J. O. & Christian, F. A. (1997). Effect of glyphosate on the development of *Pseudosuccinea columella* snails. *Arch. Environ. Contam. Toxicol.*, 33, 286-289

Tate, T.; Jackson, R. & Christian, F. (2000). Effects of glyphosate and dalapon on total free amino acid profiles of *Pseudosuccinea columella* Snails. *Bull. Environ. Contam. Toxicol.*, 64, 258–262

Turgut, C. & Fomin, A. (2002). Sensitivity of the rooted macrophyte *Myriophyllum aquaticum* (Vell.) Verdcourt to seventeen pesticides determined on the basis of EC50. *Bull. Environ. Contam. Toxicol.*, 69, 601-608

Tsui, M. T. K. & Chu, L. M. (2003). Aquatic toxicity of glyphosate-based formulations: comparison between different organisms and the effects of environmental factors. *Chemosphere*, 52, 1189- 1197

Tsui, M. T. K., & Chu, L. M. (2004). Comparative toxicity of glyphosate-based herbicides: Aqueous and sediment porewater exposures. *Arch. Environ. Contam. Toxicol.*, 46, 316–323

USEPA (United States Environmental Protection Agency). Pesticide Industry Sales and Usage 1994-2001. Marked estimates. http://www.epa.gov/opp00001/pestsales/

USDA (United States Departament of Agriculture). Agricultural Chemical Usage - Field Crops. National Agricultural Statistics Service. http://usda.mannlib.cornell.edu

Vendrell, E.; Ferraz, D. G.; Sabater, C. & Carrasco, J. M. (2009). Effect of glyphosate on growth of four freshwater species of phytoplankton: a microplate bioassay. *Bull. Environ. Contam.Toxicol.*, 82, 538-42

Vera, M. S.; Lagomarsino, L.; Sylvester, M.; Pérez, G. L.; Rodriguez, P., Mugni, H.; Sinistro, R.; Ferraro, M.; Bonetto, C.; Zagarese, H. & Pizarro, H. (2010). New evidences of Roundup® (glyphosate formulation) impact on the periphyton community and the water quality of freshwater ecosystems. *Ecotoxicology*, 19, 710-719

Wan, M. T.; Watts, R. G. & Moul, D. J. (1989). Effects of different dilution water types on the acute toxicity to juvenile Pacific salmonids and rainbow trout of glyphosate and its formulated products. *Bull. Environ. Contam. Toxicol.*, 43, 378-385

Wax, L. M.; Leibl, R. M. & Bush, D. R. (1994). Surfactant-increased glyphosate uptake into plant membrane vesicles isolated from common lambsquateres leaves. *Plant Physiol.*, 105, 1419-25

Van Gessel, M.J. (2001).Confirming glyphosate-resistant horseweed (*Conyza canadensis*) in Delaware. *Weed Sci.*, 49, 703–712

Wetzel, R. (2001). *Limnology, Lake and River Ecosystems,* Academic Press. San Diego, California.

Woodburn, A. T. (2000). Glyphosate: production, princing and use worldwide. *Pest Manage. Sci.*, 56, 309-312

Xi, Y. L. & Feng, L. K. (2004). Effects of thiophanate-methyl and glyphosate on asexual and sexual reproduction in the rotifer *Brachionus calyciflorus* Pallas. *Bull. Environ. Contam.Toxicol.*, 73, 644-651

Yamada, T.; Kremer, R. J.; de Camargo e Castro, P. R. & Wood, B. W. (2009). Glyphosate interactions with physiology, nutrition, and diseases of plants: Threat to agricultural sustainability? *Europ. J. Agronomy*, 31, 111-113

Yates, W. E., Akesson, N. B. & Bayer, D. E. (1978). Drift of glyphosate sprays applied with aerial and ground equipment. *Weed Science*, 26, 597-604

17

The Impact of Herbicides on Benthic Organisms in Flooded Rice Fields in Southern Brazil

Joele S. Baumart and Sandro Santos
Universidade Federal de Santa Maria, 97105-900 Santa Maria, RS Brazil

1. Introduction

Agricultural production is always directly associated with the use of agrochemicals to control the harmful organisms that attack the crops and reduce the harvest. In spite of their benefits, the use of agrochemicals usually causes great problems, considering that these often-toxic chemicals are used in large quantities over large areas, and generally persist in the environment for some time (Prime et al., 2005).

As a consequence of agrochemical application, water quality and aquatic biodiversity have been compromised due to the destructuring of the physical and chemical environment and alteration of the natural dynamics of the biological communities (Goulart & Callisto, 2003). According to Biggs et al. (2007), it is highly important to regulate the use of an agrochemical and its action against non-target organisms in the aquatic environment. However, specific data on the occurrence and population dynamics of aquatic organisms in agricultural areas are very limited.

Studies by Mesléard et al. (2005) indicate that the use of herbicides, insecticides, and fertilizers can modify the feeding pattern and alter the development of animal communities present in rice fields, especially the invertebrates. Because they are sedentary organisms and have relatively short life cycles (compared to fish), benthic macroinvertebrates are considered good indicators of water quality. Due to their short life cycles, they express more rapidly the changes in the environment through changes in structure of their populations and communities (Rosemberg & Resh, 1993). Because they also have great biological diversity, they tend to exhibit a greater variability of responses to different kinds of environmental impacts (Rosemberg & Resh, 1993). Another aspect refers to adaptive strategies to environmental instabilities of the environment, in general, resilience and persistence. A resilient biota can rapidly recolonize areas disturbed by flooding; and a persistent biota demonstrates a good capacity to resist disturbances (Winterbotton et al., 1997).

The use of herbicides can indirectly influence the zoobenthic community, since, as seen in experiments carried out by Moreby & Southway (1999), the use of selective herbicides against a species of weed is essential to conserve the invertebrates that feed on plants. The use of broad-spectrum herbicides risks negative effects on the food chain of these herbivores and thus causing an imbalance in the community.

The herbicide Quinclorac (3, 7 - dichloroquinoline -8-carboxylic acid) has a relatively high persistence in rice crops compared to other herbicides (Reimche et al., 2008). Marchezan et al. (2003) found Quinclorac in the water of rivers in central Rio Grande do Sul in considerable concentrations, sufficient to harm the local benthic community.

Other products, that are widely used in irrigated rice fields in southern Brazil are the herbicides Bispyribac-sodium {sodium 2,6-bis-[(4,6-dimethoxypyrimidin-2-yl]benzoate} and a formulated mixture of the herbicides Imazethapyr and Imazapic [(RS)-5-ethyl-2-(4-isopropyl-4-methyl-5-oxo-2-imidazolin-2yl) nicotinic acid and (RS)-2-(4-isopropyl-4-methyl-5-oxo-2-imidazolin-2-yl)-5-methyl nicotinic acid]. However, there are no data available about their effect on the environment (MAPA, 2009).

Considering that irrigated rice fields are large shallow lakes during the growing season, and tend to shelter a considerable aquatic fauna, studies in these areas can contribute considerably to elucidate the real impact of the agrochemicals used in this crop. The aim of this study was to investigate the impact of the use of commercial formulations of the mixture of the herbicides Imazethapyr + Imazapic (ONLY®), Bispyribac-sodium (NOMINEE 400SC), and Quinclorac (FACET PM) on the density and abundance of different groups of benthic organisms present in irrigated areas. The influence of agrochemicals on water quality as indicated by selected physical and chemical parameters, and the period of action of these agrochemicals on the community were evaluated.

2. Material and methods

The study was carried out in an experimental marshy area belonging to the Plant Technology Laboratory of the Federal University of Santa Maria, Rio Grande do Sul, Brazil, during the 2007/08 irrigated-rice growing season. The block was constituted by four plots (treatments) and each experimental unit measured 8 x 6 m, including refuge (48 m^2). In these areas the following herbicides were used (treatments): TO - Imazethapyr (75.ha^{-1}) + Imazapic (25.ha^{-1}) (ONLY®); TB - Bispyribac-sodium(50.ha^{-1}) (NOMINEE 400 SC); TQ - Quinclorac (375.ha^{-1}) (FACET PM); and TC - Control.

The experimental units were separated by embankments with irrigation and individual drainage. The rice seed IRGA 422 CL was sowed at a density of 130 kg per ha, in a direct system, using a sowing machine with rows separated by 0.17 m. The fertilizer was applied according to the soil analysis and recommendations (SOSBAI, 2005). The herbicides were applied on the dry soil, with rice in the three-to-four-leaf stage.

The area was irrigated one day after the application of the post-emergent herbicides, and the water level was maintained at 0.10 m during the entire experiment. To avoid loss of water due to lateral infiltration, a wide guard embankment was constructed, keeping a canal with water between the plots and around the experiment, with the same hydraulic load as the plot.

The benthic macroinvertebrates were collected with the help of a cylindrical PVC sampler, 0.10 m in diameter (area 0.01 m^2) and 0.10 m long. The collections were made 28 and 84 days after the entry of water into the plots; these dates correspond to the growing period and the pre-harvest (drying the area to begin the harvest). For each treatment and on each day, 12 samples were collected.

The sampling points, in each treatment, were chosen in a universe of 42 points that corresponded to one sample per m^2 of the area cultivated.

After the collection, the material was suitably packed, labeled, and taken to the laboratory, where it was washed in 0.25 mm sieves and repacked in plastic bottles, adding Rose Bengal stain. After 20 minutes in the stain, the material was fixed with absolute ethyl alcohol.

After fixing, the animals were sortted and identified to the lowest taxonomic level possible, using specialized references (Fernandez & Dominguez, 2001; Costa et al. 2006). For each sample, the population density of each taxonomic group was recorded. All the material was stored in the collection of the Carcinology Laboratory of UFSM. The species were separated into four trophic guilds: predator, detritivore, filter-collector or catcher-collector, and herbivore scraper or perforator. This classification was based on Merrit & Cummins (1996), Callisto and Esteves (1998), Marinoni (2001), and Silva et al. (2009a).

The water was collected for analyses of the physical and chemical parameters on days 3, 7, 14, 21, 28, 42, 56, and 84 after the plots were flooded, comprising the entire period of cultivation. In each sampling period, the following analyses were done: pH, with a Hanna pH meter (HI8424); total hardness, according to APHA (1992); and temperature and dissolved oxygen (YSI oximeter model Y5512). The turbidity was measured with a PoliControl turbidimeter.

During the cultivation, samples of water were collected 1, 2, 3, 7, 14, 21, 28, 42, 56, 77, 84, and 90 days after the plots were flooded; day 90 was the day of harvest. After each collection, the labeled samples were taken for chemical analysis in the Group of Research Analyses of Residues and Pesticides laboratories of UFSM. The concentrations of herbicides were determined by means of High Performance Liquid Chromatography, with detection by the arrangement of diodes (HPLC - DAD), using methanol and water as the mobile phase and a C-18 column, according to the method described by Zanella et al. (2002).

The physical and chemical data were evaluated through a two-way ANOVA (time and treatment), and the means were compared by a t test ($p<0.05$). A PCA analysis was used to assess possible correlations between the physical and chemical parameters and the pesticides applied in the field.

The density data for the main groups were submitted to a Shapiro-Wilk analysis, and after satisfying the criteria and/or were transformed for data standardization, they were submitted to a one-way ANOVA, and the means were compared by a t test ($p<0.05$). An ANOSIM was applied to assess if there was similarity in the macroinvertebrate community composition among the treatments and the sampling days. This test was also used to assess if there was similarity in the fauna composition of the trophic guilds among the treatments and the sampling days. The Bray-Curtis index was used to construct the similarity matrix. For these analyses, the programs PAST 1.82b (Hammer et al., 2001) and BioEstat 5.0 (Ayres et al., 2007) were used. To assess the period when the herbicides might affect the benthic community, a PRC (Principal Response Curve) was constructed, using the statistical program R (vegan package).

3. Results

3.1 Abiotic data

The abiotic data were measured on specific dates, which covered the entire period of rice culture. The statistical analyses showed no differences among treatments, but there were differences among the sampling days (Table 1). The PCA showed no correlation among the parameters analyzed. The highest dissolved oxygen content was recorded on day 42,

11.90mg.L^{-1}, in the treatment with Quinclorac. The pH remained between 6 and 7, during almost the entire experiment; although the variability was low, the differences among sampling days were significant (Table 1).
The water temperature varied from 34.2°C on the first day of experiment to 17.7°C on the last one, in January and April, respectively. Hardness was highest on days 21 and 28, with

	O_2D (mg.L^{-1})							
	3^{c}	7^{a}	14^{c}	21^{b}	28^{a}	42^{d}	56^{cd}	84^{a}
TO	6.90	3.50	5.43	4.50	4.10	11.40	7.40	3.90
TB	6.50	3.60	7.27	6.00	2.90	11.10	7.30	3.40
TQ	6.60	4.10	5.98	6.30	4.50	11.90	6.60	3.30
TC	7.50	4.40	6.86	4.20	5.40	11.70	9.10	2.50
	pH							
	3^{cd}	7^{e}	14^{de}	21^{ab}	28^{c}	42^{bc}	56^{bc}	84^{a}
TO	6.81	7.30	7.08	6.83	6.73	6.60	6.40	6.37
TB	6.74	6.90	7.17	6.81	6.62	6.52	6.74	6.15
TQ	6.73	7.19	7.03	6.22	6.60	6.25	6.47	6.21
TC	6.78	7.04	6.82	5.98	6.58	6.44	6.49	6.12
	Water Temperature in °C							
	3^{f}	7^{e}	14^{c}	21^{b}	28^{c}	42^{d}	56^{d}	84^{a}
TO	34.20	26.67	22.38	20.23	22.13	23.77	23.60	17.80
TB	33.53	26.57	22.42	20.23	22.10	23.57	23.33	17.80
TQ	33.67	26.63	22.56	20.70	22.27	23.67	23.50	17.75
TC	33.23	26.63	22.50	20.70	22.70	23.57	23.53	17.70
	Hardness (mg.L^{-1} de CaCO3)							
	3^{a}	7^{bc}	14^{ab}	21^{c}	28^{c}	42^{nb}	56^{a}	84^{a}
TO	17.00	32.00	28.00	40.00	40.00	36.00	24.00	36.00
TB	22.00	36.00	20.00	40.00	40.00	32.00	28.00	32.00
TQ	26.00	32.00	24.00	40.00	36.00	32.00	20.00	24.00
TC	22.00	24.00	24.00	36.00	32.00	36.00	20.00	24.00
	Turbidity (NTU)							
	3^{c}	7^{c}	14^{bc}	21^{ab}	28^{c}	42^{c}	56^{c}	84^{b}
TO	22.73	14.00	14.80	6.60	23.90	14.60	17.10	15.90
TB	17.00	18.20	17.60	4.52	23.10	24.20	28.90	8.47
TQ	18.77	17.50	18.40	4.86	18.50	17.00	15.40	6.19
TC	22.40	15.20	12.80	3.92	11.10	9.45	10.00	6.22

Table 1. Santa Maria, Brazil: Physical and chemical parameters analyzed in the irrigated rice field, on different days after flooding, during the 2007/08 growing season, in experimental plots in central Rio Grande do Sul. TO, treatment with the herbicide Only®; TB, treatment with the herbicide Bispyribac-sodium; TQ, treatment with the herbicide Quinclorac; TC, control treatment. Similar letters indicate statistical similarity.

40 mg. L $^{-1}$, decreasing from day 42 on. Water turbidity ranged from 2.53 NTU on day 84 to 28.90 NTU on day 56 (Table 1).

3.2 Agrochemical persistence

Quinclorac showed the highest persistence (Table 2), and was detected until 84 days after the plots were flooded. On this date, however, its concentration was low and it was not detected on day 90, the day of harvest (on this day, the water sample was collected in the refuge). The herbicide Only® showed the lowest persistence, and was detected until day 21 (Table 2).

	Pesticides Concentration μg. L^{-1}												
Treatments/ Days	0	1st	2 nd	3 rd	7 th	14th	21st	28th	42nd	56th	77th	84th	90th
TO	12.9	22.2	16.0	11.7	7.1	3.7	2.4	-	-	-	-	-	-
TB	8.7	20.8	12.9	10.5	8.9	4.3	1.8	1.2	0.3	0.2	-	-	-
TQ	138.8	296.7	173.3	107.6	47.0	39.2	38.0	28.2	20.6	9.8	4.9	3.8	-

Table 2. Santa Maria, Brazil: Mean concentrations of agrochemicals in irrigated rice plots during the 2007/08 growing season. TO, treatment with the herbicide Only®; TB, treatment with the herbicide Bispyribac-sodium; TQ, treatment with the herbicide Quinclorac. Day 0, flooding of the experimental plots, and then successive dates.

3.3 Benthic macroinvertebrates

A total of 3763 animals were identified, 1799 in the first collection (day 28) and 1964 in the second one (day 84). All the taxa identified in the first collection were present in the second. In the latter, some taxa were identified that were not present in the first collection (Table 3).
In both collections the control treatment (TC) showed the highest density of organisms (no. ind. m-2), followed by TO, TB, and TQ (Table 4). However, regarding the abundance of taxa recorded in the second collection, the treatment with the most abundance was TO.
In relation to fauna composition (abundance and diversity), in the first collection the ANOSIM indicated a similarity in the community composition only between TC and TB and between TB and TO. In the second collection, all the treatments diffed among themselves and also differed from the first collection.
The ANOSIM, used to assess if there was similarity in the fauna composition (abundance and diversity) in each trophic guild when comparing the treatments, showed that some were similar. The treatment showing the highest differences was TQ. The data for the day 28 collection are given in Figure 1A, and the data from day 84 in Figure 1B. The results for each guild in the treatments also differed between collections; i.e., the guilds in TC in the first treatment showed a different community composition in the second, and the same occurred for TO, TB, and TQ.
The PRC showed that, in general, all the herbicides caused an initial stress on the benthic community, and this stress decreased with time. At the end of the cultivation, the community was structured again, except the community that received TQ which continued to be affected by this herbicide until the end of the cultivation.

4. Discussion

The results showed that the pesticides did not significantly alter the abiotic factors of the water used in the experiment, considering that there were no significant differences among treatments, in the same collection. These results differ from those obtained by Faria et al. (2007), who reported that localities with high rates of contamination by pesticides and heavy metals, showed pH and O_2 values different from contamination-free areas or areas with low contamination, and attributed this difference to the presence of chemicals in the environment. Similarly, Schulz & Liess (1999), in a study of streams that supply water for the field, separated into low, medium, and high impact, also attributed to pesticides the differences found in the water quality. The influence of pesticides in reducing water quality was also reported by Molozzi et al. (2006), who evaluated irrigation and drainage water from rice fields.

In areas of rice cultivation, the oxygen concentration tends to be low, because the water depth is between 5 and 15 cm. In experiments carried out in the same area of the present study, in previous years the oxygen concentration was 0.6 to 2.2 mg. L^{-1} (Golombieski et al., 2008) and between 2.4 - 4.6 mg.L^{-1} (Reimche et al., 2008). In this experiment, the variation in the level of oxygen was 2.4 - 11.9 mg.L^{-1}, and most days it was higher than 6 mg. L^{-1}, a high level for the area. This may have favored an increase in the abundance obtained in the last collection, since some taxa are sensitive to low levels of dissolved oxygen.

The values of hardness and alkalinity were higher than in the present study compared to data from Golombieski et al. (2008) and from Reimche et al. (2008). A difference in the dissociation of ions of the compounds used in the experiments could explain this disparity. A factor that may also explain these results is that during the rice cultivation the temperature was decreasing, and the highest temperature occurred at the beginning of the study in January 2008, and the lowest temperatures in the last collection, in April. In the above-cited studies, the temperatures were lower than those recorded in this study, 16.4-23.9 °C (Reimche et al., 2008) and 17.6-25.7°C (Golombieski et al., 2008).

Another factor to consider is the relatively low persistence of the compounds used in this study. All the herbicides except Quinclorac showed lower persistence than found in some studies (USEPA, 1996; Stevens et al., 1998; Mesléard et al., 2005; Mize et al., 2008; Reimche et al., 2008). However, Quinclorac was at least 2.5 times more persistent than observed by Reimche et al. (2008).

An evaluation of the persistence of pesticides in water of rivers in the region found that the compounds forming the commercial formulation Only (Imazapic + Imazethapyr) were present in all the sampling periods (Silva et al., 2009b). These data differ from those obtained in this study, where the herbicide Only® was measurable until day 21. These results are similar to those found in the water of rivers in the region of Pelotas, Rio Grande do Sul, in a study carried out in 2007, where Quinclorac also persisted during the entire period of rice cultivation (Grutzmacher et al., 2008). It is possible that these differences in persistence are due to the management of the rice field. It may be advisable to maintain water in the field as long as possible (Marchesan et al., 2005; Grutzmacher et al., 2008). It is important to note that the studies cited here assessed the persistence of pesticides in the field water. If this persistence were analyzed in the sediment, the values might be different.

During the period of cultivation, the absolute abundance and diversity of the aquatic community tended to increase. Similar increases were found by other researchers (Schulz and Liess, 1999; Suhling et al., 2000). The most coherent explanation for this is that, as time passes, the pesticides become diluted and dissipate, thus decreasing their toxicity to the aquatic organisms (Schulz & Liess, 1999). However, Molozzi et al. (2007) did was not record temporal differences in the benthic community in relation to the stages of the rice seedling.

The benthic macroinvertebrate community is composed of many taxa, among which one of the most studied is the insect family Chironomidae. Some genera of this family are considered plagues in rice fields, and because of this pesticides have been developed to minimize their effects. In the present study, all the herbicides tested, although they do not target these organisms, negatively affected the chironomid population. This influence was perceptible for the herbicide Only® in the first collection and Bispyribac-sodium in the second, and for Quinclorac in both. This influence was also evident in other taxa of the benthic community, mainly in the first collection when the agrochemicals were present in higher concentrations in the water.

The coleopterans Curculionidae, Dytiscidae, and Hydrophilidae, the dipterans Ceratopogonidae and Chironomidae, the hemipteran Corixidae, the odonate Aeshnidae, the trichopteran Odontoceridae, as well as the Collembola, Ostracoda, Annelidae, Nematoda, and the mollusc Ampullaridae could be classified as persistent. Only the mollusc family Planorbidae were classified as resilient. The other taxa must be considered occasional, since they either were not collected in TC or were collected only in the second sampling.

Quinclorac showed a different dynamic than that studied in the laboratory by Crosby (2003), where it showed low persistence. In the study by Reimche et al. (2008), this pesticide was detected until the 31st day, with its highest concentration observed on the 7th day with 102 $\mu g.L^{-1}$. In the present study, its persistence was higher than the other pesticides investigated, being detected until day 84 in a concentration of 3.8 $\mu g.L^{-1}$ concentration. Its highest value was on the first day, with 296.7 $\mu g.L^{-1}$. The negative influence of this pesticide was perceptible in this study, since in both collections the treated plot had the lowest abundance and diversity of taxa compared to the other treatments. This possible influence was also recorded by Reimche et al. (2008), where the cladoceran assemblage showed a density variation caused by the stress of the application of this herbicide.

The use of pesticides reduces the availability of food for the benthic macroinvertebrates and changes the structure of the algae community, and the toxic effect on these food resources may also be associated with this (Gagneten, 2002). Some studies have suggested the possibility that the toxic agents to non-target organisms, in the case of herbicides, are the adjuvants and the surfactants used in the commercial formulation of each pesticide (Tatum 2004; Kitulagodage et al., 2007). However, more studies taking these compounds into consideration separately are necessary.

In relation to the community composition, it was clear that the pesticides cause alterations in their composition. This effect was most apparent over the long term, even with the community recovery. According to Berezen et al. (2005), this potential relationship between contamination by pesticides and the community structure is based mainly on physiological differences that affect the life cycle and the species' mobility. Therefore, each pesticide has a different effect on the organisms.

In studies in rice fields in France, no significant difference was found between a conventional area of cultivation and an organic one for the family Chironomidae (Mesléard et al., 2005). This may have occurred due to a decrease of predators of this group. A decrease in the numbers of predators was also observed in the present study, and may have been responsible for the alterations recorded in the proportions among the trophic guilds analyzed. This may have made possible an increase in the density of detritivores, for instance.

The guild most injured by pesticides was the scrapers, which play an important role in this ecosystem since they feed on algae, bacteria, fungi and dead organic matter adsorbed on the substrate surface (Merrit & Cummins, 1996). This result is probably due to the action of the herbicides on the algae, which serve as food for these animals. Detritivores, which start their

feeding process after the action of microorganisms that make this food more palatable (Cummins et al., 1989), were recorded in low densities in the first collection; however, at the end of the culture their densities increased. Detritivores remained in low densities even at the end of the culture only in the TQ-treated plot, probably because of the effect of this chemical on the microorganisms needed to increase the palatability of the food.
The data from this study, carried out in area of rice cultivation, reinforce the role of bioindicators performed by the benthic macroinvertebrates. In addition, as observed here, abiotic and biotic data not always give similar results when environmental matters are investigated. However, the combination of these two parameters of analyses can generate more substantial information on environmental impacts.

Taxa	Trophic guild	Sampling	TC	TO	TB	TQ
Artropoda						
Insecta						
Coleoptera						
Curculionidae	HR	28	X	X	X	X
		84	X	X	-	-
Ditiscidae	P	28	X	X	X	-
		84	X	X	-	
Hidrophilidae	P	28	X	X	-	X
		84			X	-
Psephenidae	HR	28	-	-	-	-
		84	-	-	X	-
Diptera						
Ceratopogonidae	CF/ CC/ P	28	X	X	X	X
		84	X	X	X	X
Chironomidae	CF/ CC/ P	28	X	X	X	X
		84	X	X	X	X
Tabanidae	P	28		-		
		84	-	X	-	X
Tipulidae	P/ D	28	-	-	-	-
		84	-	-	X	-
Ephemeroptera						
Baetidae	CF/ CC	28	-	-	-	-
		84	-	-	-	X
Caenidae	CF/ CC	28	-	-	-	-
		84	X	-	-	-
Leptophlebiidae	CF/ CJ	28	-	-	-	-
		84	-	-	X	-
Leptohyphidae	CF/ CC	28	-	-	-	-
		84	-	X	X	-
Hemiptera						
Belostomatidae	P	28	-	-	-	-
		84	X	-	-	-
Corixidae	HR	28	-	-	-	-
		84	X	X	-	-

Pentatomidae	P	28	-	-	-	-
		84	-	X	-	-
Odonata						
Aeshnidae	P	28	X	-	-	X
		84	X	-	X	X
Coenagrionidae	P	28	-	-	-	-
		84	X	X	-	-
Lestidae	P	28	-	-	X	-
		84	-	-	X	-
Libellulidae	P	28	-	-	-	-
		84	X	-	-	-
Perilestidae	P	28	-	-	-	-
		84	-	X	-	-
Plecoptera						
Perlidae	P	28	-	-	-	-
		84	-	X	-	-
Trichoptera						
Odontoceridae	D	28	X	X	X	-
		84	X	X	X	-
Collembola	D	28	X	X	X	X
		84	X	X	X	X
Crustacea						
Ostracoda	D	28	X	X	X	X
		84	X	X	X	-
Arachnidae						
Aranae	P	28	-	-	-	-
		84	-	X	-	-
Hidracarina	P	28	X	-	-	-
		84	-	X	-	-
Annelida						
Hirudinea	P	28	X	X	X	-
		84	X	X	X	X
Oligochaeta	D	28	X	X	X	X
		84	X	X	X	X
Nematoda	P	28	X	X	X	X
		84	X	X	X	X
Mollusca						
Ampullaridae	HR	28	X	X	X	X
		84	X	X	X	X
Planorbidae	HR	28	X	-	-	X
		84	X	X	X	X

Table 3. Santa Maria, Brazil, 2007/08 growing season: Abundance of taxa identified in the area of irrigated rice cultivation with different treatments. HR: herbivore scraper, P: predator, CC: collector catcher, CF: filter collector, D: detritivore, TC: control treatment, TO: treatment with Only®, TB: treatment with Bispyribac-sodium, TQ: treatment with Quinclorac. "X" indicates the presence of a taxon in the plot, and "-" indicates its absence.

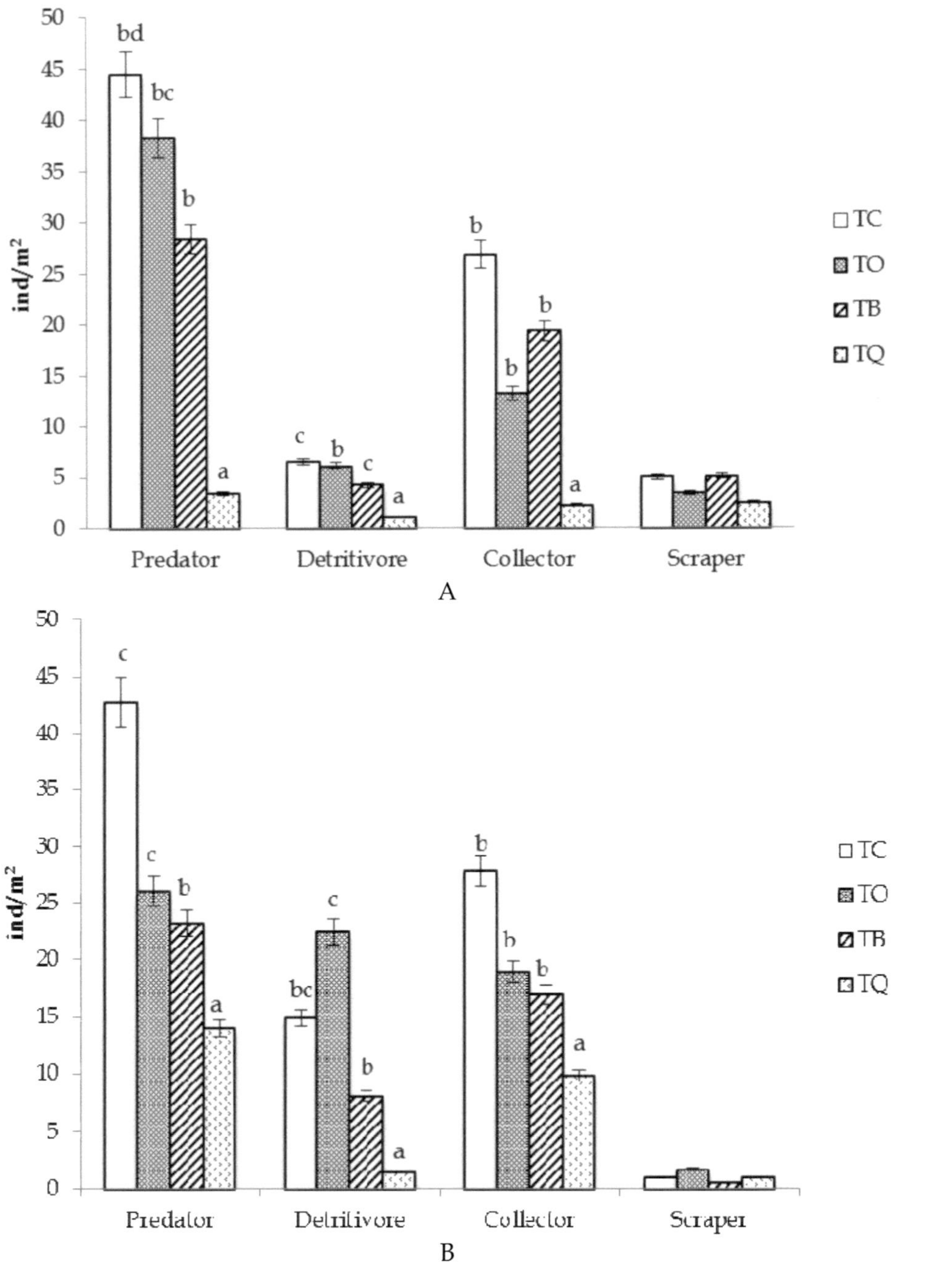

Fig. 1. Santa Maria, Brazil, 2007/08 growing season: Differences in the faunal composition in each trophic guild, evaluated by ANOSIM ($p<0.05$). Different letters indicate a statistical difference. A: collection 28 days after the experiment al plots were flooded, B: collection 84 days after flooding. TC: control treatment, TO: treatment with the herbicide Only®, TB: treatment with the herbicide Bispyribac-sodium, TQ: treatment with the herbicide Quinclorac.

	Org. average density.m-2		Abundance of the taxa collected	
	28ºd	84ºd	28ºd	84ºd
TC	147.81	157.24	15	18
TO	126.54	134.87	12	21
TB	100.66	86.84	12	17
TQ	18.42	43.20	11	11

Table 4. Santa Maria, Brazil, 2007/08 growing season: mean densities (no. ind.m^{-2}) and abundances of taxa identified in the experimental plots, by treatment.

5. Acknowledgments

To CNPq proc. nº 552546/2007-0 and Pró-Publicações Internacionais / PRPGP / UFSM by financial support. To Dr. Renato Zanella by the persistence analyses and to our colleagues for their help in the field works.

6. References

APHA (American Public Health Association). 1992. Standard methods for examination of water and wastewater. 18.ed. Sprinfield: Bru-El Graphic, 1155p.

Ayres, M.; Ayres Jr., M. Ayres, D. L. & Santos, A. S. BioEstat 5.0. 2007. Aplicações estatísticas nas áreas das ciências biológicas e médicas. Sociedade Civil Mamirauá (MCT). Imprensa Oficial do Estado do Pará.

Berenzen, N., Kumke,T., Schulz, H., Schulz, R. 2005. Macroinvertebrate community structure in agricultural streams: impact of runoff-related pesticide contamination. Ecotoxicology and Environmental Safety, 60: 37-46.

Biggs, J.; Williams, P.; Whitfield, M.; Nicolet, P.; Brown, C.; Hollis, J.; Arnold, D.; Pepper, T. 2007. The freshwater biota of British agricultural landscapes and their sensitivity to pesticides. Agriculture, Ecosystems and Environment. 122: 137-148.

Callisto, M. & Esteves, F. A. 1998. Categorização funcional dos macroinvertebrados bentônicos em quatro ecossistemas lóticos sob influência das atividades de uma mineração de bauxita na Amazônia Central (Brasil). Oecologia Brasiliensis, 5(5): 223-234.

Costa, C.; Ide, S.; Simonka, C. C. 2006. Insetos imaturos. Metamorfose e Identificação. Ribeirão Preto: Holus. 249p.

Crosby, D.R. 2003. Environmental fate of pesticides-87. Capturado em 10 mar. Online. Disponível na internet http://www.syix.com/rrb/87rpt/Enviro.htm.

Cummins, K.W., Wilzbach, M.A., Gates, D.M., Perry, J.B., Taliaferro, W.B. 1989. Shredders and riparian vegetation. Leaf litter that falls into streams influences communities of stream invertebrates. BioScience, 39 (1): 24-30.

Faria, M. F., Nogueira, A. J. A., Soares, A. M. V. M. 2007. The use of *Chironomus riparius* larvae to assess effects of pesticides from rice fields in adjacent freshwater ecosystems. Ecotoxicology and Environmental Safety, 67: 218-226.

Fernández, H. R. & Domínguez, E. 2001. Guia para La determinación de los artrópodos bentónicos Sudamericanos. Tucumán: Imprenta Central de la UNT, Argentina, 289p.

Gagneten, A.M. 2002. Efectos del herbicida paraquat sobre el zooplancton. Iheringia, 92: 47-56.

Golombieski, J. I.; Marchesan, E.; Baumart, J. S.; Reimche, G. B.; Resgalla , C. Jr.; Storck, L.; Santos, S. 2008. Cladocers, Copepods and Rotifers in rice-fish culture handled with metsulfuron-methyl and azimsulfuron herbicides and carbofuran insecticide. Ciência Rural, 38(8): 2097-2102.

Goulart, M. & Callisto, M. 2003. Bioindicadores de qualidade de água como ferramenta em estudos de impacto ambiental. Revista da FAPAM, 2(1).

Grützmacher, D. D.; Grützmacher, A. D.; Agostinetto, D.; Loeck, A. E.; Roman, R.; Peixoto, S. C.; Zanella, R. 2008. Monitoramento de agrotóxicos em dois mananciais hídricos no sul do Brasil. Revista Brasileira de Engenharia Agrícola e Ambiental, 12(6): 632-637.

Hammer, Ø; Harper, D. A. T., Ryan, P. D. PAST 1.82b. 2001. Past: Paleontological Statistics Software package for education and data analysis. Paleontologia Electronica, 4(1): 9pp.

Kitulagodage, M.; Astheimer, L.B.; Buttermer, W. A. 2007. Diacetonealcohol, a dispersant solvent, contributes to acute toxicity of a fipronil - based insecticide in a passerine bird. Ecotoxicology and Environmental Safety, 71: 597-600.

MAPA- Ministério da agricultura Pecuária e Abastecimento. Acessado em dezembro de 2009. http://www.agricultura.gov.br/

Marchesan, E.; Zanella, R.; Avila, L. A. de; Camargo, E. R.; Machado, S. L. de O.; Macedo, V. R. M. 2003. Rice herbicide monitoring in two brazilian rivers during the rice growing season. Scientia Agricola, 64 (2): 131-137.

Marchesan, E.; Santos, F. M. dos; Camargo, E. R.; Machado, S. L. de O.; Zanella, R.; Ávila, L. A. de; Gonçalves, F. F.; Primel, E. G.; Macedo, V. R. M. 2005. Monitoramento em rios e persistência de herbicidas em arroz irrigado. Santa Maria: UFSM, 8p. Informe Técnico, 05.

Marinoni, R. C. 2001. Os grupos tróficos em Coleóptera. Revista Brasileira de Zoologia, 18(1): 205-224.

Merritt, R. W. & Cummins, K. W. 1996. In: ____. An Introduction to the Aquatic Insects of North America. 3 ed., Dubuque, Kendall/Hunt, 758 p.

Mesléard, F.; Garnero, S.; Beck, N.; Rosecchi, É. 2005. Uselessness and indirect effects of an insecticide on rice field invertebrates. Comptes Rendus Biologies, 328: 955-962.

Mize, S. V.; Porter, S. D.; Demcheck, D. K. 2008. Influence of fipronil compounds and rice-cultivation land-use intensity on macroinvertebrate communities in streams of southwestern Louisiana, USA. Environmental Pollution, 152: 491-503.

Molozzi, J.; Pinheiro, A.; Silva, M. R. 2006. Qualidade da água em diferentes estádios de desenvolvimento do arroz irrigado. Pesquisa Agropecuária Brasileira, 41(9):1393-1398.

Molozzi, J.; Hepp, L. U.; Dias, A. da S. 2007. Influence of rice crop on the benthic community in Itajaí Valley (Santa Catarina, Brazil). Acta Limnologica Brasiliensia, 19(4): 383-392.

Moreby, S.J. & Southway, S. E. 1999. Influence of autumn applied herbicides on summer and autumn food available to birds in winter wheat fields in southern England. Agriculture, Ecosystems and Environment, 72: 285-297.

Primel, E.G.; Zanella, R.; Kurz, M.H.S.; Gonçalves, F.F.; Machado, S.de O.; Marchezan, E. 2005. Poluição das águas por herbicidas utilizados no cultivo do arroz irrigado na região central do estado do rio grande do sul, Brasil: predição teórica e monitoramento. Química Nova, 28(4): 605-609.

Reimche, G. B.; Machado, S. L. de O.; Golombieski, J. I.; Baumart, J. S.; Braun, N.; Marchesan, E.; Zanella, R. 2008. Persistência na água e influência de herbicidas utilizados na lavoura arrozeira sobre a comunidade zooplanctônica de Cladocera, Copepoda e Rotifera. Ciência Rural, 38(1): 7-13.

Rosenberg, D. M. & Resh, V.H. 1993. Introduction to freshwater biomonitoring and benthic macroinvertebrates. In: ____. Freshwater biomonitoring and benthic macroinvertebrates. Chapman and Hall, New York, pp. 1-9.

Schulz, R. & Liess, M. 1999. A field study of the effects of agriculturally derived insecticide input on stream macroinvertebrate dynamics. Aquatic Toxicology, 46: 155-176.

Silva, F. L. da; Pauleto, G. M.; Talamoni, J. L. B.; Ruiz, S. S. 2009a. Categorização funcional trófica das comunidades de macroinvertebrados de dois reservatórios na região Centro - Oeste do Estado de São Paulo, Brasil. Acta Scientiarum Biological Sciences, 31(1): 73-78.

Silva, D. R. O.; Avila, L. A. de; Agostinetto, D.; Dal Magro, T.; Oliveira, E. de; Zanella, R.; Noldin, J. A. 2009b. Monitoramento de agrotóxicos em águas superficiais de regiões orizícolas no sul do Brasil. Ciência Rural, 39(9): 2383-2389.

SOSBAI - Sociedade Sul-Brasileira De Arroz Irrigado. 2005. Arroz Irrigado: recomendações técnicas da pesquisa para o Sul do Brasil. Santa Maria, RS, 159p.

Stevens, M. M.; Helliwell, S.; Warren, G. N. 1998. Fipronil seed treatments for the control of chironomid larvae (Diptera:Chironomidae) in aerially-sown rice crops. Field Crops Research, 57: 195- 207.

Suhling, F., Befeld S., Hausler, M., Katzur, K., Lepkojus, S., Mesléard, F. 2000. Effects of insecticide applications on macroinvertebrate density and biomass in rice-fields in the Rhone delta, France. Hydrobiologia, 431: 69–79.

Tatum, V L. 2004. Toxicity, transport, and fate of forest herbicides. Wildlife Society Bulletin, 32(4):1042-1048.

US Environmental Protection Agency. 1996. New Pesticide Fact Sheet, PB- 96-181516. USEPA Office of Prevention, Pesticides, and Toxic Substances. EPA737-F-96-005.

Winterbotton, J.H.; Orton, S. E.; Hildrew, G.; Lancaster, J. 1997. Field experiments on flow refuge in streams. Freshwater Ecology, 37: 569-580.

Zanella, R.; Primel, E. G.; Machado, S. L. O.; Gonçalves, F. F.; Marchezan, E. 2002. Monitoring of the herbicide clomazone in environmental water samples by solid-phase extraction and high-performance liquid chromatography with ultraviolet detection. Chromatographia, 55(9/10): 573-577.

18

Structural and Functional Effects of Herbicides on Non-Target Organisms in Aquatic Ecosystems with an Emphasis on Atrazine

J.F. Fairchild
U.S. Geological Survey,
USA

1. Introduction

Herbicide use has increased dramatically around the world over the past 6 decades (Gianessi and Reigner, 2007). Few herbicides were in use in the 1950s. However, by 2001 approximately 1.14 billion kilograms of herbicides were applied globally for the control of undesireable vegetation in agricultural, silvicultural, lawncare, aquacultural, and irrigation/recreational water management activities (Kiely et al., 2004). Twenty-eight percent of the total mass of herbicides is applied in the United States, with the remaining 72 percent being applied elsewhere around the globe (Kiely et al., 2004). Herbicides represent 36% of global pesticide use, followed by insecticides (25%), fungicides (10%) and other chemical classes (Kiely et al., 2004).

Agricultural production accounts for approximately 90% of herbicide use in the U.S. (Kiely et al., 2004). Gianessi and Reigner (2007) indicated that herbicides are routinely used on more than 90% of the area designated for large commercial crops including corn, soybeans, cotton, sugar beets, peanuts, and rice. Increased farm mechanization, technological advancements in production of inexpensive sources of inorganic nitrogen fertilizer (e.g., anhydrous ammonia), and conversion of forest, grassland, and wetland habitats to cropland has led to a tremendous increase in global food production over the past half-century. Herbicides have augmented advances in large-scale agricultural systems and have largely replaced mechanical and hand-weeding control mechanisms (Gianessi and Reigner, 2007).

The wide-spread use of herbicides in agriculture has resulted in frequent chemical detections in surface and groundwaters (Gilliom, 2007). The majority of herbicides used are highly water soluble and are therefore prone to runoff from terrestrial environments. In additon, spray drift and atmospheric deposition can contribute to herbicide contamination of aquatic environments. Lastly, selected herbicides are deliberately applied to aquatic environments for controlling nuisance aquatic vegetation. Although aquatic herbicide exposure has been widely documented, these exposures are not necessarily related to adverse non-target ecological effects on natural communities in aquatic environments. This chapter evaluates the potential for effects of herbicides on the structure and function of aquatic envrionments at the population, community, and ecosystem levels of biological organization. In this manuscript I examine several critical aspects of the subject matter area: primary herbicides in use and chemical modes of action; the regulatory process used for registration and risk assessment of herbicides; data regarding non-target risks and the relative sensitivity of aquatic plants, inveretebrates, and fish to herbicides; and emerging areas of science regarding the potential for endocrine-disrupting effects of herbicides on aquatic vertebrates. Much of the focus of this paper is on atrazine due to the extensive database which exists regarding its fate and effects.

2. Herbicide production, use, and regulation in the United States

2.1 Herbicide production and use

Agricultural statistics indicate that total herbicide use in the United States has been relatively stable to declining over the past 2 decades with an approximate average usage of 250 million kilograms per year (Fig. 1) (Kieley et al., 2004). Total herbicide use in the United States has remained relatively stable over the past 25 years due to the decrease in application rates of atrazine and the increased use of the low-volume chemicals such as the acetolactate synthase (ALS) inhibitors (Reade and Cobb, 2002; Menne and Kocher, 2007).

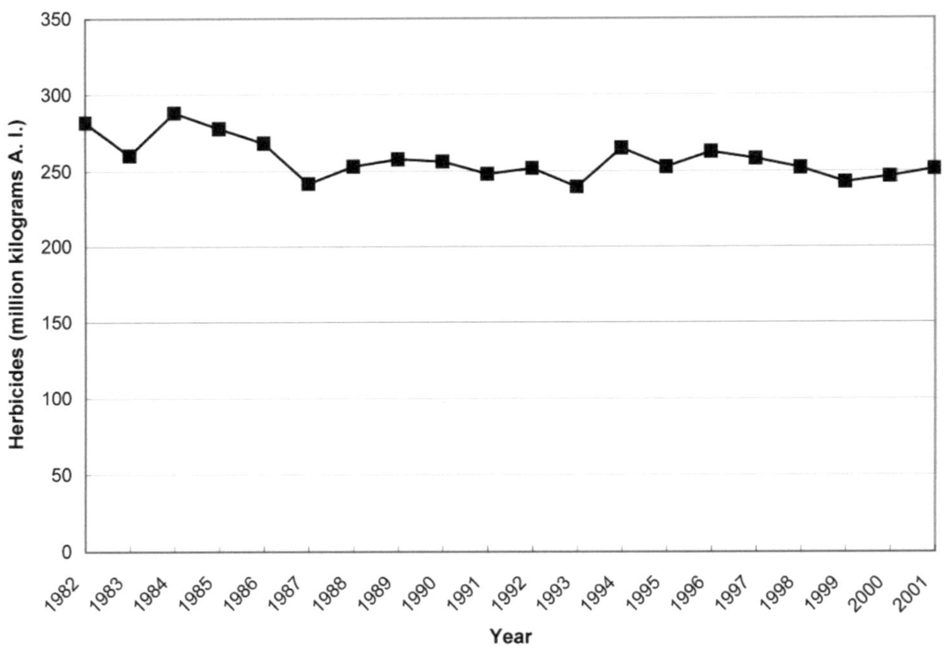

Fig. 1. Trendline in total herbicide use in the United States from years 1982 to 2001. Data from Kieley et al. (2004).

In 1987, the primary herbicides used in the United States were atrazine, alachlor, metolachlor, and 2,4-D (Table 1). The major change in herbicide use-trends has been driven by the significant increase in use of glyphosate due to the development of glyphosate-resistant crops including soybeans, corn, and cotton. In 2001, glyphosate replaced atrazine in terms of total product application, followed by the increase in use of acetochlor and 2,4-D (Table 1).

Active Ingredient	2001		1999		1997		1987	
	Rank	Range	Rank	Range	Rank	Range	Rank	Range
Glyphosate	1	39-41	2	30-33	3	15-17	9	3-4
Atrazine	2	34-36	1	34-46	1	34-37	1	32-34
Acetochlor	3	14-16	3	14-16	4	14-16	NA	NA
2,4-D	4	13-15	4	13-15	5	13-15	4	13-15
Metolachlor-S	5	9-11	8	7-9	NA	NA	NA	NA
Metolachlor	6	12-14	5	12-14	2	27-31	3	20-23
Pendimethalin	7	7-9	6	8-10	6	11-13	7	3-5
Trifluralin	8	5-7	7	8-10	7	10-11	5	11-14
Alachlor	9	3-3	9	3-5	8	6-7	2	25-27
Propanil	10	3-4	10	3-5	9	3-4	8	3-5
Dimethenamid	11	3-4	12	3-4	10	3-4	NA	NA
EPTC	12	2-4	11	3-4	11	3-4	6	8-10
Simazine	13	2-3	NA[2]	NA	NA	NA	NA	NA
Dicamba	14	2-3	13	3-4	12	3-4	10	2-3
Sulfosate	15	1-3	NA	NA	NA	NA	NA	NA

Data from Kiely et al. (2004).

Table 1. List of the major herbicides applied in the United States from 1987 to 2001 categorized by rank and amount of use (million kilograms/year).

There are approximately 220 registered herbicides that can be classified among 15 known herbicidal modes of action, whereas 48 herbicides have no identified modes of action (Cole et al., 2000; Read and Cobb, 2002; Menne and Kocher, 2007) (Table 2). The majority of

Target site or Mode of Action	Herbicide #'s	Examples of Chemical Classes
Photosystem II	59	Triazines, Phenylureas
Acetolactate synthase	43	Sulfonureas, Imidazolinone
Microtubule and cell division	29	Chloroacetamides, Dinitroanalines
Protoporphyrinogen oxidase	28	Diphenyl ethers, Triazolinones
Auxin mimics	20	Phenoxy acids, Pyridines
Acetyl CoA Carboxylase	16	Aryoxyphenoxypropionates
Phytoene desaturase	11	Pyridazinone
Hydroxyphenylpyruvate dioxygenase	3	Triketones, Isoaxazole
Oxidative phosphorylation	3	Dinitrophenols
Cellulose biosynthesis	2	Nitriles, Benzamides
Photosystem 1	2	Bypyridiliums
Auxin transport	1	Phthalamate, Semicarbazones
Dihydropteroate synthetase	1	Carbamates
Glutamine synthetase	1	Phosphinic acids
Lycopene cyclase	1	Dimethylamines
Unknown or not stated	48	

Data from Cole et al. (2000), Read and Cobb (2002), and Menne and Kocher (2007).

Table 2. Target sites/mode of action of commercial herbicides ranged by number of products.

herbicides, however, fall into a relatively small number of groups of modes of action including inhibitors of the photosystem II photosynthetic reaction; inhibitors of acetolactate synthase; inhibitors of cellular division; inhibitors of protoporhyrinogen IX oxidase; and the auxin mimics. These five modes of action are rather non-selective, which has led to proposed increases in research of herbicides using genomics to develop chemicals with new and unique modes of action that can be applied at low rates with high efficacy but that will not result in herbicidal resistance in weeds such as that recently observed with glyphosate and some acetolactate (ALS) inhibitors (Cole et al., 2000; Moss, 2002). It is anticipated that these newer modes of action would also have minimal effects on non-target aquatic organisms.

2.2 Herbicide regulation

Herbicides were first mass-produced in the early 1950s for the deliberate application to the environment for the control of weeds in agriculture, silviculture, right-of-ways, and turf lawns (Giannessi and Reigner, 2007). The history of herbicide registration and regulation has differed around the world depending on the structure of governments, the rate of scientific advancements, and the social perceptions of the need for environmental regulation. Over time we have seen major movements in developed countries toward harmonization of guidelines and requirements for herbicide registration and regulation.

United States: Herbicides and other pesticides have been regulated in the United States for over 60 years under a series of legislative mandates. Pesticide registrations were originally regulated under the Federal Insecticide, Fungicide, and Rodenticide Act (FIFRA) which was enacted in 1948 as an expansion of the Federal Water Pollution Control Act (FWPCA) of 1948. The U.S. Department of Agriculture (USDA) was responsible for regulation of pesticides in the U.S. until 1970 when responsibilities were transferred to the newly created U.S. Environmental Protection Agency (USEPA). Continued concern over the exposure and effects of pesticides on humans and non-target animals (invertebrates, fish, birds, mammals, and other wildlife) led to a total revision of the FIFRA in 1972. Substantial changes were made in 1988 to accelerate the pesticide re-registration process. The establishment of the Food Quality and Protection Act of 1996 (FQPA) brought further changes to the registration of pesticides including herbicides (Flynn, 2002; Saundry, 2006). Industrial chemicals were also first regulated under the FWPCA (1948) but requirements were subsequently revised under the Toxic Substances Control Act (TSCA) of 1976. Over the past decade testing guidelines for the effects of pesticides and industrial chemicals on non-target organisms have been harmonized and consolidated within the U.S. EPA Office of Chemical Safety and Pollution Prevention (OCSPP) (USEPA, 2010).

Europe: Prior to 1991, member states of the European Union (EU) used varying approaches in the registration and regulation of herbicides and other pesticides which collectively created a burden to agricultural trade and practices (Flynn, 2002). Therefore, the EU consolidated the regulation of herbicides and other plant protection products under Council Directive 91/414/EEC which was formally adopted on July 25, 1993 (Flynn 2002). Herbicides and other agricultural chemicals used in the European Union are now regulated under guidelines originally developed under the Organization for Economic Cooperation and Development (OECD) and more recently by the European Commission (EC; http://ecb.jrc.ec.europa.eu/). Further refinement of the pesticide registration process in Europe is expected to occur following the enactment of the Registration, Evaluation,

Authorization and Restriction of Chemicals (REACH) legislation implemented in June 2007 (REACH, 2007) with a goal of total completion by the year 2015.
Canada: Herbicide regulation in Canada was first enacted under the Agricultural Pest Act of 1927, and was subsequently amended as the Pesticide Products Act (PCPA) of 1939 (Flynn, 2002). In 1995, the Pest Management Regulatory Agency (PMRA) became a part of Health Canada which is responsible for the protection of human health and the environment including pesticide regulation of herbicides. Pesticides are regulated by the Pesticide Control Products Act (PCPA) of 2002 which was adopted in 2006 and managed by the Pest Management Regulatory Agency of Health Canada (Canadian Ministry of Justice, 2010). Tests for regulatory purposes are conducted in general accordance with U.S. and European guidelines.

2.3 Herbicide registration and the risk assessment process

Philosophically, the tiered assessment testing procedure follows the generalized framwork for ecological risk assessment that has been developed by a consortium of private, government, and academic institutions over the past 20 years (Fig. 2) (USEPA, 1998). The risk assessment process begins with a generalized problem formulation statement. For example, the problem formulation statement for the registration of an herbicide might by cast as "*The production, application, and use of herbicide X may present an unacceptable risk to non-target aquatic organisms*". Based on this problem formulation statement, the risk assessment is conducted using an iterative assessment of the toxicity of an herbicide compared to its anticipated environmental exposure. Exposure and toxicity data are then integrated in the risk characterization step. If the risk characterization reveals some degree of concern then the problem can be re-evaluated with additional research and validation of exposure and toxicity in order to refine and minimize the risk. Excessive risk, defined as risk that cannot be managed or reduced, generally results in cancellation or removal of the chemical in the registration or re-registration process.

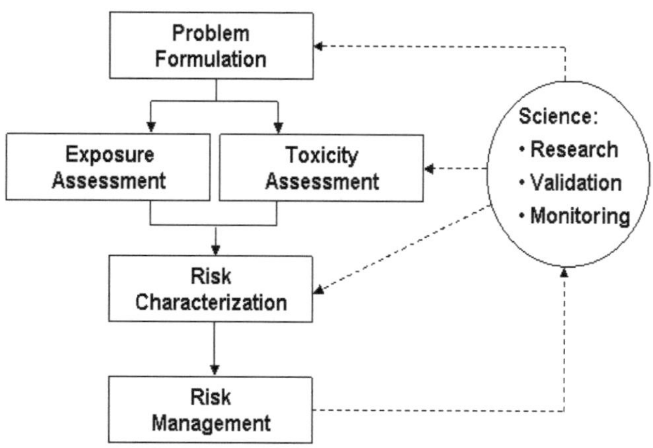

Fig. 2. Conceptual model for the framework for ecological risk assessment of herbicides and other chemicals (Source: USEPA 1998).

Registration of herbicides in the United States considers the potential for impacts on non-target aquatic organisms and is conducted within a 4-tiered process (USEPA, 2010) (Table 3).

Within each Tier, both toxicity and exposure are measured. The Tier 1 assessment consists of a comparison of the acute toxicity of an herbicide to a limited number of species potentially exposed to the maximum application rate of the chemical. The Tier 1 scenario is based on the assumption of a 10% runoff of a single application of a chemical to a 10-ha field into a 1 ha pond with a depth of 2 m. Fate predictions are made using standard exposure assessment models including the Pesticide Root Zone Model (PRZM), the Exposure Assessment Modeling System (EXAMS), and the GENeric Estimated Environmental Concentration (GENEEC2) models http://www.epa.gov/oppefed1/models/water/). These models predict a range of environmental exposures based on herbicide characteristics including water solubility, soil sorption coefficients, volatility, hydrolysis, photolysis, and biological degadation rates in soil and water. If the exposure/toxicity ratio exceeds 1 then tesing proceeds at the next highest tier. At each tier the data requirements expand to include more toxicity testing and additional chemical exposure assessment. Risk assessment at higher tiers varies depending on need, and can consist of mesocosm studies, field studies, field monitoring, and additional modeling including probablistic approaches.

Tier	Toxicity Assessment	Exposure Assessment
1	Acute testing (8 species)[1]	Deterministic models (GENEEC2)
2	Chronic testing[2]	Deterministic models (PRZM-3 and EXAMSII)
3	Mesocosm field testing[3]	Probablistic modeling
4	Field monitoring[4]	Watershed monitoring/probablistic modeling

[1]Short term (48h-96h) testing with eight species including 2 species of freshwater fish, 1 species of saltwater fish, 2 species of freshwater invertebrate, 1 species of saltwater invertebrate, and 2 species of aquatic plants.
[2]Includes fish early survival and growth; fish life cycle, and invertebrate life cycle.
[3]Experimental ecosystem testing under realistic environmental conditions.
[4]Reports of fish kills and onitoring studies at state/federal level.

Table 3. Example of the tiered assessment process used to assess risk of herbicides and other chemicals to non-target aquatic species during the pesticide registration process.

In probablistic modeling the probability of effects are compared to the probability of exposure to get a joint probability distribution of potential risk. Solomon et al. (1996) and Giddings et al. (2005) have conducted a series of probabalistic risk assessments with the herbicide atrazine. These probablistic risk assessments were possible due to the extensive datasabase available regarding the toxicity and environmental exposures to atrazine. A graphical example of the potential application of a probablistic assessment is presented in Figure 3. The data regarding probablity of exposure compared to plant and animal species sensitivity distributions indicates that the highest 20% of atrazine exposures would exceed the No Observable Effect Concentration (NOEC) values for approximately 10% of aquatic plant species (approximately 10 ug/L) whereas animals would not be affected. Similar approaches have been used to demonstrate the low risk of diquat to invertebrate and fish populations (Campbell et al., 2000). The success of probabalistic modeling is currently being evaluated and will probably be routinely used in the registration process of new chemicals in the future based on structure/activity relationships and experience with widely studied chemicals such as atrazine.

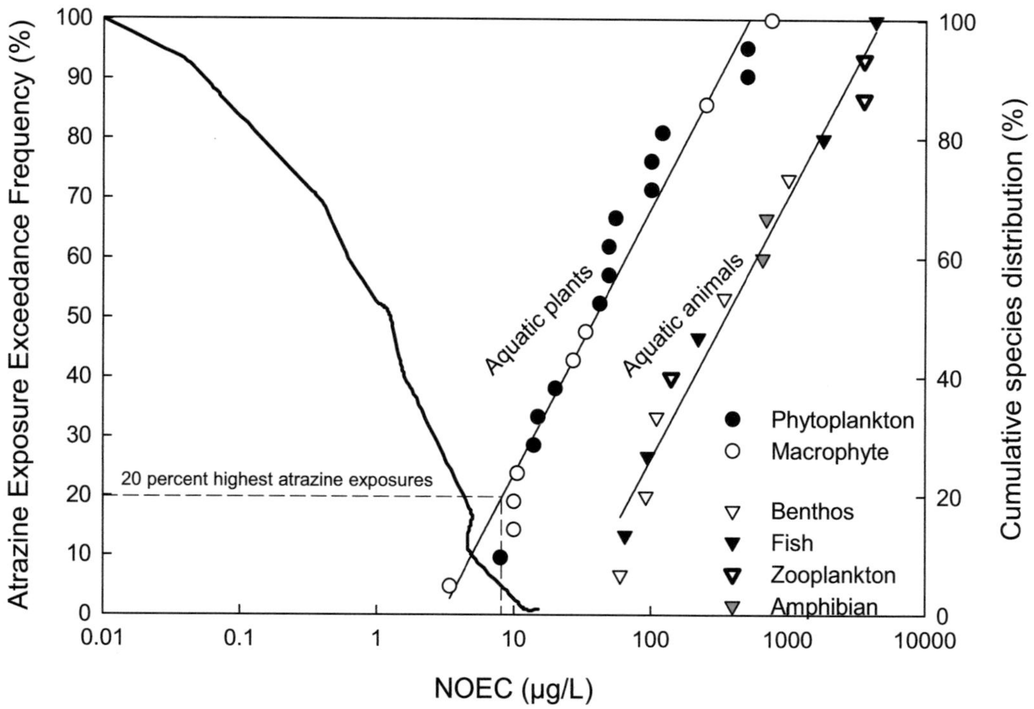

Fig. 3. Illustration of the concept of a an atrazine risk assessment based on joint probability distributions of measured environmental concentrations of atrazine in relation to the no observable effect concentrations (NOECs) determined under standardized laboratory conditions. The cumulative measured environmental exposure distribution is represented by the continuous line on the left. The other two lines with squares (aquatic plants) and triangles (benthos, zooplankton, amphibians, and fish) represent the linear cumulative species sensitivity distributions determined from chonic laboratory tests with atrazine. Plant data consists of published data from tests based on the endpoint of biomass production. Animal data consists of published data from tests on endpoints of survival, growth, or reproduction. The overlap between exposure and toxicity values represents the relative risk of exposure of plant and animal communities to atrazine. In this case, the highest 20 percent of atrazine exposures would exceed the NOEC level for aquatic plants approximately 10% of the time. No risk is anticipated for animals. Data extracted and re-drawn from Giddings et al. (2005; copyright 2005 by the Society of Environmental Toxicology and Chemistry (SETAC), Pensacola, FL, USA.) Figure reprinted with permission.

3. Herbicide exposures in aquatic ecosystems

3.1 Empirical measures of herbicide concentrations in water

Extensive use of herbicides in the United States has led to widespread detection of parent compounds and metabolites in surface and groundwaters (Gilliom, 2007). There are extensive monitoring networks for herbicides to determine spatial and temporal trends in concentrations and loads in surface waters in the U.S. Plans for implementation of monitoring networks in Europe are planned over the next 5 years in association with the

REACH legislation and the European Commission (Rabiet et al., 2010). The U.S. Geological Survey monitors herbicide concentrations at over 112 monitoring stations on a regular basis as part of its National Water Quality Assessment (NAWQA) and National Stream Quality Accounting Network (NASQAN) programs. Schnoebelen at al. (2003) monitored monthly herbicide concentrations at twelve fixed sites in the Eastern Iowa Basins NAWQA study unit from 1991-1998 and found that two herbicides, atrazine and metolachlor, were found in 100% of samples; acetachlor, cyanazine, alachlor, and bentazon were found in over 50% of samples (Fig. 4).

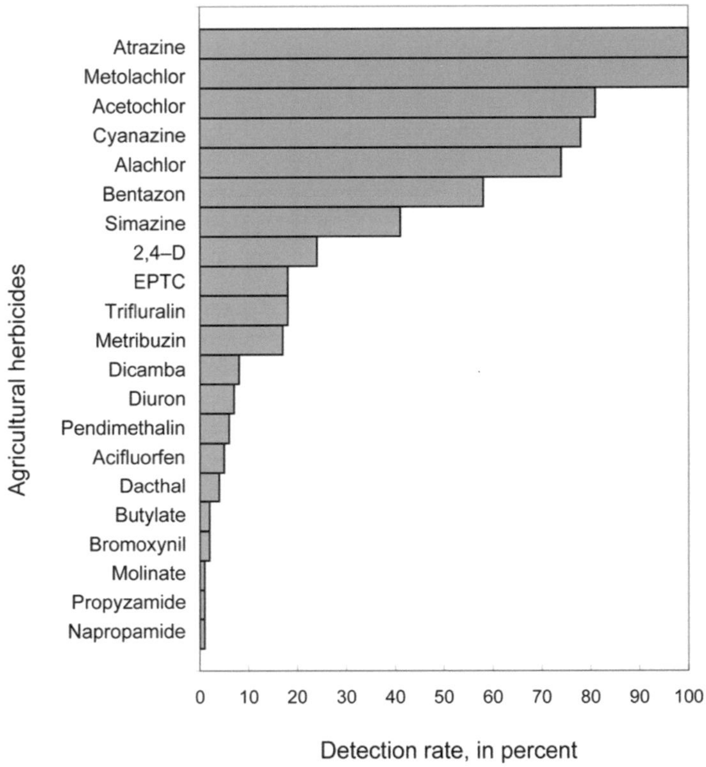

Data from Schnoebelen et al. (2003).

Fig. 4. Detection rates of herbicides monitored in Southern Iowa and Northern Missouri.

In one of the most intensive studies ever conducted, Richards and Baker (1993) monitored the concentrations of 6 herbicides daily at USGS gaging stations in 7 intensively-farmed Great Lakes tributaries from the months of April to August, 1983-1991 (Table 4). Concentrations were measured up to 3 times per day during peak exposure periods. Maximum concentrations of atrazine, alachlor, metolachlor, metribuzin, cyanazine, and linuron in the Lost Creek river basin were 68, 65, 64, 25, 23, and 13 ugL, respectively. However, these maximum concentrations were of short duration; maximum concentrations generally exceeded the 95% percentile concentrations by factors > 3 and exceeded time weighted mean concentrations by over an order of magnitude. Lerch and Blanchard (2003) monitored concentrations of atrazine, cyanazine, acetochlor, alachlor, metolachlor, and metribuzin for 3 yrs (1997-1999) at 21 streams sites in Southern Iowa and Northern Missouri on an approximate weekly basis from April 15 to July 15. Atrazine was the most frequently detected herbicide and only rarely exceeded the annual 90th percentile concentration (40 ug/L). These datasets illustrate the short duration of herbicide exposures in midwestern agricultural streams of the U.S. which are well below the majority of acute (EC50) and chronic (NOEC) levels for both algae and aquatic macrophytes and fall two orders of magnitude below levels causing effects on fish, invertebrates, and amphibians (USEPA, 2002; as summarized by Giddings et al., 2005).

River Basin (km^2) and Cropland (%)	Parameter	Herbicide and Concentration (ug/L)					
		Atrazine	Alachlor	Metolachlor	Metribuzin	Cyanazine	Linuron
Maumee (16,395 km^2) (76%)	MAX[1]	21.45	18.35	26.20	5.77	9.96	7.29
	95	7.47	3.00	5.32	1.83	1.97	0.00
	50	0.58	0.00	0.28	0.01	0.03	0.00
	TWMC[2]	1.61	0.54	1.16	0.29	0.38	0.05
Sandusky (3,240 km^2) (80%)	MAX	24.61	36.13	36.76	9.26	19.87	0.02
	95	8.84	3.76	8.59	1.68	1.73	0.29
	50	0.53	0.00	0.35	0.00	0.00	0.00
	TWMC	1.78	0.66	1.65	0.28	0.21	0.03
Honey Cr. (386 km^2) (83%)	MAX	54.04	54.87	95.75	10.52	17.47	15.50
	95	10.85	4.44	9.08	1.28	2.07	0.68
	50	0.66	0.11	0.35	0.00	0.03	0.00
	TWMC	2.33	0.89	1.80	0.24	0.40	0.17
Rock Cr. (88 km^2) (81%)	MAX	48.63	23.40	96.92	15.95	24.77	12.01
	95	6.61	2.16	8.15	1.20	0.71	0.68
	50	0.21	0.00	0.17	0.00	0.00	0.00
	TWMC	1.34	0.39	1.62	0.23	0.18	0.15
Lost Cr. (11 km^2) (83%)	MAX	68.40	64.94	63.64	25.15	22.62	13.44
	95	5.67	1.07	3.08	0.80	1.64	0.00
	50	0.27	0.00	0.00	0.00	0.00	0.00
	TWMC	1.30	0.48	0.62	0.20	0.50	0.05
Cuyahoga (1,831 km^2) (4%)	MAX	6.80	1.16	5.39	1.49	1.36	5.04
	95	0.99	0.24	0.63	0.28	0.27	0.06
	50	0.09	0.00	0.00	0.00	0.00	0.00
	TWMC	0.31	0.04	0.15	0.07	0.05	0.08
Raisin (2,699 km^2) (67%)	MAX	12.86	7.52	5.91	2.46	3.75	1.92
	95	3.91	2.02	1.50	0.37	1.11	0.18
	50	0.30	0.00	0.00	0.00	0.00	0.00
	TWMC	0.76	0.37	0.32	0.11	0.21	0.04

[1]Maximum observed concentration (MAX).
[2]Time-weighted maximum concentration (TWMC).

Table 4. Concentrations of major herbicides in 7 Lake Erie Tributaries, Apr. 1983 - Dec. 1991. Data from Richards and Baker (1993).

3.2 Modeling of herbicide concentrations in water

Exposure assessment in the pesticide registration process has primarily relied on use of deterministic models of worst-case assumptions of application and runoff. Peterson et al. (1994) modeled worst-case exposure estimates based on direct overspray of a 15-cm wetland at the maximum application rates of herbicides used in areas such as in prairie wetland areas of south-central Canada. This would rarely occur under prudent, recommended application practices. In the U.S. exposures are modeled in the herbicide registration and risk assessment process using Tier 1 (e.g., GENEEC2) and Tier 2 (e.g., PRZMII, EXAMS-3) models based on site-specific conditions including herbicide characteristics, application rates, soil types, slope, and rainfall patterns (USEPA, 2010). These Tier 1 and Tier 2 models were developed and validated using empirical data derived from edge-of-field studies conducted during the 1970s to measure the range of potential herbicide exposures in aquatic systems (e.g., Wauchope, 1978). Similar approaches are used in Europe (Huber et al., 2000).

Over the past decade more sophisticated models have been developed for higher-tier exposure assessments in the U.S. such as the Soil and Water Assessment Tool (SWAT; Gassman et al., 2007), and the Watershed Regression for Pesticides (WARP; Larson et al., 2004) and the Spatially Referenced Regression on Watershed Attributes (SPARROW; Schwarz et al., 2006) models. These higher-tier models use detailed data on soil characteristics, slope, rainfall patterns, pesticide use, and tillage patterns at various hydrologic unit scales across the nation. Recent efforts have been made to use these models to predict atrazine exposures in streams draining watersheds with highest atrazine use in order to implement site-specific monitoring programs that identify areas requiring exposure-reduction management plans (USEPA, 2006). Future development and application of these models will provide cost-effective yet sensitive methods to predict areas of highest herbicide exposures for use in both ecological and human health risk assessments.

4. Herbicide effects in aquatic systems

4.1 Direct effects on aquatic plants

The direct effects of an herbicide can be measured at either the structural or functional level. Structural endpoints include static measures of cell numbers, biomass, species composition, or community diversity. Functional endpoints are measured as changes in rates of biological processes such as carbon uptake, oxygen evolution, enzyme activity, nutrient cycling, population growth rates, or changes in system metabolism (e.g., gross production or community respiration). Both structural and functional endpoints can be measured at increasing hierarchical levels of biological organization ranging from the cell to the ecosystem to predict and assess the effects of herbicides on non-target aquatic organisms. Some functional measurements, such as the rate of enzymatic activity (e.g., peroxidase and glutathione transferase activity) are made at the plant cellular level and can be valuable in a research context because they are diagnostic of mode of action and occur rapidly at the cellular level (Field and Thurman, 1996; Ferrat et al., 2003). However, the utility and precision of these measurements is often dependent on plant species, herbicide mode of action, and other factors that make it difficult to use as an assessment endpoint for prediction of biological effects at higher levels of biological organization.

For regulatory and risk assessment purposes the direct effects of herbicides on aquatic plants are most often measured under standardized laboratory conditions using biomass production as the structural measurement endpoint. These studies are conducted under standardized conditions of lighting, temperature, and nutrient regimes using a relatively small cadre of species (e.g., algae and duckweed, etc.) (ASTM, 2009a, b). Although macrophytes are currently not routinely tested there are efforts underway in the U.S., Canada, and Europe to develop a standard toxicity test with a *Myriophyllum* sp. (Knauer et al., 2008; Kubitza and Dohmen., 2008). Single species laboratory tests are preferred for regulatory purposes due to their inherent precision, replicability, repeatability and reproducibility at relatively low monetary costs. Although no species is universally more sensitive to herbicides than another, these standard tests are commonly used to determine the toxicity of a chemical, the relative sensitivity of a number of species to a chemical, or the relative toxicity of a group of chemicals (Fairchild et al., 1997; 1998).

Herbicide effects are usually expressed as the EC50 based on regression analysis of standing crop biomass (i.e., the effective concentration of herbicide that reduces the amount of plant biomass at a fixed interval such as 4 days, 7 days, or two weeks). For example, Hughes et al. (1988) evaluated the effects of atrazine on single species aquatic plant biomass production in the laboratory over a 5-d period. Responses among species were similar with an average 120-h EC50 of 170 µg/L (95% C.I. 130-230 µg/L). Hughes et al. (1988) further compared the 120-d EC50 to several other possible endpoints including the phytostatic concentration (i.e., the concentration totally stopping plant population growth) and the phytocidcal concentration (i.e., the concentration that results in total plant populaton mortality/sterilization). The EC50 values were highly conservative compared to 120-h phytostatic concentrations (average 1,720 µg/L, range 1,450 - 4,970 µg/L) and the 120-h phytocidal concentrations (>3,200 µg/L, the highest concentration tested). Moreover, plants recovered at all concentrations once removed to clean water. This study illustrates that the most commonly used endpoint in aquatic plant ecotoxicology may be useful in comparing the relative sensitivity of various species under standardized conditions. However, for the purposes of a risk assessment, plant data must be interpreted totally differently compared to standard measurement endpoints measured in acute toxicity tests with animals such as zooplankton and fish where individual mortality is measured (e.g., 96-h LC50). While herbicides may temporarily suppress growth of non-target algae and macrophytes, populations quickly recover once exposure is reduced. Therefore, aquatic plants have intrinsic adaptive factors that allow them to tolerate herbicides in the environment where exposures are typically low and ephemeral (Table 4).

One of the greatest criticisms of single species tests is that they are ecologically unrealistic. Singles species tests do not reflect the complexity of natural systems that contain multi-species communities of algae, bacteria, and other communities in association with natural sediment and nutrient conditions which provide the biological capacity for structural and functional redundancy, resilience, and recovery. To overcome the limitations of single species tests, studies are often conducted in simulated experimental ecosystesms such as microcosms, mesocosms, and outdoor experimental ponds.

Larsen et al. (1986) examined the effects of atrazine in single species laboratory tests and experimental microcosms in order to compare the utility of single species tests for predicting community-level responses under more realistic conditions. Laboratory EC50 values were determined for 8 species of algae exposed to atrazine for 24 hr using ^{14}C uptake as a measurement endpoint. Single species EC50 values ranged from 37 - 308 ug/L with an average response at approximately 100 ug/L atrazine. These same species were studied in microcosms containing the same eight algal species along with two protozoans, an amphipod, an ostracod sp., a *Daphnia* sp., and bacterial/fungal communities in a static system containing a silica sand substrate. Microcosms were exposed for 60 days to atrazine at nominal concentrations of 0, 60, 100, 200, and 500 ug/L atrazine. Effects of atrazine were measured and compared at three intervals: early (days 10 to 20), late (days 53 to 60) and then averaged across the entire study. Short-term decreases in functional measures of ^{14}C uptake occurred in the microcosms immediately after dosing in all treatments, but recovery occurred in the 60 and 100 ug/L treatments within 10 days following treatment. ^{14}C uptake was reduced at 200 and 500 ug/L throughout the study with little recovery. Functional measures of algal community effects in the microcosms ranged from 103 - 154 ug/L for ^{14}C uptake, 126 - 165 ug/L for oxygen production, and from 106 - 164 ug/L for oxygen consumption, which approximated the average functional responses measured in single species studies. In contrast, algal biomass measured as chlorophyll *a* actually increased at 60, 100, and 200 ug/L in a hormetic fashion indicating that functional measures were more sensitive indicators of stress but not accurate in terms of estimating production of algal biomass. Single species laboratory tests based on biomass production are now known to be conservative for prediction of effects in outdoor mesocosms and most probably field effects.

DeNoyelles et al. (1982; 1986) conducted a series of experimental pond studies with atrazine over a 3-yr period in order to compare the relative structural and functional responses of natural plant communities to atrazine. Three studies were conducted during this time, with atrazine exposures at 0, 20, 100, 200, and 500 ug/L. Mesocosms exposed to environmentally relevant concentrations of 20 ug/L atrazine exhibited short term decreases in algal functional measures of ^{14}C uptake and dissolved oxygen production but quickly recovered to control values for the remainder of the studies, and no structural changes in phytoplankton were ever observed in the 20 ug/L concentration. EC50 values for ^{14}C uptake and algal biomass were 100 and 82 ug/L, indicating that community structural and functional measures were similar in sensitivity. Atrazine altered species compositions of algae at 100 ug/L, but sensitive species were replaced by more tolerant species with no changes in overall algal biomass or productivity. Both structural and functional changes in algal communities were observed at 200 and 500 ug/L atrazine; however, these high concentrations are now known to be environmentally irrelevant. Collectively, the pond studies indicated that predictions made using the most sensitive algal species under standardized laboratory conditions overestimated the predicted response of algal communities in ponds due to species replacement which provided functional redundancy at concentrations of 100 ug/L atrazine.

Fairchild et al. (1994) examined the effects of atrazine applied in combination with an insecticide (esfenvalerate) to determine if atrazine would alter the effects of the insecticide on consumer populations. The dominant macrophyte in the mesocsoms, *Najas* sp., has been shown to be sensitive to the herbicide atrazine under laboartory conditions at 24 ug/L (14-d EC50; Fairchild et al. 1998). Atrazine, applied to ponds at 50 ug/L, shifted the macrophyte community to a community dominated by *Chara* sp. Total macrophyte biomass and system metabolism did not change due to atrazine. Rapid sorption of the strongly hydrophobic esfenvalerate ($T_{1/2}$ 48 h) by aquatic plants and sediments mitigated the precicted effects of the insecticide on zooplankton dynamics and bluegill survival/growth. Therefore, alteration of one measure of ecosystem structure (macrophyte species composition) was mitigated by another measure of community structure (total biomass) and function (system metabolism) to remove any adverse effects adverse effects of an insecticide on consumer (bluegill) populations.

Fairchild et al. (2002) exposed the same experimental pond system to the triazinone herbicide metribuzin at 0, 9, 19, 38, and 75 ug/L. *Najas* sp., the dominant macrophyte species, had been shown to be sensitive to metribuzin in the laboratory at 19 ug/L (14-d EC50) which was similar to 4 other species of macrophytes (14-d EC50, range 17- 36 ug/L metribuzin) (Fairchild et al., 1998). Although metribuzin is more toxic to macrophytes (14-d EC50s ranging from 21-132 ug/L) than atrazine under laboratory conditions the relatively short half-life of metribuzin ($T_{1/2}$ = 5) days observed in the ponds resulted in no effects on water quality, periphyton biomass, macrophyte species composition, macrophyte biomass, or fish survival/growth.

Numerous other microcosm and mesocosm studies have been conducted with other herbicides including 2,4-D, hexazinone, linuron, simazine, and terbutryn. Brock et al. (2000) reviewed the results of a total of 124 microcosm and mesocosm studies of the effects of herbicides in model aquatic systems. Rigorous evaluation criteria included the following requirements: 1) the study must have had published single species toxicity data for the chemical; 2) the study must have included multiple species at different trophic levels; 3) the study was ideally conducted outdoors to allow the potential for biological recolonization; 4) the study used an appropriate experimental design (e.g. replication of treatments) ; and 5) the study must have resulted in a statistically significant lowest observable effect concentration (LOEC) of an accepted structural or functional endpoint. Very few of the studies reviewed by Brock et al. (2000) met the strict criteria for acceptance due to statistical problems, inadequate descriptions of methodologies, or concerns due to methodologies such as isolation techniques or the availability for recolonization of plants and other organisms. Significant ecological effects in studies that fully met acceptance criteria were only observed in cases where exposure concentrations exceeded those considered environmentaly relevant. Therefore, there is no compelling evidence from these studies that herbicides are likely to impact aquatic plants in the environment.

Brock et al. (2000) summarized the strengths, weaknesses, and utility of various approaches for evaluating the statistical and ecological relevance of various test systems that have historically been used to assess the risk of herbicides to non-target aquatic plants (Fig. 5). Laboratory single species tests are relatively simple to conduct and exhibit high precision and reproducibility. Single species laboratory studies are therefore useful for comparing the relative sensitivity of different species to an herbicide or the relative toxicity of several herbicides. Studies in microcosms and mesocosms have been a useful approach for validating laboratory and model predictions of the fate and effects of herbicides in aquatic environments due to their increasing realism and complexity which approximates anticipated responses under actual field conditions. However, it could be argued that additional mesocosm tests with herbicides that meet the Brock (2000) criteria are needed.

From an objective perspective the current tiered testing procedures in use by the USEPA are useful, cost-effective approach for the risk assessment of herbicides and other chemicals. Testing at higher tiers currently allows for flexibility in using either mesocosm/field testing or probablistic modeling of effects and exposure data. Current experience increasingly indicates that probablistic risk assessment of herbicide effects using single species laboratory toxicicity data and environmental exposure data similar to that described by Solomon et al. (1996) and Giddings et al. (2005) (visually illustrated in figure 3) can be used to effectively evaluate the potential for non-target effects of herbicides in aquatic systems for regulatory and risk assessment purposes. Although mesocosm tests can be used in the U.S. for registration purposes, past experience has indicated that these studies are relatively expensive due to the scarcity of available testing facilities and monetary costs associated with meeting quality assurance guidelines required by the USEPA. For these reasons, probabilistic risk assessments using single species laboratory data are increasingly being emphasized in regulatory programs for herbicide registration and re-registration in both the U.S. and Europe (Reach, 2007; USEPA, 2010).

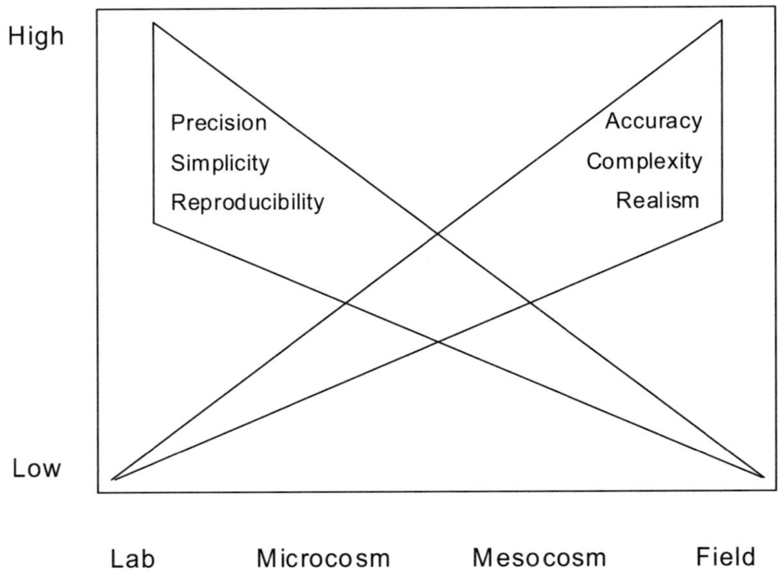

Fig. 5. Illustration of the continuum of strengths and weakness of various test systems used for the risk assessment of herbicides. Figure adapted from Brock et al. (2000).

4.2 Direct effects on invertebrates, fish, and amphibians

Direct effects of herbicides on animals are usually measured at the individual level of biological organization. As with plants, most herbicide risk assessments using invertebrates, fish, and amphibians are conducted in the laboratory using standardized species and methods. Animal tests are somewhat easier to conduct than plant tests because environmental factors such as nutrient regimes, lighting, and invasion of unwanted test organisms are less critical. Endpoints can include both structural (e.g., survival) and functional (e.g., growth and reproduction) measurements.

Few if any *environmentally relevant* concentrations have been shown to have direct effects on zooplankton, fish, or amphibians in the laboratory. This is well-illustrated in Figure 3 for the herbicide atrazine and data provided within two extensive ecological risk assessments of the chemical (Solomon et al., 1996; Giddings et al., 2005). Similar wide margins of safety have recently been shown for survival and growth of rainbow trout exposed to the herbicides picloram, 2,4-D, and clopyralid (Fairchild et al., 2009a, b, c). This is not surprising given the basic modes of action of herbicides and the phylogenetic differences between plants and animals. Field monitoring programs conducted by the USGS National Water Qaulity Assessment (NAWQA) program and the USEPA Environmental Monitoring and Assessment Program (EMAP) have found degraded aquatic communities in intensively-farmed watersheds. However, herbicides have never been shown to be associated with or causal of these effects; excessive levels of nutrients, sediments, and hydrologic alterations interact to produce these ecological impairments.

One practical way to illustrate the lack of direct effects of herbicides on invertebrates, fish, and amphibians is to examine the use of herbicides for control of aquatic nuisance plants. There are approximately 10 herbicides registered for aquatic use by the U.S. Environmental Protection Agency (Table 5). These registered herbicides represent several different chemical classes and associated modes of action. Aquatic herbicides are usually used to control native and non-native macrophytes such as Eurasian millfoil (*Myriophyllum spicatum*) and other species that interfere with desired aquatic uses including swimming, boating, and water delivery/supply. Given this intent, recommended application rates for aquatic herbicides are much higher than single species EC50s in order to reach a phytostatic or phytocidal concentration (Table 5).

Chemical	Chemical class	Recommended Application Rate (mg/L)[1]	Aqueous Half-life (d)[1]	Bluegill EC50 (mg/L)[1]
Copper	Metal	1	2 - 8	1-12
Endothall	Dicarboxylic acid	2-4	1 - 7	240
Diquat	Bipyridilium	0.1 - 0.5	1 - 4	67
Fluridone	Pyridinone	0.010 – 0.020	20 - 80	14
Glyphosate[2]	Amino-acid derivative	0.5 - 5.0	5 - 10	>1000
Imazamox	Imadizolinone	0.005 - 0.025	25 - 50	>120
Imazapyr acid	Imadizolinone	1.0	3 - 5	>100
Penoxsulam	Sulfanilamide	0.005 - 0.025	25 - 50	>103
Triclopyr acid	Pyridine	1.0	0.5 - 3	148
2,4-D acid	Phenoxy acid	1.0	2-6	2600
Atrazine[4]	*Triazine*	*1.72*[4]	*3 – 120*[4]	*28.3*

[1]Application rate and aqueous half-life data derived from University of Florida accessed on 7/28/2010 at http://plants/ifas/ufl.edu/guide/sup3herb.html.
[2]Glyphosate is a surface contact herbicide and is not phyto-toxic when applied directly to water.
[3]Note that atrazine is not registered for aquatic use; value derived from Hughes et al. (1988).
[4]Values reported in Giddings et al. (2005) as summarized from USEPA (2002).

Table 5. Chemical names, application rates, aqueous half-life, and toxicity of herbicides to fish (bluegill) registered for aquatic plant control by the USEPA.

It is reasonable to assume that phytocidal concentrations of herbicides such as those registered and used to remove aquatic plants can induce both direct and indirect effects at the population, community, and ecosystem levels of biological organization. For example, when macrophytes are deliberately removed from a system there are major structural changes in habitat when macrophyte-dominated systems are converted to phytoplankton-dominated systems. In fact, this is often the stated goal in use of herbicides for control of nuisance aquatic plants to restore native plants. Research has demonstrated that macrophyte community structure is a major factor controlling predation rates of higher level consumers such as largemouth bass on bluegill (Savino and Stein, 1982; Crowder and Cooper, 1982; Wiley et al., 1984). Loss of aquatic macrophytes following deliberate vegetation removal by herbicides can also have profound effects on ecosystem function including alteration of nutrient cycling, decreased water transparency due to wave action and bioturbation, and depletion of dissolved oxygen concentrations. Many fish kills attributed to registered aquatic-use herbicides are due to the indirect effects of oxygen depletion. Therefore, consideration must be given to the timing of application of herbicides in relation to temperature, dissolved oxygen, and standing crop of macrophyte biomass. Altered nutrient cycling and increases in turbidity following aquatic plant removal efforts are major functional indicators of the role of aquatic plants in lakes and reservoirs. However, these are ecosystem changes that are often accepted by the public in order to meet management and restoration goals in aquatic systems. These herbicide concentrations used for aquatic plant control, however, far exceed any *environmentally relevant concentrations* resulting from agricultural use.

5. Indirect effects

Indirect effects of herbicides are defined as observed effects on consumer populations such as invertebrates and fish that are not caused by direct toxicity but rather effects that have occurred due to adverse effects on primary producers such as algae and macrophytes. Previous studies have implied that the herbicide atrazine can have indirect effects on consumer populations in aquatic ecosystems at concentrations as low as 0.1 ug/L (Lampert et al. 1989). Dewey et al. (1986) reported a decrease in macroinvertebrate emergence in pond microcosms treated at 20 ug/L atrazine; in addition, Kettle et al. (1987) reported total reproductive failure in ponds treated at 20 ug/L. However, recent critical reviews of these studies by Giddings et al (2005) have revealed that the indirect effects observed by Lampert et al. (1989) on plankton communities were actually due to the effects of a solvent used to deliver atrazine which drove the system into a heterotrophic state; the indirect effects reported by Dewey et al. (1986) and Kettle et al. (1987) were not caused by atrazine, but rather the differential growth and survival of aquatic macrophytes due to the effects of grass carp (aquatic herbivores) and channel catfish (bluegill predators) that were not disclosed in the original manuscripts. Therefore, *environmentally relevant concentrations* of herbicides such as atrazine are not known to cause indirect effects on aquatic systems.

Recently, researchers have applied bioenergetic ecosystem-based models in order to predict the direct effects of herbicides on primary producers and ultimately indirect effects on consumer populations such as fish in an attempt to add ecological realism to the risk assessment process. Bartell et al. (2000) used the Comprehensive Aquatic Systems Model (CASM) which is an ecological model based on demographics and bioenergetic equations used to model the direct and indirect effects of the aquatic herbicide diquat on phytoplankton and zooplankton populations in Florida lakes. Bartell et al. (2000) showed potential for direct effects on phytoplankton but little probability for indirect effects of diquat on zooplankton consumers. Currently, the CASM model is being used to determine locations where concentrations and exposure durations of atrazine in streams may pose risks to consumer populations via indirect, dietary effects. This model is being used with intensive monitoring at selected sites in the re-registration of atrazine in order to identify areas of atrazine exposure that may require adaptive management actions to reduce risk (USEPA, 2006). However, ecological process models such as CASM are mathematically complex and require an extreme level of expertise to use and apply in a regulatory context. Therefore the utility of using ecological process models for herbicide regulatory purposes is contentious.

6. Emerging areas of research on herbicide effects on endocrine and immune function in aquatic vertebrates

Over the past decade there has been a tremendous amount of research regarding potential effects of atrazine on the endocrine and immune function in vertebrates including amphibians, fishes, and reptiles. Atrazine has been widely studied because of its frequency of detection in surface and groundwaters (Solomon et al., 1996; Schnoebelen et al., 2003) due to its widespread application (34-36 million kilograms per year in the U.S.; Kiely et al., 2004), high water solubility (33 mg/L; Wauchope et al., 1992), and long aqueous half-life (mean 159 $\pm$ 71 d; range 41-237 d; Giddings et al. 2005). Hayes et al. (2002; 2003) reported that male larval African clawed frogs (*Xenopus laevis*) exposed to atrazine at < 1.0 ug/L exhibited hermaphrodism and altered laryngeal development; in addition, male *X. laevis* suffered a 10-fold decrease in testosterone levels when exposed to 25 ug/L atrazine. The authors hypothesized that atrazine induces aromatase that promotes the conversion of testosterone to estradiol. These publications prompted numerous studies of the effects of atrazine on a vast array of physiological responses of vertebrates to atrazine. Solomon et al. (2008) conducted a critical review of over 75 atrazine studies with amphibians to evaluate reported effects on sexual differentiation, sexual development, male laryngeal development, and thyroid function; an additional 20 studies of the effects of atrazine on endocrine and behavioural functions in fishes were also examined. Solomon et al. (2008) concluded that "based on a weight of evidence of all the data, the central theory that environmentally relevant concentrations of atrazine affects reproduction, and/or reproductive development in fish, amphibians, and reptiles is not supported by the majority of observations. The same conclusions also hold for the supporting theories such as induction of aromatase, the enzyme that converts testosterone to estradiol".

Many studies have shown feminization effects in amphibians, fishes, and reptiles when exposed to the synthetic hormone 17*B* estradiol which is a standard positive control chemical for studies of endocrine function and reproductive effects; however, the metabolism, mode of action, and effects of estradiol are well-studied and known. The same cannot be said for atrazine. Solomon et al. (2008) found that the majority of studies reporting atrazine effects on endocrine and immune function contained substantial weaknesses in the areas of experimental design, methodologies, interpretation, and inferences; no studies have directly established cause and effect relationships in the laboratory or field based on established principles of epidemiology. More recently, Rohr and McCoy (2010) conducted a meta-analysis of the effects of atrazine on freshwater amphibians and fish and indicated that atrazine reduced size at metamorphosis in 15 of 17 studies in 14 species; reduced immune function in 33 of 43 studies; altered gonadal morphology in 7 of 10 studies; and altered spermatogenesis in 2 of 2 studies. However, the exact mechanisms of individual-level effects and ultimate significance in populations of amphibians, fish, and reptiles remain uncertain.

The existing concerns of the effects of atrazine and other possible chemicals on endocrine function in aquatic and semi-aquatic vertebrates remain controversial. These concerns led the USEPA to implement an amphibian metamorphosis (frog) test and a fish short-term reproduction test for assessment of potential endocrine-disrupting effects of herbicides and other chemicals on non-target aquatic organisms (USEPA, 2010). Research in this active area of ecotoxicology continues.

7. Research needs and conclusions

Research and monitoring has indicated that the widespread use of herbicides in modern agriculture has led to widespread exposures of aquatic organisms in aquatic systems. The observed high sensitivity of a few species of algae and macrophytes to herbicides has been used to infer that adverse effects on aquatic ecosystems may occur. Therefore, systematic assessment procedures that evaluate the fate and effects of herbicides on non-target aquatic organisms have been developed. Risk assessments of herbicides for registration and regulatory puroposes consist of single species laboratory toxicity tests and exposure models within a tiered testing system. Atrazine is currently one of the greatest herbicides of regulatory concern because it is a generalized inhibitor of photosynthesis, is commonly detected in aquatic systems, and has a long environmental half-life compared to other herbicides. Atrazine is occassionally observed in the environment at concentrations of 50 ug/L or higher but exposures at this level only occur over short time durations. The effects of herbicides on non-target aquatic plant communities in natural environments are likely attenuated or mitigated due to the adaptive abilities of aquatic plants through acclimation and recovery following exposures; species substitution; and functional redundancy of aquatic plant communities. Cases of direct effects of herbicides on plant communities only occur during the intentional use of registered aquatic herbicides that are deliberately applied at phytostatic or phytocidal concentrations. Studies of direct and indirect effects of herbicides on invertebrates, amphibians, and fish exposed to *environmentally relevant concentrations* using accepted measurement endpoints of survival, growth, and reproduction have not shown adverse effects in laboratory, mesocosms, or field situations.

Research regarding the effects of atrazine on endocrine and immune function continues, but remains highly controversial. Mechanistic and comphrehensive studies of atrazine need to be conducted under laboratory and field conditions in order to establish cause and effect relationships. Once these studies are conducted, additional studies will be needed to determine the ultimate ecological significance of these effects in the field at the population level of amphibians and fishes.

8. Acknowledgments

Any use of trade, product, or firm names is for descriptive purposes only and does not imply endorsement by the U.S. Government. This manuscript was produced using U.S. Geological Survey funds only; no private or outside funds were solicited or used.

9. References

American Society for Testing and Materials (ASTM). (2009a). Standard guide for conducting static 96h toxicity tests with microalgae: Practice E1218-90. In: *Annual Book of ASTM Standards: Water and Envrionmental Technology*. S.J. Baily and N.C. Baldini, (Ed), pp. 293-306, ASTM Committee E-47 on Biological Effects and Envrionmental Fate, American Society for Testing and Materials, West Conshohoken, PA. 1618 pp.

American Society for Testing and Materials (ASTM). (2009b). Standard guide for conducting toxicity tests with *Lemna gibba* G3. Practice E1415-91. In: *Annual Book of ASTM Standards: Water and Environmental Technology*. S.J. Baily and N.C. Baldini (Ed), pp.

623-632, ASTM Committee E-47 on Biological Effects and Envrionmental Fate, American Society for Testing and Materials, West Conshohoken, PA.1618 pp.

Bartell. S.M., K.R. Campbell, C.M. Lovelock, S.K Nair, and J.L Shaw. (2000). Characterizing ecological risks from pesticides using a diquat bromide case study II: Ecological process models. *Toxicol. Chem.* 19:1441-1453.

Brock, T.C.M., R.P. van Wijigaarden, and G.J. van Geest. (2000). Ecological risks of pesticides in freshwater ecosystems. Part 1: Herbicides. Aterra-Rapport 088. Alterra, Green World Research, Wageningen, the Netherlands. 127 pp.

Campbell, K.R., S.M. Bartell, and J.L. Shaw. (2000). Characterizing aquatic ecological risks from pesticides using a diquat dibromide case study II: Approaches using quotients and distributions. *Toxicol. Chem.* 19: 760-774.

Canadian Ministry of Justice. (2010). Pesticide control products: consolidation. Minister of Justice, SOR/2006-124. Accessed at http://laws-lois.justice.gc.ca, Aug. 18, 2010.

Cole, D., Pallett, K., and M. Rodgers. (2000). Discovering new modes of action for herbicides and the impact of genomics. *Pestiicde Outlook,* December, pp. 223-229.

Crowder, L.B. and W.E. Cooper. (1982). Habitat structural complexity and the interaction between bluegills and their prey. *Ecology* 63:1802-1813.

deNoyelles, F., Jr., W. D. Kettle, and D. E. Sinn. (1982). The responses of plankton communities to atrazine, the most heavily used pesticide in the United States. *Ecology* 63:1285-1293.

deNoyelles, F., Jr., and W.D. Kettle. (1985). Experimental ponds for evaluating bioassay predictions. In: *Validation and Predictability of Laboratory Methods for Assessing the Fate and Effects of Contaminants in Aquatic Systems,* T.P. Boyle, (Ed), pp. 91-103, ASTM STP 865, American Society for Testing and Materials., Philadelphia, PA. 242 pp.

Dewey, S.L.(1986). Effects of the herbicide atrazine on aquatic insect community structure and emergence. *Ecology* 67:148-162.

Fairchild, J.F., T.W. La Point, and T. Schwartz. (1994). Effects of an herbicide and insecticide mixture in experimental aquatic mesocosms. *Arch. Tox. Chem.* 27:527-533.

Fairchild, J.F., D.S. Ruessler, P.S. Haverland, and A.R. Carlson. (1997). Comparative sensitivity of *Selenastrum capricornutum* and *Lemna minor* to sixteen herbicides. *Arch. Environ. Contam. Toxicol.* 32:353-357.

Fairchild, J.F, D.S. Ruessler, and A.R. Carlson. (1998). Comparative sensitivity of five species of macrophytes and six species of algae to atrazine, metribuzin, alachlor, and metolachlor. *Toxicol. Chem.* 17:1830-1834.

Fairchild, J.F. and L.S. Sappington. (2002). Fate and effects of the triazinone herbicide metribuzin in outdoor aquatic mesocosms. *Environ. Contam. Toxicol.* 43:198-202.

Fairchild, J.F., K.P Feltz, A. Allert, L.S. Sappington, K.J. Nelson, and J. Valle. (2009a). An ecological risk assessment of the acute and chronic effects of picloram to rainbow trout (*Onchorhyncus mykiss*) and the threatened bull tourt (*Salvelinus confluentus*). *Arch. Environ. Contam. Toxicol.* 56:761-769.

Fairchild, J.F., K.P. Feltz, A. Allert, L.S. Sappington, K.J. Nelson, and J. Valle. (2009b). An ecological risk assessment of the exposure and effects of 2,4-D acid to rainbow trout (*Oncorhynhcus mykiss*). *Arch. Contam. Toxicol.* 56: 754-760.

Fairchild, J.F., A.L. Allert, K.P. Feltz, K.J. Nelson, and J. Valle. (2009c). An ecological risk assessment of the acute and chronic effects of the herbicide clopyralid to rainbow trout (*Oncorhynhcus mykiss*). *Arch. Environ. Contam. Toxicol.* 57:725-731.

Ferrat, L. C. Pergnet-Martini, and M. Romeo. (2003). Assessment of the use of biomarkers in aquatic plants for the evaluation of environmental quality: application to seagrasses. *Aquat. Tox.* 65: 187-202.

Field, J.A. and E.M. Thruman. (1996). Gultathione conjugation and contaminant transformation. *Environ. Sci. Tech.* 30: 1413-1418.

Flynn, D.J. (2002). Herbicide legislatation and regulation. In: *Weed Management Handbook*, R.E. Naylor, Ed., Pp. 114-133, British Crop Protection Council, Blackwell Publishing, Oxford, UK. 423 pp.

Gassman, P.W., M.R. Reyes, C.H. Green, and J.G. Arnold. (2007). The Soil and Water Assessment Tool: historic development, applicatioins, and future research directions. *Amer. Soc. Agricul. Biol. Eng.* 50(4): 1211-1250

Gianessi, L.P., and N.P. Reigner. (2007). The value of herbicides in U.S. crop production. *Weed Technol.* 21:559-566.

Giddings, J.M., T.D. Anderson, L.W. Hall, Jr., A.J. Hosmer, R.J. Kendall, R.P. Richards, K.R. Solomon, and W.M. Williams. (2005). *Atrazine in North American surface waters: probabalistic aquatic ecological risk asssessment.* SETAC Press, Pensacola FL. 392 pp.

Gilliom, R.J. (2007). Pesticides in streams and groundwater. *Environ. Sci. Tech.* 41:3408-3414.

Hayes, T.B, A. Collins, M. Lee, N. Mendoza, N. Noirega, A.A. Stuart, and A. Vonk. (2002). Hermaphtoditic, demasculinized frogs after exposure to the herbicide atrazine at low ecologically relevant doses. *Proc. Nat. Acad. Sci.* 99:5477-5480.

Hayes, T.B., K. Haston, M. Tsui, A. Hoang, C. Haeffele, and A. Vonk. (2003). Atrazine-induced hermaphrodism at 0.1 ppb in American leopard frogs (Rana pipiens): laboratory and field evidence. *Environ. Health Perspect.* 11:568-575.

Kubitza, J. and Dohmen, G.P. (2008). Development of a test method for the aquatic macrophyte *Myriophyllum aquaticum.* SETAC Europe 18th Annual Meeting, Warsaw, 2008.

Huber, A., M. Bach, and H.G. Frede. (2000). Pollution of surface waters with pesticides in Germany: modeling non-point source inputs. *Agricul. Ecosys.* 80:191-204.

Hughes, JS., MM Alexander, and K Balu. (1988). An evaluation of appropriate expressions of toxicity in aquatic plant bioassays as demonstrated by the effects of atrazine on algae and duckweed. In: Adams, W, GA Chapman, and WG Landis, ed. Volume 10, *Aquatic Toxicology and Hazard Assessment*, American Society for Testing and Materials, p. 531-547.

Kettle, W.D., F. deNoyelles, J.D. Heacock, and A. Kadoum. (1987). Diet and reproductive success of bluegill recovered from experimental ponds treated with atrazine. *Bull. Environ. Contam. Toxicol.* 38:47-52.

Knauer, K., Mohr, S., and U. Feiler. (2008). Comparing growth development of *Myriophyllum* spp. in laboratory and field experiments for ecotoxicological testing. *Sci. Pollut. Res.* 15:322–331.

Kiely, T., D. Donaldson, and A. Grube. (2009) Pesticide industry sales and usage: 2000 and 2001 market estimates. U.S. Environmental Protection Agency Office of Prevention, EPA-733-R-04-001, Washington, DC. 33 pp.

Kubitza, J., and G.P. Dohmen. (2008). Development of a test method for the aquatic macrophyte *Myriophyllum aquaticum*. SETAC Europe 18th Annual Meeting, Warsaw, 2008.

Lampert, W., W. Fleckner, E. Pott, U. Schober, and K.U. Storkel. (1989). Herbicide effects on planktonic systems of different complexity. *Hydrobiologia* 188/189:415-424.

Larsen, D.P., F. deNoyelles, F. Stay, and T. Shiroyama. (1986). Comparisons of single-species, microcosm, and experimental pond responses to atrazine exposure. *Toxicol. Chem.* 5:179-190.

Larson, S.J., C.G. Crawford, and R.J. Gilliom. (2004). Development and application of watershed regressions for pesticides (WARP) for estimating atrazine concentrations in streams. U.S. Geological Survey Water Resources Investigations Report 03-4047. *Environ. Toxicol. Chem.* 68 pp.

Lerch, R.N., and P.B. Blanchard. (2003). Watershed vulnerability to herbicide transport in northern Missouri and southern Iowa streams. *Environ. Sci. Technol.* 37:5518-5527.

Menne, H., and H Kocher. (2007). HRAC classification of herbicides and resistance development. In: *Modern Crop Protection Compounds*, W. Kramer and U. Schirmer, Ed., pp. 5-26, Wiley-VCH Publishing, Weingeim, Germany.

Moss, R.S. (2002). Herbicide-resistant weeds. In: *Weed Management Handbook*, R.E. Naylor, Ed., Pp. 225-252, British Crop Protection Council, Blackwell Publishing, Oxford, UK. 423 pp.

Peterson, H.G., C. Boutin, P.A. Martin, K.E. Freemark, N.R. Ruecker, and M.J. Moody (1994). Aquatic phyto-toxicity of 23 pesticides applied at expected environmental concentrations. *Aquat. Toxicol.* 28:275-292.

Rabiet, M., C. Margoum, V. Gouy, N. Carluer, and M. Coquery. (2010). Assessing pesticide concentrations and fluxes in the stream of a small vineyard catchment - effect of sampling frequency. *Environ. Pollut.* 158: 737-748.

REACH. 2007. Regulation on Registration, Evaluation, Authorization and Restriction of Chemicals. http://ec.europa.eu/environment/chemicals/reach/reach_intro.htm, accessed July 15, 2010.

Read, J.P.H., and A.H. Cobb. (2002) Herbicide modes of action and metabolism. In: *Weed Management Handbook*, R.E. Naylor, Ed., Pp. 134-170, British Crop Protection Council, Blackwell Publishing, Oxford, UK. 423 pp.

Richards, R.P., and D.B. Baker. (1993). Pesticide concentration patterns in agriculutural drainage networks in the Lake Erie Basin. *Toxicol. Chem.* 12:13-26.

Ritter, A.M, J.L. Shaw, W.M. Williams, and K.Z. Travis. (2000). Characterizing aquatic ecological risks from pesticides using a diquat dibromide case study I: Probablistic expousre estimates. *Environ. Toxicol. Chem.* 749-759.

Rohr, J.R., and K. McCoy. (2010). A qualitative metal analysis reveals consistent effects of atrazine on freshwater fish and amphibians. *Environ. Health Perspect.* 118:20-32.

Saundry, P. (2006). Federal Insecticide, Fungicide, and Rodenticide Act, Unites States. In: *Encyclopedia of Earth*, C.J. Cleveland, (Ed.), pp. 1-6, Congressional Research Service, Envrionmental Information Coalition, National Council for Science and the Envrironment, Washington, DC. Accessed Aug. 18, 2010, http://www.eoearth.org/article/Federal_Insecticide_Fungicide_and_Rodenticide_Act)_United_States.

Savino, J.F. and R.A. Stein. (1982). Predator-prey interaction between largemouth bass and bluegill as influenced byin simulated, submerged vegetation. *Trans. Amer. Fish. Soc.* 111: 255-266.

Schnoebelen, D.G., S.J. Kalkhoff, K.D. Becher, and E.M. Thurman. (2003). Water quality assesssment of the eastern Iowa basins: selected pesticides and pesticide degradates in streams, 1996-1998. U.S. Geological Survey Investigations Report 03-4075. 62 pp.

Schwarz, G.E., A.B. Hoos, R.B. Alexander, and R.A. Smith. (2006). The SPARROW surface water quality model: theory, application, and user documentation. U.S. Geological Survey Techniques and Methods Book 6, Section B, Chapter 3. 29 pp.

Solomon, K.R., D.B. Baker, R.P. Richards, K.R. Dixon, S.J. Klaine, T.W. La Point, R.J. Kendall, C.P. Weisskopf, J.M. Giddings, J.P. Giesy, L.W. Hall Jr., and W.M. Williams. (1996). Ecological risk assesssment of atrazine in North American surface waters. *Environ. Toxicol. Chem.* 15:31-76.

Solomon, K.R., J.A. Carr, L.H. DuPreez, J.P. Giesy, R.J. Kendall, E.E. Smith, and G.J. VanderKraak. (2008). Effects of atrazine on fish, amphibians, and aquatic reptiles: a critical review. *Crit. Rev. Toxicol.* 38:721-772.

USEPA. (1998). *Guidelines for Ecological Risk Assessment.* U.S. Environmental Protection Agency, Risk Assessment Forum, EPA/630/R095/002F, Washington, DD. 188 pp.

USEPA. (2002). Ecotoxicity pesticide database. http://www.epa.gov/oppefed1/general/databasesdescription.htm#ecotoxicity

USEPA. (2006). Decision documents for atrazine. Accessed Aug. 8, 2010 at http://www.epa.gov/oppsrrd1/REDs/atrazine_combined_docs.pdf. 323 pp.

USEPA. (2010). U.S. EPA Office of Chemical Safety and Pollution Prevention (OCSPP). http://www.epa.gov/ocspp/pubs/frs/home/guidelin.htm, accessed July 15, 2010.

Wauchope, R.D. (1978). The pesticide content of surface water draining from agricultural fields- a review. *Rev. Environ. Qual.* 7:459-472.

Wauchope, R.D., T.M. Buttler, A.G. Hornsby, P.W.M. Augustijn-Beckers, and J.P. Burt. (1992). The SCS/ARS/CES pesticide properties database for environmental decision-making. *Rev. Contam. Toxicol.* 123:1-164.

Weed Management Handbook (2002). Naylor, E.L., editor. Blackwell Science, 9th edition.

Wiley, M.J., R.W. Gorden, S.W. Waite, and T. Powless. (1984). The relationship between aquatic macrophytes and sport fish populations in Illinois ponds. *N. Amer. J. Fish Manage.* 4:111-119.

19

Toxicology of the Herbicide Acrolein: Risk Assessment in Aquatic Environments

Cristina Montagna, Ana María Pechen de D'Angelo and Andrés Venturino
Universidad Nacional del Comahue
Argentina

1. Introduction

Submersed weeds on irrigation systems reduce water delivery capacity, clog pumps and structures, rupture canals, increase leakages and losses of water, increase water costs, etc. (U.S.EPA, 2007). Herbicide applications are a commonly used procedure to control submersed weeds in irrigation canals because of its practicability, efficacy and cost. Acrolein, currently registered under the trade name MAGNACIDE® H by Baker Petrolite Corporation, has been used for many years in the United States (U.S.EPA, 2007), Canada (MOE, 2005), Australia (Bowmer & Sainty, 1977), and Argentina (Caldironi et al., 2004).

Acrolein, also known as acraldehyde, acrylaldehyde, acrylic aldehyde, allylaldehyde, propenal, 2-propenal, prop-2-enal, prop-2-en-1-al, is a volatile, colourless, highly flammable liquid at ordinary temperature and pressure with a pungent odour. Its Chemical Abstract Service (CAS) number is 107-02-8. The chemical formula for acrolein is C_3H_4O and the molecular weight is 56.06. Fig. 1 illustrates its chemical structure. Acrolein has a density of 0.84 g/mL, a water solubility of 206 g/L, and a vapour pressure of (kPa) 29.3 at 20°C. The log Kow (octanol/water partition coefficient) is -0.01 (high water solubility) and the log Koc (organic carbon/water partition coefficient) is 0.5 (low adsorption to soil) (WHO, 1991; U.S.EPA, 2003; ATSDR, 2007).

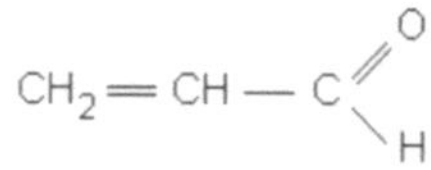

Fig. 1. Acrolein chemical structure

MAGNACIDE® H is an herbicide primarily used to remove submersed plants and algae from irrigation canals and ditches. Species of weeds such as *Potamogeton* spp., *Elodea* spp., *Najas* spp., *Zannichellia* spp., *Ceratophyllum* spp., *Zannichellia* spp. and algae such as *Anabaena* spp., *Chara* spp., *Cladophora* spp., *Selenastrum* spp., *Spirogyra* spp., are controlled by recommended label use rates of the product. According to the "Application and Safety Manual Acrolein" (Baker Petrolite, 2001), the chemical is applied by injection from the container with oxygen-free nitrogen gas into the flowing water, and travels downstream as a wave of treated water until, at some point, the chemical concentration drops to zero. Effective concentrations range from 1 mg/L to 15 mg/L and the treated area of canal is held at periods ranging from 30 minutes to 8 hours. Both the concentration and the treatment time may vary depending on the weed growth condition, water flow rate, temperature, and application time desired. The label does stipulate that "water treated with MAGNACIDE® H herbicide must be used for irrigation of fields, either crop bearing, fallow or pasture, where the treated water remains on the field or held for 6 days before being released into fish bearing waters or where it will drain into them". MAGNACIDE® H is a restricted use pesticide for retail sail to, and use only by, certified applicators or persons under their direct supervision.

2. Mode of action

Acrolein is a cell toxicant that reacts with several molecules containing sulfhydryl groups, including proteins, exerting direct cytotoxic effects or interrupting cell signalling pathways. Acrolein reacts with glutathione (GSH) to produce the adduct glutathionyl propionaldehyde, and both induce oxygen radical formation in the presence of xantine oxidase and aldehyde dehydrogenase. The depletion of GSH decreases GSH peroxidase activity, resulting in a lower level of cellular protection against oxygen radical toxicity (Adams & Klaidman, 1993). Depletion of GSH inactivates multiple enzymes in the Calvin cycle affecting the photosynthetic reactions in chloroplasts isolated from *Spinacia oleracea* L. (Mano et al., 2009). In isolated rat hepatocytes, concentrations of 0.25 and 0.5 mM of acrolein decrease GSH with a concomitant lipid peroxidation that impairs the integrity of the cell membrane. Acrolein also induces apoptosis in the Chinese hamster (Tanel & Averill-Bates, 2007) and human bronchial epithelial cell, by depletion of GSH and generation of oxidants (Nardini et al., 2002). On the other hand, Luo et al. (2005) reported that acrolein induces death of PC12 cells, mainly by necrosis. These authors suggested that the ability of acrolein to induce cell death is closely related to mitochondrial ROS production and decreased cellular ATP levels. Further, acrolein conjugation with lysine residues of low density lipoproteins has been suggested as a factor in the development of atherosclerosis (Uchida et al., 1998)

3. Toxicity on non-target species

Non-target species that are principally exposed to acrolein include aquatic organisms inhabiting the irrigation canal. Potentially exposed organisms such as fish, invertebrates, amphibians, etc., are those inhabiting natural surface water that receive treated irrigation water. Terrestrial receptors that could be exposed to acrolein include mammals, birds, reptiles, and terrestrial-phase amphibians.

3.1 Aquatic organisms

Acrolein, the active ingredient of MAGNACIDE® H, is acutely toxic to aquatic organisms (Table 1). The tadpole of the frog *Xenopus laevis* is the most sensitive aquatic species tested, with a 96-h LC_{50} of 7 μg/L (Holcombe et al., 1987). The acute toxicity (LC_{50} or EC_{50} ug/L) reviewed by the U.S.EPA (2009) ranges between 14-320 μg/L and 57-180 μg/L for freshwater fish and crustacean, respectively. Insects seem to be more tolerant to acrolein with LC_{50} values between 600 to 2,800 ug/L (Venturino et al., 2007). Acrolein is also toxic to cyanobacteria. Several species from this group showed more than 95% of growth inhibition at a concentration of 1 mg/L of acrolein (Peterson et al., 1994).
Detoxification of [^{14}C] acrolein was studied in exposed fish (*Lepomis macrochirus* and *Ictalurus punctatus*) and shellfish (*Elliptio complanata* and *Orconectes virilis*). The mayor metabolites founded were glycidol, glycerol, 1,3-propanediol, and glyceric acid (Nordone et al., 1998).

Species	Group	Toxicity (LC_{50} ug/L) - exposure time	Reference
Xenopus laevis	Amphibia	7 - 96 hours	1
Catostomus commersoni	Fish	14 - 96 hours	1
Pimephales promelas	Fish	14 - 96 hours	1
Salmo gairdneri	Fish	16 - 96 hours	1
Rhinella arenarum	Amphibia	23 - 96 hours	2
Scenedesnus subspicatus	Algae*	26 - 72 hours	3
Lepomis macrochirus	Fish	33 - 96 hours	1
Oncorhynchus mykiss	Fish	38 - 96 hours	2
Daphnia magna	Crustacea	51 - 48 hours	1
Amia calva	Fish	62 - 24 hours	3
Oncorhynchus tshawytscha	Fish	80 - 24 hours	3
Carassius auratus	Fish	80 - 24 hours	3
Penaeus aztecus	Crustacea	100 - 48 hours	3
Cladophora glomerata	Alga*	100 - 24 hours	3
Rasbora heteromorpha	Fish	130 - 48 hours	3
Tanytarsus dissimilis	Insect	151 - 48hours	1
Aplexa hypnorum	Mollusc	>151 - 96 hours	1
Micropterus salmoides	Fish	160 - 96 hours	3
Heleobia parchappii	Mollusc	210 - 96 hours	2
Hyalella curvispina	Crustacea	240 - 96 hours	2
Fundulus similis	Fish	240 - 48 hours	3
Simulium spp.	Insect	600 - 246 hours	2
Anabaena	Algae*	690- 24 hours	3
Entosiphon sulcatum	Protozoa	850 - 72 hours	3
Chilomonas paramecium	Protozoa	1,700 - 48 hours	3
Balanus ebarneus	Crustacea	2,100 - 48 hours	3
Biomphalaria glabrata	Mollusc	2,500 - 24 hours	3
Chironomus spp.	Insect	2,830 - 24 hours	3

Table 1. Acute toxicity of acrolein in aquatic organisms. *reduction in photosynthes.
References: (1) Holcombe et al. (1987), (2) Venturino et al. (2007), (3) Cited in Eisler (1994)

3.2 Terrestrial organisms

According to the review by Eisler (1994), the adverse effects of acrolein depend on the mode and concentration or dose of administration, and duration of exposure. For example, single oral doses of 4 and 28 mg/Kg of body weight resulted lethal to guinea pigs and mice, respectively. The LC_{50} reported by inhaled acrolein (mg acrolein/L air) during 30 min and 6 hours exposure were 150 and 10.5 for dog and guinea pig, respectively. Adverse effects were also observed in birds. The oral LD_{50} of 3-5 months for the mallard, *Anas platyrhynchos*, was 9.1 mg/kg body weight. Auerbach et al. (2008) reported decreased survival and toxicity to the forestomach, squamous epithelial hyperplasia in rats and mice (both sexes) exposed 5 days a week during 3 months, by gavage, to 0–10 mg/kg acrolein and 0–20 mg/kg, respectively. Studies *in vitro* (mouse embryonic fibroblasts culture) showed formation of

DNA adducts, preferentially at specific nucleotide positions, moderately resistant to DNA repair. However, the results demonstrated that acrolein was not mutagenic to these cells at doses sufficient to produce DNA adducts (Kim et al., 2007). Further, rat embryo culture system treated with acrolein (200 and 250 μM) showed a drastic inhibition of growth differentiation without teratogenic potential (Schmid et al., 1981).

Conjugation with GSH is one of the two major detoxification pathways of acrolein. The 3-hydroxypropylmercapturic acid was the principal metabolite found in urine of male Wistar rats exposed to acrolein inhalation and intraperitoneal administration. To lesser extent, the metabolite 2-carboxyethylmercapturic acid was determined (Linhart et al., 1996).

Acrolein degrades quickly in soils and in plant tissues regardless of mode of administration. Most terrestrial crop plants can tolerate 15 mg of acrolein/L of irrigation water (Eisler, 1994). Nordone et al. (1997) evaluated the accumulation of acrolein in lettuce plants receiving either a single or multiple applications of 75 ppm [^{14}C]-acrolein in irrigation water. The results showed that both treatments leave almost no radioactive residues in leaves after 53 days. This study indicates that, under normal use scenarios, irrigation of crops with MAGNACIDE® H herbicide treated water is highly unlikely to result in the accumulation of biologically significant levels of acrolein in lettuce. Further, an experiment conducted in a greenhouse, where pepper plants were irrigated with water treated with 0.25 and 0.50 mM of acrolein, showed low values of chemical concentration in the extracts of the plants (2-18 ng/gr fresh tissue to undetectable levels within the few hours). The estimated half-life of acrolein in pepper plants was of 10.3 hours (Caldironi et al., 2004).

3.3 Humans

Liquid acrolein is absorbed by the skin, and is particularly irritating to the eyes. The vapor is highly toxic and a strong irritant (lachrymator) which acts principally on the mucous membranes of the eyes, nose, throat and lungs. The vapor concentration tolerable to humans is 0.1-1 ppm in air and can cause lung injury at 2-4 ppm (Baker Petrolite Corporation, 2001). The effects of long-term atmospheric exposure of humans to acrolein at tolerable levels are not known, but the concentrations likely to be found in the environment or workplace should not affect human reproduction (WHO, 1991). There is inadequate evidence in experimental animals for the carcinogenicity of acrolein (Group 3) (IARC, 1995).

4. Fate of acrolein after direct application into irrigation canals

The high reactivity of acrolein prevents its persistence in the environment, and its transportation over long distances (WHO, 2002; U.S.EPA, 2003). Dissipation of acrolein from aquatic ecosystems includes abiotic and biotic degradation, volatilization, absorption and dilution.

Acrolein is at equilibrium with the abiotic product 3-hydroxypropanal, and the presence of both compounds is transient (Nordone et al., 1996). The decay of acrolein and its hydration product is a first order process in agricultural canals when applied at the recommended concentrations. The half-life of acrolein in weedy irrigation canals from United States, Australia, and Argentina was 10.2, 4.3, and 9.63 hours, respectively (Bowmer & Sainty, 1977; Nordone et al., 1996; Venturino et al., 2007). The U.S.EPA (2007) also reported half-lives of acrolein between 2 to 20 hours in canals from Washington and Nebraska (U.S.A). Further, Eisler (1994) summarizes the half-time persistence of acrolein in freshwaters as usually less than 50 hours, and according to ATSDR (2007) acrolein may persist for up to 6 days. Bioassays with fish and bacteria have demonstrated that acrolein losses its biocide activity in 120-180 days in different buffer systems at pH 7 and 22°C (Kissel et al., 1978). Monitoring studies in United States (U.S.EPA, 2007) showed that the compound can be transported to distances of at least 61 miles beyond the initial site of application at concentrations that are still active.

At the application rate of 15 ppm acrolein into irrigation canals, the primary microbial degradation product was 3-hydroxypropanal. Other ephemeral products such as acrlylic acid, allyl alcohol, propionic acid, propanol, and 3-hydroxipropionic acid were also identified (Smith et al., 1995).

The high water solubility of acrolein and low estimated Koc suggests that acrolein does not significantly adsorb to suspended solids and sediment (HSDB, 2010; U.S.EPA, 2007). Volatilization from water surfaces is expected to be an important fate process based upon the compound's Henry's Law constant (1.0 x 10^1 mol $atm^{-1}dm^{-3}$). Estimated volatilization half-lives for a model river and lake (1 m deep) were 7.6 hours and 4.6 days, respectively (HSDB, 2010). In the atmosphere, the primary removal mechanism for acrolein is through the reaction with hydroxyl radicals with a half-life between 15–20 hours (Faroon et al., 2008). It is unlikely that acrolein bioaccumulate or bioconcentrate significantly in aquatic organisms (WHO, 1991). Acrolein was not detected in the tissues of fish (*Lepomis macrochirus* and *Ictalurus punctatus*) and shellfish (*Elliptio complanata* and *Orconectes virilis*) exposed separately to [^{14}C]-acrolein in water (0.02 and 0.1 mg/L for fish and shellfish, respectively), over a 1-week period, and sampled 1 day after a second exposure. The presence of metabolites indicated that these species were able to rapidly metabolize acrolein (Nordone et al., 1998). An estimated Bioconcentration Factor of 3.2 suggests that the potential for bioconcentration in aquatic organisms is low ((HSDB, 2010).

5. Risk assessment of acrolein application as an herbicide in irrigation canals

We developed a risk evaluation of the use of acrolein in an irrigated valley, where applications of this compound (MAGNACIDE® H) had been currently performed for more than 20 years during the spring-summer seasons. The area of study was located in the Río Colorado valley (Argentina) between 39°10′ to 39°55′S and 62°05′ to 63°55′W. The irrigation of the area, controlled by the CORFO-Río Colorado cooperative, consists of 331 km of main canals, 3738 km of both secondary and tertiary canals, and 397 km of drainage canals which discharge directly to the Argentinean sea. The herbicide had been applied at different target concentrations and durations. The application scheme of 15 mg/L for 1 hour was used principally during the first years of use. Afterwards, it was applied at a target concentration of 4 mg/L for 12 hours. The temperature in water canals ranged from 15 to 22.5°C during the application seasons. Water flow rates were regulated currently between 0.20-0.50 m/s, according to weed proliferation status.

Keeping a tiered approach recommended by ECOFRAM (1999), we divided the ecological assessment into four tiers: (a) Literature-based screening level ecological risk assessment, (b) risk assessment with site-specific information, (c) risk assessment with native species, (d) impact of acrolein on benthic invertebrates (field study).

At tier 1, we compared the maximum predicted peak concentration in the CORFO canals with acute endpoints such as effective or lethal concentration, from the most sensitive species of freshwater fish, amphibian, molluscs and crustacean from Table 1. Even though it has been suggested that LC_5 or LC_{10} may be a more appropriate parameter (EU, 2002), EC_{50} or LC_{50} is the data generally available in the literature. Only freshwater organisms were selected to assess the aquatic risk since they are the principal ecological receptors. Hazard Quotients (HQ) were calculated using the target concentrations of acrolein in the two different application schemes. The calculated acute HQ were compared with criteria used for risk characterization in tier 1 (Urban & Cooke, 1986). The acute HQ estimated for both application schemes highly exceeded the risk criteria for all groups of organisms evaluated (Table 2). The tadpole *Xenopus laevis* and the mollusc *Aplexa hyponorum* were the most sensitive and the most tolerant species to acute exposure of acrolein, respectively.

Species	Group	Acute endpoint[a] (mg/L)	HQ[b]	
			Scheme 1	Scheme 2
Daphnia magna	Crustacea	48-h EC_{50} (0.051)	78.43	235.29
Pimephales promelas	Fish	96-h LC_{50} (0.014)	285.71	857.13
Xenopus laevis	Amphibia	96-h LC_{50} (0.0070)	571.43	1714.29
Aplexa hypnorum	Mollusc	96-h LC_{50} (0.15)	26.67	80.00
Tanytarsus dissimilis	Insect	48-h LC_{50} (0.15)	26.67	80.00

Table 2. Hazard quotients (HQ) for the most susceptible species from representative groups. HQ were calculated from endpoints and target concentrations of acrolein in two different application schemes: Scheme 1, 15 mg/L for 1 hour; Scheme 2, 4 mg/L for 12 hour. [a]Acute toxicity data cited in Holcombe et al. (1987)], [b]HQ 0.5 or greater indicates a higher risk category.

Chronic toxicity data were not used in this study because of the limited information available and the rapid dissipation of acrolein in treated canals. The risk criteria were highly exceeded for all of the species analyzed implying that progress to tier 2 was indispensable.

The environmental fate behavior of acrolein was incorporated in tier 2, to provide probabilistic expressions of the potential risk associated with its use as an herbicide. First, effects of acrolein on aquatic freshwater organisms were characterized by distribution sensitivity curves (ECOFRAM, 1999). The acute toxicity data obtained from scientific literature (Holcombe et al., 1987; WHO, 1992; Eisler, 1994) included 10 species of fish, 1 species of amphibian, 3 species of crustacean, 2 species of molluscs, 1 species of insect, 2 species of protozoan, and 3 species of algae (Table 1). According to the Guidance Document on Aquatic Ecotoxicology (EU, 2002), two algae species from different taxonomic groups should be included for herbicide risk assessment. Even though photosynthesis inhibition is not an equivalent endpoint, it was included into the list to compare algae toxicity. The species were ranked by decreasing sensitivity and the rank was transformed to a percentile $[i/(n+1)]$, where i is the species rank, and n is the total number of species listed. The probit analysis was performed to obtain the regression lines and to determine the 10th percentiles. Figure 2 represents the LC_{50} log-transformed and the percentiles converted to probabilities from the data set and the distribution profile of toxicities to organisms. From the fitted line of the distribution for all species the 10th percentile was 0.011 mg/L which means that 10% of the species have LC_{50} values lower than this concentration. On the other hand, the 10th percentile from the distribution of sensitivities to acrolein in animals, excluding the values from algae and weeds, was 0.0094 mg/L. This analysis suggests that a significant number of aquatic species may be seriously and unacceptably affected by acrolein concentrations in the canals.

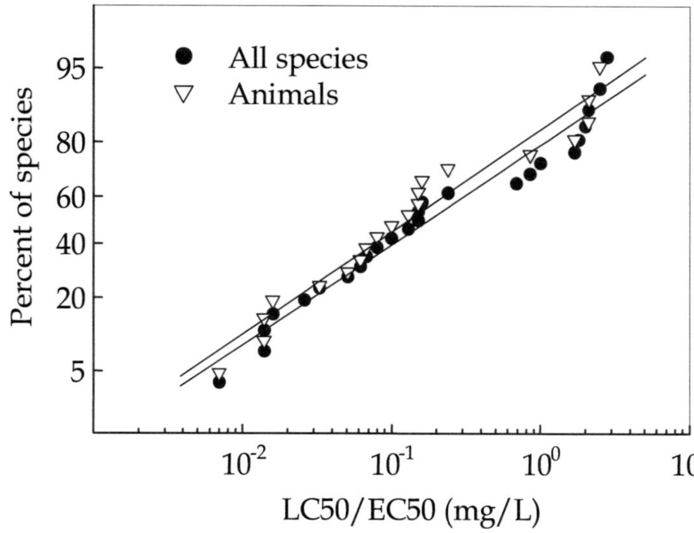

Fig. 2. Distribution of acute toxicity values for different taxonomic groups of organisms. Percentile probabilities were calculated from acrolein LC_{50} or EC_{50} values for freshwater organisms (•) (probit regression line Y = 0.8128 + 1.0861X, R=0.9783, N=25; 10th percentile 0.0118 mg/L), or animal freshwater species (∇) (probit regression line Y = 0.9794 + 1.1107X, R=0.9758, N=20; 10th percentile 0.00938 mg/L).

Next, an estimation of acrolein levels along the canals was considered for risk evaluation. On the basis of CORFO application data during several application seasons, we estimated the dissipation of acrolein in the canals. The model applied takes into account the time-space variability of acrolein concentrations within the canals, applying an exponential equation (1) that predicts the exposure at different distances from the application site:

$$[\text{Acrolein}]_d = [\text{Initial acrolein}] \times e^{-[\ln 2/t_{1/2} \times d\ (\text{km})/v\ (\text{km/h})]} \quad (1)$$

where:
$[\text{Acrolein}]_d$ = Concentration of acrolein at distance *d* from the application point.
[Initial acrolein] = Concentration of acrolein at the application point.
$t_{1/2}$ = 9.63 h according to dissipation studies performed on the canals.
v (km/h) = water flow velocity in the canal.

The pulse of acrolein passing through the canal may be visualized as a block with side heights exponentially decreasing as it is moves by water flux, and length depending on the application schedules (Fig. 3).

According to this model, the predicted concentrations were calculated at several distances from the application point in order to analyze theoretical exposures within the water body. The maximum distance calculated from the application point was 20 km since it is the approximate span wherein weed control is still effective. The predicted concentration of acrolein at different distances and the percentage of species whose LC_{50}/EC_{50} were exceeded at each concentration is summarized in Table 3. These results showed that a high percentage of the species are likely to be affected by the herbicide despite the distance from the application point in the treated canal.

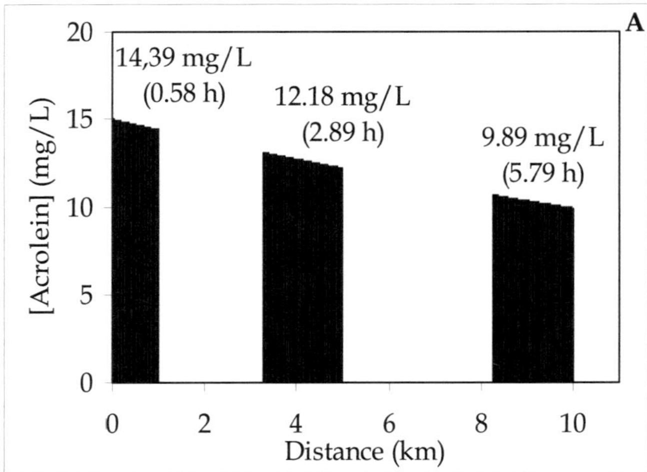

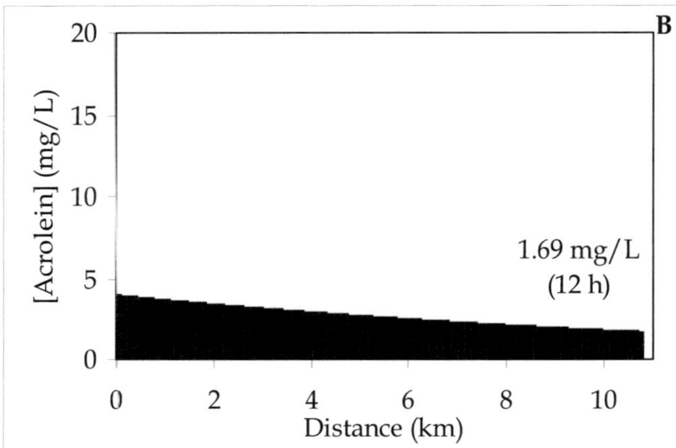

Fig. 3. Exponential dissipation model.
The exponential dissipation of acrolein was simulated for a model canal with a half-life of 9.63 h and: (A) scheme 1, flow velocity 0.48 m/s, 15 mg/L of the compound applied continuously during 1 h, or (B) scheme 2, flow velocity 0.25 m/s, 4 mg/L of the compound applied continuously during 12 hours.

Distance from the application point (km)	Predicted concentration (mg/L)	Percent of species with LC_{50}/EC_{50} exceeded
0	4.00	100
5	2.68	100
10	1.80	88.91
15	1.21	84.46
20	0.81	78.24

Table 3. Predicted concentration of acrolein and percent of species affected. Concentrations following application of acrolein were estimated for a target concentration of 4 mg/L during 12 hours at the application point and at different points along a canal with water velocity of 0.25 m/sec.

To increase the environmental realism of the tier 2 scenario, different application patterns of acrolein over six years at CORFO-Río Colorado (N=165 applications) were incorporated to the analysis. According with the dissipation model, the expected concentrations at the application point, at 10 km and at 20 km downstream were estimated. The environmental expected concentration as a cumulative exceedence curve at the above three distances, and the distribution profile of toxicities is presented in Fig. 4.

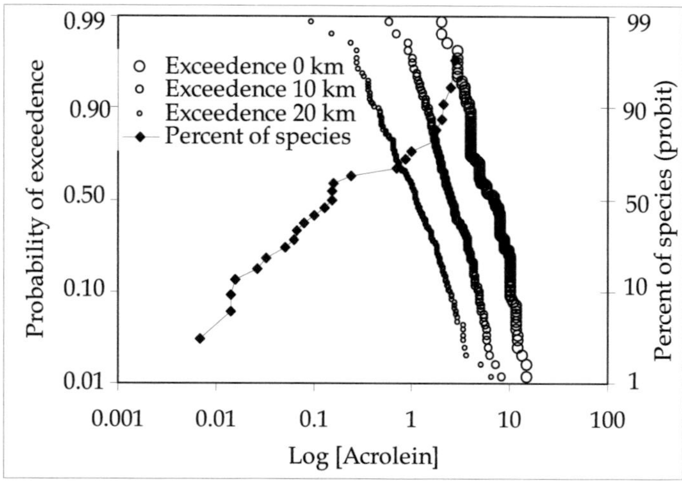

Fig. 4. Concentration exceedence probabilities and toxicity distribution profile for acrolein applications at CORFO-Río Colorado canals.

For each concentration on the X axis, this curve indicates the frequency that concentration was exceeded during the period of time analyzed, and the percentage of species affected. Therefore, each pair of points (probability and percentage of species) can be associated to a concentration. For example, during that period, the concentration causing mortality to 73% of the species (1 mg/L) was exceeded in 55%, 95% and 100% of the applications at 20, 10 and 0 km from the application point, respectively. On the basis of the results here depicted, it may be concluded that acrolein poses sufficient risk as an herbicide to require a higher level of assessment. Some of the studies proposed by ECOFRAM (1999) in tier 3 of aquatic risk assessment are: (a) Acute toxicity studies with additional species, (b) investigation of the toxicity associated with repeated exposures, (c) chronic toxicity studies, (d) sediment toxicity studies.

Additional acute toxicity studies with native species, collected nearby the potential site of exposure to acrolein or obtained from hatcheries, were then performed in tier 3. The last instar larvae of the insects *Chironomus* spp. and *Simulium* spp., the mollusc *Heleobia parchappii*, the crustacean *Hyalella curvispina*, tadpoles of the amphibia *Rhinella arenarum*, and juveniles of the fish *Oncorhynchus mykiss* were selected for the study. The different organisms were exposed to different concentrations of acrolein. The experimental conditions and the complete ecotoxicological data listing the LC_{95}, LC_{50}, Lowest Observed Effect Concentrations (LOEC), and No-Observed Effect Concentrations (NOEC) are published elsewhere (Venturino et al., 2007). The toad *R. arenarum* was the most sensitive species followed by the fish *Oncorhynchus myskiss*. However, toxicity in both species is almost three times lower (0.023 and 0.038 mg/L) than in their related counterparts *Xenopus leavis* and

Catostomus commersoni, (0.007 and 0.014 mg/L), respectively. The native species *Chironomus* spp. was the less sensitive to acrolein with the highest LC_{50} (2.83 mg/L). A probit analysis on percentile sensitivity distribution for the native species provides an estimation of 0.013 mg/L acrolein as the 10th percentile, which is the concentration eventually affecting 10 percent of native species (Fig. 5). This value is quite similar to the 10th percentile estimated from the sensitivity distribution using published data for other species (Fig. 2).

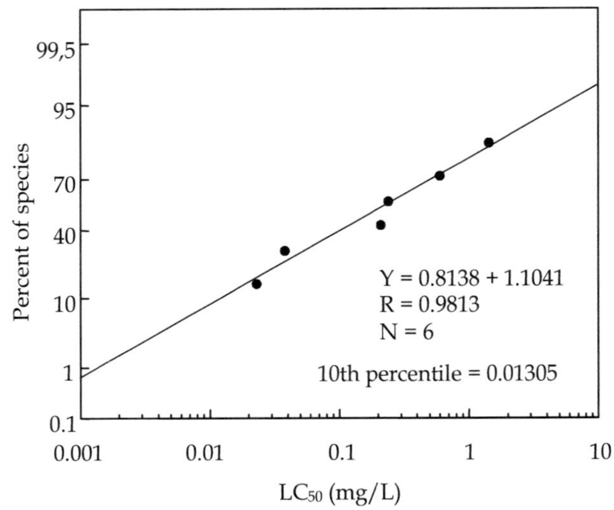

Fig. 5. Distribution of acute toxicity values for different taxonomic groups of native organisms. (Venturino et al. (2007), with permission of SETAC press).

To compare the ecotoxicological data with field predicted exposure concentrations, a first approach can be made using the percent mortality obtained at 24 hours with exposures to 0.5 and 1.0 mg/L of acrolein in laboratory tests. These levels were chosen considering the potential concentration of acrolein approaching 20 km from the application point in the low-concentration treatment schedule (0.8 mg/L), and the probable exposure time in the canals. The values obtained, shown in Table 4, indicate an unacceptable risk for fish, amphibians, and the amphipod *H. curvispina*, an intermediate risk for snails and black fly larvae, and no risk for midges.

Acrolein (mg/L)	Species (stage)					
	toad (larvae)	amphipod (adult)	snail (adult)	fish (juvenile)	black fly (larvae)	midge (larvae)
	% Death in 24 h					
0.5	100%	100%	20%	100%	36%	0%
1.0	100%	100%	n.d.	100%	98%	0%

Table 4. Percent of mortality at 24 hours of exposure to 0.5 mg/L and 1.0 mg/L acrolein in laboratory tests.

To improve the risk assessment, the sensitivity distribution along the distance from the application point was analyzed. From it, the exceedence probabilities could be assessed as percentiles of native species affected by acrolein. The two extreme application schedules of 15 mg/L-1h (scheme 1) and 4 mg/L-12h (scheme 2) were chosen, spanning the most of the alternatives used at CORFO-Río Colorado (Fig. 6). From the probit analysis, it is inferred that the distances needed for acrolein dilution and degradation to a concentration affecting no more than a 10% of native species are not physically feasible (175 km for scheme 1; 74 km for scheme 2). On the other hand, it alerts on the risks posed by both application schemes: scheme 1 produces an acute exposure (1 h) that probably affects 95% and 90% of native species at about 11 km and 30 km respectively dowstream the application point. Scheme 2 affects 90% of native species just at the application point. In both cases, the risk probability decreases linearly with the distances from the application point.

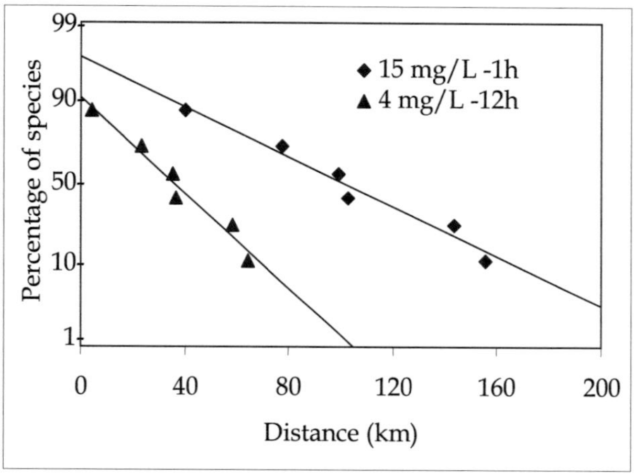

Fig. 6. Joint probability graph for native species

The use of acrolein as an herbicide against aquatic weeds currently requires periodical applications during spring-summer seasons. Acrolein exposure in aquatic ecosystems occurs in pulses, with peak concentrations in the water lasting few hours. Such pulses are applied periodically, typically every 20-30 days. Repeated exposure assays have been performed with two native species applying pulse-recovery schemes, simulating the application frequencies in CORFO-Río Colorado canals (Venturino et al., 2007). The effects were tested at 96h-LOLC and LC_{50} values for each species. *R. arenarum*, which was the most sensitive to acrolein, presented a significant increase in mortality after the first exposure-recovery cycle, and the effect was observed after acrolein removal, during the recovery period. According to this observation, a lower survival rate constant was significantly determined for acrolein treatments (Table 5). Latency effectively occurred because the onset of mortality was delayed, becoming evident during the recovery time. No cumulative effects for the repeated exposures were observed, and the final number of surviving larvae was statistically the same. Selection of tolerant individuals takes place during the first exposure, then the remaining sub-population is similar in number for both treated and control groups. Moreover, *R. arenarum* tadpoles surviving the three pulses of acrolein arrived to metamorphosis at the same time and proportion as controls. The intermediate-sensitive crustacean, *H. curvispina*, was also subjected to acrolein pulses of 1day-exposure followed by recovery. No effects were observed on this species, concluding that the short exposures did not cause cumulative damage or that the recovery time between repeated exposures (6 days) was enough to overcome the deleterious effects. So, the acute effects of acrolein on *H. curvispina* are related to peak duration since the LC_{50} determined at 96 hours of exposure (0.24 mg/L) does not cause mortality at the repeated short term exposure of 24 hours.

Treatment group	Control	LOLC	LC_{50}
		(0.010 mg/L)	(0.023 mg/L)
% Survival: *1st Exposure-Recovery*	64.2 ± 4.4	59.2 ± 4.4	53.3 ± 4.2
2nd Exposure-Recovery	58.3 ± 4.4	54.2 ± 3.0	46.7 ± 5.1
3rd Exposure-Recovery	49.2 ± 3.0	50.0 ± 5.2	41.7 ± 6.0
Latency effects		YES	YES
Mortality rate constant (d)	13.9 ± 0.1	6.8 ± 1.3*	6.9 ± 2.7*
% Metamorphosis	19.0 ± 0.2	20.1 ± 0.2	23.9 ± 5.4

Table 5. Repeated exposure effects of acrolein in *R. arenarum*. Three cycles of 1 day-exposure followed by 13 days-recovery in acrolein-free media were evaluated. LOLC: Lowest-Observed-Lethal Concentration; LC_{50}: Lethal Concentration-fifty. * denotes significant differences vs. control group, p= 0.018. Data obtained from Venturino et al. (2007), with permission of SETAC press.

The acute toxicity tests and risk assessment on native species lead to a concern about acrolein effects in the irrigation canals. Repeated exposure tests showed no cumulative effects, as population survival remained unchanged with respect to controls after the third exposure. Other studies at this stage of the evaluation are not recommended for the use of acrolein as an herbicide. The physical and chemical properties of the compound such as its high reactivity, its low tendency to partitionate to organic matter, and its low persistence in the environment do not require sediment toxicity studies as a priority. At this step, the risk evaluation needs to include field studies to broaden the analysis towards community and population levels. This category of approach in tier 4 lets the determination of effects on a variety of organisms in the ecosystem, including the interaction among species and indirect effects. In the case of the use of acrolein as an aquatic herbicide, a field study on benthic invertebrates has been designed to establish the safety of the exposure regime of MAGNACIDE® H at the CORFO-Río Colorado canals (Albariño et al., 2007). These organisms are prone to human perturbation of the ecosystem, and relatively sedentary if compared to other organisms such as fish or amphibian. A total of 34 benthic macroinvertebrates were identified in CORFO-Río Colorado canals. From the study, spanning two years, it was determined that acrolein was able to reduce community diversity and abundance during the application seasons. However, the benthic community was able to recover its ecological attributes two months after ceasing canal treatments with acrolein. Thus, the use of acrolein as an herbicide would be ecologically acceptable, taking into account that its toxic effects are reverted in a reasonable time (Campbell et al., 1999). The directional flux in the lotic systems under study probably allow the recolonization of the areas where a local perturbation has been introduced, such as the application of the herbicide, by flowing organisms from upstream sites (Winterbourn & Townsed, 1998).

6. Conclusions

There are no generally accepted quantitative criteria for evaluating ecological significance and expert judgement is always required. We have shown here evidences from literature data and from risk assessment with native species that acrolein used for weed control in irrigation canals is extremely toxic for most of the living organisms at the recommended treatment concentrations and conditions. Nevertheless, its presence in the canals is transient and it has been observed a natural recovery process, mainly operating through the introduction of species from outside the treated area that minimizes the ecological risk. Populations of species with a high intrinsic growth rate, such as zooplankton, may rapidly recover after an acute toxicity event. Species with lower intrinsic growth rate, such as amphibians and fish, will require longer periods for population recovery. One advantage in the protection of higher organisms such as birds and mammals is the irritating odour of the herbicide. The odour prevents them from getting close to the treated area, so these species are not endangered by the compound.

Taking into account the fact that most population effects derived from the use of acrolein as an aquatic herbicide are temporary, we conclude that its use is ecologically acceptable because recovery occurs within a reasonable period of time. In order to minimize the risk on the ecological receptors, a strict control on the treatment regime, concentration applied, timing and frequency of application must be ensured. Treated canals must be controlled during the applications, water release must be prevented until the product has dissipated, and it must be ensured that water is used only for irrigation purposes.

7. References

Adams, J. & Klaidman, L.K. (1993). Acrolein-induced oxygen radical formation. *Free Radical Biology and Medicine*, 15, 187-193, ISSN 0891-5849

Albariño, R.; Venturino, A.; Montagna, C. & Pechen de D'Angelo A.M. (2007). Environmental effect assessment of Magnacide H® Herbicide at Río Colorado irrigation channels (Argentina). Tier 4: In situ survey on benthic invertebrates. *Environmental Toxicology and Chemistry*, 26, 183-189, ISSN 0730-7268

ATSDR (2007). *Toxicological profile for acrolein*, Agency for Toxic Substances and Disease Registry, U.S. Department of Health and Human Services, Atlanta

Auerbach, S.S.; Mahler, J.; Travlos, G.S. & Irwin, R.D. (2008). A comparative 90-day toxicity study of allyl acetate, allyl alcohol and acrolein. *Toxicology*, 253, 79–88, ISSN: 0300-483X

Baker Petrolite Corporation (2001). *MAGNACIDE® H. Application and Safety Manual.* Baker Petrolite Corporation, Bakersfield, CA

Bowmer, K.H. & Sainty, G.R. (1977). Management of aquatic plants with acrolein. *Journal of Aquatic plants Management*, 15, (June 1977) 40-46, ISSN 0146-6623

Caldironi, H.A.; Bentivegna, D.J. & Rhea, M.R. (2004). Irrigation of pepper plant (*Capsicum* sp.) with water containing acrolein. *Agricultural Water Management* 67, 235-240, ISSN 0378-3774

Campbell, P.J.; Arnold, D.J.S.; Brock, T.C.M.; Grandy, N.J.; Heger, W.; Heinback, F.; Maund, S.J. & Streloke, M. (1999). *Guidance Document on Higher-Tier Aquatic Risk Assessment for Pesticides* (HARAP). SETAC Europe, Brussels, Belgium

ECOFRAM (1999). *Risk Assessment Methods Report of the Aquatic Workgroup.* U.S. Environmental Protection Agency, Office of Pesticide Programs, Ecological Committee on Federal Insecticide, Fungicide, and Rodenticide Act (FIFRA). Washington, DC

Eisler, R. (1994). *Acrolein hazard to fish, wildlife and invertebtrates: A synoptic review. National Biological Survey. Technical Report Series. Biological Report 23,* U.S. Department of Interior Washington, DC

EU (2002). *Guidance document on aquatic toxicology in the context of the Directive 91/414/EEC.* European Union, Working Document of the European Commission Health and Consumer Protection Directorate-General, Brussels, Belgium

Faroon, O.; Roney, N.; Taylor, J.; Ashizawa, A.; Lumpkin, M.H. & Plewak, D.J. (2008). Acrolein environmental levels and potential for human exposure *Toxicology and Industrial Health,* 24: 543-564, ISSN 1477-0393

Holcombe, G.W.; Phipps, G.L.; Sulaiman, A.H. & Hoffman, A.D. (1987). Simultaneous multiple species testing: acute toxicity of 13 chemicals to 12 diverse freshwater amphibian, fish and invertebrate families. *Archives of Environmental Contamination and Toxicology,* 16, 6, 697-710, ISSN 0090-4341

HSDB (Hazardous Substances Data Bank) (2010). National Library of Medicine's TOXNET system Available online at: (http://toxnet.nlm.nih.gov)

IARC (1995). *Acrolein, IARC Summaries and Evaluations, Vol. 63.* International Agency for Research on Cancer

Kim, S.; Pfeifer, GP. & Besaratinia, A. (2007). Lack of mutagenicity of acrolein-induced DNA adducts in mouse and human cells. *Cancer Research,* 67, 24 (December, 2008), 1164-11647, ISSN: 0008-5472

Kissel, C.L.; Brady, J.L.; Guerra, A.M.; Pau, J.K.; Rockie, B.A. & Caserio, F.F. (1978). Analysis of acrolein in aged aqueous media. Comparison of various analytical methods with bioassays. *Journal of Agriculture and Food Chemistry,* 26, 6, 1338-1343, ISSN 1520-5118

Linhart, I.; Frantík, E.; Vodičková L.; Vosmanská, M., Šmejkal, J. & Mitera, J. (1996). Biotransformation of acrolein in rat: excretion of mercapturic acids after inhalation and intraperitoneal injection. *Toxicology and Applied Pharmacology,* 136, 155-160, ISSN 0041-008X

Luo, J.; Robinson, J.P.; Shi, R. (2005). Acrolein-induced cell death in PC12 cells: role of mitochondria-mediated oxidative stress. *Neurochemistry International,* 47, 449–457, ISSN 0197-0186

Mano, J.;Miyatake, F.; Hiraoka E. & Tamoi M. (2009). Evaluation of the toxicity of stress-related aldehydes to photosynthesis in chloroplasts. *Planta,* 230, 639–648, ISSN 1432-2048

MOE (2005). *Ontario Air Standards for Acrolein,* Standards Development Branch, Ministry of Environment, Ontario

Nardini, M.; Finkelstein, E.I.; Reddy, S., Valacchi. G.; Traber, M. Cross, C.E. & van der Vliet A. (2002). Acrolein-induced cytotoxicity in cultured human bronchial epithelial cells. Modulation by alpha-tocopherol and ascorbic acid. *Toxicology*, 170, 173–185, ISSN 0300-483X

Nordone, A. J.; Kovacs, M.F. & Doane, R. (1997). [14]C Acrolein accumulation and metabolism in leaf lettuce. *Bulletin of Environmental Contamination and Toxicology*, 58, 787-792, ISSN 0007-4861

Nordone, A.; Dotson, T.A.; Kovacs, M.F.; Doane, R. & Biever, R.C. (1998). Metabolism of [14C] acrolein (MAGNACIDE® H Herbicide): nature and magnitude of residues in freshwater fish and shellfish. *Environmental Toxicology and Chemistry*, 17, 2, 276-281, ISSN 0730-7268

Nordone, A.J.; Matherly, R.; Bonnivier, B.; Doane, R.; Caravello, H.; Paakkonen, S. & Parent, R.A. (1996). The mobility and degradation of acrolein in agricultural canals treated with MAGNACIDE® H Herbicide. *Chemosphere*, 32, 5, 807-814, ISSN 0045-6535

Peterson, H.G.; Boutin, C.; Martin, P.A.; Freemark, K.E.; Ruecker, N.J. & Moody, M.J. (1994). Aquatic phyto-toxicity of 23 pesticides applied at expected environmental concentrations. *Aquatic Toxicology*, 28, 275-292, 0166-445X

Schmid, B.P.; Goulding, G.; Kitching, K. & Sanyal, M.K. (1981). Assessment of the teratogenic potential of acrolein and cyclophosphamide in rat embryo culture system. *Toxicology*, 22, 235-243, ISSN: 0300-483X

Smith, A.M.; Mao, J;. Doane, R.A. & Kovacs M.F. (1995). Metabolic fate of (14C) acrolein under aerobic and anaerobic aquatic conditions. *Journal of Agriculture and Food Chemistry*, 43, 2497–2503, ISSN 1520-5118

Tanel, A. & Averill-Bates, D.A. (2007). Activation of the death receptor pathway of apoptosis by the aldehyde acrolein. *Free Radical Biology and Medicine*, 42, 798–810, ISSN 0891-5849

U.S. EPA (1992). *Framework for ecological risk assessment. EPA/630/R-92-001*. U.S. Environmental Protection Agency, Risk Assessment Forum, Washington, DC

U.S.EPA (2003). *Toxicological review of acrolein* US Environmental Protection Agency Washington, DC

U.S.EPA (2007). *Environmental fate and ecological risk assessment for the registration of acrolein*, U.S. Environmental Protection Agency, Office of Prevention, Pesticides and Toxic substances, Washington, DC

U.S.EPA (2009). *Ambient aquatic life water quality criteria for acrolein* U.S. Environmental Protection Agency, Office of Water Office of Science and Technology, Health and Ecological Ccriteria Division, Washington, DC

Uchida, K.; Kanematsu, M.; Sakai, K.; Matsuda, T.; Hattori, N. Mizuno Y.; Suzuki, D.; Miyata, T.; Noguchii, N.ikii, E. & Osawa, T. (1998). Protein-bound acrolein: Potential markers for oxidative stress. *Proceeding of the National Academy of Science* 95, (April 1998) 4882–4887, ISSN 1091-6490

Urban, D.J. & Cook, N.J. (1986). *Standard evaluation procedure for ecological risk assessment*. U.S. Environmental Protection Agency, Hazard Evaluation Division, Office of Pesticide Programs, Washington, DC

Venturino A.; Montagna, C. & Pechen de D'Angelo, A.M. (2007). Risk assessment of MAGNACIDE H® Herbicide at Río Colorado irrigation channels (Argentina). Tier

3: Studies on native species. *Environmental Toxicology and Chemistry*, 26: 177-182, ISSN 0730-7268

WHO (1991). *Acrolein, Health and Safety Guide No 67*, World Health Organization, ISSN 0512-3038, Geneva

WHO (1992). Acrolein. *Environmental Health Criteria 127*, Health Organization, ISSN 0512-3038, Geneva

WHO (2002). *Acrolein, Document 43*, World Health Organization, ISSN 0512-3038, Geneva

Winterbourn, M.J. & Townsend, C.R. (1998). Streams and rivers: one-way flow systems, In: *Fundamentals of Aquatic Ecology*, Barnes, R.S.K. & Mann, K.H., (Ed.), 230–244, Blackwell Science, 0-632-02983-8, Cambridge, UK

20

Genetic Adaptation of Phytoplankters to Herbicides

Victoria López-Rodas[1], Eduardo Costas[1] and Antonio Flores-Moya[2]
[1]*Complutense University*
[2]*University of Málaga*
Spain

1. Introduction

One of the most important unwanted effects of herbicides is the arising of resistant non-target organisms. In fact, weeds which are resistant to almost all kinds of herbicides have been documented and the number of resistant variants is continuously increasing (http://www.weedscience.org/In.asp). Since phytoplankters (including cyanobacteria and eukaryotic microalgae) are responsible of the highest fraction of primary production in aquatic ecosystems (Falkowski & Raven, 1997), they must be highlighted among non-target herbicide organisms (Koenig, 2001). Unrelenting application of herbicides during recent decades has resulted in water pollution, with serious environmental implications and evolutionary consequences due to strong selection pressure on numerous species (Belfiore & Anderson, 2001). The problem is especially relevant in those freshwater habitats which are close to agricultural areas; these habitats are usually sinks for a large array of herbicides, so that phytoplankters are exposed to a multitude of these toxic compounds (Junghans et al., 2006). In fact, it is considered that herbicides are among the most significant human-synthesized pollutants in aquatic ecosystems (Koenig, 2001). Moreover, has been proposed that the emergence of unpredictable novelties could be a distinctive feature of the future biosphere (Tilman, 1999; Myers & Knoll 2001; Palumbi, 2001) and, consequently, the arising of resistant-herbicide phytoplankters could be considered as one of the relevant examples of human-driven selection.

The majority of studies on the effects of herbicides on phytoplankters have been focused on the degree of tolerance to the herbicides (Shehata et al., 1997; Berard et al., 1999; Kasai, 1999; Nelson et al., 1999; revised by Koenig 2001; Pinckney et al., 2002). However, adaptation includes different processes which are not usually discriminated; in particular, adaptation conferring resistance to herbicides can be achieved by three processes differing in some particular aspects (Fig. 1).

Under toxic but sub-lethal doses of herbicides, adaptation could be supported by modification of gene expression occurring in a short time (days to weeks) and within one organism's lifetime (i.e. physiological adaptation, also called acclimatization; Bradshaw & Hardwick, 1989); however, some evolutionary studies in bacteria (Cairns et al., 1988; Foster, 2000; Roth et al., 2006) and yeasts (Heidenreich, 2007) have suggested that adaptive mutations could be a process resembling Lamarckism which, in the absence of lethal selection, produces mutations that relieve selective pressure. Finally, under lethal doses of

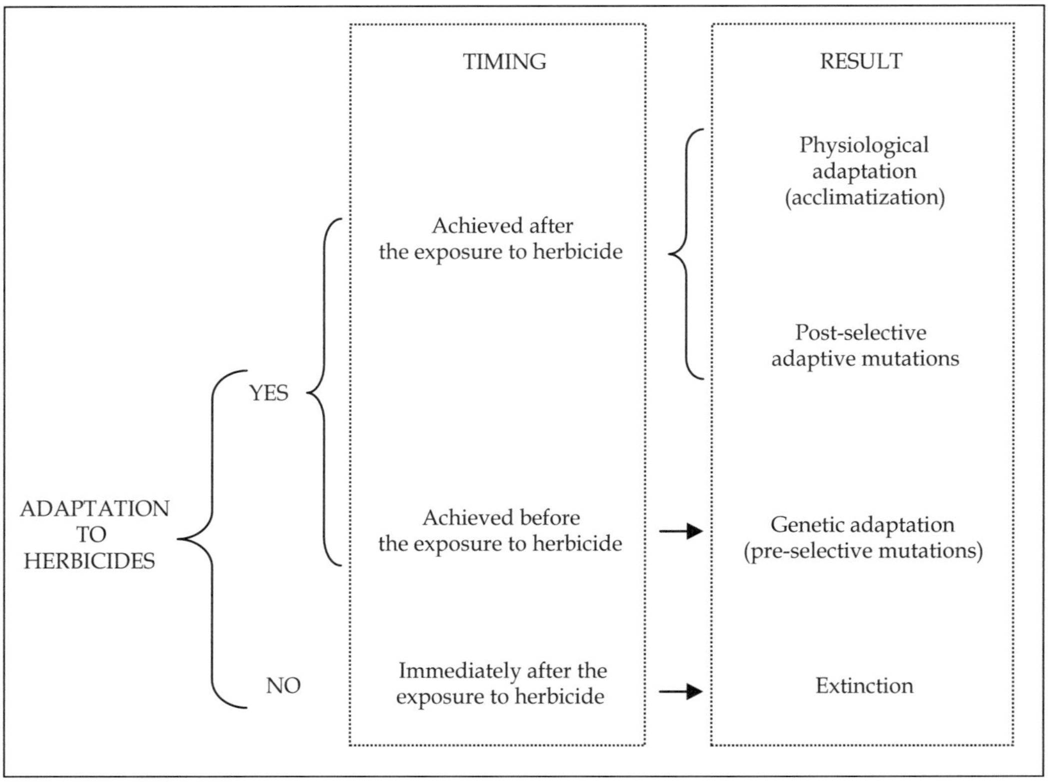

Fig. 1. Adaptation strategies of phytoplankton in herbicide-polluted waters (adapted from Marvá et al., 2010)

herbicides, genetic adaptation supported by selection of new genetic variants originated by spontaneous mutation (Sniegowski & Lenski, 1995; Belfiore & Anderson 2001; Orr, 2005) is the only possibility. Although it is assumed that this last mechanism is responsible for the arising of herbicide-resistant organisms, the empirical evidences are scarce. Thus, the discrimination of mechanisms involved in adaptation of phytoplankters to herbicides was addressed in our framework which is focused on evolutionary genetic of adaptation to anthropogenic pollution. For this purpose, we applied rigorous experimental techniques as an approach to evolutionary ecotoxicology of phytoplankters.

Here we review the studies which demonstrate the arising of new genetic-variants in phytoplankters under lethal doses of herbicides, as well as the highest concentrations of herbicides that allow genetic adaptation (an estimator of the limits of adaptation). It must be taken into account that the arising of herbicide-resistant phytoplankters is an example of adaptive evolution (Sniegowski & Lenski, 1995; Palumbi, 2001) and this allows us to hypothesize about future performance of phytoplankton communities in aquatic ecosystems.

2. Demonstrating genetic adaptation to herbicides: fluctuation analysis

2.1 Fluctuation analysis of the transformation herbicide-sensitive → herbicide-resistance

The experimental procedure called fluctuation analysis (Luria & Delbrück, 1943) is the best way to demonstrate if the adaptation to lethal doses of herbicides could take place in wild-strains of phytoplankters and, secondly, to discriminate between acquired adaptations in response to the herbicides (by acclimatization or putative adaptive mutations) and resistant cells arising from rare spontaneous mutations that appear prior to the herbicide exposure (Fig. 1). A modified fluctuation analysis for application to liquid cultures with phytoplankters (Costas et al., 2001; López-Rodas et al., 2001) has been used to investigate the origin of herbicide-resistant cells. The modification of the analysis involves the use of liquid medium containing the selective agent (different kind of herbicides) rather than plating on a solid medium, as was done by Luria & Delbrück (1943) with bacterial cultures.

Two different sets of experimental cultures are prepared. In the first set (set 1), ca. 100 culture flasks, containing non-selective culture medium, are inoculated with N_0 = 10^1-10^2 cells (a number small enough to reasonably ensure the absence of pre-existing mutants in the strain). When each culture reaches N_t = 10^5-10^6 cells, it is supplemented with a lethal dose of herbicide. The dose is previously calculated from a dose-growth rate relationship, and it is selected a dose 2-4 times higher than that originating the 100% inhibition on growth rate. For set 2 (set control), 25-50 aliquots of 10^6-10^8 cells from the same parental population are separately transferred to culture flasks containing fresh liquid medium with the herbicide at the same concentration as set 1 cultures.

In order to maintain lethal doses of herbicides in the cultures, they are centrifuged to form a pellet of cells in the tube, the medium is decanted and fresh liquid medium with the herbicide is added each 5 d. All cultures are kept under selective conditions and observed after 90 d, a period of time long enough to allow resistant cells to grow. At the end of the experiments, the number of resistant cells in both sets is counted.

Two different results can be found in the set 1 experiment when conducting a fluctuation analysis, each result being interpreted as the independent consequence of different phenomena of adaptation (Luria & Delbrück, 1943; Jones et al., 1994). In the first case, if resistant cells arose during the exposure to the herbicides (i.e. by acclimatization or putative adaptive mutations), the variance in the number of cells per culture would be low because every cell is likely to have the same chance of developing resistance (Fig. 2, set 1A). Consequently, inter-culture (flask-to-flask) variation would be consistent with the Poisson model (i.e. variance/mean ≈1). By contrast, if resistant cells arose before the exposure to the herbicides (i.e. genetic adaptation by rare spontaneous mutation occurring during the time in which the cultures grew to N_t from N_0 cells before the exposure to herbicides), a high variation in the inter-culture number of resistant cells per culture would be found (Fig. 2, set 1B). Consequently, the flask-to-flask variation would not be consistent with the Poisson model (i.e. variance/mean >1). Obviously, another result (0 resistant cells in each culture) could also be found, indicating that neither selection on spontaneous mutations that occur prior to herbicides exposure, nor specific adaptation during the exposure to the herbicides, took place.

The set 2 cultures are the experimental controls of the fluctuation analysis (Fig. 2). Variance is expected to be low, because set 2 samples the variance of the parental population. If the variance/mean ratio of set 1 is significantly greater than the variance/mean ratio of set 2 (fluctuation), this confirms that resistant cells arose by rare mutations that occurred before exposure to the herbicide. If a similar variance/mean ratio between set 1 and set 2 is found, it confirms that resistant cells arose during the exposure to the herbicide.

In addition, the fluctuation analysis allows estimation of the rate of appearance of resistant cells. There are different approaches for accomplishing this estimation (Rosche & Foster,

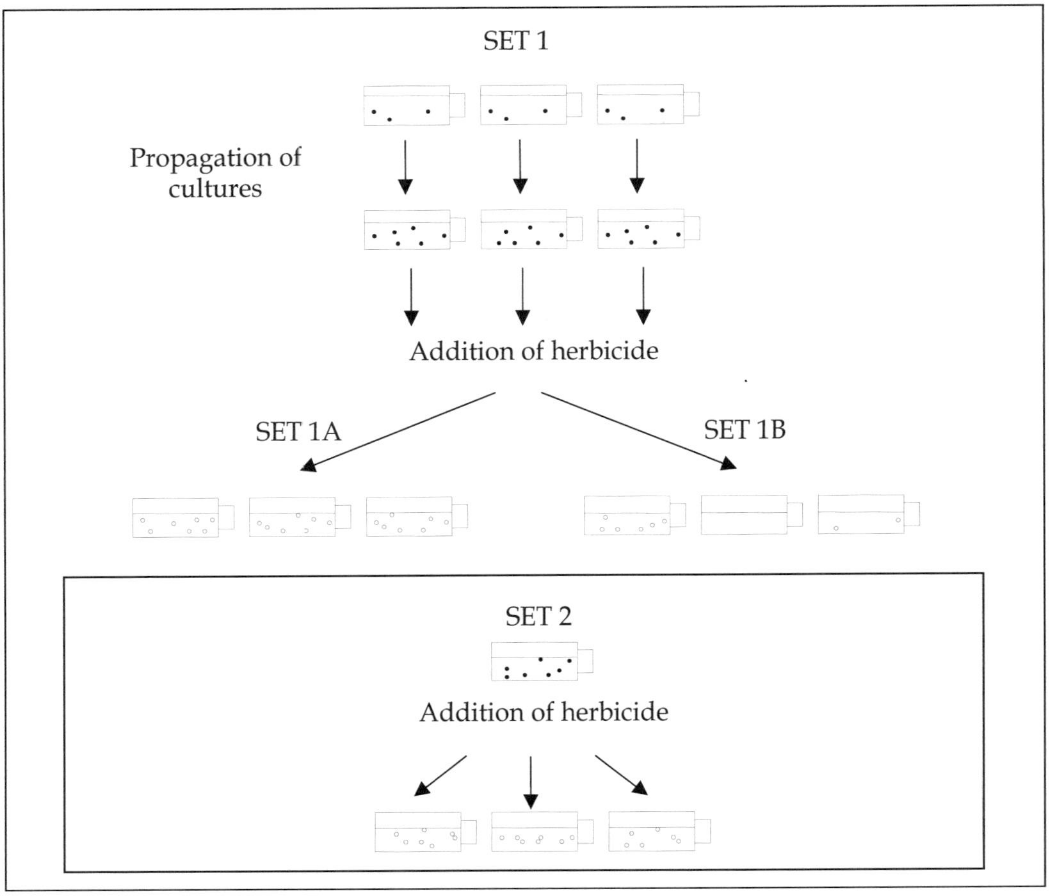

Fig. 2. Schematic diagram of the modified Luria & Delbrück (1943) fluctuation analysis. In the set 1, several cultures each inoculated with small inoculums were propagated until a high cell density was reached, and then a lethal dose of herbicide was added. If resistant cells arose by acclimatization or post-adaptive mutations, the number of resistant cells in all the cultures must be similar (set 1A). If adaptation is achieved by rare mutations (see Fig. 1) occurring in the period of the propagation of cultures the difference of the number of resistant cells in each culture must be huge (set 1B). Set 2 samples the variance of parental populations as an experimental control (adapted from Marvá et al. 2010)

2000). Due to methodological limitations imposed by a fluctuation analysis using liquid cultures, the proportion of cultures from set 1 showing no resistant cells (P_0 estimator; Luria & Delbrück, 1943) can be used to calculate the mutation rate (μ) by using the equation:

$$\mu = -\mathrm{Log}_e P_0 / \ (N_t - N_0) \quad (1)$$

2.2 Mutation-selection equilibrium

If the mutation from wild-type, herbicide-sensitive allele to herbicide-resistant allele is recurrent and, in addition, the herbicide-resistant allele is detrimental in growth rate in the absence of herbicides, most of these mutants are eliminated sooner or later by natural selection, if not by chance. At any one time, there will be a certain number of cells that are not yet eliminated. The average number of such mutants (q, frequency of the herbicide resistant allele) will be determined by the balance between μ and the rate of selective elimination, in accordance with the equation from Kimura & Maruyama (1966):

$$q = \mu / (\mu + s) \tag{2}$$

where s is the coefficient of selection, calculated as:

$$s = 1 - (m_S{}^r / m_S{}^s) \tag{3}$$

where m^r and m^s are the acclimated maximal growth rates of herbicide-resistance and herbicide-sensitive cells measured in non-selective conditions, respectively.

3. Example cases of genetic adaptation of phytoplankters to herbicides

Numerous resistant mutants to herbicides have been generated or selected in cyanobacteria and microalgae (Astier et al., 1979; Erikson et al., 1984; Golden & Haselkorn, 1985; Johanningmeier & Hallick, 1987; Chamovitz et al., 1991; Trebst et al., 1993; Singh & Singh, 1997; revised by Koenig, 2001). However, the empirical evidence of the arising of herbicide resistant-mutants in phytoplankters from wild, sensitive-cells by spontaneous mutations has only been addressed recently by using fluctuation analysis (Table 1).
In particular, when phytoplankton cultures were exposed to herbicides, they became clear after some days due to total growth inhibition and subsequent massive destruction of the cells by the lethal effect of herbicides. But after being further incubated a few cultures became green again, due to the growth of variants that were resistant to the herbicides. By using the fluctuation analysis, we demonstrated that resistance in different species of phytoplankters was due to the arising of new genetic variants caused by rare spontaneous mutation occurring randomly during propagation under non-selective conditions for the herbicides DCMU, glyphosate, simazine and diquat (Costas et al., 2001; López-Rodas et al., 2001, 2007; Marvá et al., 2010; see Table 1). Studies on adaptation to other types of herbicides are being carried out at the present.

Species	Herbicide	μ (mutants per cell per division)	q (no. of resistant cells per wild-type cell)	Ref.
Cyanobacteria				
Pseudanabaena sp.	DCMU	2.4×10^{-6}	6×10^{-4}	1
Microcystis aeruginosa	Glyphostate	3.6×10^{-7}	65×10^{-4}	2
Chlorophyta				
Dictyosphaerium chlorelloides	DCMU	2.1×10^{-6}	21×10^{-4}	3
Dunaliella tertiolecta	DCMU	3.6×10^{-6}	21×10^{-4}	1
Scenedesmus intermedius	Simazine	3.0 to 9.2×10^{-6}	11 to 30×10^{-6}	4
S. intermedius	Diquat	1.8×10^{-5}	83×10^{-6}	4
Ref., reference : 1, López-Rodas et al. (2001); 2, López-Rodas et al. (2007); 3, Costas et al. (2001); 4, Marvá et al. (2010)				

Table 1. Mutation rate (μ) and mutation-selection balance (q) of phytoplankters in the genetic adaptation to different herbicides

Table 1 shows the figures of mutation rate from herbicide-sensitivity to herbicide-resistance in different species of phytoplankters (ranging from 0.4 to 17.9 mutants per 10^6 cells per generation, depending of the species and the herbicide). Since mutation is recurrent in each generation, new mutant cells are arising continuously. One of the main characteristics of the herbicide-resistant mutants, in comparison to the wild, herbicide-sensitive organisms is that the former have significantly lower growth rate than the latter, as reflected by the q figures in Table 1. It must be taking into account that mutations usually imply an energetic cost that may affect the survival of adapting populations (Coustau et al., 2000; Vila-Auib et al., 2009). Consequently, most of resistant-mutants are eventually eliminated by natural selection (Crow & Kimura, 1970). At any given time, the balance between the continuous appearance of mutants and their selective elimination determines the number of remaining herbicide-resistant mutants in algal populations growing in the absence of herbicides. This may be the case in wild-type populations developing in non-polluted waters. Consequently, the population would be predominantly a clone line of herbicide-sensitive genotypes, accompanied by, as a very small fraction, clone lines of herbicide-resistant mutants.

Summarizing, rare spontaneous mutations conferring resistance against herbicide seems to be enough to assure survival of phytoplankters in herbicide-polluted waters. Moreover, in a hypothetical future scenario with herbicide-polluted waters, the primary production supported by phytoplankters could be significantly lower that in the present, as a consequence of the diminished growth rate of resistant mutants in comparison to wild-type cells.

4. Testing the limits of genetic adaptation: ratchet protocol

4.1 Theoretical and experimental setup

When genetic adaptation is found via fluctuation analysis, the adaptation is referred to a given lethal dose of herbicide (usually, 2-4 times higher than that originating the 100% inhibition on growth rate). However, the potential limit of adaptation to the highest concentration of the toxic is difficult to estimate via fluctuation analysis, since stronger selection pressures drastically reduce population size. This constraint can be overcome by performing experiments that include several values of selection pressure. To this end, Reboud et al. (2007) developed an experimental model aimed at evaluating the maximal potential for herbicide resistance evolution in the green microalga *Chlamydomonas*. This experiment was based on the use of different herbicide concentrations, which thereby constituted selection pressures. Furthermore, Orellana et al. (2008) provided a modified procedure that allowed for maximizing the occurrence of mutants in microalgae and their selection by applying variable selection pressures. A significant enhancement of this experimental procedure was achieved by using different replicates of each strain under each selection condition. This assures repeatability and it is referred as ratchet assays (Huertas et al. 2010). The protocol aims at reaching equilibrium between strong selection intensity, by means of ratcheting to increase herbicide dose, and at the maintenance of a population size large enough to increase probability of rare spontaneous mutations that confer adaptation. Cultures must be ratcheted only up to a dose that supports population growth. The experimental procedure is then applied in several independent replicates (Fig. 3).

During the initial phase, replicates of the control cultures containing growth medium and replicates of cultures for each of the initial doses of herbicide treatment are prepared. Each culture is transferred to the next concentration when the same net growth of the control cultures is reached; cultures that do not present net growth are maintained at the same concentration. A new ratchet cycle is concluded each time that the control cultures are transferred. The experiment ends after several cycles with net growth occurring only in the control cultures. At this point the maximal capability of adaptation corresponds to the maximal concentration of the selective agent that presents net growth (Fig. 3).

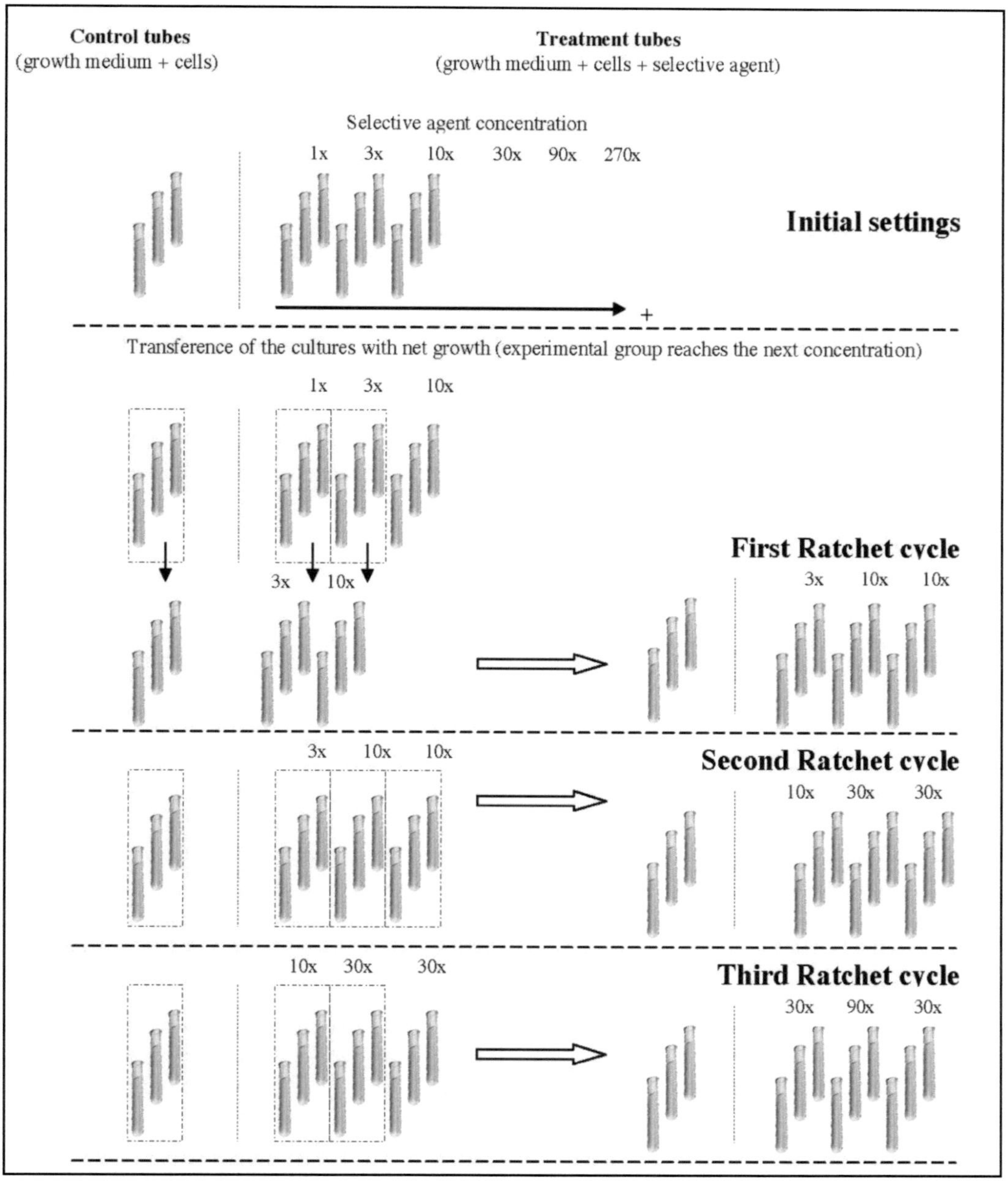

Fig. 3. Schematic representation of the ratchet experimental design (Huertas et al. 2010). Three ratchet cycles are represented but the experiment ends when net growth only occurs in controls after several ratchet cycles

4.2 Increase in adaptation to simazine in relation to concentration causing 100% growth inhibition

Huertas et al. (2010) showed differential maximal adaptation to simazine depending on taxonomic group, ploidy level (haploid and diploid vegetative cells occur in different taxonomical groups of microalgae), growth rate and habitat (Table 2).
In relation to the taxonomic group, Chlorophyta showed the greatest capacity to adapt to simazine, while Cyanobacteria did not adapt so well. It was suggested that the difference could result from the fact that prokaryotic organisms are more adversely affected by simazine than eukaryotic species (Fournadzhieva et al., 1995). Bacillariophyta and Haptophyta showed moderate and scarce ability to adapt, respectively.

Species	Adaptation increase (times)	Ploidy level	Cell division[1]	Habitat
Chlorophyta				
Scenedesmus intermedius	270	Haploid	Rapid	Continental
Dictyosphaerium chlorelloides	90	Haploid	Rapid	Continental
Tetraselmis suecica	10	Haploid	Rapid	Coastal
Cyanobacteria				
Microcystis aeruginosa (3D)	9	Haploid	Moderate	Continental
M. aeruginosa (6D)	9	Haploid	Moderate	Continental
M. aeruginosa (7D)	9	Haploid	Moderate	Continental
Bacillariophyta				
Phaeodactylum tricornutum	4.5	Diploid	Rapid	Coastal
Haptophyta				
Emiliania huxleyi (CCMP373)	3	Haploid	Slow	Oceanic
E. huxleyi (CCMP371)	1.5	Haploid	Slow	Oceanic
E. huxleyi (CCMP372)	1.5	Haploid	Slow	Oceanic
Isochrysis galbana	1.5	Haploid	Moderate	Oceanic
Monochrysis lutheri	1.5	Haploid	Moderate	Oceanic

[1]Cell division: rapid, 1 doubling every 3-4 d; moderate, 1 doubling every 4-5 d; slow, 1 doubling every 5-7 d

Table 2. Characteristics implicated in the capability of different phytoplankters to adapt to simazine. Adaptation increase column shows how much times simazine concentration causing 100% growth inhibition before the ratchet experiments is higher than that measured at the end of the experiment (data from Huertas et al., 2010)

The species with the greatest capacity for adaptation to simazine are haploid populations growing rapidly (Table 2). It is supposed that haploids will respond to selection more quickly than diploids because non-neutral mutations are expressed immediately. Moreover, growth rate is also involved in adaptation since a greater number of generations during a given time allows for more speedy adaptive evolution.
From an ecological point of view, it is very interesting to highlight that clear differences in adaptation to simazine were found depending on the habitat (Table 2). Thus, the greatest ability was found in phytoplankters from epicontinental freshwaters (usually, the sink of an array of herbicides), followed by coastal marine microalgae and, finally, the most sensitive

group was formed by oceanic microalgae (without prior exposure to simazine or other related compounds). Thus, it can be hypothesized that the capability of phytoplankters to adapt to simazine depends on a previous evolutionary history. In this way, a sudden contamination episode could be relieved by freshwater phytoplankters but not by oceanic phytoplankters.

5. Prospective

Genetic adaptation of phytoplankters to herbicides, as well as the limits of genetic adaptation, has been addressed in several species of phytoplankters representing some taxonomical and ecological groups. However, other species must be tested in order to have an extended view on the occurrence of genetic adaptation. Moreover, different herbicides are distinguished on the basis of their site and mode of action; in our framework, we tested the genetic adaptation of phytoplankters to a few of the different type of herbicides but other types will be addressed in the future.

6. Acknowledgments

The results compiled in this review were obtained in research works financially supported by the Spanish Ministry of Sciences and Innovation through grants CTM2008-05680 C02-01/MAR and CGL2008-00652/BOS, and Junta de Andalucía Research Group RNM-115.

7. References

Astier, C.; Vernotte, C.; Der-Vartanian, M. & Joset-Espardellier, F. (1979). Isolation and characterization of two DCMU-resistant mutants of the blue-green alga *Aphanocapsa* 6714. *Plant and Cell Physiology*, 20, 1501-1510, ISSN 0032-0781

Belfiore, N. M. & Anderson, S. L. (2001). Effects of contaminants on genetic patterns in aquatic organisms: a review. *Mutation Research/ Reviews in Mutation Research*, 489, 97-122, ISSN 1383-5742

Berard, A.; Leboulanger, C. & Pelte, T. (1999). Tolerante of *Oscillatoria limnetica* Lemmermann to atrazine in natural phytoplankton populations and in pure culture: influence of season and temperature. *Archives of Environmental Contamination and Toxicology*, 37, 472-479, ISSN 0090-4341

Bradshaw, A. D. & Hardwick, K. (1989). Evolution and stress –genotype and phenotype components. *Biological Journal of the Linnean Society*, 37, 137-155, ISSN 0024-4066

Cairns, J.; Overbaugh, J. & Miller, S. (1988). The origin of mutants. *Nature*, 335, 142-145, ISSN 0028-0836

Chamovitz, D.; Pecker, I. & Hirschberg, J. (1991). The molecular basis of resistance to the herbicide norflurazon. *Plant Molecular Biology*, 16, 967-974, ISSN 0167-4412

Costas, E.; Carrillo, E.; Ferrero, L. M.; Agrelo, M.; García-Villada, L.; Juste, J. & López-Rodas, V. (2001). Mutation of algae from sensitivity to resistance against environmental selective agents: the ecological genetics of *Dictyosphaerium chlorelloides* (Chlorophyceae) under lethal doses of 3-(3,4-dichlorophenyl)-1,1-dimethylurea herbicide. *Phycologia*, 40, 391-398, ISSN: 0031-8884

Coustau, C.; Chevillon, C. & Ffrench-Constant, R. (2000). Resistance to xenobiotics and parasites: can we count the cost? *Trends in Ecology & Evolution*, 15, 378-383, ISSN 0169-5347

Crow, J. F. & Kimura, M. (1970). *An Introduction to Population Genetics Theory*, Harper and Row, ISBN-10 0808729012, New York, NY, USA

Erikson, J. M.; Rahire, M.; Bennoun, P.; Delepelaire, P.; Diner, B. & Rochaix, J.-D. (1984). Herbicide resistance in *Chlamydomonas reinhardtii* results from a mutation in the chloroplast gene for the 32-kilodalton protein of photosystem II. *Proceedings of the National Academy of Sciences from the United States of America*, 81, 3617-3621, ISSN 0027-8424

Falkowski, P. G. & Raven, J. A. (1997). *Aquatic Photosyntesis*, Blackwell Science, ISBN-10 0865423873, Malden, MA, USA

Foster, P. L. (2000). Adaptive mutation: implications for evolution. *Bioessays*, 22, 1067-1074, ISSN 0265-9247

Fournadzhieva, S.; Kassabov, P.; Adnreeva, R.; Pelov, G. & Dittrt, F. (1995). Influence of the herbicide simazine on *Chlorella*, *Scenedesmus* and *Arthrospira*. *Archiv für Hydrobiologie - Supplementband*, 106, 97-109, ISSN 0342-1066

Golden, S. S. & Haselkorn, R. (1985). Mutation to herbicide resistance maps within the psbA gene of *Anacystis nidulans* R2. *Science*, 185, 1104-1107, ISSN 0036-8075

Heidenreich, E. (2007). Adaptive mutations in *Saccharomyces cerevisae*. *Critical Reviews in Biochemistry and Molecular Biology*, 42, 285-311, ISSN 1040-9238

Huertas, I. E.; Rouco, M.; López-Rodas, V. & Costas, E. (2010). Estimating the capability of different phytoplankton groups to adapt to contamination: herbicides will affect phytoplankton species differently. *New Phytologist*, ISSN 0028-646X (doi 10.1111/j.1469-8137.2010.03370.x)

Johanningmeier, U. & Hallick, C. B. (1987). The psbA gene of DCMU-resistant *Euglena gracilis* has an amino acid substitution at serine codon 265. *Current Genetics*, 12, 465-470, ISSN 0172-8083

Jones, M. E.; Thomas, S. M. & Rogers, A. (1994). Luria-Delbrück fluctuation experiments: design and analysis. *Genetics*, 136, 1209-1216, ISSN 0016-6731

Junghans, M.; Backhaus, T.; Faust, M.; Scholze, M. & Grimme, L. H. (2006). Application and validation of approaches for the predictive hazard assessment of realistic pesticide mixtures. *Aquatic Toxicology*, 76, 93-110, ISSN 0166-445X

Kasai, F. (1999). Shifts in herbicide tolerance in paddy field periphyton following herbicide application. *Chemosphere*, 38, 919-931, ISSN 0045-6535

Kimura, M. & Maruyama, T. (1966). The mutational load with epistatic gene interactions in fitness. *Genetics*, 54, 1337-1351, ISSN 0016-6731

Koenig, F. (2001). Eukaryotic algae, cyanobacteria and pesticides, In: *Algal Adaptation to Environmental Stresses. Physiological, Biochemical and Molecular Mechanisms*, Rai, L. C. & Gaur, J. P. (Eds), 389-406, Springer, ISBN 3-540-41938-1, Berlin, Germany

López-Rodas, V.; Agrelo, M.; Carrillo, E.; Ferrero, L. M.; Larrauri, A.; Martín-Otero, L. & Costas, E. (2001). Resistance of microalgae to modern water contaminants as the result of rare spontaneous mutations. *European Journal of Phycology*, 36, 179-190, ISSN 0967-0262

López-Rodas, V.; Flores-Moya, A.; Maneiro, E.; Perdigones, N.; Marva, F.; García, M. E. & Costas, E. (2007). Resistance to glyphosate in the cyanobacterium *Microcystis aeruginosa* as result of pre-selective mutations. *Evolutionary Ecology*, 21, 535-547, ISSN 0269-7653

Luria, S. & Delbrück, M. (1943). Mutations of bacteria from virus sensitivity to virus resistance. *Genetics*, 28, 491-511, ISSN 1943-2361

Marvá, F.; López-Rodas, V.; Rouco, M.; Navarro, M.; Toro, F. J.; Costas, E. & Flores-Moya, A. (2010). Adaptation of green microalgae to the herbicides simazine and diquat as result of pre-selective mutations. *Aquatic Toxicology*, 96, 130-134, ISSN 166-445X

Myers, N. & Knoll, A. H. (2001). The biotic crisis and the future of evolution. *Proceedings of the National Academy of Sciences from the United States of America*, 98, 5389-5392, ISSN 0027-8424

Nelson, K. J.; Hoagland, K. D. & Siegfried, B. D. (1999). Chronic effects of atrazine on tolerance of a benthic diatom. *Environmental Toxicology and Chemistry*, 18, 1038-1045, ISSN 0730-7268

Orellana, G.; López-Rodas, V.; Costas, E.; Maneiro, E. & Haigh, D. (2008). *Biosensores basados en microalgas para la detección de contaminantes ambientales*. CT/ES2008/000465, 2008, P200701905

Orr, H. (2005). The genetic theory of adaptation: a brief history. *Nature Reviews Genetics*, 6, 119–127, ISSN 1471-0056

Palumbi, S. R. (2001). Humans as world's greatest evolutionary force. *Science*, 293, 1786-1790, ISSN 0036-8075

Pinckney, J. L.; Ornólfsdóttir, E. B. & Lumsden, S. E. (2002). Estuarine phytoplankton group-specific responses to sublethal concentrations of the agricultural herbicide, atrazine. *Marine Pollution Bulletin*, 44, 1109-1116, ISSN 0025-326X

Reboud, X.; Majerus, N.; Gasquez, J. & Ponles, S. (2007). *Chlamydomonas reinhardtii* as a model system for pro-active herbicide resistance evolution research. *Biological Journal of the Linnean Society*, 91, 257-266, ISSN 0024-4066

Rosche, W. A. & Foster, P. L. (2000). Determining mutation rates in bacterial populations. *Methods*, 20, 4-17, ISSN 1046-2023

Roth, J. R.; Kugelberg, E.; Reams, A. B.; Kofoid, E. & Anderson, D. I. (2006). Origins of mutations under selection: the adaptive mutation controversy. *Annual Review of Microbiology*, 60: 477-501, ISSN 0066-4227

Shehata, S. A.; El-Dib, M. A. & Abou Waly, H. F. (1997). Effect of certain herbicides on the growth of freshwater algae. *Water, Air, & Soil Contamination*, 100, 1-12, ISSN 0049-6979

Singh, D. P. & Singh, N. (1997). Isolation and characterization of a metronidazole tolerant mutant of the cyanobacterium *Spirulina platensis* exhibiting multiple stress tolerance. *World Journal of Microbiology and Biotechnology*, 13, 179-183, ISSN 0959-3993

Sniegowski, P. D. & Lenski, R. E. (1995). Mutation and adaptation: the directed mutation controversy in evolutionary perspective. *Annual Review of Ecology and Systematics*, 26, 553-578, ISSN 0066-4162

Tilman, D. (1999). Global environment impacts of agricultural expansion: the need for sustainable and efficient practices. *Proceedings of the National Academy of Sciences from the United States of America*, 96, 5995-6000, ISSN 0027-8424

Trebst, A.; Hilp, U. & Draber, W. (1993). Response in the inhibitor efficiency of substituted phenols on PSII activity of six mutants of the D1 protein subunit in *Chlamydomonas reinhardtii*. *Phytochemistry*, 5, 969-977, ISSN 0031-9422

Vila-Auib, M. M.; Neve, P. & Powles, S. B. 2009. Fitness cost associated with evolved herbicide resistant alleles in plants. *New Phytologist*, 184, 751-767, ISSN 0028-646X

21

Highly Specific Biosensors to Herbicides, based on Sensitive- and Resistant-Mutants Microalgae

Javier Hernández-Allica[1], Daniel Carrera-Martínez[1],
Victoria López-Rodas[1,2], Antonio Flores-Moya[3] and Eduardo Costas[1,2]
[1]*Algasgen Biotecnología S.L.*
[2]*Complutense University of Madrid*
[3]*University of Málaga*
Spain

1. Introduction

The notion that humankind has changed the biosphere led to the Nobel prize Paul Crutzen to propose, a decade ago, the new term "Anthropocene" to denote the current interval of time in the earth dominated by human driven large-scale activities (Crutzen, 2002). The massive loss of diversity, homogenization of biotas, proliferation of opportunistic species and unpredictable emergent novelties can be considered among the distinctive features of the future biosphere (Myers & Knoll, 2001). Intensive agriculture, supported by the massive use of herbicides, pesticides and compounds with biocidal activity, is a significant cause of the biodiversity crisis (Tilman, 1999). The impact of these toxic compounds on biodiversity threatens all ecosystems, being particularly significant in those characterized by a slow response to change, such as aquatic systems. Phytoplankton are responsible for about half of the global primary production, driving essential biogeochemical cycles, exporting massive amounts of carbon to deep waters and sediments in the open ocean and strongly influencing the water–atmosphere gas exchanges (Rost et al., 2008). Since these organisms represent the basis of the aquatic food web, the repercussions of the impact on phytoplankton populations will undoubtedly affect the rest of the components of the trophic web.

Step behind of the risen up of scientific and social concern regarding the environmental pollution control, developed countries have begun to take legislative actions to protect the ecosystems from chemical pollution. For the time being, the monitoring of water quality has generally relied on the collection of spot water samples followed by extraction and laboratory-based instrumental analysis. These analytical methods usually requires the use of sophisticated equipment, skilled laboratory personnel, are time consuming, expensive and difficult to adapt for fieldwork. Besides, this provides only a snapshot of the situation at the sampling time and fails to provide more realistic information due to spatio-temporal variations in water characteristics (Rodríguez-Mozaz et al., 2006).

The European Union Water Framework Directive (WFD), one of the most important pieces of environmental legislation, is likely to transform the way that determination of water quality is undertaken. Within the next few years, the implementation of the WFD will require a considerable additional monitoring effort to be undertaken and a wide range of substances of different chemical groups to be identified. The WFD does not mandate the use of a particular set of monitoring methods, but aims at ensuring the establishment of an adequate monitoring program and encouraging the development of new technologies and more suitable methodologies allowing on-site field monitoring (Allan et al., 2006).

2. Microalgal whole cell biosensors

The need for adequate monitoring programs and early-warning procedures to detect contaminants has prompted the development of cell-based sensors as an attractive option, provided the microorganisms used as recognition bioelements are easy to isolate and manipulate (widely available, non-hazardous), their culture and maintenance is inexpensive and they provide reliable information on the presence of the target toxic agent. The essence of a whole cell biosensor is to display a cellular activity sensitive enough to stressed environments but insensitive to the physico-chemical features of the medium it operates, to the cell life cycle and to the availability of nutrients. Likewise, it should allow a smooth integration with an appropriate transducer of the biological signal.

The use of whole cells as biological recognition elements has many attractive advantages (Orellana 2008): (i) Whole-cell biosensors are usually cheap, because whole cells culturing and harvesting is easier than isolation and purification of enzymes. (ii) Whole cells are more tolerant to a significant change in pH, temperature or ionic concentration than purified enzymes. (iii) A multi-step reaction is possible because a single cell can contain all the enzymes and co-factors needed for detection of the analyte. (iv) Biosensors can easily be regenerated or maintained by letting the cells re-grow while operating in situ. (v) Extensive sample preparation is usually not required.

Green microalgae are the most preferred microorganisms to be implemented as biological element of recognition of herbicides in biosensors, since they are photosynthetic organisms. Photosynthesis inhibition estimated by chlorophyll a fluorescence of photosystem II (PSII), is an excellent indicator that rapidly reflects the toxic effect of certain pollutants. In fact, more than 50% of commercial herbicides are directed towards the direct inhibition of the electron transport at the PSII. Taking advantage of this feature, photosystem II (PSII)-based biosensors are reported to be able to detect herbicides in the environment (Giardi et al., 2001).

For example, an optical fiber based biosensor was developed for atrazine and endrine monitoring in water using *Scenedesmus subspicatus* cells, immobilized on filter paper and covered with a thin alginate layer hardened with calcium chloride (Frense et al., 1998). *Chlorella vulgaris* was immobilized at the tip of an optical fiber bundle placed inside a homemade microcell (Naessens et al., 2000), or in a rotating support holding up to five different membranes to increase the number of assays (Védrine et al., 2003), and used for the detection of herbicides affecting photosystem II (PSII) such as triazines (atrazine, simazine) or phenylureas (diuron, isoproturon) herbicides at sub-μgL^{-1} concentration level. Nguyen-Ngoc et al. (2007) investigated the response of a microalgal biosensor to the herbicide Diuron® by measuring the variation of the chlorophyll fluorescence of *Chlorella vulgaris* strain conveniently entrapped into a sol-gel translucent support. The detection limit was 1 $\mu g\ L^{-1}$ of diuron in water, a value much lower than the 115 $\mu g\ L^{-1}$ limit reported with bioassays or the 10 $\mu g\ L^{-1}$ limit reached with high performance liquid chromatography with diode array detector. The microalgae within the silica matrix kept over 95% of their initial activity after a period of 5 weeks.

To date the limiting step in the development of whole cell biosensor is the lack of specificity. Most algal biosensors have focused on reaching enough sensitivity and improving the signal measuring and the immobilization methods. Such algal biosensors are sensitive enough, although they usually present low specificity, exhibiting a summary response over a range of toxic substances. This lack of specificity represents the weak point of actual algal biosensors and is an important reason why relatively few have emerged from the laboratory to become commercially viable methods (Bengtson Nash et al., 2005). On the other hand, considering the large amount of potentially hazardous substances occurring in the environment, highly selective biosensor systems (e.g. enzyme biosensors) may also be regarded as disadvantageous (Podola et al., 2004). Ideally, combining the advantages of non-selective biosensors able to detect a variety of compounds, with a selective biosensor also allowing the identification of specific compounds would be the jump for the gap that separates academic research from field applications. In the following section we will describe recent developments directed to gain specificity in microalgae optical biosensors, focusing in the work conducted in our laboratory.

3. Theoretical and experimental setup of herbicide specific biosensors based on microalgae mutants

We have developed a new approach for increasing specificity of microalgae biosensors. This method is based in the joint use of two different genotypes from the same species, to detect a given herbicide: a highly sensitive genotype and, simultaneously, a resistant mutant to obtain high specificity.

3.1 Screening of herbicide-sensitive and –resistant genotypes

Using diverse functional and taxonomic groups, a screening of phytoplankters has been addressed to isolate strains highly herbicide-sensitive (López-Rodas et al., 2001; Costas et al., 2001; López-Rodas et al., 2007; Marvá et al., 2010; Huertas et al., 2010). The toxic effect of the herbicides has been estimated by calculating the inhibition of the acclimated maximal growth rate (*m*) in mid-log exponentially growing cells, in the presence of increasing concentrations of herbicides (diuron, simazine, diquat, glyphosate and others), by using the equation of Crow and Kimura (1970):

$$m = \mathrm{Log}_e(N_t/N_0)/t$$

where N_t and N_0 are the cell numbers at the end and at the start of the experiment, respectively, and t, is the time that cultures were exposed to different doses of herbicides.

At the same time, we have applied a selection procedure maintaining large populations of dividing cells (which ensures the occurrence of mutations that confer herbicide resistance) and a strong selection pressure (which ensures the preservation of such mutations) to isolate the most herbicide-resistant genotypes. For this purpose, an experimental system based on the ratchet protocols previously described (Reboud et al., 2007; Orellana et al., 2008; Huertas et al., 2010; López-Rodas et al., this book) was applied to estimate the maximal capability for adaptation of different phytoplankton species under increasing doses of different herbicides (Table 1). The protocol aims at reaching equilibrium between strong selection intensity, by means of ratcheting to increase herbicide dose, and the maintenance of a population size large enough to increase probability of rare spontaneous mutations that confer adaptation. These mutations occur randomly and not through specifically acquired adaptation included by herbicide (Marvá et al., 2010).

From the herbicide dose-inhibition curve it is possible to calculate the lethal dose that inhibits each microalgae ancestral population (LD_{100}). The first ratchet cycle starts with three herbicide concentrations: LD_{100}, 3 LD_{100} and 10 LD_{100}. Each herbicide ratchet cycle entails a threefold dose increase. Cultures must be ratcheted only up to a dose that supports population growth and are exposed to different selection levels. A ratchet cycle was concluded when no further cell growth was observed in a specific replicate after a period of 100 d. The number of ratchet cycles was therefore species dependent, as growth was a function of the ability to adapt to the selecting conditions. The experimental procedure was then applied in several independent experiments for each phytoplankton species.

Species	Habitat	Ancestral populations (before ratchet)	Derived populations (after ratchet)	Increase in adaptation (after ratchet)
Emiliania huxleyi (CCMP371)	Oceanic	0.10	0.15	1.5
Emiliania huxleyi (CCMP372)	Oceanic	0.10	0.15	1.5
Emiliania huxleyi (CCMP373)	Oceanic	0.05	0.15	3
Isochrysis galbana	Oceanic	0.10	0.15	1.5
Monochrysis lutheri	Oceanic	0.10	0.15	1.5
Tetraselmis suecica	Coastal marine	0.15	1.5	10
Phaeodactylum tricornutum	Coastal marine	0.10	0.45	4.5
Scenedesmus intermedius	Continental	0.15	40.5	270
Dictyosphaerium chlorelloides	Continental	0.15	13.5	90
Microcystis aeruginosa (3D)	Continental	0.05	0.45	9
Microcystis aeruginosa (6D)	Continental	0.05	0.45	9
Microcystis aeruginosa (7D)	Continental	0.05	0.45	9
Simazine concentrations (ppm) measured in triplicates of each strain inoculated with 1.5 x 10^5 cells from mid-log exponentially growing cultures and exposed to increasing simazine doses (increase interval = 0.05 ppm).				

Table 1. Simazine concentration (ppm) causing 100% growth inhibition measured in ancestral populations of various phytoplanktonic organisms before the ratchet experiments as well as in derived populations after the ratchet experiments (data from Huertas et al., 2010).

Table 1 shows the maximal capability of adaptation found by Huertas et al. (2010) using diverse functional and taxonomic groups of phytoplankters (oceanic and coastal marine microalgae, and freshwater microalgae and cyanobacteria) to isolate simazine resistant strains. A simazine concentration causing 100% growth inhibition in ancestral populations (before the ratchet experiments) is compared between the ancestral populations (before the ratchet experiments) and the derived populations (after the ratchet experiments). In spite of species diversity tested here (i.e. Cyanobacteria, Chlorophyta, Bacillariophyta, Haptophyta isolated from continental, coastal and oceanic microalgae), simazine was able to inhibit 100% growth in all ancestral populations. Adaptation of simazine at doses of 3.1 ppm (or higher) was only possible because of the occurrence of rare spontaneous simazine-resistant mutants occurring randomly during replication of organisms before exposure to simazine (Marvá et al., 2010). Consequently, adaptations obtained by Huertas et al. (2010) in the ratchet protocol (i.e. up to 40.5 ppm simazine in *S. intermedius* or 13.5 ppm simazine in *D. chlorelloides*) can only be genetically achieved by new mutations that confer resistance.

3.2 Chlorophyll *a* fluorescence of PSII as biological signal

The approach of using selected herbicide- sensitive and –resistant microalgal mutants was tested by the development of a microalgal biosensor for the specific detection of herbicides (diuron, simazine, diquat, gliphosate and others). The inhibition of chlorophyll a fluorescence of PSII by the herbicides can be used as the biological signal.

For this purpose changes in photosynthetic activity due to herbicide exposure are estimated by measuring chlorophyll a fluorescence of PSII. Immobilized algae were placed at the tip of a bifurcated fiber-optic cable (1 m long) in a homemade flow through cell, as has been described elsewhere (Orellana et al., 2007; Peña-Vazquez et al., 2009 and cites therein), for the measurements. Algae are kept in darkness for 15 min before irradiation, during that time the sample was pumped through the flow cell. Then the biosensing layers are irradiated for 5 min with the excitation light and the fluorescence measurement is carried out immediately. Experimental signals (Fig. 1) are plotted as a function of the analyte concentration (in logarithmic scale) and the experimental data fitted to a four-parameter logistic equation (Peña-Vázquez et al., 2009):

$$I = I_{min} + [(I_{max} - I_{min}) / [1 + ([\text{Analyte}]/IC_{50})^b]$$

Changes in photosynthetic activity due to herbicide can be also estimated after 30 min of exposure, by measuring fluorescence using a pulse-amplitude modulated fluorometer PAM 2000 (Walz, Effeltrich, Germany), as explained by Schreiber et al. (1986). Maximum fluorescence of light-acclimated thylakoids (F'_m) was determined after a saturating-white light pulse, which fully close all PSII reaction centres. The inhibition of F'_m was used as an estimator of the toxic effect of herbicide, according to the formula:

$$\text{Inhibition } (\%) = 100 - [100 \times (F'_m)_{herbicide} / (F'_m)_{control}]$$

where (F'_m) herbicide is the maximum fluorescence after 3 min with any given herbicide concentration, and (F'_m) control is the maximum fluorescence of the control without herbicide after the same period of time. In all the measurements in both herbicide-exposed and control cells, the irradiance was 80 µmol photons m^{-2} s^{-1}. In our previous work, we have always found that maximal fluorescence of light-adapted algae (F'_m) from herbicide resistant mutants is significantly higher than that from sensitive cells.

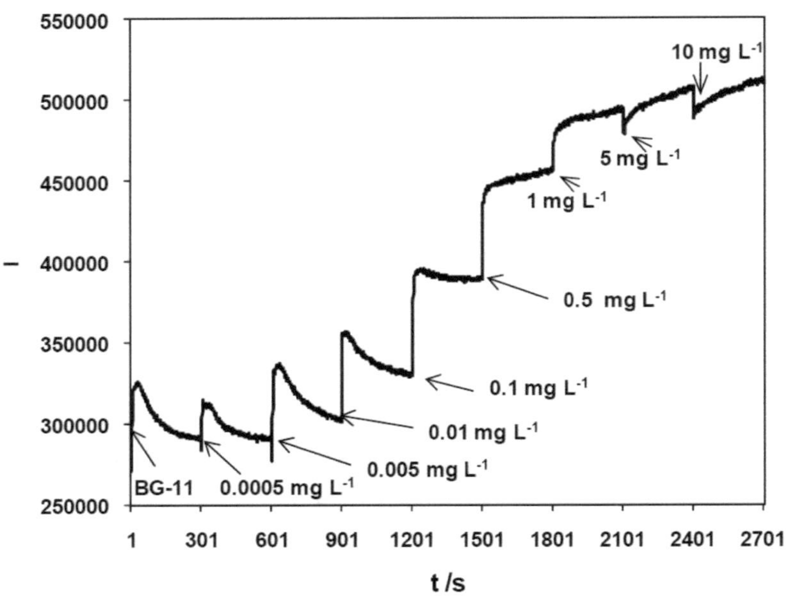

Fig. 1. Fluorescence response of a *Dictyosphaerium chlorelloides* membrane to simazine increasing concentrations in the range of 5 x 10-4 to 10 mg L^{-1}. λ_{exc}: 467 nm, λ_{em}: 699 nm. (From Peña-Vázquez et al., (2010), supplemented material).

4. Operative microalgae biosensor

Operative specific herbicide biosensors were obtained by immobilizing microalgae strains on silicone dishes (Cellon™) (Orellana et al., 2007; Orellana et al., 2008) or by encapsulating in silicate sol-gel matrices (Peña-Vázquez et al., 2009; 2010). Such procedures allow us long term stability of cells as determined by a constant F'm values for at least 3 weeks. Thus, the combined measurement of F'm from the two different genotypes (sensitive and resistant mutants to a given herbicide) allowed obtaining microalgae biosensor specificity (Table 2).

Microalgae mutant	Sensitive	Resistant
No herbicide present	+	+
Target herbicide present	-	+
Non-target herbicide present	-	-
+ and - denote high or low F'm signal respectively		

Table 2. Selective detection of a target herbicide with a dual sensing head microalgae biosensor. (Adapted from Orellana et al., 2007).

Peña-Vázquez et al. (2009) performed representative dose response curve in the 5×10^{-4} - 10 mgL^{-1} simazine range for sensitive and resistant mutants of *Dictyosphaerium chlorelloides* (D.c.) and *Scenedesmus intermedius* (S.i.) cells immobilized in a silicate sol-gel matrix. The limit of detection was lower for the D.c. (3.6 μgL^{-1}) than for S.i. (31.0 μgL^{-1}) biosensors. Further research with the D.c. biosensor showed good performance regarding important criteria that may be crucial for the implementation of a marketable biosensor (Table 3).
Despite that our D.c. biosensor described above (Peña-Vázquez et al., 2009) does not reach the European Community directive for herbicide detection (0.1 μgL^{-1}) in drinking water, when working with short irradiation times (15 min darkness/5 min irradiation), it has been demonstrated that the use of resistant strains can be a useful tool to improve biosensor specificity. In the algal biosensor wild-type sensitive cells exhibit a decrease in their biological response in presence of simazine, while resistant algae work as control exhibiting a significant smaller decrease. In contrast, under other herbicides such as atrazine or DCMU, resistant algae exhibit similar response than those of sensitive algae. This pattern demonstrates that the use of simazine-resistant cells is an appropriate procedure to improve biosensor specificity.

	D.c. biosensor
Detection limit (μgL^{-1})	3.6
IC_{50} (μgL^{-1})	125 ± 14
Dynamic range (μgL^{-1})	19 - 860
Response time (min)	20
Reversibility	Yes
Accuracy	High

Table 3. Response to simazine of an optical biosensor based on *Dictyosphaerium chlorelloides* (D.c.) immobilized in a sol-gel silicate flow-cell.

5. Prospective

There is a strong need for adequate monitoring programs and early-warning procedures allowing on-site field detection of herbicides polluting water reservoirs and ecosystems. Our approach based on the use of resistant strains can be a useful tool to improve microalgae biosensor specificity. In theory, our mutant selection method opens the possibility to obtain sensitive and resistant microalgae strain pairs for any target pollutant of interest, including of course, herbicides and other related pesticides, but also, other important environmental pollutants such as heavy metals, organics, novel emergent pollutants such as cosmetics and pharmaceuticals, etc. The way ahead is certainly long, but the ultimate goal of developing versatile and operative commercial equipments or biochips by means of immobilizing sensitive and resistant microalgae pairs on bifurcated fibre optic systems or multi-well plates, for example, appears realistic.

6. Acknowledgments

The results compiled in this review were obtained in research works financially supported by the Spanish Ministry of Sciences and Innovation through grants CTM2008-05680 C02-01/MAR and CGL2008-00652/BOS, and Junta de Andalucía Research Group RNM-115.

7. References

Allan, I.J.; Vrana, B.; Greenwood, R.; Mills, D.W.; Roig, B. & Gonzalez, C. (2006). A "toolbox" for biological and chemical monitoring requirements for the European Union's Water Framework Directive. *Talanta* 69, 302–322, ISSN 0039-9140

Bengtson Nash, S.M.; Quayle, P.A.; Schreiber, U. & Müller J.F. (2005). The selection of a model microalgal species as biomaterial for a novel aquatic phytotoxicity assay. *Aquatic Toxicology*, 72, 315–326, ISSN 0166-445X

Costas, E.; Carrillo, E.; Ferrero, L. M.; Agrelo, M.; García-Villada, L.; Juste, J. & López-Rodas, V. (2001). Mutation of algae from sensitivity to resistance against environmental selective agents: the ecological genetics of *Dictyosphaerium chlorelloides* (Chlorophyceae) under lethal doses of 3-(3,4-dichlorophenyl)-1,1-dimethylurea herbicide. *Phycologia*, 40, 391-398, ISSN 0031-8884

Crow, J.F. & Kimura, M. (1970). An introduction to population genetics theory. New York, NY, USA, Harper and Row

Crutzen, P. J. (2002). Geology of Mankind. *Nature*, 415, 23, ISSN 0028-0836

Giardi, M.T.; Koblizek M. & Masojidek J. (2001). Photosystem II-based biosensors for the detection of pollutants. *Biosensors and Bioelectronics*, 16, 1027-1033, ISSN 0956-5663

Frense, D.; Muller, A. & Beckmann, D. (1998). Detection of environmental pollutants using optical biosensor with immobilized algae cells. *Sensor and Actuators B: Chemical*, 51, 256–260, ISSN 0925-4005

Huertas, I. E.; Rouco, M.; López-Rodas, V. & Costas, E. (2010). Estimating the capability of different phytoplankton groups to adapt to contamination: herbicides will affect phytoplankton species differently. *New Phytologist*, ISSN 0028-646X (doi 10.1111/j.1469-8137.2010.03370.x)

López-Rodas, V.; Agrelo, M.; Carrillo, E.; Ferrero, L. M.; Larrauri, A.; Martín-Otero, L. & Costas, E. (2001). Resistance of microalgae to modern water contaminants as the result of rare spontaneous mutations. *European Journal of Phycology*, 36, 179-190, ISSN 0967-0262

López-Rodas, V.; Flores-Moya, A.; Maneiro, E.; Perdigones, N.; Marva, F.; García, M. E. & Costas, E. (2007). Resistance to glyphosate in the cyanobacterium *Microcystis aeruginosa* as result of pre-selective mutations. *Evolutionary Ecology*, 21, 535-547, ISSN 0269-7653

Marvá, F.; López-Rodas, V.; Rouco, M.; Navarro, M.; Toro, F. J.; Costas, E. & Flores-Moya, A. (2010). Adaptation of green microalgae to the herbicides simazine and diquat as result of pre-selective mutations. *Aquatic Toxicology*, 96, 130-134, ISSN 166-445X

Myers, N. & Knoll, A. H. (2001). The biotic crisis and the future of evolution. *Proceedings of the National Academy of Sciences from the United States of America*, 98, 5389-5392, ISSN 0027-8424

Naessens, M.; Leclerc, J.C. & Tran-Minh, C. (2000). Fiber optic biosensor using *Chlorella vulgaris* for determination of toxic compounds. *Ecotoxicology and Environmental Safety*, 46, 181–185, ISSN 0147-6513

Nguyen-Ngoc, H. & Tran-Minh, C. (2007). Fluorescent biosensor using whole cells in an inorganic translucent matrix. *Analytica Chimica Acta*, 583, 161-165, ISSN 0003-2670

Orellana, G.; Villén, L.; Haigh, D.; Maneiro, E.; Marvá, F. & Costas, E. (2007). Microalgal fiber-optic biosensors for water quality monitoring, In: Cutolo, A., Culschaw, B., López-Higuera, J.M. (Eds.), Proceedings of SPIE-volume 6619: Third European Workshop on Optical Fibre Sensors (Published on-line, article number 661921) (doi 10.1117/12.738597)

Orellana, G. & Haigh, D. (2008). New trends in fiber-optic chemical and biological sensors *Current Analytical Chemistry*, 4, 273-295, ISSN 1573-4110

Orellana, G.; López-Rodas, V.; Costas, E.; Maneiro, E. & Haigh, D. (2008). *Biosensores basados en microalgas para la detección de contaminantes ambientales.* CT/ES2008/000465, 2008, P200701905

Peña-Vázquez, E.; Maneiro, E.; Pérez-Condec, C.; Moreno-Bondi, M.C. & Costas, E. (2009) Microalgae fiber optic biosensors for herbicide monitoring using sol–gel technology. *Biosensors and Bioelectronics*, 24, 3538–3543, ISSN 0956-5663

Peña-Vázquez, E.; Pérez-Condec, C.; Costas, E. & Moreno-Bondi, M.C. (2010) Development of a microalgal PAM test method for Cu(II) in waters: comparison of using spectrofluorometry. *Ecotoxicology*, 19, 1059-1065, ISSN 0963-9292

Podola, B.; Nowack, E.C.M. & Melkonian, M. (2004). The use of multiple-strain algal sensor chips for the detection and identification of volatile organic compounds, *Biosensors and Bioelectronics*, 19, 1253–1260, ISSN 0956-5663

Schreiber, U.; Schliva, U. & Bilger,W. (1986). Continuous recording of photochemical and non-photochemical chlorophyll fluorescence quenching with a new type of modulation fluorometer, *Photosynthesis Research*, 10, 51–62, ISSN 0166-8595

Reboud, X.; Majerus, N.; Gasquez, J. & Ponles, S. (2007). *Chlamydomonas reinhardtii* as a model system for pro-active herbicide resistance evolution research. *Biological Journal of the Linnean Society*, 91, 257-266, ISSN 0024-4066

Rodríguez-Mozaz, S., López de Alda, M.J. & Barceló, D. (2006). Biosensors as useful tools for environmental analysis and monitoring. *Analytical Bioanalytical Chemistry*, 386, 1025–1041, ISSN 1618-2642

Rost, B.; Zondervan, I. & Wolf-Gladrow, D. (2008). Sensitivity of phytoplankton to future changes in ocean carbonate chemistry: current knowledge, contradictions and research directions. *Marine Ecology Progress Series*, 373, 227–237, ISSN 0028-646X

Tilman, D. (1999). Global environment impacts of agricultural expansion: the need for sustainable and efficient practices. *Proceedings of the National Academy of Sciences from the United States of America*, 96: 5995-6000, ISSN 0027-8424

Védrine, C.; Leclerc, J.; Durrieu, C. & Tran-Minh, C. (2003). Optical whole-cell biosensor using *Chlorella vulgaris* designed for monitoring herbicides. *Biosensors and Bioelectronics*, 18, 457–463, ISSN 0956-5663

22

How Early Diagenesis Reveals *In Situ* Biodegradation of Herbicides in Sediment

Devault Damien A.[1], Delmotte Sébastien[2], Macarie Hervé[3], Dolfing Jan[4] and Anschutz Pierre[5]

[1]*Cemagref, Centre de Martinique, Pôle de Recherche Agro-environnemental de la Martinique, Quartier Petit Morne, BP 214, 97 285 Lamentin*

[2]*MAD-Environnement, Allée des Demoiselles, 33170 Gradignan*

[3]*IRD, UMR IMEP (CNRS / IRD), Aix-Marseille Université, Faculté des Sciences de St. Jérôme, Boîte 441, 13397 Marseille cedex 20*

[4]*School of Civil Engineering and Geosciences, Newcastle University Newcastle upon Tyne, NE1 7RU England*

[5]*UMR Environnements et Paléo-environnements OCéaniques (EPOC) Université Bordeaux 1, 33405 Talence*

[1,2,3,5]*France*

[4]*United Kingdom*

1. Introduction

Globalisation of environment contamination is one of the actual major concerns. Concentrated in the developed countries during 1950s and 1960s, the use of pesticides, particularly the organic ones, has dramatically increased since 1970s (Alavanja, 2009). Agro chemistry expansion is parallel to diffusion of other technologies involving molecules proved to be environmental contaminants. Progressive prohibition of indiscriminate use of chemicals, considered as a threat for human and environmental health, highlights their persistence and their ubiquity. Although for partly natural contaminants like polycyclic aromatic hydrocarbons, the question is to dissociate the natural input (fires…) from the anthropogenic one; the synthesised molecules like herbicides do not give rise to such a controversy. However, understanding their flux from contaminated parcels and their fate in the environment requires historical overview that can be achieved through coring. In order to be informative coring needs to be performed on quiescent, biotic and unconsolidated growing matrix and for these reasons sediments are the most often sampled.

Coring is performed with the following five aims:

1. Defining the origin of a contamination leaning on the pollutant threat accompaniment as in source investigation (Koistinen et al., 1997) as in land use concern, distinguishing agricultural, urban or industrial input (Feng et al., 1998; Secco et al., 2005).
2. Establishing the contamination source (runoff, atmospheric deposition), especially for arctic areas due to cold condensation (Rawn et al., 2001; Macdonald et al., 2005, de Wit et al., 2006). For herbicide monitoring, such studies are scarce: even if Henry's constant are in favour to volatilization, atmospheric transport enhances photolysis.

3. Performing a retrospective analysis of dated sediment cores to evaluate the historical fluxes since the 1950s (Harrad et al., 1994; Hong et al., 2003), especially to segregate the anthropic impact to (bio) geochemical trends (Connor & Thomas, 2003). As, Nylund et al. (1992) said: *"retrospective time trends are essential in the judgement of a pollution situation"*.
4. Defining the state on the contaminant persistence in the environment. This consists to determine its half life and to estimate the time needed for the exportation of residues from an upstream watershed (Covaci et al., 2005). This aim can give information on the efficiency of a decision to ban pesticide use.
5. Estimating the fate of organic pollutants in sediment (Eganhouse et al., 2000; Devault et al., 2009).

Coring is used to answer to the major question of the fate of contaminants, in molecule as well as in watershed scale. Instead of heavy metals the presence of which can only be discussed in terms of speciation, organic pollutants may suffer profound transformations up to the point of complete mineralization (Bartha, 1971; Ficher, 1974; Pignatello, 1992). Briefly, the alteration of their chemical structure may be the result of physico-chemical reactions (photolysis, cryolysis hydrolysis…) and/or biodegradation (e.g. Grover et al., 1997 on trifluralin, Roberts et al., 2002 on paraquat, Krutz et al., 2010 on s-triazines…).

In soil and sediment, photolysis only occurs on the surface and could be so regarded as marginal in comparison to biodegradation and hydrolysis, although there are notorious exceptions such as for example in the case of the herbicide trifluralin that is strongly vulnerable to sunlight (Grover et al., 1997). Photolysis may also be promoted by the presence of MnVI and Fe(III) but their concentrations are usually quickly reduced with depth (Knights et al., 2003). In immerged sediments, the role of hydrolysis in herbicide alteration may be maximized, but generally speaking, hydrolysis is rarely fast enough to be the predominant alteration process. In such wet environment, contrarily to soil where dryness event may restrain microbial flora development, biodegradation appears as the main process responsible for the alteration of the herbicide structure although some authors have suggested that the supposed low temperatures in these environments and their negative effect on the microbial activity should result in an absence of important decrease of their herbicide stock (Mortensen and Jacobsen, 2004; Grimalt et al., 2004). This however is not supported by experimental work.

In sediment environment, diffusion of dissolved oxygen used as electron acceptor by micro organisms is limited by early compaction process and local consumption. Then, anaerobic conditions are usually dominant. In absence of oxygen, bacteria can use several alternative electron acceptors. This results in a stratification of the bacterial communities corresponding to the successive use of specific electron acceptors in order of energy yields, which in turns influences the microbial degradation of organic compounds, in particular the halogenated aromatic ones. Stratification can be significantly perturbed by benthic biota, enhancing the sediment part where early diagenesis occurs.

The aim of the present chapter is to provide an overview on a pivotal process for the fate of herbicide residues -but generally underestimate even not taken into account.

2. Early diagenesis

2.1 Introduction

Sediments record earth surface activities, such as climate or anthropogenic impact on the environment. Before burial, however, particles deposited in aqueous environments experience transformations called early diagenesis (Berner, 1980), which modify the initial recording. Biological activity of benthic micro- and macro-organisms supports the mechanisms of early diagenetic reactions. Biogeochemical processes in the first centimetres below the sediment/water interface change the chemistry and the mineralogy of sediments. Major diagenetic reactions are processes of oxidation/reduction, dissolution/precipitation, adsorption/desorption and changes in speciation. The nature and amplitude of early diagenetic processes depend mostly on physico-chemical properties of overlying waters and on organic matter flux towards the bottom. The amount of knowledge on early diagenesis processes is the result of 30 years of science, including pioneer works of Froelich et al. (1979) on the vertical sequence of red-ox reactions, Berner (1980) on the transport-reaction coupling, and Aller (1980) on the influence of macro-fauna on reactions and fluxes.

2.2 Biogeochemical processes in modern sediments

Modern sediments comprise the first decimetres below the sediment-water interface. This compartment contains organic carbon imported from the overlying water column or formed by benthic producers. Early diagenesis reactions are linked to organic matter mineralization. The mineralization of organic carbon alters the initial mineralogy and chemistry of deposited particles and interstitial waters, and determines the benthic cycle of redox species and the fate of sedimentary organic matter. Oxidation of organic matter requires an electron acceptor. The preferential use of the electron acceptor that yields the highest amount of free energy in the terminal step of the bacterially-mediated oxidation of organic matter constitutes a long-standing paradigm in biogeochemistry (Froelich et al., 1979; Postma & Jakobsen, 1996) (Table 2, Fig. 1). In undisturbed sediments, this paradigm is reflected in well documented vertical changes in pore waters and particles chemistry (Fig. 2).

The O_2 concentration decreases rapidly below the sediment surface. Nitrate increases in the oxygen containing layer and then decreases below. The disappearance of oxygen and nitrate is accompanied by a decline of the Mn-oxide content and an increase in dissolved manganese, followed deeper down by an increase in dissolved iron. Total sulphur content increases with depth, whereas sulphate concentration decreases. Deeper down, methane can reach high concentrations, particularly when sulphate reaches a zero concentration, which is common in fresh water sediments. This distribution follows the depth sequence of diagenetic reactions. Oxygen is reduced near the sediment-water interface, followed by the reduction of nitrate and manganese oxide, then reactive iron oxide (Fe_{ASC}), sulphate, and carbon dioxide (Froelich et al., 1979, Postma and Jakobsen, 1996).

Chemical changes produce vertical gradients. Gradients generate fluxes within the sediment and across the sediment/water interface. Transport mechanisms are diffusion, advection and mixing. Molecular diffusion is driven by concentration gradients and can be quantified using Fick's first law. It is a spontaneous processes carrying species downgradient, that depends on temperature, gradient, porosity and the nature of the diffusing species. In deep sediments it represents the main pathway for the transport of dissolved compounds. Bio diffusion reflects the activity of benthic organisms that adds to passive diffusion in surface sediments. Mixing by bioturbation produces a transport of particles. Then, the relative positions of reducing and oxidizing sediment components is shifted. As pointed out by Aller (1994), such disturbed systems could be better conceptualized as a spatially and temporally changing mosaic of redox reactions.

The O_2 consumption is attributed to oxic degradation of organic matter (reaction 1, Tab. 3) and the re-oxidation of products from the anaerobic degradation of organic matter (Canfield

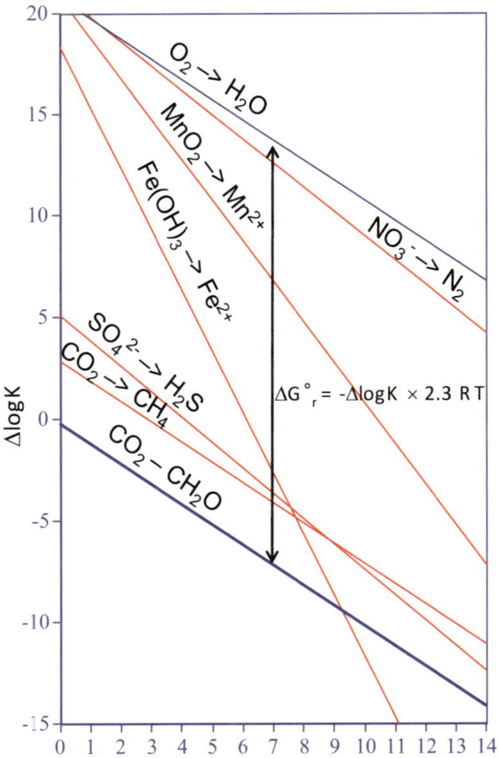

Fig. 1. Δlog K vs. pH plots for selected half redox reactions involved in organic matter oxidation. Δlog K is the total free energy at a given pH (see Table 1). The vertical distance between the line of (CH_2O) oxidation to CO_2 and any line representing reduction of electron acceptors gives the energy yield. The sequence of redox reaction can be deduced from this graph. Thermodynamically favourable secondary reactions can be deduced also from this graph. Temperature: 20°C

Element	aqueous phase	particulate phase
C	$C^{(+IV)}O_2$, $HC^{(+IV)}O_3^-$, $C^{(+IV)}O_3^{2-}$, $C^{(-IV)}H_4$, $C^{(-IV)\text{ à }(+IV)}$–org	$C^{(-IV)\text{ to }(+IV)}$-org , $CaC^{(+IV)}O_3$
N	$N^{(+V)}O_3^-$, $N^{(+III)}O_2^-$, $N^{(0)}{}_2$, $N^{(-III)}H_4^+$, $N^{(-III)}$-org	$N^{(-III)}$-org
O	$O^{(0)}{}_2$, $H_2O^{(-II)}$	
$P^{(+V)}$	PO_4^{3-}, HPO_4^{2-}, $H_2PO_4^-$	P-org, phosphates (apatite, francolite…), adsorbed-PO_4^{3-}
S	$S^{(+VI)}O_4^{2-}$, $H_2S^{(-II)}$, $S^{(-II)}H^-$	sulfides$^{(-II)}$
Fe	Fe^{2+}	$Fe^{(+III)}$-oxides, $Fe^{(+II)}$-sulfides, $Fe^{(+II)}CO_3$
Mn	Mn^{2+}	$Mn^{(+III,+IV)}$-oxides, $Mn^{(+II)}CO_3$

Table 1. Major dissolved and particulate redox compounds involved in early diagenesis processes. The redox state appears in brackets

$1/4\ CO_2 + H^+ + e^-$	=	$1/4\ CH_2O + 1/4\ H_2O$	-0.20
$1/4\ O_2 + H^+ + e^-$	=	$1/2\ H_2O$	+20.75
$1/5\ NO_3^- + 6/5\ H^+ + e^-$	=	$1/10\ N_2 + 3/5\ H_2O$	+21.05
$1/8\ NO_3^- + 5/4\ H^+ + e^-$	=	$1/8 NH_4^+ + 3/8\ H_2O$	+14.90
$1/2\ MnO_2 + 2\ H^+ + e^-$	=	$1/2\ Mn^{2+} + H_2O$	+20.80
$MnO(OH) + 3\ H^+ + e^-$	=	$Mn^{2+} + 2\ H_2O$	+25.36
$Fe(OH)_3 + 3\ H^+ + e^-$	=	$Fe^{2+} + 3\ H_2O$	+18.31
$FeO(OH) + 3\ H^+ + e^-$	=	$Fe^{2+} + 2\ H_2O$	+13.60
$1/8\ SO_4^{2-} + 5/4\ H^+ + e^-$	=	$1/8\ H_2S + 1/2\ H_2O$	+5.08
$1/8\ CO_2 + H^+ + e^-$	=	$1/8\ CH_4 + 1/4\ H_2O$	+2.87

Table 2. Redox half-reactions linked to organic carbon oxidation and their logK at 25°C. Free energy data are from Stumm & Morgan (1996), in kJ/mol : $e^- = 0$; $H^+ = 0$; $H_2O = -237.18$; $CO_2 = -394.37$; $CH_2O = -152.63$; $O_2 = 0$; $NO_3^- = -111.3$; $N_2 = 0$; $NH_4^+ = -79.37$; $IO_3^- = -128$; $I^- = -51.59$; $MnO_2 = -465.1$; $MnO(OH) = -557.7$; $Mn^{2+} = -228$; $Fe(OH)_3 = -699$; $FeO(OH) = -488.6$; $Fe^{2+} = -91.84$; $SO_4^{2-} = -744.6$; $H_2S = -27.87$; $CH_4 = -50.79$. Example of logK calculation from the first reaction: $\log K = -(\Delta G°r)/RT\ln(10) = -0.20$, with R = 8.314 J/mol/K, T = 298K, and $\Delta G°r = 1/4(-152.63)+1/4(-237.18)-1/4(-394.37) = 1140$ J/mol.

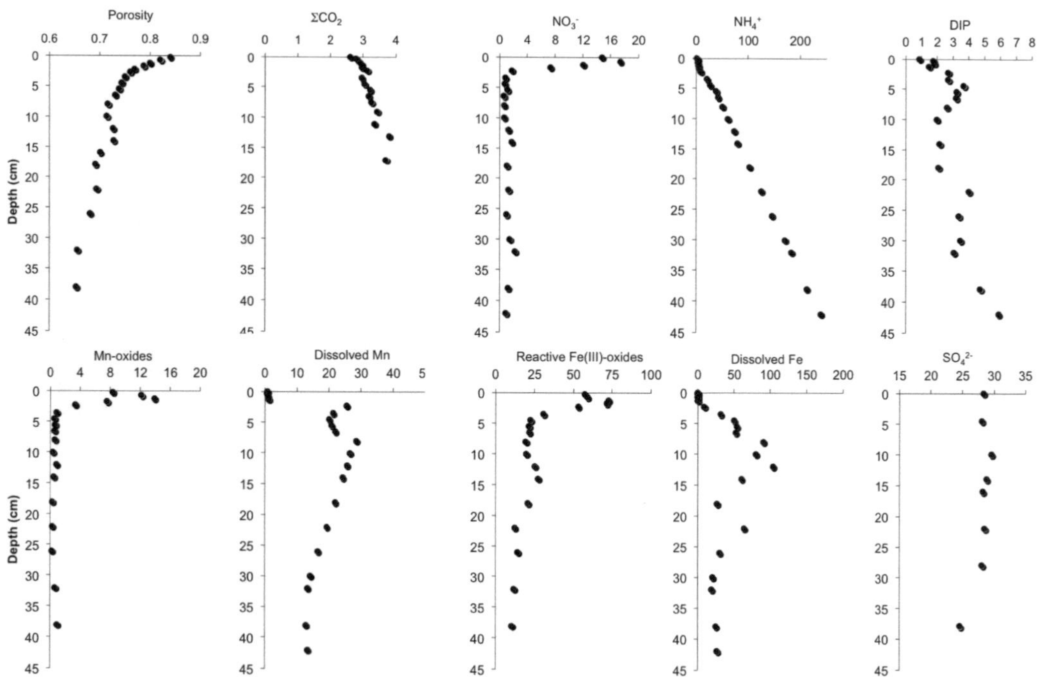

Fig. 2. (caption corrected): . Concentration vs. depth profile of dissolved and particulate compounds involved during early diagenesis and porosity. Example of a sediment collected in the Bay of Biscay at 550 m water depth. The zero depth corresponds to the sediment/water interface (modified from Hyacinthe et al., 2001). DIP = dissolved inorganic phosphorus/ Mn-oxides and reactive Fe(III)-oxides have been extracted from the solid fraction with an ascorbate reagent (Anschutz et al., 2005). Dissolved species are in µM, except sulphate in mM. Particulate oxides are in µmol/g.

et al., 1993). The observed thickness of the oxic layer depends directly on flux of labile organic carbon in the surface sediment.

The peak of nitrate in the oxic layer is attributed to the succession of reactions that lead to the bacterial nitrification of organic N (reaction 1, Tab. 3) or ammonia that diffuses from below (reaction 7, Tab. 3). The consumption of nitrate below the oxic layer is due to the bacterial denitrification (reaction 2, Tab. 3). NH_4^+ is produced from anaerobic mineralization of organic-N (reactions 4, 5 and 6, Tab. 3). The production of dissolved Mn^{2+} and Fe^{2+} in anaerobic sediments is attributed to the dissimilate reduction of manganese and iron oxides by bacteria (reactions 3 and 4, Tab. 3). The peak of Mn-oxides can be attributed to detrital Mn-oxides and authigenic Mn-oxides precipitate from the oxidation of dissolved Mn^{2+} that diffuses from below (Sundby, 1977). The concentration of Mn-oxides decreases abruptly below the oxic front, reaching values close to zero, which indicate that the Mn-oxides are totally reduced. The distribution of Fe-oxides shows highest concentrations near the oxic layer and decreases below. Amorphous manganese and iron oxide phases may be used as terminal electron acceptors in bacterial oxidation of organic carbon or they may react with reduced sulphur to form FeS.

The presence of particulate sulphur in continental margin sediments can be mostly attributed to authigenic iron-sulphide minerals that precipitate during the degradation of organic matter by sulphate-reduction (reaction 5, Tab. 3) (Berner, 1971; Jørgensen, 1982). The first compound that generally forms is amorphous FeS. It is subsequently converted to more crystalline pyrite (Berner, 1970; Schoonen and Barnes, 1991).

Within all possible reactions, recent studies attach more and more importance to metal oxides, particularly for their role in the benthic cycle of nitrogen (Luther et al., 1997; Aller et al., 1998; Anschutz et al., 2000). It has been pointed out by several authors that the reduction of Mn-oxide by diffusing ammonia could be an important process of dinitrogen production (e.g. Hulth et al., 1999; Anschutz et al., 2000). This pathway could interfere with dissimilatory Mn-reduction. Dissolved Mn(II) can be oxidized by nitrate (Aller, 1990; Schulz et al., 1994; Murray et al., 1995; Luther et al., 1997) or by iodate (Anschutz et al., 2000). Anaerobic nitrate production has been observed in several marine sediments (Aller et al., 1998; Anschutz et al., 1998; Mortimer et al., 2002 ; Deflandre et al., 2002; Chaillou et al., 2007). This process is due to oxidation of ammonia by Mn(III)-oxides. Iron oxide presents a wide speciation variety. The most labile fraction serves as oxidant for reduced compounds like dissolved ones (Canfield et al., 1992). All these alternatives or secondary reactions are thermodynamically favourable.

2.3 Interactions between benthic organisms and the sediment

In broad definition, bioturbation is the whole spectrum of physical, chemical and biological modifications of soils and sediments induced by the biological activities of all kind of living organisms (fauna, microbes, plants roots). In its strict sense, as it was employed in the first studies of bioturbation in marine sediments (Gray, 1974; Rhoads, 1974; Self and Jumars, 1978; Aller, 1977, 1980, 1982), it represents the sediment reworking and particularly the particles dispersal induced by the activities of burrowing animals. The bioturbation counterpart for solutes is called bio-irrigation, which designates the enhancement of solute transport in sediments by animal activities as gallery flushing. Even if Darwin had already shown the importance of worms activities in soils formation and functioning (Darwin, 1881), the primordial role of bioturbation and bio-irrigation in biogeochemical functioning of aquatic sediments was recently put forward in 1970'.

Bioturbation is now considered as a major process which drives diagenetic reactions, benthic landforms formation and biogeochemical cycles at the global scale although local effects and the scale of the living organisms (Meysman et al., 2006 a).

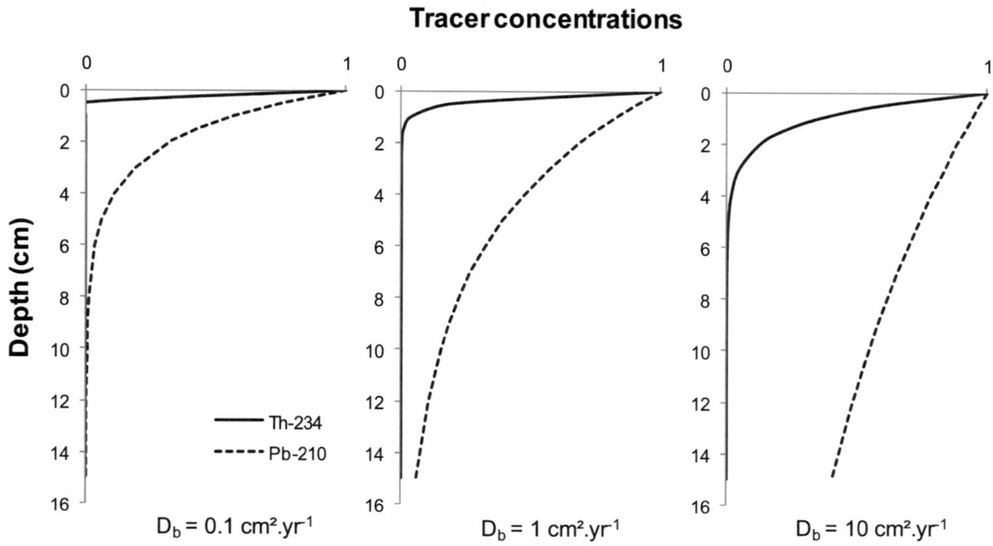

Fig. 3. Short-lived (Th-234) and long-lived (Pb-230) tracers profiles under various bioturbation intensities. These profiles are obtained in applying the biodiffusion model with a sedimentation rate of 0.1 cm.yr^{-1}.

In a general way, biological mechanisms of sediment reworking results from the activities of feeding, protection and reproduction of benthic animals. They burrow, excavate sediment and they dig galleries which communicate with overlying water. They often produce biogenic structures as mucus secretion and faecal pellets deposits resulting from ingestion-egestion of particles. They also ventilate their galleries to renew water. As consequences:

i. bioturbation generates biological transport of particles between different reactive zones in sediment on distances varying in the range from the millimetre to the metre (a classical average depth of bioturbation is about 10 cm, Boudreau, 2004);
ii. as a result of (i), bioturbation generates transport of micro-organisms associated with sediments between different reactive zones, modifies the bio-accessibility of nutrients and organic compounds and stimulates competition between bacterial groups by selective grazing (Aller, 1982; Stief and De Beer, 2002; Mermillod Blondin et al., 2003; Thullner et al., 2005);
iii. it modifies the sediment composition, e.g. in enriching the faecal pellets with organic matter as a result of the selective feeding (Brinkhurst et al., 1972; Davis, 1974);
iv. it increases exchange surface with water column and modifies sediment physical properties as porosity, permeability, compaction and stability (Rhoads, 1974; McCall and Fisher, 1980; Meadow and Tait, 1989; Sandness et al., 2000; DeDekere et al., 2001; Ciutat et al., 2006; Rasmussen et al., 1998). As they often actively irrigate their galleries by flushing, they enhance the exchange of solutes and particles between the sedimentary compartment and the free overlying water (Ciarelli et al., 1999, Meysman et al., 2006 b).

Depth sequence of bacterially-mediated oxidation of organic matter (Froelich et al., 1979; De Lange, 1986)
(O.M. = $C_{106}\,H_{263}\,O_{110}\,N_{16}\,P$)
Oxygen consumption by oxic respiration and nitrate production
$138\,O_2 + O.M. + 18HCO_3^- \rightarrow 124CO_2 + 16NO_3^- + HPO_4^{2-} + 140H_2O$ (1)
Nitrate consumption by denitrification
$94.4NO_3^- + O.M. \rightarrow 13.6CO_2 + 92.4HCO_3^- + 55.2N_2 + 84.8H_2O + HPO_4^{2-}$ (2)
Reduction of Mn-oxides by anaerobic respiration
$236MnO_2 + O.M. + 364CO_2 + 104H_2O \rightarrow 470HCO_3^- + 8N_2 + 236Mn^{2+} + HPO_4^{2-}$ (3)
Reduction of Fe-oxides and production of ammonia
$424FeOOH + O.M. + 740CO_2 + 104H_2O \rightarrow 846HCO_3^- + 424Fe^{2+} + 16NH_3 + HPO_4^{2-}$ (4)
Production of sulphide and ammonia by sulphatoreduction
$53SO_4^{2-} + O.M. \rightarrow 39CO_2 + 67HCO_3^- + 16NH_4^+ + 53HS^- + 39H_2O + HPO_4^{2-}$ (5)
Production of methane and ammonia by methanogenesis
$O.M. \rightarrow 53CO_2 + 53CH_4 + 16NH_4^+ + HPO_4^{2-}$ (6)
Production of nitrate by nitrification
$NH_4^+ + 2O_2 \rightarrow NO_3^- + 2H^+ + H_2O$ (7)
Oxidation of Mn^{2+} with oxygen
$2Mn^{2+} + O_2 + 2H_2O \rightarrow 2MnO_2 + 4H^+$
$4Mn^{2+} + O_2 + 6H_2O \rightarrow 4MnOOH + 8H^+$ (8)
Oxidation of Mn2+ with nitrate
$5Mn^{2+} + 2NO_3^- + 4H_2O \rightarrow 5MnO_2 + N_2 + 8H^+$
$10Mn^{2+} + 2NO_3^- + 14H_2O \rightarrow 10MnOOH + N_2 + 18H^+$ (9)
Oxidation of Fe^{2+} with nitrate
$5Fe^{2+} + NO_3^- + 12H_2O \rightarrow 5Fe(OH)_3 + 1/2N_2 + 9H^+$ (10)
Oxidation of Fe^{2+} with Mn-oxides
$Fe^{2+} + MnOOH + H_2O \rightarrow Fe(OH)_3 + Mn^{2+}$
$2Fe^{2+} + MnO_2 + 4H_2O \rightarrow 2Fe(OH)_3 + Mn^{2+} + 2H^+$ (11)
Reduction of Mn-oxide with ammonia to give dinitrogen
$3/2\,MnO_2 + NH_4^+ + 2\,H^+ \rightarrow 3/2\,Mn^{2+} + 1/2\,N_2 + 3H_2O$
$3\,MnOOH + NH_4^+ + 5\,H^+ \rightarrow 3Mn^{2+} + 1/2\,N_2 + 6\,H_2O$ (12)
Reduction of Mn-oxide with ammonia, production of nitrate
$8MnOOH + NH_4^+ + 14H^+ \rightarrow 8Mn^{2+} + NO_3^- + 13H_2O$
$4MnO_2 + NH_4^+ + 6H^+ \rightarrow 4Mn^{2+} + NO_3^- + 5H_2O$ (13)
Other reactions: Precipitation and dissolution of FeS, $CaCO_3$ and $MnCO_3$

Table 3. List of reactions considered in this work

The biogeochemical effects of bioturbation and bio-irrigation are multiples and often opposite: as an example, metallic contaminants tend to accumulate in sediment under the effect of particles reworking and ingestion whereas they are released to overlying water by the bio-irrigation activities (Banta and Anderson, 2003; Delmotte et al., 2007).

These effects also depend on: (i) the physico-chemical environmental conditions (pH, oxygen, sediment composition, grain size, temperature…) although they modify them; (ii) the type of benthic species present that generates different reworking and exchange effects. The complex interplays between bioturbation and physico-chemical processes make the multiplicity of effects on biogeochemical functioning not straightforward to study.

Quantification of sediment mixing is classically done using particle-tracers methods. They are based on the measurement of the vertical distribution of particle tracers within the sediment column. The underlying principle is that tracers initially deposited at the water-sediment interface, or placed at some horizon within the sediment column (Gilbert et al. 2003b), are displaced due to action of benthic fauna. The establishment of vertical profiles of tracers concentration using different methods (sediment cores slicing, image analysis) allows to monitor the tracers displacements. Sediment reworking coefficients are then computed using mathematical models that are fitted to the vertical tracer profile. (Goldberg and Koide, 1962; Guinasso and Schink, 1975; Aller, 1982; Boudreau 1986a ; Boudreau 1986b ; Wheatcroft et al. 1990 ; Gerino, 1990; Delmotte et al., 2007 ; Meysman et al. 2007). Vertical mixing was traditionally evaluated using particle-reactive radionuclide such as Pb-210 and Cs-137 in quantifying long-term (years to decade) and short-term mixing (week to month) (Robbins et al., 1977, Rice, 1986, Maire et al., 2008). More recently, conservative fluorescent particles were used to monitor the biological-induced displacements of sediment particles (Gerino, 1990; Ciutat et al., 2005). Other tracers have also been used such as pollen or Chlorophyll-a (Gerino et al., 1998; Maire et al. 2008). Some tracers are naturally deposited with a constant flux or a pulse (radionuclide, Chlorophyll-a) at the sediment-water interface whereas others are deposited by experimenters.

In the face of complexity and diversity of biological mixing processes, it is often assumed that when a sufficient number and variety of small-scale mixing events occur within a sediment interval they can be treated as random particle eddies and quantified in a mixing coefficient analogous to the physical eddy diffusion coefficient in a fluid. If we consider vertical mixing over a depth layer of thickness L, the governing tracer equation becomes:

$$\frac{\partial C}{\partial t} = D_b \frac{\partial^2 C}{\partial x^2} - \omega \frac{\partial C}{\partial x} - \lambda C$$

where x represents the depth into the sediment, C is the concentration or activity of the selected tracer, D_b is the bio diffusion coefficient, ω the burial velocity, and λ the decay constant of the tracer ($\lambda=0$ with a conservative tracer). Suitable solutions of this equation are fitted to tracer depth profiles, and from this, an optimal estimate for the bio diffusion coefficient D_b is obtained.

A major inconsistency in this analogy is that the path length of particle motion is often both specific in orientation and large in size scale with respect to chemical or physical property gradients in the sediment. For instance, sediment particles from depth may be placed directly at the oxidized sediment-water interface without mixing with material in between. Moreover, there is an inaccuracy in the time length of sediment reworking. Meysman et al. (2003) concluded that the bio diffusion analogy can be theoretically justified when the intrinsic time and length scales of sediment reworking are much smaller than their observational counterparts. These authors concluded that in many cases, the application of the bio diffusion model is questionable from a theoretical point of view. Still, in practice, tracer profiles often look diffusive and the bio diffusion analogy is widely employed to analyze these profiles. Meysman et al. (2003) refer to this apparently contradictory situation as the 'bio diffusion paradox'.

Despite such conceptual inaccuracies, some insight into the consequences of reworking as compared to simple sedimentation can be gained from the analogy. The cases of a short-lived tracer (Th-234, $t_{1/2}$=24.1 days) and a long-lived tracer (Pb-210 $t_{1/2}$=22.3 years) under various bio diffusion intensities are illustrated in Fig 3. Steady-state profiles are represented. Regarding the model coefficients, we retained values in the range of the reported ones in literature. Values of D_b varies from 0.01 to 30 $cm^2.yr^{-1}$ for marine environments (Aller, 1982), with values often in the order of 1 to 10 $cm^2.yr^{-1}$. Values for freshwater environments vary in the same range (Delmotte, 2007). Sedimentation rates can be lower than 0.1 $cm.yr^{-1}$ in marine environments and greater than 10 $cm.yr^{-1}$ in freshwater environments: a value of 0.1 $cm.yr^{-1}$ is retained.

With a small value of D_b (0.1 $cm^2.yr^{-1}$), the short-lived tracer profile shows a high decrease in the first millimetres whereas the tracer reaches a depth of 5 cm with a value 100 times greater. The long-lived tracer reaches 8 cm with the smaller bio diffusion value; its profile tends to homogenize on the first 10 cm with a high value of bio diffusion. These simulations show the importance of bioturbation in term of sediment mixing and displacement of reactive particles. They can be used to estimate the fate of diverse herbicides (labile or persistent) in bioturbated sediments.

Note that more complex models of bioturbation were developed to account for the transport of particles on long distances and simulate the effects of specific modes of bioturbation (e.g. Boudreau, 1986b; Wheatcroft, 1990; Delmotte et al., 2007, 2008).

Measurements of bio irrigation are based on tracer-methods too. Reactive or conservative tracers are often injected in the overlying water, and both the concentrations of the tracer in this compartment and in the sediments are monitored to be modelled. Application of solute transport models gives coefficients that reflect the intensity of bio irrigation. The simplest model consists in using the diffusion model with an enhancement factor of the molecular diffusion of the tracer. For instance, Wang and Matisoff (1997) reported a diffusion enhancement factor in the range from 2 to 5 by freshwater tubificids, which is a classical range of value for many irrigating organisms. Freshwater tubificids generate a water exchange of 25 $L.m^{-2}.d^{-1}$ (Woods, 1975), which is important but not massively compared to *Nereis diversicolor* (Muller, 1776) which may induce 1500 $L.m^{-2}.d^{-1}$ (Kristensen and Kostka, 2005) or *Chironomus riparius* (Meigen, 1804) with 80-800 1500 $L.m^{-2}.d^{-1}$ (Leuch, 1986). Bio irrigation greatly enhances the flux of dissolved components at the water-sediment interface and their depth of penetration: as for bioturbation, this process seems primordial in the cycles of herbicides.

In order to understand the role of fauna activities in the degradation and storage of herbicides in sediment, the proper model would be *a priori* the early diagenesis of organic matter. Aller (1982) early proposed a conceptual model about the role of bioturbation and bio-irrigation on early diagenesis of organic matter (an adaptation is presented in Fig 4). This model fits well with the observed results of numerous studies.

Starting from the classical idealization of primary reactions of organic matter degradation, two major effects of benthos activities can be distinguished: (i) the geometry of solute diffusion is altered by the presence of burrows and the irrigation of the galleries. As a consequence, solute reactants and especially oxygen and nitrates are transported deeper in the sediments and an horizontal gradient of metabolic successions takes place in the walls of the galleries; (ii) burrow and faecal pellet formation alters reaction and solute diffusion geometries, creating a mosaic of biogeochemical microenvironments rather than a vertically stratified distribution. Moreover, it has been shown that ventilation of galleries creates a dynamic succession of aerobic and anaerobic conditions in the interstitial water (Aller, 1994, Sun et al., 2002; Gilbert et al, 2003a): in a first step, worms bring oxygen in their galleries by flushing with water from the overlying water column; in a second step, this oxygen is consumed by micro-organisms (aerobic oxidation) creating an oxygen depletion and an

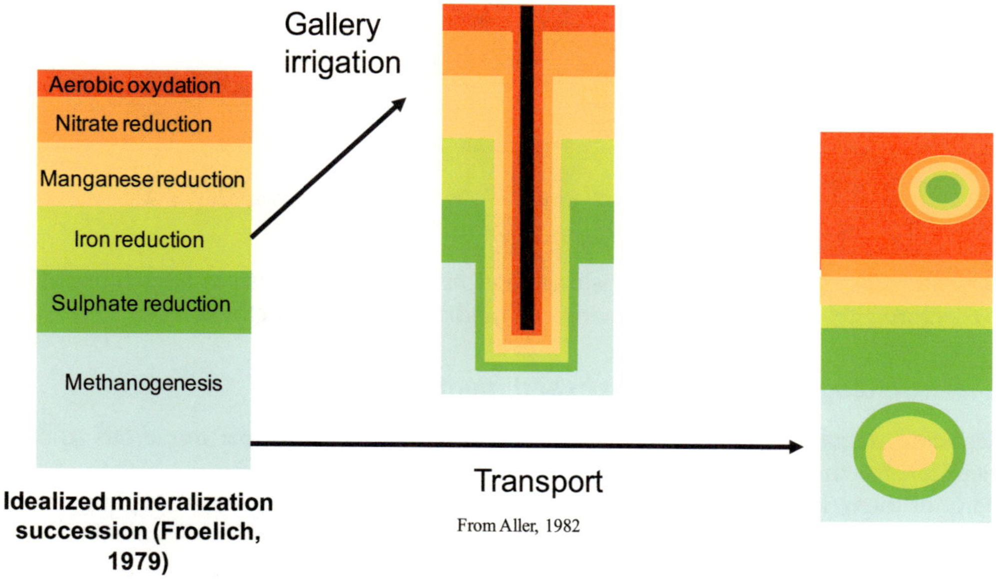

Fig. 4. Conceptual view of the effects of bioturbation and bio-irrigation on early diagenesis of organic matter.

enrichment of water with nitrates; in a third step, nitrates are consumed by micro-organisms in anaerobic conditions (denitrification); then, worms renew water bringing again oxygen. This redox oscillation in galleries enhances the organic matter mineralization by a factor 7.
Bioturbation and bio-irrigation deeply modify the biogeochemical functioning of sediments and hence the global geochemical cycles. As far as we know, the effects of bioturbation on herbicides geochemistry in sediments were not specifically addressed. Taking into account the living organisms bioturbating effects, on *in-situ* fate of this kind of contaminants, represents a challenge for the future studies and surely improves our comprehension of geochemical mechanisms of herbicides biodegradation. Numerical modelling of diagenesis of organic matter including the effects of bioturbation and bio irrigation really started in 1995 with a study of Van Cappellen and Wang. Since this paper, numerous models were presented (Boudreau, 1996; Soetaert et al., 1996; Meysman, 2001; Wijsman et al., 2002; Canavan et al., 2006; Delmotte, 2007). Such models allow exploring the influence of particular and solute transport on the numerous reactive processes, using bioturbation and bio irrigation coefficients obtained with tracer-methods. This represents a promising way to understand the fate of herbicides in bioturbated sediments.

2.4 Conclusion

The first decimetre of sediment below the sediment/water interface is an ordered and dynamic ecosystem. Benthic biogeochemical mechanisms are linked to the mineralisation of organic matter exported toward the sediment. Reactions and macro benthic activity determine the sedimentary cycle of carbon, redox species and all reactive components, including contaminants. All these phenomenon are related and depends on environmental properties, which vary in space and time. Therefore, studies on early diagenesis must take into account the complexity of the sedimentary environment and they need a multidisciplinary approach.

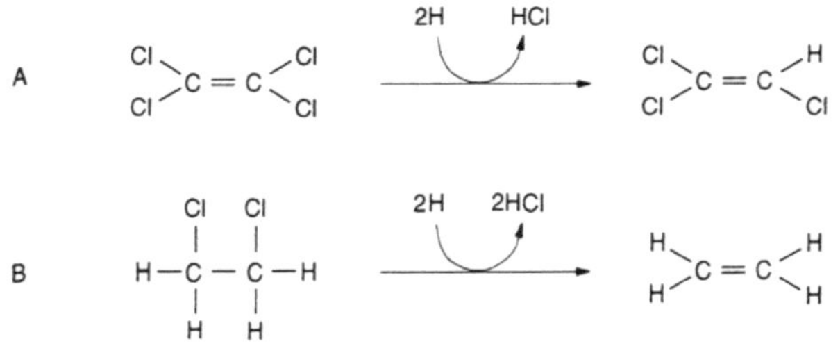

Fig. 5. Examples of two types of dehalorespiration: A. hydrogenolysis of tetrachlorethylene to trichloroethylene; B. dihaloelimination of 1,2-dichloroethane to ethene.

3. Microbiological degradation in sediment condition

Herbicide degradation pathways are as various as herbicides are numerous and their chemical structure different.

Many of them, however frequently contain chlorine or bromine groups. Typical examples are synthesized plant hormone (2,4-dichlorophenoxyacetic acid, 2,3,6-trichlorobenzoic acid…), pentachlorophenol, triazines, chloroacetanilides, hydroxy-benzonitriles, some substituted ureas (chlortoluron…) and sulfonylureas (chlorsulfuron…), diphenyl-ethers (bromofénoxime, diclofop-methyl) and pyridylphenylethers (clodinafop-propargyl) (Häggblom and Bossert, 2003; BCPC, 2007). Microorganisms have evolved a variety of metabolic strategies for cleaving the carbon halogen bond, including reductive dechlorination (Dolfing and Tiedje, 1986; Mohn and Tiedje, 1992). Reductive dechlorination is considered the predominant process in the anaerobic transformation of halogenated compounds. Although it is also a strategy used by aerobic microorganisms, reductive dehalogenation is almost exclusively observed in anaerobic environments where both aliphatic and aromatic organohalides can function as terminal electron acceptors in an anaerobic respiration process termed (de)halorespiration (Dolfing 1990; Maphosa et al. 2010). Two types of dehalorespiration can be distinguished: hydrogenolysis and dihaloelimination (Fig. 5). Both require molecular hydrogen or some other source of reducing equivalents, and both are highly exergonic. Values for the $\Delta G^{o'}$ of reductive dechlorination (hydrogenolysis) of halogenated compounds range from -130 to -180 $kJ.mol^{-1}$ of halogen removed (Dolfing, 2003), corresponding to redox potentials (E'_o) of +260 to +480 mV. This is comparable to the value for the redox couple NO_3^-/NO_2^- (E'_o = +433 mV), and substantially higher than the values for sulphate (SO_4^{2-}/HS^-; E'_o = -217 mV) and bicarbonate (HCO_3^-/CH_4; E'_o = -238 mV) (Dolfing et al. 2008), which suggests that dehalogenating organisms will outcompete sulphate reducers and methanogens for reducing equivalents when these are rate limiting. Thus there are no thermodynamic reasons why halogenated pesticides and other organohalides should not be biodegradable in the sulphate reducing and methanogenic layers of anaerobic sediments. There may well be ecophysiological reasons though. One of them could be that microorganisms simply not yet have had enough time to evolve the appropriate enzyme systems to catalyze and benefit from the dechlorination of halogenated pesticides. Although halogenated compounds per se are not necessarily xenobiotic, many chlorinated pesticides are. Chlorinated compounds tend to be fairly hydrophobic, and usually adsorb strongly to the sediment matrix. Thus the impetus for microbes to develop mechanisms to harness the energy locked up in these compounds is weak. Under those conditions fortuitous, co-metabolic conversion may well be the most prevalent route (van Eekert et al. 1999). Methanogenic *archaea,* which obtain energy for growth by converting hydrogen and acetate into methane, are for example able to degrade many organochlorines including pesticides (e.g. Jablonski & Ferry, 1992; Jablonski et al. 1996), because some of the co-factors involved in methane formation can fortuitously dechlorinate them. The methanogens have however no mechanism to harness the energy of the latter reaction. Co-metabolic transformation of halogenated compounds is typically much slower than (de)halorespiration (Löffler et al. 1999).

The $\Delta G^{o'}$ values quoted above are for standard conditions: pH = 7, 25°C, concentrations of substrates and products at 1 mol per litre, and gases at a partial pressure of 1 atm (100 000 Pa). Environmental conditions are generally different from that. This is considered in ΔG values. The rule of thumb for the conversion from $\Delta G^{o'}$ to ΔG values in reductive dechlorination reactions is that every 10-fold change in concentration results in an adjustment of 5.7 kJ/reaction. Thus, for example, a 100 fold lower product concentration (e.g. [Cl^-] is 10 mM rather than 1 M) results in a 11.4 kJ lower ("more exergonic") ΔG value, while a pH of 6 rather than 7 results in a 5.7 higher ("less exergonic") ΔG value. Molecular hydrogen levels in methanogenic sediments are generally in the range of 10 to 100 Pa; this results in adjustments of ~ 17-23 kJ per reaction. It seems reasonable to assume equimolar concentrations of chlorinated pesticides (e.g. hexachlorobenzene) and their less halogenated dechlorination product (e.g. pentachlorobenzene), but even if the concentration of the product would be 1000 fold higher than the concentration of the substrate the adjustment would still be only 17 kJ.mol^{-1}. Given the fact that for example $\Delta G^{o'}$ for the reaction hexachlorobenzene + H_2 → pentachlorobenzene + H^+ + Cl^- is -171.4 kJ per reaction (Dolfing and Harrison 1992) such adjustments are relatively minor, and it is safe to conclude that reductive dechlorination would remain exergonic under environmentally relevant conditions. Microorganisms in anoxic methanogenic sediments function at ΔG values of -10 to -20 kJ per reaction (Hoehler et al 2001).

Whereas microbiologists traditionally evaluate the thermodynamics of environmentally relevant reactions in terms of $\Delta G^{o'}$ values, that is at pH = 7, chemists generally express redox half-reactions in terms of logK values (e.g. Table 2) at pH =0. Following the latter convention Table 5 lists the logK values of all reductive dechlorination reactions involved in the sequential dechlorination pathways of hexachlorobenzene to benzene. The logK values, which range between 8.7 and 11.5, decrease with the number of chlorine atoms carried by the benzene ring, meaning that dechlorination becomes (slightly) less favourable when the number of chlorine substituents decreases. *Dehalococcoides* sp. strain CBDB1 is an example of an organism that can grow with chlorinated benzenes as electron acceptor (Adrian et al. 2000; Lorah and Olsen, 1999). Presently the only chlorinated benzene congener that appears to resist being dehalorespired is monochlorobenzene. There is however no thermodynamic reason why microorganisms should not be able to grow with this compound as electron acceptor: the hunt is on (Fung et al. 2009).

Dichloroelimination is a special case of reductive dechlorination, where two chloral substituents are removed from adjacent carbon atoms, while the aliphatic C-C bond is converted into an unsaturated C=C bond (Fig. 5, Dolfing, 2000). This reaction is of particular environmental relevance: the first step in the degradation of the organochlorine lindane (γ- hexachlorocyclohexane) is a dihaloelimination to γ-3,4,5,6-tetrachloro-1-cyclohexene (Heritage and Mac Rae 1997; Lal et al. 2010). Also, Lorah and Olsen (1999) have observed that in a freshwater tidal wetland degradation of 1,1,2,2-tetrachloroethane involved formation of 1,2-dichloroethylene rather than only classical reductive

½ hexachlorobenzene + ½ H^+ + e^-	=	½ pentachlorobenzene + ½ Cl^-	11.53
½ pentachlorobenzene + ½ H^+ + e^-	=	½ 1,2,3,4-tetrachlorobenzene + ½ Cl^-	10.63
	=	½ 1,2,3,5-tetrachlorobenzene + ½ Cl^-	11.21
	=	½ 1,2,4,5-tetrachlorobenzene + ½ Cl^-	10.83
½ 1,2,3,4-tetrachlorobenzene + ½ H^+ + e^-	=	½ 1,2,3-trichlorobenzene + ½ Cl^-	10.11
	=	½ 1,2,4-trichlorobenzene + ½ Cl^-	11.10
½ 1,2,3,5-tetrachlorobenzene + ½ H^+ + e^-	=	½ 1,2,3-trichlorobenzene + ½ Cl^-	9.54
	=	½ 1,3,5-trichlorobenzene + ½ Cl^-	10.84
	=	½ 1,2,4-trichlorobenzene + ½ Cl^-	10.53
½ 1,2,4,5-tetrachlorobenzene + ½ H^+ + e^-	=	½ 1,2,4-trichlorobenzene + ½ Cl^-	10.91
½ 1,2,3-trichlorobenzene + ½ H^+ + e^-	=	½ 1,2-dichlorobenzene + ½ Cl^-	10.41
	=	½ 1,3-dichlorobenzene + ½ Cl^-	10.64
½ 1,2,4-trichlorobenzene + ½ H^+ + e^-	=	½ 1,2-dichlorobenzene + ½ Cl^-	9.42
	=	½ 1,3-dichlorobenzene + ½ Cl^-	9.65
	=	½ 1,4-dichlorobenzene + ½ Cl^-	9.96
½ 1,3,5-trichlorobenzene + ½ H^+ + e^-	=	½ 1,3-dichlorobenzene + ½ Cl^-	9.33
½ 1,2-dichlorobenzene + ½ H^+ + e^-	=	½ monochlorobenzene + ½ Cl^-	9.94
½ 1,3-dichlorobenzene + ½ H^+ + e^-	=	½ monochlorobenzene + ½ Cl^-	9.71
½ 1,4-dichlorobenzene + ½ H^+ + e^-	=	½ monochlorobenzene + ½ Cl^-	9.40
½ monochlorobenzene + ½ H^+ + e^-	=	½ benzene + ½ Cl^-	8.75

Table 5. Redox half-reactions for reductive dechlorination of hexachlorobenzene to benzene and their logK at 25°C. Free energy (G_f^o) data are from Dolfing and Harrison (1992) and Stumm and Morgan (1996), in $kJmol^{-1}$: hexachlorobenzene = 46.0 ; pentachlorobenzene = 45.7 ; 1,2,3,4-tetrachlorobenzene = 55.8 ; 1,2,3,5-tetrachlorobenzene = 49.2; 1,2,4,5-tetrachlorobenzene = 53.5; 1,2,3-trichlorobenzene = 71.7 ; 1,2,4-trichlorobenzene = 60.5 ; 1,3,5-trichlorobenzene = 56.8 ; 1,2-dichlorobenzene = 84.3 ; 1,3-dichlorobenzene = 81.8; 1,4-dichlorobenzene = 78.3 ; monochlorobenzene = 102.3 ; benzene = 133.8 ; Cl^- = -131.3 ; logK = $-(\Delta G^o r)/RTln(10)$. For sample calculation see footnote to Table 2.

dechlorination to lesser chlorinated ethanes. This is unfortunate as subsequent anaerobic degradation of dichloroethylene will give rise to the formation of vinyl chloride, a notorious carcinogen. The energy yield per mole of reducing equivalents used for dihaloeliminations is higher than the yield for the corresponding classical hydrogenolysis reactions (Dolfing, 2000) which suggests that where possible this pathway is especially important under conditions where competition for reducing equivalents is fierce.

A prime example of anaerobic *in situ* degradation of a pesticide in anaerobic sediments has been reported for Lake Ketelmeer in the Netherlands (Beurskens et al. 1993). This lake acts as a sedimentation area for the suspended solids from the river Rhine. Comparison of recent and archived core layers revealed a 80% loss of hexachlorobenzene and increases in 1,3,5-trichlorobenzene and 1,3-dichlorobenzene during the 20 years of "in situ" incubation. Laboratory incubations with these sediments demonstrated that the native microbial population was catalyzing this reaction and that the organisms involved selectively mediated the thermodynamic most favourable reactions (Beurskens et al. 1994). Indeed, thermodynamics has been used successfully to rationalize dechlorination and dechlorination patterns of organohalogens, but this approach does not always work (Dolfing, 2003). Thus there is a need for more than "just" thermodynamics to explain and predict the dechlorination (patterns) or the lack thereof in Nature. With some unfortunate exceptions pesticides concentrations in the environment are relatively low. Hence it seems plausible that Nature needs time to respond to the challenge to use these compounds efficiently, which is as energy source. Meanwhile co-metabolism may well be the route that nature uses to cleanse itself slowly but steadily. Even for the relative success story of hexachlorobenzene degradation in Rhine sediments its estimated half life was seven years (Beurskens et al. 1993).

4. *In situ* observations of consortia biodegradation efficiency

Euphemistically, margin growth for herbicides in situ biodegradation knowledge remains important. Bromoxynil, dicamba, 2-4D and atrazine are the only herbicides whose degradation has been substantially studied *in vitro*, and data illustrate the pertinence of electron-acceptor consortium segregation approach.

Bromoxynil is aerobically degraded into 3,5-dibromo-4-hydroxybenzonitrile. Knight et al. (2003) estimated the microbial degradation under denitrifying, Fe(III)-reducing, sulfidogenic while 140 days are not enough to observe a significant degradation under denitrifying conditions, bromoxynil is degraded much faster (40d) under more reduced conditions.

Dicamba anaerobic metabolisation produces 3,6-dichlorosalicylate that is in turn transformed under methanogenic conditions by O-demethylation in 6-chlorosalycilate. Unfortunately, those dicamba metabolites accumulate and, as for bromoxynil, their degradation seems to be drastically less under denitrifying conditions. However, in sediment, 3- and 5-chlorosalycilate are degraded under denitrifying conditions, revealing that the position of the chlorine substituent is key for chlorosalicylates degradation in anoxic environments (Milligan & Häggblom, 2001). In sediment, methanogenic consortia are able to degrade all the halogenated dicamba metabolites, but under sulphate- and Fe(III)- reducing conditions micro-organisms are not able to do so, except for the final dechlorined metabolite (salycilate). Milligan and Häggblom (2001) highlight that sediment ability is due to specific consortia: the same experiment on agricultural soil did not show any dehalogenation.

Not all sediment floras are the same, though. For atrazine for example, Crawford et al. (1998) use the presence of indigenous organisms capable of anaerobic atrazine biodegradation as explanation for the observation that inoculation of a denitrifying bacterium cleaving the triazine ring enhance the anaerobic mineralization of atrazine compared to a non inoculated sediment. However, Devault et al. (2009, 2010), comparing *in situ* early diagenesis and herbicide vertical profile did not find any evidence of atrazine degradation under the considered environmental conditions.

Briefly, notwithstanding consortia abilities, biodegradation is theoretically proportional to energy yield. However, as explained above, microbial degradation of persistent pesticides tends to increase with decreasing redox potential when dehalogenation processes involving anaerobic enzymes are involved: thus biotransformation of organohalide contaminants *in situ* will vary as a function of the redox conditions within the electron acceptor profile (Häggblom et al., 2003b; Kuhn et al., 1990; Colberg, 1991).

Biodegradation efficiency depends on:

- Presence of organic carbon. Enrichment of sediment induces an early diagenesis increase (Crawford et al., 1998, Walker et al., 2001).
- Local flora. Consequently, each sediment consortia should be characterised to estimate the degradation efficiency. Inoculating is an alternative to give ability to the local biota (Goux et al., 2000).
- The enzymatic set of each electron-acceptor segregated consortia. By the way, some consortia could be the less effective for a chemical family and being the most effective for another.
- Molecule characteristics as conformation influences the degradation efficiency (Milligan & Häggblom, 2001).

Even if early diagenesis is underestimated process, *in situ* relating metabolisation could lead to a metabolic deadlock and metabolite accumulation (harmlessness to proven). Moreover, degradation by consortia involved in early diagenesis could be effective for molecules considered as non-degradable as 6-chlorosalycilate -but metabolic pathways are mostly unknown due the lack of early diagenesis consideration.

-pH and redox modification that could modify the sorption of molecules: even if chemicals are not degraded, their sorption could be modified directly (Morillo et al., 1997, for glyphosate and Hermosin & Cornejo, 1993, for 2-4D) or indirectly -due to desorption of desorbing promoters like Fe(II) (Klupinski & Chin, 2003) or Cu(II) for glyphosate (Morillo et al., 1997). Desorption could facilitate biodegradation too (Walker et al., 2001) and, therefore has a major effect on herbicide fate (Daniels et al, 2000).

In the field, early diagenesis is influenced by electron acceptor abundance. Nitrogen fertilisers are widely used in the same agricultural areas where herbicides are spread. Denitrifying activity is so reported as a major anaerobic condition in fresh water sediments (Camargo & Alonso, 2006), which may be beneficial for atrazine removal (Crawford et al. 1998). Inversely, since sulphate reducers competitively scavenge reducing equivalents, reductive dehalogenation is often inhibited in the presence of sulphate; as this has been observed in practice in methanogenic systems (Townsend et al., 1997; Kuhn, 1990). However, microbial degradation process of organohalogens under sulphate-reduction conditions has also been reported, as for example in the case of 2-4 dichlorophenoxyacetic acid (2,4 D) (Boyle et al., 1999). Such occurrences depend on the autochthonous microbial populations, and the degree to which these populations leave space and opportunities for incoming foreign microbes to handle contaminants (Goux et al., 2000). Inversely, herbicide presence could inhibit flora as reported by Isolda & Hayasaka (1991) for simazine or Crawford et al., (1998) for atrazine. However, those observations could be propitious for a feed-back effect (Laursen & Carlton, 1999): herbicide presence could inhibit consortia activities.

The previous observations however should not give rise to premature optimism. For instance, early diagenesis consortia had no observable effect on herbicide degradation in the sediments of an intensive agricultural area where s-triazines, substituted ureas and chloroacetanilides where widely spread (Devault et al., 2009). Considering sedimentation and erosion events, coring needs an optimal knowledge on the sampling place. Interpolating half-life of herbicides in soils and herbicide profiles could inform on the degradation length but important sedimentation rate induces a rapid evolution of lain matters from aerobic condition to methanogenesis ones. By the way, if degradation is not obvious, it inform about the longer remanence of studied herbicides in sediment than in soils. Moreover, it illustrates the threat of enhanced erosion: under the methanogenic condition reigns abiotic degradation and desorption (Arildskov et al., 2001). If degradation is observed in the lab for some consortia, field studies are still too scarce to ensure representativeness and the kinetics of *in-situ* degradation remains to establish.

5. Perspectives

Scientific literature about the *in situ* biodegradation of herbicides in sediments is still limited but the increase of their release in the environment should highlight sediment as biodegrading matrix. Coring in order to extrapolate the ancient fluxes, as it was used for the nowadays banned persistent organic pollutants (POPs), is becoming obsolete considering the emergence of new molecules with short half-life, especially herbicides. Instead of being considered as an inert matrix of POPs settling and archiving, sediment role into biodegradation of emergent pollutants should be increasingly taken into account. Herbicide degradation in sediment with depth could be considered as a screening of metabolisation efficiency of segregated consortia. Generalisation of interpretation of vertical profile or in-lab metabolisation testing could allow to find consortia able to degrade herbicides and to use them as inocula for the bioremediation of contaminated areas or to determine the fate of chemicals in lake, lentic and marine area, where contaminated sediment management becomes of increasing concern.

6. References

Adrian, L., Szewzyk, U., Wecke, J. & Gorisch, H. (2000). Bacterial dehalorespiration with chlorinated benzenes. *Nature* Vol. 408, 580-583.

Alavanja, M.C.R. (2009). Introduction: Pesticides use and exposure extensive worldwide. *Reviews on Environmental Health* Vol. 24, No. 4, pp. 303-309

Aller, R.C. (1977). The influence of macrobenthos on the chemical diagenesis of marine sediments. Ph.D. thesis. Yale University.

Aller, R.C., (1980). Quantifying solute distributions in the bioturbated zone of marine-sediments by defining an average micro-environment. *Geochimica et Cosmochimica Acta*, Vol. 44, 1955-1965.

Aller, R.C. (1982). The effects of macrobenthos on chemical properties of marine sediment and overlying water. In: Animal-Sediment relations. Eds P.L. McCall and M.J.S. Tevest, Plenum, N.Y., 53-102.

Aller R.C. (1990) Bioturbation and manganese cycling in hemipelagic sediments. *Philosophic Transactions of Royal Society of London*. A, Vol. 331, 51-68.

Aller, R.C. (1994). Bioturbation and remineralization of sedimentary organic matter: effects of redox oscillation. *Chemical Geology*, Vol. 114, 331-345.

Aller R.C., Hall P.O.J., Rude P.D., Aller J.Y. (1998) Biogeochemical heterogeneity and suboxic diagenesis in hemipelagic sediments of the Panama Basin. *Deep-Sea Research* part 1, 45, 133-165.

Anschutz, P., Zhong, S., Sundby, B., Mucci, A., Gobeil, C. (1998). Burial efficiency of phosphorus and the geochemistry of iron in continental margin sediments. *Limnology and Oceanography*, Vol. 43, No 1, 53-64.

Anschutz, P., Sundby, B., Lefrançois, L., Luther III, G.W., Mucci, A. (2000). Interactions between metal oxides and species of nitrogen and iodine in bioturbated marine sediments *Geochimica et Cosmochimica Acta*, Vol. 64, No 16, 2751-2763.

Anschutz P., Dedieu K., Desmazes F. Chaillou G. (2005) Solid speciation, oxidation state, and reactivity of manganese in marine sediments. *Chemical Geology*, Vol. 281, 265-279.

Arildskov, N.P., Pedersen, P.G. & Albrechtsen, H.-J., (2001). Fate of the herbicides 2,4,5-T, atrazine, and DNOC in a shallow, anaerobic aquifer investigated by in situ passive diffusive emitters and laboratory batch experiments, *Ground Water*, Vol. 39, No. 6, 819-830

Banta, G.T. & Andersen O. (2003). Bioturbation and the fate of sediment pollutants. *Vie Milieu*, Vol. 53, 233-248.

Bartha, R. (1971). Metabolism of the herbicide pronamide in soil. *Journal of Agricultural And Food Chemistry*, Vol. 19, No. 2, 385-387.

Berner R.A. (1971). Principles of Chemical Sedimentology. Mc Graw-Hill Book Co. 240p.

Berner, R.A. (1980). Early Diagenesis: A Theoretical Approach. Princeton University Press. 260pp.

Beurskens, J.E.M., Dekker, C.G.C., Jonkhoff, J. & Pompstra, L. (1993) Microbial dechlorination of hexachlorobenzene in a sedimentation area of the Rhine river. *Biogeochemistry* Vol. 19, 61-81.

Beurskens, J.E.M., Dekker, C.G.C., van den Heuvel, H., Swart, M., de Wolf, J. & Dolfing, J. (1994). Dechlorination of chlorinated benzenes by an anaerobic microbial consortium that selectively mediates the thermodynamic most favorable reactions. *Environmental Science & Technology*, Vol. 28, 701-706.

Boudreau, B.P. (1986a). Mathematics of tracer mixing in sediments: I. Spatially-dependent, diffusive mixing. *American Journal of Sciences*, Vol. 286, 161-198.

Boudreau, B.P. (1986b). Mathematics of tracer mixing in sediments: II. Nonlocal mixing and biological conveyor-belt phenomena. *American Journal of Sciences*, Vol. 286, 199-238.

Boudreau, B.P. (1996). A method-of-lines code for carbon and nutrient diagenesis in aquatic sediments. Comp. Geosc., 22, 479-496.

Boudreau, B.P. (2004). What controls the mixed-layer depth in deep-sea sediments? The importance of POC flux - COMMENT. *Limnology and Oceanography*, Vol. 49,620-622.

Boyle, A.W., Knight, V.K., Häggblom, M.M. & Young, L.Y. (1999). Transformation of 2,4-dichlorophenoxyacetic acid in four different marine and estuarine sediments: Effects of sulphate, hydrogen and acetate on dehalogenation and side-chain cleavage. *FEMS Microbiology Ecology*, Vol. 29, No1, 105-113.

Brinkhurst, R.O., Chua, K.E. & Kaushik, N.K. (1972). Interspecific interactions and selective feeding by tubificid oligochaetes. *Limnology and Oceanography* Vol. 17, 122-133.

Camargo, J.A. & Alonso, Á. (2006). Ecological and toxicological effects of inorganic nitrogen pollution in aquatic ecosystems: A global assessment. *Environment International*. Vol. 32, No. 6 831-849.

Canavan, R.W., Slomp, C.P., Jourabchi, P., Van Cappelen, P., Laverman, A.M. & Van Den Berg, G.A. (2006). Organic matter mineralization in sediment of a coastal freshwater lake and response to salinization. *Geochimica Cosmochimica Acta*, 90, 2836-2855.

Canfield, D.E., Raiswell, R., Bottrell, S. (1992). The reactivity of sedimentary iron minerals toward sulfide. *American Journal of Science*, Vol. 292, No. 9, 659-683.

Canfield, D.E., Thamdrup, B., Hansen, J.W., (1993). The anaerobic degradation of organic matter in Danish coastal sediments: Iron reduction, manganese reduction and sulfate reduction. *Geochimica Cosmochimica Acta* 57 (16), 3867-3883.

Chaillou, G., Anschutz, P., Dubrulle, C., Lecroart, P. (2007). Transient states in diagenesis following the deposition of a gravity layer: Dynamics of O2, Mn, Fe and N-species in experimental units. *Aquatic Geochemistry*, Vol. 13, No 2, 157-172.

Ciarelli, S., Van Straalen, N.M., Klap, V.A. & Van Wezel, A.P. (1999). Effects of sediment bioturbation by the estuarine amphipod *Corophium volutator* on fluoranthene resuspension and transfer into the mussel (*Mytilis edulis*). *Environmental. Toxicology and Chemistry*, Vol. 18, 318-328.

Ciutat, A., Anschutz P., Gerino, M., & Boudou, A. (2005). Effects of bioturbation on cadmium transfer and distribution into freshwater sediments. *Environmental Toxicology and Chemistry*, Vol. 24, 1048–1058.

Ciutat, A., Weber, O., Gerino, M. & Boudou, A. (2006). Stratigraphic effects of tubificids in freshwater sediments: a kinetic study based on X-ray images and grain-size analysis. *Acta Oecologica*, Vol. 30, 228-237.

Colberg, 1991

Connor, S.E., Thomas, I. (2003). Sediments as archives of industrialisation: evidence of atmospheric pollution in coastal wetlands of southern Sydney, Australia. *Water, Air and Soil Pollution*, Vol. 149, 189-210.

Covaci, A., Gheorghe, A., Voorspoels, S., Maervoet, J., Steen Redeker, E., Blust, R. & Schepens, P. (2005). Polybrominated diphenyl ethers, polychlorinated biphenyls and organochlorine pesticides in sediment cores from the Western Scheldt river (Belgium): analytical aspects and depths profiles. *Environmental International*, Vol. 31, 367-375.

Crawford, J.J., Sims, G.K., Mulvaney, R.L. Radosevich, M. (1998). Biodegradation of atrazine under denitrifying conditions. *Applied Microbiology and Biotechnology* Vol. 49, No. 5, 1998, 618-623

Daniels, W.M., House, W.A., Rae, J.E. & Parker, A. 2000. The distribution of micro-organic contaminants in river bed-sediment cores. *Science of the Total Environment*, Vol. 253, No. 1-3, 15, 81-92.

Darwin, C. (1881). The formation of vegetable mould through the action of worms with observation of their habits. John Murray.

Davis, R.B. (1974). Stratigraphic effects of tubificids in profundal lake sediments. *Limnogy and Oceanography*, Vol. 19, 466-488.

De Deckere, E. M. G. T, Tolhurst, T. J. & De Brouwer, J. F. C. (2001). Destabilization of Cohesive Intertidal Sediments by Infauna. *Estuarine and Coastal Shelf Sciences*, Vol. 53, 665-669.

Deflandre, B., Mucci, A., Gagné, J.-P., Guignard, C., Sundby, B.jorn (2002). Early diagenetic processes in coastal marine sediments disturbed by a catastrophic sedimentation event. *Geochimica et Cosmochimica Acta*, Vol. 66, No 14, 2547-2558.

De Lange, G.J. (1986). Early diagenetic reactions in interbedded pelagic and turbiditic sediments in the Nares Abyssal Plain (western North Atlantic): Consequences for the composition of sediment and interstitial water. *Geochimica et Cosmochimica Acta* , Vol. 50, No 12, 2543-2561.

Delmotte, S. (2007). Role of bioturbation in the biogeochemical functioning of the sediment-water interface : Modelling the diversity of biological transports and the effects on the early diagenesis within the sediments of a freshwater reservoir. Ph.D. Dissertation, Toulouse University, France, 2007.

Delmotte, S., Meysman, F.J.R, Ciutat, A., Sauvage, S., Boudou, A. & Gerino, M. (2007). Cadmium transport in sediments by tubificid bioturbation: an assessment of model complexity. *Geochimica Cosmochimica Acta* , Vol. 71, 844-862.

Delmotte, S. Gerino, M., Thebault, J-M. & Meysman, F.J.R. (2008). Modeling effects of patchiness and biological variability on transport rates within bioturbated sediments. *Journal of Marine Research*, Vol. 66, 191-218.

Devault, D.A., Delmotte, S., Merlina, G., Lim, P., Gerino & M., Pinelli, E. (2009). Influence of *in situ* biological activity on the vertical profile of pre-emergence herbicides in sediment. *Journal of Environmental Monitoring*, Vol. 11, No 6, 1206-1215.

Devault, D.A., Ith, C., Merlina, G., Lim, P. & Pinelli, E. (2010). Study of a vertical profile of pre-emergence herbicide contamination in middle Garonne sediments. *International Journal of Environmental Analytical Chemistry*, Vol. 90, No. 3-6, 311-320.

de Wit C.A., Alaee M., Muir D.C.G. 2006. Levels and trends of brominated flame retardants in the Arctic. *Chemosphere* 64, pp. 209-233.

Dolfing, J. & Tiedje, J.M. (1986). Hydrogen cycling in a three-tiered food web growing on the methanogenic conversion of 3-chlorobenzoate. FEMS *Microbiology Ecology*, Vol.38, 293-298.

Dolfing, J. (1990). Reductive dechlorination of 3-chlorobenzoate is coupled to ATP production and growth in an anaerobic bacterium, strain DCB-1. *Archives of Microbiology*, Vol. 153, 264-266.

Dolfing, J. & Harrison , B. K. (1992). Gibbs free energy of formation of halogenated aromatic compounds and their potential role as electron acceptors in anaerobic environments. *Environmental Science & Technology*, Vol. 26, 2213-2218.

Dolfing, J. (2000). Energetics of anaerobic degradation pathways of chlorinated aliphatic compounds. *Microbial Ecology*, Vol. 40, 2-7.

Dolfing, J. (2003). Thermodynamic considerations for dehalogenation. In: Dehalogenation: Microbial Processes and Environmental Applications (M.M. Häggblom and I.D. Bossert eds.), 89-114, Kluwer Academic Publishers, Boston, MA, USA,.

Dolfing, J., van Eekert, M., Seech, A., Vogan, J. & Mueller, J. (2008). *In situ* chemical reduction technologies: Significance of low Eh reactions. *Soil & Sediment Contamination* Vol. 17, 63-74.

Eganhouse, R.P., Pontolillo, J. & Leiker, T.J. (2000). Diagenetic fate of organic contaminants on the Palos Verdes Shelf, California. *Marine Chemistry* , Vol. 70, No. 4, 289-315.

Feng, H., Cochran K., Lwiza H., Brownawell, B.J. & Hirschberg, D.J. (1998). Distribution of heavy metal and PCB contaminants in the sediments of an urban estuary: the Hudson river. *Marine Environmental Research*, Vol. 45, No. 1, 68-88.

Fisher, J.D. (1974). Fate of herbicide-derived chloroanilines in soil.. *Journal of Agricultural and Food Chemistry*, Vol. 22, No. 4, 606-608.

Froelich, P.N., Klinkhammer, G.P., Bender, M.L., Luedtke, N.A., Heath, G.R., Cullen, D., Dauphin, P., Hammond, D., Hartman, B. & Maynard, V. (1979). Early oxidation of organic-matter in pelagic sediments of the Eastern Equatorial Atlantic- suboxic diagenesis. *Geochimica et Cosmochimica Acta*, 43, 1075-1090.

Fung, J.M., Weisenstein, B.P., Mack, E.E., Vidumsky, J.E., Ei T.A. & Zinder, S.H. (2009). Reductive dechlorination of dichlorobenzenes and momochlorobenzene to benzene in microcosms. *Environmental Science and Technology*, Vol. 43, 2302-2307.

Gerino, M. (1990). The effects of bioturbation on particle redistribution in Mediterranean coastal sediment. Preliminary result. *Hydrobiologia*, Vol. 207, 251-258

Gerino, M., Aller, R.C., Lee, C., Cochran, J.K., Aller, J.Y, Green, M.A. & Hirschberg, D. (1998). Comparison of different tracers and methods used to quantify bioturbation during a spring bloom: 234 Thorium, luminophores and Chlorophyll a. *Estuarine and Costal Shelf Sciences*, Vol. 15, 1483-1496.

Gibson, S.A., Suflita, J.M. (1990). Anaerobic biodegradation of 2,4,5-trichlorophenoxyacetic acid in samples from a methanogenic aquifer: Stimulation by short-chain organic acids and alcohols. *Applied and Environmental Microbiology*, Vol. 56, No 6, 1825-1832.

Gilbert, F., Hulth, S. & Aller, R.C. (2003a). The influence of macrofaunal burrow spacing and diffusive scaling on sedimentary nitrification and denitrification: An experimental and model approach. *Journal of Marine Research*, Vol. 61, 101-125.

Gilbert, F., Hulth, S., Stroemberg, N., Ringdahl, K. & Poggiale J.C. (2003b) 2-D optical quantification of particle reworking activities in marine surface sediments. *Journal of Experimental Marine Biology and Ecology*, Vol. 285/286, 251-263.

Goldberg, E. D. & Koide, M. (1962). Geochronological studies of deep-sea sediments by the Io/Th method. *Geochimica et Cosmochimica Acta*, Vol. 26, 417-450.

Goux, S.J., Ibanez, M., Van Hoorick, M. Debongnie, P., Agathos, S.N., Pussemier, L. 2000. Biodegradation of atrazine in sand sediments and in a sand-filter. *Applied Microbiology and Biotechnology*, Vol. 54, No. 4, 589-596.

Gray, J.S. (1974). Animal sediment relationship. *Oceanographic and Marine Biology Revue*, Vol. 12, 223-261.

Grimalt, J.O., van Drooge, B.L., Ribes A., Vilanova, R.M., Fernandez, P. & Appleby, P. (2004). Persistent organochlorine compounds in soil and sediments of European high altitude mountain lakes. *Chemosphere*, Vol. 54, 1549-1561.

Grover, R., Wolt, J.D., Cessna, A.J. & Schiefer, H.B. (1997). Environmental fate of trifluralin. *Reviews of Environmental Contamination and Toxicology*, Vol. 153, 1-64.

Guinasso, N.L. & Schink, D.R. (1975). Quantitative estimates of biological mixing rates in abyssal sediments. *Journal of Geophysical Research*, Vol. 80, 3032-3043.

Häggblom, M.M. and Bossert, I.D. (2003). Halogenated compounds – A global perspective. In: Dehalogenation: Microbial Processes and Environmental Applications (M.M. Häggblom and I.D. Bossert eds.). Kluwer Academic Publishers, Boston, MA, USA, pp. 3-29.

Häggblom, M.M., Ahn, Y.-B., Fennell, D.E., Kerkhof, L.J. & Rhee, S.-K. (2003b). Anaerobic Dehalogenation of Organohalide Contaminants in the Marine Environment. *Advances in Applied Microbiology*, Vol. 53, 61-84.

Harrad, S.J., Sewart, A.P., Alcock, R., Boumphrey, R., Burnett, V., Duatrte-Davidson, R., Halsall, C., Sanders, G., Waterhouse, K., Wild, S.R. &, Jones K.C. (1994).

Polychlorinated biphenyls (PCB) in the British environment: sinks, sources and temporal trends. *Environmental Pollution*, Vol. 85, 131-146.

Heritage, A.D. & Mac Rae, I.C. (1977). Identification of intermediates formed during the degradation of hexachlorocyclohexanes by *Clostridium sphenoides*. *Applied and Environmental Microbiology*, Vol. 33, 1295-1297.

Hermosin, M.C. & Cornejo, J. 1993. Binding mechanism of 2,4-dichlorophenoxyacetic acid by organo-clays. *Journal of Environmental Quality*, Vol. 22, No. 2, 325-331

Hoehler, T.M., Alperin, M.J., Albert, D.B. & Martens, C.S. (2001). Apparent minimum free energy requirements for methanogenic *archaea* and sulphate-reducing bacteria in an anoxic marine sediment. *FEMS Microbiology Ecology*, Vol 38, 33-41.

Hong, S.H., Yim U.H., Shim, W.J., Oh, J.R. & Lee, I.S. (2003). Horizontal and vertical distribution of PCBs and chlorinated pesticides in sediments from Masan Bay, Korea. *Marine Pollution Bulletin*, Vol. 46, 244-253.

Hulth, S., Aller, R.C., Gilbert, F. (1999). Coupled anoxic nitrification/manganese reduction in marine sediments. *Geochimica et Cosmochimica Acta*. Vol. 63, No 1, 49-66.

Hyacinthe, C., Anschutz, P., Carbonel, P., Jouanneau, J.-M., Jorissen, F.J. (2001). Early diagenetic processes in the muddy sediments of the Bay of Biscay. Marine Geology, Vol. 177, No 1-2, 111-128.

Isolda, A. & Hayasaka, S.S. 1991. Effect of herbicide residues on microbial processes in pond sediment. *Archives of Environmental Contamination and Toxicology*. 20, 1, 81-86

Jablonski, P.E. & Ferry, J.G. (1992). Reductive dechlorination of trichloroethylene by the CO-reduced CO dehydrogenase enzyme complex from Methanosarcina thermophila. *FEMS Microbioloy Letters*, Vol. 96, 55-60.

Jablonski, P.E., Phaesant, D.J. & Ferry, J.G. (1996). Conversion of Kepone by *Methanosarcina thermophila*. *FEMS Microbioloy Letters*, Vol. 139, 169-173.

Jayachandran, G., Görisch, H. & Adrian, L. (2003). Dehalorespiration with hexachlorobenzene and pentachlorobenzene by *Dehalococcoides* sp. strain CBDB1. *Archives of Microbiology*, Vol. 180, 411-416.

Jorgensen 1982

Klupinski, T.P. & Chin, Y.-P. (2003). Abiotic degradation of trifluralin by Fe(II): kinetics and transformation pathways. *Environmental Science & Technology*, Vol. 37, 1311-1318.

Knight, V.K., Berman, M.H. & Häggblom, M.M. 2003. Biotransformation of 3,5-dibromo-4-hydroxybenzonitrile under denitrifying, Fe(III)-reducing, sulfidogenic, and methanogenic conditions. *Environmental Toxicology and Chemistry*, Vol. 22, No. 3, 540-544.

Koistinen, J., Stenman, O., Haahti, H., Suonperä, M. & Paasivirta, J. (1997). Polychlorinated diphenyl ethers, dibenzo-p-dioxins, dibenzofurans and biphenyls in seals and sediment from the Gulf of Finland. *Chemosphere*, Vol. 35, No. 6, 1249-1269.

Kristensen, E. & Kostka, J.E. (2005). Macrofaunal burrows and Irrigation in marine sediment: microbiological and biogeochemical interactions. In Macro - and Microorganisms in Marine Sediments, Eds Kristensen E., Haese R.E., and Kostka J. E., *American geophysical Union*, p. 125-157.

Krutz, J.L., Shaner, D.L., Weaver, M.A., Webb, R.M.T., Zablotowicz, R.M., Reddy, K.N., Huang, Y. & Thomson, S.J. (2010). Agronomic and environmental implications of enhanced s-triazine degradation. *Pest Management Science*, Vol. 66, No. 5, 461-481.

Lal, R., Pandey, G., Sharma, P., Kumari, K., Malhotra, S., Pandey, R., Raina, V., Kohler, H.-P., Holliger, C., Jackson, C. & Oakeshott, J.G. (2010). Biochemistry and microbial degradation of hexachlorocyclohexane and prospects for bioremediation. *Microbiology and Molecular Biology Reviews*, Vol. 74, 58-80.

Laursen, A.E., Carlton, R.G. (1999). Responses to atrazine of respiration, nitrification, and denitrification in stream sediments measured with oxygen and nitrate microelectrodes. *FEMS Microbiology Ecology*. Vol. 29, No. 3, 229-240.

Leuchs, V.H. (1986). The ventilation activity of *Chironomus larvae* (Diptera) from shallow and deep lakes and the resulting water circulation in correlation to temperature and oxygen conditions, *Archives of Hydrobiology*, Vol. 108, 281-299.

Lorah, M.M., Olsen, L.D. (1999). Natural attenuation of chlorinated volatile organic compounds in a freshwater tidal wetland: Field evidence of anaerobic biodegradation. *Water Resources Research*, Vol. 35, No 12, 3811-3827.

Löffler, F.A., Tiedje, J.M. & Sanford, R.A. (1999). Fraction of electrons consumed in electron acceptor reduction and hydrogen thresholds as indicators of halorespiratory physiology. *Applied and Environmental Microbiology*, Vol. 64, 4049-4056.

Luther III G.W., Sundby B., Lewis B.L., Brendel P.J., Silverberg N., 1997. Interactions of manganese with the nitrogene cycle: Alternative pathways to dinitrogen. *Geochimica Cosmochimica Acta*, Vol. 61, 4043-4052.

Macdonald, W, Harner, T. & Fyfe, J. (2005). Recent climate change in the Arctic and its impact on contaminant pathways and interpretation of temporal trend data. *Science of the Total Environment*, Vol. 342, 5-86.

Maphosa, F., de Vos, W.M. & Smidt, H. (2010). Exploiting the ecogenomics toolbox for environmental diagnostics of organohalides-respiring bacteria. *Trends in Biotechnology*, Vol. 28, 308-316.

Maire, O., Lecroart, P., Meysman, F.J.R., Rosenberg, R., Duchene, J.C. & Gremare, A. (2008). Methods of sediment reworking assessment in bioturbation research: a review. *Aquatic Biology*, Vol. 2, 219-238.

Matisoff, G. & Wang, X. (1998). Solute transport in sediments by freshwater infaunal bioirrigators. *Limnology and Oceanography*, Vol. 43, 1487-1499.

McCall, P.L. & Fisher, J.N. (1980). Effects of tubificids oligochaetes on physical and chemical properties of Lake Erie sediments. In *Aquatic Oligochaete Biology*. Edited by Brinkhurst R.O. and Cook D.G., pp253-318, Plenum, New York, 1980.

Meadows, P.S. & Tait, J. (1989). Modification of sediment permeability and shear strength by two burrowing invertebrates. *Marine Biology*, Vol. 101, 75-82.

Mermillod-Blondin, F., Gaudet, J.P., Gerino, M., Desrosiers, G & Des Chatelliers, M.C. (2003). Influence of macroinvertebrates on physico-chemical and microbial processes in hyporheic sediments. *Hydrology Proceedings*, Vol. 17, 779-794.

Meysman, F.J.R. (2001). Modelling the influence of ecological interactions on reactive transport processes in sediments. Ph.D. Dissertation, Ghent University, 2001.

Meysman, F.J.R., Boudreau, B.P. & Middelburg, J.J. (2003). Relations between local, nonlocal, discrete and continuous models of bioturbation. *Journal of Marine Research*, Vol. 61, 391-410.

Meysman, F.J.R., Middelburg, J.J. & Heip, C.H.R. (2006a). Bioturbation: a fresh look at Darwin's last idea. *Trends in Ecology and Evolution*, Vol. 21, No. 12, 688-695.

Meysman, F.J.R., Galaktionov, O.S., Gribsholt, B. & Middleburg, J.J. (2006b). Bio irrigation in permeable sediments: Advective pore-water transport induced by burrow ventilation. *Limnology and Oceanography*, Vol. 51, 142-156.

Meysman, F.J.R., Galaktionov, O.S., Cook, P.L.M., Janssen, F., Huettel, M., Middelburg, J.J. (2007). Quantifying biologically and physically induced flow and tracer dynamics in permeable sediments. *Biogeosciences*, Vol. 4, No. 4, 627-646.

Milligan, P.W. & Häggblom, MM. (2001). Anaerobic degradation and dehalogenation of chlorosalicylates and salicylate under four reducing conditions. *Biodegradation*, Vol. 12, 159-167.

Mohn, W.W. & Tiedje, J.M. (1992). Microbial reductive dehalogenation. *Microbiology Revue*, Vol. 56, 482-507.

Morillo, E., Undabeytia, T. & Maqueda, C. (1997). Adsorption of glyphosate on the clay mineral montmorillonite: Effect of Cu(II) in solution and adsorbed on the mineral. *Environmental Science and Technology*, Vol. 31, No. 12, 3588-3592.

Mortensen, S.K. & Jacobsen, C.S. (2004). Influence of frozen storage on herbicide degradation capacity in surface and subsurface sandy soils. *Environmental Science and Technology*, Vol.38, No. 24, 6625-6632.

Mortimer, R.J.G., Krom, M.D., Harris, S.J., Hayes, P.J., Davies, I.M., Davison, W., Zhang, H. (2002). Evidence for suboxic nitrification in recent marine sediments. *Marine Ecology Progress Series*, Vol. 236, 31-35.

Murray J.W., Codispoti L.A., Friederich G.E. (1995) Oxidation-reduction environments: The suboxic zone of the Black Sea. In Aquatic chemistry: Interfacial and Interspecies Processes (ed. C.P. Huang et al.) Vol. 244, 157-176. American Chemical Society.

Nylund, K, Asplund, L., Jansson, B. & Jonnson, P. (1992). Analysis of some polyhalogenated organic pollutants in sediment and sewage sludge. *Chemosphere*, Vol. 24, No. 12, 1721-1730.

Petty, D.G., Getsinger, K.D., Woodburn, K.B. (2003). A review of the aquatic environmental fate of triclopyr and its major metabolites. *Journal of Aquatic Plant Management*, Vol. 41, No 2, 69-75.

Pignatello, J.J. (1992). Dark and photoassisted Fe3+-catalyzed degradation of chlorophenoxy herbicides by hydrogen peroxide. *Environmental Science & Technology*, Vol. 26, No. 5, 944-951.

Rasmussen, A.D., Banta, G.T. & Andersen, O. (1998). Effects of bioturbation by the lugworm *Arenicola marina* on cadmium uptake and distribution in sandy sediments. *Marine Ecology Progress Series*, Vol.164, 179-188.

Rawn, D.F.K., Lockhart, W.L., Wilkinson, P., Savoie, D.A., Rosenberg, G.B. & Muir, D.C.G. (2001). Historical contamination of Yukon Lake sediment by PCBs and organochlorine pesticides: influence of local sources and watershed characteristics. *The Science of the Total Environment*, Vol. 280, 17-37.

Rhoads, D.C. (1974). Organism-sediment relations on the muddy sea floor. *Oceanographic and Marine Biology Annual Revue*, Vol. 12, 263-300.

Rice D.L. (1986). Early diagenesis in bioadvective sediments: Relationships between the diagenesis of beryllium-7, sediment reworking rates, and the abundance of conveyor-belt deposit-feeders. *Journal of Marine Research*, Vol. 44, 149-184.

Roberts, T.R., Dyson, J.S., Lane & M.C.G. (2002). Deactivation of the biological activity of paraquat in the soil environment: A review of long-term environmental fate. *Journal of Agricultural and Food Chemistry*, Vol. 50, No. 13, 3623-3631.

Robbins J.A., Kresoski J.R. & Mozley S.C. (1977). Radioactivity in sediments of the Greats Lakes: post depositional redistribution by deposit-feeding organisms. *Earth and Planetary Science Letters*, Vol. 36, 325-333.

Schoonen, M.A.A., Barnes, H.L. (1991). Reactions forming pyrite and marcasite from solution: II. Via FeS precursors below 100°C *Geochimica et Cosmochimica Acta*, Vol. 55, No 6, 1495-1514.

Schulz H.D., Dahmke A, Schnizel U., Wallmann K., Zabel M. (1994) Early diagenetic processes, fluxes, and reaction rates in sediments of the South Atlantic. *Geochimica et Cosmochimica Acta*, Vol. 58, 2041-2060.

Secco T., Pellizzato F., Sfriso A., Pavoni B. (2005). The changing of contamination in the Lagoon of Venice. Part 1: organic pollutants. *Chemosphere* , Vol. 58, 279-290.

Self, R.F. and Jumars, P.A. (1978). New resource axes for deposit feeders. *Journal of Marine Research*, Vol. 36, 627-641.

Soetaert, K., Herman, P.M.J., Middleburg, J.J. (1996). A model of early diagenetic processes from the shelf to abyssal depths. *Geochimica Cosmochimica Acta*, Vol. 60, 1019-1040.

Stief, P. & De Beer, D. (2002). Bioturbation effects of *Chironomus riparius* on the benthic N-cycle as measured using microsensors and microbiological assays. *Aquatic Microbiology and Ecology*, Vol. 27, 175-185.

Stumm W. and Morgan J. J. (1996). Aquatic chemistry. 3rd edition, J. Wiley and Sons, New York, 780p.

Sun, M.-Y., Aller, R.C., Lee, C. & Wakeham, S.G. (2002). Effects of oxygen and redox oscillation on degradation of cell-associated lipids in surperficial marine sediments. *Geochimica Cosmochimica Acta*, Vol. 66, 2003-2012.

Sundby, B. (1977). Manganese-rich particulate matter in a coastal marine environment. *Nature*, Vol. 270, No 5636, 417-419.

Thullner, M., Van Cappellen, P. & Regnier, P. (2005). Modeling the impact of microbial activity on redox dynamics in porous media. *Geochimica Cosmochimica Acta*, Vol. 21, 5005-5019.

Townsend, G.T., Ramanand, K. & Suflita, JM. (1997). Reductive dehalogenation and mineralization of chlorobenzoate in the presence of sulphate by microorganisms from a methanogenic aquifer. *Applied and Environmental Microbiology*, Vol. 63, 2785-2791.

Van Cappellen, P. & Wang, Y. (1995). Metal cycling in surface sediments: modeling the interplay of transport and reaction. In: Allen, H.E. (Ed.), Metal Contaminated Sediments. Ann Arbor Press, Chelsea, MI, 21-64.

van Eekert, M.H.A., Stams, A.J.M., Field, J.A. & Schraa, G. (1999). Gratuitous dechlorination of chloroethanes by methanogenic granular sludge. *Applied Microbiology and Biotechnology*, Vol. 51: 46-52.

Walker, A., Jurado-Exposito, M., Bending, G.D. & Smith, V.J.R. (2001). Spatial variability in the degradation rate of isoproturon in soil. *Environmental Pollution*. Vol 111, No. 3, 407-415.

Wheatcroft, R.A., Jumars, P.A., Smith, C.R. & Nowell, A.R.M. (1990). A mechanistic view of the particulate bio diffusion coefficient: step lengths, rest period and transport directions. *Journal of Marine Research*, Vol. 48, 177-207.

Wijsman, J.W.M., Herman, P.M.J., Middelburg, J.J. & Soetaert, K. (2002). A model for early diagenetic processes in sediments of the continental shelf of the Black Sea. *Estuarine and Coastal Shelf Science*, Vol. 54, 403-421.

Wood, L.W. (1975). Role of oligochaetes in the circulation of water and solutes across the mud water interface. *Verhandlungen der Internationalen Vereinigung fur Theoretische und Angewandte Limnologie*, Vol. 19, 1530-1533.

Part 5

Toxicological Aspects

23

Risk Estimate of Water Contamination and Occurrence of Pesticides in the South of Brazil

Sergiane Souza Caldas[1], Renato Zanella[2] and Ednei Gilberto Primel[1]
[1]*Federal University of Rio Grande (FURG)*
[2]*Federal University of Santa Maria (UFSM)*
Brazil

1. Introduction

Worldwide pesticide usage has increased dramatically during the last two decades, coinciding with changes in farming practices and the increasingly intensive agriculture (Konstantinou et al., 2006). Developing countries represent approximately 30% of the global pesticide consumer market. Within this group, Brazil is the most important individual market, with an estimated 6.5 billion dollars per year spent on pesticides, making the country the second largest consumer of pesticides in the world (Peres et al., 2006). Rio Grande do Sul State, located in the south of Brazil, is responsible for 10.4% of pesticide consumption in Brazil (SINDAG, 2006).

The higher the agricultural productivity, the more contamination with pesticides. Pesticides are a group of artificially synthesized substances which have been used in agriculture to control pests and to increase production. These substances protect the agricultural crops, but overuse and incorrect use can pose risks to human health and the environment (Caldas et al., 2009). The impact of pesticide uses on human health is especially evident in developing countries, where pesticides are widely used in agriculture. According to the 2005 annual report of the Brazilian Toxico-pharmacological Information System (Sinitox), 8,000 notifications of cases of human pesticide poisoning had been received annually. The Brazilian Ministry of Health estimates that for each reported case there are at least 50 non-reported ones, it increases the annual pesticide poisoning rates to 400,000 cases per year, with 2,000 deaths (Peres et al., 2006). This human contamination can occur in different ways, and one of them is through contaminated waters.

Brazil has approximately 10% of the world's fresh water, with an annual average river flow of 182,600 $m^3 s^{-1}$. When all areas of the Amazon region are considered, including the territory in neighboring countries, the outflow is around 272,000 $m^3 s^{-1}$. Although Brazil is considered rich in water due to its extensive landmass, there is a high disparity in distribution of this resource among its regions. Moreover, the rapid process of urban development and increases in agricultural productivity in the last few decades have been affecting the quality and availability of water (Marques et al., 2007).

The surface water is considered the water stored or flowing on the earth's surface: natural bodies of water such as rivers, lakes, and wetlands, as well as constructed (artificial) water reservoirs such as canals, man-made lakes, and drainage ditches. The quantity and quality

of surface water is important for many activities: consumption, recreation, transportation, waste assimilation, agricultural production, and industrial use (Whitford et al., 2010).
Groundwater, in the broadest sense, refers to all subsurface water which is also important for many activities, and represents an important source of drinking water supply in many countries.
Pesticide residues reach the aquatic environment through direct run-off, leaching, careless disposal of empty containers, equipment washing, etc; and the water contamination depends on the physico-chemical characteristics of the compound and the interaction of the pesticides with soil, surface water and groundwater.
Moreover, the pesticide fate is controlled by numerous simultaneous biological, physical, and chemical reactions (Whitford et al., 2010). When a pesticide is applied to a field, certain reactions follow. Foliar-applied pesticides stick to leaves, where they are absorbed. But rainfall inevitably washes some of the chemical off the leaf surface onto the soil below; and some may be transformed by sunlight. For instance, pesticides that are tightly sorbed to soil particles have decreased mobility and are less likely to contaminate groundwater. Pesticides strongly sorbed to soil particles may travel primarily with eroded soil and enter surface water, while weakly sorbed pesticides that are more water soluble may be released into soil water solution and enter surface water as runoff (Whitford et al., 2010). Chemicals, which are sufficiently resistant to degradation and are adequately soluble to be transported in water, may reach the water bodies in significant amounts (Konstantinou et al., 2006).
The surface and groundwater contamination by pesticides has been documented in many papers around the world (Carabias-Martínez et al., 2002; Cerejeira et al., 2003; Hernández-Romero et al., 2004; D'Archivio et al., 2007; Baugros et al., 2008; Barrek et al., 2009). In some areas in Brazil, where pesticides are widely used, mainly due to agricultural activities, some pesticides have been found. In Table 1, results from researches that analyzed sample waters in Rio Grande do Sul (RS) State, in the south of Brazil, are shown in a summarized way. All studies found most of the investigated compounds. Normally, the compounds analyzed are used in rice and tobacco cultures. The herbicide clomazone was investigated in all studies and found in all, showing the tendency of this compound to be present in water samples, and the high use of this herbicide in the region. Only one study with pesticides determination in groundwater was found; it emphasizes the need for more studies in this compartment.
The amount of pesticide that will reach the environment compartments varies with environmental changes, and the estimation is difficult. But, because of the lack of analytical facilities and the high cost of analyses, one strategy to identify, from among the long list of chemicals applied to the area, those pesticides which are likely to be present in surface and groundwater is to focus in their physico-chemical properties, using some model environmental partitioning indexes such as Groundwater Ubiquity Score (GUS), the screening criteria recommended by the U.S. Environmental Protection Agency (US-EPA) and GOSS (Cabrera et al., 2008).

2. Objective

The aim of this chapter is to present an evaluation of the contamination of surface, drinking and groundwater by herbicides, insecticides and fungicides used in different crop cultures in Rio Grande do Sul State, Brazil.

Water sample	Compounds	Culture related	Pesticides found (maximum concentration found - µg L^{-1})	Reference
surface	clomazone	rice	clomazone (1.72)	Zanella et al., 2002
surface	2,4-D, bentazone, clomazone, propanil, quinclorac	rice	2,4-D (>2.0) bentazone (>2.0) clomazone (>2.0) propanil (>2.0) quinclorac (>2.0)	Primel et al., 2005
surface	atrazine, clomazone, chlorpyrifos, flumetralin, imidacloprid, iprodione, simazine	tobacco	atrazine (0.63) clomazone (1.72) imidacloprid (2.18)	Bortoluzzi et al., 2006
surface and groundwater	atrazine, chlorpyrifos, flumetralin, imidacloprid, iprodione, simazine, clomazone	tobacco	atrazine (ground – 0.69 - surface – 0.82) clomazone(ground – 12.84 - surface – 15.69) chlorpyrifos (ground-0.14 – surface – 0.13) imidacloprid (ground –2.33 - surface – 2.59) simazine (ground – 0.81)	Bortoluzzi et al., 2007
surface	clomazone, propanil, quinclorac	rice	clomazone (8.85) propanil (12.9) quinclorac (6.6)	Marchesan et al., 2007
surface	beta cyfluthrin, carbofuran, clomazone, fipronil, quinclorac	rice	carbofuran (14.99) clomazone(6.51) fipronil (1.14) quinclorac (5.34)	Grützmacher et al., 2008
surface	carbofuran 3-hydroxy, carbofuran, clomazone, fipronil, imazapic, imazethapyr, penoxsulam, quinclorac, tebuconazole	rice	carbofuran 3-hydroxy (0.008) carbofuran (1.4) clomazone (0.064) fipronil (3.45) imazapic (0.014) imazethapyr (0.33) Penoxsulam (0.15) quinclorac (0.12) tebuconazole. (0.015)	Silva et al., 2009
surface	2,4-D, bentazone, carbofuran, clomazone, fipronil, imazethapyr, propanil, quinclorac	rice	2,4-D (3.4) bentazone (3.6) carbofuran (0.8) clomazone (3.4) fipronil (26.2) imazethapyr (1.2) propanil (5.4) quinclorac (4.1)	Marchesan et al., 2010

Table 1. Studies that analyzed pesticides in sample waters in RS state after 2002

This region has special characteristics because of its agricultural production. Pesticides of different classes are widely used and some of them, due to their physico-chemical properties, can reach the water systems. Firstly, a theoretical evaluation was made using the approaches suggested by US-EPA, the GUS index and the Goss method to estimate the contamination possibilities. Afterwards, a monitoring program was established for the surface and groundwater of the area to investigate the presence of pesticide residues.

3. Methods

3.1 Characterization of the physico-chemical properties

The knowledge of most relevant physico-chemical parameters is important to set up the analytical scheme and to understand the environmental behavior of the compounds. It helps in the selection of the analytes that can be present in the area under study, the sample extraction and the technique for the analyses. A review about the physico-chemical characteristics was carried out and all information was based on Barceló & Hennion (1997).

3.1.1 Water solubility

Water solubility is a fundamental, specific chemical property defined as the concentration of a chemical dissolved in water when that water is both in contact and in equilibrium with the pure chemical. Water solubility indicates the tendency for a pesticide to be removed from soil by runoff or irrigation water and to reach the surface water. It also indicates its tendency to precipitate on the surface soil.

3.1.2 Octanol-Water partition coefficient

This parameter is usually reported as a logarithm log K_{ow} or log P. It is defined as the ratio of the equilibrium concentrations of the two-phase system consisting of water and *n*-octanol. This parameter is characteristic of the liphophility of the molecule and gives an indication of the tendency the compound has to accumulate in biological membranes and living organisms. It is generally considered that substances with a log K_{ow} value higher than 3 can show accumulation. The polarity of a molecule is strongly related with K_{ow}. As a rough rule, non-polar analytes are characterized by log K_{ow} values above 4.5, whereas polar analytes have K_{ow} values below 1 or 1.5. Between these two values, compounds are classified as moderately polar. K_{ow} has proved valuable for prediction of mobility and persistence in soils, and of soil sorption. However, K_{ow} alone cannot be considered to be an indicator of soil affinity because it merely represents a partition between two well-defined nonmiscible phases, whereas the real soil sorption process involves other mechanisms such as partition, adsorption, ion-exchange, complexation and precipitation.

3.1.3 Acid-base ionization constants

Ionic pesticides behave differently from non-ionic pesticides. It is therefore important to know which pesticides are capable of ionization within the normal soil/water environmental pH range of 5-8. Since soils have a tendency to be negatively charged, anionic compounds may be potential leachers whereas cationic compounds will be strongly retained. This knowledge is also important for perform trace analysis, especially to extractions from water, because it is easier to extract a non-ionic compound than an ionic one. A simple pH adjustment can help greatly in extraction recovery, which requires a knowledge of the ionization constant values. The water solubility of ionic species is always much higher than that of their non-ionic form. The ionization constant is usually expressed as pKa. The higher the pKa value, the weaker is the acid and the tendency to be ionized.

3.1.4 Vapour pressure

The vapour pressure is defined as the partial pressure of a chemical, in the gas phase, in equilibrium, with the pure solid or liquid chemical. Vapour pressures are very temperature-dependent. This parameter governs the distribution between liquid and gas phase or between solid and gas phase.

3.1.5 Henry's Law constant

Henry's Law constant, denoted H or K_H is a partition coefficient defined as the ratio of a chemical concentration in air to its concentration in water in equilibrium. This parameter is important in several aspects. The tendency for the pesticides to volatilize from water solution into air is largely determined by their H values, a high value favouring volatilization.

3.1.6 Normalized soil sorption coefficient (K_{oc})

One of the most critical factors for assessing the potential mobility of most pesticides in the soil compartment is the distribution between the solid and liquid phases of soil. This portioning presents a difficult problem since the types of soil in the environment vary enormously. The first characterization is the measurement of the simple "sorption" coefficient, K_d, defined as the ratio of the concentration of the chemical adsorbed on soil to the concentration of pesticide in the soil solution. Therefore, taking into account the different soil organic matter or organic carbon content, the adsorption constant is normalized, and K_d values are expressed per unit of organic carbon as K_{oc}. They are referred to as "soil organic carbon sorption coefficients" and expressed in $cm^3\ g^{-1}$. The environmental relevance of this parameter is important to the leaching properties in groundwater. Pesticides with K_{oc} values below 50 are considered to be highly mobile; values of 150-500 signify moderately mobile, and above 2000, slightly mobile compounds.

3.1.7 Field half-life

The half-life ($t_{1/2}$) is defined as the time required for the pesticide to undergo dissipation or degradation to half of its initial concentration. However, there is not a single half-life for pesticides, and measurements depend strongly on the environmental conditions (soil, site, climate, soil microbial activity, etc).

3.1.8 Chemical structures of the studied pesticides

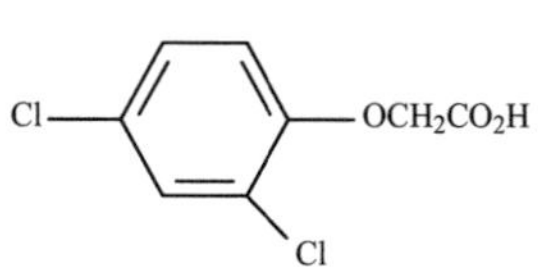

Name: 2,4-D
Mw: 221.0
CAS number: 94-75-7

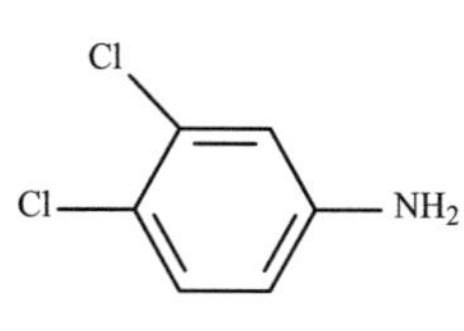

Name: 3,4-DCA
Mw: 162.0
CAS number:95-76-1

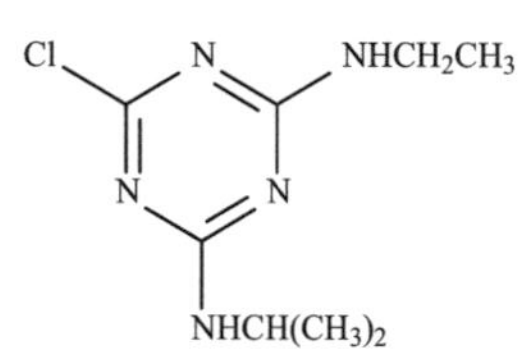

Name: atrazine
Mw: 215.7
CAS number: 1912-24-9

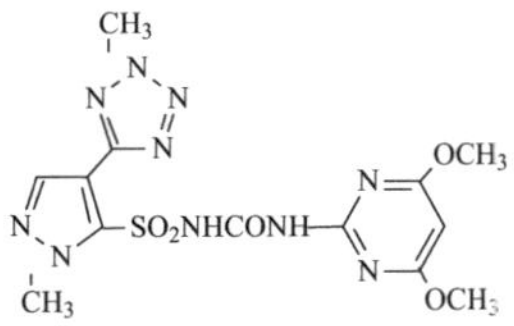

Name: azimsulfuron
Mw: 424.4
CAS number: 120162-55-2

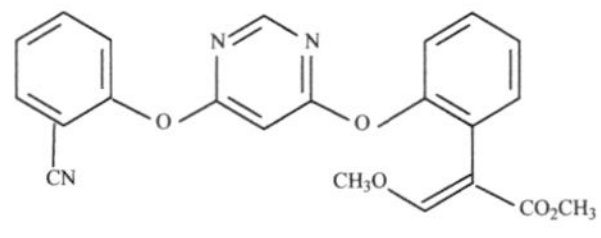

Name: azoxystrobin
Mw: 403.4
CAS number: 131860-33-8

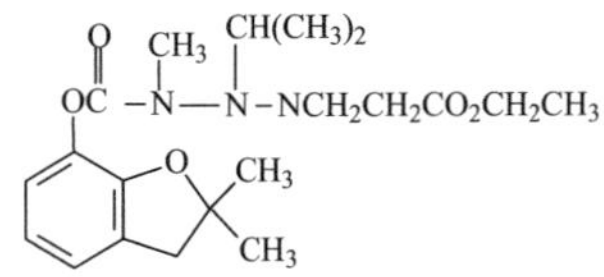

Name: benfuracarb
Mw: 410.5
CAS number: 82560-54-1

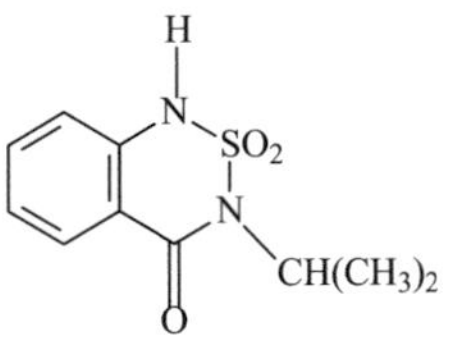

Name: bentazone
Mw: 240.3
CAS: 25057-89-0

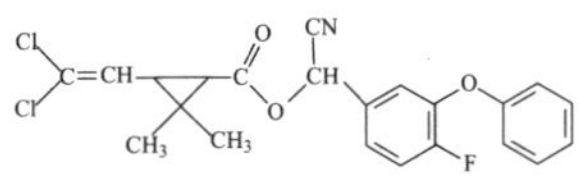

Name: beta-cyfluthrin
Mw: 434.3
CAS number: 68359-37-5

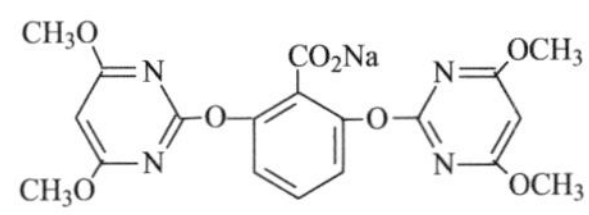

Name: bispyribac-sodium
Mw: 452.4
CAS number: 125401-92-5

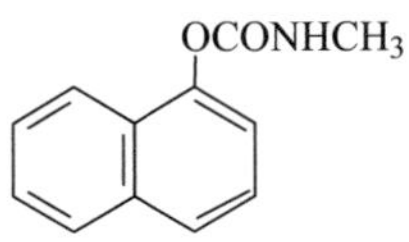

Name: carbaryl
Mw: 201.2
CAS number: 63-25-2

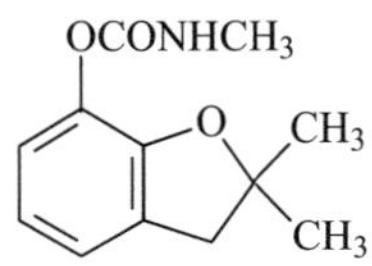

Name: carbofuran
Mw: 221.3
CAS number: 1563-66-2

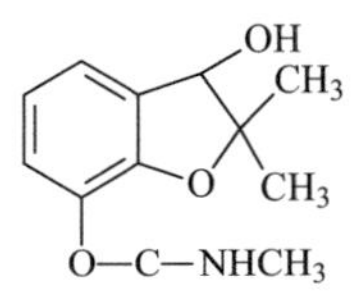

Name: carbofuran 3-hydroxy
Mw: 237.3
CAS number: 16655-82-6

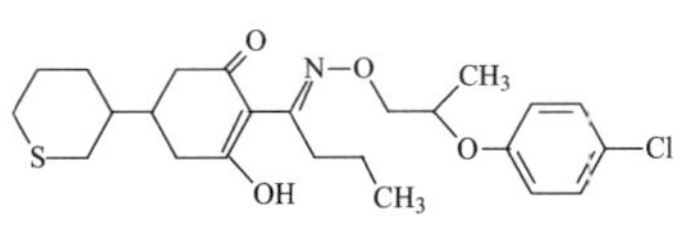

Name: clefoxydim
Mw: 466.0
CAS number: 139001-49-3

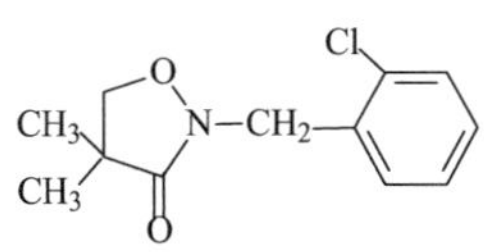

Name: clomazone
Mw: 239.7
CAS number: 81777-89-1

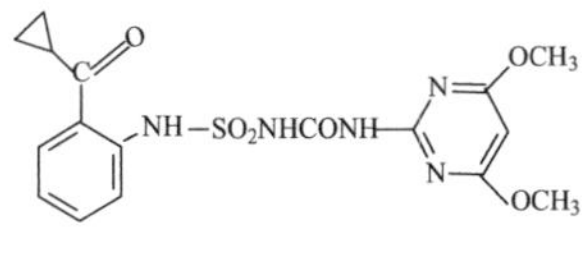

Name: cyclosulfamuron
Mw: 421.4
CAS number: 136849-15-5

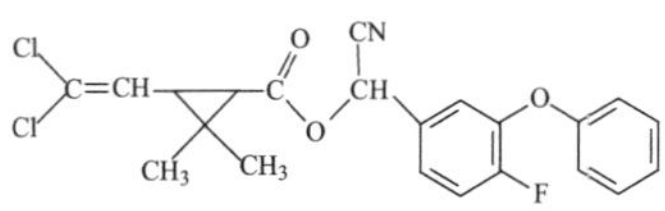

Name: cyfluthrin
Mw: 434.3
CAS number: 68359-37-5

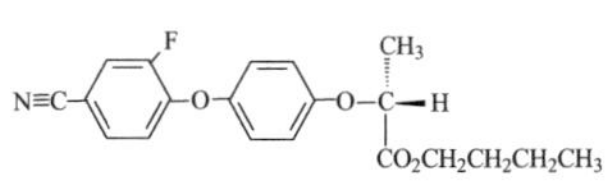

Name: cyhalofop-butyl
Mw: 357.4
CAS number: 122008-85-9

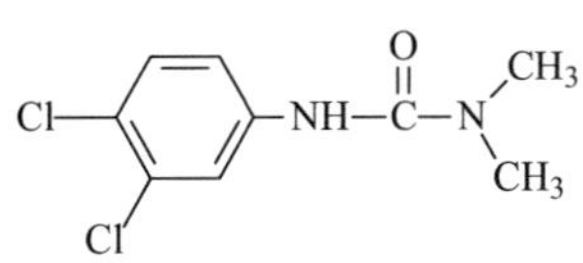

Name: diuron
Mw: 233.1
CAS number: 330-54-1

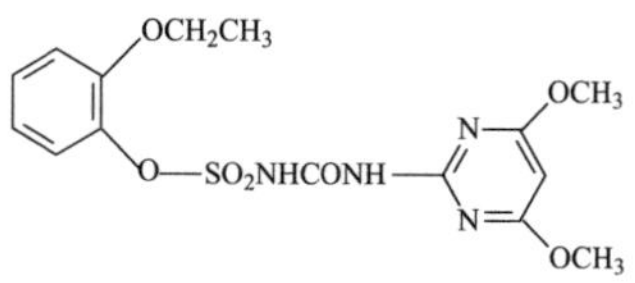

Name: ethoxysulfuron
Mw: 398.4
CAS number: 126801-58-9

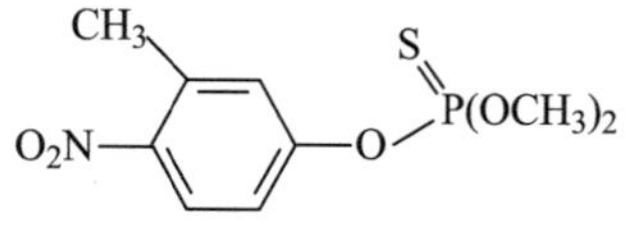

Name: fenitrothion
Mw: 277.2
CAS number: 122-14-5

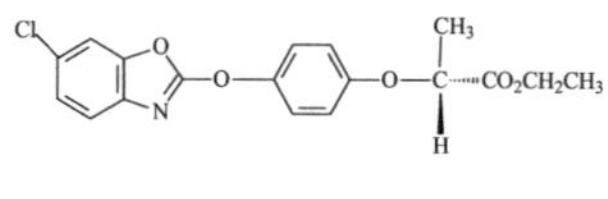

Name: fenoxaprop-p-ethyl
Mw: 361.8
CAS number: 71283-80-2

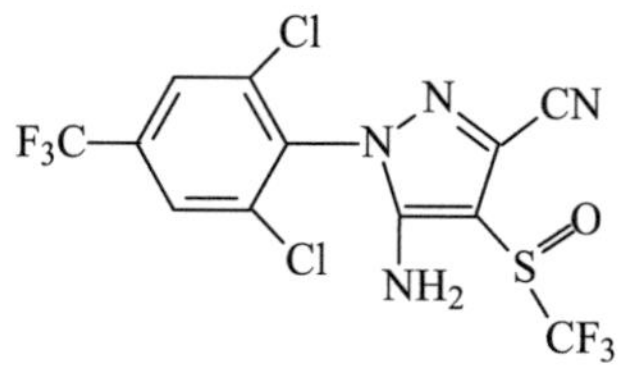

Name: fipronil
Mw: 437.2
CAS number: 120068-37-3

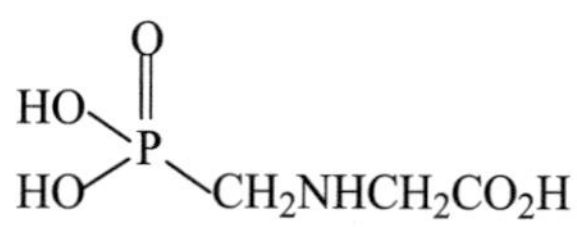

Name: glyphosate
Mw: 169.1
CAS number: 1071-83-6

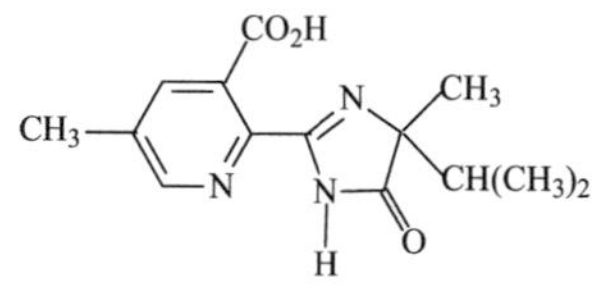

Name: imazapic
Mw: 275.3
CAS number: 104098-48-8

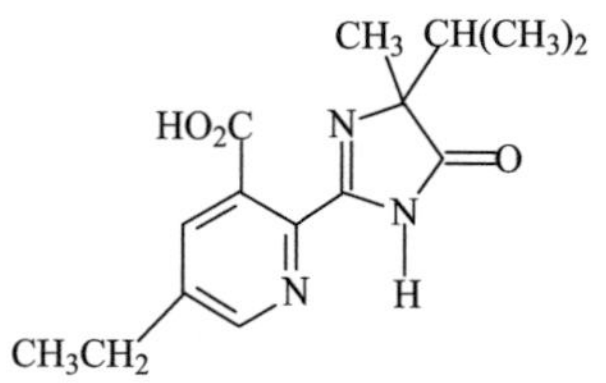

Name: imazethapyr
Mw: 289.3
CAS number: 81335-77-5

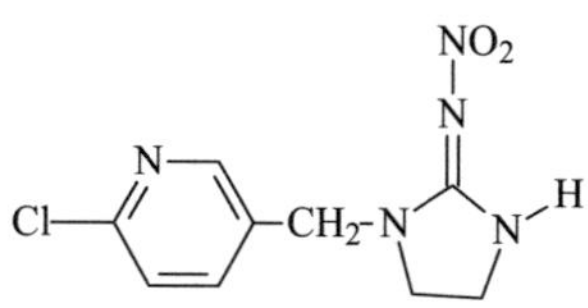

Name: imidacloprido
Mw: 255.7
CAS number: 138261-41-3

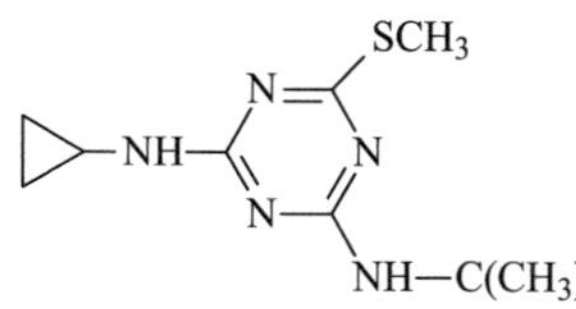

Name: Irgarol
Mw: 253.4
CAS number: 28159-98-0

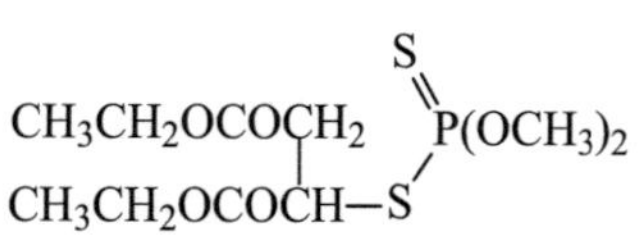

Name: malathion
Mw: 330.4
CAS number: 121-75-5

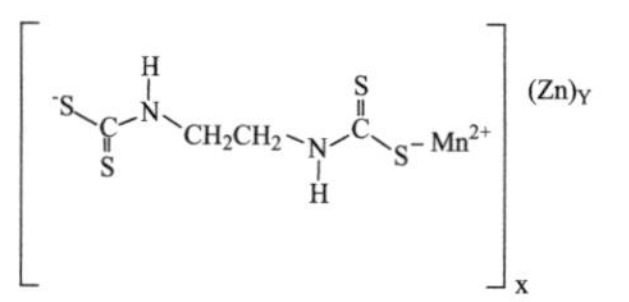

Name: mancozeb
Mw: 271.2
CAS number: 8018-01-7

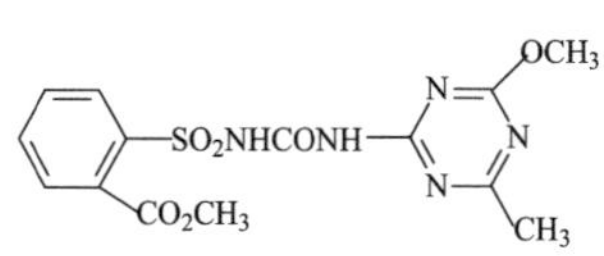

Name: metsulfuron-methyl
Mw: 381.4
CAS number: 74223-64-6

Name: molinate
Mw: 187.3
CAS number: 2212-67-1

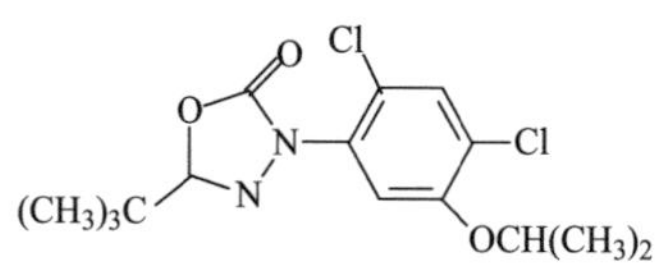

Name: oxadiazon
Mw: 345.2
CAS number: 19666-30-9

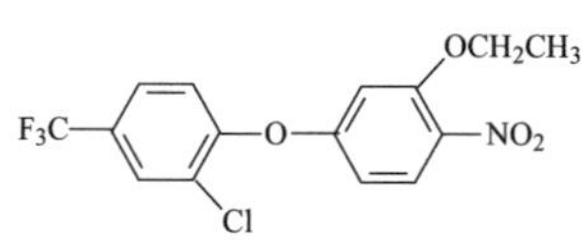

Name: oxyfluorfen
Mw: 361.7
CAS number: 42874-03-3

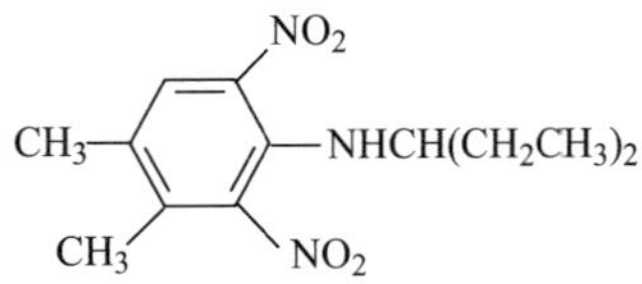

Name: pendimethalin
Mw: 281.3
CAS number: 40487-42-1

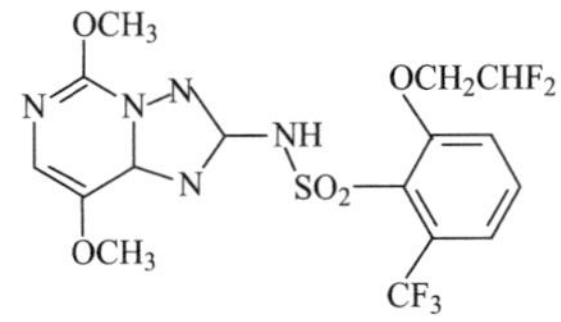

Name: penoxsulam
Mw: 483.4
CAS number: 219714-96-2

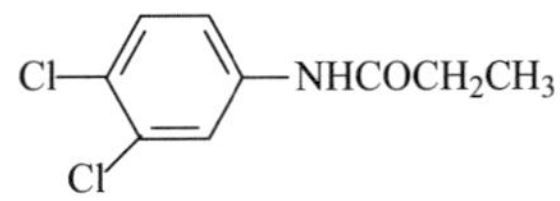

Name: propanil
Mw: 218.1
CAS number: 709-98-8

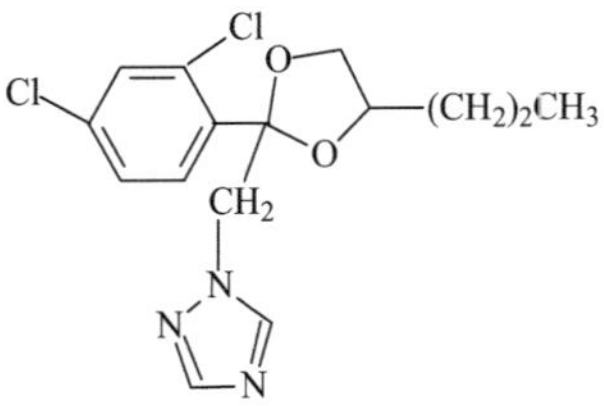

Name: propiconazole
Mw: 342.2
CAS number: 60207-90-1

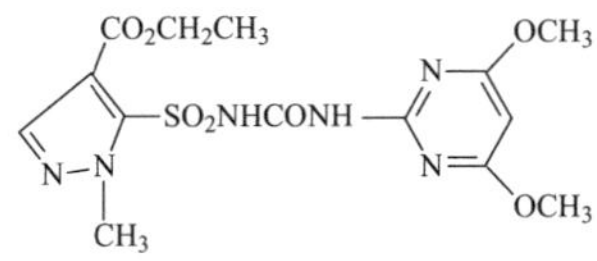

Name: pyrazosulfuron-ethyl
Mw: 414.4
CAS number: 93697-74-6

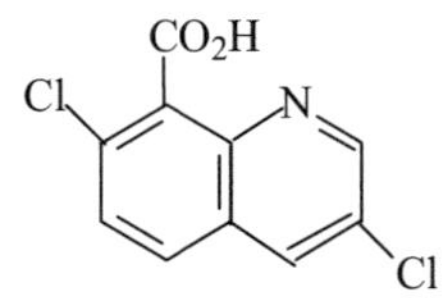

Name: quinclorac
Mw: 242.1
CAS number: 84087-01-4

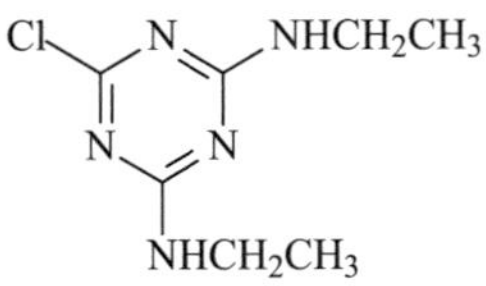

Name: simazine
Mw: 201.7
CAS number: 122-34-9

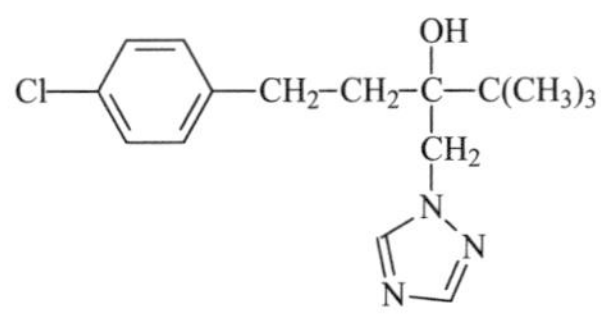

Name: tebuconazole
Mw: 307.8
CAS number: 107534-96-3

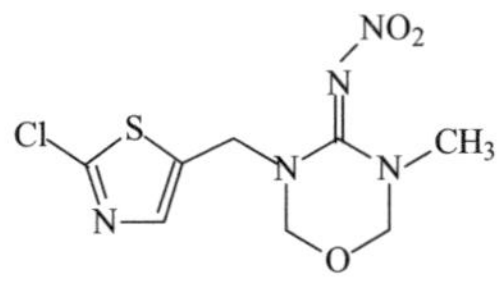

Name: thiamethoxam
Mw: 291.7
CAS number: 153719-23-4

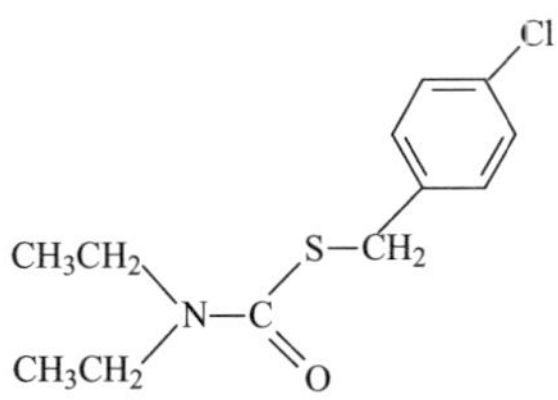

Name: thiobencarb
Mw: 257.8
CAS number: 28249-77-6

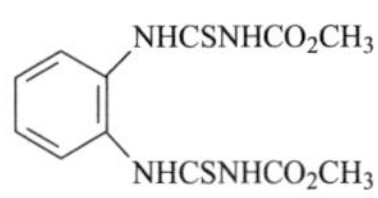

Name: thiophanate-methyl
Mw: 342.4
CAS number: 23564-05-8

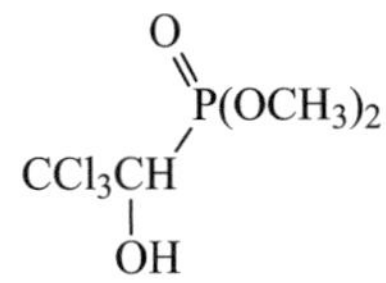

Name: trichlorphon
Mw: 257.4
CAS number: 52-68-6

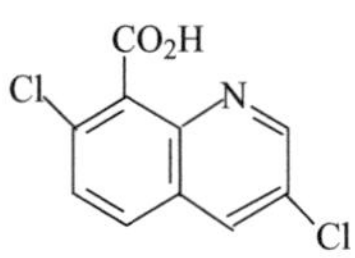

Name: triclopyr
Mw: 242.1
CAS number: 84087-01-4

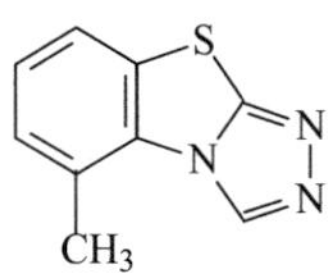

Name: tricyclazole
Mw: 189.2
CAS number: 41814-78-2

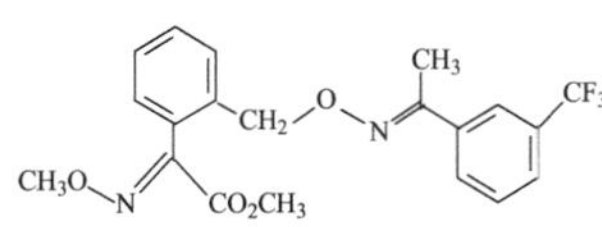

Name: trifloxystrobin
Mw: 342.2
CAS number: 60207-90-1

Fig. 1. Chemical structures, molecular weight (Mw) and CAS number of the studied compounds

3.2 Risk Assessment for groundwater

3.2.1 GUS

The leaching potential was determined by calculating the GUS which is based on the partition coefficient between soil organic carbon and water—sorption coefficient (K_{oc}) and the half-life ($t_{1/2}$) in soil for each pesticide compound.
The GUS index was calculated according to equation 1.

$$GUS = \log t_{1/2} \times (4 - \log K_{oc}) \quad (1)$$

where K_{oc} is the soil sorption coefficient (mL g^{-1}) and $t_{1/2}$ is the half-life in soil (days). GUS values lower than 1.8 and higher than 2.8 indicate, respectively, non-leacher and leacher pesticide compounds; for GUS values between 1.8 and 2.8 the pesticide is considered in a transition zone (Gustafson, 1989).

3.2.2 US-EPA

The US-EPA screening criteria was also used for predicting the leaching potential. It is based on the following: water solubility > 30 mg L^{-1}; K_{oc} < 300-500; Henry's Law constant - $K_H < 10^{-2}$ Pa m^3 mol^{-1}; speciation - negatively charged (either fully or partially) at room temperature pH (5-8); soil $t_{1/2}$ > 2-3 weeks; hydrolysis $t_{1/2}$ > 25 weeks; vulnerable field conditions (annual precipitation > 250 mm, aquifer not confined, porous soil).

3.3 Risk assessment for surface waters

3.3.1 GOSS

For the evaluation of which pesticides might contaminate the surface waters, a criterion proposed by Goss was used (Goss, 1992).
This criterion considers the half-life in soil ($t_{1/2}$), water solubility and the soil sorption coefficient (K_{oc}).
According to this criteria the compounds are divided in groups: High Sediment-Transport Runoff Potential (HSTRP); Low Sediment-Transport Runoff Potential (LSTRP); High Water-Phase-Transport Runoff Potential (HWTRP) and Low Water-Phase-Transport Runoff Potential (LWTRP). The values considered for each parameter are: HSTRP - $t_{1/2}$ soil ≥ 40 d and K_{oc} ≥ 1000 cm^3 g^{-1}, or $t_{1/2}$ soil ≥ 40 days, K_{oc} ≥ 500 cm^3 g^{-1} and water solubility ≤ 0.5 mg L^{-1}; LSTRP - $t_{1/2}$ soil < 1 d, or $t_{1/2}$ soil ≤ 40 d, K_{oc} ≤ 500 cm^3 g^{-1} and water solubility ≥ 0.5 mg L^{-1}, or $t_{1/2}$ soil ≤ 2 days and K_{oc} ≤ 500 cm^3 g^{-1} or $t_{1/2}$ soil ≤ 4 days, K_{oc} ≤ 900 cm^3 g^{-1} and water solubility ≥ 0.5 mg L^{-1} or $t_{1/2}$ soil ≤ 40 d K_{oc} ≤ 900 cm^3 g^{-1} and water solubility ≥ 2 mg L^{-1}; HWTRP - $t_{1/2}$ soil > 35 d, K_{oc} < 1000000 cm^3 g^{-1} and water solubility > 1 mg L^{-1}, or $t_{1/2}$ soil > 35 d, K_{oc} ≤ 700 cm^3 g^{-1} and water solubility between 10 and 100 mg L^{-1}; LWTRP - K_{oc} > 1000000 cm^3 g^{-1}, or $t_{1/2}$ soil ≤ 1 d and K_{oc} ≥ 1000 cm^3 g^{-1}, or $t_{1/2}$ soil < 35 d and water solubility < 0.5 mg L^{-1}. The compounds which do not meet any of these criteria are considered with medium potential, in other words, Medium Water-Phase-Transport Runoff Potential (MWTRP) or associated with sediment (MSTRP).

3.4 Reagents and chemicals

In this study, high purity standards of eighteen pesticides were selected: clomazone, tebuconazole, diuron, irgarol, atrazine, simazine, metsulfuron-methyl, quinclorac, 2,4-D, pyrazosulfuron-ethyl, bentazone, propanil, carbofuran and the two metabolites, 3,4-DCA and carbofuran 3-hydroxy , which derive from propanil and carbofuran, respectively, were purchased at Sigma Aldrich (São Paulo, Brazil). Imazethapyr, imazapic, fipronil, bispyribac-sodium and penoxsulam were purchased at Dr. Ehrenstorfer GmbH (Augsburg, Germany). Methanol and acetonitrile of chromatographic grade were supplied by Mallinckrodt (Phillipsburg, NJ, USA). Phosphoric acid (85%) and formic acid (98%) of analytical grade were purchased at Merck (Darmstadt, Germany). Ultra pure water was produced by a Direct-Q UV3® system (Millipore, Bedford, MA, USA). The SPE extraction tubes were Chromabond C18ec, 500 mg/3 mL (Macherey-Nagel, Düran, Germany).

3.5 Equipment

All analyses were performed on a Waters Alliance 2695 Separations Module HPLC, equipped with a quaternary pump, an automatic injector and a thermostatted column compartment, and detection was carried out on a Quattro micro API (triple quadrupole) mass spectrometer, equipped with a Z-spray electrospray (ESI) ionization source, from Micromass (Waters, Milford, MA, USA), according to previous studies (Caldas et al., 2010; Demoliner et al., 2010).

3.6 Sampling

Samples were collected directly in 1 L amber glass bottles in each sampling site. These bottles had been cleaned prior to sampling by rinsing them three times in the water to be sampled. The bottle was filled to the top with as little remaining air as possible, and sealed tightly. All samples were properly labeled with details of the source and sampling date, and stored at 4 °C until the solid-phase extraction, which was carried out on the same day of the sampling.

3.6.1 Groundwater

The studies of groundwater contamination with pesticides were carried out in Quitéria, a rural area located near Rio Grande, a city in the southeast of Brazil that has about 185,000 inhabitants. The region of Quitéria is an agricultural area where there are different kinds of production, mainly onion, fruits and vegetables in general. The intense agricultural production, and consequently the use of pesticides, can cause the contamination of groundwaters.

Samples from drinking water wells (Figure 2) with depths ranging from 2.5 up to 37 m were collected in ten sampling points from September 2007 to August 2008, and analyzed to determine the concentration of carbofuran, 2,4-D, clomazone and tebuconazole. Sampling was carried out once per month in each well.

3.6.2 Drinking and surface water

The sampling of the drinking and surface water was carried out monthly at CORSAN, the Water Treatment Station in Rio Grande, from January, 2008 to April, 2010. Two different samples were collected: a surface water sample, collected at the entrance of the water channel (the São Gonçalo Channel), and a drinking water sample, collected after the water treatment, in the output of the station (Figure 2).

3.7 Sample preparation

The samples were acidified at pH 3.0 with phosphoric acid. Afterwards, they were pre-concentrated and extracted by SPE tubes containing 500 mg of octadecylsilane

(Chromabond C18ec) with an average particle size of 45 µm. Cartridges were conditioned with 3 mL of methanol, 3 mL of ultrapure water and 3 mL of ultrapure water pH 3.0, acidified with phosphoric acid 1:1 (v/v). After the conditioning step, aliquots of 250 mL of water samples, acidified at pH 3.0 with phosphoric acid (to increase the pesticide retention) were loaded through the cartridges with a flow rate of 6 mL min^{-1}. Then, the analytes were eluted with 1 mL (2 x 500 µL) methanol, volume adjusted in 1 mL and injected into the chromatographic system. This extraction procedure had been previously developed and optimized in our laboratory (Caldas et al., 2009).

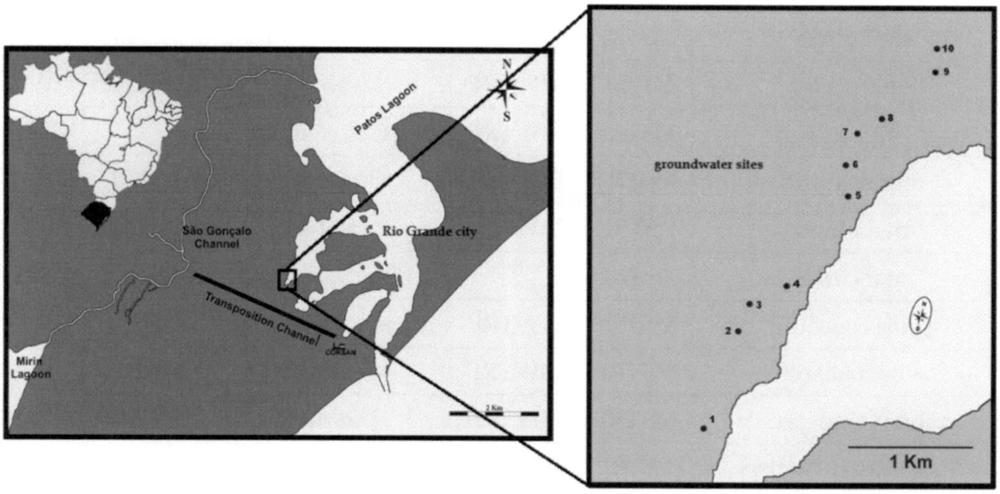

Fig. 2. Regions of surface and groundwater collection

4. Results

The pesticides evaluated by Goss, GUS and US-EPA criteria were chosen because of their high use in RS state, mainly in the rice cultivation (Câmara dos Agrotóxicos, 2008; SIA, 2003).

4.1 Risk assessment for surface water

The results for the risk assessment for surface water are shown in Table 2. According to the GOSS criteria, regarding the water-phase-transport runoff potential, the compounds fipronil, imidacloprid, propiconazole, atrazine, cyclosulfamuron, diuron, fenoxaprop-p-ethyl, imazethapyr, metsulfuron-methyl, molinate, oxadiazon, triclopyr, glyphosate and thiamethoxam show high potential. Clomazone, imazapic and thiobencarb besides presenting HWTRP showed LSTRP, which is a high indicative that these compounds have good chance to be found in surface waters.

Some compounds such as beta-cyfluthrin, cyfluthrin, thiophanate, cyhalofop-butyl and penoxsulam show LWTRP showing less tendency to be found in surface waters.

Some others due to their physico-chemical properties, presented intermediate probabilities, showing MSTRP and MWTRP. They are benfuracarb, tebuconazole, trifloxystrobin, bispyribac, oxyfluorfen, pendimethalin, quinclorac and simazine.

Among the compounds under study, only carbofuran showed HSTRP.

The contamination potential for the herbicide clefoxidim, mancozeb and tricyclazole was not evaluated because of the lack of data about these compounds.

Fungicides and Insecticides	GOSS classification		Herbicides	GOSS classification	
azoxystrobin	LSTRP	MWTRP	2,4-D	LSTRP	MWTRP
benfuracarb	MSTRP	MWTRP	atrazine	MSTRP	HWTRP
beta-cyfluthrin	MSTRP	LWTRP	azimsulfuron	LSTRP	MWTRP
carbaryl	LSTRP	MWTRP	bentazone	LSTRP	MWTRP
carbofuran	HSTRP	MWTRP	bispyribac-sodium	MSTRP	MWTRP
cyfluthrin	MSTRP	LWTRP	cyhalofop-butyl	LSTRP	LWTRP
fenitrothion	LSTRP	MWTRP	cyclosulfamuron	MSTRP	HWTRP
fipronil	MSTRP	HWTRP	clefoxydim	I	MWTRP
imidacloprid	MSTRP	HWTRP	clomazone	LSTRP	HWTRP
malathion	LSTRP	MWTRP	diuron	MSTRP	HWTRP
mancozeb	I	I	ethoxysulfuron	LSTRP	MWTRP
propiconazole	MSTRP	HWTRP	fenoxaprop-p-ethyl	MSTRP	HWTRP
tebuconazole	MSTRP	MWTRP	glyphosate	MSTRP	HWTRP
thiamethoxam	MSTRP	HWTRP	imazapic	LSTRP	HWTRP
thiophanate-methyl	LSTRP	LWTRP	imazethapyr	MSTRP	HWTRP
tricyclazole	I	I	metsulfuron-methyl	MSTRP	HWTRP
trichlorphon	LSTRP	MWTRP	molinate	MSTRP	HWTRP
trifloxystrobin	MSTRP	MWTRP	oxadiazon	MSTRP	HWTRP
			oxyfluorfen	MSTRP	MWTRP
			pendimethalin	MSTRP	MWTRP
			penoxsulam	MSTRP	LWTRP
			pyrazosulfuron-ethyl	LSTRP	MWTRP
			propanil	LSTRP	MWTRP
			quinclorac	MSTRP	MWTRP
			simazine	MSTRP	MWTRP
			thiobencarb	LSTRP	HWTRP
			triclopyr	MSTRP	HWTRP

Table 2. Goss classification for the compounds under study (HSTRP= High Sediment-Transport Runoff Potential; LSTRP= Low Sediment-Transport Runoff Potential; HWTRP= High Water-Phase-Transport Runoff Potential; LWTRP= Low Water-Phase-Transport Runoff Potential; MWTRP= Medium Water-Phase-Transport Runoff Potential; MSTRP= Medium Sediment-Transport Runoff Potential and I = non-conclusive)

4.2 Risk assessment of groundwaters

For the risk of contamination of the groundwater, according to GUS and US-EPA, the herbicides atrazine, bentazone, clomazone, imazethapyr, imazapic, metsulfuron-methyl, quinclorac, simazine and triclopyr, besides the fungicides and the insecticides carbofuran, imidacloprid, thiamethoxam, tricyclazole and trichlorphon are classified as potential contaminants of groundwaters (Table 3) by both methods. The compounds azoxystrobin, benfuracarb, beta-cyfluthrin, cyfluthrin, fenitrothion, malathion, mancozeb, thiophanate, trifloxystrobin, bispyribac-sodium, cyhalofop, cyclosulfamuron, fenoxaprop-p-ethyl and oxyfluorfen did not show any leaching tendency by both methods.
Some compounds show different classifications by the EPA-US and GUS methods, or due to the lack of physico-chemical parameters the prediction was not possible. These compounds were 2,4-D, azimsulfuron, diuron, ethoxysulfuron, glyphosate, oxadiazon, pendimethalin, penoxsulam, pyrazosulfuron, propanil, thiobencarb, carbaryl, fipronil, propiconazole and tebuconazole. These differences between the methods can occur because the methods consider different physico-chemical characteristics.

4.3 Surface Waters

4.3.1 Choice of the pesticides for the surface waters monitoring

For the surface water monitoring, eighteen pesticides and two metabolites were chosen. The choice was based on the pesticide usage in the region and on the possibility of contamination of the surface water by these compounds. The sources of contamination near the water source were also considered. The company that supplies water for the Rio Grande city takes water from the Sao Goncalo channel. This channel is an important connection between the Patoos Lagoon and the Mirim Lagoon, and in the margins there are many agricultural activities; such as horticultural, grains and rice irrigation, besides the industrial activities and the navigation. The compounds 2,4-D, bispyribac-sodium, propanil, bentazone, carbofuran, pyrazosulfuron-ethyl and quinclorac besides being used for the rice culture showed MWTRP. The compounds atrazine, clomazone, diuron, fipronil, imazethapyr, imazapic, metsulfuron-methyl showed HWTRP and penoxsulam showed LWTRP. Tebuconazole and simazine showed MWTRP. And, besides these compounds, two metabolites (3,4-DCA and carbofuran 3-hydroxy) and the biocide irgarol were also analyzed.

4.3.2 Monitoring studies of surface water

In Figure 3, the sum of the monthly results of the monitoring analysis of the surface water from the São Gonçalo Channel and of the drinking water in Rio Grande are presented.
Taking into account the 18 pesticides and 2 metabolites under analysis, concentrations above the LOQ were detectable, at least once, for carbofuran 3-hydroxy, 3,4-DCA, atrazine, clomazone, bentazone, bispyribac-sodium, carbofuran, clomazone, diuron, fipronil, imazethapyr, imazapic, irgarol, pyrazosulfuron-ethyl, propanil, quinclorac, simazine and tebuconazole. It is noticed that, except in February 2009, March and April 2010 the sum of concentrations of pesticides was always lower than 0.5 $\mu g\ L^{-1}$, the value established by the European legislation (EU Council Directives, 1998).
Some compounds were found in more than 60% of the collections. In surface water, they were atrazine, clomazone, diuron, imazapic, imazethapyr, irgarol and tebuconazole; and in drinking waters they were clomazone, diuron, imazapic, imazethapyr and tebuconazole.

The presence of diuron and irgarol in almost all the monitoring period indicates that in the region of the São Gonçalo Channel, concerning the contamination by pesticides, is influenced not only by the agricultural process but also by other sources. The occurrence of diuron can be due to the fact that this herbicide is largely used in other kinds of cultures such as lettuce, citrus fruits, and onion (Camara dos Agrotóxicos, 2008). Besides, diuron and irgarol are used as antifouling in paints for vessels (Buma et al., 2009; Avila et al., 2005). The association of these uses increases the contamination by diuron in waters, especially in portuary regions such as Rio Grande where there is intensive navigation. The presence of 3,4-DCA, the main metabolite of diuron and of propanil, although lower than LOQ, confirms the use of these compounds.

Whereas diuron is used in the culture of rice, the herbicide irgarol is not, but both are used as antifouling (Sapozhnikova et al., 2007), indicating that the contamination by these compounds has the same source, the vessels.The other compounds found in the samples are probably due to irrigated rice, because all compounds are recommended for this culture and this is the agricultural practice that dominates the southern region in RS state, besides the cultures of onion and tomato (Camara dos Agrotóxicos, 2008).

The herbicides imidazolinones analyzed in this study, imazethapyr and imazapic, were detected during all the sampling period, both in the surface water and in potable water. The highest concentrations found for imazapic were 0.34 $\mu g\ L^{-1}$ for drinking water and 0.2 $\mu g\ L^{-1}$ for surface waters. For imazethapyr the highest concentrations were 0.27 $\mu g\ L^{-1}$ for drinking water and 0.35 $\mu g\ L^{-1}$ for surface waters. The fact that they are frequently detected in the samples can be explained, for example, by the system of rice cultivation, which works with a variety of rice that is very resistant to the pesticides of the imidazolinones class; they are effective against red rice, the main pest in the irrigated rice plantation (Avila et al., 2005). Most of the time, the compounds were found in the same month probably indicating the use of a mixture, that was sold commercially and is composed by both compounds.

The fungicide tebuconazole and the herbicide atrazine were detected in the sampling in many months. Triazines are used in the world as pre and post-emerging selective herbicides to control weeds in many cultures, such as corn, wheat, sugar cane and barley.

The insecticide fipronil was detected in all samples, since it is recommended for several cultures (agriculture and silviculture). This compound was found in surface waters in RS, in other studies. In the study developed by Marchesan et al. (2010) it was found in concentration until 26.2 $\mu g\ L^{-1}$ and with high frequency of detection. The herbicide pyrazossulfuron-ethyl was also detected in some samples since this compound is recommended for rice cultivation.

The detection of the metabolite carbofuran 3-hydroxy, from carbosulfan and carbofuran, indicates that one of these compounds had been used on farms near the São Gonçalo Channel.

The herbicide clomazone was detected in concentrations lower than the LOQ in all samples, except in July, when it was detected in surface and potable waters. Marchesan et al. (2007) detected clomazone in river waters in RS, with high frequency. They comment the relation of the frequency with the rainfall regime and with the rice water management used in the fields.

4.4 Groundwater

4.4.1 Choice of the pesticides for the grondwater monitoring

The selection of the analytes included in this study was based on their extensive use as pesticides in agricultural areas. Moreover, these pesticides are some of the mostly used pesticides in the area under study; according to the US Environmental Protection Agency (US EPA), they have potential to reach the groundwater systems (Table 3).

Herbicides	US-EPA Criteria						GUS index
	Water solubility	K_{oc}	K_H	$t_{1/2}$ soil	$t_{1/2}$ water	Result	
2,4-D	A	A	A	N	N	PC	TL
atrazine	A	A	A	A	N	PC	PL
azimsulfuron	A	A	A	A	N	PC	TL
bentazone	A	A	A	A	A	PC	PL
bispyribac-sodium	A	N	N	N	A	NC	IL
cyhalofop-butyl	N	N	A	N	N	NC	IL
cyclosulfamuron	N	N		A	N	NC	IL
clefoxydim	N		N		N	NC	
clomazone	A	A	A	A	N	PC	PL
diuron	A	A	A	A		PC	TL
ethoxysulfuron	A	A	A	A	N	PC	TL
fenoxaprop-p-ethyl	N	N	A	N	N	NC	IL
glyphosate	A	N	A	A	N	PC	IL
imazapic	A	A		A	N	PC	PL
imazethapyr	A	A	A	A	N	PC	PL
metsulfuron-methyl	A	A	A	A	A	PC	PL
molinate	A	A	N	A	A	PC	PL
oxadiazon	N	N	A	A	N	NC	TL
oxyfluorfen	N	N	A	A	N	NC	IL
pendimethalin	N	N	N	A	N	NC	PL
penoxsulam	N	A	A	A	N	PC	TL
pyrazosulfuron-ethyl	N	A		A	N	I	TL
propanil	A	A	A	N	N	PC	IL
quinclorac	N	A	A	A		PC	PL
simazine	N	A	A	A	N	PC	PL
thiobencarb	A	N	A	A	A	PC	IL
triclopyr	N	A	A	A	N	PC	PL
fungicides and insecticides							
azoxystrobin	N	N	A	N	N	NC	IL
benfuracarb	N	N	A	N	N	NC	IL
beta-cyfluthrin	N	N	A	N	N	NC	IL
carbaryl	A	A	A	N	N	PC	TL
carbofuran	A	A	A	A	N	PC	PL
cyfluthrin	N	N	N	A	N	NC	IL
fenitrothion	N	N	A	N	N	NC	IL
fipronil	N	N	A	A	N	NC	PL
imidacloprid	A	A	A	A	N	PC	PL
malathion	A	A	N	N	N	NC	IL

Fungicides and Insecticides	US-EPA Criteria						GUS index
	Water solubility	K_{oc}	K_H	$t_{1/2}$ soil	$t_{1/2}$ water	Result	
mancozeb	N	N	A	N		NC	IL
propiconazole	A	N	A	A	N	PC	TL
tebuconazole	A	N	A	A		PC	TL
thiamethoxam	A	A	A	A	N	PC	PL
thiophanate-methyl	N	A	A	N	N	NC	IL
tricyclazole	A	A	A	A		PC	PL
trichlorphon	A	A	A	A		PC	PL
trifloxystrobin	A	N	A	N	N	NC	IL

Table 3. Risk assesment of groundwater contamination based on US-EPA criteria and GUS index (US-EPA, N= not meet the criteria; A= meet the criteria; I= non-conclusive; PC = potential contaminant NC = non-contaminant; blank - data not available; GUS IL = not leachable, TL = transition zone and PL= probably leachable)

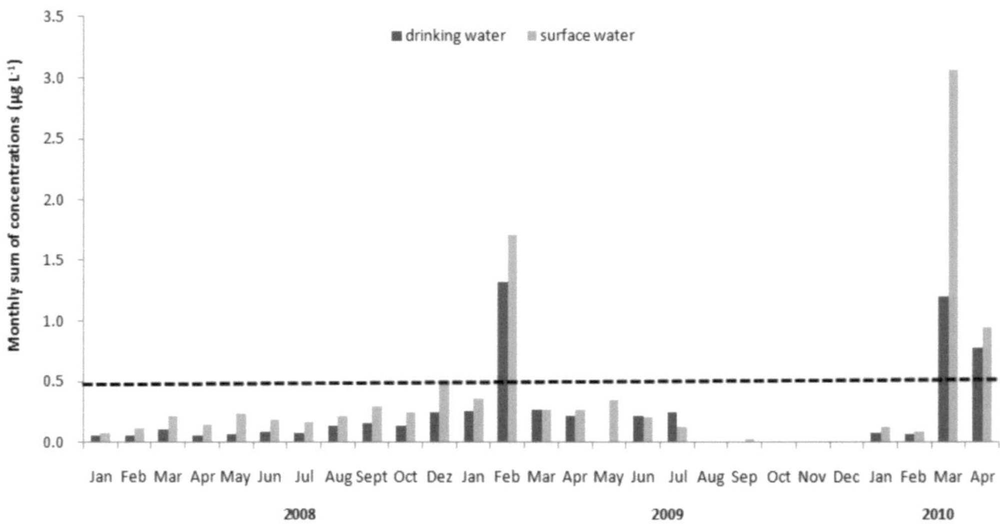

Fig. 3. Sum of pesticide concentrations determined from January 2008 to April 2010.

4.4.2 Pesticide determination in groundwater

Figure 4 show the pesticide levels detected at each location in various sampling campaigns, carried out from September 2007 until August 2008. Results are not surprising at all, since the agricultural practice in the area is intense and has been going on for years. Carbofuran, clomazone and tebuconazole were the compounds found. In general, significant differences were observed in the levels and the profile of the pesticides detected in distinct sampling periods.

The behavior of the samples was strongly different. 2,4-D was the only pesticide that was not detected, probably because it is less used in Quitéria than the others. It is mainly used in the cultures of rice, corn, soy and wheat (Camara dos Agrotóxicos, 2008), which are not raised in Quitéria. This compound has lower half life than others and, moreover, has the

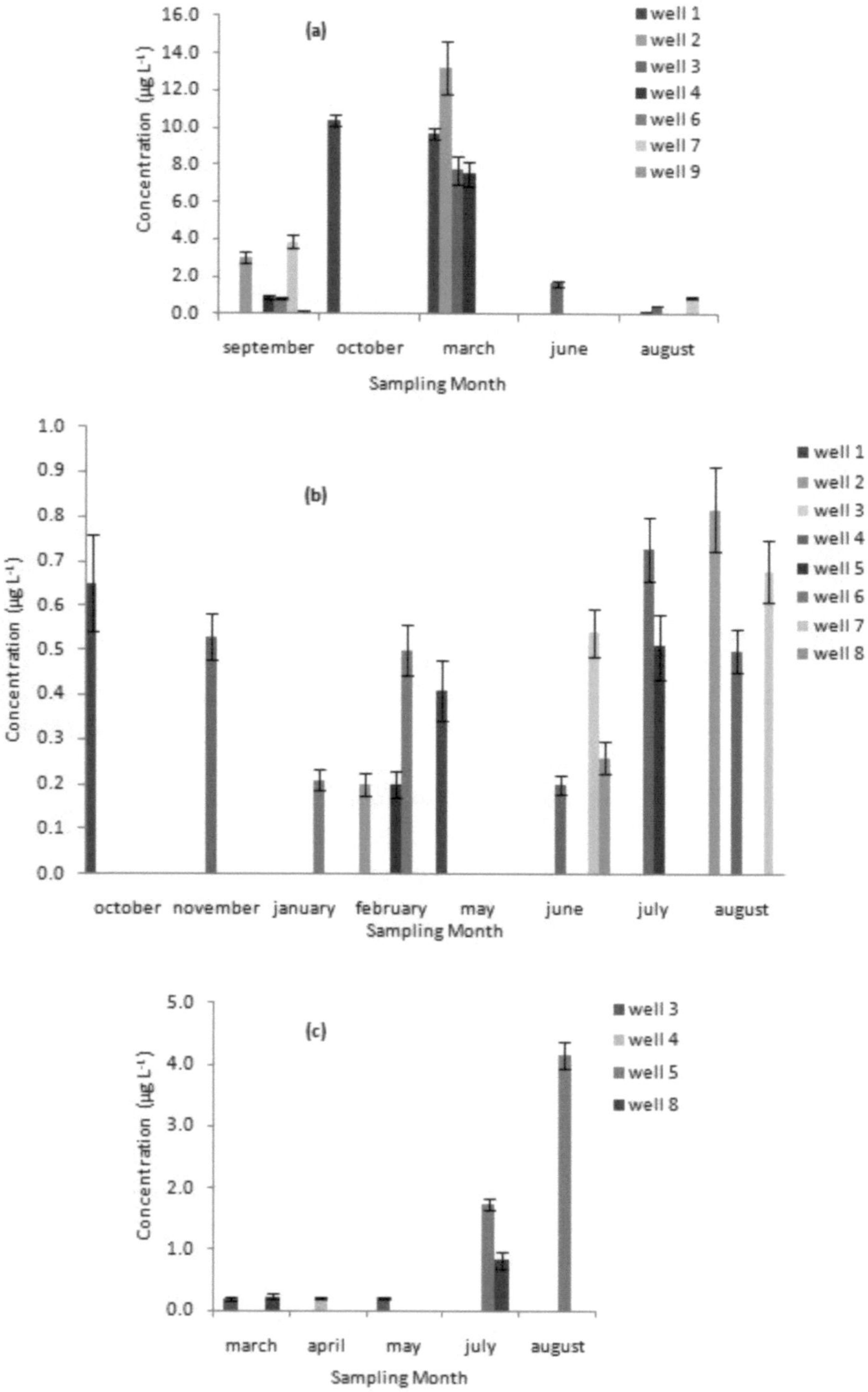

Fig. 4. Concentrations of carbofuran (a), clomazone (b), and tebuconazole (c) (µg L^{-1}) in water samples from wells in different months (Caldas et al., 2010)

highest solubility in water and the lowest half life in soil, probably being degraded rapidly before reaching the groundwater. In surface waters, it has been found in many studies (Laganà et al., 2002; Palma et al., 2004; Primel et al., 2005).
The pesticide carbofuran showed the highest concentrations. October and March were the months that showed the highest contamination, 10.4 and 9.75 μg L^{-1}, respectively; exceeding the maximum value permitted by the Brazilian legislation for groundwaters. Wells 1 (9 m) and 2 (37 m) presented the major occurrence. Tariq et al. (2004) detected carbofuran residues in well waters with a detection frequency of 59% in July and 43% in October. This is one of the factors related to the high solubility of the compound in water. The highest concentration was 23.1 μg L^{-1}. Bacigalupo & Meroni (2007) analyzed water from wells in an agricultural area in the south of Milan for 11 months and found residues of carbofuran in 90% of the samples at concentrations below 5.0 μg L^{-1}. Hernández et al. (2001) detected carbofuran in groundwater in around 25% of samples, but never in levels higher than 0.1 μg L^{-1}.
Clomazone has high water solubility and water half life higher than 30 days and showed its highest levels in August in well 2 (9 m), 0.82 μg L^{-1}. The high frequency of contamination of clomazone in groundwaters was detected by Bortoluzzi et al. (2007); they detected the compound in 50% of well samples, at a mean concentration of 6.76 μg L^{-1}. In this study, the herbicide clomazone was detected in 70 % of wells. This is the only work found that analyze pesticides in groundwater in RS state.
The high detection frequency of the pesticides carbofuran and clomazone in groundwater, both in shallow and deep wells can be explained by their high potential for leaching as a result of their low soil sorption coefficient and high solubility in water, as well as relatively high half life in the soil.
The systemic tebuconazole fungicide is used to control a wide range of fungi on fruit and vegetables. The concentrations of tebuconazole were higher in July and August, and well 5 (6 m) showed the highest levels, 1.73 μg L^{-1} in July and 3.65 μg L^{-1} in August.
The compound tebuconazole was detected in wells 29 m deep. Although it has occurred with low frequency, it confirms the high potential for leaching that this compound has. This compound was found in groundwaters by Baugros et al. (2008). The amounts ranged from 0.03 to 0.89 μg L^{-1}.

5. Conclusion

The use of models/parameters to predict the pesticide behavior in the environment is important to obtain an estimate about the contamination risk. The evaluation of the data presented in the list of compounds reveals that among the pesticides most commonly used in agriculture in Rio Grande do Sul State, most of those investigated presented mobility in the environment, and risk to contaminate the surface and groundwaters.
The results of the monitoring show that surface and groundwater contamination by agriculture is still an important issue that cannot be ignored.
In an attempt to minimize risks associated with pesticides, technological advances that allow pesticide detections at very low concentrations, combined with research findings help to increase the scientific understanding of the potential of pesticides to contaminate water resources.

The public interest is well served by a cooperative effort among regulators, university researchers, and industry to establish reasonable use restrictions. These efforts should ensure that, when used appropriately, pest control products will not pose a threat to water quality.

6. References

Avila, L.; Lee, D.; Senseman, S. McCauley G.; Chandler, J. & Cothren, J. (2005). Assessment of acetolactate synthase (ALS) tolerance to imazethapyr in red rice ecotypes (*Oryza* spp) and imidazolinone tolerant/

Bacigalupo, M. & Meroni, G. (2007). Quantitative Determination of Diuron in Ground and Surface Water by Time-Resolved Fluoroimmunoassay: Seasonal Variations of Diuron, Carbofuran, and Paraquat in an Agricultural Area *Journal of Agricultural and Food Chemistry*, Vol. 55, No. 10, 3823-3828. ISSN 1520-5118

Barceló, D. & Hennion, M.C. (1997) *Trace determination of pesticides and their degradation products in water*. 3. ed. The Netherlands: Elsevier. 542 p.

Barrek, S.; Cren-Olivé, C.; Wiest, l.; Baudot, R.; Arnaudguilhem, C. & Grenier-Loustalot, M. (2009). Multi-residue analysis and ultra-trace quantification of 36 priority substances from the EuropeanWater Framework Directive by GC–MS and LC-FLD-MS/MS in surface Waters, *Talanta*, Vol. 79, No. 3, 712–722, ISSN 0039-9140

Baugros, J.B.; Giroud, B.; Dessalces, G.; Grenier-Loustalot, M. & Cren-Olivé, C. (2008) Multiresidue analytical methods for the ultra-trace quantification of 33 priority substances present in the list of REACH in real water samples, *Analityca Chimica Acta*, Vol. *607*, No. 28, 191-203, ISSN 0003-2670

Bortoluzzi, E.; Rheinheimer, D.; Gonçalves, C.; Pellegrini, J.; Zanella, R. & Copetti, A. (2006). Contaminação de águas superficiais por agrotóxicos em função do uso do solo numa microbacia hidrográfica de Agudo, RS, *Revista Brasileira de Engenharia Agrícola e Ambiental*, Vol.10, No.4, 881-887, ISSN 1807-1929

Bortoluzzi, E.; Rheinheimer, D.; Gonçalves, C.; Pellegrini, J.; Maroneze, A.; Kurz, M.; Bacar, N. & Zanella, R. (2007). Investigation of the occurrence of pesticide residues in rural wells and surface water following application to tobacco, *Química Nova*, Vol. 30, No. 8, 1872-1876, ISSN 0100-4042

Buma, A.; Sjollema, S.; van de Poll, W.; Klamer, H. & Bakker, J. (2009) Impact of the antifouling agent Irgarol 1051 on marine phytoplankton species, *Jounal of Sea Research*, Vol. 61, No. 3, 133-139, ISSN 1385-1101

Cabrera, L.; Costa, F. & Primel, E. (2008) Estimativa de risco de contaminação das águas por pesticidas na região sul do estado do RS, *Química Nova*, Vol. 31, No. 8, 1982-1986, ISSN 0100-4042

Caldas, S.; Demoliner, A. & Primel, E. (2009). Validation of a Method using Solid Phase Extraction and Liquid Chromatography for the Determination of Pesticide Residues in Groundwaters, *Journal of the Brazilian Chemical Society*, Vol. 20, No. 1, 125-132, ISSN 0103-5053

Caldas, S.; Demoliner, A.; Costa, F.; D'Oca, M. & Primel, E. (2010) Pesticide Residue Determination in Groundwater using Solid-Phase Extraction and High-Performance Liquid Chromatography with Diode Array Detector and Liquid

Chromatography-Tandem Mass Spectrometry, *Journal of the Brazilian Chemical Society*, Vol. 21, No. 4, 642-650, ISSN 0103-5053

Câmara dos agrotóxicos. (2008). I mapeamento de agrotóxicos utilizados no Rio Grande do Sul. *CRQ - Conselho Regional de Química da 5º região,* 1.ed., CRQ-V: Porto Alegre.

Carabias-Martínez, R.; Rodríguez-Gonzalo, E.; Herrero-Hernández, E.; Román, F. & Flores, M. (2002). Determination of herbicides and metabolites by solid-phase extraction and liquid chromatography Evaluation of pollution due to herbicides in surface and q groundwaters, *Journal of Chromatography A*, Vol. 950, No. 1-2, 157–166, ISSN 0021-9673

Cerejeira, M.; Viana, P.; Batista, S.; Pereira, T.; Silva, E.; Valério, M.; Silva, A.; Ferreira, M. & Silva-Fernandes, A. (2003). Pesticides in Portuguese surface and ground Waters, *Water Research*, Vol. 37, No. 5, 1055–1063, ISSN 0043-1354

EU Council Directives 98/83/EC (1998). Directive on the Quality of Water Intended for Human Consumption, *Official Journal of the European Community,* L 330, 32-54

D'Archivio, A.; Fanelli, M.; Mazzeo, P. & Ruggieri, F. (2007). Comparison of different sorbents for multiresidue solid-phase extraction of 16 pesticides from groundwater coupled with high-performance liquid chromatography, *Talanta*, Vol. 71, No. 1, 25–30, ISSN 0039-9140

Demoliner, A.; Caldas, S.; Costa, F.; Gonçalves, F.; Clementin, R.; Milani, M. & Primel, E. (2010). Development and Validation of a Method Using SPE and LC-ESI-MS-MS for the Determination of Multiple Classes of Pesticides and Metabolites in Water Samples, *Journal of the Brazilian Chemical Society,* Vol. 21, No. 8, 1-10, ISSN 0103-5053

Goss, D.W. (1992) Screening Procedure for Soils and Pesticides for Potential Water Quality Impacts. *Weed Technology*, Vol. 6, No. 3, 701-708, ISSN 0890-037X.

Grützmacher, D.; Grützmacher, A.; Agostinetto, D.; Loeck, A.; Roman, R.; Peixoto, S. & Zanella, R. (2008). Monitoramento de agrotóxicos em dois mananciais hídricos no sul do Brasil, *Revista Brasileira de Engenharia Agrícola e Ambiental,* Vol.12, No.6, 632–637, 1807-1929

Gustafson, D. (1989) Groundwater ubiquity score: a simple method for assessing pesticide leachability, *Environmental Toxicology and Chemistry,* Vol. 8, No. 4, 339-357, ISSN 1552-8618

Hernández, F.; Sancho, J.; Pozo, O.; Lara, A. & Pitarch, E. (2001). Rapid direct determination of pesticides and metabolites in environmental water samples at sub-µg/l level by on-line solid-phase extraction-liquid chromatography–electrospray tandem mass spectrometry, *Journal of Chromatography A,* Vol. 939, No. 1, ISSN 0021-9673

Hernández-Romero, A.; Tovilla-Hernández, C.; Malo, E. & Bello-Mendoza, R. (2004). Water quality and presence of pesticides in a tropical coastal wetland in southern Mexico, *Marine Pollution Bulletin*, Vol. 48, No. 11-12, 1130–1141, ISSN 0025-326X

Konstantinou, I.; Hela, D. & Albanis, T. (2006). The status of pesticide pollution in surface waters (rivers and lakes) of Greece. Part I. Review on occurrence and levels, *Environmental Pollution*, Vol. 141, No. 3, 555-570, ISSN 0269-7491

Laganà, A.; Bacaloni, A.; Leva, I.; Faberi, A.; Fago, G. & Marino, A.; (2002). Occurrence and determination of herbicides and their major transformation products in environmental waters, *Analytica Chimica Acta, Vol. 462,* No. 2, 187-198, ISSN 0003-2670

Marchesan, E.; Zanella, R.; Avila, L.; Camargo, E.; Machado, S. & Macedo, V. (2007). Rice herbicide monitoring in two brazilian rivers during the rice growing season, *Scientia Agricola,* Vol. 64, No. 2, 131-137, ISSN 0103-9016

Marchesan, E.; Meneghetti, G.; Sartori, S.; Avila, L.; Machado, S.; Zanella, R.; Primel, E.; Macedo, V. & Marchezan, M. (2010). Resíduos de agrotóxicos na água de rios da Depressão Central do Estado do Rio Grande do Sul, Brasil, *Ciência Rural,* Vol. 40, No. 5, 1053-1059, ISSN 0103-8478

Marques, M.; Badiru, A.; Beltrame, M. & Pires, M. (2007). Pesticide Leaching and Run-off Hazard in the Ribeira de Iguape River Basin in São Paulo State, Brazil, *Journal of the Brazilian Society of Ecotoxicology,* Vol. 2, No. 2, 179-185, ISSN 1809-8401

Palma, G.; Sánchez, A.; Olave, Y.; Encina, F.; Palma, R. & Barra, R. (2004). Pesticide levels in surface Waters in an agricultural-forestry basin in Southern Chile, *Chemosphere,* Vol. *57,* No. 8, 763-770. ISSN 0045-6535

Peres, F.; Moreira, J.; Rodrigues, K. & Claudio, L. (2006). Risk Perception and Communication Regarding Pesticide Use in Rural Work: A Case Study in Rio de Janeiro State, Brazil, *International Journal of Occupational and Environmental Health,* 12, 400–407, ISSN 1077-3525

Primel, E.; Zanella, R.; Kurz, M.; Gonçalves, F.; Machado, S. & Marchezan, E. (2005). Poluição das águas por herbicidas utilizados no cultivo do arroz irrigado na região central do estado do Rio Grande do Sul, Brasil: predição teórica e monitoramento, *Química Nova,* Vol. 28, No. 4, 605-609, ISSN 0100-4042

Sapozhnikova, Y.; Wirth, E.; Schiff, K.; Brown, J. & Fulton, M. (2007). Antifouling pesticides in the coastal waters of Southern California, *Marine Pollution Bulletin, Vol. 54,* No. 12, 1962-1989, ISSN 0025-326X

SIA - SISTEMA DE INFORMAÇÃO SOBRE AGROTÓXICOS (2003). Available at: http://www4.anvisa.gov.br/agrosia/asp/default.asp, accessed in 10 jul 2010.

Silva, D.; Avila, L.; Agostinetto, D.; Dal Magro, T.; Oliveira, E.; Zanella, R. & Noldin, J. (2009). Monitoramento de agrotóxicos em águas superficiais de regiões orizícolas no sul do Brasil, *Ciência Rural,* Vol. 39, No. 9, p.2383-2389, ISSN 0103-8478

Sindag (2006) In: *Sindicato Nacional da Indústria de Produtos para Defesa Agrícola,* available at http://www.sindag.com.br, accessed in August 2007.

Tariq, M.I.; Afzal, S. & Hussain, I. (2004). Pesticides in shallow groundwater of Bahawalnagar, Muzafargarh, D.G. Khan and Rajan Pur districts of Punjab, Pakistan, *Environonmental International,* Vol. *30,* No. 4, 471-479 ISSN 0160-4120

Whitford, F.; Wolt, J.; Nelson, D.; Barrett, M.; Brichford, S. & Turco, R; Pesticides and Water Quality Principles, Policies, and Programs, *Purdue University Cooperative Extension Service,* available at http://www.btny.purdue.edu/Pubs/PPP/PPP-35.pdf, accessed in July 7 2010.

Zanella, R.; Primel, E.; Machado, S.; Gonçalves, F. & Marchezan, E. (2002). Monitoring of the Herbicide Clomazone in Environmental Water Samples by Solid-Phase Extraction and High-Performance Liquid Chromatography with Ultraviolet Detection, *Chromatographia, Vol.* 55, No.9/10 ,573-577, ISSN 1612-1112

24

Saccharomyces cerevisiae as a Tool to Evaluate the Effects of Herbicides on Eukaryotic Life

Daniela Braconi, Giulia Bernardini, Lia Millucci, Gabriella Jacomelli, Vanna Micheli and Annalisa Santucci
Dipartimento di Biologia Molecolare, Università degli Studi di Siena
Italy

1. Introduction

1.1 Herbicides and their toxicity

Agricultural herbicides play undoubted beneficial roles in preserving crop yields, although they may pose serious concerns for the environment and humans because of their widespread and intensive use/misuse (due to careless applications, high and repeated application rate, accidental spillage). It is also evident the potential to injure non-target cultivars and microorganisms, in particular those contributing to soil quality, and to cause adverse side-effects in mammals, including humans (Cabral et al.,2003; Cabral et al.,2004).

Herbicide-related toxicity may be often underestimated because little information is provided on the effects of herbicides as complete formulations. As a matter of fact, acute or chronic toxicity tests are generally carried out using pure active ingredients (AIs). Herbicide-induce damaging effects are proportional to the absorbed substance and to the time of exposure, but also to the inherent characteristics of each compound. Agriculturists are most likely exposed to acute intoxication, by inhalation or direct contact with toxic substances, though each person can be subjected to possible toxic effects due to accumulation of pesticide residues in the body, being the diet the most important source of exposure.

Typically, pure herbicide molecules are of limited value to end users. To give them practical value and make them usable, most herbicides are combined with appropriate solvents or surfactants to obtain a formulation. A given chemical may be formulated in a variety of differing formulations and sold under different trade names. For this reason, pure AIs are mixed with coformulants, also called 'inert' or auxiliary substances, to allow their use in common and convenient vehicles, such as water, and to obtain uniform and effective distributions. Several substances fall within the definition of coformulant, including: carrier substances, solvents, surfactants, dispersing agents, adhesives, absorption-promoting agents, antioxidants, bactericides, dyes, fillers, and perfumes. Coformulants can have various technical and physico/chemical properties in relation to their function in the pesticide formulation. These chemicals can be expected to have various toxicological profiles, some of them harmless (*e.g.*, water) and some with serious toxicological properties (*e.g.*, the organic solvent isophorone, which is a suspected carcinogen). Because

coformulants represent the highest amount in pesticide mixtures, even a minor toxicological concern could become significant in relation to their use (Tobiassen et al., 2003) especially considering the fact that, besides being toxic themselves, they can also increase the toxic effects of AIs (Surgan, 2005; Séralini, 2005). However, for the majority of herbicides that are currently used, no information on the identity of coformulants is publically made available.
Consequently, adjuvants can be included in herbicide formulations as a part of the total product which is sold by the manufacturer or as an additive to be mixed with pesticide products in the spray tank. Adjuvants can directly increase the toxicity of pesticide formulations. For instance, they can promote their penetration through clothes and skin and increase their persistence in the environment (Cox & Surgan, 2006). Consequently, different problems might be posed when manipulating different formulations. As an example, in the case of emulsifiable concentrates, problems of dosing and mixing could be encountered and the absorption through skin could be facilitated. Granules, since they usually do not adhere to foliage and are not intended for foliar applications, may have more serious consequences for soil quality and microbial populations commonly associated (http://www.ext.colostate.edu/ pubs/crops/00558.html).

1.2 Yeast as an eukaryotic model system

The yeast *Saccharomyces cerevisiae* is an optimal eukaryotic model system to study toxic effects and mammalian biological responses upon exposure to exogenous and endogenous perturbations. The high degree of evolutionary conservation of stress pathways between yeast and higher eukaryotes means that yeast can serve as a suitable model system for the characterization of stress responses in more complex organisms (Estruch, 2000). Particularly important advantages deriving from the use of *S. cerevisiae* as a well consolidated eukaryotic model are the non-pathogenicity and the feasibility of such a model, thus avoiding risks for manipulators. Being unicellular, it is a simple and easily accessible system allowing to avoid ethical and safety problems arising from the use of more complex eukaryotic cells, representing a consolidated, appropriate and cost-effective alternative to animal testing. At the same time, the obtained response is easier to decipher and, once extrapolated to humans, it could produce a better understanding of molecular mechanisms of toxicity. Being the first eukaryotic organism whose genome was completely sequenced in 1996 (Goffeau et al., 1996), yeast offers also tremendous opportunities for genome/proteome/metabolome and other 'omic' studies.
Yeast cells have remarkable similarities to mammalian cells at the macromolecular and organelle level, and a number of yeast proteins have been shown to be functionally interchangeable with the highly homologous human proteins. Thus it is not surprising the use of yeast as a model system with relevant contributions to the understanding of molecular mechanisms underlying oxidative stress. The involvement of oxidative stress in ageing, apoptosis, and a significant number of diseases led to the characterization of the antioxidant systems and the elucidation of their functional physiological role. For example, yeast cells have contributed in recent years to clarify the role reactive oxygen species (ROS) in diseases such as amyotrophic lateral sclerosis or Friedereich's ataxia (Costa & Moradas-Ferreira, 2001).
Studies on the evaluation of pesticides still depend extensively on the use of animals, but in the last years several publications have dealt with the development and appropriateness of alternative methods for assessing toxicity, that do not depend on animal utilization but explore the use of rapid and cost-effective alternatives (Ribeiro et al., 2000).

Yeast cells have already been proposed as a tool for assessing toxicity of environmental pollutants (Cabral et al., 2004; Cabral et al., 2003; Simoes et al., 2003), as shown by the few following examples:

i. *S. cerevisiae* can be a bio-accumulator of metals such as Cu, Cr, Cd, Ni, Zn, and Pb (Malik, 2004).
ii. *S. cerevisiae* shows affinity for different pesticides, and in particular can partially degrade and significantly adsorb the fungicides quinoxyfen (Cabras et al., 2000) and fenhexamid (Cabras et al., 2003). Moreover, yeast can degrade and adsorb several pyrethroid pesticides (Cabras et al., 1995).
iii. Wild-type yeasts can be used as biosensors and a valid tool for preliminary evaluations of xenobiotic toxicity (Baronian, 2004) (Campanella et al., 1995).

2. Aim of the work

In the scientific literature, little attention has been paid to the understanding of toxic effects of pesticides as commercial formulations. In this context, the most important exception is represented by the herbicide glyphosate, for which recent toxicity studies have been carried out using some of its formulations (Peixoto, 2005; Mann & Bidwell, 1999; Chan et al., 2007; Pieniazek et al., 2004). On the contrary, for many other herbicides (and pesticides in general) little information is currently available.

In order to fill this gap in the knowledge of herbicide-related toxicity, we undertook a comparative analysis on an autochthonous *S. cerevisiae* strain, isolated during the spontaneous fermentation of grapes and selected as a potential 'starter' for the production of high quality wines (Vagnoli et al., 1993; Trabalzini et al., 2003a; Trabalzini et al., 2003b) testing in parallel three herbicides and their corresponding pure AIs. The three tested herbicides were chosen among those used in the same geographical areas where the yeast strain was isolated. They were: Pointer (P), water dispersible granules containing tribenuron methyl (T); Silglif (S), a soluble concentrate containing glyphosate (G), and Proper Energy (PE), an oil/water emulsion containing fenoxaprop-P-ethyl (F). In order to clarify the mechanisms underlying the toxicity of herbicides as commercial formulations, we moved towards the analysis of parameters related to oxidative stress and undertook a proteomic and redox-proteomic study on the effects of the tested compound on yeast protein repertoires.

Our analyses might contribute to elucidate response mechanisms in more complex and experimentally less accessible eukaryotes while avoiding the complexity of higher eukaryotic cells and consequent ethical problems. Importantly, the use of a *S. cerevisiae* strain used in the wine-making industry provided us with an improved understanding of the perturbing effects of herbicides on fermentation, a process making yeast a milestone in the production of wine, bread, beer and many other foods and beverages.

3. Materials and methods

3.1 General materials

All high-purity reagents were from Oxoid (Garbagnate Milanese, Milan, Italy), Sigma-Aldrich (Milan, Italy) or J. T. Baker (Deventer, Holland).

Commercial grade herbicides, namely PE and P (Aventis CropScience, Milan, Italy) and S (Siapa, Milan, Italy) were from commercial sources. High purity AIs (F, T and G) were analytical standards from Riedel de Haën (Schweiz, Germany).

All of the experiments were performed using Milli-Q (Millipore, Bedford, MA, U.S.A.) water.

3.2 Yeast cell cultures

The yeast strain used in this work was *S. cerevisiae* K310. Yeast cells were pre-cultured in yeast peptone dextrose (YPD) medium [1% (w/v) yeast extract, 2% (w/v) peptone, 2% (w/v) glucose] at 30°C with rotary shaking for 10 hours. Then, an aliquot of the cell culture was inoculated in 150 mL of a modified YPD medium (YPDm) adjusted to a final pH of 4.5 by adding 0.2 M citrate/phosphate buffer and containing 100 g/L glucose, to obtain an initial concentration of 1×10^4 cells/mL. The new cell suspension was incubated at 28 °C in the dark without shaking and allowing semi-anaerobic growth.
Commercial grade herbicides were singularly added to the culture medium at the beginning of exponential growth phase (16th hour of cell culture, about 3×10^6 cells/mL). P was added as a water dispersion to a final concentration of 100 mg/L; S was added as a water solution to a final concentration of 1 g/L; PE was added as a water emulsion to a final concentration of 500 mg/L.
In parallel, single AIs were assayed and added to yeast cultures at the same time and the same concentration obtained with commercial formulations. Whereas G was prior dissolved in water, F and T were dissolved in DMSO (1 mg/mL and 4 mg/mL, respectively). Consequently, vehicle controls were prepared for F [2.62% (v/v) DMSO] and T [1.87% (v/v) DMSO]. Final concentrations of AIs in culture medium were as follows: 75 mg/L T, 304 mg/L G and 26.2 mg/L F).
At various moment during growth, cells were harvested by centrifugation, washed and cell-free protein extracts prepared as already described (Trabalzini et al., 2003a) for further analyses. Protein concentration was determined according to the Bradford's method (Bradford, 1976).

3.3 Monitoring of yeast colony forming ability and fermentative performance

K310 colony forming ability was monitored by plating on YPD-agar proper dilutions (ranging from 1:10 up to 1:100000) of the cell suspension; plates were then incubated at 28°C for three days before counting colony forming units (CFU).
Ethanol levels in culture medium were determined using an enzymatic assay (kit code 10 176 290 Boehringer Mannheim, Germany) as already described (Braconi et al., 2006b). Briefly, samples taken from cell suspensions were rapidly cooled and centrifuged (centrifuge 1515R, Eppendorf, Hamburg, Germany). The supernatants were then filtered through a 0.2 μm pore size membrane and determination of ethanol concentrations was performed spectrophotometrically (Agilent 8453 UV-visible spectroscopy system, Waldbronn, Germany) on the obtained filtered, properly diluted in accordance to manufacturer's instructions.

3.4 HPLC analysis of K310 nucleotide patterns

HPLC analysis of nucleotides in K310 cells was carried out on protein extracts with a System Gold Beckman (San Ramon, CA, USA) with a programmable Solvent Module 126 and Dual Channel Scanning Detector Module (mod. 168), equipped with Gold.8 software.
All separations were performed on a ODS Phenomenex Luna C18 column (7.5 cm × 4.6 mm, 3 μm particle size) equipped with a Phenomenex Security Guard pre-column (4 mm L × 3

mm ID) and filter. Analyses were carried out by gradient elution with 0.1 M KH_2PO4, containing 6 mM pH 5.5 tetrabutylammonium hydrogen phosphate (TBA) (eluent A) and methanol (eluent B).

The chromatographic conditions (total run time 22 min) were set as follows:

initial condition: 95% A and 5% B;

i. 5 minutes isocratic in the initial conditions;
ii. 2 minutes gradient to reach 80% A and 20% B;
iii. 5 minutes isocratic in the new conditions;
iv. sudden increase to 80% A and 30% B;
v. 2 minutes isocratic in these conditions;
vi. restoring of the initial conditions in 8 minutes.

Flow rate was 1 mL/minute at room temperature. Both UV absorbances at 260 and 280 nm were monitored and used to record sample chromatographic spectra. On individual peaks, absorbance spectra were recorded from 194 to 354 nm. Compound of interest were identified on the basis of their retention time or with the co-elution with proper internal standards. Single compounds were quantified according to calibration curves obtained with standard solutions.

3.5 Catalase and superoxide dismutase (SOD) activity assays

For enzyme activity determination, 50 µg of proteins were resolved by 10% discontinuous native PAGE according to Ornstein (Ornstein, 1964) and stained as follows, performing all operations in the dark.

For catalase activity, gels were incubated 5 minutes in 5% (v/v) methanol, briefly washed three times with water and incubated 10 minutes in 10 mM H_2O_2. Gels were again rinsed with water and soaked in a 1:1 mixture of freshly prepared 2% (w/v) potassium ferric cyanide and 2% (w/v) ferric chloride. Gels turned to blue except in the zones where H_2O_2 was decomposed by active catalase. Colour development was blocked soaking the gels in 10% (v/v) acetic acid and 5% (v/v) methanol (Agarwal et al., 2005).

For SOD activity, gels were first soaked in 2.5 mM NBT[1] for 20 minutes, then briefly washed with water and soaked in 500 mM PBS[2] containing 2.8 mM TEMED[3] and 28 µM riboflavin. For revelation, gels were illuminated on a light box for 20 minutes until the appearance of white bands on a dark background (Teixeira et al., 2004).

Images of gels were acquired (Image Scanner, Amersham Biosciences) and analysed with Image Master™ Platinum (Amersham Biosciences) choosing as a reference parameter the intensity of bands, which is automatically normalized against the surrounding background.

3.6 Proteomics and redox-proteomics

3.6.1 Two-dimensional gel electrophoresis (2D-PAGE)

K310 proteins were first mixed with a buffer containing 8 M urea, 35 mM CHAPS[4], 10 mM DTE[5], and a trace of Bromophenol Blue. Proteins were adsorbed onto Immobiline Dry Strips (IPG 18 cm, non linear 3-10 pH range, Bio-Rad, Milan - Italy) and allowed to stand at room

---

[1] Nitro blue tetrazolium

[2] Phosphate buffered saline

[3] *N,N,N',N'*-Tetramethylethylenediamine

[4] 3-[(3-Cholamidopropyl)dimethylammonio]-1-propanesulfonate

[5] Dithioerythritol

temperature for 10 hours. 100 µg (2D gels to be transferred onto nitrocellulose (NC) membranes or 1000 µg (preparative 2D gels) of proteins were used, respectively. Isoelectric focusing (IEF) was carried out with a Protean IEF cell (Bio-Rad) according to manufacturer's instruction and then IPG strips were equilibrated in 6 M urea, 30% (v/v) glycerol, 2% (w/v) SDS[6], 0.05 M Tris-HCl pH 6.8 containing first 2% (w/v) DTE and later 2.5% (w/v) IAA[7]. Sodium Dodecyl Sulfate PolyAcrylamide Gel Electrophoresis (SDS-PAGE) was carried out applying 40 V per gel until the dye front reached the bottom of gels. Preparative gels were stained with Coomassie Brilliant Blue as described (Candiano et al., 2004).

3.6.2 Western blot analysis of carbonylated proteins

After the IEF, strips for the Western blot detection of carbonylated proteins were briefly rinsed with water and incubated at room temperature for 20 minutes with 10 mM DNPH[8] in 5 % (w/v) trifluoroacetic acid (Reinheckel et al., 2000). Subsequently, strips were rapidly washed twice with a solution containing 8M urea, 20% (v/v) glycerol, 9 M SDS and 150 mM Tris-HCl pH 6.8 and then subjected to equilibration procedures and SDS-PAGE as described previously. The obtained 2D-gels were washed and equilibrated in a transfer buffer [50 mM Tris, 40 mM glycine, 1.3 mM SDS, 20% (v/v) methanol] and protein transfer from 2D gels onto NC membranes was carried out using a semidry Novablot transblot cell (Bio-Rad) applying 0.7 mA/cm^2 for 75 minutes. Protein transfer was checked with staining of membranes with 0.2% (w/v) Ponceau S in 3% (v/v) trichloroacetic acid. Carbonylated proteins were detected with anti-DNP antibodies (Sigma-Aldrich) (at the dilution of 1:10000) and secondary horseradish peroxidase-linked antibodies (at the dilution of 1:7000), followed by a chemiluminescence reaction using the Immun-star™ HRP kit (Bio-Rad).

3.6.3 Western blot detection of protein thiols labelled with biotinylated iodoacetamide (BIAM)

Before the IEF, K310 proteins were derivatized in MES-Tris buffer (pH 6.5) containing 200 µM BIAM for 15 minutes in the dark. The labelling reaction was quenched by the addition of 2 mM β-mercaptoethanol (final concentration) (Kim et al., 2000). Then proteins were resolved through 2D-PAGE and transferred onto NC membranes as already described (see 3.6.2).
BIAM-labelled proteins were detected with horseradish peroxidase (HRP)-linked streptavidin (dilution 1:10000), followed by a chemiluminescence reaction using the Immun-starTM HRP kit (Bio-Rad).

3.6.4 Image analysis

Digitalized images of Coomassie stained gels and Western Blot films were acquired (Image Scanner, Amersham Biosciences) and analyzed with Image Master™ Platinum (Amersham Biosciences). For comparative proteomic analysis, spot % relative abundance was used to compare herbicide-treated samples with the untreated control. In case of multiple spots identified as different molecular species of a same protein, the average % relative abundance was calculated.

[6] Sodium dodecyl sulfate
[7] Iodoacetamide
[8] 2,4-Dinitrophenylhydrazine

3.6.5 Protein identification

Protein spots were identified by gel matching with previously produced and characterized proteomic maps of K310 strain (Trabalzini et al., 2003a; Trabalzini et al., 2003b). Alternatively, spots from 2D gels were excised, minced, washed with water, in-gel reduced, S-alkylated, digested with trypsin and identified by MALDI-ToF[9] mass spectrometry as described previously (Braconi et al., 2009).

4. Results and discussion

4.1 Yeast colony forming ability and fermentative performance

In a first phase of work, we aimed at investigating how the herbicides P, S and P E, with AIs belonging to different classes (T, a sulfonylurea; G, an organophosphate, and F, an aryloxyphenoxy-propionate, respectively) can affect colony-forming ability and ethanol production in a wild-type wine yeast strain. We decided to use commercial grade herbicides in order to reproduce, as far as possible, the same conditions during application in crops, and to compare results with those obtained in parallel with single high-purity AIs.
The tested herbicides were chosen among those authorized and employed in central regions of Italy during the last years. At the same time, we chose, as the eukaryotic cell model, a *S. cerevisiae* strain adopted for oenological applications instead of a conventional commercial or laboratory-adapted/mutated baker's yeast strain. This type of yeast may reside on grapes in vineyards of the same geographical areas where herbicides are used (Vagnoli et al., 1993). Therefore, the use of a locally isolated wine wild-type *S. cerevisiae* strain is advantageous with respect to other previous approaches since this unmanipulated strain can reveal biological effects closer to physiological ones. Additionally, K310 strain is well-characterized physiologically, for its protein repertoire and stress response (Trabalzini et al., 2003a; Trabalzini et al., 2003b; Santucci et al., 2000; Martini et al., 2006; Martini et al., 2004; Ricci, et al., 2004).
In Fig. 1, results obtained for colony forming ability assays are reported. Only in the case of 100 mg/L P and 75mg/L T a good homology was found. As a matter of fact, both the tested compound were found to negatively affect yeast colony forming ability up to the mid-log phase of growth (Fig. 1 A,B). On the contrary, the commercial herbicides S and PE (Fig. 1 A) were found to be more cytotoxic than their pure AIs since they had negative effects in colony forming ability whereas no significant differences were found for G- and F- treated cells with respect to controls (Fig. 1 B).
Fermentative performances of herbicide- treated yeast cells were evaluated measuring ethanol levels in culture medium (Fig. 2). As observed for colony forming ability, only in the comparison between the effects of P and its AI T a good homology was found, since we detected in both the samples a similar delay in ethanol production with respect to the untreated control (Fig. 2 A, D). On the contrary, whereas G- and F- treated cells showed no differences with respect to their controls in ethanol levels (Fig. 2B, C), their commercial formulations S and PE had significant negative effects for K310 fermentative performance (Fig. 2 E, F). Importantly, in the case of PE- treated cells, fermentation showed a 100 hours delay *versus* the untreated control (Fig. 2 C).
The finding that the herbicide S, differently from its pure AI G, had clear negative effects on K310 was in accordance with findings by Low and colleagues. They reported that bakers'

---

[9] Matrix-assisted laser desorption/ionization - time of flight

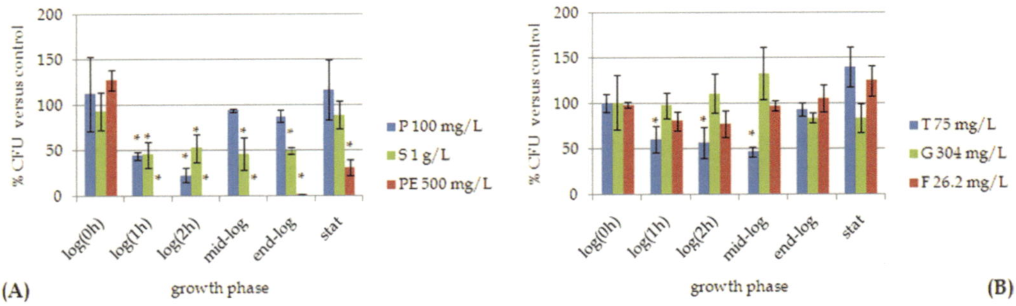

Fig. 1. Colony forming ability assay. K310 cells were treated with the tested herbicides as commercial formulations (A) or their pure AIs (B) and colony forming ability was evaluated by plate assay. Results are reported as % CFU *versus* control (cells grown in standard medium for P-, S-, PE, and G- treated cells; DMSO vehicle for T- and F- treated cells, as detailed under Material and Methods). Data are presented as average values of three independent experiments carried out under identical conditions; standard deviation is indicated with vertical bars. *P <0.05 compared with control.

yeast can metabolize up to 20% of G during the bread making process, probably with detoxification mechanisms which can result in the production of new degradation compounds, whose action is still unknown (Low et al., 2005).

Additionally, our finding is also in good agreement with a consistent number of reports indicating that commercial formulations containing G are much more toxic than G alone in several organisms/cells (dos Santos et al., 2005; Tsui & Chu, 2003; Peixoto, 2005; Benedetti et al., 2004).

Treatment of yeast cells with the herbicide PE induced a period of latency with a significant loss of colony-forming ability, followed by a restoration of the exponential phase of growth, presumably attributable to a cell population adapted to the chemical stress. This finding is in accordance with previously reported works on 2,4-D herbicide, to which yeast cells were proven to adapt before restoring growth (Viegas et al., 2005).

In the overall, our data strongly indicated how the effects produced by commercial herbicide formulations should be distinguished by those produced by pure AIs. This pointed out that coformulants can substantially contribute to cell damage. Such a damage cannot be revealed if, as usual, single AIs are tested. Because AIs are never applied alone in crops but always with several coformulants, our data reinforced the hypothesis that, for eco-toxicological considerations, commercial compounds are the most appropriate ones to be tested. Additionally, our data for S are particularly important considering that the introduction in the market of crops genetically tolerant to G will probably result in an increased use of this herbicide.

Our results should be read considering yeast not only as an eukaryotic model, but also and especially as a microorganism used for the production of wine, nowadays considered in all respects as a food. We found that significant amounts of herbicides can inhibit initial wine-making steps. Additionally, considering yeast ability to metabolize several compounds during alcoholic fermentation for the production of H_2S and other sulphur compounds (for example, in physiological conditions, intermediates of methionine biosynthesis), the possibility that yeast strains could utilize sulphur compounds contained in commercial formulations should not be underestimated. This could have important consequences on productivity of wine-making processes, from either a quantitative (fermentation yield and final ethanol production) or a qualitative point of view (organoleptic and sanitary properties of produced wines), thus with relapses for both the economic profile and the consumers' health. Altogether, results of this comparative analysis strongly encouraged us to deepen our studies on the effects of commercial herbicides on yeast cells.

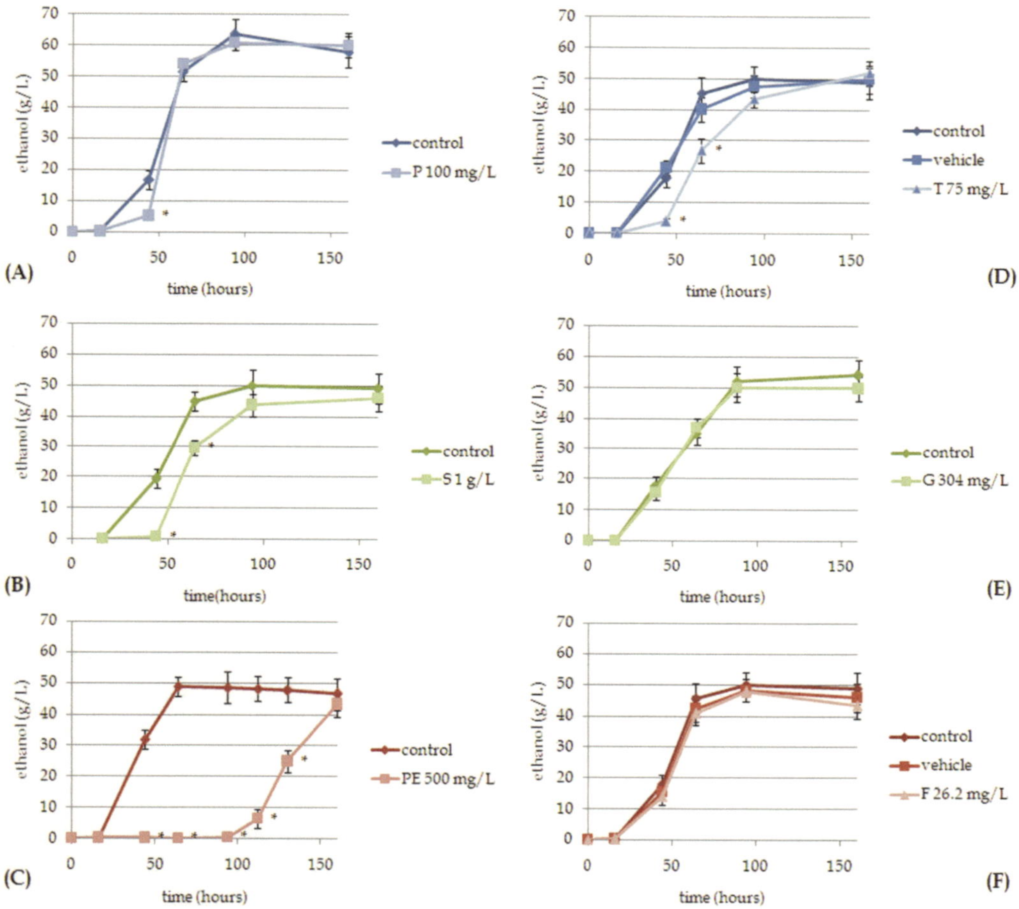

Fig. 2. Fermentative performance assay. K310 cells were treated with the tested herbicides as commercial formulations (A-C) or their pure AIs (D-F) and fermentative performance was evaluated by measuring ethanol levels in culture medium. Results are reported as g/L ethanol. Cells grown in standard medium were used as a control for P-, S-, PE, and G-treated cells; a DMSO vehicle for T- and F- treated cells, as detailed under Material and Methods, was used as well. Data are presented as average values of three independent experiments carried out under identical conditions; standard deviation is indicated with vertical bars. **P* <0.05 compared with control.

4.2 Analysis of nucleotide patterns

On the basis of our preliminary comparative investigations (Braconi et al., 2006a; Braconi et al., 2006b), from this point on we carried out our analyses only in cells treated with the commercial herbicides P, S and PE. Since we found that significant negative effects could be induced in a very short time after the addition of herbicides to yeast culture, our analyses were limited to 2 hours of treatment with the tested compounds. First, we investigated through an HPLC analysis if the tested herbicides could impair the K310 intracellular pools of some triphosphate nucleotides, their hydrolysis products, as well as NAD e NADP (Fig. 3).

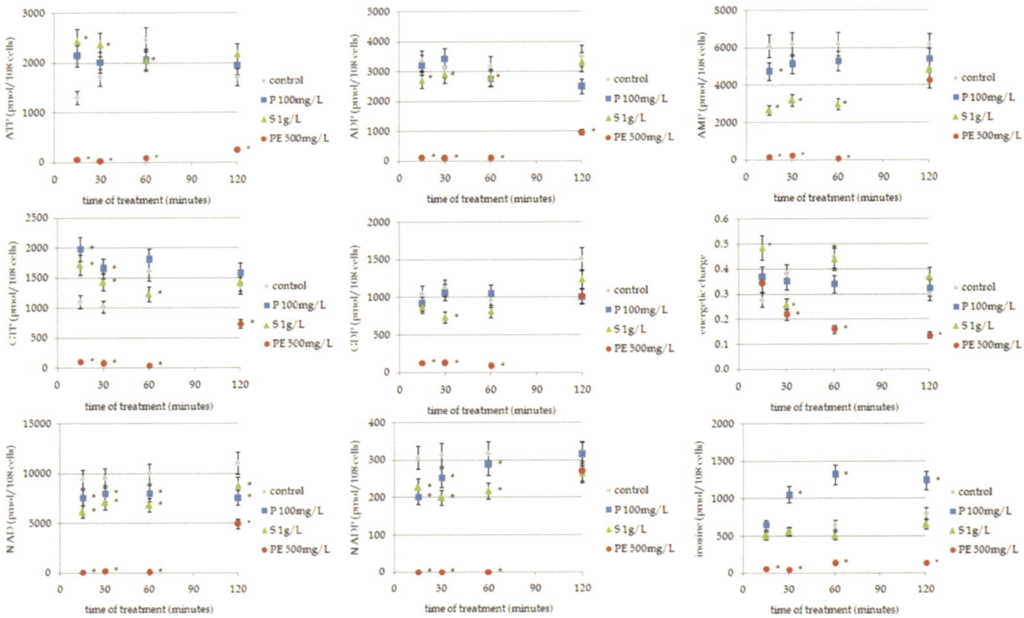

Fig. 3. HPLC analysis of nucleotide patterns in K310 cells treated with commercial herbicide formulations (P 100 mg/L, S 1 g/L or PE 500 mg/L) or in the absence of herbicides (control). Data are presented as average values of three independent experiments carried out under identical conditions; standard deviation is indicated with vertical bars. *P <0.05 compared with control.

When grown in the presence of 100 mg/L P, K310 cells exhibited a sudden raise in ATP and GTP levels; conversely, ADP, GDP and total energetic charge were quite similar between P-treated and control cells. Inosine levels were higher in the presence of P, whereas AMP, NAD and NADP levels were lower in P-treated cells *versus* the control.

Responses obtained treating K310 with 1 g/L S were partially superimposable with those obtained with P. The treatment with S made ATP and GTP levels raise, concomitantly with the decrease in AMP, NAD and NADP levels. The total energetic charge was significantly lowered after 30 minutes treatment in S-treated cells, while similar to the control at the other samplings.

When exposed to 500 mg/L PE, K310 cells showed the most important alterations in nucleotide patterns. All the parameters under evaluation were decreased by PE treatment during the whole observed period, though values registered after 120 minutes indicated a raise in AMP, GDP and NADP levels.

Stress responses involve many adaptive mechanisms in a number of metabolic processes to allow cell homeostasis and maintenance of the integrity of those structures necessary to survive. Yeast cells acquired, through the evolution, the ability to cope and survive to sudden environmental changes, and developed a peculiar ability in changing the internal *milieu* to adapt to new growth conditions. One of the most relevant aspects of this response is cell energetic balance, and for this reason we decided to investigate, through an HPLC analysis, if P, S and PE could alter nucleotide patterns in K310 cells.

Globally, our results suggested that both P- and S-treated cells had to face an increased energy demand to cope with the herbicide-induced stress, phenomenon that had been already reported for yeast cells exposed to the herbicide 2,4-D (Teixeira et al., 2005). On the other hand, they seemed to indicate a sudden impairment in the redox balance, as indicated by the lower levels of NAD and NADP, which are indirect indexes of a decreased reducing power generated by oxidative stress. Additionally, the increased levels of inosine represent an indirect index that cyototoxic effects of the tested herbicides might be exerted through oxidative mechanisms (Carvalho et al., 2003).

Our results also showed that PE-mediated cytotoxicity in K310 cells could be the result of important alterations of the cell redox homeostasis and/or imbalances of vital cell metabolic processes that cells could not counteract. Nevertheless, the trend showed by various parameters after 120 minutes of treatment with PE suggested the existence of a population of yeast cells which can adapt and survive to such stressing growth conditions, like already indicated by our previous experiments (Braconi et al., 2006a) (Braconi et al., 2006b).

4.3 Monitoring of herbicide-induced oxidative stress

Possible oxidative damage induced by G-containing herbicides has been already postulated (Beuret et al., 2005; Gehin et al., 2005; Pieniazek et al., 2004; Richard et al., 2005). Considering these findings in the light of our previous investigations on nucleotide patterns, which suggested a redox imbalance in herbicide treated cells, we decided to investigate if herbicide treatments could affect the enzymatic activity of catalase and superoxide dismutase (SOD), two key enzymes in the cell response to oxidative stress. In order to avoid the production of interfering substances during a spectrophotometric assay due to the presence of herbicide residues, we chose to evaluate catalase and SOD enzymatic activities by specific staining after native PAGE electrophoresis of K310 proteins (Braconi et al., 2008). Images of gels were acquired and quantitatively evaluated selecting the intensity of unstained active protein bands, and reported these values as arbitrary units in Fig. 4.

During a 2 hours treatment, only S was able to promptly induce catalase activity even when such an enzymatic activity was not detectable under control conditions or in the presence of P or PE (Fig. 4A). It is known, in fact, that a high glucose concentration in culture medium (such as it can be found in the conditions we tested) can exert a negative pressure on antioxidant systems, catalase included (Martínez-Pastor et al., 1996). The activation of catalase is one of the most common cellular responses to redox alterations, since this enzyme is easily induced by a wide range of *stimuli* often related to the energetic status of the cell. In turn, we could hypothesize that catalase activation in S-treated cells might represent an adaptation mechanism to the herbicide-induced oxidative stress.

Results obtained for SOD activity (Fig. 4B) showed quite a different situation. In the case of S, SOD activity was initially lowered (30 minutes treatment) then restored (60 minutes) and increased (120 minutes) in respect to control. In the presence of P and PE, SOD activity was suddenly decreased by both herbicides with effects that were observable throughout the whole observed period.

The general inactivation of SOD observed after short treatments with herbicides could be the result of oxidative damage of this enzyme, which is known to be inactivated by various peroxides (Pigeolet et al., 1990). This can probably reflect the decreased ability of yeast cells to adapt efficiently to the oxidative stress encountered. A sudden decrease in SOD activity had been already associated to brief treatments of human erythrocytes with the herbicide 2,4,5-trichlorophenoxyacetic acid and to one of its metabolites (2,4,5-trichlorophenol) (Bukowska, 2004). Whereas SOD is considered an essential anti-oxidant enzyme, at the same time it can have pro-oxidant effects *in vivo* (Lushchak et al., 2005), and thus SOD inactivation can be read as a defence mechanism as well.

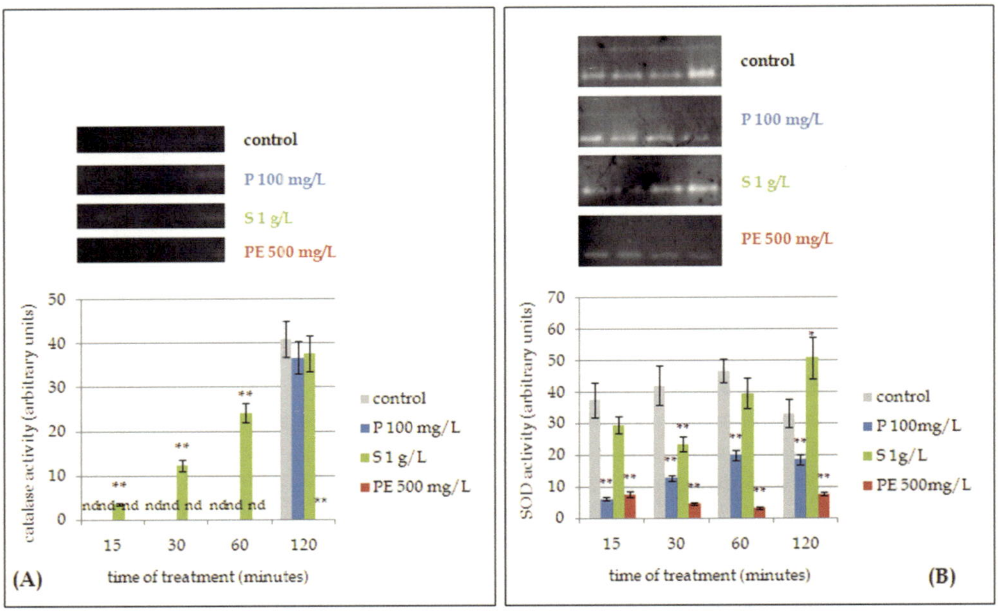

Fig. 4. Native PAGE and staining of active catalase (A) and SOD (B) in K310 cells treated with commercial herbicide formulations (P 100 mg/L, S 1 g/L or PE 500 mg/L) or in the absence of herbicides (control). Quantitative evaluation was performed on band intensities, as detailed under Material and Methods. Data reported are average values of three independent experiments carried out under identical conditions; standard deviations are indicated with vertical bars. *$P < 0.05$ compared with the control culture.; nd: not detectable

Literature reports about the activity of antioxidant enzymes after herbicide treatments are quite heterogeneous, probably reflecting different experimental conditions and the specificity of stress responses in different systems. Other factors that must be taken into account are the different ROS types that can be generated, their toxicity for the cells, their detoxification rate as well as the involvement of enzymatic and non-enzymatic defense mechanisms. Only as a few examples, pure G did not alter catalase and SOD activity in the liver of rats (Beuret et al., 2005) while SOD activity in roots and leaves of pea plants was not affected by the treatment with imazethapyr, an imidazolinone herbicide (Zabalza et al., 2007). Moreover, while catalase activation in yeast had been already associated to the herbicide 2,4-D (Teixeira et al., 2004; Teixeira et al., 2007), either a commercial formulation containing 2,4-D or pure G did not alter catalase activity in mice (Dinamarca et al., 2007). Nevertheless, our data for S are in good agreement with a recent report for another G-based formulation (Roundup Ultra 360 SL) that induced catalase activity in human erythrocytes after very short treatments (Pieniazek et al., 2004).

4.3 Proteomics and redox-proteomics

4.3.1 Comparative proteomic analysis of herbicide-induced effects

In order to gain a deeper insight into the effects of P, S and PE on K310 protein profile, we carried out a comparative proteomic analysis combining 2D-PAGE and protein identification by gel-matching or, alternatively, MALDI-ToF mass spectrometry.

Proteome analysis is conceptually attractive because it fits with the concept that determination of protein rather than mRNA levels has major advantages as it is proteins that carry out functions within the cells (Jamnik & Raspor, 2005). In our work, we found that the herbicides P, S and PE could affect the expression of several yeast proteins, which were identified, classified according to their cell function, and reported in detail in a previous paper of ours (Braconi et al., 2009). Among the differently expressed proteins, many were found to be involved in cell rescue and defence (Fig. 5A). This group includes a heterogeneous set of molecular chaperones, such as heat shock proteins (HSPs), devoted to guarantee the correct protein folding and to avoid protein aggregation (Young, 2004). HSPs are induced in response to a wide variety of *stimuli* and many of them may help in regulating cell metabolism under stressing conditions.

Additionally, the comparative proteomic analysis showed very clearly the under-expression of several enzymes belonging to the carbohydrate metabolism group in herbicide-treated cells (Braconi et al., 2009). Our data were, again, in accordance with works studying cellular responses to oxidative stress (Godon et al., 1998; Magherini et al., 2007; Perzov et al., 2000; Sales et al. 2000; Weeks et al., 2006) and underlined the susceptibility of glycolysis to oxidative insults. A decrease in the activity of glycolytic enzymes is thought to be beneficial during exposure to oxidants (Costa et al., 2002; Shanmuganathan et al., 2004). This finding suggested, in turn, the need of cells to save energy to face the herbicide-induced

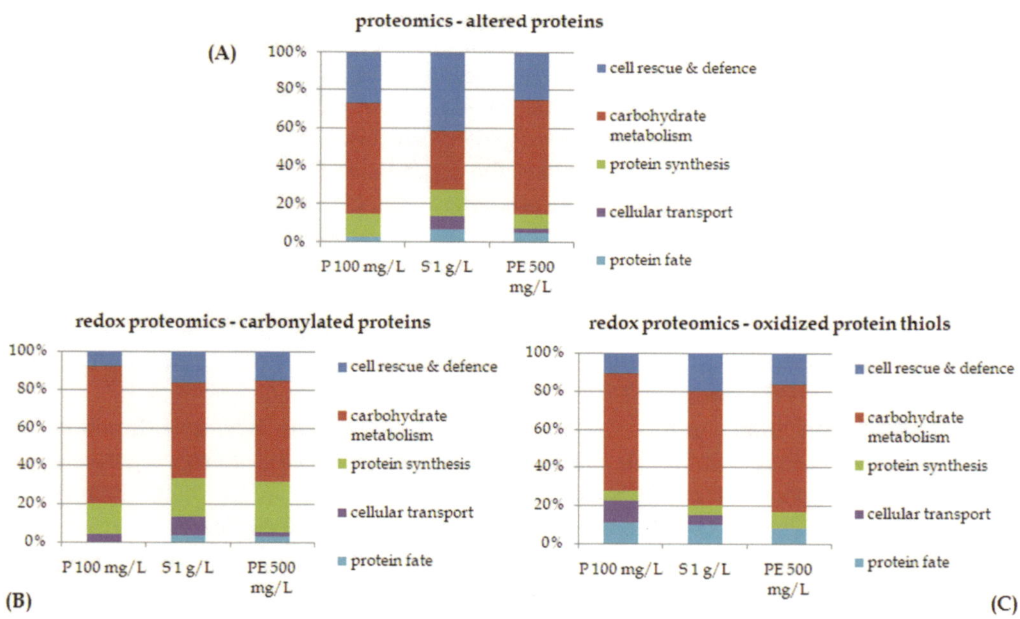

Fig. 5. Functional classification of: proteins differently expressed in K310 cells treated with the herbicides P, S or PE *versus* the untreated control (A), and proteins carbonylated (B) or with oxidized thiols following herbicide treatment (C).

stress as well as the redirection of cell metabolic fluxes to set up appropriate defence mechanisms. Alterations of carbohydrate metabolism for the generation of NADPH is a fundamental cellular response to the oxidative stress, since this factor is required for many defence mechanisms such as glutathione reductase and thioredoxine reductase, which have a critical role in maintaining the proper redox-balance of thiol groups. At the same time, the inhibition of the glycolytic route could support the production of protective substances, such as trehalose.

4.3.2 Redox-proteomic analysis of herbicide-induced oxidation of proteins

Oxidative stress occurs when the rate at which ROS are generated exceeds the capacity of the cell to remove them by anti-oxidants. When an increased ROS production is concomitant with a reduction of anti-oxidant systems, cells have to cope with oxidative damage of proteins, DNA and lipids. A large body of evidence suggests that herbicides may promote toxic events via the intermediate release of ROS which, because of their extreme reactivity, lead to the formation of lesions in target molecules. Although this notion was established for an array of herbicides, including some of those employed in the present study (Soltaninejad. & Abdollahi, 2009; Muniz et al., 2008), very little information is available on the specific molecular targets of such a damage.

Since proteins are considered the most important targets of oxidative stress (Sheehan, 2006), we evaluated through redox-proteomic techniques two commonly investigated oxidative post-translational modifications (PTMs) of proteins: carbonylation and thiol oxidation.

The introduction of carbonyls groups into proteins is an irreversible, easy detectable and non-enzymatic PTM often accompanied by loss of function. For these reasons, it is universally accepted as a good indicator of oxidative stress (Dalle-Donne et al., 2003a; Dalle-Donne et al., 2003b; Dalle-Donne et al., 2006; Dalle-Donne et al., 2005). Carbonylation of mistranslated or otherwise aberrant proteins points to an important physiological role of carbonylation in the control of protein quality. Being irreversible, carbonylation can be a signal for a protein degradation pathway rather than the chaperone-assisted repair. This can in turn act as a mechanism providing amino acids for *de novo* protein synthesis by targeting proteins that are no longer required or have become damaged to the degradation (Nyström, 2005).

In our work we found that carbonylation is a rapid and dynamic event, mainly affecting proteins involved in the cell rescue and defence or in the carbohydrate metabolism (Braconi et al., 2009) (Fig. 5B), with good homology with previous studies on yeast cells treated with oxidizing agents (Cabiscol et al., 2002; Shanmuganathan et al., 2004; Dirmeier et al., 2002; Reverter-Branchat et al., 2004; Sumner et al., 2005). Nonetheless, if the herbicide P and S induced the carbonylation of specific patterns of K310 proteins, in the case of PE protein carbonylation was nearly random (Braconi et al., 2009).

Since carbonylation often induces alterations in protein structure and function, the carbonylation of proteins with fundamental cellular roles might explain the negative effects observed in the analysis of herbicide-induced toxicity in K310 cells. Hence, the carbonylation of HSPs may counteract their beneficial effect in protecting cells and provide at the same time a pivotal decisional checkpoint to determine the fate of cells subjected to oxidative damage. Additionally, carbonylation of glycolytic enzymes might account for herbicide-induced alteration of yeast fermentative performances and have negative relapses in the wine-making process.

Thiol oxidation has a pivotal role during oxidative stress, since it has an increasingly recognized role in protein structure/function and redox signalling (Biswas et al., 2006). If on

the one hand thiol oxidation can activate specific protein functions or protect critical residues from irreversible oxidation, on the other it can generate a wide range of chemically reactive sulphur species propagating the oxidative imbalance in the form of pro- anti-oxidant redox cascades (Jacob et al., 2006).
Either in the case of treatment of K310 cells with P, S or PE, we were able to point out herbicide-mediated oxidative damage towards yeast protein thiols (Braconi et al., 2010). Once classified according to their functions, we found that many of the proteins containing oxidized thiols were involved in carbohydrate metabolism (Braconi et al., 2010) (Fig. 5C), confirming what recently observed by other authors (Le Moan et al., 2006; Magherini et al., 2009; McDonagh et al., 2009). Our results globally indicated that thiol oxidation could coordinate the metabolic response to herbicide-induced oxidative stress and regulate fluxes of carbohydrates (McDonagh et al., 2009); this could eventually have, in turn, negative relapses on the fermentative performance of yeasts. Thiol specific oxidation was observed also for some proteins involved in the stress response or covering specific important functions in yeast cells (Braconi et al., 2010) (Fig. 5C), in good agreement with results from Le Moan (Le Moan et al., 2006).

5. Conclusions and future perspectives

The importance of evaluating the potential toxicity of complete pesticide formulations, rather than just testing their AIs, began to be appreciated only recently. In this context, the yeast *S. cerevisiae* is an optimal model system to study herbicide-induced toxicity and, especially, herbicide-induced oxidative stress. Cell stress studies are usually carried out in baker's yeast strains, often genetically manipulated, under extreme environmental conditions very unlike the physiological ones. Whereas baker's *S. cerevisiae* strains have been widely studied both by genome sequencing and other types of characterization, brewing or wine yeasts are still more neglected. Nevertheless, oenological *S. cerevisiae* strains are much more interesting cell models for eukaryote stress studies, since they are constitutively stress-resistant, adapted to grow under the stressful conditions of grape must (glucose excess and then rapid exhaustion, low assimilable nitrogen, low pH, low content of oxygen, high concentration of ethanol) and thus can represent a good example of tolerance/adaptation response.
In our work, we used a wild-type wine yeast strain, namely K310, for the evaluation of toxic effects induced by three herbicides. Hence, K310 strain should be considered not only a good model cell, but also a microorganism with important biotechnological applications in the wine making process. Globally our findings reinforced the hypothesis that, for eco-toxicological considerations, commercial pesticides should be tested in parallel with their AIs. Our investigations allowed us to highlight how significant amounts of herbicides can inhibit initial wine-making steps and have negative impacts on yeast fermentative ability, having in turn negative consequences for the biological processes that oenological *S. cerevisiae* strains are supposed to properly carry out. Protein oxidation, and specifically carbonylation and thiol oxidation of enzymes involved in the fermentative processes as well as in the cell rescue and defence, was suggested to play a key role in herbicide-induced toxicity.

6. References

Agarwal, S., Sharma, S., Agrawal, V. & Roy, N. (2005). Caloric restriction augments ROS defense in *S. cerevisiae*, by a Sir2p independent mechanism, *Free Radical Research* Vol. 39(No.1):55-62.

Baronian, K. H. (2004). The use of yeast and moulds as sensing elements in biosensors, *Biosensensors & Bioelectronics* Vol. 19(No. 9):953-962.

Benedetti, A. L., de Lourdes Vituri, C., Trentin, A. G., Custódio Domingues, M. A. & Alvarez-Silva, M. (2004). The effects of sub-chronic exposure of Wistar rats to the herbicide Glyphosate-Biocarb®, *Toxicology Letters* Vol. 153:227-232.

Beuret, C. J., Zirulnik, F., & Giménez, M. S. (2005). Effect of the herbicide glyphosate on liver lipoperoxidation in pregnant rats and their fetuses, *Reproductive toxicology* Vol. 19(No.4):501-504.

Biswas, S., Chida, A. S. & Rahman, I. (2006). Redox modifications of protein-thiols: emerging roles in cell signaling, *Biochemical pharmacology* Vol. 71(No. 5):551-564.

Braconi D., Bernardini G., Fiorani M., Azzolini C., Marzocchi B., Proietti F., Collodel G, Santucci A. (2010). Oxidative damage induced by herbicides is mediated by thiol oxidation and hydroperoxides production, *Free Radical Research*, 44(08):891-906

Braconi, D., Bernardini, G., Possenti, S., Laschi, M., Arena, S., Scaloni, A., Geminiani, M., Sotgiu, M. & Santucci, A. (2009). Proteomics and redox-proteomics of the effects of herbicides on a wild-type wine *Saccharomyces cerevisiae* strain, *Journal of proteome research* Vol. 8(No. 1):256-267.

Braconi, D., Sotgiu, M., Bernardini, G., Paffetti, A.,. Tasso, F., Alisi, C., Martini, S., Rappuoli, R., Sprocati, A.R., Rossi, C. & Santucci A. (2006a). Wild-type wine *Saccharomyces cerevisiae* as a tool to evaluate the effects on eukaryotic life of locally used herbicides, *International Journal of Ecodynamics* Vol. 1(No. 3):266-283.

Braconi, D., Sotgiu, M., Millucci, L., Paffetti, A., Tasso, F., Alisi, C., Martini, S., Rappuoli, R., Lusini, P., Sprocati, A. R., Rossi, C. & Santucci, A. (2006b). Comparative analysis of the effects of locally used herbicides and their active ingredients on a wild-type wine *Saccharomyces cerevisiae* strain, *Journal of Agricultural and Food Chemistry* Vol. 54(No. 8):3163–3172.

Braconi, D., Possenti, S., Laschi, M., Geminiani, M., Lusini, P., Bernardini, G. & Santucci A. (2008). Oxidative damage mediated by herbicides on yeast cells. Journal of Agricultural and Food Chemistry Vol. 56(No. 10):3836-3845.

Bradford, M. M. (1976). A rapid and sensitive method for the quantitation of microgram quantities of protein utilizing the principle of protein-dye binding, *Analitical Biochemistry* Vol. 72:248-254.

Bukowska, B. (2004). 2,4,5-T and 2,4,5-TCP induce oxidative damage in human erythrocytes: the role of glutathione, *Cell biology international* Vol. 28(7):557-563.

Cabiscol, E., Bellí, G., Tamarit, J., Echave, P., Herrero, E., & Ros, J. (2002). Mitochondrial Hsp60, resistance to oxidative stress, and the labile iron pool are closely connected in Saccharomyces cerevisiae. *The Journal of Biological Chemistry* Vol. 277(No. 46):44531-44538.

Cabral, M. G., Sá-Correia, I. & Viegas, C. A. (2004). Adaptative responses in yeast to the herbicide 2-methyl-4-chlorophenoxyacetic acid at the level of intracellular pH homeostasis, *Journal of applied microbiology* Vol. 96(No. 3):603-612.

Cabral, M. G., Viegas, C. A., Teixeira, M. C. & Sá-Correia I. (2003). Toxicity of chlorinated phenoxyacetic acid herbicides in the experimental eukaryotic model *Saccharomyces cerevisiae*: role of pH and of growth phase and size of the yeast cell population, *Chemosphere* Vol. 51(No. 1):47-54.

Cabras P., Farris G. A., Fiori M. G. & Pusino A. (2003). Interaction between fenhexamid and yeasts during the alcoholic fermentation of Saccharomyces cerevisiae, *Journal of Agriculture and Food Chemistry* Vol. 51(No. 17):5012-5015.

Cabras P., Garau V. L., Angioni A., Farris G. A., Budroni M.& Spanedda L. (1995). Interactions during fermentation between pesticides and oenological yeasts producing H_2S and SO_2, *Applied microbiology and biotechnology* Vol. 43(No.2):370-373.

Cabras, P., Angioni, A., Garau, V. L., Pirisi, F. M., Cabitza, F., Pala, M. & Farris G. A. (2000). Fate of quinoxyfen residues in grapes, wine, and their processing products, *Journal of Agriculture and Food Chemistry* Vol. 48(No. 12):6128-6131.

Campanella, L., Favero, G. & Tomassetti M. (1995). Immobilised yeast cells biosensor for total toxicity testing, *The Science of the total environment* Vol. 171(No. 1-3):227-234.

Candiano, G., Bruschi, M., Musante, L., Santucci, L., Ghiggeri, G. M., Carnemolla, B., Orecchia, P., Zardi, L. & Righetti, P. G. (2004). Blue silver: a very sensitive colloidal Coomassie G-250 staining for proteome analysis, *Electrophoresis* Vol.25(No. 9):1327-1333.

Carvalho, E., del Valle, M., Günther Sillero, M. A., Moradas-Ferreira, P. & Sillero, A. (2003). H_2O_2, but not menadione, provokes a decrease in the ATP and an increase in the inosine levels in *Saccharomyces cerevisiae*. An experimental and theoretical approach, *European Journal of Biochemistry* Vol. 270(No. 7):1578-1589.

Chan, Y. C., Chang, S. C., Hsuan, S. L., Chien, M. S., Lee, W. C., Kang, J. J., Wang, S. C., & Liao, J. W. (2007). Cardiovascular effects of herbicides and formulated adjuvants on isolated rat aorta and heart, *Toxicology in vitro* Vol. 21(No. 4):595-603.

Costa, V. & Moradas-Ferreira P. (2001). Oxidative stress and signal transduction in *Saccharomyces cerevisiae*: insights into ageing, apoptosis and diseases, *Molecular Aspects of Medicine* Vol. 22(No. 4-5):217-246.

Costa, V. M., Amorim, M. A., Quintanilha, A. & Moradas-Ferreira, P. (2002). Hydrogen peroxide-induced carbonylation of key metabolic enzymes in *Saccharomyces cerevisiae*: the involvement of the oxidative stress response regulators Yap1 and Skn7, *Free radical biology & medicine* Vol. 33(No. 11):1507-1515.

Cox, C. & Surgan, M. (2006). Unidentified inert ingredients in pesticides: implications for human and environmental health, *Environmental health perspectives* Vol. 114(No. 12):1803-1806.

Dalle-Donne, I., Aldini, G., Carini, M., Colombo, R., Rossi, R., & Milzani, A. (2006). Protein carbonylation, cellular dysfunction, and disease progression, *Journal of cellular and molecular medicine* Vol. 10(No. 2):389–406.

Dalle-Donne, I., Giustarini, D., Colombo, R., Rossi, R., & Milzani, A. (2003b). Protein carbonylation in human diseases, *Trends in molecular medicine* Vol. 9(No. 4):169–176.

Dalle-Donne, I., Rossi, R., Giustarini, D., Milzani, A., & Colombo, R. (2003a). Protein carbonyl groups as biomarkers of oxidative stress, *Clinica chimica acta; international journal of clinical chemistry* Vol. 329(No.1-2):23–38.

Dalle-Donne, I., Scaloni, A., Giustarini, D., Cavarra, E., Tell, G., Lungarella, G., Colombo, R., Rossi, R., Milzani, A. (2005). Proteins as biomarkers of oxidative/nitrosative stress in diseases: the contribution of redox proteomics, *Mass spectrometry reviews* Vol. 24(No. 1), 55–99.

Dinamarca, V. M., Hidalgo, M. E. & Cavieres, M. F. (2007). Lack of effects of 2,4-dichlorophenoxyacetic acid administration on markers of oxidative stress during early pregnancy in mice, *Toxicology* Vol. 237(No. 1-3):104-110.

Dirmeier, R., O'Brien, K. M., Engle, M., Dodd, A., Spears, E., & Poyton, R. O. (2002). Exposure of yeast cells to anoxia induces transient oxidative stress. Implications for the induction of hypoxic genes. *The Journal of Biological Chemistry* Vol. 277(No.38): 34773-34784.

dos Santos, J., Ferreira, E. A., Kasuya, M. C. M., da Silva, A. A. & de Oliveira Procópio, S. (2005). Tolerance of *Bradyrhizobium* strains to glyphosate formulations, *Crop Protection* Vol. 24(No. 6):543-547.

Estruch F. (2000). Stress-controlled transcription factors, stress-induced genes and stress tolerance in budding yeast, *FEMS Microbiology Reviews* Vol. 24(No. 4):469-486.

Gehin, A., Guillame, Y. C., Millet, J., Guyon, C., & Nicod, L. (2005). Vitamins C and E reverse effect of herbicide-induced toxicity on human epidermal cells HaCaT: A biochemometric approach, *International journal of pharmaceutics* Vol. 288(No. 2):219-226.

Godon, C., Lagniel, G., Lee, J., Buhler, J. M., Kieffer, S., Perroti, M., Boucherie, H., Toledano, M. B. & Labarre, J. (1998). The H_2O_2 stimulon in *Saccharomyces cerevisiae*, *The Journal of biological chemistry* Vol. 273(No. 35):22480-22489.

Goffeau, A., Barrell, B. G., Bussey, H., Davis, R. W., Dujon, B., Feldmann, H., Galibert, F., Hoheisel, J. D., Jacq, C., Johnston, M., Louis, E. J., Mewes, H. W., Murakami, Y., Philippsen, P., Tettelin, H. & Oliver, S. G. (1996). Life with 6000 genes, *Science* Vol. 274(No. 5287):546,563-567.

Jacob, C., Knight, I. & Winyard, P. G. (2006). Aspects of the biological redox chemistry of cysteine: from simple redox responses to sophisticated signalling pathways, *Biological chemistry* Vol. 387(No. 10-11):1385-1397.

Jamnik, P. & Raspor, P. J. (2005). Methods for monitoring oxidative stress response in yeasts, *Journal of biochemical and molecular toxicology* Vol. 19(No. 4):195-203.

Kim, J. R., Yoon, H. W., Kwon, K. S., Lee, S. R. & Rhee, S. G. (2000). Identification of proteins containing cysteine residues that are sensitive to oxidation by hydrogen peroxide at neutral pH, *Analytical Biochemistry* Vol. 283(No.2):214-221.

Le Moan, N., Clement, G., Le Maout, S., Tacnet, F. & Toledano, M. B. (2006). The *Saccharomyces cerevisiae* proteome of oxidized protein thiols: contrasted functions for the thioredoxin and glutathione pathways, *The Journal of biological chemistry* Vol. 281(No. 15):10420-10430.

Low, F. L., Shaw, I. C. & Gerrard, J. A. (2005). The effect of *Saccharomyces cerevisiae* on the stability of the herbicide glyphosate during bread leavening, *Letters in Applied Microbiology* Vol. 40(No. 2):133-137.

Lushchak, V., Semchyshyn, H., Mandryk, S. & Lushchak, O. (2005). Possible role of superoxide dismutases in the yeast *Saccharomyces cerevisiae* under respiratory conditions, *Archives of biochemistry and biophysics* Vol. 441(No. 1):35-40.

Magherini, F., Carpentieri, A., Amoresano, A., Gamberi, T., De Filippo, C., Rizzetto, L., Biagini, M., Pucci, P. & Modesti, A. (2009). Different carbon sources affect lifespan and protein redox state during *Saccharomyces cerevisiae* chronological ageing, *Cellular and molecular life sciences : CMLS* Vol. 66(No. 5):933-947.

Magherini, F., Tani, C., Gamberi, T., Caselli, A., Bianchi, L., Bini, L., & Modesti, A. (2007). Protein expression profiles in *Saccharomyces cerevisiae* during apoptosis induced by H_2O_2, *Proteomics* Vol. 7(No. 9):1434-1445.

Malik, A. (2004). Metal bioremediation through growing cells, *Environment International* Vol. 30(No. 2):261-278.

Mann, R. M. & Bidwell, J. R. (1999). The toxicity of glyphosate and several glyphosate formulations to four species of southwestern Australian frogs, *Archives of environmental contamination and toxicology* Vol. 36(No. 2):193-199.

Martínez-Pastor, M. T., Marchler, G., Schüller, C., Marchler-Bauer, A., Ruis, H. & Estruch, F. (1996). The *Saccharomyces cerevisiae* zinc finger proteins Msn2p and Msn4p are required for transcriptional induction through the stress response element (STRE), *The EMBO Journal* Vol.15(No. 9):2227–2235.

Martini, S., Ricci, M., Bartolini, F., Bonechi, C., Braconi, D., Millucci, L., Santucci, A. & Rossi, C. (2006). Metabolic response to exogenous ethanol in yeast: an in vivo NMR and mathematical modelling approach, *Biophysical Chemistry* Vol. 120 (No.2):135-142.

Martini, S., Ricci, M., Bonechi, C., Trabalzini, L., Santucci, A. & Rossi, C. (2004). In vivo ^{13}C-NMR and modelling study of metabolic yield response to ethanol stress in a wild-type strain of *Saccharomyces cerevisiae, FEBS Letters* Vol. 23(No.564):63-68.

McDonagh, B., Ogueta, S., Lasarte, G., Padilla, C. A. & Bárcena, J. A. (2009). Shotgun redox proteomics identifies specifically modified cysteines in key metabolic enzymes under oxidative stress in *Saccharomyces cerevisiae, Journal of proteomics* Vol. 72(No. 4):677-689.

Muniz, J. F., McCauley, L., Scherer, J., Lasarev, M., Koshy, M., Kow, Y. W., Nazar-Stewart V. & Kisby, G. E. (2008). Biomarkers of oxidative stress and DNA damage in agricultural workers: a pilot study, *Toxicology and applied pharmacology* Vol. 227(No. 1):97-107.

Nyström, T. (2005). Role of oxidative carbonylation in protein quality control and senescence, *The EMBO Journal* Vol. 24(No.7):1311-1317.

Ornstein, L. (1964). Disc electrophoresis I: background and theory, *Annals of the New York Academy of Sciences* Vol. 121:321-349.

Peixoto, F. (2005). Comparative effect of the Roundup and glyphosate on mitochondrial oxidative phosphorylation, *Chemosphere* Vol. 61(No. 8):1115-1122.

Perzov, N., Nelson, H., & Nelson, N. (2000). Altered distribution of the yeast plasma membrane H^+-ATPase as a feature of vacuolar H^+-ATPase null mutants, *The Journal of biological chemistry* Vol. 275(No. 51):40088-40095.

Pieniazek, D., Bukowska, B., & Duda, W. (2004). Comparison of the effect of Roundup Ultra 360 SL pesticide and its active compound glyphosate on human erytrocytes, *Pesticide biochemistry and physiology*, Vol. 79(No.2):58-63.

Pigeolet, E., Corbisier, P., Houbion, A., Lambert, D., Michiels, C., Raes, M., Zachary, A. D. & Remacle, J. (1990). Glutathione peroxidase, superoxide dismutase, and catalase inactivation by peroxides and oxygen derived free radicals, *Mechanisms of ageing and development* Vol. 51(No. 3):283–297.

Reinheckel, T., Körn, S., Möhring, S., Augustin, W., Halangk, W., & Schild, L. (2000). Adaptation of protein carbonyl detection to the requirements of proteome analysis demonstrated for hypoxia/reoxygenation in isolated rat liver mitochondria, *Archives of biochemistry and biophysics* Vol. 376(No. 1):59-65.

Reverter-Branchat, G., Cabiscol, E., Tamarit, J., & Ros, J. (2004). Oxidative damage to specific proteins in replicative and chronological-aged *Saccharomyces cerevisiae*: common targets and prevention by calorie restriction. *The Journal of Biological Chemistry* Vol. 279(No. 30):31983-31989.

Ribeiro I. C., Veríssimo I., Moniz L., Cardoso H., Sousa M. J., Soares A. M. & Leão C. (2000). Yeasts as a model for assessing the toxicity of the fungicides Penconazol, Cymoxanil and Dichlofluanid, *Chemosphere* Vol. 41(No. 10):1637-1642.

Ricci, M., Martini, S., Bonechi, C., Trabalzini, L., Santucci, A. & Rossi, C. (2004). Inhibition effects of ethanol on the kinetics of glucose metabolism by *S. cerevisiae*: NMR and modelling study, *Chemical Physics Letters* Vol. 387(No. 4-6):377-382.

Richard, S., Moslemi, S., Sipahutar, H., Benachour, N., & Seralini, G. E. (2005). Differential effects of glyphosate and Roundup on human placental cells and aromatase, *Environmental health perspectives* Vol. 113(No. 6):716-720.

Sales, K., Brandt, W. F., Rumbak, E., & Lindsey, G. G. (2000). The LEA-like protein HSP 12 in *Saccharomyces cerevisiae* has a plasma membrane location and protects membranes against desiccation and ethanol-induced stress, *Biochimica et biophysica acta* Vol. 1463(No. 2):267-278.

Santucci, A., Trabalzini, L., Bovalini, L., Ferro, E., Neri, P., & Martelli, P. (2000). Differences between predicted and observed sequences in *Saccharomyces cerevisiae, Electrophoresis* Vol. 21(No. 17):3717-3723.

Séralini G-E. (2005). "Inert" and active ingredients, *Environmental health perspectives* Vol. 113(No. 10):A658.

Shanmuganathan, A., Avery, S. V., Willetts, S. A., & Houghton, J. E. (2004). Copper-induced oxidative stress in *Saccharomyces cerevisiae* targets enzymes of the glycolytic pathway, *FEBS Letters* Vol. 556(No.1-3):253-259.

Sheehan, D. (2006). Detection of redox-based modification in two-dimensional electrophoresis proteomic separations, *Biochemical and biophysical research communications* Vol. 349(No.2):455-462.

Simões, T., Teixeira, M. C., Fernandes, A. R. & Sá-Correia, I. (2003). Adaptation of *Saccharomyces cerevisiae* to the herbicide 2,4-dichlorophenoxyacetic acid, mediated by Msn2p- and Msn4p-regulated genes: Important role of SPI1, *Applied and Environmental Microbiology* Vol. 69(No. 7):4019-4028.

Soltaninejad, K. & Abdollahi M. (2009). Current opinion on the science of organophosphate pesticides and toxic stress: a systematic review, *Medical science monitor : international medical journal of experimental and clinical research* Vol. 15(No. 3):RA75-90.

Sumner, E. R., Shanmuganathan, A., Sideri, T. C., Willetts, S. A., Houghton, J. E., & Avery, S. V. (2005). Oxidative protein damage causes chromium toxicity in yeast. *Microbiology* Vol. 151(No.6):1939-1948.

Surgan, M. H. (2005). Toxicity tests: "inert" and active ingredients, *Environmental health perspectives* Vol. 113(No. 10):A657-658.

Teixeira, M. C., Duque, P. & Sá-Correia I. (2007). Environmental genomics: mechanistic insights into toxicity of and resistance to the herbicide 2,4-D, *Trends in biotechnology* Vol. 25(No. 8):363-370.

Teixeira, M. C., Santos, P. M., Fernandes, A. R. & Sá-Correia, I. (2005). A proteome analysis of the yeast response to the herbicide 2,4-dichlorophenoxyacetic acid, *Proteomics* Vol. 5(No. 7):1889-1901.

Teixeira, M. C., Telo, J. P., Duarte, N. F. & Sá-Correia, I. (2004). The herbicide 2,4-dichlorophenoxyacetic acid induces the generation of free-radicals and associated oxidative stress responses in yeast, *Biochemical and biophysical research communications* Vol. 324(No. 3):1101-1107.

Tobiassen, L. S., Nielsen, E., Nørhede, P., & Ladefoged, O. (2003). Report on the Health Effects of Selected Pesticide Coformulants, http://www.mst.dk/.

Trabalzini, L., Paffetti, A., Scaloni, A., Talamo, F., Ferro, E., Coratza, G., Bovalini, L., Lusini, P., Martelli, P., Santucci, A. (2003). Proteomic response to physiological fermentation stresses in a wild-type wine strain of *Saccharomyces cerevisiae, Biochemical Journal* Vol. 370(No. 1):35-46.

Trabalzini, L.; Paffetti, A.; Ferro, E., Scaloni, A.; Talamo, F.; Millucci, L.; Martelli, P.; Santucci, A. (2003b). Proteomic characterization of a wild-type wine strain of *Saccharomyces cerevisiae, The Italian Journal of Biochemistry* Vol. 52(No. 4):145-153.

Tsui, M. T. & Chu, L. M. (2003). Aquatic toxicity of glyphosate-based formulations: comparison between different organisms and the effects of environmental factors, *Chemosphere* Vol. 52(No. 7), 1189-97.

Vagnoli, P., Musmanno, R. A., Cresti, S., Di Maggio, T., Coratza, G. (1993). Occurrence of killer yeasts in spontaneous wine fermentations from the Tuscany region of Italy, *Applied & Environmental Microbiology* Vol. 59(No. 12):4037-4043.

Viegas, C. A., Cabral, M. G., Teixeira, M. C., Neumann, G., Heipieper, H. J. & Sá-Correia, I. (2005). Yeast adaptation to 2,4-dichlorophenoxyacetic acid involves increased membrane fatty acid saturation degree and decreased *OLE1* transcription, *Biochemical and biophysical research communications* Vol. 330(No. 1):271-278.

Weeks, M. E., Sinclair, J., Butt, A., Chung, Y., Worthington, J. L., Wilkinson, C. R., Griffiths, J, Jones, N, Waterfield, M. D. & Timms, J. F. (2006). A parallel proteomic and metabolomic analysis of the hydrogen peroxide- and Sty1p-dependent stress response in *Schizosaccharomyces pombe*, *Proteomics* Vol. 6(No. 9):2772-2796.

Young, J. C. (2004). Pathways of chaperone-mediated protein folding in the cytosol, *Nature reviews. Molecular cell biology* Vol. 5(No. 10):781-791.

Zabalza, A., Gaston, S., Sandalio, L.M., del Río, L.A., Royuela, M. (2007). Oxidative stress is not related to the mode of action of herbicides that inhibit acetolactate synthase, *Environmental and experimental botany* Vol. 59(No. 2):150-159.

25

Herbicides in Argentina. Comparative Evaluation of the Genotoxic and Cytotoxic Effects on Mammalian Cells Exerted by Auxinic Members

Sonia Soloneski and Marcelo L. Larramendy
Faculty of Natural Sciences and Museum, National University of La Plata
Argentina

1. Introduction

In epidemiological and in experimental biology studies, the existence of an increasing interest in biomonitoring markers to achieve both a measurement and an estimation of biologically active/passive exposure to genotoxic pollutants, is nowadays a real fact.

Significant contributions to the advancement of pesticide toxicology came and continue to come from many sources, e.g., academic, governmental/regulatory, and industrial. Regulatory agencies, private sector, and academia worldwide combine expertise to assess pesticide safety and risk potential demanding adequate data of high quality to serve as the basis for establishing safe exposure levels. The extent of testing was and is often determined by the depth of the science, as well as the chemical and physical properties of the agent and the extent of exposure. The importance of pesticide toxicology has evolved from listing poisons to protecting the public from the adverse effects of chemicals, from simply identifying effects (qualitative toxicology), to identifying and quantifying human risks from exposure (quantitative toxicology), and from observing phenomena to experimenting and determining mechanisms of action of pesticide agents and rational management for intoxication. Humans and living species may, therefore, be exposed to a number of different chemicals through dietary and other routes of exposure.

Nonetheless, there continues to be concern that the presence of multiple chemical residues in foods may cause adverse health side effects, including effects that would not be predicted from consideration of single exposures to individual compounds. It is known that the regulatory system for pesticide products found in foods does not routinely address the toxic effects of different substances in combination. The implications, both for risk assessment and for approval processes, of exposure to mixtures of pesticides are among the topics examined by different international agencies, e.g., World Health Organization (WHO, www.who.int), International Agency for Research on Cancer (IARC, www.iarc.fr), United States Environmental Protection Agency (EPA, www.epa.gov), European Chemicals Agency (ECHA, www.echa.auropa.eu), Health Canada Pest Management Regulatory Agency (PMRA, www.pmra-arla.gc.ca), among others. These international agencies, particularly WHO and EPA, have contributed a great deal in their attempts to control pesticide poisoning. They continue their efforts, with particular emphasis on safety in the use of

pesticides and applied research activities, playing the role of intermediary for the involvement of agrochemical industries in safety activities.
It has been strongly recommended that the nature and extent of combined exposure to pesticides and related chemicals, together with the likelihood of any adverse effects that might result, should be evaluated, when carrying out risk assessment. Furthermore, a scientific and systematic framework should be established to decide when it is appropriate to carry out combined risk assessments of exposures to more than one pesticide. Finally, it has been also recommended that groups of pesticides having common targets of toxicological action should be identified (www.food.gov.uk).
Pesticides are ubiquitous on the planet and they are employed to control or eliminate a variety of agricultural and household pests that can damage crops and livestock and reduce the productivity. Despite the many benefits of the use of pesticides in crops field and its significant contribution to the lifestyles we have come to expect, pesticides can also be hazardous if not used appropriately and many of them may represent potential hazards due to the contamination of food, water, and air, which can result in severe health problems not only for humans but also for ecosystems (WHO-FAO, 2009). The actual number of pesticide-related illnesses is unknown, since many poisonings go unreported. It has been estimated that at least three million cases of pesticide poisoning occur worldwide each year (www.who.int). The majority of these poisonings occur in developing countries where less protection against exposure is achieved, knowledge of health risks and safe use is limited or even unknown. Studies in developed countries have demonstrated the annual incidence intoxication in agricultural workers can reach values up to 182 per million and 7.4 per million among full time workers (Calvert et al., 2004) and schoolchildren (Alarcon et al., 2005), respectively. However, the number of poisonings increases dramatically in emerging countries where the marketing of pesticides is often uncontrolled or illicit and the misbranded or unlabelled formulations are sold at open stands (www.who.int). Yet, cases of pesticide intoxication may be the result of various causes in different regions of the world. In emerging countries, where there is insufficient regulation, lack of surveillance systems, less enforcement, lack of training, inadequate or reduced access to information systems, poorly maintained or nonexistent personal protective equipments, and larger agriculturally based populations, the incidences are expected, then, to be higher (IFCS, 2003). Despite the magnitude of the problem of pesticide poisoning, there have been very few detailed studies around the world to identify the risk factors involved with their use. The use of pesticides banned in industrialized countries, in particular, highly toxic pesticides as classified by WHO, EPA, and IARC, obsolete stockpiles and improper storage techniques may provide unique risks in the developing world, where 25% of the global pesticide production is consumed (WHO-FAO, 2009). Particularly, the impact of increased deregulation of agrochemicals in Latin America threatens to increase the incidence of pesticide poisoning, which has already been termed a serious public health problem throughout the continent by the WHO. Many of the pesticides used in Latin America are United States exports and the companies can make a number of changes to ensure the "safe" use of their products. However, the social, economic and cultural conditions under which they are used, pesticides acutely poison hundreds of thousands each year, including many children. In the majority of Latin American countries, poisoning registries are so inadequate that most acute poisoning cases never get recorded. Meanwhile, health effects of chronic or long-term pesticide exposures such as cancer or birth defects are not available, omissions that serve to hide the epidemic proportion of pesticide-related illness in the region. In Argentina, e.g.,

available official data revealed that 79% of the intoxications due to pesticides are related with the use of herbicides followed by insecticides and fungicides (www.msal.gov.ar), values that correlate with the evolution of the phytosanitary market demonstrating that herbicides accounted for the largest portion of total use (69%), followed by insecticides (13%), and fungicides (11%) (www.casafe.org). Consequently, Argentina a larger producer of cereals, including soy, is actually the world eight-largest agrochemical market. The country has seen an explosion in genetically modified soybean production with soy exports topping $16.5 billion in 2008 (www.casafe.org). The fertile South American nation is now the world's third largest producer of soy, trailing behind the United States and Brazil.

2. Herbicides. Auxinic herbicides

The most widely applied agrochemicals around the world are herbicides and consequently the environment is inevitably exposed to these chemicals. Such large amount of herbicides released into the environment may present an impending hazard to living organisms. Exposure to some of these herbicides may lead to alterations in the genetic material thereby causing mutagenicity, carcinogenicity, teratogenicity, and immunotoxicity among other side effects (IARC, 1977, 1999; Dearfield et al., 1999).

The auxinic herbicides have been around since World War II and were the first selective herbicides developed. Herbicides are classified as auxinic based on their growth-promoting effects observed in plant cell cultures, specific tissue systems (coleoptiles, roots), and in whole plants (Pipke et al., 1987; Liu et al., 1999). Generally, the auxinic herbicides are used to selectively control broadleaf weeds in grass crops such as cereal grains and turfgrass swards (Pipke et al., 1987; Reinbothe et al., 1996). These agrochemicals are usually applied as foliar treatments but at higher doses can be used as pre-emergent treatments (Reinbothe et al., 1996). The general susceptibility of dicotyledonous species and tolerance of monocotyledonous species to these herbicides is primarily determined by differences in plant morphology, rate of herbicide translocation and metabolism. For instance, the destruction of the phloem of dicotyledonous species results from abnormal tissue proliferation after exposure to auxinic herbicides. Monocotyledonous species are tolerant since the phloem is scattered in bundles surrounded by protective sclerenchyma tissue. Broadleaf species can be tolerant because they metabolize the herbicide to a less toxic form. These herbicides are considered mimics of the natural plant auxins and are thought to induce changes in gene expression leading to plant death (Reinbothe et al., 1996; Liu et al., 1999). Although they continue to be a very important class of herbicides, their precise mode of action is still unknown. In plants, as it has been stated, these chemicals mimic the action of auxins, hormones that stimulates growth, but in mammals and other species no mimic hormonal activity has been reported (Osterloh et al., 1983).

Among this family of herbicides, the 2,4–dichlorophenoxyacetic acid, commonly known as 2,4-D, and the 3,6-dichloro-2-methoxybenzoic acid, commonly known as Dicamba, are two post-emergent auxinic herbicides released in large amount daily into the environment worldwide. This family of herbicides includes many very effective broadleaf weed killers employed in lawns, golf courses, rights-of-way, and agricultural fields.

2,4-D is an herbicide from the phenoxy acid family that is used post-emergence for selective control of a wide variety of broadleaf and aquatic weeds and forestry applications. It is produced in a variety of forms, including: acid, salt, amine and ester. While at low concentrations 2,4-D acts as an auxin analogue promoting plant growth, increasing cell-wall

plasticity, biosynthesis of proteins and the production of ethylene, at high concentrations it is lethal and is employed as herbicide against broad-leafed and woody plants (Sinton et al., 1986; Devine, 1993; Tripathy et al., 1993). Worldwide, it is the most extensively used herbicide, and third most widely employed in the United States (www.epa.gov).
Dicamba, member of the benzoic acid family, is a chlorinated benzoic acid-derivative compound registered in the United States as a post-emergent herbicide in 1967 (EPA, 1983). It is produced in a variety of forms, including acid and different kinds of salts, e.g., dimethlylammonium salt, potassium salt, and sodium salt, among others (FAO, 2001). This compound is used in different crops, e.g. cereals, maize, sorghum, sugar cane, asparagus, perennial seed grasses, turf, pastures, rangeland, and non-crop land against annual and perennial broad-leaved weeds and brush species (FAO, 2001).

3. Genotoxicity and cytotoxicity of 2,4-D

On the basis of its acute toxicity, 2,4-D has been classified as a class II member (moderately hazardous) by WHO (http://www.who.int/ipcs/publications/pesticides hazard/en/) and slightly to moderately toxic (category II-III) by EPA (EPA, 1974).
Genotoxicity and cytotoxicity studies have been conducted with this auxinic member using several end-points on different cellular systems. A summary of the results reported so far is presented in Table 1. On bacterial systems, either the Ames test or reverse mutation tests performed on both *Salmonella typhimurium* and *Bacillus subtilis* gave negative results regardless of the presence or absence of a rat liver metabolic activation system (Charles et al., 1999; Grabinska-Sota et al., 2002). Whereas the herbicide induced DNA adducts on *Saccharomyces cerevisiae* (Teixeira et al., 2004), negative results were obtained for the induction of unscheduled DNA synthesis in primary rat hepatocytes (Charles et al., 1999). When tested for its carcinogenic potential, the transformation assay in Syrian hamster embryo assay gave positive results (Maire et al., 2007). Induction of DNA single-strand breaks estimated by the alkaline comet assay was evaluated in normal and transformed cells exposed *in vitro* to 2,4-D. González et al. (2005), Maire et al. (2007), and Sandal and Yilmaz (2010) observed an increased frequency of DNA primary lesions in CHO and SHE cells as well as in human lymphocytes. On the other hand, negative results were also revealed when this end-point was assayed on the same cell type by others researchers (Sorensen et al., 2005; Sandal & Yilmaz, 2010). However, Maire and co-workers (2007) showed that 2,4-D was unable to induce DNA fragmentation in SHE cells. Both González et al. (2005) and Soloneski et al. (2007) demonstrated the ability of the herbicide to induce sister-chromatid exchanges (SCEs) in CHO cells and human lymphocytes treated *in vitro*, respectively. An increased frequency of chromosomal aberrations was reported in V79 cells and human lymphocytes-treated *in vitro* in the presence/absence of rat liver metabolic activation system (Pavlica et al., 1991; Zeljezic & Garaj-Vrhovac, 2004) but not when the S9 fraction was absent (Mustonen et al., 1986). Zeljezic and Garaj-Vrhovac (2004) reported the induction of micronuclei in human lymphocytes regardless of the presence or absence of S9 fraction. The induction of alterations in the cell-cycle progression of different cellular systems including plant and V79 cells, human lymphocytes and bovine cells were reported to occur after *in vitro* exposure to 2, 4-D (Basrur et al., 1976; Bayliss, 1977; Pavlica et al., 1991, 2005; Soloneski et al., 2007). However, González and co-workers (2005) were unable to demonstrate such cytotoxic effect in CHO cells. Finally, controversial results were reported for the cell viability assay on yeast and mammalian cells (Sorensen et al., 2004; Teixeira et al., 2004). Similar end-points for both genotoxicity and cytotoxicity were also applied in *in vivo*

End-point System	Concentration[a]	Results	Referencies
In vitro assays			
Ames test			
Salmonella typhimurium + S9	96.1 - 9610 μg/plate	-	Charles et al., 1999
H17 Rec$^+$, M45 Rec$^-$ reverse mutation			
Bacillus subtilis	3x10^{-5} - 90 kg m^{-3}	-	Grabinska-Sota et al., 2002
Transformation assay			
SHE cells	11.5 - 23 μM	+	Maire et al., 2007
DNA adducts			
Saccharomyces cerevisiae	0.45 - 0.65 mM	+	Teixeira et al., 2004
UDS			
Primary rat hepatocytes	2.42 - 96.9 μg/ml	-	Charles et al., 1999
Alkaline comet assay			
CHO cells	200 μM - 4 mM	-	Sorensen et al., 2005
CHO cells	2 - 10 μg/ml	+	González et al., 2005
SHE cells	11.5 - 23 μM	+	Maire et al., 2007
Non-smokers HL	1 - 10 μM	-	Sandal & Yilmaz, 2010
Smokers HL	10 μM	+	Sandal & Yilmaz, 2010
DNA fragmentation analysis			
SHE cells	4.5 - 34 μM	-	Maire et al., 2007
SCE assay			
CHO cells	2 - 10 μg/ml	+	González et al., 2005
Non-smokers HL	10 - 50 μg/ml	+	Soloneski et al., 2007
Chromosomal aberrations			
V79 cells	10 μg/ml	+	Pavlica et al., 1991
Non-smokers HL +/- S9	0.4 - 4μg/ml	+	Zeljezic & Garaj-Vrhovac, 2004
Non-smokers HL	0.125 – 0.350 mM	-	Mustonen et al., 1986
Micronuclei assay			
Non-smokers HL +/- S9	0.4 - 4 μg/ml	+	Zeljezic & Garaj-Vrhovac, 2004
Alteration in CCP			
Daucus carota cells	15 - 30 μg/ml	+	Bayliss, 1977
CHO cells	2 - 10 μg/ml	-	González et al., 2005
V79 cells	10 μg/ml	+	Pavlica et al., 1991
Bovine cells	2 - 20 mg/L	+	Basrur et al., 1976
Non-smokers HL	25 - 50 μg/ml	+	Soloneski et al., 2007
Cell viability			
Saccharomyces cerevisiae	0.45 - 0.65 mM	+	Teixeira et al., 2004
CHO cells	100 - 750 μM	-	Sorensen et al., 2004
In vivo assays			
Root tip assay			
Allium cepa	25 - 100 ppm	+	Kumari & Vaidyanath,

End-point System	Concentration[a]	Results	Referencies
Chlorophyll mutation, specific locus			1989
Oryza sativa	25 - 100 ppm	+	Kumari & Vaidyanath, 1989
Wing spot test			
Wing spot and SLRL test	1 - 10 mM	+	Kaya et al., 1999
Wing spot and white-ivory eye	NA	+	Tripathy et al., 1993
spot test	5 mM	+	Graf & Wurler, 1996
Drosophila melanogaster			
TCRG-TCRB recombination			
Mice thymocytes	0-100 mg/Kg/day	-	Knaap et al., 2003
Alkaline comet			
Clarias batrachus erythrocytes	25 - 75 ppm	+	Ateeq et al., 2005
Non-smokers HL*		+	Garaj-Vrhovac & Zeljezic, 2001
SCE assay			
Chick embryo cells	4 mg/embryo	+	Arias, 2003, 2007
Mouse bone marrow and spermatogonial cells	100 - 200 mg/Kg bw	+	Madrigal-Bujaidar et al., 2001
Non-smokers HL*		+	Garaj-Vrhovac & Zeljezic, 2001; Zeljezic & Garaj-Vrhovac, 2002
Chromosomal aberrations			
Allium cepa cells	NA	+	Ateeq et al., 2002a
Shallot root-tip cells	45 - 450 μM	+	Pavlica et al., 1991
Mouse bone marrow cells	NA	+	Venkov et al., 2000
Mouse bone marrow and spermatogonial cells	3.3 - 33 mg/Kg bw	+	Amer & Aly, 2001
Rat bone marrow cells	NA	+	Adhikari & Grover, 1988
Non-smokers HL*		+	Garaj-Vrhovac & Zeljezic, 2001
Hair follicle nuclear aberration			
Mouse bone marrow cells	1/32 LD50	+	Schop et al., 1990
Micronuclei			
Clarias batrachus and *Channa*	25 - 75 ppm	+	Ateeq et al., 2002b
punctatus erythrocytes	25 - 75 ppm	+	Farah et al., 2003. 2006
Mouse bone marrow			
Non-smokers HL*	NA	-	Schop et al., 1990
		+	Garaj-Vrhovac & Zeljezic, 2001
Alteration in CCP			
Allium cepa	NA	+	Ateeq et al., 2002a
Shallot root-tip cells	45 - 450 μM	+	Pavlica et al., 1991

End-point System	Concentration[a]	Results	Referencies
Chick embryos	2 mg/embryo	+	Arias, 2003, 2007
Mouse bone marrow and spermatogonial cells	50 - 200 mg/Kg bw	-	Madrigal-Bujaidar et al., 2001
Mouse bone marrow	NA	+	Venkov et al., 2000
Non-smokers HL*		-	Zeljezic & Garaj-Vrhovac, 2002

Table 1. Evaluation of 2,4-D-induced genotoxicity and cytotoxicity on different target systems. [a], expressed as reported by authors; *, from agricultural workers occupationally exposed to several pesticides, including 2,4-D. UDS, unscheduled DNA synthesis; HL, human lymphocytes; CCP, cell-cycle proliferation; NA, data not available.

systems. 2,4-D has been reported to induce mutations in plants (Kumari & Vaidyanath, 1989) as well as in insects (Tripathy et al., 1993; Graf & Wurler, 1996; Kaya et al., 1999) but not in mice exposed *in vivo* (Knaap et al., 2003). Ateeq and co-workers (2005) reported an increased frequency DNA single-strand breaks in piscine erythrocytes and in the peripheral lymphocytes of a group of agricultural workers occupationally exposed to the herbicide (Garaj-Vrhovac & Zeljezic, 2001). It should be noted that this later positive result could not be totally committed to the 2,4-D but to other pesticides, since the cohort of donors included in the study was exposed to a panel of diverse pesticides. Several reports were able to revealed that 2,4-D increased the frequency of SCEs in chick embryo and mammalian cells (Garaj-Vrhovac & Zeljezic, 2001; Madrigal-Bujaidar et al., 2001; Zeljezic & Garaj-Vrhovac, 2002; Arias, 2003, 2007), and chromosomal aberrations in plants, mouse, rat and human cells, including human lymphocytes from occupationally exposed workers (Adhikari & Grover, 1988; Schop et al., 1990; Pavlica et al., 1991; Venkov et al., 2000; Amer & Aly, 2001; Garaj-Vrhovac & Zeljezic, 2001; Ateeq et al., 2002a). When the micronuclei induction end-point was employed, whereas positive results were found in the piscine system (Ateeq et al., 2002b; Farah et al., 2003, 2006) and human lymphocytes (Garaj-Vrhovac & Zeljezic, 2001), no induction was found in mouse bone marrow cells (Schop et al., 1990). Finally, non-congruent results were reported when the analysis of the cell-cycle progression was used as and end-point for cytotoxicity. Alterations in the progression of the cell-cycle was reported to occur after 2,4-D exposure of plants, chick embryo, and mouse bone marrow cells (Pavlica et al., 1991; Venkov et al., 2000; Ateeq et al., 2002a; Arias, 2003, 2007). However, others authors were unable to revealed such alterations after *in vivo* exposure to the herbicide in bone marrow and spermatogonial mouse cells as well as in non-smokers human lymphocytes (Madrigal-Bujaidar et al., 2001; Zeljezic & Garaj-Vrhovac, 2002).

4. Genotoxicity and Cytotoxicity of Dicamba

Based on its acute toxicity, Dicamba has been classified as a class II member (moderately hazardous) by WHO (http://www.who.int/ipcs/publications/pesticides hazard/en/) and slightly to moderately toxic (category II-III) by EPA (EPA, 1974).

Genotoxicity and cytotoxicity studies have been conducted with this auxinic member using several end-points on different cellular systems. A summary of the results reported so far is presented in Table 2. When mutagenic activity was assessed in bacterial systems with the *Salmonella typhimurium* Ames test either positive or negative results have been reported (Simmon, 1979; Plewa et al., 1984; Kier et al., 1986). Furthermore, similar situation were observed in *Escherichia coli* and *Bacillus subtilis* when the reverse mutation assay was applied (Simmon, 1979; Leifer et al., 1981; Waters et al., 1981). Whereas the herbicide was unable to induce mitotic recombination in *Saccharomyces cerevisiae* (Zimmermann et al., 1984), negative and positive results were obtained for the induction of unscheduled DNA synthesis in human primary cells regardless of the presence or absence of a rat liver metabolic activation system (Simmon, 1979; Perocco et al., 1990). Induction of DNA single-strand breaks, estimated by the alkaline comet assay, was evaluated in CHO cells exposed *in vitro* to Dicamba. González et al. (2007) demonstrated an increase in the frequency of DNA lesions in this cell line. Similar observations were reported by Sorensen et al. (2004, 2005) on Dicamba-treated CHO cells cultured in the presence of reduced-clay smectites but not when the clay system was not included within the culture protocol. Both González et al. (2006, 2007, 2009) and Perocco et al. (1990) demonstrated the ability of the herbicide to induce SCEs in CHO cells and human lymphocytes with and without S9 fraction treated *in vitro*, respectively. The induction of alterations in the cell-cycle progression of different cellular systems including CHO cells and human lymphocytes were reported to occur after *in vitro* exposure to Dicamba (González et al., 2006, 2007, 2009). Finally, similar results were reported for the cell viability assay in CHO cells (Sorensen et al., 2004; González et al., 2009). In genotoxic and cytotoxic studies *in vivo*, Dicamba was able to induce different types of lesions. It has been reported the ability of the herbicide to give positive results by using the gene mutation and recombination assays when *Arabidopsis thaliana* was used as experimental model (Filkowski et al., 2003). However, both negative and inconclusive results were reported for the sex-linked recessive lethal mutation end-point on Dicamba-exposed *Drosophila melanogaster* (Waters et al., 1981; Lee et al., 1983). Perocco and co-workers (1990) reported an increased frequency of DNA unwinding rate in rat hepatocytes. It has been also reported that the herbicide is able to enhance the frequency of chromosomal aberrations in the root- and hoot-tip cells of barley (*Hordeum vulgare*) and in rat bone marrow cells (Hrelia et al., 1994). On the other hand, no increased frequency of chromosomal rearrangements has been observed in the durum wheat *Triticum turgidum* by Satyavathi and co-workers (2004). Finally, when the micronuclei induction end-point was employed, positive results were reported in *Tradescantia* sp (clone 03) by Mohamed and Ma (1999).

End-point System	Concentration[a]	Results	Referencies
In vitro assays			
Ames test			
Salmonella typhimurium +/- S9	0 - 5000 μg/plate	+	Plewa et al., 1984
		-	Simmon, 1979; Kier et al., 1986
Rec A- reverse mutation			
Bacillus subtilis	0.01 - 5.0 mg/disk	+	Leifer et al., 1981
Pol A reverse mutation			
Escherichia coli	0 - 5000 μg/plate	+	Waters et al., 1981
	0 - 5000 μg/plate	-	Simmon, 1979; Leifer et al., 1981
Mitotic recombination/Gene conversion			

End-point System	Concentration[a]	Results	Referencies
Saccharomyces cerevisiae	0.1 - 5.0 % (w/v)	-	Zimmermann et al., 1984
UDS			
Human diploid fibroblasts +/-S9	0.1 - 3000 μg/ml	-	Simmon, 1979
Non-smokers HL +/- S9			
	0.1 - 0.8 mg/ml	+	Perocco et al., 1990
Alkaline comet			
CHO cells	50 - 500 μg/ml	+	González et al., 2007
CHO cells	10 μM - 10 mM	-	Sorensen et al., 2004, 2005
			Sorensen et al., 2004, 2005
CHO cells + reduced clay	10 μM - 10 mM	+	
SCE assay			
CHO cells	1 - 500 μg/ml	+	González et al., 2007, 2009
			González et al., 2006
Non-smokers HL	200 μg/ml	+	Perocco et al., 1990
Non-smokers HL +/- S9	0.1 - 0.8 mg/ml	+	
Alteration in CCP			
CHO cells	200 - 500 μg/ml	+	González et al., 2007, 2009
			González et al., 2006
Non-smokers HL	100 - 200 μg/ml	+	
Cell viability			
CHO cells	500 μg/ml	+	González et al., 2009
CHO cells	>1000 μM	+	Sorensen et al., 2004
In vivo assays			
A → G/T → G mutation and recombination assay			
Arabidopis thaliana	120 μg/L	+/-	Filkowski et al., 2003
Sex-linked recessive lethal mutations			
Drosophila melanogaster	NA	-	Waters et al., 1981
	3 - 2000 ppm	IN	Lee et al., 1983
DNA unwinding rate			
Rat hepatocytes	NA	+	Perocco et al., 1990
Chromosomal aberrations			
Hordeum vulgare root- and hoot-tip cells, microsporocytes	NA	+	Hrelia et al., 1994
Triticum turgidum	2 mg/L	-	Satyavathi et al., 2004
Rat bone marrow cells	NA	+	Hrelia et al., 1994
Micronuclei assay			
Tradescania sp. Clone 03	50 - 200 mg/L	+	Mohammed & Ma, 1999

Table 2. Evaluation of Dicamba-induced genotoxicity and cytotoxicity on different target systems. [a], expressed as reported by authors; CCP, cell cycle proliferation; NA, data no available; IN, inconclusive results.

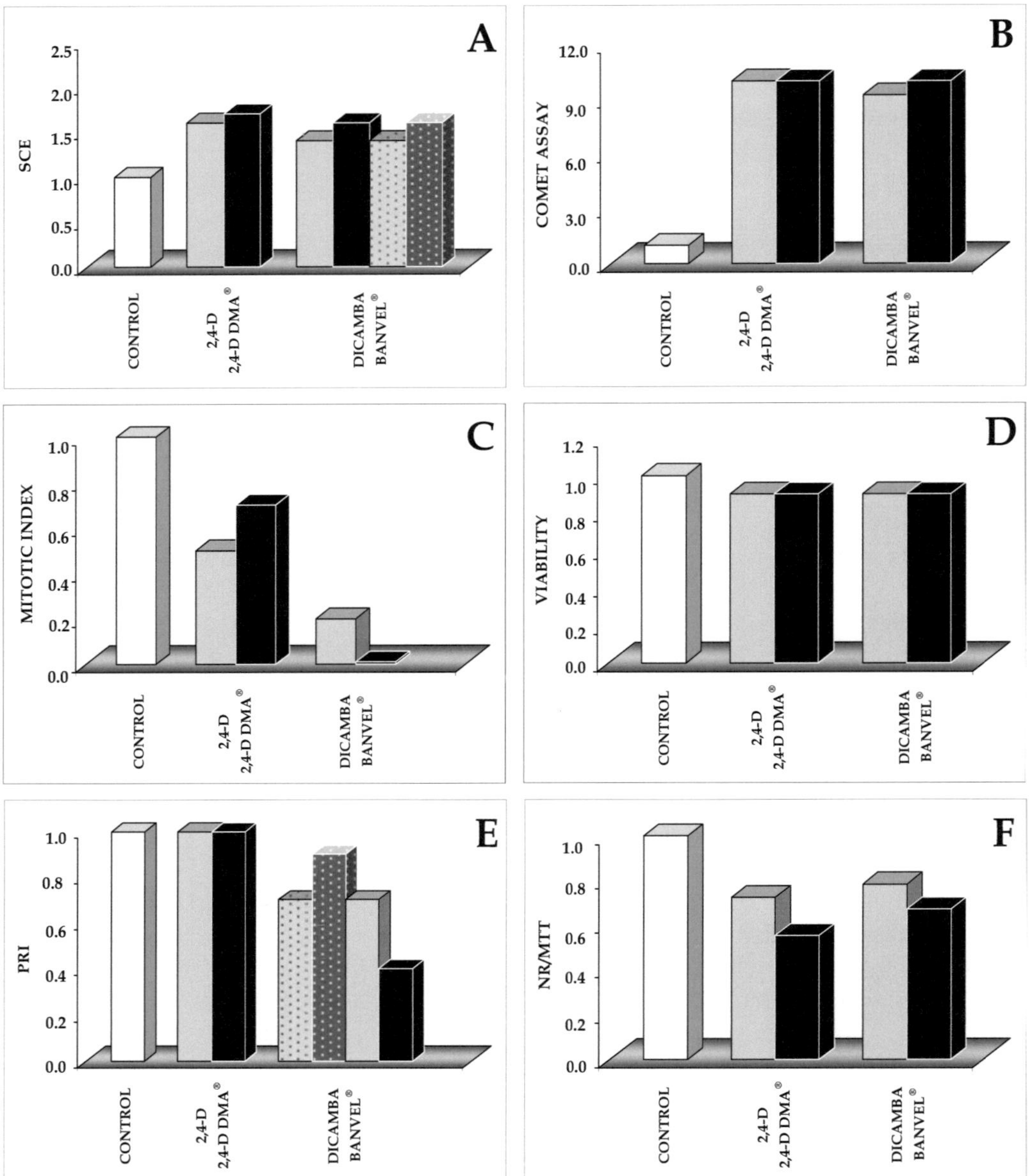

Fig. 1. Comparative genotoxicity and cytotoxicity effects induced by 2,4-D and Dicamba pure herbicides Pestanal® analytical standards (grey bars) and their technical formulations (black bars) commonly used in Argentina on mammalian cells *in vitro* (plain bars, CHO-K1 cells; dotted bars, human lymphocytes). Results are expressed as fold-time values over control data. Evaluation was performed using end-points for genotoxicity [Sister Chromatid Exchanges frequency (A), Comet Assay (B)] and cytotoxicity [Mitotic Index (C), Viability (D), Proliferative Rate Index (E), 3(4,5-dimethylthiazol-2-yl)-2,5-diphenyltetrazolium bromide (MTT) and Neutral Red (NR) (F)].

5. Comparison of the genotoxicity and cytotoxicity of 2,4-D and Dicamba and some Argentinean technical formulations

One of the major goals of our research laboratory is to evaluate comparatively the genotoxic and cytotoxic effects exerted by several pure pesticide Pestanal® analytical standards (Riedel-de Haën, Germany) and their technical formulations commonly used in Argentina on mammalian cells *in vitro*. In this section we evaluate comparatively the genotoxic and cytotoxic effects induced in CHO cells and human lymphocytes from non-smoker donors exposed *in vitro* to the auxinic pure herbicides 2,4-D (CAS 94-757) and Dicamba (CAS 1918-00-9) and their technical commercial formulations commonly used in Argentina 2,4-D DMA® (60.2% 2,4-D, Delente Laboratories SRL, Buenos Aires, Argentina) and Banvel® (57.7% Dicamba, Syngenta Agro S.A., Buenos Aires, Argentina), respectively. Evaluation was performed using end-points for genotoxicity [Sister chromatid exchanges frequency and Comet assay] and cytotoxicity [Mitotic index, Cell viability, Proliferative rate index, and 3(4,5-dimethylthiazol-2-yl)-2,5-diphenyltetrazolium bromide (MTT) and Neutral Red assays] (González et al., 2005; González et al., 2006, 2007, 2008, 2009; Soloneski et al., 2007).
A summary of the results obtained is presented in Fig. 1. The figure clearly reveals that all compounds assayed are able to inflict DNA damage in CHO cells and human lymphocytes when analyzed at chromosomal and DNA level. We observed that 2,4-D/2,4-D DMA® and Dicamba/Banvel® caused SCEs in mammalian cells indicating that they have clastogenic activity (Fig. 1A). It has been suggested that at the chromosomal level, the induction of SCEs is a reliable indicator for the screening of clastogens, since the bioassay is more sensitive than the analysis of clastogen-induced chromosomal aberrations (Palitti et al., 1982). The results also demonstrate the ability of 2,4-D/2,4-D DMA® and Dicamba/Banvel® to induce DNA single-strand breaks quali- and quantitative analyzed by the comet assay (Fig. 1B). The analysis of the mitotic (Fig. 1C) and the proliferative replication indexes (Fig. 1D) demonstrated that both 2,4-D/2,4-D DMA® and Dicamba/Banvel® are able to exert a marked reduction of the cellular mitotic activity as well as to delay the cell-cycle progression *in vitro* with a concomitant reduction of the proliferative rate index in both cell types. Besides, 2,4-D/2,4-D DMA® and Dicamba/Banvel® are able to induced a clear cellular cytotoxicity, estimated by means of the ethidium bromide/acridine orange assay in CHO cells (Fig 1.E). Finally, a loss of lysosomal activity, indicated by a decrease in the uptake of neutral red, as well as alteration in energy metabolism induced by 2,4-D/2,4-D DMA® and Dicamba/Banvel®, measured by mitochondrial succinic dehydrogenase activity in the MTT assay, were clearly revealed in herbicides-treated CHO cells (Fig. 1F) which corroborate the results obtained applying different end-points for cytotoxicity. Overall, the results clearly demonstrated that the damage induced by the commercial formulations of both herbicides is in general greater than that produced by the pure pesticides, suggesting the presence of deleterious components in the excipients with a toxic additive effect over the pure agrochemicals (Fig. 1). Unfortunately, the identity of the components present within the excipient formulations was not made available by the manufactures. Moreover, though almost improbable, the possibility that the amount of the active ingredient incorporated into the technical Argentinean formulations could be higher than that officially registered cannot be discarded.

6. Final remarks

In agriculture, agrochemicals are generally not used as a single active ingredient but as part of a complex commercial formulation. An active ingredient is a substance that prevents,

kills, or repels a pest or acts as a plant regulator, among others. In addition to the active component, the formulated products contain different solvents, carriers and adjuvants, some of which have been reported to induce damage in mammalian cells, among other cellular systems (Lin & Garry, 2000; Zeljezic et al., 2006; González et al., 2007, 2009; Soloneski et al., 2008; Molinari et al., 2009; Soloneski & Larramendy, 2010). Hence, risk assessment must also consider additional toxic effects caused by the excipient/s. Thus, both the workers as well as non-target organisms are exposed to the simultaneous action of the active ingredient and a variety of other chemical/s contained in the formulated product.
Finally, the results highlight that a complete knowledge of the toxic effect/s of the active ingredient of a pesticide is not enough in biomonitoring studies as well as that agrochemical/s toxic effect/s should be evaluated according to the commercial formulation available in the market. Furthermore, the deleterious effect/s of the excipient/s present within the commercial formulation should be neither discarded nor underestimated.

7. References

Adhikari, N., Grover, I.S. (1988). Genotoxic effects of some systemic pesticides: *in vivo* chromosomal aberrations in bone marrow cells in rats. *Environmental and Molecular Mutagenesis,* 12, 2, 235-242, ISSN 0893-6692

Alarcon, W.A., Calvert, G.M., Blondell, J.M., Mehler, L.N., Sievert, J., Propeck, M., Tibbetts, D.S., Becker, A., Lackovic, M., Soileau, S.B., Das, R., Beckman, J., Male, D.P., Thomsen, C.L., Stanbury, M. (2005). Acute illnesses associated with pesticide exposure at schools. *JAMA : The Journal of the American Medical Association,* 294, 10, 455-465, ISSN 0098-7484

Amer, S.M., Aly, F.A. (2001). Genotoxic effect of 2,4-dichlorophenoxy acetic acid and its metabolite 2,4-dichlorophenol in mouse. *Mutation Research,* 494, 1-2, 1-12, ISSN 0027-5107

Arias, E. (2003). Sister chromatid exchange induction by the herbicide 2,4-dichlorophenocyacetic acid in chick embryos. *Ecotoxicology and Environmental Safety* 55, 3, 338-343, ISSN 0147-6513

Arias, E. (2007). Cytogenetic effects of short- and long-term exposure of chick embryos to the phenoxyherbicide 2,4-D. *Environmental and Molecular Mutagenesis,* 48, 6, 462-466, ISSN 0893-6692

Ateeq, B., Abul Farah, M., Ahmad, W. (2005). Detection of DNA damage by alkaline single cell gel electrophoresis in 2,4-dichlorophenoxyacetic-acid- and butachlor-exposed erythrocytes of *Clarias batrachus. Ecotoxicology and Environmental Safety* 62, 3, 348-354, ISSN 0147-6513

Ateeq, B., Abul Farah, M., Niamat Ali, M., Ahmad, W. (2002a). Clastogenicity of pentachlorophenol, 2,4-D and butachlor evaluated by *Allium* root tip test. *Mutation Research,* 514, 1-2, 105-113, ISSN 0027-5107

Ateeq, B., Abul farah, M., Niamat Ali, M., Ahmad, W. (2002b). Induction of micronuclei and erythrocyte alterations in the catfish *Clarias batrachus* by 2,4-dichlorophenoxyacetic acid and butachlor. *Mutation Research,* 518, 2, 135-144, ISSN 0027-5107

Basrur, S.V., Fletcher, R.A., Basrur, P.K. (1976). *In vitro* effects of 2,4-dichlorophenoxy acetic acid (2,4-D) on bovine cells. *Canadian Journal of Comparative Medicine (Gardenvale, Québec)* 40, 4, 408-415, ISSN 0846-8389

Bayliss, M.W. (1977). The effects of 2,4-D on growth and mitosis in suspension cultures of *Daucus carota Plant Science Letters,* 8, 2, 99-103, ISSN 0168-9452

Calvert, G.M., Plate, D.K., Das, R., Rosales, R., Shafey, O., Thomsen, C., Male, D., Beckman, J., Arvizu, E., Lackovic, M. (2004). Acute occupational pesticide-related illness in the US, 1998-1999: surveillance findings from the SENSOR-pesticides program. *American journal of Industrial Medicine,* 45, 1, 14-23, ISSN 0271-3586

Charles, J.M., Cunny, H.C., Wilson, R.D., Ivett, J.L., Murli, H., Bus, J.S., Gollapudi, B. (1999). *In vivo* micronucleus assays on 2,4-dichlorophenoxyacetic acid and its derivatives. *Mutation Research,* 444, 1, 227-234, ISSN 0027-5107

Dearfield, K.L., McCarroll, N.E., Protzel, A., Stack, H.F., Jackson, M.A., Waters, M.D. (1999). A survey of EPA/OPP and open literature on selected pesticide chemicals. II. Mutagenicity and carcinogenicity of selected chloroacetanilides and related compounds. *Mutation Research,* 443, 1-2, 183-221, ISSN 0027-5107

Devine, M. (1993). *Physiology of herbicide action,* Englewood Cliffs, ISBN 0133690679, New Jersey

EPA (1974). *Compendium of Registered Pesticides,* U.S. Government Printing Office, Washington, DC

EPA (1983). Pesticide Fact Sheet Number 8: Dicamba. U.S. Environmental Protection Agency, Office of Pesticide Programs, U.S. Government Printing Office, Washington, DC

FAO (2001). FAO Specifications and evaluation for plant protection product Dicamba. 1-20,

Farah, M.A., Ateeq, B., Ahmad, W. (2006). Antimutagenic effect of neem leaves extract in freshwater fish, *Channa punctatus* evaluated by cytogenetic tests. *The Science of the Total Environment,* 364, 1-3, 200-214, ISSN 0048-9697

Farah, M.A., Ateeq, B., Ali, M.N., Ahmad, W. (2003). Evaluation of genotoxicity of PCP and 2,4-D by micronucleus test in freshwater fish *Channa punctatus. Ecotoxicology and Environmental Safety* 54, 25-29, ISSN 0147-6513

Filkowski, J., Besplug, J., Burke, P., Kovalchuk, I., Kovalchuk, O. (2003). Genotoxicity of 2,4-D and dicamba revealed by transgenic *Arabidopsis thaliana* plants harboring recombination and point mutation markers. *Mutation Research,* 542, 1-2, 23-32, ISSN 0027-5107

Garaj-Vrhovac, V., Zeljezic, D. (2001). Cytogenetic monitoring of croatian population occupationally exposed to a complex mixture of pesticides. *Toxicology,* 165, 2-3, 153-162, ISSN 0300-483X

González, M., Soloneski, S., Reigosa, M.A., Larramendy, M.L. (2005). Genotoxicity of the herbicide 2,4-dichlorophenoxyacetic and a commercial formulation, 2,4-dichlorophenoxyacetic acid dimethylamine salt. I. Evaluation of DNA damage and cytogenetic endpoints in Chinese Hamster ovary (CHO) cells. *Toxicology In Vitro,* 19, 2, 289-297, ISSN 0887-2333

González, N.V., Soloneski, S., Larramendy, M. (2008). Dicamba-induced genotoxicity in Chinese hamster ovary (CHO) cells is prevented by vitamin E. *Journal of Hazardous Materials,* 163, 1, 337-343, ISSN 0304-3894

González, N.V., Soloneski, S., Larramendy, M.L. (2006). Genotoxicity analysis of the phenoxy herbicide dicamba in mammalian cells *in vitro. Toxicology In Vitro,* 20, 8, 1481-1487, ISSN 0887-2333

González, N.V., Soloneski, S., Larramendy, M.L. (2007). The chlorophenoxy herbicide dicamba and its commercial formulation banvel induce genotoxicity in Chinese hamster ovary cells. *Mutation Research,* 634, 1-2, 60-68, ISSN 0027-5107

González, N.V., Soloneski, S., Larramendy, M.L. (2009). Dicamba-induced genotoxicity in Chinese hamster ovary (CHO) cells is prevented by vitamin E. *Journal of Hazardous Materials,* 163, 1, 337-343, ISSN 0304-3894

Grabinska-Sota, E., Wisniowska, E., Kalka, J., Scieranka, B. (2002). Genotoxicological effects of some phenoxyherbicides and their metabolites on *Bacillus subtilis* M45 Rec- and H17 Rec+ strains. *Chemosphere,* 47, 1, 81-85, ISSN 0045-6535

Graf, U., Wurler, F. (1996). The somatic white-ivory eye spot test does not detect the same spectrum of genotoxic events as the wing somatic mutation and recombination test in *Drosophila melanogaster*. *Environmental and Molecular Mutagenesis,* 27, 3, 219-226, ISSN 0893-6692

Hrelia, P., Vigagni, F., Maffei, F., Morotti, M., Colacci, A., Perocco, P., Grilli, S., Cantelli-Forti, G. (1994). Genetic safety evaluation of pesticides in different short-term tests. *Mutation Research,* 321, 4, 219-228, ISSN 0027-5107

IARC (1977). *Some Fumigants, the Herbicides 2,4-D and 2,4,5-T, Chlorinated Dibenzodioxins and Miscellaneous Industrial Chemicals,* ISBN 978-9283212157, Lyon

IARC (1999). *Some halogenated hydrocarbons and pesticide exposures,* International Agency for Research on Cancer, ISBN 978-9283212416, Lyon

IFCS (2003). Acutely toxic pesticides: initial input on extent of problem and guidance for risk management. Fourth session of the Intergovernmental Forum on Chemical Safety. Doc number: IFCS/FORUM-IV/10, Bangkok

Kaya, B., Yanikoglu, A., Marcos, R. (1999). Genotoxicity studies on the phenoxyacetates 2,4-D and 4-CPA in the *Drosophila* wing spot test. *Teratogenesis, Carcinogenesis and Mutagenesis,* 19, 4, 305-312, ISSN 0270-3211

Kier, L.D., Brusick, D.J., Auletta, A.E., Von Halle, E.S., Brown, M.M., Simmon, V.F., Dunkel, V., McCann, J., Mortelmans, K., Prival, M., Rao, T.K., V, R. (1986). The *Salmonella typhimurium*/mammalian microsomal assay: A report of the U.S. Environmental Protection Agency Gene-Tox Program. *Mutation Research,* 168, 2, 69-240, ISSN 0027-5107

Knaap, G.W., Setzer, R.W., Fuscoe, J.C. (2003). Quantitation of aberrant Interlocus T-cell receptor rearrangements in mouse thymocytes and the effect of the

herbicide 2,4-dichlorophenoxyacetic acid. *Environmental and Molecular Mutagenesis,* 42, 1, 37-43, ISSN 0893-6692

Kumari, T.S., Vaidyanath, K. (1989). Testing of genotoxic effects of 2,4-dichlorophenoxyacetic acid (2,4-D) using multiple genetic assay systems of plants. *Mutation Research,* 226, 4, 235-238, ISSN 0027-5107

Lee, W.R., Abrahamson, S., Valencia, R., von Halle, E.S., Würgler, F.E., Zimmering, S. (1983). The sex-linked recessive lethal test for mutagenesis in *Drosophila melanogaster*. A report of the U.S. Environmental Protection Agency Gene-Tox Program. *Mutation Research,* 123, 2, 183-279, ISSN 0027-5107

Leifer, Z., Kada, T., Mandel, M., Zeiger, E., Stafford, R., Rosenkranz, H.S. (1981). An evaluation of tests using DNA repair-deficient bacteria for predicting genotoxicity and carcinogenicity. A report of the U.S. EPA's Gene-Tox Program. *Mutation Research,* 87, 3, 211-297, ISSN 0027-5107

Lin, N., Garry, V.F. (2000). *In vitro* studies of cellular and molecular developmental toxicity of adjuvants, herbicides, and fungicides commonly used in Red River Valley, Minnesota. *Journal of Toxicology and Environmental Health,* 60, 6, 423-439, ISSN 1528-7394

Liu, C.M., McLean, P.A., Sookdeo, C.C., Cannon, F.C. (1999). Degradation of the herbicide glyphosate by members of the family Rhizobiaceae. *Applied and Environmental Microbiology,* 57, 6, 1799-1804, ISSN 0099-2240

Madrigal-Bujaidar, E., Hernandez-Ceruelos, A., Chamorro, G. (2001). Induction of sister chromatid exchanges by 2,4-dichlorophenoxyacetic acid in somatic and germ cells of mice exposed in vivo. *Food and Chemical Toxicology*, 39, 9, 941-946, ISSN 0278-6915

Maire, M.A., Rast, C., Landkocz, Y., Vasseur, P. (2007). 2,4-Dichlorophenoxyacetic acid: effects on Syrian hamster embryo (SHE) cell transformation, c-Myc expression, DNA damage and apoptosis. *Mutation Research*, 631, 2, 124-136, ISSN 0027-5107

Mohammed, K.B., Ma, T.H. (1999). Tradescantia-micronucleus and -stamen hair mutation assays on genotoxicity of the gaseous and liquid forms of pesticides. *Mutation Research*, 426, 2, 193-199, ISSN 0027-5107

Molinari, G., Soloneski, S., Reigosa, M.A., Larramendy, M.L. (2009). *In vitro* genotoxic and cytotoxic effects of ivermectin and its formulation ivomec® on Chinese hamster ovary (CHO_{K1}) cells. *Journal of Hazardous Materials*, 165, 1-3, 1074-1082, ISSN 0304-3894

Mustonen, R., Kangas, J., Vuojolahti, P., Linnainmaa, K. (1986). Effects of phenoxyacetic acids on the induction of chromosome aberrations *in vitro* and *in vivo*. *Mutagenesis*, 1, 4, 241-245, ISSN 0267-8357

Osterloh, J., Lotti, M., Pond, S.M. (1983). Toxicologic studies in a fatal overdose of 2,4-D, MCPP, and chlorpyrifos. *Journal of Analytical Toxicology*, 7, 3, 125-129, ISSN 0146-4760

Palitti, F.C., Tanzarella, C., Cozzi, R., Ricondy, E., Vitagliana, A., Fiori, M. (1982). Comparison of frequencies of SCEs induced by chemical mutagens in bone-marrow spleen and spermatogoneal cell of mice. *Mutation Research*, 103, 2, 191-195, ISSN 0027-5107.

Pavlica, M., Papes, D., Nagy, B. (1991). 2,4-Dichlorophenoxyacetic acid causes chromatin and chromosome abnormalities in plant cells and mutation in cultured mammalian cells. *Mutation Research*, 263, 2, 77-81, ISSN 0027-5107

Perocco, P., Ancora, G., Rani, P., Valenti, A.M., Mazzullo, M., Colacci, A., Grilli, S. (1990). Evaluation of genotoxic effects of the herbicide dicamba using *in vivo* and *in vitro* test systems. *Environmental and Molecular Mutagenesis*, 15, 3, 131-135, ISSN 0893-6692

Pipke, R., Amrhein, N., Jacob, G.S., Schaefer, J., Kishore, G.M. (1987). Metabolism of glyphosate in an *Arthrobacter sp*. GLP-1. *European Journal of Biochemistry* 165, 2, 267-273, ISSN 0014-2956

Plewa, M.J., Wagner, E.D., Gentile, G.J., Gentile, J.M. (1984). An evaluation of the genotoxic properties of herbicides following plant and animal activation. *Mutation Research*, 136, 3, 233-245, ISSN 0027-5107

Reinbothe, S., Reinbothe, C., Neumann, D., Apel, K. (1996). A plastid enzyme arrested in the step of precursor translocation *in vivo*. *Proceedings of the National Academy of Sciences of the United States of America*, 93, 21, 12026-12030, ISSN 0027-8424

Sandal, S., Yilmaz, B. (2010). Genotoxic effects of chlorpyrifos, cypermethrin, endosulfan and 2,4-D on human peripheral lymphocytes cultured from smokers and nonsmokers. *Environmental Toxicology*, In Press, ISSN 1520-4081

Satyavathi, V.V., Jauhar, P.P., Elias, E.M., Rao, M.B. (2004). Effects of growth regulators on *in vitro* plant regeneration in durum wheat. *Crop Science*, 44, 5, 1839-1846, ISSN 0011-183X

Schop, R.N., Hardy, M.H., Goldberg, M.T. (1990). Comparison of the activity of topically applied pesticides and the herbicide 2,4-D in two short-term *in vivo* assays of genotoxicity in the mouse. *Fundamental and Applied Toxicology: Official Journal of the Society of Toxicology*, 15, 4, 666-675, ISSN 0272-0590

Simmon, V.F., 1979. *In vitro* microbiological mutagenicity and unscheduled DNA synthesis studies of eighteen pesticides. EPA-600/1-79-041. EPA, Research Triangle Park, pp. 1-79

Sinton, G.L., Fan, L.T., Erickson, L.E., Lee, S.M. (1986). Biodegradation of 2,4-D and related xenobiotic compounds. *Enzyme and Microbial Technology,* 8, 7, 395-403 ISSN 0141-0229

Soloneski, S., González, N.V., Reigosa, M.A., Larramendy, M.L. (2007). Herbicide 2,4-D dichlorophenoxyacetic acid (2,4-D)-induced cytogenetic damage in human lymphocytes *in vitro. Cell Biology International,* 31, 11, 1316-1322, ISSN 1065-6995

Soloneski, S., Larramendy, M.L. (2010). Sister chromatid exchanges and chromosomal aberrations in Chinese hamster ovary (CHO-K1) cells treated with the insecticide Pirimicarb. *Journal of Hazardous Materials,* 174, 1-3, 410-415, ISSN 0304-3894

Soloneski, S., Reigosa, M.A., Molinari, G., González, N.V., Larramendy, M.L. (2008). Genotoxic and cytotoxic effects of carbofuran and furadan® on Chinese hamster ovary (CHO_{K1}) cells. *Mutation Research,* 656, 1-2, 68-73, ISSN 0027-5107

Sorensen, K.C., Stucki, J.W., Warner, R.E., Plewa, M.J. (2004). Alteration of mammalian-cell toxicity of pesticides by structural iron(II) in ferruginous smectite. *Environmental Science & Technology,* 38, 16, 4383-4389, ISSN 0013-936X

Sorensen, K.C., Stucki, J.W., Warner, R.E., Wagner, E.D., Plewa, M.J. (2005). Modulation of the genotoxicity of pesticides reacted with redox-modified smectite clay. *Environmental and Molecular Mutagenesis,* 46, 3, 174-181, ISSN 0893-6692

Teixeira, M.C., Telo, J.P., Duarte, N.F., Sá-Correia, I. (2004). The herbicide 2,4-dichlorophenoxyacetic acid induces the generation of free-radicals and associated oxidative stress responses in yeast. *Biochemical and Biophysical Research Communications,* 324, 3, 1101-1107, ISSN 0006-291X

Tripathy, N., Routray, P., Sahu, G., Kumar, A. (1993). Genotoxicity of 2,4-dichlorophenoxyacetic acid tested in somatic and germ-line cells of *Drosophila. Mutation Research,* 319, 3, 237-242, ISSN 0027-5107

Venkov, P., Topashka-Ancheva, M., Georgieva, M., Alexieva, V., Karanov, E. (2000). Genotoxic effect of substituted phenoxyacetic acids. *Archives in Toxicology,* 74, 9, 560-566, ISSN 0340-5761

Waters, M.D., Nesnow, S., Simmon, V.F., Mitchell, A.D., Jorgenson, T.A., Valencia, R. (1981). *The pesticide chemist and modern toxicology,* American Chemical Society, ISBN 9780841206366, Washington, DC

WHO-FAO, 2009. Pesticides residues in food-2009. FAO Plant Production and Protection paper World Health Organization and Food and Agriculture Organization of the United Nations, Rome, pp. 1-426

Zeljezic, D., Garaj-Vrhovac, V. (2002). Sister chromatid exchange and proliferative rate index in the longitudinal risk assessment of occupational exposure to pesticides. *Chemosphere,* 46, 2, 295-303, ISSN 0045-6535.

Zeljezic, D., Garaj-Vrhovac, V. (2004). Chromosomal aberrations, micronuclei and nuclear buds induced in human lymphocytes by 2,4-dichlorophenoxyacetic acid pesticide formulation. *Toxicology* 200, 1, 39-47, ISSN 0300-483X

Zeljezic, D., Garaj-Vrhovac, V., Perkovic, P. (2006). Evaluation of DNA damage induced by atrazine and atrazine-based herbicide in human lymphocytes *in vitro* using a comet and DNA diffusion assay. *Toxicology In Vitro* 20, 1, 923-935, ISSN, 0887-2333

Zimmermann, F.K., von Borstel, R., von Halle, E.S., Parry, J.M., Siebert, D., Zetterberg, G., Barale, R., Loprieno, N. (1984). Testing of chemicals for genetic activity with *Saccharomyces cerevisiae*: a report of the U.S. Environmental Protection Agency Gene-Tox Program. *Mutation Research Reviews,* 133, 3, 199-244, ISSN 1383-5742

26

A Study on Dioxin Contamination in Herbicide Sprayed Area in Vietnam by GIS

Dang Duc Nhu[1], Teruhiko Kido[2], Nguyen Ngoc Hung[1], Phung Tri Dung[1], Le Thi Hong Thom[1], Rie Naganuma[2], Nobuhiro Sawano[3], Le Ke Son[4], Kenji Tawara[5], Hideaki Nakagawa[5] and Le Vu Quan[6]

[1]*Division for Mitigation of the Consequences of the Chemicals Used During the War on Human Health (10-80 Division), Hanoi Medical University,*

[2]*Division of Health Sciences, Graduate School of Medical Science, Kanazawa University, Ishikawa,*

[3]*Seiryo Women's Junior College, Tori-1, Goshomachi, Kanazawa, Ishikawa 920-0813*

[4]*Vietnam Environment Administration, Ministry of Natural Resources and Environment*

[5]*Department of Public Health, Kanazawa Medical University, 1-1 Daigaku, Uchinada, Ishikawa 920-0293,*

[6]*Department of Economics, Seattle University,*

[1,4]*Vietnam,*

[2,3,5]*Japan*

[6]*USA*

1. Introduction

During the Vietnam war (1961 - 1971), the United States military forces carried out the operation named Ranch Hand that sprayed over 19.5 million gallons of herbicide for defoliation over wide areas of forests and crops in Vietnam, Laos and Cambodia to deny their use by opposition force (Westing, 1984; Stellman et al., 2003a). Two thirds of the chemical herbicides used was Agent Orange, a 50:50 mixture of 2,4,5-trichlorophenoxyacetic acid (2,4,5-T) and 2,4-dichlophenoxyacetic acid (2,4-D) herbicides (IOM 2002). The defoliant 2,4,5-T was contaminated with an extremely toxic substance, 2,3,7,8-tetrachlorodibenzo-p-dioxin (TCDD) known to have adverse effects on human health.

The Vietnam war ended over 35 years ago. There are still adverse effects of herbicide residues on people who lived in sprayed area and also its ecosystems (Schecter et al., 2001, 2003; Cau et al., 2003; Mai et al., 2007). In environmental health research, there is a recognized need to develop methodologies to carry out epidemiologic research. The geographic information system (GIS) is a technology which can improve the study of dioxin in Vietnam to estimate people's herbicide exposure. GIS can also combine both extensive and intensive databases on dispersal of herbicides, locations of US army military units and bases, locations of civilian population centers in Vietnam, then provide a unique basis to integrate and improve epidemiologic studies (Stellman et al., 2003b; Waring et al., 2005; Viel et al., 2008). GIS-based exposure assessment has been used in a small number of epidemiologic environmental studies (Rushton, 2003). They can efficiently integrate records of where herbicides were used.

Results of investigations in a sprayed site, A Luoi Valley in southern Vietnam, demonstrate the apparent food chain transfer of TCDD from contaminated soil to cultured fish pond sediments to fish and duck tissues, then to humans as measured in whole blood and breast milk. A Luoi Valley is considered a microcosm of southern Vietnam, where the soils have been contaminated by numerous reservoirs of TCDD (Dwernychuk et al., 2002). Another study made clear that TCDD was increased in the blood of 19 of 20 persons taken from an Agent Orange contaminated site, Bien Hoa in southern Vietnam. TCDD levels among these persons, one of whom reached 271 ppt, was 135-times higher than among persons living in non-contaminated areas in northern Vietnam (Schecter et al., 2001). A project for searching dioxin hot spots was carried out in provinces in southern Vietnam. The results showed that dioxin contamination in soil and sediment are higher than standard, and indentified specific US bases as hot spots (797 pg/g TCDD in Bien Hoa, 227 pg/g TCDD in Da Nang, 194 pg/g TCDD in Phu Cat) (Dwenychuk et al., 2005, 2006).

In 2001, a Japanese medical research team initiated studies on human impacts of herbicide spraying. The study has shown significantly higher dioxins levels of serum, breast milk and adipose tissues in inhabitants of sprayed areas (Quang Tri province) than those in non-sprayed area (Ha Tinh province), while no significant difference was found on early indicators of adverse health effects such as liver or thyroid function and immunological activities. Dioxin levels in breast milk of people in the sprayed area are significantly higher than those in non-sprayed areas (Tawara et al., 2006). The study on mothers revealed high dioxin levels in breast milk. Kido et al.(2006) has shown visual acuities of both eyes of people in sprayed areas were also significantly lower than those in non-sprayed area at every condition with change of contrast from 100 to 2.5 % except for 2.5 % contrast of the left eye, the simple relationships between dioxin levels and visual acuity were shown as mild in both eyes.

In the world and Vietnam there have been numerous studies. They have shown the hazards of dioxin on the environmental and human health. However, some studies remain limited with respect to subject of study, sample size, method of study and difficulties in chemical analysis, etc. In addition, there were not many studies applying GIS to the study of dioxin levels in soil and breast milk in Vietnam.

The main objective of this study was to assess the correlation between dioxin concentration in soil, sediment and breast milk of females in Cam Chinh commune, Quang Tri province, Vietnam. Herbicides that contained high concentration of dioxin were sprayed during the Vietnam war.

2. Materials and methods

2.1 Study area

This study was implemented in two areas where herbicides were sprayed and not sprayed. A herbicide sprayed area in Cam Chinh commune in Quang Tri province was chosen as a case site. Quang Tri province received 150 herbicide missions with a quantity of 6602 l/hectare (10-80 committee, 1999; Vietnam Investment Review, 1999). Quang Tri borders the demilitarized zone along the 17th parallel that once divided the north from South Vietnam. The control site was Cam Phuc commune in Ha Tinh province, which did not experience herbicide operations during the war.

2.2 Estimation of spatial distribution of polychlorinated dibenzodioxins (PCDDs) and polychlorinated dibenzofurans (PCDFs)

If a mother's house or residential area was more polluted with PCDD/Fs, the mother will have higher level of dioxins in her body. With 16 soil samples and one sediment sample, spatial distribution of PCDD/Fs was estimated by Kriging interpolation method. There are some detailed methods in Kriging interpolation (Armstrong et al., 1988), however, Saito et al., (2000) referred "log-normal Kriging" consistently yields the best result for concentration of dioxin.

Kriging interpolator is a specialized interpolation method that assumes the distance or direction between sample points and shows spatial correlation that helps describe the surface (Largueche, 2006). The initial step, a list of mothers whose breast milk was collected from lactating females aged between 20-40 years old, was made at Quang Tri in 2002-2003; these breast milk samples were analyzed in the laboratory of Kanazawa Medical University, Japan. All mother's house locations were checked by GPS (Global Positioning System). GPS data have been coded as latitude (Y) and longitude (X) then located on the map in accordance with the position of their residence (Fig 2). Soil and sediment samples were taken in the same area (Fig1) and also analyzed in the Japanese laboratory. These data have been used to create a surface contour of dioxin of this area by using Geostatistical Analyst, one of the extension systems of ArcGIS.

2.3 Soil and sediment sample sites

The soil and sediment samples were taken randomly around Cam Chinh commune of Quang Tri province. Sixteen soil samples and one sediment sample were collected throughout Cam Chinh commune with a stainless steel core sampler. Cores were of the 0-10 cm depth fraction (Dwernychuk et al., 2002). At any give site, ten cores were collected, composited, and thoroughly mixed to represent a single sample for laboratory analysis. The location of each sample was geo-referenced using global positioning system (GPS). The locations of sampling sites are shown in Fig 1.

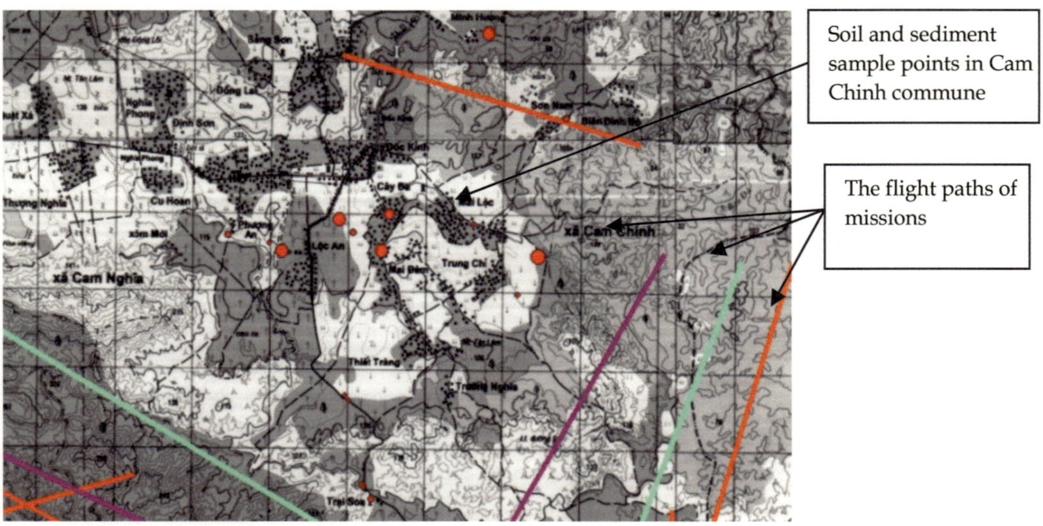

Fig. 1. Location of soil and sediment sample sites in Cam Chinh commune

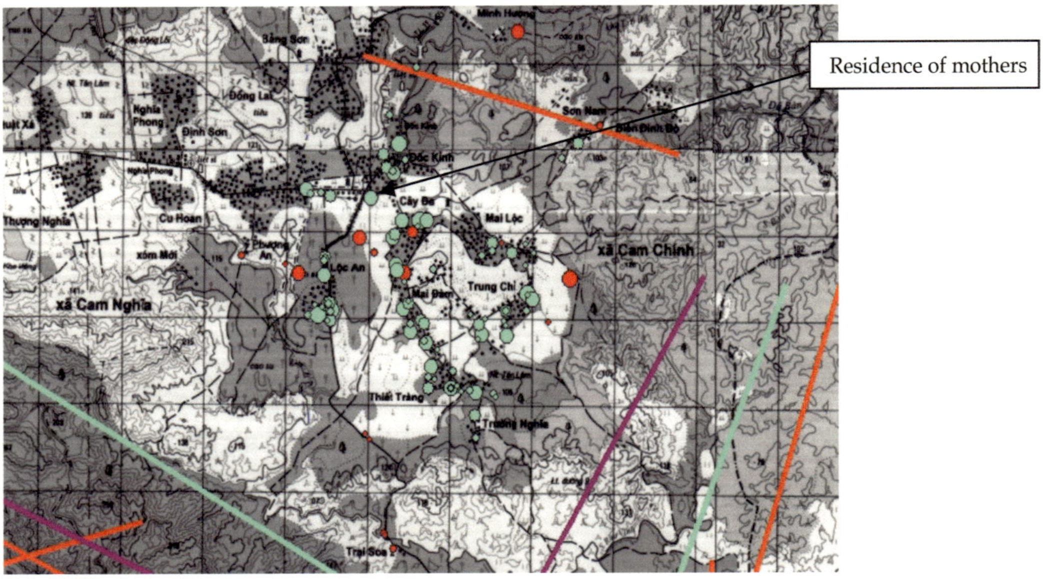

Fig. 2. Overlaid GPS data of breast milk samples in Cam Chinh map

2.4 Breast milk sampling site

In September 2002 and July 2003, breast milk samples were taken from lactating mothers aged between 20-40 years old in two communes. The study purpose was explained to the eighty-six lactating mothers in Cam Chinh commune and seventy-one lactating mothers in Cam Phuc commune by local authorities and medical staff. In sprayed area, 64 mothers lived in Cam Chinh commune, and 12 mothers were born in Cam Nghia commune but got married in Cam Chinh commune. Cam Nghia commune is near Cam Chinh commune; it is also one of the communes that was sprayed during the war. In the non-sprayed area, 71 mothers lived in Cam Phuc commune. All lactating mothers consented to cooperate to donate milk samples. The mothers were breast-feeding their infants aged from 20 days to one year. Donors provided 10-20 ml volume of milk. In each local clinic, samples were collected by the mothers themselves, and local medical staff. All samples were frozen immediately after collection. The residence of mothers was confirmed by GPS. The GPS data were coded as latitude (Y) and longitude (X) by an excel program and saved database. Dots were put on the map in accordance with position of their residence (Fig 2).

2.5 Survey method

A retrospective cohort survey, based on epidemiological interviewing was implemented. All lactating mothers in both communes who donated milk samples after participated in the survey participated. These mothers were interviewed directly by researchers of the 10-80 Division and Kanazawa University using a standard questionnaire to acquire information. Collected information during interview included personal habits like smoking, alcohol drinking, contraceptive drug use, history of pesticide contacting, disease history, number of pregnancies, age of each pregnancy and kinds of pregnant failures. Data were entered into computers using JMP® 6 software for analysis.

2.6 Analytic method

Statistic comparisons were made using chi-square and Wilcoxon signed rank test statistics for categorical variables and Turkey-Kramer HSD test for continuous variables (α=0.05 level). Odds ratios (OR) and 95% confidence intervals (95%) based on the chi-square and Wilcoxon signed rank test were calculated on epidemiology data.

All soil and sediment samples from Vietnam were frozen and brought to Ishikawa Prefectural Institute of Public Health and Environmental Science to analyze. Analysis methods of PCDD/Fs levels described in paper (Kakimoto et al., 2004; Oka et al., 2007). Breast milk samples were brought to the Center of High Technology of Kanazawa Medical University, Japan. All samples were spiked with ^{13}C-labelled surrogate standards prior to analysis. All extracts were subject to a series of chromatographic cleanup steps prior to analysis for PCDDs and PCDFs by a high resolution mass spectrometer (HRMS, HR-GC/MS; MA Station-JMS700, JOEL, Japan) equipped with a gas chromatograph (HP-6980), and measurements were performed by selected ion monitoring (SIM) method (Tawara et al., 2003; Nishijo et al., 2008). Concentration levels of dioxins were shown by actual measurement values and ones were converted to 2,3,7,8-TCDD toxic equivalents (TEQ), submitting the international Toxicity Equivalent Factor (TEF) of WHO-TEF in 1997 (Van den Berg et al., 1998), and WHO-TEF in 2005 (Van den Berg et al., 2006). For non-detectable (ND) and NDR (chromatographic peak was detected but did not meet quantification criteria) were not designated.

Maps representing summary findings of the study were generated using ArcGIS. The database was merged into the GIS software using a Kriging interpolation method to Geostatistical Analysis (Cattle et al., 2002; Wu et al., 2003). Distribution of dioxin contamination was estimated based on the soil samples. Dioxin levels in the soil of each mother's house were estimated and relation between the value of dioxin level in the soil of each mother's house and breast milk were analyzed by Kriging.

3. Results and discussions

The mean concentration of dioxin in soil of the sprayed area is 1.9 pg-TEQ/g; this value is significantly higher than 0.38 pg-TEQ/g in the non-sprayed area.

The reproductive cohort survey on mothers was carried out in two communes, one area sprayed with herbicide and another area not sprayed. The results on Table 1 show that mean dioxin level of breast milk in the sprayed area was significantly ($p<0.001$) higher than the non-sprayed area. A statistically significant difference was showed in mother's height and weight in non-sprayed area, this being higher than in sprayed areas, but BMI is not significant. The mothers have used pesticides in the non-sprayed area is significantly ($p<0.001$) higher than sprayed area, but other chemicals are not significant between the two areas. All mothers in the two areas do not use alcohol, and three mothers in the sprayed area smoke (3.6%). Cousins who have birth defects in the sprayed area were 15.2% and non-sprayed area 10.6%, and is not significant. Present disease of mothers in sprayed area (22.1%) was significantly ($p<0.05$) higher than non-sprayed area (8.6%). Reproductive failure history in sprayed area (22.1%) was significantly ($p<0.01$) higher than non-sprayed area (5.6%).

This study is to assess the correlation between dioxin contamination levels in soil and dioxin exposure levels in breast milk of females in Cam Chinh commune, Quang Tri province of Vietnam. One of the GIS technologies of Kriging interpolation method was used to check the

		Sprayed area				Non-sprayed area				p-value
		N	Mean ±SD or N (%)			N	Mean ±SD or N (%)			
Age	(years)	86	29.3	±	5.7	71	26.1	±	4.3	***b)
Height	(cm)	79	152.4	±	4.6	54	154.1	±	5.5	*b)
Weight	(kg)	85	43.6	±	3.9	66	45.8	±	4.0	**b)
BMI	(kg/m^2)	78	18.8	±	1.5	54	19.4	±	1.8	n.sb)
Contact with herbicides during the war	Yes	82	4(4.9%)			71	0 (%)			-
	No		71(86.6%)				71 (100%)			
	Don't know		7(8.5%)				0 (%)			
Using pesticides	Yes	82	7(8.5%)			68	31 (45.6%)			***c)
Using other chemicals	Yes	84	2(2.4%)			65	0 (%)			-
Alcohol habit	Yes	83	83(100%)			69	69(100%)			-
Smoke status	Yes	83	3(3.6%)			71	0 (%)			-
Using contraceptive	Yes	80	3(3.8%)			65	1 (1.5%)			n.sc)
Cousin who is birth defect people	Yes	79	12(15.2%)			66	7 (10.6%)			n.sc)
Cousin have birth defect	(People)	12	1.3	±	0.9	7	1	±	0	n.sb)
Present disease	Yes	86	19(22.1%)			70	6 (8.6%)			*c)
Reproductive failure history	Yes	86	19(22.1%)			71	4 (5.6%)			**c)
TEQ-PCDDs	[pgTEQ/gFat]a)	86	5.1	±	3.2	71	2.1	±	0.9	***b)
TEQ-PCDFs	[pgTEQ/gFat]a)	86	6.0	±	4.0	71	2.0	±	0.8	***b)
TEQ-TOTAL	[pgTEQ/gFat]a)	86	11.3	±	7.0	71	4.2	±	1.6	***b)

a)Data of dioxin levels were log transformed
b) Wilcoson signed rank test, ***:p<0.001
c) Chi-square
*:p<0.05; **: p<0.01; ***: p<0.001, n.s: not significant

Table 1. Comparison of characteristics and dioxin levels in breast milk of the mothers between herbicide sprayed and non-sprayed areas.

correlation; however, no significant correlation appeared (Fig 4, 6). We suspect the reason for the lack of correlation was the extended period since the end of the war (35 years passed). The half-life of dioxin elimination in the human body, which is estimated at 7 - 11 years (Mukerjee 1998; Kerger et al., 2007), and in the soil environment is 28.5 - 274 years (Sinkkonen et al., 2000). The second reason was the rather low number of the samples; 16 for soil, one for sediment and 86 for breast milk are highly imbalanced. In contrast, no significant relationship means present environment status represented by dioxin levels of soil is not consistent with human exposure that is represented by breast milk. There is one possibility that another exposure route such as exposure due to herbicides during Vietnam war might affect dioxin levels of breast milk.
In addition, we also found that the distributions of PCDDs and PCDFs in soils were quite different by the Kriging interpolation analysis. More details in differences are presented below. On the Kriging map of PCDDs, no correlation was shown between dioxin in soil and breast milk (fig 4). We can see red points (locations of soil samples) which are dioxin levels in soil, and blue points (locations where mothers are living) which are dioxin levels in breast milk. Around the red big points (high dioxin level) are dark colors and red small points (low dioxin level) are light. PCDDs were thought to come from Agent Orange, given Agent Orange in the Quang tri area was sprayed 1965 - 1970 (IOM 1998) (Fig 3).

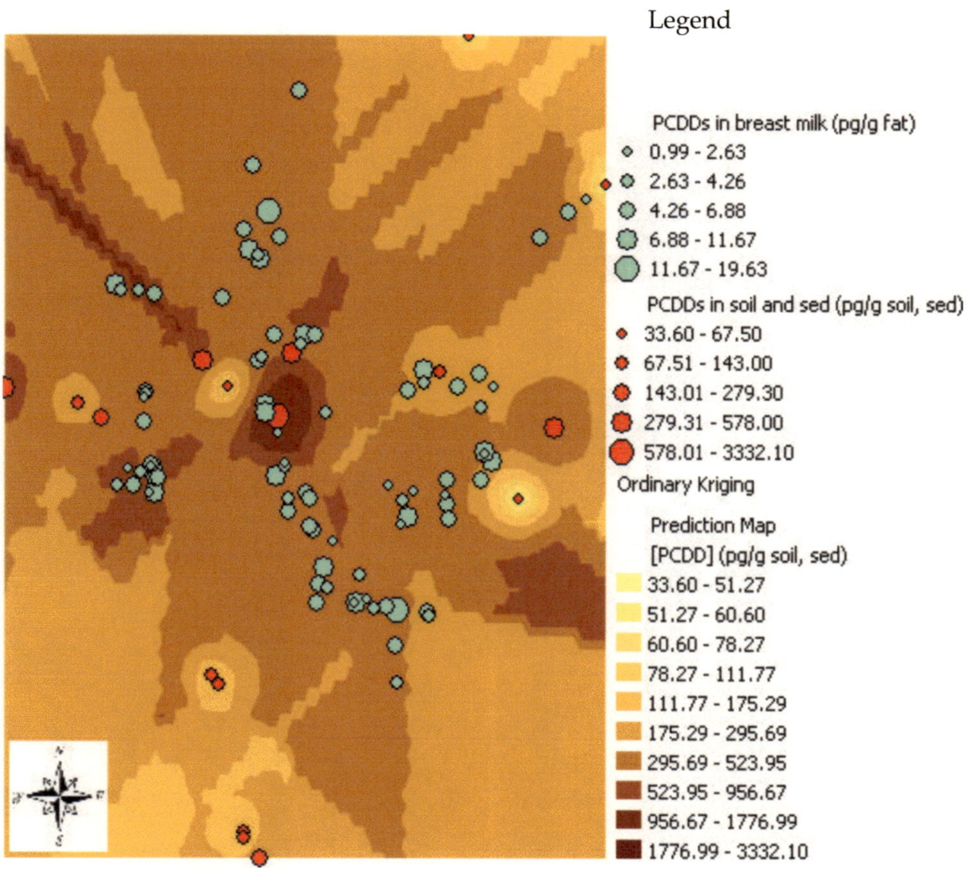

Fig. 3. Estimated distribution of PCDDs from soil samples by Kriging

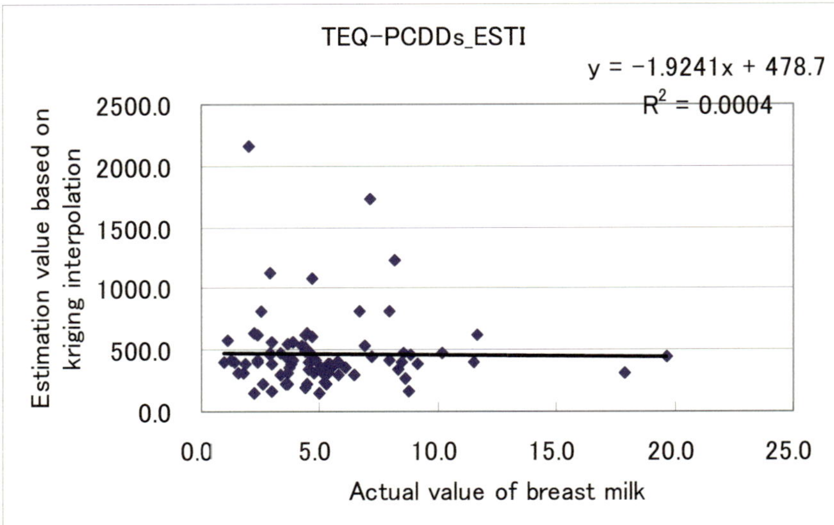

Fig. 4. Estimation value based on Kriging interpolation and actual value of breast milk

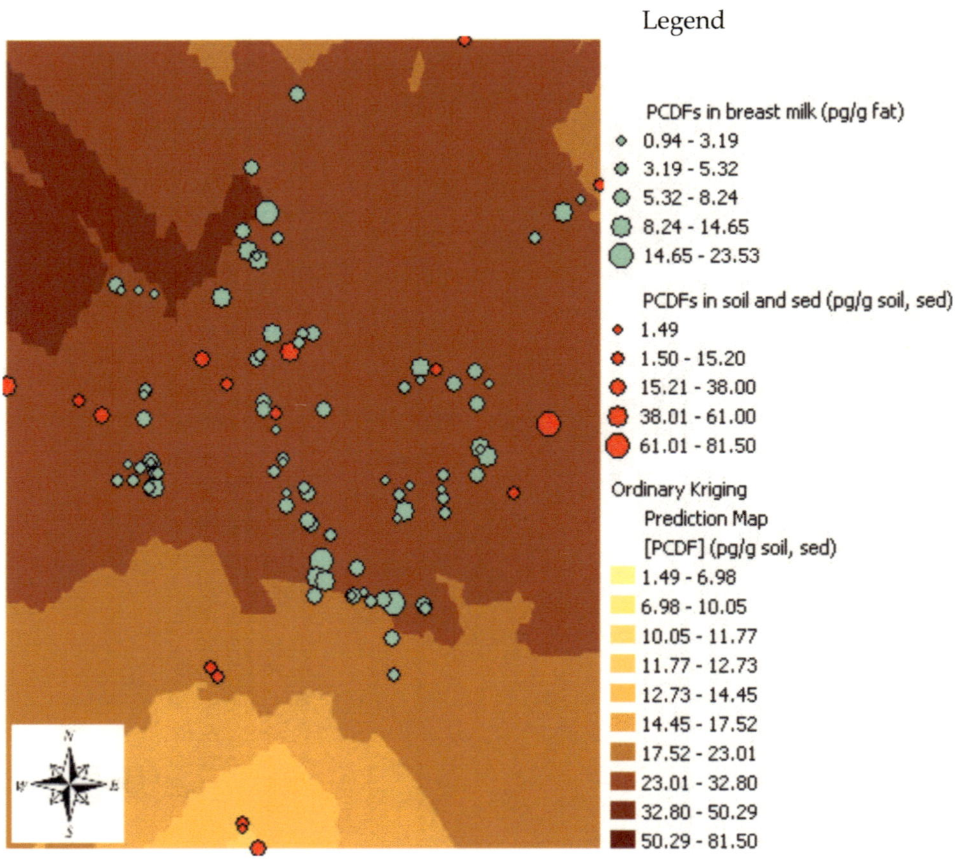

Fig. 5. Estimated distribution of PCDFs from soil samples by Kriging

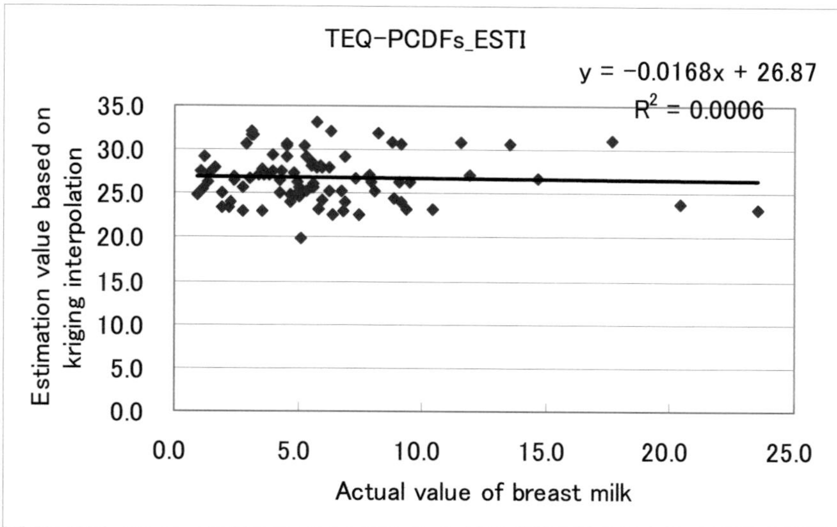

Fig. 6. Estimation value based on Kriging interpolation and actual value of breast milk

Results of the analysis show that PCDFs are not correlated with dioxin in soil and breast milk (Fig 6). On the Kriging map of PCDFs we can see red points which are dioxin levels in soil and blue points which are dioxin levels in breast milk (Fig 5). But location of red points is different with Kriging map of PCDDs and the dark color is also different, it is not around the high dioxin level in soil. PCDFs in this area are thought to come from herbicides other than Agent Orange sprayed during the war plus other chemical and pesticides applied after the war.

Reasons why we cannot detect clear correlation between estimated value of soil and breast milk follow: 1) duration after the war, 2) imbalanced alignment of soil samples, 3) larger area will be tested if people's immigration tendencies are taken into account. Then, more studies will be required to take a larger number of soil samples, which will be well-distributed in the study site including dioxin hot spots.

4. Conclusions

The concentrations of PCDDs, PCDFs in soil, sediment and breast milk samples collected in Cam Chinh commune, Quang Tri, Vietnam to assess correlations using Geostatistic algorithms of log-normal Kriging. Results showed that mean dioxin level of breast milk in sprayed areas was significantly higher than that in non-sprayed area. Significant correlations did not appear between estimated dioxin levels in soil by Kriging method and those in breast milk. There is one possibility that another exposure route such as exposure due to herbicides during Vietnam war might affect dioxin levels of breast milk. The distribution pattern of PCDDs and PCDFs is also quite different each other. More soil data should be needed to make more reliable geographical estimations.

5. Acknowledgements

This study was supported by grants from Japan Society for the Promotion of Science (Grant-in-Aid for Scientific Research, (B)no.17406016 and Grant-in-Aid for Scientific

Research (A) no.19209021) and Yoshida Scholarship Foundation, Japan. The authors would like to thank Dr. Wayne Dwernychuk at Hatfield Consultant in Vancouver, Canada; Dr. Nguyen Ngoc Hung and officers of 10-80 Division, Hanoi Medical University, Vietnam for making this possible.

6. References

10-80 Committee. *Herbicides Sprayed Map*. 1999.

Armstrong, M., Boufassa, A., 1988. Comparing the robustness of Ordinary Kriging and Lognormal Kriging: Outlier resistance. Math. Geol. 20, 447-457.

Cattle, J.A., McBratney, A.B., Minasny, B., 2002. Kriging method evaluation for assessing the spatial distribution of urban soil lead contamination. J. Environ. Qual. 31, 1576-1588.

Cau, H.D., 2003. Environment and human health in Vietnam: thirty years after the Ranch Hand Operation, Hanoi, 10-80 Committee.

Dwernychuk, L.Wayne., Hung, T.M, Boivin, T.C, Bruce, G.S., Dung, P.T., Son, T.K., Hatfield, C.T., Dung, N.T., Allan, J.A., Nhu, D.D., Thuc, P.V., Moats, D.J., Borton, L., 2006. The Agent Orange dioxin issue in Vietnam: A manageable problem. Organohalogen Compd. 68, 312-315.

Dwernychuk LW, 2005. Dioxin hot spots in Vietnam. Chemosphere. 60, 998-999.

Dwernychuk, L.Wayne., Cau, H.D., Hatfield, C.T., Boivin, T.G., Hung, T.M., Dung, P.T., Thai, N.D., 2002. Dioxin reservoirs in southern Viet Nam - a legacy of Agent Orange. Chemosphere. 47, 117-137.

Rushton, G., 2003. Public health, GIS, and spatial analytic tools. Annu. Rev. Public Health 24, 43-56.

IOM (Institute of Medicine). Veterans and Agent Orange - Health effects of herbicides used in Viet Nam. Update 2002. Washington, D.C: National Academy Press.

IOM (Institute of Medicine). Veterans and Agent Orange - Health effects of herbicides used in Viet Nam. Update 1998. Washington, D.C: National Sciences Academy Press.

Kakimoto, H., Oka, H., Harada, Y., Ushijima, S., Toriba, A., Kizu, R., Hayakawa, K., 2004. Comparison of compositions of polychlorilated-dipenzo-p-dioxins (PCDDs) and dipenzofurans (PCDFs) in air and soil samples collected in Ishikawa. J. Health Sci. 50, 58-65.

Kerger, B.D., Leung, H.W., Scott, P.K., Paustenbach, D.J., 2007. An adaptable internal dose dodel for risk assessment of dietary and soil dioxin exposures in young children. Toxicol. Sci. 100, 224–237.

Kido, T., Suzuki, H., Naganuma, R., Tawara, K., Nishijo, M., Nakagawa, H., Hung, T.M., Thom, L.H.T., Dung, P.T., Nhu, D.D., 2006. An epidemiological study on health effects by Dioxin in Vietnam; Comparison of contrast acuity between inhabitants of herbicide sprayed and non-sprayed areas. The 26th International Symposium on Halogenated Persistent Organic Pollutants. Organohalogen Compd. 68, 1990-1993.

Largueche, F.B., 2006. Estimating soil contamination with Kriging interpolation method. American J. of Applied Sciences. 3, 1894-1898.

Mai, T.A., Doan, T.V., Tarradellas, J., de Alencastro, L.F., Grandjean, D., 2007. Dioxin contamination in soils of Southern Vietnam, Chemosphere. 67, 1802-1807.

Mukerjee, D., 1998. Health impact of polychlorinated dibenzo-p-dioxins: a critical review. J. Air Waste Manag. Assoc. 48, 157-165.

Nishijo, M., Tawara, K., Nakagawa, H., Honda, R., Kido, T., Nishijo, H., Saito, S., 2008. 2,3,7,8-Tetrachlorodibenzo-p-dioxin in maternal breast milk and newborn head circumference. J. Expo. Sci. Environ. Epidemiol. 18, 246-251.

Oka, H., Miyakawa, S., Noguchi, K., Watanuki, F., Tsukabayashi, H., Yada, M., 2007. Dioxin levels in soil and sediment of Vietnam. J. Ishikawa Prefectural Institute of Public Health and Environ. Sci. 44, 79-81 (in Japanese).

Rushton, G., 2003. Public health, GIS, and spatial analytic tools. Annu. Rev. Public Health. 24, 43-56.

Saito, H., Goovaerts, P., 2000. Geostatistical interpolation of positively skewed and censored data in a dioxin contaminated site. Environ. Sci. Technol. 34, 4228 - 4235.

Scheter, A., Quynh, H.T., Papke, O., Malisch, R., Constable, J.D., Tung, K.C., 2003. Halogenated organics in Vietnamese and in Vietnam food: Dioxin, dibenzofurans, PCBs, polybrominated diphenyl ethers and selected pesticides. J. Occup. Environ. Med. 45, 781-788.

Scheter, A., Dai, L.C., Papke, O., Prange, J., Constable, J.D., Matsuda, M., Thao, V.D., Piskac, A.L., 2001. Recent dioxin contamination from Agent Orange in residents of a Southern Vietnam city. J. Occup. Environ. Med. 43, 435-443.

Sinkkonen, S., Paasivirta, J., 2000. Degradation half-life times of PCDDs, PCDFs and PCBs for environmental fate modeling. Chemosphere. 40, 943–949.

Stellman, J.M., Stellman, S.D., Christian, R., Weber, T., Tomassalo, C., 2003a. The extent and patterns of usage of Agent Orange and other herbicides in Viet Nam. Nature. 422, 681-687.

Stellman, J.M., Stellman, S.D., Weber, T., Tomasallo, C., Stellman, A.B., Christian, R., 2003b. A geographic information system for characterizing exposure to Agent Orange and other herbicides in Vietnam. Environ. Health Perspect. 111, 321-328.

Tawara, K., Nishijo, M., Nakagawa, H., Kido, T., Naganuma, R., Hung, T.M., Thom, L.T.H., Dung, P.T., Nhu, D.D., 2006. Areal differences of concentration levels of polychlorinated dibenzo-p-dioxins and dibenzofurans in human breast milk from Vietnam and Japan. Organohalogen Compd. 68, 1655-1658.

Tawara, K., Honda, R., Nishijo, M., Nakagawa, H., 2003. Pretreatment procedure of dioxin analysis for a small volume of human breast milk. J. Kanazawa Med. Univ. 28, 17-25 (in Japanese).

Van den Berg, M., Birnbaum, L., Denison, M., De Vito, M., Farland, W., Feeley, M., Fiedler, H., Hakansson, H., Hanberg, A., Hwas, L., Rose, M., Safe, S., Schrenk D., Toyama, C., Tritscher, A., Tuomisto, J., Tysklind, M., Walker, N., Peterson, R.E., 2006. The 2005 World Health Organization re-evaluation of human and mammalian toxic equivalency factors for dioxins and dioxin-like compounds. Toxicol. Sci. 93, 223-241.

Van den Berg, M., Birnbaum, L., Bosveld, A.T.C., Brunstrom, B., Cook, P., Feeley, M., Giesy, J.P., Hanberg, A., Hasegawa, R., Kennedy, S.W., Kubiak, T., Larsen, J.C., van Leeuwen, F.X., Liem, A.K., Nolt, C., Peterson, R.E., Poellinger, L., Safe, S., Schrenk D., Tillitt, D., Tysklind, M., Younes, M., Waern, F., Zacharewski, T., 1998. Toxic equivalency factors (TEFs) for PCBs, PCDDs, PCDFs for humans and wildlife. Environ. Health Perspect. 106, 77-792.

Viel, J.F., Clément, M.C., Hägi, M., Grandjean, S., Challier, B., Danzon, A., 2008. Dioxin emissions from a municipal solid waste incinerator and risk of invasive breast cancer: a population - based case-control study with GIS-derived exposure. Int. J. Health Geogr. 2008, 7:4.

Vietnam Investment Review, Ministry of Planning and Investment, Socialist Republic of Vietnam, 1999. War border province's Agent Orange tragedy. Issue: 0410, August.

Waring, S.C., Zakos-Feliberti, A., Wood, R., Stone, M., Padgett, P., Arafat, R., 2005. The utility of geographic information systems (GIS) in rapid epidemiological assessments following weather-related disasters: Methodological issues based on the Tropical Storm Allison experience. Int. J. Hyg. Environ. Health. 208, 109-116.

Westing, A.H., 1984. Herbicides in War: the long-term ecological and human consequences. Taylor and Francis, London and Philadenphia. Pg 3-24.

Wu, J., Norvell, W.A., Hopkins, D.G., Smith, D.B., Ulmer, M.G., Welch, R.M., 2003. Improved prediction and mapping of soil copper by Kriging with auxiliary data for cation-exchange capacity. Soil Sci. Soc. Am. J. 67, 919-927.

27

Developmental Toxicity of Nitrophenolic Herbicide Dinoseb, 2-sec-butyl-4,6-dinitrophenol

Mariko Matsumoto, Akihiko Hirose and Makoto Ema
Division of Risk Assessment, Biological Safety Research Center,
National Institute of Health Sciences
Japan

1. Introduction

Dinoseb (2-*sec*-butyl-4,6-dinitrophenol; CAS No. 88-85-7), a dark reddish-brown solid or dark orange viscous liquid, depending on the temperature (melting point: 32-42 °C) (Kidd & James, 1991), was approved for sale in the US in 1948 as a nitrophenolic herbicide in soybeans, vegetables, fruits, nuts, citrus and other field crops for the selective control of grass and broadleaf weeds (EXTOXNET, 1996; Schneider, 1986). Dinoseb is also used as an insecticide for grapes and as a seed crop drying agent (EXTOXNET, 1996). Dinoseb is one of the chemicals available on the market on the basis of safety tests conducted by Industrial Bio-Test Laboratory, a concern later found to have submitted many flawed and even fraudulent reports on its procedures and results (Shabecoff, 1986). Subsequently, several studies showed that dinoseb has the potential to produce developmental toxicity including teratogenicity in rats, mice and rabbits (Giavini et al., 1986; Johnson, 1988; Preache & Gibson, 1975a; Preache & Gibson, 1975b).

Dinoseb as a pesticide was banned in the US in 1986 and the EU in 1991 owing to the potential risk of adverse health effects in humans (EXTOXNET, 1996; Rotterdam Convention, 2006), but dinoseb and its salts are still widely used as other agricultural products (PAN, 2006). Dinoseb as a pesticide is also banned in Japan, but its import is permitted (PAN, 2006). The volumes of dinoseb imported into Japan were estimated to be 615 tons in fiscal year 2008 and 726 tons in fiscal year 2009 (NITE, 2009). Dinoseb is a high-volume chemical with production or importation exceeding 1,000 tons per year in Organisation for Economic Co-operation and Development (OECD) member countries (OECD, 2004).

Dinoseb is well absorbed from the gastrointestinal tract by the oral route and can pass through the placenta into the fetus in mice (Gibson & Rao, 1973). A dermal study showed that in six hours young and adult female rats absorbed about 44% of the dose, while at 120 hours 75.9% was absorbed in young and 92.5% in adults (Hall et al., 1992). Dinoseb shows relatively strong acute toxicity with an oral LD_{50} of 5-50 mg/kg in female rats (MHLW, 2005), an intraperitoneal LD_{50} of 14.1-22.5 mg/kg in mice (US EPA, 2003b) and a dermal LD_{50} of 40 mg/kg in rabbits (US EPA, 2003b). Inhalation LC_{50} is 33-290 mg/m^3 for 4-hour exposure in rats (US EPA, 2003b). The basic mechanism of toxicity is thought to be stimulation of oxidative metabolism in cell mitochondria by the uncoupling of oxidative phosphorylation (Leftwich et al., 1982). Toxicity of dinoseb is enhanced by physical activity and high ambient temperature such as in an outdoor agricultural environment (Leftwich et al., 1982; US EPA, 2007). Early symptoms of dinoseb exposure include hyperthermia, sweating, headache and confusion. Severe exposure may result in restlessness, seizures, coma and death (Leftwich et al., 1982; US EPA, 2006; US EPA, 2007).

Exposure to dinoseb may occur by direct contact, ingestion or inhalation by users and producers, but indirect exposure to dinoseb via the environment is also anticipated. The microbial breakdown of dinoseb has been demonstrated in soils, but dinoseb persists for about two to four weeks after application (Health Canada, 1991). A soil persistence of 24 to 42 months was also observed in potato fields in Canada (O'Neill et al., 1989). It has been reported that dinoseb was detected in water supplies in Canada and the US, and dinoseb residues were found in a cotton meal sample (Health Canada, 1991).

Developmentally toxic effects of chemicals are influenced by the susceptibility of animal species and strains, the developmental stages of offspring and administration doses (Schardein, 2000; Wilson & Warkany, 1965). Teratogenicity is governed by dose-effect relations, but there are many variable factors such as the duration of chemical treatment (Wilson, 1966), frequency of dosing (Isaacson & Chaudhry, 1962), routes or modes of administration (Hansen & Billings, 1986; Kavlock et al., 1982; Kimmel, 1977; Staples et al., 1976), the vehicle/suspending agent (Anderson & Morse, 1966) or a combination of chemicals (Wilson, 1964). Dinoseb is one of the chemicals that show differences in developmental toxicity according to these variable factors. We have already reviewed studies on the developmental effects of dinoseb exposed prenatally in experimental animals (Matsumoto et al., 2008c). In the following sections, available literature including new information concerning the developmental toxicity of dinoseb is introduced by focusing on the variable factors for risk assessment of dinoseb. It should be noted that the term dinoseb has been used in the literature to refer to several related chemicals based on 2-*sec*-butyl-4,6-dinitrophenol (CAS: 88-85-7). In this chapter, dinoseb refers to the parent molecule only.

2. Developmental toxicity in rabbits

Table 1 shows the results of developmental toxicity of dinoseb in rabbits. There are gavage and dermal dose studies. The data are shown by routes of administration, in order of the most likely route of human intake. Only statistically significant effects are summarized unless noted otherwise.

2.1 Gavage study in rabbits

In a teratology study, 16 Chinchilla rabbits were exposed by gavage to dinoseb at 0 (vehicle: corn oil), 1, 3 or 10 mg/kg bw/day on gestation days **(GDs)** 6-18 (Research and Consulting Company, 1986). There were no differences in fetal body weight and sex ratio between the dinoseb-treated and control groups. In the highest dose group, there were neural tube defects including dyscrania associated with hydrocephaly, scoliosis, kyphosis, malformed or fused caudal or sacral vertebrae and/or encephalocele in a total of 40 fetuses in 11/16 litters. Eleven fetuses showed only hydrocephalus and/or micro- or anophthalmia, and 4 fetuses showed only skeletal abnormalities. No maternal death occurred. Body weight gain and food consumption in dams and number of implantations were not affected.

2.2 Dermal study in rabbits

In a New Zealand white rabbit study, 16-17 pregnant rabbits were dermally given dinoseb at 0, 1, 3, 9 or 18 mg/kg bw/day on GDs 7-19 (Johnson, 1988). The dinoseb (no vehicle was used) was dermally applied to rabbits wearing Elizabethan collars for 6 hours, and the application site was wiped and then dried. Because overt maternal toxicity was observed at 18 mg/kg bw/day and animals were also dying in the 9 mg/kg bw/day group, animals treated with dinoseb at the high dose were reassigned to the 9 mg/kg bw/day dose group and did not contribute to the evaluation. There were increased incidences of anophthalmia and hydrocephaly at 3 and 9 mg/kg bw/day. Dead and resorbed fetuses and fetuses with cleft palate, microphthalmia and microcephaly were increased at 9 mg/kg bw/day. At 3 mg/kg bw/day and higher, hyperthermia and reduced body weight were observed in maternal rabbits.

Species (Reference)	Dose	Exposure time	Developmental effect
Gavage			
Chinchilla rabbit	10 mg/kg	GDs 6-18	External, internal and skeletal defects
(Research and Consulting Company, 1986)			
Dermal			
NZ white rabbit	3 mg/kg	GDs 7-19,	Hydrocephaly, anophthalmia
	9 mg/kg	6 h/day	Dead and resorbed fetuses, cleft palate, microcephaly, microphthalmia
(Johnson, 1988)			

Table 1. Developmental toxicity of dinoseb in rabbits
GDs: gestation days

3. Developmental toxicity of dinoseb in mice

Tables 2.1-2.3 show the results of developmental toxicity studies of dinoseb in mice. There are gavage, intraperitoneal (i.p.) and subcutaneous (s.c.) administration studies. The data are shown by routes of administration, in order of the most likely route of human intake. Only statistically significant effects are summarized unless noted otherwise.

3.1 Gavage studies in mice

Pregnant CD-1 mice were administered dinoseb in corn oil on GDs 8-12 at 15 mg/kg bw/day, the expected maximum tolerated dose level of dinoseb. No effects were observed in reproductive and developmental parameters (Chernoff & Kavlock, 1982). Pregnant CD-1 mice were given dinoseb in corn oil by gavage at 26 or 33 mg/kg bw on GD 7. Two out of 40 pregnant animals died at 33 mg/kg bw, but percent mortality and body weight of pregnant mice were not changed. An increased incidence of supernumerary ribs was observed in both dinoseb-treated groups. The authors noted that increased incidence of supernumerary ribs may be a response to a non-specific disruption in maternal status (Kavlock et al., 1985).
Administration of dinoseb to pregnant CD-1 mice by gavage on GDs 7-8 at 50 mg/kg bw/day in NaOH produced reduced fetal weight and increased incidence of fetuses with supernumerary ribs (71% in litters) without maternal death. The authors suggested that

supernumerary ribs are indicative of basic alterations in the development of the axial skeleton (Branch et al., 1996). A similar study conducted by Rogers et al. (2004) revealed a dose-related increased incidence of mouse fetuses with supernumerary ribs following maternal administration of dinoseb in NaOH at 50 mg/kg bw/day on GDs 7-8 and suggested that increased incidence of supernumerary ribs in fetuses is toxicologically significant. Skeletal anomalies such as sternum or vertebral centrum defects and fused ribs were also detected in fetuses of mice given dinoseb on GDs 7-8 at 50 mg/kg bw/day in NaOH. Although the treatment regimes of Branch et al. (1996) and Rogers et al. (2004) were essentially the same, they obtained different developmental effects in fetuses of mice given dinoseb at 50 mg/kg bw/day. Rogers et al. (2004) used 25 pregnant mice. On the other hand, Branch et al. (1996) used only two pregnant mice, which is too few to evaluate the developmental toxicity. Therefore, it appears that a gavage dosing of dinoseb on GDs 7-8 at 50 mg/kg bw/day can induce teratogenic effects without maternal toxicity in CD-1 mice.
Dinoseb was administered to pregnant Swiss-Webster mice during GDs 7-15, GDs 9-11 or GDs 13-15 by gavage up to 50 mg/kg bw/day in NaOH. Gavage dosing of dinoseb produced no increased incidence of gross or soft-tissue anomalies. When dinoseb was given by gavage on GDs 9-11, six out of eight pregnant animals died at 50 mg/kg bw/day, but no effects were observed on developmental parameters. Skeletal variations such as supernumerary ribs and vertebrae were observed after doses of 20 and/or 32 mg/kg bw/day during GDs 7-15. The fetal crown-rump length (CRL) was also reduced at 32 mg/kg bw/day after administration of dinoseb on GDs 7-15. A dose of 32 mg/kg bw/day dinoseb during GDs 13-15 induced absent or not ossified sternebrae. The dose levels that caused these adverse effects in fetuses were also lethal to some dams (Gibson, 1973).

Species (Reference)	Dose	Exposure time	Developmental effect
CD-1 mouse (Chernoff & Kavlock, 1982)	15 mg/kg	GDs 8-12	No effects
CD-1 mouse (Kavlock et al., 1985)	26, 33 mg/kg	GD 7	Supernumerary ribs
CD-1 mouse (Branch et al., 1996)	50 mg/kg	GDs 7-8	Supernumerary ribs, ↓fetal weight
CD-1 mouse (Rogers et al., 2004)	50 mg/kg	GDs 7-8	Supernumerary ribs, sternum and vertebral centrum defects, fused ribs
SW mouse	50 mg/kg	GDs 9-11	No effects (6/8 dams died)
	20 mg/kg 32 mg/kg	GDs 7-15	Supernumerary ribs and vertebrae ↓fetal crown-rump length
(Gibson, 1973)	32 mg/kg	GDs 13-15	Absent or not ossified sternebrae

Table 2.1. Developmental toxicity of dinoseb administered by gavage in mice
GDs: gestation days

3.2 Intraperitoneal studies in mice

No adverse effects were observed in reproductive and developmental parameters after an i.p. administration of dinoseb on GDs 7-15 at 5 mg/kg bw/day in Swiss-Webster mice; however, teratogenicity was obtained after i.p. administration of dinoseb on GDs 13-15 and GDs 9-11 (Gibson, 1973). An increased incidence of soft tissue malformation such as internal hydrocephalus was observed at 10-15.8 mg/kg bw/day in NaOH after i.p. treatment of dinoseb on GDs 9-11. At these doses, no maternal toxicity was observed. Increased incidences of defects in the limbs, tail, ribs, sternebrae and vertebrae, internal hydrocephaly and hydronephrosis were also induced at 17.7 mg/kg bw/day. Fetal body weight and number of fetuses were decreased at 18.8 mg/kg bw/day, and fetal CRL was decreased at 20.0 mg/kg bw/day. At 17.7-20.0 mg/kg bw/day, dinoseb produced hyperthermia and death in dams. Dinoseb at 12.5 and 17.7 mg/kg bw/day on GDs 13-15 caused increased resorptions and decreased fetal body weight, but not maternal toxicity. Unlike administration of dinoseb on GDs 9-11, teratogenicity was not observed after administration of dinoseb on GDs 13-15 up to 17.7 mg/kg bw/day.

In a later review study for perinatal nephropathies, Gibson (1976) stated that an incidence of 30-40% of fetuses with hydronephrosis was observed at cesarean section owing to i.p. administration of dinoseb on GDs 9-11; however, no grossly observable hydronephrosis was evident in pups at 1 or 2 weeks of age. Renal alteration observed in offspring of mice given dinoseb seems to be a transient dilatation of the renal pelvis, which is also suggested by studies in rats (Daston et al., 1988; McCormack et al., 1980), as described in 4.3. On the other hand, i.p. treatment of dinoseb on GDs 9-11 at 15.8 mg/kg bw/day caused a impairment in p-aminophippuric acid (PAH) uptake into renal cortical slices of offspring at one and two weeks of age, and this effect was also evident at seven weeks of age (Gibson, 1976).

Effects of food deprivation, phenobarbital (an inducer of chemical metabolism) and 2-diethylaminoethyl-2,2-diphenylvalerate hydrochloride (SKF-525A; an inhibitor of chemical metabolism) on the developmental toxicity of dinoseb were evaluated in Swiss-Webster mice (Preache & Gibson, 1975a). Pregnant mice were treated i.p. with dinoseb at doses of 0-18.8 mg/kg bw/day on GDs 9-11. These treatments were preceded by 24 or 48 h food deprivation or by pretreatment with phenobarbital or SKF-525A. Dinoseb-induced external and skeletal anomalies were increased by 24 h food deprivation. Effects of phenobarbital pretreatments on dinoseb-induced developmental toxicity were inconsistent at 17.7 and 18.8 mg/kg bw/day. At these doses, maternal death was not observed. Pretreatment with SKF-525A in combination with dinoseb at 15.8 mg/kg bw/day caused fetal anomalies, potentiated dinoseb-induced resorptions and produced maternal mortality. SKF-525A in combination with dinoseb at 17.7 mg/kg bw/day was markedly lethal maternally; however, developmental parameters could not be analyzed because of the small number of litters surviving. Therefore, it is likely that the proximate toxicant for maternal toxicity was dinoseb itself.

Swiss-Webster mice were treated with dinoseb on GDs 9-11 and maintained at an increased environmental temperature (32 °C) for 24 h or a decreased temperature (0-6 °C) for 1.5-4 h (Preache & Gibson, 1975b). Exposure to 32 °C enhanced adverse effects of dinoseb; it increased maternal mortality, decreased fetal body weight and increased the incidence of fetal anomalies at 7.5 mg/kg bw/day. Fetal body weight and the frequency of malformations were generally the same in groups exposed to low temperature and maintained at room temperature at 15.8-17.7 mg/kg bw/day. Maternal mortality was observed at doses that caused fetal toxicity. On the basis of these results, higher temperature enhanced the maternal and developmental toxicity of dinoseb.

Species (Reference)	Dose	Exposure time	Developmental effect
SW mouse	10 mg/kg	GDs 9-11	Soft-tissue malformation
	17.7 mg/kg		Gross and skeletal malformations
	18.8 mg/kg		↓Fetal body weight, no. of fetuses, resorption
	20 mg/kg		↓Fetal crown-rump length
	12.5-17.7 mg/kg	GDs 13-15	↓Fetal body weight, ↑resorption
	5 mg/kg	GDs 7-15	No effects
(Gibson, 1973)			
SW mouse (Gibson, 1976)	15.8 mg/kg	GDs 9-11	↓PAH uptake by renal cortical slices
SW mouse	15.8 mg/kg	GDs 9-11	Hydronephrosis, ectrodactyly, resorption
	17.7, 18.8 mg/kg (combination with SKF-525A)	GDs 9-11	Hydronephrosis
	14.1 mg/kg	GDs 9-11	Delayed ossification
	15.8 mg/kg		External malformations
	17.7 mg/kg		Hydronephrosis
	18.8 mg/kg (combination with phenobarbital)		↓Fetal body weight, resorption
	14.1 mg/kg	GDs 9-11	Delayed ossification, ↓fetal body weight
	15.8 mg/kg (24 h deprivation)		Hydronephrosis, ectopic kidney, internal hydrocephalus, External and skeletal malformations
	14.1 mg/kg	GDs 9-11	↓Fetal body weight,
	15.8 mg/kg (48 h deprivation)		External and skeletal malformations
(Preache & Gibson, 1975a)			
SW mouse	7.5 mg/kg (32°C)	GDs 9-11	↓Fetal body weight, external, soft-tissue and skeletal malformations, delayed ossification
	15.8 mg/kg (Room temp; wet)	GDs 9-11	↓Fetal body weight, external and soft-tissue malformations, resorption
	15.8 mg/kg (6°C; wet)	GDs 9-11	↓Fetal body weight, external and soft-tissue malformations
	15.8 mg/kg	GDs 9-11	↓Fetal body weight
	17.7 mg/kg (Room temp; dry)		External and soft-tissue malformations, skeletal retardation, variation and malformation
	17.7 mg/kg (6°C; dry)	GDs 9-11	↓Fetal body weight, external and soft-tissue malformations, skeletal retardation, variation and malformation
(Preache & Gibson, 1975b)			

Table 2.2. Developmental toxicity of dinoseb administered by intraperitoneally in mice
GDs: gestation days

3.3 Subcutaneous study of dinoseb

Dinoseb was subcutaneously administered to pregnant Swiss-Webster mice during GDs 8-16, 10-12 or 14-16 at 0, 10 or 17.7 mg/kg bw/day (Gibson, 1973). Adverse effects were observed only at 17.7 mg/kg bw/day. Dinoseb on GDs 14-16 induced increases in resorption rate and the incidence of cleft palate and decreases in the number of live fetuses, fetal CRL and fetal body weight. At this dose, one out of eight dams died. Dinoseb on GDs 10-12 induced an increase in the incidence of fused ribs/vertebrae and absent or not ossified sternebrae, and on GDs 8-16 induced supernumerary ribs/vertebrae, absent or not ossified sternebrae, decreased fetal body weight and decreased fetal CRL without maternal toxicity. The authors concluded that an s.c. dose of dinoseb was not teratogenic and cleft palate induced by treatment of dinoseb was not considered as a toxicological response because this anomaly was not found in any i.p. treated groups, as described in 3.2, or in other s.c. treatment groups given 17.7 mg/kg bw/day. However, this anomaly can be considered as a toxic effect because the incidence of cleft palate was statistically significant, and other later studies showed that i.p. dose of dinoseb induced cleft palate in mice (Preache & Gibson, 1975b) and in rabbits (Johnson, 1988). Moreover, a recent survey by international experts in the field of reproductive/developmental toxicology resulted in strong agreement that fused ribs and vertebrae can be considered as malformations (Solecki et al., 2001). Therefore, it can be concluded that an s.c. dosing of dinoseb in mice may have the potential to produce teratogenic effects in the same way as i.p. dosing of dinoseb.

Species (Reference)	Dose	Exposure time	Developmental effect
SW mouse	17.7 mg/kg	GDs 10-12	Fused ribs and vertebrae, absent or not ossified sternebrae
	17.7 mg/kg	GDs 14-16	Cleft palate, ↑resorption, ↓no. of fetuses, ↓fetal body weight, ↓fetal crown-rump length
(Gibson, 1973)	17.7 mg/kg	GDs 8-16	Skeletal variations, ↓fetal body weight, ↓fetal crown-rump length, absent or not ossified sternebrae

Table 2.3. Developmental toxicity of dinoseb administered by subcutaneously in mice
GDs: gestation days

4. Developmental toxicity in rats

Tables 3.1-3.3 show the results of developmental toxicity studies of dinoseb in rats. There are oral (gavage and diet) and i.p. administration studies. The data are shown by routes of administration, in order of the most likely route of human intake. Only statistically significant effects are summarized unless noted otherwise.

4.1 Gavage studies in rats

In our previous study, male Crj:CD(SD)IGS rats were administered dinoseb by gavage for a total of 42 days beginning 14 days before mating and females underwent this treatment for a total of 44-48 days beginning 14 days before mating to day 6 of lactation at 0 (vehicle: corn oil) , 0.78, 2.33 or 7.0 mg/kg bw/day (Matsumoto et al., 2008a). As for the developmental parameters, no changes attributable to the chemical were noted in the 0.78 and 2.33 mg/kg bw/day dose groups. Eight of twelve females died and two animals were moribund during late pregnancy at 7.0 mg/kg bw/day. Developmental toxicity of dinoseb was not precisely estimated because only one dam with live pups was obtained at the highest dose, and newborn rats were only examined externally. No increased incidence of pups with an external malformation was noted in the dinoseb-treated groups.

In teratology studies in rats, skeletal variation, delayed ossification and/or decreased fetal body weight was commonly observed in fetuses of dams treated with dinoseb. Giavini et al. (1986) administered dinoseb to pregnant CD rats by gavage in corn oil either once a day on GDs 5-14 at 0, 2.5, 5, 10 or 15 mg/kg bw/day or twice a day on GDs 5-12 at 15 (7.5 x 2) or 20 (10 x 2) mg/kg bw/day. Dinoseb was also administered to pregnant rats on GDs 5-12 at 15 mg/kg bw/day in NaOH. This vehicle was selected to conform to a vehicle used in a study by Gibson (1973) in which dinoseb in NaOH showed teratogenicity in mice when administered i.p. but not by gavage. An increased incidence of supernumerary ribs was observed at 10 mg/kg bw/day and higher, and fetal weight was decreased at 15 and 20 mg/kg bw/day regardless of frequency of dosing or vehicle. Delayed ossification of caudal vertebrae, metacarpals or sternebrae was observed at a single dose of 15 mg/kg bw/day (in both corn oil and NaOH). These doses also caused maternal toxicities such as mortality and decrease in body weight gain. No malformations were observed in fetuses of dams treated with dinoseb under the test condition regardless of the dosing regimen or vehicle used in the experiment.

Fetal body weight was decreased when pregnant Crl:CD rats were given dinoseb at 15 mg/kg bw/day with diet A (protein 21%, fat 3.5%, fiber 6.5%, ash 7.5% and N-free extractives 61.5%; Italiana Mangimi, Settimo Milanese, Italy) and diet B (protein 21%, fat 4.8%, fiber 4.2%, ash 8.5% and N-free extractives 61.5%; Mangimi Piccioni, Gessate, Italy) on GDs 5-13 (Giavini et al., 1989). Dinoseb induced microphthalmia in fetuses of animals fed diet B but did not induce maternal toxicity. Maternal mortality and decreased maternal body weight gain were observed when dinoseb was given with diet A. Although developmental toxicity was different according to the type of diet, there were no differences in dinoseb concentrations in maternal plasma and in embryos between the two dietary groups.

Wistar/Han rats were administered dinoseb by gavage on GDs 6-15 at 0, 1, 3 or 10 mg/kg bw/day (Health Canada, 1991). No information on the vehicle was presented in this study. Only slight depressions were observed in food consumption and body weight gain of dams at 10 mg/kg bw/day. Fetuses at the highest dose showed a slight decrease in body weight, and increases in the incidence of delayed ossification and incidence of skeletal variations, especially supernumerary ribs. At 3 mg/kg bw/day and higher, absence of thoracic vertebrae was observed. No further information is available for this study, but the result indicates that dosing of dinoseb by gavage is hazardous in Wistar/Han rats.

The details of our new findings (Matsumoto et al., 2010) shown in Table 3.1 are described below (see 4.4).

4.2 Feeding studies in rats

Feeding of dinoseb to CD rats on GDs 5-14 produced a specific teratogenic effect, increased incidence of fetuses with microphthalmia, reduced fetal weight and increased incidence of fetuses with supernumerary ribs at 200 ppm (15 mg/kg bw/day) accompanied by decreased maternal body weight gain (Giavini et al., 1986). An increased incidence of fetuses

Species (Reference)	Dose	Exposure time	Developmental effect
Crj:CD(SD) IGS rat (Matsumoto et al., 2008a)	7 mg/kg	44-48 days	↓ No. of dams delivered, ↓no. of dams with live pups at delivery
CD rat	10 mg/kg	GDs 5-14	Skeletal variations
	15 mg/kg (in corn oil)		Delayed ossification, ↓fetal body weight
	7.5, 10 mg/kg (twice/day)	GDs 5-12	Skeletal variations, ↓fetal body weight
(Giavini et al., 1986)	15 mg/kg (in NaOH)	GDs 5-12	Skeletal variations, delayed ossification ↓fetal body weight
Crl:CD rat	15 mg/kg (with diet B)	GDs 5-13	↓Fetal body weight, microphthalmia
(Giavini et al., 1989)	15 mg/kg (with diet A)	GDs 5-13	↓Fetal body weight
Wistar/Han rat [a] (Health Canada, 1991)	3 mg/kg	GDs 6-15	Absence of thoracic vertebrae,
	10 mg/kg		Absence of thoracic vertebrae, skeletal variations
SPF Crl:CD (SD) rat [b]	8.0 mg/kg	GDs 6-15	↓Fetal body weight, skeletal variations, delayed ossifications
(Matsumoto et al., 2010)	10 mg/kg		↓ Fetal body weight, skeletal variations, delayed ossifications, microphthalmia

Table 3.1. Developmental toxicity of dinoseb administered by gavage in rats
a: only secondary literature or abstract is available.
b: the details are described in 4.4
GDs: gestation days

with microphthalmia and reduced fetal weight were also observed when pregnant Crl:CD rats were given dinoseb in diet B at 200 ppm on GDs 5-13. At this dose, maternal food consumption and body weight gain were decreased compared with those of control groups (Giavini et al., 1989). When dinoseb was fed with diet A, maternal food consumption and body weight gain were reduced, but no effects were found in fetuses (Giavini et al., 1989). These findings indicate that the developmental toxicity, including teratogenicity, of dinoseb in rats was influenced by diet composition (see 4.1 for the compositions of diet A and diet B).

Following feeding of dinoseb on GDs 5-14 at 0, 50, 100, 150, 200, 250, 300 and 350 ppm (0, 3.26, 6.9, 9.23, 10.86, 9.38, 9.49 and 8.6 mg/kg bw/day) in SD rats, the number of resorptions at 200-350 ppm, early embryo loss at 200-350 ppm, and total intra-uterine loss at 150-350 ppm were increased in a dose-related manner (Spencer & Sing, 1982; US EPA, 2003b). Body weight gain in dams was decreased at 150-350 ppm. At 200 ppm, hypoplastic tail was observed in 8 out of 62 fetuses and fetal weight was decreased. In decidualized females given dinoseb on days 7-10 of pseudopregnancy, uterine protein and glycogen concentrations were decreased at 200 ppm and higher in a dose-related manner. The authors suggested a toxic role of dinoseb in the uterine environment.

Hall et al. (1978) provided a brief summary of a subchronic feeding study in which Sherman male and female rats were fed a diet containing dinoseb at 0, 50, 100, 150, 200, 300, 400 and 500 ppm for 153 days. The 300, 400 and 500 ppm groups were terminated at day 21 of administration owing to mortality of 14, 100 and 100%, respectively, and only animals fed dinoseb up to 200 ppm were evaluated. Fertility, fecundity, neonate survival, weight gain, viability and lactation were depressed. No further details are available.

In an unpublished five-generation study, decreased body weight gains were observed in parents during the pre-mating period (F0, F1 and F2) at 10 mg/kg bw/day dinoseb in the diet and in pups on postnatal day **(PND)** 21 (F1, F2 and F3) at 1, 3 and 10 mg/kg bw/day, but weights at birth were similar to the controls. Body weight gain in F4 and F5 pups was increased and absolute and relative gonadal weights in F4 pups were decreased at all dose levels. A low viability index was obtained (from F4 to F5) at all dose levels. No detailed information is available for this study (Health Canada, 1991; US EPA, 2003a).

The details of our new findings (Matsumoto et al., 2010) shown in Table 3.2 are described below (see 4.4).

4.3 Intraperitoneal studies in rats

Two i.p. studies in rats showed similar results on developmental toxicity. When dinoseb was given to SD rats on GDs 9-11 at doses up to 15.8 mg/kg bw/day in NaOH, all pregnant rats given dinoseb at 11.2 mg/kg bw/day and higher and three of the 16 pregnant rats at 9.0 mg/kg bw/day died. There were dilated renal pelvis and ureters in fetuses, decreased body weight in fetuses, and pathological changes in the liver and kidney in both fetal and neonatal rats at 8.0 mg/kg bw/day without maternal toxicity. At 9.0 mg/kg bw/day, fetal CRL was decreased, and neonatal body weight was decreased on PNDs 1 and 7 but not on PND 42. In surviving dams, dinoseb did not affect the number of live fetuses or the resorption rate in surviving dams (McCormack et al., 1980).

When dinoseb was administered i.p. to pregnant SD rats on GDs 9-11 or 10-12 at 0-18.0 mg/kg bw/day in NaOH, fetal body weight was decreased at 7.5 mg/kg bw/day and higher, but weights at birth and on PND 6 were not affected. Maternal death was observed at 8.0 mg/kg bw/day and higher, and 10.5 mg/kg bw/day was an approximate LD_{50} in pregnant rats. Postnatal observation on PND 30 revealed that there was a body weight reduction and an increase in relative kidney weight at 10.5 mg/kg bw/day. On PND 6, there were a deficit in urinary concentrating ability in pups of dams given dinoseb on GDs 9-11 at 10.5 mg/kg bw/day (Daston et al., 1988).

As described above, dinoseb produced suggestive renal damage in rat offspring following maternal administration. However, pathological changes in the kidney observed in prenatal rats were reduced in incidence or not detected at 42-day postpartum in the study of

Species (Reference)	Dose	Exposure time	Developmental effect
CD rat	200 ppm (15 mg/kg)	GDs 5-14	↓Fetal body weight, microphthalmia, skeletal variations
(Giavini et al., 1986)			
Crl:CD rat	200 ppm (diet B)	GDs 5-13	↓Fetal body weight, microphthalmia
	200 ppm (diet A)	GDs 5-13	No effects
(Giavini et al., 1989)			
SD rat	150 ppm (9.23 mg/kg)	GDs 5-14	↑Total intra-uterine loss
	200 ppm (10.86 mg/kg)		↑Early embryonic loss, resorptions, ↓fetal body weight, hypoplastic tail
(Spencer & Sing, 1982; US EPA, 2003b)			
Sherman rat [a]	< 200 ppm	153 days	↓ Fertility, ↓fecundity, ↓neonate survival, ↓body weight gain, ↓viability, ↓ lactation
(Hall et al., 1978)			
CD (SD) rat [a]	1, 3, 10 mg/kg	3-generation	↓Body weight gain in pups (F1, F2, F3)
		Next 2-generation	↓Body weight gain in pups (F4, F5), ↓absolute/relative gonadal weight (F4), ↓viability index (F5)
(Health Canada, 1991; US EPA, 2003a)			
SPF Crl:CD (SD) rat [b]	120 ppm (6.52 mg/kg)	GDs 6-16	↓Fetal body weight
	10 mg/kg (8.0 mg/kg)		↓Fetal body weight, skeletal variations, delayed ossifications
(Matsumoto et al., 2010)			

Table 3.2. Developmental toxicity of dinoseb administered in diet in rats
a: only secondary literature or abstract is available
b: the details are described in 4.4
GDs: gestation days

McCormack et al. (1980). In the study of Daston et al. (1988) a deficit in urinary concentrating ability observed during postnatal development also disappeared after functional maturation (PND 30). Prenatal incidence of dilated renal pelvis was not dose-dependent. Moreover, Woo and Hoar (1972) noted that the renal parenchyma increased in weight rapidly, but that the renal papilla increased in length solely during late pregnancy, and they suggested that this discrepancy in growth rate frequently resulted in the kidney with an enlarged renal pelvis. Taken together, these renal effects appear to be a developmental delay, but not a permanent functional impairment.

Species (Reference)	Dose	Exposure time	Developmental effect
SD rat	8.0 mg/kg	GDs 9-11	↓Fetal body weight, dilated renal pelvis and ureters in fetuses Pathological changes in liver and kidney in fetuses and neonates
	9.0 mg/kg		↓Fetal crown-rump length, ↓ neonatal body weight
(McCormack et al., 1980)			
SD rat	7.5 mg/kg	GDs 9-11, 10-12	↓Fetal body weight
	10.5 mg/kg		Functional defect of kidney (PND 6), ↓body weight in pups (PND 30), ↑relative weight of kidney (PND 30)
(Daston et al., 1988)			

Table 3.3. Developmental toxicity of dinoseb administered by intraperitoneally in rats
GDs: gestation days
PND: postnatal day

4.4 Further clarification of teratogenicity in rats

Several studies including our previous study did not demonstrate the teratogenicity of dinoseb in rats (Daston et al., 1988; Matsumoto et al., 2008a; McCormack et al., 1980), but we considered that the teratogenic potential of dinoseb in rats was unclear because of the influence of variable factors. Because detailed test conditions were not described in the studies of Giavini et al. (Giavini et al., 1989; Giavini et al., 1986), adequate experimental conditions for the production of fetal malformations by the administration of dinoseb to pregnant rats remained unknown. Therefore, we recently conducted gavage and feeding studies to clarify the experimental conditions that produce fetal malformations when dinoseb is given to pregnant rats (Matsumoto et al., 2010).

Pregnant rats (12 animals/group) were given dinoseb by gavage at 0, 8.0 or 10 mg/kg bw/day on GDs 6-15 or in the diet (CRF-1; protein 22%, fat 5.7%, fiber 2.9%, ash 6.3% and N-free extractives 55.3%; Oriental Yeast Co., Ltd., Tokyo, Japan) at 0, 120 or 200 ppm on GDs 6-16 (Figure 1). The feeding dose groups were expected to consume similar amounts of dinoseb to those in the gavage groups. Dinoseb induced dose-dependent decreases in maternal body weight gain and food consumption during pregnancy in all the dinoseb-treated groups. The decrease in food consumption was greater in the feeding dose groups than the gavage dose groups; therefore, the decreased food consumption may be related to a reduced palatability of the diet in the feeding groups. Intakes of dinoseb by feeding dose were estimated to be 0, 6.52 and 8.50 mg/kg bw/day (0, 120 and 200 ppm).

Significantly decreased body weights of fetuses were observed in all the dinoseb-treated groups, except for the group fed dinoseb at 120 ppm. Skeletal examinations of fetuses revealed an increased incidence of fetuses with skeletal variations in all the dinoseb-treated groups and delayed ossification at 8.0 and 10 mg/kg bw/day and at 200 ppm. An increased incidence of fetuses with microphthalmia was observed at 10 mg/kg bw/day, but there was no increased incidence of fetuses with external, internal or skeletal malformations in the groups given dinoseb at 8.0 mg/kg bw/day by gavage or 120 or 200 ppm by feeding (Table 3.1 and 3.2).

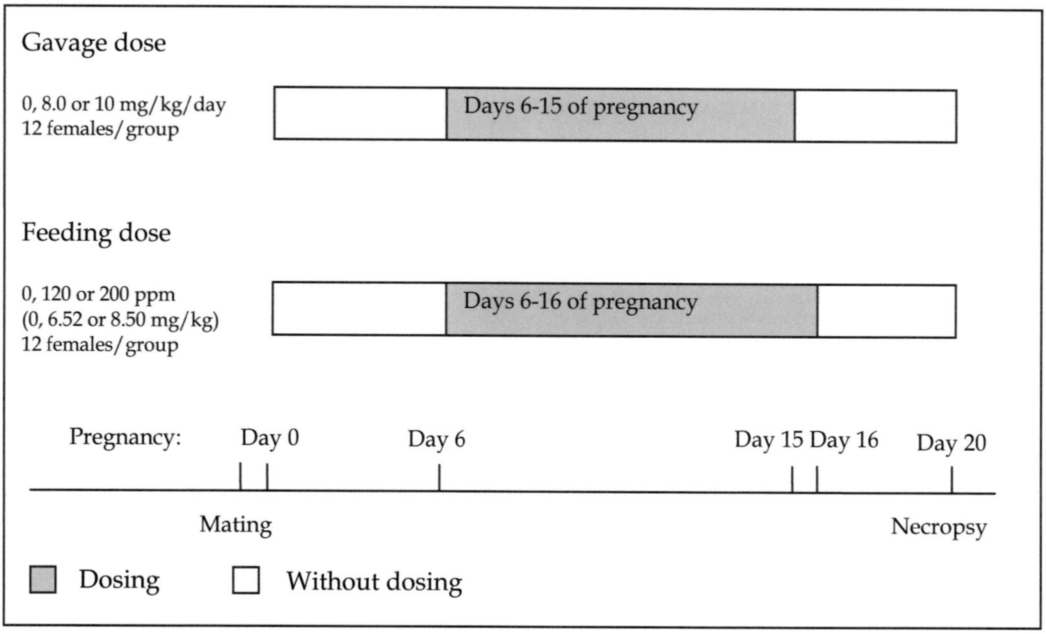

Fig. 1. A study design of prenatal developmental toxicity of gavage or feeding doses of dinoseb in rats (Matsumoto et al., 2010)

Although the feeding dose of dinoseb at 200 ppm (15 mg/kg bw/day) was previously reported to be teratogenic in rats (Giavini et al., 1986), the feeding dose of dinoseb up to 200 ppm (8.5 mg/kg bw/day) did not induce teratogenicity in our study. The diets used in the studies of Giaini et al. did not meet the current nutrient requirement of rats for fat (more than 5%) (ILAR, 1995; Suckow et al., 2005) while the diet used in our study is a standard rat diet; however, fat concentration seems unrelated to dinoseb-induced teratogenicity, and it seemed impossible to identify the definitive dietetic factor involved. Dose levels of dinoseb in our study might not have been sufficiently high to induce teratogenicity; however, pregnant rats did not consume sufficiently high amounts of dinoseb to produce fetal malformations because food consumption was reduced in the feeding groups. It seems unlikely that a feeding study is appropriate to evaluate the toxicity of dinoseb.

Microphthalmia, which was found in rats after exposure to dinoseb by gavage or feeding (Giavini et al., 1989; Giavini et al., 1986) and in rabbits by gavage (Research & Consulting Company, 1986) or dermal application (Johnson, 1988), was predominantly observed after administration of dinoseb at 10 mg/kg bw/day by gavage. As a rule, the administration of a suitable dosage of a teratogen generally results in the production of some normal offspring, some malformed offspring and some dead or resorbed offspring (Schardein, 2000). In our study, the increased incidence of malformed fetuses was not accompanied by an increased incidence of intrauterine deaths of offspring after the administration of dinoseb. This phenomenon was also observed in the previous studies of Giavini et al. (Giavini et al., 1989; Giavini et al., 1986). One possible explanation for this is that microphthalmia itself is not lethal in utero. Because maternal death was observed after the gavage dose of dinoseb at 10 mg/kg bw/day, the exposure range of dinoseb where malformations are observed seems to be narrow in rats. The findings of our study confirmed the experimental condition that could induce malformation in rats fed a standard diet.

5. Discussion and conclusions

A difficulty lies in the risk assessment of chemical compounds for developmental toxicity because there are many variable factors in the manifestation of developmental toxicity of chemicals. The administration route is one of the definitive factors for risk assessment of chemicals. The data obtained from animal experiments by oral administration are the most important for risk assessment of chemicals because the oral route is the most relevant route for human exposure to dinoseb.

Gavage dosing of dinoseb during organogenesis in rabbits produced external, internal and skeletal malformations in fetuses without maternal toxicity at 10 mg/kg bw/day (Health Canada, 1991; US EPA, 2003a). In mice, gavage dosing of dinoseb during organogenesis induced skeletal variations and growth retardation at or above maternally toxic levels (26-50 mg/kg bw/day) (Branch et al., 1996; Kavlock et al., 1985). Teratogenic effects were observed without maternal toxicity at 50 mg/kg bw/day by gavage in CD-1 mice (Rogers et al., 2004). Doses of dinoseb in rats during organogenesis induced skeletal variations and growth retardation at maternally toxic levels (8.0-20 mg/kg bw/day) by gavage and (6.52-15 mg/kg bw/day) by feeding (Giavini et al., 1986; Matsumoto et al., 2010). Malformations such as microphthalmia or hypoplastic tail were observed when dinoseb was given in the diet (10.86-15 mg/kg bw/day) with maternal toxicity (Giavini et al., 1989; Giavini et al., 1986; Spencer & Sing, 1982), but not in our study (Matsumoto et al., 2010). Microphthalmia was also observed when dinoseb was given by gavage (8.0-15 mg/kg bw/day) in CD rats with maternal toxicity (Giavini et al., 1989; Giavini et al., 1986; Matsumoto et al., 2010). In Wistar/Han rats, absence of thoracic vertebrae was observed by gavage dose of dinoseb at 3 mg/kg bw/day and higher without maternal toxicity. No detailed test condition is available for this study, but genetic difference in strains of rats may also influence the teratogenic potential of dinoseb. Although there are differences in susceptibility of developmental toxicity by the oral route among rabbits, mice and rats, namely susceptibility to developmental toxicity caused by dinoseb was greater in rabbits than in rats and mice, teratogenicity was noted at some doses without maternal toxicity in these animal species. More precisely, dinoseb can be a selective teratogen in these animal species.

Dermal exposure is the next most likely route of exposure to dinoseb in humans, especially in users and producers. A dermal teratology study in rabbits showed a markedly increased incidence of dead and resorbed fetuses (Johnson, 1988). The survivors exhibited a high incidence of external and soft tissue malformations at application levels of dinoseb, but these dose levels were also maternally toxic.

Prenatal i.p. and s.c. doses of dinoseb induced growth retardation, embryolethality and/or teratogenicity at or over the maternally toxic dose levels (10-20 mg/kg bw/day) in Swiss-Webster mice (Gibson, 1973; Preache & Gibson, 1975a; Preache & Gibson, 1975b). Prenatal i.p. dose of dinoseb did not induce teratogenicity but induced growth retardation at or above the maternally toxic level in rats (Daston et al., 1988; McCormack et al., 1980). The teratogenic effects were observed with or without maternal toxicity in rats and mice, but the maternal toxicity of dinoseb seems greater in rats than in mice because dinoseb treatment (i.p.) during GDs 10-12 at 9.0 mg/kg bw/day caused 3/16 maternal deaths in rats (McCormack et al., 1980) while no maternal toxicity was observed at 15.8 mg/kg bw/day after i.p. dosing of dinoseb during GDs 10-12 in mice (Gibson, 1973). This may explain why teratogenicity was induced in mice, but not in rats, after i.p. dosing of dinoseb. It can be considered that maternal mice were tolerant to dose levels that can produce fetal malformations. Prenatal i.p. and s.c. doses of dinoseb also showed teratogenic potentials; however, these exposure routes are not likely to be relevant to human exposure to dinoseb and may not be important for risk assessment of dinoseb.

The developmental toxicity of dinoseb was also influenced by administration methods. These effects are considered to be related to differences in absorption due to the concentration of the chemical, duration of exposure and rate of release or to differences in metabolic fate and the nature of the metabolites reaching the embryo (Kalter, 1968). In fact, food deprivation for 24 h that enhanced external, soft-tissue and skeletal malformations slowed the disappearance of dinoseb from the plasma, but phenobarbital, which reduced developmental toxicity, hastened the disappearance of dinoseb from the plasma. SKF-525A pretreatment, which enhanced both maternal and developmental toxicity, decreased the rate of disappearance from the liver (Preache & Gibson, 1975a). When pregnant mice were administered dinoseb, either i.p. at 17.7 mg/kg bw or by gavage at 32 mg/kg bw, the amount of dinoseb and its metabolites present in the embryo was greater after i.p. than oral administration, and peak levels were reached much earlier after i.p. administration (8 min vs. 12 h for oral) (Gibson & Rao, 1973). Developmental effects of i.p. dosing of dinoseb in mice can be related to rapid and relatively extensive uptake of the compound or its metabolites by the embryo.

Over the years, many investigations have been conducted using laboratory animals to assess the risk to humans. We here reiterate the importance of the administration method to extrapolate laboratory results to humans. We showed that fetal malformations by dinoseb were produced by the anticipated routes of human exposure (oral and dermal exposure) in laboratory animals. These results for routes/modes of administration relevant to human intake should be used for human risk assessment.

There is no clear understanding of the fundamental mechanism of developmental toxicity of dinoseb, although an energy-deficient intrauterine environment due to uncoupling of cellular oxidative phosphorylation may explain dinoseb-induced developmental toxicity. A prenatal dose of thiabendazole, an ATP-synthesis inhibitor, induced a deformity involving reduced limb size in mice fetuses (Ogata et al., 1984), and ATP levels in fore- and hindlimb buds of fetuses were related to the incidence of this deformity (Tsuchiya & Tanaka, 1985). Dinoseb-induced teratogenicity may be related to the degree of reduction in ATP expression influenced by variable factors.

Recent studies have investigated the role that mitochondria play in mediating apoptotic signals (Green & Kroemer, 2004; Linsinger et al., 1999; Little & Mirkes, 2002). Programmed cell death **(PCD)** is an essential component of normal physiological processes such as embryogenesis and normal tissue development (Vaux & Korsmeyer, 1999). Altering normal patterns of PCD could be teratogenic because areas of the body with a high incidence of malformations coincide with areas where PCD occurs (Knudsen, 1997; Sulik et al., 1988). Some studies showed a positive correlation between mitochondrial uncoupling activity and PCD (Maccarrone et al., 2001; Maccarrone et al., 2003), and 2,4-dinitrophenol, an uncoupling agent, enhanced the Fas apoptotic signal in Jurkat Bcl-2 cells (Linsinger et al., 1999). These findings imply that the enhanced uncoupling of oxidative phosphorylation in mitochondria may alter normal patterns of PCD. However, the link between malformations and mitochondrial uncoupling activity is still poorly understood. In addition, we previously showed that these apoptotic activities could also involve in testicular toxicity of dinoseb in rats and mice (Matsumoto et al., 2008b). Further mechanistic studies are necessary to clarify the toxicity of dinoseb.

6. References

Anderson, I. & Morse, L. M. (1966). The influence of solvent on the teratogenic effect of folic acid antagonist in the rat. *Exp Mol Pathol* 5. 2. 134-145.

Branch, S., Rogers, J. M., Brownie, C. F. & Chernoff, N. (1996). Supernumerary lumbar rib: manifestation of basic alteration in embryonic development of ribs. *J Appl Toxicol* 16. 2. 115-119.

Chernoff, N. & Kavlock, R. J. (1982). An in vivo teratology screen utilizing pregnant mice. *J Toxicol Environ Health* 10. 4-5. 541-550.

Daston, G. P., Rehnberg, B. F., Carver, B., Rogers, E. H. & Kavlock, R. J. (1988). Functional teratogens of the rat kidney. I. Colchicine, dinoseb, and methyl salicylate. *Fundam Appl Toxicol* 11. 3. 381-400.

EXTOXNET. (1996). Extension Toxicology Network Pesticide Information Profiles Dinoseb Retrieved February 26, 2007, from http://extoxnet.orst.edu/pips/dinoseb.htm.

Giavini, E., Broccia, M. L., Prati, M., Cova, D. & Rossini, L. (1989). Teratogenicity of dinoseb: role of the diet. *Bull Environ Contam Toxicol* 43. 2. 215-219.

Giavini, E., Broccia, M. L., Prati, M. & Vismara, C. (1986). Effect of method of administration on the teratogenicity of dinoseb in the rat. *Arch Environ Contam Toxicol* 15. 4. 377-384.

Gibson, J. E. (1973). Teratology studies in mice with 2-sec-butyl-4,6-dinitrophenol (dinoseb). *Food Cosmet Toxicol* 11. 1. 31-43.

Gibson, J. E. (1976). Perinatal nephropathies. *Environ Health Perspect* 15. 121-130.

Gibson, J. E. & Rao, K. S. (1973). Disposition of 2-sec-butyl-4,6-dinitrophenol (dinoseb) in pregnant mice. *Food Cosmet Toxicol* 11. 1. 45-52.

Green, D. R. & Kroemer, G. (2004). The pathophysiology of mitochondrial cell death. *Science* 305. 5684. 626-629.

Hall, L., Linder, R., Scotti, T., Bruce, R., Moseman, R., Heiderscheit, T., Hinkle, D., Deferton, T., Chaney, S., Goldstein, M., Gage, M., Farmer, J., Bennett, L., Stevens, J., Durham, W. & Furley, A. (1978). Subchronic and reproductive toxicity of dinoseb. *Toxicol Appl Pharmacol* 45. 235.

Hall, L. L., Fisher, H. L., Sumler, M. R., Hughes, M. F. & Shah, P. V. (1992). Age-related percutaneous penetration of 2-sec-butyl-4,6-dinitrophenol (dinoseb) in rats. *Fundam Appl Toxicol* 19. 2. 258-267.

Hansen, D. K. & Billings, R. E. (1986). Effect of route of administration on phenytoin teratogenicity in A/J mice. *J Craniofac Genet Dev Biol* 6. 2. 131-138.

Health Canada. (1991). Dinoseb. Retrieved March, 2007, from http://www.hc-sc.gc.ca/ewh-semt/pubs/water-eau/doc_sup-appui/dinoseb/index_e.html#ref_22.

ILAR (1995). Nutrient Requirements of the Laboratory Rat. *Nutrient Requirements of Laboratory Animals, Fourth Revised Edition*, NATIONAL ACADEMY PRESS. 11-79.

Isaacson, R. J. & Chaudhry, A. P. (1962). Cleft palate induction in strain a mice with cortisone. *The Anatomical Record* 142. 4. 479-484.

Johnson, E. M., Bellet, E.M., Christian, M.S. and Hoberman, A.M. (1988). The hazard identification and animal NOEL phases of developmental toxicity risk estimation: a case study employing dinoseb. *Advances in modern environmental toxicology* 15. 123-132.

Kalter, H. (1968). Teratology of the Central Nervous System. Chicago, University of Chicago Press.

Kavlock, R. J., Chernoff, N., Gray, L. E., Jr., Gray, J. A. & Whitehouse, D. (1982). Teratogenic effects of benomyl in the Wistar rat and CD-1 mouse, with emphasis on the route of administration. *Toxicol Appl Pharmacol* 62. 1. 44-54.

Kavlock, R. J., Chernoff, N. & Rogers, E. H. (1985). The effect of acute maternal toxicity on fetal development in the mouse. *Teratog Carcinog Mutagen* 5. 1. 3-13.

Kidd, H. & James, D. R. (1991). The Agrochemicals Handbook, Third Edition. Cambridge, Royal Society of Chemistry Information Services.

Kimmel, C. A. (1977). Effect of route of administration on the toxicity and teratogenicity of EDTA in the rat. *Toxicol Appl Pharmacol* 40. 2. 299-306.

Knudsen, T. (1997). Cell death. *Drug toxicity in embryonic development I.* R. J. Kavlock and G. Dalton. New York, Springer-Verlag. 211-244.

Leftwich, R. B., Floro, J. F., Neal, R. A. & Wood, A. J. (1982). Dinitrophenol poisoning: a diagnosis to consider in undiagnosed fever. *South Med J* 75. 2. 182-184.

Linsinger, G., Wilhelm, S., Wagner, H. & Hacker, G. (1999). Uncouplers of oxidative phosphorylation can enhance a Fas death signal. *Mol Cell Biol* 19. 5. 3299-3311.

Little, S. A. & Mirkes, P. E. (2002). Teratogen-induced activation of caspase-9 and the mitochondrial apoptotic pathway in early postimplantation mouse embryos. *Toxicol Appl Pharmacol* 181. 2. 142-151.

Maccarrone, M., Bari, M., Battista, N., Di Rienzo, M., Falciglia, K. & Finazzi Agro, A. (2001). Oxidation products of polyamines induce mitochondrial uncoupling and cytochrome c release. *FEBS Lett* 507. 1. 30-34.

Maccarrone, M., Taccone-Gallucci, M. & Finazzi-Agro, A. (2003). 5-Lipoxygenase-mediated mitochondrial damage and apoptosis of mononuclear cells in ESRD patients. *Kidney Int Suppl* 84. S33-36.

Matsumoto, M., Fujii, S., Hirose, A. & Ema, M. (2010). Prenatal developmental toxicity of gavage or feeding doses of 2-sec-butyl-4,6-dinitrophenol in rats. *Reprod Toxicol* 29. 3. 292-297.

Matsumoto, M., Furuhashi, T., Poncipe, C. & Ema, M. (2008a). Combined repeated dose and reproductive/developmental toxicity screening test of the nitrophenolic herbicide dinoseb, 2-sec-butyl-4,6-dinitrophenol, in rats. *Environ Toxicol* 23. 2. 169-183.

Matsumoto, M., Hirose, A. & Ema, M. (2008b). Review of testicular toxicity of dinitrophenolic compounds, 2-sec-butyl-4,6-dinitrophenol, 4,6-dinitro-o-cresol and 2,4-dinitrophenol. *Reprod Toxicol* 26. 3-4. 185-190.

Matsumoto, M., Poncipe, C. & Ema, M. (2008c). Review of developmental toxicity of nitrophenolic herbicide dinoseb, 2-sec-butyl-4,6-dinitrophenol. *Reprod Toxicol* 25. 3. 327-334.

McCormack, K. M., Abuelgasim, A., Sanger, V. L. & Hook, J. B. (1980). Postnatal morphology and functional capacity of the kidney following prenatal treatment with dinoseb in rats. *J Toxicol Environ Health* 6. 3. 633-643.

MHLW, Japan (2005). Single Dose Oral Toxicity Test of 2-*sec*-Butyl-4,6-dinitrophenol in rats.

NITE (2009). Chemical Risk Information Platform (CHRIP) by the National Institute of Technology and Evaluation (NITE).

O'Neill, H. J., Pollock, T. L., Bailey, H. S., Milburn, P., Gartley, C. & Richards, J. E. (1989). Dinoseb presence in agricultural subsurface drainage from potato fields in northwestern New Brunswick, Canada. *Bull Environ Contam Toxicol* 43. 6. 935-940.

OECD. (2004). The 2004 OECD List of High Production Volume Chemicals. Retrieved February, 2007, from http://www.oecd.org/dataoecd/55/38/33883530.pdf.

Ogata, A., Ando, H., Kubo, Y. & Hiraga, K. (1984). Teratogenicity of thiabendazole in ICR mice. *Food Chem Toxicol* 22. 7. 509-520.

PAN (2006). PAN (Pesticide Action Network) Pesticides Database-Pesticide Registration Status.

Preache, M. M. & Gibson, J. E. (1975a). Effect of food deprivation, phenobarbital, and SKF-525A on teratogenicity induced by 2-sec-butyl-4,6-dinitrophenol (dinoseb) and on disposition of [14C]dinoseb in mice. *J Toxicol Environ Health* 1. 1. 107-118.

Preache, M. M. & Gibson, J. E. (1975b). Effects in mice of high and low environmental temperature on the maternal and fetal toxicity of 2-sec-butyl-4,6-dinitrophenol (dinoseb) and on disposition of [14C]-dinoseb. *Teratology* 12. 2. 147-156.

Research and Consulting Company (1986). Embryotoxicity study with dinoseb technical grade in the rabbit (oral administration). , Research and Consulting Co.

Rogers, J. M., Setzer, R. W., Branch, S. & Chernoff, N. (2004). Chemically induced supernumerary lumbar ribs in CD-1 mice: size distribution and dose response. *Birth Defects Res B Dev Reprod Toxicol* 71. 1. 17-25.
Rotterdam Convention (2006). PIC Circular XXIV-December 2006. 92.
Schardein, J. L. (2000). Principles of Teratogenesis Applicable to Drug and Chemical Exposure. *Chemically Induced Birth Defects*. New York, Marcel Dekker, inc., 1-65.
Schneider, K. (1986). Some older pesticides yield a harvest of ugly surprises. The New York Times.
Shabecoff, P. (1986). Emergency order bans much-used pesticide. The New York Times.
Solecki, R., Burgin, H., Buschmann, J., Clark, R., Duverger, M., Fialkowski, O., Guittin, P., Hazelden, K. P., Hellwig, J., Hoffmann, E., Hofmann, T., Hubel, U., Khalil, S., Lingk, W., Mantovani, A., Moxon, M., Muller, S., Parkinson, M., Paul, M., Paumgartten, F., Pfeil, R., Platzek, T., Rauch-Ernst, M., Scheevelenbos, A., Seed, J., Talsness, C. E., Yasuda, M., Younes, M. & Chahoud, I. (2001). Harmonisation of rat fetal skeletal terminology and classification. Report of the Third Workshop on the Terminology in Developmental Toxicology. Berlin, 14-16 September 2000. *Reprod Toxicol* 15. 6. 713-721.
Spencer, F. & Sing, L. T. (1982). Reproductive responses to rotenone during decidualized pseudogestation and gestation in rats. *Bull Environ Contam Toxicol* 28. 3. 360-368.
Staples, R. E., Kellam, R. G. & Haseman, J. K. (1976). Developmental toxicity in the rat after ingestion or gavage of organophosphate pesticides (Dipterex, Imidan) during pregnancy. *Environ Health Perspect* 13. 133-140.
Suckow, M. A., Weisbroth, S. H. & Franklin, C. L. (2005). The laboratory rat, ACADEMIC PRESS.
Sulik, K. K., Cook, C. S. & Webster, W. S. (1988). Teratogens and craniofacial malformations: relationships to cell death. *Development* 103 Suppl. 213-231.
Tsuchiya, T. & Tanaka, A. (1985). In vivo inhibition of adenosine triphosphate (ATP) synthesis associated with thiabendazole-induced teratogenesis in mice and rats. *Arch Toxicol* 57. 4. 243-245.
US EPA. (2003a, 2007). Dinoseb (CASRN 88-85-7). Integrated Risk Information System (IRIS) Retrieved August 3, 2007, from http://www.epa.gov/iris/subst/0047.htm.
US EPA (2003b). High Production Volume Challenge Program (HPV), Robust Summaries and Test Plans: phenol, 2-(1-methylpropyl)-4,6-dinitro-.
US EPA. (2006). Recognition and Management of Pesticide Poisonings, 5th Edition. Section III Herbicides: 11 Nitrophenolic and Nitrocresolic Herbicides Retrieved February 26, 2007, from http://www.epa.gov/pesticides/safety/healthcare/handbook/contents.htm.
US EPA. (2007, February, 2007). Chemical Emergency Preparedness and Prevention Emergency First Aid Treatment Guide (Dinoseb:88-85-7) Retrieved February 26, 2007, from http://yosemite.epa.gov/oswer/CeppoEHS.nsf/firstaid/88-85-7?OpenDocument.
Vaux, D. L. & Korsmeyer, S. J. (1999). Cell Death in Development. *Cell* 96. 2. 245-254.
Wilson, J. G. (1964). Teratogenic Interaction of chemical agents in the rat. *J Pharmacol Exp Ther* 144. 3. 429-436.
Wilson, J. G. (1966). Effects of acute and chronic treatment with actinornycin D on pregnancy and the fetus in the rat. *Harper Hops. Bull.* 24. 109-118.
Wilson, J. G. & Warkany, J. (1965). Teratology. Chicago, The University of Shicago Press.
Woo, D. C. & Hoar, R. M. (1972). "Apparent hydronephrosis" as a normal aspect of renal development in late gestation of rats: the effect of methyl salicylate. *Teratology* 6. 2. 191-196.

28

New Concept for Evaluating the Toxicity of Herbicides for Ecological Risk Assessment

Munir Mohammad and Kazuhito Itoh
Life and Environmental Science, Shimane University
Japan

1. Introduction

The use of agricultural pesticides is increasing day by day for controlling pests and weeds in crop production, and among these pesticides, more than 65% of total amount are herbicides (USDA, 1998). Unfortunately, their exposure is usually not limited to the location where they are applied, and the pesticides reach aquatic environmental locations and compartments through various physical transport processes, such as spray drift, leaching, runoff or accidental spill, and affect the organisms living in the locations (Thurman et al., 1991; Squillace & Thurman, 1992). The ecotoxicological impact of pesticides has been usually measured by their effects on non-target organisms. Among the non-target aquatic organisms, aquatic plants received less attention for assessing the impact of pesticides, compared with the other aquatic organisms (i.e. algae, fish, daphnia etc.), however, aquatic plants play important roles in the environmental conditions of stagnant and flowing waters. They produce organic matter and oxygen, and provide food, shelter and substrate for a variety of aquatic organisms (Sand-Jensen, 1997), therefore, toxicity of herbicides to the non-target photosynthetic aquatic organisms is of concern. Peterson et al. (1997) showed that there can be several order of variation in sensitivity to herbicides between animals and aquatic plants. Huxley (1984) suggested that if one plant species becomes extinct from aquatic ecosystem, 10-30 other non-plant organisms may also become extinct. Therefore, it is of great important to understand the adverse effects by herbicides on non-target aquatic plants in the ecosystem. There is thus need of a convenient method to assay the toxicity of herbicides. For this purpose, test guidelines for ecotoxicology have been set in many countries (EU, US EPA, Japan, etc). For hazard prediction, two types of information are required: the exposure levels of non-target organisms to the chemicals, and the toxic effects of the chemicals on the non-target group under consideration. The expected environmental concentration (EEC) for the agricultural usage, which is a concentration calculated based on the input of the maximum proposed application rate, is used for the estimation of the exposure levels in an aquatic habitat (Boutin et al., 1993). The toxicity is expressed as the EC50 value, which causes 50% reduction of growth, and NOEC (no observed effect concentration), which is the maximum concentration that does not harm the test organisms. If the relationship between these two pieces of information suggests a hazard, the next step of risk assessment is to refine the assumptions to accurately predict risk. Although a short-term exposure test is required in the ecotoxicological test guidelines using an aquatic plant, duckweed, the results are not enough when considering environmentally because of the three reasons: 1) the toxicity though a long-term exposure to a pesticide might be higher than that obtained by the short-term exposure test, 2) toxicological indexes does not indicate the lethality of the test material, and the recovery potential of the test species from the damage should be considered due to the rapid growth rate of duckweed, and 3) the exposure to several herbicides is usual in the aquatic environments, and the joint toxic effects could be affected by the combination of the chemicals.

2. Current toxicity test guidelines

In the past, the EPA relied extensively on the results of simulated and actual aquatic field studies to make final recommendations concerning risk to aquatic organisms for pesticides that trigger acute and/or chronic regulatory risk criteria (Urban & Cook, 1986). As a result of the controversy associated with the design and interpretation of the field studies in the 1980's and in the early 1990's, the EPA in 1992 decided to de-emphasize and limit its use of such studies in the aquatic risk assessment process. Instead, greater emphasis was placed on the traditional laboratory-derived toxicity test and comparison of the toxicity with EEC. Presently, the EPA requires field studies only under special circumstances, and post-registration monitoring studies are used to verify the mitigation of pesticides (Touart & Maciorowski, 1997).

In the ecological risk assessment for aquatic plants, most guidelines have focused their attention on short-term exposure toxicity, and the toxicity is usually expressed as the EC50 values (OECD, 2006; U.S. EPA, 1996). Then, the toxicity and environmental concentrations are compared to evaluate the magnitude and probabilities of the possible hazard. In the actual field, however, recovery of the reproduction capability of organisms after exposure to chemicals is another important factor that must be considered. Hughes et al. (1988) suggested that the determination of the EC50 alone does not indicate the lethality of the test material or the recovery potential of the test species. Due to the environmental significance, they recommended that if substantial inhibition is observed from a 4- or 5-day exposure to the test material, the long-term exposure toxicity and the recovery phase should be conducted, and the phytostatic and phytocidal concentrations should be determined as the primary responses, because the test procedure provides a better assessment of toxic effects on an aquatic plant population. In a study with an aquatic plant, *Lemna gibba*, the phytocidal concentrations were 2.6 to >36 times higher than the corresponding EC50 values depending on the type of the herbicides tested (Mohammad et al., 2006, 2010). In another study with *Scenedesmus quadricauda*, the EC50 of paraquat decreased with an increasing of the exposure period, and paraquat caused algistatic rather than algicidal effects at the higher concentration (Saenz et al., 2001). These findings suggest that it is important to establish a different model for understanding the potential impact of chemicals in aquatic ecosystems other than the model typically used, in which only short-term effects are considered.

In addition, misconceptions arise concerning the use of the toxicity tests. There has been some debate as to whether effects should be based on the EC50 value or the realistic exposure scenario of chemicals in the aquatic systems adjacent to the agricultural areas. To some cases, 50% reduction in growth is not considered ecologically significant due to the rapid growth rate of algae or duckweed, and the algistatic (phytostatic) and algicidal (phytocidal) concentrations are considered to be more relevant (Payne & Hall, 1979; Hughes et al., 1988). The choice of the effect parameters and calculations can impact the test results. The NOEC values based on several growth parameters varied by up to 10-fold for several chemicals (Adams & Dobbs, 1984). The similar effect was seen between using a pigment content and a dry weight as the effect parameters (Sirois, 1990).

The way in which exposure is calculated also varies among regulatory agencies. An estimate of exposure is generally based on crop application rates. In the US, Canada and the UK guidelines, EEC in the aquatic environment is calculated from a hypothetical overspray of a water body at the maximum recommended label rate applications (Boutin et al., 1993; Holst & Ellwanger, 1982). In the calculation of the resulting concentrations in the water body, the US and Canada use 15 and 30 cm of water depth for forestry and agriculture, respectively, while in the UK, the concentration is calculated using a 100 cm-deep water body. The U.S. calculates its EEC value from 60% overspray, while Boutin et al. (1993) recommended a 100% overspray. Therefore, the concentration used to estimate hazard could be >10-fold difference among the guidelines.

3. Aquatic test organisms

Phytotoxicity data for aquatic plants have served a relatively minor role in regulatory decisions concerning the environmental hazard of most potential contaminants. A variety of phytotoxicity tests have been conducted with freshwater green algae, blue-green algae and diatoms (OECD, 2002; US EPA, 1996), and duckweed (OECD, 2006; US EPA, 1996). One of the important issues that needs to be resolved in toxicity testing is the great variability among organisms. Most aquatic toxicological research with chemicals has been conducted on algae as the standard organism, and the current scientific understanding concerning the phytotoxic effects of the contaminants is based mostly on results of algal test. The greatest limitation of these results is their uncertain environmental relevance due to the large variation among organisms in response of standard algal test species.

In addition, the interspecies variation of algae in sensitivity to a toxicant has been reported on many occasions. The sensitivities of different strains and geographical races of algae have varied as much as 200-fold (Blanck et al., 1984). Due to these differences in the algal sensitivity, there is an inability to extrapolate toxicity from one algal species to another. To improve this situation, a species battery approach needs to be used in laboratory phytotoxicity tests where several taxonomically different algae are exposed to the test substance. Swanson et al. (1991) provide a list of possible species.

Aquatic macrophytes are used less frequently than algae in the toxicity tests. Research with aquatic macrophytes has centered in the past on determining effective eradication techniques for nuisance growths of several species such as *Elodea canadensis* and *Ceratophyllum demersum* (Nichols, 1991). In addition, considerable research has been conducted to determine the usefulness of macrophytes as biomonitors of polluted environments (Haslam, 1982; Sortkjaer, 1984), and as bioremediative agents in wastewater treatment (Tripathi & Shukla, 1991). When macrophytes have been used in toxicity tests, the duckweeds (*Lemna* spp.) have been the species of choice, and they are often used as a representative species for all other vascular plants.

Although vascular aquatic plants are not as cosmopolitan as algae, *Lemna* sp. have been used as a test organism in various ecotoxicological test guidelines (ASTM, 1993; OECD, 2006; US EPA, 1996), and it has been reported that *Lemna* sp. is more sensitive than algae to some herbicides (Fairchild et al., 1997; Peterson et al., 1994; Mohammad et al., 2005). *Lemna* sp. is a relatively new bioindicator species, and commonly used in phytotoxicity tests because of its small size, high reproductive rate, ease of cultivation and ease of growth measurement without specialized instruments (Wang, 1990). Due to unique in its floating structure, the exposure to herbicides can be both aerial and aquatic. There appears to be little difference in the sensitivities of the two more widely used species, *L. minor* and *L. gibba,* based on the results of Cowgill et al. (1991) and King & Coley (1985).

4. Effects of long-term exposure

4.1 Backgroud

Application of herbicides several times in the season are common in the actual field, therefore, non-target organisms are exposed to chemicals for longer periods than expected from their dissipation rates, and also in the case of slow degradation in the aquatic environment. It is believed that the toxic effects depend on both the duration and the concentration of the chemical. Davies et al. (2003) reported that the exposure to sulfosulfuron at 3.33 ppb for up to 21 days was tolerated by *Lemna* sp., but adverse effects were observed when the plants were exposed for 70 days at the same concentration. The toxicity tests for *Lemna* sp. are typically conducted for 7 days of exposure to pesticides, and toxicity usually evaluated by determining EC50 (OECD, 2006; US EPA, 1996). But, evaluation based on the short-term toxicity alone is not environmentally significant for risk assessment, because the organisms might be exposed to herbicides for longer periods as mentioned above. Therefore, it is necessary to examine long-term exposure effects on non-target organisms.

To obtain the basic information of toxicity to *Lemna gibba* of several herbicides with different mode of action, the short-term exposure tests were conducted (Mohammad et al., 2008). The herbicides used and their mode of action are listed in Table 1. The inhibitory effects were expressed by relative growth rate (RGR) at the seventh day of exposure compared with the control according to the equation (1) below.

$$\mathrm{RGR}(\%) = \frac{\text{Numberof new fronds in the test vessel at 7th day}}{\text{Numberof new fronds in the control vessel at 7th day}} \times 100 \qquad (1)$$

The frond number of *L. gibba* in the control cultures increased almost exponentially during exposure and the fronds remained green and healthy throughout the experiment. When herbicides were added, growth was affected depending on the type and concentration of the chemicals. Although growth was inhibited, no visible changes in appearance and no lethal effects were observed at any concentrations of any chemicals, except for paraquat. Higher concentrations of paraquat (100 and 1000 ppb) caused plant death with a bleaching effect. RGRs of *L. gibba* during exposure to herbicides are summarized in Fig. 1.

Five typical patterns were observed as follows: (1) cyhalofop-butyl and thiobencarb were relatively week. These chemicals inhibited growth moderately even at 1000 ppb, (2) atrazine showed moderate toxicity among the herbicides and inhibited growth completely at 1000 ppb, (3) simetryn, alachlor and diuron inhibited growth less than 16 % RGR at 100 ppb, (4) paraquat with 86% RGR in exposure at 10 ppb caused death at 100 ppb, and (5) bensulfuron-methyl and cyclosulfamuron showed higher toxicity with 24% RGR at 10 ppb and 48% RGR at 1 ppb, respectively.

Based on the results, long term exposure effects were examined for the representative herbicides, atrazine, alachlor, paraquat and cyclosulfamuron with different mode action (Mohammad et al., 2006, 2010).

Name	Chemical Family	CAS number	Mode of Action
Alachlor	Chloroacetamide	15972-60-8	Inhibition of very-long-chain fatty acid biosynthesis
Atrazine	Triazine	1912-24-9	Inhibition of photosynthesis at photosystem II
Bensulfuron-methyl	Sulfonylurea	83055-99-6	Inhibition of acetolactate synthase
Cyclosulfamuron	Sulfonylurea	136849-15-5	Inhibition of acetolactate synthase
Cyhalofop-butyl	Aryloxyphenoxy propionate	122008-85-9	Inhibition of acetyl CoA propionate carboxylase
Diuron	Urea	330-54-1	Inhibition of photosynthesis at photosystem II
Paraquat	Bipyridylium	1910-42-5	Photosystem-I-electron diversion
Simetryne	Triazine	1014-70-6	Inhibition of photosynthesis at photosystem II
Thiobencarb	Thiocarbamate	28249-77-6	Inhibition of very-long-chain fatty acid biosynthesis

Table 1. Herbicides used in this study

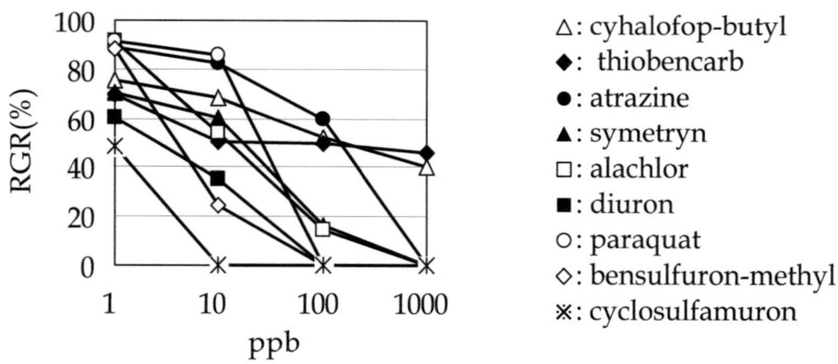

Fig. 1. Relative growth rate (RGR) of *Lemna gibba* with 7-day exposure to nine herbicides at 0, 1, 10, 100, and 1000 ppb.

4.2 Long-term exposure experiment

The long-term toxicity to *L. gibba* was tested according to the draft OECD guidelines for the testing of chemicals (OECD, 2006). Fronds of *L. gibba* were collected from the pond in front of Lake Shinji Nature Museum, Izumo, Shimane prefecture, Japan. After collection, steps were taken to eliminate the contaminating organisms. A sample of plant materials was taken and the roots were cut off. The fronds were then shaken vigorously in clean water, followed by immersion in a 0.5% (v/v) sodium hypochlorite solution for 1 minute. The fronds were then rinsed with sterile water and placed on agar medium containing 1% saccharose to confirm the sterility. Visibly contamination-free fronds were then transferred to the same agar, and cultured for eight weeks. Sufficient colonies were transferred aseptically from the stock culture into fresh sterile medium and cultured for 10 days under the test condition before starting the test. *L. gibba* was cultivated using light conditions of 12:12 light:dark cycle, cool white fluorescent lighting at 85 $\mu E^{-2}s^{-1}$ and temperature conditions of 24 $\pm$ 2 ^{o}C.

Tests were conducted under static conditions using 9 fronds in each 100 mL test beaker containing 50mL growth medium. The beakers were covered by transparent wrapping paper with some pores for aeration. Stock solutions were prepared in either acetone or water, and different concentrations of test solution were prepared by mixing with 20X-APP growth medium based on OECD guidelines. The final concentration of acetone in the test solution was less than 0.01%. All stock solutions were prepared just before the experiments.
Frond numbers were counted at the third, fifth and seventh days of the test period. Inhibition of growth was estimated on the basis of frond number, which was calculated on the basis of frond area with a fraction of 0.2 compared with the standard mother frond. Each concentration was tested in triplicate. RGR was determined at the seventh day to evaluate the capacity of mother fronds to produce new ones.
The experiment was conducted with different exposure periods of 1, 2, 3 and 4 weeks at 200-3200 ppb for atrazine, 6.25-400 ppb for alachlor, 2.5-80 ppb for paraquat, and 1-100 ppb for cyclosulfamuron. Exposure to all chemicals were conducted under static-renewal conditions every 7 days.
Toxicity data, expressed as EC50, were determined by Ecotox-Statics 2.4 (Japanese Society of Environmental Toxicology). Multiple comparisons among the treatments in each week were analyzed by analysis of variance (ANOVA) with Duncan's test ($p>0.05$) using SPSS 12.0.

4.3 Long-term exposure effects of atrazine

When atrazine was tested at concencentrations of 0, 200, 400, 800, 1600 and 3200 ppb, the inhibition patterns with different exposure periods are shown in Fig. 2. Growth was

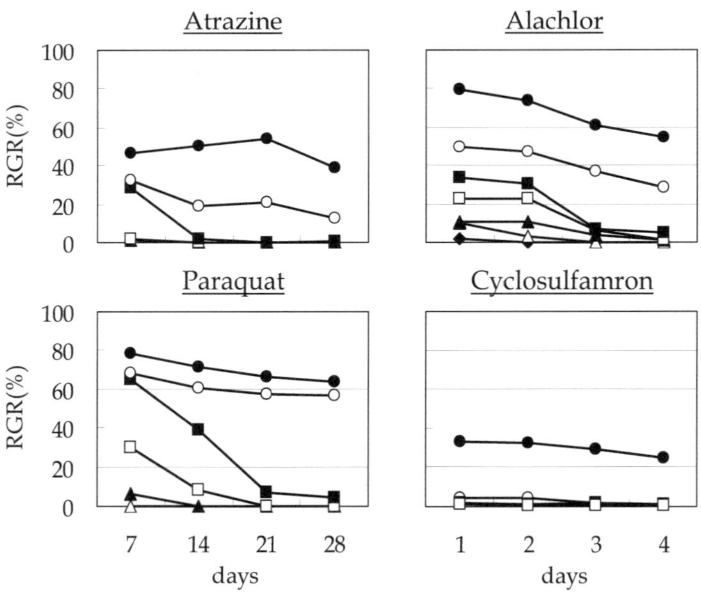

atrazine; ●:200, ○:400, ■:800, □:1600, ▲:3200 ppb, alachlor; ●:6.25, ○:12.5, ■:25, □:50, ▲:100, △:200, ◆:400 ppb, paraquat; ●:2.5, ○:5, ■:10, □:20, ▲:40, △:80 ppb, cyclosulfamron; ●:1, ○:10, ■:50, □:100 ppb

Fig. 2. Relative growth rate (RGR) of *Lemna gibba* fronds in exposure to herbicides for 28 days.

significantly inhibited ($p>0.05$) after a 7-day exposure at 200 ppb, and the comparable inhibition continued during 28 days of exposure. The RGR at 400 ppb slowly decreased during 28 days, from 32% to 12%. With exposure at 800 ppb, the RGR was 29% at 7 days but decreased to 2% at 14 days, and no growth was observed after 14 days of exposure. No growth was observed at 1600 or 3200 ppb after a 7-day exposure. There were no significant changes in the color of the fronds at any concentrations at any stages of exposure (Mohammad et al., 2010).

Atrazine disrupts photosynthesis, the most basic functionin the plant kingdom. It blocks the electron transport of photosynthesis, leading to a reduction in photosynthetic oxygen production and, finally, reducing the RGR. It has been assumed that chloroplast membranes can be damaged by this type of chemical (Corre et al., 1996).

4.4 Long-term exposure effects of alachlor

When the experiment was designed with alachlor at 0, 6.25, 12.5, 25, 50, 100, 200, 400, 800, 1600 and 3200 ppb for 7, 14, 21 and 28 day exposure, the results provide evidence that the growth of *L. gibba* was significantly affected at 6.25 ppb and almost stopped at 400 ppb for a 7-day exposure. There was a decrease in RGR as the exposure period and concentrations of alachlor increased (Fig. 2). A slowly decreasing tendency of RGR was observed at concentrations lower than the EC50, in which the RGR decreased from 80% to 55% at 6.25 ppb and from 50% to 30% at 12.5 ppb during a 28-day exposure. However, at concentrations higher than the EC50 level of 25 and 50 ppb, a rapidly decreasing tendency was observed for the RGR after 14 days (Mohammad et al., 2010).

The effects of 21-day exposure to alachlor on an algal community showed that a significant negative effect on algal biomass was observed at >10 pbb and approximately half the dominant algal taxa were affected, suggesting different sensitivity among algal species (Spawn et al., 1997). These researchers concluded that alachlor altered both algal community composition and biomass in agricultural streams. Therefore, it is necessary to examine the toxicity at the community level in duckweed for ecological risk assessment.

4.5 Long-term exposure effects of paraquat

Growth was significantly affected at 2.5 ppb and almost stopped at 40 ppb for a 7-day exposure (Fig. 2). The toxicological response varied after different exposure durations and concentrations of the compound paraquat. At the end of each contact test period of 28 days, there was some population growth with exposure to concentrations at <10 ppb, but the RGR decreased drastically at 10 ppb, which was lower than the EC50. The fronds were severely affected and appeared to be dead because of the bleaching effect at >20 ppb at the test duration of 28 days (Mohammad et al., 2010).

During photosynthesis, paraquat disrupts photosynthetic electron transfer by accepting electrons from photosystem I, and produces highly destructive superoxide radicals (Tomlin, 2000). Therefore, photosynthetic organisms are severely affected by exposure to paraquat and often die. In a previous study, it was found that freshwater algae generally died at paraquat concentrations between 0.25 and 0.5 ppm (Eisler, 1990), but the exposure period was not mentioned. A study with different exposure periods, 1, 2, 3, and 4 days, showed that the EC50 decreased from 0.89 to 0.22 ppm with an increasing exposure period with *Scenedesmus quadricauda* (Saenz et al., 2001).

4.6 Long-term exposure effects of cyclosulfamuron

Effects of exposure period (1, 2, 3, and 4 weeks) and concentration (1, 10, 50, and 100 ppb) on growth inhibition were examined using cyclosulfamuron. Growth was inhibited at 1 ppb, and completely stopped at 10 ppb in the first week of exposure (Fig 2). When the exposure period was prolonged by transferring the mother fronds once a week to new media, no change was observed in inhibition at 1 ppb even in the fourth week of exposure. But at higher concentrations (10-100 ppb), a bleaching effect was observed with longer exposure (3-4 weeks), during which color of fronds turned yellow to white. Sulfonylureas temporally inhibited the growth at less than the EEC of 3-20 ppb, and a longer exposure, beyond 21 days, caused severe damage, such as the death of fronds, at the EEC level (Mohammad et al., 2006).

Important points in risk assessment of sulfonylureas to *Lemna* sp. are presented in this study. Sulfonylureas are inhibiting the enzyme, acetolactete synthase, which is necessary in the first step for plants to synthesize the branched amino acids, valine, leucine and isoleucine (Brown, 1990; Schloss, 1994). Sulfonylureas inhibited only cell division on short-term exposure, but prolonged exposure resulted in lethality at the same concentration.

4.7 Conclusion

The toxic effects were affected by the exposure period and concentration, depending on the type of herbicide. All the tested herbicides showed stronger toxicity with the increasing exposure period than the toxicity of the standard exposure period suggested by guidelines. These characteristics of herbicides required a different model than typically used, where only short-term exposure is usually assumed, for understanding the potential impact of herbicides in aquatic ecosystems, e.g. comparison of toxicity at different concentrations with different exposure periods.

5. Recovery potential from damage

5.1 Background

In the actual field, recovery of the reproduction capability of *Lemna* sp. after exposure to herbicides is another important factor which must be considered. Hughes et al. (1988) suggested the importance of examining the recovery potential and determining phytostatic and phytocidal concentrations for a better assessment of toxic effects on an aquatic plant population. However, very few studies assessing the recovery potentials of *Lemna* sp. have been conducted. Nathalie et al. (2008) found on the algae *Scenedesmus vacuolatus* that the delay in recovery subsequent to S-metolachlor exposure contrasted with the fast recovery upon exposure to triazines and phenylureas (photosystem II inhibitors). While the effects following exposure to the photosynthesis inhibitors were readily reversible, exposure to herbicides that impaired cell division induced a delayed recovery. The results suggest that the mode of action of chemicals, the reversibility of their binding at the target site, and the degree of damage during exposure, all influence the potential recovery following exposure.

5.2 Recovery experiment

Recovery potential of *L. gibba* from the damage by several herbicides listed in Table 1 with different mode of action was examined (Mohammad et al., 2008). After each exposure period for 7 days was conducted according to the draft OECD guidelines (OECD, 2006), the nine mother fronds were collected from each beaker, washed in sterilized distilled water, and transplanted to fresh medium for recovery. The tests were done under static conditions using 100 mL test beaker containing 50 mL growth medium. Frond numbers were counted at the third, fifth and seventh days of the recovery periods, and at the tenth day when the recovery was slow.

The effect of longer-term exposure on the recovery potential was examined for the herbicides, atrazine, alachlor, paraquat and cyclosulfamuron with different mode action (Mohammad et al., 2006, 2010). The basic test conditions were the same as those of the above mentioned experiment. After exposure for 1, 2, 3 and 4 weeks, the nine mother fronds were transplanted to fresh medium for recovery, and RGR was determined at the third, fifth and seventh days of the recovery periods. The phytostatic and phytocidal concentrations of the tested chemicals for *L. gibba* were determined according to the definition described by Hughes et al. (1988).

5.3 Recovery potential from damage by herbicides with different mode of action

When the fronds were transferred to fresh medium after 7 day-exposure for recovery, *L. gibba* started to grow again even in plots where they did not grow during the exposure period. RGRs of *L. gibba* during recovery are shown in Fig. 3. Patterns of RGRs in the recovery periods showed a tendency corresponding to the mode of action of the herbicides. Cyhalofop-butyl and thiobencarb exhibited rapid recovery as well as the untreated control even at 1000 ppb and growth recovered to more than 70% RGR. Results from the recovery test with alachlor, having the same mode of action with thiobencarb, showed a slow recovery tendency for all the concentration tested. RGR in recovery for 1000ppb was 15% and for 100ppb was 32%. Triazine and urea herbicides showed moderate recovery. Although the growth was inhibited completely at 1000ppb in exposure (Fig. 1), 76% RGR was observed in recovery in case of atrazine. In case of simetryn, the RGR was >20% in recovery for 1000ppb. Therefore, the chemicals which act as inhibitors of photosystem II are moderately toxic to *L. gibba*, and moderate recovery (RGR, 76%) was observed with exposure for 7 days. Paraquat showed no recovery above the critical concentration. Recovery potential of *L. gibba* from inhibition by the sulfonylureas herbicides was greater than with other types of herbicides and recovery was possible even at 1000 ppb with 57 % RGR for bensulfuron-methyl and with 71% RGR at 10 days during the recovery period (data not shown) for cyclosulfamuron. In risk assessment, the expected environmental concentrations of the sulfonylureas were reported as 3-20 ppb (Peterson et al., 1994), which

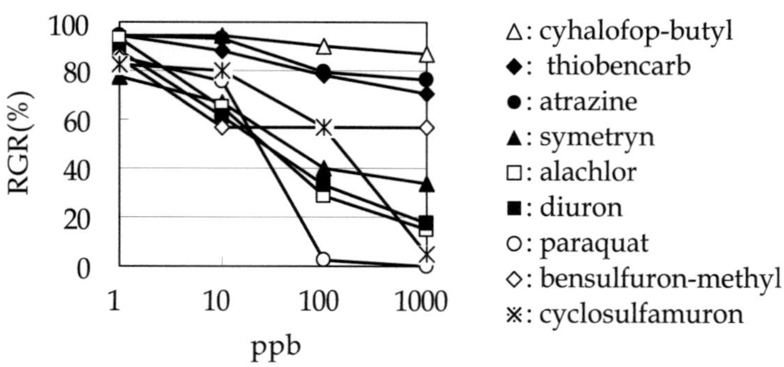

Fig. 3. Relative growth rate (RGR) of *Lemna* sp. In recovery in fresh medium after exposure to nine herbicides at 0, 1, 10, 100, and 1000 ppb for 7 days.

are greater than EC50 of *Lemna* sp. for some sulfonylureas, but recovery of growth is possible when the chemicals are dissipated by degradation in the environment.

5.4 Effect of long-term exposure on recovery potential from damage by atrazine

Fig. 4 shows the recovery after prolonged exposure of *L. gibba* at concentrations of 0, 200, 400, 800, 1600 and 3200 ppb. There was an apparent recovery of the population, even from no growth for 28 days at 3200 ppb. The RGRs were higher than those in the exposure for all concentrations tested (Fig. 2), although the significant difference was observed even at 200 ppb. The RGR in recovery depended on the concentration of atrazine in the exposure. The RGR decreased slightly after 7 days of exposure and were almost constant between 14 and 28 days of exposure. There were no significant changes in the color of the fronds at any concentrations at any stages of exposure and recovery. Under the experimental conditions, phytostatic concentrations of atrazine to *L. gibba* were 1600 and 800 ppb in the exposure periods of 14 and 28 days, respectively, and the phytocidal concentration was >3200 ppb for a 28-day exposure (Table 2) (Mohammad et al., 2010). At the highest concentration of 3200 ppb with exposure for 28 days, the RGR was 43% in recovery. The results suggest that even after a 28-day exposure to atrazine at an EEC level of 2667 ppb (Peterson et al., 1994), *L. gibba* might have capability to re-grow in the environment after the removal of atrazine by degradation.

5.5 Effect of long-term exposure on recovery potential from damage by alachlor

Recovery of alachlor after different exposure period and concentrations in fresh growth medium was according to the exposure duration and concentrations. The RGR was almost the same for 28 days when *L. gibba* was exposed at <12.5 ppb, but it constantly decreased at >12.5 ppb with exposure for longer than 14 days. The growth of populations exposed to 400 ppb for 14 days indicated a phytostatic response, while populations exposed for 21 days at ≥200 ppb showed a phytocidal response. The phytostatic concentration of alachlor to *L. gibba* was 400 ppb for a 14-day exposure, and the phytocidal concentration was >400 ppb within 14 days of exposure, but it decreased to 200 ppb for 21- and 28-day exposures (Fig. 4, Table 2) (Mohammad et al., 2010).

Alachlor interferes with metabolism and inhibits the synthesis of fatty acids (Weisshaar et al., 1993; Couderchet & Boger, 1993). It has nearly the same scenario as sulfonylurea, which disrupts amino acid biosynthesis and affects processes essential to all photosynthetic organisms (Moberg & Cross, 1990). Mohammad et al. (2006) showed that short exposures to cyclosulfamuron at higher concentrations caused longer lag periods for the initiation of growth in recovery, and a longer exposure period caused a slower growth rate without the lag period. The same tendency was observed in the case of alachlor (data not shown).

A recovery study with an algal community after exposure to alachlor at >10 ppb for 21 days showed that some algal taxa recovered after exposure, while others took longer or did not recover (Spawn et al., 1997). Therefore, it is assumed that recovery of other species of duckweed should be examined as there are differences in recovery among algal species.

5.6 Effect of long-term exposure on recovery potential from damage by paraquat

The RGR was higher in recovery than in exposure, but the difference was slight. Recovery was smooth at <10 ppb within a 14-day exposure, but it decreased when the exposure was longer than 21 days. The RGR constantly decreased at concentrations >20 ppb, and no recovery was observed because of death in the case of an exposure longer than 7 days. The phytostatic concentration of paraquat was not determined because all phytostatic fronds could not grow in the recovery period. The phytocidal concentration decreased with exposure period from 80 ppb for a 7-day exposure to 20 ppb for 21- and 28-day exposures (Fig. 4, Table 2) (Mohammad et al., 2010).

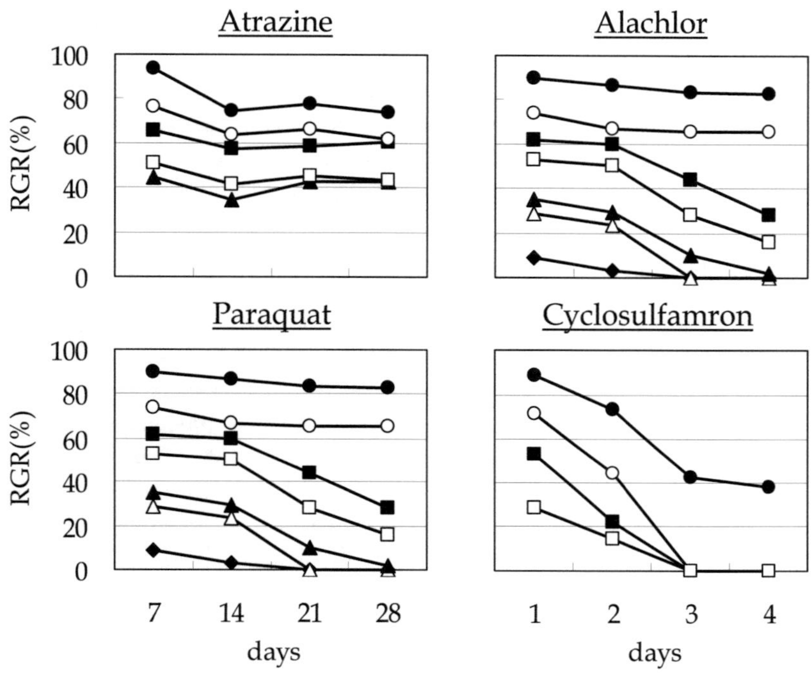

atrazine; ●:200, ○:400, ■:800, □:1600, ▲:3200 ppb, alachlor; ●:6.25, ○:12.5, ■:25, □:50, ▲:100, △:200, ◆:400 ppb, paraquat; ●:2.5, ○:5, ■:10, □:20, ▲:40, △:80 ppb, cyclosulfamron; ●:1, ○:10, ■:50, □:100 ppb

Fig. 4. Relative growth rate (RGR) of *Lemma gibba* fronds in recovery in fresh medium after the exposure to herbicides.

Chemicals	Phytostatic concentrations (ppb)				Phytocidal concentrations (ppb)			
	Exposure period (days)				Exposure period (days)			
	7	14	21	28	7	14	21	28
Atrazine	1600	1600	800	800	>3200	>3200	>3200	>3200
Alachlor	>400	400	nd[a]	nd	>400	>400	200	200
Paraquat	nd	nd	nd	nd	80	40	20	20
Cyclosulfamuron	100	50	nd	nd	>100	>100	10	10

Table 2. Phytostatic and phytocidal concentrations of atrazine, alachlor, paraquat and cyclosulfamuron for *Lemna gibba* in different exposure periods. [a] not determined

Another study with different exposure periods, 1, 2, 3, and 4 days, showed that *Scenedesmus quadricauda* could recover its reproduction capability even with exposure at the maximum concentration in their test, 3.2 ppm, for 4 days, but the EC50 decreased from 0.89 to 0.22 ppm with an increasing exposure period (Saenz et al., 2001). Those researchers also explained that an extended lag phase was required for recovery, and that it was more extended for the population exposed at 3.2 ppm. However, they did not check the lethal concentrations with a longer exposure to paraquat. Similar results were also obtained in our study, as the RGR varied from 72% to 13% in recovery from a concentration of 10 ppb with exposure periods from 7 to 28 days. Therefore, growth was possible in recovery with exposure at 10 ppb, but the recovery potential drastically decreased with a longer exposure. Phytocidal concentrations were 80 and 20 ppb for exposure for 7 and 28 days, respectively. This shows remarkable variation in the sensitivity between duckweed and algae to paraquat.

5.7 Effect of long-term exposure on recovery potential from damage by cyclosulfamron

In the recovery period after different length of exposure, RGR decreased with longer exposure as shown in Fig. 4. Reproduction was observed within two weeks of exposure at 100 ppb, but no recovery occurred after exposure for three weeks at more than 10 ppb. Cyclosufamuron, with an EC50 of 0.91 ppb, was phytostatic at 100 and 50 ppb for 7- and 14-day exposures, respectively, indicating no lethal effects at more than 50 times the concentration of EC50 within an exposure period of 14 days. In the case of exposure for longer than 21 days, however, it exhibited phytocidal activity at 10 ppb (Table 2). The results suggest that the recovery potential of *L. gibba* drastically changes depending on both chemical concentration and exposure period. *L. gibba* can reproduce again at the same rate as that before the exposure if cyclosufamuron is removed within two weeks, even after complete inhibition at 100 ppb (Mohammad et al., 2010).

5.8 Conclusion

Recent studies demonstrated that a longer period of exposure caused more serious adverse effects on *Lemna* sp. and the exposure period could affect on recovery. When the relationship of RGRs between exposure and recovery periods was examined, the RGR in recovery from the damage by atrazine was not affected much by the RGR in exposure. In case of alachlor and paraquat, the RGR in recovery was dependent on the RGR in exposure. For cyclosulfamuron, RGR decreased along with exposure period, therefore, the potential for recovery was dependent on the exposure period (Mohammad et al., 2010). When considering phytostatic and phytocidal scenario, the phytostatic and phytocidal concentrations decreased with exposure period. In some cases, phytocidal concentration became lower than the EC50 value when exposure was prolonged. Therefore, incorporation of the both concentrations associated with the exposure period would be important for ecotoxicological risk assessment of herbicides.

6. Combined effect of herbicides

6.1 Background

It is common to find combinations of several herbicides in the surface water in agricultural areas, with the exact type of substance depending on the dominant crops in the area.

Herbicides in the environment rarely occur alone. In a large US monitoring program, more than 50% of all stream samples contained 5 or more pesticides, and about 15% contained more than 10 compounds (Gilliom et al., 1999). Therefore, herbicide toxicity in natural ecosystems is not generally the result from exposure to a single toxicant, but rather exposure to mixture of toxicants. Therefore, mixture toxicity has been a subject of ecotoxicological interest for several decades.

A small number of studies have reported the potential threat to macrophytes exposed to pesticide mixtures in aquatic model ecosystems (Fairchild et al., 1994; Lytle & Lytle, 2002, 2005; Wendt-Rasch et al., 2004). Recent laboratory studies with the standard OECD plant species *Lemna minor* addressed the joint toxicity of pesticides by applying the two common models of mixture toxicity: concentration addition (CA) and independent action (IA) (Belz et al., 2008; Cedergreen et al., 2007a, b; Munkegaard et al., 2008). The concept of CA is based on the assumption that any component of a mixture can be replaced by another without altering the overall effect of the mixture, and applied for toxicants with similar molecular target sites (Loewe & Muischnek, 1926). The concept of IA is based on the assumption that all mixture components independently contribute to a given effect by different modes of action (Bliss, 1939). However, how the joint toxicity of such combinations of pesticides should be estimated is still a matter of debate in the case of considering effects of long term exposure and recovery potential.

In this section, the results of our recent study are presented on the combined effects of the mixtures of herbicides with dissimilar modes of action. Based on the results of our previous studies (Mohammad et al., 2006, 2010), we selected three combinations: paraquar + atrazine, paraquat + alachlor, and paraquat + cyclosulfamuron. Because the combinations of the herbicides with different modes of action were used, the expected joint effects were calculated based on the IA model from the individual effects, and the actual joint effects were evaluated by comparing with the expected effects. Deviation from the prediction is thus an indication of antagonism (weaker effects) or synergism (stronger effects). The mixture effects were also evaluated on a basis of the effects of long term exposure and recovery potential from the damage.

6.2 Herbicides' combined effects experiment

The long-term toxicity to *L. gibba* and the recovery potential from the damage was tested. The experiments of the mixture of herbicides were conducted for 7, 14, 21 and 28 days exposure, followed by a 7-day recovery test in a fresh medium for each length of the exposure. The experimental conditions, test procedures, measurement of the number of fronds, determination of RGR, phytostatic and phytocidal concentrations were the same as described in the previous sections. The concentrations for mixture were set based on the results of our previous study as follows (Mohammad et al., 2006, 2010). In order to observe the combined effects during the long-term exposure to chemicals, the concentrations were basically set at lower than the EC50 values.

The EC50 of paraquat was found to be 31 ppb, but during a 28-day exposure, RGR drastically decreased at <10 ppb, and the bleaching effect was observed at >20 ppb. Recovery was smooth at <10 ppb within a 14-day exposure, but it decreased when the exposure was longer than 21 days. Therefore, the concentrations for paraquat were set at 2.5, 5 and 10 ppb.

The EC50 of atrazine was found to be 89 ppb, and its phytotostatic concentrations were 1600 and 800 ppb for exposure periods of 14 and 28 days, respectively, and no phytocidal effects were observed up to 3200 ppb for a 28-day exposure. Therefore, the concentration of atrazine for mixture with paraquat was set at 100 ppb, which is near the EC50 value of atrazine.

The EC50 of alachlor was found to be 31 ppb, and RGR slowly decreased during the exposure at concentrations lower than EC50 level of 6.25 and 12.5 ppb, but rapidly decreased at 25 and 50 ppb for 14 days of exposure. Therefore, the concentration of alachlor was set at 10 ppb for mixture with paraquat.

The EC50 of cyclosulfamuron was found to be 0.91 ppb for a 7-day exposure, and the phytostatic concentrations were 100 and 50 ppb, for a 7- and 14-day exposures, respectively, and the phytocidal activity was 10 ppb when the exposure was longer than 21 days. Considering the results of our study with cyclosulfamuron, the concentration of cyclosulfamuron for mixture with paraquat was set at 0.15ppb.

The expected RGR in exposure to combined herbicides was calculated based on the RGR values in exposure to each herbicide according to the equation (2) below.

$$\begin{aligned}&\text{ExpectedRGR}(\%)\text{ in exposure to compounds A and B mixture}\\&=\text{RGR }(\%)\text{ in exposure to compound A xRGR }(\%)\text{ in exposure to compoundB} / 100\end{aligned} \quad (2)$$

The expected RGR in recovery from the damage by combined herbicides was calculated based on the RGR values in recovery from the damage by each herbicide according to the equation (3) below.

$$\begin{aligned}&\text{ExpectedRGR}(\%)\text{ in recovery for compounds A and B mixture}\\&=\text{RGR }(\%)\text{ in recovery for compound A xRGR }(\%)\text{ in recovery for compoundB} / 100\end{aligned} \quad (3)$$

6.3 Mixture effects of paraquat and atrazine

The influence of mixtures of paraquat and atrazine on the growth and recovery potential of *L. gibba* are shown in Fig. 5. When *L. gibba* was exposed to mixtures of paraquat and atrazine, the growth inhibition increased with all the mixture concentrations, compared with individual paraquat and atrazine. The RGR value decreased from 78% to 32% for 7 days exposure, and 70% to 0% for 28 days exposure by the addition of 100 ppb of atrazine to 2.5 ppb of paraquat. The expected RGR calculated from the RGR for each herbicide was larger than the observed RGR, therefore, stronger synergistic effects than the expected ones were indicated. At the highest test concentration of mixture (paraquat 10 ppb + atrazine 100 ppb), the RGR decreased 62% to 5% for 7 days exposure, but there was no change in colour of any fronds at the end of the exposure period.

In the recovery phase, the reproduction was very slow in the case of mixture compared with the individual herbicide. The RGR in recovery ranged from 90% to 73% after a 7-day exposure to only paraquat at from 2.5 to 10 ppb, whereas the RGR was from 47% to 32% when 100 ppb of atrazine was added. The RGR in recovery in atrazine alone at 100 ppb was above 82% even after exposure for 28 days, and the observed RGR in recovery for the mixed herbicides were smaller than the corresponding expected RGR, suggesting that there was the synergistic effect also in the recovery phase.

Although no growth was observed at the mixture of 10 ppb of paraquat and 100 ppb of atrazine after 21-day exposure, there was no phytocidal effect in appearance. Phytostatic effect was found at the mixture of 5 and 10 ppb of paraquat and 100 ppb of atrazine for 21-day exposure, and at the mixture of 2.5 ppb of paraquat and 100 ppb of atrazine for 28-day exposure. The phytostatic concentration of atrazine was 800 ppb for the same period of exposure (Table 2), therefore, atrazine showed eight times stronger phytostatic effects by adding paraquat at 10 ppb. On the other hand, paraquat did not show any phytostatic effect at this concentration, but paraquat showed this type of character when mixed with atrazine. This is an interesting phenomenon in this combination.

Atrazine is a common contaminant of surface waters, as a result of agricultural non point surface and subsurface runoff, and is usually detected in levels from less than 0.5 ppb (Albanis et al., 1995; Squillace & Thurman, 1992) up to 100 ppb (Thurman et al., 1992). The toxic effects of a mixture of atrazine and metolachlor were examined in unialgal cultures of *Chlorella fusca var-fusca* using a bioassay system. In concentrations lower than the EC50, the combination resulted in reduced toxicity (antagonism) in comparison with the toxicity caused by the sum of toxic actions of the same levels of concentration from single chemicals (Kotrikla et al., 1999). Another study analyzed the toxicity of two mixtures (atrazine and the insecticide chlorpyrifos; atrazine and the fungicide chlorothalonil) to the marine

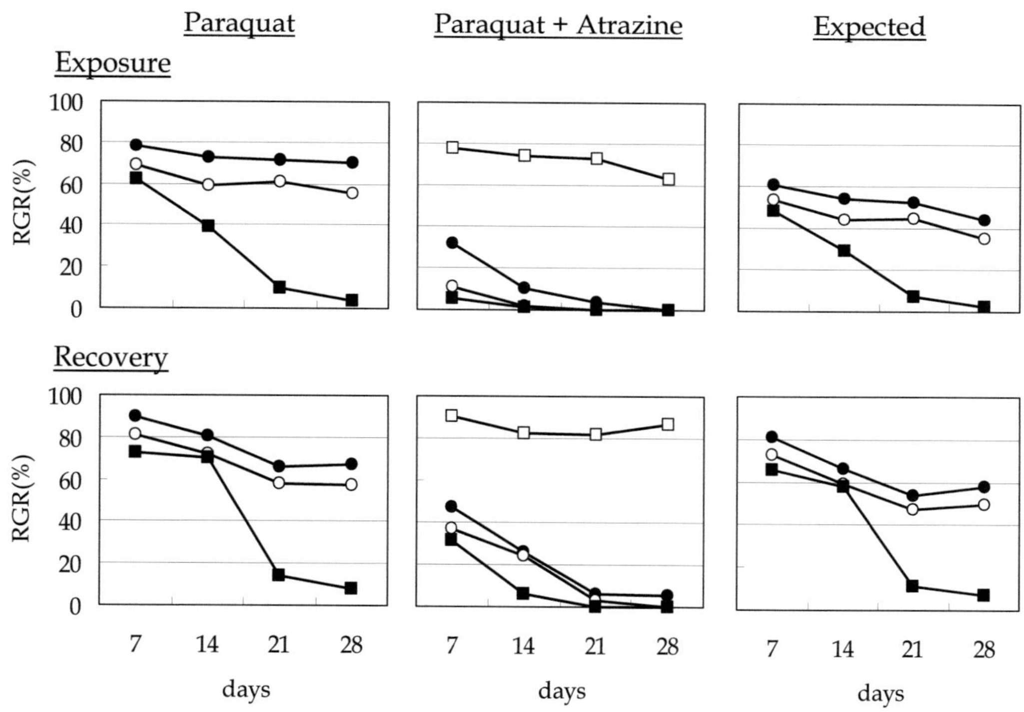

Fig. 5. Relative growth rate (RGR) of Lemna giba in exposure and in recovery in fresh medium after the exposure to paraquat with and without atrazine. Expexted RGR in exposure and in recovery were calculated as (paraquat exposure RGR) x (atrazine exposure RGR)/100, and (paraquat recovery RGR) x (atrazine recovery RGR)/100, respectively.

phytoplankton species *Dunaliella tertiolecta* (Chlorophyta). Atrazine and chlorpyrifos in mixture displayed additive toxicity, whereas atrazine and chlorothalonil in mixture had a synergistic effect. The toxicity of atrazine and chlorothalonil combined was approximately 2 times greater than that of the individual chemicals (DeLorenzo & Serrano, 2003).
Our study using *L. gibba* showed that the sensitivity increased in presence of atrazine with paraquat for 7 to 28-day exposure. Fig. 5 shows that there are large difference between the expected and actual inhibition scenario in both exposure and recovery phases The results suggest the importance of examining combined effects of herbicides for ecotoxicological risk assessment.

6.4 Mixture effects of paraquat and alachlor

After 7 days of exposure, RGR significantly decreased from 70% to 29% by adding 10 ppb alachlor with the lowest paraquat concentration of 2.5 ppb (Fig. 6). The RGR in exposure to only alachlor for the same period was 50%, therefore, the observed RGR value was almost the same as the expected RGR (35%). When the exposure prolonged up to 28 days and the concentration of paraquat increased up to 10 ppb, the RGR decreased from 34% to 3% and

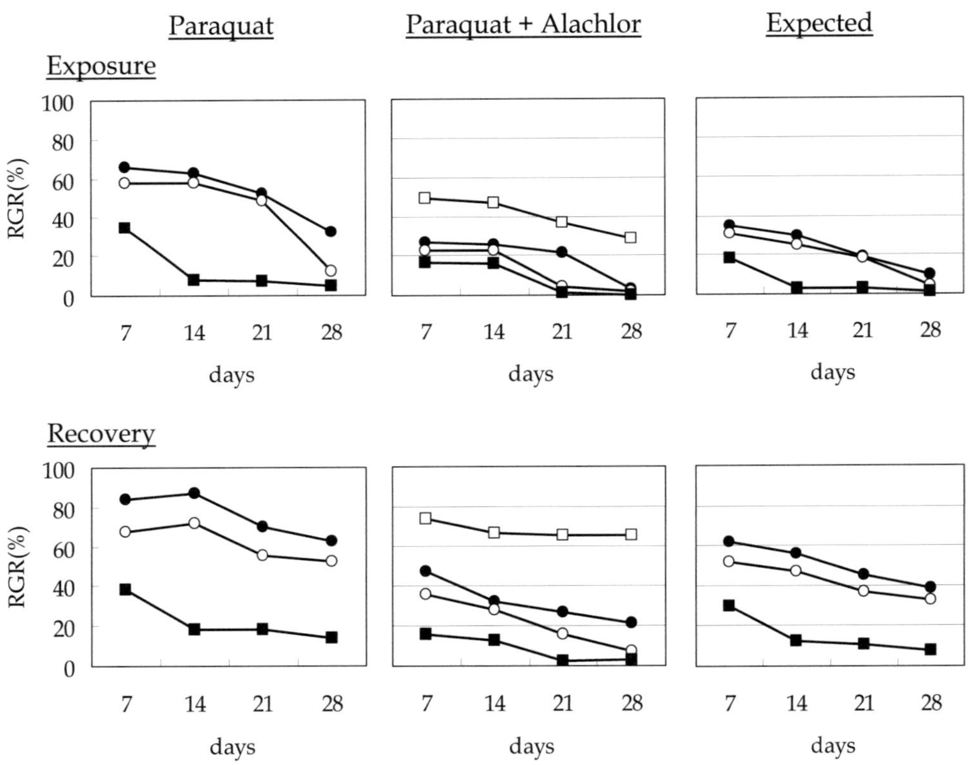

paraquat; ●:2.5, O:5, ■:10, alachor alone; □:100 ppb

Fig. 6. Relative growth rate (RGR) of *Lemna gibba* fronds in exposure and in recovery in fresh medium after the exposure to paraquat with and without alachor. Expexted RGR in exposure and in recovery were calculated as (paraquet exposure RGR) x (alachor exposure RGR)/100, and (paraquet recovery RGR) x (alachor recovery RGR)/100, respectively.

from 36% to 17%, respectively, by adding alachlor. These decreases of RGR were also comparable to the expected RGR. The results indicated that the effects of a mixture of paraquat and alachlor could be predicted from the individual toxicity. Although the RGR becomes 0% at the highest mixture concentration, no discoloration of frond was seen.

Higher RGR values were observed in recovery than those in exposure for all selected mixture concentrations, even from complete inhibition at the highest mixture of 10 ppb of paraquat and 10ppb of alachlor, but the recovery was slow compared with each corresponding individual concentration of paraquat and alachlor. RGR was 47% at the lowest mixture concentration of 2.5 ppb of paraquat and 10 ppb of alachlor for 7 days exposure, while 84% and 74% RGR was observed in the individual corresponding concentrations of paraquat and alachlor, respectively. Moreover, although there was not a marked difference between the expected and actual RGRs in the exposure phase, in the recovery phase, the actual RGR was lower than the expected RGR at all combinations of the mixture. Therefore, the mixture of paraquat and alachlor showed stronger synergistic effects on *L. gibba* in recovery than in exposure.

6.5 Mixture effects of paraquat and cyclosulfamuron

The RGR in exposure was significantly affected by all the selected mixture of paraquat and cyclosulfamuron (Fig. 7). RGR decreased from 58% to 23% by adding 0.15 ppb of cyclosulfamuron to 2.5 ppb of paraquat for 7 days exposure. When the exposure was prolonged up to 28 days, the RGR did not change as in the case of individual herbicide. The RGR at this set of concentration were similar to the corresponding expected RGR, while at 5 ppb of paraquat, lower RGR values were observed compared with the expected RGR values throughout the exposure period. With the highest mixture concentrations of 10 ppb paraquat and 0.15 ppb cyclosulfamuron, the RGR became 0% and strong discoloration (chlorosis) of total fronds was seen after 21 days exposure.

At the end of each test period, the fronds were transferred to fresh medium to observe the recovery potential of individual and mixture of chemical treatment up to 7 days. The recovery of fronds, which were exposed to individual chemical, was clearly possible, as RGR in recovery was 77% and 90% after individual exposure for 7 days to paraquat at 2.5 ppb and cyclosulfamuron at 0.15 ppb, respectively. And the mixture of the herbicides at these concentrations showed 67% RGR which was the almost same as the expected RGR (69%). But when the exposure period or concentration increased, the actual RGR were lower than the corresponding expected RGR. Bleached fronds at the height mixture concentration, exposed for 21 and 28 days, were supposed to be dead because no growth was observed in recovery. Therefore, the phytocidal concentration of the mixture was found to be a mixture of 10 ppb of paraquat and 0.15 ppb of cyclosulfamuron for 21 and 28 days exposure, and phytostatic scenario was identified for 14 days exposure at the same concentrations. Phytocidal concentrations of individual paraquat and cyclosulfamuron were 20 ppb and 10 ppb, respectively, for the same period of exposure (Table 2). Therefore, the results showed that a mixture of paraquat and cyclosulfamuron at lower concentration than their phytocidal level caused phytocidal effects on *L. gibba*. The results also showed that the mixture toxicity of paraquat and cyclosulfamuron not only increased sensitivity of *Lemna gibba,* but also lethal if exposure prolonged beyond 14 days with concentration of paraquat 10 ppb and cyclosulfamuron 0.15 ppb, which showed nether phytostatic nor phytocidal effect when act individually.

Deneer et al. (2000) found that the concentration addition was quite common for narcotic acting compounds. They demonstrated in experiments with 50 nonreactive chemicals that even chemicals present at very low concentrations, equivalent to 0.01, 0.005, and 0.0025 toxic units, contributed to the overall toxicity.

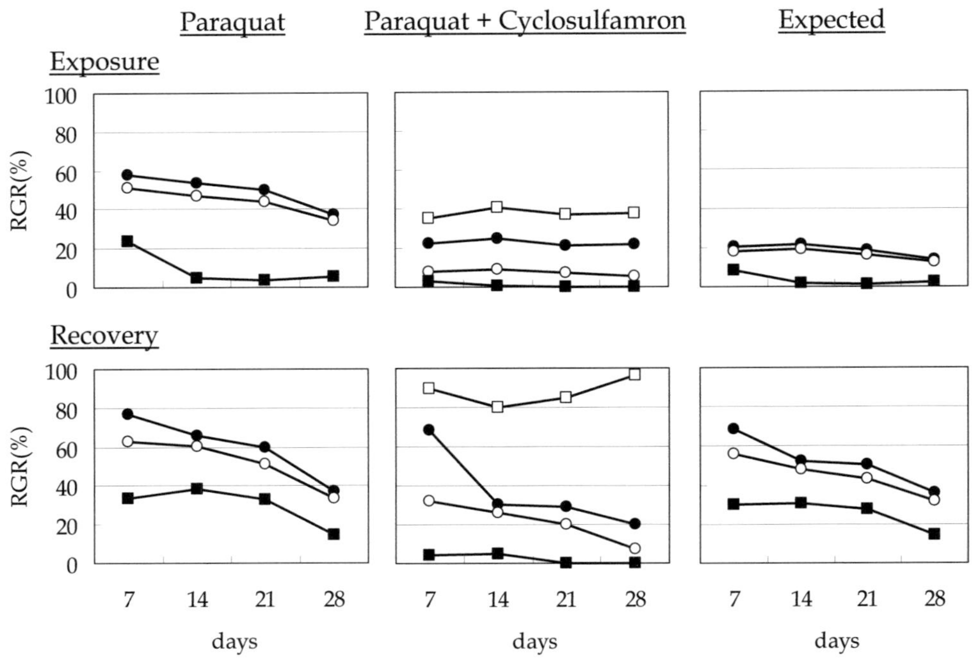

paraquat; ●:2.5, ○:5, ■:10, cyclosulfamron alone; □:0.15 ppb

Fig. 7. Relative growth rate (RGR) of *Lemna gibba* fronds in exposure and in recovery in fresh medium after the exposure to paraquat with and without cyclosulfamron. Expexted RGR in exposure and in recovery were calculated as (paraquet exposure RGR) x (cyclosulfamron exposure RGR)/100, and (paraquet recovery RGR) x (cyclosulfamron recovery RGR)/100, respectively.

6.6 Conclusion

When focusing on ecotoxicological studies and risk assessments of mixture effects of herbicides on aquatic plants, synergism is the most important effect to protect against, since it can not be predicted and results in an increase of toxicological effects, as it is the worst interaction between components of mixtures. If one, two or more chemicals were present at low levels in the same ecosystems, each of them would be poorly deleterious to non-target species if considered separately, but their addition increase significantly the ecotoxicological risk by the accumulation of low level risks. Therefore, increased toxicity due to the synergistic nature of the herbicides could results in detrimental effects to primary producers at concentrations lower than expected from the individual toxicity. The joint effects of the herbicides to *L. gibba* presented in this section suggested that they often appeared stronger than the expected ones, therefore, they could not be predicted from the standard toxicity test using a single herbicide. In addition, the effects of long-term exposure to herbicides and recovery potential of duckweed were also affected by their combination, indicating further understanding of mechanisms how mixtures of herbicides affect non-target aquatic species is necessary.

7. Future perspective

Current scientific knowledge concerning the phytotoxicities of potential contaminants is based largely on results from laboratory toxicity tests for a few freshwater green algal species. The available results are used sometimes, with little scientific justification, as surrogates for other types of aquatic plants and organisms. Our knowledge in regard to how different organisms respond to herbicides is simply insufficient to be able to speculate about cause and effect scenarios. In addition to the regulatory testing that needs to be developed, there is a need for complimentary research that will expand our knowledge beyond the level given by these standard regulatory tests. The ultimate goal of any phytotoxicity test should be to provide results for a battery of relevant surrogate species. As a result, a composite picture can be obtained to estimate the short- and long-term influence of contaminants on the condition of an exposed plant community and ecosystem. With this in mind, there is a need not only to increase use of the available phytotoxicity test methods but also to continue to develop their ability to provide useful results.

Our results presented that the relative risk of a variety of scenarios of exposure and recovery with an aquatic vascular plant *Lemna* sp. exposed to individual and mixtures of herbicides are significant from both regulatory and research perspectives. To address actual environmental situations, the application of this approach would be a good solution for a better understanding of the ecological significance of the end points used in toxicity testing and how they are interpreted and applied in ecological risk assessment.

8. References

Adams, N. & Dobbs, A.J. (1984). A comparison of results from two test methods for assessing the toxicity of aminothiazole to *Selenastrum capricornutum*. *Chemosphere*, 13, 965-971.

Albanis, T.A., Danis, T.G. & Hela, D.G. (1995). Transportation of pesticides in estuaries of Louros and Aracthos rivers (Amvrakikos Gulf, N.W. Greece), *Sci Total Environ*, 171, 85-93.

American Society for Testing and Materials (ASTM) (1993). Standard guide for conducting static toxicity tests with *Lemna gibba* G3. Annual Book of ASTM Standards, Section 11, vol 11.04, Designation E1415-91. Philadelphia, pp. 1137-1146.

Belz, R.G., Cedergreen, N. & Sorensen, H. (2008). Hormesis in mixtures - can it be predicted?, *Sci Total Environ*, 404, 77-87.

Blanck, H., Wallin, G. & Wangberg, S. (1984). Species-dependent variation in algal sensitivity to chemical compounds. *Ecotoxicol Environ Saf*, 8, 339-351.

Bliss, C.I. (1939). The toxicity of poisons applied jointly. *Ann Appl Biol*, 26, 585-615.

Boutin, C, Freemark, K.E. & Keddy, C.J. (1993). Proposed guidelines for registration of chemical pesticides in Canada: nontarget-plant testing and evaluation. Canadian Wildlife Service Technical Report Series 145, Canadian Wildlife Service, Environment Canada, Ottawa, ON.

Brown, H.M. (1990) Mode of action, crop selectivity, and soil relations of the sulfonylurea herbicides. *Pestic Sci* 29, 263-281.

Cedergreen, N., Abbaspoor, M., Sorensen, H. & Streibig, J.C. (2007a). Is mixture toxicity measured on a biomarker indicative of what happens on a population level? a study with *Lemna minor*. *Ecotoxicol Environ Saf*, 67, 323-332.

Cedergreen, N., Kudsk, P., Mathiassen, S.K. & Streibig, J.C. (2007b). Combination effets of herbicides on plants and algae: do species and test system matter?. *Pest Manage Sci*, 63, 282-295.

Corre, G., Templier, J. & Largeau, C. (1996). Influence of cell wall composition on the resistance of two *Chlorella* species (Chlorophyta) to detergents. *J Phycol*, 32, 584-590.

Couderchet, M. & Boger, P. (1993). Chloroacetamide-induced reduction of fatty acid desaturation. *Pestic Biochem Physiol*, 45, 91-97.

Cowgill, U.M., Milazzo, D.P. & Landenberger, B.D. (1991). The sensitivity of *Lemna gibba G-3* and four clones of *Lemna minor* to eight common chemicals using a 7-day test. *J Water pollt Contr Fed*, 63, 991-998.

Davies, J., Honegger, J.L., Tencalla, F.G., Meregalli, G., Brain, P., Newman, J.R. & Pitchford, H.F. (2003). Herbicide risk assessment for non-target aquatic plants: sulfosulfuron--a case study. *Pest Manage Sci*, 59, 231-237.

DeLorenzo, M.E. & Serrano, L. (2003). Individual and mixture toxicity of three pesticides; atrazine, chlorpyrifos, and chlorothalonil to the marine phytoplankton species *Dunaliella tertiolecta. J Environ Sci Health, Part B*, B38, 529-538.

Deneer, J.W. (2000). Toxicity of mixtures of pesticides in aquatic systems. *Pest Manage Sci*, 56, 516–520.

Eisler, R. (1990). Paraquat hazards to fish, wildlife and invertebrates: a synoptic review. Contaminant Hazard Reviews. U.S. Fish Wildlife Service Biological Report 85 (1.22) pp 28.

Fairchild, J.F., La Point, T.W. & Schwartz, T.R. (1994). Effects of an herbicide and insecticide mixture in aquatic mesocosms. *Arch Environ Contam Toxicol*, 27, 527–533.

Fairchild, J.F., Ruessler, D.S., Haverland, P.S. & Carlson, A.R. (1997) Comparative sensitivity of *Selenastrum capricornutum* and *Lemna minor* to sixteen herbicides. *Arch Environ Contam Toxicol*, 32, 353-357.

Gilliom, R.J., Barbash, J.E., Kolpin, D.W. & Larson, S.J. (1999). Testing water quality for pesticide pollution. *Environ Sci Technol*, 33, 164A–169A.

Grace, J.B. & Wetzel, R.G. (1978). The production biology of Eurasian watermilfoil (*Myriophyllum spicatum* L.): a review. *J Aquat Plant Manage*, 16, 1-11.

Haslam, S.M. (1982). A proposed method for monitoring river pollution using macrophytes. *Environ Technol*, 3, 19-34.

Holst, R.W. & Ellwanger, T.C. (1982). Pesticide assessment guidelines. Subdivision J. Hazard evaluation: nontarget plants. US EPA, Washington, DC, EPA-54019-82-020.

Hughes, J.S., Alexander, M.M. & Balu, K. (1988). An evaluation of appropriate expressions of toxicity in Aquatic bioassays as demonstrated by the effects of atrazine on algae and duckweed. In: *Aquatic Toxicology and Hazard Assessment*, vol.10, ASTM STD971, Adams, W.J., Chapman, G.A. & Landis, W.G. (Ed.) American Society for Testing and Materials, Philadelphia, pp 531-547.

Huxley, A. (1984). *Green inheritance*, University of California Press, CA.

King, J.J. & Coley, K.S. (1985). Toxicity of aqueous extracts of natural and synthetic oils to three species of *Lemna*. ASTM STP 891, American Society for Testing and Materials, Philadelphia, pp. 302-309.

Kotrikla, A., Gatidou, G. & Lekkas, T.D. (1999). Toxic effects of atrazine, dethyl-atrazine, deisopropyl-atrazine and metolachlor on *Chlorella fusca var-fusca. Global Nest: Int J*, 1, 39-45.

Loewe, S. & Muischnek, H. (1926). Effect of combinations: mathematical basis of problem. *N-S. Arch Ex Path Ph*, 114, 313-326.

Lytle, J.S. & Lytle, T.F. (2002). Uptake and loss of chlorpyrifos and atrazine by *Juncus effusus* L. in a mesocosm study with a mixture of pesticides. *Environ Toxicol Chem*, 21, 1817–1825.

Lytle, T.F. & Lytle, J.S. (2005). Growth inhibition as indicator of stress because of atrazine following multiple toxicant exposure of the freshwater macrophyte, *Juncus effusus* L. *Environ Toxicol Chem*, 24, 1198–1203.

Moberg, W.K. & Cross, B. (1990). Herbicides inhibiting branched-chain amino acid biosynthesis. *Pesticide Sci*, 29, 241-246.

Mohammad, M., Kishimoto, T., Itoh, K., Suyama, K. & Yamamoto, H. (2005). Comparative sensitivity of *Pseudokirchnetiella subcapitata* vs. *Lemna* sp. to eight sulfonylurea herbicides. *Bull Environ Contam Toxicol*, 75, 866-872.

Mohammad, M., Itoh, K., Suyama, K. & Yamamoto, H. (2006). Recovery of *Lemna* sp. after exposure to sulfonylurea herbicides. *Bull Environ Contam Toxicol*, 76, 256-263.

Mohammad, M., Itoh, K. & Suyama, K. (2008). Comparative effects of different families of herbicides on recovery potentials in *Lemna* sp. *J Pestic Sci*, 33, 171-174.

Mohammad, M., Itoh, K. & Suyama, K. (2010). Effects herbicides on *Lemna Gibba* and recovery from damage after prolonged exposure. *Arch Environ Contam Toxicol*, 58, 605-612.

Munkegaard, M., Abbaspoor, M. & Cedergreen, N. (2008). Organophosphorous insecticides as herbicide synergists on the green algae *Pseudokirchneriella subcapitata* and the aquatic plant *Lemna minor*. *Ecotoxucol*, 17, 29-35.

Nathalie, V., Daya, M., Rik, I.L.E., Marion, J. & Nathalie, C. (2008). *S*-metolachlor pulse exposure on the alga *Scenedesmus vacuolatus*: Effects during exposure and the subsequent recovery. *Chemosphere*, 73, 395–400.

Nichols, S.A. (1991). The interaction between biology and the management of aquatic macrophytes. *Aquat Bot*, 41, 225-252.

OECD (2002). OECD guidelines for the testing of chemicals, proposal for updating guideline 201, Freshwater algae and cyanobacteria, growth inhibition test. OECD, Paris, France.

OECD (2006). OECD guidelines for the testing of chemicals, revised proposal for a new guideline 221, *Lemna* sp. growth inhibition test. OECD, Paris, France.

Payne, A.G. & Hall, R.H. (1979). A method for measuring algal toxicity and its application to the safety assessment of new chemicals. In: *Aquatic Toxicology*, ASTM STP 667, Marking, L.L. & Kimerle, R.A. (Ed.) American Society for Testing and Materials, Philadelphia, pp. 171-180.

Peterson, H.G., Boutin, C., Martin, P.A., Freemark, K.E., Ruecker, N.J. & Moody, M.J. (1994). Aquatic phyto-toxicity of 23 pesticides applied at expected environmental concentrations. *Aquat Toxicol*, 28, 275-292.

Peterson, H.G., Boutin, C., Freemark, K.E. & Martin, P.A. (1997). Toxicity of hexazinone and diquat to green algae, diatom, cyanobacteria and duckweed. *Aquat Toxicol*, 39, 111-134.

Sand-Jensen, K. (1997). Macrophytes as biological engineers in the ecology of Danish streams. In: *Freshwater biology: priorities and development in Danish research*, Sand-Jensen, K. & Pedersen, O. (Ed.) University of Copenhagen and G.E.C. Gad Publishers, Copenhagen, pp. 74-101.

Saenz, M.E., Marzio, W.D., Di Alberdi, J.L. & Tortorelli, M.C. (2001). Algal growth recovery studies after paraquat exposure. *Bull Environ Contam Toxicol*. 66, 263-268.

Schloss, J.V. (1994). Recent advances in understanding the mechanism and inhibition of acetolactate synthase. In: *Chemistry of Plant Protection*, vol.10, Stetter. J. (Ed). Springer-Verlag, Berlin, pp 3-14.

Sirois, D.L. (1990). Evaluation of protocols for the assessment of phytotoxicity. In: *Plants for Toxicity Assessment*, ASTM STP 1091, Wang, W., Gorsuch, J.W. & Lower, W.R. (Ed.) American Society for Testing and Materials, Philadelphia, pp. 225-234.

Sortkjaer, O. (1984). Macrophytes and macrophyte communities as test systems in ecotoxicological studies of aquatic systems. *Ecol Bull*, 36, 75-80.

Spawn, R.L., Kyle, D.H. & Blair, D.S. (1997). Effects of alachlor on an algal community from a midwestern agricultural stream. *Environ Toxicol Chem*, 16, 785-793.

Squillace, P.J. & Thurman, E.M. (1992). Herbicide transport in rivers: importance of hydrology and geochemistry in nonpoint-source contamination. *Environ Sci Technol*, 26, 538-545.

Swanson, S.M., Rickard, C.P., Freemark, K.E. & Mac-Quarrie, P. (1991). Testing for pesticide toxicity to aquatic plants: Recommendations for test species. In: *Plants for Toxicity Assessment*, vol.2, ASTM STP 1115, Gorsuch, J.W., Lower, W.R., Wang, W. & Lewis, M.A. (Ed.) American Society for Testing and Materials, PA, pp. 77-97.

Thurman, E.M., Goolsby, D.A., Meyer, M.T. & Kolpin, D.W. (1991). Herbicides in surface waters of the midwestern United States: the effect of spring flush. *Environ Sci Technol*, 25, 1794-1796.

Thurman, E.M., Goolsby, D.A., Meyer, M.T., Mills, M.S., Pomes, M.L. & Kolpin, D.W. (1992). A reconnaissance study of herbicides and their metabolites in surface water of the midwestern United States using immunoassay and gas chromatography/mass spectrometry. *Environ Sci Technol*, 26, 2440-2447.

Tomlin, C.D.S. (Ed.) (2000). *The Pesticide Manual*, 12th edition, British Crop Protection Council, UK.

Touart, L.W. & Maciorowski, A.F. (1997). Information needs for pesticide registration in the United States. *Ecol Applicat*, 7, 1086-1093.

Tripathi, B.D. & Shukla, S.C. (1991). Biological treatment of wastewater by selected aquatic plants. *Environ Pollut*, 69, 69-78.

Urban, D.J. & Cook, N.J. (1986). Hazard evaluation division standard evaluation procedure: Ecological risk assessment. EPA-540/9-85-001. US EPA, Washington, DC.

USDA (1998). Agicultural chemical Usage: Field Crop Summary. US Department of Agriculture, July 99, Washington, DC.

U.S. Environmental Protection Agency (1996). OPPTS 850.4400 Aquatic plant toxicity test using *Lemna* sp., tiers I and II, EPA 712-C-96-156, EPA, Washington DC.

U.S. Environmental Protection Agency (1996). OPPTS 850.5400 Algal toxicity, tiers I and II, EPA 712-C-96-164, EPA, Washington DC.

Wang, W. (1990). Literature review on duckweed toxicity testing. *Environ Res*, 52, 7-22.

Weisshaar, H., Retzlaff, G., Boger, P. (1988). Chloroacetamide inhibition of fatty acid synthesis. *Pestic Biochem Physiol*, 32, 212-216.

Wendt-Rasch, L., Van den Brink, P.J., Crum, S.J.H. & Woin, P. (2004). The effects of a pesticide mixture on aquatic ecosystems differing in trophic status: responses of the macrophyte *Myriophyllum spicatum* and the periphytic algal community. *Ecotoxicol Environ Saf*, 57, 383–398.

Part 6

Herbicide Resistance

29

Resistance of Weeds to Herbicides

William Vencill, Timothy Grey, and Stanley Culpepper
Department of Crop & Soil Sciences, University of Georgia
USA

1. Introduction

Herbicides are the most widely used group of pesticides worldwide. The widespread use of herbicides has allowed tremendous gains in agricultural productivity worldwide. Since the 1950's herbicides have progressively replaced mechanical weed control because herbicides are more cost effective (Gianessi & Reigner, 2007). In 2009 over 95% of all the major agronomic crops grown in the U. S. were treated with herbicides (USDA-NASS, 2009). Transgenic herbicide-resistant crops were commercially introduced in the U. S. in 1996 when glyphosate-resistant (Roundup Ready®) soybean was released. Use of the very broad-spectrum herbicide, glyphosate, provided outstanding weed control (Dill, 2005; Dill et al. 2008). The most recent data indicates that the percent of the total acres of each of the following crops planted with glyphosate-resistant cultivars is soybeans 91%, canola 91%, cotton 71%, and corn 68% (Brookes & Barfoot 2009). Herbicides are used on >90% of arable farmland in the U.S. and herbicide-resistant crops has been used widely since the mid-1990's. Herbicide resistance in weeds was first discovered in 1968 (Ryan 1970) and there are currently 347 confirmed weed biotypes worldwide (Heap 2010).

When discussing pest resistance, whether it is weeds, pathogens, or insects, it is important to define the resistance. Some of the basic differences in the definitions of pest resistance depend on the basic definitions. The most basic unit of biological classification is the species, defined as a group of individual organisms displaying common characteristics and having the ability to mate and produce fully viable progeny. A population is a group of organisms within a species that co-occur in time and space (Radosevich et al. 1997) and share a distinct range of genetic variation. A species is usually composed of several to many populations. A genotype is the sum of the genetic coding or the genome of an individual. A biotype may not be fully coincident with genotype, as an individual has many genes. Certain genes may be expressed or unexpressed and not pertain to the phenotype associated with the biotype. A biotype is a phenotype that consistently expresses or exhibits a specific trait or set of traits. Weed scientists tend to refer to a biotype as a group of individuals with distinctive biochemical or morphological traits (e.g. resistance to a specific herbicide mechanism of action; growth and morphological traits). A phenotype refers to the physiological and morphological profile of the expressed genes in an individual. A single genotype can produce different phenotypes in response to environmental conditions present. This fundamental property of organisms is known as phenotypic plasticity. The alteration of phenotype (morphological or biochemical) without change in either the coding sequence of a gene or the upstream promoter region is classified as epigenetic change (Rapp & Wendel 2005). There is some controversy over whether

epigenetic changes can be inherited. The enhanced expression of EPSP synthase gene in glyphosate-resistant Palmer amaranth may be such a change.
The Weed Science Society of America's (WSSA) (1998) published its approved definitions for terms as follows: "Herbicide resistance (HR) is the inherited ability of a plant to survive and reproduce following exposure to a dose of herbicide normally lethal to the wild type. In a plant, resistance may be naturally occurring or induced by such techniques as genetic engineering or selection of variants produced by tissue culture or mutagenesis." In herbicide-resistant crops, the resistance trait allows the use of a herbicide that would otherwise injure or kill the crop. Herbicide tolerance: "Herbicide tolerance (HT) is the inherent ability of a species to survive and reproduce after herbicide treatment. This implies that there was no selection or genetic manipulation to make the plant tolerant; it is naturally tolerant." In crops, herbicide-tolerance allows the use of herbicides that control weeds but do not injure the crop.
Whereas there are many individual herbicide products, the Herbicide Resistance Action Committee (HRAC) recognizes only 16 unique modes of action (Senseman et al. eds. 2007), excluding those that are unclassified. The mode of action of an herbicide is the way the chemical controls the weed, thus it characterizes the selection factor. From the beginning of large-scale herbicide use, there were concerns about the potential for herbicide resistance (Appleby 2005). Like bacteria, fungi, and arthropods, weed populations adapt to selection; the most susceptible individuals are eliminated by exposure, while the less susceptible reproduce and present a succeeding generation that is more difficult to control than the former. The first case of herbicide resistance was to the triazine herbicide, simazine, in 1968 (Ryan 1970). Since then over 347 resistant weed biotypes have been reported; virtually all major modes of action of herbicide have certain weeds that have developed resistance to them (Heap 2010). During the 1970s and 80s different agronomic crops tended to use different combinations of herbicides, because the crops tolerated different herbicide modes of action, and generally more than one mode of action was needed in each crop to control the several species of weeds that might infest them. Since glyphosate had such broad activity against weeds, it was often used alone. Initially the argument was advanced that glyphosate resistance was highly improbable (Bradshaw et al. 1997). Nevertheless, a resistant biotype of rigid ryegrass (*Lolium rigidum* L.) was confirmed in Australia in 1996 (Heap 2010). There are now 18 reported instances of weed species that are resistant to glyphosate; they are found on all agriculturally productive continents.

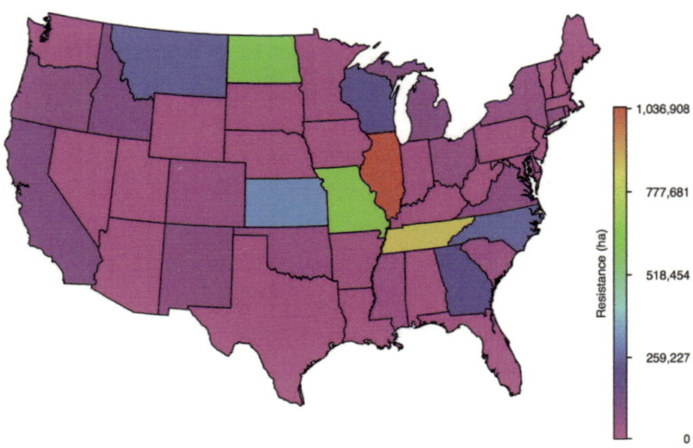

Fig. 1. Hectares of herbicide-resistant weeds in the US (Heap 2009).

Agronomic weed management is increasingly difficult and costly due to the apparent increase in the rate of development of weed resistance to herbicides and the lack of development of new modes of herbicide action. No new class of herbicides has been registered in the U.S. since mesotrione, an hydroxyphenyl pyruvate dioxygenase inhibitor in 1993. In contrast, the number of herbicide resistant weeds continues to increase, as have those specifically resistant to glyphosate.

Herbicide resistance in weeds occurs via target site resistance, enhanced metabolism, sequestration, reduced uptake, and over-production of the herbicide target site. Herbicide resistance has been confirmed to ten specific herbicide mechanisms of action. The most widespread resistance is to photosystem II-inhibitors, photosystem I inhibitors, acetolactate synthase (ALS)-inhibitors, acetyl-CoA carboxylase (ACCase) inhibitors, protoporphyrinogen oxidase (PPO or PROTOX) inhibitors, carotenoid synthesis inhibitors, EPSP synthase inhibitors (e.g. glyphosate), mitotic inhibitors, and auxinic herbicides. Within each of these herbicide mechanisms of action, there are multiple amino acid changes within the herbicide-binding domain. For many herbicide mechanism of action, there are multiple mechanisms of resistance possible. The specific mechanism can affect the level of cross-resistance observed. There are many factors such as herbicide rate can affect the type of resistance mechanism that occurs in the field. The presence of a fitness penalty associated with the resistance mechanism can also determine some dynamics of the herbicide resistance phenomenon.

The rapid adoption of herbicide-resistant crops has lead to a high dependence on a small range of herbicide mechanisms of action for weed management while suppressing the introduction of novel herbicide mechanisms of action. This increases the impact of weed resistance to one or two herbicide mechanisms of action can be economically devastating because of the paucity of alternative herbicide choices. Weed resistance to glyphosate in glyphosate-tolerant crops has become particularly problematic in areas of concentrated glyphosate-tolerant crop production. To minimize the spread of herbicide-resistance in weeds, growers will have to emphasize integrated weed management techniques of using cultural weed control, mechanical weed control, and using more than one herbicide mechanism of action to control targeted weed problems.

Of the weedy Amaranths, herbicide resistance has been reported in eleven species (Table 1). The first reported incidence of herbicide resistance in an agronomic crop in North America was in *Amaranthus hybridus* to the triazine herbicide atrazine in 1970 (Ryan 1970). *Amaranthus tuberculatus* biotype has been shown to have multiple resistance across three herbicide sites of action (ALS, PPO, PSII) (Patzoldt, et al., 2005).

Glyphosate- and ALS-resistant *Amaranthus palmeri* and *rudis* are of most concern and potential to disrupt current weed management systems in soybean, maize, and cotton in the United States.

2. PS II resistance

The first case of herbicide resistance in a row crop situation was *A. hybridus* to triazine herbicides in 1970 (Ryan 1970). Currently, triazine-resistant *Amaranthus* infests greater than 500,000 ha in North America. Resistance to photosystem II inhibitors is via target site resistance and enhanced metabolism. Target based resistance in the classical change in the Qb protein. The Qb protein is the site where electron transfer from chlorophyll to an initial electron acceptor, pheophytin, occurs in photosynthetic electron flow. Although many point mutations have been documented in cyanobacteria conferring resistance to triazine

Species	HRAC Group	Ha infested (worldwide)
Amaranthus albus	C1	250
Amaranthus blitoides	C1, B, C1 and B	4500
Amaranthus cruetus	C1	50
Amaranthus hybridus	C1, B	>75,000
Amaranthus lividus	B, C1, D	300
Amaranthus palmeri	B, C1, G, K1, B and G	1,000,000
Amaranthus quitensis	B	830000
Amaranthus retroflexus	B, C1, C2, B and C1, C1 and C2	>70000
Amaranthus rudis	B, C1, E, G, B and C1, B and E and G	>2,000,000
Amaranthus tuberculatus	B	250

Table 1. Herbicide Resistance in Amaranthus worldwide (Heap, 2010).

herbicides; in higher plants, only Ser-264-Gly, Val-219-Ile, and Ans-266-Thr have been documented (Patzoldt et al. 2003). However, the vast majority of triazine-resistance has been the Ser-264-Gly mutation. Other point mutations are less frequent. The Ser-264-Thr mutation confers resistance to triazine and substituted urea herbicides (Masabni & Zandstra, 1999).

In a survey of *A tuberculatus* in Illinois, 14 out of 59 randomly sampled populations were segregating for atrazine resistance, with only one of the 59 populations having site-of-action resistance (Patzoldt et al. 2003). The *A tuberculatus* population with site-of-action resistance, which was used in this study (UniR population), was also identified to have a second, non-site-of-action mediated mechanism. Thus, this novel triazine resistance mechanism may already be prevalent in *A tuberculatus* populations (Patzoldt et al. 2003). Similarly, an atrazine-resistant population of Amaranthus palmeri has been described in Georgia. This population seems to have enhanced glutathione conjugation of atrazine (Vencill 2008) and is not cross-resistant to other triazines such as ametryn.

The rate of CO_2 reduction in the *S*-triazine-resistant biotype of smooth pigweed (*Amaranthus hybridus* L.) was lower at all levels of irradiance than the rate of CO_2 reduction in the susceptible biotype. The intent of this study was to determine whether or not the lower rates of CO_2 reduction are a direct consequence of the same factors which confer triazine resistance. The quantum yield of CO_2 reduction was 23 ± 2% lower in the resistant biotype of pigweed and the resistant biotype of pigweed had about 25% fewer active photosystem II centers on both a chlorophyll and leaf area basis. This quantum inefficiency of the resistant biotype can be accounted for by a decrease in the equilibrium constant between the primary and secondary quinone acceptors of the photosystem II reaction centers that in turn would lead to a higher average level of reduced primary quinone acceptor in the resistant biotype. Thus, the photosystem II quantum inefficiency of the resistant biotype appears to be a direct consequence of those factors responsible for triazine resistance but a caveat to this conclusion is discussed. The effects of the quantum inefficiency of photosystem II on CO_2 reduction should be overcome at high light and therefore cannot account for the lower light-saturated rate of CO_2 reduction in the resistant biotype. Chloroplast lamellar membranes isolated from both triazine-resistant and triazine-susceptible pigweed support equivalent rates of whole chain electron transfer and these rates are sufficient to account for the rate of light-saturated CO_2 reduction. This observation shows that the slower transfer of electrons from the primary to the secondary quinone acceptor of photosystem II, a trait which is characteristic of the resistant biotype, is nevertheless still more rapid than subsequent reactions of photosynthetic CO_2 reduction. Thus, it appears that the lower rate of light-saturated CO_2 reduction of the resistant biotype is not limited by electron transfer capacity and therefore is not a direct consequence of those factors that confer triazine resistance.

3. ALS resistance

Acetolactate synthase (ALS) is the first enzyme in the biosynthetic pathway leading to the synthesis of the branch-chain amino acids isoleucine, leucine, and valine. The branch-chain amino acids comprise part of the amino acid pool essential to protein synthesis and other plant functions. Inhibition of the ALS enzyme results in a cessation of growth followed by purpling of younger foliage then older foliage, as essential proteins cannot be synthesized. There are five chemical classes (sulfonylureas, imidazolinones, pyrimidinylthiobenzoates, triazolopyrimidines, and sulfonylaminocarbonyltriazolinones) that are confirmed to inhibit and ALS and these are used worldwide in numerous weed control situations in row crops and non-cropping situations. ALS resistance in is widespread in eight *Amaranthus* species (see Table 1). Of these, *A. hybridus* a *A. rudis* are the most widespread. There are documented cases of eight point mutations to the ALS gene conferring resistance to ALS-inhibiting herbicides. The Trp-574-Leu seems to be the most common and provides resistance to the greatest range of ALS inhibiting herbicides.
ALS-resistance in *A. rudis* had become so widespread in the midwestern US that ALS-inhibiting herbicides are not recommended (Syngenta press release). One of the reasons that glyphosate-resistant crops were adopted in the mid-1990's in the US so quickly and to such a great extent was because of ALS-resistance in the *Amaranthus* spp.

Point Mutation	Species	Resistance[a]
Ala-122 - Thr	*retroflexus, powelli*	IMI (SCT not tested)
Ala-205-Val	*retroflexus*	IMI (PTB, TP, SCT not tested)
Asp-376-Glu	*hybridus*	All groups
Pro-197-Leu	*retroflexus*	IMI, SU, PTB, TP (SCT not tested)
Pro-197-Ser	*blitoides*	PTB, SU, TP (SCT not tested)
Ser-653-Thr	*A.powelli, retroflexus, rudis*	IMI (PTB, TP, SCT not tested)
Ser-653-Asn	*rudis, hybridus*	IMI (PTB, SCT not tested)
Trp-574-Leu	*A.rudis, blitoides, retroflexus, powelli*	IMI, SU, PTB, TP, SCT

[a]IMI = imidazolinone, SU = sulfonylurea, PTB = pyrimidinylthiobenzoates,
TP = triazolopyrimidine, SCT = sulfonylaminocarbonyltriazolinone.

Table 2. Point mutations leading to ALS-resistance in Amaranthus spp. (Tranel et al. 2007)

There are no reported cases in Amaranthus where the ALS-resistance trait has lead to reductions in ecological fitness. *A. retroflexus* and *A. blitoides* were specifically examined and none were found.

4. PPO resistance

Amaranthus tuberculatus is only one of three species worldwide to develop resistant to PPO-inhibiting herbicides. Evaluation of a PPO-inhibitor-resistant *A. tuberculatus* biotype revealed that resistance was a (incompletely) dominant trait conferred by a single, nuclear gene. In plants, chlorophyll synthesis occurs exclusively in the plastids, while heme synthesis occurs in the plastids and mitochondria (Patzoldt et al. 2005). There are two nuclear genes to encode

PPO isozymes in the plastid and mitochondria. These are called PPX1 and PPX2 for the plsatid and mitochondria, respectively. Protogen IX accumulates in sensitive plants treated with PPO inhibitors. Protogen IX exported to the cytoplasm is converted to proto IX that in the presence of light causes the formation of singlet oxygen that results in membrane damage and eventual plant death. One gene from the resistant biotype, designated PPX2L, contained a codon deletion (G210) (Patzoldt et al 2005). PPX2L is predicted to encode both plastid- and mitochondria-targeted PPO isoforms, allowing a mutation in a single gene to confer resistance to two herbicide target sites. Resistant biotypes of *A. tuberculatus* have robust resistant to most PPO-inhibiting herbicides (lactofen, sulfentrazone, flumioxazin). Deletion of a codon rather than substitution is a unique formation of target site resistance to herbicides. There have been no studies to determine if there is a fitness costs to PPO resistance in weeds.

5. Glyphosate resistance

Glyphosate-resistance was first confirmed in *Lolium rigidum* in 1996 from Australia (Heap 2010). There are nineteen biotypes of weeds that have confirmed glyphosate-resistance worldwide. The most widespread resistance in from *Conyza canadensis*, first cofirmed in Delaware in 2001. It is estimated to infest more than three million hectares in the US alone. The first reported case of glyphosate-resistance in an in-season row crop was in *Amaranthus palmeri* in 2005. Currently, glyphosate-resistance has been confirmed in *A. palmeri, A. rudis,* and *A. tuberculatus.* Culpepper et al (2006) showed that the mechanism of resistance differs from that described in *Conyza candadensis* and *Lolium spp.* Glyphosate-resistance in a Amaranthus palmeri is due to increased EPSPS expression (Gaines et al 2010). While increased expression of EPSPS as a molecular glyphosate resistance mechanism has been reported to endow relatively low level glyphosate resistance in lab studies, this is the first report in a field weed population. It is likely that glyphosate selection pressure over several years in the Georgia cotton field (3) either selected plants with previously existing *EPSPS* gene amplification, or *EPSPS* gene amplification occurred during a period of less than seven years over which glyphosate was repeatedly applied. If we examine glyphosate-resistant *Amaranthus palmeri,* we see at least two mechanisms of resistance (reduced translocation and a target site change) and perhaps biotypes with both types of resistance as well as individuals that are resistant to glyphosate and ALS-inhibitors. Other collections of Palmer amaranth that seem to have very low levels (<2 x) of glyphosate resistance that have been difficult to characterize may be a third type of resistance. Before weed scientists can effectively manage glyphosate-resistant Palmer amaranth as well as other glyphosate-resistant weed species, we will need to better characterize at the genetic level whether individual plants are resistant via translocation mechanism, target site, combinations of these, and whether they are resistant to other herbicide mechanisms of action. Sammons et al. (2007) suggest that there are three primary mechanisms which confer herbicide selectivity among plants: 1) differences in herbicide target sites, 2) inactivation of an herbicide by chemical modification (i.e. metabolism), and 3) exclusion mechanisms which either reduce herbicide uptake or sequester the herbicide away from the target site. To clarify the exclusion mechanism, Ge et al. (2010) reports that glyphosate-resistant *Conyza canadensis* actively transports glyphosate to the vacuoles of the cell compared to the cytoplasm preventing it from getting to the target site.

Greenhouse data indicate that the glyphosate-resistant *A. palmeri* may have a fitness cost. The GS biotype grew at an 11% faster rate than the GR biotype, and the GR biotype assimilated carbon at 60.2% the rate that the GS biotype assimilated carbon. Measurements of photosystem I activity, chlorophyll content, and branching help to characterize the GR biotype of Palmer amaranth, and suggest a mechanism of resistance different from that of *Conyza canadensis* and some other confirmed glyphosate-resistant weed biotypes, but did not correlate with relative fitness differences.

Glyphosate resistance has been particularly troublesome in the central U. S. including the states of Illinois, Missouri, Arkansas, and Tennessee. Glyphosate-resistant horseweed was first discovered in Delaware (van Gressel, 2001), but quickly spread to Indiana (Davis et al., 2007, Davis et al. 2008), Tennessee (Steckel & Gwathmey, 2009), and Arkansas. Glyphosate resistant horseweed increased the cost of weed management by about $13/acre (Mueller et al. 2005). While troublesome, glyphosate-resistant horseweed is primarily a problem at pre-plant before crop establishment. The emergence of glyphosate-resistant Palmer amaranth (*Amaranthus palmeri*) and water hemp (*Amaranthus rudis, A. tuberculatus*) have caused severe and well-documented management problems for in-season weed management in cotton and soybeans (Culpepper et al. 2008; Legleiter et al. 2008, Legleiter et al., 2009; Norsworthy et al., 2008a; Norsworthy et al. 2008b; Patzoldt et al. 2002; Patzoldt et al. 2005; Steckel & Sprague, 2004a; Steckel & Sprague, 2004b; Steckel et al. 2007; Steckel et al., 2008; Volenberg et al. 2007).

6. Mitotic inhibitor resistance

There are a number of herbicide classes that inhibit mitosis via disruption of microtubule formation. These include dinitroaniline herbicides such as trifluralin, pendimethalin, and ethalfluralin as well as some pyridine, carbamate, and phosphoroamidate herbicides. Microtubules are an integral part of mitosis as well as other cellular process such as cytokinesis and vesicular transport (Powles and Yu, 2010). These herbicides bind to one of the α- and β-tubulin dimers. Sensitive plants symptoms include malformed root areas that come in contact with the herbicide. Resistance occurs through a Thr-239_Ile substitution in the α-tubulin gene resulting in reduced binding of the herbicide. Resistance to mitotic inhibiting herbicides is not widespread with evolved resistance reported in 10 species worldwide (Heap, 2010). Resistance has been reported in South Carolina in *Amaranthus palmeri* in 1994 (Heap 2010) and a population was found with resistance in Georgia in 2010 (Vencill, personal communication).

7. HPPD-inhibitor resistance

Three classes of chemistry (triketones, isoxazoles, and callistemones) are bleaching herbicides that inhibit 4-hydroxyphenyl pyruvate dioxygenase (HPPD), a key enzyme required for the formation of carotenoids. The inhibition of carotenoid synthesis by the inhibition of the HPPD enzyme leads to white foliage because the carotenoid pigments protect chlorophyll pigment in plant tissues. Carotenoid synthesis can be inhibited by two other herbicide mechanisms of action, the inhibition of phytoene desaturase (e.g. norflurazon and fluridone) and the inhibition of deoxyxylulose 5-phosphate synthase (DXP) by clomazone. Resistance has been confirmed for all three bleaching herbicide mechanisms of action. Fluridone (phytoene desaturase inhibition) resistance is widespread in hydrilla in Florida and clomazone-resistant barnyard grass is reported in rice production in Arkansas and Louisiana. Resistance has been reported in a population of Amaranthus rudis in Illinois (Ag News, 19 July 2010). The mechanism of resistance is not understood, but resistance seems limited to foliar applications of HPPD-inhibiting herbicides while soil applications of the same herbicides seem to still provide control.

8. Muliple resistance

In the United States, the only documented case of resistance to multiple herbicide mechanisms of action has been in the Amaranthaceae. Cases of multiple resistant to ALS and PSII as well as ALS and glyphosate. Describe the ALS/PSII. In Georgia, there are biotypes resistant to ALS and glyphosate but little is known about specifics.

There are several populations of *A. tuberculatus* that have evolved multiple herbicide resistances. An Illinois biotype has resistance to PSII, ALS, and PPO inhibitors while a population from Missouri has evolved resistance to ALS, PPO, and EPSPS inhibitors (Patzoldt et al. 2003 According to Mueller (2005), there are >150000 ha of PSII/PPO/ALS-resistant common waterhemp in Illinois. In Georgia, populations of *Amaranthus palmeri* have been documented to be resistant to ALS and EPSP inhibitors. There are populations of A. palmeri that are reported to be resistant to mitotic inhbitors, ALS, and EPSP inhibitors.

9. Conclusion

In Europe, *Alopecurus* has been documented to a weed of serious agronomic potential to have evolved widespread resistance to commonly used herbicides and to multiple mechanisms of action in some cases (Delye 2005). In Australia, the niche is occupied by *Lolium* where resistance is documented to several groups of herbicides (Neve et al. 2004). In the United States, *Amaranthus* has long been one of the most common and troublsome weeds in agronomic crops and has been of the first weeds to develop resistance to herbicides in many situations. They were the first weeds to develop resistance to triazine herbicides, ALS-resistance in A. tubercualtus was widespread in the mid-1990's before the introduction of glyphosate-resistant crops, and glyphosate-resistance has been found in three species of *Amaranthus* and is growing rapidly. PPO-inhibiting herbicides have become the standard recommendation for glyphosate-resistant *Amaranthus* spp. However, we now see PPO-resistant *A. tuberculatus*. There are unconfirmed reports of resistance in A. palmeri in the southeastern US. The first case of multiple herbicide resistance in the US was in *Amaranthus tuberculatus* and *palmeri.*

In the past, herbicide resistance in *Amaranthus* caused growers to shift to another herbicide mechanism of action. There has only been one new herbicide mechanism of action introduced since 1990 so we are to a crisis point where growers may not have another herbicide mechanism of action to go to when resistance to PSII, ALS, PPO, and EPSPS inhibitors become more widespread in one of our most common and troublesome weed species. Without the introduction of new herbicide mechanisms of action or better herbicide-resistance management, a technology that has allowed tremendous increases in agricultural productivity is at risk.

10. References

Appleby, A.P. 2005. A history of weed control in the United States and Canada - a sequel. Weed Sci. 53:762-768.

Bradshaw, L.D., S.R. Padgette, S.L. Kimball, and B.H. Wells. 1997. Pesrpectives on glyphosate resistance. Weed Technol. 11:189-198.

Brookes, G. and P. Barfoot. 2009. GM crops - global socio-economic and environmental impacts 1996-2007. PG Economics Ltd., Dorchester, UK.

Culpepper, A.S., J.R. Whitaker, A.W. McRae, and A.C. York. 2008. Distribution of glyphosate-resistant Palmer amaranth (Amaranthus palmeri) in Georgia and North Carolina during 2005 and 2006. J. Cotton Sci. 12:306-312.

Culpepper, A.S., T.L. Grey, W.K. Vencill, J.M. Kichler, T.M. Webster, S.M. Brown, A.C. York, J.W. Davis, and W.W. Hanna. 2006. Glyphosate-resistant Palmer amaranth (*Amaranthus palmeri*) confirmed in Georgia. Weed Sci. 54:620-626.

Davis, V.M., K.D. Gibson, T.T. Bauman, S.C. Weller, and W.G. Johnson. 2007. Influence of weed management practices and crop rotation on glyphosate-resistant horseweed population dynamics and crop yield. Weed Sci. 55:508-516.

Davis, V.M., K.D. Gibson, and W.G. Johnson. 2008. A field survey to determine frequency of glyphosate-resistant horseweed (*Conyza canadensis*) in Indiana. Weed Technol. 22:331-338.

Délye, C. 2005. Weed resistance to acetyl-coenzyme A carboxylase inhibitors: an update. Weed Sci. 53:728-746.

Dill, G.M., C.A. Jacob, and S.R. Padgette. 2005. Glyphosate-resistant crops, history, status, and future. Pest Mgmt. Sci. 61:219-224.

Dill, G.M., C.A. Jacob, and S.R. Padgette. 2008. Glyphosate-resistant crops: adoption, use, and future considerations. Pest Mgmt. Sci. 64:326-331.

Gaines, T.A., W. Zhang, D. Wang, B. Bukun, S.T. Chisholm, D.L. Shaner, S.J. Nissen, W.L. Patzoldt, P.J. Tranel, A.S. Culpepper, T.L. Grey, T.M. Webster, W.K. Vencill, R.D. Sammons, J. Jiang, C. Preston, J.E. Leach, and P. Westra. 2010. Gene amplification confers glyphosate resistance in *Amaranthus palmeri*. Proc. National Acad. Sci. 107:955-956.

Ge, X., D. d'Avignon, J.H. Ackerman, and R.D. Sammons. 2010. Rapid vacuolar sequestration: the horseweed glyphosate resistance mechanism. Pest Mgmt. Sci. 66:345-348.

Gianessi, L.P. and N.P. Reigner. 2007. The value of herbicides in U.S. crop production. Weed Technol. 21:559-566.

Haider, J.B., W.K. Vencill, A.S. Culpepper, and T.L. Grey. 2007. Physiological aspects of glyphosate-resistant Palmer amaranth (*Amaranthus palmeri*). Proc. South. Weed Sci. 60:180.

Heap, I.M. 2010. International Survey of Herbicide Resistant Weeds. www.weedscience.com.

Ligleiter, T.R. and K.W. Bradley. 2008. Glyphosate and multiple herbicide resistance in waterhemp (*Amaranthus rudis*) populations from Missouri. Weed Sci. 56:582-587.

Ligleiter, T.R., K.W. Bradley, and R.E. Massey. 2009. Glyphosate-resistant waterhemp (*Amaranthus rudis*) control and economic returns with herbicide programs in soybean. Weed Technol. 23:54-61.

Norsworthy, J.K., G.M. Griffith, R.C. Scott, K.L. Smith, and L.R. Oliver. 2008a. Confirmation and control of glyphosate-resistant Palmer amaranth (*Amaranthus palmeri*) in Arkansas. Weed Technol. 22:108-113.

Norsworthy, J. K., R. C. Scott, K. L. Smith, and L. R. Oliver. 2008b. Response of northeastern Arkansas Palmer amaranth (*Amaranthus palmeri*) accession to glyphosate. Weed Technology 22:408-413.

Massabni, J.G. and B.H. Zandstra. 1999. A serine to threonine mutation in linuron-resistant *Portulaca oleracea*. Weed Sci. 47:393-400.

Mueller, T. M., P.D. Mitchell, B.G. Young, and A.S. Culpepper. 2005. Proactive versus reactive management of glyphosate-resistant or –tolerant weeds. Weed Technol. 19:924-933.

Neve, P., J. Sadler, and S. B. Powles. 2004. Multiple herbicide resistance in a glyphosate-resistant rigid ryegrass (*Lolium rigidum*) population. Weed Sci. 52:920-928.

Patzoldt, W.L., B.S. Dixon, and P.J. Tranel. 2003. Triazine resistance in *Amaranthus tuberculatus* (Moq) Sauer that is not site-of-action mediated. Pest Mgmt. Sci. 59:1134-1142.

Patzoldt, W.L., P. J. Tranel, and A. G. Hager. 2005. A waterhemp (*Amaranthus tuberculatus*) biotype with multiple resistanceacross three herbicide sites of action. Weed Sci. 53:30-36

Patzoldt, W.L., A. G. Hager, J. S. McCormick, and P. J. Tranel. 2005. A codon deletion confers resistance to herbicides inhibiting protoporphyrinogen oxidase. Proc. Natl. Acad. Sci. 103:12329-12334.

Powles, S.B. and Q. Yu. 2010. Evolution in Action: Plants Resistant to Herbicides. Annu. Rev. Plant Biol. 61:317-347.

Rapp, R.A. and J.F. Wendell. 2005. Epigenetics and plant evolution. New Phytol. 168:81-91.

Radosevich, S.R., J. Holt, and C. Ghersa. 1997. Weed Ecology: Implications for Management, 2nd Ed. John Wiley & Sons, New York. 589 pp.

Ryan, G.F. 1970. Resistance of common groundsel to simazine and atrazine. Weed Sci. 18:614-616.

Sammons, R.D., D.C. Heering, N. Dinicola, H. Glick, and G.A. Elmore. 2007. Sustainability and stewardship of glyphosate and glyphosate-resistant crops. Weed Technol. 21:347-354.

Senseman, S. 2007. Herbicide Handbook of the Weed Science Society of America. Weed Science Society of America, Champaign, IL, 458 pp.

Sibony, M. and B. Rubin. 2003. The ecological fitness of ALS-resistant *Amaranthus retroflexus* and multiple-resistant *Amaranthus blitoides*. Weed Res. 43:40-47.

Steckel, L.E. and C. O. Gwatheny. 2009. Glyphosate-resistant horseweed (*Conyza canadensis*) growth, seed production, and interference in cotton. Weed Sci. 57:346-350.

Steckel, L. E. and C.L. Sprague. 2004a. Common waterhemp (*Amaranthus rudis*) interference in corn (*Zea mays*). Weed Sci. 52:359-364.

Steckel, L.E. and C.L. Sprague. 2004b. Common waterhemp (*Amaranthus rudis*) interference in soybean (*Glycine max*). Weed Technol. 18:947-952.

Steckel, L.E., C.L. Main, A.T. Ellis, and T.C. Mueller. 2008. Palmer amaranth (*Amaranthus palmeri*) in Tennessee has low level glyphosate resistance. Weed Technol. 22:119-123.

Tranel, P.J., T.R. Wright, and I.M. Heap. 2007. ALS-mutations from herbicide-resistant weeds. Online. Internet. September 25, 2007. Available http://www.weedscience.com .

Tranel, P.J., J.J. Wassom, , M.R. Jeschke, and A.L. Rayburn. 2002. Transmission of herbicide resistance from a monoecious to a dioecious weedy Amaranthus species. Theor. Appl. Genet. 105: 674-679.

Trucco, F. T., M. R. Jeschke, A. L. Rayburn, and P. J. Tranel. 2005. Promiscuity in weedy amaranths: high frquency of female tall waterhemp (*Amaranthus tuberculatus*) x smooth pigweed (*A. hybridus*) hybridization under field conditions. Weed Sci. 53:46-54.

USDA-NASS (U.D. Department of Agriculture - National Agricultural Statistics Service). 2009. Acreage. June 30. Cr Pr 2-5 (6-09) Washington, DC. Available online at: http://usda.mannlib.cornell.edu/usda/current/Acre/Acre-06-30-2009.pdf .

Van Gessel, M. 2001. Glyphosate-resistant horseweed from Delaware. Weed Sci. 49:703-705.

Vencill, W.K., E. Prostko, P.J. Tranel, and J. Kichler. 2008. Mechanism of triazine resistance in *Amaranthus palmeri* and *A. rudis*. 2009 Proc. South. Weed Sci. Soc. 62:262.

Volenberg, D.S., W.L. Patzoldt, A.G. Hager, and P.J. Tranel. 2007. Response of contemporary and historical waterhemp (*Amaranthus tuberculatus*) accessions to glyphosate. Weed Technol. 55:327-333.

Wassom, J.J. and P.J. Tranel. 2005. Amplified fragment length polymorphism-based genetic relationships among weedy *Amaranthus* species. J. Hered. 96:410-416.

Weed Science Society of America. 1998. http://www.weedscience.net.

30

Brief Approach of Herbicide Resistance in Context of Crops Development Worldwide

Ioan Gh. OROIAN
University of Agricultural Sciences and Veterinary Medicine Cluj-Napoca
Romania

1. Introduction

The review aims to emphasize the state of art of herbicide resistance mechanisms in weeds, when ourdays realities confront us with crop productions in continuous expansion in EU and worldwide. Regarding sustainable agriculture and plant biodiversity maintenance as premise for successful development in the future, the virtual environmental threat produced by a large diversity of herbicides in use appears as entitled concern. This reality is exacerbated by another phenomenon: the development of herbicide resistant weeds reported more and more often in last four decades. In this way, even herbicides still represent the most important and effective tool in controlling weeds, in many areas, their field value decreased. The alteration produced, often as punctiform mutations, in weed genotypes in order to counteract the herbicide attacks, led to development of resistant weed biotypes that may produce huge harmful environmental effects, besides yield loss. One of the most important reasons of this phenomenon is represented by administering the same herbicide for many years to the same culture in the same cultivation field. The situation is aggravated when the herbicide has only one site of action in herbicide genome (Sigematsu et al., 1989; Kremer et al., 1999; Owen et al., 2005; Powels et al., 2010). This underlines the importance of obtaining a better understanding of the mechanism that represents the basis of resistance, in order to successful fighting against this phenomenon. It was elucidated by emphasizing the biochemical processes and genetic mechanisms that occur in plant in order to develop defending mechanisms. The genetic processes involved in weed genome react to environmental stimuli, as the attack of active principle from herbicides. Usually, the gene/genes that metabolize the active substance/substances from the herbicide are naturally multiplied of several hundred times, compared to the genotype of a plant that is not resistant. Now, gene amplification is one of the well known mechanisms of resistance against herbicides (Vila-Aiub et al., 2003; Oroian, I., 2008; Ge et al., 2010).
These mechanisms once understood can be counteracted by suitable measures taken at different levels. Some of the most important crop management strategies that could contribute to reduce the appearance and/or decrease of the frequency of this phenomenon consist in: using herbicides only when necessary, rotate crops and herbicides site of action, using combined herbicide administration system, combine mechanical weed control with chemical treatment (Paoletti et al., 1995; Berca M., 2007).

2. The development of crop production in EU and world

The production of the **most important cultures** according to the importance of their place in human society is presented, in order to underline the importance of the weed resistance control in these plants. The cultivated area in both EU and the world is also emphasized.

The wheat (Kingd. *Plantae*, Ord. *Poales*, Fam. Poaceae; ***Triticum spp.***) orginary from the Near East is the main human food crop. According to FAO last updating (December 2008), the greatest production was obtained in 2008 in EU and the same orientation was reported by entire world (Fig. 1.a). The area harvested was bigger in 2008 compared to former year, but lower compared to year 2000, and 2002 when the biggest area was wheat cultivated in EU, while in 2008 was recorded the biggest area cultivated worldwide (Table 1).

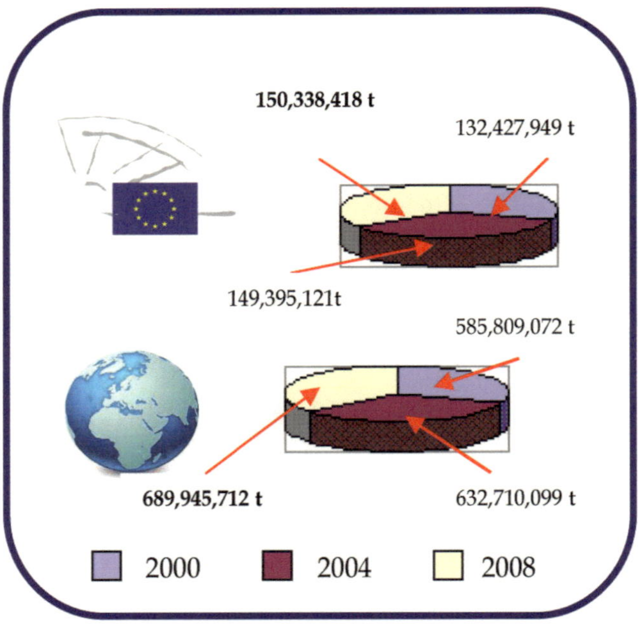

Fig. 1. a. The wheat production in EU and in the world in three reference years of the last decade

The barley (Kingd. *Plantae*, Ord. *Poales*, Fam. Poaceae; ***Hordeum vulgare***) was cultivated from the second millennium b.Ch. in Mesopotamia. It is a cereal grain of major importance as animal fodder. According to FAO (December 2008), the greatest production was obtained in 2008 in EU and the same orientation was reported by entire world (Fig. 1.b). The area harvested was bigger in 2008 compared to 2000, while in EU increased in 2004 compared to 2000, and the same tendency was recorded worldwide, since 2004 when the biggest value was recorded (Table 1).

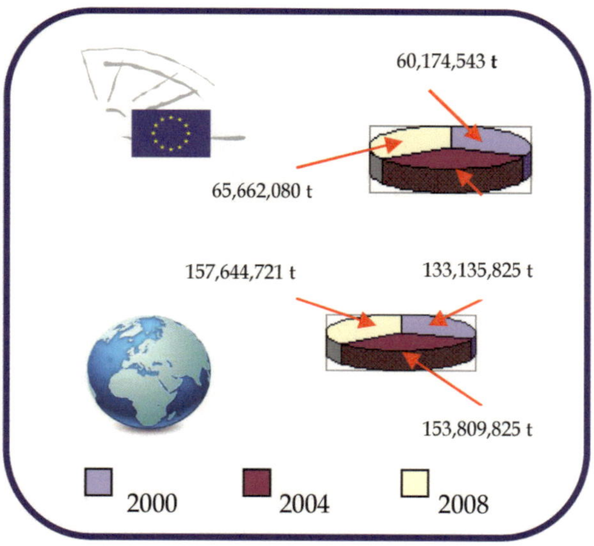

Fig. 1.b. The barley production in EU and in the world in three reference years of the last decade

The oats (Kingd. *Plantae*, Ord. *Poales*, Fam. Poaceae; ***Avena sativa***), similarly to barley, has the most common use as livestock fodder. According to FAO last updating (December 2008), the greatest production was obtained in 2008 in EU. In 2001 was obtained the biggest barley production by entire world, while in 2008 was bigger compared to 2000, but still smaller than in 2001 (Fig. 1.c). The area harvested was bigger in 2001 compared to 2000 and 2008 worldwide, while in EU increased in 2001 compared to 2000, but in 2008 decreased by 10.64% compared to 2000 and by 13.65% compared to 2001 (Table 1).

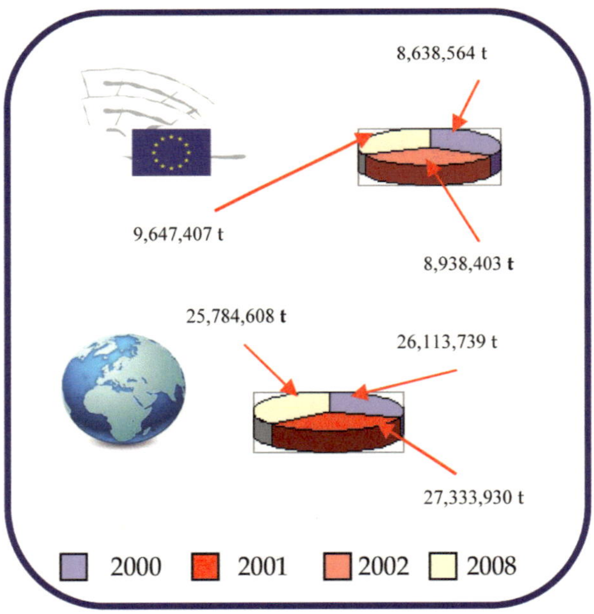

Fig. 1.c. The oats production in EU and in the world in three reference years of the last decade

The maize (Kingd. *Plantae,* Ord. *Poales,* Fam. Poaceae; **Zea mays**) is a known since prehistoric times, but now is spread all over the world due to its large production potential and ability to grow in diverse climates. According to FAO (December 2008), the greatest production was obtained in 2004 in EU and in 2008 worldwide (Fig. 1.d). The area harvested was 4.93% smaller in 2008 compared to 2000 in EU, while worldwide it increased by 17.53% in 2008 compared to 2000 (Table 1).

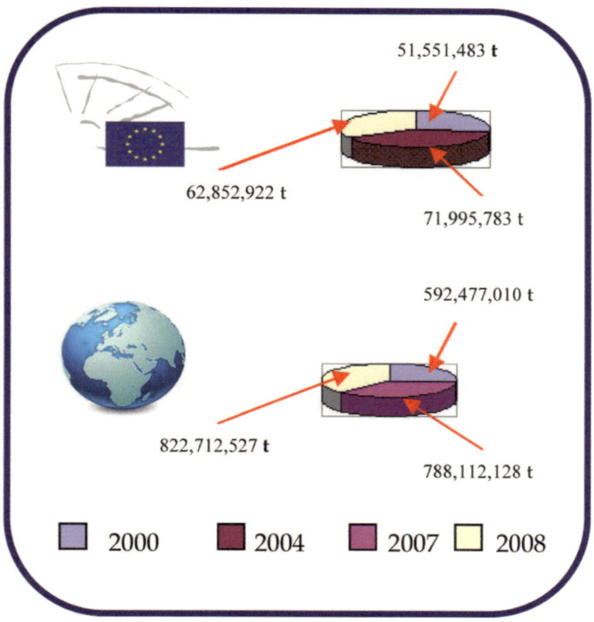

Fig. 1.d. The maize production in EU and in the world in three reference years of the last decade

The soya bean (Kingd. *Plantae*, Ord. *Fabales*, Fam. Fabaceae; **Glycine max**.) native to Asia, is an annual species of legume, rich source of oil, carbohydrates and protein. It is also considered one of the most versatile crops. According to FAO last updating (December 2008), the greatest production was obtained in 2001 in EU, and worldwide in 2008 (Fig. 1.e). The smallest production (almost one half of that obtained in 2001) was obtained in 2008 in EU. The area harvested was 52.89% smaller in 2008 compared to 2000 in EU, while worldwide, it increased by 30.26% in 2008 compared to 2000 (Table 1).

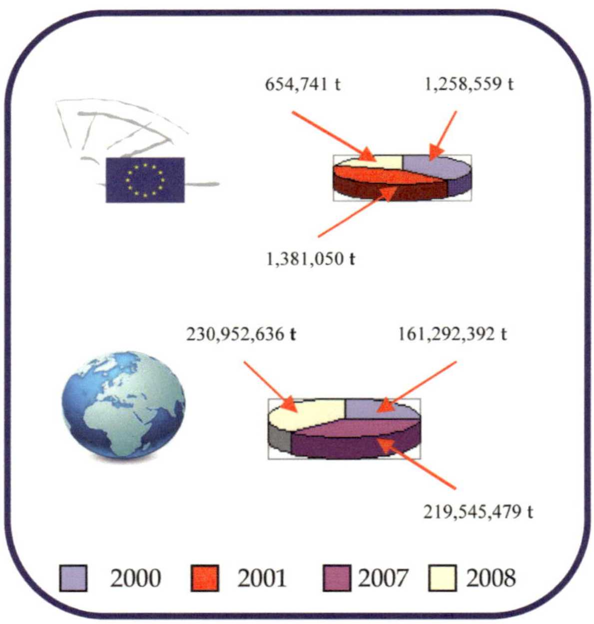

Fig. 1. e. The soya bean production in EU and in the world in three reference years of the last decade

The sunflower (Kingd. *Plantae*, Ord. *Asterales*, Fam. Asteraceae; ***Helianthus***) is orginary from Central America. It is of great importance in food industry. According to FAO (December 2008), the greatest production was obtained in 2008 in EU and the same orientation was reported by entire world (Fig. 1.b). The area harvested was bigger in 2008 compared to 2000 worldwide. It increased by 18.47%. In EU, the area cultivated increased by 1.31% in 2008 compared to 2000, but decreased by 11.87% compared to 2003, when the biggest area was cultivated during the analyzed interval, 2000 - 2001, respectively (Table 1).

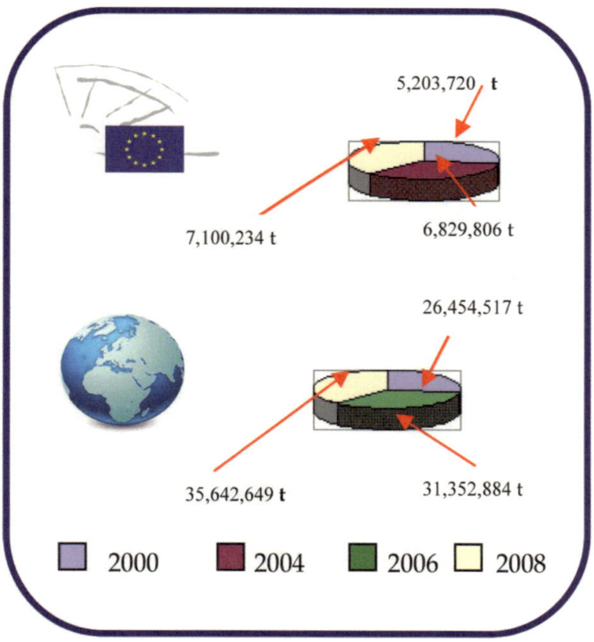

Fig. 1. f. The sunflower production in EU and in the world in three reference years of the last decade

The rapeseed (Kingd. *Plantae*, Ord. *Brassicales*, Fam. Brassicaceae; ***Brassica napus***) is a valuable source of animal feed and vegetable oil for human consumption and biodiesel. Certain varieties of oilseed rape are commonly named canola. According to FAO last updating (December 2008), the greatest production was obtained in 2008 in EU and the same orientation was reported by entire world (Fig. 1.a). The area harvested was 17,31% bigger in 2008 compared to 2000 worldwide, while in EU it increased by 48,01% in 2008 compared to 2000 (Table 1).

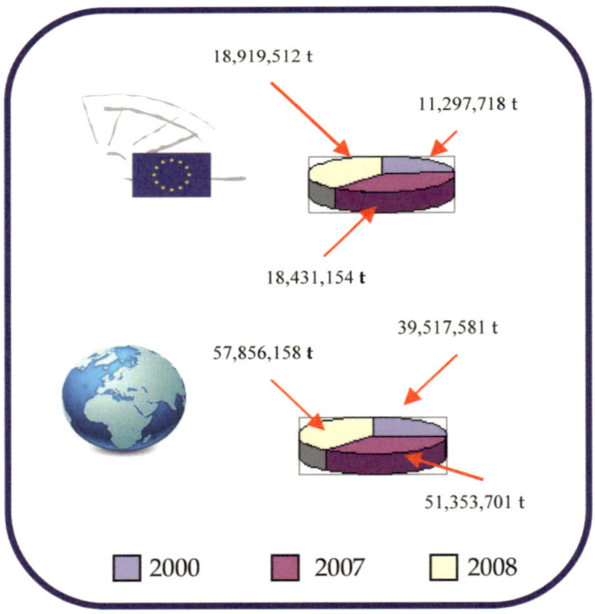

Fig. 1. g. The rapeseed production in EU and in the world in three reference years of the last decade

Vineyard (Kingd. *Plantae*, Ord. *Vitales*, Fam. Vitaceae; ***Vitis vinifera***). The wine producing is known since millenniums b.Ch According to FAO (December 2008), the greatest production was obtained in 2008 in EU (by 36.44% compared to production obtained in 2000) and the same orientation was reported by entire world, by 34.74% bigger in 2008 compared to the production obtained in 2000, respectively (Fig. 1.b). The area harvested was 9.34% bigger in 2008 compared to 2000 in EU, while it increased by 0,91% in 2008 compared to 2000, worldwide (Table 1).

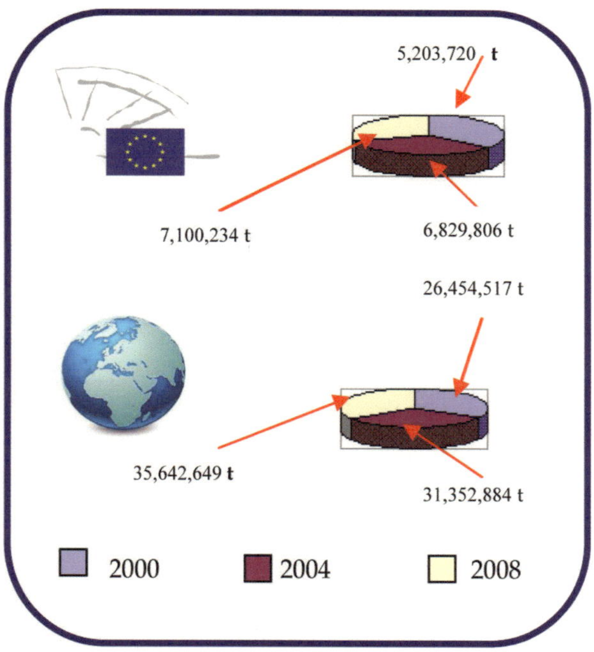

Fig. 1. h. The wheat production in EU and in the world in three reference years of the last decade

Vegetables include a series of plants (potatoes, eggplants, tomatoes, cucumbers, cabbages, onions, garlic, peas, cauliflowers, spinach, pumpkins, etc.) often intensively managed and with high input costs. They are important both agricultural economy of countries, food industry, and food consumption, being essential to a healthy diet. In 2005, according to FAO, there were 66 million Ha of vegetables grown in the world. This figure includes of potatoes (19 million Ha) but excludes nearly 5 million Ha of watermelons and other melons which are sometimes included with vegetables. If we include melons, the biggest production was obtained in 2005 in EU (Fig. 1.i.).

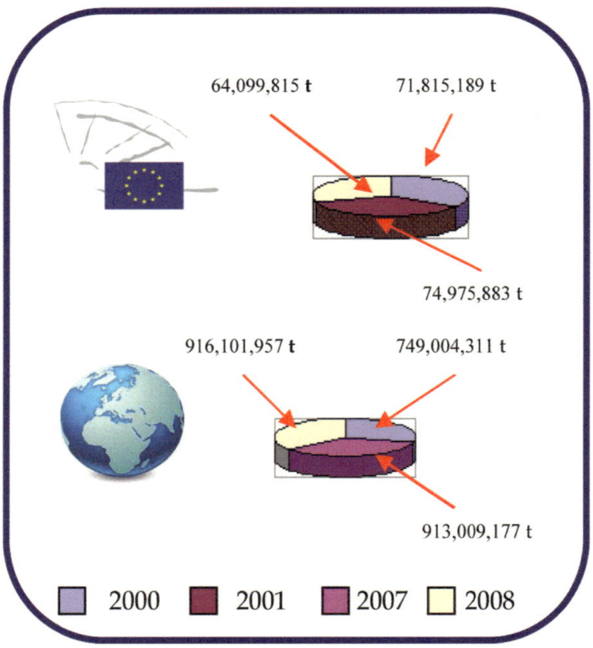

Fig. 1. i. The vegetables production in EU and in the world in three reference years of the last decade

Issue/Year		2000	2001	2002	2003	2004	2005	2006	2007	2008
Wheat	1)	26560780	26412836	26888507	24329061	26597861	26441014	24920478	24859885	26503083
	2)	215472863	214597539	213806041	207658164	216894341	219736189	211815280	213894296	223564097
Maize	1)	9337537	9618975	9262990	9737571	10059931	8991942	8553137	8028447	8877660
	2)	136999105	137485945	137293647	144673388	147472375	147441686	148607870	158606742	161016542
Barley	1)	14202011	14330156	14269355	14027708	13704760	13824772	13758706	13685865	14473752
	2)	54520176	56173845	55272435	57730908	57535566	55342779	56361192	55430143	56774297
Oats	1)	3063167	3055184	3240881	3178923	2912364	2879314	2925499	2966607	2997924
	2)	12682536	13124757	12452273	12141204	11696607	11269443	11641613	11583196	11333331
Soya beans	1)	501604	452795	354185	422878	386726	419319	487491	342772	236317
	2)	74366760	76799819	78962290	83663393	91602610	92506171	95248048	90111139	96870395
Sunflower	1)	3693610	3484659	3473261	4245848	3716073	3599920	3922551	3270569	3742142
	2)	21120631	17811931	19448548	23417226	21436622	23025310	23790326	21172596	25023511
Rapeseed	1)	4139854	4172497	4246639	4161306	4557222	4867009	5399346	6528779	6127566
	2)	25835280	22560773	22914090	23468954	25316569	27693540	27440505	29673836	30308662
Grapes	1)	3920220	3876898	3871812	3860016	3737869	3716264	3667345	3628630	3665097
	2)	7341221	7406923	7438751	7497740	7343476	7345439	7389298	7349718	7408127
Vegetables	1)	3031888	2902895	2848534	2940864	2927856	2826443	2627700	2531619	2486497
	2)	45074440	46808658	47805184	50200877	49881448	50992160	52234017	53042269	54091889

1) European Union, 2) World

Table 1. Area Harvested (Ha) in European Union and in the world

Worldwide, the biggest production was obtained in 2008. The cultivated area decreased by 18% in 2008 compared to 2000, while worldwide it increased by 20% in 2008 compared to 2000 (Table 1).
Discrepancies are recorded in both production and cultivated areas in EU and worldwide. The production (tonnes) of some cultures dramatically decreased in EU in 2008 compared to previous years, but it considerably increased worldwide during the same time interval. The same tendency is observed concerning the area cultivated (Ha).
The oats production increased in 2008 compared to 2000 in EU, while de production decreased I worldwide during the same time interval. The maize production decreased in 2008 compared to 2004, even it is bigger compared to 2000. Worldwide, it is bigger in 2008 compared to 2000. The soy bean production dramatically decreased in 2008 compared to 2000 in EU, while it increased in 2008 compared to 2000 worldwide. The sunflower production increased in 2008 compared to 2000, but it is smaller compared to production obtained in 2004, when the biggest production of the analyzed time interval was recorded. In the same culture, the production increased in 2008 compared to 2000 worldwide. The vegetable production, including melons, decreased in 2008 compared to 2000 in EU, while it increased during the same time interval (2000 - 2008) worldwide (Fig. 1).
Differences in tendencies concerning area cultivated with analyzed crops were also recorded between EU and world during the time interval 2000 - 2008. Smaller area was cultivated with oats, maize, soya beans and vegetables in 2008 compared to 2000 in EU, while bigger field areas were cultivated with the same cultures in 2008 compared to 2000 worldwide (Table 1).

3. Common herbicides in use that confront weed resistance

A series of large scale used herbicides in use confront with weed herbicide resistance phenomenon (Practical Guide for Farmers, Alcedo 2010). They are commonly administered to most important crops, in fight against a large variety of annual and/or perennial dicotyledonous and monocotyledonous weeds. The resistance developed against these herbicides is based on the activity of specific enzymatic systems (Table 2).

Commercial name of herbicide and producer	Active principle/s	Enzymatic system developing resistance/target class of weeds	Chemical formula
		Wheat	
BUCTRIL UNIVERSAL (bromoxynil 280 g/l + acid 2,4-D 280 g/l) Bayer Cropscience Ag., Germany	Bromoxynil 3,5-dibromo-4-hydroxybenzonitrile MCPA (4-Chloro-*o*-tolyloxy-acetic acid)	Photosystem II/ Annual and perennial dicotyledonous weeds	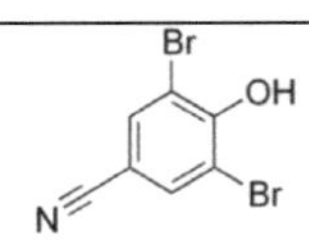 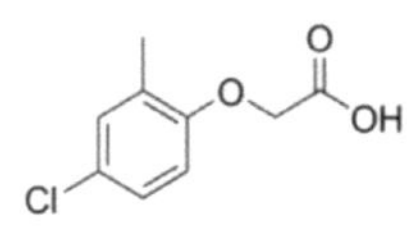

Table 2. Herbicides in use and enzymatic system of weeds susceptible to develop resistance

Commercial name of herbicide and producer	Active principle/s	Enzymatic system developing resistance/target class of weeds	Chemical formula
		Maize	
BROMOTRIL 40 EC (bromoxynil 400 g/l) Makhteshim Agan, Israel	Bromoxynil 3,5-dibromo-4-hydroxybenzonitrile	Photosystem II/ Annual dicotyledonous weeds	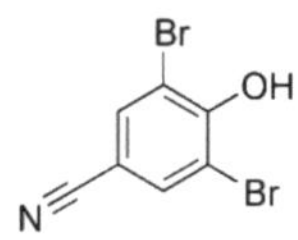
BUCTRIL UNIVERSAL			
CAMBIO (bentazone 320 g/l + dicamba 90 g/l) BASF SE, Germany	Bentazone 2,2-dioxo-3-propan-2-yl-1H-benzo[1,2,6] thiadiazin-4-one Dicamba 3,6-dichloro-2-methoxybenzoic acid	Photosystem II/ Annual dicotyledonous weeds	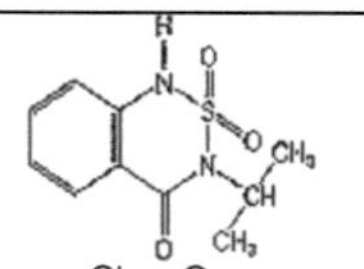 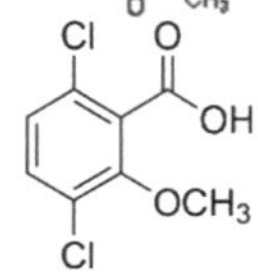
SURDONE (metribuzin 70%) Feinchemie Schwebda GmbH, Germany	Surdone, 4-amino-6-*tert*-butyl-4,5-dihydro-3-methylthio-1,2,4-triazin-5-one	Photosystem II/ Annual dicotyledonous and some mono-cotyledonous weeds	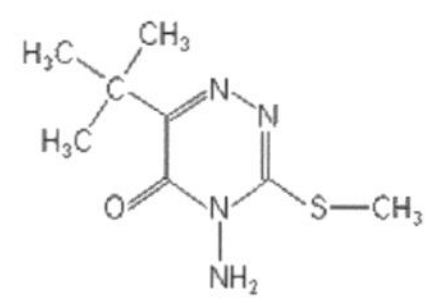
ZEAGRAN 340 SE (bromoxinil 90 g/l + terbutilazine 250 g/l) Nufarm GmbH & Co. Kg, Austria	Bromoxynil 3,5-dibromo-4-hydroxybenzonitrile Terbutilazine *N*2-tert-butyl-6-chloro-*N*4-ethyl-1,3,5-triazine-2,4-diamine	Photosystem II/ Annual dicotyledonous weeds and partially perennial	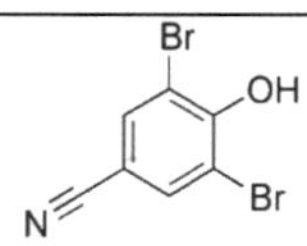 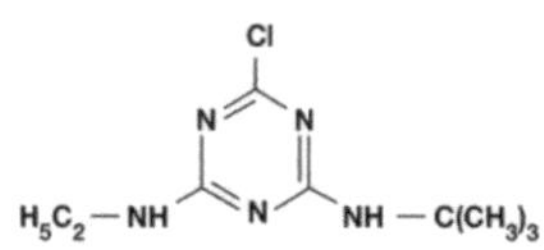
		Sunflower	
FOCUS ULTRA (cicloxidim 100 g/l) BASF Ag., Germany	Cicloxidim (5*RS*)-2-[(*EZ*)-1-(ethoxyimino)butyl]-3-hydroxy-5-[(3*RS*)-thian-3-yl]cyclohex-2-en-1-one	Acetyle CoA carboxylase/ *Sorghum halepense*	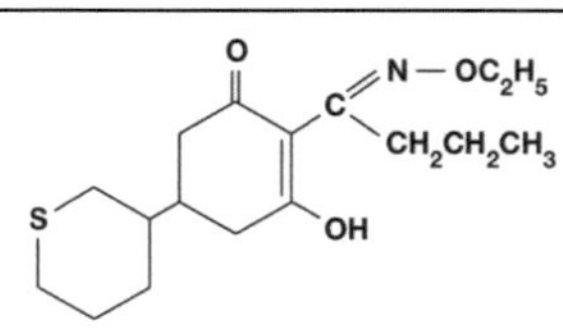

Table 2. Herbicides in use and enzymatic system of weeds susceptible to develop resistance – continued

Commercial name of herbicide and producer	Active principle/s	Enzymatic system developing resistance/target class of weeds	Chemical formula
SURDONE			
BROMOTRIL 40 EC			
Soya bean			
BASAGRAN FORTE (bentazon 480 g/l + Wettol 150 g/l) BASF Ag., Germany	Bentazon 3-isopropyl-(IH)-benzo-2,1,3-thiadiazine-4-Retone-2,2-dioxide Wettol Water dispersible granules of liquid pesticides	Photosystem II/ Annual dicotyledonous weeds	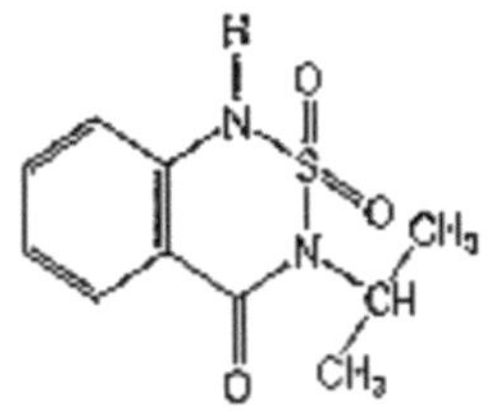
SENCOR 70 WG (metribuzin 70%) Bayer CropScience Ag., Germany	Metribuzuin (1, 1-dimethylethyl)-3-(methylthio)-1, 2, 4-triazin-5(4H)-one	Photosystem II/ Annual dicotyledonous weeds	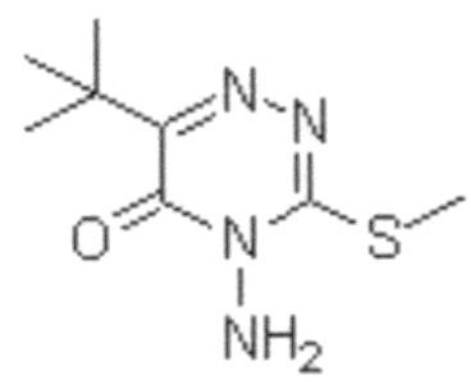
SELECT SUPER (cletodim 120 g/l) Arvesta Lifescience Corporation USA	Cletodim (RS)-2-[(E)-1-[(E)-3-chloroallyloxyimino] propyl]-5-[2-(ethylthio)propyl]-3-hydroxycyclohex-2-enone	Acetyle CoA carboxylase/ Annual dicotyledonous weeds and *Sorghum halepense*	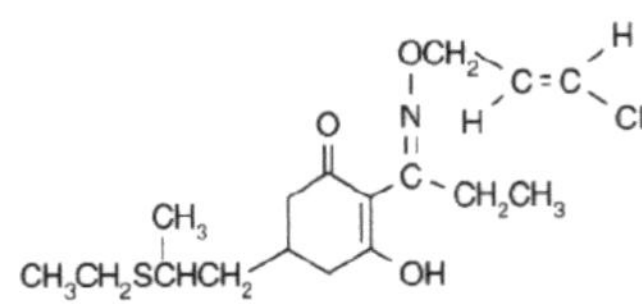
FOCUS ULTRA			
Vegetables			
SENCOR 70 WG (potato, tomato)			
SELECT SUPER (potato, beans)			
SURDONE (potato)			
FOCUS ULTRA (tomato, beans)			
BASAGRAN FORTE (beans)			

Table 2. Herbicides in use and enzymatic system of weeds susceptible to develop resistance - continued

3. Mechanism of herbicide resistance

The herbicide resistance is a concept, which could regard both, culture plants and weeds. According to AgBiosafety - University of Nebraska "**Herbicide resistance is when plants become resistant to the negative effects of a particular herbicide formulation or are genetically engineered in order to have resistance to a particular herbicide. Herbicide-resistant crops can be sprayed with a particular herbicide, killing weeds growing in and around the crop plants, without being damaged.**"

Concerning weeds, the Illinois Agricultural Pest Management Handbook (2008) adopts the following definition "**Resistance is the inherited ability of a plant to survive and reproduce following exposure to a dose of herbicide normally lethal to the wild type.**"

Resistance of weeds to herbicides is not a unique phenomenon. In fact, resistance to herbicides is a worldwide problem that became more and more acute in last 20 years, due to increased production capacity.

According to Prather et al. (2000), resistance may appear in weeds as consequence of random punctiform mutation occurred in plant genome. If we deal with a herbicide selection pressure the resistant plants will survive, while susceptible plant will disappear.

The first report concerning resistance against pest, fungi, herbicide, etc. was reported in 1908, when insects resistant to insecticides were identified. Plant pathogens resistant to fungicides were emphasized for the first time in 1940, and weeds resistant to herbicides beginning with 1950s, when dandelion and wild carrot biotypes were reported to be resistant to 2,4-D (MCPA or 4-Chloro-*o*-tolyloxy-acetic acid), and then groundsel resistant to triazines was reported in 1968 (Kremer et al., 1999; Oroian, I., 2008).

By 2008,183 weed biotypes that were resistant to triazine herbicides and 15 other herbicide families were reported and well documented. Results of a 1992 North Central Weed Science Society survey of the north central United States and Canada reflect a world wide trend of increasing appearance of herbicide resistance (Yamasue et al., 1992; Oroian, I., 2008).

Besides intrinsic weed traits, a series of culture practices may favourize the installation of herbicide resistance phenomenon (Fig. 2)

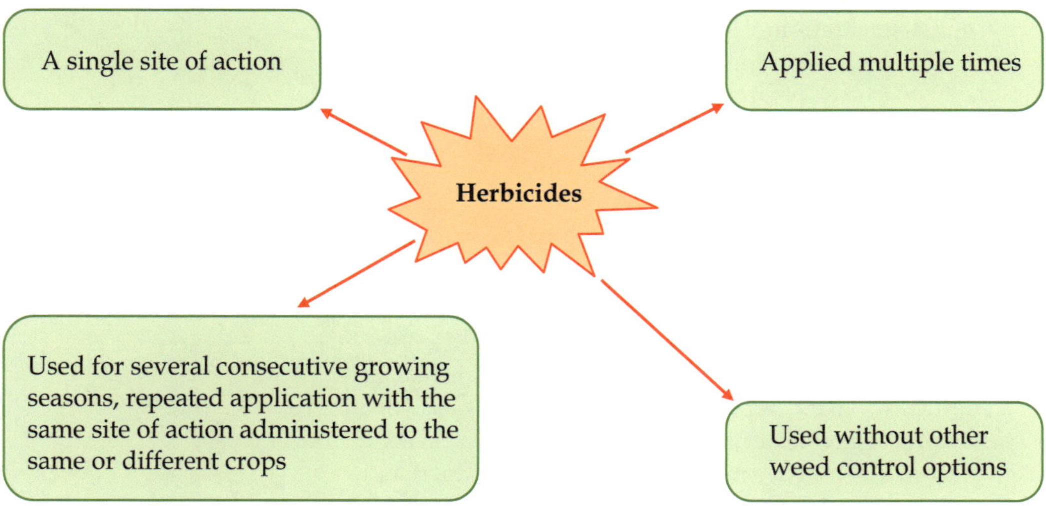

Fig. 2. The herbicide administration practices and/or intrinsic traits that contribute to herbicide resistance

Two mechanisms are at the origin of the herbicide resistance. One is the so-called "exclusionary resistance" and the other, "the site action resistance" (Cummins et al., 2010).

3.1 The exclusionary resistance

The mechanism consists in the exclusion of the herbicide molecule from the specific site located in target plant where it induces the characteristic toxic response (Fig. 3).

Several well known mechanisms, elucidated by now, can explain the processes involved in resistance developed by weeds against herbicides. Among them, we can enumerate: differential herbicide uptake, differential translocation, compartmentation in different specific regions of the weed, and metabolic detoxification (Oroian, I., 2008; Cummins et al., 2010).

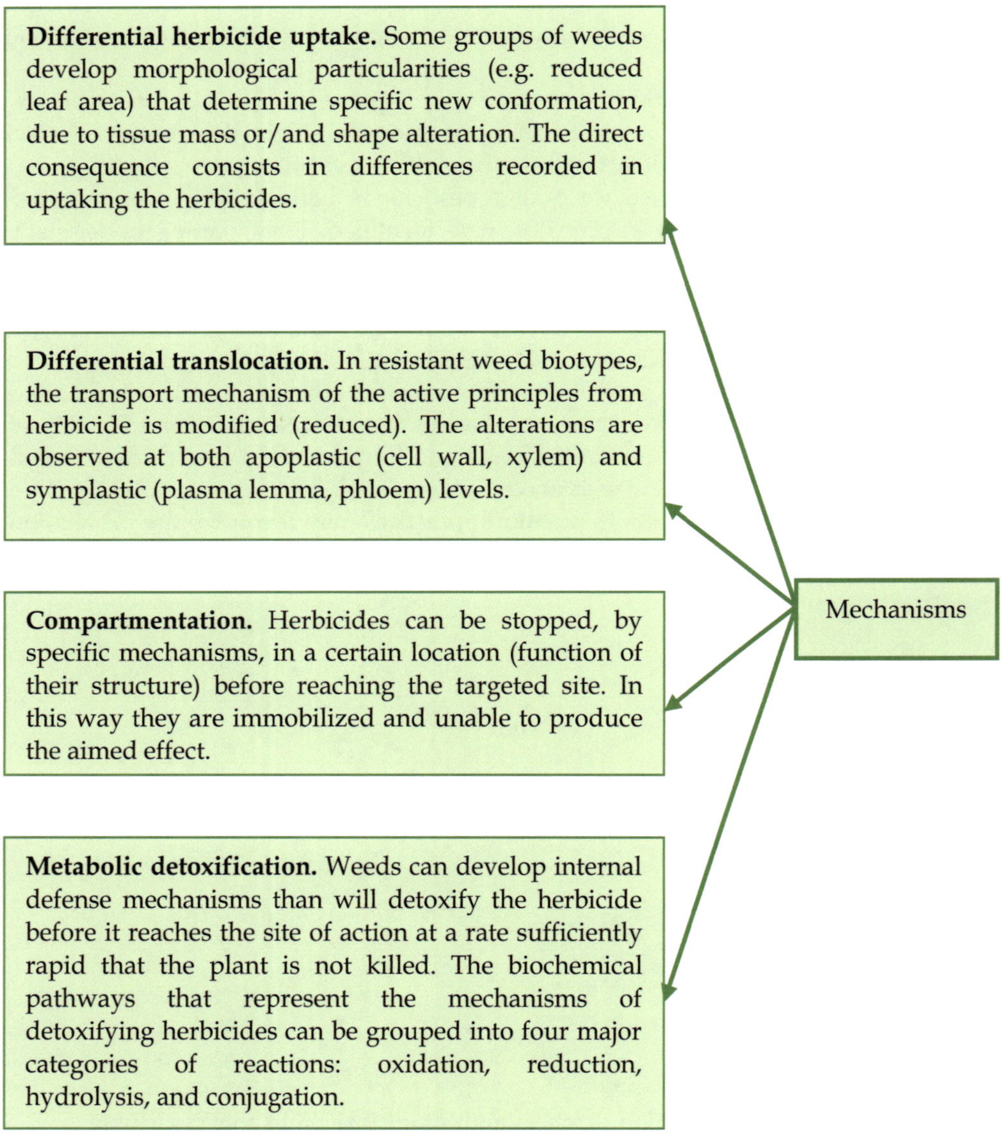

Fig. 3. The exclusionary resistance mechanisms

The metabolic detoxification involves three enzyme systems (Hakala et al., 2006; Kurth et al., 2009; McCourt et al., 2006) that catalyze the specific reactions that represent the basis of herbicide resistance in weeds (table 3).

Enzyme	Global process	Action	Target substance
Glutathione-s-transferase	Resistance to herbicides	Detoxifies	Atrazine
Aryl-acylamidase	Resistance to herbicides	Detoxifies	Propanil
Cytochrome P450 monoxygenase	Resistance to herbicides	Increased herbicide metabolism	Inhibitors of ACCase, ALS and PSII

Table 3. The enzymes and their role in herbicide resistance of weeds

3.2 Resistance due to the site of action

From this point of view, two types of resistance are known: due to the altered site of action and site of action overproduction (Cummins et al., 2010).

Altered site of action. The site of action is modified in such a way that it cannot be affected by the herbicide action. The molecular mechanisms of resistance are mainly based on impossibility of herbicide binding to the target protein from weed specific tissue. This is the consequence of a mutation, usually punctiform (resulted form a single nucleotide change) in the gene encoding the protein to which the herbicide normally binds.

The direct consequence is the change produced in the amino acid sequence of the target protein. And this is the process that represents the basis of destroying the possibility of the herbicide to interact with the specific protein.

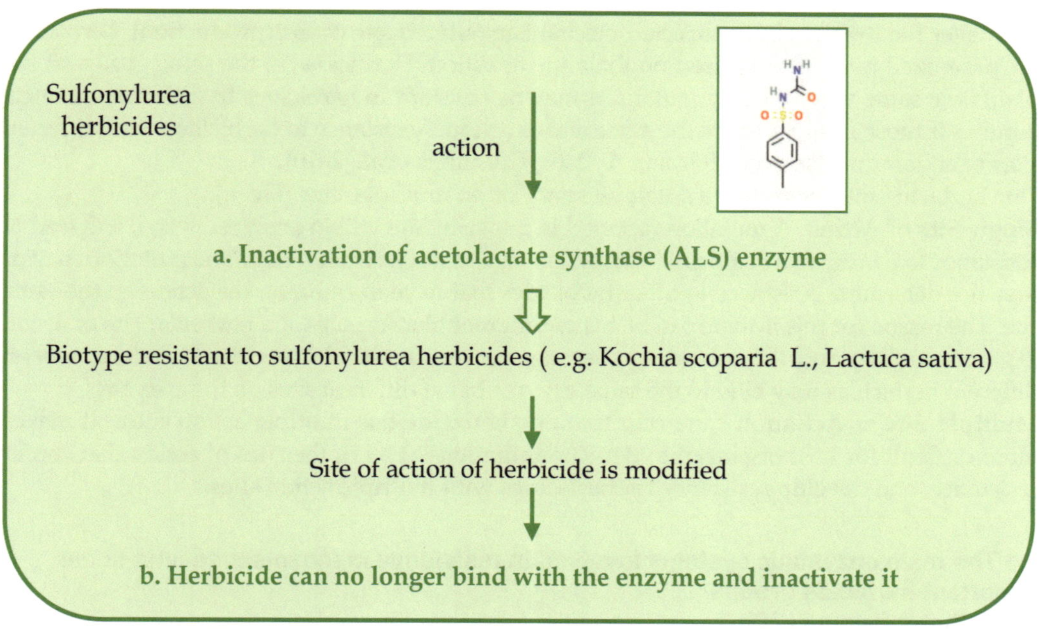

Fig. 4. Mechanism of herbicide action (a) and resistance (b) in biotypes resistant to sulfonylurea herbicides (e.g. *Kochia scoparia* L., *Lactuca sativa*)

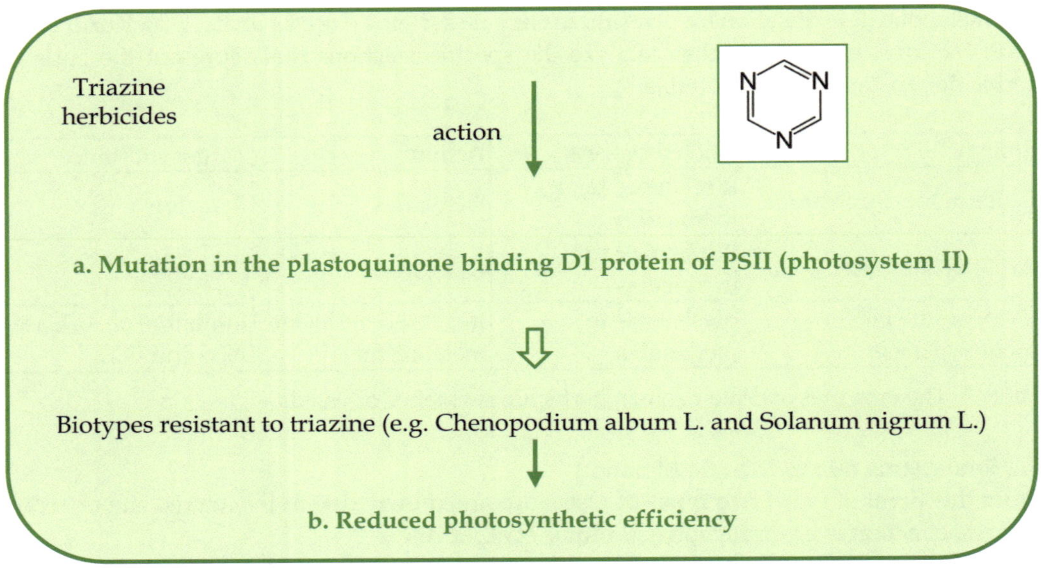

Fig. 5. Mechanism of herbicide action (a) and resistance (b) in biotypes resistant to triazine herbicides (e.g. *Chenopodium album* L. and *Solanum nigrum* L.)

Site of action overproduction. The first consequence of this overproduction will be a decreased intensity of the herbicide effect. An overproduction of the herbicide action site being recorded, the active principle of the weed control product will be less effective and finally it will not be able to inactivate the entire enzyme produced.

3.2.1 Herbicides and site of action

Whatever the resistance site directed mechanism (altered site or overproduction), herbicides are organized into groups based on their site of action. Herbicides in the same group all kill plants the same way. Weed populations may be resistant to herbicides in different chemical families if those families share the same site of action. Resistance to herbicides with different sites of action can also occur (Oroian, I., 2008; Cummins et al., 2010).
One herbicide may action on a single site and/or on multiple sites (Fig. 6).
Single Site of Action. A mutation reported in a specific site within analyzed weed, will lead to resistance to a particular herbicide with action on the site with mutation. The mutation may or may not determine resistance to other herbicides that have in common the action on the same site. The reason for this is there can be many different binding sites at a particular site of action (e.g. an enzyme) and those binding sites can be very herbicide specific. Therefore, several different herbicides may bind to the same enzyme but at different sites on the enzyme.
Multiple Site of Action. It can occur that one herbicide has multiple action sites. It makes more difficult for host organism to develop mutations at all of the sites of action that confer resistance, and develop resistance for herbicides with multiple action sites.

3.3 The main enzymatic systems involved in resistance mechanism against some important herbicide groups

3.3.1 Photosystem I (PSI)

Photosystem I (plastocyanin: ferredoxin oxidoreductase) is characteristic for the photosynthetic light reactions of algae, plants, and some bacteria, being part of photosynthetic

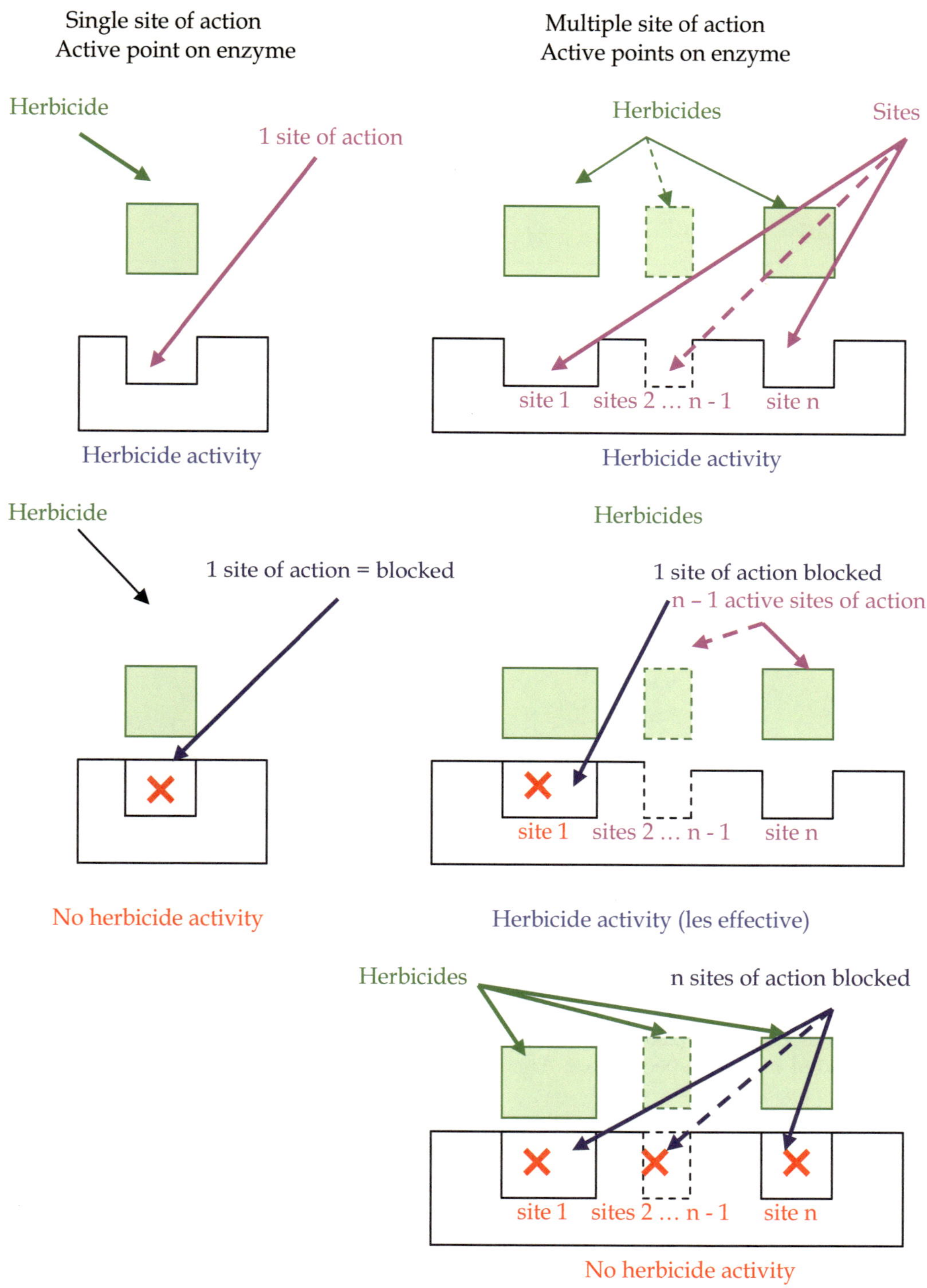

Fig. 6. Herbicide activity on one ore more sites

electron transport located in thylakoid membranes and an iron-sulfur type (Fe-S) reaction centre (Fig. 7).
It is involved in electron flow in the chloroplast thylakoid membrane of plants and in cyanobacteria, using light energy. It is excited best by light at about 700 nm, and is thus sometimes called P700 (Stockel et al., 2004; Hakala et al., 2006; Sharon et al., 2009). It contains eleven protein subunits, the cytochrome b_6f complex, lipids, pigments (chlorophylls and β-carotene and cofactors (early electron acceptors vitamin K_1 phylloquinone in electron transport chain, Q_K A and B, Ferredoxin-$NADP^+$ oxidoreductase enzyme, FNR respectively, Mg^+ and Ca^+).

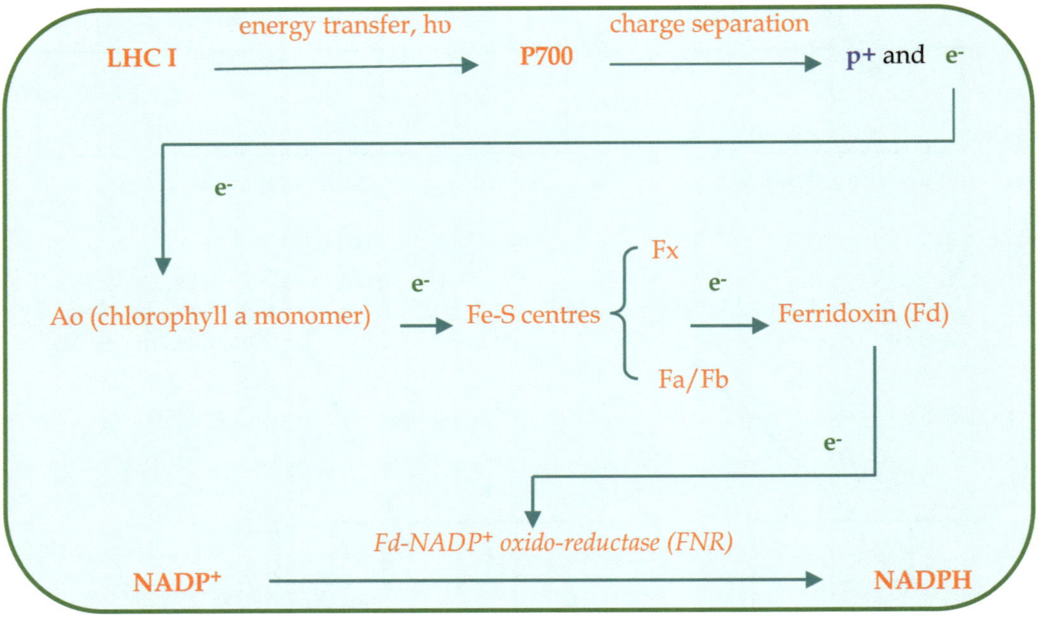

Fig. 7. The mechanism of PSI function in driving energy

The main inhibitors of photosystem I belong to post-emergence non-selective contact herbicides. They are represented by the chemical family Bipyridillum with herbicides Paraquat and Diquat.
According to Oroian I. (2008) herbicide resistance which involves the enzymatic PSI has been reported in weed species like: *Amaranthus lividus, Bidens pilosa, Conyza spp., Eleusine indica, Solanum nigrum.*
Mechanism of resistance to PSI inhibitors. The strategies against inhibitors of PSI may be grouped in two classes, detoxification of the toxic products formed and rapid sequestration of the herbicide mechanisms, respectively (fig. 8).

3.2.2 Photosystem II (PSII)

PSII (water-plastoquinone oxidoreductase) is a protein complex located in chloroplast thylakoid membrane (Guskov et al., 2009; Watanabe et al., 2009). It is characteristic for photosynthetic organisms and has important role in driving light energy to split H_2O into

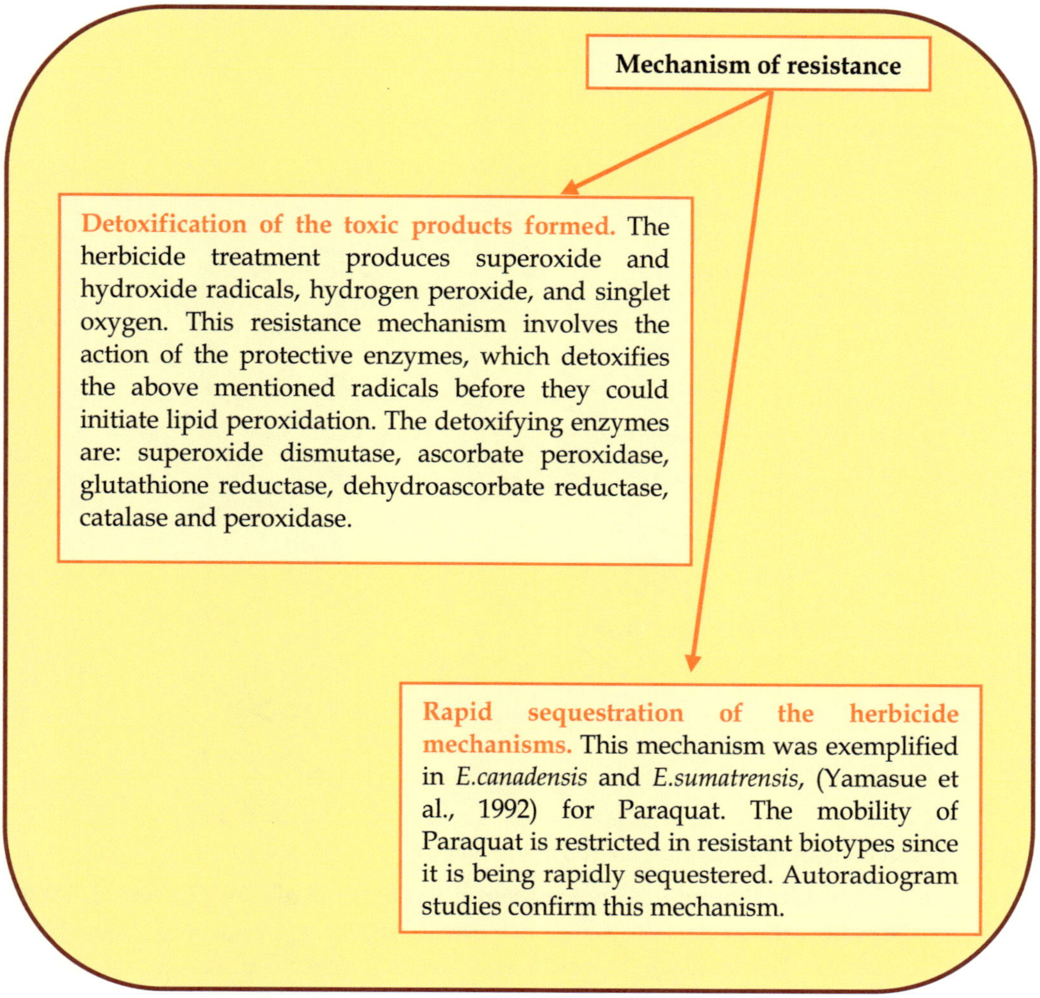

Fig. 8. The mechanisms involved in weed resistance against PSI inhibitors

O_2, protons and electrons (Fig. 9). It consist of 20 subunits and light-harvesting proteins and more than 99 cofactors. Among subunits, we can enumerate:

- light harvesting complex - LHC -,
- P680 reaction centre,
- protein D1 - reaction centre protein, quinone and manganese centre, binds Chlorophyll P 680 and beta-carotene,
- protein D2 - reaction centre protein -,
- PsbO – managnsese stabilizing protein – etc.

The cofactors are represented by:

- chlorophyll a, which absorbs light,
- beta-carotene with function in quenching excess photoexcitation energy,
- plastoquinones as intra-thylakoid membrane mobile electron carriers - two mobile electron carriers-plastoquinone-A (PQA), and plastoquinone-B (PQB),
- pheophytin as primary electron acceptor,

- heme,
- bicarbonate,
- n-dodecyl-beta-D-maltoside detergent molecules,
- managanese centre (Guskov et al., 2009).

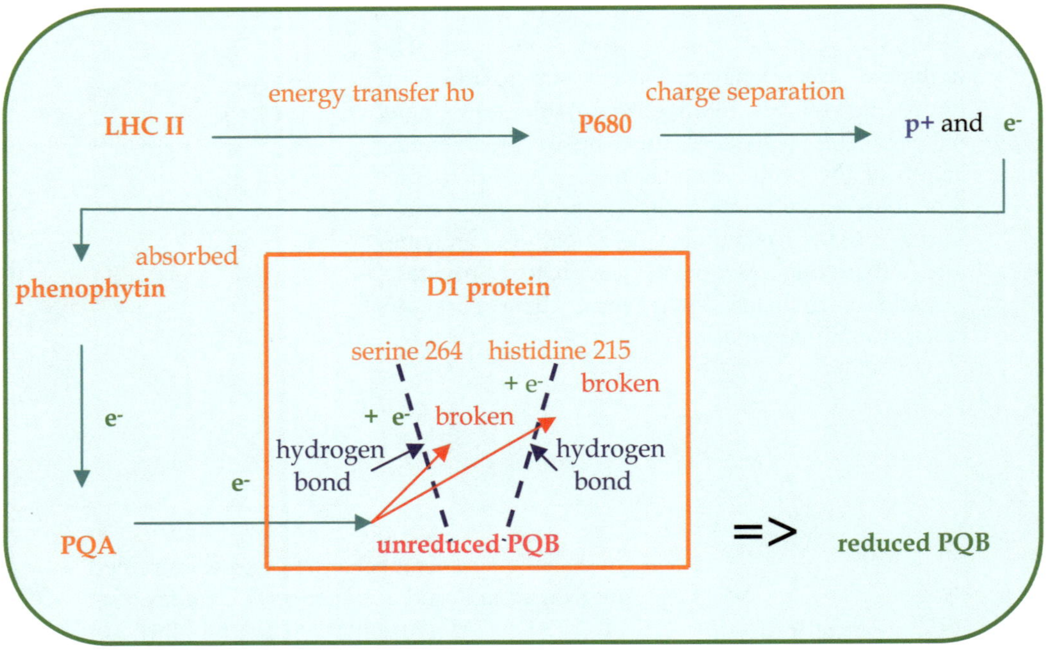

Fig. 9. The mechanism of PSII function in driving energy

The chemical herbicides that inhibit photosystem II are (Saari et al, 1990; Oroian, I., 2008): Atrazine, Cynazine, Simazine, Propazine (chemical family of Triazines), Metribuzin (chemical family of Triazinones), Bromacil and Terbacil (chemical family of Uracils), Bromoxinil (chemical family of Nitriles), Diuron and Fenuron (chemical family of Phenylureas), Pyrazon (chemical family of Pyridazinones), and Bentazon (chemical family of Benzothiadiazole).

According to Oroian I. (2008) herbicide resistance which involves the enzymatic PSI has been reported in weed species like: *Amaranthus hybridius, Solanum nigrum, Chenopodium album, Phalaris paradoxa.*

Mechanism of resistance to PSII inhibitors. Three ways are known to explain the weed internal defence strategies against characteristic inhibitors (fig. 10).

3.2.3 Acetyle CoA carboxylase (ACCase)

ACCase is a multifunctional, biotinylated protein located in stroma of plastids. It catalyzes the ATP dependent carboxylation of Acetyl CoA to form malonyl CoA, which is the precursor of fatty acids (Sellwood et al., 2000).

It is present in prokaryotes, chloroplasts of most plants and algae, endoplasmic reticulum of most eukariotes, and even mycobacteria (Kurth et al., 2009). ACCase catalyses two partial reaction occurring at two different sites (Fig. 11).

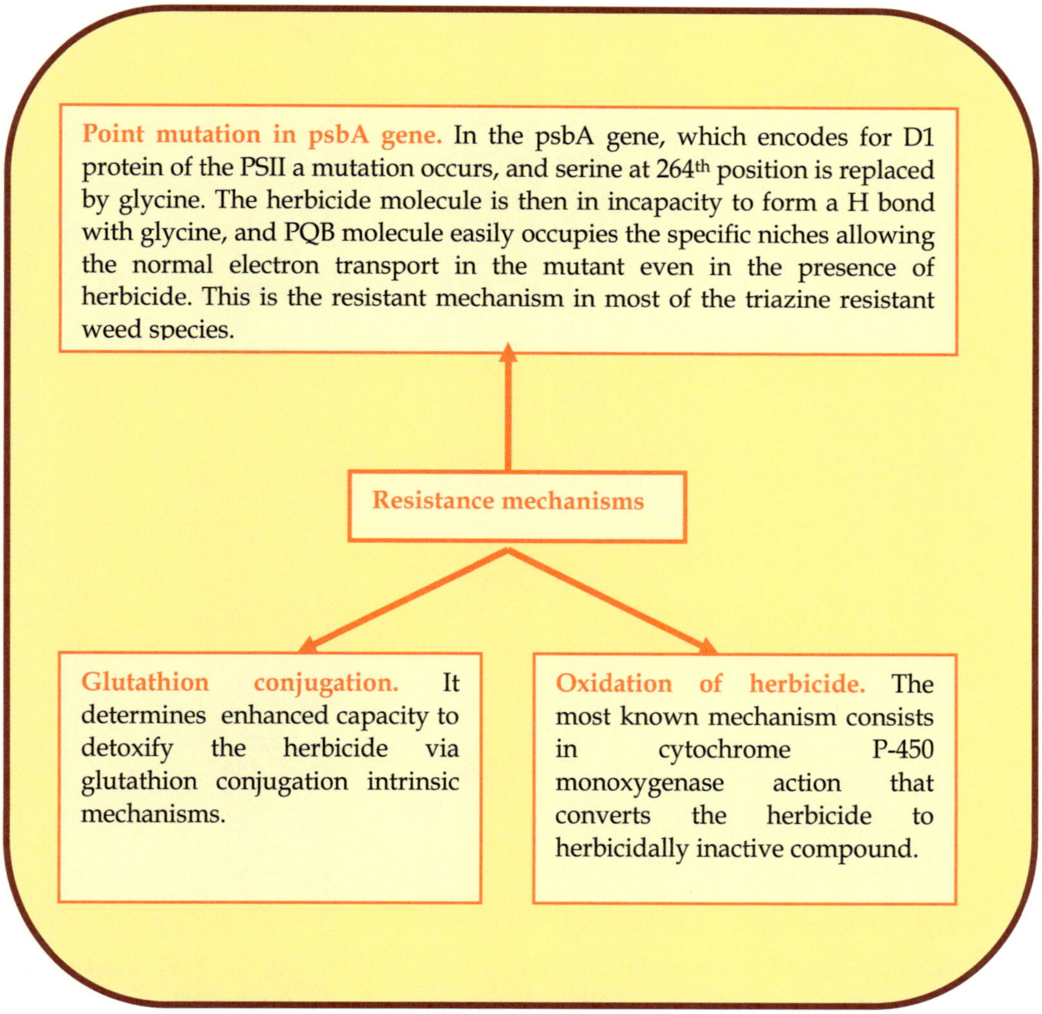

Fig. 10. The mechanisms involved in weed resistance against PSII inhibitors

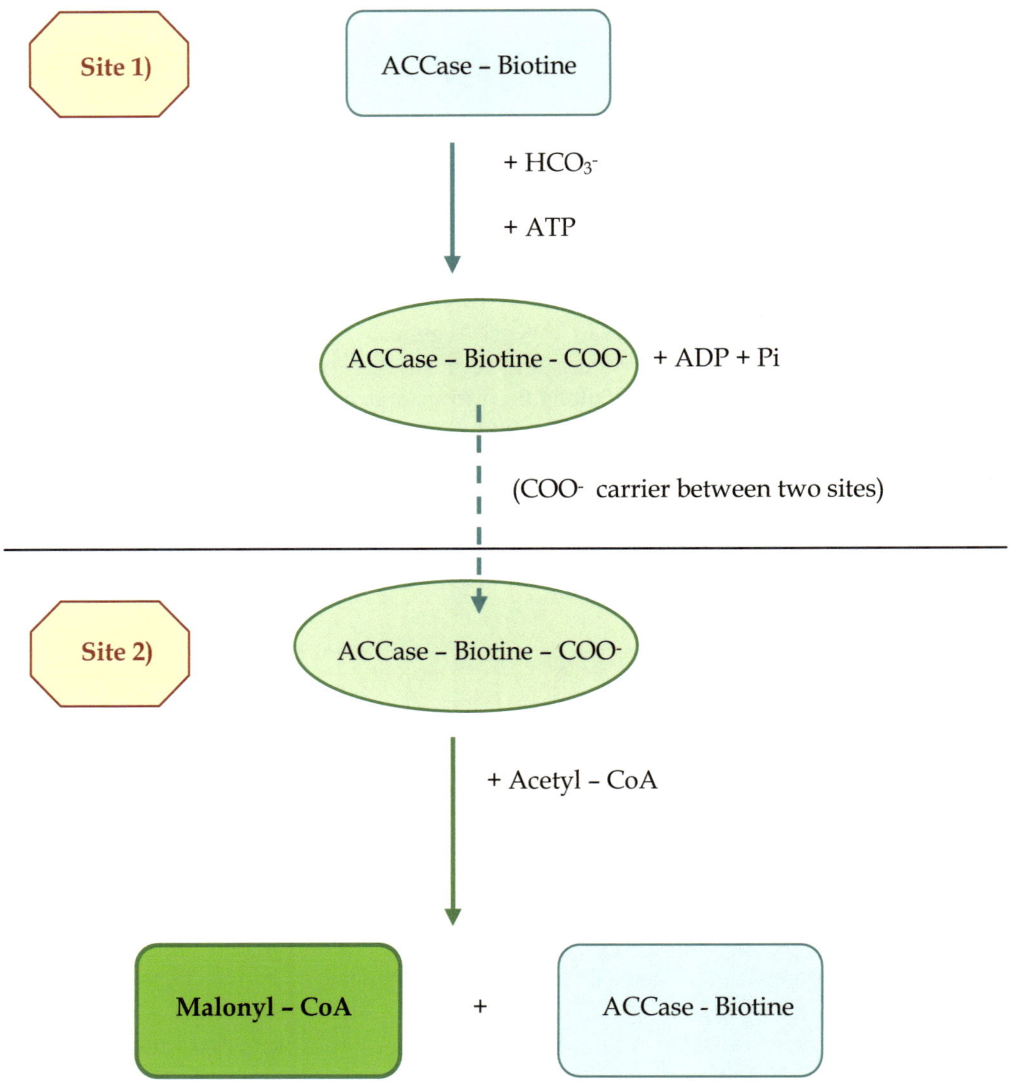

Fig. 11. The mechanism of ACCase activity

The chemical herbicides that inhibit ACCase activity are (Oroian, I., 2008): Clodinafop, Diclofop, Fenoxaprop (chemical family of Aryloxyphenoxypropionates), Sethoxydim, Cycloxidim, and Clethodim (chemical family of Cyclohexanediones).

According to Oroian I. (2008) herbicide resistance which involves the enzymatic ACCase activity have been reported in weed species like: *Avena fatua, Digitaria sangunalis, Echinochloa crusgali, Echinochloa colona, Lolium* spp.

Mechanism of resistance to ACCase inhibitors. Up to date, three mechanisms of herbicide resistance, in which ACCase activity is involved, are described. They are: Presence of tolerant form of ACCase (alteration of target site enzyme), detoxification mechanism, and overproduction of ACCase (fig. 12).

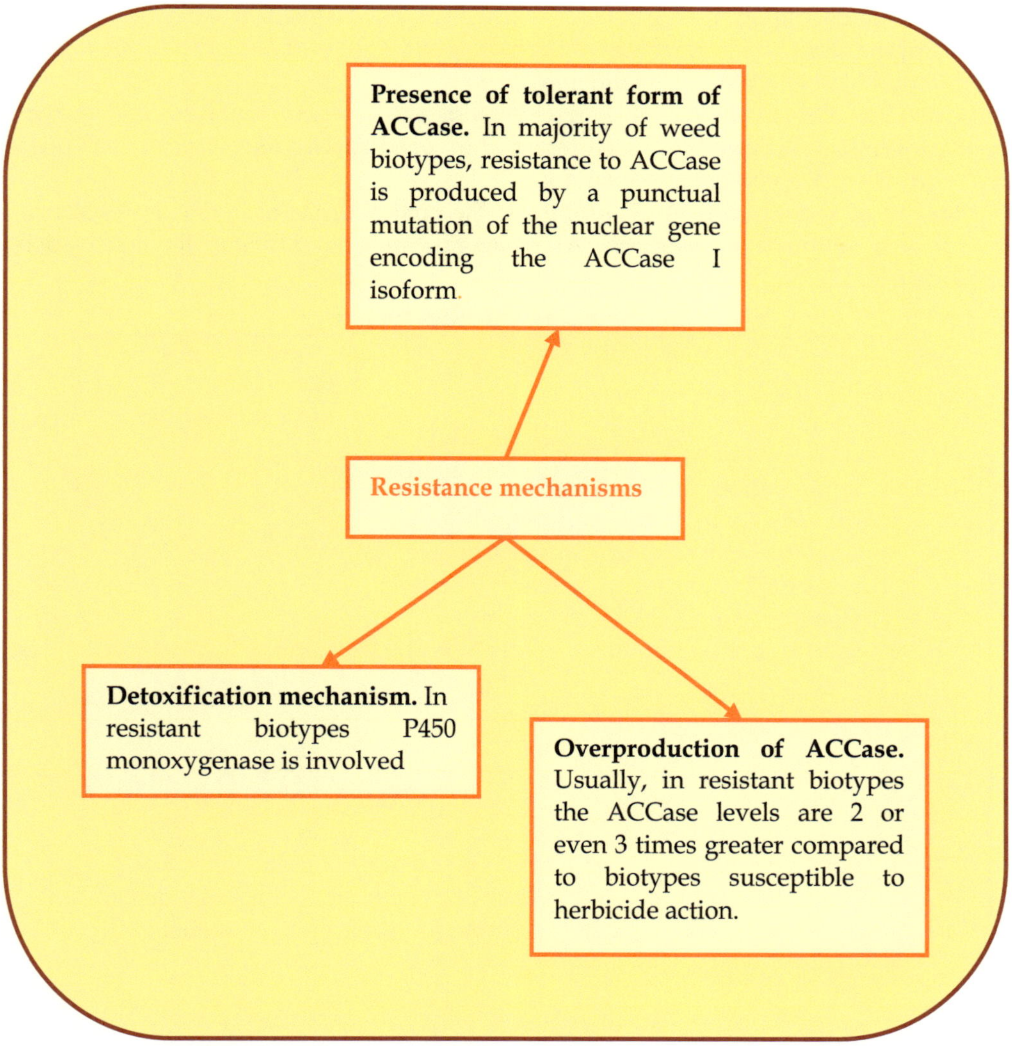

Fig. 12. The mechanisms involved in weed resistance against ACCase inhibitors

3.2.4 Acetolactate synthase

Two groups of enzymes are referred as Acetolactate synthase: Acetohydroxyacid synthase (AHAS), and Acetolactosynthase (ALS), respectively. It is a protein consisting of 590 residues. These residues are classified into three separate subunits: d1yhya1, d1yhya2 and 1yhya3 (Chipman et al., 1998; Preston et al., 2002; Scarabel et al., 2004). AHAS/ALS is the first enzyme common to biosynthesis of branched chain amino acids leucin, valine and isoleucine (McCourt et al., 2006) It is found in plants and micro-organisms.

It is involved in the reaction of ketobutyrate conjugation with pyruvate to form acetohydroxybutyrate, which develops in parallel with reaction of conjugation of 2 molecules of pyruvate to form acetolactate, reaction also catalyzed by this enzyme.

The chemical herbicides that inhibit the enzyme activity are (Oroian, I., 2008): Chlorosulfuron, Sulfosulfuron (chemical family of Sulfonylureas), Imazapyr (chemical family of Imidazolinones), Diclosulam, Flumetsulam, Metosulam (chemical family of Triazolopyrimidines), Pyriminobac-methyl, Bispyribac, Pyriftalid (chemical family of Pyrimidinyl-thio-benzoate Imidazolinones).

According to Oroian I. (2008) herbicide resistance which involves the enzymatic acetolactosynthase activity have been reported in weed species like: *Amaranthus* sp. *Avena fatua, Conyza* sp., *Eleusine indica,Lolium* spp.

Mechanism of resistance to ALS inhibitors. Two mechanisms of herbicide resistance are known: less sulfonylurea sensitive ALS enzyme and rapid metabolic inactivation of herbicide (fig. 13).

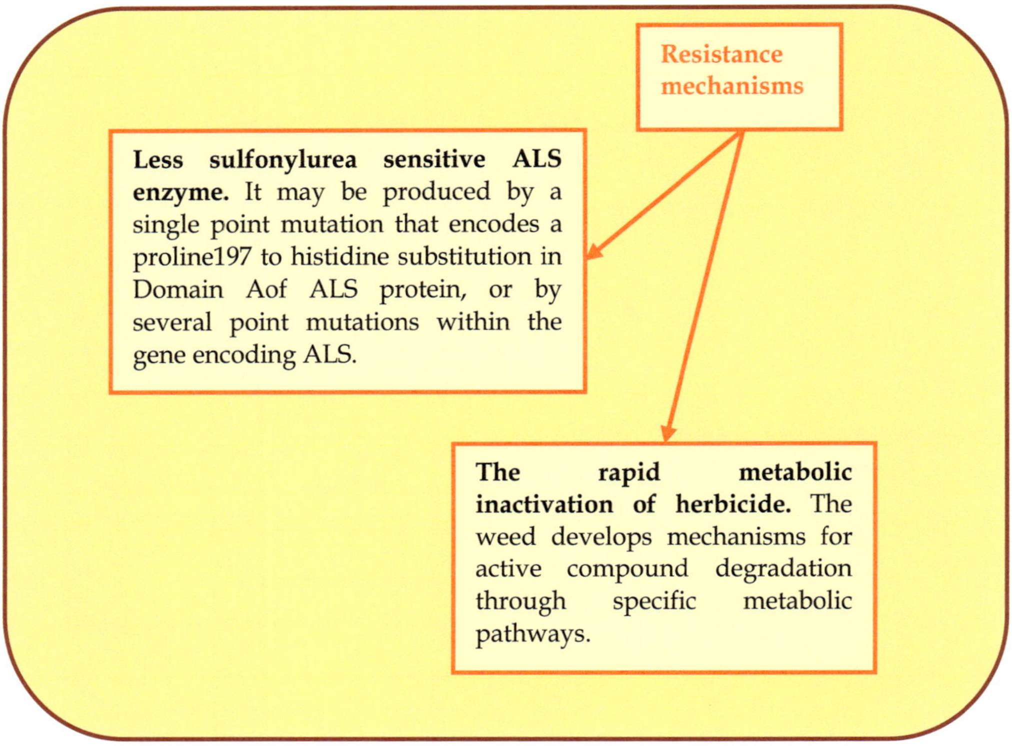

Fig. 13. The mechanisms involved in weed resistance against ALS inhibitors

4. Conclusion

Even the concern of weed resistance against herbicides appeared a few decades ago, it still remains a serious concern and a continuous challenge for both farmers and producing industry.

To date, the most valuable information concerning the weed resistance against herbicides came from weed genetics. This capacity of auto defence is the result of the plant ability to auto generates gene variability. Studying the plant intrinsic molecular processes, valuable answers were obtained.

Understanding herbicide mechanisms of action in context of permanent evolution of weed resistance represents the main way of fighting against this phenomenon, and this approach delivers the most valuable tool for research, development and practice.

Mastering the state of art in the field will supply the appropriate basis to efficient fight against new and more and more developed herbicide resistance mechanisms.

5. References

Berca, M. (2007). Fungicide Resistance in Agricultural Cultures from European Perimeter, *ProEnvironment*, Pilot number, pp. 8 - 20

Chipman, D., Barak,, Z., Schloss, J.V. (1998). Biosynthesis of 2-aceto-2-hydroxy acids: acetolactate synthases and acetohydroxyacid synthases, *Biochim Biophys Acta.*, 1385 (2), pp 401-419

Cummins, I., Edwards, R. (2010). The Biochemistry of Herbicide Resistance in Weeds, *Outlooks on Pest Management*, Vol. 21, No.2, pp. 73-77

Ge., X., d'Avignon, D.A., Ackerman, J.J.H., Sammons, R.D. (2010). Rapid vacuolar sequestration: the horseweed glyphosate resistance mechanism, Pest Management Science, *Vol. 66, Issue 4,* pp 345–348

Guskov, A., Kern, J., Gabdulkhakov, A., Broser, M., Zouni, A., Saenger, W. (2009). Cyanobacterial photosystem II at 2.9-A resolution and the role of quinones, lipids, channels and chloride, *Nature Structural & Molecular Biology*, 16(3), pp 334-42.

Hakala, M., Rantamaki, S., Puputti, E.-M., Tyystjarvi, T., Tyystjarvi, E. (2006). Photoinhibition of manganese enzymes: insights into the mechanism of photosystem II photoinhibition, *Journal of Experimental Botany*, Vol. 57, No. 8, pp. 1809–1816

Kremer, E. & Kropff, M.J. (1999). Comparative Growth of Triazine-susceptible and -resistant Biotypes of *Solanumnigrum* at Different Light Levels, *Annals of Botany*, 83, pp. 637-644

Kurth, D.G., Gago, G.M., de la Iglesia, A., Lyonnet, B.B., Lin, T.W., Morbidoni, H.R., Tsai, S.-C., Gramajo, H. (2009). ACCase 6 is the essential acetyl-CoA carboxylase involved in fatty acid and mycolic acid biosynthesis in mycobacteria, *Microbiology*, 155, pp 2664-2673

McCourt, J. A. & Duggleby, R. G. (2006). Acetohydroxyacid synthase and its role in the biosynthetic pathway for branched-chain amino acids, *Amino Acids*, Vol. 31, No. 2, pp 173-210

Oroian, I. (2008). Plant protection and environment, 325 pg., Todesco Cluj – Napoca, ISBN 978-973-7695-47-5 [In Romanian]

Owen, M.D.K.& Zelaya, I.A. (2005). Herbicide-resistant crops and weed resistance to herbicides, *Pest Management Science*, 61, pp. 301 - 311

Paoletti, M.G. & Pimentel, D. (1995). The environmental and economic costs of herbicide resistance and host-plant resistance to plant pathogens and insects, *Technological Forecasting and Social Change*, Vol. 50, Issue 1, pp 9-23

Powles, S.B. & Yu, Q. (2010). Evolution in Action: Plants Resistant to Herbicides, *Annual Review of Plant Biology*, Vol. 61, pp. 317-347

Prather, S.T., Ditomaso, J.M., Holt, J.S. (2000). Herbicide resistance: Definition and Management strategies, Division of Agriculture and Natural Resources, University of California, http://anrcatalog.ucdavis.edu

Preston, C. & Powles, S.B. (2002). Evolution of herbicide resistance in weeds: initial frequency of target site-based resistance to acetolactate synthase-inhibiting herbicides in *Lolium rigidum*, *Heredity*, 88, pp. 8 - 13

Saari, L.L., Cotterman, J.C., Primiani M.M. (1990). Mechanism of Sulfonylurea Herbicide Resistance in the Broadleaf Weed, *Kochia scoparia*, *Plant Physiology*, 93, 55 -61

Scarabel, L., Carraro, N., Sattin, M., Varotto, S. (2004).Molecular basis and genetic characterisation of evolved resistance to ALS-inhibitors in *Papaver rhoeas*, *Plant Science*, Vol., 166, Issue 3, pp. 703 - 709

Sellwood, C., Slabas, A.R., Rawsthorne S. (2000). Effects of manipulating expression of acetyl-CoA carboxylase I in *Brassica napus* L. embryos, *Biochemical Society Transactions*, Vol. 28, part 6598, pp. 600 - 615

Sharon, I., Alperovitch, A., Rohwer, F., Haynes, M., Glaser, F., Atamna-Ismaeel, N., Pinter, R.Y., Partensky, F., Koonin, E.V., Wolf, Y.I., Nelson, N., Béjà, O. (2009). Photosystem I gene cassettes are present in marine virus genomes, *Nature*, 461, pp 258-262

Sigematsu, Y., Sato, F., Yamada, Y. (1989). The Mechanism of Herbicide Resistance in Tobacco Cells with a New Mutation in the QB Protein, *Plant Physiology*, 89, pp. 986-992

Stockel, J.& Oelmuller, R. (2004). Novel Protein for Photosystem I Biogenesis, *The Journal of Biological Chemistry*, Vol. 279, No. 11, pp. 10243–10251

Vila-Aiub, M. M., Martinez-Ghersa, M.A., Ghersa, C. M. (2003). Evolution of herbicide resistance in weeds: vertically transmitted fungal endophytes as genetic entities, *Evolutionary Ecology*, Vol. 17, Nos 5 – 6, pp. 441-456

Watanabe, M., Iwai, M., Narikawa, R., Ikeuchi, M. (2009). Is the Photosystem II Complex a Monomer or a Dimer ?,*Plant Cell Physiology*, 50 (9), pp. 1674–1680

Yamasue, Y., Kamiyama, K., Haniokaand, Y., Kusanagi, T. (1992). Paraquat resistance and its inheritance in seed germination of the foliar-resistant biotypes of *Erigeron canadensis* L. and *E. sumatrensis* Retz, *Pesticide Biochemistry and Physiology*, Vol. 44, Issue 1,pp 21 - 27

*** (2008). Illinois Agricultural Pest Management Handbook, http://ipm.illinois.edu/pubs/iapmh/12chapter.pdf, pp. 272 – 276

*** (2010). Practical Guide for Farmers, Alcedo 2010, pp. 3 – 142, ISBN 978-973-7982-31-5 [In Romanian]

*** (2010). FAOSTAT, http://faostat.fao.org/site/567/default.aspx#ancor

*** (2010). http://agbiosafety.unl.edu/

31

Resistance to Herbicides in the Model Organisms *Saccharomyces cerevisiae* and *Arabidopsis thaliana*: the Involvement of Multidrug Resistance Transporters

Tânia R Cabrito[1], Estelle Remy[2], Miguel C Teixeira[1], Paula Duque[2] and Isabel Sá-Correia[1]
[1]*IBB – Institute for Biotechnology and BioEngineering, Centre for Biological and Chemical Engineering, Instituto Superior Técnico, Technical University of Lisbon* [2]*Instituto Gulbenkian de Ciência*
Portugal

1. Introduction

Herbicides are agrochemicals that control the growth of undesired weeds, bringing about a significant overall increase in crop productivity. Herbicide resistance, the inherited ability of a plant to survive and reproduce following exposure to a dose of herbicide normally lethal to the wild type, is a weed physiological characteristic (Twonson, 1997). This trait has the potential to cause not only large economic losses in agricultural production, forestry and landscaping, but also deleterious effects on the environment and human health as a result of rising herbicide application rates (Hayes & Wolf, 1997). On the other hand, crops of agronomic relevance exhibiting chemical stress resistance are highly desirable and can be obtained through genetic manipulation, based on the knowledge gathered on the underlying mechanisms.

Emergence of multidrug resistance (MDR), that is, the simultaneous acquisition of resistance to a wide range of structurally and functionally unrelated cytotoxic chemicals, is found in a wide variety of organisms, from bacteria to mammals. The widespread use of herbicides has led, since the early 1970s, to a growing number of resistant weed species and biotypes that possess multiple resistance to various classes of herbicides (Hayes & Wolf, 1997). Indeed, herbicide application is one of the most important selective forces acting on a weed community in an agroecosystem (Owen & Zelaya, 2005).

To date, and particularly during the past two decades, numerous cases of herbicide resistant weed species have been described, in at least 60 different countries (www.weedscience.org). Table 1 shows the number of cases of single and multiple herbicide resistance development in weed species registered in the last 30 years. Among them there are several weed strains that have acquired simultaneous resistance to herbicides with different modes of action (Table 1). For example, in 1982 in South Australia a strain of *Lolium rigidum* (rigid ryegrass), a monocot weed that infests barley and wheat fields, was registered as having developed resistance to multiple herbicides, exhibiting seven diverse modes of action (from the following HRAC[1]/WSSA[2] groups: ACCase inhibitors - A/1, ALS inhibitors - B/2, Triazoles, ureas, isoxazolidiones - F3/11, Dinitroanilines and others - K1/3, Mitosis inhibitors - K2/23, Chloroacetamides and others - K3/15, Thiocarbamates and others - N/8). Owing to its capacity to rapidly acquire multiple herbicide resistance, *L. rigidum* bursts exert the highest negative economic impact in Australian crops (Owen et al., 2007). Another case of multiple herbicide resistance is *Alopecurus myosuroides* (blackgrass), a monocot weed that infests winter wheat plantations. In 1996, multiple herbicide resistant biotypes of this weed were found in Belgium, exhibiting simultaneous resistance to atrazine, chlorotoluron, clodinafop-propargyl, fenoxaprop-P-ethyl, flupyrsulfuron-methyl-Na, pendimethalin, and propaquizafop (from the following HRAC groups: ACCase inhibitors - A/1, ALS inhibitors - B/2, Photosystem II inhibitors - C1/5, Ureas and amides - C2/7, and Dinitroanilines and others - K1/3)(www.weedscience.org; Eelen et al., 1997).

Mode of action (HRAC/WSSA group)	Herbicide example	Number of resistant weed species	Number of MDR weed species*
ACCase inhibitors (A/1)	Diclofop-methyl	38	16
ALS inhibitors (B/2)	Chlorsulfuron	107	33
Photosystem II inhibitors (C1/5)	Atrazine	68	9
Ureas and amides (C2/7)	Chlorotoluron	21	11
Nitriles and others (C3/6)	Bromoxynil	3	0
Bipyridiliums (D/22)	Paraquat	24	4
PPO inhibitors (E/14)	Oxyfluorfen	4	3
Carotenoid biosynthesis inhibitors (F1/12)	Flurtamone	2	1
Triazoles, ureas, isoxazolidiones (F3/11)	Amitrole	4	1
Glycines (G/9)	Glyphosate	19	10
Dinitroanilines and others (K1/13)	Trifluralin	10	3
Mitosis inhibitors (K2/23)	Propham	1	1
Chloroacetamides and others (K3/15)	Butachlor	4	4
Cellulose inhibitors (L/27)	Dichlobenil	1	1
Thiocarbamates and others (N/8)	Triallate	8	4
Synthetic auxins (O/4)	2,4-D	28	8
Arylaminopropionic acids (Z/25)	Flamprop-methyl	2	1
Organoarsenicals (Z/17)	MSMA	1	0
Unknown (Z/27)	(chloro) - flurenol	1	1

* Number of weed species that have biotypes described as resistant to herbicides with different modes of action.

Table 1. Number of resistant weed species and MDR weed species according to the herbicide mode of action (adapted from www.weedscience.com).

[1] Herbicide Resistance Action Committee

[2] Weed Science Society of America

Some herbicide families seem to be particularly prone to induce herbicide resistance in their target weeds. Indeed, approximately 70% of the registered herbicide resistant weeds exhibit tolerance to only four out of 19 herbicide families. These four herbicide families include ALS inhibitors, photosystem II (PSII) inhibitors, ACCase inhibitors and synthetic auxins.

The great majority of herbicides act by inhibiting a specific plant enzyme essential for metabolism, whereas the remainder, including auxinic herbicides, act as general inhibitors. Two types of resistance can be distinguished: target-site and non-target-site herbicide resistance (Powles & Yu, 2010). Target-site herbicide resistance is sustained by modifications of the target site that impair the action of the herbicide. One well described case of such a mechanism involves resistance to the PSII-inhibiting triazine herbicide, shown to be due in most cases to a point mutation in the chloroplastic psbA gene, modifying the site at which triazine competes with plastoquinone within the PSII. Interestingly, this single resistance mutation has evolved independently worldwide (Arntzen et al., 1982). In contrast, non-target-site herbicide resistance implies usually a reduction in the concentration of active herbicide reaching the target site (Powles & Yu, 2010), including often the action of membrane pump-catalyzed active efflux or compartmentalization of the herbicide in plant cells (Windsor et al., 2003).

The present chapter aims to summarize current knowledge on the function of plant plasma membrane transporters in the efflux of herbicides during non-target-site herbicide resistance. Particular emphasis will be given to data obtained through the exploitation of gene-by-gene and genome-wide approaches applied to the single cell eukaryotic model, *Saccharomyces cerevisiae*, and the plant model, *Arabidopsis thaliana* (Cabrito et al., 2009a; Teixeira et al., 2007), and extrapolated, when possible, to plants with relevance in agriculture. Finally, a combined approach using the two model organisms aimed at deciphering the role of membrane transporters in auxinic herbicide resistance will be presented. The integration of the current knowledge on plant herbicide resistance in delineating rationales to design crops with increased herbicide tolerance, leading to the development of important strategies to overcome multiple herbicide resistance, is discussed.

2. *S. cerevisiae* and *A. thaliana* as model organisms to study herbicide resistance

Saccharomyces cerevisiae is widely accepted as a single cell eukaryotic model, being non-pathogenic, easy to manipulate genetically, well characterized, and exhibiting fast and inexpensive growth. The paradigm of research using *S. cerevisiae* has changed following the publication of its genome sequence more than a decade ago (Goffeau et al., 1996). Research on *S. cerevisiae* pioneered the development of several post-genomic experimental approaches and computational tools, which has allowed the gathering of large amounts of biological information easily accessible through public databases (SGD, Saccharomyces Genome Database - www.yeastgenome.com; and YEASTRACT - http://www.yeastract.com, among others). Large-scale biological material has also been produced, including a collection of mutants in which each yeast gene was individually deleted (EUROSCARF - http://web.uni-frankfurt.de/fb15/mikro/euroscarf) (Kelly et al., 2001), facilitating quick, easy and high-throughput search for genes involved in the resistance or susceptibility to any given environmental stressor. Genome-wide analyses in yeast have been successfully used to identify the genes responsible for yeast response and resistance to environmental stress, in particular those induced by xenobiotic compounds of agricultural interest, such as the herbicide sulfometuron methyl (Jia et al., 2000), the dithiocarbamate fungicides mancozeb (Dias et al., 2010; Santos et al., 2009), thiuram, zineb and maneb (Kitagawa et al., 2003), the benzimidazole fungicide benomyl (Lucau-Danila et al., 2005), the pesticide lindane (Parveen et al., 2003), and the herbicide 2,4-D (Teixeira et al., 2007; Teixeira et al., 2006; Teixeira et al., 2005). Even though many cytotoxic compounds used in agriculture may act in their target organisms via physiological mechanisms that do not exist in yeast, many of the basic mechanisms underlying resistance and adaptation to chemical and environmental stresses are apparently conserved among phylogenetically distant organisms (Landis & Yu, 1999). Therefore, the characterization of mechanisms of resistance to herbicides in yeast cells is expected to contribute to the understanding of these mechanisms in more complex and less easily accessible eukaryotes, such as weeds and crops.

The small flowering plant *Arabidopsis thaliana* is extensively used as a model organism in plant biology. *Arabidopsis* has been the focus of intense genetic, biochemical and physiological study for more than 40 years because of its rapid life cycle, small size and prolific seed production, offering important advantages for laboratory manipulations and cultivation. The complete *Arabidopsis* genome sequence (e.g. http://www.arabidopsis.org, http://www.tigr.org/tdb/e2k1/ath1) has been available since 2000 (The Arabidopsis Genome, 2000), and genome-wide expression approaches have been applied to it with success. Transformation protocols using *Agrobacterium tumefaciens* have also been optimized and large sets of mutant lines are available (e.g. T-DNA Express database - http://signal.salk.edu), making *Arabidopsis* the system of choice for molecular and system-wide studies of plant responses to chemical stress.

3. Multiple herbicide resistance in yeast and plants

3.1 Non-target site mechanisms of resistance to herbicides in plants

The understanding of the molecular mechanisms underlying acquired herbicide resistance is crucial to deal with the emergence of resistant weeds. As stated above, there are two mechanisms involved in weed herbicide resistance: target-site and non-target-site herbicide resistance (Powles & Yu, 2010). The mechanisms underlying non-target-site herbicide resistance are usually the cause for unexpected multiple herbicide resistance (Yuan et al., 2007; Preston, 2004; Preston et al., 1996).

Non-target-site herbicide resistance has been proposed to be caused by a plant detoxification process with three phases (Martinoia et al., 1993; Bartholomew et al., 2002; Sandermann, 2004; Yuan et al., 2007). Phase I consists in metabolic changes of the herbicide within the plant, typically oxidation, peroxydation, or reduction mainly by P450 monooxygenases, resulting in the formation of less toxic metabolites. During phase II, herbicides or their metabolites are directly conjugated with glutathione, sugars or amino acids to produce hydrophilic molecules that are easily handled by the plant cell. Phase III consists in the translocation of the resulting conjugated metabolites into the vacuole or the extracellular space through the action of transporters. Further degradation of the resulting molecules may then occur in the vacuole or the extracellular space. Non-target-site herbicide resistance has been described to involve mainly P450 monooxygenases (phase I), glutathione S-transferases and glycosyltransferases (phase II) and ABC (ATP-binding cassette) transporters (phase III) (Reade et al., 2004; Yun et al., 2005; Bowles et al., 2005; Klein et al., 2006).

Although this three-step mechanism of non-target-site herbicide resistance is largely recognised, it is more than probable that it is not sustaining the resistance to some classes of herbicides. Such is the case of glyphosate and paraquat, two herbicides for which mutation of the target site or phase I metabolism cannot explain entirely the multiple herbicide resistant phenotype observed (Coupland, 1985; Feng et al., 1999; Lorraine-Colwill et al., 2002). In these cases, decreased herbicide influx into the plant, decreased level of translocation within the plant, and/or increased rate of sequestration/efflux could be the mechanisms underlying the reduction of the amount of herbicide reaching the target site. All of these three mechanisms involve the action of membrane transport systems. Therefore, and owing to their high relevance in herbicide resistance, this chapter will focus on the action of membrane pumps, mainly those putatively involved in efflux during MDR.

3.2 Multidrug efflux pumps involved in herbicide resistance

Several membrane transport systems have been shown to be involved in drug resistance, both in prokaryotes and eukaryotes. One of the first systems of this nature to be described was the mammalian P-glycoprotein efflux pump (Kartner et al., 1983); cell lines overexpressing this transporter were shown to display multiple resistance to drugs. As a matter of fact, many of these membrane transporters play an important role in conferring MDR, presumably due to the catalysis of energy-dependent extrusion of a large number of structurally and functionally unrelated compounds out of the cells or within cell compartments, against their concentration gradient (Balzi & Goffeau 1995; Bolhuis et al., 1997; Hayes & Wolf 1997; Roepe et al., 1996).

On average, the predicted multidrug pumps represent more than 10% of all transporters in a cell. The increasing number of multidrug transporters identified so far is a clear indication that MDR by efflux pumps is not an exceptional phenomenon, but a highly conserved defence mechanism. Significantly, most multidrug transporters share significant homology with transporters with more specific substrates (Paulsen, 2003). Based on bioenergetic and structural criteria, drug transporters are classified as ATP-dependent or secondary transporters (Bolhuis et al., 1997). The latter can be driven by the electrochemical proton gradient or proton motive force (Δp), composed of an electrical potential ($\Delta\psi$; negative interior) and a chemical proton gradient (ΔpH; higher interior). Currently, two large ubiquitous superfamilies, the major facilitator (MFS) and the ATP binding cassette (ABC) superfamilies (Bolhuis et al., 1997; Jack et al., 2001; Sá-Correia et al., 2009; Crouzet et al., 2006), are recognized to include multidrug efflux pumps in eukaryotes. Structures have now been obtained for multidrug transporters of each of these families (see Higgins, 2007). Further discussion will be focused on these two types of transporters and their involvement in MDR and herbicide resistance.

3.2.1 ABC transporters

ABC transporters constitute one of the largest gene families in all living organisms, using the free energy of ATP hydrolysis to drive substrate transport across cell membranes (Bolhuis et al., 1997). These proteins mediate diverse cellular transport processes, having been reported to transport a wide range of substances, including ions, carbohydrates, lipids, xenobiotics, antibiotics, drugs and heavy-metals. They can also work as ion channels or display antigen activity (Martinoia et al., 2002). The mammalian P-glycoprotein is probably the best characterised of all ABC transporters, but many others have been described in bacteria and fungi. The minimal functional unit of all ABC transporters consists of four domains: two hydrophobic transmembranar domains, each with six transmembrane α-helical segments (TMS), plus two cytoplasmic, nucleotide-binding domains containing the ATP-binding cassette. The four can be fused into a single polypeptide, as in the mammalian P-glycoprotein. This typical 'two times two' domain organisation (Bolhuis et al., 1997) probably arose from an internal gene duplication event. On the other hand, most bacterial ABC transporters comprise only one TMS and one nucleotide-binding domain.

The *S. cerevisiae* genome encodes a total of 22 ABC proteins (Paumi et al., 2009). Among these, 16 are about twice as long as the remainder, the former having 12 TMS (indicative of a putative duplication of the half-size ABC transporters). The 22 identified yeast ABC transporters belong to four families, recently re-designated as ABCB, ABCC, ABCD and ABCG (formerly named MDR (Multi Drug Resistance), MRP (Multidrug Related Protein), ALDP (AdrenoLeukoDystrophy protein) and PDR (Pleiotropic Drug Resistance)) (Paumi et al., 2009). The ABCB family includes four members, three of which are half transporters that dimerize to form full transporters localized to the mitochondrial inner membrane (Mdl1p, Mdl2p and Atm1p) and Ste6p, the plasma membrane a-factor pheromone exporter. The ABCC family includes five vacuolar membrane transporters, two of which, Ycf1p and Bpt1p, are glutathione conjugate transporters involved in cytosolic detoxification from metal ions, and Yor1, a plasma membrane multidrug transporter. The ABCD family is composed of Pxa1p and Pxa2p that transport acyl-CoA across the peroxisomal membrane. Finally, the PDR family in yeast comprises nine transporters (Pdr5p, Pdr10p, Pdr11p, Pdr12p, Pdr15p, Pdr18p, Snq2p, Aus1, Adp1 and Yol075c), which can be seen as the cell's primary line of defence: PDR function not only decreases the cell's sensitivity to many unrelated chemical stresses, but also protects it against endogenous toxic metabolites (Paumi et al., 2009; Jungwirth & Kuchler 2006). Interestingly, PDR genes are only found in plants and fungi.

In *S. cerevisiae* only one ABC transporter, Pdr5p, was described as a determinant of resistance to herbicides (Teixeira & Sá-Correia, 2002). Transient activation (two fold) of *PDR5* transcription takes place during the adaptation period preceding cell division under stress induced by the auxin-like herbicides 2-methyl-4-chlorophenoxiacetic acid (MCPA) or 2,4-dichlorophenoxiacetic acid (2,4-D). *PDR5* induction is mediated by both Pdr1p and Pdr3p, two transcription factors controlling MDR in yeast and, as soon as adapted cells start duplication under herbicide stress, mRNA levels are drastically reduced to basal values (Teixeira & Sá-Correia, 2002). Differently from what has been seen for fungicides and drugs, resistance to herbicides has not been extensively studied in *S. cerevisiae*. Thus, it is expectable that many other of the yeast ABC multidrug transporters may also render the cell herbicide resistant..

The *Arabidopsis* genome was found to encode about 600 predicted membrane transporters, a large number of which belong to the ABC superfamily. In fact, this superfamily alone comprises 131 members, with 53 encoding full-size and the remainder half-size transporters. *Arabidopsis* genes encoding full-size ABC transporters are significantly more numerous than those reported in other eukaryotes such as yeast and humans, but the reasons for such diversification in the plant kingdom are unknown. The great majority of plant ABC genes remains uncharacterised, with those encoding full-size transporters being the best studied to date. These can be subdivided into three main groups: the MDR or PGP (P-glycoproteins), the MRP and the PDR gene families (Martinoia et al., 2002).

The MDR-PGP family is the largest, consisting of 22 *Arabidopsis* genes coding for full-size ABC transporters. The first demonstration of an MDR-like mechanism in plants, able to confer tolerance to different herbicides, stemmed from functional studies on *AtPGP1*. Overexpression of this P-glycoprotein in *A. thaliana* conferred increased resistance to herbicides from different chemical classes, such as dicamba, pendimethalin, oryzalin or monosodium acid methanearsonate (MSMA), suggesting a resistance mechanism related to decreased retention or increased active toxin efflux from plant cells (Windsor et al., 2003). Loss of function approaches later showed that both *AtPGP1* and the closely related *AtMDR1* participate in auxin efflux required for polar auxin transport in *Arabidopsis* (Lin & Wang, 2005). In fact, all plant PGP transporters characterized so far appear to be involved in cellular and long-distance auxin transport (Terasaka et al., 2005; Geisler & Murphy 2006; Lewis & Muday 2009), pointing to a role of these plant transporters in growth and development in addition to detoxification processes.

The best characterized plant ABC transporters to date belong to the MRP gene family. The *Arabidopsis* genome includes a small redundant family of 15 *AtMRP* isogenes, which encode vacuolar sequesters of glutathione conjugates potentially involved in phase III of the detoxification process during non-target-site herbicide resistance. Five of the *Arabidopsis* MRPs (*AtMRP1-5*) have been biochemically characterized and shown to encode *bona fide* functional transporters of model glutathione conjugates after heterologous expression in *S. cerevisiae*. In addition, several are able to transport other amphipatic organic anions, such as linear tetrapyrroles derived from chlorophyll catabolism and folates (Lu et al., 1998; Raichaudhuri et al., 2009). Functional analysis of four *Arabidopsis* MRP isogenes (*AtMRP1, 2, 11* and *12*) has shown that their contribution to detoxification is marginal, as the corresponding loss-of-function mutants showed no drastic changes in sensitivity towards different herbicides. Only a knockout allele for *AtMRP2* exhibited reduced sensitivity to 1-chloro-2,4-dinitrobenzene, a classical glutathione *S*-transferase substrate (Frelet-Barrand et al., 2008). These results may be explained by putative high levels of functional redundancy among MRP *Arabidopsis* transporters. On the other hand, it is also likely that the contribution of vacuolar sequestration during phase III of the detoxification process has been largely overestimated, with plasma membrane efflux pumps potentially playing a more preponderant role than recognised so far.

The third group of plant ABC transporters was defined based on similarity to the yeast *PDR5* gene and, in *Arabidopsis*, includes 15 members encoding putative efflux transporters for cytotoxic compounds. Expression of *SpTUR2* from the water plant *Spirodella polyrhiza*, the first plant PDR gene identified, is induced by abscisic acid, cold and salt stresses, suggesting a role for the encoded transporter in the response to abiotic stress (Smart & Fleming, 1996). Interestingly, the herbicide 2,4-D also has an inductive effect on the transcription of *SpTUR2*, and expression of *SpTUR2* in *Arabidopsis* confers increased resistance to the antifungic sclareol and to lead (van den Brule & Smart, 2002). Expression of another member of the PDR gene family, *NpABC1* from wild tobacco, is also strongly induced by sclareol and has been suggested to be involved in its excretion from the tobacco cell (Jasinski et al., 2001). In *Arabidopsis*, three PDR transporters have been characterized at the functional level. These include *AtPDR8*, which is a key factor controlling the extent of cell death upon pathogen infection (Kobae et al., 2006) and contributes to non-host resistance (Stein et al., 2006), having also been identified as a plasma membrane cadmium extrusion pump (Kim et al., 2007). *AtPDR12* is regulated by multiple plant defence signalling pathways and, as its homologue *SpTUR2*, confers resistance to the diterpenoid sclareol (Campbell et al., 2003) and lead (Lee et al., 2005). In addition, *AtPDR12* was recently reported to be a plasma membrane ABA uptake transporter (Kang et al., 2010). Importantly, another *Arabidopsis* PDR member, *AtPDR9*, a homologue of *S. cerevisiae PDR5*, has recently been shown to encode a 2,4-D efflux facilitator localized in the plasma membrane; while overexpression of *AtPDR9* leads to 2,4-D resistance and hypoaccumulation, a loss-of-function mutant for this gene displays increased sensitivity and hyperaccumulation of the herbicide (Ito & Gray, 2006). Results from a more recent study suggest that *AtPDR9* also facilitates the efflux of the auxin precursor IBA from root cells (Strader & Bartel, 2008). Although none of the 23 *PDR* genes identified in *Oryza sativa* has been functionally analysed so far, expression of many of these genes is induced in response to diverse abiotic stresses (Moons, 2008), suggesting that these transporters may also play a role in conferring resistance to chemical stress agents in rice.

3.2.2 MFS-MDR transporters

The MFS is a very large and ancient superfamily found throughout nature, from bacteria to mammals, and is energised by the electrochemical gradient across membranes. Diverse substrates show affinity to these transporters in a proton motive force dependent mechanism of symport, antiport or uniport. The MFS proteins are involved in drug resistance, among other cellular functions (Paulsen et al., 1996). Among the eleven known families of MFS transporters in yeast (www.tcdb.org), MFS multidrug resistance (MFS-MDR) efflux pumps are proposed to be drug:H^+ antiporters, belonging to the DHA1 family (12 predicted spanners) or the DHA2 family (14 predicted spanners). These transporters comprise at least 23 proteins that have largely escaped characterisation by classical approaches, as most of the present knowledge on these putative drug:H+-antiporters was driven by the disclosure of the *S. cerevisiae* genome sequence in April 1996 (Goffeau et al., 1996; Sá-Correia et al., 2009).

Among the MFS-MDR, Tpo1p, which belongs to the DHA1 family, is the transporter with the broadest described range of substrates. In addition to polyamine excretion, *TPO1* is involved in resistance to cytotoxic compounds ranging from antimalarials to herbicides and fungicides (Alenquer et al., 2006; Markovich et al., 2004; Sá-Correia et al., 2009; Tomitori et al., 2001; Tomitori et al., 1999; Tucker & Fields, 2001; Kennedy & Bard, 2001). This transporter is a yeast determinant of resistance to the auxin-like herbicides 2,4-D and MCPA (Teixeira & Sá-Correia 2002). In addition, the DHA1 transporters Qdr1 (Vargas et al., 2004), Qdr2 (Vargas et al., 2004) and Qdr3 (Tenreiro et al., 2005) were found to be determinants of yeast resistance to the herbicide barban.

Although several *S. cerevisiae* MFS transporters have been linked to MDR, this is not the case in plants. The *Arabidopsis* genome encodes a wide array of metabolite transporters, around 100 within the MFS family alone. Available sequence data indicate that a similar situation exists in other plant species as well (Williams et al., 2000). The best characterised subfamily of MFS transporters in plants is the monosaccharide-like transporters one, whose several members are indeed involved in sugar transport (Reinders et al., 2006; Buttner 2007). Another subfamily has been implicated in phosphate transport (Stefanovic et al., 2007; Mudge et al., 2002). To date, only two plant MFS transporters have been related to chemical stress resistance: *A. thaliana ZIF1* is a tonoplast-localized protein that influences zinc tolerance and accumulation (Haydon & Cobbett, 2007) and, in maize, Zm-mfs1 was identified as a transcript induced by a range of defence-related conditions, including fungal pathogenic infection (Simmons et al., 2003). By direct *in silico* methods few MFS-MDR family members have been predicted in *A. thaliana* or in any higher eukaryote, including humans.

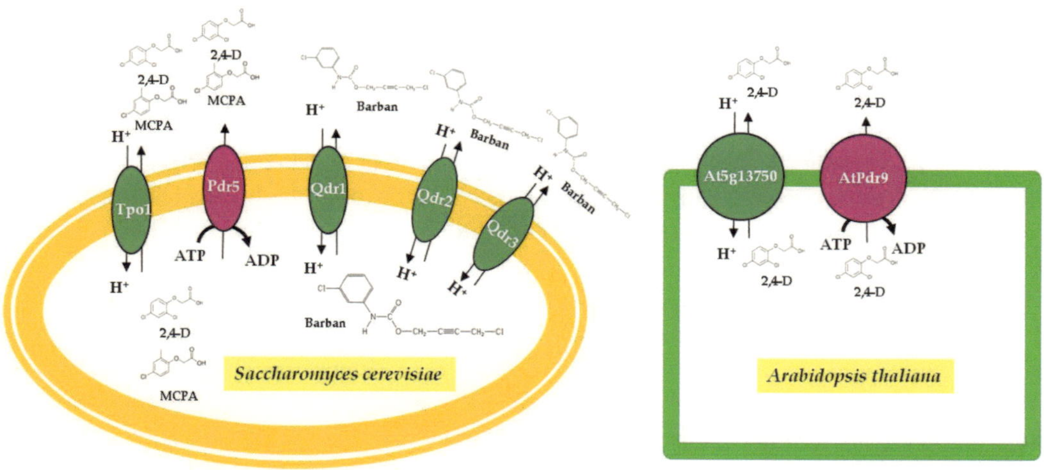

Fig. 1. Proposed model for the role of the so far described *S. cerevisiae* and *A. thaliana* ABC efflux pumps (purple) and MFS-MDR transporters (green) involved in herbicide stress resistance. The role of At5g13750 in 2,4-D stress resistance in plants, proposed based on the results obtained through heterologous expression in yeast, remains to be clarified.

Mima *et al.* (2007) identified a human orthologue of the yeast MFS-MDR gene *TPO1*, *TETRAN*. Although sequence homology was relatively low, the authors were able to show that both *TPO1* and *TETRAN* are capable of conferring NSAIDs (non-steroidal anti-inflammatory drug) resistance. There are also reports that plant pathogenic fungi utilise MFS drug:H^+ antiporters to export their own toxins, thus rendering themselves resistant, while delivering toxins to the plant (Del Sorbo et al., 2000).

The data linking MFS transporters and MDR in higher eukaryotes are less numerous and conclusive than those available for ABC transporters, and the role of the plant MFS is just beginning to be unravelled. This could reflect increased functional redundancy among the MFS transporter family, but also perhaps decreased interest in analysing their involvement in MDR. It is consequently of utmost importance to further investigate the potential role of MFS transporters in the MDR phenomenon in higher plants. Results stemming from the combined study described below strongly strengthen the hypothesis that a subfamily of plant MFS transporters may be involved in MDR. Furthermore, it is expected that genome-wide expression analysis applied to the study of plant response to stress induced by herbicides may bring additional clues to the functional analysis of MDR transporters in plants. For example, the observed up-regulation of the putative MFS encoding ORF *At1g79410* in *A. thaliana* exposed to herbicidal concentrations of 2,4-D suggests that this transporter may play a role in the adaptation to this stress (Raghavan et al., 2005).

4. Mechanisms underlying auxin-like herbicide resistance in *S. cerevisiae* and *A. thaliana*: a case study

2,4-dichlorophenoxyacetic acid (2,4-D) is the most commonly used member of the auxin-like herbicide family and, having been introduced in 1946, is still one of the most widely used herbicides in the world. The effects of 2,4-D in weeds include epinastic bending and growth abnormalities, its toxicity depending mainly of the acid form (Twonson, 1997). Recent genome-wide gene expression studies focused on the *Arabidopsis thaliana* response to herbicidal concentrations of 2,4-D report the remodelling of its transcriptional repertoire at the level of genes involved in the auxin response, ethylene signalling and ABA biosysthesis, signalling and response (Raghavan et al., 2005; Raghavan et al., 2006). *S. cerevisiae* has been intensively used as a model to investigate the mechanisms underlying herbicide resistance, focusing on 2,4-D. In low pH environments, the highly lipophilic weak acid 2,4-D exists in its undissociated lipophilic toxic form (RCOOH), which can readily cross the plasma membrane by passive diffusion. In the neutral cytosol, the 2,4-D molecule dissociates leading to internal acidification (Fernandes et al., 2003; Simões et al., 2003) and to accumulation of the toxic counter-ion ($RCOO^-$) that cannot easily cross the plasma membrane lipid bilayer. Therefore, at low pH (e.g. acidic soils, the alimentary canal of animals) the toxic potential of the herbicide increases dramatically (Cabral et al., 2003). In acidified growth medium, suitable for fungal growth, yeast cells challenged with the herbicide 2,4-D suffer a strong reduction in their cytosolic and vacuolar pH (Fernandes et al., 2003; Simões et al., 2003), which is counteracted by the activation of the plasma and vacuolar membrane H^+-ATPases (Fernandes et al., 2003; Teixeira et al., 2005). Significantly, auxins were also shown to induce the activity of the *Arabidopsis* plasma membrane H^+-ATPase, contributing to maintain the intracellular pH in plant roots (Shen et al., 2006).

Based on the participation of the MDR transporters Tpo1p and Pdr5p (belonging to the MFS and ABC superfamily, respectively) in yeast resistance to 2,4-D, the active export of the 2,4-D counter-ion catalysed by these plasma membrane transporters was postulated (Teixeira & Sá-Correia, 2002). Interestingly, the *Arabidopsis PDR5* homologue *AtPDR9* was recently shown to encode a 2,4-D plasma membrane efflux facilitator contributing to 2,4-D resistance in this plant (Ito & Gray, 2006). The expression of the *S. cerevisiae TPO1* gene was shown to decrease the accumulation of the herbicide 2,4-D counter-ion in yeast cells, indicating that this gene is also, directly or indirectly, involved in 2,4-D export (Cabrito et al., 2009a). Other details on the yeast adaptive response to the mechanisms underlying 2,4-D toxic effects were reviewed by Teixeira *et al.* (2007) and are summarized in Figure 2.

Since the *TPO1* gene was previously found to confer resistance to 2,4-D in yeast (Teixeira & Sá-Correia, 2002), to be transcriptionally activated in response to the herbicide (Teixeira & Sá-Correia, 2002; Teixeira et al., 2006), and required to reduce the intracellular concentration of the 2,4-D counter-ion (Cabrito et al., 2009a), ScTpo1p homologs encoding putative plasma membrane MFS transporters from the plant model *A. thaliana* were analysed by Cabrito *et al.* (2009) for a possible role in 2,4-D resistance. *At5g13750* transcript levels were found to increase in 2,4-D stressed plants. The functional heterologous expression of this plant ORF in yeast was found to confer increased resistance to the herbicide in wild-type and *Δtpo1* cells, through the reduction of the intracellular concentration of 2,4-D counter-ion. Heterologous expression of *At5g13750* in yeast also leads to increased resistance to Al^{3+} and Tl^{3+}. Hence, *At5g13750* gene-encoded protein is the first plant putative MFS transporter to be suggested as possibly involved in MDR (Cabrito et al., 2009a). These new insights suggesting a role for higher eukaryotic MFS transporters in multidrug resistance may open an entirely new field of research with promising repercussions not only in agriculture but also in medicine and biotechnology.

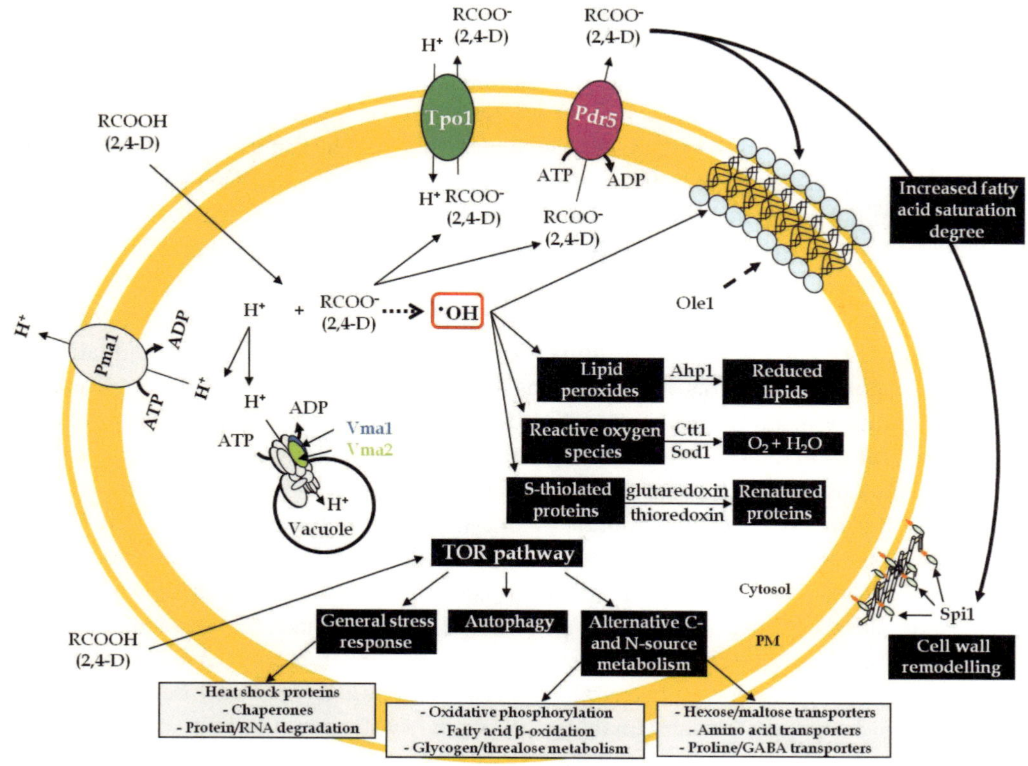

Fig. 2. Model for the adaptive yeast response to 2,4-D.

5. Conclusions and perspectives

Multiple herbicide resistance is a widespread phenomenon among weed species and poses severe agro-economical and environmental problems (Hayes & Wolf, 1997). At the same time, herbicide-resistant crop strains have been developed during the past decades, altering profoundly agricultural practices as this approach allows the use of specific herbicides in previously sensitive crops. The use of genetic engineering to express multidrug efflux pumps, identified as specific for a given herbicide, has been suggested as a method to increase crop resistance to that particular herbicide (Windsor et al., 2003). Thus, the identification of new MDR plant transporters playing a role in herbicide resistance may prove an invaluable asset in modern and future agricultural practices. Such studies are facilitated by the use of *S. cerevisiae* as a eukaryotic model system in the search for herbicide resistance determinants. Similar extrapolations have been made for agricultural fungicide resistance in phytopathogenic fungi (Cabrito et al., 2009b). The use of yeast as a host for the heterologous expression of plant transporters has also been considered. Although a number of reports have found this task extremely difficult mainly due to the inhibitory effect that the overexpression of foreign membrane proteins appears to have on yeast viability (Crouzet et al., 2006; Ito & Gray, 2006; van den Brule & Smart, 2002), Cabrito *et al.* (2009a) managed to optimize heterologous expression conditions in order to functionally express the plant gene *At5g13750* in the yeast cell. These results reinforce the notion that *S. cerevisiae* is highly suitable not only as a eukaryotic model system but also as an experimental platform for the study of the MDR phenomenon in higher eukaryotes. Furthermore, approaches combining *S. cerevisiae* and the plant *A. thaliana* as a model for functional and comparative genome-wide studies in weed species can help us understand better the mechanisms underlying herbicide resistance.

Based on the herbicide resistance mechanisms identified so far, different approaches aimed at improving the resistance of crops against various herbicides have been developed in the last decade. Initial practices relied on the type and application rate of the herbicide used in the agrosystem. For instance, herbicide tolerance can be improved by the use of safeners, synthetic compounds that enhance herbicide tolerance in selected monocot crops without impairing herbicide susceptibility in target weeds. They are suggested to reduce toxicity in crops by inducing the expression of glutathione-S-transferases capable of conjugating the herbicide to glutathione. This approach has been successful in different crops, such as maize, sorghum, wheat, rice and barley (DeRidder & Goldsbrough, 2006). The second and most successful contribution of modern agricultural biotechnology was the introduction of herbicide resistance traits in crops. Traditional methods involved plant breeding, but in the last decade the most promising approach consists in the generation of transgenic crops exhibiting cross resistance to various classes of herbicides. In 2005, only three years after the introduction of the first transgenic crop in the field, over 52 million hectares of transgenic crops were planted worldwide, and from these 41 million hectares were planted with herbicide-resistant crops (Owen & Zelaya, 2005). The vast majority of transgenic crops consists of soybean containing a bacterial gene that encodes a glyphosate-insensitive form of the enzyme 5-enolypyruvylshikimate-3-phosphate synthase (EPSPS), catalyzing the penultimate step of the shikimate pathway, the glyphosate target. This transgenic soybean enables glyphosate, a non selective herbicide active on nearly all plant species, to be used selectively after soybean emergence (Dill, 2005). Similar approaches are currently being pursued for further development. For example, expression of the human cytochrome P450 monooxygenase encoded by the *CYP1A1* gene in rice plants has been recently shown to confer broad cross resistance towards various herbicides with different structures and mode of action, via an increased and quicker metabolism of these herbicides (Kawahigashi et al., 2007).

Similarly, the identification of an MDR transporter functioning as a multi-component detoxification system in plants could open the possibility of developing engineered plants, thus achieving useful phenotypes from the agricultural point of view. Plants grow in diverse environments, in which their roots are exposed to toxic or inhibitory chemical substances. Toxins can be pollutants, such as herbicides or the xenobiotic products of neighbouring plants or endemic microorganisms. Phytoremediation – i.e. the use of plants and their associated microbes for the remediation of soils contaminated with organic and inorganic pollutants – can also represent a positive outcome of the MDR phenomenon. The phytoremediation efficiency of plants can also be substantially improved by engineering technologies, including overexpression of genes involved in the transport and sequestration of toxic compounds. The development of transgenic plants in heavy metal and metalloid soil decontamination has been largely prospected and revealed their usefulness and suitability (Kotrba et al., 2009). Again, identification of MDR efflux pumps involved in herbicide detoxification could also have an interest in phytoremediation approaches of herbicide-contaminated soils.

6. Acknowledgements

"Research carried out by the authors in this context is supported by FEDER and "Fundação para a Ciência e a Tecnologia" (FCT) (Contracts PTDC/AGR-AAM/67858/2006, PTDC/BIA-MIC/72577/2006 and PTDC/AGR-AAM/102967/2008 and PhD and post-doctoral grants to TRC and ER, respectively).

7. References

Alenquer, M., S. Tenreiro & I. Sá-Correia (2006) Adaptive response to the antimalarial drug artesunate in yeast involves Pdr1p/Pdr3p-mediated transcriptional activation of the resistance determinants *TPO1* and *PDR5*. *FEMS Yeast Res,* 6, 1130-9.

Arntzen, C. J., K. Pfister & K. E. Steinback. 1982. The mechanism of chloroplast triazine resistance: Alterations in the site of herbicide action. In *Herbicide Resistance in Plants,* 185–214. New York: Wiley: ed. HM LeBaron, J Gressel.

Balzi, E. & A. Goffeau (1995) Yeast multidrug resistance: the PDR network. *J Bioenerg Biomembr,* 27, 71-6.

Bartholomew, D. M., D. E. Van Dyk, S. M. Lau, D. P. O'Keefe, P. A. Rea & P. V. Viitanen (2002) Alternate energy-dependent pathways for the vacuolar uptake of glucose and glutathione conjugates. *Plant Physiol,* 130, 1562-72.

Bolhuis, H., H. W. van Veen, B. Poolman, A. J. Driessen & W. N. Konings (1997) Mechanisms of multidrug transporters. *FEMS Microbiol Rev,* 21, 55-84.

Bowles, D., J. Isayenkova, E.-K. Lim & B. Poppenberger (2005) Glycosyltransferases: managers of small molecules. *Current Opinion in Plant Biology,* 8, 254-263.

Buttner, M. (2007) The monosaccharide transporter(-like) gene family in *Arabidopsis*. *FEBS Lett,* 581, 2318-24.

Cabral, M. G., C. A. Viegas, M. C. Teixeira & I. Sá-Correia (2003) Toxicity of chlorinated phenoxyacetic acid herbicides in the experimental eukaryotic model *Saccharomyces cerevisiae*: role of pH and of growth phase and size of the yeast cell population. *Chemosphere,* 51, 47-54.

Cabrito, T. R., M. C. Teixeira, A. A. Duarte, P. Duque & I. Sá-Correia (2009a) Heterologous expression of a Tpo1 homolog from *Arabidopsis thaliana* confers resistance to the herbicide 2,4-D and other chemical stresses in yeast. *Appl Microbiol Biotechnol,* 84, 927-936.

Cabrito, T. R., M. C. Teixeira & I. Sá-Correia. 2009b. Global adaptive response and resistance to agricultural fungicides: lessons from yeast and phytopathogenic fungi. In *Fungicides: Chemistry, Environmental Impact and Health Effects,* ed. F. Columbus, 228-253. New York, USA: Nova Science Publishers.

Campbell, E. J., P. M. Schenk, K. Kazan, I. A. Penninckx, J. P. Anderson, D. J. Maclean, B. P. Cammue, P. R. Ebert & J. M. Manners (2003) Pathogen-responsive expression of a putative ATP-binding cassette transporter gene conferring resistance to the diterpenoid sclareol is regulated by multiple defence signaling pathways in *Arabidopsis*. *Plant Physiol,* 133, 1272-84.

Coupland, D. 1985. Metabolism of glyphosate in plants. In *The Herbicide Glyphosate,* eds. E. Grossbard & D. Atkinson, 25–34. London/Boston: Butterworth.

Crouzet, J., T. Trombik, A. S. Fraysse & M. Boutry (2006) Organization and function of the plant pleiotropic drug resistance ABC transporter family. *FEBS Lett,* 580, 1123-30.

Del Sorbo, G., H. Schoonbeek & M. A. De Waard (2000) Fungal transporters involved in efflux of natural toxic compounds and fungicides. *Fungal Genet Biol,* 30, 1-15.

DeRidder, B. P. & P. B. Goldsbrough (2006) Organ-specific expression of glutathione S-transferases and the efficacy of herbicide safeners in *Arabidopsis*. *Plant Physiol,* 140, 167-75.

Dias, P. J., M. C. Teixeira, J. P. Telo & I. Sá-Correia (2010) Insights into the mechanisms of toxicity and tolerance to the agricultural fungicide mancozeb in yeast, as suggested by a chemogenomic approach. *OMICS: A Journal of Integrative Biology,* 14, 211-227.

Dill, G. M. (2005) Glyphosate-resistant crops: history, status and future. *Pest Manag Sci,* 61, 219-24.

Eelen, H., R. Bulcke & D. Callens (1997) Herbicide resistance in blackgrass (*Alopecurus myosuroides* Huds.) in Belgium. *Mededelingen Faculteit Landbouwkundige en Toegepaste Biologische Wetenschappen,* 62, 777-784.

Feng, P. C. C., J. E. Pratley & J. A. Bohn (1999) Resistance to glyphosate in *Lolium rigidum*. II. Uptake, translocation, and metabolism. *Weed Science,* 47, 412-415.

Fernandes, A. R., P. J. Durão, P. M. Santos & I. Sá-Correia (2003) Activation and significance of vacuolar H^+-ATPase in *Saccharomyces cerevisiae* adaptation and resistance to the herbicide 2,4-dichlorophenoxyacetic acid. *Biochem Biophys Res Commun,* 312, 1317-24.

Frelet-Barrand, A., H. U. Kolukisaoglu, S. Plaza, M. Ruffer, L. Azevedo, S. Hortensteiner, K. Marinova, B. Weder, B. Schulz & M. Klein (2008) Comparative mutant analysis of *Arabidopsis* ABCC-type ABC transporters: *AtMRP2* contributes to detoxification, vacuolar organic anion transport and chlorophyll degradation. *Plant Cell Physiol,* 49, 557-69.

Geisler, M. & A. S. Murphy (2006) The ABC of auxin transport: the role of p-glycoproteins in plant development. *FEBS Lett,* 580, 1094-102.

Goffeau, A., B. G. Barrell, H. Bussey, R. W. Davis, B. Dujon, H. Feldmann, F. Galibert, J. D. Hoheisel, C. Jacq, M. Johnston, E. J. Louis, H. W. Mewes, Y. Murakami, P. Philippsen, H. Tettelin & S. G. Oliver (1996) Life with 6000 genes. *Science,* 274, 546, 563-7.

Haydon, M. J. & C. S. Cobbett (2007) A novel major facilitator superfamily protein at the tonoplast influences zinc tolerance and accumulation in *Arabidopsis*. *Plant Physiol,* 143, 1705-19.

Hayes, J. D. & C. R. Wolf (1997) Molecular genetics of drug resistance. *In: Modern Genetics,* v. 3, Amsterdam, the Netherlands: Harwood Academic.

Higgins, C. F. (2007) Multiple molecular mechanisms for multidrug resistance transporters. *Nature,* 446, 749-57.

Ito, H. & W. M. Gray (2006) A gain-of-function mutation in the *Arabidopsis* pleiotropic drug resistance transporter *PDR9* confers resistance to auxinic herbicides. *Plant Physiol,* 142, 63-74.

Jack, D. L., N. M. Yang & M. H. Saier, Jr. (2001) The drug/metabolite transporter superfamily. *Eur J Biochem,* 268, 3620-39.

Jasinski, M., Y. Stukkens, H. Degand, B. Purnelle, J. Marchand-Brynaert & M. Boutry (2001) A plant plasma membrane ATP binding cassette-type transporter is involved in antifungal terpenoid secretion. *Plant Cell,* 13, 1095-107.

Jia, M. H., R. A. Larossa, J. M. Lee, A. Rafalski, E. Derose, G. Gonye & Z. Xue (2000) Global expression profiling of yeast treated with an inhibitor of amino acid biosynthesis, sulfometuron methyl. *Physiol Genomics,* 3, 83-92.

Jungwirth, H. & K. Kuchler (2006) Yeast ABC transporters - a tale of sex, stress, drugs and aging. *FEBS Lett,* 580, 1131-8.

Kang, J., J. U. Hwang, M. Lee, Y. Y. Kim, S. M. Assmann, E. Martinoia & Y. Lee (2010) PDR-type ABC transporter mediates cellular uptake of the phytohormone abscisic acid. *Proc Natl Acad Sci U S A,* 107, 2355-60.

Kartner, N., Riordan, J.R., & Ling, V. (1983) Cell surface P-glycoprotein associated with multidrug resistance in mammalian cell lines. *Science* 221, 1285–1288.

Kawahigashi, H., S. Hirose, H. Ohkawa & Y. Ohkawa (2007) Herbicide resistance of transgenic rice plants expressing human *CYP1A1*. *Biotechnol Adv,* 25, 75-84.

Kelly, D. E., D. C. Lamb & L. Kelly (2001) Genome-wide generation of yeast deletion strains. *Comp Funct Genom,* 2, 236-242.

Kennedy, M. A. & M. Bard (2001) Positive and negative regulation of squalene synthase (*ERG9*), an ergosterol biosynthetic gene, in *Saccharomyces cerevisiae*. *Biochim Biophys Acta,* 1517, 177-89.

Kim, D. Y., L. Bovet, M. Maeshima, E. Martinoia & Y. Lee (2007) The ABC transporter *AtPDR8* is a cadmium extrusion pump conferring heavy metal resistance. *Plant J,* 50, 207-18.

Kitagawa, E., Y. Momose & H. Iwahashi (2003) Correlation of the structures of agricultural fungicides to gene expression in *Saccharomyces cerevisiae* upon exposure to toxic doses. *Environ Sci Technol,* 37, 2788-93.

Klein, M., B. Burla & E. Martinoia (2006) The multidrug resistance-associated protein (MRP/ABCC) subfamily of ATP-binding cassette transporters in plants. *FEBS Letters,* 580, 1112-1122.

Kobae, Y., T. Sekino, H. Yoshioka, T. Nakagawa, E. Martinoia & M. Maeshima (2006) Loss of *AtPDR8*, a plasma membrane ABC transporter of *Arabidopsis thaliana,* causes hypersensitive cell death upon pathogen infection. *Plant Cell Physiol,* 47, 309-18.

Kotrba, P., J. Najmanova, T. Macek, T. Ruml & M. Mackova (2009) Genetically modified plants in phytoremediation of heavy metal and metalloid soil and sediment pollution. *Biotechnol Adv,* 27, 799-810.

Landis, W. G. & M. H. Yu (1999) Introduction to Environmental Toxicology. *In: Impacts of chemicals upon ecological systems.* , Lewis, Boca Raton, FL, USA.

Lee, M., K. Lee, J. Lee, E. W. Noh & Y. Lee (2005) *AtPDR12* contributes to lead resistance in *Arabidopsis*. *Plant Physiol,* 138, 827-36.

Lewis, D. R. & G. K. Muday (2009) Measurement of auxin transport in *Arabidopsis thaliana*. *Nat Protoc,* 4, 437-51.

Lin, R. & H. Wang (2005) Two homologous ATP-binding cassette transporter proteins, AtMDR1 and AtPGP1, regulate *Arabidopsis* photomorphogenesis and root development by mediating polar auxin transport. *Plant Physiol,* 138, 949-64.

Lorraine-Colwill, D. F., S. B. Powles, T. R. Hawkes, P. H. Hollinshead, S. A. J. Warner & C. Preston (2002) Investigations into the mechanism of glyphosate resistance in *Lolium rigidum*. *Pesticide Biochemistry and Physiology,* 74, 62-72.

Lu, Y. P., Z. S. Li, Y. M. Drozdowicz, S. Hortensteiner, E. Martinoia & P. A. Rea (1998) AtMRP2, an *Arabidopsis* ATP binding cassette transporter able to transport glutathione S-conjugates and chlorophyll catabolites: functional comparisons with Atmrp1. *Plant Cell,* 10, 267-82.

Lucau-Danila, A., G. Lelandais, Z. Kozovska, V. Tanty, T. Delaveau, F. Devaux & C. Jacq (2005) Early expression of yeast genes affected by chemical stress. *Mol Cell Biol,* 25, 1860-8.

Markovich, S., A. Yekutiel, I. Shalit, Y. Shadkchan & N. Osherov (2004) Genomic approach to identification of mutations affecting caspofungin susceptibility in *Saccharomyces cerevisiae. Antimicrob Agents Chemother,* 48, 3871-6.

Martinoia, E., M. Klein, M. Geisler, L. Bovet, C. Forestier, U. Kolukisaoglu, B. Muller-Rober & B. Schulz (2002) Multifunctionality of plant ABC transporters - more than just detoxifiers. *Planta,* 214, 345-55.

Martinoia, E., R. Locher & E. Vogt (1993) Inositol Trisphosphate Metabolism in Subcellular Fractions of Barley (*Hordeum vulgare* L.) Mesophyll Cells. *Plant Physiol,* 102, 101-105.

Mima, S., H. Ushijima, H. J. Hwang, S. Tsutsumi, M. Makise, Y. Yamaguchi, T. Tsuchiya, H. Mizushima & T. Mizushima (2007) Identification of the *TPO1* gene in yeast, and its human orthologue *TETRAN,* which cause resistance to NSAIDs. *FEBS Lett,* 581, 1457-63.

Moons, A. (2008) Transcriptional profiling of the PDR gene family in rice roots in response to plant growth regulators, redox perturbations and weak organic acid stresses. *Planta,* 229, 53-71.

Mudge, S. R., A. L. Rae, E. Diatloff & F. W. Smith (2002) Expression analysis suggests novel roles for members of the Pht1 family of phosphate transporters in *Arabidopsis. Plant J,* 31, 341-53.

Owen, M. D. & I. A. Zelaya (2005) Herbicide-resistant crops and weed resistance to herbicides. *Pest Manag Sci,* 61, 301-11.

Owen, M. J., M. Walsh, R. Llewellyn & S. B. Powles (2007) Widespread occurrence of multiple herbicide resistance in western australia annual ryegrass (*Lolium rigidum*). *Aust J Agric Res,* 58, 711-718.

Parveen, M., Y. Momose, E. Kitagawa, S. Kurita, O. Kodama & H. Iwahashi (2003) Bioassay of pesticide Lindane using yeast-DNA microarrays technology. *Chem-Bio Informatics Journal,* 3, 12-29.

Paulsen, I. T. (2003) Multidrug efflux pumps and resistance: regulation and evolution. *Curr Opin Microbiol,* 6, 446-51.

Paulsen, I. T., R. A. Skurray, R. Tam, M. H. Saier, Jr., R. J. Turner, J. H. Weiner, E. B. Goldberg & L. L. Grinius (1996) The SMR family: a novel family of multidrug efflux proteins involved with the efflux of lipophilic drugs. *Mol Microbiol,* 19, 1167-75.

Paumi, C. M., M. Chuk, J. Snider, I. Stagljar & S. Michaelis (2009) ABC transporters in *Saccharomyces cerevisiae* and their interactors: new technology advances the biology of the ABCC (MRP) subfamily. *Microbiol Mol Biol Rev,* 73, 577-93.

Powles, S. B. & Q. Yu (2010) Evolution in action: plants resistant to herbicides. *Annu Rev Plant Biol,* 61, 317-47.

Preston, C. (2004) Herbicide resistance in weeds endowed by enhanced detoxification: complication for management *Weed Sci,* 52, 448–53.

Preston, C., F. J. Tardif, J. T. Christopher & S. B. Powles (1996) Multiple Resistance to Dissimilar Herbicide Chemistries in a Biotype of *Lolium rigidum* Due to Enhanced

Activity of Several Herbicide Degrading Enzymes. *Pesticide Biochemistry and Physiology*, 54, 123-134.

Raghavan, C., E. K. Ong, M. J. Dalling & T. W. Stevenson (2005) Effect of herbicidal application of 2,4-dichlorophenoxyacetic acid in *Arabidopsis*. *Funct Integr Genomics*, 5, 4-17.

Raghavan, C., E. K. Ong, M. J. Dalling & T. W. Stevenson (2006) Regulation of genes associated with auxin, ethylene and ABA pathways by 2,4-dichlorophenoxyacetic acid in *Arabidopsis*. *Funct Integr Genomics*, 6, 60-70.

Raichaudhuri, A., M. Peng, V. Naponelli, S. Chen, R. Sanchez-Fernandez, H. Gu, J. F. Gregory, 3rd, A. D. Hanson & P. A. Rea (2009) Plant Vacuolar ATP-binding Cassette Transporters That Translocate Folates and Antifolates in Vitro and Contribute to Antifolate Tolerance in Vivo. *J Biol Chem*, 284, 8449-60.

Reade, J. P. H., L. J. Milner & A. H. Cobb (2004) A role for glutathione S-transferases in resistance to herbicides in grasses. *Weed Sci*, 52, 468–74.

Reinders, A., A. B. Sivitz, A. Hsi, C. P. Grof, J. M. Perroux & J. M. Ward (2006) Sugarcane ShSUT1: analysis of sucrose transport activity and inhibition by sucralose. *Plant Cell Environ*, 29, 1871-80.

Roepe, P. D., L. Y. Wei, M. M. Hoffman & F. Fritz (1996) Altered drug translocation mediated by the MDR protein: direct, indirect, or both? *J Bioenerg Biomembr*, 28, 541-55.

Sá-Correia, I., S. Santos, M. Teixeira, T. Cabrito & N. Mira (2009) Drug:H^+ antiporters in chemical stress response in yeast. *Trends Microbiol*, 17, 22-31.

Sandermann, H., Jr. (2004) Molecular ecotoxicology of plants. *Trends Plant Sci*, 9, 406-13.

Santos, P. M., T. Simões & I. Sá-Correia (2009) Insights into yeast adaptive response to the agricultural fungicide mancozeb: A toxicoproteomics approach. *Proteomics*, 3, 657-670.

Shen, H., J. Chen, Z. Wang, C. Yang, T. Sasaki, Y. Yamamoto, H. Matsumoto & X. Yan (2006) Root plasma membrane H^+-ATPase is involved in the adaptation of soybean to phosphorus starvation. *J Exp Bot*, 57, 1353-62.

Simmons, C. R., M. Fridlender, P. A. Navarro & N. Yalpani (2003) A maize defence-inducible gene is a major facilitator superfamily member related to bacterial multidrug resistance efflux antiporters. *Plant Mol Biol*, 52, 433-46.

Simões, T., M. C. Teixeira, A. R. Fernandes & I. Sá-Correia (2003) Adaptation of *Saccharomyces cerevisiae* to the herbicide 2,4-dichlorophenoxyacetic acid, mediated by Msn2p- and Msn4p-regulated genes: important role of *SPI1*. *Appl Environ Microbiol*, 69, 4019-28.

Smart, C. C. & A. J. Fleming (1996) Hormonal and environmental regulation of a plant *PDR5*-like ABC transporter. *J Biol Chem*, 271, 19351-7.

Stefanovic, A., C. Ribot, H. Rouached, Y. Wang, J. Chong, L. Belbahri, S. Delessert & Y. Poirier (2007) Members of the *PHO1* gene family show limited functional redundancy in phosphate transfer to the shoot, and are regulated by phosphate deficiency via distinct pathways. *Plant J*, 50, 982-94.

Stein, M., J. Dittgen, C. Sanchez-Rodriguez, B. H. Hou, A. Molina, P. Schulze-Lefert, V. Lipka & S. Somerville (2006) *Arabidopsis PEN3/PDR8*, an ATP binding cassette transporter, contributes to nonhost resistance to inappropriate pathogens that enter by direct penetration. *Plant Cell*, 18, 731-46.

Strader, L. C. & B. Bartel (2008) A new path to auxin. *Nat Chem Biol,* 4, 337-9.

Teixeira, M. C., P. Duque & I. Sá-Correia (2007) Environmental genomics: mechanistic insights into toxicity of and resistance to the herbicide 2,4-D. *Trends Biotechnol,* 25, 363-70.

Teixeira, M. C., A. R. Fernandes, N. P. Mira, J. D. Becker & I. Sá-Correia (2006) Early transcriptional response of *Saccharomyces cerevisiae* to stress imposed by the herbicide 2,4-dichlorophenoxyacetic acid. *FEMS Yeast Res,* 6, 230-48.

Teixeira, M. C. & I. Sá-Correia (2002) *Saccharomyces cerevisiae* resistance to chlorinated phenoxyacetic acid herbicides involves Pdr1p-mediated transcriptional activation of *TPO1* and *PDR5* genes. *Biochem Biophys Res Commun,* 292, 530-7.

Teixeira, M. C., P. M. Santos, A. R. Fernandes & I. Sá-Correia (2005) A proteome analysis of the yeast response to the herbicide 2,4-dichlorophenoxyacetic acid. *Proteomics,* 5, 1889-901.

Tenreiro, S., R. C. Vargas, M. C. Teixeira, C. Magnani & I. Sá-Correia (2005) The yeast multidrug transporter Qdr3 (Ybr043c): localization and role as a determinant of resistance to quinidine, barban, cisplatin, and bleomycin. *Biochem Biophys Res Commun,* 327, 952-9.

Terasaka, K., J. J. Blakeslee, B. Titapiwatanakun, W. A. Peer, A. Bandyopadhyay, S. N. Makam, O. R. Lee, E. L. Richards, A. S. Murphy, F. Sato & K. Yazaki (2005) PGP4, an ATP binding cassette P-glycoprotein, catalyzes auxin transport in *Arabidopsis thaliana* roots. *Plant Cell,* 17, 2922-39.

The Arabidopsis Genome, I. (2000) Analysis of the genome sequence of the flowering plant *Arabidopsis thaliana. Nature,* 408, 796-815.

Tomitori, H., K. Kashiwagi, T. Asakawa, Y. Kakinuma, A. J. Michael & K. Igarashi (2001) Multiple polyamine transport systems on the vacuolar membrane in yeast. *Biochem J,* 353, 681-8.

Tomitori, H., K. Kashiwagi, K. Sakata, Y. Kakinuma & K. Igarashi (1999) Identification of a gene for a polyamine transport protein in yeast. *J Biol Chem,* 274, 3265-7.

Tucker, C. L. & S. Fields (2001) A yeast sensor of ligand binding. *Nat Biotechnol,* 19, 1042-6.

Twonson, J. K. 1997. Herbicide Resistance. In *Molecular Genetics of Drug Resistance.,* ed. J. D. Hayes, Wolf, C.R. , 139-174. Amsterdam: Harwood Academic Publishers.

van den Brule, S. & C. C. Smart (2002) The plant PDR family of ABC transporters. *Planta,* 216, 95-106.

Vargas, R. C., S. Tenreiro, M. C. Teixeira, A. R. Fernandes & I. Sá-Correia (2004) *Saccharomyces cerevisiae* multidrug transporter Qdr2p (Yil121wp): localization and function as a quinidine resistance determinant. *Antimicrob Agents Chemother,* 48, 2531-7.

Williams, L. E., R. Lemoine & N. Sauer (2000) Sugar transporters in higher plants - a diversity of roles and complex regulation. *Trends in Plant Science,* 5, 283-290.

Windsor, B., S. J. Roux & A. Lloyd (2003) Multiherbicide tolerance conferred by AtPgp1 and apyrase overexpression in *Arabidopsis thaliana. Nat Biotechnol,* 21, 428-33.

Yuan, J. S., P. J. Tranel & C. N. Stewart, Jr. (2007) Non-target-site herbicide resistance: a family business. *Trends Plant Sci,* 12, 6-13.

Yun, M. S., Y. Yogo, R. Miur, Y. Yamasue & A. J. Fischer (2005) Cytochrome P-450 monooxygenase activity in herbicide-resistant and -susceptible late watergrass (*Echinochloa phyllopogon*). *Pestic. Biochem. Physiol.* , 107–14.

Part 7

Herbicide Applications and Modeling

32

Herbicides Applications: Problems and Considerations

Qasem, Jamal R.
University of Jordan
Jordan

1. Introduction

Methods of weed control are many and varied including prevention, mechanical, cultural, physical, biological and chemical (Qasem, 1992), each may be used alone or all are usually integrated for any successful weed control program (Qasem, 2003; Singh et al., 2006). However, weed control is jointly used with other farm operations for efficient crop production. Herbicides are chemicals inhibit or kill weeds, do not harm crops if properly handled and selectively used. They are organic or inorganic (contain no carbon) chemicals (California Weed Conference, 1985) and can be easily designated from botanical or micoherbicides (Rice, 1983, Mo & He, 2005). Although the use of herbicides for weed control create public concern and receives much criticisms nowadays at which most naturalists and environments oppose their use and other pesticides, recommend alternative weed control methods or natural eco-friendly chemicals, but synthetic herbicides remained widely used, considered for weed control and intensively applied in developed and many developing countries. Herbicides however, represent 44% of the world's pesticide markets while out of which 57% is for the USA (Kiely et al., 2004). In developing countries, chemical weed control is not widely practiced because of relatively cheap labor, high chemicals prices and lack of technical extension and experience in herbicides application which leads in most cases to the misuse of these chemicals and crop injury (Su, 2006), failure of selective herbicides and weed control operation, soil and air pollution and limitation in crop rotation options. Different factors are occasionally involved in herbicides failure under field conditions which most due to human mistakes lead to improper application technique (Ross & Lembi, 1999). In addition, lack of correct diagnosis of the weed problem, selection of incorrect herbicide and prevalence of unsuitable weather conditions for chemical weed control are all reasons behind failure of the herbicide in controlling existing weeds (Gwynne & Murray, 1985). In this chapter, proper handling, precautions and some considerations in herbicides application in the field are considered and discussed as factors contribute in the success of chemical weed control; enable farmers to effectively control weeds in crop fields with low cost and more economical, less hazardous and low environmental pollutions.

2. Important

The chapter deals with problems and considerations in herbicides applications mainly facing farmers in the field. Factors affecting herbicides application and performance under field conditions including herbicidal, plant and environmental factors are discussed. Technical problems that lead to improper application and failure of the herbicide are covered. Aspects related to sprayer calibration, herbicide labels, methods of application and requirements, surfactants, chemical drift, formulations, selectivity, mixtures, development of herbicide resistance, herbicides stability and effectiveness in the field, integration of chemical control with other methods of weed control are all dealt with. The chapter also covered the interaction between weeds and crop plants and the responses of both to herbicides and environmental conditions. Impact of the interaction between the three components on the success of chemical weed control operation, crop production and clean environment was discussed. Measures aimed to reduce crop injuries, herbicide persistence, development of weed tolerance/resistance, weed races and crop relatives, and shift in weed population are suggested. Therefore the chapter addressed the proper handling and application of herbicides as an effective weed control technology and their proper application in the field.

3. Information

3.1 Considerations

3.1.1 Weed factors

3.1.1.1 Weed identification

Diagnosis of the weed species found in the field is the first step toward any successful weed control program. Generally, weeds are narrow- or broad-leaved, annuals, biennials or perennials. Biennial weeds are few in number, most are prolific seed producers and many are ruderals (Grime, 1986); annuals reproduce from seeds, mostly short-lived terminate their life cycles before crop mature and can be easily controlled by other weed control methods or with contact herbicides (Qasem, 2003). Weeds, however, are different in their competitive abilities that determine the extent of yield loss and other negative effects of weeds on crop plants, the need for control and time required (Zimdahl, 1980). They are also widely differing in their physiology, biology and biochemistry. Yield losses although are different from one weed to another since differ in requirements and growth habits but competition is greatly influenced by weed density (Ziomdahl, 1980). Therefore, considering the critical density at which the weed must be controlled is important if to avoid any significant yield loss (Qasem, 2009). Although weed growth in certain important cash crops, is not allowed at any density eventhough for a short period. However, as a general rule weeds of similar requirements and growth habit to crop plants are known to be more competitive and cause greater yield loss and higher damage to crop plants than weeds of dissimilar requirements or morphology (Ross & Lembi, 1999).

3.1.1.2 Weed morphology and herbicide application

It is generally approved that narrow- leaved weeds intercept and retain less amount of spray solution on their vegetative parts than broad-leaved weeds , growing points or shoot apexes in many species of these are encased (e.g. cereals) by coleoptiles and thus protected against spray solution or herbicide droplets. In contrast, shoot apexes of broad-leaved weeds are directly exposed to herbicide solution; have horizontally extended leaves and thus better exposed to herbicide and subsequently intercept and retain higher amounts of spray solution.

Early weed (< 10 cm in height) control is always recommended. Weeds at early growth stage are small, short, their competition effect not started yet or minimum and thus weed management by herbicides is quite possible at the recommended or even lower rates of herbicide application beside they require low volume of spray solution for thorough coverage with contact herbicides. Annual weeds and to some extent biennials may be easily controlled using contact herbicides since reproduce by seeds, while creeping perennials are generally difficult to control by such chemicals since vegetatively reproduce from the below or sometimes above ground organs, and thus need translocated herbicides while repeat application may be mostly necessary. However, full coverage of the weed vegetative parts with contact herbicide solution is essential since only plant parts that became in contact with spray solution are affected while others are not, these may remain photosynthesizing, recover from herbicide injuries, then weed resume growth and successfully complete its life cycle.

However, morphological differences between the two groups may not be always the base since depend on ecology at which these weeds are growing (Grime, 1986; Holt, 2006). Broad-leaved weeds may have hairy, leathery small leaves with a low number of stomata on the upper surface or have some morphological modification under certain conditions (e.g. in arid and semi-arid regions) which may challenge chemical control at normal rates. The opposite is also true at which narrow-leaved weeds could be easily controlled at the recommended or lower rates of herbicides when found in moist, fertile soils or humid regions. Weeds under such conditions may lack or have a very thin waxy layer on leaf surface, absence of hairs with a high number of stomata projected on the upper surface and high physiological activity in general. It is well established that weeds are more susceptible to herbicides at early than at late growth stages which may be largely due to morphological and/or physiological sensitivity to herbicides.

Annual weed control should be carried out from after emergence to the full vegetative or pre-flowering stage. Weed control at this period prevents seed production and competition effect of tall growing weeds. However, this could be easily achieved using contact general herbicides such as glufosinate- ammonium, paraquat, bromoxynil, oxyfluorfen and others.

Perennial weed control using herbicides must be timed at three main important and critical growth stages of weeds, these are as follows:

1. Before establishment, at which weed seedlings can be easily controlled by a nonselective contact herbicide, but sometimes repeat application may be necessary.
2. After establishment, at which weeds are treated with translocated herbicides at full vegetative/pre-flowering growth stage, during which most of the reserved food in regenerative organs is directed toward the above ground parts for flowering and thus stored food left in the below-ground organs at this time is at low ebb. Although the symplast translocation of herbicides from shoots to the below ground parts with photosynthate materials is opposing the upward movement of stored food at this stage which may be more rapid and occurs at a higher rates. Killing or harming the above ground vegetative parts could result in a serious damage to weeds or even their death because of deficient food in perennating organs.
3. At late growth stage, at which perennial weeds may be effectively controlled by translocated herbicides late in the growing season and before plants loose their green color. At this stage the herbicide easily translocates with carbohydrates to the below ground parts and accumulates where carbohydrates are finally stored. Herbicide movement from the above to the below ground parts normally occurs and possible re-vegetation of treated weeds is slim.

Sometimes translocated herbicide may be followed with a contact herbicide after a while. This complementary treatment may be effective in controlling difficult perennial weeds by allowing transloacted herbicides reaching the below ground parts and then prevention of any supply of photosynthate materials using foliage contact applied herbicide.

Chemical weed control however, is generally not recommended at late growth stages because of the following:

1. Weed competition effects on crop plants and thus on yield is irreversible at late growth stages.
2. Weed physiology (absorption, translocation and metabolism) is low, leaf surface may be heavily covered by wax and hairs and thus herbicide treatment may not be effective in most cases.
3. Weed morphology and anatomy is well differentiated and mostly do not facilitate herbicide penetration through thick cuticle layer on leaf surface and hard tissues on other foliage parts. Movement of the herbicide within plant tissues is becoming more difficult due to deposition of cutin, wax, cellulose and other byproducts or secondary metabolites in cell sap and on tissues themselves.
4. Weed vegetative mass may be thick enough, mostly not possible to fully cover with spray solution and/or needs a high volume of spray solution, and therefore possible recovery from herbicide treatment is high.
5. Spray retention is lower on mature than young plants since cuticle and thus waxy layer on leaves in the latter are not complete.
6. In most cases weeds reached seeding stage or finished their life cycle, therefore, any disturbance may help their seed dissemination, and herbicide application may not be effective in preventing weed seed production or maturation.
7. Herbicide possible residue in crop materials (mainly straw and grains) at harvest is high and the same is the run-off herbicide solution into the soil.
8. Cost of weed control may much exceeds the return yield value, considering that more crop plants are destroyed or crushed by labors and spray machines.

Soil applied herbicides should be used at a certain time when weed seeds are readily germinating and most sensitive to herbicides. At pre-germination stage, seeds may easily absorb herbicide with soil water; imbibed seeds take up an amount of the herbicide solution enough to kill the embryo. Therefore application of herbicides to control dormant weed seeds or during drought conditions should be avoided since of low or no response of the physiologically inactive weed seeds. It is suggested that these herbicides may be applied after seed bed preparation and before sowing or planting and may be followed by irrigation or rain that activate both the herbicide molecules (Friesen & Dew, 1966), and seed absorption, imbibition and germination. It is usually recommended that soil applied herbicides should be selected according to the type of weed to be treated, cultivated crop and soil type, while most of these chemicals are susceptible to soil microorganisms. However, when shallow-root annual weeds present in a perennial deep rooted crop it becomes important to select a soil surface active, none-leachable herbicide and the opposite is true for deep-rooted weeds in shallow-rooted perennial crops. The adsorptive relationship between the herbicide and treated soil, the size, diversity and activity of soil microbes' population, soil organic matter content, pH, and soil type are all factors determining the rate of herbicide application and its degradation. While leachability of the herbicide through the soil is determined by the adsorptive relationship between herbicide molecules and soil colloids and the amount of water passing through the soil.

3.1.1.3 Weed height and vegetative mass

Large growing weeds require higher volume of spray solution especially from contact herbicides. In this case weed control may becomes more expensive, time and labor consuming. Tall growing weeds could escape chemical control because of location of their growing points and shoot apexes which are not possibly hit by herbicide droplets since placed at a higher level above spryer boom that is usually fixed at a standard regular height in tractor-mounted, pull-type and self-propelled sprayers. Uniform spray is also difficult to achieve having tall and large growing weeds.

3.1.2 Crop factors

3.1.2.1 Crop species /varieties

Crop species are different in their sensitivity and responses to herbicides. Differences have been also reported between different crop cultivars in response to the same or different herbicides (Duwayri & Saghir, 1983; Felix et al., 2007; Abit at al., 2009; Kong et al., 2009 & Jin et al., 2010) hence differ in morphology, physiology, growth habit and phenology. In addition, cultivars are different in germination, emergence, growth development and duration, physiological and biochemical responses (Grime, 1986). All greatly affect herbicides performance in plants, their effectiveness and thus herbicides physiological, physical and biochemical degradation.

3.1.2.2 Crop growth and herbicide tolerance

As early mentioned, tolerance of crop plants to herbicides determines herbicide selectivity and its safe uses. Crop plants similar to weeds in that are more susceptible to herbicides early in the growth at which crop injury is commonly observed. Anatomy, morphology, succulence and physiology of crop seedlings are all factors determine the extent of herbicide effectiveness and crop injuries. It is generally approved that crop plants grown from seeds are more susceptible to herbicides than those from transplants and therefore, herbicides sensitive crops are recommended to be grown from seedlings instead of seeds when soil applied herbicides are used. Foliage applied herbicides are well effective against weeds at early (1-5 leaf) growth stages while crop seedlings at this stage may also show some injury symptoms which could recover from otherwise a more selective and safe herbicide must be used. This depends on time and rate of herbicide application (Friesen, 1967; Friesen et al., 1968; Carter et al., 2007) and prevailing environmental conditions (Coupland, 1987). Selectivity of the herbicide is highly dependent on crop internal factors as well as climatic conditions that activate/ inactivate herbicide molecules. For example, it is well established that crop plants have certain period/s in the life cycles during which physiological activity is low and thus absorption and translocation of herbicide may become very limited. Conversion of the herbicide from a toxic to a non- or less toxic form or the opposite is also affected and varied at different growth stages. Active growing weeds are required for effective chemical control. For example, application of Diclofop-methyl, Barban, Diallate, Triallate, Difezoquat, Bromoxynil in cereals, and many others are usually recommended at early growth of weeds and before reaching 5-leaf stage (Virginia Polytechnic Institute and State University, 1999a & b; Mu et al., 2007). The same is also true with other herbicides in vegetable crops such as oxadiazon (Fig. 1) and oxyfluorfen for weed control in onions

Fig. 1. Selective control of *Ammi majus* L. in onion with post-planting treatment of oxadiazon. Treated onion rows (left) and untreated rows (right).

(Qasem, 2005). On the other hand, 2,4-D and MCPA are best applied at full tillering stage of cereal (wheat and barley) crops and when crop plants are at resting stage and exhibit low physiological activities. Weeds at the same time are physiologically active; their cell division and cell elongation rates are high and thus accumulate lethal doses from these herbicides in embryonic tissues (Fig. 2). Differences in selectivity between legume crops and their associated broad-leaved weeds to 2,4-DB depends mainly on the ability of these to accumulate and convert large amounts from this herbicide to 2,4-D (more toxic) and the rapid conversion process. Weeds accumulate and convert larger amount within a relatively short period and thus affected more rapidly than crop plants that escape injuries because of

Fig. 2. Selective control of broad-leaved weeds by 2, 4-D in wheat field.

slow conversion process and the non-lethal level accumulated in their tissues. Therefore, herbicide application is a skill that allows farmers take an advantage from morphological, physiological or biochemical differences between weeds and crop plants in response to herbicides and thus may be considered as an opportunistic operation should be implemented on time and when conditions permit.

3.1.2.3 Crop morphology

Crops are different from above ground morphology; some are narrow-leaved while others are broad-leaved. The same as for weeds leaf arrangements on the stem and their display are also different for different crop species, in addition to differences in morphology size and heights. All affect herbicide spray retention on crops vegetative parts, herbicide selectivity and safe use on these crops. Narrow- leaved crops with somehow vertically arranged and less displayed leaves receive less volume of foliage-applied spray solution and the opposite is true for broad-leaved crops. In addition, the deposited waxy layer on leaves of different crop species or varieties is widely different which in turn affects the amount of herbicide intercept and retained on leaves and subsequently their absorption rates. Crucifers and onions are examples on crops of wax covered leaves and thus their wettabilty with herbicide solution may not be possible without surfactants and the run-off into the soil surface is quite high. Crop species are different in location of their growing points, amounts of embryonic tissues, presence or absence of cambium, physiological and biochemical activities, depth of their root systems, root distribution, surface area, capacity and efficiency in taking up herbicides molecules from soil solution. All these factors could affect herbicide sensitivity or tolerance of crop plants. However, differences were detected on families, species and cultivars levels in response to herbicides and therefore any selective herbicide on a specific crop cultivar may confer injury to other cultivars (Felix et al., 2007; Abit at al., 2009; Kong et al., 2009; Jin et al., 2010). However, production of cuticular wax tends to be reduced under warm humid conditions which may resulted in a higher retention of spray solution on crop leaves and more injury under such conditions.

3.1.3 Herbicide factors

3.1.3.1 Herbicide status and type

Herbicides are chemicals negatively affect weeds normal growth, they quickly act and may be applied to a specific area or where other methods of weed control are not possible, or may be integrated with other weed control methods. However, chemical weed control has many advantages among which is the use of pre-emergence herbicides which provides control of strong competing weeds during crop establishment and thus help get a weed control job that otherwise cannot be done.

Herbicides are classified based on different factors among which is the selectivity. Based on that, herbicides kill or inhibit weeds and don't harm crop plants beyond point of economic recovery. The nonselective herbicides referred to those chemicals kill all plants when applied at enough rates and mainly used where no vegetation is required. Selectivity however, depends on rate and formulation of the herbicide applied, time of and method of application, environmental conditions and stage of weeds and crop plants. Herbicides are also categorized based on their method of action to contact which only kill plant parts that became in contact with and most are effective against annual weeds, but if used on perennials another applications may be necessary. On the other hand, systemic (translocated) herbicides are those absorbed by leaves, shoots or roots, transport from the site of application to where toxic action take place. These are useful against established perennials and unlike contact herbicides uniform coverage is less important. These chemicals kill slowly (toxicity symptoms may take weeks and up to a year to be developed for certain herbicides such as amitrole) while contact herbicides are rapidly acting. Herbicides are also classified based on method of application to soil-applied absorbed by roots of large plants mainly affect seed germination, germinated seeds or small seedlings and thus used to eliminate weeds before emerge and can selectively eliminate germinating weeds in established crops. These may or not be foliage active.

Foliage applied herbicides are contact or translocated and may or not be soil active. If translocated, they move passively through the symplast system with photosythate materials.

Considering time of application, herbicides are either preplanting applied at which crop should resist herbicide treatment and herbicide must be short- lived; pre-emergence used after crop is planted (vegetative parts) or sown (seeds) but before emergence and crop tolerance is necessary; post-emergence applied after crop emerged, may be soil or foliage applied but crop resistance is essential. If soil applied, they kill newly germinated seedlings while foliage applied herbicides kill emerged weeds. Persistence of soil-applied herbicide may be required since offer a long period of weed control and may be through out the whole growing season or until harvest. However, herbicide carry-over problem should be taken into account using such chemicals (Schnoor, 1992).

3.1.3.2 Time of application

Time of herbicides application is important in determining the effectiveness and length of weed control duration (Carter et al., 2007; James et al., 2007). Herbicides are applied at preplanting in crops grown from seedlings such as of most vegetables; pre-sowing in case of seed-sown grain field crops and some vegetables such as many cucurbits; and as post-emergence. The first two groups are soil applied, control germinated seeds or weed seedlings either by a contact or systemic action, prevent weed seed germination or seedlings growth and thus prevent early weed competition and protect crops from planting or sowing date until a good canopy is developed. This period in crop life cycle is most vulnerable to weed competition and crop recovery may not be possible if weeds escape control during this period. In all cases crop tolerance and herbicide selectivity are essential factors.

3.1.3.3 Herbicide formulations

Herbicides formulations are many including water soluble (WS, S, SL) liquids require wetting agents, water soluble powders (SP) need stirring or agitation in preparation, water emulsions (E, EC) require some agitation and held together by an emulsifier, wettable powders (W, WP) need continuous agitation and most used on soil, water dispersal liquids (WDL, L, F), water dispersed granules (WDG, DF), granules (G) need water to leach them down into the soil, and pellets usually used in spot treatments (Foy & Pritchard, 1996). Herbicide formulation affect selectivity and thus crop growth. The most used herbicide formulations are aqueous and granules. For each formulation there are certain equipments to apply, these are sprayers or spreaders, respectively. However, both forms should be uniformly applied with the second most commonly used in horticultural crops and is least selective. It is well known that forms such as powders (dust), granules, ester forms are less stable than others while water soluble forms are more stable than emulsions.

3.1.3.4 Labels and technical pamphlets

Herbicide label and technical bulletin must be carefully read prior to herbicide application. Label shows herbicide trade and common names and carry important information on herbicide selectivity, formulation, active ingredient, volume of spray solution or carrier required per unit area, method of application, time of herbicide application, rate of application, post-application treatments, persistence (carry- over) (Felix et al., 2007), weed species affected, and crops in which the herbicide is recommended, weed control spectrum and tolerant weeds, volatility and conditions under which the herbicide is selectively used and any possible crop injures and precautions. All these information are of high value for the farmer and must be consulted and strictly followed if a successful weed control operation is the ultimate farmers' goal. Considering all of these would help in achieving effective chemical weed control with no hazards or environmental pollution.

3.1.3.5 Herbicide resistance and selection pressure

One of the causes of herbicide tolerance/resistance in weeds is the continuous application of the same herbicide or herbicides of the same mechanism of action year after another in the same field (Caseley et al., 1991; Duke, 1996). This practice imbalances weed population at which certain group of weeds (broad or narrow-leaved) is affected while the other is not harmed (Fig. 3). This is usually resulted from use of selective rather than general herbicides. However, within the affected weed group sensitive species are the first harmed and disappear later. The less affected species may include some individuals within populations that withstand the rate of herbicide applied or may be partially affected but could recover late in the season and set seeds. Continuous use of the same herbicide impose a selection pressure that force some individuals within the weed population to physiologically, morphologically and may be biochemically distinct and adapt themselves to tolerate the rate of herbicide used (Powles & Holtum, 1994; Pline et al., 2003; Qasem, 2003). This will lead later to development of individuals within the same weed population that resists chemical

Fig. 3. Weed population shift to narrow-leaved weeds (Mainly bulbous barley, *Hordeum bulbosum* L.) resulted from continuous application of broad-leaf killers in olive orchard.

treatments (e.g. *Senecio vulgaris, Chenopodium album* and *Amaranthus retroflexus* to traizines). At present, the total number of resistant weeds may exceed 668 species of more than 346 biotypes (htpp://www.weedscience.org/summary/MOASummary.asp). However, continuous application of the same herbicide or group of herbicide with the same mechanism of action may leads to the following problems:

1. Imbalance natural weed population
2. Cause shift in weed population toward perennation.
3. Encourage dominance and existence of certain weed species
4. Promote development of weed tolerance followed by weed resistance
5. Lead to development of weed races or crop- relative weeds that become irresponsive to agricultural practices employed and dominate cultivated fields.

However, these problems can be avoided by the following measures:

1. Rotation of chemical control method with other methods of weed control
2. Rotation of the herbicide with other herbicides of different mechanism of action.
3. Adoption of crop rotation system which is usually accompanied with herbicide rotation and thus leads to brake down the build up of dominant weed population.
4. Alternate selective chemical control treatment with fallow years during which general (nonselective) herbicides may be used to control all weed species in the field including resistant species.

3.1.3.6 Sprayers and calibration

The main and serious problem with herbicides application or uses is the people apply these chemicals. Wrong application is commonly resulted from failure in sprayer calibration.

Sprayer calibration aims at uniform herbicide spray distribution and coverage of treated weeds/surface that means the receipt of the same amount of spray solution per each unit area of treated surface. Sprayer calibration is the first step to be carried out before herbicide application. Herbicide application in the field should be properly carried out since failure of uniform distribution of any herbicide may result ineffective weed control or crop injures and thus herbicide residues (Badowski et al., 2008). Therefore, sprayer calibration is mainly conducted to determine the volume of the carrier (mostly water) required to dissolve in the amount of the herbicide calculated (based on recommended rate) and required for specific area. In other words, it aimed to determine exactly the volume of water required to hold/dissolve the amount of the formulated herbicide specified to cover a measured area infested by weeds. It is also required to know how much of the herbicide is applied per treated area (Kansas State University, 2010). Failure to calibrate the sprayer can injure crop, cause pollution, and waste time and money. Therefore, calibration should be carried out regularly and nozzles must be cleaned at each spray time and before herbicide application. Therefore, several factors must be considered which influence this operation, these are as follows:

1. Solution/ tank pressure. It is well known that pressure inside sprayer tank may range between 15 to 50 pounds per square inch (psi) and application rates can vary from less than 15 to more than 250 gallons per hectare. The higher the pressure imposed on spray solution the more fine are the spray droplets and the higher is their liability to drift. The volume of spray solution ejected from the nozzle is linearly correlated with the pressure imposed on spray solution and the greater is the nozzle discharge. In addition, the higher pressure the wider is the spray swath and the larger the area covered per unit time. Therefore, pressure must be calibrated and adjusted to be the same as that used in calibration and must be kept constant (labor carried sprayers) through out the whole time of herbicide application. Any

deviation from a constant pressure may result failure in obtaining uniform coverage of the weeds vegetative parts or differences in coverage of the treated soil surface when pre-emergence herbicides are used.

2. Sprayer ground speed. The volume of spray solution applied is affected by sprayer speed. The higher the speed the larger is the area covered by herbicide solution and the lower is the volume used pre unit time and *vice versa*. Doubling the ground speed (mph) of the sprayer reduces the gallons of spray applied by one-half. Sprayer speed should be determined through calibration directly in the field to be sprayed. However, this could be affected by weed species, weed size, height, population thickness and soil structure and topography. Small and short grown weeds may be easily controlled using small spray volume. In contrast, large and tall growing weeds require more coverage by herbicide solution, thus large volume of spray solution is applied at low sprayer speed. This however, is in link with herbicide method of action whether is a contact or translocated herbicide. Application of the former requires slow movement, low speed and full coverage of the treated surface. However, both pressure and nozzle size are factors controlling the delivered spray volume.

3. Sprayer boom height. Nozzle height has obvious effect on the area covered by herbicide solution. The higher the boom height the wider is the spray swath and the larger is the area covered with spray solution of different nozzles mounted on the sprayer boom. Changing boom height from above the soil level during spray operation may easily result in a non-uniform coverage leads to overlapping between spray swath of different nozzles fixed on the boom and thus crop may be injured or killed since received a double dose of the herbicide solution which could be easily observed in the field . When nozzles are placed at low height from the soil level possible gaps left uncovered by the herbicide solution are mostly resulted and thus weeds in these gaps may continue growing in regularly spaced strips in the field (Fig. 4). In both situations lack of uniform coverage by the herbicide solution could result in equally spaced strips of growing weeds or killed crop plants under

Fig. 4. Wild mustard (*Sinapis arvensis* L.) grown in strips after escaped from control by 2, 4-D in wheat due to low boom height level.

low and high levels of sprayer boom height, respectively. Therefore, one of the objectives of calibration is to exactly determine the optimum boom height that provides uniform coverage (a single coverage treatment) of the sprayed surface or treated weeds with herbicide solution.

4. Nozzle type and discharge of spray solution

Nozzle type determines the spray discharge, the uniformity of spray application and spray pattern, the coverage of sprayed surface of a target plant and the drift rate. It is generally agreed that the same type of nozzles must be used on the sprayer and these should be carefully checked for correct number and against tip damage or orifice clogging. All should be mounted and directed similarly on the sprayer boom. However, nozzle type is different between different nozzles but generally two main types are commonly employed. These are cone type and flat fan type nozzles. In the first, orifice is rounded gives a cone type spray while the orifice opening in the second is straight flattened. Each type includes hollow or solid spray patterns. Cone type is used to produce a coarse spray for aerial and other low drift purposes and is more commonly used for woody shrub or tree control but its use in field crops is limited or rare. Turbo flat-fan, venture flat fans and drift reduction pre-orifice flat-fans are commonly employed in crop protection products. For herbicide application, the flat fan type of a straight orifice opening is usually used since gives a flat spray swath in a straight line (sheet of spray) which could thoroughly cover all weeds within the spray swath limits. Each nozzle type could carry a number (mostly not found in knapsack sprayers or other compressed manual operated sprayers) fixed on the nozzle tip that represents spray angle and spray discharge. The higher the spray angle the larger the area covered and the delivered spray volume. Nozzle discharge is measured as gallon per minute. The higher the number the higher is the nozzle discharge. Spray angle and solution discharge are in link with type of the herbicide used and the volume of spray solution and influenced by tank pressure and nozzle size. It is advised that low spray solution volume is better used for systemic herbicide application while high volume is more relevant to thorough coverage of weed vegetation using contact herbicide solution. All thus depend on the nozzle type and could be determined from the type of weeds found in the field. Annual weeds may require contact herbicide, full coverage treatment, high volume of spray solution and subsequently higher spray angle and solution discharge. The opposite is true for perennial weeds and systemic herbicides at which may be absorbed and translocated through the whole plant system once an amount of the spray solution is absorbed from any part of the plant. In summary, nozzles are different in specifications, spray angle and discharge, orifice or tip type and size, materials which are made from, all these determine with the density of the spray liquid the spray volume applied, the type of the herbicide used, and the type of weeds to be treated by each nozzle type.

4. Problems

Like any weed control method, chemical weed control has advantages and disadvantages. Among problems encountered are the following:

1. Herbicides may cause shift in weed population (Owen et al., 2008) because of continuous use of certain chemical, chemicals of the same mechanism of action or family create selection pressure and development of weed races.
2. Imbalance weeds population mainly when selective herbicides are used. Broad-leaf killers imbalance weed population toward dominance of narrow-leaved weeds and the opposite

is true for narrow-leaved killers and broad-leaved proportions. Therefore, both types of herbicide groups should be occasionally complimented or alternated in the field.

3. Herbicide residues in plant materials. Herbicides are absorbed by crop plants as well as by weeds. Certain herbicides are completely metabolized while others are not (Hatzios & Penner, 1982; Pline et al., 2003). The second group mostly is not recommended on food or feed plants. Examples are paraquat that not permitted on food crops, and dinoseb-acetate, another contact herbicide used in alfalfa fields. Because toxicity is extended on harvested alfalfa plants therefore is always advised that first harvest should not be fed to the cattle but rather burned and people should stay away from the smoke of burned plant materials. Residues of herbicides in plant materials are most commonly found when herbicides were applied late in weed and crop growth.
4. Herbicide carry- over problem and limitation in crop rotation options. This problem may be easily observed with some long persistent herbicides. Some examples are urea and sufonyl-urea herbicides. Planting of non cereal crops in soil treated with these chemicals may result in serious crop injuries. Cucurbits and some legumes are good examples on sensitive crops to herbicides. Injury effect on the same crops may be also observed after metribuzin and many other soil applied herbicides.
5. Herbicide leaching to ground water. Pollution of ground water is possible when leachable herbicides are used (Smith et al., 1984; Vizantinopoulos & Lolos, 1994). Leachability may be high with salt formulated herbicides used or in treated sandy soils under heavy irrigation or rain.
6. Herbicide chemical drift and crop injuries. This is an important problem encountered with certain herbicides and usually take place as spray-drift occurs when small droplets of spray liquid are carried away from target plants by wind currents, vapour-drift resulted from application of high volatile chemicals enhanced by high temperature or wind currents, and the blow-off which is associated with light soils and resulted from movement of tiny particles of soil treated by herbicide.
7. Herbicide persistence in environment. This may be in relation to half life of the herbicide and its rate of degradation in the environment. Certain herbicides have long half-life and persist in the environment for along period of time that limits agricultural land uses. These however, may be traced in the soil, air, surrounding environment, ground water and even agricultural products and may become a serious threat to public health. Therefore, these chemicals should be avoided, highly restricted and only applied in absence of other alternatives and when are extremely needed.
8. Herbicide storage and handling. Herbicides and pesticides in general should be properly handled and safely stored since are toxic to humans and other living organisms. Stores should be properly located, designed specifically and locked to prevent contamination, pollution and protect chemicals from any degradation and loss in activity before use. Loss in any form must be prevented. Therefore, temperature, humidity, and aeration must be strictly controlled to prevent any loss in chemicals activities. Chemicals however, should be stored in their original containers, kept away and separately stored from other stored agricultural materials and preferably used within a short period where possible.

   Empty glass, metal, or plastic made containers should be thoroughly washed with water that to be added to the spray solution in the spryer tank. Containers must be punctured, destroyed and then deeply buried (deeper than 1m) in the soil or disposed in a sanitary licensed area. Herbicides containers made from cartoon or paper bags should not be burned in a residential area or where smoke of these may becomes in contact with animals or cause air pollution and toxicity.

9. Herbicide and toxicity to mammals. It should be clearly declared that all pesticides are toxic to humans and mammalians in general but toxicity is different between these chemicals. Herbicides are the least toxic among all agricultural pesticides. This however, may be determined by considering the LD_{50} or LC_{50}. Among herbicides with relatively low LD_{50} are paraquat, diaquat, dinoseb, diclofop-methyl and some others. It should be known that toxicity may occurs through absorption, smoking, inhalation, swallow through contaminated hands, food and drink. Certain highly toxic chemicals should be avoided unless no alternatives are available.
10. Herbicide resistant weeds. This is a serious problem that could totally cause shift in weed population from susceptible to resistant. The main cause of this is the high selection pressure imposed by use of only a single herbicide or herbicides having the same mode or mechanism of action for a relatively long period. Farmers should keep watching changes occur in weed population in response to chemicals used and take measures when observe scattered individuals of certain weed species start tolerating the herbicide normal rate of application. This is the first sign on development of herbicide resistance in that species.
11. Crop injury. This may be due to direct or indirect exposure of crop plants to herbicide molecules. Herbicide treated crop may show signs of herbicide injuries that may or may not recover from. This is mostly related to herbicide rate of application, form used, surfactant added, time of application relative to crop physiological or morphological stages or prevailing climatic conditions at spray time. Indirect injury may resulted from spryer tank contamination through herbicide traces left unwashed and used to apply insecticides or fungicides using the same sprayer (Fig. 5), drift of volatile herbicides applied in the nearby fields, leached herbicides to ground water, running water used to irrigate sensitive crops or from carry-over problem of certain previously applied herbicides.

Fig. 5. Toxicity symptoms on Jute (*Chorchorus olitorius* L.) resulted from 2,4-D traces left in the sprayer tank used for insecticide application. Plants showed abnormal growth (malformed leaves, twisting and yellowing).

12. Herbicide application and technicality required. This is an important factor determines the success or failure of weed control operation. Some of the problems detected including herbicide application without sprayer calibration but rather addition of a measured/unmeasured amount of the selected herbicide to the tank (unmarked) filled by unmeasured volume of water, spray weeds thoroughly, heavily and deeply covered by dew droplets which may dilute the spray solution, failure to thorough mixing of herbicide solution, failure of continuous stirring or agitating spray solution (suspensions) during spray time, failure to maintain a fixed pressure or a constant ground speed during spraying (in case of knapsack or shoulders-hilled sprayers) this resulted in uneven spray distribution, partial coverage of sprayed surface or low volume of spray solution received per unit area, failure in maintaining straight forward travel in manual operated sprayers with no overlapping or area left uncovered by the spray solution.
13. Cost of chemical weed control and precautions required. In absence of any studies on the economic use of herbicides compared with other methods of weed control, chemical weed control may be thought by farmers as an expensive tool, needs a lot of preparations and accuracy, unsafe to crop plants if not properly and carefully considered, cause soil and environmental pollutions if not well designed and carefully selected, and doesn't give complete solution to all weeds present in the field and thus other methods of weed control should be also incorporated.
14. Herbicides toxicity and public concern. With the present trend toward public concern from synthetic chemicals and tendency in crop cultivation and food production through an organic farming system, use of agrochemicals in general including herbicides became much restricted. People world over preferred, chemical untreated and naturally produced crops despite high prices and payments for this staff of organic food.
15. Human failure to properly apply the herbicide on time, conducting sprayer calibration, after herbicide application operations and uniform distribution and application.

5. General rules, recommendations and concluding remarks

Before start considering chemical weed control method think carefully in the following:

1. Look first for weed control options other than herbicides.
2. There is no best weed control method but rather integrated methods of weed management.
3. Consider the economic return value from weed control using herbicides.
4. Weed control in general, and using herbicides in particular may be of no value after certain period of weed/crop competition since crop recovery may not possible and more expensive without crop yield return.
5. Weeds as well as crops are different in their competitive abilities and these are different from region to another within the same country.
6. Critical period of weed control is different for the same crop and weed species from area to another.
7. Always search for an eco-friendly herbicide suit weed control in the specified crop, cheap, with low rate, short life and non residual, environmentally compatible and safe for mammals and biological system in general.
8. Identify weed species and predicted weed population in the field, determine if the critical density of weeds is reached, therefore weed population prediction is important

before start any weed control program. Weed population under critical density may doesn't need immediate control.

9. Always look for differences between crop plants and weeds in growth, requirements, development and responses, benefit from these to favor crop plants over weeds. Therefore, field managements may be of high value for crop plants the same as the use of herbicide.
10. Weed control aimed at keeping weed population below the level cause economic crop injury. Eradication of weeds commonly found in the area may not be possible or economically unfeasible. Therefore, weed control should be implemented when weed density in the field is becoming critical. Sometimes it is strongly advised to use a spot treatment rather than an overall spraying of the herbicide when weeds are found growing in patches or forming scattered colonies in the field.
11. Chose registered herbicide recommended for use in the grown crop.
12. Obtain the herbicide from local chemical agent with full detailed information on its uses and supplement of technical bulletin.
13. Be sure that the purchased herbicide is a new valid (not or nearly expired) product; handle carefully its storage until being used.
14. Think always that both crop plants and weeds take up herbicide and the differences in susceptibility between both may be marginal and easily affected by plant, herbicide and environmental factors. These marginal differences determine herbicide selectivity.
15. Weeds grown under harsh environmental conditions (drought, high temperature, salinity, highly disturbed habitats) have different morphological modifications and are more difficult to control by herbicides compared with those grown under favorable growth conditions.
16. Early weed control is always recommended and for each crop there is a critical period of weed control at which weeds must be removed to avoid any significant yield loss. Weeds competed with crop plants before or after this period cause no significant effect on crop yield providing weeds are controlled during this period. This however, is greatly influenced by crop species and cultivars, weed species and density, and agricultural practices followed. At the critical period chemical weed control may not the best method of weed control considering crop growth stage and status, crop physiology and possible injury or weather conditions that may lead to more toxic herbicide even though the recommended rate is used.
17. Determine the best time for herbicide application and chose conditions favored crops over weeds. Consider environment, herbicide and plant factors before herbicide application. Plant morphology, physiology, anatomy and biochemical activities are important factors in determining the ultimate effect of herbicide on crop and weed plants.
18. Be sure that there are certain periods in crop and weed growth cycle at which herbicide by any is tolerated/resisted or these are more susceptible. Look always for the differences in weed and crop plants tolerance to herbicides treatment.
19. Before preparing herbicide solution, carefully consult the technical bulletin provided by the manufacturer with herbicide, read carefully all information in herbicide label and consider all factors affect herbicide performance in the field and weed control operation.
20. Sprayer inspection, preparation and calibration must be considered. Sprayer check out, tank or nozzle leakage, pump working order, type of nozzles and fixation on spray

boom, nozzle clogging, nozzle spacing and direction, boom height, tank pressure, ground speed of the sprayer and spray volume, all should be considered and optimized.

21. Conduct sprayer calibration directly in the field to be treated by the herbicide. Always keep in mind that spray volume is affected by sprayer ground speed, pressure, boom height and nozzle type, weed growth and population thickness.
22. Calibration of the sprayer mainly aims to determine the volume of water needed to carry in the amount of herbicide required for weed control in a specific area. It is conducted always at fixed pre-determined ground speed, pressure, boom height, same nozzle type and at suitable climatic conditions.
23. Once the volume of water needed to dissolve or carry with the amount of herbicide required per intended area for spray is determined start preparations for herbicide application. Prepare the special dress/apron designed for spray operation.
24. Calculate the amount of herbicide required for that specific area based on the recommended rate of herbicide application by the manufacturer, don't exceed the specified rate or go lower than that because in both cases wrong results may be obtained. In the first the crop may be killed/injured while in the second failure of weed control. In both cases the required amount of herbicide per unit area is not received on the sprayed surface.
25. Herbicide selectivity is not an absolute value or character of a herbicide rather than resulted from the interaction between plant, herbicide and environmental factors.
26. Herbicide mixtures must be compatible, beneficial and used mainly to widen the weed control spectrum. However, these may be already available and prepared in advance by the manufacturer or used as tank mixtures at spray time.
27. Weed population is either active or passive, each of which needs certain herbicides determined based on their method of application. Passive weed populations is controlled using soil applied herbicides while active weed population may be treated by foliage or soil applied herbicides. Both weed populations must be considered in any weed control program.
28. Before herbicide application, consider crop grown and tolerance, crop rotation followed, herbicide cost, duration of weed control required and herbicide persistence in the environment. Don't exceed or lower the recommended rate of the herbicide unless based on experimental results obtained under local conditions and no residues or pollution are left in the environment.
29. Avoid the use of growth regulator herbicides (such as phenoxy or benzoic herbicides) during flowering time of field crops (e.g. tomato) or fruit trees (e.g. citrus). This may result in flower abortion in most cases.
30. Chemical weed control may cause shift in weed population toward perennation therefore this practice must be changed by time, rotated or integrated with other weed control methods to maintain weed population balance.
31. Selective herbicides may imbalance weed population thus herbicide mixtures of broad-leaf and narrow leaf killers are recommended or may be alternately applied.
32. Wettable-powder formulated herbicides may need continuous agitation of spray solution through out the whole spray operation to prevent settling at tank bottom, separation and precipitation. Liquid herbicides may need emulsifying agent when dissolved in water especially for emulsifiable concentrate (EC) formulated herbicides
33. Use surfactants when necessary especially wetting, spreading, sticking and drift control agents.

34. Certain materials may act as surfactants by modifying leaf surface and increase herbicide wettability to leaf surface or penetration through increasing compatibility between herbicide and waxy leaf layer on leaf surface, these include ammonium sulfate, urea, soap, washing or cleaning powders, liquid detergents and acidic plant extracts.
35. Different soil types require different rates of application from the same herbicide based on their mechanical and structural analysis.
36. Herbicide susceptible weeds may become tolerant at certain period after emergence or when reaching certain morphological stage since resulted from changes in weed physiology and/ or biochemistry.
37. Herbicide rotation is essential to keep weed population under control; any deviation from this system may result in weed population shift, development of weed races, crop-relative weeds and herbicide resistance.
38. High volatile herbicides should not be applied during hot weather conditions, since chemical drift in vapor form or crop injury is highly possible. Instead less or nonvolatile forms are recommended.
39. Photosensitive herbicides as well as volatile soil applied herbicides need soil-incorporation or light irrigation after application.
40. In preparing herbicide solution for field application, the amount of herbicide calculated to treat a specific area and volume of water needed must be added consequently (as water-herbicide-water) to the sprayer tank, thus herbicide is first dissolved in certain amount of water then spray solution may be then completed by adding water up to the final volume consumed in calibration. The herbicide-water mixture must then be thoroughly mixed by hand shaking of knapsack sprayers or spray solution must be kept agitated during spraying operation. The pressure, sprayer ground speed, boom height and the same nozzle type used in calibration must all be set the same as were in calibration. If all factors are kept the same then the herbicide solution must run-out by finishing the sprayed area without any shortage or spray solution left in the sprayer tank. In this case distribution of the herbicide solution is considered as uniform and equally applied over the sprayed surface; otherwise some problems in weed control may be resulted.
41. Consider that no herbicide gives 100% killing of all weeds, some weed species specified on the label as sensitive to certain herbicide may be left uncontrolled and the opposite is also true. Therefore, selective herbicides may show some overlapping up to certain extents in their weed control spectrum of broad- and narrow-leaved weeds.
42. Soil applied herbicides are influenced by edaphic factors including; soil pH or soil reaction, mechanical and structural analysis, leachability, microorganism population, organic matter content and height of soil water table.
43. Activity of soil-applied herbicides is affected by seed dormancy, hardness of weed seed coat, weed seed population and richness of seed bank, seed bank species composition and microbe's population...
44. Nonselective herbicides may be selectively used if directed only toward weeds and contact between these and crop plant is prevented. In this case the herbicide is not selective but rather than selectively applied.
45. Consider that certain herbicides working on photosynthesis need high light intensity associated with high temperature (De Vleeschauwer et al., 1992) for effective weed control (e.g. paraquat, diuron, cycloxidim) while others exhibit photo degradations (e.g.

glyphosate) and need to be applied in cloudy or foggy days, early in the morning or late in the evening.

46. Keep observing treated weed population, switch to other herbicide of different mechanism of action and family or change weed control method once observed that some plants of one or more weed species start tolerating the herbicide used.
47. Weed races or herbicide resistance or persistence and dominance of certain weed species are results of selection pressure of the agricultural practices employed. To prevent development of such conditions frequently change the existing agricultural practices.
48. Any herbicide traces left in the sprayer tank may cause toxicity to sensitive crops; therefore it is always recommended that herbicides sprayers must not be used for application of other pesticides even though were thoroughly washed or carefully cleaned.
49. Higher rate of application than that specified on herbicide label may cause changes in method of herbicide action, instead of giving a systemic action, tanslocated herbicide kills by contact action since high application rate kills phloem tissues and then no more translocation to the below ground regenerative organs occurs and thus perennials are quickly re-vegetate after treatment with translocated herbicides.
50. Chemical drift could be substantially reduced by selecting the nozzle type that deliver large droplet size (more than 100 micron) but providing adequate coverage at the intended application rate and sprayer pressure, avoidance of herbicide application where wind current speed exceeding 10 miles per hour, and temperature is high, avoid application of volatile herbicides and ester forms of phenoxy herbicides when air temperature is high, avoid using of high vapor pressure herbicides, lower the nozzle height to be closer to or above weed vegetation, incorporation of volatile soil applied herbicides, smoking, timing of spray operation either early in the morning or late in the evening since high relative humidity result in slow evaporation of droplets and decrease drift, selection of nozzles work at low pressure (swirl chamber, flood jet), allow enough time for herbicide absorption before rain occurs, and by adding drift control agents.
51. Failure in herbicide application may be resulted from loss of herbicide solution by heavy rain or irrigation, high rainfall occurs after application and wash-off the herbicide from leaf surface, failure of uniform coverage, herbicide precipitation (suspensions) in sprayer tank during spraying, wrong time of herbicide application, wrong physiological or morphological growth stage at which weeds or crop plants are treated and no or wrong sprayer calibration.
52. Highly volatile herbicides should not be applied when air temperature is high, temperature may increase loss of herbicide through volatiles, and at the same time may activates other herbicides to become more toxic and cause crop injuries even though are used at normal recommended rates of application.
53. There should be enough time allowed for foliage applied herbicide absorption after its application. This period is different from one herbicide to another but generally six hours are at least required for most herbicides to be absorbed through foliage parts and thus farmers should ensure that no rain occurs during this period when apply herbicides.

6. Conclusion

Herbicide application in the field is a skill needs full consideration of all plant, herbicide and environmental factors if a successful weed control is the ultimate farmer aim. Failure in weed control or crop injury is mainly due to human errors at which failure in or wrong sprayer calibration is a common mistake. Weed identification, size of weed problem exist, critical period of weed competition, weed density, selection of proper herbicide, rate and time of herbicide application, herbicide persistence, development of herbicide resistant weeds and environmental pollution are among factors to be considered and determine the success or failure of weed control operation. However, chemical control is usually integrated with other methods of weed control in any successful weed management program.

7. References

Abit, M. J. M.; Al-Khatib, K.; Regehr, D. L.; Tuinstra, M. R.; Claassen, M. M.; Geier, P. W.; Stahlman, P. W.; Gordon, Barney W. & Currie, R. S. (2009) Differential response of grain sorghum hybrids to foliar-applied mesotrione *Weed Technology*, 23(1), 28-33.

Badowski, Marek & Sadowski, Jerzy (2008). The level of residues of 2,4-D in plants on grassland. *Progress in Plant Protection*, 48(4), 1185-1189.

California Weed Conference. (1985). *Principles of Weed Control in California.* Thomson Publications. Fresno, California.

Carter, A. H.; Hansen, J.; Koehler, T.; Thill, D. C. & Zemetra, R. S. (2007). The effect of imazamox application timing and rate on imazamox resistant wheat cultivars in the Pacific Northwest. *Weed Technology*, 21(4), 895-899

Caseley, J.C., Cussans, G.W. & Atkin, R.K. (Eds.). (1991). *Herbicide Resistance in Weeds and Crops*. Butterworth-Heinemann, Oxford, England.

Coupland, D. (1987). Influence of environmental factors on the performance of sethoxydim against *Elymus repens* (L.). *Weed Research*, 27(5), and 329-336.

De Vleeschauwer, J.; Jennrich, H. & Defloor, K. (1992). Experience with cycloxidim for the control of grass weeds in Europe. Mededelingen van de Faculteit Landbouwwetenschappen, Universiteit Gent 57(3b), 1021-1030.

Duke, S.O. (1996). *Herbicide-Resistant Crops. Agricultural, Environmental, Economic, Regulatory, and Technical Aspects*. CRC Press, Boca Raton, Florida.

Duwayri, M. & Saghir, A.R. (1983). Effect of herbicides on weeds and winter cereals in Jordan. *Dirasat*, 10, 115-122.

Felix, J.; Doohan, D. J. & Bruins, D. (2007). Differential vegetable crop responses to mesotrione soil residues a year after application. *Crop Protection*, 26(9), 1395-1403

Foy, C.L. & Pritchard, D.W. (Eds.). (1996). *Pesticide Formulation and Adjuvant Technology*. CRC Press, Boca Raton, Florida.

Friesen, G. (1967). The efficiency of barban as influenced by growth stages of wild oats and spring wheat. *Weeds*, 15, 160-163.

Friesen, H. A. & Dew, D. A. (1966).The influence of temperature and soil moisture on the phytotoxicity of dicamba, picloram, bromoxynil, and 2, 4-D ester. *Canadian Journal of Plant Science*, 46(6), 653-660.

Friesen, H.A., Born, W.H.V., Keys, C.H., Dryden, R.D., Molbery, E.S. & Simons, B. (1968). Effect of time of application and dosage of dicamba on the tolerance of wheat, oats and barley. *Canadian Journal of Plant Science*, 48, 213-215.

Grime, J.P. (1986). *Plant Strategies and Vegetation Processes*. John Wiley and Sons, Chichester, England.

Gwynne, D.C. & Murray, R.B. (1985). *Weed Biology and Control in Agriculture and Horticulture.* Batsford Academic and Educational. London.

Hatzios, K.K. & Penner, D. (1982). *Metabolism of Herbicides in Higher Plants.* Burgess Publishing, Minneapolis.

Holt, J. (2006). Ecology and evolution of herbicide resistant weeds. Abstracts of Papers, 232nd ACS National Meeting, San Francisco, CA, United States, Sept. 10-14, 2006, AGRO-191.

htpp://www.weedscience.org/summary/MOASummary.asp

James, T. K.; Rahman, A. & Trolove, M. (2007). Optimizing time of planting and herbicide application for control of problem weeds in maize. *New Zealand Plant Protection*, 60, 183-188.

Jin, Z. L.; Zhang, F.; Ahmed, Z. I.; Rasheed, M.; Naeem, M. S.; Ye, Q. F. & Zhou, W. J. (2010). Differential morphological and physiological responses of two oilseed Brassica species to a new herbicide ZJ0273 used in rapeseed fields. *Pesticide Biochemistry and Physiology*, 98 (1), 1-8.

Kansas State University. (2010). *Chemical Weed Control for Field Crops, Pastures, Rangeland, and Noncropland. Report of Progress 1027.* Kansas State University, Agricultural Experimental Station and Cooperative Extention Service. Kansas.

Kiely, T., Donaldson, D. and Grube, A. (2004). *Pesticides Industry Sales and Usage: 2000 and 2001 Market Estimates.* EPA's Biological and Economic Analysis Division, Office of Pesticide Programs, and Office of Prevention, Pesticides, and Toxic Substances. http://www.epa.gov/pesticides.

Kong, L., Si, J., Feng, B.; Li, S., Wang, F., Sayre, K.. (2009). Differential responses of two types of winter wheat (*Triticum aestivum* L.) to autumn- and spring-applied mesosulfuron-methyl. *Crop Protection*, 28(5), 387-392

Mo, Wen-yan & He, Hong-wu. (2005). Development of herbicidal-active materials from plants, *Huaxue Yu Shengwu Gongcheng*, 22(8), 7-9.

Mu, Juan-wei; Li, Peng & Li, De-ping (2007). Studies on herbicide application by stage to control barnyard grass in paddy field of cold region. *Beifang Shuidao*, 6, 40-42.

Owen, M. D. K. (2008). Weed species shifts in glyphosate-resistant crops. *Pest Management Science*, 64(4), 377-387.

Pline, W. A. & Hatzios, K. K. (2003). Physiological factors affecting resistance to herbicides in crop and weed species . Abstracts of Papers, 226th ACS National Meeting, New York, NY, United States, September 7-11, 2003, AGRO-088.

Powles, S.B. & Holtum, J.A.M. (eds.). (1994). *Herbicide Resistance in Plants. Biology and Biochemistry*. Lewis Publishers, Boca Raton, Florida.

Qasem, J.R. (1992). Suggested strategy for weed control. *The First Arab Symposium for Weed Control in Orchards and Vegetable Fields.* Federation of the Arab Scientific Research Councils. 11-13 October, 1992. Amman- Jordan.

Qasem, J. R. (2003). *Weeds and their Control.* University of Jordan Publications. Amman, Jordan. 628 pp.

Qasem, J.R. (2005). Chemical control of weeds in onion (*Allium cepa* L.). *Journal of Horticultural Science and Biotechnology*, 80 (6), 721-726.

Qasem, J.R. (2009). Weed competition in cauliflower (*Brassica oleracea* L. var. *botrytis*) in the Jordan Valley. *Scientia Horticulturae*, 121, 255-259.

Rice, E.L. (1983). *Pest Control with Nature's Chemicals: Allelochemics and Pheromones in Gardening and Agriculture.* The University of Oklahoma Press, Oklahoma. Publishing Division of the University. Norman, Oklahoma.

Ross, M.A. & Lembi, C.A. (1999). *Applied Weed Science.* 2nd Edition. Prentice-Hall Inc. New Jersey 07458.

Schnoor, J.L. (1992). *Fate of Pesticides and Chemicals in the Environment.* John Wiley-Interscience, New York.

Singh, H.P., Batish, D.R. & Kohli, R.K. (eds.). (2006). *Weed Management Handbook.* The Haworth Press, USA. 892 pp.

Smith, K. W. & Stephenson, G. R. (1984). Fate of [14C]-HOE 00736 in soil. Eds. Chow, P.N. P.; Grant, C. A. H. *Act. Plants Soils, Proc. Workshop Chem. Biochem. Herbic., 11th (1984)*, 113-126.

Su, Shao-quan (2006). Crop injury from herbicides and its prevention in China. *Xiandai Nongyao*, 5(4), 1-4, 12.

Virginia Polytechnic Institute and State University. (1999a). *Pest Management Guide for Horticulture and Forest Crops,* 1999. Virginia State University, Publication 456-017.

Virginia Polytechnic Institute and State University. (1999b). *Pest Management Guide for Field Crops,* 1999. Virginia State University, Publication 456-016.

Vizantinopoulos, S. & Lolos, P. (1994). Persistence and leaching of the herbicide imazapyr in soil. *Bulletin of Environmental Contamination and Toxicology*, 52(3), 404-410.

Zimdahl, R. (1980). *Weed Crop Competition, a Review.* International Plant Protection Center. Oregon State University. Corvallis.

33

Impacts, Efficacy and Economics of Bushwacker Sc (Bromacil) In Controlling *Acacia* Invasion in South Africa

Sikhalazo Dube[1], Fatunbi A. Oluwole[2] and 'Mota, S. Lesoli[3]

[1]*Natural Resources and the Environment, CSIR, Ecosystems Processes & Dynamics Research Group, PO Box 395, Pretoria 0001,*

[2]*Forum for Agricultural Research in Africa (FARA), PMB 173, Cantonment, 2 Gowa Close Roman Ridge, Accra*

[3]*Departnment of Livestock and Pasture Science, University of Fort Hare, P Bag X1314, Alice 5700,*

[1,3]*South Africa*

[2]*Ghana*

1. Introduction

The widespread transition of grasslands and savannas to shrub-lands during the last 50-100 years has elicited significant debate concerning the causes of these changes and has given rise to a number of investigations (Rollins *et. al.*, 1997). Often the encroaching species suppress the growth of palatable grasses and herbs, as they grow into impenetrable thickets (Wiegand et al., 2006). This often results in reduced grazing capacities in livestock farms. As a result, livestock farmers have regarded bush encroachment as a major problem and have resorted to various control measures. The causes of bush encroachment is not clear but it suffice to say that they are diverse and complex (Smit et al., 1999), consequently, it is difficult to devise complete control measures for all encroaching species. On the bases of some comprehensive reviews of the literature (Archer 1994, Van Auken 2000, Dube et al., 2009), it can be concluded that the primary mechanism behind the increase in shrub cover has been a dramatic shift in patterns of herbivory and fire frequency, although shifts in climate and carbon dioxide (CO_2) concentrations have also been cited as possible factors.

According to Dube *et al.*, (2009), the encroachment of rangelands by *Acacia karroo* bush, in South Africa is known to greatly reduce rangeland productivity with immense economic implications, especially in systems where grazers are preferred to browsers. Bush encroachment is defined as the invasion and/or thickening of aggressive undesired woody species resulting in an imbalance of the grass: bush ratio, a decrease in biodiversity, and a decrease in carrying capacity, causing severe economic losses in both the commercial and communal farming areas. The phenomenon of bush encroachment in savannas is seen to be part of the process of desertification (Tainton, 1999). *Acacia karroo* is one of the main species causing the encroachment problem in the Amathole Montane grassland, Bhisho Thronveld and Eastern Cape Escapement Thicket vegetation.

Various control methods viz., cultural, mechanical, chemical and biological uses have been tested and advanced to control the growth and spread of encroaching plants (Fatunbi et al. 2008). Burning, browsing with goats, cutting and application of herbicides are some of the methods that are widely used to control encroaching species.

The use of herbicide containing bromacil (5-Bromo-3-sec-butyl-6-methyluracil) for control of *A. karoo* has been observed in a number of commercial holdings in South Africa. Bromacil herbicide act by interfering with the photosynthetic pathway of the plant. It achieves a gradual kill that could span over two years. The use of herbicides have however brought about various environmental concerns (EXTONET, 1993; Rosner et al.,1999; Singh et al., 2003). This has resulted in the need to investigate the effects of different types of herbicides on components of the ecosystem (Dube et al. 2009).

A number of herbicides in South African markets are currently being used in controlling encroaching species, of particular note are herbicides containing bromacil (5- Bromo-3-sec-butyl-6-methyluracil) as the active ingredient (a.i), for example Bushwacker SC(Enviro Weed Control Systems (Pty) Ltd), Bushwacker GG (Enviro Weed Control Systems (Pty) Ltd), Rinkhals 400 PA (Dow AgroSciences LLC) e.t.c. These herbicides differ mainly in their bromacil concentration, for instance Bushwacker SC contains 500 g bromacil per litre, Bushwacker GG contains 200 g bromacil per kilogram and Rinkhals 400 PA contains 400 g bromacil per kilogram. The different concentrations of bromacil determine the specific use of the herbicide, coupled with the concentration of other reactive ingredients. Herbicides are usually selective within certain application rates, environmental conditions, and methods of application (Masters and Sherley, 2001).

Bromacil belongs to the uracil family of herbicides (Arteca, 1994). It can be used to selectively control annual and perennial weeds, broad leaved and woody plants on cropland and non-cropland areas (EXTOXNET, 1993; Meister, 1998; Zhu and Li, 2002). It is also widely used for selective weed control in pineapple and citrus crops (EXTONET 1993). Bromacil works by interfering with the photosynthetic pathway of plants (EXTOXNET, 1993). One herbicide that is gaining importance in bush control in South Africa is Bushwacker SC. It has been reported to be an effective herbicide for general weed and bush controls in agricultural and non-agricultural areas (Zhu and Li, 2002). This herbicide can be sprayed on the plants or spread dry. It quickly dissolves in soil water and may stay in the soil for several years (EXTOXNET, 1993). Its application is usually done just before the active growth stage of plants thus, before the wet season stabilizes. Bromacil is readily absorbed through the root system (Gangstad, 1989) and is a specific inhibitor of photosynthesis. In the soil, there is little adsorption of bromacil to soil colloids therefore it moves (leaches) through the soil and it can contaminate groundwater (EXTOXNET, 1993); however, it is highly susceptible to microbial degradation (Arteca, 1995). When used as a selective herbicide it can persist in the soil for one year, however if it is applied at high concentrations it can persist for more than one year (Arteca, 1995).

There is the speculation that bromacil can destroy some grasses if it stays too long on the upper horizon of the soil profile. Grasses are assumed to extract water from the top soil layer (0 - 15 cm) due to their shallow rooting characteristics, while trees and bushes derive their nutrients from the lower layers (Wiegand et al., 2006). The movement rate of bromacil when applied is therefore important to its economic use and ecological suitability. There is the perception that phytotoxicity could occur when animals ingest plants that may have taken up bromacil from soil water. Furthermore, the relative effects of bromacil on soil microbial activity and dynamics need to be investigated.

This aim of this paper is to clear doubts about the effectiveness and safety of the use of bromacil for the control of invasive species in South African rangelands. It also aim to identify the economic implication of Bromacil use and determine the best application method at the farmers' level. To provide sufficient scientific discussion in this chapter, the authors represents a substantial proportion their comprehensive review work (Dube et al., 2009) and also report a field research.

2. The chemical basis for the use of bromacil as an herbicide

Bromacil (Figure 1) falls under the substituted uracil family; other members of the family include terbacil and isocil Terbacil is an effective herbicide for the control of annual and perennial weeds. The general characteristic of the uracil family is the presence of a methyl group located at the sixth position on the ring. The members of the substituted uracil herbicide family differ one from another by substituents at the third or fifth position of the ring, or both. The bromacil molecule consists of a uracil nucleus containing bromine, methyl and a secondary butyl substituent (Canadian Council of Ministers of the environment 1999).

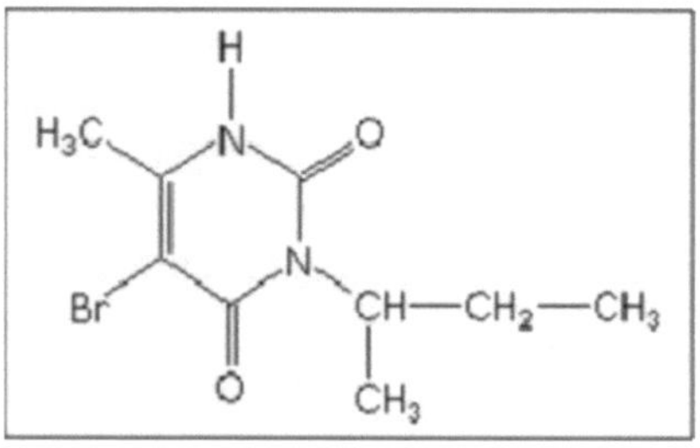

(a)

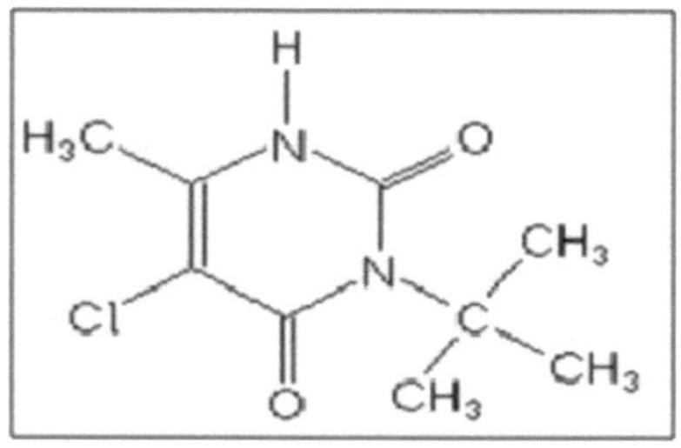

(b)

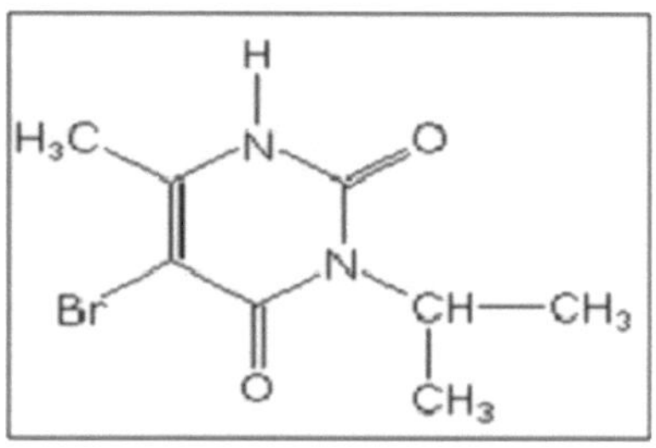

(c)

Fig. 1. (a). Bromacil (5-bromo-3-sec-butyl-6-methyluracil). (b). Terbacil (3-tert-butyl-5-chloro-6-methyluracil). (c). Isocil (5-bromo-3-isopropyl-6-methyluracil). Adapted from the International Organization for Standardization (ISO).

The family as a whole posses a broad toxicity to many plant species, however specific compounds differ significantly in their toxicity to plants, solubility in water, persistence in soil and other economically significant characteristics (Lakoski et al., 1993). Firstly introduction of this family of herbicides was intended for general vegetation control, mainly because the different compounds have broad spectrum activity over a wide range of plant species and they also persist long in the soil, therefore, having long residual activity for weed control in industrial areas.

The principal use of these herbicides are the selective control of many annual and perennial weed species in certain crops and general weed control in non-crop areas, such as railroads, highways, pipeline right-of-ways, lumber- yards, storage areas and industrial sites. The uracils are formulated as both wettable powders and as water soluble preparations (Dube et al,. 2009). All of the uracil herbicides in pure form are white, crystalline solids and are temperature stable up to their melting point of 335° C. The uracils are characteristically low in mammalian toxicity and are non-volatile. Their long persistence in soil, however, does create problems in crop rotations. In South Africa, bromacil is traded as Bushwhacker and it is applied as a spray, spread dry just before or during the period of active growth, preferable when rain is expected for soil activation or aerial application of granules.

The mode-of-action refers to the manner in which an herbicide affects a plant at the tissue or cellular level. Bromacil is a powerful mobile inhibitor of photosynthesis (Prostko, 2001). The target plant must be undergoing active photosynthesis for the herbicide to be effective. It is readily absorbed through the root system (Gangstad, 1989); the leaves and stems can also absorb some bromacil. It is translocated upward via the xylem to foliage and interferes with light-harvesting complexes (Prostko, 2001). It inhibits photosynthesis by blocking the photosystem II reaction; thereby preventing the conversion of sunlight into chemical energy (Prostko, 2001), thus it blocks the photosynthetic electron transport (Prostko, 2001). Bromacil blocks electron transport from QA to QB in the chloroplast thylakoid membranes by binding to the D-1 protein at the QB binding niche. The electrons that are blocked from passing through photosystem II are transferred through a series of reactions to other reactive toxic compounds. These compounds disrupt cell membranes and cause chloroplast swelling, membrane leakage, and ultimately cellular destruction (Tu et al., 2001). Inhibition of photosynthesis thus results in slow starvation of the target plant and eventual death.

Bromacil is readily absorbed through the plant root system (Bovey, 2001; Gangstad, 1989). Little or no bromacil moves from the apex downward toward the base of a treated leaf via the phloem. The early symptom of bromacil kill activities in a plant is leaf chlorosis concentrated around the veins, this is often noticed at the lower leaves and it gradually moves up the plant. The structure of the leaves' chloroplasts is altered while further cell wall development will cease. Chlorosis will then appear first between leaf veins and along the margins which is later followed by necrosis of the tissue and eventual death of the plant (Prostko, 2001).

The control of undesirable species in rangelands is a basic maintenance activity in livestock production. This could be carried out using methods that range from cultural, biological, chemical and a combination of these methods. The amount of drudgery involved in administering these methods makes some of them practically undesirable. Bromacil has many benefits in this case; it is used as an herbicide for general weed or bush control in non-croplands; it is also particularly useful against perennial grasses (Meister, 1998). The current use of bromacil in agriculture is necessary so as to sustain high productivity, reduce cost and drudgery and give high profit margins. It is used on rail road rights of way and other industrial, non-cropland areas.

Bromacil is one of the most commonly used herbicides to control weeds in citrus orchards. It is used in citrus and pineapple fields for selective control of weeds (Turner, 2003). Bromacil is also effective in the control of deep-rooted perennial broadleaf and grass weeds. Other commonly used herbicides are glyphosate, diuron, diquat, simazine, linuron, terbuthylazine and terbumeton (Gomez-Barreda et al., 1991).

3. Effectiveness of different application methods for bromacil based herbicides

Bromacil can be applied in a variety of ways. Application can be made as broadcast, band or spot treatments (Gangstad, 1989). The most appropriate application method is determined by the weed being treated, the herbicide being applied, the skills of the applicator and the application site (Tu et al., 2001). The bromacil application method is of considerable importance; since it determines the extent of contact with the target plant and its movement within the soil. Three conventional methods of application are known, these are aerial spraying with the use of an aircraft and direct application to the soil near the target plant with the use of a backpack sprayer (liquid application) or the placement of the granular form at a close proximity to the target plant.
Bromacil and lithium bromacil are often applied by ground application. Aerial application is allowed in areas where it is too dense or dangerous such as military firing ranges. Fixed boom sprayers are typically used for broadcast treatment of perennial and annual weeds; for brush control, basal spot applications are made to the specific shrubs and trees to be controlled. In citrus groves, it is often applied only in bands between the tree rows (Turner, 2003). Bromacil needs to be watered into the soil to be effective, and it is best applied to soil that is already damp. For rights-of-way use, which is not amenable to irrigation, and for un-irrigated citrus applications, this generally means that bromacil would be applied during the rainy season in the winter (Turner, 2003).
Bromacil can be applied on its own as a selective herbicide or it can be applied in mixture with other herbicides to control a broad spectrum of weeds (Gomez de Barreda Jr. et al., 1998). Generally bromacil is applied at rates of 2 - 4 kg/ha (Meister 1998), depending on soil properties and persisting environmental conditions. In citrus, pineapple, and non-crop areas, Bromacil can be applied at rates of 5 - 7, 1.5 - 3, and 1.5 - 5 kg/ha respectively (Raov, 2000). Differences in soils could affect the overall performance of bromacil and these differences must be taken into consideration, i.e. soils with low clay or organic matter content and lower application rates must be used so as to avoid high rates of leaching of the chemical (Gangstad, 1989). The type, diversity and height of the vegetation are also important factors to be considered for the effective application of bromacil (Gangstad, 1989). Other methods of bromacil application are the foliar application, where bromacil is directly applied to the leaves and stems of a plant (Tu et al., 2001). Bromacil can be applied with a backpack sprayer or a hand-held bottle to the basal bark of the target plant (Tu et al., 2001).

4. Effects of bromacil on the environment

Bromacil toxicity to mammals and birds is described by its LD50, which is the dose received either as oral or dermal that kills half the population of studied animals. The LD_{50} is typically reported in grams of bromacil per kilogram of animal body weight (Tu et al., 2001). Tests conducted by the United States Environmental Protection Agency (USEPA) for bromacil mammalian toxicity revealed an LD_{50} of 3998 mg kg^{-1} when administered on acute oral basis; this showed that bromacil is practically non toxic to mammals. A similar result was obtained for birds (LD_{50} of 2250 mg kg^{-1}) and reptiles.
Bromacil's toxicity to aquatic organisms is quantified with LC_{50}, which is the concentration of the herbicide in water that is required to kill half of the study animals. The LC_{50} is typically measured in micrograms of bromacil per liter of water (Tu et al., 2001). The USEPA

test with rainbow trout and bluegill sunfish resulted in an LC_{50} of 36 and 127 ppm of the ai. This suggests that bromacil is slightly toxic to rainbow trout and non-toxic to the bluegill sunfish. Consequently, bromacil is viewed to be slightly toxic to fishes and amphibians. Determining the implication of this toxicity on the secondary component of the aquatic food chain will constitute an interesting endeavor.

Bromacil is mainly degraded by micro-organisms in the soil and several forms of micro-organisms are involved in the process such as the bacteria Psuedomans spp. Which can use bromacil as a source of carbon (Chaudhry & Cortez, 1988). Bromacil has varying effects on soil microbial populations depending on herbicide concentrations and the microbial species present. Low residue levels can enhance populations while higher levels can cause population declines (Tu et al., 2001).

Water bodies can be contaminated by direct overspray, or when herbicides drift, volatilize, leach through soils to groundwater or are carried in surface or subsurface runoffs. Amounts of leaching and runoff are largely dependent on total rainfall in the first few days after an application (Tu et al., 2001). Most environmental fate and impact concerns linked to the use of herbicides are related to offsite movement into aquatic ecosystems (Zhu & Li, 2002). Bromacil rapidly moves through the soil, as a result it has the potential to be a ground water contaminant (Gomez de Barreda Jr. et al., 1998). Bromacil may degrade in natural waters through microbial degradation and photo-sensitized degradation. Bromacil is moderately soluble in water (0.815 gl^{-1} at 25° C) (Gomez de Barreda Jr. et al., 1998; Zhu & Li, 2002). Bromacil is one of the most commonly found herbicides in groundwater; it is usually detected at higher concentrations than those of terbuthylazine and simazine (de Paz and Rubio, 2006).

Bromacil in the atmosphere is mainly degraded by light, in a process known as photo-oxidation. The hydroxyl radicals and superoxide radicals are the primary oxidizing species in the photocatalytic oxidation process of bromacil (Singh et al., 2003). Oxygen has no pronounced effect on the initiation of the photolytic process of bromacil, as compared to that of metribuzin where oxygen has a pronounced effect and hydrogen peroxide has a lesser effect (Muszkat et al., 1998). The photolytic process in bromacil is initiated by hydroxyl radicals generated by hydrogen peroxide photolysis (Muszkat et al., 1998). In a study carried out by Singh et al. (2003), the immediate photolytic products in the presence of titanium dioxide were 5-hydroxy-3-secbutyl-6-methyl uracil and diisopropyl urea.

5. Persistence and degradation of bromacil in the environment

Increasingly, herbicides are continually being applied onto the environment. Ideally a herbicide should control or eradicate the targeted species selectively, remain stationary at the site of application and degrade rapidly once its purpose is achieved, however, their persistence in the environment together with their low degradability rates have become a cause of concern especially the ecological risks they might possess (Dowd et al., 1998, Muszkat et al., 1998, Singh et al., 2003, Rosner et al., 1999, Girotti et al., 2008). The degree of bromacil persistence and mobility (Hornsby et al., 1995) is mainly dependent on soil properties and environmental conditions such as water availability.

Any herbicide's persistence in soils is often described by its half-life (also known as the DT_{50}). The half-life is the time it takes for half of the herbicide applied to the soil to be dissipated (Tu et al., 2001). Bromacil has a lengthy half-life. Its soil half-life ranges from 2 to

8 months depending upon the patterns of use and other environmental factors such as temperature and availability of water (Fishel, 2005; Meister, 1998). Bromacil activity, movement and persistence in the soil depend on the interaction of the bromacil molecule with the soil's colloids adsorption capacity (Paterson & Mackay, 1994). Soil organic carbon-water partitioning coefficient (Koc) is the the ratio of the mass of a chemical that is adsorbed in the soil per unit mass of organic carbon in the soil per the equilibrium chemical concentration in solution. Koc value of less than 100 indicates that a pesticide is very mobile in soils (Branham et al., 1995). Bromacil moves quite readily through the soil (EXTONET, 1993, Rosner et al., 1999); this is because bromacil adsorbs only slightly soil particles, with a Koc value of 32 g/ml (de Paz & Rubio, 2006; EXTONET, 1993; Gomez de Barreda Jr. et al., 1998). Bromacil is a good candidate for leaching and therefore, a groundwater contaminant (Gomez de Barreda Jr. et al., 1998).

Due to its ability to move readily through the soil and its solubility in water, concerns on the use of Bromacil arise as it is able to contaminate groundwater (EXTONET, 1993; Rosner et al., 1999, Singh et al., 2003). Relatively Bromacil behaves differently on different types of soils with different constituents. Thus Bromacil is more strongly adsorbed to by organic matter colloids rather than clay particles; as a result it is more persistent and less mobile in soils with high organic matter content (5% or more) (James & Lauren, 1995). Soils with moderate to high organic matter content may retain bromacil residues for 1 to 2 years, thus, a soil half-life of 3 to 7 months is more likely in soils with low organic matter content (less than 5%) (EXTOXNET, 1993). A soil with high organic matter content will also bind bromacil and prevent it from being available in soil solution, this obviously will affect it effectiveness on plant. In a study carried out by de Paz and Rubio (2006), involving eight of the most frequently applied herbicides in citrus orchards (glyphosate, diuron, diquat, bromacil, simazine, linuron, terbuthylazine, and terbumeton), a ranking according to the potential to leach was obtained. The leaching potential of the herbicides was as follows, from highest to potential to least; terbumeton > bromacil > simazine > terbuthylazine > diuronNlinuron > glyphosate > diquat.

On relative terms, bromacil is one of the polluters of groundwater that should be given considerable attention (EXTONET 1993, Rosner et al., 1999; Singh et al., 2003). Other report by Sanders et al., (1996) showed that bromacil was degraded within 4 to 6 months when it was applied once compared to when it was applied twice in the same season; it was also reported that Bromacil persisted in the top 75 mm of soil for nearly a year (Alavi et al., 2008). Soil with no previous bromacil use had higher chemical residue levels in lower depths and slower degradation rates than soils with a 10 year history of asparagus management and associated bromacil use.

6. The economic implications of bromacil application methods on rangelands encroached by *Acacia karroo*

The encroachment of woody plants into grassland is a global problem. Different methods have been used to control bush encroachment and most of them have been ineffective in the total elimination of bush encroachment. There are a number of chemicals that have been used in controlling bush encroachment. Consequently, large volumes of potentially hazardous chemicals, produced by various industries and agricultural operations, are entering the ecosystem. Bromacil, a broad spectrum herbicide, is used to control undesirable woody plants on noncropland so as to increase the carrying capacity of the veld. The active ingredient (bromacil) is carried into the root zone by rain. It is readily absorbed through the root system and is then translocated to foliage. The leaves then become yellow and abscise. When new leaves are formed, they also turn yellow and abscise. This process continues until the tree no longer has reserves to initiate re-growth and so it dies. Figure 2 and 3 show the efforts of bromacil application on Acacia species in the Eastern Cape Province of South Africa.

7. Encroachment of South Africa rangelands by *Acacia karoo*

(a) (b)

Fig. 2. (a). Stand of *Acacia karoo* (b). Dead stands of *Acacia karoo* form plots treated with Bromacil

Adelaide: I week before spraying **Adelaide: I year before spraying**

Fig. 3. Effect of bromacil application on Acacia *Karoo* over time

Box 1: Field Research on the Economics and Effects of bromacil Application Method

The cost implication of Bromacil application methods is vital to decisions on methods which are suitable for specific vegetation density and field size.

Methodology: The bromacil experiment was conducted at the Honeydale Section of the University of Fort Hare Research farm located at 32°47'58.78" S, 26°52'25.59" E at an altitude of 517.86 m asl in Bhisho Thornveld (Mucina and Rutherford 2006). Mean annual rainfall is 480 mm with the average summer temperature of 18.7 °C. The soil is generally characterized as a silty loam of the Glenrosa form.

The two camps were identified; one of low and the other high density *Acacia karroo* were identified. In each paddock 6 x 200m² plots were marked. Bromacil is applied to *Acacia karroo* plants in the marked plots; 2 of the plots were subjected to liquid spraying; the other 2 to granular spreading and the remaining 2 were treated as controls as follows: Treatment 1: Ground application (liquid) 1Lt Bushwhacker in 5Lt water. 2ml was applied for every 0.5m in height of trees. Treatment 2: Ground application (granules) about 3g - 2m, 6g- 4m and 12g ≥ 8m tree. Treatment 3: Control, no bromacil applied.

After a year following a bromacil application, herbaceous and tree height were measured. A disc pasture meter was used to measure the grass height for herbaceous biomass estimation. Tree height and canopy width was measured using a 2m rod. A tuft-to-tuft point system was used to determine species composition.

The partial budget technique was used to analyze the data for the time spent on application, volume and weight of the herbicide used time spent on cutting for penetrex and access. Teams working were monitored as was the amount of herbicide used in each plot. The cost of aerial application was known before hand. Total cost of each method per hectare was calculated, it included labour and chemical costs, furthermore, it was ascertained that the minimum area that can be aerially sprayed is 50 ha. The data on time spent on application, and volume and weight of herbicide used between treatments with the analysis of variance (ANOVA) with SAS (1999).

Low density : < 500 stands/ha

High density : > 1000 stands/ha

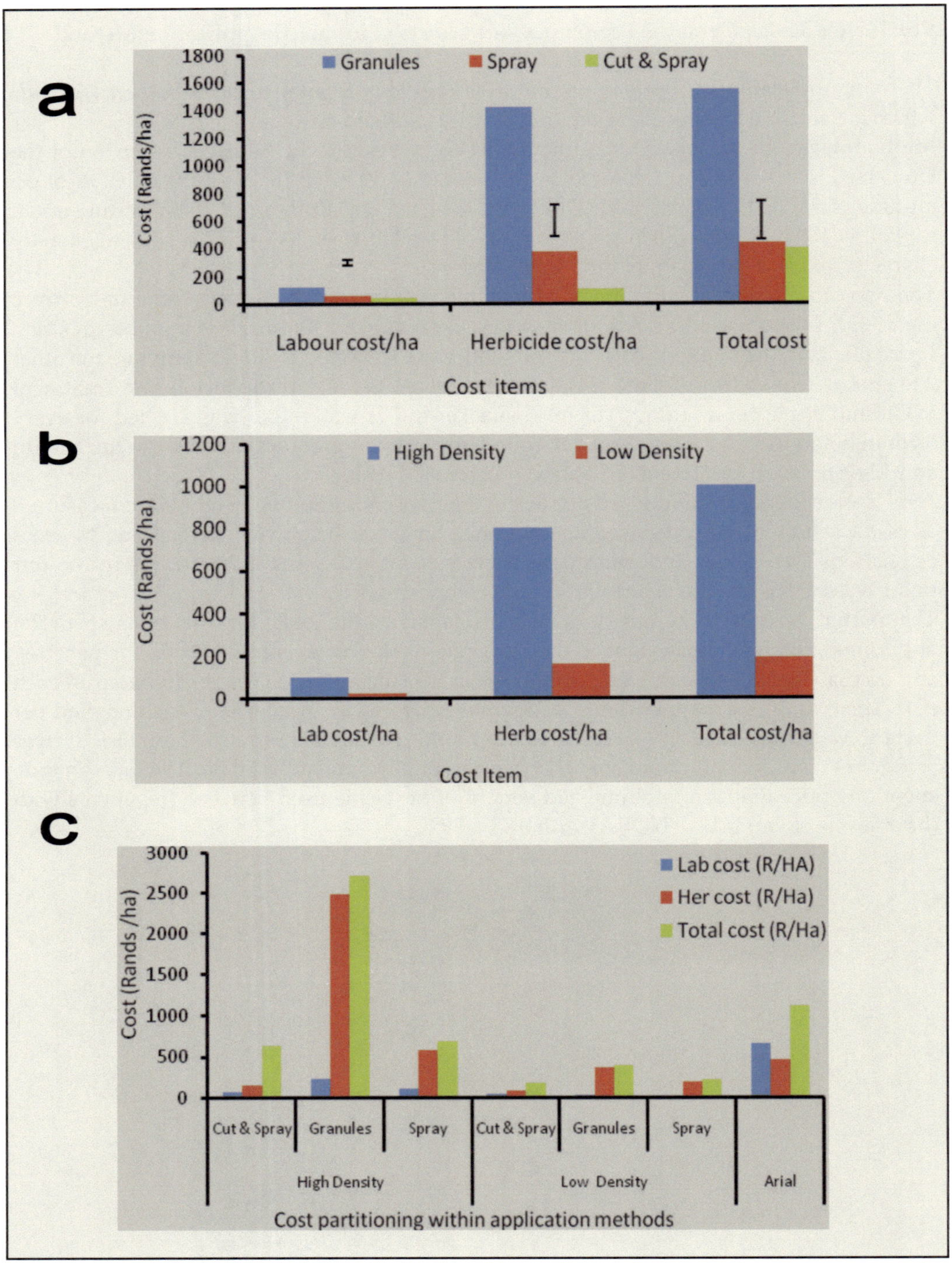

Fig. 4. (a). Absolute cost of Bromacil herbicide type and application method for the control of *Acacia Karoo* bush. (b). Absolute cost of Bromacil herbicide application on different densities of *Acacia Karoo* bush invasion. (c). A comparison aerial application cost with hand application

Box 2: Result of Field Research
Economic Implications of bromacil Application Method

The cost of bromacil was affected by bush density between the density levels. Labour was slightly higher at the high density. This could be due to large number of trees on which each has to be treated with bromacil.

The total cost of bromacil applied as granules was higher than bromacil spray and use of Penetrex and access. The time spent and labor was not different between application methods (Figure 4a). This implies that the use of granules is more expensive than the use of spray bromacil.

The time spent on bromacil application was determined by the bush density, the higher the bush density the more the time spent (Figure 4b). The amount of herbicide used was affected by the bush density. The higher density bush utilised more herbicide than the low density bush (Figure 3c). This could be due to the difficulty in movement and large number of trees and each had to get the bromacil. This implies that with hand application the high density bush will require large number of labourers for a short time or small number of labourers for a long time. In the case where labour is paid per hour per person the expenses will be high as well as herbicide purchase expenses.

The result indicated that, on a per hectare basis, it was more expensive to hand apply granules in the high density plots (R2700.ha^{-1}) compared to either spraying liquid (R675.ha^{-1}) or aerial spraying (R600.ha^{-1}). This could be due to the unease of movement for the person applying the granules, while the spray application was faster. Under low tree density aerial spraying, the cost of which does not change with density, was more expensive compared to hand application of granules (R389.ha^{-1}) compared to either spraying liquid (R207.ha^{-1}). This implies that in the low density bush, the hand application is less expensive compared with the aerial application. The cost of water used in the mixing was minimal and was not included in the calculations.

Box 3: Result of Field Research

Effects of Bromacil on Forage Productivity of Herbaceous Vegetation

Herbaceous biomass production was not affected by bromacil herbicide after one year of application. The forage yield between the plot where two forms of bromacil (granules and liquid) were applied does not vary. Rollins *et. al.*, (1988) indicated that forage yields and availability, particularly of herbaceous vegetation, is often low in dense shrub communities. Furthermore this study shows that the forage yield was similar between high and low bush densities after one year of bromacil application. The forage yield was not significantly affected by bromacil treatment after the first year of application, the pre-bromacil treatment forage yield 2083 $kgDM.ha^{-1}$ while the post treatment forage yield was 2163 $kgDM.ha^{-1}$.

After the application of bromacil the forage yield under high bush density showed a higher rate of accumulation (2604 $kgDM.ha^{-1}$) compared to the low bush density site (1642 $kgDM.ha^{-1}$). Both bromacil in a granular form and in a liquid form have shown the difference in forage yield when compared with the control sites (Table 3). The bromacil treated sites have shown a great forage yield accumulation within the first year of application. The results from this study implies that there the bromacil herbicide does not affect the forage yield, regardless of the method of application.

Effects of Bush Density on the Herbaceous Vegetation and Ecological Condition Index for Herbaceous Vegetation

The grasses species are classified into their ecological status determined by their pecieved acceptability to animals and grazing. (i) Decreaser species (highly desirable species)- those species which occur in rangeland in good condition and decrease with over grazing (Sisay and Baars 2002), increaser I species (undesirable species)- those species which occur in rangeland in good condition and increase with under utilisation and increaser II species (undesirable) occur in rangeland in poor condition and increase with overgrazing. The response of grasses according to the mentioned classes the decreaser species were not affected by bromacil application, the pre-bromacil treatment and post-bromacil treatment were different after one year. Under the different bush densities the ecological index values of decreaser species was different (Table 2). The bromacil application methods (granula, spray and control) (Table 3), did not affect the ecological value of decreaser species between the granula and spray methods. However, there was a significance difference ($P<0.05$) in the ecological value index between both granular and spray application methods and control. There was no significance ($P>0.05$) in the ecological index value of increaser Ia, Ib, IIa, IIc and other species (Table 1) between pre-treatment and post-treatment. In contrast the ecological index values of increaser IIb were significantly different ($P<0.05$) between the pre-treatment and post-treatment.

Looking at the data of all species the ecological values of all species were higher in pre-treatment than in post- treatment ($P<0.05$). The ecological index values of decreasers, increaser IIa, increaser IIb, (Table 2) and other species were higher in high bush density than in low bush density sites ($P<0.05$). All the increaser I species and increaser IIc ecological values were not affected by bush density ($P>0.05$). Pooling the ecological value of all the species there were variations ($P<0.05$) between high bush density and low bush density, ecological values were higher in high bush density than in low density site. The ecological index of decreasers were significantly different ($P<0.05$) between the three bromacil treatment methods (granula, spray and control) (Table 3). There was no significant difference ($P>0.05$) in the ecological values of increaser (Ib, IIa, IIc and others) between different bromacil treatment application methods. In contrast there were variations ($P<0.05$) in the ecological index values of increaser (Ia, IIb). Concerning ecological index of the total species there were no variation ($P>0.05$) between the three treatment application methods.

Vegetation response to bromacil treatment		Bromacil treatments	
		Pre-treatment	One year Post-treatment
Biomass production (kg/ha)		2083±89.7^{a}	2163±89.7^{a}
Ecological status (%)	Decreaser	30.3±2.2^{a}	24.3±2.2^{a}
	Increaser 1a	8.2±1.5^{a}	3.8±1.5^{a}
	Increaser 1b	0.2±0.4^{a}	0.7±0.4^{a}
	Increaser IIa	15.2±1.8^{a}	15.5±1.8^{a}
	Increaser IIb	8.2±1.8^{a}	14.3±1.8^{b}
	Increaser IIc	15.8±2.0^{a}	10.8±2.0^{a}
	Others	2.2±0.4^{a}	1.0±0.4^{a}
Bush density (Plants.ha^{-1})	*A. karroo*	3221±896.9^{a}	2100±896.9^{a}
	Others	1442±272.3^{a}	1375±272.3^{a}

(a)

Herbaceous vegetation variables		High Bush Density	Low Bush Density
Biomass production.(kg-ha^{-1})		2604±89.7^{a}	1641±89.7^{b}
Ecological status %	Decreaser	32.7±2.2^{a}	22.0±2.2^{b}
	Increaser 1a	4.0±1.5^{a}	8.0±1.5^{a}
	Increaser 1b	0.8±0.4^{a}	0.0±0.4^{a}
	Increase IIa	18.5±1.8^{a}	12.2±1.8^{b}
	Increaser IIb	14.7±1.8^{a}	7.8±1.8^{b}
	Increaser IIc	15.2±2.0^{a}	11.5±2.0^{a}
	Others	1.8±0.0^{a}	1.3±0.0^{b}

(b)

Vegetation parameter		Bromacil application methods		
		Granular	Spray	Control
Biomass production. (kg-ha^{-1})		2159±109.8^{ab}	2345±109.8^{a}	1864±109.8^{b}
Ecological status (%)	Decreaser	31.0±2.7^{a}	32.8±2.7^{a}	18.3±2.7^{b}
	Increaser 1a	4.8±1.8^{ab}	3.3±1.8^{b}	10.0±1.8^{a}
	Increaser 1b	0.5±0.5^{a}	0.8±0.5^{a}	0.0±0.5^{a}
	Increaser IIa	16.8±2.3^{a}	14.3±2.3^{a}	15.0±2.3^{a}
	Increaser IIb	5.8±2.2^{b}	6.5±2.2^{b}	21.5±2.2^{a}
	Increaser IIc	12.0±2.2^{a}	14.8±2.2^{a}	13.3±2.2^{a}
	Others	2.0±2.4^{a}	1.3±2.4^{a}	1.5±2.4^{a}
Bush density (Plants.ha^{-1})	A. karroo	1706±1098.1^{a}	3288±1098.1^{a}	2988±1098.1^{a}
	Others	101±333.5^{a}	1881±333.5^{a}	1325±333.5^{a}

Means in the same row followed by same superscripts are not significantly different ($P > 0.05$).

(c)

Table 1. **(a)** Herbaceous and woody (species composition, biomass production and bush density) vegetation status before and after bromacil treatment. **(b).** Effect of bush density on herbaceous production and species composition (Mean±SE). **(c).** Effects of bromacil application methods on herbaceous biomass production, species composition and bush density (Mean ± SE).

8. Conclusion

Bromacil based herbicides are effectively used to control annual and perennial weeds. The family possesses a broad toxicity to many plant species, however, the specific compounds differ in their toxicity to plants, solubility in water and persistence in soil. The chemical is absorbed through the root system and translocated upwards via the xylem vessels to the leaves where it interferes with light harvesting complexes by blocking the photosystem II reaction. These compounds disrupt cell membranes and cause chloroplast swelling and cellular destruction. The early symptoms of bromacil kill activities in a plant is leaf chlorosis concentrated around the veins, this is often noticed at the lower leaves and gradually moves up the plant.

The current use of bromacil in agriculture is necessary so as to sustain high productivity, reduce cost, reduce drudgery and give high profit margins. There are various ways of applying bromacil; the most appropriate method is determined by the weed being treated, the herbicide being applied, the skills of the applicator and the application site. The application method determines the extent of contact with target plant and its movement within the soil. The conventional methods of application include aerial spraying and direct application to the soil. The performance of bromacil is influenced by soil characteristics, thus soils with low clay or organic matter content are highly leachable, therefore require lower application rates. The vegetation structure and composition are also very important factors to consider.

Bromacil is non-toxic to mammals, however, it is slightly toxic to fish and amphibians. The effect of Bromacil on microbial populations depends on herbicide concentration and microbial species present. Most environmental fate and impact concerns linked to the use of herbicides are related to offsite movements into aquatic ecosystems. Bromacil is mainly degraded by micro-organisms in the soil and in natural waters. The chemical is degraded mainly by light in the atmosphere through photo-oxidation. Bromacil provides for sustained weed control because of its persistence in the environment and its low degradability rates, however, this has become the cause of environmental concern especially the ecological risks. We therefore concluded that the use of bromacil in areas with important aquatic ecosystems should be carefully undertaken and monitored.

Generally granules application by hand is higher than other application methods. Bush density class had a significant effect on the cost of Bromacil application. The time and herbicide cost constitute the main items where intervention to reduce cost may be targeted. Under low tree density condition, the cost of aerial application was higher compared to hand application of granules. Full economic implication of application methods will be better assessed with; assessment of biophysical component viz., mortality rate of *Acacia karoo* plant, rate and volume of grass biomass accumulation, grass species diversity, prevalence of decreaser species and ecological benefits

In the current study bromacil herbicide did not show any effect on *Acaccia karroo* in short term. It is clear from other research that bromacil achieves its total kill in a minimum of two years under favourable conditions. The current use of bromacil in agriculture is necessary so as to sustain high productivity. The vegetation structure and composition are also very important factors to consider.

9. References

Alavia G, Sandab M, Looc B, Greend RE & Raya C (2008). Movement of bromacil in a Hawaii soil under pineapple cultivation – a field study. *Chemosphere* 72: 45-52.

Archer, S. (1994) Woody plant encroachment into southwestern grasslands and savannas: rates, patterns and proximate causes. In: *Ecological Implications of Livestock Herbivory in the West* (Eds M. Vavra, W. Laycock and R. Pieper), Society for Range Management, Denver, CO, pp. 13–68.

Arteca RN (1995). *Plant growth substances; principles and substances.* Chapman and Hall.

Bovey RW (2001). *Woody plants and woody plant management; Ecology, safety and environmental impact.* Marcel Decker, Inc, New York. p. 185.

Branham B, Milnert E & Rieke P (1995). Potential groundwater contamination from pesticides and fertilizers used on golf courses.Michigan State University. *USGA* 33: 33-39.

Chaudhry GR & Cortez L (1988). Degradation of Bromacil by a *Pseudomonas sp. Appl. Environ. Microbiol.* pp. 2203-2207.

De Paz JM & Rubio JL (2006). Application of a GIS–AF/RF model to assess the risk of herbicide leaching in a citrus-growing area of the Valencia Community, Spain. *Sci. Total Environ.* 37: 44-54.

Dowd RM, Anderson MP & Johnson ML (1998). *Proceedings of the second national outdoor action conference on aquifer restoration, groundwater, monitoring geophysical methods.* National Water Well Association, Dublin, OH., pp. 1365-1379.

Dube S, M.S Lesoli & A.O Fatunbi (2009). The efficacy and safety of bromacil based herbicide for the control of the invasive bush species in South African rangelands. *African Journal of Biotechnology* Vol. 8 (9), pp. 1776-1781.

EXTONET (1993). *Extension Toxicology Network.* A Pesticide Information Project of Cooperative Extension Offices of Cornell University, Michigan State University, Oregon State University, and University of California at Davis. (URL:http://pmep.cce.cornell.edu/profiles/extoxnet) (Accessed 13 May 2010)

Fatunbi Abiodun Oluwole, Johan Mackson Sambo and Dube Sikhalazo. 2008. Long term effects of differing burning frequencies on the dry savannah grassland in South Africa. *African Journal of Agricultural Research* 3(2): 147-153

Gangstad EO (1989). *Woody brush control.* CRC Press, Boca Ralon, Florida. p. 103

Girotti S, Ferri EN. Fumo MG & Maiolini E (2008). Monitoring of environmental pollutants by bioluminescent bacteria. *Anal. Chim. Acta* 608(1): 2-29

Gomez-Barreda D, Lorenzo E, Gamon M, Monteaguado E, Saez A. & DE LA Cuadra JD (1991). Survey of herbicide residues in soil and wells in three citrus orchards in Valencia, Spain. *Weed Res.* 31: 143-151.

Gomez DE, Barreda DJR, Vila MG, Rueda EL, Olmo AS, Gomez DE, Barred AD, Garcia De La Cuadra J, Ten A. & Peris C (1998). Dissipation of some citrus selective residual herbicides in an irrigation well. *J. Chromatogr.* 795: 125-131.

James TK. & Lauren DR (1995). Determination of bromacil in groundwater and in high organic matter soils. *J. Agric. Food Chem*, 43: 684-690.

Lakoski, J.M., M.I. Arentsen, N.L. Kneisley, W.W. Au. &M.S. Legator. 1993. Evaluation of toxic and genotoxic effects of bromacil: Part II. Open-field behavioural assessment of locomotor effects in the rat. *J. Occup. Med. Toxicol.* 2(2):173–187.

Meister R (1998). *Farm chemicals handbook.* Meister publishing company, Willoughby, OH. pp. 101-116.

Muszkat L, Feigelson L, Bir L. & Muszkat, KA (1998). Reaction patterns in photooxidative degradation of 2 herbicides. *Chemosphere* 36(7): 1485-1492.

Prostko EP (2001). *Herbicide mode of action*. Extension Weed Specialist. University of Georgia Tifton, GA.
http://www.cropsoil.uga.edu/weedsci/slides/newmode/slide21.html#menuX000.

Raov S (2000). *Principles of weed science* (second edition). Science Publishers, Enfield, NE, USA, p. 107.

Rollins, D., D.N. Ueckert. & C.G. Brown (eds.). 1997. *Proceedings: Brush Sculptors Symposium, Sept. 17-18, 1997,* Abilene, Texas. Texas Agr. Exp. Sta., San Angelo, Tex.

Rosner M, Yaasur IT, Hadas A, Russo D & Yaron B (1999). Leaching of tebuthylazine and bromacil through field soils. *Water Air Soil Pollut*. 113: 319-335.

Sanders P, Wardle D & Rahman A (1996). Persistence of bromacil in soils with different management histories. A paper from the *49th conference proceedings of the New Zealand plant protection society incorporated.*

Singh HK, Muneer M. & Bahneman N (2002). Photocatalyzed degradation of a herbicide derivative, bromacil in aqueous suspension of titanium dioxide. The royal society of chemistry and owner societies. *Photochem. Photobiol. Sci.* 2: 151-156.

Smit, G.N., Aucamp, A. and Richter, C.G.F., 1999. Bush encroachment: an approach to understanding and managing the problem. In: Tainton, N.M., Editor, , 1999. *Veld Management in Southern Africa,* University of Natal Press, Pietermaritzburg.

Tu M, Hurd C & Randall JM (2001). *Weed Control Methods Handbook: Tools and Techniques for Use in Natural Areas*. The Nature Conservancy, http://tnc weeds.ucdavis.edu, version: April 2001.

Turner L (2003). *Bromacil and Lithium Bromacil: Analysis of Risks from Herbicide Use to Ten Evolutionarily Significant Units of Pacific Salmon and Steelhead*. Ph.D Environmental Field Branch Office of Pesticide Programs.

Van Auken, O.W. (2000) Shrub invasions of North American semiarid grasslands. *Annual Review of Ecological Systems* 31, 197–215.

Wiegand K, Saltz D. & Ward D (2006). A patch-dynamics approach to savanna dynamics and woody plant encroachment - Insights from an arid savannah. Perspectives in Plant Ecology, *Evol. Syst.* 7: 229-242.

34

Pyrimidinylsalicylic Based Herbicides: Modeling and Prediction

Eduardo J. Delgado
Theoretical and Computational Chemistry Group (QTC)
Department of Physical-Chemistry, Faculty of Chemical Sciences
Universidad de Concepción, Concepción,
Chile

1. Introduction

The necessity of increasing the productivity per cultivation area is a peremptory demand since, on hand, the available surface is limited, even worse it has diminished due to the degradation of the soil; and on the other hand, it is necessary to supply the food demand of a steadily increasing population.

To supply this demand of the current world population, about six thousand million, it is required to produce more and more. To do this is necessary to use massively chemicals, known generically as agrochemicals (insecticides, fungicides, acaricides, nematocides and herbicides). The use of these chemicals has allowed for significant reduction of the agriculture plagues and consequently increased the productivity. Among the pesticides, the herbicides deserve special attention since, due to the resistance developed by weeds, new products have to be steadily introduced to market.

Plants and many microorganisms are able to synthesize from inorganic precursors all the metabolites needed for their survival. In contrast, animals must obtain many compounds, such as vitamins, essential fatty acids and certain amino acids, from their diet. This is because they lack the full biosynthetic machinery, so there are metabolic pathways and their component enzymes that are not found in animals. The branched-chain amino acids (BCAAs) are synthesized by plants, algae, fungi, bacteria and archae, but no by animals. Therefore, the enzymes involved in the BCAA biosynthetic pathway are potential targets for the development of herbicides, fungicides, and antimicrobial compounds.

Pyrimidinylsalicylic acid (PSA) based compounds show potent herbicidal activity. This activity has been identified as a result of the inhibition of acetohydroxyacid synthase, AHAS. Unfortunately, this family of compounds has been poorly characterized from the physical-chemical point of view. This lack of information has prevented the assessment of their impact in the environment. The difficulty to obtain accurate experimental values arise mainly from limitations of analytical techniques, cost, safety and time. For this reason, it is very useful to be able to predict these properties. Such a prediction may be important additionally for the design of novel herbicides since their properties could be predicted prior to synthesis and consequently the design may, in this way, be guided by the results of calculations. Once these properties are known the effect of these chemicals on the environment could be evaluated in advance, reaching in this way the desired compromise between productivity and environment protection. In this chapter, classical and quantum chemical methods are applied to predict and calculate important properties of PSA based herbicides.

2. Partitioning of a chemical in the environment

When a chemical is released in the environment it distributes among the diverse compartments which comprise it, Figure 1, (Stangroom et al., 2000; Ballschmiter, 1992). A knowledge of chemical partitioning is needed to assess pathways of pollutant transport and transformation. Hence, a detailed understanding of chemical partitioning behaviour is a fundamental requirement for the development of environment models for the assessment of contaminant fate. An improved understanding of chemical partitioning would facilitate the estimation of exposure levels in the various environment compartments (soil, water, air and biota). Being the acid dissociation constant pK_a one of the most important of such parameters because ionization alters the macro properties such as solubility and lipophilicity.

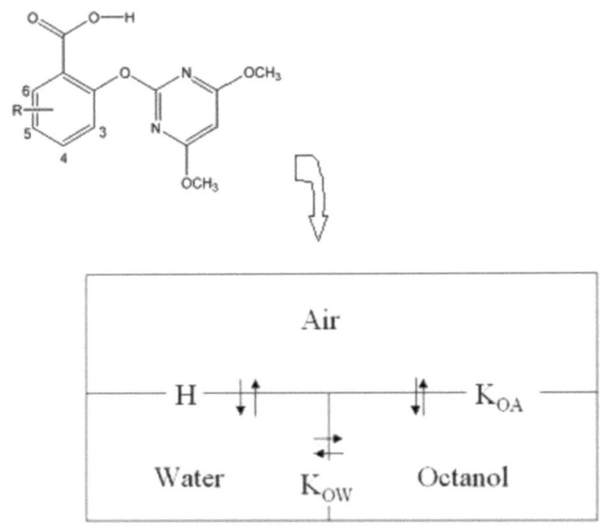

Fig. 1. Distribution of PSA compounds in the environment.

3. Pyrimidinyl salicylic acids

The synthesis and herbicidal activity of pyrimidinylsalicylic acid (PSA) based compounds, Figure 2, was reported for first time by Nezu *et al.* (Nezu et al., 1996; Nezu et al., 1998). These compounds show highly potent herbicidal activity characteristic of AHAS inhibition. The effect of substituents R introduced into various positions of the benzene ring is striking. Some substituents at the 6-position increase the AHAS inhibitory activity enormously, whereas others, as well as most substituents at the other positions, have the reverse effect. The replacement of O, in the X position, with S does not much affect the AHAS inhibition except for a few pairs of analogs. The phytotoxicity in the oxy-series is, however, reduced, in general, in the corresponding thio-series. The extent of the toxicity reduction varies depending on the crop/weed species, leading to selectively herbicidal thiosalicylic acids.

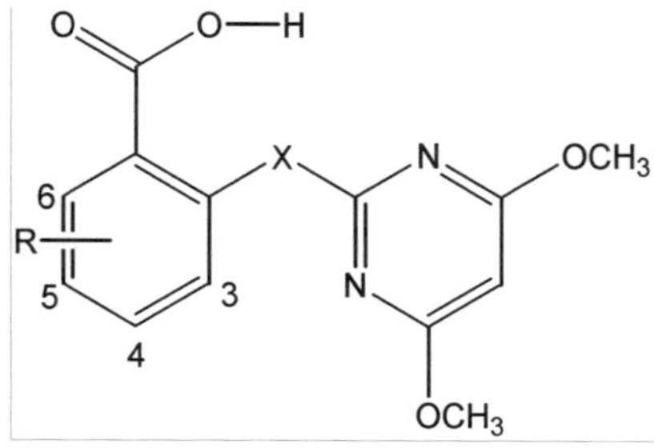

Fig. 2. General structure of PSA compounds: X=O, S

4. Acetohydroxyacid synthase.

Acetohydroxyacid synthase, AHAS, the first common enzyme in the biosynthetic route to the branched chain amino acids, valine, leucine and isoleucine, Figure 3, has been identified as the target of action of several structurally different types of chemicals (sulfonylureas, sulfonamides, imidazolinones and pyrimidylsalicylates) with high herbicidal activity. These four classes of herbicides, all obtained by traditional screening methods, have the attributes of low application rates, good crop selectivity, environmental safety and low mammalian toxicity. These herbicides act inhibiting AHAS leading to the starvation of the plant for these amino acids, it is this starvation that is thought to be the primary mechanism by which these chemicals cause plant death (Singh & Shaner, 1995; Zohar et al., 2003; Pang et al., 2002, Pang et al., 2003; McCourt et al., 2005; MCCourt et al., 2006; Duggleby & Pang, 2000; MCCourt & Duggleby, 2006).

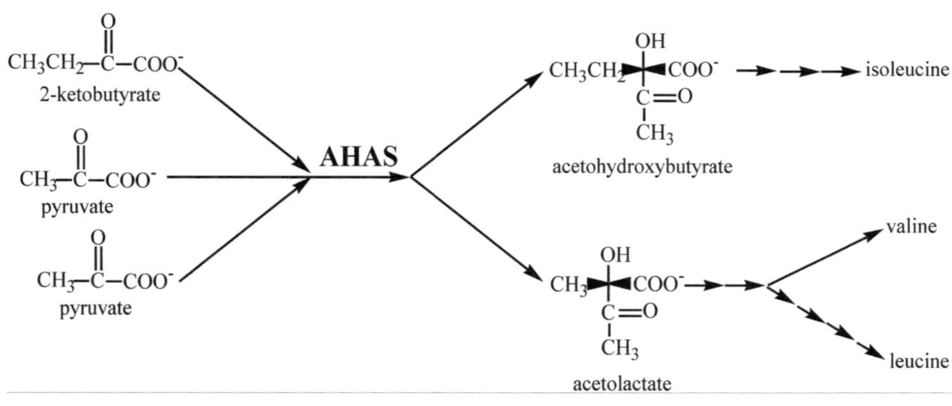

Fig. 3. Branched-chain amino acid biosynthesis pathway

Valine, leucine and isoleucine are synthesized by a common pathway in micro organisms and plants. One unusual feature of this pathway is the employment of parallel steps leading to the formation of valine and isoleucine. These parallel steps involve four enzymes, namely the anabolic AHAS, ketol-acid reductoisomerase, dihydroxyacid dehydratase, and a transaminase, each of which is capable of catalyzing two slightly different reactions.

The anabolic AHAS catalyzes the first of the parallel steps and is at branch point in the pathway because its reactions will determine the extent of carbon flow through to the branched-chain amino acids. The reactions involve the irreversible decarboxylation of pyruvate and the condensation of the acetaldehyde moiety with a second molecule of pyruvate to give 2-acetolactate or with a molecule of 2-ketobutyrate to yield 2-aceto-2-hydroxybutyrate.

Each of the products in then converted further in three reactions to give valine and isoleucine; for leucine biosynthesis, four additional enzymes are required using valine precursor 2-ketoisovalerate as the starting point for synthesis.

The regulation of the biosynthesis of the branched–chain amino acids is complex and carefully controlled. This regulation is essential, not only to ensure a balanced supply of the amino acids within cells, but also because its intermediates interact with other cellular metabolic pathway. Even through microbes and plants share the common branched-chain amino acid pathway, its regulation may vary among organisms and is not fully understood.

AHAS belongs to a super family of thiamine diphosphate, ThDP, dependent enzymes that are capable of catalyzing a variety of rections, including both the oxidative and non-oxidative decarboxylation of 2-ketoacids. This cofactor is bound by a divalent metal ion such as Mg^{+2}, which coordinates to the diphosphate group of ThDP and two highly conserved residues in these proteins (Mc Court et al., 2006). AHAS also binds a molecule of flavin adenine dinucleotide, FAD. There is no intrinsic feature of the reaction catalyzed that demands this requirement, and is demonstrated by the fact that there is an FAD-independent form of the enzyme in some bacteria. The hypothesis that FAD plays a purely structural role is supported by comparison of the structures of the FAD-dependent and FAD-independent enzyme (Pang et al., 2004).

5. Molecular approaches to studying chemistry in solution.

Modern chemistry is oriented more and more towards elucidating in detail how the macroscopic properties are determined by the microscopic properties of matter, thus enabling subsequent experimentation to be concentrated in the most promising directions. There exist several approaches for the prediction of chemical properties in solution, which may be classified as: discrete, classical and quantum chemical methods.

Discrete methods based on statistical mechanics, link those two pictures through the probabilistic treatment of particle ensembles. The most popular are molecular dynamics (MD) and the Monte Carlo (MC) method. In both cases, the condensed system is represented by an assembly of interacting particles, the statistical distribution of any property, or its evolution in time, is obtained as a sum over all particles, with appropriate rules. Applications of such techniques to study phase equilibria have been reported widely in literature (Frenkel & Smit, 1996; Siepmann, 1999; Duffy & Jorgensen, 2000; Kollman, 1993). Although some simple hydrocarbons can nowadays be reasonably well described by MD and MC methods, water and especially water mixtures, still represent challenges for such simulations techniques despite 30 years of active parameterization of appropriate force-fields. This is due to the extremely strong and complicated electrostatic and hydrogen-bonding interactions (Estrada et al., 2007).

The best developed classical method is the multivariate quantitative structure-activity/property relationship (QSAR/QSPR) methodology (Katritzky at al., 1997; Karelson et al., 1996; Hansch at al., 1996; Hansch, 1993, Cramer at al., 1988; Klebe at al., 1994). The underlying assumption in this methodology is that the molecular structure of an organic compound contains, in principle, coded within it all of the information which predetermines the chemical, biological and physical properties of that compound. If we could elucidate in detail how these properties are determined by structure, we could predict such properties simply from a knowledge of the molecular structure. A major goal of a QSAR/QSPR study is to find a mathematical relationship between a certain property and one or more descriptive parameters, known as descriptors, related to the structure of the molecule. These descriptors are numerical representations of structural features of molecules that attempt to encode important information that causes structurally different compounds to have different property values. These models so developed are important for the design of novel compounds since properties can be predicted prior to synthesis, and in this way the design of new chemicals, with specific properties, may be guided by the calculation results.

Quantum mechanical continuum solvation methods (Tomasi & Persico, 1994; Cramer, 2002; Young, 2001) are based on the self-consistent reaction field (SCRF) approach which considers the solvent as a structureless and continuous dielectric medium, characterized with a dielectric constant ε, with the solute placed in a suitable shaped hole within it. The SCRF method is an adaptation of the Poisson method for *ab initio* calculations. There are quite a number of variations on this method. One point of difference is the shape of the solvent cavity. Various models use spherical cavities, spheres for each atom, or an isosurface of electron density. The second difference is the description of the solute, which could be a dipole, multipole expansion, or numerical integration over the charge density.

The quantum mechanical based methods have been developed to the point that they are useful tools for predicting thermodynamic properties and phase behaviour of some substances to an accuracy useful in engineering calculations (Sandler, 2003).

6. Prediction of PSA properties: three case studies

6.1 Quantitative prediction of AHAS inhibition by PSA compounds

We have studied AHAS inhibition by PSA based herbicides within the framework of quantitative structure-activity relationship (QSAR) methodology (Diaz & Delgado, 2009). A general model for this family of herbicides has been developed to predict molar pI_{50}, i.e, the logarithm of the reciprocal molar concentration of herbicide required for 50% inhibition of the AHAS activity. The model involves only four descriptors: two geometric and two quantum chemical, accounting for the steric, electrostatic and hydrogen bonding interactions responsible for the binding of the herbicide to the enzyme.

6.1.1 Chemical data

The data set of the pI_{50} was taken from the data reported by Nezu *et al.* (1998). The set contains 46 structures of substituted *O*-(4,6-dimethoxypyrimidin-2-yl)salicylic acids and thio analogs inhibiting AHAS, including 6-substituted(thio)-, 5- and 6- substituted salicylic acids; covering a pI_{50} range from about 3 to 8 units.

6.1.2 Methodology

Empirical evidences show that the acidic carboxyl group of these pyrimidylsalicylates is indispensable for AHAS inhibition; moreover, it has been suggested the carboxylate group is responsible for the binding of the inhibitor molecule to the enzyme (Nezu at al., 1998). Enzymes are proteins which are actives under relative mild reactions conditions: temperature below 100ºC, atmospheric pressure and nearly neutral pH. Therefore, at these conditions of pH the PSA compounds will be in their anionic form, since their pK_a values go from about 3.3 to 4.4 (Delgado, 2009).

Modeling was performed in order to set the anions in their lowest energy 3D conformations. To achieve this goal, initial three-dimensional geometries of the chemical structures were generated using Hyperchem 7.0 molecular modeling package. These 3D structures were refined later using Ampac 5.0, a semiempirical molecular modeling program, using AM1 parameterization. To determine the lowest-energy conformations for each molecule, geometry optimizations were carried out allowing one or more torsional angles to vary systematically. The keyword CHARGE= -1 was always used in all cases. The Ampac output files, containing the lowest–energy structures and the respective electron wave functions of individual compounds, were loaded into the Codessa program to calculate the molecular descriptors. This pool of descriptors was reduced by removing descriptors that could not be calculated for every structure in the data set, and by eliminating one descriptor from those pairs highly correlated. Afterwards, from this reduced pool of descriptors the best multilinear correlation QSAR model was searched using the Sigmastat statistical package

6.1.3 Results

A total of 184 descriptors, 12 geometrical and 172 quantum chemical, were calculated for all compounds. The best regression equation found involves only four descriptors, two geometrical and two quantum chemical: S_M, the molecular surface area; S_{XY}, the normalized shadow area of the molecule projected on the XY plane; HOMO, the energy of the highest occupied molecular orbital; and FNSA, the fractional partial negatively charged surface area. It is noteworthy that these descriptors individually correlate poorly with the property (R^2 = 0.22, 0.27, 0.11, 0.01 for S_M, S_{XY}, HOMO and FNSA, respectively), however they collectively correlate pretty well with the property, R^2 = 0.89. This is an interesting result since, from the respective individual correlation coefficient, these descriptors are seemingly not relevant for predicting inhibition. Nevertheless, the relevance of these descriptors it is made evident only when the correlation with the collective is considered. This means that the interaction of information among the descriptors add an important additional predictive value which goes further from the simple sum of the information contained in the individual descriptors. This finding is in agreement with the widely accepted idea that inhibition is driven by several interactions (steric, electrostatic and hydrogen-bonding) which occur simultaneously and synergically. The best regression equation found is the following:

$$pI_{50} = 13.70 - 0.04 S_M + 18.77\, S_{XY} + 0.79 HOMO - 4.99 FNSA \quad (1)$$

and its respective statistics is shown in Tables 1 and 2.

In these tables the statistical parameters have the usual meaning. The p-values indicate that all descriptors are statistically significant at the 99% confidence level. On the other hand, the VIF values, about 1, indicate there is no serious collinearity between the involved descriptors (Ott & Longnecker, 2001). Therefore, it is concluded that all the independent variables included in the model are relevant.

The experimental and calculated values of pI_{50} along with the values of the descriptors involved in the model are shown in Table 3. The respective scatter plot is shown in Figure 4.

To check the predictive capability of the model, it was tested with an external set of chemicals not contains in the training set. The validation data set included 13 chemicals including 5- and 6- substituted salicylic acids as well as 6- substituted thio analogs. In Table 4, the values of the molecular descriptors along with the experimental and calculated values of pI_{50} for the validation set are shown. The statistics for the validation is as follows: $R^2 = 0.84$, s = 0.33, F= 59. These results confirm the prediction capability of the model.

	Coefficient	Std. Error	p-value	VIF
Constant	13.70	2.76	< 0.001	
S_M	-0.04	0.003	< 0.001	1.31
S_{XY}	18.77	2.39	< 0.001	1.81
HOMO	0.79	0.26	0.006	1.99
FNSA[(1)]	-4.99	1.30	< 0.001	1.11
	$R^2 = 0.89$, $R^2_{CV} = 0.85$, $R^2_{df} = 0.88$, s = 0.43			

Table 1. Statistical parameters for the best QSAR model

	Analysis of Variance				
	DF	**SS**	**MS**	**F**	**P**
Regression	4	37.38	9.34	52	< 0.001
Residual	25	4.53	0.18		
Total	29	41.91	1.44		

Table 2. Analysis of variance of the best QSAR model

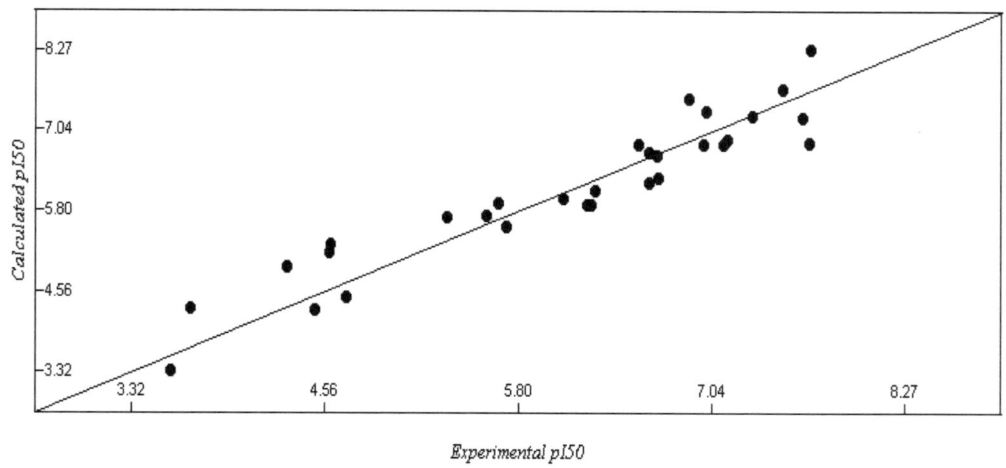

Fig. 4. Scatter plot of the calculated vs. experimental pI_{50}

Y	S_M	S_{XY}	HOMO	FNSA[(1)]	pI_{50} (exp)	pI_{50} (calc)	Diff.
5-substituted Pyrimidinylsalicilates							
F	279.23	0.57	-5.35	0.47	6.27	5.86	-0.41
Cl	313.83	0.63	-4.92	0.53	5.35	5.67	0.32
Br	318.11	0.57	-4.93	0.54	4.50	4.25	-0.25
C_2H_5	310.27	0.56	-5.17	0.35	4.59	5.15	0.56
OCH_3	300.19	0.55	-5.21	0.39	4.60	5.25	0.65
OC_6H_5	357.88	0.60	-5.40	0.42	3.58	3.32	-0.26
SCH_3	338.00	0.61	-4.86	0.38	4.32	4.93	0.61
CN	301.27	0.56	-5.51	0.53	3.71	4.29	0.58
NH_2	285.19	0.57	-5.17	0.42	6.09	5.98	-0.11
CCH	309.95	0.57	-5.28	0.50	4.70	4.45	-0.25
6-substituted Pyrimidinylsalicilates							
H	268.10	0.57	-5.16	0.42	6.64	6.69	0.05
F	267.34	0.61	-5.24	0.46	7.30	7.23	-0.07
Cl	273.62	0.63	-5.18	0.50	7.62	7.21	-0.41
I	281.59	0.60	-5.21	0.40	7.66	6.83	-0.83
CH_3	280.47	0.62	-5.13	0.38	6.89	7.51	0.62
C_2H_5	294.75	0.62	-5.12	0.37	6.57	6.81	0.24
OC_3H_7	309.91	0.60	-5.25	0.33	6.24	5.87	-0.37
$OCH(CH_3)_2$	322.80	0.62	-5.22	0.37	5.73	5.54	-0.19
SC_2H_5	316.95	0.62	-4.31	0.33	7.11	6.79	-0.32
SC_3H_7	324.40	0.60	-4.32	0.32	6.29	6.10	-0.19
NO_2	292.87	0.63	-5.65	0.48	6.64	6.21	-0.43
CO_2CH_3	307.35	0.62	-5.56	0.41	5.68	5.89	0.21
NH_2	280.47	0.62	-5.24	0.39	7.00	7.32	0.32
6-substituted Pyrimidinyl(thio)salicylates							
F	301.31	0.70	-4.35	0.47	7.67	8.27	0.60
Cl	308.83	0.69	-4.32	0.49	7.49	7.65	0.16
I	294.59	0.62	-5.11	0.39	6.99	6.80	-0.19
OC_2H_5	341.24	0.65	-4.19	0.36	6.70	6.30	-0.40
OC_6H_5	364.64	0.70	-4.37	0.44	5.60	5.69	0.09
NO_2	324.48	0.69	-4.71	0.51	6.69	6.63	-0.06
$COCH_3$	326.72	0.68	-4.43	0.44	7.14	6.87	-0.27

Table 3. Molecular descriptors and values of experimental and calculated pI_{50} for the training set.

Y	S_M	S_{XY}	HOMO	FNSA(1)	pI_{50} (exp)	pI_{50} (calc)	Diff.
5-substituted Pyrimidinylsalicilates							
H	268.10	0.57	-5.16	0.42	6.64	6.69	0.05
I	321.03	0.55	-4.93	0.39	5.05	4.50	-0.55
OH	308.43	0.69	-4.87	0.48	7.20	7.29	0.09
6-substituted Pyrimidinylsalicylates							
C_3H_7	310.23	0.61	-5.15	0.35	5.89	6.17	0.28
OC_2H_5	302.35	0.66	-5.22	0.33	7.05	7.37	0.32
OC_4H_9	328.96	0.64	-5.15	0.32	6.21	5.98	-0.23
CF_3	289.31	0.68	-5.51	0.52	6.96	7.15	0.19
6-substituted Pyrimidinyl(thio)salicylates							
Br	288.99	0.63	-5.08	0.40	7.40	7.23	-0.17
CH_3	286.51	0.67	-4.98	0.41	7.53	8.05	0.52
OCH_3	323.96	0.67	-4.20	0.39	7.05	7.18	0.13
SC_2H_5	345.36	0.65	-4.26	0.35	6.67	6.13	-0.54
CF_3	326.52	0.69	-4.52	0.55	6.02	6.48	0.46
COC_6H_5	366.52	0.72	-4.46	0.45	5.19	5.75	0.56

Table 4. Molecular descriptors and values of experimental and calculated pI_{50} for the validation set.

6.1.4 Discussion

From the explanations suggested in literature, it seems to be logical to think the first requirement that the inhibitors must meet is the steric factor, since the chemicals must fit in the active site channel. To this the chemical has to have the adequate size and the shape. Consequently, the more relevant descriptors in the model are the molecular surface area and the shadow area of the molecule projected on the XY plane. These descriptors encode the size and the geometrical shape of the molecule, respectively. The normalized shadow indices, introduced by Jurs as molecular shape descriptors (Stanton & Jurs, 1990), are calculated as the ratio of the areas of three orthogonal projections to the maximum dimensions along the respective axes, taking the X coordinate along the main axis of inertia and so on.

The correlation coefficients for the above two descriptors have opposite sign, indicating size and shape have antagonistic effects on inhibition. Thus, on one hand, the pI_{50} value decreases as the surface area increases; on the other hand, the normalized shadow area in the XY plane, S_{XY}, increases the value of pI_{50}. Therefore, both descriptors have inverse effect on inhibition, disfavoring and favoring inhibition, respectively. This inverse effect is presumably due to the inhibitors must accommodate in a size-limited cavity in the enzyme, on one hand; and to enter into the enzyme the inhibitors should have the adequate conformation to facilitate the entry and to favor the diverse intermolecular interaction with amino acid side chains, on the other hand. The conformations which maximize the shadow XY area are those in which the benzene ring and the pyrimidinyl ring are aligned in the

same plane in such a way the main moment of inertia lies in this plane. This conformation, which is observed in the thio-derivates, facilitates the entry of the inhibitor into the active site channel and set the ring atoms in a favorable position to make interactions with the amino acids side chains. In the other two families instead the benzene and the pyrimidinyl ring lies in planes almost orthogonals each other, adopting a nearly L-shaped structure hindering the entry into the active site tunnel.
The correlation coefficient for the HOMO energy is positive indicating the property, pI_{50}, and this descriptor vary in symmetrical way, i.e., the higher the value of the Homo energy the higher the value of pI_{50}, which means a lower molar concentration for 50% inhibition of the enzyme activity. Therefore, the HOMO energy is a key descriptor for an increased inhibitory potency. The energy of the HOMO characterizes the susceptibility of the molecule toward attack by electrophiles. Thus, it is expected this descriptor is involved in hydrogen bonding interactions between partners with complementary properties, i.e. hydrogen acceptors on the ligand and hydrogen donors on the receptor. It has been well established that this type of interaction is one of the factors responsible for the binding of the inhibitor to the enzyme. The carboxylate group is expected to be the key for the extent of these interactions due to its ability to act as hydrogen acceptor because of the high electron density on the oxygen atoms of this group. This could explain the empirical finding about the carboxylate group is indispensable for AHAS inhibition because of its crucial role in the binding of the inhibitor to the enzyme.
The fourth descriptor in relevance is the fractional partial negatively charged surface area, $FNSA^{(1)}$, i.e. the ratio of the partial negatively charged surface area to the total molecular surface area (Stanton & Jurs, 1990); encoding features related to polar interactions. Its respective correlation coefficient is negative indicating this electrostatic factor disfavors the inhibition. This effect is similar to that observed in the inhibition of AHAS by sulfonylurea herbicides (Wang at al., 2005), wherein the chemicals need contributions from positively charged groups to achieve enhanced inhibition, and only small areas of the active site channel of AHAS have preference for negatively charged groups. This analog trend between these two families of herbicides seems to suggest they share the same binding site in the enzyme or partially overlapping sites. This finding has already been observed for imidazolinones, as well. These results are in agreement with recently reported results obtained by the integration of molecular docking, CoMFA, CoMSIA and DFT calculations (He at al., 2007). Nevertheless, our model predicts pI_{50} with fewer descriptors and similar statistics than those models reported in the just mentioned article. In QSAR modeling, the Parsimony Principle (Occam´s Razor Principle) calls for using models and procedures that contain all that is necessary for the modeling but nothing more, i.e. given a number of models with nearly the same predictive error, that containing fewer parameters should be preferred because simplicity is desirable in itself (Estrada et al., 2004).

6.2 DFT calculation of pK_a´s for PSA based herbicides

The acid-dissociation constant, pK_a, of a compound influences many characteristics of the compound such as its reactivity. In biochemistry, the pK_a values of pesticides are of major importance for the activity of the related enzymes. In environment chemistry, this property is of general importance because ionization of a compound alters its physical behavior and macro properties such as solubility and lipophilicity. The nature and location of substituents may induce drastic changes in the values of the acid-dissociation constants, and consequently important changes in the physical and chemical properties. The interest in determine the

pK_a values of diversely substituted chemicals is, on one hand, to correlate the substituent effect and the observed pK_a value; and, on the other hand, to evaluate its effect on derived physical-chemical properties. We have calculated the acid-dissociation constants for 39 PSA derived herbicides by using density functional theory (DFT) methods at B3LYP/6-31G(d,p) level of theory (Delgado, 2009).

6.2.1 Computational methods

The quantum chemical calculations were carried out using Jaguar and its graphic interface Maestro. Gas-phase molecular geometries and electronic energies were computed at DFT B3LYP/6-31G(d,p) level of theory. To determine the lowest-energy conformations for each molecule, geometry optimizations were carried out allowing one or more torsional angles to vary systematically. Free energies of solvation in water were computed by single point calculation on the gas-phase optimized geometry using the Poisson-Boltzmann Solvation Model (PB) (Marten at al., 1996). The pK_a values were calculated with the Jaguar pK_a prediction module using the following thermodynamic cycle, scheme 1, and subsequent equation:

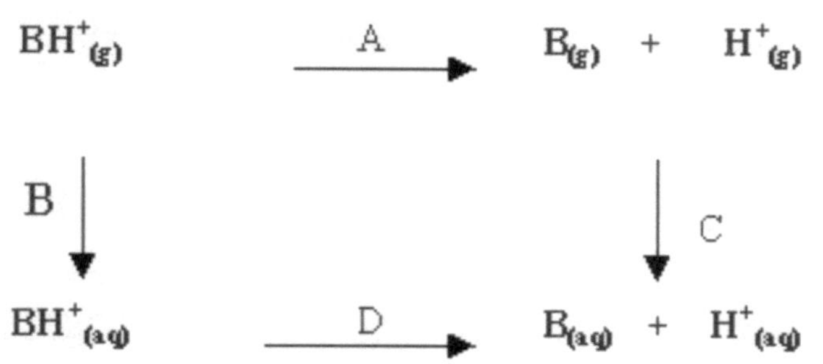

Scheme 1. Thermodynamic cycle used to calculate the pK_a.

$$pK_a = \frac{1}{2.3RT} D \tag{2}$$

where D is the free energy change involved in the step D, and may be calculated from the free energy changes of the other cycle steps by the following relation: D = A+ C - B. In this cycle, the BH^+ species refers to the adduct formed from the reaction between the electron-pair donor B species, Lewis base, and the electron-pair acceptor H^+ species, Lewis acid.

6.2.2 Results and discussion

The calculated values of pK_a's, Table 5, fall in a very narrow range going from 3.3 to 4.4, as it was conjectured by Nezu et al. (1998). However, small differences in the pK_a scale may entail large changes in the degree of ionization and related properties such as enzyme inhibition and environmental fate. Therefore, the knowledge of the pK_a for each compound, rather than to take an average value for the all family, is of fundamental importance to get a better comprehension of the behavior of these compounds since the average value does not account for the differences observed in the properties of different substituted compounds.

Unfortunately, as mentioned above, for these compounds there is no experimental pK_a data available to validate the calculated figures. However, considering that the observed

differences in the values of the pK_a's have to be effect of the R-substituent group, since the O-dimethoxypyrimidynil (ODMP) group keeps constant for all compounds, an estimation of the reliability of the calculated figures can be reached quantifying the effect of the R-substituents on the pK_a values. From the mid-1930's the primary means to assess the effect of a substituent in the meta or para position of the benzene ring have been the Hammett σ constants (Hammett, 1937). The σ constants were originally formulated from the logarithm of the ratios of the acid-dissociation constants of substituted benzoic acids relative to the acid-dissociation constant of benzoic acid itself. Since the definition of σ considers a reference compound respect to it the effect of the substituent is measured, we hypothesize that for a same substituent, the σ constant value should be the same for benzoic acids and PSA acids, once we take the 6-H substituted PSA, (compound 1 in table 5) as reference structure to isolate the effect of the R substituent group.
The results of this comparison are shown in Table 6. It is possible to observe that the calculated σ values of this study for the selected substituents, not only, do exhibit the correct qualitative trend, but also show quantitative accuracy, within the error reported by Hammett. This finding validates the quality of the pK_a's values calculated in this study.
In Fig. 5, the pK_a values are plotted as function of σ for benzoic acids and PSA compounds with the same substituents. The figure shows a strong correlation between the pK_a's and the

Compound	R	pK_a	Compound	R	pK_a
1	6-H	4.0	**21**	6-$OCH(CH_3)_2$	4.3
2	6-F	3.4	**22**	6-OC_4H_9	4.2
3	6-Cl	3.4	**23**	6-$OCHF_2$	3.8
4	5-F	3.7	**24**	6-OC_6H_5	4.0
5	5-Cl	3.6	**25**	5-OCH_3	3.8
6	3-F	3.9	**26**	5-OC_6H_5	3.9
7	5,6-$(Cl)_2$	3.3	**27**	6-SCH_3	4.1
8	6-C_6H_5	4.4	**28**	6-SC_2H_5	4.2
9	6-CH_3	4.0	**29**	6-SC_3H_7	4.2
10	6-C_2H_5	4.0	**30**	5-SCH_3	4.0
11	6-C_3H_7	4.0	**31**	6-CO_2CH_3	4.1
12	6-CF_3	3.5	**32**	6-COC_6H_5	4.2
13	5-CH_3	4.0	**33**	6-$COCH_3$	3.6
14	5-C_2H_5	4.0	**34**	6-NO_2	3.7
15	5-CN	3.4	**35**	6-CH_3SO_2	3.9
16	5-CCH	3.7	**36**	6-NH_2	4.3
17	3-CH_3	4.0	**37**	5-NH_2	3.7
18	6-OCH_3	4.3	**38**	5-NO_2	3.4
18	6-OC_2H_5	4.1	**39**	5-OH	3.8
20	6-OC_3H_7	4.2			

Table 5. Calculated acid-dissociation constants for R-substituted PSA compounds.

Substituent	Benzoic acids [42]		DMPS acids	
	pK_a	σ	pK_a	σ
m-chloro	3.83	0.37 ± 0.04	3.6	0.4
m-cyano	3.60	0.678	3.4	0.6
m-fluor	3.87	0.34 ± 0.08	3.7	0.3
m-hydroxy	4.08	0.12	3.8	0.2
m-methoxy	4.09	0.12 ± 0.12	3.8	0.2
m-methyl	4.27	-0.07 ± 0.04	4.0	0.0
m-nitro	3.49	0.71 ± 0.07	3.4	0.6

Table 6. pK_a's and Hammet σ constants for selected substituents of benzoic acids and PSA compounds.

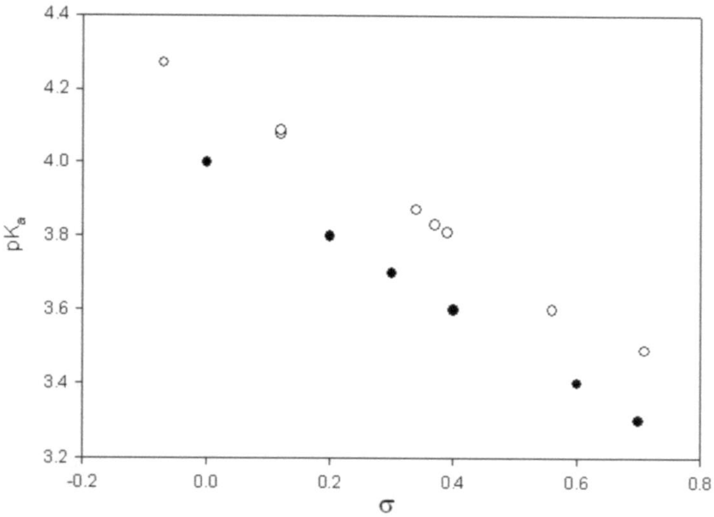

Fig. 5. pK_a vs the Hammett σ constants (open circles: benzoic acids; solid circles: PSA compounds.

σ constants for the two families of compounds. Both straight lines are shifted in an extent that we identify as a consequence of the substituent effect of the ODMP group.

Additionally to the above checking procedure, we propose a second checking test in order to discard the presence of potential systematic errors in the method. The reported correlation equation of pK_a as function of σ for benzoic acids is (Hollingsworth at al., 2002):

$$pK_a = 4.20 - \sigma \tag{3}$$

This equation allows us to predict the pK_a's of the compounds of this study if we consider the ODMP group as a substituent of the respective benzoic acids, once the value of the Hammett σ constant for this group is known. This value can be determined by taking the logarithm of the ratios of the acid-dissociation constants of the R-substituted PSA compounds to the respective R-substituted benzoic acids shown in Table 6, in analogy to the

original definition of Hammett for σ for benzoic acids. This procedure allows to isolate the substituent effect of the ODMP group.
The results of these calculations are shown in Table 7; the resulting mean is 0.21. With this value of σ for the ODMP group and the respective value for the other substituent, we are able to predict the pK_a´s for the family of compounds of this study using the empirical equation (3) obtained for substituted benzoic acids. The predicted pK_a's coincide exactly with those values calculated in this study using DFT methodology, Figure 6.

Substituent	σ_{ODMP}
6-H	0.189
5-F	0.169
5-Cl	0.228
5-CH_3	0.270
5-CN	0.200
5-OCH_3	0.288
5-NO_2	0.089
5-OH	0.278

Table 7. Hammett σ constants for the O-dimethoxypyrimidinyl group (ODMP)

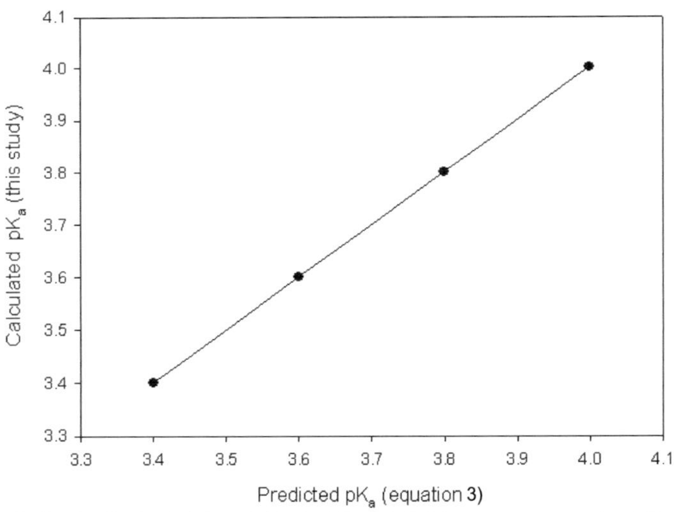

Fig. 6. Calculated pK_a (this study) vs. predicted pK_a (eqn. 3) for PSA compounds.

6.3 Theoretical calculation of partition coefficients pf PSA compounds

Continuing with the physicochemical characterization of this family of herbicides, we have reported environmentally important partition coefficients, Henry´s law constant, H, octanol/water, K_{OW}, and octanol/air, K_{OA}, partition coefficients for 39 PSA compounds (Delgado, 2010). These coefficients are calculated using density functional theory (DFT) at B3LYP/6-31G(d,p) level of theory using the Poisson-Boltzmann solvation model. These properties have not been reported previously for this family of herbicides, neither experimentally nor theoretically.

6.3.1 Thermodynamics

Free energy of solvation ΔG_S^0, in addition to its fundamental interest, may combined with other thermodynamic data to predict a variety of equilibrium constants, being one of the most important the partitioning of a solute between two immiscible phases. Therefore, ΔG_S^0 is a key property to estimate the fate of a chemical once it is released in the environment. From an environmental point of view, three of the most important partition coefficients are the Henry´s law constant (H), the octanol/air (K_{OA}) and the octanol/water (K_{OW}) partition coefficients which can be calculated straightforwardly from the free energies of solvation in water and octanol by means of the well known equations:

$$\ln H = \ln(RT) + \Delta G_s^0(water) / RT \tag{4}$$

$$\log K_{OA} = -\Delta G_s^0(oct) / 2.303RT \tag{5}$$

$$\log K_{OW} = \left\{\Delta G_s^0(water) - \Delta G_s^0(oct)\right\} / 2.303RT \tag{6}$$

6.3.2 Computational methods

Initial three-dimensional geometries of the chemical structures were generated using Hyperchem 7.0 molecular modeling package. These 3D structures were refined later using the Jaguar suite and its graphic interface Maestro. Gas-phase molecular geometries and electronic energies were obtained by density functional theory (DFT) calculations at the same level of theory, basis set including polarization functions on all atoms in conjunction with the hybrid functional B3LYP, which uses a combination of the three-parameter Becke exchange functional along with the Lee-Yang-Parr nonlocal correlation functionals: B3LYP/6-31G(d,p). To determine the lowest-energy conformation for each molecule, geometry optimizations were carried out allowing one or more torsional angles to vary systematically. Free energies of solvation in water and octanol were computed by single point calculations, including implicit solvation, on the gas-phase optimized geometry using the Poisson-Boltzmann Solvation Model (Marten et al., 1996).

6.3.3 Results

The calculated free energies of solvation in water and octanol, along with the respective values of the Henry´s law constants, the octanol/air partition and the octanol/water partition coefficients are shown in Table 8.

From this table it is possible to observe that all these compounds have low values of gas/solvent partition coefficients, consequently they show low volatility and a clear preference for the condensed phases. The preferred condensed phase, either water or octanol, is determined by the value of K_{OW}. The calculated values of log K_{OW} range from -0.50 to about 3.0, breaking down in the following way: two cases, compounds 35 and 39, with negative values; three cases, compounds 24, 26 and 38, with values greater than 2; and the remaining compounds having intermediate values between 0 and 2.

	R	$\Delta G_S^0(water)$ $(kJ\,mol^{-1})$	$\Delta G_S^0(octanol)$ $(kJ\,mol^{-1})$	H (Pa m^3 mol^{-1})	K_{OA}	Log K_{OW}
1	6-H	-44.22	-49.20	4.44E-05	4.17E+08	0.87
2	6-F	-42.80	-48.79	7.88E-05	3.53E+08	1.05
3	6-Cl	-43.39	-47.82	6.21E-05	2.39E+08	0.78
4	5-F	-42.47	-48.20	9.00E-05	2.78E+08	1.00
5	5-Cl	-40.46	-46.74	2.02E-04	1.54E+08	1.10
6	3-F	-43.89	-48.41	5.08E-05	3.03E+08	0.79
7	5,6-$(Cl)_2$	-40.79	-46.11	1.77E-04	1.20E+08	0.93
8	6-C_6H_5	-49.75	-55.10	4.77E-06	4.51E+09	0.94
9	6-CH_3	-38.99	-45.69	3.66E-04	1.01E+08	1.17
10	6-C_2H_5	-40.25	-45.77	2.20E-04	1.05E+08	0.97
11	6-C_3H_7	-39.96	-45.35	2.48E-04	8.84E+07	0.94
12	6-CF_3	-42.09	-47.99	1.05E-04	2.56E+08	1.03
13	5-CH_3	-43.64	-48.33	5.61E-05	2.93E+08	0.82
14	5-C_2H_5	-42.30	-47.78	9.64E-05	2.35E+08	0.96
15	5-CN	-52.93	-64.22	1.32E-06	1.79E+11	1.98
16	5-CCH	-44.35	-53.76	4.22E-05	2.63E+09	1.65
17	3-CH_3	-44.89	-45.40	3.39E-05	8.99E+07	0.09
18	6-OCH_3	-43.43	-52.43	6.11E-05	1.53E+09	1.58
18	6-OC_2H_5	-42.89	-51.76	7.60E-05	1.17E+09	1.55
20	6-OC_3H_7	-42.47	-51.55	9.00E-05	1.07E+09	1.59
21	6-$OCH(CH_3)_2$	-48.74	-54.77	7.17E-06	3.94E+09	1.06
22	6-OC_4H_9	-41.63	-51.34	1.26E-04	9.87E+08	1.70
23	6-$OCHF_2$	-45.06	-55.23	3.17E-05	4.74E+09	1.78
24	6-OC_6H_5	-37.74	-54.35	6.07E-04	3.33E+09	2.91
25	5-OCH_3	-45.98	-54.81	2.18E-05	4.01E+09	1.55
26	5-OC_6H_5	-39.08	-56.74	3.53E-04	8.71E+09	3.09
27	6-SCH_3	-45.06	-50.12	3.17E-05	6.05E+08	0.89
28	6-SC_2H_5	-44.98	-50.58	3.27E-05	7.29E+08	0.98
29	6-SC_3H_7	-44.77	-50.25	3.56E-05	6.37E+08	0.96
30	5-SCH_3	-43.64	-48.33	5.61E-05	2.93E+08	0.82
31	6-CO_2CH_3	-52.26	-59.41	1.73E-06	2.57E+10	1.25
32	6-COC_6H_5	-55.31	-61.21	5.07E-07	5.30E+10	1.03
33	6-$COCH_3$	-55.61	-59.12	4.49E-07	2.28E+10	0.61
34	6-NO_2	-50.25	-61.42	3.90E-06	5.77E+10	1.96
35	6-CH_3SO_2	-76.36	-74.64	1.04E-10	1.20E+13	-0.30
36	6-NH_2	-53.76	-57.24	9.47E-07	1.07E+10	0.61
37	5-NH_2	-65.73	-66.32	7.58E-09	4.16E+11	0.10
38	5-NO_2	-48.41	-61.04	8.20E-06	4.96E+10	2.21
39	5-OH	-69.50	-66.40	1.66E-09	4.30E+11	-0.54

Table 8. Calculated free energy of solvation in water and octanol, Henry´s law constant, octanol-air partition coefficient and octanol-water partition coefficient for R-substituted O-pyrimidinylsalicylic acids.

Those with negative values show preference for the aqueous phase. This behavior is explained, in the case of compound 35, by the high polar nature of the sulfonyl group which provides an highly hydrophilic character. Moreover, it is well known that sulfones are able to stabilize negative charges on neighbor atoms, such as those of the carboxylic group. On the other hand, in the compound 39, the hydroxyl group in meta position with respect to the carboxylic group strengthens the acidity by inductive effect, favoring in consequence its water solubility.

On the other hand, those compounds with values of log $K_{OW} > 2$ will have preference for the octanol phase. This behavior is explained in terms of the high hydrophobic character provided by the OC_6H5 and NO_2 substituent groups present in compounds 24, 26 and 38. This finding is in agreement with the empirical evidence of the surprisingly low solubility in water of these compounds, therefore in the ambient they will be preferably found in the lipids of aquatic and animal biota and potentially they could scale in the food chain.

6.3.4 Discussion

The validation of the calculated figures unfortunately can not be made in a direct way since for this family of compounds there is no experimental data available to check the calculated values. However, indirectly we can have an estimation of their reliability. For instance, a recent paper reports the following model, based on the fragmental method, to predict the log K_{OW} of halogenated benzoic acids in terms of certain group and factor values (Qiao et al., 2008):

$$\log K_{OW}^{HBz} = \sum_{i=1}^{11} n_i g_i + \sum_{j=1}^{2} l_j f_j + 1.117 \qquad (7)$$

where n_i is the number of i-type groups, g_i is the value of the group i, l_j is the number of j-type factors, f_1 and f_2 denote the factors for *ortho* and *para* substituents, respectively. The model reproduces the experimental data of log K_{OW} for halogenated benzoic acids with an average absolute error of 0.22 log units. Since, the DMPS-compounds of our study can be viewed as benzoic acids substituted in the ortho position with a O-dimethoxypyrimidynil (ODMP) group, we hypothesize this model should be also valid to our compounds. If it is so, we could apply the model to determine the group value for the ODMP group using our calculated values of log K_{OW} and the group and factor values reported by Qiao for the analog R-substituted benzoic acids. The results of these calculations are shown in Table 9; the resulting mean is -0.768. If this value for the ODMP group is correct, we should be able to reproduce the experimental values of log K_{OW} for R-substituted benzoic acids from the calculated log K_{OW} for the analog R-substituted PSA compound, since structurally benzoic acids can be viewed as DMPS acids without the ODMP group; therefore the application of eqn. (7) leads to the following equation:

$$\log K_{OW}^{HBz} = \log K_{OW}^{PSA} - g_{ODMP} - f_2 - \Delta g \qquad (8)$$

where g_{ODMP} is the value for the ODMP group, f_2 is the factor value for ortho substituents, and Δg accounts for the difference in the group value of the benzene ring with two and three substituents. The results of these calculations are shown in Table 10. The average absolute error of log K_{OW} is 0.26, quite comparable with the reported error (0.22) for the original data set of halogenated benzoic acids. This result allows to confirm the quality of the values of

K_{OW} calculated in this study, and also it validates the applicability of eqn. (7) for this family of compounds.

Substituent	ODMP group value
6-H	-0.679
6-F	-0.260
6-Cl	-1.039
5-F	-0.726
5-Cl	-1.135

Table 9. Group values for the *O*-dimethoxypyrimidinyl group (ODMP).

Compound	**Exp. log K_{OW}** (Qiao et al., 2008)	**Calc. log K_{OW}** (eqn. 8)
Benzoic acid	1.87	2.06
o-chlorobenzoic acid	2.05	1.87
m-chlorobenzoic acid	2.68	2.19
o-fluorobenzoic acid	1.77	2.14
m-fluorobenzoic acid	2.15	2.09

Table 10. Experimental and calculated log K_{OW} for benzoic acids.

The value of the Henry´s law constant is determined fundamentally by the free energy of solvation in water according to eqn. (4). The reliability of the methodology used in this article to calculate ΔG_S^0 in water was confirmed in an earlier article (Delgado, 2010), and consequently the values of H derived from it.
Since the calculated values of K_{OW} and H are supported by the above checking procedures, then the calculated values of K_{OA} should be also checked considering that these three coefficients are related by the well known equation (Meylan & Howard, 2005):

$$K_{OA} = RT\frac{K_{OW}}{H} \tag{9}$$

Thus, the plot of K_{OA}, calculated according to eqn. (5), vs (K_{OW} / H) should give rise to a straight line whose slope equals to RT. The plot, shown in Figure 7, confirm this fact and therefore supports the calculated values of K_{OA}. Moreover, the value of the slope corresponds to the value of RT at 298 K, within the error of the methodology.

7. Concluding remarks.

Modern chemistry is oriented more and more towards the elucidation in detail of how the macroscopic properties are determined by the microscopic properties of matter, this current tendency is motivated by both academic and applied reasons. From academic point of view, it allows to have a detailed picture of the intermolecular interactions that determine the macro properties; and from an applied point of view, it allows systematize the search of a compound with desired properties in base of its molecular structure, selecting, in this way, the most promising compounds to be synthesized on a rational base. Quantum chemical

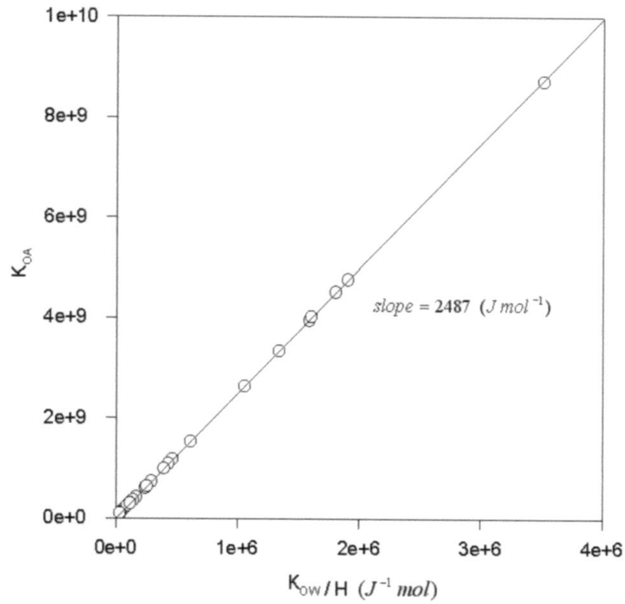

Fig. 7. Calculated K_{OA} (eqn. 2) vs (K_{OW} /H)

calculations allow the most accurate description of the electronic and geometric structure of molecules, as well as their interactions. These methods, which range from semi-empirical to *ab initio* approaches, have advantages and drawbacks which are necessary to evaluate before their use, since there exists a compromise among accuracy, computation time, physical interpretation and applicability.

Several computer-assisted quantum chemical approaches have been successfully applied to a variety of chemical systems, ranging their applicability from biological chemistry (Lie & Schiott, 2008) to chemical engineering (Sandler, 1999, 2003), passing by pesticide (Wan et al., 2004) and environmental chemistry (Delgado & Alderete, 2002). Thus, theoretical studies often may be considered not only as other option, but also as the only option in those cases in which the empirical information is not ready available, like those showed in this chapter.

In closing, we believe that computer-assisted quantum chemical studies represent one of the most important approaches in the present and future of chemistry, since they, on one hand, allow to obtain information not available from other techniques, and, on the other hand, the phenomena can be understood at molecular level, this is essential for the design of novel herbicides since their properties may be predicted prior to synthesis and consequently the design may, in this way, be guided by the results of calculations.

8. Acknowledgement.

The author acknowledges financial support from Fondecyt under grant N⁰ 1100064.

9. References

Ballschmiter, K. (1992). Transport and fate of organic compounds in the global environment. *Angew. Chem. Int. Ed. Engl.* 31, 487-515.

Cramer, C.J. (2002). *Essentials of Computational Chemistry. Theories and Models*, John Wiley & Sons Ltd., Chichester.
Cramer, R.D.; Patterson, D.E.; Bunce, J.D. (1988). Comparative molecular field analysis (CoMFA). 1. Effect of shape on binding of steroids to carrier proteins. *J. Am. Chem. Soc.* 110, 5959-5967.
Delgado, E.J. (2009). DFT calculation of pK_a`s for dimethoxypyrimidinylsalicylic based herbicides. *Chem. Phys. Lett.* 471, 133-135.
Delgado, E.J. (2010). Theoretical calculation of partition coefficients of dimethoxypyrimidinylsalicylic Acids. *J. Mol. Model.* (in press), DOI 10.1007.
Delgado, E.J; Alderete, J. (2002). On the calculation of Henry´s law constants of chlorinated benzenes in water from semiempirical quantum chemical methods. *J. Chem. Inf. Comput. Sci.* 42, 559-563.
Diaz, G.A.; Delgado, E.J. (2009). Quantitative prediction of AHAS inhibition by pyrimidinylsalicylate based herbicides. *Pestic. Biochem. Physiol.* 95, 33-37.
Duffy, E.M.; Jorgensen, W.L. (2000). Prediction of properties from simulations: free energies of solvation in hexadecane, octanol and water. *J. Am. Chem. Soc.*, 122, 2878-2888.
Duggleby, R.G.; Pang, S-S. (2000). Acetohydroxyacid synthase. *J. Biochem. Mol. Biol.* 33, 1-36.
Estrada, E.; Delgado, E.J.; Alderete, J.B.; Jaña, G.A. (2004). Quantum-connectivity descriptors in modeling solubility of environmentally important organic compounds. *J. Comput. Chem.* 25, 1787-1796.
Estrada, E.; Delgado, E.J.; Simon-Manso, Y. (2007). Modeling the solubility in water of environmentally important organic compounds, In: *Thermodynamics, Solubility and Environmental Issues*, Trevor L. Letcher, (Ed.), 17-29, Elsevier, Amsterdam.
Frenkel, D.; Smit, B. (1996). *Understanding Molecular Simulation*, Academic Press, San Diego, CA.
Hansch, C. (1993). Quantitative structure–activity relationships and the unnamed science. *Acc. Chem. Res.* 26, 147-153.
Hansch, C.; Hoekman, D.; Gao, H. (1996). Comparative QSAR: toward a deeper understanding of chemicobiological interactions. *Chem. Rev.* 96, 1045-1075.
Hammett, L.P. (1937). The effect of structure upon the reactions of organic compounds. Benzene derivates. *J. Am. Chem. Soc.* 59, 96.
He, Y.Z.; Li, Y.X.; Zhu, X.L.; Xi, Z.; Niu, C.; Wan, J.; Zhang, L.; Yang, G.F. (2007). Rational design on bioactive conformation analysis of pyrimidinylbenzoates as acetohydroxyacid symthase inhibitors by integrating molecular docking, CoMFA, CoMSIA, and DFT calculations. *J. Comput. Chem.* 47, 2335-2344.
Hollingsworth, C.A.; Seybold, P.G.; Hadad, C.M. (2002). Substituent effects on the electronic structure and pK_a of benzoic acid. *Int. J. Quantum. Chem.* 90, 1396-1403.
Karelson, M.; Lobanov, V.S.; Katritzky, A.R. (1996). Quantum-Chemical descriptors in QSAR/QSPR studies. *Chem. Rev.* 96, 1027-1043.
Katritzky, A.L.; Karelson, M.; Lobanov, V.S. (1997). QSPR as a means of predicting and understanding chemical and physical properties in terms of structure. *Pure & Appl. Chem.* 69, 245-248.
Klebe, G.; Abraham, U.; Mietzner, T. (1994). Molecular similarity indices in a comparative analysis (CoMSIA) of drug molecules to correlate and predict their biological activity. *J. Med. Chem.* 37, 4130-4146.

Kollman, P. (1993). Free-energy calculations - applications to chemical and biochemical phenomena. *Chem. Rev.* 93, 2395- 2417.

Lie, M.A.; Schiott, B. (2008). A DFT study of solvation effects on the tautomeric equilibrium and catalytic ylide generation of thiamin models. *J. Comput. Chem.* 29, 1037-1047.

Lyman Ott, R.; Longnecker, M. (2001). *An Introduction to Statistical Methods and Data Analysis,* Duxbury, Pacific Grove.

Marten, B.; Kim, K.; Cortis, C.; Friesner, R.A; Murphy, R.B.; Ringnalda, M.N.; Sitkoff, D.; Honig, B. (1996). New model for calculation of solvation free energies: correction of self-consistent reaction field continuum dielectric theory for short-range hydrogen-bonding effects. *J. Phys. Chem.* 100, 11775-11788.

Meylan, W.M., Howard, P.H. (2005). Estimating octanol-air partition coefficients with octanol-water partition coefficients and Henry´s law constants. *Chemosphere* 61,640-644.

McCourt, J.A.; Pang, S.S.; Guddat, L.W.; Duggleby, R.G. (2005). Elucidating the specifity of binding of sulfonylurea herbicides to Acetohydroxyacid synthase. *Biochemistry* 44, 2330-2338.

McCourt, J.A.; Duggleby, R.G. (2006). Acetohydroxyacid synthase and its role in the biosynthetic pathway for branched-chain amino acids. *Amino Acids* 31, 173-210.

McCourt, J.A.; Pang, S.S.; King-Scott, J.; Guddat, L.W.; Duggleby, R.G.(2006). Herbicide-binding sites revealed in the structure of plant Acetohydroxyacid synthase. *Proc. Natl. Acad. Sci. U.S.A.* 103, 569-573.

Nezu, Y.; Miyasaki, M.; Sugiyama, K.; Wada, N.; Kajiwara, I.; Miyazawa, T. (1996). Dimethoxypyrimidines as novel herbicides. Part 2.Synthesis and herbicidal activity of O-Pyrimidinylsalicylates and analogues. Pest. Sci. 47, 115-124.

Nezu, Y.; Wada, N.; Yoshida, F.; Miyazawa, T.;. Shimizu, T.; Fujita, T. (1998). Dimethoxypyrimidines as novel herbicides. Part 4. Quantitative structure-activity relationships of dimethoxypyrimidinyl(thio)salicylic acids. *Pestic. Sci.* 52, 343.-353

Pang, S.S.; Duggleby, R.G.; Guddat, L.W. (2002). Crystal structure of yeast Acetohydroxyacid synthase: a target for herbicidal inhibitors. *J. Mol. Biol.* 317, 249-262.

Pang, S.S.; Guddat, L.W.; Duggleby, R.G. (2003). Molecular basis of sulfonylurea herbicide inhibition of Acetohydroxyacid synthase. *J. Biol. Chem.* 278, 7639-7644.

Pang, S.S.; Duggleby, R.G.; Schowen, R.L.; Guddat, L.W. (2004). The crystal structures of *Klebsiella pneumoniae* acetolactate synthase with enzyme-bound cofactor and with an unusual intermediate. *J. Biol. Chem.* 279, 2242-2253.

Qiao, Y., Xia, S., Ma, P. (2008). Octanol/water partition coefficienrs of substituted benzene derivates containing halogens and carboxyls: determiantion using the shake-flask method and estimation using the fragment method. *J. Chem. Eng. Data* 53, 280-282.

Sandler, S.I. (1999). Unusual chemical thermodynamics. *J. Chem. Theromodynamics* 31, 3-25.

Sandler, S.I. (2003). Quantum Mechanics: a new tool for engineering thermodynamics, *Fluid Phase Equilibria* 210, 147-160.

Siepmann, J.I. (1999). Monte Carlo methods for simulating phase equilibria of complex fluids. *Adv. Chem Phys.* 105, 443-460.

Singh, B. K.; Shaner, D.L. (1995). Biosynthesis of branched chain amino acids: from test tube to field. *Plant Cell* 7, 935-944.

Stangroom, S.J.; Lester, J.N.; Collins, C.D. (2000). Abiotic behaviour of organic micropollutants in soils and the aquatic environment. A review: I. Partitioming. *Environ. Technol.* 21, 845-863.

Stanton, D.T.; Jurs, P.C. (1990). Development and use of charged partial surface area structural descriptors in computer-assisted quantitative structure-property relationship studies. *Anal. Chem.* 62, 2323- 2329.

Tomasi, J.; Persico, M. (1994). Molecular interactions in solution: an overview of methods based on continuous distributions of the solvent. *Chem. Rev.* 94, 2027-2094.

Wan, J.; Zhang, L.; Yang, G. (2004). Quantitative structure-activity relationships for phenyl triazolinones of protoporphyrinogen oxidase inhibitors: a density functional theory study. *J. Comput. Chem.* 25, 1827-1832.

Wang, J.G.; Li, Z.M.; Ma, N.; Wang, B.L.; Jiang, L.; Pang, S.S.; Lee, Y.T.; Guddat, L.W.; Duggleby, R.G. (2005). Structure-activity relationships for a new family of sulfonylurea herbicides. *J. Comput. Aid. Mol. Des.* 19, 801-820.

Young, D. (2001). *Computational Chemistry. A practical guide for applying techniques to real world problems,* John Wiley & Sons, Inc., New York.

Zohar, Y.; Einav, M.; Chipman, D.M.; Barak, Z. (2003). Acetohydroxyacid synthase from Mycobacterium avium and its inhibition by sulfonylureas and imidazolines. *Biochim. Biophys. Acta* 1649, 97-105.

Part 8

Alternative Sources for Herbicidal Compounds

35

Marine and Freshwater Microalgae as a Potential Source of Novel Herbicides

John Berry
Florida International University
U. S. A.

Introduction

Despite the considerable contribution of synthetic herbicides to modern agriculture, lingering concerns remain - and, in some cases, have grown - with respect to most commonly used agents. In particular, concerns have included documented toxic effects on human, animal and ecosystem health, as well as associated potential for bioaccumulation, and growing evidence of acquired resistance (and cross-resistance) by target species with consequent implications for both cost and sustainability (Horrigan et al., 2002; Powles et al., 1996). As such, there is a continuing interest in the identification of novel herbicidal agents, specifically toward discovery of compounds that may address these concerns (e.g. reduced toxicity and/or potential for bioaccumulation, new molecular targets, high production capacity and low-cost of production).

In particular, numerous studies have explored the potential of naturally occurring secondary metabolites (i.e. "natural products"), particularly from plants and microorganisms, as a source of such compounds. This approach has included both purified and characterized metabolites, as well as mixtures (i.e. extracts or preparations), and "semi-synthetic" compounds based on natural product leads. Accordingly, in recent years, several good reviews have emerged on various topics related to bioactive natural products as a source of potential herbicides (e.g. Duke et al., 2002; Copping & Duke, 2007; Dayan et al., 2009; Lindell et al., 2009), and the reader would be directed to these resources for further information outside the scope of the current chapter.

This chapter focuses specifically on the marine and freshwater microalgae as a source of herbicidal compounds. As a group, the microalgae are recognized to produce a remarkable diversity of biologically active metabolites (see *Marine and Freshwater Microalgae: A Rich Source of Bioactive Compounds*, below). Furthermore, in addition to this recognized chemical diversity, a growing body of knowledge supports the notion that many of these bioactive metabolites from algae - as a aquatic photoautotrophs - may specifically have evolved in the capacity of *allelopathy* providing a competitive advantage via interspecific, and particularly negative (i.e. inhibition), effects on growth, survival and reproduction of photoautotrophic competitor species (see *Algal Allelopathy*, below). Moreover, given the overlapping targets of action between allelopathic metabolites and herbicides (e.g. inhibition of photosynthesis), there is an emerging literature on the potential - including possible commercial potential - of algal metabolites as herbicides (as discussed in the final two sections of the chapter, below).

2. Marine and freshwater microalgae: a rich source of bioactive compounds

Algae, although not a taxonomically valid classification, encompass oxygenic photosynthetic organisms, exclusive of the true plants, spanning the prokaryotic and eukaryotic kingdoms of the Monera (bacteria) and Protista (protists). The group itself is frequently subdivided functionally into the so-called "microalgae" and "macroalgae" with the latter classification generally reserved for macrophytic representatives (i.e. "seaweeds"), and the former for strictly unicellular forms. Considerable overlap, however, can exist (e.g. colonial forms), as all algae are, strictly speaking, unicellular. Accordingly, the macroalgae classically include the protistan phyla of the Chlorophyta ("green algae"), Rhodophyta ("red algae") and Phaeophyta ("brown algae"), whereas the microalgae include several photoautotrophic protistan taxa, particularly including the dinoflagellates, raphidophytes, diatoms (Bacilliarophyta), haptophytes and Chrysophyta ("golden algae"), as well as the prokaryotic cyanobacteria ("blue-green algae"). Alongside obvious morphological differences between the macroalgae and microalgae, there are notable differences in the secondary metabolites of the groups (see below) with the latter arguably representing a more chemically diverse repertoire. The current chapter will largely focus on the microalgae, however, references to relevant cases of herbicidal metabolites from macroalgae will be included as appropriate.

Equally diverse as their phylogeny is the ecology of the microalgae. Perhaps most conspicuous in aquatic systems, microalgae are, in fact, ubiquitous in their distribution, and include free-living, terrestrial representatives, as well as numerous symbiotic species (e.g. cyanobacterial symbionts of lichens, dinoflagellate zooxanthellae of corals, root nodule symbionts). Although the current chapter primarily focuses on marine and freshwater microalgae, examples of potentially herbicidal compounds from non-aquatic representatives will also be briefly mentioned.

In terms of bioactive compounds, the secondary metabolites of the microalgae include nearly every chemical class of natural product - ranging from fatty acids to alkaloids, and even a particularly interesting cabal of amino acids - but are particularly characterized by an array of metabolites with polyketide and non-ribosomal peptide biosynthetic origins (Dittman et al., 2001; Snyder et al., 2003; Jones et al., 2009). Respectively, the chemical repertoires of the cyanobacteria and eukaryotic microalgae are perhaps best typified by a myriad of peptides derived from the non-ribosomal peptide synthetase (NRPS), and polyketides derived from a pathway of sequential condensations (of malonyl- or acetyl-CoA units) orchestrated by the polyketide synthase (PKS) family of genes. Most notably, both biosynthetic pathways are specifically characterized by a family of modular genes (and associated gene products) such that nearly all biosynthetic steps, in both cases, are largely coordinated within so-called "modules" of a single, assembly-line like "megasynthases". These megasynthases are responsible for several core biosynthetic steps (e.g. adenylation, condensation and peptide transfer in the case of NRPSs, and ketosynthesis and acyl transfer in the case of PKSs), as well as various secondary modification steps (e.g. ketone reduction, dehydration, methylation). In fact, metabolites presumably derived from both pathways are seen in prokaryotic and eukaryotic algae alike, and it is, moreover, becoming increasingly clear that many metabolites may be derived from mixed (PKS/NRPS) pathways (and mixed modular genes/enzymes). As an example, the mixed NRPS/PKS pathway (as proposed by Moffitt and Neilan, 2004) for the cyanobacterial metabolite, nodularin, is shown in Fig. 1.

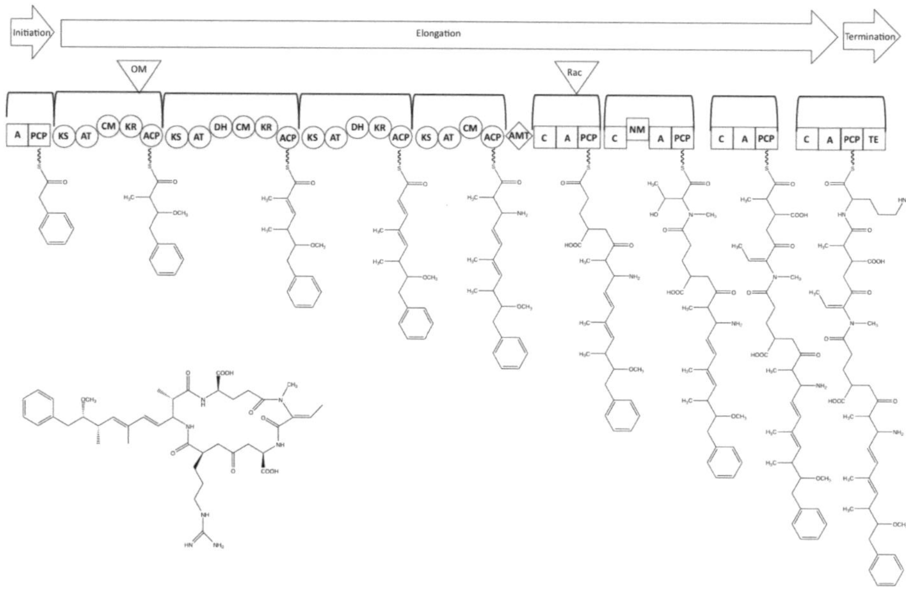

Fig. 1. Modular non-ribosomal peptide synthetase (NRPS)/polyketide synthase (PKS) biosynthetic pathway of the cyanobacterial toxin, nodularin. Modified from Moffitt and Neilan (2004). Each module, as indicated in the figure, contains essential domains - condenstation (C), adenlyation (A) and peptide carrier protein (PCP) for NRPS, and ketosynthase (KS), acyltrasferase (AT) and acyl carrier protein (ACP) for PKS – as well as additional domains for modifications including methylation (C-methyltransferase, CM; *N*-methyltransferase), ketoreductases (KR), dehydratases (DH). A thioesterase (TE), in the final NRPS module, releases and cyclizes the peptide. In addition to these NRPS and PKS modules, several other domains are incorporated, including *O*-methyltransferase (OM) for transfer of methyl to hydroxyl of C-9, racemase (Rac) which epimerizes the *D*-glutamate, aminotransferase (AMT).

This shared biosynthetic origin of algal secondary metabolites is, perhaps, what makes these organisms such particularly compelling candidates as a source of bioactive compounds. The modular - and presumably "interchangeable"- nature of these biosynthetic enzymes is likely, at its core, responsible for the high diversity of bioactive compounds found within the group. As simply put by Welker & von Döhren (2006), these biosynthetic pathways likely represent "nature's own combinatorial chemistry," resulting in an unequalled diversity of secondary metabolites, and presumably directed by largely unknown selective pressures (e.g. chemical defense against micrograzers, allelopathy). Moreover, from a compound discovery and development (and even sustainability) point-of-view, such a biosynthetic mechanism presents several exciting opportunities. For example, "semi-synthetic" modifications (i.e. via genetic modification of PKS/NRPS genes) can enable improvement of activity or other relevant parameters (e.g. reduced toxicity, increased bioavailability). Co-opting of biosynthetic genes, and subsequent heterologous expression, enables synthesis of relevant pharmacophores and scaled-up production of candidate compounds (discussed further in *Commercial Potential of Herbicides from Microalgae*).

Accordingly, the resulting diversity of toxic or otherwise bioactive metabolites has received considerable attention (see, for example, reviews by Codd et al., 1999; van Dolah, 2000; Gerwick et al., 2001; Cardozo et al., 2007; Tan, 2007; Berry et al., 2008) both as a leads to compounds with potential biomedical (i.e. drug discovery) or other commercially relevant (e.g. pesticides, herbicides) applications, as well as a growing relevance to public and environmental health (i.e. as "toxins"). In terms of the former, several bioactive compounds from microalgae, including both eukaryotic and prokaryotic representatives, are currently being investigated as potential "drug leads." Several metabolites from cyanobacteria (e.g. cryptophycins, dolastatins, curacin A) have, for example, been investigated with respect to anticancer activities (i.e. inhibition of cell-division; Trimurtulu et al., 1994; Verdier-Pinard et al., 1998; Luesch et al., 2001). Dinoflagellate toxins (e.g. saxitoxins and brevetoxins) have recognized applications as biomedical research tools (i.e. inhibitors and activators, respectively, of voltage-gated sodium channels) and emerging applications as drugs (e.g. brevenal as an antagonist of brochoconstriction and related pulmonary effects, such as those found in cystic fibrosis; Potera, 2007). On the other hand, as toxic compounds (see Fig. 2), and specifically in relation their role in "harmful algal blooms" (HABs), microalgal compounds have been well documented (Codd et al., 1999; van Dolah, 2000) to pose threats to human, animal and environmental health via contamination of drinking water, bioaccumulation in fish, shellfish and other seafood, and various other routes (e.g. recreational exposure).

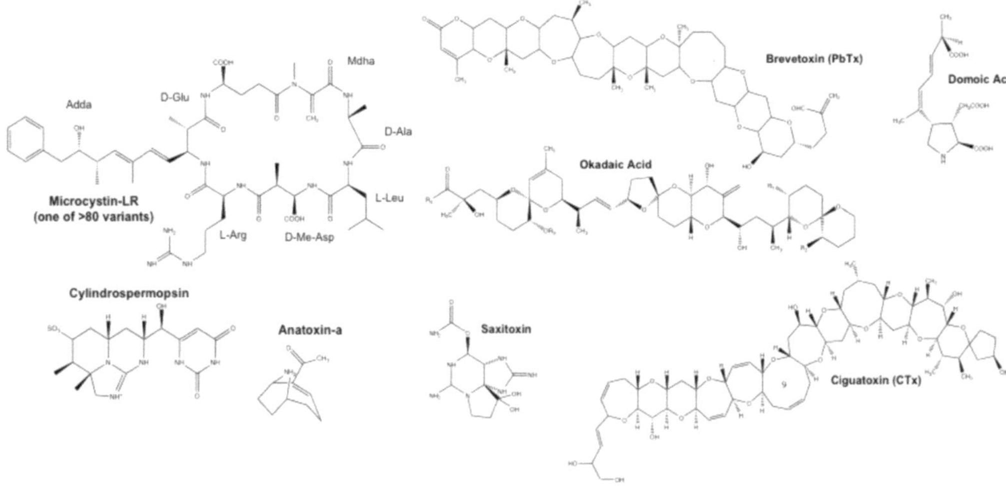

Fig. 2. Toxins from marine and freshwater microalgae with established concerns for human, animal and environmental health.

Interestingly, the role of these compounds in the life history of the algae remains largely unknown. However, it has been frequently proposed (e.g. Berry et al., 2008) that bioactive secondary metabolites of microalgae may be involved in the chemical ecology of these organisms, particularly including a role in allelopathy (discussed further below) and chemical defense (e.g. feeding deterrents, qualitative toxins) against potential micrograzers in aquatic habitats. Indeed, with respect to the current discussion, the specifically suited role of these compounds as allelochemicals (i.e. allelopathic inhibitors of competing photoautotrophic plants and microbes), that makes them a particularly compelling candidate as a source of potential herbicidal agents.

3. Algal allelopathy

Given the likely intense competition for physical (e.g. space, light) and chemical (e.g. micro- and macronutrient) resources in aquatic habitats, it is not surprising that a growing number of studies have documented apparent allelopathy between photoautotrophic species - as well as phyla and kingdoms - in these systems. In fact, several good reviews on the topic (e.g. Irfanullah & Moss, 2005; Erhard, 2006; Graneli et al., 2008; Macias et al., 2007), specifically emphasizing aquatic plants, macroalgae and microalgae, have appeared over the past several years. Not only have these studies supported the possible role of these compounds as inhibitors of sympatric autotrophs, but have, in some cases, even provided convincing evidence that interplay of these interspecific inhibitors may support ecological succession in these habitats. Moreover, given an overlap - and presumptive evolutionary convergence or divergence - of targets shared by algae and true plants (e.g. oxygenic photosynthesis), the repertoire of such allelochemicals provides an especially rich source of compounds with activity directly relevant to possible development as herbicides and/or identification of novel herbicide targets.

Although the term "allelopathy" was not coined (Molisch, 1937) until the 1930s, scientific observation of apparent allelopathy in aquatic habitats dates back to the turn of the previous century. In particular, studies by Apstein (1896) and Pütter (1908) that detailed an apparent role of allelochemicals in determining community structure in these systems. A number of studies subsequently followed including, perhaps most notably, the seminal studies by Keating et al. (1977, 1978) which examined the role of extracellular metabolites from cyanobacteria with respect to interannual succession patterns of algal blooms in a closed aquatic system (i.e. Linsley Pond). These studies specifically demonstrated that filtrates prepared from axenic (or unialgal) cultures of "dominant" bloom strains (from a particular year) negatively affected (i.e. inhibited) growth of "predecessor" strains - but not "successor" strains on which filtrates, instead, exerted a positive or neutral effect - indicating an apparent role of extracellular metabolites in structuring the community from year to year. Interestingly, even these early studies (e.g. Keating, 1977) - although focused on ecological aspects - recognized the potential implications of these interactions as a means to control populations of nuisance algae. The subsequent decades witnessed numerous studies on the ecological role of allelopathy in both freshwater and marine habitats (see, for example, review by Gross, 2003). However, the vast majority of these studies examined extracts, filtrates or other crude preparations. With respect to the potential applications for novel herbicides (and/or algicides), identification and subsequent chemical characterization is an obvious necessity.

Indeed, numerous algal metabolites (Fig. 3) with algicidal activity, and accordingly suggested roles in allelopathy, have been isolated and characterized. The reader is directed to previous reviews of these compounds (Smith and Doan, 1999; Berry et al., 2008). The first such compound, cyanobacterin, was reported by Mason et al. (1982). The compound was isolated from cellular extracts of *Scytonema hofmanni* (UTEX 1581), specifically based on inhibition of cyanobacteria, and chemically characterized as a chlorinated aromatic γ-lactone (see Fig. 3). Extracts of *S. hofmanni* were shown to inhibit (presumably due to cyanobacterin) a wide range of microorganisms, particularly including a diversity of cyanobacteria and green algae, as well as rhodophytes, and was suggested, as such, to play a role in allelopathy of the species (Mason et al., 1982). Subsequently, species of the cyanobacterial genus, *Nostoc*, were found (Gromov et al., 1991; Vepritskii et al., 1991) to produce antialgal metabolites, termed cyanobacterin LU-1 and LU-2, which were, likewise, shown to inhibit a variety of cyanobacteria and other microalgae. Although the structures of LU-1 and LU-2 (not completely characterized) were generally considered to be unrelated to those isolated from *S. hofmanni*, it was found that both cyanobacterins from *S. hofmanni* and *Nostoc* (i.e. LU-1 and LU-2) seemed to act via inhibition of photosystem II (PSII), a target of several commercial herbicides (discussed further below).

Based on this, several screening studies (Flores & Wolk, 1986; Shlegel et al., 1998; Volk, 2005; Gantar et al., 2008) followed, using algicidal assays to identify metabolites with potential allelopathic roles. The presumably high importance of such compounds is underscored by the fact that up to 40% of cyanobacterial isolates evaluated in these screening studies were found to produce algicidal metabolites (Volk et al., 2005). Of particular note, it was found by Flores and Wolk (1986) and Schlegel et al. (1998), who screened sixty-five and approximately 200 strains of cyanobacteria, respectively, that algicidal metabolites were primarily restricted to heterocystous, nitrogen-fixing filamentous taxa (of Sections IV and V of the standard classification scheme of Rippka et al., 1979). In fact, this was somewhat confirmed by subsequent screening by Volk et al. (2005) who, likewise, found that the majority of algicide producers belong to these, or closely akin filamentous groups (e.g. *Oscillatoria, Arthrospira, Phormidium* from Section III). Moreover, it was found in this latter study that members of these same groups were, in fact, the more susceptible (versus unicellular cyanobacterial or green algal representative) to the apparent algicidal metabolites, suggesting a role in the allelopathy toward most related (and perhaps ecologically similar) photoautotrophic competitors. This is further supported by recent screening studies (Gantar et al., 2008) that used co-cultivation of sympatric isolates cyanobacteria and chlorophytes (specifically from the Florida Everglades), and found apparent allelopathic activity among all ecologically co-occurring cyanobacterial, and 4 of 6 green algal, strains examined the study.

Within Section IV and V cyanobacteria, two groups - the genus, *Nostoc*, and several members of the family Stigonemataceae - are particularly notable producers of apparent allelochemicals. Perhaps the most frequently identified producer of algicidal metabolites is the cyanobacterial genus, *Nostoc*. In addition to being consistently represented, among algicide producers in screening studies (Flores & Wolk, 1986; Shlegel et al., 1998; Volk, 2005), several apparently allelopathic metabolites (see Fig. 3) have been characterized from the genus, including alkaloids (e.g. nostocarboline), phenolics (e.g. 4.'-dihydroxybiphenyl) and peptides (e.g. nostocyclamide). Although perhaps not as chemically diverse, members of the family of branched filamentous cyanobacteria, Stigonemataceae, particularly including the genera *Hapalosiphon* and *Fischerella*, have been, likewise, widely cited as producers of allelopathic compounds (e.g. Doan et al., 2000 and 2001; Leflaive and Ten-Hage, 2006; Gantar et al., 2008; Graneli et al., 2008; Leão et al., 2009).

In addition to an apparent phylogenetic conservation, there is some evidence to support chemical conservation within the algicidal chemistry of microalgae. In particular, a number of metabolites belonging to the indole class of alkaloids have been found to possess anti-algal activity, and consequently associated with allelopathic interactions. Perhaps the most frequently cited allelochemicals from cyanobacteria are the hapalindoles and related alkaloids (e.g. 12-*epi*-hapalindole E isonitrile, Fig. 3), including ambiguines, welwitindolinones and fischerindoles, that have been isolated from both marine and

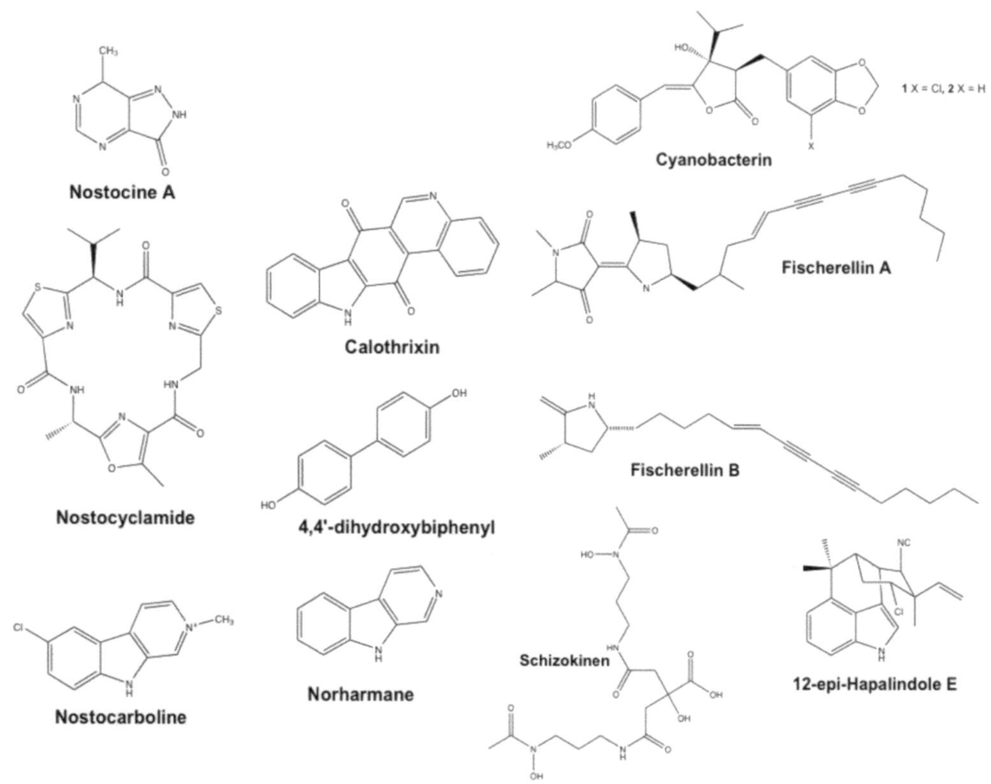

Fig. 3. Algicidal metabolites from microalgae with proposed role in allelopathy.

freshwater representatives of the family Stigonemataceae, and particularly the genera *Hapalosiphon* and *Fischerella*. Indeed, the first member of the class, hapalindole A, was initially identified, in part, based on algicidal acivity (Moore et al., 1984). Several subsequent studies pointed to the apparent role of hapalindoles, as algicidal metabolites, in allelopathy (Doan et al., 2000; Etchegaray et al., 2004; Gantar et al., 2008). Doan et al. (2000), for example, purified 12-epi-hapalindole E isonitrile (Fig. 3) as an algicidal metabolite of an isolate of *Fischerella*, previously identified from screening studies (Schlegel et al., 1999), and specifically demonstrated inhibition of RNA synthesis. Although only a few hapalindole alkaloids have been specifically characterized with respect to algicidal activity, and possible allelopathy, it is likely that other members of the class, and related indole alkaloids in the taxonomic group (e.g. welwitindolinones, fischerindoles), may have similar roles and biological activities, particularly due to structural similarity, and generally widespread antimicrobial and antimitotic activity among these compounds. In fact, Etchegaray et al. (2004), in their identification of the 12-epi-hapalindole isothiocyanate, identified several other uncharacterized, algicidal metabolites from a strain of *Fischerella*, including those with chemical similarity (e.g. molecular weight) to the hapalindoles.

Moreover, other taxonomically distinct groups have, likewise, been found to produce compounds (Fig. 3 and 6) with a chemically similar indole core, including calothrixins isolated from *Calothrix*, and β-carbolines, and the chlorinated nostocarboline and norharmane (and related harmane alkaloids), isolated from species of *Nostoc* and *Nodularia*, respectively (Doan et al., 2000; Becher et al., 2005; Blom et al., 2006; Volk, 2005). Interestingly an indole metabolite, harmane (or 1-methyl norharmane) - chemically related to one of these apparent algal allelochemicals (i.e. norharmane) - was also identified separately (Kodani et al., 2002) from cultures of a non-algal bacterium, *Pseudomonas*, isolated from a eutrophic freshwater system, specifically based on apparent algicidal activity (i.e. against several cyanboacterial species). Furthermore, in some case (e.g. calothrixin), there is also shared biological activity (e.g. inhibition of RNA synthesis) with other cyanobacterial indole alkaloids (e.g. hapalindoles, see above). The possibility of an indole "pharmacophore" is particularly intriguing, in the context of possible herbicides, since the same heterocyle is found among naturally occurring auxins, including the major - and most potent - congener, indole-3-acetic acid (IAA), and synthetic auxins (e.g. 2,4-dichlorophenoxyacetic acid, or "2,4-D") which mimic these plant growth regulators (see section 4.2 *Herbicidal Target-Based Approach*, below), and consequently represent an important and widely used class of commercial herbicides.

In fact, algicidal metabolites from cyanobacteria and other microalgae encompass a fairly diverse array of secondary metabolites. In addition to the previously discussed indole alkaloids, algicidal metabolites so far identified (see Fig. 3) include various other alkaloids (e.g. the aminoacylpolyketide fischerellins; nostocine A), phenolics (e.g. 4,4′-dihydroxybiphenyl), hydroxamate chelators (e.g. schizokinen) and fatty acids (e.g. 2,5-dimethyldodecanoic acid, polyunsaturated fatty acids), as well as peptides (e.g. nostocyclamide). The latter is perhaps of particular interest. This group of metabolites is, as discussed above, a particularly characteristic class of secondary metabolites from cyanobacteria (and, to a lesser extent, other microalgae). Moreover, this class of compounds presents a number of compelling attributes toward the development of commercially viable candidate compounds (e.g. modification via semi-synthesis, heterologous expression; see *Commercial Potential of Herbicides from Microalgae,* below). Moreover, the biosynthetic genes (NRPSs), associated with this class of metabolites, appear to be especially widespread in the Section IV/V cyanobacteria that (as noted above) are most frequent producers of algicidal metabolites. Specifically, it was found (Christiansen et al., 2001) using sequence-specific primers that nearly all (97% and 100%, respectively) of the cyanobacteria in these two sections were positive for NRPS genes (compared to only 52%, 80% and 64%, respectively, for Sections I, II and III).

In a very recent study, Leão et al. (2010) used bioassay-guided fractionation, specifically based on inhibition of the green alga, *Chlorella vulgaris,* to isolate a series of cyclic peptides, the portoamides (Fig. 4) as apparent allelopathic agents from of the blue-green alga, *Oscillatoria.* In addition to *C. vulgaris*, as well as other chlorophyte species, these metabolites differentially inhibited cyanobacteria, including *Cylindrospermopsis raciborskii*, but not several other cyanobacterial species (e.g. *Microcystis, Aphanizomenon* and *Anabaena*), nor other microalgae (e.g. the diatom, *Cyclotella menenghiniana*) tested. Interestingly, allelopathic activity was found to specifically peak during the early stages of exponential growth of the cyanobacterial culture, suggesting an ecological role (i.e. to "open" a niche for colonization), and consequently providing, in this case, an opportunity to purify a sufficient quantity of the otherwise low concentration metabolite for chemical characterization (Leão et al., 2010). Even more interesting, the investigators documented an apparent synergism of these metabolites. Specific mixtures of the most abundant congeners, portoamides A and B, in the ratios of 2:1 and 1:2.6 inhibited *C. vulgaris* at 30 ppm, but not in other ratios (e.g. 4.4:1) tested (Leão et al., 2010). This finding is particularly stunning given the rather minimal difference between the structure of the two (differing only in the presence of a methoxy group), and is further underscored by the apparent lack of activity associated with other chemically related congeners (e.g. metabolites specifically lacking an esterified *N*-acetyl-*N*-methyl tyrosine found in portoamides A and B).

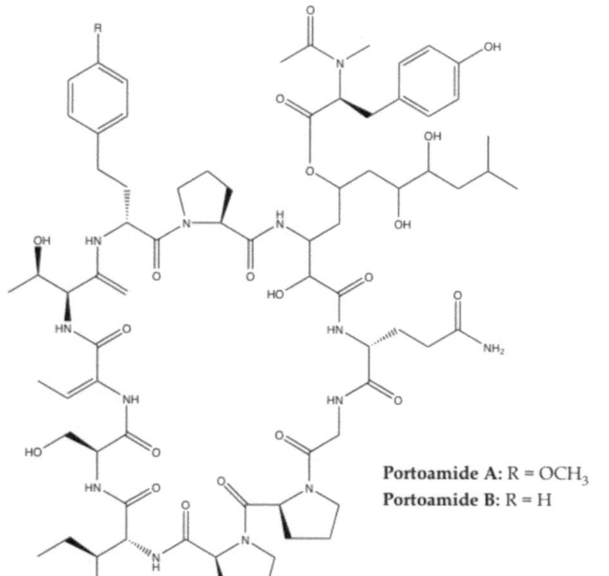

Fig. 4. Portoamides A and B isolated from a species of *Oscillatoria* as synergistic allelopathic metabolites (Leão et al., 2010).

In addition to metabolites identified based on their algicidal activity, it has been suggested that various 'HAB toxins,' largely recognized based on their effects on human health, may also play an ecological role in allelopathy. In fact, it has been suggested (e.g. Graneli et al., 2008) that nutritional imbalance (and consequent limitation) associated with eutrophication in aquatic systems may drive (via competitions for these resources) production of allelochemicals, and consequently production of these toxins. Elegant work by Kearns and Hunter (2000, 2001), for example, investigated allelopathic interactions between the cyanobacterial species, *Anabaena flos-aquae*, and green alga, *Chlamydomonas reinhardtii*, and found that well characterized toxins, microcystin-LR and anatoxin-a, from *A. flos-aquae* inhibited the green alga specifically via apparent paralysis, and subsequent settling. Interestingly, production of the latter of the two toxins was found to be stimulated by extracellular products of *C. reinhardtii*, while microcystin production was seemingly inhibited by the same (Kearns and Hunter, 2000). Subsequent studies (e.g. Pflugmacher, 2002) have, likewise, shown an allelopathic role of microcystins. Interestingly, both microcystin and anatoxin-a have been also shown (Mitrovic et al., 2004; Jang et al., 2007) to inhibit growth of plants, specifically using an aquatic plant model, "duckweed" (*Lemna japonica*).

Moreover, allelopathic interactions are not limited to the freshwater cyanobacteria, and a possible role of various metabolites from marine microalgae, including dinoflagellates, diatoms, haptophytes and raphidophytes, as well as from microalgal chlorophytes, have been, likewise, documented (see, for example, a good review by Graneli et al., 2008). In some cases, allelopathy has been associated with known HAB-associated toxins. For dinoflagellate, haptophytes and diatom HAB species, in particular, studies have suggested a possible role of various polyethers - characteristic metabolites of these groups - including brevetoxins (Kubanek and Hicks, 2005), okadaic acid (Sugg and Van Dolah, 1999), prymnesin (Graneli and Johansson, 2003), gymnomidine (cf. Legrand et al., 2003) and cooliatoxin (cf. Legrand et al., 2003). However, in several cases, it has been suggested that these metabolites, although algicidal, likely are not solely responsible for allelopathic activity, and that other extracellular metabolites (which remain to be characterized) may play a larger role (e.g. Sugg and Van Dolah, 1999; Kubanek and Hicks, 2005). Indeed, although apparent allelopathy has been demonstrated for a wide range of HAB species, specifically using culture-based experiments and/or extracellular preparations (e.g. culture filtrates), the allelochemicals responsible largely remain, in most cases, to be characterized.

Spanning eukaryotic and prokaryotic microalgae, one particularly interesting class of compounds that has been suggested to play a role in allelopathy are fatty acids - and specifically a polyunsaturated fatty acids (PUFAs) - either in free acid form, or as part of acylated sugars (i.e. glycolipids). A thorough review of these metabolites, and their possible role in allelopathy, has been previously presented by Ikawa (2004). Indeed, a growing number of studies have generally pointed to the biological activity of fatty acids. However, it is becoming generally clear that desaturation - as well as other modifications (e.g. oxidations, aldehydes) - may be particularly important to the observed toxicity or other bioactivity, and accordingly their possible role in allelopathy. Table 1 illustrates this trend, specifically based on several prior studies, in which algicidal activity has been consistently associated with relative desaturation of fatty acids isolated from microalgae. As an example of the importance of PUFAs in allelopathy, Arzul et al. (1993, 1995) investigated the apparent inhibition of the marine diatom, *Chaetoceros gracile*, by blooms of the dinoflagellate, *Gyrodinium*, and identified a high concentration of octadecapentaenoic aid (OPA, C18:5n3) and docosahexaenoic acid (DHA, C22:6n3) in the fatty acid profiles of *Gymnodinium*. These studies identified further found that these metabolites specifically inhibited the growth of *C. gracile*. Similarly, Uchida et al. (1988) identified relatively large amounts of DHA, along with eicosapentaenoic acid (C20:5n3), and small amounts of C18:2 and C18:3 fatty acids, from *Peridinium bipes*, and characterized their inhibition of cyanobacteria.

Chemically related to fatty acids are a family of mono- and diacylglycerides in which acyl groups are esterified to the 2′ and/or 3′ position of glycerol (typically with either a mono- or disaccharide, and particularly galactose, at the 1′ position). As for free fatty acids, a variety of biological activities (e.g. inhibition of DNA polymerase, antimicrobial activity) have been described for these compounds (e.g. Kurihara et al., 1996; Ohta et al., 1998; Hanashima et al., 2000; Eitsuka et al., 2004; Mizushina et al., 2005; Cantillo-Ciau et al., 2010). Of particular note are the sulfated galactosyl (i.e. sulfoquinovosyl) substituted mono- and diacylglycerides which are found alongside non-sulfated glycolipids in the thylakoid membranes of prokaryotic and eukaryotic photoautotrophs (including higher plants). In fact, it has been proposed that, in addition to other possible mechanisms (e.g. acting as secondary messengers for biochemical pathways, phospholipases), the importance of both free fatty acids and acylglycerides in cellular membranes - and the consequent disruption and destabilization of these membranes by non-endogenous variants - may be responsible for the widespread biological activity of these compounds (Ikawa, 2004). Given the obvious and unique importance of subcellular plastids in photosynthetic organisms, including algae and plants alike, further exploration of these compounds as selective inhibitors of photoautotrophs, and specifically as a novel target for herbicidal agents, is warranted.

	Cyanobacteria	**Green Algae (Chlorophyta)**			
	Phormidium tenue	*Chlamydomonas*	*Haematococcus*	*Pandorina*	*Ankistrodesmus*
Myristic Acid (14:0)	>100	>100	>100	>100	>100
Palmitic Acid (C16:0)	>100	>100	>100	>100	>100
Palmitoleic Acid (C16:1)	**2.5**	**12.5**	**50**	**50**	**50**
Stearic Acid (C18:0)	n.d.	>100	>100	>100	>100
Oleic Acid (C18:1 cis-9)	**1**	**50**	100	100	100
Cis-Vaccenic Acid (C18:1 cis-11)	**5**	n.d.	n.d.	n.d.	n.d.
Linoleic Acid (C18:2)	**0.5**	**50**	**12.5**	**25**	**25**
Linolenic Acid (C18:3)	**0.5**	**50**	**<12.5**	**12.5**	**50**

Table 1. Algicidal activity of saturated and unsaturated fatty acids against cyanobacteria and green algae (Chlorophyta). Data are taken from McCracken et al. (1980) and Yamada et al. (1993) for chlorophyte and cyanobacteria data, respectively. Given are the minimum concentration (mg/L) that inhibits algal culture growth >50%. "n.d." indicates evaluation not done. Apparent increase in algicidal activity, associated with desaturation, highlighted in **bold** text.

Finally, allelopathy is not, by any means, limited to microalgae, and similar interspecific, chemically mediated interactions have been, likewise, suggested in marine and freshwater macroalgae. For example, several studies (Friedlander et al., 1996; Jin & Dong, 2003; Nan et al., 2004; Jin et al., 2005; Wang et al., 2007; Nan et al., 2008) have documented the apparent allelopathy of the macroalgal chlorophye, *Ulva*, toward various microalgae and other macroalgae. Friedlander et al. (1996) identified apparently lipophilic metabolites from culture media of *Ulva lactuca* that inhibited the growth of the red alga, *Gracilaria conferta*, and suggested that these exometabolites may play a role in the observed growth inhibition of the latter species when grown together, under otherwise controlled conditions, in culture. Based on these studies, and citing a documented reciprocal relationship between growth of micro- and macroalgae, Nan et al. (2004) evaluated allelopathy of *U. pertusa* against eight species of microalgae, including dinoflagellates, haptophytes, diatoms and microalgal chlorophytes. These studies found that all eight species were inhibited in co-cultivation with the macroalga. However, only one of the species (i.e. the haptophyte, *Chroomonas placoidea*) was inhibited by culture filtrates. This was taken, by the authors, citing several related examples, to be due to instability of released metabolites (Nan et al., 2004). This observation was confirmed by several subsequent studies (Jin et al., 2005; Wang et al., 2007) which additionally showed that continuous release of allelochemicals by *Ulva*, either by semi-continuous addition of filtrate, or presence of fresh algal material or dried powdered biomass, was sufficient to inhibit growth of microalgae. Such effects, moreover, are not limited to *Ulva*, and similar studies (e.g. Kim et al., 2004; Wang et al., 2006; Wang et al., 2007; An et al., 2008; Wang et al., 2009) have, likewise, shown inhibition of microalgae by other green algae (e.g. *Enteromorpha*), as well as coralline red algae (e.g. *Lithophyllum, Corallina, Porphyra, Gracilaria*) and brown algae (e.g. *Sargassum, Undaria, Laminaria*).

By-and-large, algicidal metabolites from macroalgae remain to be well characterized, however, several studies suggest a role of PUFAs, and associated metabolites (e.g. acylglycerides), as discussed above for microalgae. Work by Alamsjah et al. (2005, 2007) screened a collection of thirty-seven macroalgae, including representatives of the Chlorophyta, Rhodophyta and Phaeophyta, and specifically found extracts of the genus, *Ulva*, to be the most potently algicidal against the HAB organism, *Heterosigma akashiwo*. The investigators subsequently identified (Alamsjah et al., 2005, 2007) a series of PUFAs - particularly including unsaturated C16 and C18 acids - from several species of *Ulva* (*U. fasciata, U. pertusa, U. arasakii, U. conglobota*) with algicidal activity toward a range of microalgal representatives. Based on the relatively higher production of these by the most algicidal species (*U. fasciata* and *U. pertusa*) it was argued that they likely play a role in observed allelopathy. Fatty acids, including PUFAs, have similarly been identified as apparent allelopathic agents of other macroalgae, including brown algae (e.g. inhibition of various microalgae by *Cladosiphon,* Kakisawa et al., 1988), red algae (e.g. *Neodilsea, Chondrus* and *Ptilota;* Macias et al., 2007) and Charophytes (e.g. inhibition of the cyanobacterium, *Microcystis aeruginosa*, by *Chara vulgaris*, Zhang et al., 2009).

PUFAs, moreover, are not the only class of bioactive metabolites from macroalgae, and various bioactive metabolites, including alkaloids (e.g. Gross et al., 2006; Arunkumar et al., 2010; Güven et al., 2010), phenolics (e.g. Hassan and Ghareib, 2009) and terpenoids (e.g. Fenical & Paul, 1984; Paul & Fenical, 1984; Lane et al., 2007; Arunkumar et al., 2010), have been characterized. Although the potential of these compounds as herbicides remains to well-studied, based on the apparent importance of allelopathy among photoautotrophs in aquatic habitats, the chemical diversity of the macroalgae, like the microalgae, represents a wealth of secondary metabolites to be explored to this end.

4. Microalgal metabolites as herbicides

In contrast to algicidal metabolites - specifically in relation to allelopathy - relatively scant studies have investigated biological activity of microalgal toxins with respect to higher plants (i.e. herbicides). The current body of knowledge with regards to potential herbicidal metabolites from microalgae currently remains limited. However, given the importance of allelopathy among aquatic photoautotrophs, and presumptive homologies of the biochemistry and physiology between both algae and higher plants, as photosynthetic organisms, the chemical diversity of bioactive secondary metabolites represents a likely rich source of such compounds. As such three approaches are proposed as a means to explore the, otherwise well recognized, chemical repertoire of these taxa with respect to potential herbicidal agents. These inclue: (1) a *in vivo* bioassay approach, based on identifying phytotoxicity in representative plant species; (2) a target-based approach, specifically incorporating assays based on recognized herbicide targets; and (3) a pharmacophore-based approach, specifically surveying algal metabolites to identify those with recognized structural features associated with herbicidal compounds.

4.1 *In vivo* bioassay approach

Perhaps the primary means of identifying potentially herbicidal metabolites from microalgae has been a variety of *in vivo* biological assays, relevant to herbicidal activity, which are readily available (and adaptable to most laboratory environments). Such assays, typically utilize a variety of endpoints relevant to herbicidal activity, including inhibition of seed germination and seedling growth, and inhibition of photosynthesis and/or "bleaching" of photosynthetic pigments, as well as various biochemical markers (e.g. oxidative stress).
In particular, the aquatic monocotyledonous angiosperm, "duckweed", including *Lemna minor* and *L. gibba,* and related species, have been widely utilized as a model system for algae-derived herbicides due, in part to its rapid growth (and presumptively its role of these metabolites in aquatic allelopathy). *Lemna* has, in fact, been adopted by the U.S. Environmental Protection Agency (EPA) as a general model for aquatic phytotoxicity (EPA 712-C-96-156). As an early example of this, Entzeroth et al. (1985) used *Lemna* to identify herbicidal compounds from ethanol extracts of the cyanobacterial species, *Lyngbya aesturii*. Using the species for bioassay-guided fractionation, herbicidal activity was found to be associated with an unusual fatty acid, 2,5-dimethyldodecanoic acid, which inhibited growth at concentrations as low as 200 ng/mL. (Entzeroth et al., 1985). Although not an algal metabolite in the strict sense, *Lemna* was also used to evaluate the phytotoxic activity of usnic acid, a metabolite of lichens (composite organisms comprised of a fungal and algal symbiont).
More recenty, several studies (Weiss et al., 2000; Leblanc et al., 2005; Mitrovic et al., 2004 and 2005; Jang et al., 2007; Saqrane et al., 2007; Yi et al., 2009) have utilized *Lemna* to investigate the apparent effects of the widespread cyanobacterial toxin, microcystin, on plant growth. Such studies have indicated that microcystins can inhibit growth and photosynthesis, as well as oxidative stress, in *Lemna* (e.g. Weiss et al., 2000; Mitrovic et al., 2005). Interestingly, not all studies have equally shown inhibitory effects of microcystins on *Lemna,* likely due to the lack of a standardized genetic background (and associated variability). Leblanc et al. (2005), for example, showed that - contrary to various other studies in the same species (e.g. Saqrane et al., 2007) - microcystin-LR had no apparent effect on *L. gibba* growth or photosynthesis in the species.
Aside from studies in *Lemna,* as well as a number of other aquatic plant species (e.g. *Myriophyllum spicatum, Ceratophyllum demersum, Phragmites australis*; Pflugmacher et al., 2001; Pflugmacher, 2002; Yi et al., 2009) studies have used a range of other plant models, including agriculturally important ones, to investigate phytotoxicity of algal metabolites. For example, Sanevas et al. (2006) evaluated aqueous methanolic extracts from a species of the cyanobacterial genus, *Hapalosiphon,* in a range of monocotyledonous and dicotyledenous agricultural plant species, including radish, cabbage, carrot, lettuce, wheat, onion, rice and maize, and specifically demonstrated a dose-dependent inhibition of roots and, to a lesser extent, shoot elongation, apparently due to inhibition of cell division. Hassan and Ghareib (2009) more recently utilized tomato (*Lycopersicon esculentum*) and lettuce (*Lactuca sativa*) to demonstrate inhibition of seed germination and seedling growth by apparent phenolic compounds extracted from the green alga, *Ulva lactuca*. Similarly, extracts of a strain of *Nostoc,* isolated from an agricultural pond, was found to specifically inhibit root and shoot growth of seedling, but not affect seed germination, in rice (*Oryza sativa*). Phytotoxicity of the lichen-derived usnic acid has been evaluated and found (Lasceve and Gaugain, 1990;

Ozturk et al., 1999) to inhibit growth in several monocot (e.g. maize, onion) and dicot (e.g. sunflower) species. Gleason and Case (1986) showed that cyanobacterin from *Scytonema hofmanni* equally inhibited crop plant species, including maize (*Zea mays*) and peas (*Pisum sativum*), as well as wild plant species, including dock (*Rumex crispus*), wild buckwheat (*Polygonium convolvulus*) and wild oats (*Avena fatua*). Moreover, several studies (McElhiney et al., 2001; Pflugmacher et al., 2007; Saqrane et al., 2008) investigated the possible impacts of microcystin on crop plants by evaluating phytotoxicity in a range of species including potato (*Solanum tuberosum*), mustard (*Synapis alba*), bean (*Phaseolus vulgaris*), maize (*Z. mays*), lentils (*Lens esculentum*), peas (*P. sativum*) wheat (*Triticum durum*) and spinach (*Spinacia oleracea*). In particular, microcystin-LR inhibited both epicotyl and root length, in addition to seed germination in maize (Saqrane et al., 2008).

Here we propose - and briefly present preliminary data on - the novel use of the model angiosperm, *Arabidopsis thaliana*, as the basis of a biological assay for investigation of herbicides from microalgae (and potentially other sources). *A. thaliana* ("thale cress"), a member of the mustard family (Brassicaceae), has been well studied – dating back to the work of George Rédei in the 1950s. However it took nearly four decades before it would be widely accepted as a model organism in plant biology. Specific advantages of the model include small size and ease of cultivation (e.g. ability to grow on agar plates), a small (ca. 120 Mb) and now completely sequence genome, and prolific seed production and rapid life cycle (ca. 6 weeks from seed germination to mature seed), as well as a large collection of described genetic mutants and various features that make it amenable to transgenic manipulations. An early, seminal review of the species as a model organism was presented by Meinke et al. (1998), following its acceptance by the Security Council of Model Genetic Organisms in 1998 (Fink, 1998). *The Arabidopisis Information Resource* (TAIR; www.arabidopisis.org) currently maintains a database, specifically relevant the genetic and molecular biology of *Arabidopsis*.

Owing, in particular, to various practical advantages (e.g. small size, laboratory cultivation, rapid germination and growth), *A. thaliana* represents an ideal candidate for investigation of herbicides generally, and specifically as a means to rapidly assess herbicidal compounds including, as relevant to the current topic, those from microalgae. Moreover, as a model organism, and particularly one with well described genetic background, as well as general amenability to various molecular biological methodologies, herbicidal activity can, in principle, be readily investigated via, for example, comparison to available genetic mutants, genetic manipulation, etc toward the goal of target identification. As such, we have developed a phytotoxicity assay based on *A. thaliana*, and specifically applied this to the screening of metabolites from a collection of freshwater cyanobacteria.

As illustrated in Fig. 5, extracts of cyanobacterial culture biomass were assayed in agar plates seeded with *A. thaliana*, and grown in a standard light- and temperature-controlled environmental chamber. Biomass (ca. 100 mg) from eleven strains of freshwater cyanobacteria, specifically isolated from the Florida Everglades (see Berry et al., 2007; Gantar et al., 2008), was sequentially extracted in non-polar (chloroform) and polar (30% ethanol) solvents. Prior to seeding, small "treatment wells" were made into the agar medium (i.e Murashige-Skoog medium with 0.8% agar) in the approximate center of each "sector" of a standard square Petri dish, specifically using an autoclave-sterilized Pasteur pipet, and low vacuum (see Fig. 5). Each well was filled with aliquots (20 μL) of prepared extracts. Subsequently, seeds of the wild-type *Columbia* ecotype of *A. thaliana* (*Col-0*), obtained from commercial sources (e.g. Lehle Seeds, Round Rock, TX, U.S.A.), were seeded into assay plates, as per standard techniques. Briefly, an appropriate amount of seeds (ca. 50 per milligram, or approximately 8 mg per assay plate) were weighed into microcentrifuge tubes, and sterilized (for 5 minutes) with 30% bleach, followed by repeated (3-4 times) rinsing with nanopure water, and subsequent centrifugation. Rinsed, sterilized seeds were re-suspended into 0.1% agar, and seeded into appropriate sectors of the test plate (ca. 10-15 seeds). Seeded plates were kept for approximately two weeks (to germinate and grow) in a environmental growth chamber (Therm Scientific Precision Model 818) at 28° C with a 14:10 light/dark cycle.

As seen in Fig. 5, the assay specifically demonstrated apparent inhibition of *A. thaliana* seedling growth by a polar (30% ethanol) extract of the isolate, *Microcystis* 95-13. Although germination was not apparently inhibited, the growth of hypocotyls was clearly reduced in seedlings exposed to the extract (Fig. 5). None of the other polar or non-polar extracts evaluated showed any apparent inhibition of seed germination or seedling growth in the two-week test period. Likewise, no effect of solvent (neither 30% ethanol or chloroform) was detected relative to untreated (i.e. "no solvent") controls. As referenced above, *Microcystis* is, as a genus, recognized to produce the potently toxic microcystins that have been associated with human and other animal intoxication events, as well as phytotoxicity. These results support the latter, but the recognized toxicity of the microcystins - if they are, in fact, responsible for phytoxicity against *A. thaliana* - would, as further discussed below, likely preclude their application as herbicides. We are currently investigating the phytotoxic metabolites identified here, but these results suggest the assay presented here may represent an effective means to rapidly screen - and subsequently characterize targets of - microalgal metabolites with respect to herbicidal activity.

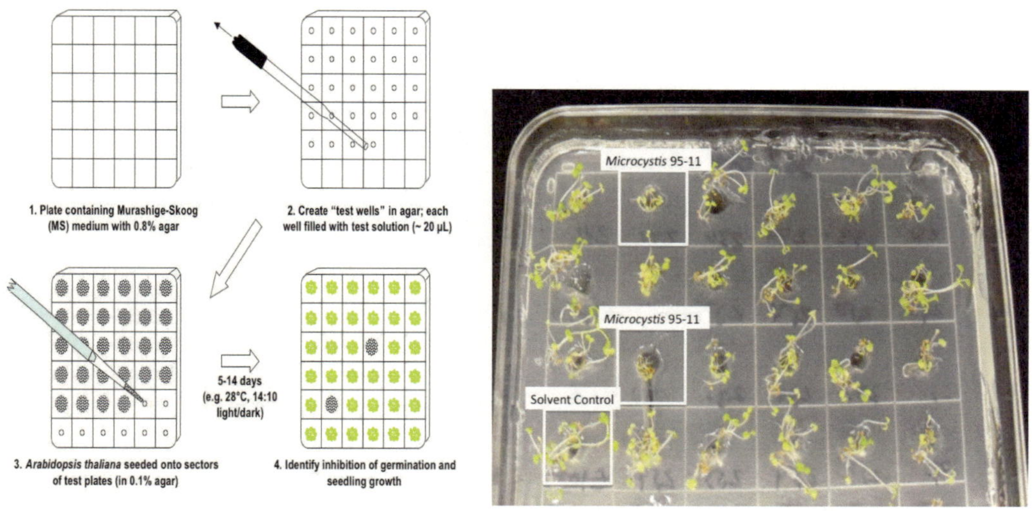

Fig. 5. *A. thaliana* phytotoxicity assay of cyanobacterial culture extracts showing inhibition of seedling growth by the isolate *Microcystis* 95-11. Shown are the polar (30% extracts).

4.2 Herbicidal target-based approach

With respect to the target-based approach, commercial herbicide generally fall into several categories based on their biochemical, molecular or cellular "mechanisms of actions" (MOA), including: (1) inhibition of primary metabolism, particularly including biosynthesis of lipids and amino acids, and specifically "branched" and aromatic amino acids, e.g. inhibitors of acetolactate synthase (ALS) and enolpyruvulshikimate 3-phosphate synthase (ESPS), respectively; (2) inhibition of oxygenic photosynthesis, including photosystems I and II (PSI/II), as well as biosynthesis of associated pigments, e.g. protoporphoryrinogen oxidase (PPO), a key enzyme in the synthesis of chloropyll; (3) synthetic "mimics" of plant growth regulators, e.g. auxins, cytokinins; and (4) inhibitors of microtubules or other components of cell-division.

By far, the most well investigated target of phytotoxic metabolites from microalgae has been inhibition of oxygenic photosynthesis, and associated molecular/biochemical targets. A good review of photosynthesis inhibitors from microalgae has been previously presented by Smith & Doan (1999). Recent studies by Gantar et al. (2008), for example, used pulse amplitude-modulated (PAM) fluorescence to show that lipophilic extracts of the cyanobacterial strain, *Fischerella* 52-1 (isolated from the Florida Everglades), specifically inhibited photosystem II (PSII), in addition to apparent degeneration of thylakoids. This finding is particularly notable as the strain was found to produce hapalindoles (Gantar et al., 2008; Walton et al., in press) previously associated with allelopathy (see *Algal Allelopathy* above). Within the same genus, the fischerellins A and B (Fig. 3) were shown (Srivastava et al., 1998) to potently inhibit PSII, specifically acting at multiple sites of PSII, distinct from that of the photosynthesis-inhbiting herbicide, 3-(3,4-dichlorophenyl)-1,1-dimethylurea (DCMU). Likewise, cyanobacterin, originally isolated from the cyanobacterial species, *Scytonema hofmanni*, inhibited PSII at a site distinct from - and with a potency nearly five times - that of DCMU, and specifically at the oxidizing site of quinone-B electron acceptor (Gleason & Paulson, 1984; Gleason & Case, 1986; Gleason et al., 1986). On the other hand, the peptide, nostocyclamide (Fig. 3), from the genus, *Nostoc*, seems to inhibit PSII by uncoupling electron transport (Jüttner et al., 1997 cf. Smith & Doan, 1999).

A second important class of herbicides are those that mimic plant growth regulators ("hormones") including auxins, cytokinins and gibberellins. Both macroalgae and microalgae - and particularly cyanbacteria and chlorophytes - have been found to produce extracellular compounds that act as all three of these, including metabolites chemically identical (e.g. the auxin, indole-3-acetic acid) to the plant hormones (see review by Tsavkelova et al., 2006). The realization that microalgae may produce metabolites with plant growth regulator activity extends back to Zulpa de Caire et al. (1979) who demonstrated that culture medium of the cyanobacterium, *Nostoc*, exhibited auxin-like activity. Zaccaro et al. (1996) subsequently showed that extracellular products from *Scytonema hofmanni* stimulated growth of *Lilium alexandrae* in a manner similar to the synthetic auxin, 1-napthaleneacetic acid (NAA), and implied that metabolites might represent a non-toxic alternative to this agent. Sergeeva et al. (2002) screened thirty-four strains of cyanobacteria, and not only identified auxin-like activity in 21 of the 34 strains, but confirmed (using GC-MS, and analytical standards) the presence of the main auxin, indole-3-acetic acid (IAA), in two species of *Nostoc*. Interestingly, a higher amount of auxin-like activity was found for symbiotic (83%), versus free-living (38%), species of cyanobacteria. Similarly, isopentanyl adenine cytokinins, including zeatin, and specifically *cis* isomers, as well as aromatic cytokinins, including benzyladenine (chemically similar to the synthetic cytokinin,) have been previously isolated from several species of macroalgae (Stirk et al., 1999; Stirk et al., 2002; Stirk et al., 2003). More recently, Hussain et al. (2010) identified both IAA and zeatin-like cytokinins in a variety of cyanobacterial species. Cyanobacteria have been similarly shown to produce gibberellin-like metabolites (Gupta and Agarwal, 1973; Rodriguez et al., 2006).

Although plant growth regulators are, obviously, involved in the positive growth of higher plants (e.g. cell division, elongation), directed application of agonists can lead to uncontrolled growth, and consequently act as herbicides. Perhaps the best example of this strategy is the widely used synthetic auxin, 2,4-dichlorophenoxyacetic acid (2,4-D). In fact, several studies (Jäger et al., 2005; Manickavelu et al., 2006) have shown that cyanobacterial metabolites can replace 2,4-D in plant cell culture. In accordance with the dose-dependent effects of plant growth regulators, Hassan and Ghareib (2009) evaluated phenolics extracted from *Ulva lactuca,* and found moderate concentration (30 ppm) stimulated germination of lettuce and tomato seeds. However, at higher concentrations (300 ppm), the same extracts reduced both seed germination and seedling growth.

A third target of several herbicides is cell division. Acting in some ways like synthetic plant growth regulators (e.g. effects on elongation, cell swelling), such compounds differ in affecting cell division via direct mechanisms (e.g. inhibition of tubulin/microtubule assembly) rather than indirect routes (e.g. regulation of gene expression). Perhaps the best described are the pre-emergence dinitroaniline herbicides (DNHs). DNHs, such as the widely used trifluralin, inhibit mitosis (as well as other effects, e.g. cell swelling, induction of multiple nuclei) by interaction with tubulin subunits of microtubules, in a fashion similar to the well described "spindle poison," colchine (Upadhyaya & Noodén, 1977, 1980).

Indeed, microalgae produce a wide diversity of cell division inhibitors, although these metabolites have been largely investigated with respect to their potential as antibiotics or anticancer drugs. Perhaps the best investigated microalgal metabolite in this regard, are the cryptophycins (Trimurtulu et al., 1994), comprised of more than twenty-five congeners specifically isolated from the cyanobacterial genus, *Nostoc,* that have been shown (Smith et al., 1994) to act via depolymerization of microtubules, and consequently investigated as both antifungal and antitumor compounds (including, in the latter case, Phase I clinical trials). Likewise, other cyanobacterial metabolites, including curacin A (from the marine cyanobacterium, *Lyngbya majuscula*) and the dolastatins (from the marine cyanobacterium, *Symploca*) are inhibitors of microtubule assembly and tubulin polymerization (Verdier-Pinard et al., 1998; Luesch et al., 2001), and have also been investigated with respect to potential anticancer drug development. Indeed, many such compounds - and particularly an array of non-ribosomal peptides - have been isolated and characterized (to some extent) from cyanobacteria and other microalgae, and a number of good reviews on these are available (Moore, 1996; Gerwick et al., 2001; Tan, 2007). Moreover, aside from cyanobacteria, other algae have been, likewise, found to produce inhibitors of mitosis. For example, various species of brown algae have been found to produce terpenoids, including the mediterraneols and bifurcarenone (Sun et al., 1980; Francisco et al., 1986) that inhibit cell-division.

Despite this focus on cell division inhibitors from microalgae with respect to drug development, limited studies have also shown apparent cell division in plant tissues. Extracts from the cyanobacterium, *Hapalosiphon,* for example, were shown (Sanevas et al., 2006) to inhibit root and shoot growth in a wide range of crop plants (see above), and subsequently found to inhibit mitosis, as evidenced by a dose-dependent reduction in the mitotic index (in the onion root model). In a study (Baskin and Wilson, 1997) using the *A. thaliana* model (see above), the effect of the protein phosphatase inhibitors, okadaic acid and microcystin-LR - metabolies from dinoflagellate and cyanobacterial species, respectively -

were evaluated with respect to cortical microtubules of roots. Interestingly, although both toxins are recognized inhibitors of type 1/2A serine/threonine protein phosphatases, they exerted markedly different effects. Okadaic acid affects both cell elongation and radial expansion (at higher concentrations), as well as microtubule disorganization, whereas microcystin-LR only minimally inhibited elongation. Although inherent toxicity of both metabolites may obviously limit the potential of these compounds themselves, as commercial herbicides, the differential activity of these metabolites suggests an otherwise uncharacterized difference in their effects (via protein phosphatase inhibition) on cell division, and supports a need to further investigate their specific targets, particularly in relation to phytotoxic effects such as those observed.

A final important target of commercial herbicides is the inhibition of biosynthetics pathways of primary metabolites, including lipids and amino acids. Among important examples of these are the amino acid biosynthesis inhibitors (see review by Kishore and Shah, 1988), including specific inhibitors of acetolactate synthase (ALS, e.g. sulfonylureas) and enopyruvylskikimate 3-phosphate synthase (ESPS, e.g. glyphosate), key steps in the synthesis of branched chain, and aromatic, amino acids, respectively. In particular, the ALS pathway is one that exists only in plants (and not animals) making it a particularly good target for herbicides. Similarly, inhibitors of the enzyme acetyl coenzyme A carboxylase (ACCase) act via inhibition of a key step in lipid biosynthesis, and show selective activity, specifically between monocots and dicots.

To date, there are no known (to the author's knowledge) metabolites from microalgae that have been found to target either amino acid or lipid biosynthesis. That said, the unique diversity of both lipid and amino acids in the secondary metabolites of microalgae makes them seemingly rife with opportunities to explore possible effects of these as, for example, potential inhibitors of their respective primary metabolic pathways. In particular, the cyanobacteria are recognized to produce and utilize a wide range of "unusual amino acids" including D-isomers, β-hydroxy, N-methylated or otherwise modified version of essential amino acids, as well as completely unique representatives (see, for example, review by Gerwick et al., 2001). Also along these lines, it is worth noting that several studies (e.g. Powell et al., 1991; Forlani et al., 2008) have shown an apparently widespread tolerance of cyanobacteria to the ESPS inhibitor, glyphosate, specifically, it seems, via insensitivity of this enzyme in these microalgae. Interestingly, although such a possible role of these small molecules with respect to amino acid metabolism remains to be investigated, at least two study have shown inhibition, by algal (or related) metabolites, of enzymes involved in catabolism of amino acids. Specifically, the lichen-derived toxin, usnic acid, was found to inhibit hydroxyphenylpyruvate dioxygenase (HPPD), similar to several commercial herbicides (e.g. triketones), leading to blockage of plastoquinone synthesis, and consequent "bleaching" of plant cells (Duke et al., 2002). Likewise, norharmane (isolated from *Nostoc* and several other cyanobacteria; Volk, 2008) has been shown to inhibit the equivalent enzyme - indoleamine 2,3-dioxygenase - involved in the catabolism of the amino acid, tryptophan. In general, the paucity of studies represents an open area for research.

4.3 Pharmacophore-based approach

In addition to their classification based on target or MOA, commercially important herbicides also span a wide range of chemical classes. Based on established structure-activity relationships of either herbicides - or, alternatively, endogenous/exogenous

regulators – the identification of structurally related metabolites (i.e. shared "pharmacophores") serves as a means to identify potential herbicidal candidates. To be sure, there have been, to date, very limited use of this approach as a systematic means to screen micralgal metabolites for herbicidal compounds. However, at least one study arguably has. Specifically, a recent study by Volk (2008) utilized a combination of high-performance thin-layer chromatography, and high-performance liquid chromatography (HPLC), to screen thirty-three species of microalgae for norharmane, based on prior (Volk, 2006) identification of this metabolite as an algicidal metabolite of *Nostoc*. Interestingly, this study identified an additional seven strains of cyanobacteria that produced norharmane or related compounds (Volk, 2008). Similarly, the pharmacophore-based approach has been used direct synthetic studies of microalgal metabolites. As an example, Blom et al. (2006) used norharmane (as a synthetic starting material), and the structurally related metabolite, nostocarboline (Fig. 3), to prepare several synthetic analogs of the latter, and demonstrate structure-activity relationships, as well as (in, at least, one case) enhanced algicidal activity.
Moreover, several cases of structural similarity between microalgal metabolites and established herbicides exist. For example, as discussed above (see *Algal Allelopathy*) - and as shown in Fig. 6 - several of the purportedly allelopathic metabolites from microalgae contain an indole core, structurally similar to that of the naturally occurring auxins (e.g. IAA, IBA, CI-IAA). Although bioactive indole alkaloids are widespread in the natural world, including marine and freshwater habitats (see, for example, review by Gul and Hamann, 2005), and the algicidal activity of these microalgal metabolites, as currently understood, do not parallel the effects of auxins, this supports the importance of indoles as a target for pharmacophore-based approaches.

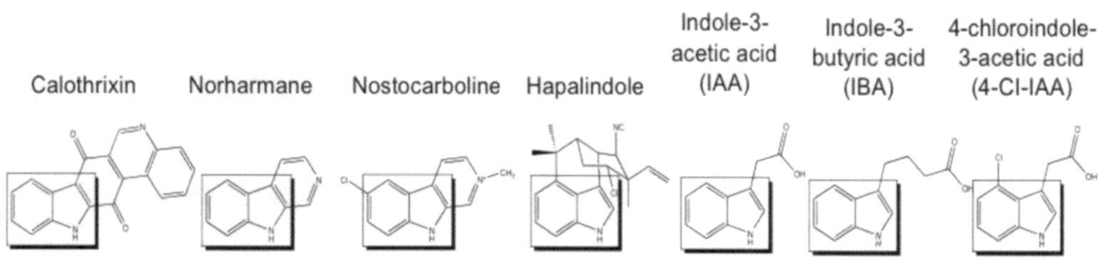

Fig. 6. **Indole-containing allelochemicals from cyanobacteria, and structural comparison to plant auxins.** From left to right are calothrixin A, norharmane, nostocarboline and 12-eip-hapalindole E isonitrile, and the naturally-occurring auxins, indole-3-acetic acid (IAA), indole-3-butyric acid (IBA) and 4-chloroindole-3-acetic acid.

Likewise, various other chemical classes, represented by commercially important herbicides, are found in marine and freshwater microalgae. For example, oxazole rings – characteristic of a number of herbicides (e.g.) - are commonly found among bioactive non-ribosomal peptides from cyanobacteria (e.g. muscoride from *Nostoc muscorum*, Nagatsu et al., 1995; microcyclamides from *Microcystis* sp., Raveh et al., 2010; raocyclamides from *Oscillatoria raoi*, Admi et al., 1996). Indeed, the recognized allelopathic metabolite, nostocyclamide, contains this heterocyclic ring system (Fig. 3). Likewise, various marine algae are known to produce chlorinated hydrocarbons similar to the halogenated aliphatic class of herbicides (e.g. trihaloacetic acids, such as dalapon), and which have been associated with possible allelopathy (see, for example, review by Kladi et al. (2004).

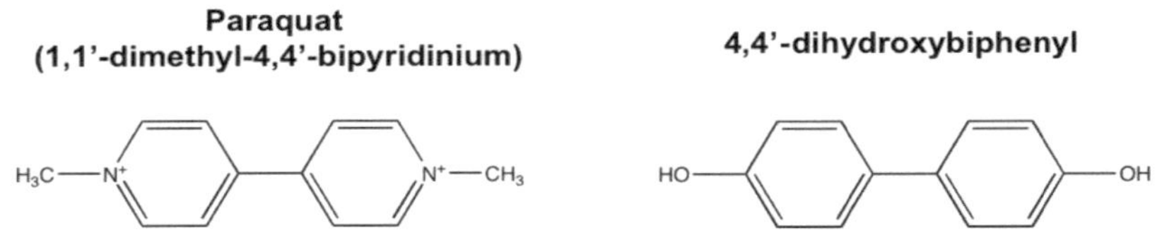

Fig. 7. General structural resemblance between the algicidal exometabolite, 4,4'-dihydroxybiphenyl, from cyanobacteria, and the herbicide, paraquat.

In other cases, the structure of microalgal metabolites, rather than containing identical structural features, closely resemble, in a more general way, recognized herbicides. For example, the algicidal phenolic compound, 4,4'-dihydroxybiphenyl, isolated (Volk, 2006) as an "exometabolite" of cyanobacterial cultures, bears general structural resemblance to the commercially important herbicide, paraquat (Fig. 7). Similarly, bioactive metabolites structurally resembling (Fig. 8) halogenated diphenyl ether herbicides (e.g. oxyfluorfen, acifluorfen, diclofop), have been isolated from algae. In particular, bioactive polybrominated diphenyl ethers, including hydroxylated and methoxylated variants, have been isolated from rhodophytes, chlorophytes and cyanobacteria (Malmvärn et al., 2008). One such example is shown in Fig. 8.

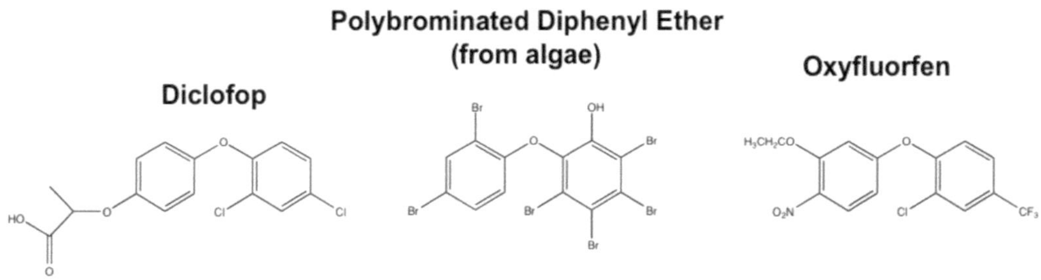

Fig. 8. Comparison of polybrominated diphenyl ethers from microalgae, and the diphenyl ether herbicide, oxyfluorfen).

5. Commercial potential of herbicides from microalgae

As demonstrated in this chapter, microalgae represent a rich repository of bioactive metabolites, including numerous compounds with biological activity either indirectly (e.g. algicidal) or directly (e.g. photosynthesis inhibiting, antimitotic) related to herbicidal potential. Moreover, a number of features make these metabolites particularly compelling as a source of potential herbicides.

As discussed, the cyanobacteria, in particular, are recognized producers of an array of non-ribosomal peptides, while the secondary metabolism of eukaryotic microalgae is particularly characterized by a diversity of metabolites derived from the polyketide synthesis pathway. As modular enzymes, in both cases, these megasynthases (e.g. NRPSs, PKSs) are highly amenable to "*in vivo* combinatorial" approaches (through directed manipulation of gene, and consequently enzyme, modules), as well as heterologous expression. The former, in principle, enables a means for rapid biosynthetic modification, and essentially "design," of molecules for improved activity and uptake potential, as well as reduced toxicity, and other characteristics germane to effective and safe herbicides. As one example, Christiansen et al. (2003) demonstrated, using targeted mutation of one module (*mcyJ*, an *O*-methyltransferase) of the NRPS/PKS gene responsible for microcystin biosynthesis chemical modification specifically resulting in a novel variant with modified activity. Heterologous expression, on the other hand, enables the potential for large-scale production of these metabolites in suitable hosts (e.g. *E. coli*) which can be readily cultured with high yields on a large-scale (e.g. biofermentation). Such an approach has, in fact, been demonstrated for a very limited number of algal or other microbial metabolites. In one example, the biosynthetic NRPS gene cluster for the algal metabolite, patellamide A, naturally produced by an endosymbiotic algae (from didemnid ascidians), was heterologously expressed in *E. coli* (Schmidt et al., 2005), and proposed (Long et al., 2005) as a means of sustainable production of the bioactive metabolite.

Other compelling feature of these algal metabolites as herbicides include a presumptive "biodegradability" of many of these natural products, as well as water solubility of many (e.g. peptides) which might minimize concerns of bioaccumulation, and associated environmental concerns. Although these are not universally true of algal metabolites, both water solubility and lability have been described in at least some of the allelopathic metabolites from algae (e.g. apparent algicides from the macroalga, *Ulva,* as discussed above).

In addition to potential application as herbicides, particularly with regards to, for example, crop pests (i.e. weeds), microalgal metabolites hold an equally important potential as algicides and related "antifouling" agents. As pointed-out by Duke et al. (2002) aquaculture represents "one of the fastest growing areas of agriculture in the world," and although not traditionally grouped with herbicides, the diversity algicides represent a "niche market." Presence of algae in aquaculture facilities is associated with a number of concerns related to both health (e.g. toxic contaminants) and quality (e.g. "off-flavors" associated with the cyanobacterial metabolite, 2-methylisoborneol, MIB) that are consequently responsible for substantial economic losses. Currently there are rather limited approved agents for control of noxious algae in aquaculture, and as such, discovery of natural products, as potentially "environmentally safe" algicides, from microalgae (or other sources) holds tremendous - albeit largely untapped - promise.

Although a number of microalgal metabolites have been shown to have either directly, or potentially, herbicidal or algicidal activity, there is an obvious need for further study. In many cases, compounds, studied with respect to herbicidal or algicidal activity, have recognized potential for human or animal toxicity which would presumably limit their direct use as herbicides. This is perhaps particularly true for a number of so-called "HAB toxins" which despite demonstrated allelopathic, and even herbicidal, activity have well documented human toxicity, as well as impacts on wildlife and domestic animals. That said, although such toxicity would likely preclude any direct commercial application of these compounds, their continued investigation would provide a means of identify possible novel targets for development of (less toxic) herbicides based on their MOAs. Furthermore, aside from continued identification of metabolites with herbicidal activity, evaluation of taxa specific (e.g. monocots versus dicots, crop plant versus weed" species) differences in activity, as well as further elucidation of MOAs, will be required for potential of these compounds to be realized in any commercially relevant sense.

Finally, despite the obvious potential of the microalgal secondary metabolites with respect to discovery and development of herbicides, there remains a relatively limited commercial exploration of this resource. In fact, to the author's knowledge, in only one case - specifically based on the work of Gleason et al. (1986, 1990) with respect to the algicidal and phototoxic activity of cyanobacterin – has a patent (United States Patent 4626271) been submitted for explicit use of the compound (specifically in conjunction with a surfactant) as a green plant herbicide. It is, of course, the hope that continued exploration of this chemical diversity will lead to future realization of its tremendous potential.

6. References

Admi, V.; Afek, U. & Carmeli, S. (1996) Raocyclamides A and B, novel cyclic hexapeptides isolated from the cyanobacterium *Oscillatoria raoi*. *Journal of Natural Products* 59, 396-399.

Alamsjah, M. A.; Hirao, S.; Ishibashi, F. & Fujita, Y. (2005) Isolation and structure determination of algicidal compounds from *Ulva fasciata*. *Bioscience, Biotechnology and Biochemistry* 69, 2186-2192.

Alamsjah, M. A.; Hirao, S.; Ishibashi, F.; Oda, T. & Fujita, Y. (2007) Algicidal activity of polyunsaturated fatty acids derived from *Ulva fasciata* and *U. pertusa* (Ulvaceae, Chlorophyta) on phytoplankton. *Journal of Applied Phycology* 20, 713-720.

An, Z.; Wang, Z.; Li, F.; Tian, Z. & Hu, H. (2008) Allelopathic inhibition on red tide microalgae *Skeletonema costatum* by five microalgal extracts. *Frontiers of Environmental Science & Engineering in China* 2, 297-305.

Apstein, C. (1896) *Das Sübwwasserplankton*. Kiel, Leipzig, Germany.

Arunkumar, K.; Sivakumar, S. R. & Rengasamy, R. (2010) Review on bioactive Potential in seaweeds (marine macroalgae): a special emphasis on bioactivity of seaweeds against plant pathogens. *Asian Journal of Plant Sciences* 9, 227-240.

Arzul, G.; Erard-Le Denn, E.; Videau, C.; Jegou, A. M. & Gentien, P. (1993) Diatom growth repressing factors during an offshore bloom of *Gyrodinium* cf. *aureolum*. In: *Toxic Phytoplankton Blooms in the Sea*, T. J. Smayda & Y. Shimizu (Eds.), pp. 719-724, Elsevier, Amsterdam.

Arzul, G.; Gentien, P.; Bodennec, G.; Toularastel, F.; Youenou, A. & Crassous, M. P. (1995) Comparison of toxic effects of *Gyrodinium* cf. *nagasakiense* polyunsaturated fatty acids. In: *Harmful Marine Algal Blooms*, P. Lassus, G. Arzul, E. Erard, P. Gentien & C. Marcaillou (Eds.), Lavoisier, Paris.

Baskin, T. I. & Wilson, J. E. (1997) Inhibitors of protein kinase and phosphatase alter root morphology and disorganize cortical microtubules. *Plant Physiology* 113, 493-502.

Becher, P. G.; Beuchat, J.; Gademann, K. & Jüttner, F. (2005) Nostocarboline: isolation and synthesis of a new cholinesterase inhibitor from *Nostoc* 78-12A. *Journal of Natural Products* 68, 1793-1795.

Berry, J. P.; Gantar, M.; Perez, M. H.; Berry, G. & Noriega, F. G. (2008) Canobacterial toxins as allelochemicals with potential applications as algaecides, herbicides and insecticides. *Marine Drugs* 15, 117-146.

Blom, J. F.; Brütsch, T.; Bethuel, B. D.; Locher, H. H.; Hubschwerien, C. & Gademann, K. (2006) Potent algicides based on the cyanobacterial alkaloid nostocarboline. *Organic Letters* 16, 737-740.

Cardozo, K. H. M.; Guaratini, T.; Barros, M. P.; Falcão, V. R.; Tonon, A. P.; Lopes, N. P.; Campos, S.; Torres, M. A.; Souza, A. O.; Colepicolo, P. & Pinto, E. (2007)

Metabolites from algae with economical impact. *Comparative Biochemistry and Physiology, Part C* 146, 60-78.

Cantillo-Ciau, Z.; Moo-Puc, R.; Quijano, L. & Freile-Pelegrin, Y. (2010) The tropica brown alga *Lobophora variegata*: a source of antiprotozoal compounds. *Marine Drugs* 16, 1291-1304.

Christiansen, G. E.; Dittmann, L. V., Ordorika, R.; Rippka, M; Herdman, T.; Börner, T. (2001) Nonribosomal peptide synthetase genes occur in most cyanobacterial genera as evidenced by their distribution in axenic strains of the PCC. *Archives of Microbiology* 176, 452-458.

Christiansen, G.; Fastner, J.; Erhard, M.; Börner, T. & Dittman, E. (2003) Microcystin biosynthesis in *Planktothrix*: genes, evolution and manipulation. *Journal of Bacteriology* 185, 564-572.

Codd, G. A.; Bell, S. G.; Kaya, K.; Ward, C. J.; Beattie, K. A. & Metcalf, J. S. (1999) Cyanobacterial toxins, exposure routes and human health. *European Journal of Phycology* 34, 405-415.

Copping, L. G. & Duke, S. O. (2007) Natural products that have been used commercially as crop protection agents. *Pest Management Science* 63, 524-554.

Dayan, F. E.; Cantrell, C. L. & Duke, S. O. (2009) Natural products in crop protection. *Biorganic & Medicinal Chemistry* 17, 4022-4034.

Dittmann, E.; Neilan, B. & Börner, T. (2001) Molecular biology of peptide and polyketide biosynthesis in cyanobacteria. *Applied Microbiology and Biotechnology* 57, 467-473.

Doan, T. N.; Rickards, R. W.; Rothschild, J. M. & Smith, G. D. (2000) Allelopathic actions of the alkaloid 12-*epi*-hapalindole E isonitrile and calothrixin A from cyanobacteria of the genera *Fischerella* and *Calothrix*. *Journal of Applied Phycology* 12, 409-416.

Doan, N. T.; Stewart, P. R. & Smith, G. D. (2001) Inhibition of bacterial RNA polymerase by the cyanobacterial metabolite 12-*epi*-hapalindole E isonitrile and calothrixin. *FEMS Microbiology Letters* 196, 135-139.

Duke, S. O.; Dayan, F. E.; Rimando, A. M.; Schrader, K. K.; Aliotta, G.; Oliva, A. & Romagni, J. (2002) Chemicals from nature for weed management. *Weed Science* 50, 138-151.

Eitsuka, T.; Nakagawa, K.; Igarishi, M. & Miyazawa, T. (2004) Telomerase inhibition by sulfoquinovosyldiacylglycerol from edible purple laver (*Porphyra yezoensis*). *Cancer Letters* 212, 15-20.

Entzeroth, M.; Mead, D. J.; Patterson, G. M. L. & Moore, R. E. (1985) A herbicidal fatty acid produced by *Lyngbya aestuarii*. *Phytochemistry* 24, 2875-2876.

Erhard, D. & Gross, E. M. (2006) Allelopathic activity of *Elodea Canadensis* and *Elodea nuttallii* against epiphytes and phytoplankton. *Aquatic Botany* 85, 203-211.

Etchegaray, A.; Rabello, E.; Dieckmann, R.; Moon, D. H.; Fiore, M. F.; von Döhren, H.; Tsai, S. M. & Neilan, B. A. (2004) Algicide production by the filamentous cyanobacterium *Fischerella* sp. CENA 19. *Journal of Applied Phycology* 16, 237-243.

Fenical, W. & Paul, V. J. (1984) Antimicrobial and cytotoxic terpenoids from tropical green algae of the family Udoteaceae. *Hydrobiologia* 116-117, 135-140.

Fink, G. R. (1998) Anatomy of a revolution. *Genetics* 149, 473-477.

Flores, E. & Wolk, C. P. (1986) Production by filamentous, nitrogen fixing cyanobacteria of a bacteriocin and of other antibiotics that killed related strains. *Archives of Microbiology* 145, 215-219.

Forlani, G.; Pavan, M.; Gramek, M.; Kafarski, P. & Lipok, J. (2008) Biochemical bases for a widespread tolerance of cyanobacteria to the phosphonate herbicide glyphosate. *Plant and Cell Physiology* 49, 443-456.

Francisco, C.; Banaigs, B.; Valls, R.; Codomier, L. (1985) Mediterraneol a, a novel rearranged diterpenoid-hydroquinone from the marine alga *Cytoseira mediterranea*. *Tetrahedron Letters* 26, 2629-2632.

Friedlander, M.; Gonen, Y.; Kashman, Y. & Beer, S. (1996) *Gracilaria coferta* and its epiphytes: 3. Allelopathic inhibition of the red seaweed by *Ulva* cf. *lactuca*. *Journal of Applied Phycology* 8, 21-25.

Gantar, M.; Berry, J. P.; Thomas, S.; Wang, M.; Perez, R.; Rein, K. S. & King, G. (2008) Allelopathic activity among cyanobacteria and microalgae isolated from Florida freshwater habitats. *FEMS Microbiology Ecology* 64, 55-64.

Gerwick, W. H., Tan, L. T. & Sitachitta, N. (2001) Nitrogen-containing metabolites from marine cyanobacteria. *The Alkaloids: Chemistry and Biology* 57, 75-184.

Gleason, F. K. & Paulson, J. L. (1984) Site of action of the natural algicide, cyanobacterin, in the blue-green alga, *Synechococcus* sp. *Archives of Microbiology* 138, 273-277.

Gleason, F. K. & Case, D. E. (1986) Activity of the natural algicide, cyanobacterin, on angiosperms. *Plant Physiology* 80, 834-837.

Gleason, F. K.; Case, D. E.; Sipprell, K. D. & Magnuson, T. S. (1986) Effect of the natural algicide, cyanobacterin, on a herbicide-resistant mutant of *Anacystis nidulans* R2. *Plant Science* 46, 5-10.

Granéli, E. & Johansson, N. (2003) Increase in the production of allelopathic substances by *Prymnesium parvum* cells grown under N- or P-deficient conditions. *Harmful Algae* 2, 135-143.

Granéli, E.; Weberg, M. & Salomon, P. S. (2008) Harmful algal blooms of allelopathic microalgal species: the role of eutrophication. *Harmful Algae* 8, 94-102.

Gromov, B. V.; Vepritskii, A. A.; Titota, N. N.; Mamkayeva, K. A. & Alexandrova, O. V. (1991) Production of the antibiotic cyanobacterin LU-1 by *Nostoc linckia* CALU 892. *Journal of Applied Phycology* 3, 55-59.

Gross, E. M. (2003) Allelopathy in aquatic autotrophs. *Critical Reviews in Plant Sciences* 22, 313-339.

Gross, H.; Goeger, D.; Hills, P.; Mooberry, S. L.; Ballantine, D. L.; Murray, T. F.; Valeriote, F. A. & Gerwick, W. H. (2006) Lophocladines, bioactive alkaloids from the red alga *Lophocladia* sp. *Journal of Natural Products* 69, 640-644.

Gul, W. & Hamann, M. T. (2005) Indole alkaloid natural products: an established source of cancer drug leads with considerable promise for the control of parasitic, neurological and other diseases. *Life Science* 78, 442-453.

Gupta, A. B. & Agarwal, P. R. (1973) Extraction, isolation and bioassay of a gibberellin-like substance from *Phormidium foveolarum*. *Annals of Botany* 37, 737-741.

Güven, K. C.; Percot, A. & Sezik, E. (2010) Alkaloids in marine algae. *Marine Drugs* 8, 269-284.

Hanashima, S.; Mizushina, Y.; Ohta, K.; Yamazaki, T.; Sugawara, F. & Sakaguchi, K. (2000) Structure-activity relationship of a novel group of mammalian DNA polymerase inhibitors, synthetic sulfoquinovosylacylglycerols. *Japanese Journal of Cancer Research* 91, 1073-1083.

Hassan, S. M. & Ghareib, H. R. (2009) Bioactivity of *Ulva lactuca* acetone extracts on germination and growth of lettuce and tomato plants. *African Journal of Biotechnology* 8, 3832-3838.

Horrigan, L.; Lawrence, R. S. & Walker, P. (2002) How sustainable agriculture can address the environmental and human health harms of industrial agriculture. *Environmental Health Perspectives*, 110, 445-456.

Hussain, A.; Krischke, M.; Roitsch, T.; Hasnain, S. (2010) Rapid determination of cytokinins and auxin in cyanobacteria. *Current Microbiology*, in press.

Ikawa, M. (2004) Algal polyunsaturated fatty acids and effects on plankton ecology and other organisms. *UNH Center for Freshwater Biology Research* 6, 17-44.

Irfanullah, H. M. & Moss, B. (2005) Allelopathy of filamentous green algae. *Hydrobiologia* 543, 169-179.

Jäger, K.; Ördög, V.; Barnabás, B. (2005) Effect of cyanobacterial and microalgal biomass on anther culture response of wheat (*Triticum aestivum*). *Acta Agronomica Hungarica* 53, 99-107.

Jang, M.-H.; Ha, K. & Takamura, N. (2007) Reciprocal allelopathic responses between toxic cyanobacteria (*Microcystis aeruginosa*) and duckweed (*Lemna japonica*). *Toxicon* 49, 727-733.

Jin, Q. & Dong, S. L. (2003) Comparative studies on allelopathic effects of two different strains of *Ulva pertusa* on *Heterosigma akashiwo* and *Alexandrium tamarense*. *Journal of Experimental Marine Ecology and Biology* 293, 41-45.

Jin, Q.; Dong, S. & Wang, C. (2005) Allelopathic growth inhibition of *Prorocentrum micans* (Dinophyta) by *Ulva pertusa* and *Ulva linza* (Chlorophyta) in laboratory cultures. *European Journal of Phycology* 40, 31-37.

Jones, A. C.; Gu, L.; Sorrels, C. M.; Sherman, D. H. & Gerwick, W. H. (2009) New tricks from ancient algae: natural products biosynthesis in marine cyanobacteria. *Current Opinions in Chemical Biology* 13, 216-223.

Kakisawa, H.; Asari, F.; Kusumi, T.; Toma, T.; Sakurai, T.; Oohusa, T.; Hara, Y. & Chiharai, M. (1988) An allelopathic fatty acid from the brown alga *Cladosiphon okamuranus*. *Phytochemistry* 27, 731-735.

Kearns, K. D. & Hunter, M. D. (2000) Green algal extracellular products regulate antialgal toxin production in a cyanobacterium. *Environmental Microbiology* 2, 291-297.

Kearns, K. D. & Hunter, M. D. (2001) Toxin-producing *Anabaena flos-aquae* induces settling of *Chlamydomonas reinhardtii*, a competing motile alga. *Microbiol Ecology* 42, 80-86.

Keating, K. L. (1977) Allelopathic influences on blue-green sequence in a eutrophic lake. *Science* 196, 885-887.

Keating, K. L. (1978) Blue-green algal inhibition of diatom growth: transition from mesotrophic to eutrophic community structure. *Science* 199, 971-973.

Kim, M.-J.; Choi, J.-S.; Kang, S.-E.; Cho, J.-Y.; Jin, H.-J., Byung-Soo, C. & Hong, Y.-K. (2004) Multiple allelopathic activity of the crustose coralline alga *Lithophyllum yessoense* against settlement and germination of seaweed spores. *Journal of Applied Phycology* 16, 175-179.

Kishore, G. M. & Shah, D. M. (1988) Amino acid biosynthesis inhibitors as herbicides. *Annual Reviews in Biochemistry* 57, 627-663.

Kodani, S.; Imoto, A.; Mitsutani, A. & Murakami, M. (2002) Isolation and identification of the antialgal compound, harmane (1-methyl-b-carboline), produced by the algicidal bacterium, *Pseudomonas* sp. K44-1. *Journal of Applied Phycology* 14, 109-114.

Kubanek, J. & Hicks, M. K. (2005) Does the red tide dinoflagellate *Karenia brevis* use allelopathy to outcompete other phytoplankton? *Limnology and Oceanography* 50, 883-895.

Kurihara, H.; Tada, S.; Takahashi, K. & Hatano, M. (1996) Digalactosyldiacylglycerol suppression of inhibition by sulfoquinovosyldiacylglycerol of alpha-glucosidase. *Bioscience, Biotechnology and Biochemistry* 60, 932-933.

Lane, A. L.; Stout, E. P.; Hay, M. E.; Prusak, A. C.; Hardcastle, K.; Fairchild, C. R.; Franzblau, S. G.; Le Roch, K.; Prudhomme, J.; Aalbersberg, W. & Kubanek, J. (2007) Callophycoic acids and callophycols from the Fijian red alga *Callophycus serratus*. *Journal of Organic Chemistry* 72, 7343-7351.

Lascéve, E. & Gaugain, F. (1990) Effects of usnic acid on sunflower and maize plantlets. *Journal of Plant Physiology* 136, 723-727.

Leão, P. N.; Vasconcelos, M. T. & Vasconcelos, V. M. (2009) Allelopathy in freshwater cyanobacteria. *Critical Reviews in Microbiology* 35, 271-282.

Leão, P. N.; Pereira, A. R.; Liu, W.-T.; Ng, J.; Pevzner, P. A.; Dorrestein, P. C.; König, G. M.; Vasconcelos, V. M. & Gerwick, W. H. (2010) Synergistic allelochemicals from a freshwater cyanobacterium. *Proceedings of the National Academy of Sciences U. S. A.* 107, 11183-11188.

Leblanc, S.; Pick, F. R. & Aranda-Rodriguez, R. (2005) Allelopathic effects of the toxic cyanobacterium *Microcystis aeruginosa* on duckweed, *Lemna gibba*. *Environmental Toxicology* 20, 67-73.

Leflaive, J. & Ten-Hage, L. (2006) Algal and cyanobacterial secondary metabolites in freshwaters: a comparison of allelopathic compounds and toxins. *Freshwater Biology* 52, 199-214.

Legrand, C.; Rengefors, K.; Fistarol, G. O. & Granéli, E. (2003) Allelopathy in phytoplankton - biochemical, ecological and evolutionary aspects. *Phycologia* 42, 406-419.

Lindell, S. D.; Pattenden, L. C. & Shannon, J. (2009) Combinatorial chemistry in the agrosciences. *Bioorganic & Medicinal Chemistry* 17, 4035-4046.

Long, P. F.; Dunlap, W. C.; Battershill, C. N. & Jaspers, M. (2005) Shotgun cloning and heterologous expression of the patellamide gene cluster as a strategy to achieving sustained metabolite production. *Chembiochem* 6, 1760-1765.

Luesch, H.; Yoshida, W. Y.; Moore, R. E.; Paul, V. J.; Mooberry, S. L. & Corbett, T. H. (2001) Symplostatin 3, a new dolastatin 10 analogue from the marine cyanobacterium *Symploca* sp. VP 452.

Macías, F. A.; Galindo, J. L. G.; Garcia-Diaz, M. D. & Galindo, J. C. G. (2007) Allelopathic agents from aquatic ecosystems: potential biopesticides models. *Phytochemistry Reviews* 7, 155-178.

Malmvärn, A.; Zebühr, Y.; Kautsky, L.; Bergman, A. & Asplund, L. (2008) Hydroxylated and methoxylated polybrominated diphenyl ethers and polybrominated dibenzo-*p*-dioxins in red alga and cyanobacteria living in the Baltic Sea. *Chemosphere* 72, 910-916.

Manickavelu, A.; Nadarajan, N.; Ganesh, S. K.; Ramalingam, R., Raguraman, S. & Gnanamalar, R. P. (2006) Organogenesis induction in rice callus by cyanobacterial extracellular product. *African Journal of Biotechnology* 5, 437-439.

Mason, C. P.; Edwards, K. R.; Carlson, R. E., Pignatello, J.; Gleason, F. K. & Wood J. M. (1982) Isolation of a chlorine-containing antibiotic from the freshwater cyanobacterium, *Scytonema hofmanni*. *Science* 215, 400-402.

McCracken, M. D.; Middaugh, R. E. & R. S. Middaugh (1980) A chemical characterization of an algal inhibitor obtained from *Chlamydomonas*. *Hydrobiologia* 70: 271-276.

McElhiney, J.; Lawton, L. A. & Leifert, C. (2001) Investigations into the inhibitory effects of microcystins on plant growth, and the toxicity of plant tissues following exposure. *Toxicon* 39, 1411-1420.

Meinke, D. W.; Cherry, J. M.; Dean, C.; Rounsley, S. D. & Koornneef, M. (1998) *Arabidopsis thaliana:* a model plant for genome analysis. *Science* 282, 662-682.

Mitrovic, S. M.; Pflugmacher, S.; James, K. J. & Furey, A. (2004) Anatoxin-a elicits an increase in peroxidase and glutathione-S-transferase activity in aquatic plants. *Aquatic Toxicology* 10, 185-192.

Mitrovic, S. M.; Allis, O.; Furey, A. & James, K. J. (2005) Bioaccumulation and harmful effects of microcystin-LR in the aquatic plants *Lemna minor* and *Wolffia arrhiza* and the filamentous alga *Chladophora fracta*. *Ecotoxicology and Environmental Safety* 61, 345-352.

Mizushina, Y.; Kasai, N.; Iijima, H.; Sugawara, F.; Yoshida, H. & Sakaguchi, K. (2005) Sulfoquinovosylacylglycerol, a eukaryotic DNA polymerase inhibitor and anti-cancer agent. *Current Medicinal Chemistry and Anticancer Agents* 5, 613-625.

Moffitt, M. C. & Neilan, B. A. (2004) Characterization of the nodularin synthetase gene cluster and proposed theory of evolution cyanbacterial hepatotoxins. *Applied and Environmental Microbiology* 70, 6353-6362.

Molisch H. (1937) *Der Einflub einer Pflanze auf die andere Pflanze, Allelopathie* . Gustav Fischer, Jena, Germany.

Moore, R. E.; Cheuk, C. & Patterson, G. M. L (1984) Hapalindoles: new alkaloids from the blue-green alga *Hapalosiphon fontinalis*. *Journal of the American Chemical Society* 106, 6456-6457.

Moore, R. E. (1996) Cyclic peptides and depsipeptides from cyanobacteria: a review. *Journal of Industrial Microbiology* 16, 134-143.

Nagatsu, A.; Kajitani, H. & Sakakibara, J. (1995) Muscoride A: a new oxazole peptide alkaloid from freshwater cyanobacterium *Nostoc muscorum*. *Tetrahedron Letters* 36, 4097-4100.

Nan, C. R.; Zhang, H. Z. & Zhao, G. Q. (2004) Allelopathic interactions between the macroalga *Ulva pertusa* and eight microalgal species. *Journal of Sea Research* 52, 259-268.

Nan, C.; Zhang, H.; Lin, S.; Zhao, G. & Liu, X. (2008) Allelopathic effects of *Ulva lactuca* on selected species of harmful bloom-forming microalgae in laboratory cultures. *Aquatic Botany* 89, 9-15.

Ohta, K.; Mizushina, Y.; Hirata, N.; Takemura, M.; Sugawara, F.; Matsukage, A.; Yoshida, S. & Sakaguchi, K. (1998) Sulfoquinovosyldiacylglycerol, KM043, a new potent inhibitor of eukaryotic DNA polymerase and HIV-reverse transcriptase type 1 from a marine red alga, *Gigartina tenella*. *Chemical Pharmacology Bulletin* 46, 684-686.

Öztürk, S.; Güvenc, S.; Arikan, N.; Yilmaz, Ö (1999) Effect of usnic acid on mitotic index in root tips of *Allium cepa*. *Lagascalia* 21, 47-52.

Paul, V. J. & Fenical, W. (1984) Bioactive terpenoids from Caribbean marine algae of the genera *Penicillus* and *Udotea* (Chlorophyta). *Tetrahedron* 40, 2913-2918.

Pflugmacher, S.; Wiegand, C.; Beattie, K. A.; Krause, E., Steinberg, C. E. W. & Codd, G. A. (2001) Uptake, effects, and metabolism of cyanobacterial toxins in the emergent reed plant *Phragmites Australia*. *Environmental Toxicology and Chemistry* 20, 846-852.

Pflugmacher, S. (2002) Possible allelopathic effects of cyanotoxins, with reference to microcystin-LR, in aquatic ecosytems. *Environmental Toxicology* 17, 407-413.

Pflugmacher, S.; Aulhorn, M. & Grimm, B. (2007) Influence of a cyanobacterial crude extract containing microcystin-LR on the physiology and antioxidative defence systems of different spinach variants. *New Phytologist* 175, 482-489.

Potera, C. (2007) Florida Red Tide brews up drug lead for cystic fibrosis. *Science* 15, 1561-1562.

Powles, S. B.; Preston, C.; Bryan, I. B. & Jutsum, A. R. (1996) Herbicide resistance: impact and management. *Advances in Agronomy* 58, 57-93.

Powell, H. A.; Kerbby, N. W. & Rowell, P. (2006) Natural tolerance of cyanobacteria to the herbicide glyphosate. *New Phytologist* 119, 421-426.

Raveh, A.; Moshe, S.; Evron, Z.; Flescher, E. & Carmeli, S. (2010) Novel thiazole and oxazolee containing cyclic hexapeptides from a waterbloom of the cyanobacterium *Microcystis* sp. *Tetrahedron* 66, 2705-2712.

Rippka, R.; Deruelles, J.; Waterbury, J. B.; Herdman, M. & Stanier, R. Y. (1979) Generic assignments, strain histories and properties of pure cultures of cyanobacteria. *Journal of General Microbiology* 111, 1-61.

Rodriguez, A. A.; Stella, A. M.; Storni, M. M.; Zulpa, G. & Zaccaro, M. C. (2006) Effects of cyanobacterial extracellular products and gibberellic acid on salinity tolerance in *Oryza sativa*. *Saline Systems* 2, 7-11.

Saqrane, S.; El Ghazali, I.; Ouahid, Y.; El Hassni, M.; El Hadrami, I.; Bourarab, L.; del Campo, F. F.; Oudra, B. & Vasconcelos, V. (2007) Phytotoxic effects of cyanobacteria extract on the aquatic plant *Lemna gibba*: microcystin accumulation, detoxification and oxidative stress induction. *Aquatic Toxicology* 83, 284-294.

Saqrane, S.; El Ghazali, I.; Oudra, B.; Bouarab, L. & Vasconcelos, V. (2008) Effects of cyanobacteria producing microcystins on seed germination and seedling growth of several agricultural plants. *Journal of Environmental Science and Health B* 43, 443-451.

Schmidt, E. W.; Nelson, J. T.; Rasko, D. A.; Sudek, S.; Elsen, J. A.; Haygood, M. G. & Ravel, J. (2005) Patellamide A and C biosynthesis by a microcin-like pathway in *Prochloron didemni*, the cyanobacterial symbiont of *Lissoclinum patella*. *Proceedings of the National Academy of Sceinces U.S.A.* 102, 7315-7320.

Sanevas, N. Sunohara, Y. & Matsumoto, H. (2006) Crude extract of the cyanobacterium, *Hapalosiphon* sp., causes a cessation of root elongation and cell division in several species. *Weed Biology and Management* 6, 25-29.

Sergeeva, E.; Liaimer, A. & Bergman, B. (2002) Evidence for production of phytohormone indol-3-acetic acid by cyanobacteria. *Plant* 215, 229-238.

Schlegel, I.; Doan, N. T.; de Chazal, N. & Smith, G. D. (1999) Antibiotic activity of new cyanobacterial isolates from Australia and Asia against green algae and cyanobacteria. *Journal of Applied Phycology* 10, 471-479.

Smith, C. D.; Zhang, Z.; Mooberry, S. L.; Patterson, G. M. L. & Moore, R. E. (1994) Cryptophycin: a new microtubule agent active against drug-resistant cells. *Cancer Research* 54, 3779-3784.

Smith, G. D. & Doan, N. T. (1999) Cyanobacterial metabolites with bioactivity against photosynthesis in cyanobacteria, algae and higher plants. *Journal of Applied Phycology* 11, 337-344.

Snyder, R. V.; Gibbs, P. D.; Palacios, A.; Ably, L.; Dickey, R.; Lopez, J. V. & Rein, K. S. (2003) Polyketide synthase genes from marine dinoflagellates. *Marine Biotechnology* 5, 1-12.

Srivastava, A.; Jüttner, F. & Strasser, R. J. (1998) Action of the allelochemical, fischerellin A, on photosystem II. *Biochimica et Biophysica Acta* 1364, 326-336.
Stirk, W.A.; Ördög, V. & Van Staden, J. (1999) Identification of the cytokinin isopentenyladenin in a strain of *Arthronema africanum*. *Journal of Phycology*, 35 89-92.
Stirk, W. A.; Ördög, V.; Van Staden, J. and Jäger, K. (2002) Cytokinin- and auxin-like activity in cyanophyta and microalgae. *Journal of Applied Phycology* 14, 215-221.
Stirk, W. A.; Novak, O.; Strnad, M. & van Staden, J. (2003) Cytokinins in macroalgae. *Plant Growth Regulation* 41, 13-24.
Sugg, L. M. & Van Dolah, F. M. (1999) No evidence for allelopathic roles of okadaic acid among ciguatera-associated dinoflagellates. *Journal of Phycology* 35, 93-103.
Sun, H. H.; Ferrara, N. M.; McConnell, O. J. & Fenical, W. (1980) Bifurcarenone, an inhibitor of mitotic cell division from the brown alga. *Tetrahedron Letters* 3123-3126.
Tan, L. T. (2007) Bioactive natural products from marine cyanobacteria for drug discovery. *Phytochemistry* 68, 954-979.
Trimurtulu, G.; Ohtani, I.; Patterson, G. M. L.; Moore, R. E.; Corbett, T. H.; Valeriote, F. A. & Demchik, L. (1994) Total structures of cryptophycins, potent antitumor depsipeptides from the blue-green alga *Nostoc* sp. strain GSV 224. *Journal of the American Chemical Society* 116, 4729-4737.
Tsavkelova, E. A.; Klimova, S. Y.; Cherdyntseva, T. A. & Netrusov, A. I. (2006) Microbial producers of plant growth stimulators and their practical use: a review. *Applied Biochemistry and Microbiology* 42, 117-126.
Upadhyaya, M. K. & Noodén, L. D. (1977) Uptake and binding of [^{14}C]oryzalin in corn root segments. *Plant Physiol.* 59, 5-46.
Upadhyaya, M. K. & Noodén, L. D. (1980) Mode of dinitroaniline herbicide action. *Plant Physiology* 66, 1048-1052.
Uchida, A.; Shimada, A.; Ishida, Y. (1988) Antibacterial and antialgal substance produce by the dinoflagellate *Peridinium bipes*. *Nippon Suisan Gakkaishi* 54, 1941-1945.
Van Dolah, F. M. (2000) Marine algal toxins: origins, health effects, and their increased occurrence. *Environmental Health Perspectives* 108, 133-141.
Vepritskii, A. A.; Gromov, B. V.; Titota, N. N. & Mamkaeva, K. A. Production of the antibiotic-algicide cyanobacterin LU-2 by a filamentous cyanobacterium *Nostoc* sp. *Mikrobiologiia* 60, 21-25.
Verdier-Pinard, P.; Lai, J.-Y.; Yoo, H.-D.; Yu, J.; Marquez, B.; Nagle, D. G.; Nambu, M.; White, J. D.; Falck, J. R.; Gerwick, W. H.; Day, B. W. & Hamel, E. (1998) Structure-activity analysis of the interaction of curacin A, the potent colchicines site antimitotic agent, with tubulin and effects of analogs on the growth of MCF-7 breast cancer cells. *Molecular Pharmacology* 53, 62-76.
Volk, R. B. (2005) Screening of microalgal culture media for the presence of algicidal compounds and isolation and identification of two bioactive metabolites, excreted by the cyanobacteria *Nostoc insulare* and *Nodularia harveyana*, respectively. *Journal of Applied Phycology* 17, 339-347.
Volk, R. B. (2008) Screening of microalgae for species excreting norharmane, a manifold biologically active indole alkaloid. *Microbiology Research* 163, 307-313.
Wang, R.; Xiao, H.; Zhang, P.; Qu, L.; Cai, H. & Tang, X. (2007) Allelopathic effects of *Ulva pertusa*, *Corallina pilulifera* and *Sargassum thunbergii* on the growth of the dinoflagellates *Heterosigma akashiwo* and *Alexandrium tamarense*. *Journal of Applied Phycology* 19, 109-121.

Wang, Y.; Bin, Z. & Xuexi, T. (2009) Effects of two species of macroalgae - *Ulva pertusa* and *Gracilaria lemaneiformis* - on growth of *Heterosigma akashiwo* (Raphidophyceae). *Journal of Applied Phycology* 21, 375-385.

Weiss, J.; Libert, H.-P. & Braune, W. (2000) Influence of microcystin-RR on growth and photosynthetic capacity of the duckweed *Lemna minor*. *Journal of Applied Botany* 74, 100-105.

Welker, M. & von Döhren, H. (2006) Cyanobacterial peptides - nature's own combinatorial biosynthesis. *FEMS Microbiology Reviews* 30, 530-563.

Yamada, N.; Murakami, N.; Morimoto, T. & Sakakibara, J. (1993) Autogrowth inhibitory substances from the freshwater cyanobacterium *Phormidium tenue*. *Chemical Pharmaceutical Bulletin* 41, 1863-1865.

Yi, D.; Yijun, Z.; Xue, B.; Zhihui, F. & Kai, C. (2009) Cytotoxic effects of cyanobacteria extract on *Lemna minor* and *Myriophyllum spicatum* phyto-tolerance and superoxide dismutase activity. *Environmental Toxicology* 24, 304-308.

Zaccaro, M. C.; Kato, A.; Zulpa, G.; Storni, M. M.; Steyerthal, N.; Lobasso, K & Stella, A. M. (2006) Bioactivity of *Scytonema hofmanni* (Cyanobacteria) in *Lillium alexandrae in vitro* propogation. *Electronic Journal of Biotechnology* 9, 211-214.

Zhang, T. T.; He, M.; Wu, A. P. & Nie, L. W. (2009) Allelopathic effects of submerged macrophyte *Chara vulgaris* on toxic *Microcystis aeruginosa*. *Allelopathy Journal* 23.

Zulpa de Caire, G.; Zaccaro De Mulé, M. C..; Storni De Cano, M. (1979) Productos extracelulares de *Nostoc muscorum* obtenidos en medios con y sin nitrógenocombinado. I: Sus efectos sobre plantulas de arroz. *Phyton* 37, 1-13.

36

Herbicidal Potentiality of Fusel Oil

Andréa Aparecida de Padua Mathias Azania[1],
Carlos Alberto Mathias Azania[1], Marcos Omir Marques[2]
and Maria do Carmo Morelli Damasceno Pavani[2]
[1]*Instituto Agronômico, Centro de Cana, Ribeirão Preto,*
[2]*Faculdade de Ciências Agrárias e Veterinárias, Jaboticabal,*
Brasil

1. Introduction

During the process of manufacturing sugar, various by-products are generated. Some, such as bagasse and vinasse, which are generated in large quantities, have the potential for use as fertilizers or soil conditioners and thus for reducing the environmental impact of sugar processing. Others, because they are produced in smaller quantities, are mixed with other materials and taken to the field or marketed as raw material for other processes, as is done with fusel oil.

In alcohol distilleries, fusel oil is a by-product produced during alcohol fermentation and is taken from the rectifying columns. Average production ranges from 0.1 to 0.5 liters per 100 liters of alcohol produced. Although there is no mention of its role in agriculture or distilleries, it has commercial value and is usually marketed according to the amount of isoamyl alcohol present.

Generally, the use of fusel oil is restricted to perfume industries, where it can be employed as a fixative in cosmetics manufacturing, and for preparing artificial flavors or flavoring.

A study from Azania (2003) suggested the possible use of this byproduct as a plant desiccant. When applied directly into the soil of pots planted with sugarcane, the plants wilted, progressively dried out and later died.

It is worth noting that up to this time, this work was unique, given the lack of studies using this approach.

2. Utilization of by-products in mills and distilleries

Reducing the cost of operations in the area of sugarcane reform can be substantial when by-products generated in the manufacturing of sugarcane are employed. Vinasse is the best known by-product in terms of reuse. Due to its pollutant potential and the large quantity in which it is produced, where vinasse is utilized merits attention.

There are several alternatives for vinasse reuse, but none is used to the same magnitude as fertigation. Fertigation of sugarcane may improve the physical, chemical and biological properties of the soil (Ferreira & Monteiro, 1987). Moreover, through fertigation, one can fully or partially replace mineral fertilization, increase the availability of macro and micronutrients and raise the pH of the soil. These effects, along with increased productivity

of sugarcane, may also contribute to an increase in the overall production of grains on former cane plantations.
Another by-product of manufacturing alcohol is flegmass, which is mixed with vinasse and applied in the field despite its effects on soil not being fully known. Flegmass is obtained from the rectifying column of phlegm during the process of alcohol production.
Fusel oil, another byproduct of this process, is a result of alcohol redistillation. It consists of a mixture of high-grade alcohols, usually stored in separate tanks. It is often used as a raw material with low commercial value for chemical industries recovering isoamyl alcohol, whose reuse in agriculture was not mentioned in the literature until a decade ago.
When a sufficient amount of by-product is accumulated in the storage tanks, it is sold to cosmetics or perfume companies or as a solvent, hydraulic fluid or defoamer (Patil et al., 2002). For Vauclair et al. (1997), the presence of high-grade alcohols increases the commercial value of fusel oil. However in Brazil, the sale price depends almost exclusively on the amount of amyl alcohol present, which is the largest constituent of fusel oil.
According to Nascimento et al. (2003), for each 1000 L of ethanol produced, about 2.5 L are fusel oil. Taking this estimate and the Brazilian production of 18 billion liters of alcohol in the harvest of 2005/2006 into account, approximately 42.5 million gallons of fusel oil were produced. According to data from Agrianual (2008), total production of ethanol from sugarcane in Brazil in the 2007/2008 harvest was approximately 21.3 billion gallons, which means that roughly 53.2 million liters of fusel oil were produced.
The data show that the trend for ethanol production is always increasing, and the production of fusel oil follows this trend as well. In the most current calculation by Conab (2010), the 2010/2011 crop will produce 28.5 billion liters of ethanol, which generates approximately 71 million gallons of fusel oil without direct use by production entities.
In an unpublished study, Azania (2003) mentioned a possible desiccant potential of fusel oil, indicating a possible use in the fields for chemical eradication of sugarcane to renew the sugarcane plantations because the product has contact action: it destroys the tissues of the plant surface. The only products that have contact action are acids, salts, oils and detergents (Deuber, 2003). Many oils can be highly phytotoxic, and some have been used as herbicides (Kissmann, 1985). This contact action is a result of the oil breaking down the tissues of the leaf surface and destroying the cells.
If studies demonstrate the economic feasibility of using fusel oil for weed control or eradication of sugarcane, this would be an option to reduce the expense of using certain herbicides, as it costs very little to produce fusel oil in the distilleries.

3. General characteristics of fusel oil

In general, fusel oil is a dark-colored liquid that may also have yellow and green tones. Its odor is unpleasant, distinctive and responsible for the characteristic taste and aroma of aguardentes (Lima, 1964). It is produced during the processing of amino acids by yeasts during fermentation in distilleries, and its components are high-grade alcohols (Botelho, 1945). According to Rasovsky (1973), its specific weight is 0.83, with a boiling point between 75 and 134°C. Fusel oil is flammable and burns with a bright blue flame. It mixes easily with other products such as alcohol, chloroform and ether.
Fusel oil is composed of a mixture of alcohols such as ethyl, amyl, isoamyl, propyl and butyl. According to Almazan et al. (1998), amyl alcohol, isoamyl alcohol, n-butanol and other compounds can be separated by another distillation process, which has some

economic advantages. However, this other distillation process is performed not by the distilleries, but by the companies that buy the product to market it. Besides the compounds listed above, Nascimento et al. (2003) report that sec-butanol, esters, alkanes and terpenes may also be present. Once separated and purified, these substances are useful in the preparation of artificial flavors or flavoring. Souza & Llistó (1978) reported that fusel oil may also contain hexyl, heptyl and octyl alcohols in small quantities.

However, the main constituent of fusel oil is isoamyl alcohol (Vauclair et al. 1997; Kuçuk & Ceylan, 1998; Pérez et al. 2001; Ceccato-Antonini & Silva, 2002; Azania, 2003, Nascimento et al. 2003), which comes from the decomposition of iso-leucine, an amino acid derived from the hydrolysis of proteins in yeast (Codistil, 1978). According to Nascimento et al. (2003), isoamyl alcohol has been investigated by chemical industries as a reagent in organic synthesis or as a solvent in the extraction of pharmacological compounds such as esters. According to Rasovsky (1973), fusel oil mostly consists of methanol, superior alcohols in the series of greases, isobutyl and propyl alcohols and 8-10% ethyl alcohol. Therefore, it is believed that the composition of fusel oil is dependent on the raw material used, the fermentation conductions and the process of distillation and decanting of the oil (Rasovsky, 1973).

In Brazil, the raw material used is sugarcane. In other countries, other types of raw materials also produce fusel oil as a byproduct. Accordingly, Brinker (2000) noted that alcohol distilled from potatoes has a greater amount of fusel oil, which is almost pure isoamyl alcohol, whereas alcohol produced from grain has more fusel oil than alcohol distilled from grapes. Grain fusel oil consists mainly of isoamyl alcohol, whereas the grapes have more butyl alcohol and other volatile fatty acids.

Table 1 describes the main components of fusel oil according to analysis using gas chromatography. Analyses performed in two different years showed higher percentages of isoamyl alcohol, followed by ethanol and butanol as principal products.

Components	Fusel oil (%)	
	2003	2007
Ethanol	11.70	14.46
n-propanol	0.83	1.69
i-butanol	8.47	10.30
n-butanol	0.21	0.67
i-amyl	28.66	61.60
n-amyl	0.12	0.49

Table 1. Chemical composition of fusel oil, analyzed by gas chromatography. (Azania, 2003 and 2007)

The fusel oil used by Azania (2003 and 2007) was from the same distillery but was collected in different seasons, which shows that its composition may vary according to time collected, method of conducting the process and characteristics of the raw material (Rasovsky, 1973). Kuçuk & Ceylan (1998) also showed this order of products in terms of percentage, but their results differ, as all of the values were lower than those of this work.

In the process of fusel oil production, Rasovsky (1973) states that the phase of oil extracted from the column is of paramount importance, without which there will not be a high quality product. To have properly-washed oil of good market value, one should measure the alcohol content of wash water, which should range between 8 and 12° GL (Codistil, 1978).

The most widely used system is the "settle-wash," in which crude oil is washed countercurrently with water, leaving the alcohol that was carried and deposited spontaneously, having then been extracted. When the fusel oil is not taken from the column, it is taken away by either flegmass or its own alcohol, which inevitably causes a fall in temperature at the base of the column (Codistil, 1978).
By presenting a difference in the type of raw materials and processes used, the quantity of fusel oil produced varies greatly between producing units. In Brazilian literature, two references state that 2.5 L of fusel oil are produced for every 1000 L of ethanol produced (Nascimento et al. 2003) or between 0.1 to 0.5 L per 100 liters of alcohol (Codistil, 1978).
To understand the technological differences between countries, see Turkey as an example, where 5 liters of alcohol are produced for every 1,000 liters of alcohol distillation, and approximately 30 million gallons of fusel oil are produced per year (Kuçuk & Ceylan, 1998) from beets.

4. Results from research on the applications of fusel oil in agriculture

4.1 Application of fusel oil in sugarcane and effects on the soil

Until recently, there was no mention of fusel oil use in agriculture, for this potential was first described by Azania (2003) to test different products derived from alcohol (vinasse, flegmass and fusel oil) and their impacts on soil fertility in sugarcane plants and weeds. Sugarcane plants treated with fusel oil at four concentrations (12.5, 25.0, 50.0 and 100.0% v/v) applied at a volume corresponding to 150 m^3 ha^{-1} wilted in a progressive and irreversible process of drying of the leaves and branches in the first 24 hours after application.
One of the conclusions of this study pointed to a possible desiccant potential of the by-product toward the plants because the drying and death of the sugarcane plant was observed a short time after application of the by-product. In the soil, the chemical attributes were altered with increased levels of calcium, acid potential and aluminum.

4.2 Effects of fusel oil application in *Sida rhombifolia* and *Brachiaria decumbens*

The study was conducted under laboratory conditions by Azania et al. (2003) to evaluate the germination of *Sida rhombifolia* and *Brachiaria decumbens* after applications of vinasse, flegmass and fusel oil. These by-products at concentrations of 12.5, 25.0, 50.0 and 100.0% (v/v) and controls (water with pH and osmolality adjusted depending on the characterization of the by-products and their dilutions) were applied directly to 100 seeds in plastic boxes using paper as a substrate.
The seeds of *Sida rhombifolia* that were treated with fusel oil did not germinate and had reduced viability, especially with higher applications. *Brachiaria decumbens* seeds that were treated with a higher concentration of flegmass had reduced viability and germination speed index. In the presence of fusel oil, *Brachiaria decumbens* seeds did not germinate and were completely unviable.
In a similar work, Azania et al. (2004) evaluated the effects of fusel oil, compared to flegmass and vinasse, on the growth of plants (*Sida rhombifolia* and *Brachiaria decumbens*) grown simultaneously in the greenhouse. Concentrations of 12.5, 25.0, 50.0 and 100.0% (v/v) of each by-product and the control (water) were applied (at a rate equivalent to 150 m^3 ha^{-1}) in pots (22 L) containing 100 seeds of each weed. Fusel oil inhibited the emergence of *Sida rhombifolia* and *Brachiaria decumbens*. Vinasse and flegmass reduced the emergence and development of *B. decumbens* and *S. rhombifolia*.

4.3 Mixing fusel oil in the spray solution

After publication of the first results, several studies were conducted to understand better the way that by-product behaves, its characteristics and the application rates that were most effective in the eradication of sugarcane and weeds. However, one of the problems encountered during this research was how to blend the fusel oil spray because the product does not homogenize easily with water.

To remedy this problem, work was undertaken by Azania (2007), whose objective was to study the effectiveness of surfactants and/or adhesive spreaders and alcohol on the stability of the spray solution, which consisted of fusel oil and water. A range of products already used for this purpose were tried.

The results indicated that of all the products tested, a non-ionic wetting agent, trademarked Energic (226 g L of nonyl phenoxy poly (ethyleneoxy) ethanol + 226 g L of sodium salt of dodecyl benzene sulfonic acid to 548 g inert L) and a neutral detergent were those that presented the best results and were able to maintain the emulsion for 30 minutes without separation. Alcohol was unfeasible due to the amount that would be needed to achieve good results.

4.4 Application of fusel oil for the eradication of sugarcane plants

While Azania (2003) found a possible desiccant potential of fusel oil in sugarcane, the tests were performed by placing different concentrations of product directly into the potted soil. When used with manual spray application with the appropriate bar and nozzles, the isolated product was not satisfactory for use as a desiccant in any of the tested concentrations because the effects ranged from mild to moderate and some plants recovered (Azania, 2007).

When the product hit the leaves, a yellow color appeared in the location of the drops in just a few hours, and the tissue generally began wilting and posterior necrosis. In this study, were used XR Teejet 110.02 VS nozzles for the preliminary results. However, the results were not satisfactory, so another application was made, this time with TT 110.02 nozzles, which provided better spreading of the product.

Based on previous research results, Azania (2007) evaluated the effect of fusel oil, glyphosate and a mixture of the two products in the eradication of sugarcane. Due to the lack of knowledge about the use of fusel oil in agriculture, it became necessary to test different dosages compared to glyphosate. Dosages for fusel oil application were determined by the cost of the individual products, so that the final cost would not exceed that of the glyphosate application (4 L ha^{-1} commercial dose) usually used by farmers for sugarcane eradication.

The experiment was conducted in 22 L pots maintained until 60 days after treatment (DAT). The treatments were (1) 4 L ha^{-1} glyphosate; (2) 150 L ha^{-1} fusel oil; (3) glyphosate 2.5 + 6.25 L ha^{-1} fusel oil; (4) glyphosate 2.0 + 25.0 L ha^{-1} fusel oil; (5) glyphosate 1.5 + 43.75 L ha^{-1} fusel oil; (6) glyphosate 1.0 + 62.50 L ha^{-1} fusel oil; (7) 100 L ha^{-1} fusel oil; (8) glyphosate 1.50 + 18.75 L ha^{-1} fusel oil; (9) glyphosate 1.0 + 37.50 L ha^{-1} fusel oil; (10) 75.0 L ha^{-1} fusel oil; (11) glyphosate 1.0 + 12.50 L ha^{-1} fusel oil; and (12) 50.0 L ha^{-1} fusel oil applied with pressurized backpack equipment, fitted with four Teejet 110.02 Turbo TT nozzles, at a pressure of 30 lb $in.^2$ (2.1 kgf cm^2) and spray volume of 200 L ha^{-1}. Also, 0.2% neutral detergent was added to all treatments involving fusel oil to facilitate mixing.

For application, the developmental stage of the culture was considered, which was an average of 28 cm in height measured from the first visible ligule to the ground. Moreover, the plants were in full vegetative development and not subjected to any kind of stress.
The fusel oil alone did not promote drying of sugarcane. In treatments with a mixture of glyphosate and fusel oil, the injuries were not as severe initially but were enough to dry out the main tiller until 60 DAT. Another important factor was the sprouting of suckers, which may mean that the results were not satisfactory from an agronomic point of view, as the expected result would be that the whole plant suffered stress resulting in the desiccation process. Overall, the soil chemical properties did not suffer alterations with the application of glyphosate, fusel oil or a mixture of the two.

4.5 Application of fusel oil in late post-emergence in a natural community of weeds

Azania et al. (2008) evaluated the effectiveness of fusel oil applied singly or in combination with desiccant herbicides during the late post-emergence stage in a natural community of weeds.
The experiment was conducted in the field in 3x3 m plots. The least satisfactory results of the experiment were those in which only the fusel oil was applied. In the treatments (1.50 + 43.75, 1.00 + 62.50, 1.50 + 18.75, 1.00 + 37.5, 1.00 + 12.5 L ha^{-1} glyphosate and fusel oil, respectively) that contained glyphosate, fusel oil allowed a reduction of the herbicide dosage and provided maximum control over a natural community of weeds consisting of *Amaranthus spp.*, *Brachiaria spp.* and *C. echinatus*, but not *Commelina* and *Cyperus* genera. These species, even with yellow spots, showed tolerance to the products and did not desiccate (Figure 1).

4.6 Application of fusel oil for weed management in no-till farming

Osipe et al. (2009) proved the effectiveness of fusel oil for weed management in no-till systems, for they obtained positive results in controlling *Digitaria insularis* and *Commelina benghalensis*.
The authors concluded that fusel oil was effective in controlling weeds in the doses tested (220, 440, 660 and 880 L ha^{-1}) and that it has the potential to become a product for use in no-till management. However, they stressed that further studies are necessary to design an application technique that reduces the volume of product applied per hectare.

4.7 Eradication of sugarcane with application of fusel oil spray and soil treatment and evaluation of subsequently-planted sunflowers

The eradication of sugarcane ratoon was also studied by Azania et al. (2010a) to test different concentrations of fusel oil which were pulverized and applied to the soil, as well as the development of subsequently-planted sunflowers.
The rationale for performing this work was that in the reform of a sugarcane plantation, the use of fusel oil could be another option for producers, which consequently would have economic and environmental benefits because there would be less application of desiccant herbicides. However, the use of fusel oil should be studied further because the residue only highlights its desiccant potential.
The planting of sugarcane, which is costly and consumes considerable amounts of herbicides that may even harm the environment, could benefit from the use of this

byproduct. From this standpoint, Pérez et al. (2001) thought that the low commercial value of fusel oil coupled with high production volume of crops are factors that should encourage proposals to develop technology to use fusel oil. Consequently, the choice of desiccant herbicides for of sugarcane reform is essential for the selection of both the sugarcane and the species that can be cultivated after eradication of the ratoon.

The cultivation of legumes (Bolonhezi, 2007), sunflowers and peanuts (Ramos et al., 2009) is common after ratoon eradication and before new sugarcane planting. However, Pires et al. (2003) stated that it is essential that there be no herbicide residues that are selective for these plants in the soil to ensure the development of new plants.

Sunflowers are cultivated to obtain green mass to be incorporated into the soil to improve the physical and chemical properties. In the case of fusel oil, it is interesting to identify portions that enable the development of sunflowers or other crops after eradication of sugarcane ratoons.

Therefore, to highlight the use of fusel oil, the objective was to test the hypothesis that fusel oil applied by spraying or directly into the soil promotes the eradication of the ratoon without harming the development of successive sunflower plantings.

The authors found that fusel oil applied by spraying the sugarcane did not eradicate the ratoon. However, applying fusel oil directly to the soil was effective in eradication of ratoons, and it did not alter the chemical attributes of the soil that are essential for plant development (except phosphorus values, which were less than those of the control due to the 0.1 m^3 ha^{-1} of fusel oil). It also did not jeopardize sunflower development afterward. The failure of fusel oil to alter the majority of soil chemical properties, which was found in the work of Azania (2003 and 2007) and corroborated by the results of this research, is a positive signal to continue this type of research. One should not forget that agro-industrial residues such as fusel oil are generally chemically unbalanced and should be applied in the environment with caution.

4.8 Efficacy of herbicide management in association with fusel oil on species of weeds

In a more recent publication on the subject, Pizzo et al. (2010) warned that it is common for producers to mix different herbicides directly into the same spray tank in the field, with standard formulations for two or more molecules, according to Rodrigues & Almeida (2005). However for technical purposes, studies that demonstrate the efficiency of control of two or more herbicides mixed in a single application are important for the subsequent possibility of registering new standard formulations.

Considering the technical need to verify the effectiveness of the mixtures made directly in the tank, the herbicide potential and low commercial value of fusel oil, Pizzo et al. (2010) evaluated the efficiency of herbicide control in combination with fusel oil on *Panicum maximum*, *Amaranthus deflexus*, *Ipomoea quamoclit*, *Euphorbia heterophylla* and *Brachiaria decumbens*.

The treatments consisted of: diuron + hexazinone (1170 + 330 g ha^{-1}); diuron + hexazinone (1170 + 330 g ha^{-1}) + fusel oil (25.0 L ha^{-1}); diuron + hexazinone (819 + 231 g ha^{-1}) + fusel oil (25.0 L ha^{-1}); metribuzin (1920 g ha^{-1}); metribuzin (1920 g ha^{-1}) + fusel oil (25.0 L ha^{-1}); metribuzin (1344 g ha^{-1}) + fusel oil (25.0 L ha^{-1}); amicarbazone (1400 g ha^{-1}); amicarbazone (1400 g ha^{-1}) + fusel oil (25.0 L ha^{-1}); amicarbazone (980 g ha^{-1}) + fusel oil (25.0 L ha^{-1}) and no

herbicide. The treatments were applied with a pressurized backpack sprayer at a volume corresponding to 250 L ha-1 in post-crop emergence (30 cm) and weeds (up to 20 cm).
The herbicides diuron + hexazinone, metribuzin and amicarbazone isolates were effective in controlling all species, but full dosage and 70% of the dose plus fusel oil only showed satisfactory control in species of *I. quamoclit* and *E. heterophylla*.

4.9 Fusel oil applied in early and late post-emergence of weeds

Azania et al. (2010b) studied the response of weeds to doses of fusel oil applied in early and late post-emergence. For this study, the species *Ipomoea hederifolia, Ipomoea quamoclit, Euphorbia heterophylla, Digitaria spp. Cenchrus echinatus* and *Panicum maximum* were used. Fusel oil was applied in early and late post-emergence at doses of 50, 125, 250, 375 and 500 L ha^{-1} plus the untreated control. The plots were made of polyethylene pots with a capacity of 3 L and contained soil from the topsoil of a fallow field. The percentage of intoxication at 7 and 30 days after application (DAA) were visually observed.
These species were only susceptible to the use of 500 L ha^{-1} of fusel oil applied in early or late post-emergence. *Digitaria spp.* was susceptible, *E. Heterophylla* was tolerant, and all of the other species were moderately tolerant to fusel oil applied in early post-emergence. *E. heterophylla* was susceptible, *Digitaria spp., C. echinatus* and *P. maximum* were moderately tolerant and *I. hederifolia* and *I. quamoclit* were tolerant to fusel oil applied in late post-emergence.

4.10 Influence of fusel oil on the development of peanut crops planted in rotation with sugarcane

Azania et al. (2009) evaluated the influence of by-products (vinasse, flegmass and fusel oil) that had previously been applied to sugarcane on the chemical attributes of the soil and the development of peanut crops (*Arachis hypogaea*) in a greenhouse.
The experiment was set up in 22 L pots with treatments of a combination of the three by-products at four concentrations (12.50, 25.00, 50.00 and 100.00% v/v), as well as a control that was irrigated with water and fertilized. Forty days after treatment, the pots that the sugarcane was taken from were planted with seeds of the peanut cultivar Tatu vermelho at 10 seeds per vase, and two plants were maintained per pot after thinning (20 days after planting - DAP).
Fusel oil at 50.0 and 100.0% reduced the soil pH and increased the levels of toxic aluminum. The vinasse increased the pH, freeing up more nutrients. The initial development and subsequent peanut production were influenced by the fusel oil at the two highest concentrations (50.0 and 100.0%), where there was little plant development.

4. Conclusion

The residuals and/or by-products generated by different industrial processes most often have no practical application, and when stored, they may contaminate the environment. Research dedicated to exploring possible uses for these materials is of great importance. In this way, research so far shows that the fusel oil applied in combination with herbicides or herbicide alone has potential, and it only needs to be further explored scientifically to adjust doses and expand application technology.

Fig 1. A) Weed community before the application of treatments; B) lack of control of *Cyperus* sp. and *Commelina benghalensis* (1.50 + 43.75, 1.00 + 62.50, 1.50 + 18.75, 1.00 + 37.5, 1.00 + 12.5 L ha^{-1} glyphosate and fusel oil, respectively); C) *Commelina benghalensis* with yellow leaves and D) total control of weeds (1.50 + 43.75, 1.00 + 62.50, 1.50 + 18.75, 1.00 + 37.5, 1.00 + 12.5 L ha^{-1} glyphosate and fusel oil, respectively). Ribeirão Preto, Brazil, (Azania, 2007).

5. References

Agrianual - *Anuário da agricultura brasileira* (2008). Instituto FNP, São Paulo. ISSN 1807157-0

Almazan, O.; Gonzales, L.; Galvez, L. (1998). The sugarcane, its by-products and co-products. *Food and Agricultural Research Council,* pp.13-25. Réduit, Mauritius

Azania, A. A. P. M. (2003). Influência de subprodutos da indústria alcooleira nos atributos químicos do solo e em plantas de cana-de-açúcar, guanxuma e capim-braquiária. Dissertação de Mestrado em Produção Vegetal da Faculdade de Ciências Agrárias e Veterinárias, Universidade Estadual Paulista, Jaboticabal-Brasil

Azania, A. A. P. M. (2007). Potencialidade herbicídica do óleo fúsel. Tese de Doutorado em Produção Vegetal da Faculdade de Ciências Agrárias e Veterinárias, Universidade Estadual Paulista, Jaboticabal-Brasil

Azania, A. A. P. M.; Azania, C. A. M.; Marques, M. O.; Pavani, M. C. M. D. (2004). Emergência e desenvolvimento de guanxuma (*Sida rhombifolia*), capim-braquiária (*Brachiaria decumbens*) e cana-de-açúcar (*Saccharum* spp.) influenciados por subprodutos da destilação do álcool. *Planta Daninha,* Vol. 22, No.3, 331-336, 0100-8358

Azania, A. A. P. M.; Marques, M. O.; Pavani, M. C. M. D.; Azania, C. A. M. (2003). Germinação de sementes de *Sida rhombifolia* e *Brachiaria decumbens* influenciada por vinhaça, flegmaça e óleo de fúsel. *Planta Daninha,* Vol. 21, No.3, 443-449, 0100-8358

Azania, A.A.P.M; Azania, C.A.M.; Marques, M.O.; Pavani, M.C.M.D.; Furtado, D.E.; Rodrigues, D. (2008). Aplicação de óleo fúsel isolado e em mistura com glifosato na pós-emergência tardia de plantas daninhas. *Planta Daninha,* Vol.26, No.1, 231-236, 0100-8358

Azania, C. A. M.; Azania, A. A. P. M.; Rolim, J. C.; Schiavetto, A. R.; Pizzo, I. V.; Zera, F. S. (2010a). Aplicação de óleo fúsel na erradicação química da cana-de-açúcar e seu efeito na cultura do girassol em sucessão. *Bragantia,* Vol. 69, No.3, 0006-8705

Azania, C. A. M.; Azania, A. A. P. M.; Pizzo, I. V.; Schiavetto, A. R. (2010b). Tolerância de plantas daninhas em pós-emergência inicial e tardia na aplicação do óleo fúsel. Planta Daninha, Vol. 28, n.3, p. 541-549 0100-8358

Azania, A. A. P. M.; Azania, C. A. M., Marques. M. O.; Pavani, M. C. M. D. (2009). Alterações em atributos químicos do solo e crescimento de amendoim em função de doses de subprodutos da cana-de-açúcar. *Ecossistema,* Vol.34, n.1/2 , p. , 0100-4107

Bolonhezi, D. (2007). Uso de adubos verdes na renovação de canavial. In: *Expansão e renovação de canavial.* Segato, S. V. et al. (Ed.), 159-174, ISBN 978-85-60409-03-7, Piracicaba, Vol.28, n.2, p.347-357

Botelho, J. (1945). Impurezas do álcool etílico. *Brasil Açucareiro,* Vol. 25, 99-101

Brinker, F. N. D. The importance of Alcohol in medicine - related health world articles. (2000) http://www.healthy.net/scr/article.asp?ID=772

Ceccato-Antonini, S. R.; Silva, P. C. (2002). Produção de pseudo-hifas e extensões semelhantes a hifas em linhagens industriais de leveduras induzida por álcool isoamílico. *Brazilian Journal of Microbiology,* Vol. 33, No. 3, 1-8, 1517-8382

Codistil (1978). Fermentação alcoólica - destilação. Departamento de Engenharia. 110p. Piracicaba

CONAB - Companhia Nacional de Abastecimento. Avaliação da Safra Agrícola de Cana-de-Açúcar. (2010) http://www.conab.gov.br/conabweb/download/safra/1_cana_10.pdf

Deuber, R. (2003). *Ciência das plantas infestantes: Fundamentos.* FUNEP, p.218, ISBN 85-87632-58-2, Jaboticabal

Ferreira, E. S.; Monteiro, A. O. (1985). Efeitos da aplicação da vinhaça nas propriedades químicas, físicas e biológicas do solo. (Boletim Técnico Copersucar, 36), pp. 3-7, São Paulo

Kissmann, K. G. (1985). Óleos: valiosos auxiliares para tratamentos fitossanitários. Basf. 7 pp.

Kuçuk, Z.; Ceylan, K. (1998). Potential utilization of fusel oil: A kinetic approach for production of fusel oil esters through chemical reaction. *Turkish Journal of Chemistry,* Vol. 22, 289-300, 1300-0527

Lima, U. A. (1964). Estudo dos principais fatores que afetam os componentes do coeficiente não-alcoólico das aguardentes de cana. Tese de Cátedra em Tecnologia do Açúcar e do Álcool da Escola Superior de Agricultura Luiz de Queiroz, Universidade de São Paulo, Piracicaba-Brasil

Nascimento, E. A.; Morais, S. L.; Castro, A. M. de. (2003). Análise dos constituintes do óleo fúsel. http://www.destilariaslenzi.com.br

Osipe, R.; Baldini, V.; Gandolfo, A.M.; Braido, A.(2009). Avaliação da eficiência de diferentes doses de óleo fúsel no controle de *Digitaria insularis* (capim-amargoso) e *Commelina benghalensis* (trapoeraba). *Revista Brasileira de Agroecologia,* Vol. 4, No. 2, 1042-1045, 1980-9735

Patil, A. G.; Koolwal, S. M.; Butala, H. D. (2002). Fusel oil: composition, removal and potential utilization. *International Sugar Journal,* Vol. 104, No. 1238, 51-63, 0020-8841

Pérez, E. R.; Cardoso, D. R.; Franco, D.W. (2001). Análise dos álcoois, ésteres e compostos carbonílicos em amostras de óleo fúsel. *Química Nova,* Vol. 24, No.1, 10-12, 0100-4042

Pires, F. R.; Souza, C. M.; Silva, A. A.; Queiroz, M. E. L. R.; Procópio, S. O.; Santos, J. B.; Santos, E. A.; Cecon, P. R. (2003). Seleção de plantas com potencial para fitorremediação de tebuthiuron. *Planta Daninha,* Vol.21, 451-458, 0100-8358

Pizzo, I. V.; Azania, C. A. M.; Azania, A. A. P. M.; Schiavetto, A. R. (2010). Seletividade e eficácia de controle da associação óleo fúsel e herbicidas em cana-de-açúcar. *Planta Daninha,* Vol.28, n.2, p.347-357, 0100-8358

Ramos, N. P.; Novo, M. C. S. S.; Lago, A. A.; Ungaro, M. R. G. (2009). Girassol: emergência e crescimento inicial de plantas sob resíduos de cana-de-açúcar. *Ciência Rural,* Vol.39, 45-51, 0103-8478

Rasovsky, E. M. (1973). Álcool: destilarias. (Coleção Canavieira, 12). Instituto do Açúcar e do Álcool. 384p. Rio de Janeiro

Rodrigues, B. N.; Almeida, F. S. de (2005). *Guia de herbicidas.* 5ª ed. Londrina

Souza, L. G. de; Llistó, A. M. S. M. (1978). Alguns componentes do coeficiente não-álcool das aguardentes de cana - Determinação por cromatografia em fase gasosa. *Brasil Açucareiro,* Vol.3, 13-16.

Vauclair, C.; Tarjus, H.; Schaetzel, P. (1997). Permselective properties of PVA-PAA blended membrane used for dehydration of fusel oil by pervaporation. *Journal of Membrane Science*, Vol. 125, 293-301, 0376-7388